ELECTRICAL MEASUREMENT, SIGNAL PROCESSING, and DISPLAYS

Edited by
JOHN G. WEBSTER

CRC PRESS

Boca Raton London New York Washington, D.C.

Library of Congress Cataloging-in-Publication Data

Electrical measurements, signal processing, and displays / John G. Webster, editor-in-chief.
p. cm. -- (Principles and applications in engineering)
Includes bibliographical references and index.
ISBN 0-8493-1733-9 (alk. paper)
1. Electronic measurements. 2. Electric measurements. 3. Signal processing. I. Webster, John G., 1932- II. Series.

TK7878.E435 2003
621.3815′48—dc21 2003048530

International Standard Book Number 0-8493-1733-9
Library of Congress Card Number 2003048530
Printed in the United States of America 1 2 3 4 5 6 7 8 9 0
Printed on acid-free paper

Preface

Introduction

The purpose of *Electrical Measurement, Signal Processing, and Displays* is to provide a reference that is both concise and useful for engineers in industry, scientists, designers, managers, research personnel and students, as well as many others who have measurement problems. The book covers an extensive range of topics that comprise the subject of measurement, instrumentation, and sensors.

The book describes the use of instruments and techniques for practical measurements required in electrical measurements. It includes sensors, techniques, hardware, and software. It also includes information processing systems, automatic data acquisition, reduction and analysis and their incorporation for control purposes.

Chapters include descriptive information for professionals, students, and workers interested in measurement. Chapters include equations to assist engineers and scientists who seek to discover applications and solve problems that arise in fields not in their specialty. They include specialized information needed by informed specialists who seek to learn advanced applications of the subject, evaluative opinions, and possible areas for future study. Thus, *Electrical Measurement, Signal Processing, and Displays* serves the reference needs of the broadest group of users — from the advanced high school science student to industrial and university professionals.

Organization

The book is organized according to the measurement problem. Section I covers electromagnetic variables measurement such as voltage, current, and power. Section II covers signal processing such as amplifiers, filters, and compatibility. Section III covers displays such as cathode ray tubes, liquid crystals, and plasma displays.

John G. Webster
Editor-in-Chief

Editor-in-Chief

John G. Webster received the B.E.E. degree from Cornell University, Ithaca, NY, in 1953, and the M.S.E.E. and Ph.D. degrees from the University of Rochester, Rochester, NY, in 1965 and 1967, respectively.

He is Professor of Electrical and Computer Engineering at the University Wisconsin-Madison. In the field of medical instrumentation he teaches undergraduate and graduate courses, and does research on RF cardiac ablation and measurement of vigilance.

He is author of *Transducers and Sensors*, an IEEE/EAB Individual Learning Program (Piscataway, NJ: IEEE, 1989). He is co-author, with B. Jacobson, of *Medicine and Clinical Engineering* (Englewood Cliffs, NJ: Prentice-Hall, 1977), with R. Pallás-Areny, of *Sensors and Signal Conditioning* (New York: Wiley, 1991), and with R. Pallás-Areny, of *Analog Signal Conditioning* (New York: Wiley, 1999). He is editor of *Encyclopedia of Medical Devices and Instrumentation* (New York: Wiley, 1988), *Tactile Sensors for Robotics and Medicine* (New York: Wiley, 1988), *Electrical Impedance Tomography* (Bristol, UK: Adam Hilger, 1990), *Teaching Design in Electrical Engineering* (Piscataway, NJ: Educational Activities Board, IEEE, 1990), *Prevention of Pressure Sores: Engineering and Clinical Aspects* (Bristol, UK: Adam Hilger, 1991), *Design of Cardiac Pacemakers* (Piscataway, NJ: IEEE Press, 1995), *Design of Pulse Oximeters* (Bristol, UK: IOP Publishing, 1997), *Medical Instrumentation: Application and Design, Third Edition* (New York: Wiley, 1998), and *Encyclopedia of Electrical and Electronics Engineering* (New York, Wiley, 1999). He is co-editor, with A.M. Cook, of *Clinical Engineering: Principles and Practices* (Englewood Cliffs, NJ: Prentice-Hall, 1979) and *Therapeutic Medical Devices: Applications and Design* (Englewood Cliffs, NJ: Prentice-Hall, 1982), with W.J. Tompkins, of *Design of Microcomputer-Based Medical Instrumentation* (Englewood Cliffs, NJ: Prentice-Hall, 1981) and *Interfacing Sensors to the IBM PC* (Englewood Cliffs, NJ: Prentice-Hall, 1988, and with A.M. Cook, W.J. Tompkins, and G.C. Vanderheiden, *Electronic Devices for Rehabilitation* (London: Chapman & Hall, 1985).

Dr. Webster has been a member of the IEEE-EMBS Administrative Committee and the NIH Surgery and Bioengineering Study Section. He is a fellow of the Institute of Electrical and Electronics Engineers, the Instrument Society of America, and the American Institute of Medical and Biological Engineering. He is the recipient of the AAMI Foundation Laufman-Greatbatch Prize and the ASEE/Biomedical Engineering Division, Theo C. Pilkington Outstanding Educator Award.

Advisory Board

Contributors

A. Roberto Ambrosini
Institute of Radioastronomy
Bologna, Italy

Jeff P. Anderson
LTV Steel Corporation
Independence, Ohio

Pasquale Arpaia
Università di Napoli Federico II
Naples, Italy

Francesco Avallone
Università di Napoli Federico II
Naples, Italy

Aldo Baccigalupi
Università di Napoli Federico II
Naples, Italy

William A. Barrow
Planar Systems
Beaverton, Oregon

Cipriano Bartoletti
Institute of Radioastronomy
Bologna, Italy

David M. Beams
University of Texas at Tyler
Tyler, Texas

K. Beilenhoff
Institut für
Hochfrequenztechnik,
Technische Universität
Darmstadt, Germany

Richard J. Blotzer
Cleveland, Ohio

C. Bortolotti
Institute of Radioastronomy
Bologna, Italy

Arnaldo Brandolini
Politecnico di Milano
Milan, Italy

Saps Buchman
Stanford University
Stanford, California

Barrett S. Caldwell
Purdue University
West Lafayette, Indiana

Robert B. Campbell
Sandia National Laboratories
Albuquerque, New Mexico

Robert M. Crovella
Plano, Texas

N. D'Amico
Institute of Radioastronomy
Bologna, Italy

Claudio de Capua
Università di Napoli Federico II
Naples, Italy

Alfons Dehé
Siemens AG
Munich, Germany

Achim Dreher
German Aerospace Center
Wessling, Germany

Halit Eren
Curtin University of Technology
Bentley, Australia

Alessandro Ferrero
Politecnico di Milano
Milan, Italy

K. Fricke
Institut für
Hochfrequenztechnik,
Technische Universität
Darmstadt, Germany

Alessandro Gandelli
Politecnico di Milano
Milan, Italy

Daryl Gerke
Kimmel Gerke Associates, Ltd.
Mesa, Arizona

W. A. Gillespie
University of Abertay
Dundee, Scotland

James Goh
Curtin University of Technology
Perth, Australia

G. Grueff
Institute of Radioastronomy
Bologna, Italy

H. L. Hartnagel
Institut für Hochfrequenztechnik, Technische Universität
Darmstadt, Germany

Michael B. Heaney
Palo Alto Research Center
Palo Alto, California

Albert D. Helfrick
Embry-Riddle Aeronautical University
Daytona Beach, Florida

David A. Hill
National Institute of Standards and Technology
Boulder, Colorado

Rahman Jamal
National Instruments Germany
Munich, Germany

Motohisa Kanda
National Institute of Standards and Technology
Boulder, Colorado

Mohammad A. Karim
City College of New York
New York, New York

William Kimmel
Kimmel Gerke Associates, Ltd.
Mesa, Arizona

H. Klingbeil
Institut für Hochfrequenztechnik, Technische Universität
Darmstadt, Germany

V. Krozer
Institut für Hochfrequenztechnik, Technische Universität
Darmstadt, Germany

Carmine Landi
Università de L'Aquila
L'Aquila, Italy

W. Marshall Leach, Jr.
Georgia Institute of Technology
Atlanta, Georgia

Yufeng Li
Samsung Information Systems
San Jose, California

E. B. Loewenstein
National Instruments
Austin, Texas

Michael Z. Lowenstein
Harmonics Limited
Mequou, Wisconsin

Albert Lozano-Nieto
The Pennsylvania State University
Lehman, Pennsylvania

Steven A. Macintyre
Macintyre Electronic Design
Herndon, Virginia

Allan M. MacLeod
University of Abertay
Dundee, Scotland

Sergio Mariotti
Institute of Radioastronomy
Bologna, Italy

P. F. Martin
University of Abertay
Dundee, Scotland

Edward McConnell
National Instruments
Austin, Texas

Robert T. McGrath
The Pennsylvania State University
University Park, Pennsylvania

Douglas P. McNutt
The MacNauchtan Laboratory
Colorado Springs, Colorado

John Mester
Stanford University
Stanford, California

Jeffrey P. Mills
Illinois Institute of Technology
Chicago, Illinois

Devendra K. Misra
University of Wisconsin
Milwaukee, Wisconsin

William C. Moffatt
Sandia National Laboratories
Albuquerque, New Mexico

Stelio Montebugnoli
Institute of Radioastronomy
Fontano, Italy

Jerry Murphy
Hewlett Packard Company
Geneva, Switzerland

Steven A. Murray
California State University
Fullerton, California

A. Orfei
Institute of Radioastronomy
Bologna, Italy

Peter O'Shea
Royal Melbourne Institute of Technology
Melbourne, Australia

Ramón Pallás-Areny
Universitat Politecnica de Catalyna
Barcelona, Spain

Ronney B. Panerai
University of Leicester
Leicester, England

Luca Podestà
University of Rome La Sapienza
Rome, Italy

Rodney Pratt
University of South Australia
Adelaide, Australia

Gordon W. Roberts
McGill University
Montreal, Canada

Giancarlo Sacerdoti
University of Rome La Sapienza
Rome, Italy

Kalluri R. Sarma
Honeywell, Inc.
Phoenix, Arizona

Christopher J. Sherman
Merrimack, New Hampshire

Robert Steer
Frequency Devices
Haverhill, Massachusetts

Timothy J. Sumner
Imperial College
London, England

Peter H. Sydenham
University of South Australia
Adelaide, Australia

Michał Szyper
University of Mining and Metallurgy
Cracow, Poland

G. Tomassetti
University of L'Aquila
L'Aquila, Italy

Michael F. Toner
Northern Telecom Ltd.
Nepean, Ontario, Canada

Ramanapathy Veerasingam
The Pennsylvania State University
University Park, Pennsylvania

Herman Vermariën
Vrije Universiteit Brussel
Brussels, Belgium

Contents

SECTION I Electromagnetic Variables Measurement

SECTION II Signal Processing

SECTION III Displays

I

Electromagnetic Variables Measurement

1

Voltage Measurement

Alessandro Ferrero
Politecnico di Milano

Jerry Murphy
Hewlett Packard Company

Cipriano Bartoletti
Institute of Radioastronomy

Luca Podestà
University of Rome La Sapienza

Giancarlo Sacerdoti
University of Rome La Sapienza

1.1 Meter Voltage Measurement

Alessandro Ferrero

Instruments for the measurement of electric voltage are called *voltmeters*. Correct insertion of a voltmeter requires the connection of its terminals to the points of an electric circuit across which the voltage has to be measured, as shown in Figure 1.1. To a first approximation, the electric equivalent circuit of a voltmeter can be represented by a resistive impedance Z_v (or a pure resistance R_v for dc voltmeters). This means that any voltmeter, once connected to an electric circuit, draws a current I_v given by:

$$I_v = \frac{U}{Z_v} \tag{1.1}$$

where U is the measured voltage. The higher the value of the internal impedance, the higher the quality of the voltmeter, since it does not significantly modify the status of the electric circuit under test.

Different operating principles are used to measure an electric voltage. The mechanical interaction between currents, between a current and a magnetic field, or between electrified conductors was widely adopted in the past to generate a mechanical torque proportional to the voltage or the squared voltage to be measured. This torque, balanced by a restraining torque, usually generated by a spring, causes the instrument pointer, which can be a mechanical or a virtual optical pointer, to be displaced by an angle proportional to the driving torque, and hence to the voltage or the squared voltage to be measured. The value of the input voltage is therefore given by the reading of the pointer displacement on a graduated scale. The thermal effects of a current flowing in a conductor are also used for measuring electric voltages, although they have not been adopted as widely as the previous ones. More recently, the widespread diffusion of semiconductor devices led to the development of a completely different class of voltmeters: *electronic* voltmeters. They basically attain the required measurement by processing the input signal by means of electronic semiconductor devices. According to the method, analog or digital, the input signal is processed, the electronic voltmeters can be divided into *analog* electronic voltmeters and *digital*

0-8493-1733-9/04/$0.00+$1.50

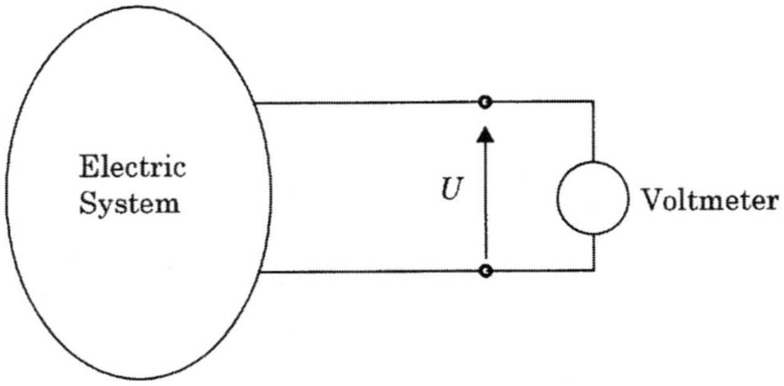

FIGURE 1.1 Voltmeter insertion.

TABLE 1.1 Classification of Voltage Meters

Class	Operating principle	Subclass	Application field
Electromagnetic	Interaction between currents and magnetic fields	Moving magnet	Dc voltage
		Moving coil	Dc voltage
		Moving iron	Dc and ac voltage
Electrodynamic	Interactions between currents	—	Dc and ac voltage
Electrostatic	Electrostatic interactions	—	Dc and ac voltage
Thermal	Current's thermal effects	Direct action	Dc and ac voltage
	Indirect action		Dc and ac voltage
Induction	Magnetic induction	—	Ac voltage
Electronic	Signal processing	Analog	Dc and ac voltage
		Digital	Dc and ac voltage

electronic voltmeters. Table 1.1 shows a rough classification of the most commonly employed voltmeters, according to their operating principle and their typical application field.

This chapter section briefly describes the most commonly employed voltmeters, both electromechanical and electronic.

Electromechanical Voltmeters

Electromechanical voltmeters measure the applied voltage by transducing it into a mechanical torque. This can be accomplished in different ways, basically because of the interactions between currents (*electrodynamic voltmeters*), between a current and a magnetic field (*electromagnetic voltmeters*), between electrified conductors (*electrostatic voltmeters*, or *electrometers*), and between currents induced in a conducting vane (*induction voltmeters*). According to the different kinds of interactions, different families of instruments can be described, with different application fields. Moving-coil electromagnetic voltmeters are restricted to the measurement of dc voltages; moving-iron electromagnetic, electrodynamic, and electrostatic voltmeters can be used to measure both dc and ac voltages; while induction voltmeters are restricted to ac voltages.

The most commonly employed electromechanical voltmeters are the electromagnetic and electrodynamic ones. Electrostatic voltmeters have been widely employed in the past (and are still employed) for the measurement of high voltages, both dc and ac, up to a frequency on the order of several megahertz. Induction voltmeters have never been widely employed, and their present use is restricted to ac voltages.

Therefore, only the electromagnetic, electrodynamic, and electrostatic voltmeters will be described in the following.

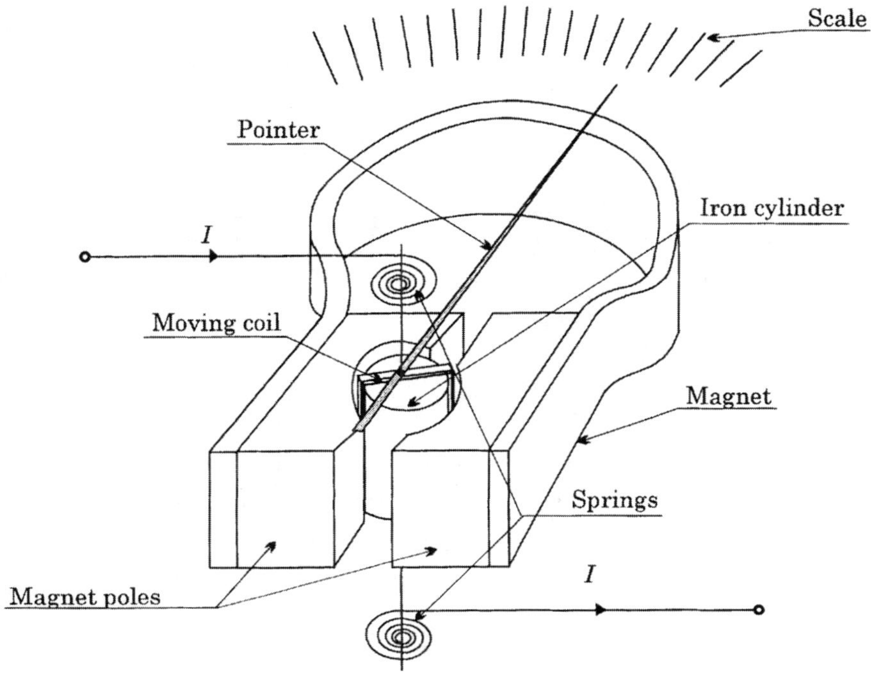

FIGURE 1.2 Dc moving-coil meter.

Electromagnetic Voltmeters

Dc Moving-Coil Voltmeters.

The structure of a dc moving-coil meter is shown in Figure 1.2. A small rectangular pivoted coil is wrapped around an iron cylinder and placed between the poles of a permanent magnet. Because of the shape of the poles of the permanent magnet, the induction magnetic field B in the air gap is radial and constant.

Suppose that a dc current I is flowing in the coil, the coil has N turns, and that the length of the sides that cut the magnetic flux (active sides) is l; the current interacts with the magnetic field B and a force F is exerted on the conductors of the active sides. The value of this force is given by:

$$F = NBlI \tag{1.2}$$

Its direction is given by the right-hand rule. Since the two forces applied to the two active sides of the coil are directed in opposite directions, a torque arises in the coil, given by:

$$T_i = Fd = NBldI \tag{1.3}$$

where d is the coil width. Since N, B, l, d are constant, Equation 1.3 leads to:

$$T_i = k_i I \tag{1.4}$$

showing that the mechanical torque exerted on the coil is directly proportional to the current flowing in the coil itself.

Because of T_i, the coil rotates around its axis. Two little control springs, with k_r constant, provide a restraining torque T_r. The two torques balance when the coil is rotated by an angle δ so that:

$$k_i I = k_r \delta \tag{1.5}$$

which leads to:

$$\delta = \frac{k_i}{k_r} I \tag{1.6}$$

Equation 1.6 shows that the rotation angle of the coil is directly proportional to the dc current flowing in the coil. If a pointer with length h is keyed on the coil axes, a displacement $l = h\delta$ can be read on the instrument scale. Therefore, the pointer displacement is proportional to the current flowing in the coil, according to the following relationship:

$$\lambda = h\frac{k_i}{k_r} I \tag{1.7}$$

This instrument is hence intrinsically a current meter. A voltmeter can be obtained by connecting an additional resistor in series with the coil. If the coil resistance is R_c, and the resistance of the additional resistor is R_a, the current flowing in the coil when the voltage U is applied is given by:

$$I = \frac{U}{R_a + R_c} \tag{1.8}$$

and therefore the pointer displacement is given by:

$$\lambda = h\delta = h\frac{k_i}{k_r} I = h\frac{k_i}{k_r\left(R_a + R_c\right)} U \tag{1.9}$$

and is proportional to the applied voltage. Because of this proportionality, moving-coil dc meters show a proportional-law scale, where the applied voltage causes a proportional angular deflection of the pointer.

Because of the operating principle expressed by Equation 1.3, these voltmeters can measure only dc voltages. Due to the inertia of the mechanical part, ac components typically do not cause any coil rotation, and hence these meters can be also employed to measure the dc component of a variable voltage. They have been widely employed in the past for the measurement of dc voltages up to some thousands volts with a relative measurement uncertainty as low as 0.1% of the full-scale value. At present, they are being replaced by electronic voltmeters that feature the same or better accuracy at a lower cost.

Dc Galvanometer.

General characteristics. A galvanometer is used to measure low currents and low voltages. Because of the high sensitivity that this kind of measurement requires, galvanometers are widely employed as null indicators in all dc balance measurement methods (like the bridge and potentiometer methods) [1, 2].

A dc galvanometer is, basically, a dc moving-coil meter, and the relationship between the index displacement and the current flowing in the moving coil is given by Equation 1.7. The instrument constant:

$$k_a = h\frac{k_i}{k_r} \tag{1.10}$$

is usually called the galvanometer *current constant* and is expressed in mm μA^{-1}. The galvanometer *current sensitivity* is defined as $1/k_a$ and is expressed in μA mm^{-1}.

According to their particular application field, galvanometers must be chosen with particular care. If k_a is taken into account, note that once the full-scale current and the corresponding maximum pointer displacement are given, the value of the ratio hk_i/k_r is also known. However, the single values of h, k_i,

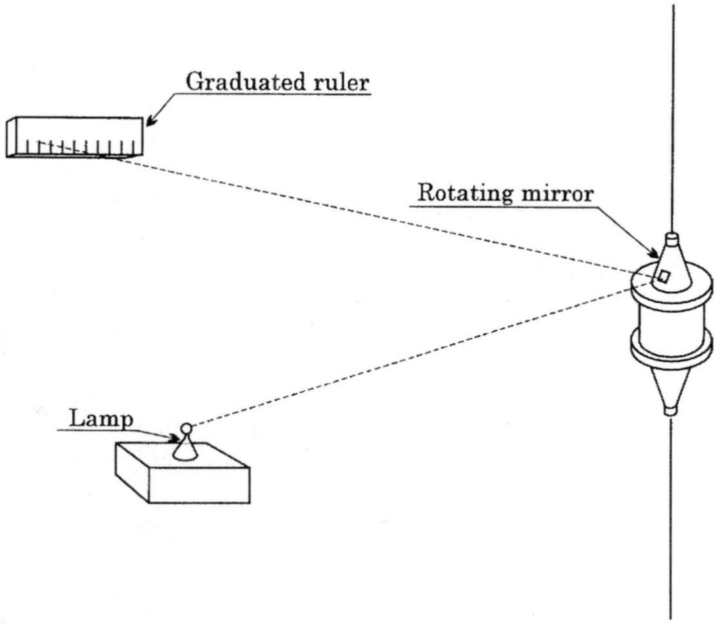

FIGURE 1.3 Virtual optical pointer structure in a dc galvanometer.

and k_r can assume any value and are usually set in order to reduce the friction effects. In fact, if the restraining friction torque T_f is taken into account in the balance equation, Equation 1.5 becomes:

$$k_i I = k_r \frac{\lambda}{h} \pm T_f \quad (1.11)$$

where the ± sign shows that the friction torque does not have its own sign, but always opposes the rotation.

The effects of T_f can be neglected if the driving torque $hk_i I$ and the restraining torque $k_r \lambda$ are sufficiently greater than T_f. Moreover, since the galvanometer is employed as a null indicator, a high sensitivity is needed; hence, k_a must be as high as possible. According to Equations 1.10 and 1.11, this requires high values of hk_i and low values of k_r. A high value of h means a long pointer; a high value of k_i means a high driving torque, while a low value of k_r means that the inertia of the whole moving system must be low.

The pointer length can be increased without increasing the moving system inertia by employing virtual optical pointers: a little, light concave mirror is fixed on the moving-coil axis and is lit by an external lamp. The reflected light hits a translucid, graduated ruler, so that the mirror rotation can be observed (Figure 1.3). In this way, a virtual pointer is obtained, whose length equals the distance between the mirror and the graduated ruler.

The reduction of the moving system inertia is obtained by reducing the weight and dimension of the moving coil, and reducing the spring constant. This is usually done by suspending the moving coil with a thin fiber of conducting material (usually bronze). Thus, the friction torque is practically removed, and the restraining spring action is given by the fiber torsion.

According to Equations 1.3 and 1.4, the driving torque can be increased by increasing the coil flux linkage. Three parameters can be modified to attain this increase: the induction field B, the coil section ld, and the number of turns N of the coil winding.

The induction field B can be increased by employing high-quality permanent magnets, with high coercive force, and minimizing the air gap between the magnet's poles. This minimization prevents the use of moving coils with a large section. Moreover, large coil sections lead to heavier coils with greater inertia, which opposes the previous requirement of reduced inertia. For this reason, the coil section is usually rectangular (although a square section maximizes the flux linkage) and with $l > d$.

If the galvanometer is used to measure a low voltage U, the *voltage sensitivity*, expressed in $\mu V\ mm^{-1}$ is the inverse of:

$$k_v = \frac{\lambda}{U} \tag{1.12}$$

where k_v is called the galvanometer's *voltage constant* and is expressed in mm μV^{-1}.

Mechanical characteristics. Due to the low inertia and low friction, the galvanometer moving system behaves as an oscillating mechanical system. The oscillations around the balance position are damped by the electromagnetic forces that the oscillations of the coil in the magnetic field exert on the coil active sides. It can be proved [1] that the oscillation damping is a function of the coil circuit resistance: that is, the coil resistance r plus the equivalent resistance of the external circuit connected to the galvanometer.

In particular, the damping effect is nil if the coil circuit is open, and maximum if the coil is short-circuited. In practical situations, a resistor is connected in series with the moving coil, whose resistance is selected in such a way to realize a critical damping of the coil movement. When this situation is obtained, the galvanometer is said to be *critically damped* and reaches its balance position in the shortest time, without oscillations around this position.

Actual trends. Moving-coil dc galvanometers have been widely employed in the past when they represented the most important instrument for high-sensitivity measurements. In more recent years, due to the development of the electronic devices, and particularly high-gain, low-noise amplifiers, the moving-coil galvanometers are being replaced by electronic galvanometers, which feature the same, or even better, performance than the electromagnetic ones.

Electrodynamic Voltmeters

Ac Moving-Coil Voltmeters.

The structure of an ac moving-coil meter is shown in Figure 1.4. It basically consists of a pivoted moving coil, two stationary field coils, control springs, a pointer, and a calibrated scale. The stationary coils are series connected and, when a current i_f is applied, a magnetic field B_f is generated along the axis of the stationary coils, as shown in Figure 1.5. A magnetic flux is therefore generated, whose instantaneous values are given by:

$$\varphi_f(t) = k' m_f i_f(t) \tag{1.13}$$

where m_f is the number of turns of the stationary coil and k' is a proportionality factor. When a current i_m is applied to the moving coil, a torque arises, whose instantaneous values are proportional to the product of φ_f and i_m instantaneous values:

$$T_i(t) = k'' \varphi_f(t) i_m(t) = k i_f(t) i_m(t) \tag{1.14}$$

The driving torque is therefore proportional to the instantaneous product of the currents flowing in the two coils. Due to this driving torque, the moving element is displaced by an angle (δt), until the spring restraining torque $T_s(t) = k_s \delta(t)$ balances the driving torque. The moving element rotation is thus given by:

$$\delta(t) = \frac{k}{k_s} i_f(t) i_m(t) \tag{1.15}$$

and, if the pointer length is h, the following pointer displacement can be read on the scale:

$$\lambda(t) = h \frac{k}{k_s} i_f(t) i_m(t) \tag{1.16}$$

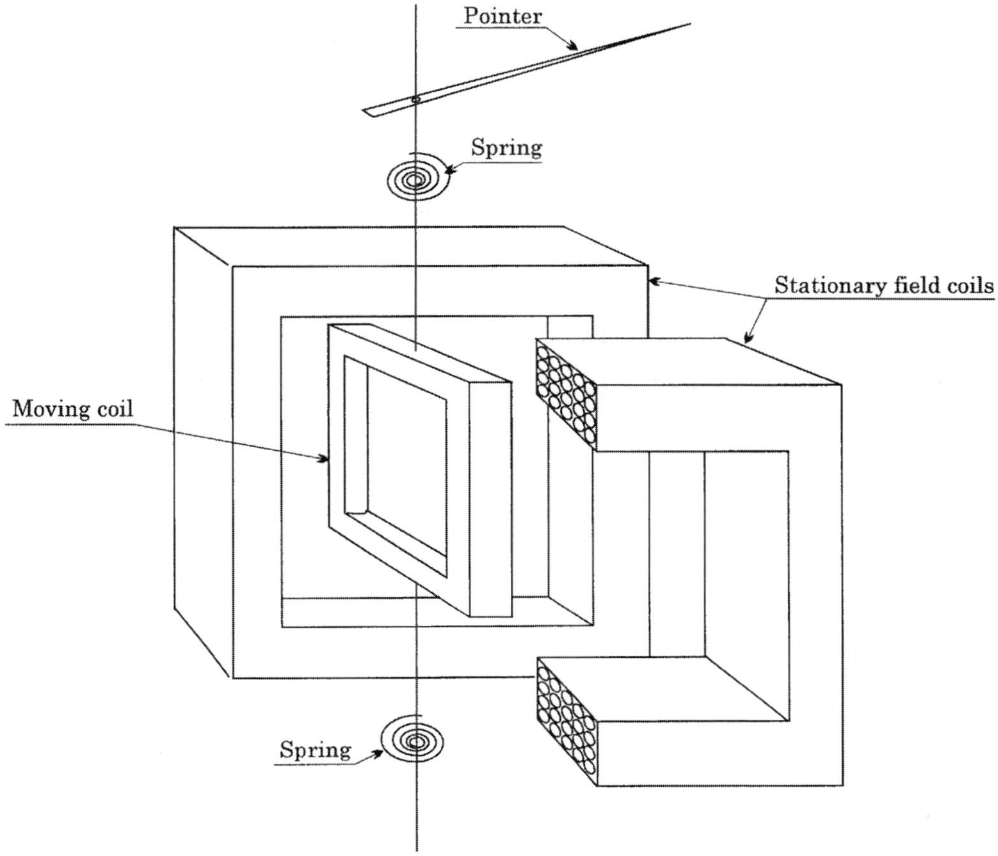

FIGURE 1.4 Ac moving-coil meter.

The proportionality factor k is generally not constant, since it depends on the mutual inductance between the two coils, and thus on their number of turns, shape, and relative position. However, if the two coils are carefully designed and placed, the magnetic field can be assumed to be constant and radial in the rotation area of the moving coil. Under this condition, k is virtually constant.

Because the bandwidth of the moving element is limited to a few hertz, due to its inertia, the balance position is proportional to the average value of the driving torque when the signal bandwidth exceeds this limit. If i_f and i_m currents are sinusoidal, with I_f and I_m rms values, respectively, and with a relative phase displacement β, the driving torque average value is given by:

$$\overline{T_i} = kI_f I_m \cos\beta \tag{1.17}$$

and thus, the pointer displacement in Equation 1.16 becomes:

$$\lambda = h\frac{k}{k_S} I_f I_m \cos\beta \tag{1.18}$$

In order to realize a voltmeter, the stationary and moving coils are series connected, and a resistor, with resistance R, is also connected in series to the coils. If R is far greater than the resistance of the two coils, and if it is also far greater than the coil inductance, in the frequency operating range of the voltmeter, the rms value of the coils' currents is given by:

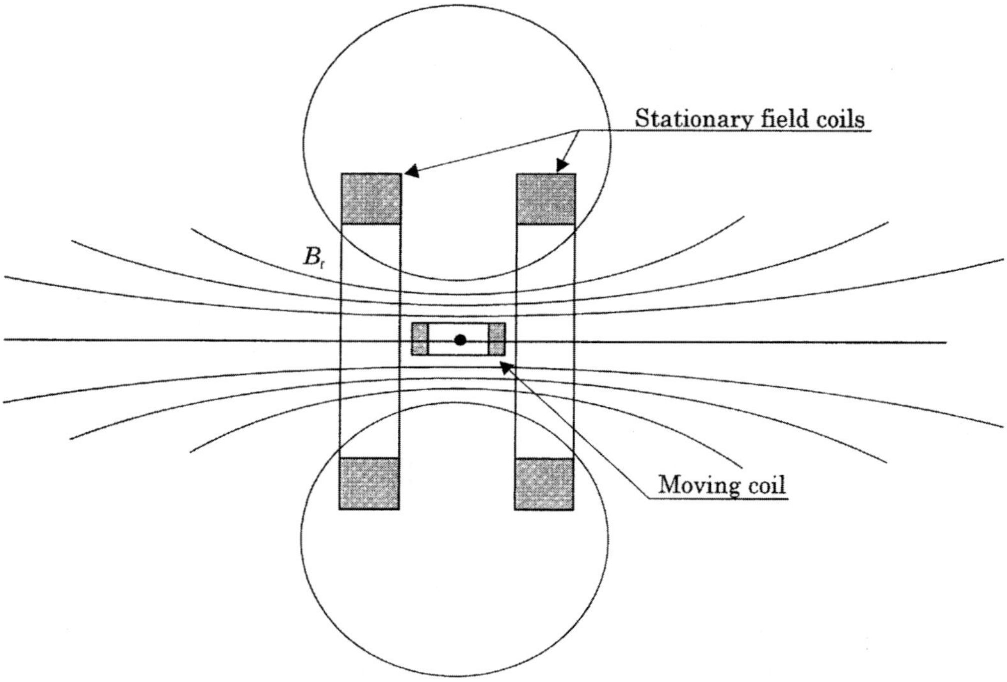

FIGURE 1.5 Magnetic field generated by the field coils in an ac moving-coil meter.

$$I_f = I_m = \frac{U}{R} \tag{1.19}$$

U being the applied voltage rms value. From Equation 1.18, the pointer displacement is therefore given by:

$$\lambda = h\frac{k}{k_S}\frac{U^2}{R^2} = k_v U^2 \tag{1.20}$$

Because of Equation 1.20, the voltmeter features a square-law scale, with k_v constant, provided that the coils are carefully designed, and that the coils' inductance can be neglected with respect to the resistance of the coils themselves and the series resistor. This last condition determines the upper limit of the input voltage frequency.

These voltmeters feature good accuracy (their uncertainty can be as low as 0.2% of the full-scale value), with full-scale values up to a few hundred volts, in a frequency range up to 2 kHz.

Electrostatic Voltmeters

The action of electrostatic instruments is based on the force exerted between two charged conductors. The conductors behave as a variable plate air capacitor, as shown in Figure 1.6. The moving plate, when charged, tends to move so as to increase the capacitance between the plates. The energy stored in the capacitor, when the applied voltage is U and the capacitance is C, is given by:

$$W = \frac{1}{2}CU^2 \tag{1.21}$$

This relationship is valid both under dc and ac conditions, provided that the voltage rms value U is considered for ac voltage.

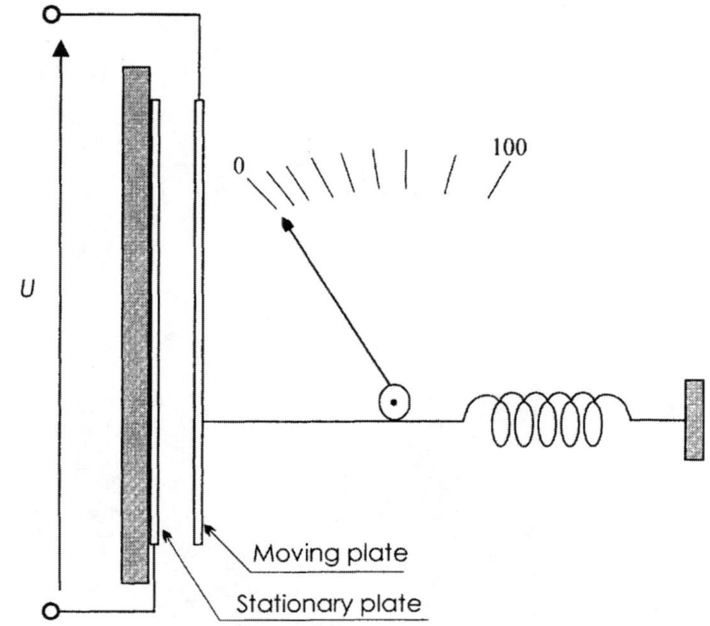

FIGURE 1.6 Basic structure of an electrostatic voltmeter.

When the moving plate is displaced horizontally by d*s*, while the voltage is held constant, the capacitor energy changes in order to equal the work done in moving the plate. The resulting force is:

$$F = \frac{dW}{ds} = \frac{U^2}{2}\frac{dC}{ds} \tag{1.22}$$

For a rotable system, Equation 1.21 leads similarly to a resulting torque:

$$T = \frac{dW}{d\vartheta} = \frac{U^2}{2}\frac{dC}{d\vartheta} \tag{1.23}$$

If the action of a control spring is also considered, both Equations 1.22 and 1.23 show that the balance position of the moving plate is proportional to the square of the applied voltage, and hence electrostatic voltmeters have a square-law scale. These equations, along with Equation 1.21, show that these instruments can be used for the measurement of both dc and ac rms voltages. However, the force (or torque) supplied by the instrument schematically represented in Figure 1.6 is generally very weak [2], so that its use is very impractical.

The Electrometer.

A more useful configuration is the quadrant electrometer, shown in Figure 1.7. Four fixed plates realize four quadrants and surround a movable vane suspended by a torsion fiber at the center of the system. The opposite quadrants are electrically connected together, and the potential difference ($U_1 - U_2$) is applied. The moving vane can be either connected to potential U_1 or U_2, or energized by an independent potential U_3.

Let the zero torque position of the suspension coincide with the symmetrical X-X position of the vane. If $U_1 = U_2$, the vane does not leave this position; otherwise, the vane will rotate.

Let C_1 and C_2 be the capacitances of quadrants 1 and 2, respectively, relative to the vane. They both are functions of ϑ and, according to Equation 1.23, the torque applied to the vane is given by:

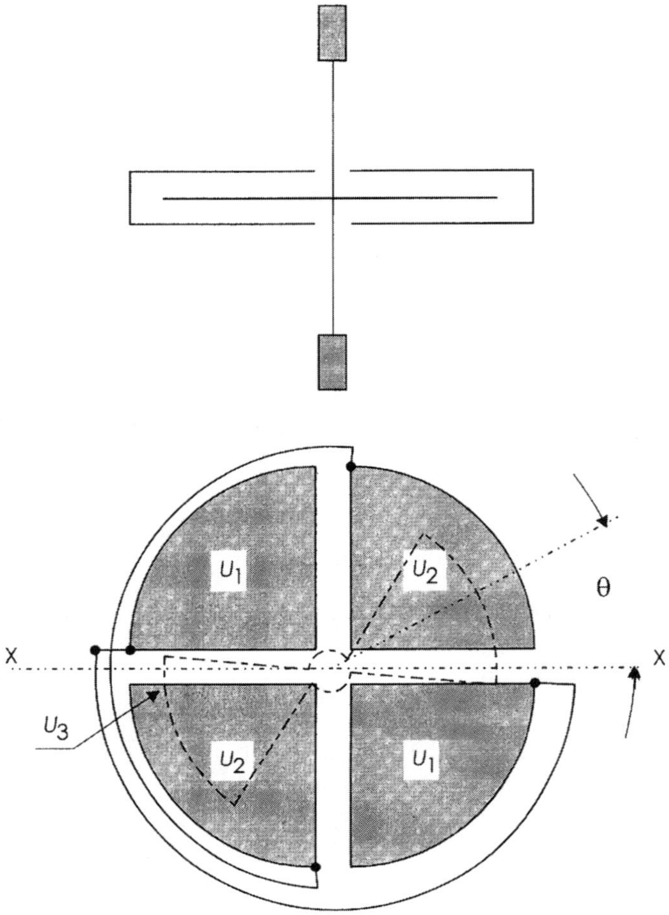

FIGURE 1.7 Quadrant electrometer structure.

$$T = \frac{\left(U_3 - U_1\right)^2}{2}\frac{dC_1}{d\vartheta} + \frac{\left(U_3 - U_2\right)^2}{2}\frac{dC_2}{d\vartheta} \tag{1.24}$$

Since the vane turns out of one pair of quadrants as much as it turns into the other, the variations of C_1 and C_2 can be related by:

$$-\frac{dC_1}{d\vartheta} = \frac{dC_2}{d\vartheta} = k_1 \tag{1.25}$$

Taking into account the suspension restraining torque $T_r = k_2\vartheta$, the balance position can be obtained by Equations 1.24 and 1.25 as:

$$\vartheta = \frac{k_1}{2k_2}\left[\left(U_3 - U_2\right)^2 - \left(U_3 - U_1\right)^2\right] \tag{1.26}$$

If the vane potential U_3 is held constant, and is large compared to the quadrant potentials U_1 and U_2, Equation 1.26 can be simplified as follows:

$$\vartheta = \frac{k_1}{k_2}U_3\left(U_1 - U_2\right) \tag{1.27}$$

Equation 1.27 shows that the deflection of the vane is directly proportional to the voltage difference applied to the quadrants. This method of use is called the *heterostatic* method.

If the vane is connected to quadrant 1, $U_3 = U_1$ follows, and Equation 1.26 becomes

$$\vartheta = \frac{k_1}{2k_2}\left(U_1 - U_2\right)^2 \tag{1.28}$$

Equation 1.28 shows that the deflection of the vane is proportional to the square of the voltage difference applied to the quadrants, and hence this voltmeter has a square-law scale. This method of use is called the *idiostatic* method, and is suitable for the direct measurement of dc and ac voltages without an auxiliary power source.

The driving torque of the electrometer is extremely weak, as in all electrostatic instruments. The major advantage of using this kind of meter is that it allows for the measurement of dc voltages without drawing current by the voltage source under test. Now, due to the availability of operational amplifiers with extremely high input impedance, they have been almost completely replaced by electronic meters with high input impedance.

Electronic Voltmeters

Electronic meters process the input signal by means of semiconductor devices in order to extract the information related to the required measurement [3, 4]. An electronic meter can be basically represented as a three-port element, as shown in Figure 1.8.

The input signal port is an input port characterized by high impedance, so that the signal source has very little load. The measurement result port is an output port that provides the measurement result (in either an analog or digital form, depending on the way the input signal is processed) along with the power needed to energize the device used to display the measurement result. The power supply port is an input port which the electric power required to energize the meter internal devices and the display device flows through.

One of the main characteristics of an electronic meter is that it requires an external power supply. Although this may appear as a drawback of electronic meters, especially where portable meters are concerned, note that, this way, the energy required for the measurement is no longer drawn from the signal source.

The high-level performance of modern electronic devices yields meters that are as accurate (and sometimes even more accurate) as the most accurate electromechanical meters. Because they do not require the extensive use of precision mechanics, they are presently less expensive than electromechanical meters, and are slowly, but constantly, replacing them in almost all applications.

Depending on the way the input signal is processed, electronic meters are divided into *analog* and *digital* meters. Analog meters attain the required measurement by analog, continuous-time processing of the input signal. The measurement result can be displayed both in analog form using, for example, an electromechanical meter; or in digital form by converting the analog output signal into digital form.

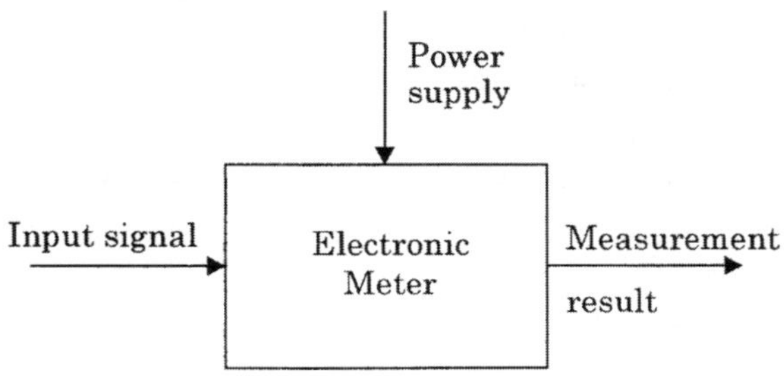

FIGURE 1.8 Electronic meter.

Digital meters attain the required measurement by digital processing of the input signal. The measurement result is usually displayed in digital form. Note that the distinction between analog and digital meters is not due to the way the measurement result is displayed, but to the way the input signal is processed.

Analog Voltmeters

An electronic analog voltmeter is based on an electronic amplifier and an electromechanical meter to measure the amplifier output signal. The amplifier operates to make a dc current, proportional to the input quantity to be measured, flow into the meter. This meter is hence a dc moving-coil milliammeter.

Different full-scale values can be obtained using a selectable-ratio voltage divider if the input voltage is higher than the amplifier dynamic range, or by selecting the proper amplifier gain if the input voltage stays within the amplifier dynamic range.

The main features of analog voltmeters are high input impedance, high possible gain, and wide possible bandwidth for ac measurements. The relative measurement uncertainty can be lower than 1% of full-scale value. Because of these features, electronic analog voltmeters can have better performance than the electromechanical ones.

Dc Analog Voltmeters

Figure 1.9 shows the circuit for an electronic dc analog voltmeter. Assuming that the operational amplifier exhibits ideal behavior, current I_m flowing in the milliammeter A is given by:

$$I_m = I_o + I_2 = \frac{U_o}{R_o} + \frac{U_o}{R_2} = -U_i \frac{R_2}{R_1} \frac{R_2 + R_o}{R_2 R_o} = -\frac{U_1}{R_1}\left(1 + \frac{R_2}{R_o}\right) \tag{1.29}$$

If $R_1 = R_2$, and the same resistances are far greater than R_o, Equation 1.29 can be simplified to:

$$I_m = -\frac{U_1}{R_o} \tag{1.30}$$

Equation 1.30 shows that the milliammeter reading is directly proportional to the input voltage through resistance R_o only. This means that, once the milliammeter full-scale value is set, the voltmeter full-scale value can be changed, within the dynamic range of the amplifier, by changing the R_o value. This way, the meter full-scale value can be changed without changing its input impedance.

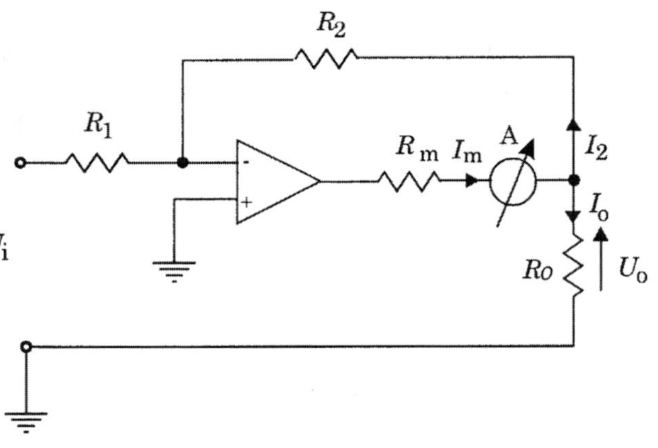

FIGURE 1.9 Electronic dc analog voltmeter schematics.

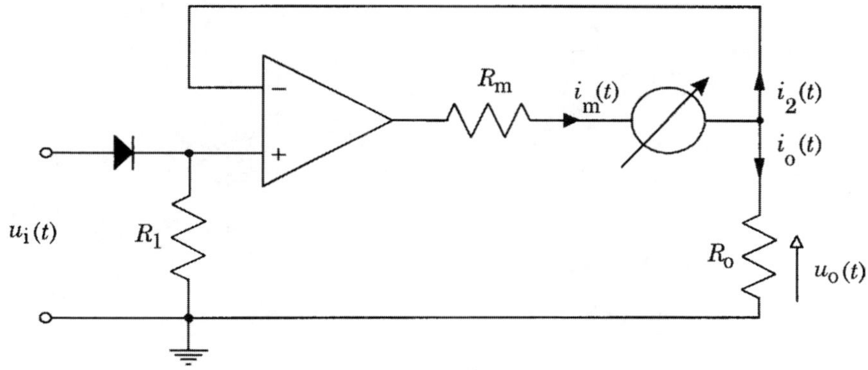

FIGURE 1.10 Electronic, rectifier-based ac analog voltmeter schematics.

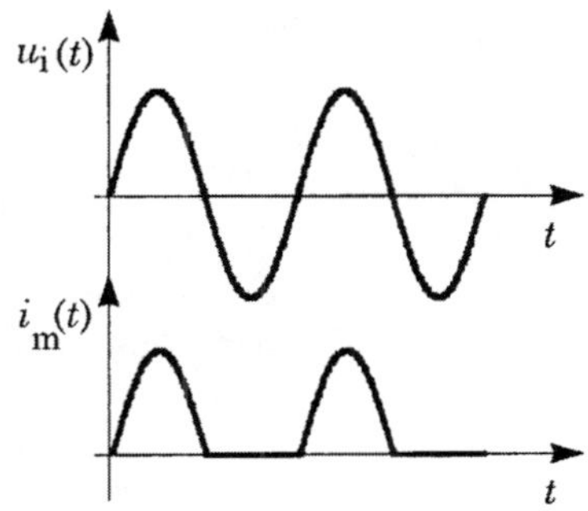

FIGURE 1.11 Signal waveforms in a rectifier-based ac analog voltmeter when the input voltage is sinusoidal.

Rectifier-Based Ac Analog Voltmeters.

Analog meters for ac voltages can be obtained starting from the dc analog voltmeters, with a rectifying input stage. Figure 1.10 shows how the structure in Figure 1.9 can be modified in order to realize an ac voltmeter.

Because of the high input impedance of the electronic amplifier, $i_2(t) = 0$, and the current $i_m(t)$ flowing in the milliammeter A is the same as current $i_o(t)$ flowing in the load resistance. Since the amplifier is connected in a voltage-follower configuration, the output voltage is given by:

$$u_o(t) = u_i(t) \tag{1.31}$$

Due to the presence of the input diode, current $i_m(t)$ is given by:

$$i_m(t) = \frac{u_i(t)}{R_o} \tag{1.32}$$

when $u_i(t) > 0$, and

$$i_m(t) = 0 \tag{1.33}$$

when $u_i(t) \le 0$. If $u_i(t)$ is supposed to be a sine wave, the waveform of $i_m(t)$ is shown in Figure 1.11.

The dc moving-coil milliammeter measures the average value $\bar{I}_m$ of $i_m(t)$, which, under the assumption of sinusoidal signals, is related to the rms value U_i of $u_i(t)$ by:

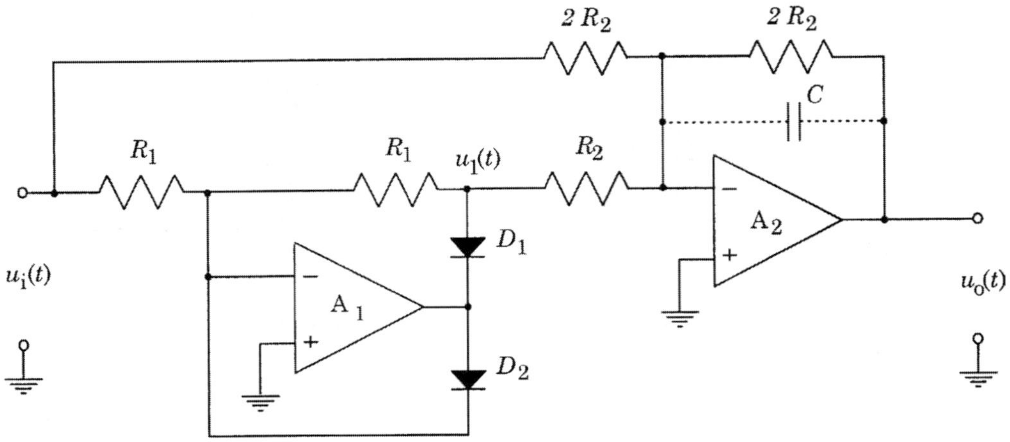

FIGURE 1.12 Electronic, full-wave rectifier-based ac analog voltmeter schematics.

$$\bar{I}_m = \frac{2\sqrt{2}}{\pi R_o} U_i \tag{1.34}$$

The performance of the structure in Figure 1.10 can be substantially improved by considering the structure in Figure 1.12 which realizes a full-wave rectifier. Because of the presence of diodes D_1 and D_2, the output of amplifier A1 is given by:

$$u_1(t) = \begin{cases} -u_i(t) & \text{for } u_i(t) \geq 0 \\ 0 & \text{for } u_i(t) < 0 \end{cases} \tag{1.35}$$

where $u_i(t)$ is the circuit input voltage.

If capacitor C is supposed to be not connected, amplifier A_2 output voltage is:

$$u_o(t) = -\left[u_i(t) + 2u_1(t)\right] \tag{1.36}$$

which gives:

$$u_o(t) = \begin{cases} u_i(t) & \text{for } u_i(t) \geq 0 \\ -u_i(t) & \text{for } u_i(t) < 0 \end{cases} \tag{1.37}$$

thus proving that the circuit in Figure 1.12 realizes a full-wave rectifier.

If $u_i(t)$ is a sine wave, the waveforms of $u_i(t)$, $u_1(t)$, and $u_o(t)$ are shown in Figure 1.13.

Connecting capacitor C in the feedback loop of amplifier A2 turns it into a first-order low-pass filter, so that the circuit output voltage equals the average value of $u_o(t)$:

$$\bar{U}_o = \overline{\left|u_i(t)\right|} \tag{1.38}$$

In the case of sinusoidal input voltage with rms value U_i, the output voltage is related to this rms value by:

$$\bar{U}_o = \frac{2\sqrt{2}}{\pi} U_i \tag{1.39}$$

$\bar{U}_o$ can be measured by a dc voltmeter.

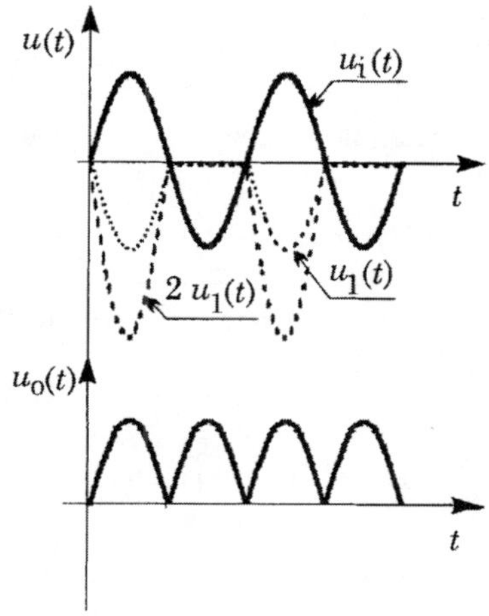

FIGURE 1.13 Signal waveforms in a fullwave rectifier-based ac analog voltmeter when the input voltage is sinusoidal.

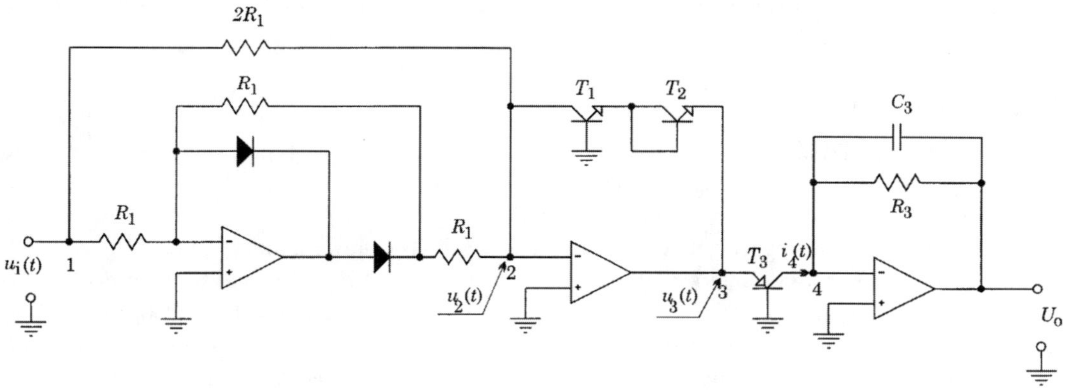

FIGURE 1.14 True rms electronic ac voltmeter schematics.

Both meters in Figures 1.10 and 1.12 are actually average detectors. However, due to Equations 1.34 and 1.39, their scale can be labeled in such a way that the instrument reading gives the rms value of the input voltage, provided it is sinusoidal. When the input voltage is no longer sinusoidal, an error arises that depends on the signal form factor.

True rms Analog Voltmeters.

The rms value U_i of a periodic input voltage signal $u_i(t)$, with period T, is given by:

$$U_i = \sqrt{\frac{1}{T}\int_0^T u_i^2(t)dt} \tag{1.40}$$

The electronic circuit shown in Figure 1.14 provides an output signal U_o proportional to the squared rms value of the input signal $u_i(t)$. The circuit section between nodes 1 and 2 is a full-wave rectifier. Hence, node 2 potential is given by:

$$u_2(t) = |u_i(t)| \tag{1.41}$$

The circuit section between nodes 2 and 4 is a log multiplier. Because of the logarithmic characteristic of the feedback path due to the presence of T1 and T2, node 3 potential is given by:

$$u_3(t) = 2k_1 \log[u_2(t)] = k_1 \log[u_2^2(t)] = k_1 \log\left[|u_i(t)|^2\right] = k_1 \log[u_i^2(t)] \quad (1.42)$$

and, due to the presence of T_3, the current flowing in node 4 is given by:

$$i_4(t) = k_2 \exp[u_3(t)] = k_3 u_i^2(t) \quad (1.43)$$

The circuit section after node 4 is a low-pass filter that extracts the dc component of the input signal. Therefore, the circuit output voltage is given by:

$$U_o = \frac{k}{T}\int_0^T u_i^2(t)dt = kU_i^2 \quad (1.44)$$

thus providing an output signal proportional to the squared rms value of the input signal $u_i(t)$ in accordance with Equation 1.40. Quantities k_1, k_2, and k depend on the values of the elements in the circuit in Figure 1.14. Under circuit operating conditions, their values can be considered constant, so that k_1, k_2, and k can be considered constant also.

If carefully designed, this circuit can feature an uncertainty in the range of ±1% of full scale, for signal frequencies up to 100 kHz.

Digital Voltmeters

A digital voltmeter (DVM) attains the required measurement by converting the analog input signal into digital, and, when necessary, by discrete-time processing of the converted values. The measurement result is presented in a digital form that can take the form of a digital front-panel display, or a digital output signal. The digital output signal can be coded as a decimal BCD code, or a binary code.

The main factors that characterize DVMs are speed, automatic operation, and programmability. In particular, they presently offer the best combination of speed and accuracy if compared with other available voltage-measuring instruments. Moreover, the capability of automatic operations and programmability make DVMs very useful in applications where flexibility, high speed, and computer controllability are required. A typical application field is therefore that of automatically operated systems.

When a DVM is directly interfaced to a digital signal processing (DSP) system and used to convert the analog input voltage into a sequence of sampled values, it is usually called an analog-to-digital converter (ADC).

DVMs basically differ in the following ways: (1) number of measurement ranges, (2) number of digits, (3) accuracy, (4) speed of reading, and (5) operating principle.

The basic measurement ranges of most DVMs are either 1 V or 10 V. It is however possible, with an appropriate preamplifier stage, to obtain full-scale values as low as 0.1 V. If an appropriate voltage divider is used, it is also possible to obtain full-scale values as high as 1000 V.

If the digital presentation takes the form of a digital front-panel display, the measurement result is presented as a decimal number, with a number of digits that typically ranges from 3 to 6. If the digital representation takes the form of a binary-coded output signal, the number of bits of this representation typically ranges from 8 to 16, though 18-bit ADCs are available.

The accuracy of a DVM is usually correlated to its resolution. Indeed, assigning an uncertainty lower than the 0.1% of the range to a three-digit DVM makes no sense, since this is the displayed resolution of the instrument. Similarly, a poorer accuracy makes the three-digit resolution quite useless. Presently, a six-digit DVM can feature an uncertainty range, for short periods of time in controlled environments, as low as the 0.0015% of reading or 0.0002% of full range.

The speed of a DVM can be as high as 1000 readings per second. When the ADC is considered, the conversion rate is taken into account instead of the speed of reading. Presently, the conversion rate for

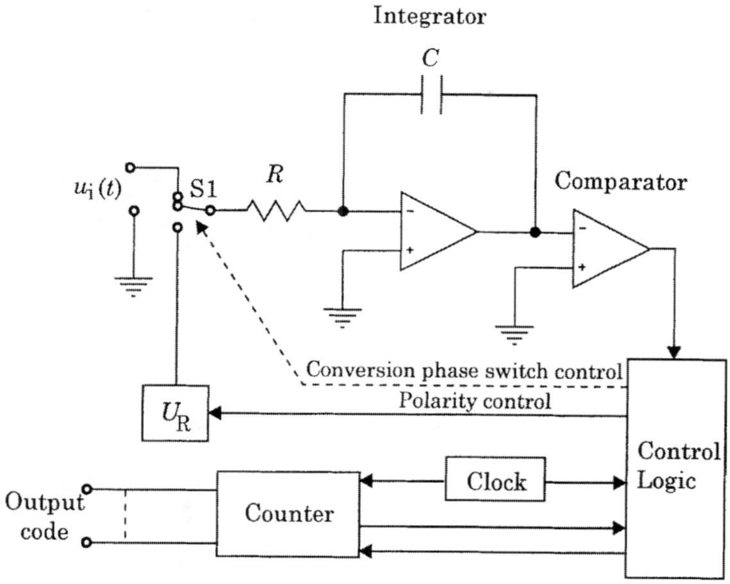

FIGURE 1.15 Dual slope DVM schematics.

12-bit, successive approximation ADCs can be on the order of 10 MHz. It can be on the order of 100 MHz for lower resolution, flash ADCs [5].

DVMs can be divided into two main operating principle classes: the *integrating* types and the *non-integrating* types [3]. The following sections give an example for both types.

Dual Slope DVM.

Dual slope DVMs use a counter and an integrator to convert an unknown analog input voltage into a ratio of time periods multiplied by a reference voltage. The block diagram in Figure 1.15 shows this operating principle. The switch S1 connects the input signal to the integrator for a fixed period of time t_f. If the input voltage is positive and constant, $u_i(t) = U_i > 0$, the integrator output represents a negative-slope ramp signal (Figure 1.16). At the end of t_f, S1 switches and connects the output of the voltage reference U_R to the integrator input. The voltage reference output is negative for a positive input voltage. The integrator output starts to increase, following a positive-slope ramp (Figure 1.16). The process stops when the ramp attains the 0 V level, and the comparator allows the control logic to switch S1 again. The period of time t_v the ramp takes to increase to 0 V is variable and depends on the ramp peak value attained during period t_f.

The relationship between the input voltage U_i and the time periods t_v and t_f is given by:

$$\frac{1}{RC}\int_0^{t_f} U_i dt = \frac{t_v}{RC} U_R \tag{1.45}$$

that, for a constant input voltage U_i, leads to:

$$U_i = U_R \frac{t_v}{t_f} \tag{1.46}$$

Since the same integrating circuit is used, errors due to comparator offset, capacitor tolerances, long-term counter clock drifts, and integrator nonlinearities are eliminated. High resolutions are therefore possible, although the speed of reading is low (in the order of milliseconds).

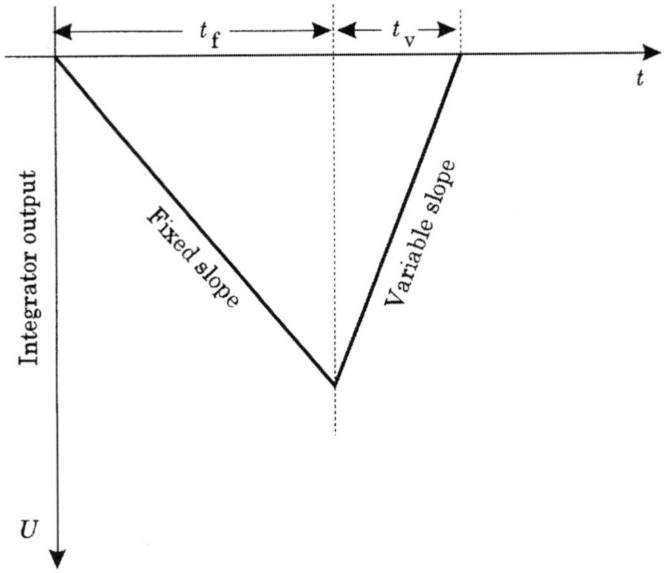

FIGURE 1.16 Integrator output signal in a dual slope DVM.

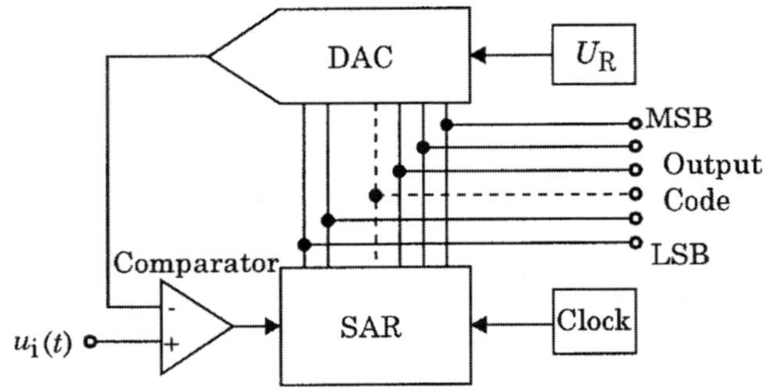

FIGURE 1.17 Successive approximation ADC schematics.

Slowly varying voltages can be also measured by dual slope DVMs. However, this requires that the input signal does not vary for a quantity greater than the DVM resolution during the reading time. For high-resolution DVMs, this limits the DVM bandwidth to a few hertz.

Successive Approximation ADC.

The successive approximation technique represents the most popular technique for the realization of ADCs. Figure 1.17 shows the block diagram of this type of converter. The input voltage is assumed to have a constant value U_i and drives one input of the comparator. The other comparator's input is driven by the output of the digital-to-analog converter (DAC), which converts the binary code provided by the successive approximation register (SAR) into an analog voltage. Let n be the number of bits of the converter, U_R the voltage reference output, and C the code provided by the SAR. The DAC output voltage is then given by:

$$U_c = \frac{C}{2^n} U_R \tag{1.47}$$

When the conversion process starts, the SAR most significant bit (MSB) is set to logic 1. The DAC output, according to Equation 1.47, is set to half the reference value, and hence half the analog input

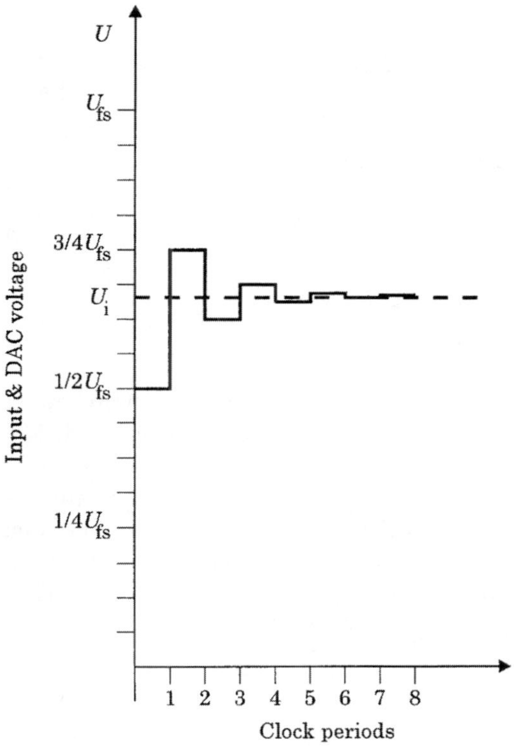

FIGURE 1.18 DAC output signal in a successive approximation ADC.

full-scale range. The comparator determines whether the DAC output is above or below the input signal. The comparator output controls the SAR in such a way that, if the input signal is above the DAC output, as shown in Figure 1.18, the SAR MSB is retained and the next bit is set to logic 1.

If now the input signal is below the DAC output (Figure 1.18), the last SAR bit set to logic 1 is reset to logic 0, and the next one is set to logic 1. The process goes on until the SAR least significant bit (LSB) has been set. The entire conversion process takes time $t_c = nT_c$, where T_c is the clock period. At the end of conversion, the SAR output code represents the digitally converted value of the input analog voltage U_i.

According to Equation 1.47, the ADC resolution is $U_R/2n$, which corresponds to 1 LSB. The conversion error can be kept in the range ±1/2 LSB. Presently, a wide range of devices is available, with resolution from 8 to 16 bits, and conversion rates from 100 ms to below 1 ms.

Varying voltages can be sampled and converted into digital by the ADC, provided the input signal does not vary by a quantity greater than $U_R/2n$ during the conversion period t_c. The maximum frequency of an input sine wave that satisfies this condition can be readily determined starting from given values of n and t_c.

Let the input voltage of the ADC be an input sine wave with peak-to-peak voltage $U_{pp} = U_R$ and frequency f. Its maximum variation occurs at the zero-crossing time and, due to the short conversion period t_c, is given by $2\pi f t_c U_{pp}$. To avoid conversion errors, it must be:

$$2\pi f t_c U_{pp} \leq \frac{U_R}{2^n} \tag{1.48}$$

Since $U_{pp} = U_R$ is assumed, this leads to:

$$f \leq \frac{1}{2^n 2\pi t_c} \tag{1.49}$$

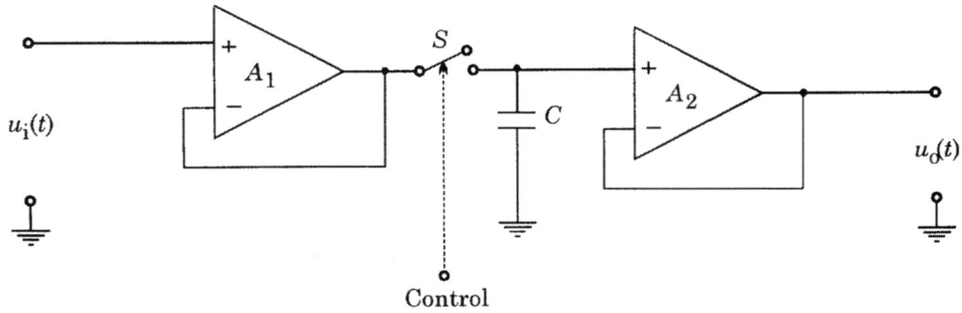

FIGURE 1.19 Sample and Hold schematics.

If $t_c = 1$ ms and $n = 12$, Equation 1.49 leads to $f \leq 38.86$ Hz. However, ADCs can still be employed with input signals whose frequency exceeds the value given by Equation 1.49, provided that a *Sample and Hold* circuit is used to keep the input voltage constant during the conversion period.

The *Sample and Hold* circuit is shown in Figure 1.19. When the electronic switch S is closed, the output voltage $u_o(t)$ follows the input voltage $u_i(t)$. When switch S is open, the output voltage is the same as the voltage across capacitor C, which is charged at the value assumed by the input voltage at the time the switch was opened. Due to the high input impedance of the operational amplifier A_2, if a suitable value is chosen for capacitor C, its discharge transient is slow enough to keep the variation of the output voltage below the ADC resolution.

Ac Digital Voltmeters.

True rms ac voltmeters with digital reading can be obtained using an electronic circuit like the one in Figure 1.14 to convert the rms value into a dc voltage signal, and measuring it by means of a DVM. However, this structure cannot actually be called a digital structure, because the measurement is attained by means of analog processing of the input signal.

A more modern approach, totally digital, is shown in Figure 1.20. The input signal $u_i(t)$ is sampled at constant sampling rate f_s, and converted into digital by the ADC. The digital samples are stored in the memory of the digital signal processor (DSP) and then processed in order to evaluate Equation 1.40 in a numerical way. Assuming that the input signal is periodic, with period T, and its frequency spectrum is upper limited by harmonic component of order N, the sampling theorem is satisfied if at least $(2N + 1)$ samples are taken over period T in such a way that $(2N + 1)T_s = T$, $T_s = 1/f_s$ being the sampling period [6, 7]. If $u_i(kT_s)$ is the kth sample, the rms value of the input signal is given by, according to Equation 1.40:

$$U^2 = \sqrt{\frac{1}{2N+1}\sum_{k=0}^{2N} u_i^2(kT_s)} \tag{1.50}$$

This approach can feature a relative uncertainty as low as ±0.1% of full scale, with an ADC resolution of 12 bits. The instrument bandwidth is limited to half the sampling frequency, according to the sampling theorem. When modern ADCs and DSPs are employed, a 500 kHz bandwidth can be obtained. Wider bandwidths can be obtained, but with a lower ADC resolution, and hence with a lower accuracy.

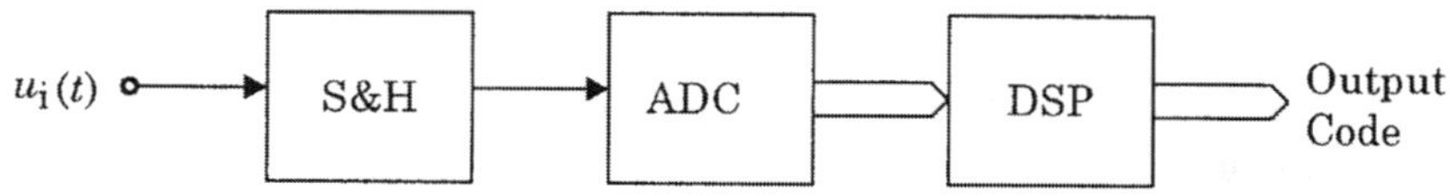

FIGURE 1.20 Block diagram of a modern digital meter.

Frequency Response of Ac Voltmeters.

When the frequency response of ac voltmeters is taken into account, a distinction must be made between the analog voltmeters (both electromechanical and electronic) and digital voltmeters, based on DSP techniques.

The frequency response of the analog meters is basically a low-pass response, well below 1 kHz for most electromechanical instruments, and up to hundreds of kilohertz for electronic instruments.

When digital, DSP-based meters are concerned, the sampling theorem and aliasing effects must be considered. To a first approximation, the frequency response of a digital meter can be considered flat as long as the frequency-domain components of the input signal are limited to a frequency band narrower than half the sampling rate. If the signal components exceed this limit (the so-called Nyquist frequency), the aliasing phenomenon occurs [6]. Because of this phenomenon, the signal components at frequencies higher than half the sampling rate are folded over the lower frequency components, changing them. Large measurement errors occur under this situation.

To prevent the aliasing, a low-pass filter must be placed at the input stage of any digital meter. The filter cut-off frequency must ensure that all frequency components above half the sampling rate are negligible. If the low-pass, anti-aliasing filter is used, the digital DSP-based meters feature a low-pass frequency response also.

References

1. M.B. Stout, *Basic Electrical Measurements,* Englewood Cliffs, NJ, Prentice-Hall, 1960.
2. I.F. Kinnard, *Applied Electrical Measurements,* New York, John Wiley & Sons, Chapman & Hall, Ltd. London, 1956.
3. B.M. Oliver and J.M. Cage, *Electronic Measurements and Instrumentation,* London, McGraw-Hill, Inc. 1975.
4. T.T. Lang, *Electronics of Measuring Systems,* New York, John Wiley & Sons, 1987.
5. Analog Devices, *Analog-Digital Conversion Handbook,* Englewood Cliffs, NJ, Prentice-Hall, 1986.
6. A.V. Oppenheim and R.W. Schafer, *Digital Signal Processing,* Englewood Cliffs, NJ, Prentice-Hall, 1975.
7. A. Ferrero and R. Ottoboni, High-accuracy Fourier analysis based on synchronous sampling techniques. *IEEE Trans. Instr. Meas.,* 41(6), 780-785, 1992.

1.2 Oscilloscope Voltage Measurement

Jerry Murphy

Engineers, scientists, and other technical professionals around the world depend on oscilloscopes as one of the primary voltage measuring instruments. This is an unusual situation because the oscilloscope is not the most accurate voltage measuring instrument usually available in the lab. It is the graphical nature of the oscilloscope that makes it so valued as a measurement instrument — not its measurement accuracy.

The oscilloscope is an instrument that presents a graphical display of its input voltage as a function of time. It displays voltage waveforms that cannot easily be described by numerical methods. For example, the output of a battery can be completely described by its output voltage and current. However, the output of a more complex signal source needs additional information such as frequency, duty cycle, peak-to-peak amplitude, overshoot, preshoot, rise time, fall time, and more to be completely described. The oscilloscope, with its graphical presentation of complex waveforms, is ideally suited to this task. It is often described as the "screwdriver of the electronic engineer" because the oscilloscope is the most fundamental tool that technical professionals apply to the problem of trying to understand the details of the operation of their electronic circuit or device. So, what is an oscilloscope?

The oscilloscope is an electronic instrument that presents a high-fidelity graphical display of the rapidly changing voltage at its input terminals.

The most frequently used display mode is voltage vs. time. This is not the only display that could be used, nor is it the display that is best suited for all situations. For example, the oscilloscope could be called on to produce a display of two changing voltages plotted one against the other, such as a Lissajous display. To accurately display rapidly changing signals, the oscilloscope is a high bandwidth device. This means that it must be capable of displaying the high-order harmonics of the signal being applied to its input terminals in order to correctly display that signal.

The Oscilloscope Block Diagram

The oscilloscope contains four basic circuit blocks: the vertical amplifier, the time base, the trigger, and the display. This section treats each of these in a high-level overview. Many textbooks exist that cover the details of the design and construction of each of these blocks in detail [1]. This discussion will cover these blocks in enough detail so that readers can construct their own mental model of how their operation affects the application of the oscilloscope for their voltage measurement application. Most readers of this book have a mental model of the operation of the automatic transmission of an automobile that is sufficient for its successful operation but not sufficient for the overhaul or redesign of that component. It is the goal of this section to instill that level of understanding in the operation of the oscilloscope. Those readers who desire a deeper understanding will get their needs met in later sections.

Of the four basic blocks of the oscilloscope, the most visible of these blocks is the display with its *cathode-ray tube* (CRT). This is the component in the oscilloscope that produces the graphical display of the input voltage and it is the component with which the user has the most contact. Figure 1.21 shows the input signal is applied to the vertical axis of a cathode ray tube. This is the correct model for an analog oscilloscope but it is overly simplified in the case of the digital oscilloscope. The important thing to learn from this diagram is that the input signal will be operated on by the oscilloscope's vertical axis circuits so that it can be displayed by the CRT. The differences between the analog and digital oscilloscope are covered in later sections.

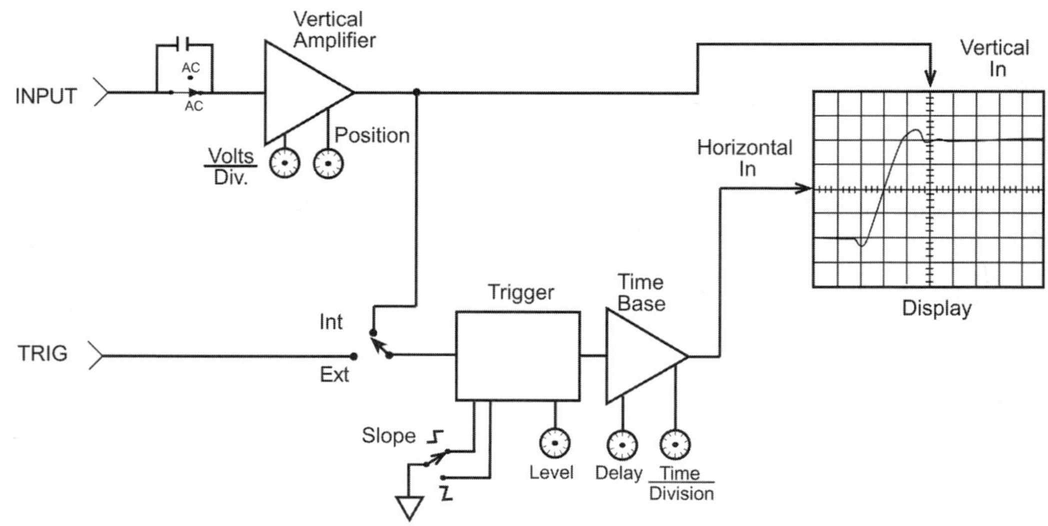

FIGURE 1.21 Simplified oscilloscope block diagram that applies to either analog or digital oscilloscopes. In the case of the digital oscilloscope, the vertical amplifier block will include the ADC and high-speed waveform memory. For the analog scope the vertical block will include delay lines with their associated drivers and a power amplifier to drive the CRT plates.

The *vertical amplifier* conditions the input signal so that it can be displayed on the CRT. The vertical amplifier provides controls of volts per division, position, and coupling, allowing the user to obtain the desired display. This amplifier must have a high enough bandwidth to ensure that all of the significant frequency components of the input signal reach the CRT.

The *trigger* is responsible for starting the display at the same point on the input signal every time the display is refreshed. It is the stable display of a complex waveform that allows the user of an oscilloscope to make judgments about that waveform and its implications as to the operation of the device under test.

The final piece of the simplified block diagram is the *time base.* This circuit block is also known as the horizontal system in some literature. The time base is the part of the oscilloscope that causes the input signal to be displayed as a function of time. The circuitry in this block causes the CRT beam to be deflected from left to right as the input signal is being applied to the vertical deflection section of the CRT. Controls for time-per-division and position (or delay) allow the user of the oscilloscope to adjust the display for the most useful display of the input signal. The time-per-division controls of most oscilloscopes provide a wide range of values, ranging from a few nanoseconds (10^{-9} s) to seconds per division. To get a feeling for the magnitude of the dynamic range of the oscilloscope's time base settings, keep in mind that light travels about 1 m in 3 ns.

The Oscilloscope as a Voltage Measurement Instrument

That the oscilloscope's vertical axis requires a wide bandwidth amplifier and its time base is capable of displaying events that are as short as a few nanoseconds apart, indicates that the oscilloscope can display rapidly changing voltages. Voltmeters, on the other hand, are designed to give their operator a numeric readout of steady-state or slowly changing voltages. Voltmeters are not well suited for displaying voltages that are changing levels very quickly. This can be better understood by examination of the operation of a voltmeter as compared to that of an oscilloscope. The analog voltmeter uses the magnetic field produced by current flowing through a coil to move the pointer against the force of a spring. This nearly linear deflection of the voltmeter pointer is calibrated by applying known standard voltages to its input. Therefore, if a constant voltage is applied to the coil, the pointer will move to a point where the magnetic force being produced by the current flowing in its coil is balanced by the force of the spring. If the input voltage is slowly changing, the pointer will follow the changing voltage. This mechanical deflection system limits the ability of this measurement device to the measurement of steady-state or very low-frequency changes in the voltage at its input terminals. Higher-frequency voltmeters depend on some type of conversion technique to change higher frequencies to a dc signal that can be applied to the meter's deflection coil. For example, a diode is used to rectify ac voltages to produce a dc voltage that corresponds to the average value of the ac voltage at the input terminals in average responding ac voltmeters.

The digital voltmeter is very much like the analog meter except that the mechanical displacement of the pointer is replaced with a digital readout of the input signal. In the case of the digital voltmeter, the input signal is applied to an analog-to-digital converter (ADC) where it is compared to a reference voltage and digitized. This digital value of the input signal is then displayed in a numerical display. The ADC techniques applied to voltmeters are designed to produce very accurate displays of the same signals that were previously measured with analog meters. The value of a digital voltmeter is its improved measurement accuracy as compared to that of its analog predecessors.

The oscilloscope will display a horizontal line displaced vertically from its zero-voltage level when a constant, or dc voltage is applied to its input terminals. The magnitude of this deflection of the oscilloscope's beam vertically from the point where it was operating with no input being applied is how the oscilloscope indicates the magnitude of the dc level at its input terminals. Most oscilloscopes have a graticule as a part of their display and the scope's vertical axis is calibrated in volts per division of the graticule. As one can imagine, this is not a very informative display of a dc level and perhaps a voltmeter with its numeric readout is better suited for such applications.

There is more to the scope–voltmeter comparison than is obvious from the previous discussion. That the oscilloscope is based on a wide-bandwidth data-acquisition system is the major difference between

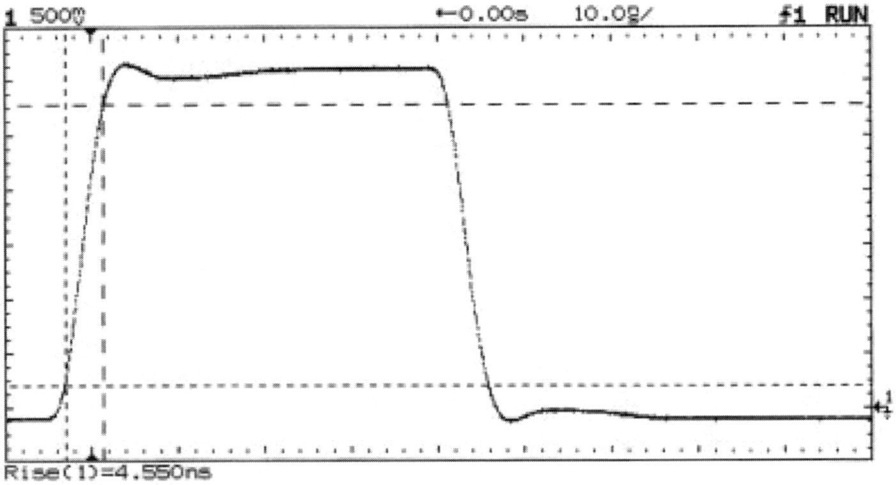

FIGURE 1.22 A typical complex waveform. This waveform is described by measurements of its amplitude, offset, risetime, falltime, overshoot, preshoot, and droop.

these two measurement instruments. The oscilloscope is designed to produce a high fidelity display of rapidly changing signals. This puts additional constraints on the design of the oscilloscope's vertical system that are not required in the voltmeter. The most significant of these constraints is that of a constant group delay. This is a rather complex topic that is usually covered in network analysis texts. It can be easily understood if one realizes the effect of group delay on a complex input signal.

Figure 1.22 shows such a signal. The amplitude of this signal is a dc level and the rising edge is made up of a series of high-frequency components. Each of these high-frequency components is a sine wave of specific amplitude and frequency. Another example of a complex signal is a square wave with a frequency of 10 MHz. This signal is made up of a series of odd harmonics of that fundamental frequency. These harmonics are sine waves of frequencies of 10 MHz, 30 MHz, 50 MHz, 70 MHz, etc. So, the oscilloscope must pass all of these high-frequency components to the display with little or no distortion. Group delay is the measure of the propagation time of each component through the vertical system. A constant group delay means that each of these components will take the same amount of time to propagate through the vertical system to the CRT, independent of their frequencies. If the higher-order harmonics take more or less time to reach the scope's deflection system than the lower harmonics, the resulting display will be a distorted representation of the input signal. Group delay (in seconds) is calculated by taking the first derivative of an amplifier's phase-vs.-frequency response (in radians/(l/s)). If the amplifier has a linearly increasing phase shift with frequency, the first derivative of its phase response will be a horizontal line corresponding to the slope of the phase plot (in seconds). Amplifier systems that have a constant group delay are known as Gaussian amplifiers. They have this name because their pass band shape resembles that of the bell curve of a Gaussian distribution function (Figure 1.23). One would think that the oscilloscope's vertical amplifier should have a flat frequency response, but this is not the case because such amplifiers have nonconstant group delay [1].

The oscilloscope's bandwidth specification is based on the frequency where the vertical deflection will be –3 dB (0.707) of the input signal. This means that if a constant 1 V sine wave is applied to the oscilloscope's input, and the signal's frequency is adjusted to higher and higher frequencies, the oscilloscope's bandwidth will be that frequency where its display of the input signal has been reduced to be 0.707 V. Noticable errors in ampitude measurements will start at 20% of the scope's bandwidth. The oscilloscope's error-free display of complex waveforms gives it poor voltage accuracy. For the measurement of dc and single frequency signals such as sine waves, other instruments can produce more accurate measurements.

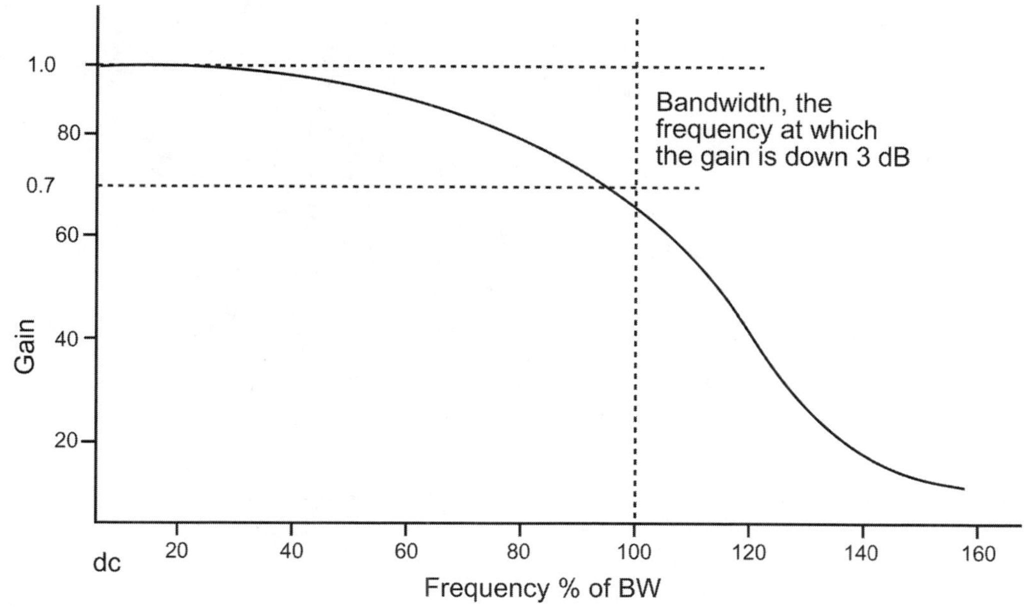

FIGURE 1.23 The Gaussian frequency response of the oscilloscope's vertical system which is not flat in its pass band. Amplitude measurements made at frequencies greater than 20% of the scope's bandwidth will be in error.

Conclusion: The voltmeter makes the most accurate measurements of voltages that are dc, slowly changing, or can be converted to a dc analog of their ac content. The oscilloscope is not the most accurate voltage measurement instrument, but it is well suited to measurements of voltages that are changing very rapidly as a function of time. Oscilloscopes are the instrument of choice for observing and characterizing these complex voltages.

Analog or Digital

The world of oscilloscopes is divided into two general categories: analog and digital. The first oscilloscopes were analog. These products are based on the direct-view vector cathode-ray tube (DVVCRT or CRT for short). The analog oscilloscope applies the input signal to the vertical deflection plates of the CRT where it causes the deflection of a beam of high-energy electrons moving toward the phosphor-coated faceplate. The electron beam generates a lighted spot where it strikes the phosphor. The intensity of the light is directly related to the density of the electrons hitting a given area of the phosphor. Because this analog operation is not based on any digitizing techniques, most people have little trouble creating a very accurate and simple mental model in their minds of its operation.

The analog oscilloscope produces a display of the input signal that is bright and easy to see under most conditions. It can also contain as many as 16 shades of gray-scale information. This means that an event that occurs less frequently will appear at a lower intensity in the display than another event that occurs more frequently. This oscilloscope does not produce a continous display of the input signal. It is blind during retrace and trigger hold-off times. Because the display depends on the production of visible light from the phosphor being excited by an electron beam, the display must be refreshed frequently. This makes the analog oscilloscope a low-dead-time display system that can follow rapidly changing signals. Also, there is little lag time in front panel control settings.

The analog oscilloscope is not without its shortcomings. The strength of the analog oscilloscope is its CRT, but this is also the source of its weaknesses. The biggest problem with analog scopes is their dependence on a display that is constantly being refreshed. This means that these scopes do not have any waveform storage. If the input signal fails to repeat frequently, the display will simply be a flash of light

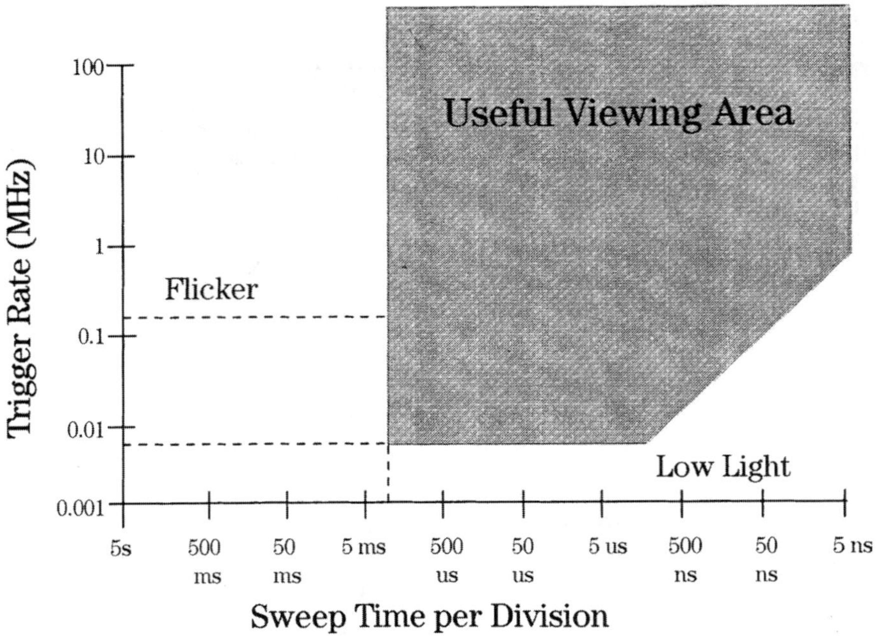

FIGURE 1.24 The operating range of the analog oscilloscope. This is a plot of input signal repetition rate from the lower limit of single shot to the full bandwidth of the scope plotted against sweep speed. The shaded area is the area where the analog oscilloscope will produce a usable display.

when the beam sweeps by the phosphor. If the signal's repetition rate falls below 100 Hz, the display will flicker annoyingly. Figure 1.24 shows a plot of the range of an input signal's repetition frequency range from a single-shot event to the full bandwidth of a scope vs. the scope's sweep speeds. The result is a map of the scope's operational area. Figure 1.24 shows that the analog oscilloscope fails to map onto the full range of possible input signals and sweep speeds.

Another problem of the analog oscilloscope is its inability to display information ahead of its trigger. This is a problem in applications where the only suitable trigger is at the end of the event of interest. Another limitation of analog scopes is their timing accuracy. The time base of the analog scope is based on the linearity of a voltage ramp. There are other sources of errors in the analog oscilloscope's horizontal axis, but the sweep nonlinearity is the major contributor. This results in these scopes having a timing accuracy of typically ±3% of their full-scale setting. Therefore, if the time base is set to 100 ns/div, in order to view a 100 ns wide pulse, the full scale will be 1000 ns or 1 ms. The accuracy of this pulse width measurement will be ±30 ns or ±30% of the pulse width!

The digital oscilloscope or digital storage oscilloscope (DSO) differs from its analog counterpart in that the input signal is converted to digital data and therefore it can be managed by an embedded microprocessor. The waveform data can have correction factors applied to remove errors in the scope's acquisition system and can then be stored, measured, and/or displayed. That the input signal is converted from analog to digital and manipulations are performed on it by a microprocessor results in people not having a good mental model of the digital oscilloscope's operation. This would not be a problem except for the fact that the waveform digitizing process is not totally free from errors, and a lack of a correct mental model of the scope's operation on the part of its user can increase the odds of a measurement error. To make matters worse, various manufacturers of these products make conflicting claims, making it easy to propagate incorrect mental models of the digital scope's operation. It is the intention of this presentation to give the information needed to create a mental model of the operation of these devices that will enable the user to perform error-free measurements with ease.

The digital storage oscilloscope offers many advantages over its analog counterpart. The first is accuracy. The voltage measurement accuracy of the digital oscilloscope is better than that of an analog scope because the microprocessor can apply correction factors to the data to correct for errors in the calibration of the scope's vertical system. The timing accuracy of a digital oscilloscope is an order of magnitude better than that of an analog scope. The digital scope can store the waveform data for comparison to other test results or uploading to a computer for analysis or project documentation. The digital oscilloscope does not depend on the input signal being continuously updated to produce an easily viewable display. A single-shot event is displayed at the same brightness level as a signal that repeats in time periods corresponding to the full bandwidth of the scope.

The disadvantages of the digital oscilloscope are its more complex operation, aliasing, and display performance. The analog-to-digital conversion process [1] is used to convert the input signal into a series of discrete values, or samples, uniformly spaced in time, which can be stored in memory. Voltage resolution is determined by the total number of codes that can be produced. A larger number permits a smoother and more accurate reproduction of the input waveform but increases both the cost and difficulty in achieving a high sample frequency. Most digital oscilloscopes provide 8-bit resolution in their ADC. As the ADC's sampling speed is increased, the samples will be closer together, resulting in smaller gaps in the waveform record.

All digital scopes are capable of producing an aliased display. Some models are more prone to this problem than others, but even the best will alias under the right conditions. An alias is a lower frequency false reproduction of the input signal resulting from under-sampling, i.e., sampling less than the Nyquist frequency. The display of the digital scope is based on computer display technology. This results in a display that is very bright and easy to see, even under conditions where an analog scope would have difficulty in producing a viewable display. The disadvantage of the digital scope's display is its lower horizontal resolution. Most of the scopes on the market have a raster scan display with a resolution of 500 lines, less than half the resolution of an analog scope's display. This is not a problem in most applications. It could become a factor where very complex waveforms, such as those found in TV systems, are being analyzed. Many digital scopes have display systems that exhibit large dead- or blind-times. Scopes based on a single CPU will be able to display their waveform data only after the CPU has finished all of its operations. This can result in a display that is unresponsive to front panel control inputs as well as not being able to follow changes in the input signal.

Table 1.2 shows that both analog and digital oscilloscopes have relative advantages and disadvantages. All the major producers of oscilloscopes are pushing the development of digital scopes in an attempt to overcome their disadvantages. All the major producers of these products believe that the future is digital. However, a few manufacturers produce scopes that are both analog and digital. These products appear to have the best of both worlds; however, they have penalties with respect to both cost and complexity of operation.

TABLE 1.2 A Comparison of Analog and Digital Oscilloscopes

	Analog Oscilloscope	Digital Oscilloscope
Operation	Simple	Complex
Front panel controls	Direct access knobs	Knobs and menus
Display	Real-time vector	Digital raster scan
Gray scales	>16	>4
Horizontal resolution	>1000 lines	500 lines
Dead-time	Short	Can be long
Aliasing	No	Yes
Voltage accuracy	±3% of full scale	±3% of full scale
Timing accuracy	±3% of full scale	±0.01% of full scale
Single shot capture	None	Yes
Glitch capture	Limited	Yes
Waveform storage	None	Yes
Pretrigger viewing	None	Yes
Data out to a computer	No	Yes

One of the driving forces making scope manufacturers believe that the future of the digital oscilloscope is bright is that modern electronic systems are becoming ever more digital in nature. Digital systems place additional demands on the oscilloscope that exceed the capabilities of the analog scope. For example, often in digital electronic systems, there is a need to view fast events that occur at very slow or infrequent rates. Figure 1.24 shows that these events fail to be viewable on analog scopes. Another common problem with digital systems is the location of trigger events. Often the only usable trigger is available at the end of the event being viewed. Analog scopes can only display events that occur after a trigger event. The rapid growth of digital electronics that occurred in the late 1990s is being attributed to the lowering of the cost of single-chip microcontrollers. These devices, which contain a complete microprocessor on one integrated circuit, are responsible for the "electronics everywhere" phenomenon, where mechanical devices are becoming electronic as well as those devices that were previously electrical in nature. In 1996, Hewlett Packard introduced a new class of oscilloscope designed to meet the unique needs of the microcontrollerbased applications. This new class of oscilloscope is known as the mixed signal oscilloscope or MSO [2].

Voltage Measurements

Voltage measurements are usually based on comparisons of the waveform display to the oscilloscope's graticule. Measurements are made by counting the number of graticule lines between the end-points of the desired measurement and then multiplying that number by the sensitivity setting. This was the only measurement available to most analog scope users, and it is still used by those performing troubleshooting with their digital scope as a time-saving step. (Some late-model analog oscilloscopes incorporate cursors to enhance their measurement ability.) For example, a waveform that is 4.5 divisions high at a vertical sensitivity of 100 mV/div would be 450 mV high.

Switching the scope's coupling between ac and dc modes will produce a vertical shift in the waveform's position that is a measurement of its dc component. This technique can be applied to either analog or digital scopes. Simply note the magnitude of the change in waveform position and multiply by the channel's sensitivity.

Additional measurements can be performed with an analog oscilloscope but they usually require more skill on the part of the operator. For example, if the operator can determine the location of the top and base of a complex waveform, its amplitude can be measured. Measurements based on percentages can be made using the scope's vernier to scale the waveform so that its top and bottom are 5 divisions apart. Then, each division represents 20% of the amplitude of the waveform being studied. The use of the vernier, which results in the channel being uncalibrated, prevents performance of voltage measurements. Many analog scopes have a red light to warn the operator that the scope is uncalibrated when in vernier mode.

The digital oscilloscope contains an embedded microprocessor that automates the measurement. This measurement automation is based on a histogramming technique, where a histogram of all the voltages levels in the waveform are taken from the oscilloscope's waveform data. The histogram is a plot of the voltage levels in the waveform plotted against the number of samples found at each voltage level. Figure 1.25 shows the histogramming technique being applied to the voltage measurements of complex waveforms.

Understanding the Specifications

The oscilloscope's vertical accuracy is one place that a person's mental model of the scope's operation can lead to measurement trouble. For example, the oscilloscope's vertical axis has a frequency response that is not flat across its pass band. However, as noted above, the scope has a Gaussian frequency response to produce the most accurate picture of complex signals. This means that the oscilloscope's accuracy specification of ±3% is a dc-only specification. If one were to attempt to measure the amplitude of a signal whose frequency is equal to the bandwidth of the scope, one would have to add another 29.3% to the error term, for a total error of ±32.3%. This is true for both analog and digital oscilloscopes. This limitation can be overcome by carefully measuring the frequency response of the oscilloscope's vertical

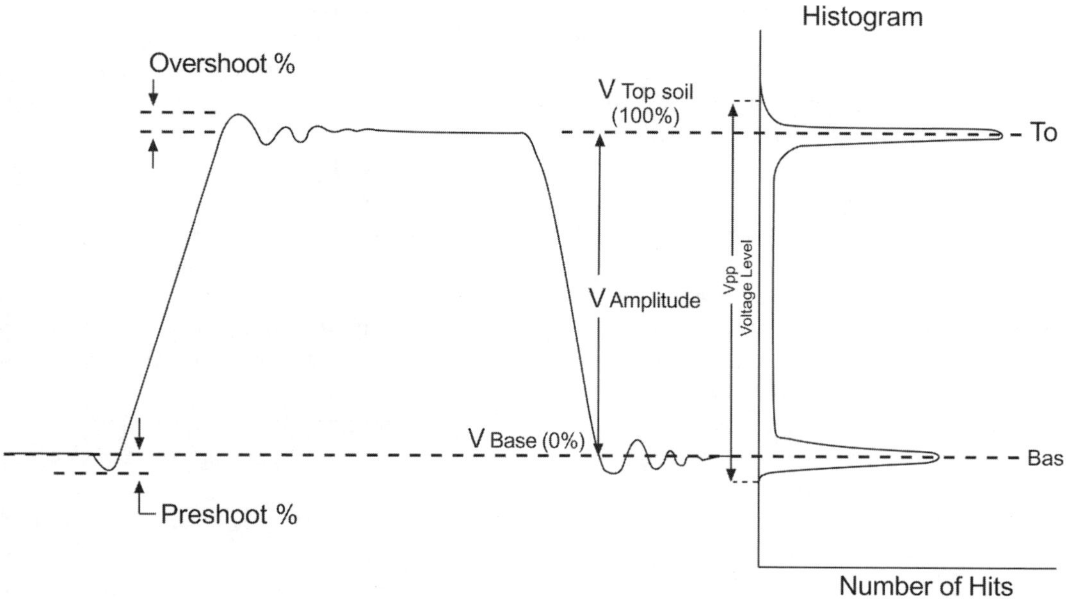

FIGURE 1.25 Voltage histograms as applied by a digital oscilloscope. The complex waveform is measured by use of the voltage histogram. This histogram is a plot of each voltage level in the display and the number of data points at that level.

channels. One will need to repeat this process every time the scope is serviced or calibrated, because the various high-frequency adjustments that may need to be made in the scope's vertical axis will affect the scope's frequency response. One is probably asking, why don't the manufacturers do this for me? The answer is twofold. The first is cost, and the second is that this is not the primary application of an oscilloscope. There are other instruments that are much better suited to the measurement of high-frequency signals. The spectrum analyzer would be this author's first choice.

Additionally, the vertical accuracy is a full-scale specification. This means that at 1 V/div, the full-scale value is typically 8 V. The measurement error for a scope with a ±3% specification under these conditions will be ±0.24 V. If the signal being measured is only 1 V high, the resulting measurement will be ±24% of reading. Check the manual for the scope being used, as some manufacturers will specify full-scale as being 10 or even 10.2 divisions. This will increase the error term because the full-scale term is larger.

In digital oscilloscopes, the vertical accuracy is often expressed as a series of terms. These attempt to describe the analog and digital operations the scope performs on the input signal. Terms might include digitizing resolution, gain, and offset (sometimes called as position). They also might be called out as single and dual cursor accuracies. The single cursor accuracy is a sum of all three terms. In the dual cursor case, where the voltage measurement is made between two voltage cursors, the offset term will cancel out, leaving only the digitizing resolution and gain errors. For example, the Hewlett Packard model 54603B has a single cursor accuracy specification of ±1.2% of full scale, ±0.5% of position value, and a dual cursor specification of ±0.4% of full scale.

HINT: Always try to make the voltage measurements on the largest possible vertical and widest possible display of the signal.

The horizontal accuracy specifications of analog and digital scopes are very different; however, both are based on a full-scale value. In the analog scope, many manufacturers limit accuracy specifications to only the center eight divisions of their display. This means that a measurement of a signal that starts or ends in either the first or ninth graticule, will be even more error prone than stated in the scope's specifications. To the best of this author's knowledge, this limitation does not apply to digital scopes. The horizontal specifications of digital scopes are expressed as a series of terms. These might include the crystal accuracy, horizontal display resolution, and trigger placement resolution. These can be listed as cursor accuracy. For example, the Hewlett Packard model 54603B has a horizontal cursor accuracy specification of ±0.01% ±0.2% full-scale ±200 ps. In this example, the first term is the crystal accuracy, the second is the display resolution (500 lines), and the final term is twice the trigger placement error. By comparing the analog and digital scopes' horizontal specifications, it can be seen that in either case, the measurement is more accurate if it can be made at full screen. The digital scope is more accurate than its analog counterpart.

Digital scopes also have acquisition system specifications. Here is another place where the operator's mental model of the operation of a digital scope can produce measurement errors. All manufacturers of digital scopes specify the maximum sampling speed of their scope's acquisition system as well as its memory depth and number of bits. The scope's maximum sampling speed does not apply to all sweep speeds, only memory depth and number of bits applies to all sweep speeds. The scope's maximum sampling speed applies only to its fastest sweep speeds.

The complexity of the digital scope results from the problem of having to sample the input. There is more to be considered than Nyquist's Sampling Theorem in the operation of a digital scope. For example, how does the scope's maximum sampling rate relate to the smallest time interval that the scope can capture and display? A scope that samples at 100 MSa s^{-1} takes a sample every 10 ns; therefore, in principle, it cannot display any event that is less than 10 ns wide because that event will fall between the samples. In practice, however, this limit can — under certain circumstances — be extended. If the scope is operating in an "equivalent time" or "random repetitive" mode and if the signal is repetitive, even if very infrequently, the scope will be able to capture any event that is within its vertical system bandwidth. Figure 1.26 shows an infrequently occurring pulse that is 25 ns wide embedded into a data stream being captured and displayed on an oscilloscope with a maximum sampling speed of 20 MSa s^{-1} (sampling interval of 50 ns). Figure 1.26(b) shows this pulse at a faster sweep speed. An analog scope would produce a similar display of this event, with the infrequent event being displayed at a lower intensity than the rest of the trace. Notice that the infrequent event does not break the baseline of the trace.

The correct mental model of the digital scope's ability to capture signals needs to be based on the scope's bandwidth, operating modes, and timing resolution. It is the timing resolution that tells the operator how closely spaced the samples can be in the scope's data record.

The most common flaw in many mental models of the operation of a digital scope is related to its maximum sampling speed specification. As noted, the maximum sampling speed specification applies only to the scope's fastest sweep speeds. Some scope manufacturers will use a multiplex A/D system that operates at its maximum sampling speed only in single-channel mode. The scope's memory depth determines its sampling speed at the sweep speed being used for any specific measurement. The scope's memory depth is always equal to the scope's horizontal full-scale setting. For scopes with no off-screen memory, this is 10′ the time base setting. If the scope has off-screen memory, this must be taken into account. For example, assume that one has two scopes with a maximum sampling speed of 100 MSa s^{-1}. One scope has a memory depth of 5 K points and the other only 1 K. At a sweep speed of 1 ms per division, both scopes will be able to store data into their memory at their full sampling speed, and each will be storing 100 data points per division, for a total of 1000 data points being stored. The scope with the 5 K memory will have a data point in one of every 5 memory locations, and the scope with the 1 K memory will have a data point in every memory location. If one reduces the sweep speed to 5 ms/div, the deeper memory scope will now fill every one of its memory locations with data points separated by 10 ns. The scope with only 1 K of memory would produce a display only 2 divisions wide if its sampling speed is not reduced. Scope designers believe that scope users expect to see a full-length sweep at every sweep speed. Therefore, the 1 K scope must reduce its sampling speed to one sample every 50 ns, or 20 MSa s^{-1}, to be able to fill its memory with a full sweep width of data. This 5:1 ratio of sampling speeds between these two scopes will be maintained as their time bases are set to longer and longer sweeps. For example, at 1 s/div, the 5 K scope will be sampling at 500 samples per second, while the 1 K scope will be sampling at only 100 samples per second. One can determine a scope's sampling speed for any specific time base setting from Equation 1.51.

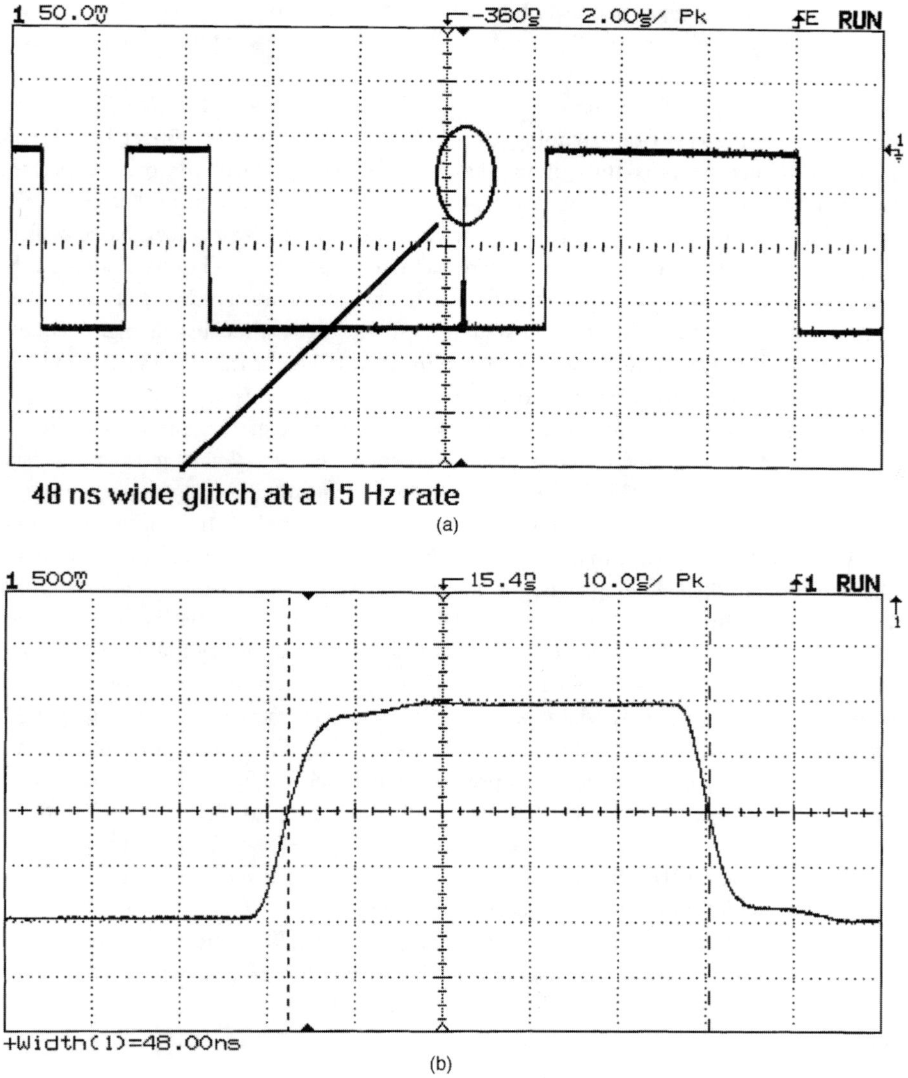

FIGURE 1.26 An infrequently occurring event as displayed on a digital oscilloscope with random repetitive sampling. (a) The event embedded in a pulse train. (b) Shows the same event at a faster sweep speed. The fact that the waveform baseline is unbroken under the narrow pulse indicates that it does not occur in every sweep. The width of this pulse is less than half the scope's sampling period in (b). Both traces are from a Hewlett Packard model 54603B dual channel 60 MHz scope.

$$S\left(\text{samples/second}\right) = \frac{\text{memory depth}\left(\text{samples}\right)}{\text{full-scale time base}\left(\text{seconds}\right)}, \tag{1.51}$$

or the scope's maximum sampling speed, whichever is less

One must look closely at the application to determine if a specific scope is best suited to that application. As a rule, the deeper the memory, the faster the scope will be able to sample the signal at any given time base setting. Memory depth is not free. High-speed memory required to be able to store the data out of the scope's A/D is costly, and deeper memory takes longer to fill, thus reducing the scope's display update rate. Most scopes that provide memory depths of 20 K or more will also give the user a memory depth selection control so that the user can select between fast and deep. (In 1996, Hewlett Packard Co.

introduced two scopes based on an acquisition technology known as MegaZoom (TM) [10] that removes the need for a memory depth control.) A correct mental model for the sampling speed of a digital scope is based on Equation 1.51 and not just on the scope's maximum performance specifications.

Some digital oscilloscopes offer a special sampling mode known as *peak detection*. Peak detection is a special mode that has the effect of extending the scope's sampling speed to longer time records. This special mode can reduce the possibility of an aliased display. The performance of this special mode is specified as the minimum pulse width that the peak detection system can capture. There are several peak detection systems being used by the various manufacturers. Tektronix has an analog-based peak detection system in some of its models, while Hewlett Packard has a digital system in all of its models. Both systems perform as advertised, and they should be evaluated in the lab to see which system best meets one's needs. There is a downside to peak detection systems and that is that they display high-frequency noise that might not be within the bandwidth of the system under test. Figure 1.27 shows a narrow pulse being captured by peak detection and being missed when the peak detection is off.

What effect does display dead-time have on the oscilloscope's voltage measurement capabilities? Display dead-time applies to both analog and digital oscilloscopes, and it is that time when the oscilloscope is not capturing the input signal. This is also a very important consideration in the operation of a digital scope because it determines the scope's ability to respond to front-panel control commands and to follow changing waveforms. A digital scope that produces an incorrect display of an amplitude-modulated signal is not following this rapidly changing signal because its display update rate is too low. Sampling speed is not related to display update rate or dead-time. Display dead-time is a function of the scope's ability to process the waveform data from its A/D and plot it on the display. Every major oscilloscope manufacturer has been working on this problem. Tektronix offers a special mode on some of its products known as InstaVu (TM) [4]. This special mode allows these scopes to process up to 400,000 waveforms per second to their display. Hewlett Packard has developed a multiple parallel processor technology [5] in the HP 54600 series of benchtop scopes that provides a high-speed, low dead-time display in a lowcost instrument. These instruments can plot 1,500,000 points per second to their display and they have no dead-time at their slower sweep speeds. LeCroy has been applying the Power PC as an embedded processor for its scopes to increase display throughput. There are other special modes being produced by other vendors, so be sure to understand what these can do before selecting an oscilloscope. Figure 1.28 shows the effect of display update rate on a rapidly changing waveform. An amplitude-modulated signal is displayed with a high-speed display and with the display speed reduced by the use of hold-off.

Triggering

The trigger of the oscilloscope has no direct effect on the scope's ability to measure a voltage except that the trigger does enable the oscilloscope to produce a stable display of the voltage of interest. Ref. [6] presents a thorough discussion of this subject.

Conclusion

The mental model that oscilloscope users have created in their minds of the oscilloscope's operation can be helpful in reducing measurement errors. If the operator's mental model is based on the following facts, measurement errors can be minimized:

- Oscilloscopes have a frequency response that affects measurement accuracy.
- Digital scopes are more accurate than analog scopes.
- Analog scopes do not have continuous displays.
- Oscilloscope accuracy specifications always contain a percent of full-scale term.
- Measurements should be made at the largest possible deflection in order to minimize errors.
- Maximum sampling speed is available only at the scope's fastest sweep speeds.
- Deeper memory depth allows faster sampling at more sweep speeds.

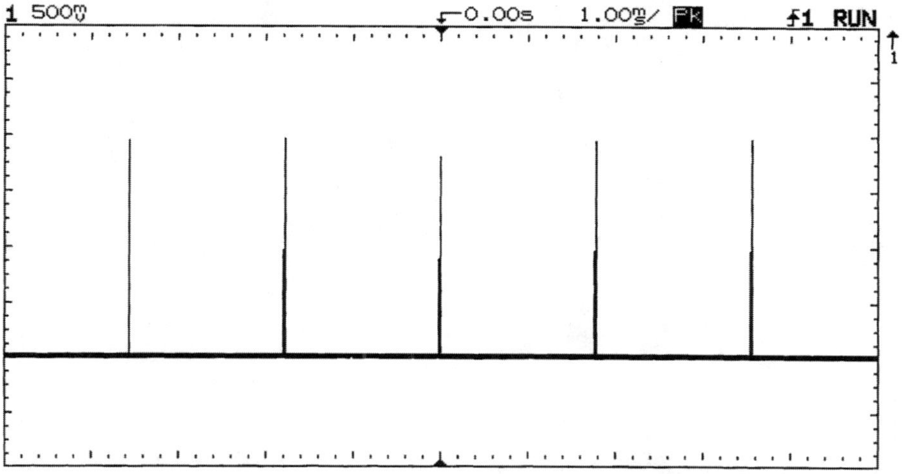

(a)

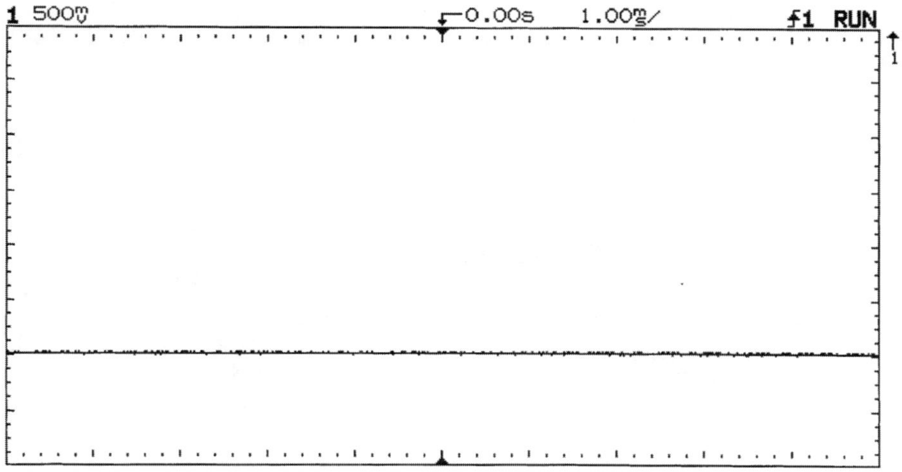

(b)

FIGURE 1.27 Peak detection. This special mode has the effect of increasing the scope's sampling speed at time base settings where it would be decimated. In operation, each memory location contains either the maximum or minimum value of the waveform at that location in time. (a) A series of 300 ns wide pulses being captured at a slow sweep speed; (b) the same setup with peak detection disabled. These narrow pulses would appear as intermittent pulses if the scope could be seen in operation with peak detection disabled.

- All digital scopes can produce aliases, some more than others.
- Display dead-time is an important characteristic of digital scopes that is often not specified.
- Display dead-time affects measurement accuracy because it can cause a distorted display.
- The scope with the highest maximum sampling speed specification might not be the most accurate or have the lowest display dead-time.
- The operator must have some knowledge of the signals being measured to be able to make the best possible measurements.

The person who has the mental model of the oscilloscope that takes these factors into account will be able to purchase the scope that is best suited to his/her application and not spend too much money on unnecessary performance. In addition, that person will be able to make measurements that are up to the full accuracy capabilities of the scope.

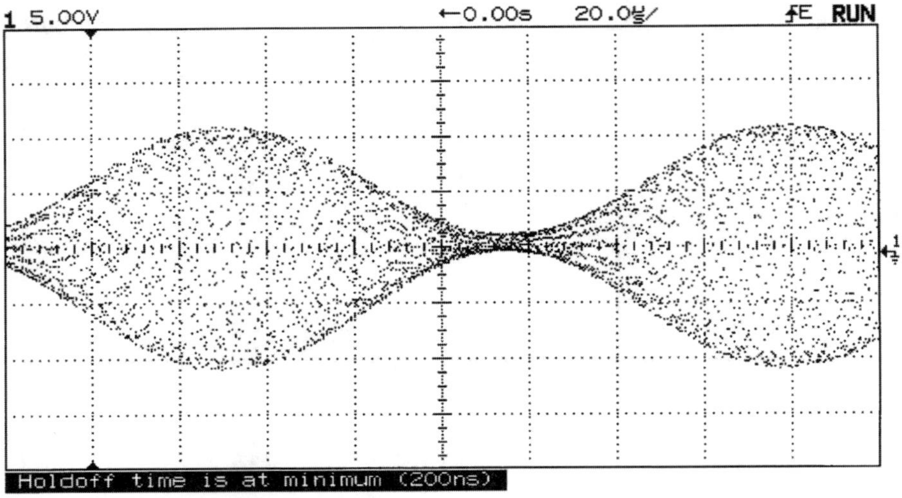

(a)

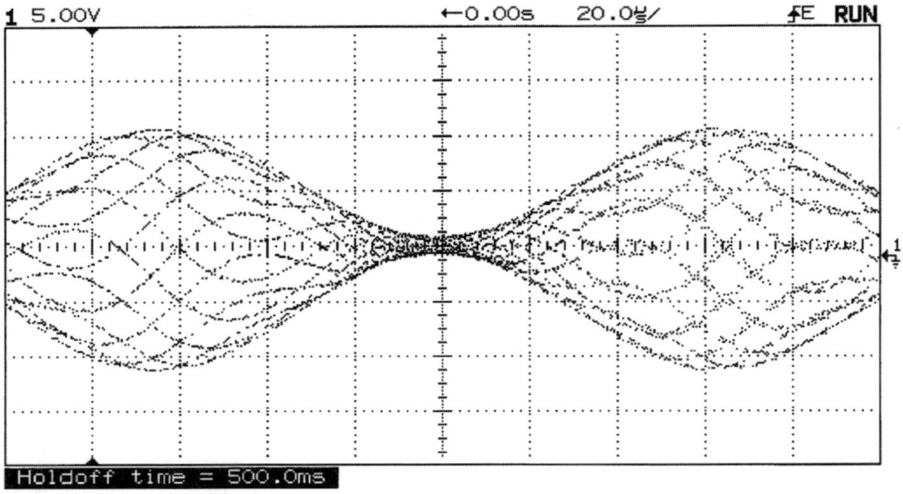

(b)

FIGURE 1.28 Display dead-time. The time that an oscilloscope is blind to the input signal has an effect on the scope's ability to correctly display rapidly changing signals. (a) An amplitude-modulated signal with a high-speed display; (b) the same signal with the dead-time increased by use of hold-off.

Selecting the Oscilloscope

There are ten points to consider when selecting an oscilloscope. This author has published a thorough discussion of these points [7] and they are summarized as follows:

1. **Analog or Digital?** There are a few places where the analog scope might be the best choice, and the reader can make an informed selection based on the information presented here.
2. **How much bandwidth?** This is a place where the person selecting an oscilloscope can save money by not purchasing more bandwidth than is needed. When analog oscilloscopes were the only choice, many people were forced to purchase more bandwidth than they needed because they needed to view infrequent or low repetition signals. High-bandwidth analog scopes had brighter CRTs so that they were able to display high-frequency signals at very fast time base settings. At a sweep speed of 5 ns/div, the phosphor is being energized by the electron beam for 50 ns, so the

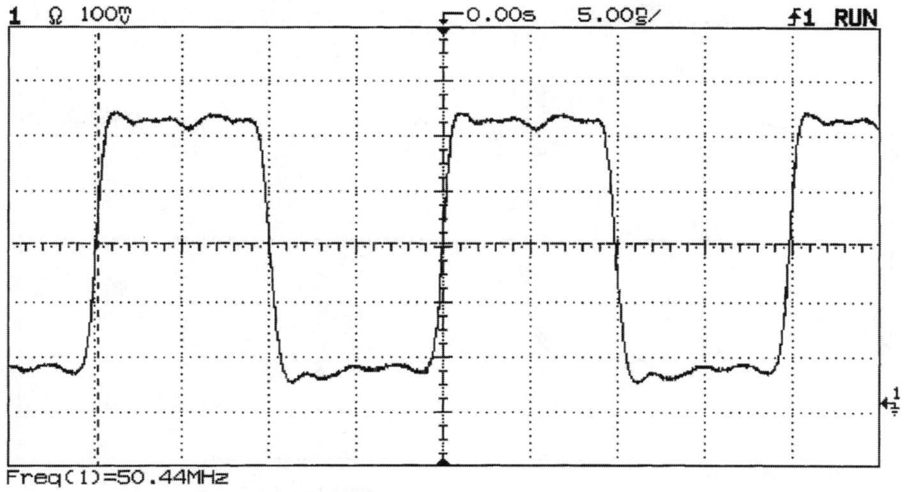

(a) **BW = 500 Mhz**

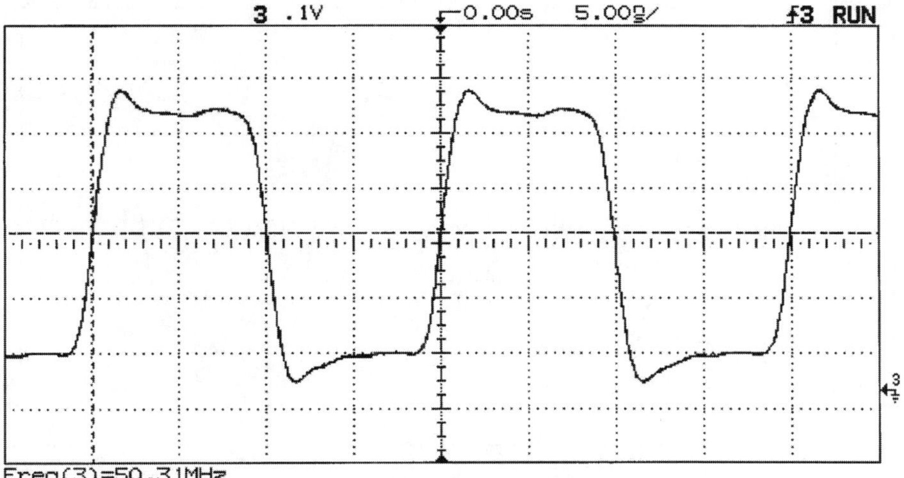

(b) **BW = 250 Mhz**

FIGURE 1.29 The effect of the scope's bandwidth is shown in this set of waveforms. The same 50 MHz square wave is shown as it was displayed on scopes of 500 MHz in Figure 1.28(a) all the way down to 20 MHz in Figure 1.29(e). Notice that the 100 MHz scope produced a usable display although it was missing the high-frequency details of the 500 MHz display. The reason that the 100 MHz scope looks so good is the fact that its bandwidth is slightly greater than 100 MHz. This performance, which is not specified on any data sheet, is something to look for in any evaluation.

electron beam had to be very high energy to produce a visible trace. This situation does not apply to digital scopes. Now, one needs to be concerned only with the bandwidth required to make the measurement. Figure 1.29 shows the effect of oscilloscope bandwidth on the display of a 50 MHz square wave.

The oscilloscope's bandwidth should be >2× the fundamental highest frequency signal to be measured.

The bandwidth of the scope's vertical system can affect the scope's ability to correctly display narrow pulses and to make time interval measurements. Because of the scope's Gaussian frequency response, one can determine its ability to correctly display a transient event in terms of risetime with Equation 1.52.

$$t_r = 0.35/\text{BW} \tag{1.52}$$

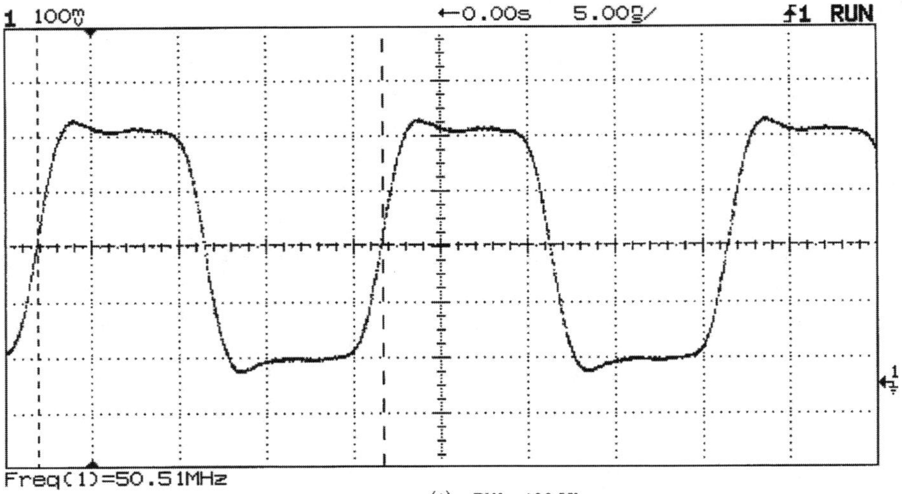

(c) BW = 100 Mhz

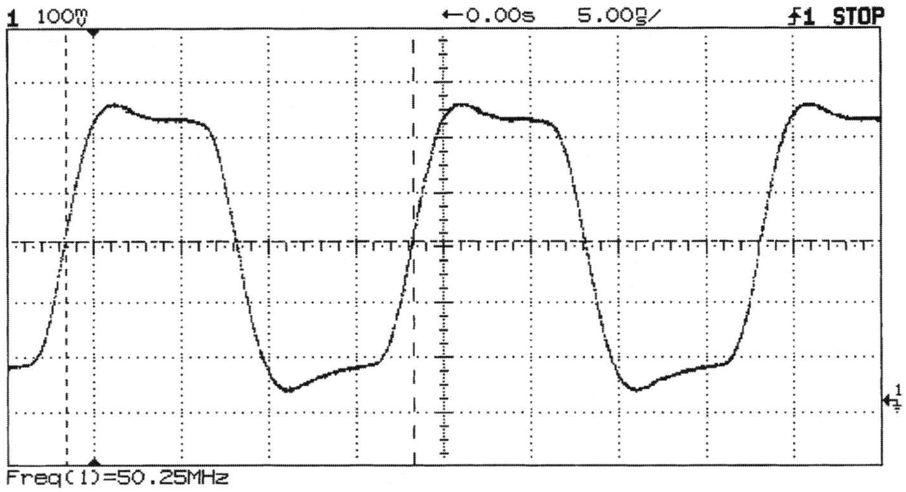

(d) BW = 60 Mhz

FIGURE 1.29 (continued)

Therefore, a 100 MHz scope will have a risetime of 3.5 ns. This means that if the scope were to have a signal at its input with zero risetime edges, it would be displayed with 3.5 ns edges. This will affect the scope's measurements in two ways. First is narrow pulses. Figure 1.30 shows the same 5 ns wide pulse being displayed on oscilloscopes of 500 MHz and 60 MHz bandwidths, and the effect of the lower bandwidth on this event that is closest to the risetime of the slower scope is apparent.

The second is fast time interval measurements. A measurement of signal risetime is an example. The observed risetime on the scope's display is according to Equation 1.53.

$$t_{\text{observed}} = \left(t_{\text{signal}}^{\ 2} + t_{\text{scope}}^{\ 2}\right)^{1/2} \tag{1.53}$$

If a 10 ns risetime were to be measured with a 100 MHz scope, one would obtain a measurement of 10.6 ns based on Equation 1.53. The scope would have made this measurement with a 6% reading error before any other factors, such as time base accuracy, are considered.

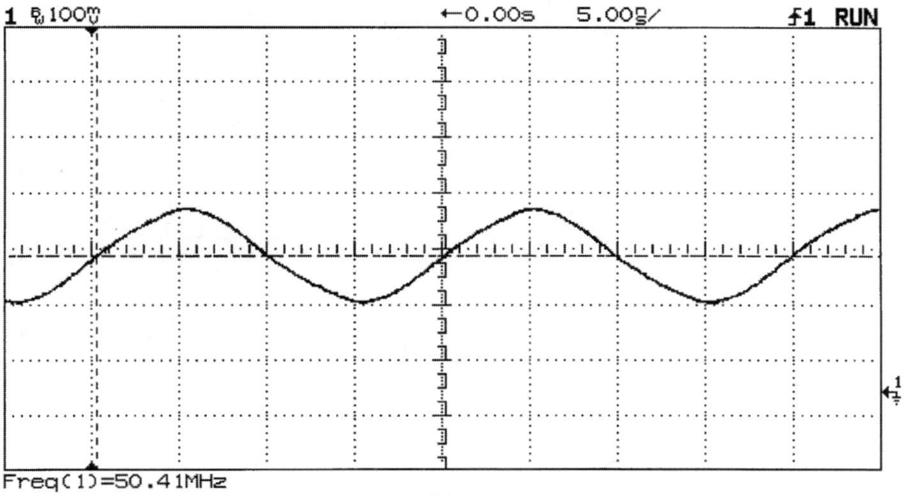

(e) **BW = 20 Mhz**

FIGURE 1.29 (continued)

The scope's risetime should be at least no more than 1/5 of the shortest time interval to be measured. For time interval measurements, this should be >1/10.

3. **How many channels?** Most oscilloscopes in use today are dual-channel models. In addition, there are models described as being 2+2 and four channels. This is one time where 2+2 is not equal to 4. The 2+2 models have limited features on two of their channels and cost less than 4-channel models. Most oscilloscope suppliers will hold the 4-channel description only for models with four full-featured channels, but the user should check the model under consideration so as to be sure if it is a 4- or 2+2 model. Either of the four channel classes is useful for applications involving the testing and development of digital-based systems where the relationship of several signals must be observed.

   Hewlett Packard introduced a new class of oscilloscopes that is tailored for the applications involving both analog and digital technologies, or mixed-signal systems. The mixed signal oscilloscope (MSO) [4] provides 2 scope channels and 16 logic channels so that it can display both the analog and digital operation of a mixed-signal system on its display.
4. **What sampling speed?** Do not simply pick the scope with the highest banner specification. One needs to ask, what is the sampling speed at the sweep speeds that my application is most likely to require? As observed in Equation 1.51 the scope's sampling speed is a function of memory depth and full-scale time base setting. If waveforms are mostly repetitive, one can save a lot of money by selecting an oscilloscope that provides equivalent time or random repetitive sampling.
5. **How much memory?** As previously discussed, memory depth and sampling speed are related. The memory depth required depends on the time span needed to measure and the time resolution required. The longer the time span to be captured and the finer the resolution required, the more memory one will need. High-speed waveform memory is expensive. It takes time to process a longer memory, so the display will have more dead-time in a long memory scope than a shallow memory model. All the suppliers of deep memory scopes provide a memory depth control. They provide this control so that the user can choose between a high-speed display and deep memory for the application at hand. Hewlett Packard introduced MegaZoom (TM) technology [3] in 1996; it produces a high-speed low dead-time display with deep memory all the time.
6. **Triggering?** All scope manufacturers are adding new triggering features to their products. These features are important because they allow for triggering on very specific events. This can be a valuable troubleshooting tool because it will let the user prove whether a suspected condition

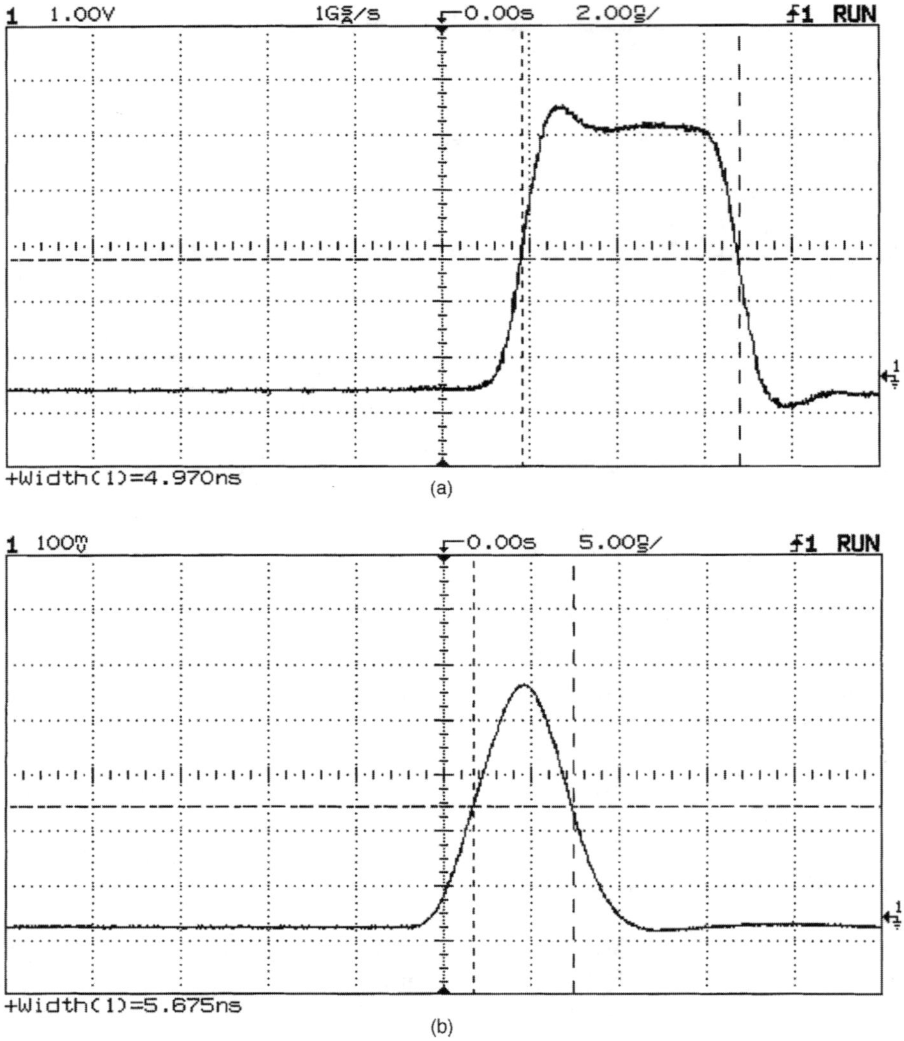

FIGURE 1.30 Bandwidth and narrow events. (a) A 5 ns wide pulse as displayed on a 500 MHz scope; (b) the same pulse displayed on a 60 MHz scope. The 60 MHz scope has a risetime of 5.8 ns, which is longer than the pulse width. This results in the pulse shape being incorrectly displayed and its amplitude being in error.

exists or not. Extra triggering features add complexity to the scope's user interface; so be sure to try them out to make sure that they can be applied.

7. **Trustworthy display?** Three factors critically affect a scope's ability to display the unknown and complex signals that are encountered in oscilloscope applications. If the user loses confidence in the scope's ability to correctly display what is going on at its probe tip, productivity will take a real hit. These are display update rate, dead-time, and aliasing.

   Because all digital scopes operate on sampled data, they are subject to aliasing. An alias is a false reconstruction of the signal caused by under-sampling the original. An alias will always be displayed as a lower frequency than the actual signal. Some vendors employ proprietary techniques to minimize the likelihood of this problem occurring. Be sure to test any scope being considered for purchase on your worse-case signal to see if it produces a correct or aliased display. Do not simply test it with a single-shot signal that will be captured at the scope's fastest sweep speed because this will fail to test the scope's ability to correctly display signals that require slower sweep speeds.

TABLE 1.3 Major Suppliers of Oscilloscopes and their Web Addresses

Vendor	Description	Web address
B&K Precision 6460 W. Cortland St. Chicago, IL 60635	Analog and digital scopes and Metrix scopes in France	http://bkprecision.com
Boonton Electronics Corp. 25 Estmans Road P.O. Box 465 Parsippany, NJ 07054-0465	U.S. importer for Metrix analog, mixed analog, and digital scopes from France	http://www.boonton.com
Fluke P.O. Box 9090 Everett, WA 98206-9090	Hand-held, battery-powered scopes (ScopeMeter), analog scopes, and CombiScopes(R)	http://www.fluke.com
Gould Roebuck Road, Hainault, Ilford, Essex IG6 3UE, England	200 MHz DSO products	http://www.gould.co.uk
Hewlett Packard Co. Test & Measurement Mail Stop 51LSJ P.O. Box 58199 Santa Clara, CA 95052-9952	A broad line of oscilloscopes and the Mixed Signal oscilloscope for technical professionals	http://www.tmo.hp.com/tmo/pia search on "oscilloscopes"
LeCroy Corp. 700 Chestnut Ridge Road Chestnut Ridge, NY 10977	Deep memory oscilloscopes for the lab	http://www.lecroy.com
Tektronix Inc. Corporate Offices 26600 SW Parkway P.O. Box 1000 Watsonville, OR 97070-1000	The broad line oscilloscope supplier with products ranging from hand-held to high-performance lab scopes	http://www.tek.com/measurement search on "oscilloscopes"
Yokogawa Corp. of America Corporate offices Newnan, GA 1-800-258-2552	Digital oscilloscopes for the lab	http://www.yca.com

8. **Analysis functions?** Digital oscilloscopes with their embedded microprocessors have the ability to perform mathematical operations that can give additional insight into waveforms. These operations often include addition, subtraction, multiplication, integration, and differentiation. An FFT can be a powerful tool, but do not be misled into thinking it is a replacement for a spectrum analyzer. Be sure to check the implementation of these features in any scope being considered. For example, does the FFT provide a selection of window functions? Are these analysis functions implemented with a control system that only their designer could apply?
9. **Computer I/O?** Most of the digital scopes on the market today can be interfaced to a PC. Most of the scope manufacturers also provide some software that simplifies the task of making the scope and PC work together. Trace images can be incorporated into documents as either PCX or TIF files. Waveform data can be transferred to spreadsheet applications for additional analysis. Some scope models are supplied with a disk drive that can store either waveform data or trace images.
10. **Try it out?** Now one has the information to narrow oscilloscope selection to a few models based on bandwidth, sampling speed, memory depth, and budget requirements. Contact the scope vendors (Table 1.3) and ask for an evaluation unit. While the evaluation unit is in the lab, look for the following characteristics:

- Control panel responsiveness: Does the scope respond quickly to inputs or does it have to think about it for a while?
- Control panel layout: Are the various functions clearly labeled? Does the user have to refer to the manual even for simple things?
- Display speed: Turn on a couple of automatic measurements and check that the display speed remains fast enough to follow changing signals.
- Aliasing: Does the scope produce an alias when the time base is reduced from fast to slow sweep speeds? How does the display look for the toughest signal?

The oscilloscope is undergoing a period of rapid change. The major manufacturers of oscilloscopes are no longer producing analog models and the digital models are evolving rapidly. There is confusion in the oscilloscope marketplace because of the rapid pace of this change. Hopefully, this discussion will prove valuable to the user in selecting and applying oscilloscopes in the lab in the years to come.

References

1. A. DeVibiss, Oscilloscopes, in C.F. Coombs, Jr. (ed.), *Electronic Instrument Handbook,* 2nd ed., New York, McGraw-Hill, 1995.
2. R. A. Witte, A family of instruments for testing mixed-signal circuits and systems, *Hewlett Packard J.,* April 1996, Hewlett Packard Co., Palo Alto, CA.
3. M.S. Holcomb, S.O. Hall, W.S. Tustin, P.J. Burkart, and S.D. Roach, Design of a mixed signal oscilloscope, *Hewlett Packard J.,* April 1996, Hewlett Packard Co., Palo Alto, CA.
4. InstaVu acquisition mode, *Tektronix Measurement Products Catalog,* Tektronix Inc., Beaverton, OR, 1996, 69.
5. M.S. Holcomb and D.P. Timm, A high-throughput acquisition architecture for a 100 MHz digitizing oscilloscope, *Hewlett Packard J.,* February 1992, Hewlett Packard Co., Palo Alto,CA.
6. R.A. Witte, *Elecronic Test Instruments, Theory and Applications,* Englewood Cliffs, NJ, Prentice-Hall, 1993.
7. J. Murphy, Ten points to ponder in picking an oscilloscope, *IEEE Spectrum,* 33(7), 69-77, 1996.

1.3 Inductive and Capacitive Voltage Measurement

Cipriano Bartoletti, Luca Podestà, and Giancarlo Sacerdoti

This chapter section addresses electrical measurements where the voltage range to be measured is very large — from 10^{-10} V to 10^{7} V. The waveform can be continuous, periodic, or impulsive. If it is periodic, the spectrum components can vary for different situations, and within the same electric power network there may be subharmonic components. In impulsive voltage measurement, it is often important to get maximum value, pulse length, etc. Capacitive and inductive voltage sensors are mainly utilized in low-frequency electric measurements.

Capacitive Sensors

The voltage to be measured can be reduced by means of capacitive dividers (Figure 1.31). Capacitive dividers are affected by temperature and frequency and therefore are not important, at least in Europe. Capacitive sensors detect voltage by different methods:

1. Electrostatic force (or torque)
2. Kerr or Pockels effect
3. Josephson effect
4. Transparency through a liquid crystal device
5. Change in refractive index of the optic fiber or in light pipe

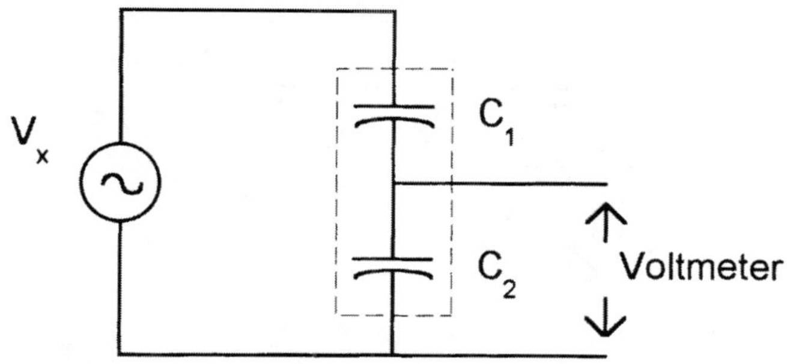

FIGURE 1.31 Schematic arrangement of a capacitive divider.

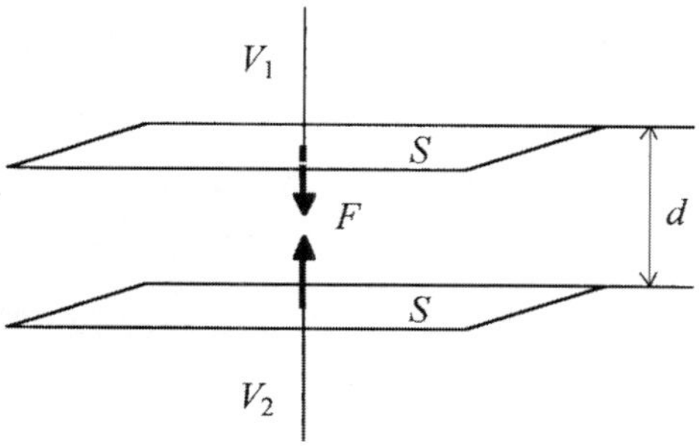

FIGURE 1.32 Force between two electrodes with an applied voltage.

1. The relations that rule the listed capacitive voltage sensors are reported below. The force between two electrodes is (Figure 1.32):

$$F = \varepsilon_0 \frac{S}{d}(V_1 - V_2)^2 \tag{1.54}$$

where ε_0 = Dielectric constant
S = Area of the electrode
d = Distance
V_1, V_2 = Potentials of the electrodes

The *torque* between electrostatic voltmeter quadrants (Figure 1.33) is given by:

$$T = \frac{1}{2}\frac{\partial C}{\partial \theta}(V_1 - V_2)^2 \tag{1.55}$$

where C = Capacitance
θ = Angle between electrodes

To get the torque from the rate of change (derivative) of electrostatic energy vs. the angle is easy. Obtaining the torque by mapping the electric field is difficult and requires long and complex field computing.

2. The rotation of the polarization plane of a light beam passing through a KDP crystal under the influence of an electric field (*Pockels effect*) is expressed by (Figure 1.34):

$$\theta = k_\pi l(V_1 - V_2) \tag{1.56}$$

where k_π = Electro-optic constant
l = Length of crystal

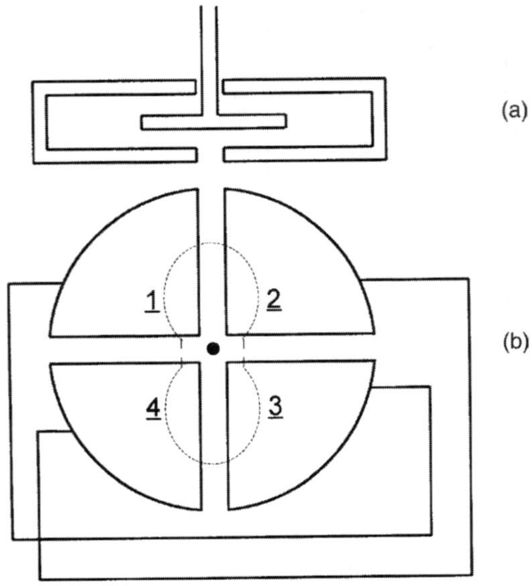

FIGURE 1.33 Scheme of an electrostatic voltmeter. (a) Lateral view; (b) top view: (1), (2), (3), (4) are the static electrodes; the moving vane is shown in transparency.

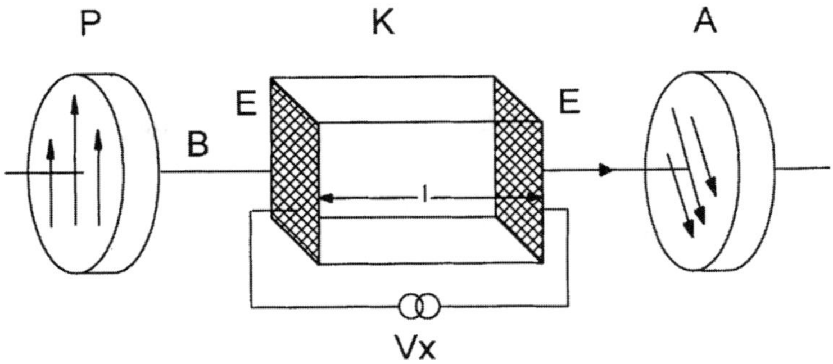

FIGURE 1.34 Scheme of an electrooptic KDP device. The parts are labeled as: (B) a light beam, (P) a polarizer, (A) an analyzer, (K) a KDP crystal, with the voltage to be measured *Vx* applied to its (E) transparent electrodes.

One obtains a rotation of $\pi/2$ by applying a voltage of the order of 1 kV to a KDP crystal of a few centimeters in length.

If a light beam passes through a light pipe that performs the *Kerr effect*, one observes a quadratic dependence of the rotation vs. *V*.

$$\theta \equiv kE^2 \equiv k'V^2 \tag{1.57}$$

3. The *Josephson effect* consists of translation of a voltage into a periodical signal of a certain frequency, carried out by a special capacitive sensor. There is an array of *N* layers of Josephson superconducting junctions; the frequency of emitted signal, when a voltage *V* is applied, is given by:

$$\nu = \frac{2eV}{Nh} \tag{1.58}$$

4. The *transparency* of a liquid crystal device depends on the difference of potential applied. There are liquid crystal devices working in transmission or in reflection. A change in transparency is obtained when a difference of potential of a few volts is applied.

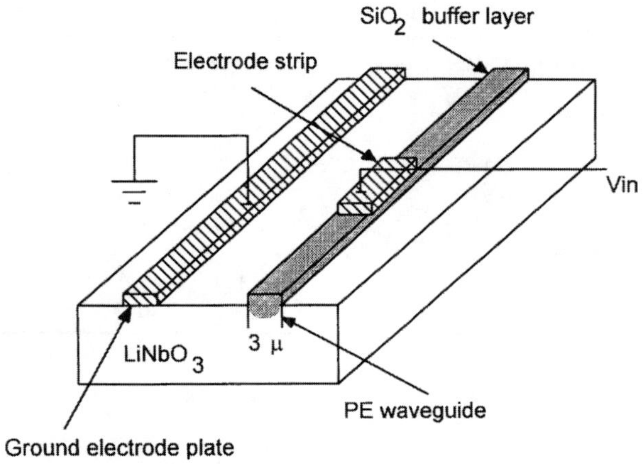

FIGURE 1.35 Li-Nb optical wave guide device.

5. The *change in refractive index* due to the presence of an electric field can be detected by:
   - Interferometric methods (where the velocity of light is equal to c/n)
   - Change in light intensity in a beam passing through an optical wave guide device like Li-Nb (Figure 1.35).

By means of method 1, many kinds of instruments (voltmeters) can be realized. Methods 2 through 5 are used in research laboratories but are not yet used in industrial measurements.

Inductive Sensors

Voltage Transformers (VTs)

Voltage transformers have two different tasks:

- Reduction in voltage values for meeting the range of normal measuring instruments or protection relays
- Insulation of the measuring circuit from power circuits (necessary when voltage values are over 600 V)

Voltage transformers are composed of two windings — one primary and one secondary winding. The primary winding must be connected to power circuits; the secondary to measuring or protection circuits. Electrically, these two windings are insulated but are connected magnetically by the core.

One can define:

$$\textit{Nominal ratio} = K_n = \frac{V_{1n}}{V_{2n}} \tag{1.59}$$

as the ratio between the magnitude of primary and secondary rated voltages.

$$\textit{Actual ratio} = K = \frac{V_1}{V_2} \tag{1.60}$$

as the ratio between the magnitudes of primary and secondary actual voltages.

Burden is the value of the apparent power (normally at $\cos\varphi = 0.8$) that can be provided on the secondary circuit (instruments plus connecting cables).

Burden limits the maximum value of secondary current and then the minimum value of impedance of the secondary circuit is:

TABLE 1.4 Angle and Ratio Error Limit Table Accepted by CEI-IEC Standards

Class	Percentage voltage (ratio) error (±)	Phase displacement Minutes (±)	Centiradians (±)
0.1	0.1	5	0.15
0.2	0.2	10	0.3
0.5	0.5	20	0.6
1	1	40	1.2
3	3	—	—
3P	3	120	3,5
6P	6	240	7

$$Z_{min} = \frac{V_{2n}^2}{A_n} \tag{1.61}$$

where A_n = VT burden

For example, if A_n = 25 VA and V_{2n} = 100 V, one obtains:

$$Z_{min} = \frac{100}{0.25} = 400\ \text{W} \tag{1.62}$$

There are two kinds of errors:

1. *Ratio error* = $\text{Ratio error} = h_{\%} = \dfrac{K_n - K}{K}$ (1.63)
2. *Angle error* = the phase displacement between the primary voltage and the secondary voltage (positive if the primary voltage lags the secondary one).

Voltage transformers are subdivided into accuracy classes related to the limits in ratio and angle error (according to CEI and IEC normative classes 0.1, 0.2, 0.5, 1, 3; see Table 1.4). To choose the voltage transformer needed, the following technical data must be followed:

- Primary and secondary voltage (rated transformation ratio). Normally, the secondary value is 100 V.
- Accuracy class and rated burden in VA: e.g., cl. 0.5 and A_n = 10 VA.
- Rated working voltage and frequency
- Insulation voltage
- Voltage factor: the ratio between maximum operating voltage permitted and the rated voltage. The standard voltage factor is 1.2 V_n (i.e., the actual primary voltage) for an unlimited period of time (with VT connected with phases), and is 1.9 V_n for a period of 8 h for VT connected between phase and neutral.
- Thermal power is the maximum burden withstood by VT (errors excluded).

For extremely high voltage values, both capacitive dividers and voltage transformers are normally used, as shown in Figure 1.36. The capacitive impedance must compensate for the effect of the transformer's internal inductive impedance at the working frequency.

Other Methods

The ac voltage inductive sensors act by interaction between a magnetic field (by an electromagnet excited by voltage to be measured) and the eddy current induced in an electroconductive disk, producing a force or a torque. This can be achieved by the scheme shown in Figure 1.37. The weight of many parts of the

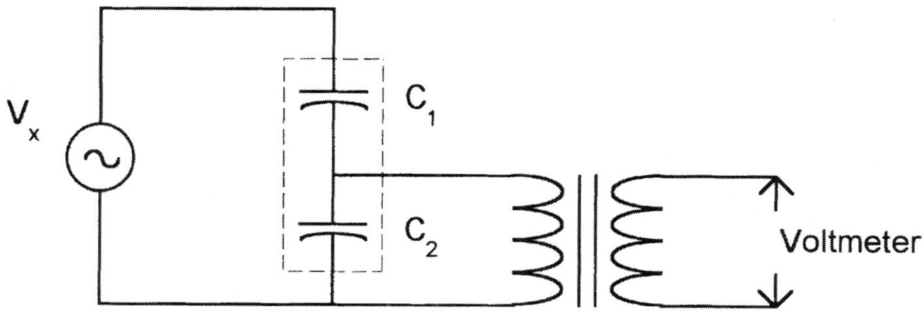

FIGURE 1.36 Capacitive divider and voltage transformer device for extremely high voltage.

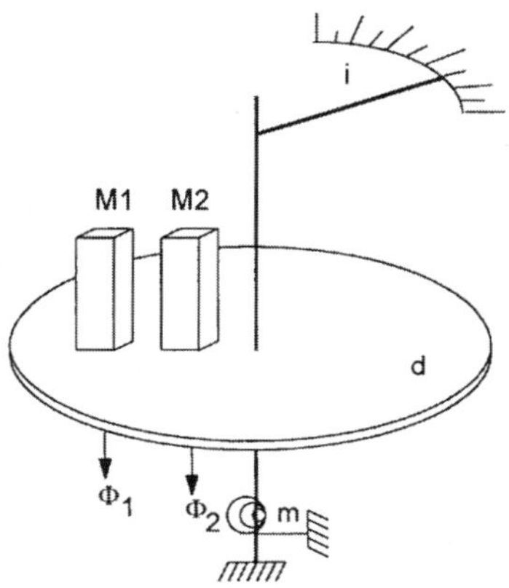

FIGURE 1.37 Schematic inductive voltmeter. The parts are labeled as: (i) index, (d) metallic disk, (M1) and (M2) electromagnets, (m) spring, (Φ1) and (Φ2) generated fluxes.

indicator can be some tens of grams. The power absorbed is on the order of a few watts. The precision is not high, but it is possible to get these sensors or instruments as they are similar to the widely produced induction energy meters. They are quite robust and are priced between \$50 and \$100, but they are not widely used. The relation between torque and voltage is quadratic:

$$T = k_i V^2 \tag{1.64}$$

The proportionality factor k_i depends on magnet characteristics and disk geometry.

G.E.C., Landys & Gyr, A.B.B., Schlumberger, etc. are the major companies that furnish components and instruments measuring voltage by inductive and capacitive sensors.

Defining Terms

CEI: Comitato Elettrotecnico Italiano.
IEC: International Electric Committee.
KDP: Potassium dihydrogen phosphate.
Li-Nb: ($LiNbO_3$) lithium niobate.

Further Information

J. Moeller and G. Rosenberger, Instrument Transformers for HV Systems, Siemens Power Engineering III (1981) Special Issue, *High-Voltage Technology.*

G. Sacerdoti, O. Jappolo, and R. Paggi, *Misure Elettriche, Vol. I Strumenti*, Bologna, Zanichelli, 1994.

G. Zingales, *Metodi e Strumenti per le Misure Elettriche*, Bologna, UTET, 1976.

2

Current Measurement

Douglas P. McNutt
The MacNauchtan Laboratory

Current measuring devices are selected for:

- Precision
- Stability
- Frequency response, including dc
- Galvanic isolation
- Presentation of the data
- Effect on measured circuit

A few of the most common sensors will be introduced in this chapter. Details and some less common sensors will follow after definitions and a bit of magnetic theory, which can be skipped. Any magnetic field sensor can be used as a current sensor and there are some exotic examples, such as quantum effects in low-temperature superconductors used to measure currents in neurons within the brain. This discussion is limited to measurement of currents in wires with commercially practical devices.

An isolated current sensor is free of any metallic connection to the circuit being measured. It is also essentially free of capacitive coupling so that it is safe to use with grounded amplifiers and other equipment. The quality of the isolation is measured in volts and is usually the breakdown potential of an insulator. 5 kV is typical for personal safety around light industrial power.

0-8493-1733-9/04/$0.00+$1.50

By far, the simplest current-to-voltage converter is the resistor. In current measuring service, it is called a shunt although it is typically placed in series with the load. That is because shunts are sometimes used to increase the range of another current-measuring device using a connection that bypasses part of the current around a meter. Frequency response of a shunt is good and includes dc. Shunts produce a voltage output that can be presented by a variety of secondary meters, including analog meters, digital meters, oscilloscopes, and 4 to 20 mA converters. Shunts provide no isolation and have a potentially unacceptable effect on the circuit being measured. Shunts used for dc are as accurate as the resistance and the associated voltmeter.

The common moving-coil meter, the D'Arsonval movement [1, 2], probably with an internal shunt and/or rectifier, is an easily used device and still available. Its isolation is by means of the human eye since that is the only way to read out the result. It is useful for power panels where an operator needs quick data. Accuracy is no better than 2%.

For power frequency, 25 to 400 Hz service, the current transformer, called a "donut" transformer, or CT in the trade, is commonly employed. The current-carrying conductor is passed through the hole in a toroid of magnetic material. A shorted secondary winding of n turns carries current, which is $1/n$ times the measured current, and is typically passed to another ammeter or used as the current input to a power measuring device. Isolation is as good as the insulation on the primary conductor; frequency response is fair but does not include dc; there is minimal effect on the measured circuit, and the cost is low. Operational safety is an issue; see the note below.

A variety of noncontact sensors is available for dc sensing. Most depend on the Hall effect and all require a source of operating power. Frequency response from dc to 200 kHz is advertised. Because operating power is available, output to later processing can be voltage, current, or digital. Accuracy is whatever one wants to pay for. Long-term stability depends on dc-coupled operational amplifiers and can exhibit zero drift. Externally, these look much like CTs.

CTs, Hall devices, and other similar noncontact schemes are available in wrap-around form so they can be installed without disconnecting power. The wrapping process always involves breaking a magnetic path, and consistency of reassembly becomes a limit on precision. Everything from current-sensing probes for oscilloscopes to CTs for 10000 A circuits can be found as wrap-arounds.

2.1 Definition of the Ampere

There is perpetual argument about the number of "fundamental" quantities required to describe our environment. Is there a fundamental unit of electricity or are the electrical units derived from more basic things such as mass, length, and time? Standards laboratories such as the U.S. National Institute of Standards and Technology, NIST, measure the force between current-carrying wires to compare the ampere to the meter, the kilogram, and the second, and provide a standard stated as "that constant current which, if maintained in two straight parallel conductors of infinite length, of negligible circular cross-section, and placed 1 m apart in vacuum, would produce between these conductors a force equal to 2×10^{-7} newton per meter of length" [3]. For the rest of this discussion, it is assumed that instrumentation is "traceable to the NIST," meaning that one way or another, an electrical unit as measured is compared to a standard ampere maintained by NIST. Absolute calibrations using the quantum properties of the Hall effect and the Josephson junction are not practical in the field.

For practical reasons, it is often easier to distribute a voltage standard than a current standard. Chemical cells and, more recently, semiconductor voltage references are quite stable and do not depend on the length of the wires used to connect them together during calibration and use. As a result, current measuring is usually a matter of conversion of current to an equivalent voltage.

An exception is current measurement by comparison to magnetic forces provided by a permanent magnet and that might be why the older unrationalized centimeter-gram-second, the cgs units, with no separate electrical unit is still found in specifications of magnetic quantities. The gauss and the oersted are particular examples of these now deprecated units. It is this confusion of units that frightens many who would measure current away from even attempting calculations involving magnetic devices.

2.2 Magnetics

Magnetic current sensors have advantages over shunts. To understand them, one should delve into the interaction between currents and magnetic fields. Following Maxwell by way of Sommerfeld [4], it is convenient to describe magnetic effects in terms of two vector fields, **B** and **H**. **H** is the field created by an electric current, and **B** is the field that acts on a moving charge or a current-carrying wire. **B** and **H** are related by characteristics of the material in which they coexist. Strictly speaking, **B** is the flux density and **H** is the field, but they are both called the field in less than precise usage.

The SI units of **B** and **H** are the tesla (T) and the ampere per meter (A m^{-1}). They are called rationalized because a ubiquitous 4π has been suppressed in the underlying equations.

For practical engineering, it is still necessary to understand the unrationalized equivalents, the gauss and the oersted, because they are universally used in specifications for magnetic materials. To convert from gauss to tesla, divide by 10^4. To convert from oersted to amperes per meter, multiply by $1000/(4\pi)$, a number that is commonly approximated as simply 80; but strictly speaking, the units of **H** in the two systems are dimensionally different and cannot be converted.

The relationship between **B** and **H** is most generally a tensor that reflects spatial anisotropy in the material, but for common magnetic materials used in current sensing, a scalar value μ applies. In SI units, **B** and **H** have different physical dimensions and are not equal in vacuum. The "permeability of free space" μ_0 is defined so that the μ for a material is a dimensionless constant. Thus,

$$\mathbf{B} = \mu\mu_0\mathbf{H} \tag{2.1}$$

where μ_0 is exactly $4\pi \times 10^{-7}$ H m^{-1}. (H here is the henry, the SI unit of inductance, base units m^2 kg s^{-2} A^{-2}) . For many problems, μ will be a constant of the magnetic material, but when magnetic saturation needs to be considered it will be a variable. For engineering calculations, μ is often treated as though it depends on frequency. Values of μ range from less than 100 for high-frequency ferrite to 5000 for transformer iron, to 10^5 for carefully annealed magnetic alloys.

The field due to a long straight wire is shown in Figure 2.1 and Equation 2.2. The **H** vector obeys the right-hand rule and is everywhere perpendicular to the wire. The amplitude of **H** is proportional to the current and falls off linearly with the radial distance from the wire. This is the field that makes noncontact sensing of current possible.

$$\mathbf{H} = \frac{\mathbf{I}}{2\pi r} \tag{2.2}$$

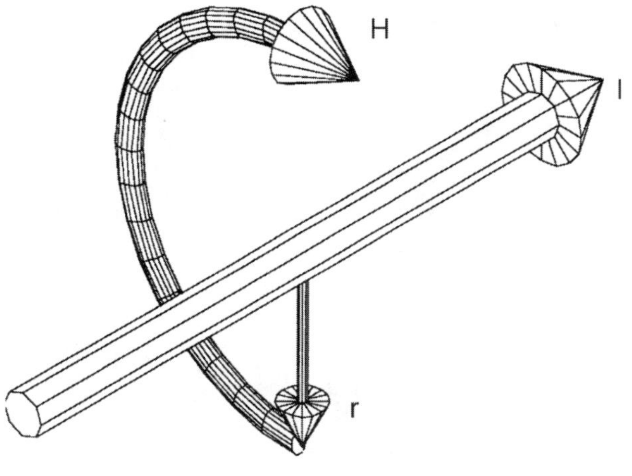

FIGURE 2.1 The magnetic field associated with a current-carrying long wire. It is this field that makes contactless sensing of current possible.

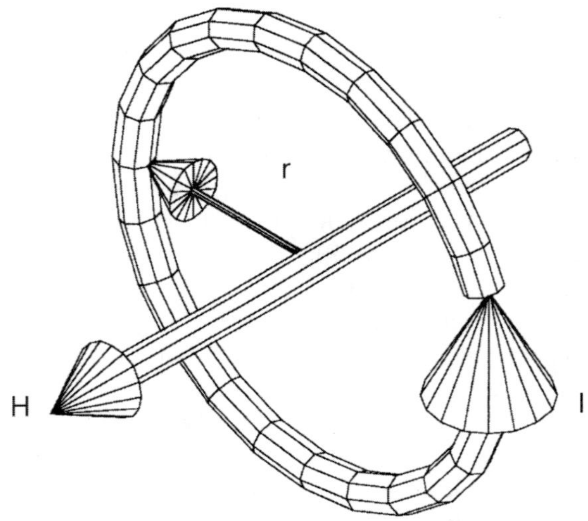

FIGURE 2.2 The magnetic field at the center of a loop of wire. This is the starting point for many calculations involving inductors and transformers.

The field at the center of a circular loop of wire is shown in Figure 2.2 and Equation 2.3. It obeys the right-hand rule, is proportional to the current, and inversely proportional to the radius of the loop.

$$\mathbf{H} = \frac{\mathbf{I}}{r} \tag{2.3}$$

A magnetic field can often be calculated by direct application of Maxwell's relations in integral form [4]. The line integral over any closed path of **H•dl** is equal to the current passing through a surface delimited by the closed path.

Magnetic flux passing through a surface, usually denoted by ϕ, is the integral of **B•n** dA over the surface with normal vector **n.** The SI unit of ϕ is the weber. The unrationalized unit, the line of force, is best relegated to history except that the lines, which form continuous loops, dramatize the fact that it does not matter how a surface is drawn to close a bounding line. The flux is the same. It is convenient to think of magnetic lines, forming closed loops, that can be bent but not interrupted or removed.

A carrier of charge, an electron in a wire, a particle in vacuum, or a hole in a semiconductor, moving in a magnetic field is acted on by a force that is perpendicular to the field and the velocity. For positive charge, the force obeys the right-hand rule. The magnitude of the force is proportional to the magnitude of **B**, the velocity **V**, and the sine of the angle between them.

$$\mathbf{F} = \mathbf{V} \times \mathbf{B} \tag{2.4}$$

A carrier, moving or not, is affected in a similar way by a changing magnetic field. The result is an electromotive force, *EMF*, in a loop of wire through which a changing magnetic flux passes. The *EMF* is equal to the rate of change of the flux enclosed by the loop with a change of sign. That is Faraday's law of induction.

$$\mathrm{EMF} = -\frac{d\phi}{dt} \tag{2.5}$$

Most magnetic materials exhibit hysteresis. That is, the relationship between **B** and **H** depends on the history of the applied **H**. A plot of **B** vs. **H** is called a hysteresis loop and a sample is shown as Figure 2.3. The area of the loop represents an energy. If a magnetic core is repeatedly driven around its hysteresis loop, there is a power loss due to hysteresis that is proportional to the area of the loop and the frequency. A material with very large hysteresis is a permanent magnet.

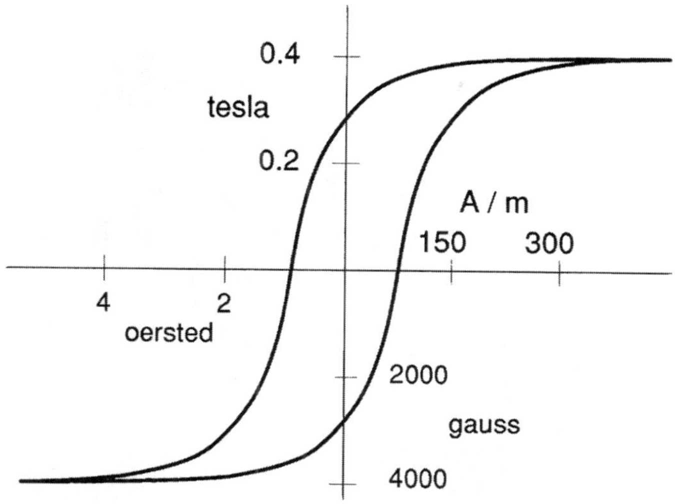

FIGURE 2.3 A hysteresis curve for some rather poor transformer iron. Unrationalized units are shown below and left of the origin because that remains standard practice in the industry.

Magnetic materials which are also electric conductors have free carriers which are affected by alternating magnetic fields. As they move back and forth they encounter electric resistance and dissipate energy in the form of heat. These eddy currents can be minimized by use of high-resistance ferrites, powdered iron, by laminating cores, or by keeping the flux small.

2.3 Shunts

Shunts were introduced above. They dissipate power as heat and the resistance changes in response to the rising temperature. The dissipation is proportional to the voltage across the shunt and a design compromise must be made because low voltage implies less accuracy in the voltmeter. A standard of 50 mV has evolved. Shunts do not provide galvanic isolation between the measured circuit and the measurement device.

Measurement of alternating current using shunts is also affected by skin effect and the inductance of the shunt. Skin effect can be minimized in the design of the shunt by the use of several parallel sheets of thin metal, Figure 2.4, a feature that also improves heat dissipation. There is not much to be done about inductance except to minimize size.

Safety is enhanced if shunts are placed in the ground leg of a circuit; that way, the output leads, usually only at 50 mV, are near ground. However, that introduces a resistance in the ground path and can interfere with common mode requirements of interdevice signal connections. If the shunt is placed in the high side, care must be taken to protect the wiring and the meter to which it is attached from accidental grounds.

Sometimes, sufficient accuracy can be obtained by measuring the voltage drop along a length of conductor that is otherwise required in the installation. In vehicular service, it is common to sense the voltage drop in a battery cable using a millivoltmeter. Such installations are always high side and should be protected with fuses installed near the points of measurement. It is also wise to provide a connection means that is independent of contact resistance where the current-carrying conductor is installed — a Kelvin connection. Including one lead of an in-line fuse holder in the terminals before they are crimped is one such technique.

2.4 The Moving Magnet Meter

The simplest current indicator balances the force on a permanent magnet created by current in a wire against a spring. A magnetic material is usually placed around the conductor to concentrate the field and

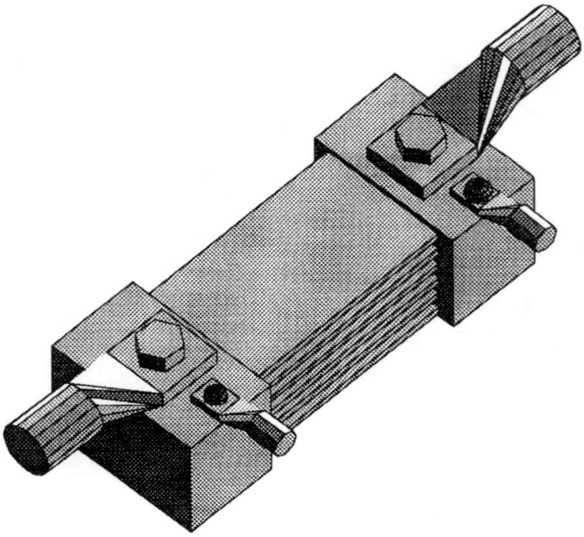

FIGURE 2.4 Multiple sheets of conductor are provided in this shunt to reduce skin effect and allow air cooling. Kelvin connections are provided for the voltmeter so that the voltage drop in the high-current connectors is not inadvertently included in the measurement.

reduce the effect of the magnetic field of the Earth. Use is limited to low-precision indicators such as a battery charging meter for a vehicle. It is a dc device.

2.5 The D'Arsonval Meter

This indicator balances the force on a current-carrying wire due to a permanent magnet against a spring. The measured current flows in a coil of wire supported in bearings. It is described in elementary texts [1, 2]. It is generally a dc instrument, but chart recorders have been built with frequency response in the kilohertz range using mirrors on the moving coil. They are then called galvanometers. For ac service, these meters are often equipped with internal copper oxide rectifiers and a nonlinear scale to correct for the diode drop. For current ranges above a few milliamperes, they will have an internal shunt.

The moving magnet meter and the D'Arsonval movement are the only current sensors that do not convert current to voltage and then depend on other devices to read out the voltage.

2.6 The Electrodynamometer

A variation of the D'Arsonval meter for ac service can be built by replacing the permanent magnet with an electromagnet. The force on the moving coil becomes proportional to both the current being measured and the voltage applied to the coil of the electromagnet. It is sensitive to the relative phase of the voltage and current in just the right way to be useful for measurement of power in a circuit with correction for power factor. An electrodynamometer in a power panel is often the load for a current transformer.

2.7 The RF Ammeter and True rms

Current to a radio transmitting antenna is commonly passed through a small resistor, a shunt, that is thermally connected to a thermocouple or other thermometer and mounted in a thermally insulating blanket. The rise in temperature is a measure of the current and is often sensed with a thermocouple.

This is an example of true rms indication. Root mean square current is the square root of the integral of the square of the instantaneous current over an unspecified time divided by that time. It is intended

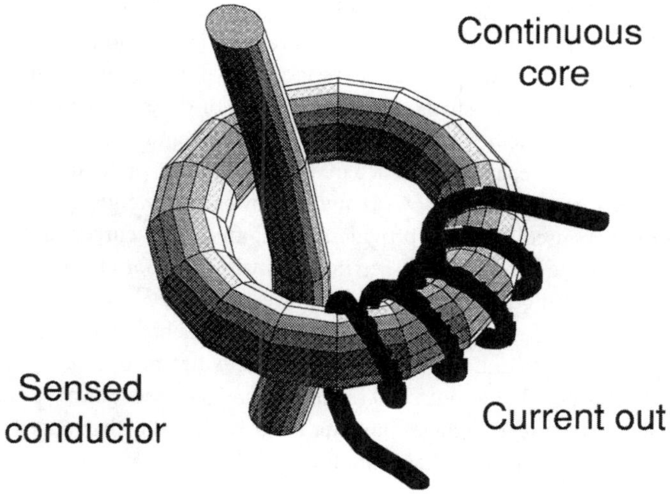

FIGURE 2.5 The ideal current transformer or CT is tightly coupled with no magnetic gap. Current in the secondary exactly balances current in the single-turn primary so that the magnetic flux in the core is zero.

to represent a stationary ac waveform by a single value that is equal to the direct current which would dissipate the same power in a resistive load. The RF ammeter does that precisely. The indication is not particularly linear, but it can easily be calibrated by applying dc to the input.

Other schemes for measuring rms current depend on analog multipliers and subsequent integration. They are limited by crest factor, the ratio of highest instantaneous peak to the rms over a measuring period. Inexpensive meters simply measure the peak, assume a sinusoidal waveform, and scale to rms.

2.8 The Current Transformer

Consider the magnetics of a toroidal core of high-μ material through which a current-carrying conductor passes. Include a secondary winding of n turns as shown in Figure 2.5. The secondary winding is connected to a low-resistance load.

In this current transformer, universally referred to as a CT, alternating current in the single-turn primary attempts to magnetize the core but in so doing, creates an emf and current in the secondary that tend to cancel the field. If the secondary truly has zero resistance, the current in it exactly cancels the field due to the primary. The result is a secondary current equal to the primary current divided by the number of secondary turns. The secondary current is in phase with the primary current. Because of the tightly closed magnetic loop, there is little effect from nearby conductors or position of the primary wire in the hole.

The secondary circuit can now be connected to a low-resistance current- or power-sensing device with assurance of calibration. But the secondary resistance is never really zero and the magnetic coupling is never perfect, so there are other considerations.

First, the concept of *secondary burden* is introduced. It is called that to avoid calling it a "load" because it behaves differently; the best burden is a short-circuit. Burden is sometimes expressed as a secondary resistance in ohms, but more often as an equivalent power in kVA for a defined current without consideration of phase. When the burden is not a perfect short-circuit, energy is dissipated and the magnetic field present in the core is no longer zero. The secondary current leads the primary current with a phase that depends on frequency.

Manufacturers of CTs have techniques to optimize accuracy of the CT when specified for a particular burden. The finished units might not have the number of secondary turns one would expect, but will nonetheless provide results accurate to a percent or so. They have laminations selected to minimize heating of the core. One will see ratings like 100:5, meaning 100 A in the primary will produce 5 A in the secondary rather than "20 turns." They should be installed in a circuit that provides the burden for which they were calibrated. The voltage across a burden resistor is commonly amplified and passed to a data collection device. CTs for large currents need to be large to avoid magnetic saturation when burdened. Cores are prepared with laminates of silicon iron in the form of disks, concentric rings, or tape that is wound on a bobbin. Even with the best of materials, eddy current and hysteresis loss are present. When power dissipation is unacceptable, another choice of sensor might be preferable.

Most CTs are used for measurement of low-frequency power and energy. They are found at the front end of kilowatt-hour meters used by power providers. Radio frequency current in transmitting antennas can be measured with suitable core material. Ferrite cores with appropriate compensation are used for sensing pulse width modulated current in switching power supplies. Very large cores are used to sense pulsing beams of high-energy particles. Some oscilloscope probes are highly compensated CTs with a core that can be opened to allow a current-carrying wire to be introduced. With modern winding equipment for toroids, it is possible to put 2000 or more turns on a secondary [Coilcraft]. The CT then begins to look more like a current-to-voltage converter in its own right without need for very small values of the burden resistor and subsequent voltage amplification.

Safety Note

The secondary of a CT should always remain shorted while there is any possibility of current flow in the primary. With an open secondary, the core can be driven back and forth between saturation in opposite directions, resulting in high-speed changes in the internal B field. The result is very high, dangerous to life, voltages on the open secondary. Insulation in the secondary circuit can be damaged by arcing. Many CTs are made with a provision for shorting the secondary if the circuit must be opened. Use it.

2.9 Gapped Inductive Sensors

It is common practice in the design of transformers to introduce a small gap in the magnetic path. For even very small gaps, the magnetic properties of the magnetic loop become almost completely determined by the length of the gap, the rest of the material serving only to contain the lines of flux. Analysis of such a device begins with understanding that the B field is continuous in the core and through the gap. The H field is not, but it still satisfies the relation that the line integral of $\mathbf{H} \cdot \mathbf{dl}$ around the core is equal to the linked current. For a magnetic path of length s in material of permeability μ with a gap g, the ratio B/H, the effective permeability is given by:

$$\mu_{\text{eff}} = \frac{s}{g + \dfrac{s}{\mu}} \tag{2.6}$$

which applies for g much smaller than s. Note that when μ is sufficiently large, the effective permeability becomes independent of μ.

Introducing a gap into what would otherwise be a CT and drastically increasing the secondary turns count to 10,000 or more results in a current sensor that is intrinsically safe because the core cannot saturate: Figure 2.6 [SRT]. Because the B field is always small, the heating effect of eddy currents is less important than in the CT. When loaded with an appropriate resistor, the high inductance of the secondary causes the sensor to act like a current source that generates a voltage across the load proportional to the primary current with better than 1% linearity. Useful bandwidths of a sensor can be over 3 decades. The output impedance is high and requires the use of electronic voltmeters. Power dissipation is low, even for very high current models. Output voltage is high enough that simple diode rectifiers can be used to provide for dc output to further processing. In many cases, such a sensor can be used without any special electronics other than a voltmeter.

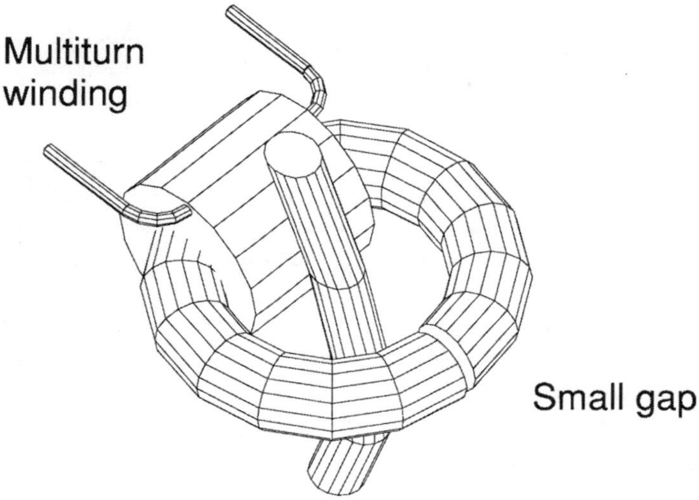

FIGURE 2.6 Placing a gap in the core and dramatically increasing the secondary turns count results in a current-to-voltage converter with high output voltage.

2.10 Hall Effect Sensor

The *Hall effect* as a sensor for magnetic fields is described in Chapter 12 of this volume and in [5]. It depends on a semiconductor crystal selected for its high carrier mobility and is placed in a magnetic field. A current is passed through the crystal along an axis perpendicular to the field. The carriers assume a mean velocity that causes them to be acted upon by the field and they move toward the other axis perpendicular to the field. The result is an emf at the faces of the crystal that can be measured. The emf is proportional to the field, the bias current, and the mobility.

In principle, such a field sensor could be placed near a current-carrying wire and oriented to sense the field created by the current, Figure 2.1, but the sensitivity is insufficient and there would always be interfering fields from currents in other nearby wires. A flux concentrator that looks like a CT with a large gap is always used. See Figure 2.7.

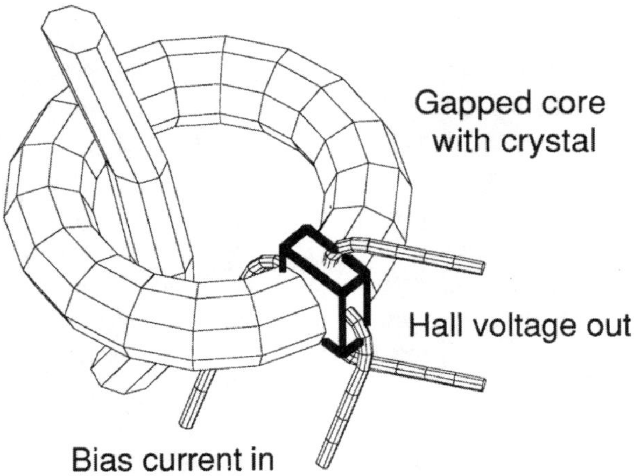

FIGURE 2.7 A semiconducting crystal is placed in the gap of a flux concentrating magnetic core. Bias current on one axis of the crystal produces a Hall voltage on the other.

The device is sensitive to direct current and the polarity is preserved. The Hall voltage is a few millivolts and amplification is always required. Zero drift in the amplifiers must be properly compensated although this is not so important for ac service. The bias current must be carefully controlled and it can be used to provide intrinsic analog multiplication and metering of power if it is made proportional to the circuit voltage.

Sensitivity is best with the smallest gap but there must be room for the crystal so gaps are larger than in the gapped inductive sensors. Fringing of the field in the larger gap reduces the natural shielding of the toroid from unwanted magnetic fields.

The accuracy and linearity of the Hall effect sensor can be improved in closed-loop mode. A feedback winding is added to the core and driven by a servo amplifier. The emf from the Hall device is used to drive the servo amplifier until the field is zero. The output is then the feedback current which is less than the sensed current by the number of turns in the feedback winding. The frequency response of the closed-loop system is surprisingly good, hundreds of kilohertz [F. W. Bell].

2.11 Clamp-On Sensors

It is often desirable to measure current in an existing system without removing power in order to install a device; thus, most of the magnetic devices are available in a clamp-on configuration. The variety ranges from tiny oscilloscope probes to clamps for 3000 A power buses.

Accuracy is always reduced in clamp-on mode because the clamp itself constitutes a gap that is uncontrollable and subject to wear. Some manufacturers provide highly polished surfaces that slide together. Others have iron fingers that interlace as the clamp is closed. Still others do not worry about it because the instrument is intended for field use where accuracy is not so critical.

Some units have handles for one-hand operation; others require a wrench or other tool. An interesting variation is the flexible core by [Flexcorp]. Installation is by bending the core and allowing it to spring back to shape.

2.12 Magnetoresistive Sensors

Most of the features of a Hall effect sensor are available if the Hall crystal is replaced by a device whose resistance changes with magnetic field. The discovery of giant magnetoresistive devices has recently made this idea attractive. [NVE]

Such devices still exhibit rather small resistance change and are sensitive to other effects such as temperature, so it is imperative that they be used in self-compensating bridge circuits in the manner of a strain gage. They are also insensitive to the polarity of the field.

Zetex has delivered magnetoresistive current sensors using thin-film permalloy in a variety of printed circuit packages. They are constructed in the form of a bridge and require bias and a differential amplifier. Measured current up to 20 A passes through the chip via its solder pins.

2.13 The Magnetic Amplifier

The efficiency of a transformer can be adjusted by a dc or low-frequency current that moves the operating point on a hysteresis curve. Excitation, a pump, is required at a higher frequency; it is passed through the transformer and then synchronously rectified and filtered into a higher power representation of the low-frequency signal. The magnetic configuration must be designed so that the pump does not put power into the signal circuit. Figure 2.8 shows one such arrangement. The pump coils are phased in series so that the associated flux cancels in the center leg of the E-E transformer core. The output coils are also in series and phased to add. When dc is applied to the center winding, it saturates both sides and reduces the signal on the output coils. More complicated arrangements can preserve polarity in the phase of the output.

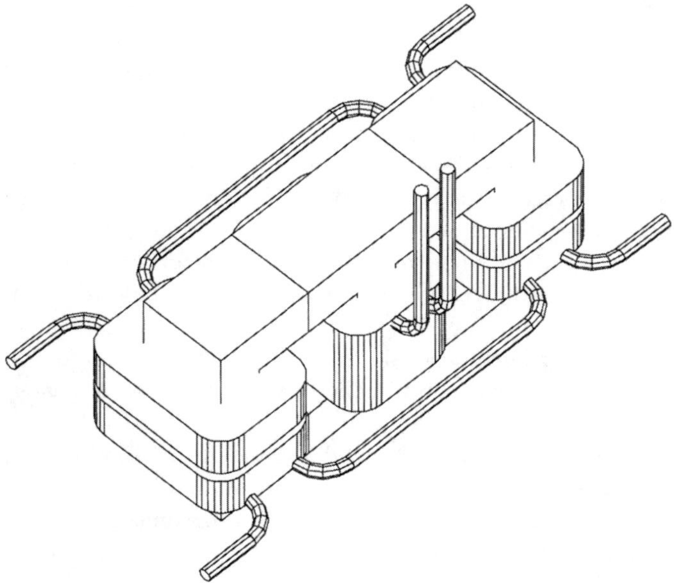

FIGURE 2.8 The windings of a magnetic amplifier. For simple current detection, the coils can be as little as two turns on ferrite toroids using an RF pump.

As a current measuring device the magnetic amplifier leaves much to be desired in linearity and frequency response, but it does provide isolation and is limited in sensitivity only by the number of turns placed on the center winding. The 20 mA dc off-hook current in a telephone is one such application.

2.14 Fluxgates

In its simplest form, a *fluxgate magnetometer* uses a driving coil to drive a high-permeability rod into saturation, first in one direction and then in the other. A second winding observes the rate of change of the B field inductively. In the absence of an external magnetic field, the observed signal is symmetric; but when an external field shifts the hysteresis loop to the right or left, the symmetry is lost. An amplifier tuned and phase-locked to the second harmonic of the driving frequency can be used to determine the amplitude of the external field.

In principle, such a fluxgate could be used to sense the field in the gap of a flux concentrator as with the Hall sensor, but it is too big and the linearity would be unacceptable. It is better to think of a way to drive the whole flux concentrator with a pumping frequency similar to that used in a magnetic amplifier.

Driving a toroid this way has the undesirable effect of coupling the pump energy into the circuit being measured, but one can pump two toroids in opposite directions. Now a current to be sensed passes through both cores and biases one in one direction and one in the other relative to the pump. A pickup winding senses the second harmonic, which is demodulated with a phase-locked sensor to preserve the direction of the sensed current.

Linearity is improved by adding one more winding through both toroids; it is driven by a servo amplifier that is satisfied only when the amplitude of the second harmonic is zero; the flux at the measured frequency is zero. This is now the ideal current transformer. At the frequency of the sensed current, which might be dc, there is zero average flux in the toroids and the output of the servo amplifier is a true representation of the current being measured, reduced by the turns count of the servo winding.

Unfortunately, complicated electronics and considerable power are required to accomplish all of this, but the resulting accuracy is significantly better than anything other than precision shunts. Two suppliers — GMW and Holec — provide equipment that uses similar principles. Many of the details are either patented or proprietary.

2.15 Optical Sensors

The Faraday effect is a rotation of the plane of polarization of light as it passes in a transparent medium parallel to a magnetic field. Optical fibers that do not randomize polarization are available with a small Faraday coefficient. Winding such a fiber on a form to be installed around a wire so that the light propagates parallel to the field produced by current in the wire, Figure 2.1, results in a sensor. A measurable rotation proportional to the current comes about because of the long optical path. When the myriad problems of preparation are solved, the result is a sensor that looks externally like a current transformer but has no wires. Using a reflecting ring made of yttrium-iron-garnet with a large Faraday coefficient, NIST reports sensitivity of 220 nA [6]. Matsushita Electric Industrial Company makes sensors using thin garnet films.

Analysis of the polarization requires a polarized light source and a polarization analyzer at the other. An advantage of such a sensor is that the fiber leading to and from can be long enough to allow isolation for very high voltages.

Winding and annealing of fiber sensors without destroying the polarization-preserving properties of the fiber and temperature sensitivity of the Faraday coefficient must be addressed. Further information on these sensors can be found in [7–9]. ABB, Sweden, reports a maximum detectable current of >23 kA, a sensitivity of about 2 A, and a relative error of ±0.15% [7].

2.16 Fault Indicators

Latching indicators which save an indication of high pulse current which was present sometime in the past are needed for power distribution systems subject to lightning strikes. Such indicators are often a bar magnet that moves into a latched position when a current pulse occurs. Readout is visual and they can be reset and installed on live wires using high-voltage tools.

2.17 Other Schemes

Cantor of Ford Aerospace has described a flux reset transformer scheme for measurement of direct current [10]. An oscillator and semiconductor switch periodically drive a core into reverse saturation. When the drive current is switched off, the flux rises due to dc in the sense winding and produces a voltage on the drive winding that is sensed synchronously with the oscillator. No commercial device based on this principle is available.

The double-balanced mixer, familiar to RF engineers, is also a sensor for low-frequency current. The IF port is usually dc-coupled through diodes in a ring configuration. dc applied there will modulate an RF pump applied to one of the RF ports and recovered on the other. There are no known products that use this principle.

2.18 Some Generalities and Warnings

Except for very simple loads, the waveform of the current drawn by a load does not resemble the voltage waveform. Besides the well-known phase shift and power factor, current flows in harmonics and subharmonics of the power frequency. Bridge rectifiers with capacitor input filters draw pulses of current near the peaks of the voltage waveform. Triacs cause a phase shift of the fundamental and introduce odd harmonics. Triacs in full-cycle mode with zero current switching can draw full current for a few cycles, followed by zero current for a few more introducing frequency components below the power frequency. Frequency changers are likely to draw current in pulse width modulated bursts.

A classic error is to measure true rms voltage and true rms current and multiply the two to get power. Even after correcting for phase, this is usually wrong. In short, accurate measurement demands some knowledge of the characteristics of the load. Beware of sensors labeled "true rms" for they can be anything but.

2.19 Current Actuated Switches and Indicators

Another form taken by current sensors is the current actuated switch. There was a time when contacts were placed on the pointer of a D'Arsonval movement to establish upper and lower limits for a process variable. The modern way is to configure a sensor so that it operates a solid-state switch.

When operating power is available, any of the current sensors can be configured as a switch; but when the switch must be controlled solely by the power that can be extracted from the current being measured, the gapped toroid is superior. The high voltage available can be rectified to control MOSFETs or the base of an open collector Darlington transistor [SRT].

One company — CR Magnetics — markets a light-emitting diode, LED, indicator that shows the presence of alternating current without any connection or external source of power.

Circuit breakers can be implemented with a magnetic coil or with a bimetallic thermal element. For completeness, there is also the fuse.

Ground fault breakers use a toroidal CT through which the line current is passed twice: once on the way to the load and again on the way back to the neutral wire in such a way as to cancel. Any fault current to ground associated with the load causes the return current to be less than the source and induces voltage on the secondary of the toroid. This is amplified and used to trip a switch. This application demands absolute insensitivity to the position of a current-carrying wire in the hole of the toroid. Gapped sensors typically are not suitable because of magnetic leakage.

2.20 Where to Get Current Sensors

CTs, shunts, and some other sensors are commodity items available from a variety of manufacturers and distributors. Well-known manufacturers of switchgear — General Electric and Westinghouse — have their own CTs for use with their power meters. Test equipment manufacturers — Simpson, Tektronix, Fluke, Extech, and the like — offer hand-held current meters. Table 2.1 shows distributors and manufacturers of current sensing equipment. There is no way that a list such as this can be complete. It has selected itself from those companies willing to contribute to the author's library.

TABLE 2.1 Selected Manufacturers of Current Sensors

AEMC Instruments/Chauvin Arnoux, Inc.
200 Foxborough Blvd.
Foxborough, MA 02035
Tel: (800) 343-1391, (508) 698-2115, Fax: (508) 698-2118
http://www.aemc.com
Makes a selection of clamp-on and flexible-core current transformers which are suitable for extending the range of other alternating current meters.

ABB
SSAC division of ABB Inc has a line of current actuated switches and sensors.
ABB has its own line of current metering equipment for the power industry and has been involved with optical current sensing using the Faraday effect.
http://www.ssac.com
Sales are through:
Entrelec Inc. (USA headquarters)
1950 Hurd Drive
Irving, TX 75038-4312
Tel: (972) 550-9025, (800) 431 2308, Fax: (972) 550-9215, (800) 862-5066
entrelec.info@us.abb.com
http://www.entrelec.com/ssacofclink

TABLE 2.1 (continued) Selected Manufacturers of Current Sensors

Amecon, Inc.
1900 Chris Lane
Anaheim, CA 92805
Tel: (714) 634-2220, (800) 394-2234, Fax: (714) 634-0905
http://www.ameconinc.com
sales@ameconinc.com
Offers a line of smaller CTs and a Hall effect device

American Aerospace Controls
570 Smith St.
Farmingdale, NY 11735
Tel: (516) 694-5100, Fax: (516) 694-6739
http://www.a-a-c.com
Has a line of ac and dc current sensors for mounting in remote locations with wiring to collect data.

Coilcraft
1102 Silver Lake Rd.
Cary IL, 60013
Tel: (800) 322-2645, Fax: (847) 639-6400, (708) 639-6400, (847) 639-1469
http://www.coilcraft.com
Makes smaller current transformers usable up to 100 amps for mounting on printed circuit boards.

CR Magnetics Inc.
544 Axminister Drive
Fenton (St. Louis)
MO 63026
Tel: (636) 343-8518, Fax: (636) 343-5119
http://www.crmagnetics.com
sales@crmagnetics.com
Provides a full line of current transformers and current actuated switches. Their catalog contains a useful discussion of accuracy classes and selection criteria.

Dranetz-BMI (Headquarters)
1000 New Durham Rd.
Edison, NJ 08818-4019
Tel: (800) 372-6832, Fax: (732) 248-1834
http://www.dranetz-bmi.com
Offers a line of equipment for analysis of electrical power and is included here even though it does not make instruments dedicated to current sensing.

Extech Instruments Corporation
335 Bear Hill Road
Waltham, MA 02154-1020
Tel: (617) 890-7440, Fax: (617) 890 7864
http://www.extech.com
Builds a line of hand-held meters suitable for current measurement to 2000 A.

F.W. Bell, A Division of Bell Technologi (sic)
6120 Hanging Moss Rd.
Orlando FL 32807
Tel: (407) 677-6900, (800) 775-2550, Fax: (407) 677-5765
http://www.fwbell.com
Offers noncontact sensors mostly using the Hall effect with a variety of output connections.

Fluke
http://www.fluke.com
Numerous sales offices and distributors, has a line of sensors for their hand-held instruments.

TABLE 2.1 (continued) Selected Manufacturers of Current Sensors

Flex-Core division of Morlan and Associates
6625 McVey Blvd.
Columbus, OH 43235
Tel: (614) 889-6152, (614) 876-8308, Fax: (614) 876-8538
http://www.flex-core.com
mailto:flexcore ail.msn.com
Makes a line of sensors and current actuated switches. Of particular interest is their flexible iron core which can be installed over an existing conductor by bending it.

GMW Danfysik
GMW Associates
955 Industrial Road
San Carlos, CA 94070
Tel: (650) 802-8292, Fax: (650) 802-8298
http://www.gmw.com/index.html
sales@gmw.com
Offers a line of laboratory grade equipment for precision noncontact measurement. Precision in the ppm range and prices to match. Up to 10000 A, dc to more than 200 kHz. Large area CTs suitable for beam current in accelerators under the Bergoz tradename.

LEM USA, Inc. is the North American subsidiary of LEM holding SA of Geneva, Switzerland
6643 W. Mill Road
Milwaukee, WI 53218
Tel: (414) 353-0711, Fax: (414) 353-0733
http://www.lemusa.com
Offers Hall effect sensors in both open- and closed-loop form for ac and dc currents up to 18000 A and up to 500 kHz.

Honeywell Inc. (Microswitch)
http://www.honeywell.com
http://content.honeywell.com/sensing
Offers current sensors mostly using the Hall effect along with their line of magnetic sensors and switches.

Magnetics Division of Spang, Inc.
P.O. Box 11422
Pittsburgh, PA 15238-0422
Tel: (412) 696-1333, (800) 245-3984, Fax: (412) 696-0333
http://www.mag-inc.com
http://www.mag-inc.com/software/transformer.asp
magnetics@spang.com
Offers core material suitable for current transformers with design data.

Metrosil unit of M&I Materials Ltd.
PO Box 136
Manchester, UK
M60 1AN
Tel: +44(0)161 875 4332, Fax: +44(0)161 875 2695
metrosilsales@metrosil.com
http://www.metrosil.com/protection_units.htm
Supplies protective devices for use on secondaries of current transformers.

Neilsen-Kuljian, Inc./NK Technologies
1400 Dell Avenue, Suite A
Campbell, CA 95008-6620
Tel: (800) 959-4014, (408) 871-7510, Fax: (408) 871-7515
http://www.nktechnologies.com
mailto:sales@nktechnologies.com
Current transformers, other current sensors.

TABLE 2.1 (continued) Selected Manufacturers of Current Sensors

NVE Corporation (Nonvolatile Electronics)
11409 Valley View Road
Eden Prairie, MN 55344-3617
Tel: (952) 829-9217, (800) 467-7141, Fax: (952) 996-1600
http://www.nve.com
info@nve.com
http://www.nve.com
Offers a giant magnetoresistive technology for current sensing.

NxtPhase Corporation
3040 East Broadway
Vancouver BC V5M 1Z4
Tel: (604) 215-9822, (604) 215-9833
http://www.nxtphase.com
http://www.nxtphase.com/sub-products-optical_current-nxvct_voltage.htm
info@nxtphase.com
Offers high kilovolt isolated optical current and voltage measurement.

Ohio Semitronics, Inc.
4242 Reynolds Dr.
Hilliard, OH 43026
Tel: (800) 537-6732, (614) 777-1005, Fax: (614) 777-4511
sales@ohiosemitronics.com
http://www.ohiosemi.com
Offers a wide range of current, voltage, and power transducers for ac and dc.

Pearson Electronics Inc.
4009 Transport Street
Palo Alto, California 94303
Tel: (650) 494-6444, (650) 494-6716
http://www.pearsonelectronics.com
Provides high-speed compensated current transformers for pulse work with oscilloscopes.

Smith Research & Technology, Inc. (SRT)
3109 S. Cascade Ave #201
Colorado Springs, CO, 80907-5190
Tel: (719) 634-2259, Fax: (719) 634-2601
Offers gapped inductive sensors with ac, dc, and 4-20 mA output up to 1000 A. Ac current actuated switches which control isolated 200 W loads using no external power for the switch are offered.

SSAC - see ABB

Tektronix
http://www.tektronix.com
Offers current sensors and hand-held instruments as well as current-sensing probes for oscilloscopes.

Zetex plc
Lansdowne Road
Chadderton
OL9 9TY
United Kingdom
Tel: +44 (0) 161 622 4444, Fax: +44 (0) 161 622 4720
Tel: (516) 543-7100, Fax: (516) 864-7630 in the USA
http://www.zetex.com
This supplier of integrated circuits and transistors offers magnetoresistive current and magnetic field sensors (ZMZ20) for use on printed circuit boards.

References

1. I. Genzer and P. Youngner, *Physics,* Chapter 10-6, Morristown, NJ: General Learning Corporation, 1973. A high school physics text.
2. H.E. White, *Modern College Physics,* Chapter 51, New York: D. Van Nostrand, 1952.
3. B.N. Taylor, *Guide for the Use of the International System of Units* (SI), NIST Special Publication 811, Washington, D.C., U.S. Government Printing Office, 1995, Appendix A.5.
4. A. Sommerfeld, *Electrodynamics, Lectures on Theoretical Physics, Vol. III,* New York: Academic Press, 1952.
5. R.C. Dorf (Ed.), *The Electrical Engineering Handbook,* Chapter 49, The Hall Effect, Boca Raton, FL, CRC Press, 1993.
6. K.B. Rochford, A.H. Rose, M.N. Deeter, and G.W. Day, Faraday effect current sensor with improved sensitivity-bandwidth product, *Optics Lett.,* 19, 1903-1905, 1994.
7. S.R. Forrest (Ed.), *JTEC Panel Report on Optoelectronics in Japan and the United States,* Baltimore, MD: International Technology Research Institute, Loyola College, February 1996. NTIS PB96-152202. Available *http://itri.loyola.edu/opto/c6_s3.htm.*
8. M.N. Deeter, Domain effects in Faraday effect sensors based on iron garnets, *Appl. Opt.,* 34, 655, 1995.
9. M.N. Deeter, A.H. Rose, and G.W. Day, Faraday effect magnetic field sensors based on substituted iron garnets, in *Fiber Optic and Laser Sensors VIII,* R.P. DePaula and E. Udd, eds., Proc. Soc. Photo-Opt. Instrumentation. Eng. 1367, 243-248, 1990.
10. S. Cantor, NASA Tech Briefs, November 1993, p. 54.

3

Power Measurement

Pasquale Arpaia
Università di Napoli Federico II

Francesco Avallone
Università di Napoli Federico II

Aldo Baccigalupi
Università di Napoli Federico II

Claudio de Capua
Università di Napoli Federico II

Carmine Landi
Università de L'Aquila

In this chapter, the concept of electric power is first introduced, and then the most popular power measurement methods and instruments in *dc*, *ac*, and *pulse* waveform circuits are illustrated.

Power is defined as the *work performed per unit time.* So, dimensionally, it is expressed as joules per second, J s^{-1}. According to this general definition, electric power is the electric work or energy dissipated per unit time and, dimensionally, it yields:

$$\mathrm{J\,s^{-1} = J\,C^{-1} \times C\,s^{-1} = V \times A} \tag{3.1}$$

where J = Joules
s = Seconds
C = Coulombs
V = Volts
A = Amperes

The product *voltage times current* gives an electrical quantity equivalent to *power.*

3.1 Power Measurements in Dc Circuits

Electric power (P) dissipated by a load (L) fed by a dc power supply (E) is the product of the voltage across the load (V_L) and the current flowing in it (I_L):

$$P = V_L \times I_L \tag{3.2}$$

Therefore, a power measurement in a dc circuit can be generally carried out using a voltmeter (V) and an ammeter (A) according to one of the arrangements shown in Figure 3.1. In the arrangement of Figure 3.1(a), the ammeter measures the current flowing into the voltmeter, as well as that into the load; whereas in the arrangement of Figure 3.1(b), this error is avoided, but the voltmeter measures the voltage drop across the ammeter in addition to that dropping across the load. Thus, both arrangements give a surplus of power measurement absorbed by the instruments. The corresponding measurement errors are generally referred to as insertion errors.

0-8493-1733-9/04/$0.00+$1.50

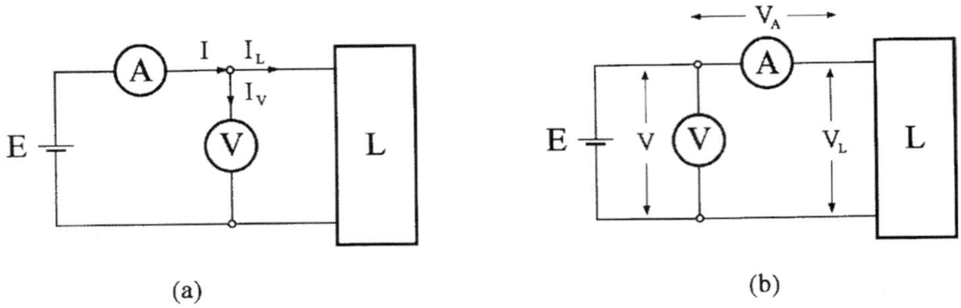

FIGURE 3.1 Two arrangements for dc power measurement circuits.

According to the notation:

- I, Current measured by the ammeter (Figure 3.1(a))
- V, Voltage measured by the voltmeter (Figure 3.1(b))
- R_V, R_A, Internal resistance of the voltmeter and the ammeter, respectively
- R_L, Load resistance
- I_V Current flowing into the voltmeter (Figure 3.1(a))
- V_A, Voltage drop across the ammeter (Figure 3.1(b))

the following expressions between the measurand electric power P and the measured power $V \times I$ are derived by analyzing the circuits of Figures 3.1(a) and 3.1(b), respectively:

$$P = V_L \times I_L = V \times I \times \left(\frac{R_V - R_L}{R_V} \right) \tag{3.3}$$

$$P = V_L \times I_L = V \times I \times \left(\frac{R_L - R_A}{R_L} \right) \tag{3.4}$$

If:

- I_V, compared with I
- V_A, compared with V

are neglected for the arrangements of Figure 3.1(a) and 3.1(b), respectively, it approximately yields:

$$\frac{I_V}{I} = \frac{R_L}{R_V + R_L} \cong \frac{R_L}{R_V} \cong 0; \quad \frac{V_A}{V} \cong \frac{R_A}{R_A + R_L} \cong \frac{R_A}{R_L} \cong 0; \tag{3.5}$$

consequently, measured and measurand power will be coincident.

On this basis, from Equations 3.3, 3.4, and 3.5, analytical corrections of the insertion errors can be easily derived for the arrangement of Figures 3.1(a) and 3.1(b), respectively.

The instrument most commonly used for power measurement is the *dynamometer.* It is built by (1) two fixed coils, connected in series and positioned coaxially with space between them, and (2) a moving coil, placed between the fixed coils and equipped with a pointer (Figure 3.2(a)).

The torque produced in the dynamometer is proportional to the product of the current flowing into the fixed coils times that in the moving coil. The fixed coils, generally referred to as *current coils,* carry the load current while the moving coil, generally referred to as *voltage coil,* carries a current that is proportional, via the multiplier resistor R_V, to the voltage across the load resistor R_L. As a consequence, the deflection of the moving coil is proportional to the power dissipated into the load.

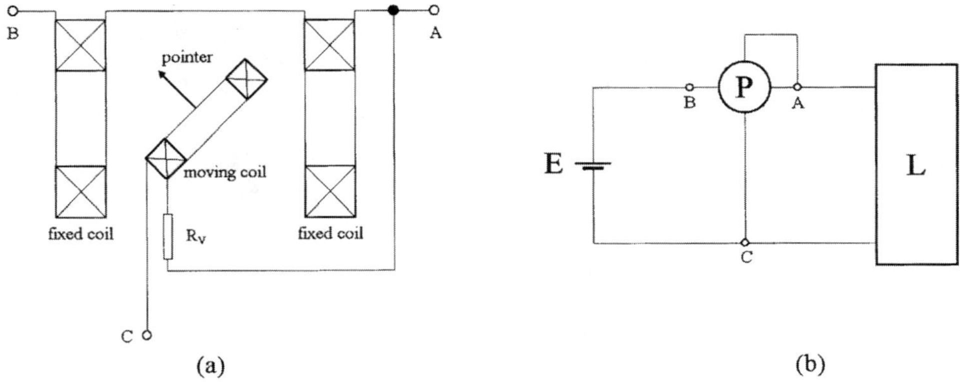

FIGURE 3.2 Power measurement with a dynamometer. (a) Working principle; (b) measurement circuit.

As for the case of Figure 3.1, insertion errors are also present in the dynamometer power measurement. In particular, by connecting the voltage coil between A and C (Figure 3.2(b)), the current coils carry the surplus current flowing into the voltage coil. Consequently, the power P_L dissipated in the load can be obtained by the dynamometer reading P as:

$$P_L = P - \frac{V^2}{R'_v} \tag{3.6}$$

where R'_v is the resistance of the voltage circuit ($R'_v = R_v + R_{vc}$, where R_{vc} is the resistance of the voltage coil). By connecting the moving coil between B and C, this current error can be avoided, but now the voltage coil measures the surplus voltage drop across the current coils. In this case, the corrected value is:

$$P_L = P - I^2 R_C \tag{3.7}$$

where R_C is the resistance of the current coil.

3.2 Power Measurements in Ac Circuits

Definitions

All the above considerations relate to *dc* power supplies. Now look at power dissipation in *ac* fed circuits. In this case, electric power, defined as voltage drop across the load times the current flowing through it, is the function:

$$p(t) = v(t) \times i(t) \tag{3.8}$$

referred to as the *instantaneous power.* In ac circuits, one is mainly interested in the mean value of instantaneous power for a defined time interval. In circuits fed by periodic ac voltages, it is relevant to define the mean power dissipated in one period T (*active power P*):

$$P = \frac{1}{T}\int_0^T p(t)\,dt \tag{3.9}$$

The simplest case is a sinusoidal power supply feeding a purely resistive load. In this case, $v(t)$ and $i(t)$ are in phase and $p(t)$ is given by:

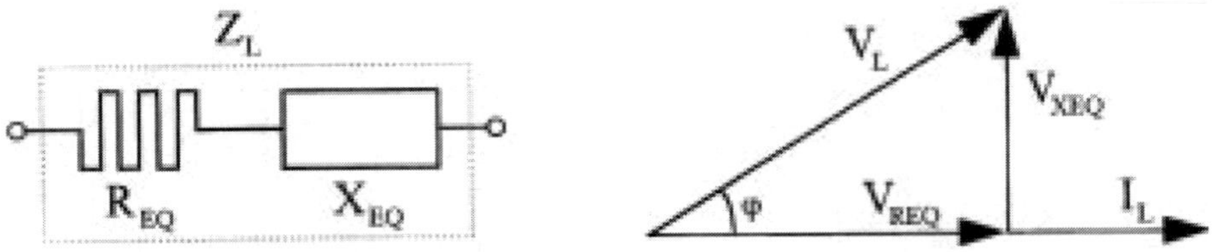

FIGURE 3.3 Voltage drop on the load and on its equivalent components.

$$p(t) = VI[1 - \cos(2\omega t)] \tag{3.10}$$

where V and I = rms values of $v(t)$ and $i(t)$, respectively
ω = power supply angular frequency

Therefore, the instantaneous power is given by a constant value VI plus the ac quantity oscillating with twice the angular frequency of the power supply; thus, the active power is simply the product VI. In this case, all the above considerations referring to active power for dc circuits are still correct, but voltages and currents must be replaced by the corresponding rms values.

The case of purely reactive loads is the opposite; the voltage drop across the load and current flowing through it are out of phase by 90°. Instantaneous power $p(t)$ is given by:

$$p(t) = VI\cos(2\omega t) \tag{3.11}$$

Thus, the active power dissipated by a reactive load is zero, owing to the phase introduced by the load itself between voltage and current.

The simplest cases of sinusoidal power sources supplying purely resistive and purely reactive loads have been discussed. In these cases, the load is expressed by a real or a pure imaginary number. In general, the load is represented by a complex quantity (the impedance value). In this case, load impedance can be represented by its equivalent circuit (e.g., a pure resistance and a pure reactance in series). With this representation in mind, the electric power dissipated in the load Z_L (Figure 3.3) can be expressed by the sum of power components separately dissipated by resistance R_{EQ} and reactance X_{EQ} of the equivalent circuit Z_L.

Considering that no active power is dissipated in the reactance X_{EQ}, it yields:

$$P = V_{REQ} I_L = V_L I_L \cos\varphi \tag{3.12}$$

The term $\cos\varphi$ appearing in Equation 3.12 is referred to as the *power factor.* It considers that only a fraction of voltage V_L contributes to the power; in fact, its component V_{XEQ} (the drop across the reactance) does not produce any active power, as it is orthogonal to the current I_L flowing into the load.

Figure 3.4 plots the waveforms of instantaneous power $p(t)$, voltage $v(t)$, and current $i(t)$. The effect of the power factor is demonstrated by a dc component of $p(t)$ that varies from a null value (i.e., $v(t)$ and $i(t)$ displaced by 90°) toward the value VI (i.e., $v(t)$ and $i(t)$ in phase).

The term:

$$P_A = V_L I_L \tag{3.13}$$

is called the *apparent power*, while the term:

$$Q = V_{XEQ} I_L = V_L I_L \sin\varphi \tag{3.14}$$

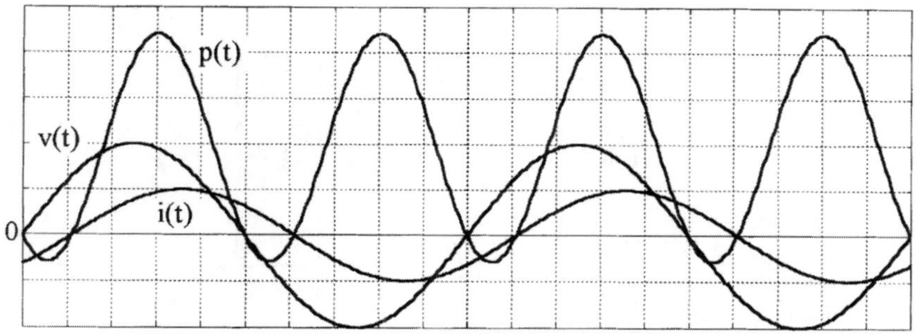

FIGURE 3.4 Waveforms of instantaneous power (p), voltage (v), and current (i).

is called the *reactive power* because it represents a quantity that is dimensionally equivalent to power. This is introduced as a consequence of the voltage drop across a pure reactance and, therefore, does not give any contribution to the active power. From Figure 3.3, the relationship existing between *apparent power, active power*, and *reactive power* is given by:

$$P_A = \sqrt{P^2 + Q^2} \tag{3.15}$$

Dynamometers working in ac circuits are designed to integrate instantaneous power according to Equation 3.9. Insertion errors can be derived by simple considerations analogous to the dc case. However, in ac, a phase uncertainty due to the not purely resistive characteristic of voltage circuit arises. In sinusoidal conditions, if ε_w (in radians) is the phase of the real coil impedance, and $\cos\varphi$ is the load power factor, the relative uncertainty in active power measurements can be shown to be equal to $-\varepsilon_w T_g \varphi$. The phase uncertainty depends on the frequency. By using more complex circuits, the frequency range of the dynamometer can be extended to a few tens of kilohertz.

The above has presented the power definitions applied to ac circuits with the restrictions of sinusoidal quantities. In the most general case of distorted quantities, obviously symbolic representation can no longer be applied. In any case, active power is always defined as the mean power dissipated in one period.

As far as methods and instruments for ac power measurements are concerned, some circuit classification is required. In fact, the problems are different, arising in circuits as the frequency of power supply increases. Therefore, in the following, ac circuits will be classified into (1) line-frequency circuits, (2) low- and medium-frequency circuits (up to a few megahertz), and (3) high-frequency circuits (up to a few gigahertz). Line-frequency circuits will be discussed separately from low-frequency circuits, principally because of the existence of problems related specifically to the three-phase power supply of the main.

Low- and Medium-Frequency Power Measurements

In the following, the main methods and instruments for power measurements at low and medium frequencies are considered.

Three-Voltmeter Method

The power dissipation in the load L can be measured using a noninductive resistor R and measuring the three voltages shown in Figure 3.5 [1]. Although one of the voltages might appear redundant on a first analysis of the circuit, in actual fact, three independent data are needed in order to derive power from Equation 3.12. In particular, from voltage drops v_{AB} and v_{BC}, the load current and load voltage can be directly derived; instead, v_{AC} is used to retrieve information about their relative phase.

If currents derived by voltmeters are neglected and the current i_L flowing into the load L is assumed to be equal to that flowing into the resistor R, the statement can be demonstrated as follows:

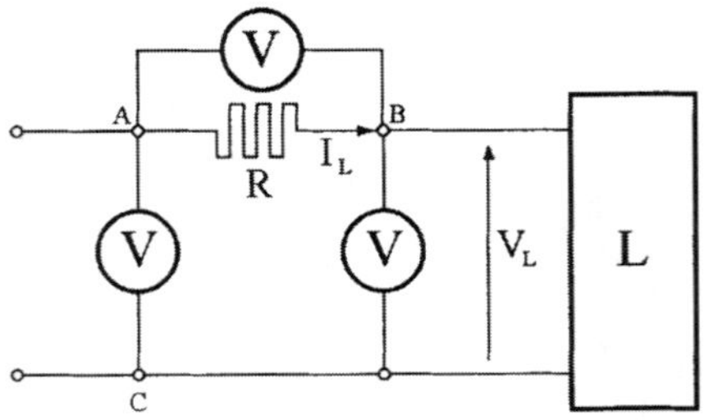

FIGURE 3.5 Three-voltmeter method.

$$\begin{aligned} v_{AC} &= v_L + Ri_L \\ v_{AC}^2 &= R^2 i_L^2 + v_L^2 + 2Rv_L i_L \end{aligned} \tag{3.16}$$

where the small characters indicate instantaneous values. By computing rms values (indicated as capital characters), one obtains the power P_L:

$$\begin{aligned} &\frac{1}{T}\int_0^T v_{AC}^2 dt = \frac{1}{T}\int_0^T R^2 i_L^2 dt = \frac{1}{T}\int_0^T v_L^2 dt + \frac{1}{T}\int_0^T 2Rv_L i_L dt \\ &V_{AC}^2 = RI_L^2 + V_L^2 + 2RP_L \\ &P_L = \frac{V_{AC}^2 - R^2 I_L^2 - V_L^2}{2R} = \frac{V_{AC}^2 - V_{AB}^2 - V_{BC}^2}{2R} \end{aligned} \tag{3.17}$$

Equation 3.17 is also the same in dc by replacing rms values with dc values. Since the result is obtained as a difference, problems arise from relative uncertainty when the three terms have about sum equal to zero.

Such a method is still used for high-accuracy applications.

Thermal Wattmeters

The working principle of thermal wattmeters is based on a couple of twin thermocouples whose output voltage is proportional to the square of the rms value of the currents flowing into the thermocouple heaters [2].

The principle circuit of a thermal wattmeter is shown in Figure 3.6(a). Without the load, with the hypothesis $S \ll r_1$ and $S \ll r_2$, the two heaters are connected in parallel and, if they have equal resistance r ($r_1 = r_2 = r$), they are polarized by the same current i_p

$$i_1 = i_2 = \frac{i_P}{2} = \frac{v}{2R + r} \tag{3.18}$$

In this case, the output voltages of the two thermocouples turn out to be equal ($e_1 = e_2$); thus, the voltage Δe measured by the voltmeter is null. In Figure 3.6(b), this situation is highlighted by the working point T equal for both thermocouples. By applying a load L with a corresponding current i_L, a voltage drop across S arises, causing an imbalance between currents i_1 and i_2. With the hypothesis that $r \ll R$, the two heaters are in series; thus, the current imbalance through them is:

$$i_1 - i_2 = \frac{Si_L}{2R} \tag{3.19}$$

This imbalance increases the current i_1 and decreases i_2. Therefore, the working points of the two thermocouples change: the thermocouple polarized by the current i_1 operates at A, and the other thermocouple

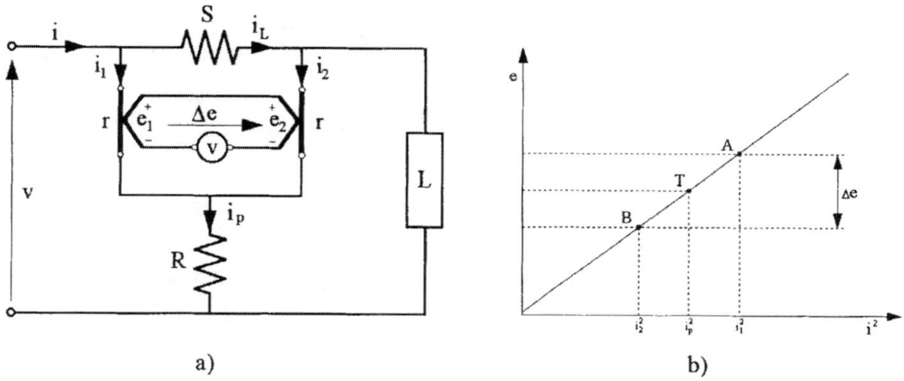

FIGURE 3.6 Thermal wattmeter based on twin thermocouples (*a*); working characteristic in ideal conditions (*b*).

operates at B (Figure 3.6(b)). In this situation, with the above hypotheses, the voltmeter measures the voltage imbalance Δe proportional to the active power absorbed by the load (except for the surplus given by the powers dissipated in R, S, r_1, and r_2):

$$\begin{aligned}\Delta e &= k\left(\left\langle i_1^2\right\rangle - \left\langle i_2^2\right\rangle\right) = k\left(\left\langle \left(i_{\mathrm{P}} + i_{\mathrm{L}}\right)^2\right\rangle - \left\langle \left(i_{\mathrm{P}} - i_{\mathrm{L}}\right)^2\right\rangle\right) \\ &= k\left\langle 4 i_{\mathrm{P}} i_{\mathrm{L}}\right\rangle = k_1\left\langle v(t) i(t)\right\rangle = k_1 P\end{aligned} \tag{3.20}$$

where the notation $\langle i \rangle$ indicates the time average of the quantity i.

If the two thermocouples cannot be considered as twins and linear, the power measurement accuracy will be obviously compromised. This situation is shown in Figure 3.7 where the two thermocouples are supposed to have two quite different nonlinear characteristics. In this case, the voltage measured by voltmeter will be Δe_{n} instead of Δe.

Wattmeters based on thermal principle allow high accuracy to be achieved in critical cases of highly distorted wide-band spectrum signals.

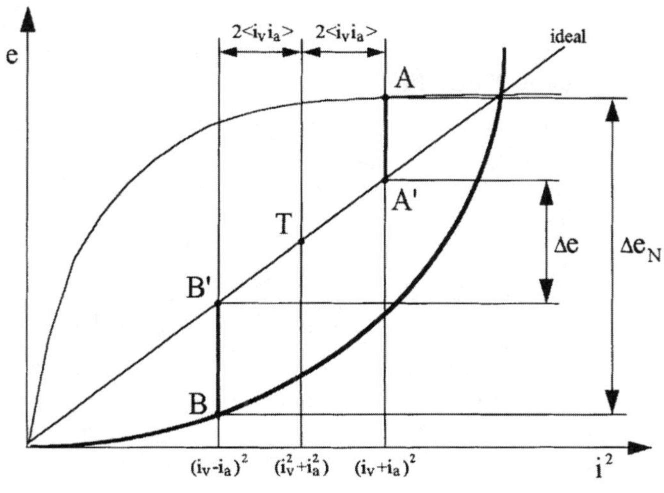

FIGURE 3.7 Ideal and actual characteristics of thermal wattmeter thermocouples.

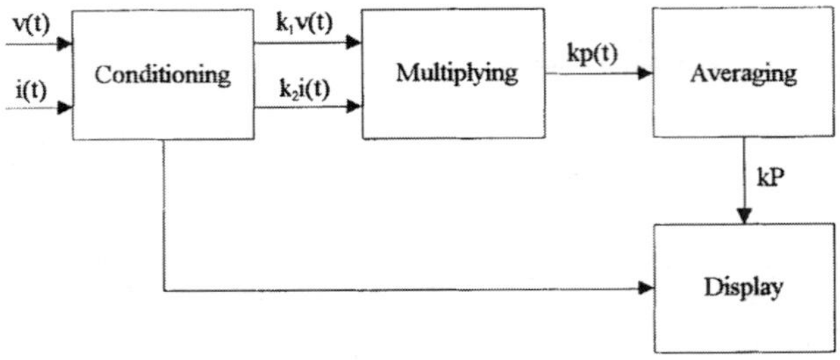

FIGURE 3.8 Block diagram of a multiplier-based wattmeter.

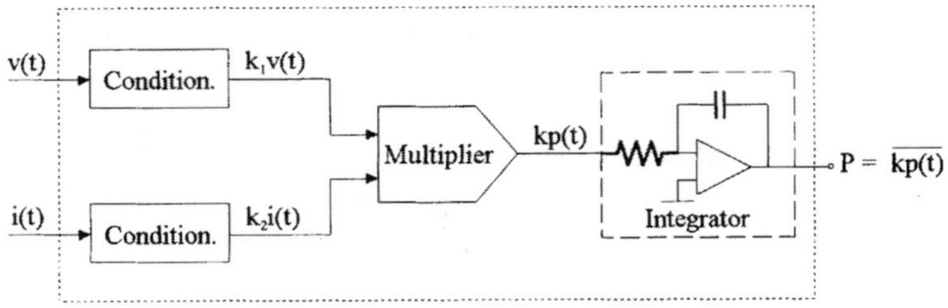

FIGURE 3.9 Block diagram of a four-quadrant, multiplier-based wattmeter.

Wattmeters Based on Multipliers

The multiplication and averaging processes (Figure 3.8) involved in power measurements can be undertaken by electronic means.

Electronic wattmeters fall into two categories, depending on whether multiplication and averaging operations are performed in a continuous or discrete way. In continuous methods, multiplications are mainly carried out by means of analog electronic multipliers. In discrete methods, sampling wattmeters take simultaneous samples of voltage and current waveforms, digitize these values, and provide multiplication and averaging using digital techniques.

Analogous to the case of dynamometers, the resistances of the voltage and current circuits have to be taken into account (see Equations 3.6 and 3.7). Also, phase errors of both current ε_{wc} and voltage ε_{wv} circuits increase the relative uncertainty of power measurement, e.g., in case of sinusoidal conditions increased at $(\varepsilon_{wc}-\varepsilon_{wv})T_g\varphi$.

Wattmeters Based on Analog Multipliers

The main analog multipliers are based on a transistor-based popular circuit such as a four-quadrant multiplier [3], which processes voltage and current to give the instantaneous power, and an integrator to provide the mean power (Figure 3.9). More effective solutions are based on (1) Time Division Multipliers (TDMs), and (2) Hall effect-based multipliers.

TDM-Based Wattmeters.

The block diagram of a wattmeter based on a TDM is shown in Figure 3.10 [4]. A square wave v_m (Figure 3.11(a)) with constant period T_g, and duty cycle and amplitude determined by $i(t)$ and $v(t)$, respectively, is generated. If T_g is much smaller than the period of measurands $v_x(t)$ and $v_y(t)$, these voltages can be considered as constant during this time interval.

The duty cycle of v_m is set by an impulse duration modulator circuit (Figure 3.10). The ramp voltage $v_g(t)$ (Figure 3.11(b)) is compared to the voltage $v_y(t)$ proportional to $i(t)$, and a time interval t_2, whose duration is proportional to $v_y(t)$, is determined. If

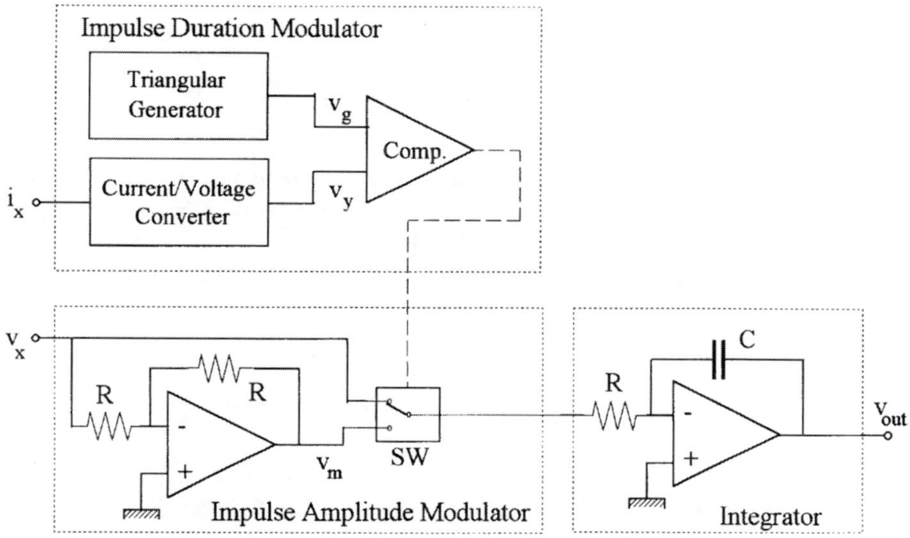

FIGURE 3.10 Block diagram of a TDM-based wattmeter.

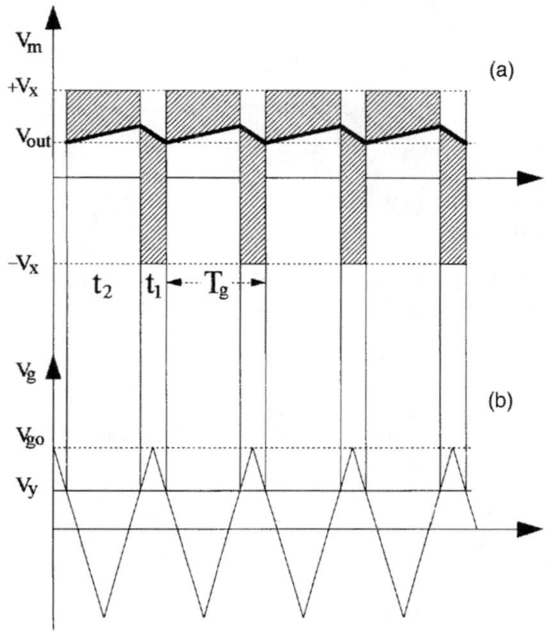

FIGURE 3.11 Waveform of the TDM-based power measurement: (*a*) impulse amplitude modulator output, (*b*) ramp generator output.

$$v_g(t) = \frac{4V_{g0}}{T_g} t \quad \text{when} \quad 0 \le t \le \frac{T_g}{4} \tag{3.21}$$

then from simple geometrical considerations, one obtains:

$$t_2 = 2\left(\frac{T_g}{4} - \frac{v_y T_g}{4V_{g0}}\right) \tag{3.22}$$

and

$$t_1 - t_2 = \frac{T_g}{V_{g0}} v_y \tag{3.23}$$

The amplitude of $v_m(t)$ is set by an impulse amplitude modulator circuit. The output square wave of the impulse duration modulator drives the output $v_m(t)$ of the switch SW to be equal to $+v_x$ during the time interval t_1, and to $-v_x$ during the time interval t_2 (Figure 3.11(a)).

Then, after an initial transient, the output voltage $v_{out}(t)$ of the low-pass filter (integrator) is the mean value of $v_m(t)$:

$$V_{out} = \frac{1}{RC}\int_0^t v_m(t)dt = K'\left(\int_0^{t1} v_x(t)dt - \int_{t1}^{t1+t2} v_x(t)dt\right) = K'v_x(t_1 - t_2) = Kv_x v_y \tag{3.24}$$

The high-frequency limit of this wattmeter is determined by the low-pass filter and it must be smaller than half of the frequency of the signal $v_g(t)$. The frequency limit is generally between 200 Hz and 20 kHz, and can reach 100 kHz. Uncertainties are typically 0.01 to 0.02% [5].

Hall Effect-Based Wattmeters.
As is well known, in a Hall-effect transducer, the voltage $v_H(t)$ is proportional to the product of two time-dependent quantities [6]:

$$v_H(t) = R_H i(t) B_t \tag{3.25}$$

where R_H = Hall constant
$i(t)$ = Current through the transducer
$B(t)$ = Magnetic induction

In the circuit of Figure 3.12(a), the power P is determined by measuring $v_H(t)$ through a high-input impedance averaging voltmeter, and by considering that $v_x(t) = ai(t)$ and $i_x(t) = bB(t)$, where a and b are proportionality factors:

$$P = \frac{1}{T}\int_0^T v_x(t)\cdot i_x(t)dt = ab\frac{1}{T}\int_0^T i(t)\cdot B(t)dt = abR_H V_H \tag{3.26}$$

where T is the measurand period, and V_H the mean value of $v_H(t)$.

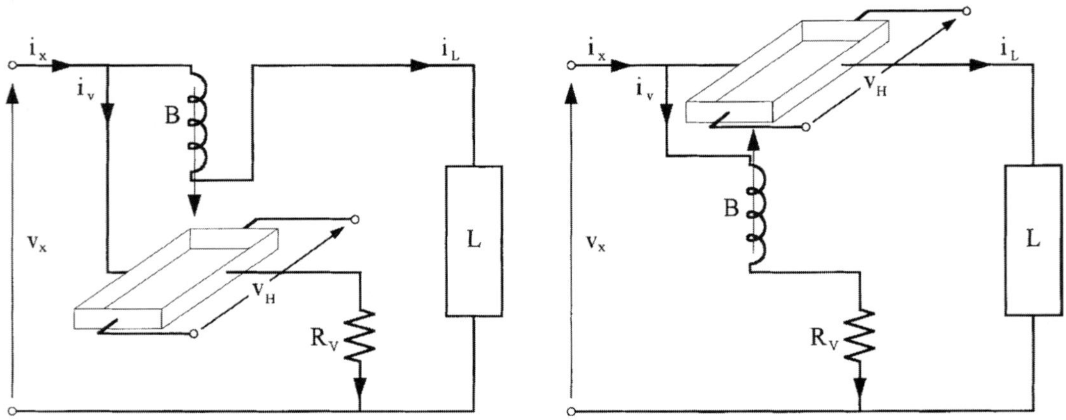

FIGURE 3.12 Configurations of the Hall effect-based wattmeter.

In the usual realization of the Hall multiplier (0.1% up to a few megahertz), shown in Figure 3.12(a), the magnetic induction is proportional to the load current and the optimal polarizing current i_v is set by the resistor R_v.

For the frequency range up to megahertz, an alternative arrangement is shown in Figure 3.12(b), in which the load current I_L flows directly into the Hall device, acting as a polarizing current, and the magnetic field is generated by the voltage v. In this same way, the temperature influence is reduced for line-frequency applications with constant-amplitude voltages and variable load currents.

In the megahertz to gigahertz range, standard wattmeters use probes in waveguides with rectifiers.

Wattmeters Based on Digital Multipliers

Sampling Wattmeters.

The most important wattmeter operating on discrete samples is the sampling wattmeter (Figure 3.13). It is essentially composed of two analog-to-digital input channels, each constituted by (1) a conditioner (C), (2) a sample/hold (S/H), (3) an analog-to-digital converter (ADC), (4) a digital multiplier (MUL), and (5) summing (SUM), dividing (DIV), and displaying units (DISP). The architecture is handled by a processing unit not shown in Figure 3.13.

If samples are equally spaced, the active power is evaluated as the mean of the sequence of instantaneous power samples $p(k)$:

$$\bar{p} = \frac{1}{N}\sum_{k=0}^{N-1} p(k) = \frac{1}{N}\sum_{k=0}^{N-1} v(k)i(k) \tag{3.27}$$

where N^* represents the number of samples in one period of the input signal, and $v(k)$ and $i(k)$ are the kth samples of voltage and current, respectively. A previous estimation of the measurand fundamental period is made to adjust the summation interval of Equation 3.27 and/or the sampling period in order to carry out a synchronous sampling [7]. The sampling period can be adjusted by using a frequency multiplier with PLL circuit driven by the input signal [8]. Alternatively, the contribution of the sampling error is reduced by carrying out the mean on a high number of periods of the input signal.

In the time domain, the period estimation of highly distorted signals, such as Pulse Width Modulation (PWM), is made difficult by the numerous zero crossings present in the waveform. Some types of digital filters can be used for this purpose. An efficient digital way to estimate the period is the discrete integration of the PWM signal. In this way, the period of the fundamental harmonic is estimated by detecting the sign changes of the cumulative sum function [9]:

$$S(k) = \sum_{i=1}^{k} p_i \qquad k = 1, 2, \ldots, N \tag{3.28}$$

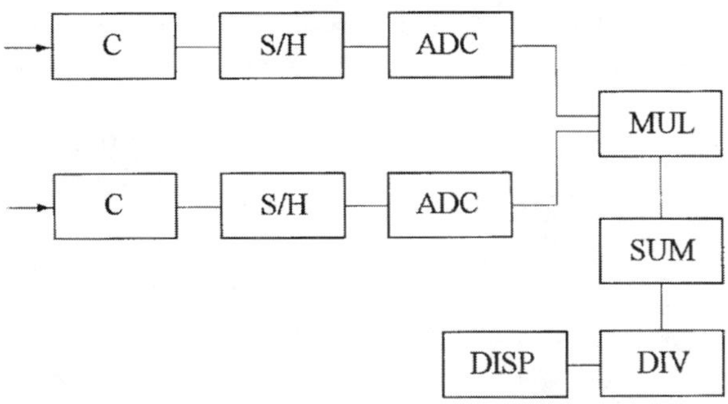

FIGURE 3.13 Block diagram of the sampling wattmeter.

If the summation interval is extended to an integer number of periods of the $S(k)$ function, a "quasisynchronous" sampling [10] is achieved through a few simple operations (cumulative summation and sign detection) and the maximum synchronization error is limited to a sampling period. Through relatively small increases in computational complexity and memory size, the residual error can be further reduced through a suitable data processing algorithm; that is, the multiple convolution in the time domain of triangular windows [9]. Hence, the power measurement can be obtained as:

$$P_{(B)} = \frac{1}{\sum_{k=0}^{2B(N^*-1)} w(k)} \sum_{k=0}^{2B(N^*-1)} w(k)\,p(k) \tag{3.29}$$

where $p(k)$ is the kth sample of the instantaneous power and $w(k)$ the kth weight corresponding to the window obtained as the convolution of B triangular windows [10].

Another way to obtain the mean power is through the consideration of the harmonic components of voltages and currents in the frequency domain using the Discrete Fourier Transform [11]. In particular, a Fast Fourier Transform algorithm is used in order to improve efficiency. Successively, a two-step research of the harmonic peaks is carried out: (1) the indexes of the frequency samples corresponding to the greatest spectral peaks provide a rough estimate of the unknown frequencies when the wide-band noise superimposed onto the signal is below threshold; (2) a more accurate estimate of harmonic frequencies is carried out to determine the fractional bin frequency (i.e., the harmonic determination under the frequency resolution); to this aim, several approaches such as zero padding, interpolation techniques, and flat-top window-based technique can be applied [12].

Line-Frequency Power Measurements

For line applications where the power is directly derived by the source network, the assumption of infinite power source can be reliably made, and at least one of the two quantities voltage or current can be considered as sinusoidal. In this case, the definition of the power as the product of voltage and current means that only the power at the fundamental frequency can be examined [13].

Single-Phase Measurements

Single-phase power measurements at line frequency are carried out by following the criteria previously mentioned. In practical applications, the case of a voltage greater than 1000 V is relevant; measurements must be carried out using voltage and current transformers inserted as in the example of Figure 3.14. The relative uncertainty is equal to:

$$\frac{\Delta P}{P} = \left(\eta_w + \eta_a + \eta_v\right) + \left(\varepsilon_w + \varepsilon_a + \varepsilon_v\right) T_g \varphi_c \tag{3.30}$$

where η_w and ε_w are the instrumental and phase uncertainty of the wattmeter, η_a and η_v are the ratio uncertainties of current (CT) and voltage (VT) transformers, and ε_a and ε_v their phase uncertainties, respectively.

If the load current exceeds the current range of the wattmeter, a current transformer must be used, even in the case of low voltages.

Polyphase Power Measurements

Three-phase systems are the polyphase systems most commonly used in practical industrial applications. In the following, power measurements on three-phase systems will be derived as a particular case of polyphase systems (systems with several wires) and analyzed for increasing costs: (1) balanced and symmetrical systems, (2) three-wire systems, (3) two wattmeter-based measurements, (4) unbalanced systems, (5) three wattmeter-based measurements, and (6) medium-voltage systems.

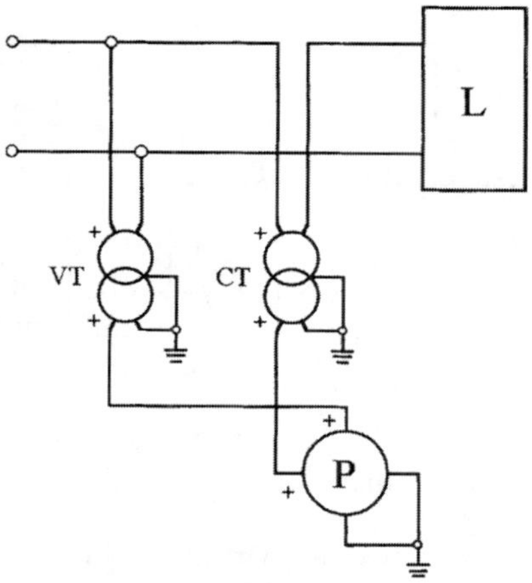

FIGURE 3.14 Single-phase power measurement with voltage (VT) and current (CT) transformers.

Measurements on Systems with Several Wires

Consider a network with sinusoidal voltages and currents composed by *n* wires. For the currents flowing in such wires, the following relation is established:

$$\sum_{1}^{n} \dot{I}_{\mathrm{i}} = 0 \tag{3.31}$$

The network can be thought as composed of $n-1$ single-phase independent systems, with the common return on any one of the wires (e.g., the sth wire). Then, the absorbed power can be measured as the sum of the readings of $n-1$ wattmeters, each one inserted with the current circuit on a different wire and the voltmeter circuit between such a wire and the sth one (Figure 3.15):

$$P = \sum_{1}^{n-1} \left(\dot{V}_{\mathrm{is}} \times \dot{I}_{\mathrm{i}} \right) \tag{3.32}$$

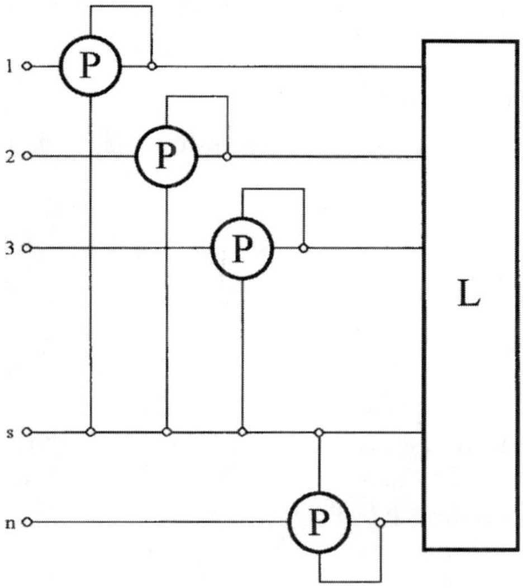

FIGURE 3.15 Power measurement on systems with several wires.

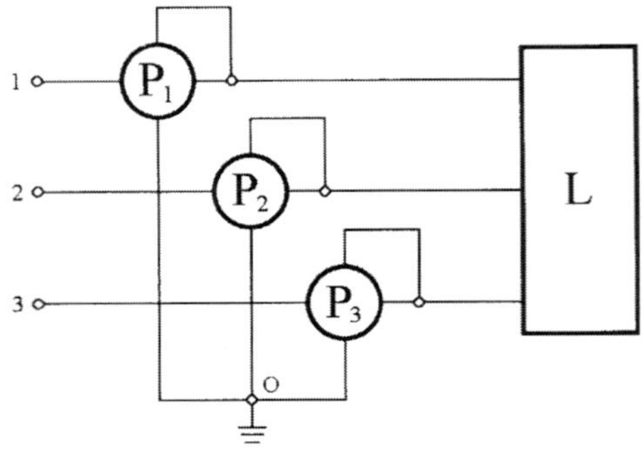

FIGURE 3.16 Power measurement on three-wire systems.

The absorbed power can be also measured by referring to a generic point O external to the network. In this case, the absorbed power will be the sum of the readings of *n* wattmeters, each inserted with the ammeter circuit on a different wire and the voltmeter circuit connected between such a wire and the point O:

$$P=\sum_{1}^{n}\left(\dot{V}_{io}\times\dot{I}_{i}\right) \tag{3.33}$$

Power Measurements on Three-Wire Systems

Active power in a three-phase power system can generally be evaluated by three wattmeters connected as shown in Figure 3.16.

For each power meter, the current lead is connected on a phase wire and the voltmeter lead is connected between the same wire and an artificial neutral point O, whose position is fixed by the voltmeter impedance of power meters or by suitable external impedances.

Under these conditions, absorbed power will be the sum of the three wattmeter indications:

$$P=\sum_{1}^{3}\left(\dot{V}_{io}\times\dot{I}_{i}\right) \tag{3.34}$$

If the three-phase system is provided by four wires (three phases with a neutral wire), the neutral wire is utilized as a common wire.

Symmetrical and Balanced Systems

The supply system is symmetrical and the three-phase load is balanced; that is:

$$\begin{cases} V_1=V_2=V_3 \\ I_1=I_2=I_3 \end{cases} \tag{3.35}$$

In Figure 3.17, the three possible kinds of insertion of an instrument S (an active power or a reactive power meter) are illustrated. The first (a in Figure 3.17) was described in the last subsection; if S is a wattmeter; the overall active power is given by three times its indication, and similarly for the reactive power if S is a reactive power meter. Notice that a couple of twin resistors with the same resistance *R* of the voltage circuit of S are placed on the other phases to balance the load.

The other two insertions are indicated by the following convention: S_{ijk} indicates a reading performed with the current leads connected to the line "*i*" and the voltmeter leads connected between the phases "*j*" and "*k*." If "*i*" is equal to "*j*", one is omitted (e.g., the notation P_{12} (b in Figure 3.17)). The active power absorbed by a single phase is usually referred to as P_1.

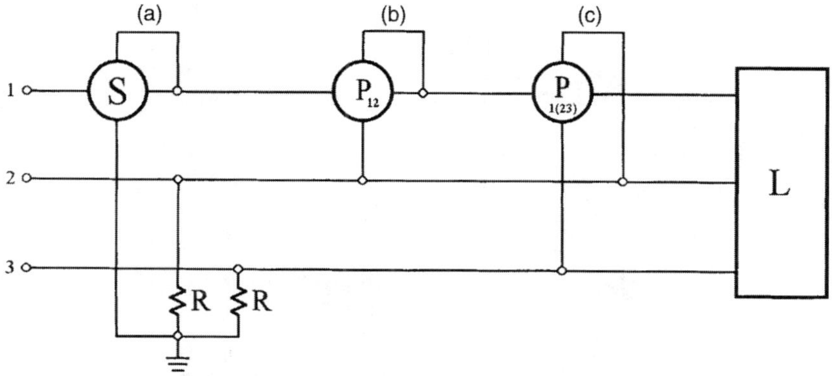

FIGURE 3.17 The three kinds of insertion of a power meter.

The wattmeter reading corresponding to the (c) case in Figure 3.17 is equal to the reactive power Q_1 involved in phase 1, save for the factor $\sqrt{3}$. Hence, in the case of symmetrical and balanced systems, the overall reactive power is given by:

$$Q = 3Q_1 = 3\,P_{1(23)}\big/\sqrt{3} = \sqrt{3}\,P_{1(23)} \tag{3.36}$$

In fact, one has:

$$P_{1(23)} = \dot{I}_1 \times \dot{V}_{23} \tag{3.37}$$

but:

$$\dot{V}_{12} + \dot{V}_{23} + \dot{V}_{31} = 0 \;\Rightarrow\; P_{1(23)} = \dot{I}_1 \times \left(-\dot{V}_{12} - \dot{V}_{31}\right)$$

$$\dot{V}_{13} = -\dot{V}_{31} \qquad \Rightarrow\; P_{1(23)} = -\dot{I}_1 \times \dot{V}_{12} + \dot{I}_1 \times \dot{V}_3$$

$$\begin{cases} \dot{I}_1 \times \dot{V}_{12} = P_{12} \\ \dot{I}_1 \times \dot{V}_{13} = P_{13} \end{cases} \qquad \Rightarrow\; P_{1(23)} = P_{13} - P_{12}$$

In the same manner, the following relationships, which are valid for any kind of supply and load, can be all proved:

$$\begin{aligned} P_{1(23)} &= P_{13} - P_{12} \\ P_{2(31)} &= P_{21} - P_{23} \\ P_{3(12)} &= P_{32} - P_{31} \end{aligned} \tag{3.38}$$

If the supply system is symmetrical, $P_{1(23)} = \sqrt{3}Q_1$.

In fact, moving from the relationship (Figure 3.18):

$$P_{1(23)} = \dot{I}_1 \times \dot{V}_{23} = I_1 V_{23} \cos\beta \tag{3.39}$$

where $\beta = 90° - \varphi_1$, one obtains $P_{1(23)} = \sqrt{3}E_1 I_1 \sin\varphi_1 = \sqrt{3}Q_1$.

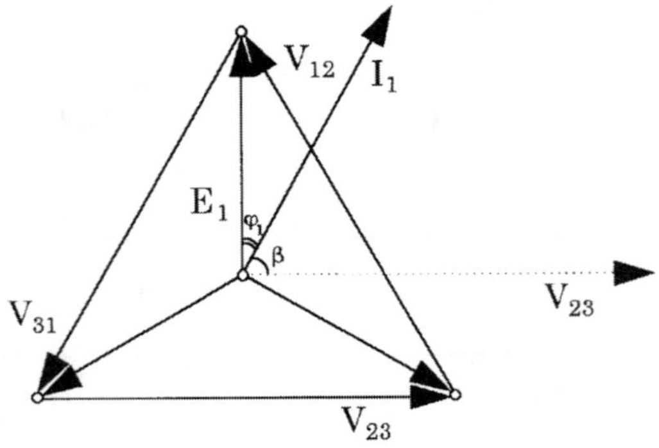

FIGURE 3.18 Phasor diagram for a three-phase symmetrical and balanced system.

In the same manner, the other two corresponding relationships for $P_{2(31)}$ and $P_{3(12)}$ are derived. Hence:

$$
\begin{aligned}
P_{1(23)} &= \sqrt{3}Q_1 = P_{13} - P_{12} \\
P_{2(31)} &= \sqrt{3}Q_2 = P_{21} - P_{23} \\
P_{3(12)} &= \sqrt{3}Q_3 = P_{32} - P_{31}
\end{aligned}
\tag{3.40}
$$

Power Measurements Using Two Wattmeters

The overall active power absorbed by a three-wire system can be measured using only two wattmeters. In fact, Aron's theorem states the following relationships:

$$
\begin{aligned}
P &= P_{12} + P_{32} \\
P &= P_{23} + P_{13} \\
P &= P_{31} + P_{21}
\end{aligned}
\tag{3.41}
$$

Analogously, the overall reactive power can be measured by using only two reactive power meters:

$$
\begin{aligned}
Q &= Q_{12} + Q_{32} \\
Q &= Q_{23} + Q_{13} \\
Q &= Q_{31} + Q_{21}
\end{aligned}
\tag{3.42}
$$

Here one of the previous statements, that is:

$$P = P_{12} + P_{32}$$

is proved. The two wattmeters connected as shown in Figure 3.19 furnish P_{12}, P_{32}:

Hence, the sum of the two readings gives:

$$
\begin{aligned}
P_{12} + P_{32} &= \dot{I}_1 \times \dot{V}_{12} + \dot{I}_3 \times \dot{V}_{32} = \dot{I}_1 \times \left(\dot{E}_1 - \dot{E}_2\right) + \dot{I}_3 \times \left(\dot{E}_3 - \dot{E}_2\right) \\
&= \dot{I}_1 \times \dot{E}_1 - \dot{I}_1 \times \dot{E}_2 + \dot{I}_3 \times \dot{E}_3 - \dot{I}_3 \times \dot{E}_2 = \dot{I}_1 \times \dot{E}_1 + \dot{I}_3 \times \dot{E}_3 - \left(\dot{I}_1 + \dot{I}_3\right) \times \dot{E}_2 \\
&= \dot{I}_1 \times \dot{E}_1 + \dot{I}_3 \times \dot{E}_3 + \dot{I}_2 \times \dot{E}_2 = P_1 + P_2 + P_3 = P
\end{aligned}
\tag{3.43}
$$

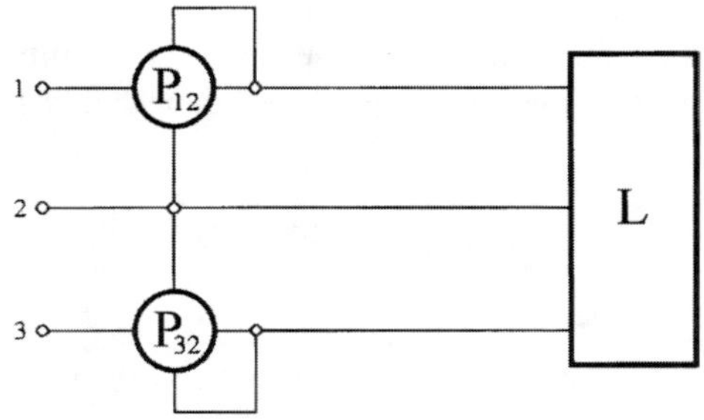

FIGURE 3.19 Power measurements using two wattmeters.

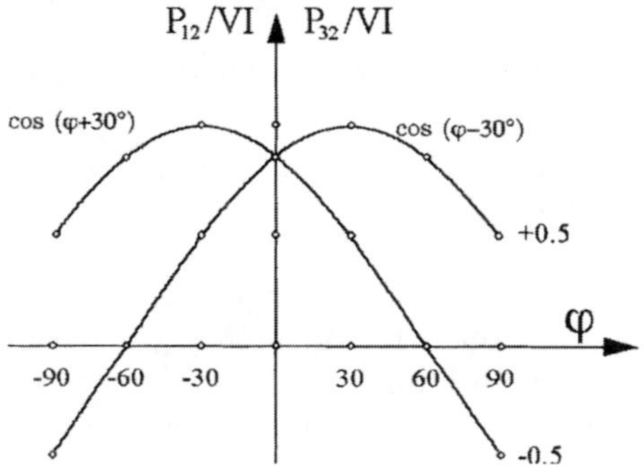

FIGURE 3.20 Sign of powers in Aron insertion.

Provided that the system has only three wires, Aron's theorem applies to any kind of supply and load. In the case of symmetrical and balanced systems, it also allows the reactive power to be evaluated:

$$Q = \sqrt{3} \cdot \left(P_{32} - P_{12}\right) \tag{3.44}$$

Using Equations 3.41 and 3.44, the power factor is:

$$\cos\varphi = \frac{P_{12} + P_{32}}{\sqrt{\left(P_{12} + P_{32}\right)^2 + 3\left(P_{32} - P_{12}\right)^2}} = \frac{P_{12} + P_{32}}{\sqrt{4P_{12}^2 + 4P_{32}^2 - 4P_{12}P_{32}}} = \frac{1 + \frac{P_{12}}{P_{32}}}{2\sqrt{\left(\frac{P_{12}}{P_{32}}\right)^2 - \left(\frac{P_{12}}{P_{32}}\right) + 1}} \tag{3.45}$$

Aron's insertion cannot be utilized when the power factor is low. In fact, if the functions:

$$\begin{aligned} \cos(\varphi + 30) &= \frac{P_{12}}{VI} \\ \cos(\varphi - 30) &= \frac{P_{32}}{VI} \end{aligned} \tag{3.46}$$

are considered (Figure 3.20), it can be argued that: (1) for $\varphi \leq 60°$, P_{12} and P_{32} are both greater than zero; (2) for $\varphi > 60°$, $\cos(\varphi - 30)$ is still greater than zero, and $\cos(\varphi + 30)$ is lower than zero.

The absolute error in the active power is:

$$\Delta P = \frac{\partial\left(P_{12} + P_{32}\right)}{\partial P_{12}} \Delta P_{12} + \frac{\partial\left(P_{12} + P_{32}\right)}{\partial P_{32}} \Delta P_{32} = \Delta P_{12} + \Delta P_{32} \tag{3.47}$$

This corresponding relative error is greater as P_{12} and P_{32} have values closer to each other and are opposite in polarity; in particular, for $\cos\varphi = 0$ ($\varphi = 90°$), the error is infinite.

If η_w and ε_w are the wattmeter amplitude and phase errors, respectively, then the error in the active power is:

$$\frac{\Delta P}{P} = \frac{\left(\eta_w + \varepsilon_w T_g \varphi_{12}\right) P_{12} + \left(\eta_w + \varepsilon_w T_g \varphi_{32}\right) P_{32}}{P_{12} + P_{32}} = \eta_w + \varepsilon_w \frac{Q}{P} \tag{3.48}$$

Let two wattmeters with nominal values V_0, I_0, $\cos\varphi_0$, and class c be considered; the maximum absolute error in each reading is:

$$\Delta P = \frac{c V_0 I_0 \cos\varphi_0}{100} \tag{3.49}$$

Therefore, the percentage error related to the sum of the two indications is:

$$\frac{\Delta P}{P} = \frac{c V_0 I_0 \cos\varphi_0}{\sqrt{3} VI \cos\varphi} = 1.11 \frac{c V_0 I_0 \cos\varphi_0}{VI \cos\varphi} \tag{3.50}$$

equal to approximately the error of only one wattmeter inserted in a single-phase circuit with the same values of I, V, and $\cos\varphi$. Consequently, under the same conditions, the use of two wattmeters involves a measurement uncertainty much lower than that using three wattmeters.

If the Aron insertion is performed via current and voltage transformers, characterized by ratio errors η_a and η_v, and phase errors ε_a and ε_v, respectively, the active power error is:

$$\frac{\Delta P}{P} = \frac{\left(\eta_{TOT} + \varepsilon_{TOT} T_g \varphi_{12}\right) P_{12} + \left(\eta_{TOT} + \varepsilon_{TOT} T_g \varphi_{32}\right) P_{32}}{P_{12} + P_{32}} = \eta_{TOT} + \varepsilon_{TOT} \frac{Q}{P} = \eta_{TOT} + \varepsilon_{TOT} T_g \Phi_c \tag{3.51}$$

where $\cos\Phi_c$ = conventional power factor

$$\left.\begin{aligned} \eta_{TOT} &= \eta_w + \eta_a + \eta_v \\ \varepsilon_{TOT} &= \varepsilon_w + \varepsilon_a + \varepsilon_v \end{aligned}\right\} \text{the error sums with } \eta_w \text{ and } \varepsilon_w \text{ being the wattmeter errors.}$$

Symmetrical Power Systems Supplying Unbalanced Loads

If the load is unbalanced, the current amplitudes are different from each other and their relative phase is not equal to 120°. Two wattmeters and one voltmeter have to be connected as proposed by Barbagelata [13] (Figure 3.21). The first wattmeter can provide P_{12} and P_{13}, and the second one gives P_{31} and P_{32}.

From the Aron theorem, the active power is:

$$P = P_{12} + P_{32} \tag{3.52}$$

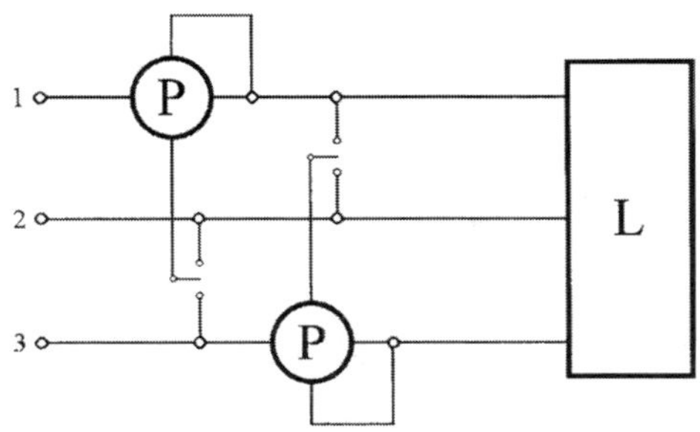

FIGURE 3.21 Barbagelata insertion for symmetrical and unbalanced systems.

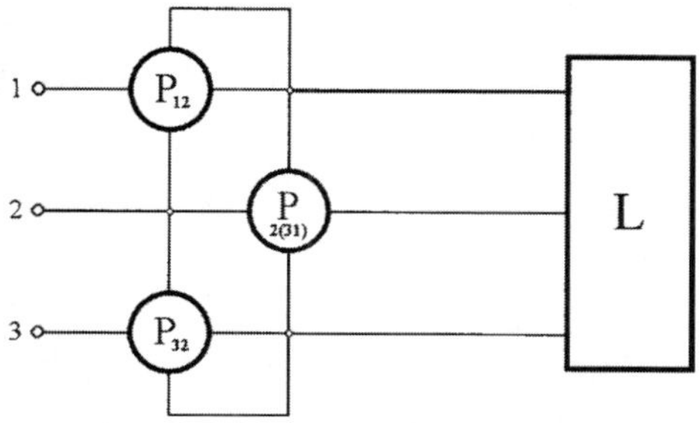

FIGURE 3.22 Righi insertion for symmetrical and unbalanced systems.

and then the reactive power Q is:

$$Q = Q_1 + Q_2 + Q_3 = \frac{1}{\sqrt{3}}\left[P_{13} - P_{12} + \underline{P_{21}} - \underline{P_{23}} + P_{32} - P_{31}\right] \tag{3.53}$$

For the underlined terms, from Aron's theorem it follows that:

$$P = P_{13} + P_{23} = P_{12} + P_{32} = P_{21} + P_{31} \tag{3.54}$$

then:

$$P_{13} + P_{23} = P_{21} + P_{31} \;\Rightarrow\; P_{21} - P_{23} = P_{13} - P_{31}$$

Thus, one obtains:

$$Q = \frac{1}{\sqrt{3}}\left[2\left(P_{13} - P_{31}\right) + P_{32} - P_{12}\right] \tag{3.55}$$

Therefore, using only four power measurements, the overall active and reactive powers can be obtained.

The main disadvantage of this method is that the four measurements are not simultaneous; therefore, any load variations during the measurement would cause a loss in accuracy. In this case, a variation proposed by Righi [13] can be used. In this variation, three wattmeters are connected as shown in Figure 3.22 and give simultaneously P_{12}, P_{32}, and $P_{2(31)}$. Reactive power is:

$$Q = \frac{1}{\sqrt{3}}\left[\underline{P_{13}} - P_{12} + P_{21} - P_{23} + P_{32} - \underline{P_{31}}\right]. \tag{3.56}$$

Analogously as above, from the Aron theorem it follows that:

$$P_{21} - P_{23} = P_{13} - P_{31} \;\Rightarrow\; P_{2(31)} = P_{21} - P_{23} = P_{13} - P_{31} \tag{3.57}$$

then:

$$Q = \frac{1}{\sqrt{3}}\left[P_{32} - P_{12} + 2P_{2(31)}\right] \tag{3.58}$$

For symmetrical and unbalanced systems, another two-wattmeter insertion can be carried out (Figure 3.23). The wattmeters give:

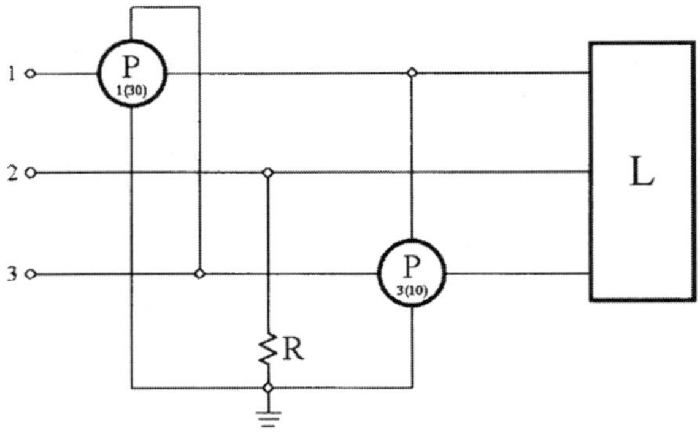

FIGURE 3.23 Two wattmeters-based insertion for symmetrical and unbalanced systems.

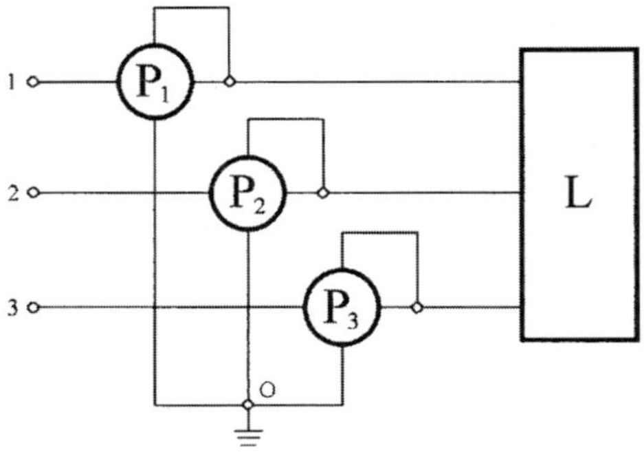

FIGURE 3.24 Three wattmeters-based insertion for three-wire, three-phase systems.

$$P_{1(30)} = \dot{E}_3 \times \dot{I}_1 = j\frac{\dot{V}_{12}}{\sqrt{3}} \times \dot{I}_1 = \frac{Q_{12}}{\sqrt{3}}$$
$$P_{3(10)} = \dot{E}_1 \times \dot{I}_3 = j\frac{\dot{V}_{23}}{\sqrt{3}} \times \dot{I}_3 = \frac{Q_{32}}{\sqrt{3}} \tag{3.59}$$

Hence, the overall reactive power is:

$$Q = Q_{12} + Q_{32} = \sqrt{3}\left[-P_{1(30)} + P_{3(10)}\right] \tag{3.60}$$

Three-Wattmeter Insertion

A three-wire, three-phase system can be measured by three wattmeters connected as in Figure 3.24. The artificial neutral point position does not affect the measurement; it is usually imposed by the impedance of the voltmeter leads of the wattmeters.

Medium-Voltage, Three-Wattmeter Insertion

Analogously to the single-phase case, for medium-voltage circuits, the three-wattmeter insertion is modified as in Figure 3.25.

Method Selection Guide

For three-wire systems, the flow chart of Figure 3.26 leads to selecting the most suitable method according to system characteristics.

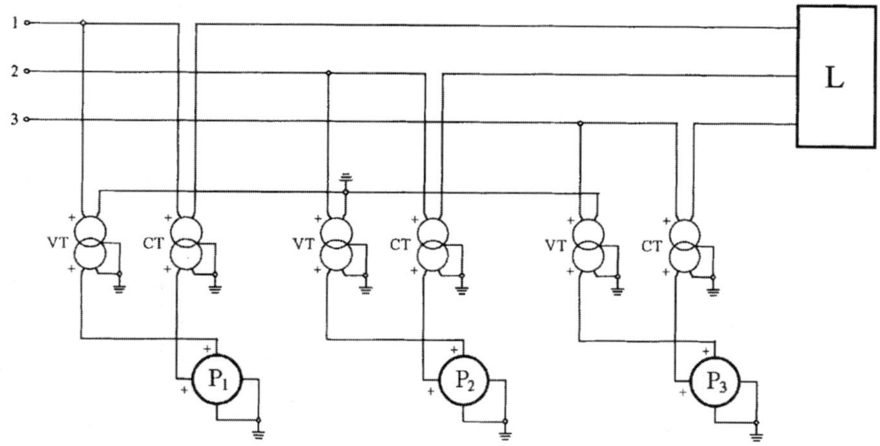

FIGURE 3.25 Medium-voltage, three-wattmeters insertion.

High-Frequency Power Measurements

Meters used for power measurements at radio or microwave frequencies are generally classified as *absorption type* (containing inside their own load, generally 50 Ω for RF work) and *transmitted* or *throughline type* (where the load is remote from the meter). Apart from the type, power meters are mainly based on thermistors, thermocouples, diodes, or radiation sensors. Therefore, to work properly, the sensor should sense all the RF power (P_{LOAD}) incoming into the sensor itself. Nevertheless, line-to-sensor impedance mismatches cause partial reflections of the incoming power (P_{INCIDENT}) so that a meter connected to a sensor does not account for the total amount of reflected power ($P_{\text{REFLECTED}}$). The relationship existing among power dissipated on the load, power incident, and power reflected is obviously:

$$P_{\text{LOAD}} = P_{\text{INCIDENT}} - P_{\text{REFLECTED}} \tag{3.61}$$

Directional couplers are instruments generally used for separating incident and reflected signals so that power meters can measure each of them separately. In Figure 3.27, the longitudinal section of a directional coupler for waveguides is sketched. It is made up by two waveguides properly coupled through two holes. The upper guide is the *primary waveguide* and connects the power source and load; the lower guide is the *secondary waveguide* and is connected to the power meter. To explain the working of directional couplers, incident and reflected waves have been sketched separately in Figure 3.27(a) and 3.27(b). In particular, section a depicts a directional coupler working as incident wave separator, whereas section (b) shows the separation of the reflected wave. The correct working is based on the assumption that the distance between the holes matches exactly *one quarter of the wave length* (λ). In fact, in the secondary waveguide, each hole will give rise to two waves going in opposite directions (one outside and the other inside the waveguide); consequently, in front of each hole, two waves are summed with their own phases. The assumption made on the distance between the holes guarantees that, in front of one hole, (1) the two waves propagating outside the waveguide will be in phase, causing an enforcing effect in that direction; (2) while, in front of the other hole, the two waves (always propagating outside) will be in opposition, causing a canceling effect in that direction. The enforcing and canceling effects for incident and reflected waves are opposite. In particular, according to the directions chosen in Figure 3.27, incident power propagates on the right side and is canceled on the left side (Figure 3.27(a)), while reflected power propagates on the left side and is canceled on the right side (Figure 3.27(b)). Therefore, directional couplers allow separate measurement of incident and reflected power by means of power meters applied, respectively, on the right and on the left side of the secondary waveguide.

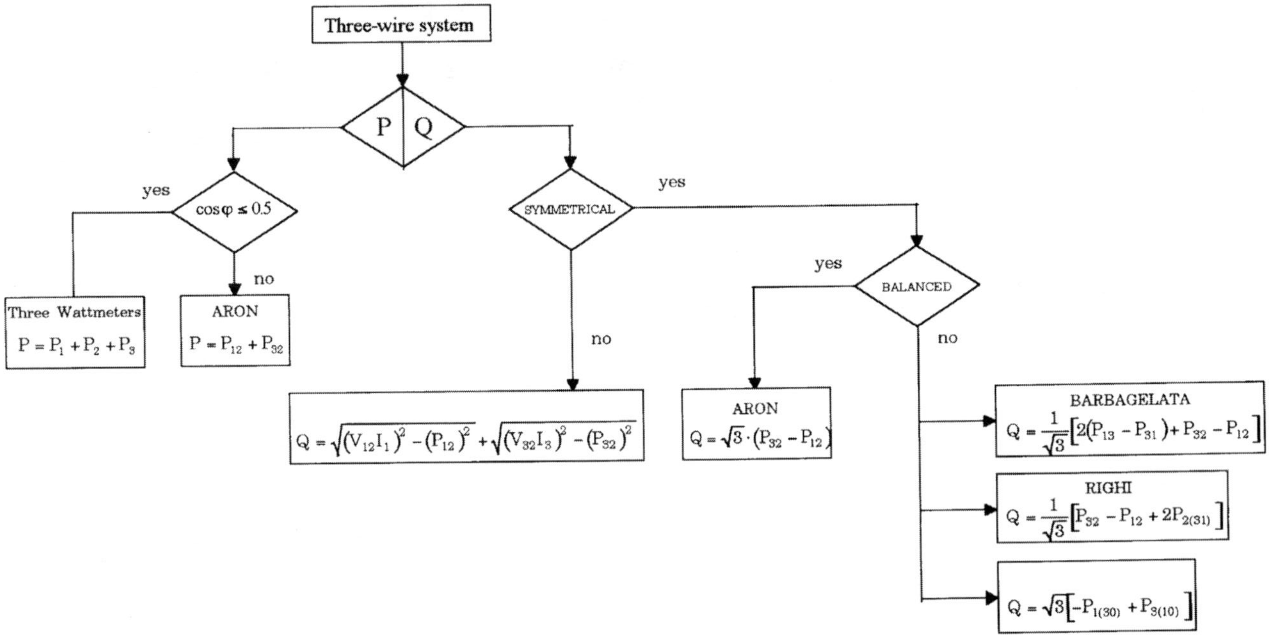

FIGURE 3.26 Method selection guide for power measurements on three-wire systems.

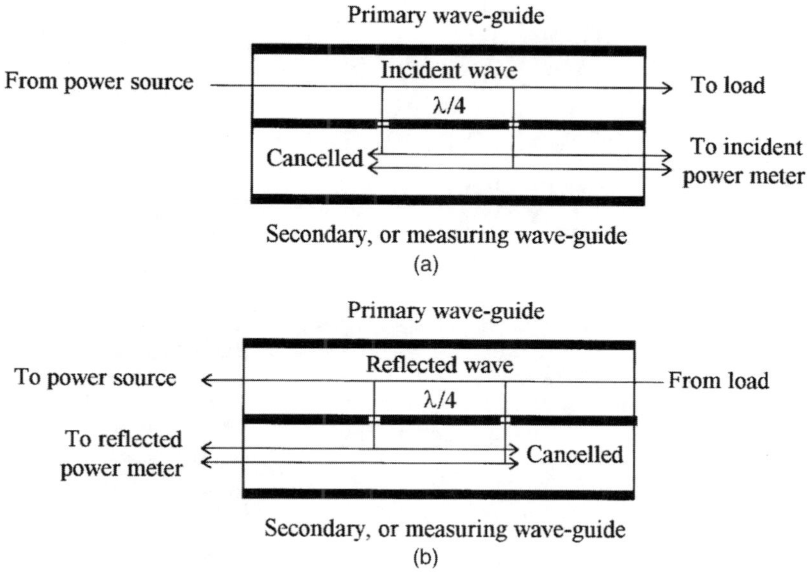

FIGURE 3.27 Directional couplers for separating incident (a) from reflected (b) power.

In any case, the secondary waveguide must be correctly matched from the impedance point of view at both sides (by adaptive loads and/or a proper choice of the power meter internal resistance) in order to avoid unwanted reflections inside the secondary waveguide.

Directional couplers are also used to determine the *reflection coefficient* ρ of the sensor, which takes into account mismatch losses and is defined as:

$$P_{\text{REFLECTED}} = \rho^2 \times P_{\text{INCIDENT}} \tag{3.62}$$

In order to take into account also the absorptive losses due to dissipation in the conducting walls of the sensor, leakage into instrumentation, power radiated into space, etc., besides the reflection coefficient, the *effective efficiency* η_C of the sensor should also be considered. Generally, the reflection coefficient and effective efficiency are included into the *calibration factor* K, defined as:

$$K = \eta_C\left(1-\rho^2\right) \times 100 \tag{3.63}$$

For example, a calibration factor of 90% means that the meter will read 10% below the incident power. Generally, calibration factors are specified by sensor manufacturers at different values of frequency.

Thermal Methods

In this section, the main methods based on power dissipation will be examined, namely: (1) thermistor-based, (2) thermocouple-based, and (3) calorimetric.

Thermistor-Based Power Meters

A thermistor is a resistor made up of a compound of highly temperature-sensitive metallic oxides [14]. If it is used as a sensor in a power meter, its resistance becomes a function of the temperature rise produced by the applied power. In Figure 3.28, typical power-resistance characteristics are reported for several values of the operating temperature.

The working principle of the thermistor power meter is illustrated in Figure 3.29 [15]: two thermistors (R_{T1} and R_{T2}) are connected (1) in parallel, for measurand signals appearing at the RF input (P_{RF}); and (2) in series, for the following measuring circuit (e.g., a bridge). The capacitance C_1 prevents the power dc component from flowing to the thermistors; the C_2 stops the RF power toward the bridge.

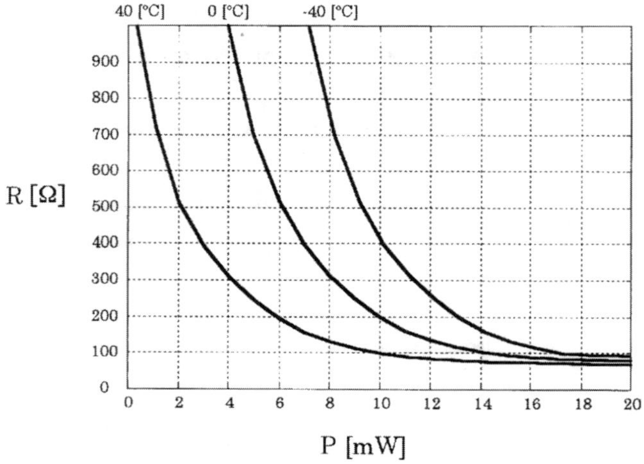

FIGURE 3.28 Typical power-resistance characteristics of commercial thermistors.

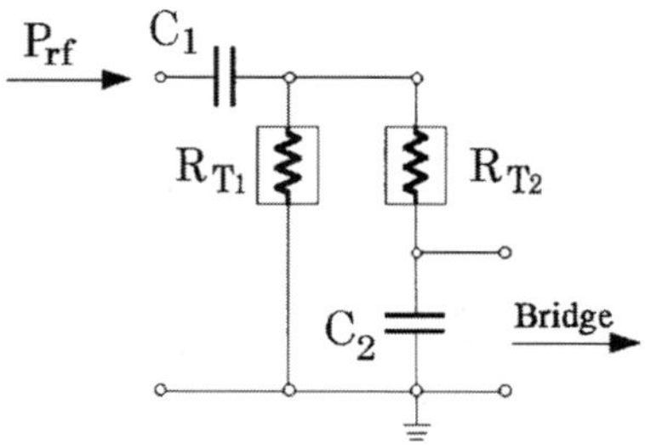

FIGURE 3.29 Working principle of the thermistor-based power meter.

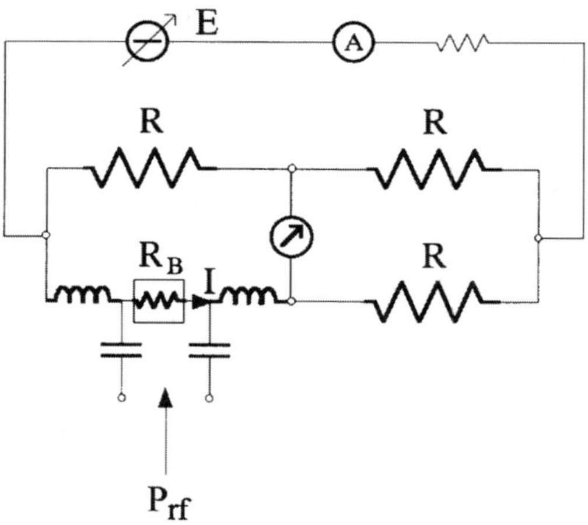

FIGURE 3.30 The manual bolometer.

A bridge with a thermistor or a barretter in one arm is called a *bolometer.* Bolometer-based measurements can be performed with (1) a manual bolometer with variation of the bias current, (2) a manual bolometer with substitution method, or (3) a self-balancing bolometer.

The *manual bolometer with a variation of the bias current* is illustrated in Figure 3.30. Its working principle consists of two steps. In the first, no RF power is applied to the sensor; the equilibrium is obtained by varying the dc power supply E until the sensor resistance R_B, related to the dc power flowing

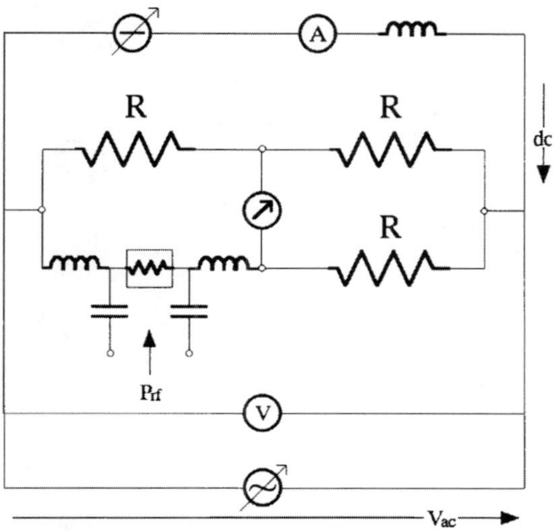

FIGURE 3.31 Manual bolometer with substitution method.

in it, is equal to R. In this condition, let the current I flowing into the sensor be equal to I_1. In the second step, an RF power P_{RF} is fed to the sensor; the power increase must be compensated by a dc power decrease, which is performed by lowering the bridge dc supply voltage E; in this case, let I be equal to I_2.

Since the power dissipated in the sensor has been maintained constant in both steps, the power P_{RF} can be evaluated as:

$$P_{RF} = \frac{R}{4}\left(I_1^2 - I_2^2\right) \tag{3.64}$$

The *manual bolometer with substitution method* (Figure 3.31) consists of two sequential steps; in the first, both RF power (P_{RF}) and dc power (P_{dc}) are present, and the power (P_d) necessary to lead the bridge to the equilibrium is:

$$P_d = P_{dc} + P_{RF} \tag{3.65}$$

During the second step, P_{RF} is set to zero and an alternative voltage V_{ac} is introduced in parallel to the dc power supply. In this case, the power P_d necessary to balance the bridge:

$$P_d = P_{dc} + P_{ac} \tag{3.66}$$

is obtained by varying v_{ac}.

Since P_d is the same in both cases, the power supplied by the alternative generator is equal to P_{RF}:

$$P_{RF} = P_{ac} = \frac{V_{ac}^2}{4R} \tag{3.67}$$

Equation 3.66 implies that the RF power can be obtained by a voltage measurement. The *self-balancing bolometer* (Figure 3.32) automatically supplies a dc voltage V to balance the voltage variations due to sensor resistance R_B changes for an incident power P_{RF}. At equilibrium, R_B is equal to R and the RF power will then be:

$$P_{RF} = \frac{V^2}{4R} \tag{3.68}$$

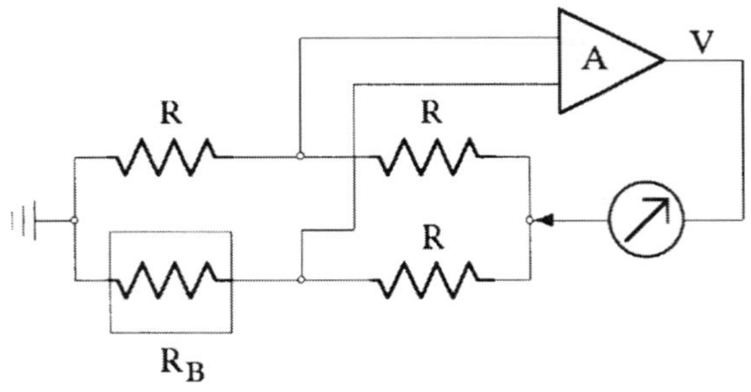

FIGURE 3.32 Self-balancing bolometer.

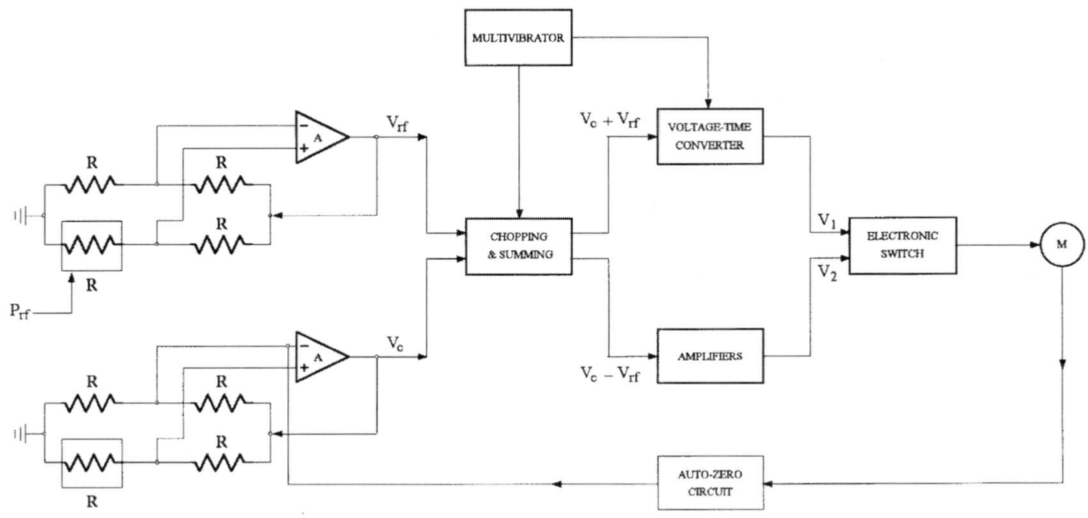

FIGURE 3.33 Power meter based on two self-balancing bridges.

As mentioned above, the thermistor resistance depends on the surrounding temperature. This effect is compensated in an instrument based on two self-balancing bridges [15]. The RF power is input only to one of these, as shown in Figure 3.33.

The equilibrium voltages V_c and V_{RF} feed a chopping and summing circuit, whose output $V_c + V_{RF}$ goes to a voltage-to-time converter. This produces a pulse train V_1, whose width is proportional to $V_c + V_{RF}$. The chopping section also generates a signal with an amplitude proportional to $V_c - V_{RF}$, and a frequency of a few kilohertz, which is further amplified. The signals V_1 and V_2 enter an electronic switch whose output is measured by a medium value meter M. This measure is proportional to the RF power because:

$$P_{RF} = \frac{(V_c + V_{RF})(V_c - V_{RF})}{4R} \Rightarrow P_{RF} = \frac{V_c^2 - V_{RF}^2}{4R} \tag{3.69}$$

Owing to the differential structure of the two bolometers, this device is capable of performing RF power measurements independent of the surrounding temperature. In addition, an offset calibration can be carried out when P_{RF} is null and V_c is equal to V_{RF}.

These instruments can range from 10 mW to 1 μW and utilize sensors with frequency bandwidths ranging from 10 kHz to 100 GHz.

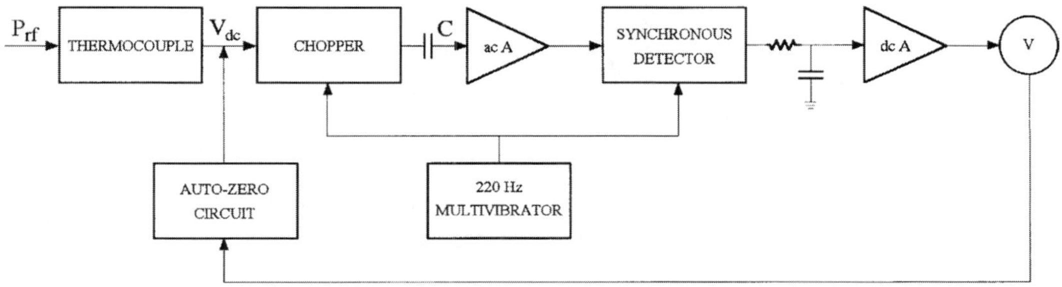

FIGURE 3.34 Power meter with thermocouple-based sensor.

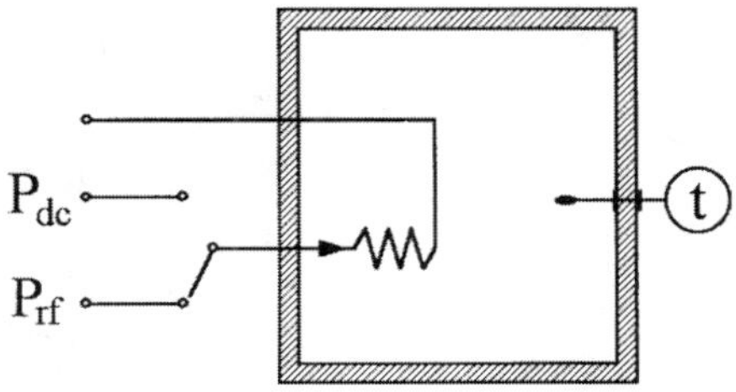

FIGURE 3.35 Calorimetric method based on a substitution technique.

Thermocouple-Based Power Meters

Thermocouples [14] can be also used as RF power meters up to frequencies greater than 40 GHz. In this case, the resistor is generally a thin-film type. The sensitivity of a thermocouple can be expressed as the ratio between the dc output amplitude and the input RF power. Typical values are 160 μV mW^{-1} for minimum power of about 1 μW.

The measure of voltages of about some tens of millivolts requires strong amplification, in that the amplifier does not have to introduce any offset. With this aim, a chopper microvoltmeter is utilized [16], as shown in the Figure 3.34.

The thermocouple output voltage V_{dc} is chopped at a frequency of about 100 Hz; the resulting square wave is filtered of its mean value by the capacitor C and then input to an ac amplifier to further reduce offset problems. A detector, synchronized to the chopper, and a low-pass filter transform the amplified square voltage in a dc voltage finally measured by a voltmeter.

Calorimetric Method

For high frequencies, a substitution technique based on a calorimetric method is utilized (Figure 3.35) [17]. First, the unknown radio frequency power P_{RF} is sent to the measuring device t, which measures the equilibrium temperature. Then, once the calorimetric fluid has been cooled to its initial temperature, a dc power P_{dc} is applied to the device and regulated until the same temperature increase occurs in the same time interval. In this way, a thermal energy equivalence is established between the known P_{dc} and the measurand P_{RF}.

A comparison version of the calorimetric method is also used for lower frequency power measurements (Figure 3.36). The temperature difference ΔT of a cooling fluid between the input (1) and the output (2) sections of a cooling element where the dissipated power to be measured P is determined. In this case, the power loss will correspond to P:

$$P = C_p \rho Q \Delta T \tag{3.70}$$

where C_p is the specific heat, ρ the density, and Q the volume flow, respectively, of the refreshing fluid.

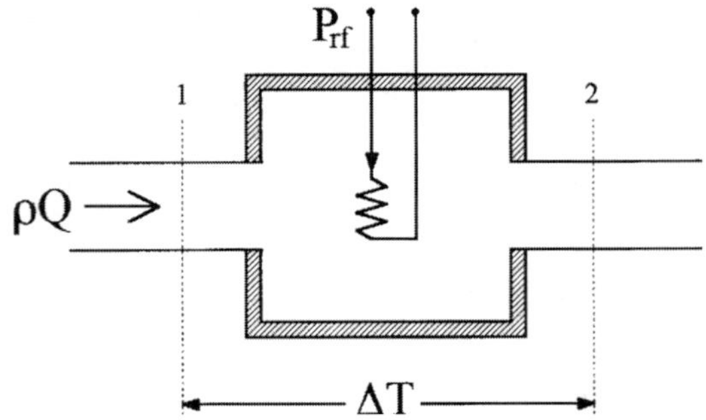

FIGURE 3.36 Calorimetric method based on a comparison technique.

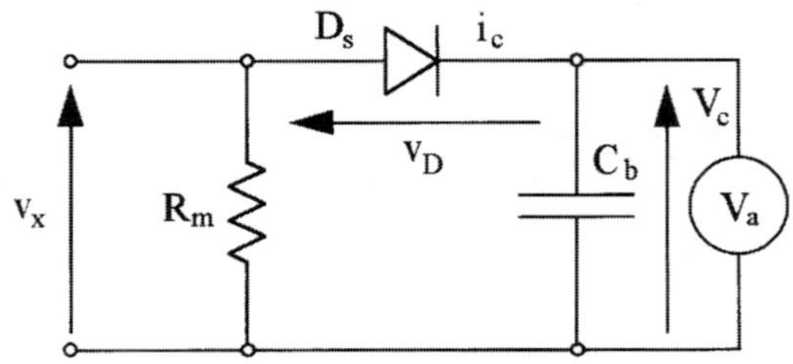

FIGURE 3.37 Circuit of the diode sensor-based power measurement.

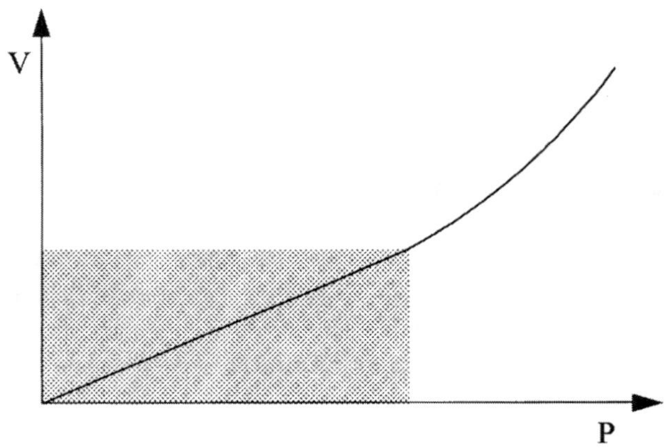

FIGURE 3.38 Characteristic of a low-barrier Schottky diode.

Diode Sensor-Based Power Measurements

Very sensitive (up to 0.10 nW, –70 dBm), high-frequency (10 MHz to 20 GHz) power measurements are carried out through a diode sensor by means of the circuit in Figure 3.37 [18]. In particular, according to a suitable selection of the components in this circuit, (1) true-average power measurements, or (2) peak power measurements can be performed.

The basic concept underlying *true-average power measurements* exploits the nonlinear squared region of the characteristic of a low-barrier Schottky diode (nondashed area in Figure 3.38). In this region, the current flowing through the diode is proportional to the square of the applied voltage; thus, the diode acts as a squared-characteristic sensor.

In the circuit of diode sensor-based wattmeters shown in Figure 3.37, the measurand v_x, terminated on the matching resistor R_m, is applied to the diode sensor D_s working in its squared region in order to produce a corresponding output current i_c in the bypass capacitor C_b. If C_b has been suitably selected, the voltage V_c between its terminals, measured by the voltmeter amplifier V_a, is proportional to the average of i_c, i.e., to the average of the squares of instantaneous values of the input signal v_x, and, hence, to the true average power.

In the true-average power measurement of nonsinusoidal waveforms having the biggest components at low frequency, such as radio-frequency AM (Amplitude Modulated), the value of C_b must also satisfy another condition. The voltage v_d on the diode must be capable of holding the diode switched-on into conduction even for the smallest values of the signal. Otherwise, in the valleys of the modulation cycle, the high-frequency modulating source is disconnected by the back-biased diode and the measurement is therefore misleading.

On the other hand, for the same signal but for a different selection of the C_b value, the circuit can act as a peak detector for *peak power measurements.* As a matter of fact, the voltage v_c on the bypass capacitor C_b during the peak of the modulation cycle is so large that in the valleys, the high-frequency peaks are not capable of switching on the diode into conduction; thus, these peaks do not contribute to the measured power level.

If higher power levels have to be measured (10 mW to 100 mW), the sensing diode is led to work out of the squared region into its linear region (dashed area in Figure 3.38). In this case, the advantage of true-average power measurements for distorted waveforms is lost; and for peak power measurements, since the diode input-output characteristic is nonlinear and the output is squared, spectral components different from the fundamental introduce significant measuring errors.

Radiation Sensor-Based Power Measurements

Very high-frequency power measurements are usually carried out by measuring a radiant flux of an electromagnetic radiation through a suitable sensor. In particular, semiconductor-based radiation microsensors have gained wider and wider diffusion [19], in that size reduction involves several well-known advantages such as greater portability, fewer materials, a wider range of applications, etc. One of the most familiar applications of radiation sensor-based power measurements is the detection of object displacement. Furthermore, they are also commonly used for low-frequency power noninvasive measurements.

Radiation sensors can be classified according to the measurand class to which they are sensitive: nuclear particles or electromagnetic radiations. In any case, particular sensors capable of detecting both nuclear particles and electromagnetic radiations, such as gamma and X-rays, exist and are referred to as nucleonic detectors.

In Table 3.1, the different types of radiation sensors utilized according to the decrease of the measurand wavelength from microwaves up to nuclear (X, gamma, and cosmic) rays are indicated.

In particular, *microwave* power radiation sensors are mainly used as noncontacting detectors relying on ranging techniques using microwaves [20]. Shorter and longer (radar) wavelength microwave devices are employed to detect metric and over-kilometer displacements, respectively.

Beyond applications analogous to microwave, power radiation *infrared* sensors also find use as contact detectors. In particular, there are two types of infrared detectors: thermal and quantum. The thermal type includes contacting temperature sensors such as thermocouples and thermopiles, as well as noncontacting pyroelectric detectors. On the other hand, the quantum type, although characterized by a strong wavelength dependence, has a faster response and includes photoconductive (spectral range: 1 μm to 12 μm) and photovoltaic (0.5 μm to 5.5 μm) devices.

The main power radiation *visible and ultraviolet* sensors are photoconductive cells, photodiodes, and phototransistors. Photodiodes are widely used to detect the presence, the intensity, and the wavelength of light or ultraviolet radiations. Compared to photoconductive cells, they are more sensitive, smaller, more stable and linear, and have lower response times. On the other hand, phototransistors are more sensitive to light.

At very low light levels, rather than silicon-based microsensors, *nuclear radiation* power microsensors are needed. In this case, the most widespread devices are scintillation counters, solid-state detectors, plastic films, and thermoluminescent devices. The scintillation counter consists of an active material that

TABLE 3.1 Operating Field of Main Radiation Power Sensors

Operating Field (Wavelength)	Microwave (1, 10^{-3} m)	Infrared (10^{-3}, 10^{-6} m)	Visible and Ultraviolet (10^{-6}, 10^{-9} m)	Nuclear Rays (10^{-8}, 10^{-15} m)
Sensors	Noncontacting displacement sensors	Pyroelectric, photoconductive, photovoltaic	Photoconductive, photovoltaic	Scintillation counters, plastic films, solid-state, thermolum

converts the incident radiation to pulses of light, and a light-electric pulse converter. The active material can be a crystal, a plastic fluorine, or a liquid. The scintillator size varies greatly according to the radiation energy, from thin solid films to large tanks of liquid to detect cosmic rays. A thin (5 μm) plastic polycarbonate film or a thermoluminescent material (e.g., LiF) can measure the radiation power falling on a surface. The film is mechanically damaged by the propagation of highly α–ionizing particles. Consequent etching of the film reveals tracks that can be observed and counted.

3.3 Pulse Power Measurements

Pulse waveforms are becoming more and more diffused in several fields such as telecommunications, power source applications, etc. The *pulse power* P_p is defined as the average power P_m in the pulse width:

$$P_p = \frac{P_m}{\tau_d} \tag{3.71}$$

where τ_d is the duty cycle of the pulse waveform (i.e., the pulse width divided by the waveform period). If the pulse width cannot be accurately defined (e.g., nonrectangular pulses in the presence of noise), the pulse power P_p becomes unmeaningful. In this case, the *peak envelope power* is introduced as the maximum of the instantaneous power detected on a time interval, including several periods of the pulse waveform (but negligible with respect to the modulation period, in the case of PWM waveforms).

Several techniques are used to measure pulse power [21]. In particular, they can be classified according to the *pulse frequency* and the necessity for *constraining real-time applications* (i.e., measuring times, including a few of the modulation periods). For real-time, low-frequency applications (up to 100 kHz), the algorithms mentioned in the above sections on wattmeters based on digital multipliers can be applied [9].

If constraining limits of real-time do not have to be satisfied, either digital or analog techniques can be utilized. As far as the digital techniques are concerned, for high-frequency applications, if the measurand pulse waveform is stationary over several modulation periods, digital wattmeters based on equivalent sampling can be applied, with accuracies increasing according to measuring times. As far as the analog techniques are concerned, three traditional methods are still valid: (1) average power per duty cycle, (2) integration-differentiation, and (3) dc/pulse power comparison.

A block diagram of an instrument measuring *average power per duty cycle* is illustrated in Figure 3.39. At first, the mean power of the measurand pulse signal, terminated on a suitable load, is measured by means of an average power meter; then, the pulse width and the pulse waveform period are measured by a digital counter. Finally, the power is obtained by means of Equation 3.71.

The *integration-differentiation* technique is based on a barretter sensor capable of integrating the measurand, and on a conditioning and differentiating circuit to obtain a voltage signal proportional to the measurand power. The signal is input to the barretter sensor, having a thermal constant such that the barretter resistance will correspond to the integral of the input. The barretter is mounted as an arm of a conditioning Wheatstone bridge; in this way, the barretter resistance variations are transformed into voltage variations, and an integrated voltage signal is obtained as an output of the bridge detecting arm. This signal, suitably integrated to reach a voltage signal proportional to the output, is detected by a peak

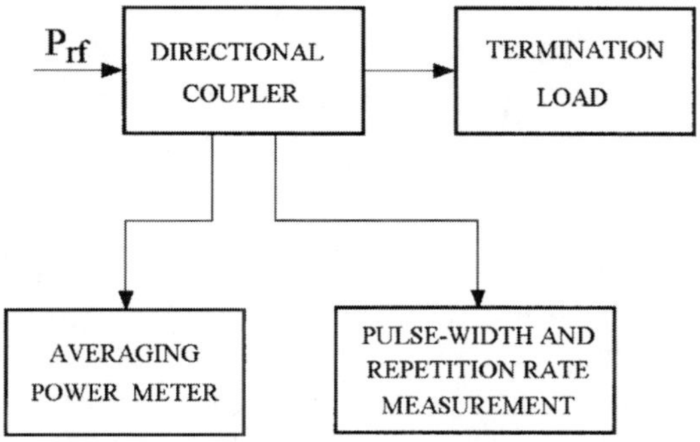

FIGURE 3.39 Block diagram of an instrument measuring average power per duty cycle.

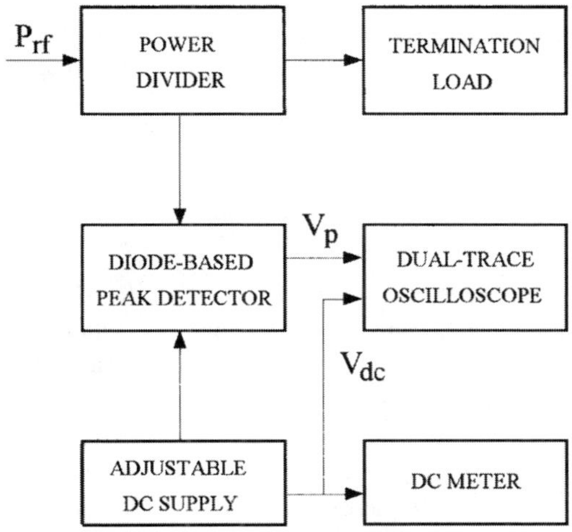

FIGURE 3.40 Block diagram of an instrument based on dc/pulse power comparison technique.

voltmeter calibrated in power. Analogously to the selection of the time constant of an *RC* integrating circuit, attention must be paid to the thermal constant selection of the barretter in order to attain the desired accuracy in the integration. With respect to the measurand pulse period, a very long thermal constant will give rise to insufficient sensitivity. On the other hand, a very short constant approaching the pulse duration will give rise to insufficient accuracy.

The *dc/pulse power comparison* technique is based on the concept of first revealing the peak envelope power through a diode sensor, and then comparing the peak to a known dc source with a dual trace scope. A block diagram of an instrument based on this concept is illustrated in Figure 3.40. The peak of the measurand pulse signal terminated on a suitable load is sensed by a peak detector by obtaining a proportional signal V_p. This signal is input to a channel of a dual trace oscilloscope and the envelope peak is displayed. An adjustable dc source is input to the other channel of the scope to obtain a signal V_{dc} to be compared to the envelope peak signal. When the two signals are made equal, a dc voltage meter directly calibrated in peak power measures the output power.

References

1. G. Zingales, Power measurements on single-phase ac circuits (Chap. VI, 6.2), in *Methods and Instruments for Electrical Measurements*, (in Italian), Torino: UTET, 1980.
2. G. Korànyi, Measurement of power and energy, in L. Schnell (Ed.), *Technology of Electrical Measurements*, Chichester: John Wiley & Sons, 1993.

3. J. Milmann and C.C. Halkias, *Integrated Electronics: Analog and Digital Circuits and Systems,* New York: McGraw-Hill, 1972.
4. F.F. Mazda, Ac analogue instruments (Chap. VI, 6.3), in *Electronic Instruments and Measurement Techniques*, Cambridge, U.K.: Cambridge University Press, 1987.
5. P.S. Filipski, A TDM wattmeter with 0.5 MHz carrier frequency, *IEEE Trans. Instrum. Meas.*, IM39, 15-18, 1990.
6. J.R. Carstens, *Electrical Sensors and Transducers*, Englewood Cliffs, NJ: Prentice-Hall, 1992.
7. Lu Zu-Liang, An error estimate for quasi-integer-period sampling and an approach for improving its accuracy, *IEEE Trans. Instrum. Meas.*, IM-23, 337-341, 1984.
8. V. Haasz, The error analysis of digital measurements of electrical power, *Measurement*, 6, 1986.
9. P. Arpaia, F. Avallone, A. Baccigalupi, and C. De Capua, Real-time algorithms for active power measurements on PWM-based electric drives, *IEEE Trans. Instrum. Meas.*, IM-45, 462-466, 1996.
10. X. Dai and R. Gretsch, Quasi-synchronous sampling algorithm and its applications, *IEEE Trans. Instrum. Meas.*, IM-43, 204-209, 1994.
11. M. Bellanger, *Digital Processing of Signals: Theory and Practice*, Chichester: John Wiley & Sons, 1984.
12. M. Bertocco, C. Offelli, and D. Petri, Numerical algorithms for power measurements, *Europ. Trans. Electr. Power*, ETEP 3, 91-101, 1993.
13. G. Zingales, Measurements on steady-state circuits (Chap. VI), in *Methods and Instruments for Electrical Measurements* (In Italian), Torino: UTET, 1980.
14. H.N. Norton, Thermometers (Chap. 19), in *Handbook of Transducers*, Englewood Cliffs, NJ: Prentice-Hall, 1989.
15. Anonymous, Thermistor mounts and instrumentation (Chap. II), in *Fundamental of RF and Microwave Power Measurements,* Application Note 64-1, Hewlett Packard, 1978.
16. R.E. Pratt, Power measurements (15.1–15.16), in C.F. Coombs (Ed.), in *Handbook of Electronic Instruments*, New York: McGraw-Hill, 1995.
17. F. F. Mazda, High-frequency power measurements (Chap. VIII, 8.4), in *Electronic Instruments and Measurement Techniques*, Cambridge: Cambridge University Press, 1987.
18. Anonymous, Diode detector power sensors and instrumentation (Chap. IV), in *Fundamental of RF and Microwave Power Measurements,* Application Note 64-1, Hewlett Packard, 1978.
19. J.W. Gardner, *Microsensors: Principles and Applications,* Chichester: John Wiley & Sons, 1994.
20. H.N. Norton, Radiation pyrometers (Chap. 20), in *Handbook of Transducers*, Englewood Cliffs, NJ: Prentice-Hall, 1989.
21. F.F. Mazda, Pulse power measurement (Chap. VIII, 8.5), in *Electronic Instruments and Measurement Techniques,* Cambridge, U.K.: Cambridge University Press, 1987.

Further Information

F.K. Harris, The Measurement of Power (Chap. XI), in *Electrical Measurements*, Huntington, NY: R.E. Krieger Publishing, 1975; a clear reference for line-frequency power measurements.

Anonymous, *Fundamental of RF and Microwave Power Measurements,* Application Note 64-1, Hewlett Packard, 1978; though not very recent, is a valid and comprehensive reference for main principles of high-frequency power measurements.

J.J. Clarke and J.R. Stockton, Principles and theory of wattmeters operating on the basis of regularly spaced sample pairs, *J. Phys. E. Sci. Instrum.*, 15, 645-652, 1982; gives basics of synchronous sampling for digital wattmeters.

T.S. Rathore, Theorems on power, mean and RMS values of uniformly sampled periodic signals, *IEE Proc. Pt. A*, 131, 598-600, 1984; provides fundamental theorems for effective synchronous sampling wattmeters.

J.K. Kolanko, Accurate measurement of power, energy, and true RMS voltage using synchronous counting, *IEEE Trans. Instrum. Meas.*, IM-42, 752–754, 1993; provides information on the implementation of a synchronous dual-slope wattmeter.

G.N. Stenbakken, A wideband sampling wattmeter, *IEEE Trans. Power App. Syst.*, PAS-103, 2919-2925, 1984; gives basics of asynchronous sampling-based wattmeters and criteria for computing uncertainty in time domain.

F. Avallone, C. De Capua, and C. Landi, Measurement station performance optimization for testing on high efficiency variable speed drives, *Proc. IEEE IMTC/96 (Brussels, Belgium)*, 1098–1103, 1996; proposes an analytical model of uncertainty arising from power measurement systems working under highly distorted conditions.

F. Avallone, C. De Capua, and C. Landi, Metrological performance improvement for power measurements on variable speed drives, *Measurement*, 21, 1997, 17-24; shows how compute uncertainty of measuring chain components for power measurements under highly distorted conditions.

F. Avallone, C. De Capua, and C. Landi, Measurand reconstruction techniques applied to power measurements on high efficiency variable speed drives, *Proc. of XIV IMEKO World Congress (Tampere, Fi)*, 1997; proposes a technique to improve accuracy of power measurements under highly distorted conditions.

F. Avallone, C. De Capua, and C. Landi, A digital technique based on real-time error compensation for high accuracy power measurement on variable speed drives, *Proc. of IEEE IMTC/97 (Ottawa, Canada)*, 1997; reports about a real-time technique for error compensation of transducers working under highly distorted conditions.

J.C. Montano, A. Lopez, M. Castilla, and J. Gutierrrez, DSP-based algorithm for electric power measurement, *IEE Proc. Pt.A*, 140, 485–490, 1993; describes a Goertzel FFT-based algorithm to compute power under nonsinusoidal conditions.

S.L. Garverick, K. Fujino, D.T. McGrath, and R.D. Baertsch, A programmable mixed-signal ASIC for power metering, *IEEE J. Solid State Circuits*, 26, 2008–2015, 1991; reports about a programmable mixed analog-digital integrated circuit based on six first-order sigma-delta ADCs, a bit serial DSP, and a byte-wide static RAM for power metering.

G. Bucci, P. D'Innocenzo, and C. Landi, A modular high-speed dsp-based data acquisition apparatus for on-line quality analysis on power systems under non-sinusoidal conditions, *Proc. of 9th IMEKO TC4 Int. Sym.* (Budapest, Hungary), 286–289, 1996; shows the strategy of measurement system design for power measurements under non-sinusoidal conditions.

K.K. Clarke and D.T. Hess, A 1000 A/20 kV/25 kHz-500 kHz volt-ampere-wattmeter for loads with power factors from 0.001 to 1.00, *IEEE Trans. Instrum. Meas.*, IM-45, 142-145, 1996; provides information on the implementation of an instrument to perform an accurate measurement of currents (1 A to 1000 A), voltages (100 V to 20 kV), and powers (100 W to 20 MW) over the frequency range from 25 kHz to 500 kHz.

P. Arpaia, G. Betta, A. Langella, and M. Vanacore, An Expert System for the Optimum Design of Measurement Systems, *IEE Proc. (Pt. A)*, 142, 330–336, 1995; reports about an Artificial Intelligence tool for the automatic design of power measuring systems.

In any case, *IEEE Transactions on Instrumentation and Measurement* and *Measurement* journals provide current research on power measurements.

4

Power Factor Measurement

Michael Z. Lowenstein
Harmonics Limited

Electricity can be thought of as a means of delivering power from one place to another to do work. The laws and relationships for delivering power were originally developed for direct current. Power delivered, expressed in *watts,* was calculated by multiplying the voltage and current as shown in Equation 4.1.

$$P_{dc} = EI \tag{4.1}$$

The situation becomes more complex when alternating current is used to deliver power. Figure 4.1 shows a sine wave representing either ac current or voltage. Since the instantaneous value of the wave is continually changing, a numerical quantity is defined to represent an average property of this wave. This quantity, the root-mean-square, or **rms** value, calculated by squaring the instantaneous value, integrating it during one cycle, dividing by the period, and taking the square root of the result, is equal to the peak value of the ac wave divided by the square root of 2, or, for ac current, $I_{rms} = i_{peak}/\sqrt{2}$. In the physical world, a sine wave ac current having an rms value of 1 A (A = ampere), passed through a resistive load, produces the same heating effect as 1 A of dc current. Thus, one might expect delivered ac power to be easily calculated in watts using Equation 4.1 and inserting rms values for current and voltage. While this simple relationship holds true for the instantaneous voltage and current values as shown in Equation 4.1a in general, it is not true for the rms quantities except for the special case when the ac current and voltage are restricted to perfect sine waves *and* the load is a pure resistance.

$$p_{inst.} = ei \tag{4.1a}$$

In real-world situations where current and/or voltage waveforms are not perfectly sinusoidal and/or the loads are other than resistive, the relationships are no longer simple and the power delivered, or *active power,* is usually less than the product of rms voltage and current, as shown in Equation 4.2.

0-8493-1733-9/04/$0.00+$1.50

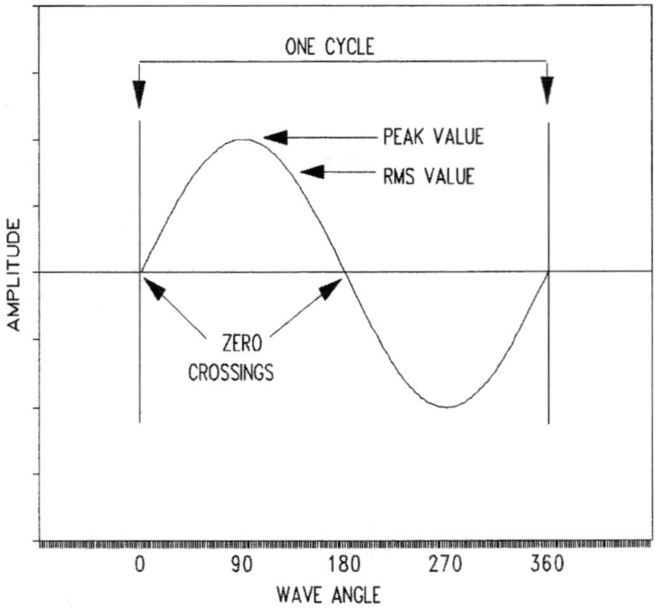

FIGURE 4.1 Sine wave characteristics. One cycle of a continuous wave is shown. The wave begins at a zero-crossing, reaches a positive peak, continues through zero to a negative peak, and back to zero. The wave repeats every 360°. The wave angle can also be expressed as a function of frequency *f* and time *t*. For a given frequency, the wave angle is related to the expression, $2\pi ft$.

$$P_{ac} \leq E_{rms} I_{rms} \tag{4.2}$$

The product of rms voltage and rms current does, however, define a quantity termed *apparent power*, U, as shown in Equation 4.3.

$$U = E_{rms} I_{rms} \tag{4.3}$$

A derived term, the *power factor*, F_p, used to express the relationship between delivered or active power, P, and apparent power, U, is defined by Equation 4.4.

$$F_p = \frac{P}{U} \tag{4.4}$$

From Equations 4.2, 4.3, and 4.4, it is clear that the value of F_p must lie in a range between zero and one.

This chapter focuses on: (1) ac power relationships and the calculation of power factor; (2) the physical meaning of these relationships; and (3) measurement techniques and instrumentation for determining these relationships and calculating power factor.

4.1 Reasons for Interest in Power Factor

Power factor is a single number that relates the active power, P, to the apparent power, U. Electric components of a utility distribution system are designed on a kVA basis; i.e., they are designed to operate at a given voltage and carry a rated current without undue temperature rise. Transformer and conductor sizes are chosen on this basis. While active power does useful work, reactive and harmonic powers do no useful work, absorb system capacity, and increase system losses; but reactive and harmonic powers are needed to provide magnetic fields or nonlinear currents. *The capacity of electric systems is limited by apparent power, not active power. Power factor expresses, with a single value, the extent to which an electrical distribution system is efficiently and effectively utilized.* A low value for the power factor means that much of the system capacity is not available for useful work. From a utility viewpoint, this means reduced ability to deliver revenue-producing active power; from a user viewpoint, a low power factor reduces the available active power or requires increased system size to deliver needed power.

4.2 Ac Electric Loads

Linear Loads

Electric loads in ac power systems with sinusoidal voltages are categorized by the way they draw current from the system. Loads that draw sinusoidal currents, i.e., the current waveshape is the same as the voltage waveshape, are defined as *linear loads.* Historically, a high percentage of electric loads have been linear. Linear loads include: (1) induction motors; (2) incandescent lighting; and (3) heaters and heating elements. Linear loads use ac electric power directly to accomplish their functions.

Nonlinear Loads

Electric loads that draw nonsinusoidal currents, i.e., the current waveshape differs from the voltage waveshape, are defined as *nonlinear loads.* As energy savings and efficient use of electricity are emphasized, an increased percentage of nonlinear electric devices, both new and replacement, are being installed. Nonlinear loads include: (1) adjustable-speed motor drives; (2) fluorescent and arc-discharge lighting; (3) computers and computerized controls; and (4) temperature-controlled furnaces and heating elements. Nonlinear loads, rather than using ac electric power directly, often convert ac power into direct current before it is used to accomplish their functions. A common element in nonlinear loads is some kind of rectifier to accomplish this ac to dc conversion. Rectifiers do not draw sinusoidal currents.

4.3 Ac Power Relationships

Sinusoidal Voltage and Current

Power calculations for sinusoidal ac electric systems require knowledge of the rms voltage, the rms current, and the phase relationships between the two. Figure 4.2 illustrates possible phase relationships between voltage and current. If the positive-going zero-crossing of the voltage is considered the reference point, then the nearest positive-going zero-crossing of the current can occur at a wave angle either less than or greater than this reference. If the current zero-crossing occurs before the reference, the current is said to *lead* the voltage. If the current zero-crossing occurs after the reference, the current is *lagging.* If the zero-crossing for the current coincides with the reference, the two waves are said to be *in phase.* The wave angle, θ, by which the current leads or lags the voltage is called the *phase angle,* in this case, 30°.

Single-Phase Circuits

Power Calculations

The power delivered to do work is easily calculated [1]. Given a sinusoidal voltage of rms magnitude E and sinusoidal current of rms magnitude I, displaced by angle θ, at time t,

$$\begin{aligned}
&\text{Instantaneous voltage} = e\sqrt{2}E\sin(2\pi ft) \\
&\text{Instantaneous current} = i = \sqrt{2}I\sin(2\pi ft - \theta)\ (\text{note that } \theta \text{ can have a positive or negative value}) \qquad (4.5) \\
&\text{Instantaneous power} = p = ei
\end{aligned}$$

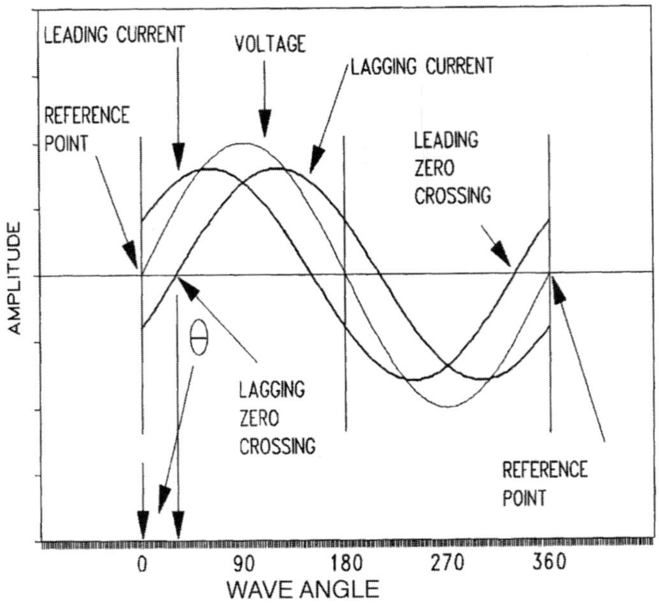

FIGURE 4.2 Sine wave phase angle. Two waves with the same zero-crossing are *in phase.* A sine wave that crosses zero before the reference wave is *leading,* and one that crosses zero after the reference wave is *lagging.* The *phase angle* θ illustrated is 30° lagging.

$$p = 2EI\sin(2\pi ft)\sin(2\pi ft - \theta)$$

$$p = EI\cos(\theta) - EI\cos(4\pi ft - \theta)$$

$$\text{Average power over an integral number of cycles} \quad P = EI\cos(\theta) \tag{4.5}$$

$$\text{Power factor } F_\text{p} = \frac{P}{U} = \frac{EI\cos(\theta)}{EI} = \cos(\theta) \tag{4.5}$$

Equation 4.5 is the fundamental equation that defines power for systems in which the current and voltage are sinusoidal. The application of this equation is illustrated for three cases: (1) the current and voltage are in phase; (2) the current and voltage are out of phase by an angle less than 90°; and (3) the current and voltage are out of phase by exactly 90°.

Ac Power Examples

Figure 4.3 shows voltage current and power when the voltage and current are in phase and the current displacement angle is zero (0). (An example would be a resistance space heater.) The power curve is obtained by multiplying together the instantaneous values of voltage and current as the wave angle is varied from 0° to 360°. Instantaneous power, the product of two sine waves, is also a sine wave. There are two zero-crossings per cycle, dividing the cycle into two regions. In region (1) both the voltage and current are positive and the resultant product, the power, is positive. In region (2) both the voltage and current are negative and the power is again positive. The average power in watts, given by Equation 4.5, $EI\cos(0°) = EI$, is the maximum power that can be delivered to do work. *When sinusoidal voltage and current are in phase, the delivered power in watts is the same as for dc and is the maximum that can be delivered.* The power factor, $F_\text{p} = \cos(0°) = 1$, or *unity.*

Figure 4.4 shows voltage, current, and power when the current lags the voltage by 60°. (An example might be a lightly loaded induction motor.) The power sine wave again is obtained by multiplying together

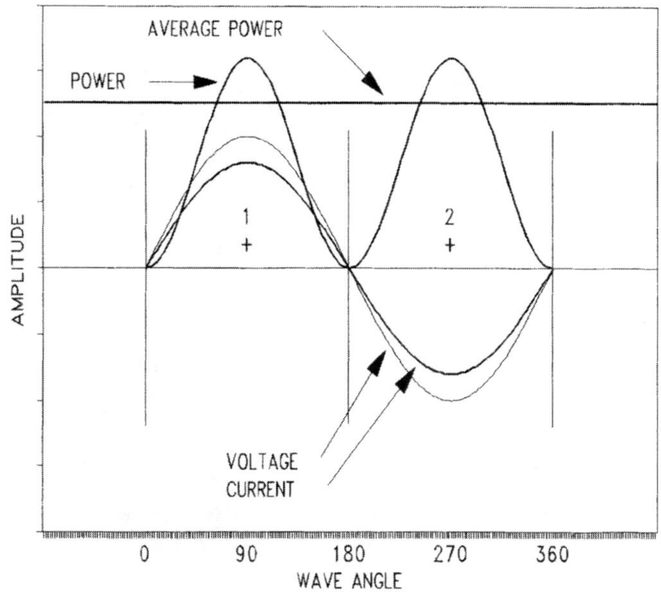

FIGURE 4.3 Voltage, current, and power for sine waves in phase. The vertical scales for voltage and current are equal. The scale for power is relative and is selected to permit the display of all three waves on a single graph. The voltage and current are in phase and both are sinusoidal. The power is everywhere positive, and average power delivered to do work is the maximum power.

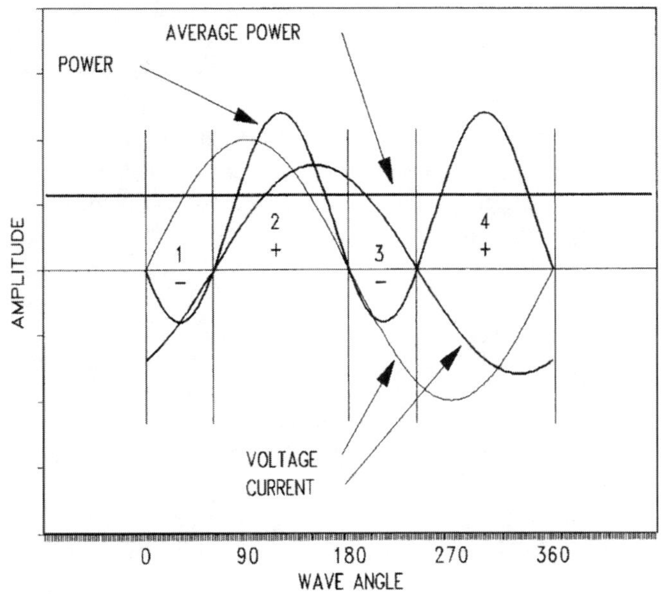

FIGURE 4.4 Voltage, current, and power for sine waves 60° out of phase. The vertical scales for voltage and current amplitudes are the same as those for Figure 4.3. Current lags voltage by 60° and both are sinusoidal. The power is positive in regions (2) and (4), and negative in regions (1) and (3). Average power delivered to do work is less than the maximum power.

the instantaneous values of voltage and current. There are now four zero-crossings per cycle, dividing the cycle into four regions. In regions (2) and (4), voltage and current have the same sign and the power is positive. In regions (1) and (3), voltage and current have opposite signs, resulting in a negative value

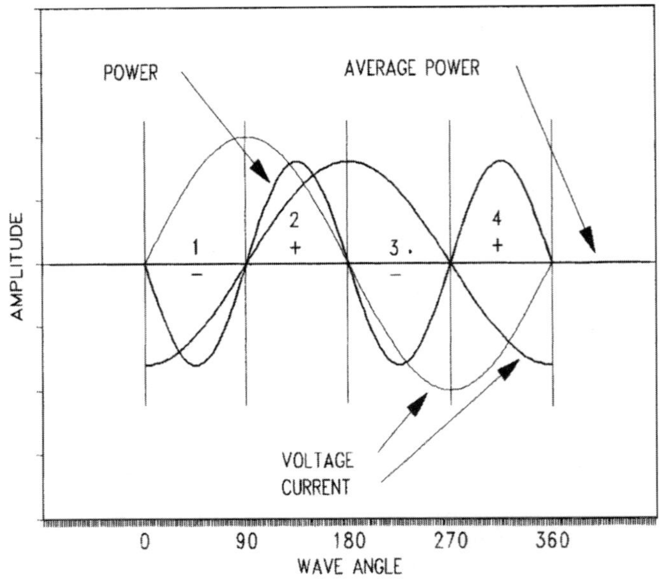

FIGURE 4.5 Voltage, current, and power for sine waves 90° out of phase. The vertical scales for voltage and current amplitudes are the same as those for Figure 4.3. Current lags voltage by 90° and both are sinusoidal. The power is positive in regions (2) and (4), negative in regions (1) and (3), and is of equal absolute magnitude in all four regions. Average power delivered to do work is zero.

for the power. The average power in watts, given by Equation 4.5, $EI\cos(60°) = EI(0.5)$, is less than the maximum that could be delivered for the particular values of voltage and current. *When voltage and current are out of phase, the delivered power in watts is always less than the maximum.* In this example, $F_p = \cos(60°) = 0.5$.

Figure 4.5 shows voltage, current, and power when the current lags the voltage by 90°. (This situation is not attainable in the real world.) The power sine wave again is obtained by multiplying together the instantaneous values of voltage and current. Again, four zero-crossings divide the cycle into four regions. In regions (2) and (4), the power is positive, while in regions (1) and (3), the power is negative. The average power in watts is given by Equation 4.5, $EI\cos(90°) = 0$. No matter what the values of voltage and current, *when voltage and current are exactly 90° out of phase, the delivered power in watts is always zero.* The power factor, $F_p = \cos(90°) =$ zero.

Power Factor

Resolving the current into orthogonal components on a phasor diagram illustrates how delivered power can vary from a maximum to zero, depending on the phase angle between the voltage and the current sine waves. Figure 4.6 shows the voltage vector along with the current resolved into orthogonal components [1]. The current I at an angle θ relative to the voltage can be resolved into two vectors: $I\cos(\theta)$ and $I\sin(\theta)$. The in-phase component $I\cos(\theta)$ multiplied by the voltage gives average power in watts. The current component that is 90° out of phase with the voltage, $I\sin(\theta)$, is not associated with delivered power and does not contribute to work. For want of a better name, this was often termed the *wattless component* of the current. Since this wattless current could be associated with magnetic fields, it was sometimes termed *magnetizing current* because, while doing no work, this current interacts through the inductive reactance of an ac motor winding to provide the magnetic field required for such a motor to operate.

Three types of power have been defined for systems in which both the voltage and current are sinusoidal. Throughout the years, a number of different names have been given to the three power types. The names in present usage will be emphasized.

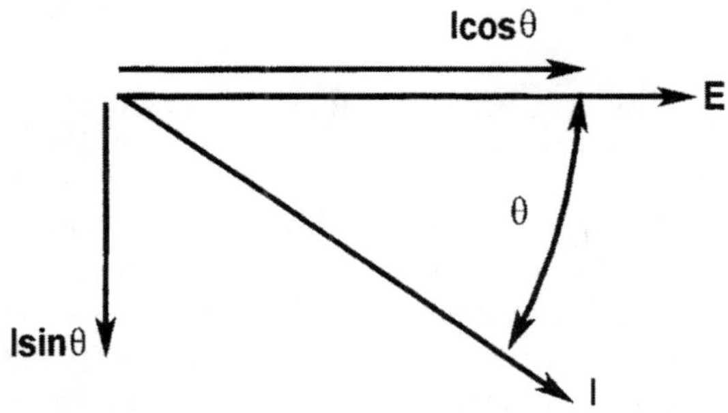

FIGURE 4.6 Phasor diagram for current and voltage. Voltage is the reference phasor. The current I has been resolved into orthogonal components $I\cos\theta$ and $I\sin\theta$. (From Reference [1].)

Active power is given the symbol P and is defined by Equation 4.5:

$$P = EI\cos(\theta) \tag{4.5}$$

Other names for active power include: (1) real power and (2) delivered power. Active power is the power that does work. Note that while all power quantities are volt-ampere products, only *active power is expressed in watts.*

Reactive power is given the symbol Q and is defined by the equation:

$$Q = EI\sin(\theta) \tag{4.6}$$

Other names for reactive power include: (1) imaginary power; (2) wattless power; (3) and magnetizing power. Reactive power is expressed in *voltamperes*$_{\text{reactive}}$ or *vars.* If the load is predominantly inductive, current lags the voltage and the reactive power is given a positive sign. If the load is predominantly capacitive, current leads the voltage and the reactive power is given a negative sign.

Phasor power is given the symbol S and is defined by the equation:

$$S = \sqrt{P^2 + Q^2} \tag{4.7}$$

Phasor power was called apparent power for many years, and it will be seen in a later section that phasor power S, *for sinusoidal voltages and currents,* is identical to what is now called apparent power U. Phasor power is expressed in *voltamperes* or *VA.*

Figure 4.7 is a phasor diagram, often called a *power triangle,* which illustrates the relationships among the three types of power defined above. Reactive power is orthogonal to active power, and is shown as positive for lagging current. It is clear that the definition of phasor power, Equation 4.7, is geometrically derived from active and reactive power.

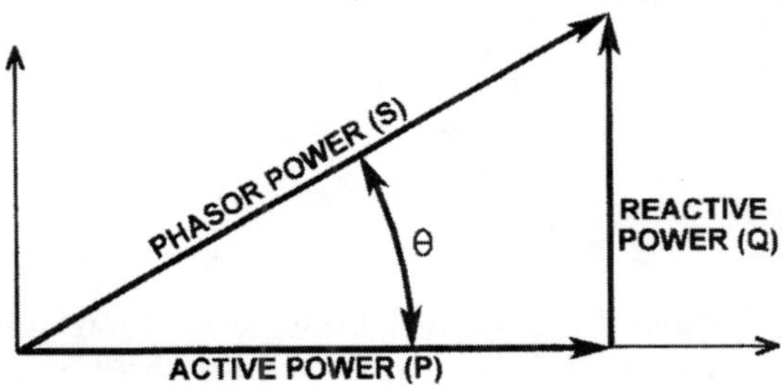

FIGURE 4.7 Power triangle showing the geometric relationships between active, reactive, and phasor power. Power factor is defined geometrically as $\cos\theta$.

Power factor is given the symbol F_p and *for sinusoidal quantities* is defined by the equation:

$$F_p = \frac{\text{ACTIVE POWER}}{\text{PHASOR POWER}} = \frac{P}{S} = \frac{\text{WATTS}}{\text{VOLTAMPS}} = \cos\theta \tag{4.8}$$

Since the power factor can be expressed in reference to the displacement angle between voltage and current, power factor so defined should be termed *displacement power factor*, and the symbol is often written $F_{p\ \text{displacement}}$. Values for displacement power factor range from one (unity) to zero as the current displacement angle varies from 0° (current and voltage in phase) to 90°. Since the cosine function is positive in both the first and fourth quadrants, the power factor is positive for both leading and lagging currents. To completely specify the voltage–current phase relationship, the words *leading* or *lagging* must be used in conjunction with power factor. Power factor can be expressed as a decimal fraction or as percent. For example, the power factor of the case shown in Figure 4.4 is expressed either as 0.5 lagging or 50% lagging.

Polyphase Circuits

Power Calculations

The power concepts developed for single-phase circuits with sinusoidal voltages and currents can be extended to polyphase circuits. Such circuits can be considered to be divided into a group of two-wire sets, with the neutral conductor (or a resistively derived neutral for the case of a delta-connected, three-wire circuit) paired with each other conductor. Equations 4.3 to 4.5 can be rewritten to define power terms equivalent to the single-phase terms. In these equations, k represents a phase number, m is the total number of phases, and α and β are, respectively, the voltage and current phase angles with respect to a common reference frame.

$$P = \sum_{k=1}^{m} E_k I_k \cos(\alpha - \beta) \tag{4.9}$$

$$Q = \sum_{k=1}^{m} E_k I_k \sin(\alpha - \beta) \tag{4.10}$$

and, restating Equation 4.7:

$$S = \sqrt{P^2 + Q^2}$$

For example, a three-phase sinusoidal power distribution service, with phases a, b, and c:

$$P = E_a I_a \cos(\alpha_a - \beta_a) + E_b I_b \cos(\alpha_b - \beta_b) + E_c I_c \cos(\alpha_c - \beta_c)$$

$$Q = E_a I_a \sin(\alpha_a - \beta_a) + E_b I_b \sin(\alpha_b - \beta_b) + E_c I_c \sin(\alpha_c - \beta_c)$$

$$S = \sqrt{P^2 + Q^2}$$

Power Factor

Power factor is defined by Equation 4.11. Note that it is no longer always true to say that power factor is equal to the cosine of the phase angle. In many three-phase balanced systems, the phase angles of all three phases are equal and the cosine relationship holds. In unbalanced systems, such as that represented

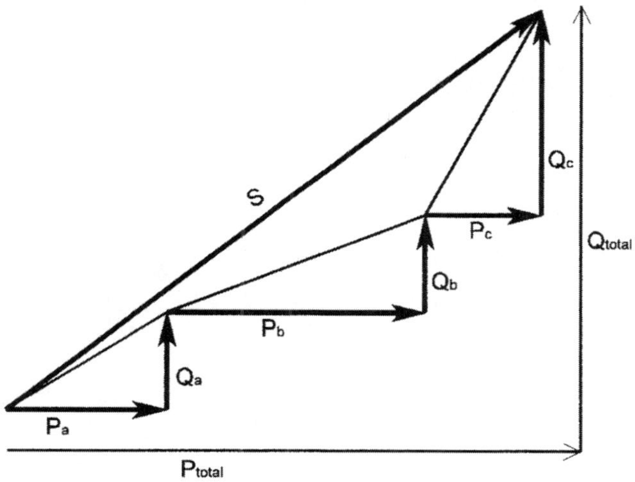

FIGURE 4.8 Phasor diagram for a sample polyphase sinusoidal service in which each phase has a different phase angle. Power factor cannot be defined as the cosine of the phase angle in this case. (From Reference [3].)

by the phasor diagram Figure 4.8, each phase has a different phase angle, the phase voltages and currents are not equal, and the cosine relationship fails [3].

$$F_p = \frac{\text{TOTAL ACTIVE POWER}}{\text{PHASOR POWER}} = \frac{P_{\text{Eq.9}}}{S} = \frac{\text{WATTS}}{\text{VOLTAMPS}} \; \textit{often} \neq \cos\theta \tag{4.11}$$

Nonsinusoidal Voltage and Current

Fourier Analysis

Figure 4.9 shows voltage, current, and power for a typical single-phase nonlinear load, a computer switch-mode power supply. Due to the nature of the bridge rectifier circuit in this power supply, current is drawn from the line in sharp spikes. The current peak is only slightly displaced from the voltage peak and the power is everywhere positive. However, power is delivered to the load during only part of the cycle and the average power is much lower than if the current had been sinusoidal. The current waveshape required by the load presents a problem to the ac power system, which is designed to deliver only sine wave current. The solution to this problem is based on mathematical concepts developed in 1807 for describing heat flow by Jean Baptiste Joseph Fourier, a French mathematician [4]. Fourier's theorem states that any periodic function, however complex, can be broken up into a series of simple sinusoids, the sum of which will be the original complex periodic variation. Applied to the present electrical problem, Fourier's theorem can be stated: *any periodic nonsinusoidal electrical waveform can be broken up into a series of sinusoidal waveforms, each a harmonic of the fundamental, the sum of which will be the original nonsinusoidal waveform.*

Harmonics

Harmonics are defined as continuous integral multiples of the fundamental waveform. Figure 4.10 shows a fundamental sine wave and two harmonic waves — the 3rd and 5th harmonics. The harmonic numbers 3 and 5 express the number of complete cycles for each harmonic wave per cycle of the fundamental (or 1st harmonic). Each harmonic wave is defined by its harmonic number, its amplitude, and its phase relationship to the fundamental. Note that the fundamental frequency can have any value without changing the harmonic relationships, as shown in Table 4.1.

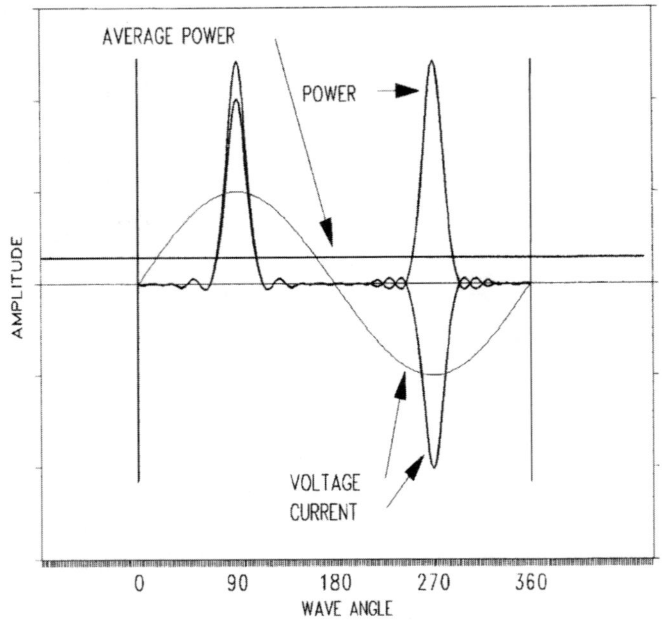

FIGURE 4.9 Voltage, current, and power for a single-phase, switch-mode computer power supply, a typical nonlinear load. The current is no longer sinusoidal.

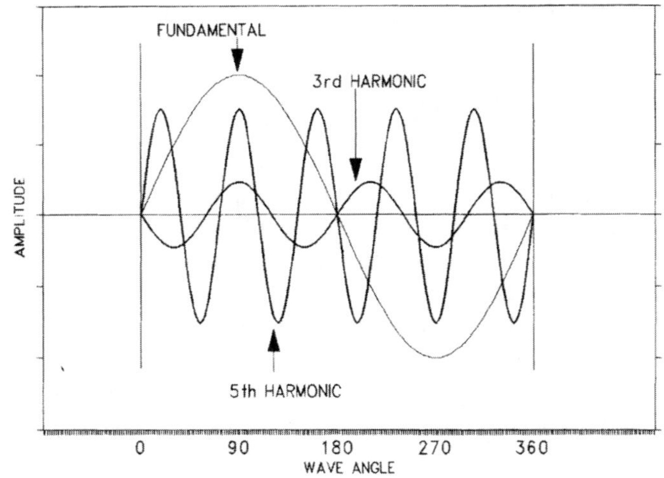

FIGURE 4.10 Harmonics are continuous integral multiples of the fundamental frequency. The 5th harmonic is shown, in phase with the fundamental, while the 3rd harmonic is 180° out of phase.

TABLE 4.1 Harmonics and Their Relationship to the Fundamental Frequency

Harmonic number	Frequency (Hz)	Frequency (Hz)	Frequency (Hz)
1	60	50	400
2	120	100	800
3	180	150	1200
4	240	200	1600
5	300	250	2000

Power Calculations

Calculating power delivered to do work for a nonlinear load is somewhat more complicated than if the current were sinusoidal. If the fundamental component of the voltage at frequency f is taken as a reference (the a-phase fundamental for a polyphase system), the subscript "1" means the fundamental, and E denotes the peak value of the voltage; then the voltage can be expressed as:

$$\varepsilon_{a1(t)} = E_{a1}\sin\left(2\pi ft + 0°\right)$$

The voltage fundamental will then have an amplitude E_{a1} and pass through zero in the positive going direction at time $t = 0$. If h = harmonic number, and E_h and I_h are peak amplitudes of the harmonic voltage and current, respectively, then general expressions for any harmonic will be:

$$e_{h(t)} = E_h \sin\left(2\pi fht + \alpha_h^\circ\right)$$

$$i_{h(t)} = I_h \sin\left(2\pi fht + \beta_h^\circ\right)$$

To compute the power associated with a voltage and current waveform, take advantage of the fact that products of harmonic voltages and currents of different frequency have a time average of zero. Only products of voltages and currents of the same frequency are of interest, giving a general expression for harmonic power as:

$$p_{h(t)} = E_h I_h \sin\left(2\pi fht + \alpha_h^\circ\right)\sin\left(2\pi fht + \beta_h^\circ\right)$$

Simplifying with trigonometric identities, evaluating over an integral number of cycles, and replacing peak voltage and current with rms values, the average power becomes:

$$P_{h(t)} = E_h I_h \cos\left(\alpha_h^\circ - \beta_h^\circ\right)$$

For a *single-phase system* where h is the harmonic number and H is the highest harmonic, the total average power or active power is given by:

$$P = \sum_{h=1}^{H} E_h I_h \cos\left(\alpha_h - \beta_h\right) \tag{4.12}$$

Total average reactive power is given by:

$$Q = \sum_{h=1}^{H} E_h I_h \sin\left(\alpha_h - \beta_h\right) \tag{4.13}$$

It should be noted that in the real world, the actual contribution of harmonic frequencies to active and reactive power is small (usually less than 3% of the total active or reactive power). The major contribution of harmonic frequencies to the power mix comes as distortion power, which will be defined later.

For a *polyphase system* wherein r is the phase identification and N is the number of conductors in the system, including the neutral conductor, total average power for a polyphase system is given by:

$$P = \sum_{r=1}^{N-1}\sum_{h=1}^{H} E_{rh} I_{rh} \cos\left(\alpha_{rh} - \beta_{rh}\right) \tag{4.14}$$

Total average reactive power is given by:

$$Q = \sum_{r=1}^{N-1} \sum_{h=1}^{H} E_{rh} I_{rh} \sin(\alpha_{rh} - \beta_{rh}) \tag{4.15}$$

Power Factor

Single-Phase Systems

For a single-phase system, phasor power is again given by Equation 4.7 and illustrated by Figure 4.7, where P is the algebraic sum of the active powers for the fundamental and all the harmonics (Equation 4.12), and Q is the algebraic sum of the reactive powers for the fundamental and all the harmonics (Equation 4.13). Therefore, phasor power is based on the fundamental and harmonic active and reactive powers. It is found, however, that phasor power S is no longer equal to apparent power U and a new power phasor must be defined to recognize the effects of waveform distortion. A phasor representing the distortion, termed *distortion power* and given the symbol D, is defined by:

$$D = \pm\sqrt{U^2 - S^2} \tag{4.16}$$

Without further definite information as to the sign of distortion power, its sign is selected the same as the sign of the total active power. The relationships among the various power terms are displayed in Figure 4.11, a three-dimensional phasor diagram. Power factor, in direct parallel with sinusoidal waveforms, is defined by the equation:

$$F_p = \frac{\text{TOTAL ACTIVE POWER}}{\text{APPARENT POWER}} = \frac{P}{U} = \frac{\text{WATTS}}{\text{VOLTAMPS}} \tag{4.17}$$

From Equations 4.7 and 4.16 we obtain:

$$S = \sqrt{P^2 + Q^2} \tag{4.7}$$

$$U = \sqrt{S^2 + D^2} = \sqrt{P^2 + Q^2 + D^2} \tag{4.18}$$

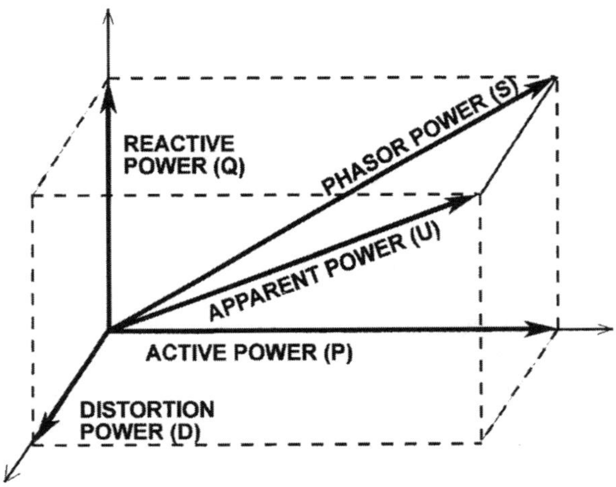

FIGURE 4.11 Phasor diagram for a single-phase, nonsinusoidal service in which the voltage and current contain harmonics. Geometric relationships are shown between active, reactive, phasor, distortion, and apparent powers.

It is clear that when waveforms are sinusoidal, i.e., linear loads are drawing current, that there is no distortion power and Equation 4.18 reduces to Equation 4.7. Likewise as shown in Figure 4.13, as the distortion power vector goes to zero, the figure becomes two-dimensional and reduces to Figure 4.7, and U becomes equal to S. When, however, nonlinear loads are drawing harmonic currents from the system, U will be greater than S. As already noted, the contribution of the harmonics to the total power quantities is small and one is frequently interested mainly in the fundamental quantities.

The power factor associated only with the fundamental voltage and current components was termed the displacement power factor $F_{\text{p displacement}}$ where Equations 4.7 and 4.8 are written [5]:

$$S_{60} = \sqrt{P_{60}^2 + Q_{60}^2}$$

and

$$F_{\text{p displacement}} = \frac{P_{60}}{S_{60}}$$

When harmonics are present, F_p is always smaller than $F_{\text{p displacement}}$.

Polyphase Systems

For a polyphase system, phasor power, S, is again given by Equation 4.7, but one must now use the total values for P and Q calculated using Equations 4.14 and 4.15. One can then define the apparent power U in one of two ways.

- *Arithmetic apparent power.* The *arithmetic apparent power* is given the symbol U_a, and is defined by Equation 4.19, where E_r and I_r are the rms values for the respective phases and M equals the number of phases. U_a is a scalar quantity.

$$U_a = \sum_{r=1}^{M-1} E_r I_r \tag{4.19}$$

- *Apparent power. Apparent power* is given the symbol U and is defined by Equation 4.18 using total values for P and Q as defined by Equations 4.14 and 4.15, and a total value for D determined using Equation 4.16 for each phase. Figure 4.12 illustrates the two alternative concepts for polyphase apparent power [6]. Note that U_a uses arithmetic addition of vector magnitudes and is equal to apparent power U only if the polyphase voltages and currents have equal magnitudes and equal angular spacings, a situation that often exists in balanced three-phase systems. The two alternative definitions of apparent power, U and U_a, give rise to two possible values for power factor: (1) $F_p = P/U$; and (2) $F_{pa} = P/U_a$. Apparent power U and power factor F_p are the preferred definitions since using U_a can give unexpected results with some nonsymmetric service arrangements such as four-wire delta, and, with extremely unbalanced resistive loads, F_{pa} can exceed 1.0. Despite these shortcomings, arithmetic apparent power has become quite widely used due to the comparative simplicity of its measurement and calculation. With the advent of sophisticated digital meters, there is no longer any advantage to using arithmetic apparent power and its use will surely decrease.

4.4 Power Factor "Measurement"

There are no instruments that measure power factor directly. (Power stations and large substations often use phase angle meters with a power factor scale representing cos (θ) to display power factor. Such meters are accurate only for sinusoidal balanced polyphase systems.) One must remember that, of all the ac power quantities discussed, the only ones that can be directly measured are voltages, currents, and their time relationships (phase angles). All other ac power quantities are derived mathematically from these measured quantities. The only one of these derived values that has physical reality is the active power P

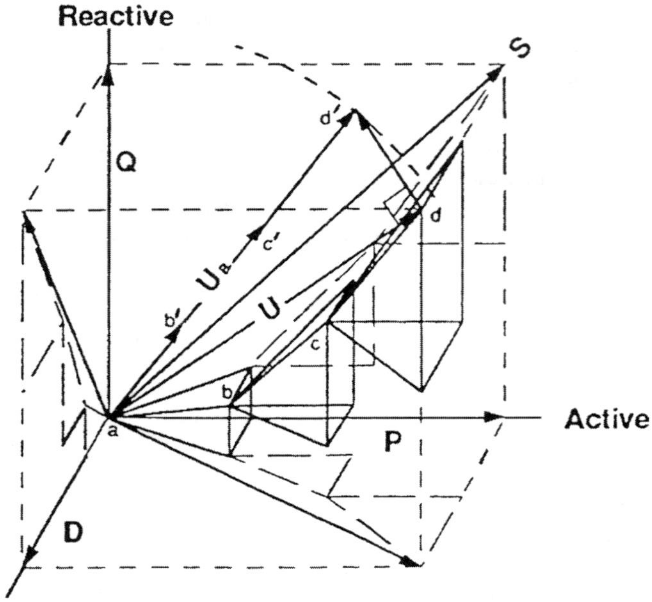

FIGURE 4.12 Phasor diagram for a three-phase, nonsinusoidal service in which the voltage and current contain harmonics. Arithmetic apparent power U_a is the length of the segmented line *abcd* and is a scaler quantity. U_a can be represented by the line *ab′c′d′*. The diagonal *ad*, a vector quantity, is the apparent power *U* [6].

(the quantity that does work); the others are mathematical constructs. Therefore, correct determination of power factor requires accurate measurement of voltage and current, and proficient mathematics.

Metering for Linear Loads

By the early 1920s, the concepts of active, reactive, and apparent (since renamed phasor) power, and power factor were known, and metering capabilities had been developed to enable their determination. Energy meters utilizing voltage and current coils driving an induction disk inherently measured active power *P* ($EI \cos\theta$), which was displayed on a mechanical register. Using the trigonometric identity $EI \sin\theta = EI \cos(90° + \theta)$, with voltage delayed 90°, a similar energy meter displayed reactive power *Q* and, with these two values, displacement power factor was calculated. Voltage delay (phase shifting) was accomplished using specially wound transformers.

Through the years, the method of obtaining the 90° phase shift has been updated. Analog electronic meters are available that provide the 90° phase shift electronically within the meter. More recently, digital meters have been developed that sample voltages and currents at regular intervals and digitize the results. Voltages and currents are multiplied as they are captured to compute active power. Past voltage samples delayed by a time equal to a quarter cycle (90°) are multiplied by present current values to obtain reactive power. Active–reactive metering of this type is a widely utilized method for determining displacement power factor for utility billing. These meters do not accurately measure the effect of harmonic currents because the delay of the voltage samples is based on the fundamental frequency and is incorrect for the harmonics. (The important 5th harmonic, which is frequently the predominant harmonic component, is accurately measured because it is delayed 450° (5 × 90°), which is equivalent to the correct 90° delay).

Metering for Nonlinear Loads

With the application of high-speed digital computing techniques for measurement of ac currents and voltages, together with digital filtering, the quantities necessary for accurate and correct calculation of power factor are susceptible to direct computation. In practice, the ac waveforms are sampled at a frequency greater than twice the highest frequency to be measured, in compliance with well-known

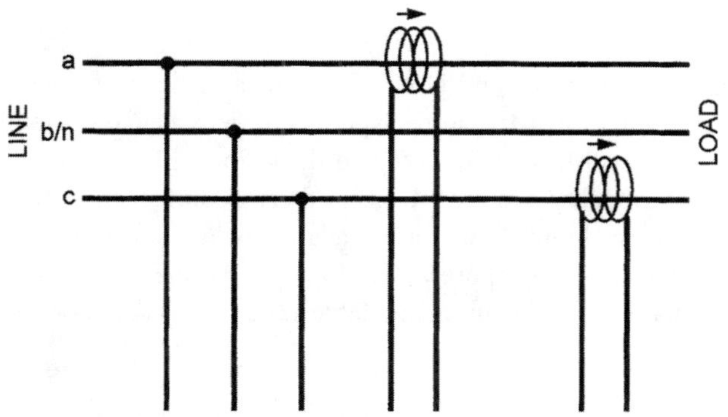

FIGURE 4.13 Meter connections for a one-, two-, or three-phase, three-wire service. Voltage leads are connected to phases a, b, and c, *or* to a, c, and neutral as the system dictates. Care must be taken to connect the voltage lead for each phase to the input corresponding to the input for the current transformer reading that phase, and directional characteristics of the transformers must be observed. Only two current connections are required.

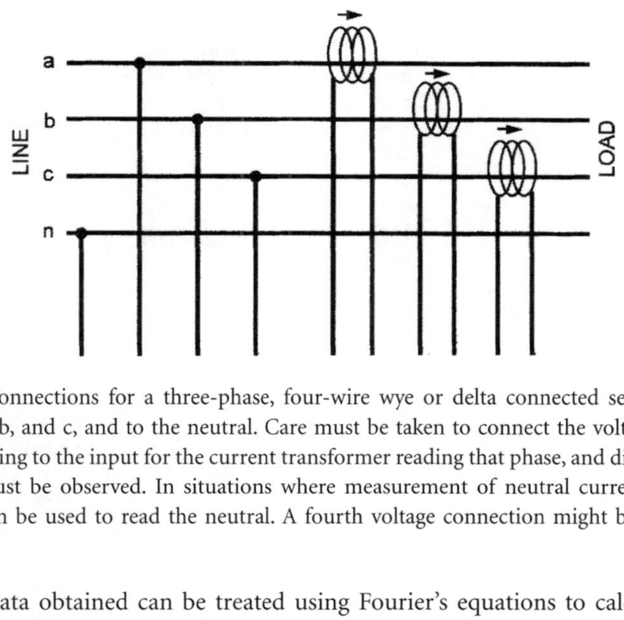

FIGURE 4.14 Meter connections for a three-phase, four-wire wye or delta connected service. Voltage leads are connected to phases a, b, and c, and to the neutral. Care must be taken to connect the voltage lead for each phase to the input corresponding to the input for the current transformer reading that phase, and directional characteristics of the transformers must be observed. In situations where measurement of neutral currents is desired, a fourth current transformer can be used to read the neutral. A fourth voltage connection might be used to read neutral-ground voltages.

sampling theories. Data obtained can be treated using Fourier's equations to calculate rms values for voltage, current, and phase angle for the fundamental and each harmonic frequency. Power quantities can be obtained with digital filtering in strict accordance with their *ANSI/IEEE STD* 100 definitions. Power quantities can be displayed for the fundamental only (displacement power factor), or for fundamental plus all harmonics (power factor for nonsinusoidal waveforms).

Metering Applications

Instruments with the capabilities described above are often called *harmonic analyzers*, and are available in both single-phase and polyphase versions. They can be portable, in which case they are often used for power surveys, or panel mounted for utility and industrial revenue metering. Polyphase analyzers can be connected as shown in Figures 4.13 and 4.14. Care must be taken to connect the instrument properly. Both voltage leads and current transformers must be connected to the proper phases, and the current transformers must also be placed with the correct polarity. Most instruments use color-coded voltage connectors. Correct polarity is indicated on the current transformers by arrows or symbols, and complete hook-up and operating instructions are included. When single-phase instruments are used, the same precautions must be followed for making connections.

4.5 Instrumentation

Table 4.2 lists a sampling of harmonic and power factor measuring instruments available from major manufacturers. All instruments listed use some type of Fourier calculations and/or digital filtering to determine power values in accordance with accepted definitions. Many can be configured to measure nonharmonic-related power quality concerns. Unless otherwise noted, all instruments require the purchase of one current probe per input. Probes are available for measuring currents from 5 A to several thousand amperes. For comparison purposes, priced probes will be those with a 600 A range. Voltage leads are usually supplied as standard equipment. Table 4.3 contains addresses of these manufacturers.

TABLE 4.2 Selected Instrumentation for Harmonic and Power Factor Measurement

Manufacturer	Model	V/I inputs	Display type	Communication	Special features	List price (US$)
			Hand-held units			
Amprobe	HA2000 1/1	Visual	RS232	Hand-held, probe included, 21 nonvolatile memories	990	
BMI	155 1/1	Visual	Optional printer, RS232	Hand-held, 1145 + 550 (printer) + 380 (probe)		
BMI	355	4/3	Visual	Optional printer, RS232	Hand-held	1995 + 550 (printer) + 380/probe
Dranetz 4300	4/4	Panel for data and graphs RS232	Hand-held, battery or ac power, optional PCMCIA memory card	5000 + 450/probe		
Fluke	39	1/1	Visual	None	Probe included	895
Fluke	41b	1/1	Visual	RS232	Probe included, 8 memories, logging with computer and supplied software	1795
			Portable units			
BMI	3030A	4/4	Built-in printer for data and graphs optional internal modem	Portable, optional built-in disk drive for storage, longterm monitoring, optional PQ configurations	6800 + 600 (modem) + 1895 (disk storage) + 590/probe	
Dranetz	PP1-R	4/4	Panel for data and graphs	PCMCIA card slot	Long-term monitoring, optional PQ configurations, remote control by software	12,000 + 450/probe

TABLE 4.2 (continued) Selected Instrumentation for Harmonic and Power Factor Measurement

Manufacturer	Model	V/I inputs	Display type	Communication	Special features	List price (US$)
			PC-based units			
BMI	7100	4/4	PC-based (not included)	PC connection cord	Portable, uses computer for storage, optional software, longterm monitoring, optional PQ configurations	5294 + PC + 395 (software) + 415/probe
Cooper	V-Monitor II	4/4	PC-based (not included)	Serial port	Portable, software and signal interface and data acquisition board, long-term monitoring	12,000 + 500/ probe
RPM	1650	4/5	PC-based, not included	Ethernet longterm monitoring, optional PQ configurations, remote control by software	6250 + 750 software + 490/ probe	
			Panel-mounted			
Cutler-Hammer/ Westinghouse	4/4	Panel	Optional IMPACC	Panel-mount to replace meters, monitoring	3690 + input devices	
General Electric	kV	Vector Electricity Meter	Socket	Programmable multifunction LCD display pulse output for measured power quantities replaces industrial revenue meters, calculates and accumulates all power and revenue data	595 + 495 (software)	
Square D	Powerlogic PM620	3/3	LCD panel	RS485	Panel meter replacement, calculates and accumulates power data	1583 + probes
Square D	Powerlogic CM2350	3/3	Panel	RS485	Connect to network, remote controller, monitoring	4290 + probes

TABLE 4.3 Manufacturers of Power Factor Measuring Harmonic Analyzers

Amprobe Instruments 630 Merrick Road Lynbrook, NY 11563 Tel: (516) 593-5600	Cutler Hammer Inc. Westinghouse & Cutler-Hammer Products Five Parkway Center Pittsburgh, PA 15220 Tel: (412) 937-6100	GE Meter 130 Main Street Somersworth, NH 03878 Tel: (603) 749-8477
BMI 3250 Jay Street Santa Clara, CA 95054 Tel: (408) 970-3700	Dranetz Technologies, Inc. 1000 New Durham Road Edison, NJ 08818-4019 Tel: (800) DRANTEC	Reliable Power Meters, Inc. 400 Blossom Hill Road Los Gatos, CA 95032 Tel: (408) 358-5100
Cooper Power Systems Division 11131 Adams Road P.O. Box 100 Franksville, WI 53126-0100 Tel: (414) 835-2921	Fluke Corporation P.O. Box 9090 Everett, WA 98206 Tel: (800) 44FLUKE	Square D Power Logic 330 Weakley Road Smyrna, TN 37167-9969 Tel: (615) 459-8552

Defining Terms

Active power: A term used to express the real power delivered to do work in an ac distribution system.

Reactive power: A term used to express the imaginary power that does no work but provides magnetization to enable work in an ac distribution system.

Phasor power: A term used to express the product of volts and amperes in an ac distribution system in which voltage and current are sinusoidal.

Harmonic power: A term used to express the power due to harmonic frequencies in an ac distribution system in which voltage and/or current are nonsinusoidal.

Apparent power: A term used to express the product of volts and amperes in an ac distribution system in which voltage and/or current are nonsinusoidal.

Power factor: A single number, calculated by dividing active power by either the phasor power or the apparent power, which describes the effective utilization of ac distribution system capacity.

References

1. D.F. Bullock, Methods of measuring apparent power, GE Meter, Missouri Valley Electrical Assoc., *Annu. Eng. Conf.,* Kansas City, MO, 1996.
2. *ANSI/IEEE Std 100, Dictionary of Electrical and Electronic Terms, 1992,* New York: IEEE, 1993.
3. D.F. Bullock, Phase Angle Relationships, Part 1: Theory, GE Meter Arkansas Electric Meter School, 1995.
4. I. Assimov, *Assimov's Biographical Encyclopedia of Science and Technology, New Rev. Ed.*, Garden City, NY: Doubleday & Co., 1972.
5. D.F. Bullock, private communication.
6. D.F. Bullock and D.D. Elmore, *MIND UR PS & QS, GE Meter EEI-AEIC Meter and Service Committee Meeting,* Dallas, TX, 1994.

Further Information

H.L. Curtis and F.B. Silsbee, Definitions of power and related quantities, *AIEE Summer Conf.*, 1935.

5 Phase Measurement

Peter O'Shea

Royal Melbourne Institute of Technology

The notion of "phase" is usually associated with *periodic* or repeating signals. With these signals, the waveshape perfectly repeats itself every time the *period* of repetition elapses. For periodic signals one can think of the phase at a given time as the fractional portion of the period that has been completed. This is commonly expressed in degrees or radians, with full cycle completion corresponding to 360° or 2π radians. Thus, when the cycle is just beginning, the phase is zero. When the cycle is half completed, the phase is half of 360°, or 180° (see Figure 5.1). It is important to note that if phase is defined as the portion of a cycle that is completed, the phase depends on where the beginning of the cycle is taken to be. There is no universal agreement on how to specify this beginning. For a sinusoidal signal, probably the two most common assumptions are that the start of the cycle is (1) the point at which the maximum value is achieved, and (2) the point at which the negative to positive zero-crossing occurs. Assumption (1) is common in many theoretical treatments of phase, and for that reason is adopted in this chapter. It should be noted, however, that assumption (2) has some benefits from a measurement perspective, because the zero-crossing position is easier to measure than the maximum.

The measurement of phase is important in almost all applications where sinusoids proliferate. Many means have therefore been devised for this measurement. One of the most obvious measurement techniques is to directly measure the fractional part of the period that has been completed on a cathode-ray oscilloscope (CRO). Another approach, which is particularly useful when a significant amount of noise is present, is to take the Fourier transform of the signal. According to Fourier theory, for a sinusoidal signal, the energy in the Fourier transform is concentrated at the frequency of the signal; the initial phase of the signal (i.e., the phase at time, $t = 0$) is the phase of the Fourier transform at the point of this energy concentration. The measurements of initial phase and frequency obtained from the Fourier transform can then be used to deduce the phase of the signal for any value of time.

0-8493-1733-9/04/$0.00+$1.50

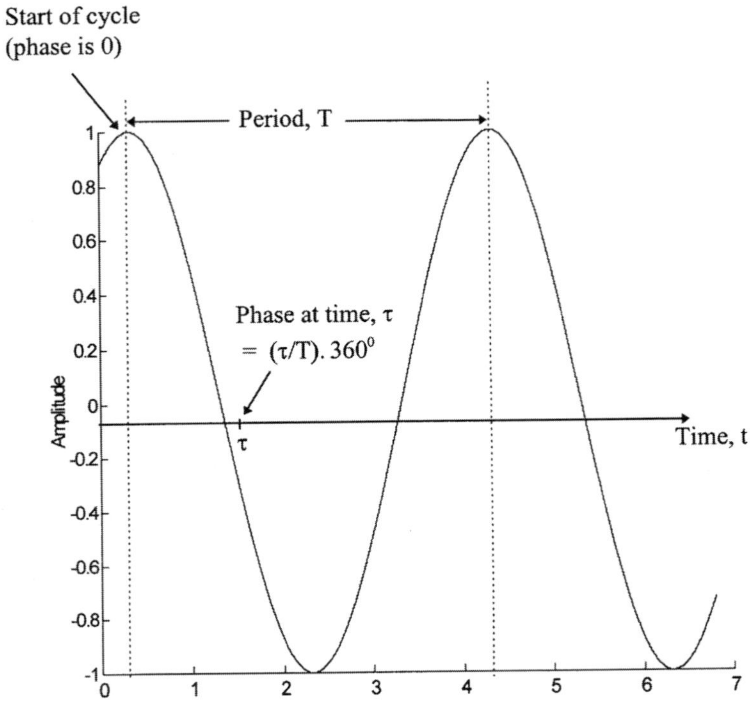

FIGURE 5.1 The phase of a periodic sinusoidal signal. The time scale is arbitrary.

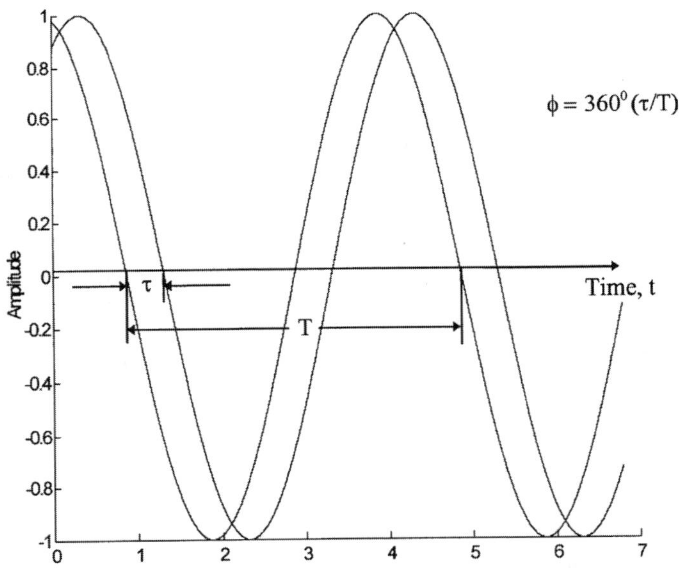

FIGURE 5.2 Two signals with a relative phase difference of ϕ between them. The time scale is arbitrary.

Frequently what is needed in practice is a measurement of the *phase difference* between two signals of the same frequency; that is, it is necessary to measure the *relative phase* between two signals rather than the *absolute phase* of either one (see Figure 5.2). Often, in the measurement of the relative phase between two signals, both signals are derived from the same source. These signals might, for example, be the current and voltage of a power system; the relative phase, ϕ, between the current and voltage would then be useful for monitoring power usage, since the latter is proportional to the cosine of ϕ.

Several techniques are available for the measurement of "relative phase." One crude method involves forming "Lissajous figures" on an oscilloscope. In this method, the first of the two signals of interest is fed into the vertical input of a CRO and the other is fed into the horizontal input. The result on the oscilloscope screen is an ellipse, the intercept and maximum height of which can be used to determine the relative phase. Other methods for determining relative phase include the crossed-coil meter (based on electromagnetic induction principles), the zero-crossing phase meter (based on switching circuitry for determining the fractional portion of the period completed), the three-voltmeter method (based on the use of three signals and trigonometric relationships), and digital methods (based on analog-to-digital conversion and digital processing).

5.1 Amplitude, Frequency, and Phase of a Sinusoidal Signal

An arbitrary sinusoidal signal can be written in the form:

$$s(t) = A\cos(2\pi ft + \phi_0) = A\cos(\omega t + \phi_0) \tag{5.1}$$

where A = Peak amplitude
f = Frequency
ω = Angular frequency
ϕ_0 = Phase at time $t = 0$

This signal can be thought of as being the real part of a complex phasor that has amplitude, A, and which rotates at a constant angular velocity $\omega = 2\pi f$ in the complex plane (see Figure 5.3).

Mathematically, then, $s(t)$ can be written as:

$$s(t) = \Re e\left\{Ae^{j2\pi ft+\phi_0}\right\} = \Re e\left\{z(t)\right\} \tag{5.2}$$

where $z(t)$ is the complex phasor associated with $s(t)$, and $\Re\{.\}$ denotes the real part. The "phase" of a signal at any point in time corresponds to the angle that the rotating phasor makes with the real axis. The initial phase (i.e., the phase at time $t = 0$) is ϕ_0. The "frequency" f of the signal is $1/2\pi$ times the phasor's angular velocity.

There are a number of ways to define the phase of a real sinusoid with unknown amplitude, frequency, and initial phase. One way, as already discussed, is to define it as the fractional part of the period that

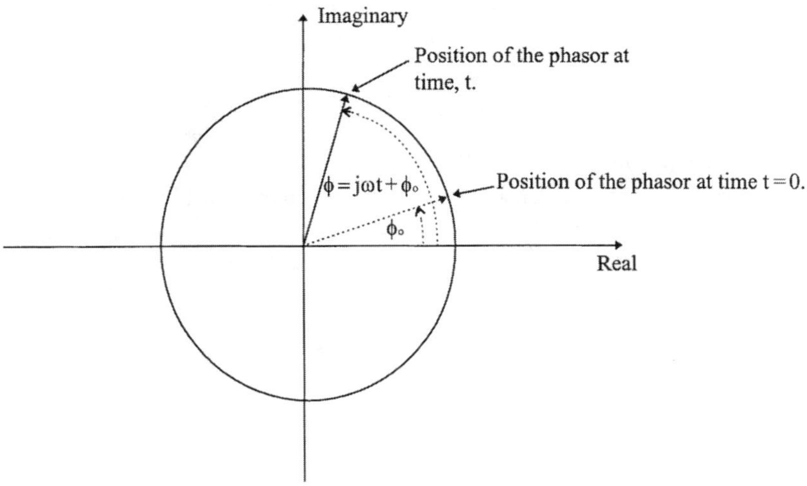

FIGURE 5.3 A complex rotating phasor, $A\exp(j\omega t + \phi_0)$. The length of the phasor is A and its angular velocity is ω. The real part of the phasor is $A\cos(\omega t + \phi_0)$.

has been completed. This is a valid and intuitively pleasing definition, and one that can be readily generalized to periodic signals that contain not only a sinusoid, but also a number of harmonics. It cannot, however, be elegantly generalized to allow for slow variations in the frequency of the signal, a scenario that occurs in communicatons with phase and frequency modulation. Gabor put forward a definition in 1946 that can be used for signals with slowly varying frequency. He proposed a mathematical definition for generating the complex phasor, $z(t)$, associated with the real signal, $s(t)$. The so-called *analytic signal* $z(t)$ is defined according to the following definition [1]:

$$z(t) = s(t) + j\mathcal{H}\{s(t)\} \tag{5.3}$$

where $\mathcal{H}\{\ \}$ denotes the *Hilbert transform* and is given by:

$$\mathcal{H}\{s(t)\} = p.v.\left[\int_{-\infty}^{+\infty} \frac{s(t-\tau)}{\pi\tau}\,\mathrm{d}\tau\right] \tag{5.4}$$

with *p.v.* signifying the Cauchy principal value of the integral [2].

The imaginary part of the analytic signal can be generated practically by passing the original signal through a "Hilbert transform" filter. From Equations 5.3 and 5.4, it follows that this filter has impulse response given by $1/\pi t$. The filter can be implemented, for example, with one of the HSP43xxx series of ICs from Harris Semiconductors. Details of how to determine the filter coefficients can be found in [2].

Having formally defined the analytic signal, it is possible to provide definitions for phase, frequency, and amplitude as functions of time. They are given below.

$$\textit{Phase:}\quad \phi(t) = \arg\{z(t)\} \tag{5.5}$$

$$\textit{Frequency:}\quad f(t) = \frac{1}{2\pi}\frac{\mathrm{d}\left[\arg\{z(t)\}\right]}{\mathrm{d}t} \tag{5.6}$$

$$\textit{Amplitude:}\quad A(t) = \mathrm{abs}\left[z(t)\right] \tag{5.7}$$

The definitions for phase, frequency, and amplitude can be used for signals whose frequency and/or amplitude vary slowly with time. If the frequency and amplitude do vary with time, it is common to talk about the "instantaneous frequency" or "instantaneous amplitude" rather than simply the frequency or amplitude.

Note that in the analytic signal, the imaginary part lags the real part by 90°. This property actually holds not only for sinusoids, but for the real and imaginary parts of all frequency components in "multicomponent" analytic signals as well. The real and imaginary parts of the analytic signal then correspond to the "in-phase (I)" and "quadrature (Q)" components used in communications systems.

In a balanced three-phase electrical power distribution system, the analytic signal can be generated by appropriately combining the different outputs of the electrical power signal; that is, it can be formed according to:

$$z(t) = v_\mathrm{a}(t) + \frac{j}{\sqrt{3}}\left[v_\mathrm{c}(t) - v_\mathrm{b}(t)\right] \tag{5.8}$$

where $v_\mathrm{a}(t)$ = Reference phase
$v_\mathrm{b}(t)$ = Phase that leads the reference by 120°
$v_\mathrm{c}(t)$ = Phase that lags the reference by 120°.

5.2 The Phase of a Periodic Nonsinusoidal Signal

It is possible to define "phase" for signals other than sinusoidal signals. If the signal has harmonic distortion components present in addition to the fundamental, the signal will still be periodic, but it will no longer be sinusoidal. The phase can still be considered to be the fraction of the period completed. The "start" of the period is commonly taken to be the point at which the initial phase of the fundamental component is 0, or at a zero-crossing. This approach is equivalent to just considering the phase of the fundamental, and ignoring the other components. The Fourier method provides a very convenient method for determining this phase — the energy of the harmonics in the Fourier transform can be ignored.

5.3 Phase Measurement Techniques

Direct Oscilloscope Methods

Cathode-ray oscilloscopes (CROs) provide a simple means for measuring the phase difference between two sinusoidal signals. The most straightforward approach to use is direct measurement; that is, the signal of interest is applied to the vertical input of the CRO and an automatic time sweep is applied to the horizontal trace. The phase difference is the time delay between the two waveforms measured as a fraction of the period. The result is expressed as a fraction of 360° or of 2π radians; that is, if the time delay is 1/4 of the period, then the phase difference is 1/4 of 360° = 90° (see Figure 5.2). If the waveforms are not sinusoidal but are periodic, the same procedure can still be applied. The phase difference is just expressed as a fraction of the period or as a fractional part of 360°.

Care must be taken with direct oscilloscope methods if noise is present. In particular, the noise can cause triggering difficulties that would make it difficult to accurately determine the period and/or the time delay between two different waveforms. The "HF reject" option, if available, will alleviate the triggering problems.

Lissajous Figures

Lissajous figures are sometimes used for the measurement of phase. They are produced in an oscilloscope by connecting one signal to the vertical trace and the other to the horizontal trace. If the ratio of the first frequency to the second is a rational number (i.e., it is equal to one small integer divided by another), then a closed curve will be observed on the CRO (see Figures 5.4 and 5.5). If the two frequencies are

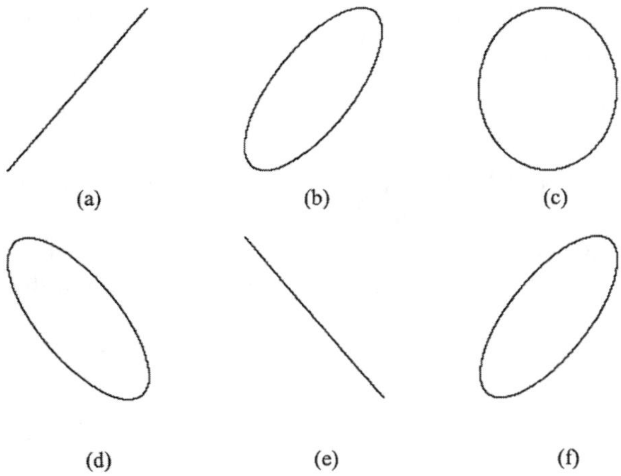

FIGURE 5.4 Lissajous figures for two equal-amplitude, frequency-synchronized signals with a relative phase difference of (a) O, (b) $\pi/4$, (c) $\pi/2$, (d) $3\pi/4$, (e) π, (f) $-\pi/4$.

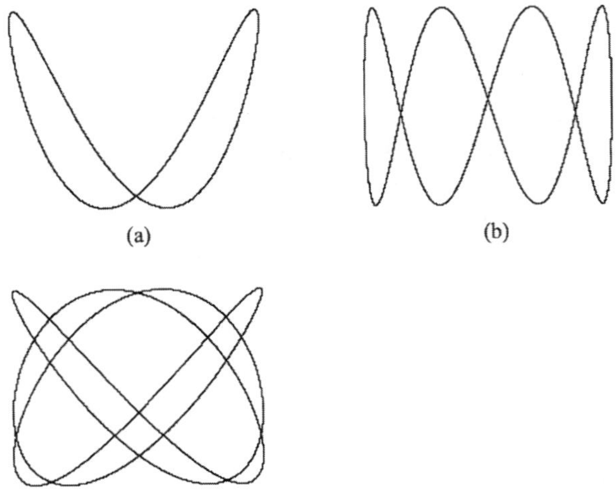

FIGURE 5.5 Lissajous figures for two signals with vertical frequency: horizontal frequency ratios of (a) 2:1, (b) 4:1, (c) 4:3.

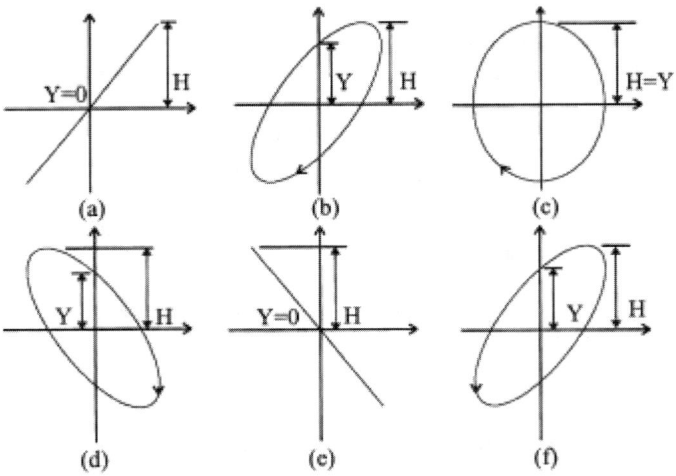

FIGURE 5.6 Lissajous figures for two signals with synchronized frequency and various phase differences: (a) phase difference = 0°, (b) phase difference = 45°, (c) phase difference = 90°, (d) phase difference = 135°, (e) phase difference = 180°, (f) phase difference = 315°.

unrelated, then there will be only a patch of light observed because of the persistance of the oscilloscope screen.

If the two signals have the same frequency, then the Lissajous figure will assume the shape of an ellipse. The ellipse's shape will vary according to the phase difference between the two signals, and according to the ratio of the amplitudes of the two signals. Figure 5.6 shows some figures for two signals with synchronized frequency and equal amplitudes, but different phase relationships. The formula used for determining the phase is:

$$\sin(\phi) = \pm \frac{Y}{H} \tag{5.9}$$

where H is half the maximum vertical height of the ellipse and Y is the intercept on the *y-axis.* Figure 5.7 shows some figures for two signals that are identical in frequency and have a phase difference of 45°, but

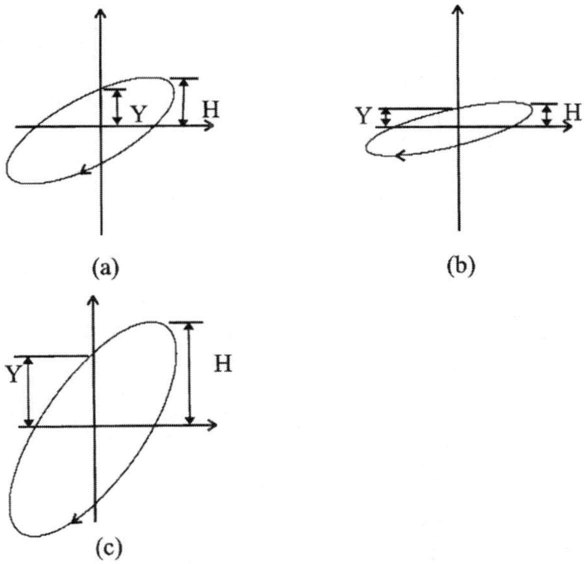

FIGURE 5.7 Lissajous figures for two signals with synchronized frequency, a phase difference of 45°, and various amplitude ratios: (a) amplitude ratio of 1, (b) amplitude ratio of 0.5, (c) amplitude ratio of 2.

with different amplitude ratios. Note that it is necessary to know the direction that the Lissajous trace is moving in order to determine the sign of the phase difference. In practice, if this is not known *a priori,* then it can be determined by testing with a variable frequency signal generator. In this case, one of the signals under consideration is replaced with the variable frequency signal. The signal generator is adjusted until its frequency and phase equal that of the other signal input to the CRO. When this happens, a straight line will exist. The signal generator frequency is then increased a little, with the relative phase thus being effectively changed in a known direction. This can be used to determine the correct sign in Equation 5.9.

Lissajous figure methods are a little more robust to noise than direct oscilloscope methods. This is because there are no triggering problems due to random noise fluctuations. Direct methods are, however, much easier to interpret when harmonics are present. The accuracy of oscilloscope methods is comparatively poor. The uncertainty of the measurement is typically in excess of 1°.

Zero-Crossing Methods

This method is currently one of the most popular methods for determining phase difference, largely because of the high accuracy achievable (typically 0.02°). The process is illustrated in Figure 5.8 for two signals, denoted *A* and *B*, which have the same frequency but different amplitudes. Each negative to positive zero-crossing of signal *A* triggers the start of a rectangular pulse, while each negative to positive zero-crossing of signal *B* triggers the end of the rectangular pulse. The result is a pulse train with a pulse width proportional to the phase angle between the two signals. The pulse train is passed through an averaging filter to yield a measure of the phase difference. It is also worth noting that if the positive to negative zero-crossings are also used in the same fashion, and the two results are averaged, the effects of dc and harmonics can be significantly reduced.

To implement the method practically, the analog input signals must first be converted to digital signals that are "high" if the analog signal is positive, and "low" if the analog signal is negative. This can be done, for example, with a Schmitt trigger, along with an *RC* stabilizing network at the output. Chapter 21 provides a circuit to do the conversion. In practice, high-accuracy phase estimates necessitate that the switching of the output between high and low be very sharp. One way to obtain these sharp transitions is to have several stages of "amplify and clip" preceding the Schmitt trigger.

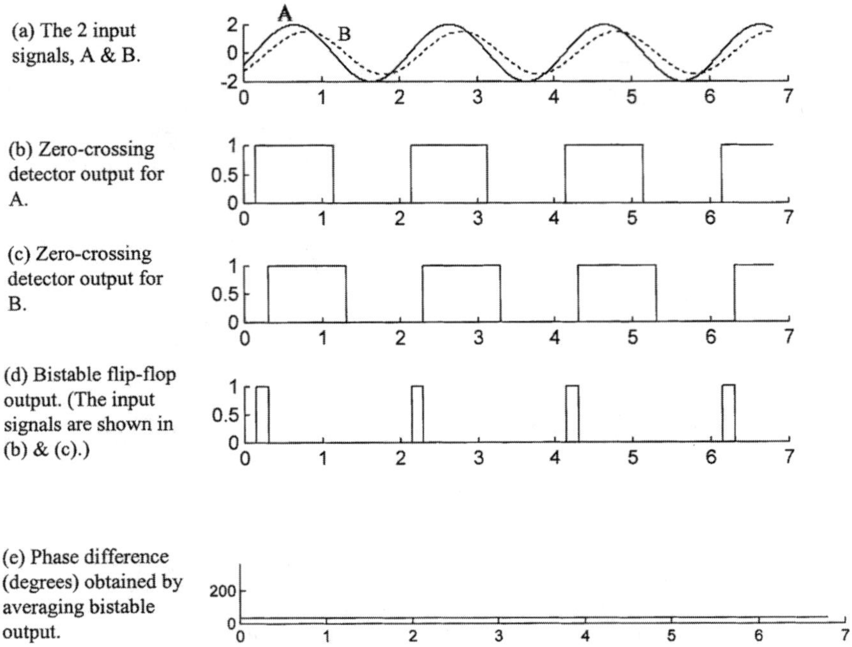

FIGURE 5.8 Input, output, and intermediate signals obtained with the zero-crossing method for phase measurement. Note that the technique is not sensitive to signal amplitude.

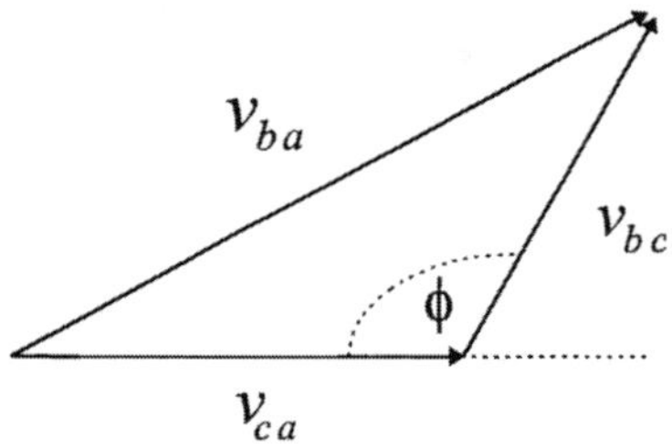

FIGURE 5.9 A vector diagram for determining the phase angle, ϕ, between two ac voltages, v_{ac} and v_{bc}, with the three-voltmeter method.

The digital portion of the zero-crossing device can be implemented with an edge-triggered RS flip-flop and some ancillary circuitry, while the low-pass filter on the output stage can be implemented with an *RC* network. A simple circuit to implement the digital part of the circuit is shown in Chapter 21.

A simpler method for measuring phase based on zero-crossings involves the use of an exclusive or (XOR) gate. Again, the analog input signals must first be converted to digital pulse trains. The two inputs are then fed into an XOR gate and finally into a low-pass averaging filter. The circuit is illustrated in Chapter 21. A disadvantage with this method is that it is only effective if the duty cycle is 50% and if the phase shift between the two signals is between 0 and π radians. It is therefore not widely used.

The Three-Voltmeter Method

The measurement of a phase difference between two voltage signals, v_{ac}, and v_{bc}, can be expedited if there is a common voltage point, *c*. The voltage between points *b* and *a* (v_{ba}), the voltage between points *b* and *c* (v_{bc}), and the voltage between points *c* and *a* (v_{ca}) are measured with three different voltmeters. A vector diagram is constructed with the three measured voltages as shown in Figure 5.9. The phase difference

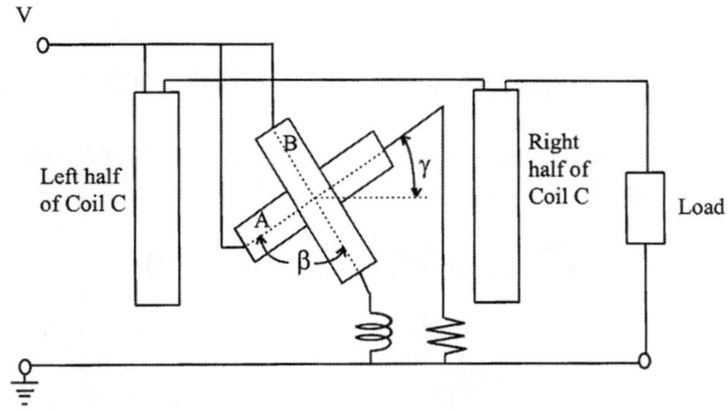

FIGURE 5.10 Diagram of a crossed-coil device for measuring phase. Coils A and B are the rotating coils. Coil C (left and right parts) is the stationary coil.

between the two vectors, v_{ac} and v_{bc}, is determined using a vector diagram (Figure 5.9) and the cos rule. The formula for the phase difference, ϕ, in radians is given by:

$$\pi - \phi = \cos^{-1}\left(\frac{v_{ca}^2 + v_{bc}^2 - v_{ba}^2}{2v_{ca}v_{bc}}\right) \tag{5.10}$$

The Crossed-Coil Method

The crossed-coil phase meter is at the heart of many analog power factor meters. It has two crossed coils, denoted A and B, positioned on a common shaft but aligned at different angles (see Figure 5.10). The two coils move as one, and the angle between them, β, never changes. There is another independent nonrotating coil, C, consisting of two separate parts, "enclosing" the rotating part (see Figure 5.10). The separation of coil C into two separate parts (forming a Helmholtz pair) allows the magnetic field of coil C to be almost constant in the region where the rotating A and B coils are positioned.

Typically the system current, I, is fed into coil C, while the system voltage, V, is applied to coil A via a resistive circuit. The current in coil A is therefore in phase with the system voltage, while the current in coil C is in phase with the system current. Coil B is driven by V via an inductive circuit, giving rise to a current that lags V (and therefore the current in coil A) by 90°. In practice, the angle between the currents in coils A and B is not quite 90° because of the problems associated with achieving purely resistive and purely inductive circuits. Assume, then, that this angle is β. If the angle between the currents in coil B and in coil C is ϕ, then the angle between the currents in coils A and C is $\beta + \phi$. The average torque induced in coil A is proportional to the product of the average currents in coils A and C, and to the cosine of the angle between coil A and the perpendicular to coil C. The average torque induced in coil A is therefore governed by the equation:

$$\overline{T}_A \propto I_A I_C \cos(\phi + \beta)\cos(\gamma) = k_A \cos(\phi + \beta)\cos(\gamma) \tag{5.11}$$

where I_A and I_C = Constants
ω = Angular frequency
$\phi + \beta$ = Relative phase between the currents in coils A and C
γ = Angle between coil A and the perpendicular to coil C
$\propto$ = Signifies "is proportional to"

Assuming that the current in coil B lags the current in coil A by β, then the average torque in coil B will be described by:

$$\overline{T}_B \propto I_B I_C \cos(\phi)\cos(\gamma+\beta) = k_B \cos(\phi)\cos(\gamma+\beta) \tag{5.12}$$

where I_B is a constant, ϕ is the relative phase between the currents in coils B and C, and the other quantities are as in Equation 5.11.

Now, the torques due to the currents in coils A and B are designed to be in opposite directions. The shaft will therefore rotate until the two torques are equal; that is, until:

$$k_A \cos(\phi+\beta)\cos(\gamma) = k_B \cos(\phi)\cos(\gamma+\beta) \tag{5.13}$$

If $k_A = k_B$, then Equation 5.13 will be satisfied when $\phi = \gamma$. Thus, the A coil will be aligned in the direction of the phase shift between the load current and load voltage (apart from errors due to the circuits of the crossed coils not being perfectly resistive/inductive). Thus, a meter pointer attached to the A plane will indicate the angle between load current and voltage. In practice, the meter is usually calibrated to read the cosine of the phase angle rather than the phase angle, and also to allow for the errors that arise from circuit component imperfections.

The accuracy of this method is limited, due to the heavy use of moving parts and analog circuits. Typically, the measurement can only be made accurate to within about 1°.

Synchroscopes and Phasing Lamps

The crossed-coil meter described above is used as the basis for *synchroscopes*. These devices are often used in power generation systems to determine whether two generators are phase and frequency synchronized before connecting them together. In synchroscopes, the current from one generator is fed into the fixed coil and the current from the other generator is fed into the movable crossed coils. If the two generators are synchronized in frequency, then the meter needle will move to the position corresponding to the phase angle between the two generators. If the generators are not frequency synchronized, the meter needle will rotate at a rate equal to the difference between the two generator frequencies. The direction of rotation will indicate which generator is rotating faster.

When frequency synchronization occurs (i.e., the meter needle rotation ceases) and the phase difference is zero, the generators can be connected together. Often in practice, the generators are connected before synchronization occurs; the generator coming on-line is deliberately made a little higher in frequency so that it can provide extra power rather than be an extra drain on the system. The connection is still made, however, when the instantaneous phase difference is zero.

Phasing lamps are sometimes used as a simpler alternative to synchroscopes. A lamp is connected between the two generators, and any lack of frequency synchronization manifests as a flickering of the lamp. A zero phase difference between the two generators corresponds to maximum brightness in the lamp.

Vector Voltmeters and Vector Impedance Methods

Alternating voltages (and currents) are often characterized as vectors consisting of a magnitude and a phase, with the phase being measured relative to some desired reference. Many instruments exist that can display the voltage amplitude and phase of a signal across a wide range of frequencies. These instruments are known as *vector voltmeters* or *network analyzers*. The phase and amplitude as a function of frequency can be obtained very simply in principle by taking the Fourier transform of the signal and simply reading the amplitude and phase across the continuum of frequencies. To achieve good accuracy, this is typically done with down-conversion and digital processing in the baseband region. The down-conversion can be analog, or it can be digital. The procedure is described more fully in the succeeding paragraphs.

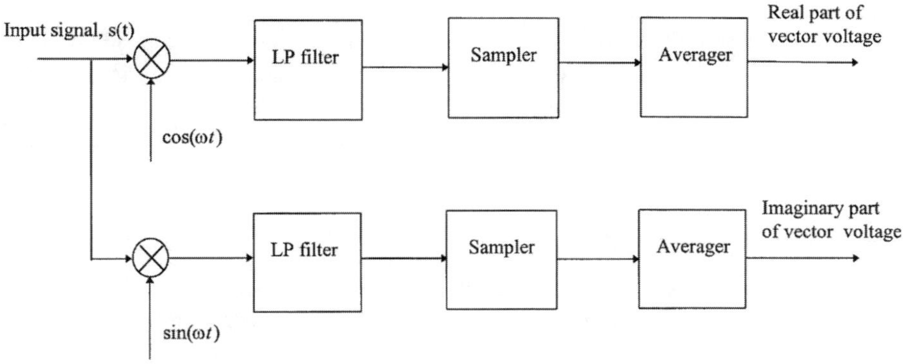

FIGURE 5.11 Vector voltmeter block diagram. The vector voltmeter determines the voltage (amplitude and phase or real and imaginary parts) of the component of the input signal at frequency *f*.

To determine the real part of the voltage vector at a given frequency *f*, the signal is first down-converted by mixing with a local oscillator signal, $\cos(2\pi ft)$. This mixing of the signal recenters the frequency component of interest at 0 Hz. The resultant signal is low-pass filtered, digitally sampled (if not in the digital domain already), and averaged. The digital sampling and averaging enables the amplitude of the newly formed 0 Hz component to be evaluated. The imaginary part is obtained in similar fashion by mixing the signal with $\sin(2\pi ft)$, low-pass filtering, digitally sampling, and again averaging the samples. The amplitude and phase of the voltage vector, *V*, are obtained from the real and imaginary parts using the standard trigonometric relationships:

$$Magnitude = Abs(V) = \sqrt{[\Re e\{V\}]^2 + [\Im m\{V\}]^2} \tag{5.14}$$

$$Phase = \arg(V) = \arctan(\Im m\{V\}/\Re e\{V\}) \tag{5.15}$$

where $\Re\{.\}$ and $\Im m\{.\}$ denote the real and imaginary parts, respectively.

The procedure for forming the vector voltage is summarized in the block diagram in Figure 5.11. In practice, the down-conversion can be carried out in more than one step. For high-frequency signals, for example, the first stage might shift a large band of frequencies to the audio region, where further down-conversion is carried out. Alternatively, the first stage might shift a band of frequencies to the intermediate frequency (IF) band, and the second stage to the audio band. More details on the physics of the down-conversion process are available in the article on "Modulation" in Chapter 21.

Just as it is possible to analyze a voltage signal and produce a magnitude and phase across any given frequency band, so it is possible to obtain a frequency profile of the magnitude and phase of a *current* signal. If current vectors and voltage vectors can be obtained for a given impedance, then it is possible to obtain a "vector impedance." This impedance is defined simply as the result of the complex division of voltage by current:

$$Z = \frac{V}{I} \tag{5.16}$$

The calculation of vector impedances are useful for such applications as impedance matching, power factor correction, and equalization.

Typically, much of the current processing for vector voltmeters and vector impedance meters is done digitally. One of the great advantages of this type of processing is the high accuracy achievable. Accuracies of 0.02° are common, but this figure is improving with developing technology. The high sampling rates that can be employed (typically beyond 1 GHz) cause the errors in the A/D conversion to be spread out over very large bandwidths. Since the ultimate measurement of a vector voltage or impedance is usually made over a very narrow bandwidth, the errors are substantially eliminated. The developments in technology that enable greater accuracy are (1) sampling rate increases, (2) word-length increases, and (3) increased A/D converter fidelity.

Phase Standard Instruments

For high-precision phase measurements and calibration, "phase standard" instruments can be used. These instruments provide two sinusoidal signal outputs, whose phase difference can be controlled with great accuracy. They typically use crystal-controlled timing to digitally synthesize two independent sinusoids with a variable user-defined phase difference. The Clarke-Hess 5500 Digital Phase Standard is one such instrument. For this standard, the sinusoids can have frequencies ranging from 1 Hz to 100 kHz, while amplitudes can vary between 50 mV and 120 V rms. The phase can be set with a resolution of 0.001°, with a typical accuracy of about 0.003°.

The Fast Fourier Transform Method

This method is one in which virtually all the processing is done in the digital domain. It operates on the pulse code modulated (PCM) digital samples of a signal. This and other similar methods are very promising means for measuring phase. This is because of the digital revolution that has resulted in cheap, fast, accurate, and highly versatile digital signal processors (DSPs). The latter are small computer chips capable of performing fast additions and multiplications, and which can be programmed to emulate conventional electronic functions such as filtering, coding, modulation, etc. They can also be programmed to perform a wide variety of functions not possible with analog circuitry. Up until the end of the 1980s, digital measurement was limited by the relatively inaccurate analog-to-digital (A/D) conversion process required before digital processing could be performed. Developments in the early 1990s, however, saw the introduction of oversampling analog-to-digital converters (ADCs), which can achieve accuracies of about one part in 100,000 [3]. ADC speeds as well as DSP chips are now running reliably at very high speeds.

In the fast Fourier transform (FFT) method, the digital signal samples are Fourier transformed with an FFT [2]. If the signal is sinusoidal, the initial phase is estimated as that value of the phase where the Fourier transform is maximized [4]. The frequency of the signal is estimated as that value of frequency where the Fourier transform is maximized. Once measurements of the frequency f and initial phase ϕ_0 have been obtained, the phase at any point in time can be calculated according to:

$$\phi = 2\pi f t + \phi_0 \tag{5.17}$$

One important practical issue in the measurement of the frequency and initial phase with an FFT arises because the FFT yields only *samples* of the Fourier transform; that is, it does not yield a continuous Fourier transform. It is quite possible that the true maximum of the Fourier transform will fall *between* samples of the FFT. For accurate measurement of frequency and initial phase, then, it is necessary to *interpolate* between the FFT samples. An efficient algorithm to do this is described in [5].

The FFT method is particularly appealing where there is significant noise present, as it is effective down to quite low signal-to-noise ratios (SNRs). Furthermore, it provides an optimal estimate of the frequency and initial phase, providing the background noise is white and Gaussian, and that no harmonic components are present [4]. If harmonics are present, the estimate of the phase is commonly taken as the phase at the FFT peak corresponding to the fundamental; this is not an optimal estimate, but serves well in many appplications. An optimal estimate in the case when harmonics are present can be obtained,

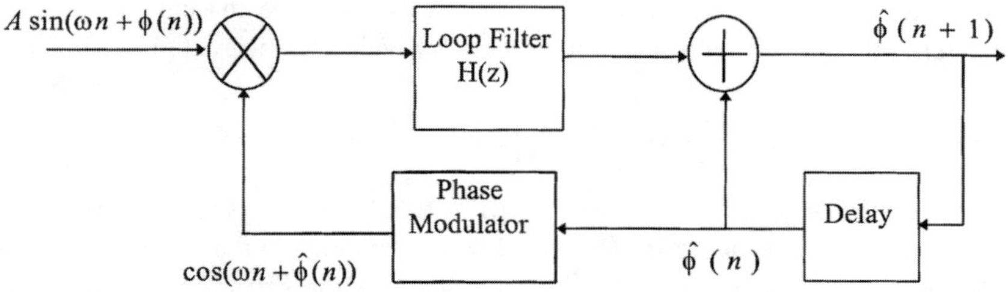

FIGURE 5.12 Block diagram of a digital phase-locked loop to implement phase demodulation.

if necessary, with the algorithms in [6], [7], and [8]. DSP chips such as the Texas Instruments TMS320C3x family, the Analog Devices ADSP21020 chip, or the Motorola MC5630x series can be used to implement the real-time FFTs.

If long word-lengths are used (say 32 bits) to perform the arithmetic for the FFTs, then determination of the phase from the samples of the FFT is virtually error-free. The only significant inaccuracy incurred in determining the phase is due to the ADC errors. Moreover, the error due to the digitization will typically be spread out over a large bandwidth, only a small amount of which will be "seen" in the phase measurement. With a high-quality ADC, accuracies of less than 0.001° are possible.

Phase-Locked Loops

If the frequency of a signal changes significantly over the period of the measurement, the FFT method described above will provide inaccurate results. If the signal's frequency does change substantially during measurement, one means to estimate the phase of the signal is to use a phase-locked loop (PLL). In this case, the signal, $s(t) = A \sin(\omega t + \phi(t))$, can be thought of as a constant frequency component, $A \sin(\omega t)$, which is phase modulated by a time-varying phase component, $\phi(t)$. The problem then reduces largely to one of demodulating a phase-modulated (PM) signal. A PLL can be used to form an estimate of the "phase-modulating" component, $\hat{\ }\phi(t)$, and the overall phase of the signal, ϕ_{oa}, can be estimated according to:

$$\phi_{oa}(t) = \omega t + \hat{\phi}(t) \tag{5.18}$$

Either analog or digital PLLs can be used, although higher accuracy is attainable with digital PLLs. Analog PLLs for demodulating a frequency-modulated (FM) signal are discussed in Chapter 21 and in [12]. The digital PLL (DPLL) was developed as an extension of the conventional analog PLL and is therefore similar in structure to its analog counterpart. The DPLL is discussed in [9] and [10]. The equation to demodulate the digital modulated signal with a first order DPLL is a simple recursive equation [9].

A block diagram of the DPLL for demodulating a PM signal is shown in Figure 5.12. In this diagram, n represents the discrete-time equivalent of continuous time t. It can be seen that there are strong similarities between Figure 5.12 and the analog PLL-based FM demodulator in Chapter 21. Both have phase comparators (implemented by the multiplier in Figure 5.12), both have loop filters, and both have modulators (either PM or FM) in the feedback loop.

The DPLL can easily be extended to measure the phase of a signal consisting not just of a fundamental component, but also of harmonically related components. Details are provided in [11]. The DPLL is near optimal for phase estimation in white Gaussian background noise down to a signal power-to-noise power ratio of about 8 dB [10].

The DPLL will function effectively whether the phase is constant or time-varying. Unlike the FFT, the DPLL is a recursive algorithm, with the feedback involved in the recursion creating a vulnerability to quantization errors. However, with proper precautions and long word-lengths, the DPLL will introduce minimal processing error. The main error would then arise from the inaccuracy of the ADC. With appropriate conditioning, one could expect the DPLL to provide accuracies approaching 0.001°.

5.4 Phase-Sensitive Demodulation

It is frequently necessary to track the phase of a carrier that "jitters" in some uncertain manner. This tracking of the carrier phase is necessary for synchronous demodulation schemes, where the phase of the demodulating signal must be made equal to the phase of the carrier. This need is explained in Chapter 21, and is briefly re-explained here. Consider, for example, double sideband (DSB) amplitude modulation. In DSB, the modulated signal is given by $f_s(t) = A[k + \mu m(t)]\cos(\omega_c t)$, where $m(t)$ is the message signal, A is the amplitude of the unmodulated carrier, μ is the modulation index, k is the proportion of modulating signal present in the modulated signal, and $\cos(\omega_c t)$ is the carrier. Demodulation is typically carried out by multiplying $f_s(t)$ by the carrier, and then low-pass filtering so that the demodulated signal is given by:

$$f_d(t) = \frac{A_c\left[k + \mu m(t)\right]}{2} \tag{5.19}$$

However, if because of carrier uncertainty, one multiplies the modulated signal by $\cos(\omega_c t + \phi)$, then the demodulated signal is given by:

$$f_d(t) = \frac{A_c\left[k + \mu m(t)\right]\cos(\phi)}{2} \tag{5.20}$$

It can be seen from Equation 5.20 that the error in the carrier phase can affect both the amplitude and the sign of the demodulated signal. The phase errors can thus yield substantial errors in system output. The following sections outline important techniques used for tracking the phase of carriers, and thus reducing phase errors.

The Phase-Locked Loop for Carrier Phase Tracking

The phase-locked loop (PLL) is well known as a means for demodulating frequency-modulated signals. It is also frequently used for tracking the phase of a carrier in noise, so that a copy of the carrier with correct phase is available for demodulation. This tracking is simple enough if a (noisy) copy of the carrier is directly available; either a digital or analog PLL can be used. In either case, the input can be assumed to have the form, $A\sin(\omega t + \phi(t))$, where $\phi(t)$ is the carrier phase. The PLL consists of a multiplier (effectively a phase comparator), a phase modulator, and a loop filter, arranged as shown in Chapter 21. The design of the loop filter is critical if noise is to be optimally removed. In proper operation, the PLL output will track the phase of the incoming signal (i.e., of the carrier). If a copy of the carrier is not available but needs to be inferred from the modulated signal, the demodulation task is more difficult. Digital PLLs using a DSP chip can be particularly helpful in this case; the carrier can be adaptively estimated using intelligent algorithms, with convergence to the "correct" signal being registered when certain desired features of the demodulated signal are observed.

The PLL is quite versatile. It can function in relatively low noise environments (typically down to about 8 dB SNR). It can be implemented digitally. It can also cope with substantial carrier frequency variations by increasing the order of the loop filter [12] (this is often necessary, for example, in satellite communications because of the Doppler effect). At very low SNR, however, the PLL fails badly. Recent developments in digital signal processing have seen the development of an alternative based on hidden Markov models (HMMs), which will function down to about –5 dB SNR [13]. The HMM method is discussed in the next section.

Hidden Markov Model-Based Carrier Phase Tracker

In the HMM method, the problem of estimating the phase and frequency of a noisy waveform is couched as a "state estimation" problem. The phase of the signal at any point in time can go from 0 to 360°. The 0 to 360° value range is divided into a finite number of intervals or "states," so that the phase at any time occupies a particular (though unknown) state. Similarly, the angular frequency normalized by the sampling frequency at any time in a digital system must be between $-\pi$ to $+\pi$. This value range is also divided into a number of states, so that the frequency at any time has a (hidden or unknown) state associated with it. The frequency is assumed to be a first-order Markov process and probabilities are assigned to the possibility of the frequency changing from one state to another for successive values of time, i.e., frequency transition probabilities are assigned. Large frequency changes are assigned low probabilities, while small changes are assigned high probabilities. The problem of estimating the true phase and frequency states underlying the noisy signal then reduces to one of estimating which states the phase and frequency occupy as time evolves, given the observed noisy signal and the transition probabilities. Computationally efficient optimal algorithms have been developed to estimate these "optimal state sequences" for both the phase and frequency. Details are provided in [13].

5.5 Power Factor

Of particular interest in many applications is the phase angle between the current and voltage of a system. This angle is important because it is a measure of the power which is dissipated in the system. The following paragraphs discuss this angle, its cosine (the system power factor), and its measurement.

In a linear electric circuit that is fed by a current of peak amplitude, I_M, with an angular frequency of ω, the current will have the form, $I_M \cos(\omega t)$. The system voltage will be given by $V_M \cos(\omega t + \phi)$, where V_M is the peak voltage and ϕ is the phase difference between the current and voltage. Then the average power dissipated in the circuit will be given by:

$$P_{av} = \frac{1}{2} V_M I_M \cos(\phi) = V_{rms} I_{rms} \cos(\phi) \tag{5.21}$$

where V_{rms} and I_{rms} are the root mean square (rms) values of the voltage and current respectively. The term $\cos(\phi)$ is known as the *power factor*. It may alternatively be expressed as the ratio of real average power to the product of the rms values of voltage and current, respectively:

$$PF = \frac{P_{av}}{V_{rms} I_{rms}} \tag{5.22}$$

The above expression is, in fact, a general definition of power factor for any current and voltage waveforms. For the special case of sinusoidal voltage and currents, PF reduces to $\cos(\phi)$.

There are a number of ways to measure the power factor. One way is to use a wattmeter to measure the real average power and a voltmeter and an ammeter to measure the rms voltage and current, respectively. The power factor is then determined according to Equation 5.22. This is probably the most effective way when the currents and/or voltages are nonsinusoidal. This procedure can easily be implemented with "digital power meters." The power is measured by time-averaging the product of the instantaneous voltage and current, while the rms values are calculated by taking the square root of the time averaged value of the square of the parameter of interest — current or voltage. Some digital power meters also provide an analysis of the individual harmonics via FFT processing. These meters are accurate and versatile, and consequently very popular.

A more direct method is based on the crossed-coil meter, the operation of which was described earlier in this chapter. Note that this meter is a "single-phase meter," which is accurately designed for one frequency only. Errors will occur at other frequencies because of the dependance of the crossed-coil meter method on a constant and known phase angle between the currents in the crossed coils.

With balanced polyphase circuits, it is possible to use a single-phase meter applied to one of the phases. Alternatively, one can use specially designed polyphase meters. In a three-phase meter, for example, one phase is connected to the fixed coil, while the other two phases are connected to the crossed coils on the rotating shaft. The crossed coils are constructed with a 60° angle between them. With four-phase systems, consecutive lines are 90° out of phase. Two of these consecutive lines are connected to the two crossed coils and the angle between the coils is made equal to 90°.

With unbalanced polyphase circuits amid the presence of harmonics, each of the harmonic components has its own power factor, and so it is likely to be misleading to use a meter that measures a single angle. These methods based on the crossed-coil meter are thus much more limited than their digital counterparts.

5.6 Instrumentation and Components

Table 5.1 lists some integrated circuits and DSP chips that can be used in the various techniques for measuring phase. The list is really only illustrative of what is available and prices are approximate costs in U.S. dollars for small quantities at the end of 1996. Table 5.2 lists some companies that manufacture these products. An extensive (and indeed rapidly increasing) product range exists for DSP chip-based products, with details being available from the companies listed in Table 5.2. Table 5.3 lists instruments used for phase measurement. These instruments include CROs, vector voltage meters, vector impedance meters, crossed-coil meters and digital power meters, zero-crossing meters, and phase standards. Again, the table is only representative, as the full range of available instruments is enormous. Addresses of some of the relevant companies are provided in Table 5.4.

TABLE 5.1 Integrated Circuits Used in Phase Measurement

Function	Designation	Manufacturer	Approximate Price
Phase-locked loop	LM566	National, Motorola, Phillips	$2.70
Phase-locked loop	74HC4046	Harris, Motorola	$2
Phase/frequency detector	MC4044P	Motorola	$18.29
Pair of retriggerable monostables (one-shot)	74HC4538	Motorola, Harris	$2
DSP Chip	TMS320C32	Texas Instruments	$50
DSP Chip	TMS320C31	Texas Instruments	$80
DSP Chip	MC56303	Motorola	$60
DSP Chip	ADSP21020	Analog Devices	$110

TABLE 5.2 Companies Making Integrated Circuits and DSP Chips Which Can Be Used for Phase Measurement

Analog Devices, Inc. One Technology Way Box 9106, Norwood, MA 02062 Tel: (617) 329-4700.	National Semiconductor Corp. 2900 Semiconductor Dr. P.O. Box 58090 Santa Clara, CA 95052-8090
Harris Semiconductor Products Division P.O. Box 883 Melbourne, FL 37902 Tel: (407) 724-3730	Texas Instruments Incorporated P.O. Box 1443 Houston, Texas 77251-1443
Motorola, Semiconductor Products Sector 3102 N. 56th St. Phoenix, AZ 85018 Tel: (602) 952-3248	

TABLE 5.3 Instruments for Measuring Phase

Description	Model Number	Manufacturer	Approximate Price
CRO	HP54600B	Hewlett Packard	$2,495
CRO	HP54602B	Hewlett Packard	$2,995
CRO	HP54616	Hewlett Packard	$5,595
CRO	TDS220	Tektronix	$1,695
CRO	TDS510A	Tektronix	$9,956
Vector signal analyzer	HP89410A	Hewlett Packard	$29,050
Vector signal analyzer	HP89440A	Hewlett Packard	$52,500
Vector signal analyzer	HP89441A	Hewlett Packard	$58,150
Gain/phase impedance meter	HP4193A	Hewlett Packard	$13,700
Zero-crossing phase meter	KH6500	Krohn-Hite	$1,300
Digital power analyzer (with power factor and phase)	NANOVIP	Elcontrol	$660
Digital analyzing vector voltmeter	NA2250	North Atlantic Instruments	
Digital power analyzer (with power factor and phase, FFT analysis)	3195	Hioki	$25,000
Crossed-coil meter	246-425G	Crompton Industries	$290
Digital phase standard	5500	Clarke-Hess	$11,000

TABLE 5.4 Instruments for Measuring Phase

Description	Model Number	Manufacturer	Approximate Price
CRO	HP54600B	Hewlett Packard	$2,495
CRO	HP54602B	Hewlett Packard	$2,995
CRO	HP54616	Hewlett Packard	$5,595
CRO	TDS220	Tektronix	$1,695
CRO	TDS510A	Tektronix	$9,956
Vector signal analyzer	HP89410A	Hewlett Packard	$29,050
Vector signal analyzer	HP89440A	Hewlett Packard	$52,500
Vector signal analyzer	HP89441A	Hewlett Packard	$58,150
Gain/phase impedance meter	HP4193A	Hewlett Packard	$13,700
Zero-crossing phase meter	KH6500	Krohn-Hite	$1,300
Digital power analyzer (with power factor and phase)	NANOVIP	Elcontrol	$660
Digital analyzing vector voltmeter	NA2250	North Atlantic Instruments	
Digital power analyzer (with power factor and phase, FFT analysis)	3195	Hioki	$25,000
Crossed-coil meter	246-425G	Crompton Industries	$290
Digital phase standard	5500	Clarke-Hess	$11,000

TABLE 5.5 Companies Making Instruments for Measuring Phase

Hewlett-Packard Co. Test and Measurement Sector P.O. Box 58199 Santa Clara, CA 95052-9943 Tel: (800) 452-4844	Krohn-Hite Corporation Bodwell St., Avon Industrial Park Avon, MA	Hioki 81 Koizumi Veda, Nagano 386-11 Japan
	Crompton Instruments Freebournes Road, Witham Essex, CM83AH England	Clarke-Hess Comm. Research Corporation 220 W. 19 Street New York, NY
Tektronix Inc. Corporate Offices 26600 SW Parkway P.O. Box 1000 Wilsonville, OR 97070-1000 Tel: (503) 682-3411, (800) 426-2200	Elcontrol Via San Lorenzo 1/4 - 40037 Sasso Marconi Bologna, Italy	North Atlantic Instruments http://www.naii.com

References

1. D. Gabor, The theory of communication, *J. Inst. Elec. Eng.*, 93(III), 429-457, 1946.
2. A.V. Oppenheim and R.W. Schafer, *Discrete-Time Signal Processing,* Engelwood-Cliffs, NJ: Prentice-Hall, 1989.
3. K.C. Pohlmann (Ed.), *Advanced Digital Audio,* Carmel, IN: Howard Sams and Co., 1991.
4. D. Rife and R. Boorstyn, Single tone parameter estimation from discrete-time observations, *IEEE Trans. Inf. Theory,* 20, 591-598, 1974.
5. T. Abotzoglou, A fast maximum likelihood algorithm for estimating the frequency of a sinusoid based on Newton's algorithm, *IEEE Trans. Acoust., Speech Signal Process.,* 33, 77-89, 1985.
6. D. McMahon and R. Barrett, ML estimation of the fundamental frequency of a harmonic series, *Proc. of ISSPA 87,* Brisbane, Australia, 1987, 333-336.
7. A. Nehorai and B. Porat, Adaptive comb filtering for harmonic signal enhancement, *IEEE Trans. Acoust., Speech Signal Process.,* 34, 1124-1138, 1986.
8. L. White, An iterative method for exact maximum likelihood estimation of the parameters of a harmonic series, *IEEE Trans. Automat. Control,* 38, 367-370 1993.
9. C. Kelly and S. Gupta, Discrete-time demodulation of continuous time signals, *IEEE Trans. Inf. Theory,* 18, 488-493, 1972.
10. B.D.O. Anderson and J.B. Moore, *Optimal Filtering,* Englewood Cliffs, NJ: Prentice-Hall, 1979.
11. P. Parker and B. Anderson, Frequency tracking of periodic noisy signals, *Signal Processing,* 20(2), 127-152, 1990.
12. R.E. Best, *Phase-Locked Loops; Theory, Design and Applications,* 2nd ed., New York: McGraw-Hill, 1993.
13. L. White, Estimation of the instantaneous frequency of a noisy signal, in B. Boashash (ed.), *Time-Frequency Signal Analysis, Methods and Applications,* Melbourne, Australia: Longman-Cheshire; New York: Halsted Press, 1992.

Further Information

A.D. Helfrick and W.D. Cooper, *Modern Electronic Instrumentation and Measurement Techniques,* Englewood Cliffs, NJ: Prentice-Hall, 1990.

McGraw-Hill Encyclopedia of Science and Technology, 8th ed., New York: McGraw-Hill, 1997.

H. Taub and D.L. Schilling, *Principles of Communication Systems,* 2nd ed., New York: McGraw-Hill, 1986.

J.D. Lenk, *Handbook of Oscilloscopes: Theory and Application,* Englewood Cliffs, NJ: Prentice-Hall, 1982.

6

Energy Measurement

Arnaldo Brandolini
Politecnico di Milano

Alessandro Gandelli
Politecnico di Milano

Energy is one of the most important physical quantities in any branch of science and engineering and especially in electrical engineering. Energy exchange processes lead to the study of electrical networks from the physical point of view and allow an in-depth knowledge of power transfer within the electrical world and between electric and other forms of energy.

The definitions of energy and power represent the starting point for any successive study.

1. Energy is the amount of work that the system is capable of doing.
2. Power is the time rate of doing work.

Energy can be mathematically defined as the definite integral of the power over a given time interval Δt.

The power available in a two-terminal section of a generic electric circuit is given by the product of the voltage across the terminals and the current flowing through the section itself ($p = vi$). The electric energy (E) flowing through the same section is defined by the integral of the power over the observation interval

$$E(\Delta t) = \int_{t_0}^{\Delta t + t_0} p \, dt \tag{6.1}$$

For this reason the energy measurement is a dynamic measurement, which means it varies with time. The energy measurement unit is the Joule [J]; but for electric energy the Watthour [Wh] is most common.

Electricity is generated starting from different forms of energy (thermal, hydraulic, nuclear, chemical, etc.); after electric transfer and distribution processes, it is converted to other forms of energy.

The main feature of electric energy is the simplicity by which one can transfer it over long distances, control the distribution, and measure energy consumption.

6.1 Electromechanical Measuring Systems

Dc Energy Measurement

The simplest way to perform this measurement is to measure voltage and current and then compute the product:

0-8493-1733-9/04/$0.00+$1.50

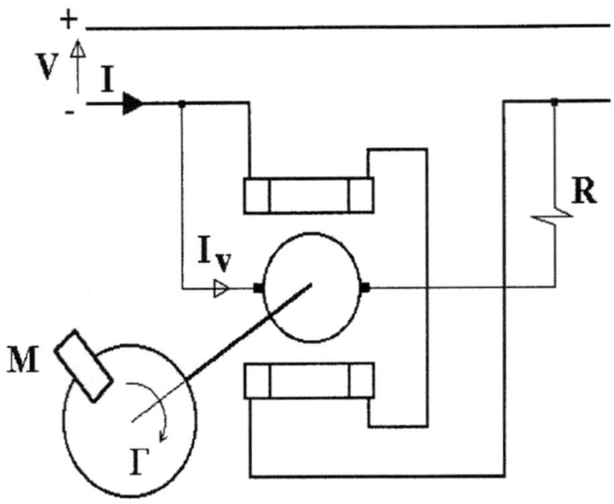

FIGURE 6.1 The Electrodynamic dc Energy Meter; M = permanent magnet.

$$E = VI\Delta t \tag{6.2}$$

where Δt is the observation interval measured by means of a chronometer or a time counter.

Note that dc power systems are limited to a restricted number of applications, as, for example: electric traction, electric drives, electrochemical power plants, and for high-voltage dc transmission systems in limited operating conditions.

Dc energy measurement has been performed in the past by means of different methodologies and instruments, such as electrodynamics measurement devices (Electrodynamics dc Energy Meter) operating as integrating wattmeters (Figure 6.1).

This measuring instrument is built using a small dc motor without iron, whose magnetic field is generated by the line current flowing through a coil arranged as the fixed part of the system. Because of the lack of iron in the magnetic circuit, the magnetic flux ϕ is strictly proportional to the current I. The rotor is connected in series with an additional resistor and is powered by the line voltage (V).

The rotor current (derived from the line voltage) is:

$$I_V = (V - E)/R \tag{6.3}$$

where $E = k_1\ \Gamma\ \phi$ is the emf induced by the angular speed Γ and R is the total resistance of the voltage circuit. It is possible to make the emf E negligible because of low angular speed Γ, limited amplitude of the flux ϕ, and a significant resistance R. In this way, Equation 6.3 becomes:

$$I_V \approx V/R \tag{6.4}$$

The torque C_m provided by the motor can be written:

$$C_m = k_2 \phi I_v \approx k_3 IV/R = k_4 P \tag{6.5}$$

C_m is, therefore, proportional to the power P flowing through the line. It is necessary, however, to remember that this torque could create a high angular speed to the rotor because of constantly incrementing speed. In order to maintain the dynamic equilibrium, a simple aluminum disk mounted on the rotor axis and placed in a constant magnetic field provided by a permanent magnet M is added to the dc motor system. In this way the induced currents in the disk introduce a damped torque proportional to the angular speed Γ, so, at equilibrium, there is a linear dependence of Γ on the power P. Thus,

$$E = \int_{\Delta t} P\,\mathrm{d}t = k_5 \int_{\Delta t} \Gamma\,\mathrm{d}t \tag{6.6}$$

A mechanical counter transfers the rotating motion into a digital representation of the total energy consumed during a specific time interval Δt in the power system.

Ac Induction Energy Meters

The most traditional and widely used ac energy meter is the *induction meter.* This device is built by means of three electrical circuits, magnetically coupled, two of them fixed and one rotating around the mechanical axis of the system. Figure 6.2 shows the two fixed circuits, (1) and (2), which are the voltage and the current coils. The third circuit is the rotating disk (3), generally made of aluminum, mounted on a rigid axis (4) transmitting the disk rotation to a mechanical counter (6), which provides the energy display.

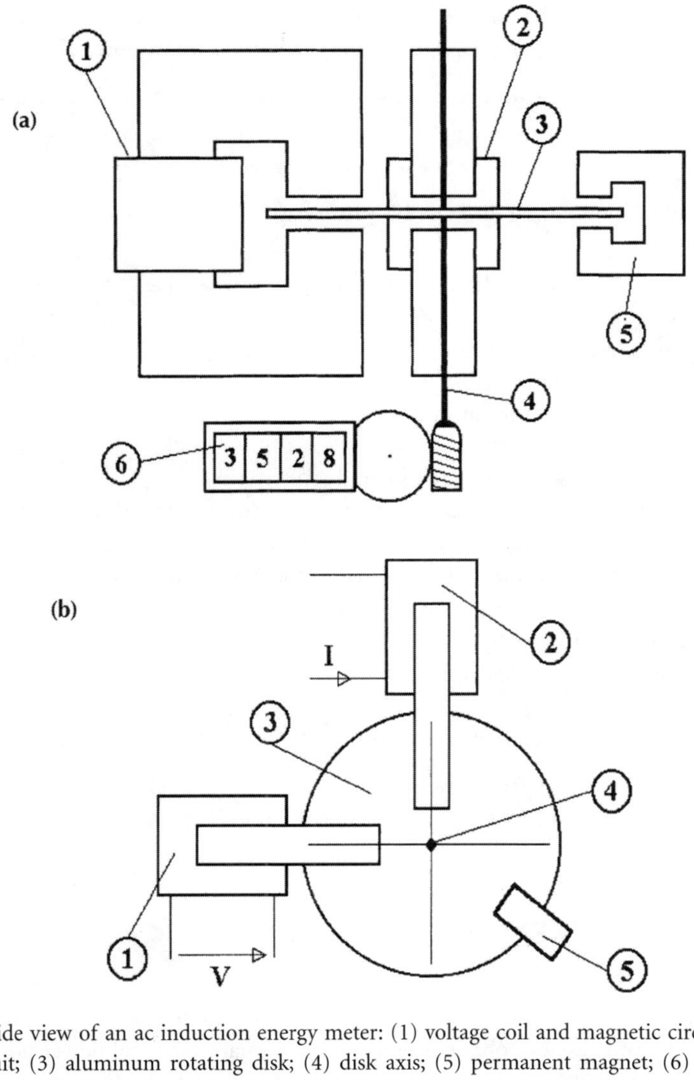

FIGURE 6.2 (a) Side view of an ac induction energy meter: (1) voltage coil and magnetic circuit; (2) current coil and magnetic circuit; (3) aluminum rotating disk; (4) disk axis; (5) permanent magnet; (6) mechanical display. (b) Top view of an ac induction energy meter: (1) voltage coil and magnetic circuit; (2) current coil and magnetic circuit; (3) aluminum rotating disk; (4) disk axis; (5) permanent magnet.

In Figure 6.2 the nuclei of fixed circuits (1) and (2) form a C shape and the disk is placed in their iron gaps. Another similar structure, arranged using a permanent magnet (5), is placed over the disk as well.

The magnetic fluxes generated by the voltage and current circuits are at the same frequency and are sinusoidal. They induce currents in the rotating disk (3) that, by means of a cross-interaction with the two generating fluxes, provide a mechanical torque acting on the disk. The torque is given by:

$$C_m = KVI\sin(\alpha) \tag{6.7}$$

where C_m = Mechanical torque
K = System constant
V = rms of the value of the applied voltage
I = rms of the value of the applied current
α = Phase angle between the fluxes generated by V and I

The acting torque causes the disk to rotate around its axis. This rotation reaches a dynamic equilibrium by balancing the torque C_m of the voltage and current coils and the reacting torque generated by the permanent magnet. The resulting angular speed, Γ, is therefore proportional to the flowing power if:

- The angular speed Γ of the disk is much smaller than the voltage and current frequency ω
- The phase difference between the voltage and current fluxes is equal to $\alpha = \pi - \varphi$, where φ is the phase difference between the voltage and current signals

The angular speed of the rotating disk can be written as:

$$\Gamma = (1/k)\omega\left(R_3/Z_3^2\right)(M_1 I)(M_2 V/Z_2)\cos(\phi) = K\,P$$

where Γ = Angular speed of the mobile circuit (conductor disk) [rad s^{-1}]
K = Instrument constant [rad s^{-1} W^{-1}]
P = Mean power in the circuit [W]
$1/k$ = Constant [Ω V^{-2} s^{-2}]
ω = Voltage and current frequency, in [rad s^{-1}]
R_3 = Equivalent resistance of the rotating disk, relative to the induced current fields [Ω]
Z_3 = Equivalent impedance of the rotating disk, relative to the induced current fields [Ω]
$(M_2\,V/Z_2)$ = rms value of the common flux related to the circuits n. 1 and 3 [Wb]
$(M_1\,I)$ = rms value of the common flux related to the circuits n. 2 and 3 [Wb]
Z_2 = Impedance of the voltage circuit (n. 1) [Ω]
V = rms value of the applied voltage [V]
I = rms value of the applied current [A]
φ = Phase difference between current and voltage signals

The integral of Γ over a defined period Δt is proportional (with enough accuracy) to the energy flowing in the power circuit. Thus, it is true that the instrument constant K is strictly related (but not proportional) to the signal frequency ω.

6.2 Electronic Energy Meters

The development of electronic multipliers led to their use in energy meters. There are many different prototypes in this class of energy meters. The first realizations were based on analog multipliers. Voltage and current signals are processed to obtain a signal proportional to the real power flowing into the line. The result is integrated over the observation time in order to calculate the *measured* energy. Even if they were not able to replace the traditional induction energy meters, they represented a good solution for all those applications where an increased accuracy was required (up to 0.2 %).

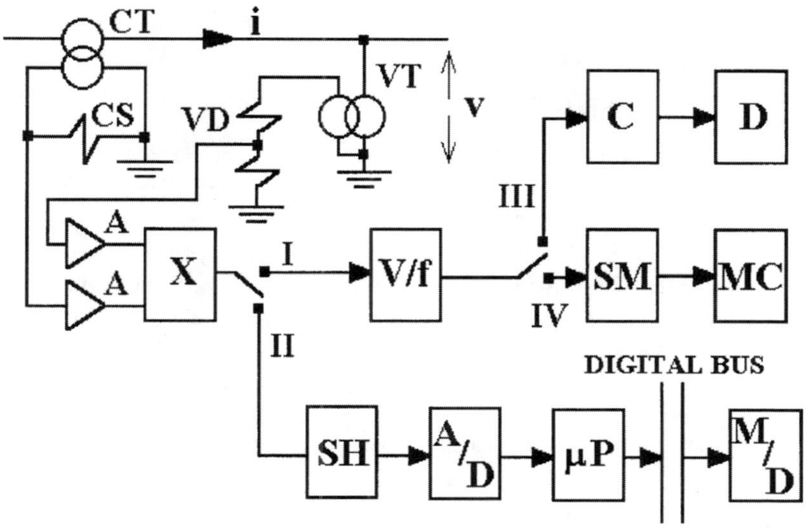

FIGURE 6.3 Electronic energy meter. Mechanical display option (I to IV). Electronic display option (I to III). Electronic display option and digital processing of the power signal (II). CT, current transformer; VT, voltage transformer; CS, current shunt; VD, voltage divider; A, analog signal processing block; X, multiplier; V/f, voltage-to-frequency converter; SM, step motor; MC, mechanical counter; C, electronic counter; D, display; SH, sample and hold; A/D, analog-to-digital converter; μP, microprocessor (CPU); M/D, memory and display.

Many of these instruments can be analyzed by means of the following functional descriptions. Figure 6.3 shows a block diagram of an electronic energy meter. The main feature of this type of instrument is the presence of voltage inputs on both voltage and current channels, because the electronic circuitry accepts only voltage signals. It has negligible current consumption from the system under measurement, due to high input impedance. The maximum amplitude level of the input signal must be limited around 5 V to 15 V. For this reason, the conditioning apparatus must guarantee the correct current-to-voltage transformation and the proper voltage reduction. Moreover, because these electronic components have a frequency range from dc to high frequencies, instruments based on them can be applied to dc, ac, or distorted power systems.

The Conditioning System for Electronic Energy Meters

The basis blocks of the conditioning system (Figure 6.3) for a dc energy meter are formed from a voltage divider for the voltage input and a current shunt for the current input. After these passive components, two preamplifiers are usually introduced before the processing system. The current preamplifier is very important because:

1. The voltage output level of the current shunt is very low, even at full scale (≤1 V).
2. Many times, the current input has to support overloaded signals; the presence of a variable gain amplifier allows acceptable working conditions for the system.
3. It can be used to implement an active filter before signal processing.

The most common devices to process ac signals for electronic energy meters are the traditional voltage and current transformers. They must be made with proper components to achieve the right amplitude of the voltage inputs (by nonreactive shunts for the current transformers, and nonreactive voltage dividers for the voltage transformers). After the transformers and related devices, a second block, based on electronic amplifiers, provides the final analog processing of the input signals as for the dc conditioning systems. It is useful to introduce this second processing element because analog filters are generally required when the input signals need to be digitally processed.

Electronic-Analog Energy Meters with Digital Output

These instruments provide the product of the two input signals (both voltages) through an analog multiplier that evaluates a voltage output proportional to the power of the input signals. This output can be followed by a filtering block.

The output signal is proportional to the instantaneous electric power flowing through the line. To calculate the energy, it is now necessary to complete the process by integrating over the observation time. This last procedure can be performed in two different ways.

1st Procedure. The power signal at the output of the analog multiplier is applied to the input of a voltage-frequency converter. Thus, the power information is converted from a voltage level to the frequency of pulse sequence, for which the counting process performs the integration of the power in the observation interval, i.e., the measurement of energy.

The final measurement can be performed by means of an electronic counter with digital display or using a dc step motor incrementing the rotor angular position every pulse by a fixed angular increment. The rotor position is shown by a mechanical counter (similar to the system mounted on the induction energy meters) indicating the total number of complete rotations performed by the system, proportional to the energy of the system under measurement. This second arrangement is normally adopted because it allows a permanent record of the energy information, which is not subject to possible lack of electric energy as in the first case.

2nd Procedure. This arrangement is based on an analog-to-digital converter (ADC) connected to the output of the analog multiplier. The sampling process is driven by an internal clock and performed by a sample and hold circuit. Thus, the ADC provides uniform sampling over the signal period and, under the condition imposed by the sampling theorem, the sum of the samples is proportional to the integral of the power signal, i.e., to the energy during the observation interval.

The calculation is performed by means of a dedicated CPU and then the results are sent to the digital memory to be stored and displayed. They can also be used to manage any other automatic process based on the energy measurement. For this purpose, data are available on a data bus (serial or parallel) connecting the measuring system with other devices.

Completely Digital Energy Meters

The most advanced solution for the energy measurement can be found in the all-digital meters (Figure 6.4), where both the voltage and current signals are sampled before any other processing. Thus, the data bus presents the sampled waveforms in digital form, giving the opportunity to perform a wide choice of digital signal processing on the power and energy information. Both the sampling devices are driven by a CPU, providing synchronized sampling signal. Filters able to meet the sampling theorem requirements, programmable gain amplifiers, and sample and hold circuits generally precede the ADCs.

Sometimes the system is equipped with a DSP capable of providing hardware resources to implement real-time evaluation of complex parameters (i.e., signal transforms) of the signal and energy measurement. Dedicated hardware and software performing instrument testing are also integrated into the meter to complete the device with the most advanced features.

Data management is arranged in two possible ways: sending the sampled data directly to the processing system for calculation, or accessing the memory using DMA procedures so the data string for a specific time period is first stored and then used for computation of energy and related parameter values. Final results of this computation are then available on the system bus to be sent to the other system resources or to be displayed.

This operational procedure introduces "artificial intelligence" into the meter capabilities. Thus, it is possible to extend the measurement ability to a cooperating environment in order to obtain articulated and controlled management of the energy fluxes in multi-managed networks.

This autonomous logic capability allows checking the accuracy of the recorded energy flow, in function of misuse of the measurement apparatus by external actors. For example, taking into account the current

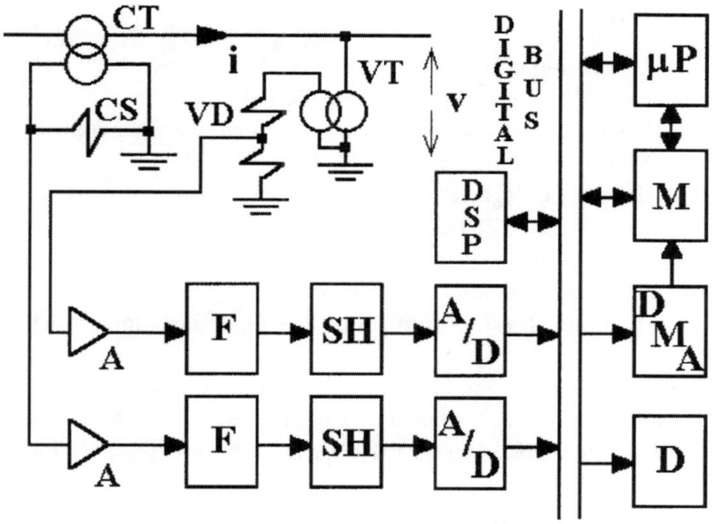

FIGURE 6.4 All-digital energy meter. CT, current transformer; VT, voltage transformer; CS, current shunt; VD, voltage divider; A, analog signal processing block; F, analog electronic filter; SH, sample and hold; A/D, analog-to-digital converter; μP, microprocessor (CPU); M, memory; DSP, digital signal processor; DMA, direct memory access circuit; D, display.

flowing in the conductors, it is possible to assign to the energy meter the capability of evaluating the correct working condition in the system. In a single-phase system this requirement leads to checking that the flowing current is the same in both conductors. On the contrary, the higher value is chosen in order to evaluate the energy, thus guaranteeing the electric company against misuses.

Accuracy of Energy Meters

Accuracy of energy meters is defined by means of relative parameters (in percent) obtained from a testing process by powering the instrument with a constant (nominal) voltage signal and a variable current signal (for example: 5, 10, 20, 50, 100, 120% of the nominal value). The testing procedures are performed by comparing the meter under test with a standard meter (Figure 6.5), or using equivalent methods.

The accuracy of commercial electromechanical (induction) energy meters is generally around 2%. Energy meters with accuracies of 1% have also been built. Electronic energy meters have a better accuracy, generally between 0.5% and 0.2%.

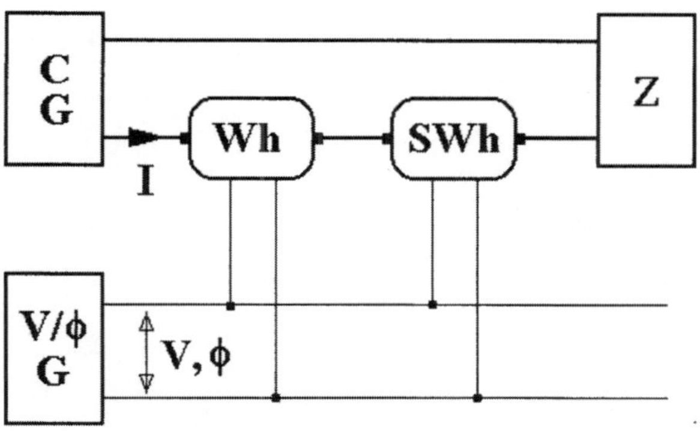

FIGURE 6.5 Testing circuit arrangement to compare an industrial meter (Wh) with a standard energy meter (SWh). CG, variable-amplitude current generator; V/ϕ G, variable-amplitude and phase voltage generator; *Z*, load impedance.

Further Information

S. Kusui and T.A. Nagai, A single-phase 3-wire watt-to-pulse frequency-converter using simple pwm and its accuracy analysis, *IEEE Trans. Instrum. Meas.*, 43, 770–774, 1994.

P. Bosnjakovic and B. Djokic, Reactive energy measurement using frequency controlled analog-to-pulse-rate converter, *Archiv fur Elektrotechnik*, 75, 131–135, 1992.

B. Djokic, P. Bosnjakovic, and M. Begovic, New method for reactive power and energy measurement, *IEEE Trans. Instrum. Meas.*, 41, 280–285, 1992.

J.C. Montano, A. Lopez, M. Castilla, and J. Gutierrez, DSP-based algorithm for electric-power measurement, *IEE Proc. A, Sci. Meas. Technol.*, 140, 485–490, 1993.

C.V. Drisdale and A.C. Jolley, *Electrical Measuring Instruments*, 2nd ed., John Wiley & Sons, New York, 1952.

F.K. Harris, *Electrical Measurements*, John Wiley & Sons, New York, 1962.

L. Schnell, Ed., *Technology of Electrical Measurements*, John Wiley & Sons, New York, 1993.

L. Finkelstein and T.V. Grattan, Eds., *Concise Encyclopedia of Measurement & Instrumentation*, Pergamon Press, New York, 1994.

K. Tam, A low-cost single-phase electricity meter using the MSP430C11x, Texas Instruments Application Report SLAA075, Oct. 1999.

Analog Devices, A power meter reference design based on the ADE7756, Analog Devices Technical Note, AN-564.

AVS Technology, Power meter IC AV7755, AVS Technology Data Sheet, Jan. 2001.

Cirrus Logic, Single phase bi-directional power/energy IC CS5460A, Cirrus Logic Data Sheet, Oct. 2001.

7

Electrical Conductivity and Resistivity

Michael B. Heaney
Palo Alto Research Center

Electrical resistivity is a key physical property of all materials. It is often necessary to accurately measure the resistivity of a given material. The electrical resistivity of different materials at room temperature can vary by over 20 orders of magnitude. No single technique or instrument can measure resistivities over this wide range. This chapter describes a number of different experimental techniques and instruments for measuring resistivities. The emphasis is on explaining how to make practical measurements and avoid common experimental errors. More theoretical and detailed discussions can be found in the sources listed at the end of this chapter.

7.1 Basic Concepts

The *electrical resistivity* of a material is a number describing how much that material resists the flow of electricity. Resistivity is measured in units of ohm·meters (Ω m). If electricity can flow easily through a material, that material has low resistivity. If electricity has great difficulty flowing through a material, that material has high resistivity. The electrical wires in overhead power lines and buildings are made of copper or aluminum. This is because copper and aluminum are materials with very low resistivities (about 20 nΩ m), allowing electric power to flow very easily. If these wires were made of high resistivity material like some types of plastic (which can have resistivities about 1 EΩ m (1×10^{18} Ω m)), very little electric power would flow.

Electrical resistivity is represented by the Greek letter ρ. Electrical conductivity is represented by the Greek letter σ, and is defined as the inverse of the resistivity. This means a high resistivity is the same as a low conductivity, and a low resistivity is the same as a high conductivity:

$$\sigma \equiv \frac{1}{\rho} \tag{7.1}$$

This chapter will discuss everything in terms of resistivity, with the understanding that conductivity can be obtained by taking the inverse of resistivity. The electrical resistivity of a material is an intrinsic

0-8493-1733-9/04/$0.00+$1.50

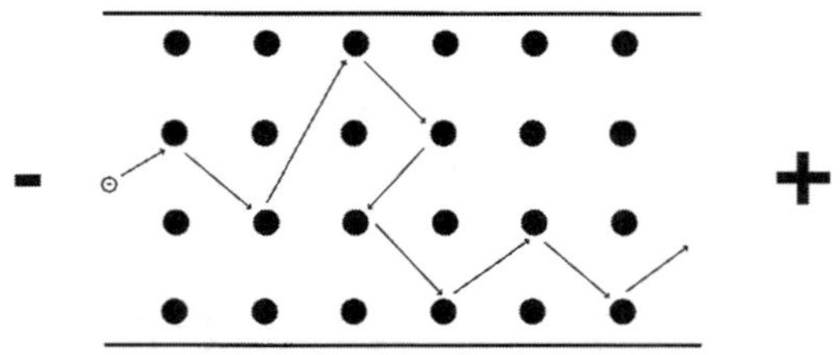

FIGURE 7.1 Simple model of electricity flowing through a material under an applied voltage. The white circle is an electron moving from left to right through the material. The black circles represent the stationary atoms of the material. Collisions between the electron and the atoms slow down the electron, causing electrical resistivity.

physical property, independent of the particular size or shape of the sample. This means a thin copper wire in a computer has the same resistivity as the Statue of Liberty, which is also made of copper.

7.2 Simple Model and Theory

Figure 7.1 shows a simple microscopic model of electricity flowing through a material [1]. While this model is an oversimplification and incorrect in several ways, it is still a very useful conceptual model for understanding resistivity and making rough estimates of some physical properties. A more correct understanding of the electrical resistivity of materials requires a thorough understanding of quantum mechanics [2].

On a microscopic level, electricity is simply the movement of electrons through a material. The smaller white circle in Figure 7.1 represents one electron flowing through the material. For ease of explanation, only one electron is shown. There are usually many electrons flowing through the material simultaneously. The electron tends to move from the left side of the material to the right side because an external force (represented by the large minus and plus signs) acts on it. This external force could be due to the voltage produced by an electrical power plant, or a battery connected to the material. As the electron moves through the material, it collides with the "stationary" atoms of the material, represented by the larger black circles. These collisions tend to slow down the electron. This is analogous to a pinball machine. The electron is like the metal ball rolling from the top to the bottom of a pinball machine, pulled by the force of gravity. The metal ball occasionally hits the pins and slows down. Just like in different pinball machines, the number of collisions the electron has can be very different in different materials. A material that produces lots of collisions is a high-resistivity material. A material that produces few collisions is a low-resistivity material.

The resistivity of a material can vary greatly at different temperatures. The resistivity of metals usually increases as temperature increases, while the resistivity of semiconductors usually decreases as temperature increases. The resistivity of a material can also depend on the applied magnetic field.

The discussion thus far has assumed that the material being measured is homogeneous and isotropic. Homogeneous means the material properties are the same everywhere in the sample. Isotropic means the material properties are the same in all directions. This is not always a valid assumption. A more exact definition of resistivity is the proportionality coefficient ρ relating a local applied electric field to the resultant current density:

$$\boldsymbol{E} \equiv \rho \boldsymbol{J} \tag{7.2}$$

where $\boldsymbol{E}$ is the electric field (V/m), $\boldsymbol{J}$ is the current density (A m^{-2}), and ρ is a proportionality coefficient (Ω m). Equation 7.2 is one form of Ohm's law. Note that $\boldsymbol{E}$ and $\boldsymbol{J}$ are vectors, and ρ is, in general, a tensor. This implies that the current does not necessarily flow in the direction of the applied electric field. In this chapter, isotropic and homogeneous materials are assumed, so ρ is a scalar (a single number).

Now consider the bar-shaped sample shown in Figure 7.2. The electric field $\boldsymbol{E}$ is given by the voltage V divided by the distance l over which the voltage is applied:

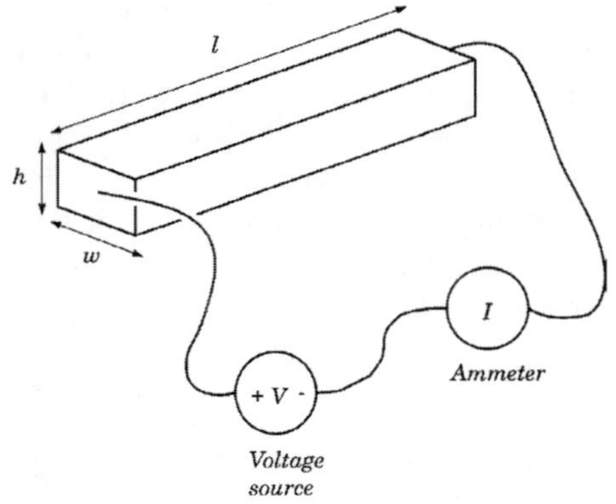

FIGURE 7.2 A two-point technique for measuring the resistivity of a bar of material. The voltage source applies a voltage across the bar, and the ammeter measures the current flowing through the bar.

$$\boldsymbol{E} \equiv \frac{V}{l} \tag{7.3}$$

The current density $\boldsymbol{J}$ is given by the current $\boldsymbol{I}$, divided by the cross-sectional area A through which the current flows:

$$\boldsymbol{J} \equiv \frac{\boldsymbol{I}}{A} \tag{7.4}$$

where the area A in Figure 7.2 is equal to the width w times the height h. Combining Equations 7.2, 7.3, and 7.4 and rearranging gives:

$$V = \frac{I\rho l}{A} \tag{7.5}$$

Now define a new quantity called "resistance" R with the definition:

$$R \equiv \frac{\rho l}{A} \tag{7.6}$$

Combining Equations 7.5 and 7.6 then gives:

$$I = \frac{V}{R} \tag{7.7}$$

where I is the current in amps (A) flowing through the sample, V is the voltage in volts (V) applied across the sample, and R is the resistance in ohms (Ω) of the sample. Equation 7.7 is another form of Ohm's law.

Note that the resistance R can depend on the size and shape of the sample, while ρ is independent of the size or shape of the sample. For example, if the length l of the sample bar is doubled, the resistance will double but the resistivity will remain constant.

The quantitative relationship between the resistivity ρ and the simple microscopic model shown in Figure 7.1 is given by:

$$\rho = \frac{m}{ne^2\tau} \tag{7.8}$$

where m is the mass of an electron, n is the number of electrons per unit volume carrying current in the material, e is the electric charge on an electron, and τ is the average time between collisions of an electron with the stationary atoms of the material. If there were more electrons per unit volume, they could carry more current through the material. This would lower the resistivity. If the electric charge on the electrons were greater, then the applied voltage would pull harder on the electrons, speeding them up. This would lower the resistivity. If the average time between collisions with the stationary atoms were longer, then the electrons could get through the material quicker. This would lower the resistivity. If electrons could be made more massive, they would move slower and take longer to get through the material. This would increase the resistivity.

7.3 Experimental Techniques for Measuring Resistivity

Two-Point Technique

The resistivity of a material can be obtained by measuring the resistance and physical dimensions of a bar of material, as shown in Figure 7.2. In this case, the material is cut into the shape of a rectangular bar of length l, height h, and width w. Copper wires are attached to both ends of the bar. This is called the two-point technique, since wires are attached to the material at two points. A voltage source applies a voltage V across the bar, causing a current I to flow through the bar. (Alternatively, a current source could force current through the sample bar, while a voltmeter in parallel with the current source measures the voltage induced across the sample bar.) The amount of current I that flows through the bar is measured by the ammeter, which is connected in series with the bar and voltage source. The voltage drop across the ammeter should be negligible. The resistance R of the bar is given by Equation 7.8a:

$$R = \frac{V}{I} \tag{7.8a}$$

where R = Resistance in Ω
V = Voltage in volts
I = Current in amps

The physical dimensions can be measured with a ruler, a micrometer, or other appropriate instrument. The two-point resistivity of the material is then:

$$\rho \equiv \frac{Rwh}{l} \tag{7.9}$$

where ρ is the resistivity in Ω m, R is the measured resistance in Ω, and w, h, and l are the measured physical dimensions of the sample bar in meters.

In practice, measuring resistivity with a two-point technique is often not reliable. There is usually some resistance between the contact wires and the material, or in the measuring equipment itself. These additional resistances make the resistivity of the material measure higher than it really is. A second potential problem is modulation of the sample resistivity due to the applied current. This is often a possibility for semiconducting materials. A third problem is that contacts between metal electrodes and a semiconducting sample tend to have other electrical properties that give wrong estimates for the actual sample resistivity. The four-point measurement technique overcomes many of these problems.

Four-Point Technique

Figure 7.3 shows the four-point measurement technique on a bar of material. Four wires are attached to the sample bar as shown. A current source forces a constant current through the ends of the sample bar.

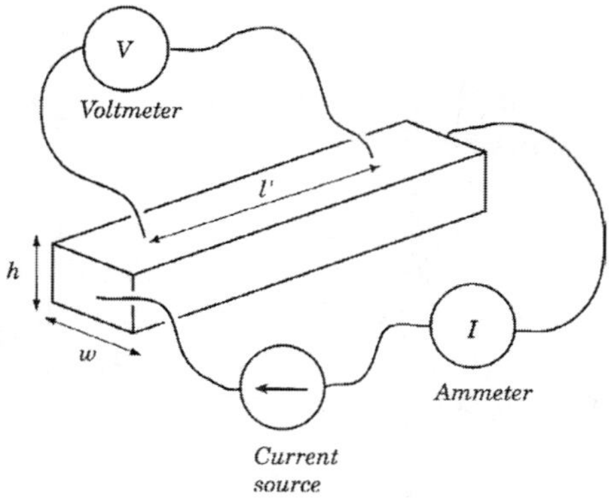

FIGURE 7.3 A four-point technique for measuring the resistivity of a bar of material. The current source forces a current through the bar, which is measured by a separate ammeter. The voltmeter measures the voltage across the middle of the bar.

A separate ammeter measures the amount of current I passing through the bar. A voltmeter simultaneously measures the voltage V produced across the inner part of the bar. (Alternatively, a voltage source could apply a voltage across the outer contacts, while an ammeter in series with this voltmeter measures the current flowing through the sample bar.)

The four-point resistivity of the material is then:

$$\rho = \frac{Vwh}{Il^1} \tag{7.10}$$

where ρ = Resistivity in Ωm
V = Voltage measured by the voltmeter in volts
w = Width of the sample bar measured in meters
h = Height of the sample bar measured in meters
I = Current the ammeter measures flowing through the sample in amperes
l^1 = Distance between the two points where the voltmeter wires make contact to the bar, measured in meters

Note that the total length l of the bar is not used to calculate the four-point resistivity: the length l^1 between the two inner contacts is used.

Common Experimental Errors

There are many experimental pitfalls to avoid when making resistivity measurements. The most common sources of error arise from doing a two-point measurement on a material that has any of the contact problems discussed earlier. For this reason, it is advisable to do four-point measurements whenever possible. This section describes experimental techniques to avoid errors in measuring resistivity:

1. The most difficult part of making resistivity measurements is often making good electric contacts to the sample. The general technique for making good electric contacts is to clean the areas of the sample where contacts are to be made with alcohol or an appropriate solvent, and then apply the contacts. If this does not work, try scraping the surface with a razor blade where contact is to be made, or cutting the sample to expose a fresh surface. Contacts can be made in many ways, such as using alligator clips, silver-paint, squeezing a wire against the material, soldering wires to the material, pressing small pieces of indium or a similar soft metal onto the contact areas, etc. Contacts can age: a good contact can become a bad contact over time. It might be necessary to make fresh contacts to a sample that has aged. There are many complications involved in the electrical

properties of contacts. Refer to the sources listed at the end of this chapter for more extensive discussions.

2. The measurement system should be calibrated before measuring any material samples. Calibration procedures are usually described in the equipment manuals.
3. The input resistance (or "impedance") of the voltmeter should be at least 10^5 higher than the resistance of the sample bar. The input impedance is usually listed in the equipment specifications. Note that some voltmeters and electrometers have a sufficiently high impedance between either of the inputs and ground, but not between the two inputs. In this case, it is necessary to use two voltmeters/electrometers (each with one input connected to ground and the other input connected to the sample bar). Measure the difference between them to obtain the voltage across the sample.
4. The measurement system should be tested before measuring any material samples. First test "short" with a thick copper wire or sheet in place of the sample. Then test "open" with nothing in place of the sample. Finally, test with a known, calibrated resistor whose resistance is within an order of magnitude of the sample resistance.
5. The geometry of the sample and electric contacts can be important. Contacts are often made by painting silver-paint or applying metal electrodes to the sample. If these contact areas are large or close to each other, this could reduce the accuracy of the resistivity measurement. It is best to make the two voltage contacts in a four-point measurement as small or thin as possible, and make the distance between inner electrodes much larger than the sample thickness. This also allows a more accurate estimate of the effective volume of the sample being probed.
6. It is critical that the four contacts to the sample bar in a four-point measurement are completely independent; there should be nothing other than the material of the bar connecting each of the four wires at the bar. For example, when pieces of indium are used to attach wires to a small sample, it is easy to smudge two adjacent indium pieces into one another. Those two contacts are no longer independent, and could easily cause errors. Visually inspect the contacts for this condition. If visual inspection is impractical, measure the resistance between the wires going to adjacent contacts. An unusually low resistance might indicate that two contacts are touching each other.
7. The applied voltage or current can cause heating of the material, which can change its resistivity. To avoid this problem, start with very small voltages or currents, and increase until the measured voltages and currents are at least 10 times larger than the random fluctuations of the meters. Then make sure the measured resistance is constant with time: the average resistance should not drift more than 10% in a few minutes.
8. Even if heating of the sample is not a problem, Ohm's law is not always obeyed. Many materials have a resistance that varies as the applied voltage varies, especially at higher voltages. Test for a linear relationship between current and voltage by measuring the resistance at several voltages on both sides of the measurement voltage. Whenever possible, make measurements in the linear (ohmic) region, where resistance is constant as voltage changes.
9. If one or both of the contacts to the voltmeter are bad, the voltmeter may effectively be disconnected from the material. In this situation, the voltmeter might display some random voltage unrelated to the voltage in the material. It might not be obvious that something is wrong, since this random voltage could accidentally appear to be a reasonable value. Check for this by setting the current source to zero amps and seeing if the voltmeter reading drops to zero volts. If it does not, try remaking the two inner contacts.
10. A critical check of a four-point measurement is to reverse the leads and remeasure the resistance. First, turn the current source to zero amps. Without disturbing any of the four contacts at the sample, swap the two sets of wires going to the voltmeter and the current source/ammeter. The two wires that originally plugged into the voltmeter should now plug into one terminal of the current source and one terminal of the ammeter. The two wires that originally plugged into the current source and ammeter should now plug into the voltmeter. Turn the current source on and remeasure the resistance. Note that current is now being forced to flow between the two inner

contact points on the sample, while the voltage is being measured between the two outer contacts on the sample. The two measured resistances should be within 10% of each other.

11. The resistivity of some materials can depend on how much light is hitting the material. This is especially a problem with semiconductors. If this is a possibility, try blocking all light from the sample during measurement.

Sheet Resistance Measurements

It is often necessary to measure the resistivities of thin films or sheets of various materials. If the material can be made into the form of a rectangle, then the resistivity can be measured just like the bar samples in Figure 7.2:

$$\rho \equiv \frac{Vwh}{Il} \tag{7.11}$$

where ρ = Sample resistivity in $\Omega\,\mathrm{m}$
V = Voltage measured by the voltmeter in volts
w = Width of the sample measured in meters
h = Thickness of the sample measured in meters
I = Current the ammeter measures flowing through the sample in amperes
l = Length of the film measured in meters

For the special case of a square film, the width w is equal to the length l, and Equation 7.11 becomes:

$$\rho(\mathit{of\ square\ film}) \equiv \frac{Vh}{I} \tag{7.12}$$

The resistivity of a square film of material is called the "sheet resistivity" of the material, and is usually represented by the symbol ρ_s. The "sheet resistance" R_s is defined by:

$$R_s \equiv R(\mathit{of\ square\ film}) = \frac{V}{I} \tag{7.13}$$

where V = Voltage measured by the voltmeter in volts
I = Current the ammeter measures flowing through the sample in amps

The units for sheet resistance are Ω, but people commonly use the units "Ω per square" or $\Omega/\square$. The sheet resistance is numerically equal to the measured resistance of a square piece of the material. Note that sheet resistance is independent of the size of the square measured, and it is not necessary to know the film thickness to measure sheet resistance. This makes sheet resistance a useful quantity for comparing different thin films of materials.

It is usually more convenient to measure thin-film samples of arbitrary shape and size. This is usually done by pressing four collinear, equally spaced contacts into a film whose length and width are both much greater than the spacing between contacts. In this situation, the sheet resistance is [3]:

$$R_s = 4.532\frac{V}{I} \tag{7.14}$$

where V = Voltage measured across the two inner contacts
I = Current applied through the two outer contacts

In many practical cases, the size of the thin-film sample will not be much greater than the spacing between the four-point contacts. In other cases, it might be necessary to measure a thin film near a corner or edge. In this situation, use geometric correction factors to accurately estimate the sheet resistance. These correction factors are available for the most commonly encountered sample geometries [3].

Instrumentation for Four-Point Resistivity Measurements

The resistivities of thin films of materials are often measured using commercial four-point probes. These probes generally have four equally spaced, collinear metal points that are pressed against the surface of the film. A current is applied between the outer two points, while the voltage is measured across the inner two points. These probes can also be used to measure the resistivity of bulk samples. Some companies that make probes and systems specifically for four-point resistivity measurements are listed in Table 7.1.

Instrumentation for High-Resistivity Measurements

Many materials such as rocks, plastics, and paper have very high resistivities, up to 1 EΩ m. The techniques described earlier for measuring resistivity are usually not reliable for these materials. In particular, it is often not possible to make a four-point measurement. One problem is that high voltages are needed to get any measurable current flowing through these materials. A second problem is that very long time constants prevent making steady-state measurements. A third problem is that the surfaces of these materials can often have significantly lower resistivity than the bulk, due to defects or other contamination. Measurements using the techniques described above then give falsely low values for the bulk resistivity. The best way to measure the resistivity of these materials is to use specialized commercial instruments. These are designed to separate out the bulk resistivity from the surface resistivity, and to minimize the many other problems encountered when measuring very high resistivities. Table 7.2 lists some companies that make high-resistivity measurement systems.

van der Pauw Technique

The four-point measurement technique described earlier has assumed the material sample has the shape of a rectangular thin film or a bar. There is a more general four-point resistivity measurement technique that allows measurements on samples of arbitrary shape, with no need to measure all the physical dimensions of the sample. This is the van der Pauw technique [4]. There are four conditions that must be satisfied to use this technique:

1. The sample must have a flat shape of uniform thickness.
2. The sample must not have any isolated holes.
3. The sample must be homogeneous and isotropic.
4. All four contacts must be located at the edges of the sample.

In addition to these four conditions, the area of contact of any individual contact should be at least an order of magnitude smaller than the area of the entire sample. For small samples, this might not be possible or practical. If sufficiently small contacts are not achievable, it is still possible to do accurate van der Pauw resistivity measurements, using geometric correction factors to account for the finite size of the contacts. See Ref. [5] for further details.

The inset illustration of Figure 7.4 illustrates one possible sample measurement geometry. A more common geometry is to attach four contacts to the four corners of a square-shaped sheet of the material.

The procedure for doing a van der Pauw measurement is as follows:

1. Define a resistance $R_{ij,kl} \ldots V_{kl}/I_{ij}$, where $V_{kl} \ldots V_k - V_l$ is the voltage between points k and l, and I_{ij} is the current flowing from contact i to contact j.
2. Measure the resistances $R_{21,34}$ and $R_{32,41}$. Define $R_>$ as the greater of these two resistances and $R_<$ as the lesser of these two resistances.
3. Calculate the ratio $R_>/R_<$ and find the corresponding value of the function $f(R_>/R_<)$ from Figure 7.4. Be careful to use the appropriate horizontal scale!

TABLE 7.1 Companies That Make Four-Point Resistivity Measurement Probes and Systems

Company and Comments
Creative Design Engineering, Inc. 20565 Elves Drive Cupertino, CA 95014 Tel: (408) 736-7273 Fax: (408) 738-3912
Creative Design Engineering makes manual and automatic four-point resistivity systems specially designed for both small and large semiconductor wafers.
Four Dimensions, Inc. 3138 Diablo Ave. Hayward, CA 94545 Tel: (510) 782-1843 Fax: (510-786-9321 http://www.4dimensions.com
Four Dimensions makes a variety of manual and automatic four-point probe systems for measurement of resistivity and resistivity mapping of flat samples such as semiconductor wafers.
Hewlett-Packard Company Test and Measurement Organization 5301 Stevens Creek Blvd. Santa Clara, CA 95052-8059 Tel: (800) 452-4844 Fax: (303) 754-4801 http://www.hp.com
Hewlett-Packard makes a variety of high-quality instruments useful for four-point measurements.
Jandel Engineering, Ltd. Grand Union House Leighton Road Linslade, Leighton Buzzard LU7 7LA U.K. Tel: (01525)-378554 Fax: (01525)-381945 http://www.getnet.com/~bridge/jandel.html
Jandel makes four-point probes useful for flat samples such as semiconductor wafers. They will build custom four-point probes for your particular needs. They also make a combined constant current source and digital voltmeter for resistivity measurements.
Keithley Instruments, Inc. 28775 Aurora Road Cleveland, OH 44139-1891 Tel: (440) 248-0400 Fax: (440) 248-6168 http://www.keithley.com
Keithley makes a wide variety of four-point measurement systems. They also have useful, free literature detailing techniques for making accurate resistivity measurements.
KLA-Tencor Corp. 1 Technology Drive Milpitas, CA 95035 Tel: (408) 875-3000 Fax: (408) 875-3030 http://www.kla-tencor.com
KLA-Tencor makes automated sheet resistance mapping systems designed for semiconductor wafers.

TABLE 7.1 (continued) Companies That Make Four-Point Resistivity Measurement Probes and Systems

Company and Comments
Lucas-Signatone Corp. 393-J Tomkins Ct. Gilroy, CA 95020 Tel: (408) 848-2851 Fax: (408) 848-5763 http://www.signatone.com
Signatone makes four-point resistivity measurement systems and a variety of four-point probe heads. They make a high-temperature, four-point probe head for temperatures up to 670 K.
Miller Design and Equipment, Inc. 2231-C Fortune Drive San Jose, CA 95131-1806 Tel: (408) 434-9544 Fax: (408) 943-1491
Miller Design makes semi-automatic resistivity probe systems, designed for semiconductor wafers.
Mitsubishi Chemicals Corp./Yuka Denshi Co., Ltd. Kyodo Bldg., 1-5 Nihonbashi Muromachi 4-chome Chuo-kyu, Tokyo 103 Japan Tel: 03-3270-5033 Fax: 03-3270-5036
Yuka Denshi makes a low-resistivity meter and a variety of four-point probe heads.
MMR Technologies, Inc. 1400 North Shoreline Blvd., # A5 Mountain View, CA 94043 Tel: (650) 962-9620 Fax: (650) 962-9647 http://www.mmr.com
MMR makes systems for four-point resistivity, Hall mobility, and Seebeck potential measurements over the temperature range 80 K to 400 K.
Napson Corporation Momose Bldg. 7F 2-3-6 Kameido Koto-kyu Tokyo 136 Japan QuadTech, Inc. 100 Nickerson Rd., Suite 3 Marlborough, MA 01752-9605 Tel: (800) 253-1230 Fax: (508) 485-0295 http://www.quadtechinc.com
QuadTech makes a four-point ohmmeter capable of measuring resistances from 1 μΩ to 2 MΩ.
Quantum Design 11578 Sorrento Valley Rd. San Diego, CA 92121-1311 Tel: (800) 289-6996 Fax: (619) 481-7410 http://www.quandsn.com
Quantum Design makes an automated system for measuring four-point resistivity, Hall mobility, and other properties over the temperature range 2 K to 400 K in magnetic fields up to 14 T.

TABLE 7.2 Companies That Make High-Resistivity Measurement Probes and Systems

Company and Comments
Hewlett-Packard Company Test and Measurement Organization 5301 Stevens Creek Blvd. Santa Clara, CA 95052-8059 Tel: (800) 452-4844 Fax: (303) 754-4801 http://www.hp.com
Hewlett-Packard makes high-resistance meters and specially designed resistivity test chambers.
Keithley Instruments, Inc. 28775 Aurora Road Cleveland, OH 44139-1891 Tel: (440) 248-0400 Fax: (440) 248-6168 http://www.keithley.com
Keithley makes special meters and resistivity test chambers for measuring high resistivities. They also have useful, free literature detailing techniques for making accurate resistivity measurements.
Mitsubishi Chemicals Corp./Yuka Denshi Co., Ltd. Kyodo Bldg., 1-5 Nihonbashi Muromachi 4-chome Chuo-kyu, Tokyo 103 Japan Tel: 03-3270-5033 Fax: 03-3270-5036
Yuka Denshi makes high-resistance meters and a variety of probes and resistivity test chambers.
Monroe Electronics, Inc. 100 Housel Avenue Lyndonville, NY 14098 Tel: (800) 821-6001 Fax: (716) 765-9330 http://www.monroe-electronics.com
Monroe Electronics makes portable and hand-held instruments for measuring surface resistivity, designed for testing antistatic materials.
QuadTech, Inc. 100 Nickerson Rd. Suite 3 Marlborough, MA 01752-9605 Tel: (800) 253-1230 Fax: (508) 485-0295 http://www.quadtechinc.com
QuadTech makes a high-resistance ohmmeter.

4. Calculate the resistivity ρ_a using:

$$\rho_a = \frac{\pi d \left(R_> + R_<\right) f\left(R_>/R_<\right)}{\ln 4} \tag{7.15}$$

where ρ_a = Resistivity in Ω m
d = Thickness of the sample in m
Resistances $R_>$ and $R_<$ are measured in Ω
ln4 = Approximately 1.3863

It is not necessary to measure the width or length of the sample.

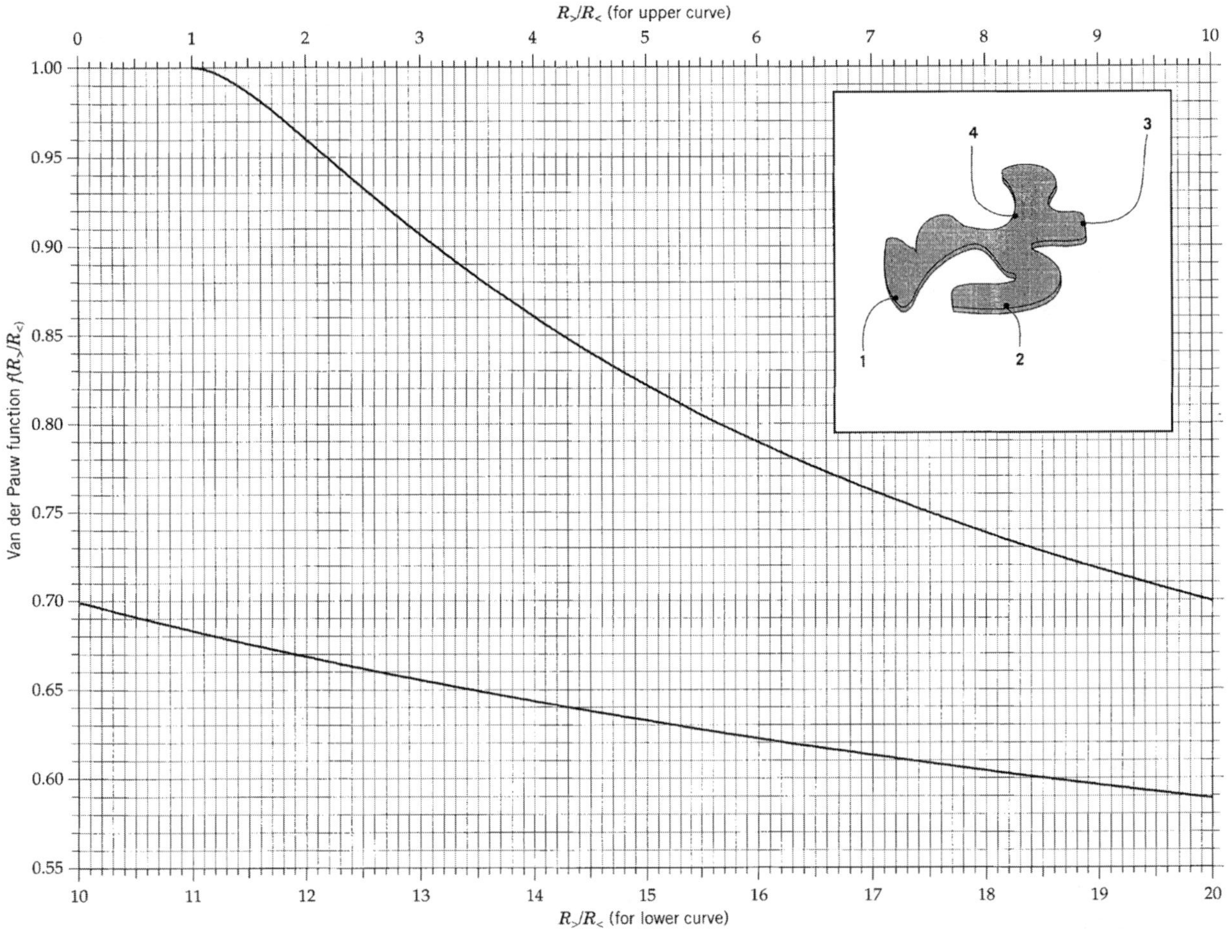

FIGURE 7.4 The van der Pauw technique. The inset shows one possible measurement geometry. The graph shows the function $f(R_>/R_<)$ needed to find the resistivity in Equation 7.15.

5. Switch the leads to measure $R_{43,12}$ and $R_{14,23}$. Repeat steps 3 and 4 to calculate ρ_b using these new values for $R_>$ and $R_<$. If the two resistivities ρ_a and ρ_b are not within 10% of each other, either the contacts are bad, or the sample is too nonuniform to measure reliably. Try making new contacts. If the two resistivities ρ_a and ρ_b are within 10% of each other, the best estimate of the material resistivity ρ is the average:

$$\rho = \frac{(\rho_a + \rho_b)}{2} \tag{7.16}$$

Note: The function $f(R_> / R_<)$ plotted in Figure 7.4 is defined by the transcendental equation:

$$f(R_>/R_<) \equiv \frac{-\ln 4(R_>/R_<)}{\left[1 + (R_>/R_<)\ln\left\{1 - 4^{-[(1+R_>/R_<)f]^{-1}}\right\}\right]} \tag{7.17}$$

Defining Terms

Conductance: The inverse of resistance.

Conductivity: The inverse of resistivity.

Contact resistance: The resistance between the surface of a material and the electric contact made to the surface.

Four-point technique: A method for measuring the resistivity of a material, using four electric contacts to the material, which avoids many contact resistance problems.

Resistance: The physical property of a particular piece of a material, quantifying the ease with which electricity can flow through it. Resistance will depend on the size and shape of the piece of material.

Resistivity: The intrinsic physical property of a material quantifying the ease with which electricity can flow through it. Resistivity will not depend on the size and shape of the piece of material. Higher resistivity means the flow of electricity is more difficult.

Sheet resistance: The resistance of a square thin film or sheet of material.

Two-point technique: A method for measuring the resistivity of a material, using two electric contacts to the material.

Van der Pauw technique: A method of measuring the four-point resistivity of an arbitrarily shaped material sample.

Acknowledgments

I thank Alison Breeze, John Clark, Kirsten R. Daehler, James M. E. Harper, Linda D. B. Kiss, Heidi Pan, and Shukri Souri for many useful suggestions.

References

1. P. Drude, Zur elektronentheorie der metalle, *Annalen der Physik,* 1, 566–613, 1900; 3, 369–402, 1900. See Ref. [2] for modern discussions of the Drude model and electrical conductivity.
2. N.W. Ashcroft and N.D. Mermin, *Solid State Physics*, Philadelphia, PA: Saunders College, 1976; C. Kittel, *Introduction to Solid State Physics,* 7th ed., New York: John Wiley & Sons, 1996.
3. L.B. Valdes, Resistivity measurements on germanium for transistors, *Proc. I.R.E.,* 42, 420–427, 1954.
4. L.J. van der Pauw, A method of measuring specific resistivity and Hall effect of discs of arbitrary shape, *Philips Res. Rep.,* 13, 1–9, 1958.
5. R. Chwang, B.J. Smith, and C.R. Crowell, Contact size effects on the van der Pauw method for resistivity and Hall coefficient measurement, *Solid-State Electron.,* 17, 1217–1227, 1974.

Further Information

H.H. Wieder, *Laboratory Notes on Electrical and Galvanomagnetic Measurements*, New York: Elsevier, 1979.

L.I. Maissel, Electrical properties of metallic thin films, 13-1 to 13-33, in L.I. Maissel and R. Glang (Eds.), *Handbook of Thin Film Technology*, San Francisco: McGraw-Hill, 1970.

D.C. Look, *Electrical Characterization of GaAs Materials and Devices*, New York: John Wiley & Sons, 1989.

D.C. Look, Bulk and contact electrical properties by the magneto-transmission-line method: application to GaAs, *Solid-State Electron.*, 30, 615–618, 1987.

8

Charge Measurement

Saps Buchman
Stanford University

John Mester
Stanford University

Timothy J. Sumner
Imperial College

Electric charge, a basic property of elementary particles, is defined by convention as negative for the electron and positive for the proton. In 1910, Robert Andrews Millikan (1868–1953) demonstrated the quantization and determined the value of the elementary charge by measuring the motion of small charged droplets in an adjustable electric field. The SI unit of charge, the *coulomb* (*C*), is defined in terms of base SI units as:

$$1 \text{ coulomb} = 1 \text{ ampere} \times 1 \text{ second} \tag{8.1a}$$

In terms of fundamental physical constants, the coulomb is measured in units of the elementary charge *e:*

$$1\text{ C} = 1.60217733 \times 10^{19}\, e \tag{8.1b}$$

where the relative uncertainty in the value of the elementary charge is 0.30 ppm [1].

Charge measurement is widely used in electronics, physics, radiology, and light and particle detection, as well as in technologies involving charged particles or droplets (as for example, toners used in copiers). Measuring charge is also the method of choice for determining the average value for small and/or noisy electric currents by utilizing time integration. The two standard classes of charge-measuring devices are the electrostatic voltmeters and the charge amplifiers.

Electrostatic instruments function by measuring the mechanical displacement caused by the deflecting torques produced by electric fields on charged conductors [2,3]. Electrostatic voltmeters also serve as charge-measurement devices, using the fact that charge is a function of voltage and instrument capacitance. This class of instruments can be optimized for a very wide range of measurements, from about 100 V to 100 kV full-scale, with custom devices capable of measuring voltages in excess of 200 kV. The accuracy of electrostatic voltmeters is about 1% of full scale, with typical time constants of about 3 s. Their insulation resistance is between 10^{10} Ω and 10^{15} Ω, with instrument capacitances in the range of 1 pF to 500 pF. Figure 8.1 gives a schematic representation of several types of electrostatic voltmeters.

Modern electronic instruments have replaced in great measure the *electrostatic voltmeters* as devices of choice for the measurement of charge. The charge amplifier is used for the measurement of charge or charge variation [4]. Figure 8.2 shows the basic configuration of the charge amplifier. The equality of charges on C_1 and C_f results in:

$$v_o = \frac{C_1}{C_f} v_i \quad \text{or} \quad \Delta v_o = \frac{C_1}{C_f} \Delta v_i \tag{8.2}$$

0-8493-1733-9/04/$0.00+$1.50

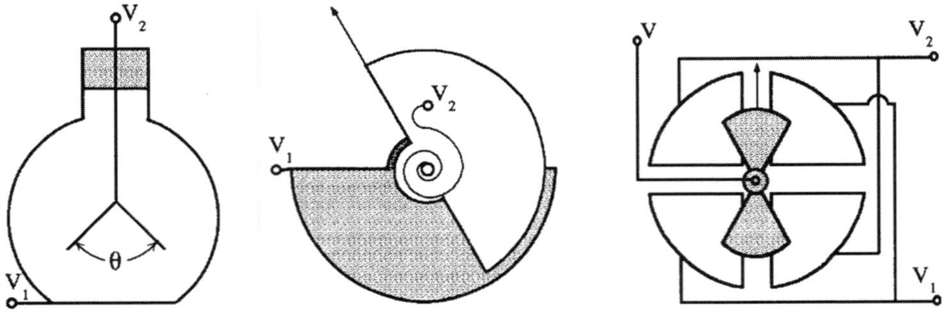

FIGURE 8.1 Examples of the repulsion, attraction, and symmetrical mechanical configurations of electrostatic voltmeters: (a) gold-leaf electroscope, (b) schematic representation of an attraction electrostatic voltmeter, (c) a symmetric quadrant electrostatic voltmeter [2].

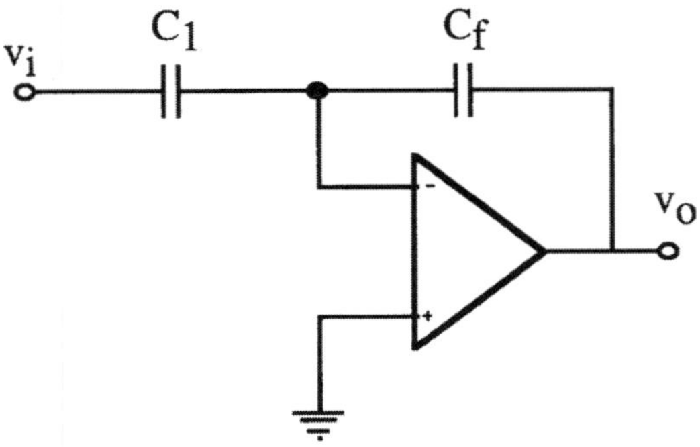

FIGURE 8.2 Basic concept of the charge amplifier. The output voltage is $v_o = C_1/C_f \times v_i$.

This same measurement principle is realized in the *electrometer.* The charge, Q, to be measured is transferred to the capacitor, C, and the value, V, of the voltage across the capacitor is measured: $Q = CV$. Figure 8.3 shows the block diagram for the typical digital electrometer [5]. Charge is measured in the coulomb mode, in which a capacitor C_f is connected across the operational amplifier, resulting in the input capacitance AC_f. Typical gain A for these amplifiers is in the range 10^4 to 10^6, making AC_f very large, and thus eliminating the problem of complete charge transfer to the input capacitor of the coulombmeter. Electrometers have input resistances in the range 10^{14} Ω to 10^{16} Ω, resulting in very long time constants, and thus minimizing the discharging of the capacitor. Typical leakage currents are 5×10^{-14} A to 5×10^{-16} A, again minimizing the variation in the charge. In the coulombmeter mode, electrometers can measure charges as low as 10^{-15} C and currents as low as 10^{-17} A.

Errors in charge-measuring instruments are caused by extraneous currents [5]. These currents are generated as thermal noise in the shunt resistance, by resistive leakage, and by triboelectric, piezoelectric, pyroelectric, electrochemical, and dielectric absorption effects. The coulombmeter function of the electrometers does not use internal resistors, thus eliminating this thermal noise source. Triboelectric charges due to friction between conductors and insulators can be minimized by using low-noise triaxial cables, and by reducing mechanical vibrations in the instrument. Under mechanical stress, certain insulators will generate electric charge due to piezoelectric effects. Judicious choices of materials and reduction of stress and mechanical motion can significantly reduce this effect.

Trace chemicals in the circuitry can give rise to electrochemical currents. It is therefore important to thoroughly clean and dry chemicals of all sensitive circuitry. Variations in voltages applied across insulators cause the separation and recombination of charges, and thus give rise to dielectric absorption

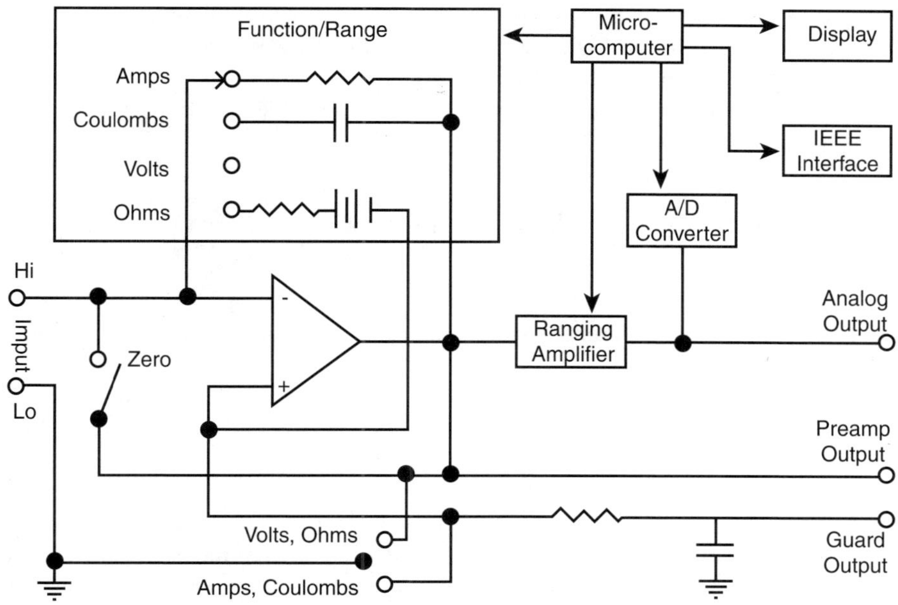

FIGURE 8.3 Conceptual block diagram of the digital electrometer. In the coulombs function, the charge to be determined is transferred to the corresponding capacitor, and the voltage across this capacitor is measured.

parasitic currents. The solution is to limit the voltages applied to insulators used for high-sensitivity charge measurements to less than about 5 V.

Dielectric materials used in sensitive charge-measurement experiments should be selected for their high resistivity (low resistive leakage), low water absorptivity, and minimal piezoelectric, pyroelectric, triboelectric, and dielectric absorption effects. Sapphire and polyethylene are two examples of suitable materials. *Guarding* is used to minimize both shunt currents and errors associated with the capacitance of cables and connectors. The block diagram in Figure 8.3 shows typical guarding arrangements for modern electrometers.

8.1 Electrostatic Voltmeters

Electrostatic voltmeters and the more sensitive mechanical electrometers use an indicator to readout the position of a variable capacitor. Depending on their mechanical configuration, the electrostatic voltmeters can be categorized into three types: repulsion, attraction, and symmetrical [2, 3]. The moving system in the high-sensitivity instruments is suspended from a torsion filament, or pivoted in precision bearings to increase ruggedness. A wide variety of arrangements is used for the capacitative elements, including parallel plates, concentric cylinders, hinged plates, and others. Motion damping of the moving parts is provided by air or liquid damping vanes or by eddy current damping.

One of the oldest devices used to measure charge is the *gold leaf electroscope,* shown in Figure 8.1a. Thin leaves of gold are suspended from a conductive contact that leads out of a protective case through an insulator. As charge applied to the contact is transferred to the leaves, the leaves separate by a certain angle, the mutual repulsion being balanced by gravity. In principle, this device can also be used to measure the voltage difference between the contact electrode and the outer case, assuming the capacitance as a function of leaf separation angle is known. The electroscope is an example of a repulsion-type device, as is the Braun voltmeter in which the electroscope leaves are replaced by a balanced needle [2]. The delicate nature and low accuracy of this class of instruments limit their use in precise quantitative measurement applications.

An example of an attraction electrostatic voltmeter used for portable applications is shown in Figure 8.1b. While the fixed sector disk is held at V_1, the signal V_2 is applied to the movable sector disk through a coil spring that supplies the balancing torque. Opposite charges on the capacitor cause the movable plate to rotate until the electric attraction torque is balanced by the spring. If a voltage $V = V_1 - V_2$ is applied, the electric torque is given by [3]:

$$\tau = \frac{dU}{d\theta} = \frac{d\left(\frac{1}{2}CV^2\right)}{d\theta} = \frac{1}{2}V^2\frac{dC}{d\theta} \tag{8.3}$$

The balancing spring torque is proportional to angular displacement, so the angle at equilibrium is given by:

$$\frac{1}{2}V^2\frac{dC}{d\theta} = K\theta \tag{8.4}$$

Since the rotation is proportional to V^2, such an instrument can be used to measure ac voltages as well.

Symmetrical instruments are used for high-sensitivity, low-voltage measurements. The voltage is applied to a mobile element positioned between a symmetrical arrangement of positive and negative electrodes. Common mode displacement errors are thus reduced, and the measurement accuracy increased. One of the first devices sensitive enough to be called an "electrometer" was the quadrant electrometer shown schematically in Figure 8.1c. As a voltage difference, $V_1 - V_2$, is applied across the quadrant pairs, the indicator is attracted toward one pair and repelled by the other. The indicator is suspended by a wire allowing the stiffness of the suspension to be controlled by the potential V, so that the displacement is given by [2]:

$$\theta = K\left[\left(V_1 - V_2\right)\left(V - \frac{1}{2}\left(V_1 - V_2\right)\right)\right] \tag{8.5}$$

where K is the unloaded spring constant of the suspension.

The advantage of electrostatic instruments is that the only currents they draw at dc are the leakage current and the current needed to charge up the capacitive elements. High-performance symmetrical electrostatic instruments have leakage resistances in excess of 10^{16} Ω, sensitivities of better than 10 μV, and capacitances of 10 pF to 100 pF. They are capable of measuring charges as small as 10^{-16} C, and are sensitive to charge variations of 10^{-19} C.

Historically, as stated above, the symmetrical electrostatic voltmeters have been called "electrometers." Note that this can give rise to some confusion, as the term *electrometer* is presently also used for the electronic electrometer. This is a high-performance dc multimeter with special input characteristics and high sensitivity, capable of measuring voltage, current, resistance, and charge.

Modern noncontacting electrostatic voltmeters have been designed for voltage measurements up to the 100 kV range. An advantage of these instruments is that no physical or electric contact is required between the instrument and test surface, ensuring that no charge transfer takes place. Kawamura, Sakamoto, and Noto [6] report the design of an attraction-type device that uses a strain gage to determine the displacement of a movable plate electrode. Hsu and Muller [7] have constructed a micromechanical shutter to modulate the capacitance between the detector electrode and the potential surface to be measured. Trek Inc. [8] electrostatic voltmeters achieve a modulated capacitance to the test surface by electromechanically vibrating the detector electrode. Horenstein [9], Gunter [10], and MacDonald and Fallone [11] have employed noncontacting electrostatic voltmeters to determine the charge distributions on semiconductor and insulator surfaces. Tables 8.1 and 8.2 contain a selection of available commercial devices and manufacturers.

TABLE 8.1 Instruments Used in Charge Measurement Applications

Instrument Manufacturer	Model #	Description	Approximate Price
Advantest	TR8652	Electrometer	$2500.00
	R8340/8340A	Electrometer	$5400.00
	R8240	Digital electrometer	$2300.00
	TR8601	Micro current meter	$3500.00
	TR8641	Pico ammeter	$2500.00
Amptek	A101	Charge preamplifier	$300.00
	A111	Charge preamplifier	$375.00
	A203	Charge preamplifier/shaper	$300.00
	A225	Charge preamplifier/shaper	$395.00
	A250	Charge preamplifier	$420.00
EIS	ESH1-33	Electrostatic voltmeter	$1650.00–$5890.00[a]
	ESD1-11	Electrostatic voltmeter	$1465.00–$1740.00[a]
	CRV	Electrostatic peak voltmeter	$2100.00
Jennings	J-1005	RF kilovoltmeter	$5266.00
Keithley	610C	Electrometer	$4990.00
	614	Digital electrometer	$2490.00
	617	Programmable electrometer	$4690.00
	642	Digital electrometer	$9990.00
	6512	Electrometer	$2995.00
	6517	Electrometer	$4690.00
Kistler	5011B	Charge amplifier	$2700.00
	5995	Charge amplifier	$1095.00
	5395A	Charge calibrator	$11655.00
Monroe	168-3	Electrostatic voltmeter	$4975.00
	174-1	Electrostatic voltmeter	$5395.00
	244AL	Electrostatic millivoltmeter	$3695.00
	253-1	Nanocoulomb meter/Faraday cup	$1765.00
Nuclear Associates	37-720FW	Digital electrometer for dosimetry	$1234.00
Trek	320B	Electrostatic voltmeter	$1930.00
	341	Electrostatic voltmeter	$6900.00
	344	Electrostatic voltmeter	$2070.00
	362A	Electrostatic voltmeter	$2615.00
	368	Electrostatic voltmeter	$2440.00–$9160.00[a]

[a] Available in a range of specifications.

8.2 Charge Amplifiers

The conversion of a charge, Q, into a measurement voltage involves at some stage the transfer of that charge onto a reference capacitor, C_r. The voltage, V_r, developed across the capacitor gives a measure of the charge as $Q = V_r/C_r$. There are two basic amplifier configurations for carrying out such measurements using the reference capacitor in either a *shunt* or *feedback* arrangement.

Shunt Amplifiers

Figure 8.4 shows a typical circuit in which the reference capacitor is used in a shunt mode. In this example, it is assumed that the charge that is to be measured is the result of the integrated current delivered by a time-dependent current source, $i(t)$. With the measurement circuit disconnected (switch in position s2), the charge on the source capacitor, C_s, at time τ will be $Q = \int_0^{c} i(t)\mathrm{d}t$ (assuming Q starts from zero at $t = 0$) and the output voltage, V_o, will be zero, as the input voltage to the (ideal) operational amplifier is zero. On closing the switch in position s1, the charge on C_s will then be shared between it and C_r and:

TABLE 8.2 Instrument Manufacturers

Advantest Corporation Shinjuku-NS Building 4-1 Nishi-Shinjuku 2-Chome, Shinjuku-ku Tokyo 163-08 Japan	Kistler Intrumente AG CH-8408 Winterthur, Switzerland
Amptek Inc. 6 De Angelo Drive Bedford, MA 01730	Monroe Electronics 100 Housel Ave. Lyndonville, NY 14098
Electrical Instrument Service Inc. Sensitive Research Instruments 25 Dock St. Mount Vernon, NY 10550	Nuclear Associates Div. of Victoreen, Inc. 100 Voice Rd. P.O. Box 349 Carle Place, NY 11514-0349
Jennings Technology Corporation 970 McLaughlin Ave. San Jose, CA 95122	Trek Incorporated 3922 Salt Works Rd. P.O. Box 728 Medina, NY 14103
Keithley Instruments Inc. 28775 Aurora Road Cleveland, OH 44139	

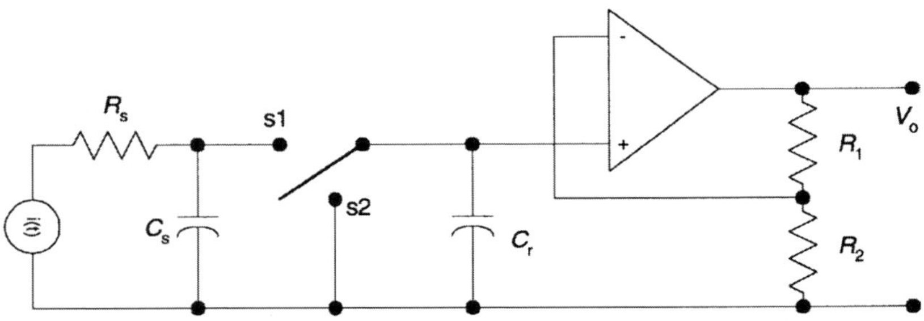

FIGURE 8.4 Schematic representation of a charge amplifier using a shunt reference capacitor. With the switch in position s1, the measurement circuit is connected and the charge is proportional to the output voltage and to the sum $C_s + C_r$. Note the significant sensitivity to C_s.

$$V_o = \left(\frac{R_1 + R_2}{R_2} \right) \frac{Q}{C_s + C_r} \tag{8.6}$$

In order to accurately relate the output voltage to the charge Q, not only does the gain of the non-inverting amplifier and the reference capacitance need to be known, which is relatively straightforward, but it is also necessary to know the source capacitance. This is not always easy to determine. The effect of any uncertainty in the value of C_s can be reduced by increasing the value of the reference capacitor to the point where it dominates the total capacitance. However, in so doing, the output voltage is also reduced and the measurement becomes more difficult. The dependence of the measurement on C_s is one of the main limitations to this simple method of charge measurement. In addition, any leakage currents into the input of the operational amplifier, through the capacitors, or back into the source circuitry during the measurement period will affect the result. For the most accurate measurements of low charge levels, *feedback amplifiers* are more commonly used.

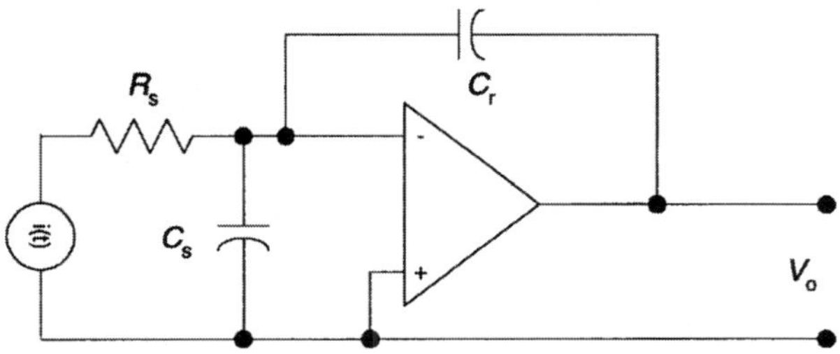

FIGURE 8.5 Schematic representation of a charge amplifier with reference feedback capacitor. The charge is proportional to the output voltage and to the sum $C_s + AC_r$, where A is the amplifier gain. Note the reduced sensitivity to C_s.

Feedback Amplifiers

Figure 8.5 shows a circuit where the reference capacitor now provides the feedback path around the operational amplifier. The output voltage from this configuration for a given charge Q transfer from the source is then:

$$V_o = \frac{AQ}{C_S + AC_r} \tag{8.7}$$

where A is the open-loop gain of the operational amplifier. For most situations, $AC_r > C_s$ and the charge measurement becomes independent of the source capacitance. In addition, the inverting input to the operational amplifier is kept close to ground potential, reducing the magnitude of leakage currents in that part of the circuit. However, in contrast to these two benefits is the new problem that the input bias current for the operational amplifier is integrated by the feedback capacitor, producing a continual drift in the output voltage. Several solutions have been used to overcome this problem, including the use of a parallel feedback resistor, R_f, which suppresses the integrating behavior at low frequencies (periods longer than $R_f C_r$), balancing the bias current with another externally provided current, and incorporating a reset switch that discharges the circuit each time the output voltage ramps beyond a set trigger level.

The sensitivity of feedback amplifiers depends on the noise sources operating within any specific application. The most impressive performance is obtained by amplifiers integrated into CCD chips (charge coupled devices) that can, under the right operational conditions, provide sensitivities measured in terms of a few electron charges. To illustrate the important parameters involved in the design of ultralow noise charge preamplifiers for CCD-type applications, consider the circuit shown in Figure 8.6. The source (detector) is now shown as a biased photodiode, which is assumed to be producing individual bursts of charge each time a photon (an X-ray, for example) interacts in it. In this example, the photodiode is coupled to the amplifier using a large value capacitor, C_c. This blocks the direct current path from the diode bias supply, V_b, but provides a low impedance path for the short-duration charge deposits. The preamplifier is a variant on that shown in Figure 8.5, in which there is now a parallel feedback resistor to provide baseline restoration on long time scales and an FET transistor to reduce the effect of the operational amplifier input bias current by virtue of its high current gain factor, β.

In practice, the dominant noise contributions in most applications of this type come from Johnson (current) noise in the bias and feedback resistors, shot noise on the photodiode bias current, voltage noise across the FET, and finally the inevitable $1/f$ component. The two resistors effectively feed thermal current noise into the input of the integrator. Similarly, the shot noise associated with the photodiode bias current feeds into the input. Together, these components are known as *parallel* noise and the total parallel noise charge is given by:

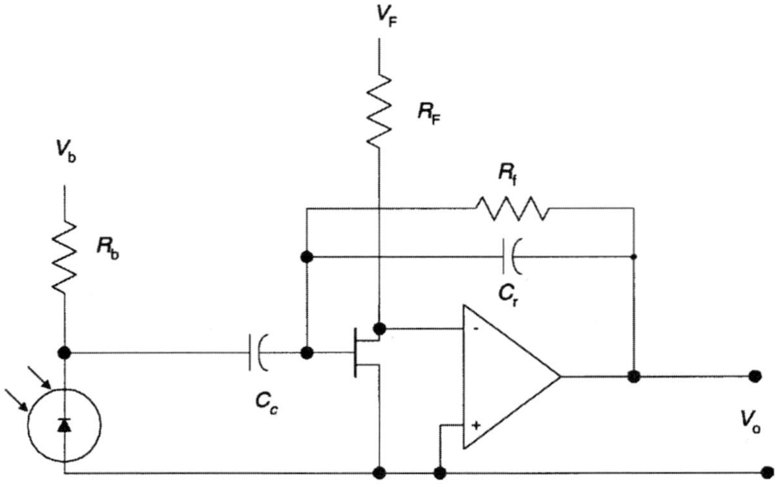

FIGURE 8.6 Typical ultralow noise charge preamplifier configuration for charge pulse readout from ionization type radiation detectors (e.g., X-ray detection using photodiodes or CCDs). The large capacitor C_c provides a low impedance path for the short charge deposit pulses, while the parallel feedback resistor provides baseline restoration on long time scales. An FET transistor reduces the effect of the operational amplifier input bias current.

$$q_p = \sqrt{\left(\frac{4kT}{R_f + R_b} + 2eI_b(T)\right)\frac{1}{\Delta B}} \tag{8.8}$$

where k is Boltzmann's constant, e is the charge on the electron, T is absolute temperature, and ΔB is the bandwidth associated with the measurement that will depend on the details of subsequent shaping amplifier stages [12]. Voltage noise across the FET (and hence operational amplifier inputs) will arise from junction noise in the FET itself and from Johnson noise in its bias resistor, R_F. In practice, the FET junction noise usually dominates, in which case this *series* noise contribution is given by:

$$q_s = \sqrt{\varepsilon_n^2 C_{in}^2 \Delta B} \tag{8.9}$$

where ε_n is the junction voltage noise for the FET in $V\sqrt{Hz^{-1}}$ and C_{in} is the total capacitance seen at the gate of the FET. This will include both the source capacitance, the gate capacitance of the FET, and any stray capacitance. The total noise is then the quadrature sum of Equations 8.8 and 8.9. The different dependencies on the bandwidth for Equations 8.8 and 8.9 imply there will be some optimum bandwidth for the measurement and this will depend on the relative contributions from each. $1/f$ noise manifests itself as a bandwidth-independent term that again, must be added in quadrature. The Johnson noise associated with the resistors and FET junction will show a temperature dependence decreasing with $\sqrt{T}$. For the FET, this reduction does not continue indefinitely and there is usually an optimum temperature for the FET around 100 K. Photodiode bias currents also fall with decreasing temperature and, for silicon devices, this is about a factor of 2 for every 10 K drop in temperature. Most ultralow noise applications thus operate at reduced temperature, at least for the sensitive components. Bias resistors and feedback resistors are kept as high as possible (typically > 100 MΩ) and FETs are specially selected for low junction voltage noise (typically 1 $nV\sqrt{Hz^{-1}}$). Ideally, photodiode capacitances should be kept as low as possible and there is also an interplay between the FET junction noise, ε_n, and the FET gate capacitance that is affected by altering the FET bias current, which can be used to fine-tune the series noise component.

Finally, there is another noise component that can often be critical and difficult to deal with. This is from microphonics. There are two effects. First, the feedback reference capacitor is typically made as small as possible (<1 pF) to reduce the effect of noise in the following shaping amplifier stages. This makes it sensitive to any stray capacitances and, if there are vibrations in the system that alter the local geometry, then this can change the feedback capacitance that changes the "gain" of the preamplifier. Second, the photodiode will be operating with some applied bias voltage (often several tens of volts) and any change in its apparent capacitance through mechanical movement of components will result in charge being moved around. These charge movements will be sensed by the charge amplifier. Ultralow noise applications that ignore mechanical stability in their design phase are in peril. Tables 8.1 and 8.2 contain a selection of available commercial devices and manufacturers.

8.3 Applications

The Millikan technique of measuring charges on particles suspended in electric fields continues to be developed for various applications. Kutsuwada et al. [13] use a modified Millikan experiment to measure the charge distribution on electrophotographic toner particles of various sizes. They show a comparison of the results obtained using an ordinary Millikan apparatus, and a modified system in which an additional ac electrode has been inserted in the hyperbolic quadrupole electrode assembly. The two methods agree to within a small multiplicative calibration factor. A different method of measuring the charge of toner particles uses the q/d meter [14] in which q and d refer, respectively, to the charge and the diameters of the particles. In the q/d meter, the charged particles are transported horizontally in a steady laminar air flow, and move vertically in an electric field until deposited on a registration electrode. The position at which the particle is deposited on the registration electrode defines the charge-to-diameter ratio. The size of the deposited particle is then measured, thus completing the determination of the charge distribution for various particle sizes.

The experiments searching for fractional charges [15] make use of superconducting niobium spheres 0.25 mm in diameter, suspended in vacuum at 4.2 K in a magnetic field. The vertical position of the spheres is modulated by an alternating electric field and measured with an ultra-sensitive magnetometer. Positrons and electrons generated by radioactive sources are used to cancel all integer charges on the spheres. Fractional charges are detected and measured as that residual charging of the niobium spheres that cannot be neutralized by the integral charges from the radioactive sources. Although this experiment is sensitive to about 0.01 electron charges (10^{-21} C), it has produced no conclusive evidence of fractional charges.

A similar approach, using force modulation, is used for the noncontact measurement of charge on gyroscopes [16]. Out-of-phase equal forces are applied to an electrostatically suspended gyroscope at a frequency well within the suspension control bandwidth. The charge of the gyroscope is then proportional to modulation frequency component of the suspension control effort. The sensitivity of this method is about 10^{-12} C gyroscope charge, limited by the allowable modulation force and position sensor noise.

Noncontact measurement of charge on liquid drops in a microgravity environment can be also performed using field mill instruments [17]. A mechanical chopper is used to modulate the electric field induced by the spherical charge on a grounded sensing plate. The resulting alternating current from sensor to ground is a measure of the charge on the drop. In a refinement of this method, the modulation is achieved by varying the distance between charge and sensor. The authors [18] claim that this system provides an increase in sensitivity of 2 to 3 orders of magnitude over the original technique.

Optical sensors based on the Pockels effect [19] are used to measure the space-charge field in gaseous dielectrics. The Pockels effect involves the change in the birefringence of certain crystalline materials on the application of an electric field. In the measuring system, a circularly polarized beam is detected after passing through the Pockels sensor and a polarizing plate. The detected intensity varies linearly with the intensity of the electric field applied to the sensor. This system is capable of performing a vector measurement of the electric field produced by the space charge, determining both intensity and direction.

Optical methods of charge measurement are also used for particles whose physical structure depends on their charge. An example is the degree of dissociation of the end groups of polystyrene particles in colloidal solutions [20]. In this application, the intensities of the Raman scattering spectrum lines depend on the degree of dissociation of the polystyrene end groups, and thus determine the charge of these particles.

A widely used application of charge measurement is as an integral element of radiation dosimetry. Radiation is passed through ionization chambers, where the generated ions are collected and the charge measured with electrometers. Ionization chambers used in dosimetry for radiation therapy have typical sensitivities in the range 0.02 nC R^{-1} to 0.2 nC R^{-1}. Coulombmeter electrometers with sensitivities of 1 pC to 10 pC are therefore required for this application.

Defining Terms

Charge, also **Electric charge:** A basic property of elementary particles, defined by convention as negative for the electron and positive for the proton. The SI unit of charge is the coulomb (**C**), defined as 1 ampere × 1 second.

Electrostatic instrument: An instrument that functions by measuring the mechanical displacement or strain caused by electric fields.

Electrostatic voltmeter: Electrostatic instrument used to measure charge. The charge is determined as a function of voltage and instrument capacitance.

Charge amplifier: Charge-measuring instrument. The charge is transferred to a reference capacitor and the resulting voltage across the capacitor is measured. Shunt and feedback versions of the charge amplifier have the reference capacitor used in shunt and feedback mode, respectively.

Electrometer: *Historic usage*, type of electrostatic voltmeter. *Modern usage*, electronic electrometer, a high-performance dc multimeter with special input characteristics and high sensitivity, capable of measuring voltage, current, resistance, and charge.

References

1. E. Cohen and B. Taylor, The 1986 Adjustment of the Fundamental Physical Constants, *Rev. Modern Phys.*, 59(4), 1121-1148, 1987.
2. F. Harris, *Electrical Measurements,* New York: John Wiley & Sons, 1952.
3. W. Michels, *Electrical Measurements and Their Applications,* Princeton, NJ: D. Van Nostrand, 1969.
4. A.J. Diefenderfer, *Principles of Electronic Instrumentation,* 3rd ed., Philadelphia: Saunders College Publishing, 1994.
5. Keithley Instruments, Inc., *Low Level Measurements,* 4th ed., Cleveland: Keithley Instruments, Inc., 1993.
6. K. Kawamura, S. Sakamoto, and F. Noto, Design and Development of New Electrostatic Voltmeter Using Strain Gauge, *IEEE Trans. Ind. Appl.,* 25, 563-568, 1989.
7. C. Hsu and R. Muller, Micromechanical Electrostatic Voltmeter, *Transducers '91. Int. Conf. Solid-State Sensors Actuators,* 1991, 659-662.
8. Trek Incorporated, 3922 Salt Works Rd., P.O. Box 728, Medina, NY 14103.
9. M.M. Horenstein, Measuring surface charge with a noncontacting voltmeter, *Proc. IEEE Industry Appl. Soc. Annu. Meeting,* Toronto, Ontario, Canada IAS'93, 3, 1811-1816, 1993.
10. P. Gunther, Determination of charge density and charge centroid location in electrets with semiconducting substrates, *IEEE Trans. Electrical Insul.,* 27, 698-701, 1992.
11. B. MacDonald and B. Fallone, Surface-charge distributions on radiation-charged electrets. *7th Int. Symp. Electrets* (ISE 7). Berlin, Germany, 1991, 798-803.
12. V. Radeka, Signal, noise and resolution in position-sensitive detectors, *20th Nuclear Science Symp. 5th Nuclear Power Syst. Symp.,* San Francisco, CA, *IEEE Trans. Nucl. Sci.,* 21, 51-64, 1974.
13. N. Kutsuwada, T. Shohdohji, N. Okada, H. Izawa, T. Sugai, Y. Nakamura, and T. Murata, Measurement of electric charge of electrophotographic toner, *J. Imaging Sci. Technol.,* 37(5), 480-484, 1993.
14. M. Mehlin and R.M. Hess, Electrical charge measurement of toner particles using the *q/d* meter, *J. Imaging Sci. Technol.,* 36, 142-150, 1992.

15. A.F. Hebard, G.S. LaRue, J.D. Phillips, and C.R. Fisel, *Search for Fractional Charge in Near Zero: New Frontiers of Physics*, New York: W. H. Freeman and Company, 1988, 511.
16. S. Buchman, T. Quinn, G.M. Keiser, D. Gill, and T.J. Sumner, Charge measurement and control for the Gravity Probe B gyroscopes, *Rev. Sci. Instrum.*, 66, 120-129, 1995.
17. M.N. Horenstein, Peak sampled vibrating-reed for the measurement of electric fields in the presence of space charge, *Rev. Sci. Instrum.*, 54, 591-593, 1983.
18. K.-C. Lin and T.G. Wang, Noncontact charge measurement, *Rev. Sci. Instrum.*, 63, 2040-2043, 1992.
19. K. Hidaka, Progress in Japan of space charge field measurement in gaseous dielectrics using a Pockels sensor, *IEEE Electrical Insul. Mag.*, 12(1), 17-28, 1996.
20. R. Kesavamoorthy, T. Sakuntala, and A.K. Arora, *In situ* measurement of charge on polystyrene particles in colloidal suspension, *Meas. Sci. Technol.*, 1(5), 440-445, 1990.

9

Capacitance and Capacitance Measurements

Halit Eren
Curtin University of Technology

James Goh
Curtin University of Technology

A capacitor is a system of two conducting electrodes, having equal and opposite charges separated by a dielectric. The capacitance C of this system is equal to the ratio of the absolute value of the charge Q to the absolute value of the voltage between bodies as:

$$C = Q/V \tag{9.1}$$

where C = Capacitance in farad (F)
Q = Charge in coulomb (C)
V = Voltage (V)

The unit of capacitance, the farad, is a large unit; practical capacitors have capacitances in mirofarads (μF or 10^{-6} F), nanofarads (nF or 10^{-9} F), and picofarads (pF or 10^{-12} F). The unit conversions are shown in Table 9.1.

The capacitance C depends on the size and shape of charged bodies and their relative positions; examples are shown in Table 9.2. Generally, capacitance is inherent wherever an electrostatic field appears. In many electronic systems, it is necessary to deal with capacitances that are designed within the circuits and also with unwanted interference and stray capacitances that are introduced externally or internally at various stages of the circuits. For example, some sensors operate on capacitive principles, giving useful signals; in others, capacitance is inherent but undesirable. In many cases, cables and external circuits introduce additional capacitances that need to be accounted for the desirable operation of the system. In these cases, Table 9.2 is useful to identify and analyze possible sources of capacitances where charged bodies are involved.

In general, the capacitance can be determined by solving Laplace's equations $\nabla^2 V(x, y, z) = 0$ with appropriate boundary conditions. One type of boundary condition specifies the electrode voltages V_1 and V_2 of the plates. Laplace's equation yields to V and the electric field $E(x, y, z) = -\nabla V(x, y, z)$ between the electrodes. The charge of each electrode can also be obtained by integration of the flux density over each electrode surface as:

0-8493-1733-9/04/$0.00+$1.50

TABLE 9.1 Capacitance Unit Conversions

microfarads (μF)	nanofarads (nF)	picofarads (pF)
10^{-6} F	10^{-9} F	10^{-12} F
0.000001 μF	0.001 nF	1.0 pF
0.001 μF	1.0 nF	1000 pF
1.0 μF	1000 nF	1,000,000 pF

$$Q = \int \varepsilon(x,y,z) E(x,y,z) \mathrm{d}A \tag{9.2}$$

If the capacitor is made from two parallel plates, as shown in Figure 9.1, the capacitance value in terms of dimensions can be expressed by:

$$C = \varepsilon A/d = \varepsilon_r \varepsilon_0 A/d \tag{9.3}$$

where ε = Dielectric constant or permittivity
ε_r = Relative dielectric constant (in air, $\varepsilon_r = 1$)
ε_0 = Dielectric constant of vacuum (8.854188×10^{-12} F m^{-1})
d = Distance of the plates in m
A = Effective area of the plates in m^2

In arriving at Equation 9.3, the fringe field is neglected for small distances, d, between the plates.

Capacitances can also be expressed in terms of dielectric properties currents and voltages. Suppose that a uniform dielectric between two parallel plates has a resistance, R, which can be written as:

$$R = d\rho/A \tag{9.4}$$

where ρ is the specific resistance of the dielectric in Ω m. Then, using Equation 9.3 gives:

$$C = \varepsilon\rho/R \tag{9.5}$$

A voltage V across the capacitor causes a leakage current $I_l = V/R$ such that:

$$C = \varepsilon\rho I_l/V \tag{9.6}$$

This indicates that the leakage current of a capacitor is proportional to its capacitance value.

As seen in Equations 9.3 and 9.4, the value of the capacitance is proportional to the permittivity of the dielectric material used. In the construction of capacitors, the permittivity of commonly used materials is given in Table 9.3.

9.1 Types of Capacitors

Commonly used fixed capacitors are constructed with air, paper, mica, polymers, and ceramic dielectric materials. A comprehensive list of common capacitors and their characteristics are given in Table 9.4 and Table 9.5 and a list of manufacturers is given in Table 9.6. Variable capacitors are generally made with air or ceramic dielectric materials. The capacitors used in electronic circuits can be classified as: low-loss, medium-loss, and high-tolerance capacitors.

TABLE 9.2 Capacitances of Various Electrode Systems

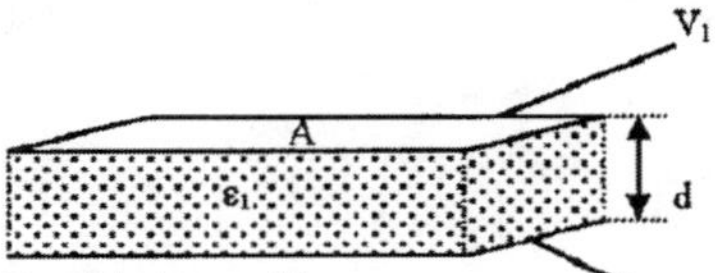

Parallel planes with homogeneous dielectric

$$C = \frac{\varepsilon A}{d}$$

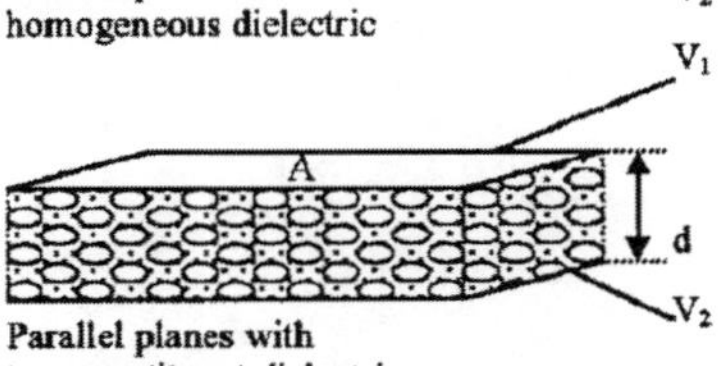

Parallel planes with two constituent dielectrics

$$C = \frac{(\varepsilon_1 r + \varepsilon_2)A}{(1+r)d}$$ where r value is the volumetric ratio

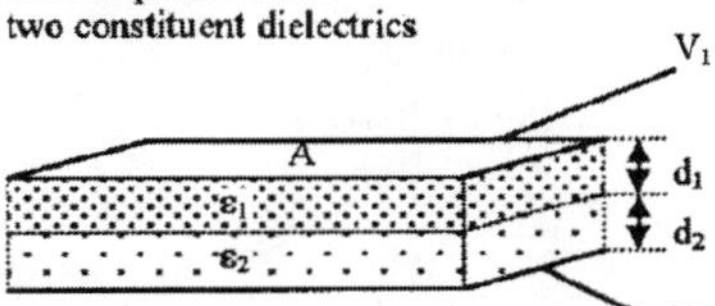

Parallel planes with two different dielectrics

$$C = \frac{\varepsilon A}{\frac{d_1}{\varepsilon_1} + \frac{d_2}{\varepsilon_2}}$$

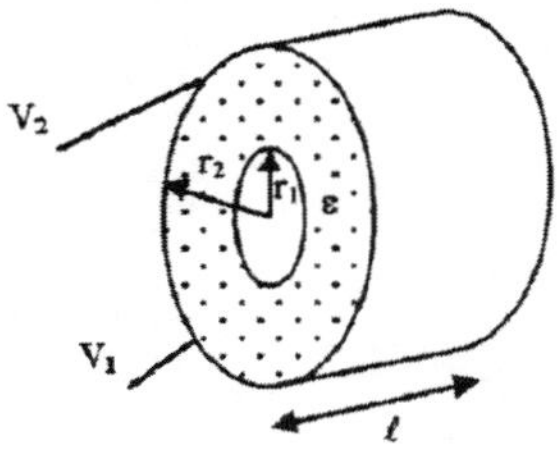

Coaxial cylinders with single dielectric

$$C = \frac{4\pi\varepsilon\ell}{\ln(r_2/r_1)}$$

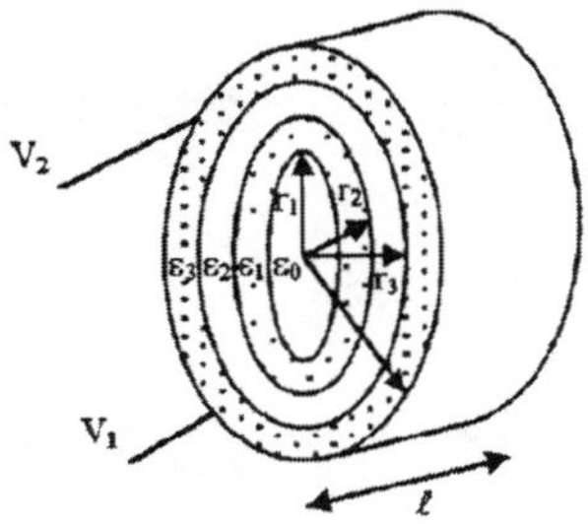

Coaxial cylinders with three dielectrics

$$C = \frac{2\pi\varepsilon_0\ell}{\ln\left[\left(\frac{r_2}{r_1}\right)^{\frac{\varepsilon_0}{\varepsilon_1}} \times \left(\frac{r_3}{r_2}\right)^{\frac{\varepsilon_0}{\varepsilon_2}} \times \left(\frac{r_4}{r_3}\right)^{\frac{\varepsilon_0}{\varepsilon_3}}\right]}$$

$$= \left(\frac{1}{C_1} + \frac{1}{C_2} + \frac{1}{C_3}\right)^{-1}$$

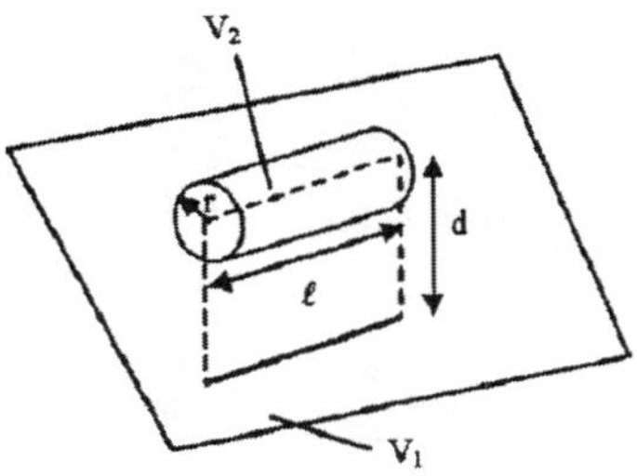

Cylinder parallel with plane

$$C = \frac{\pi\varepsilon\ell}{\ln\left[\frac{d}{r} + \sqrt{\left(\frac{d}{r}\right)^2 - 1}\right]} = \frac{2\pi\varepsilon\ell}{\mathrm{Cosh}^{-1}\left(\frac{d}{r}\right)}$$

TABLE 9.2 (continued) Capacitances of Various Electrode Systems

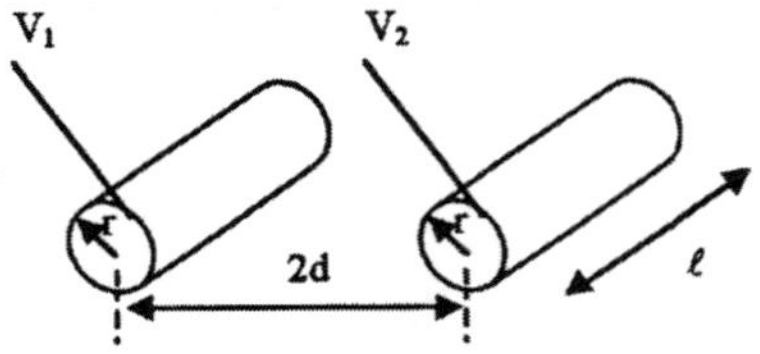

Parallel cylinders with equal radius

$$C = \frac{\pi\varepsilon\ell}{\ln\left[\frac{d}{r}+\sqrt{\left(\frac{d}{r}\right)^2-1}\right]} = \frac{\pi\varepsilon\ell}{\mathrm{Cosh}^{-1}\left(\frac{d}{r}\right)}$$

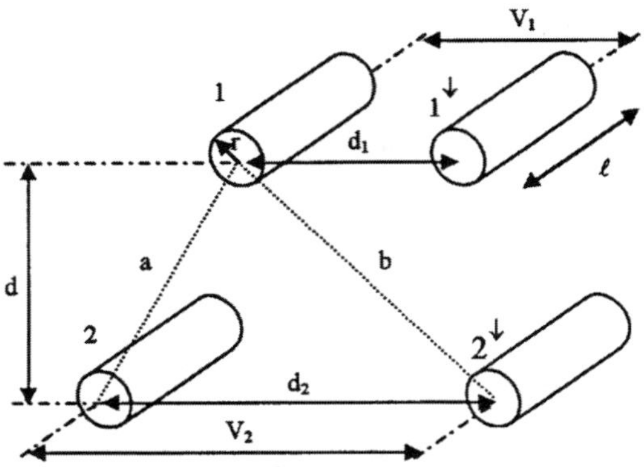

Mutual capacitance

$$C_{11} = \frac{\pi\varepsilon\ell}{\ln(d_1/r)} + \frac{\pi\varepsilon}{2}\frac{\ln(b/a)}{[\ln(a/r)]\ln(b/r)}$$

$$C_{21} = C_{12} = \frac{\pi\varepsilon\ell}{2}\frac{\ln(b/a)}{[\ln(a/r)\ln(b/r)]}$$

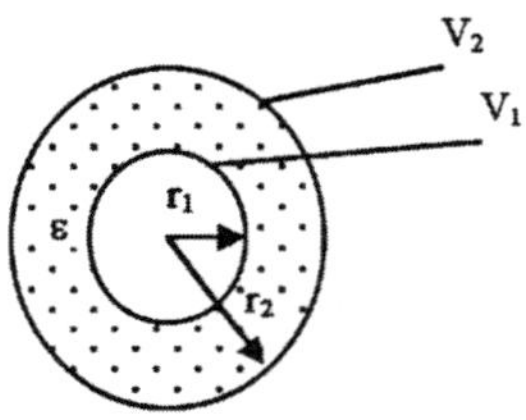

Concentric spheres

$$C = \frac{4\pi\varepsilon r_1 r_2}{r_2 - r_1}$$

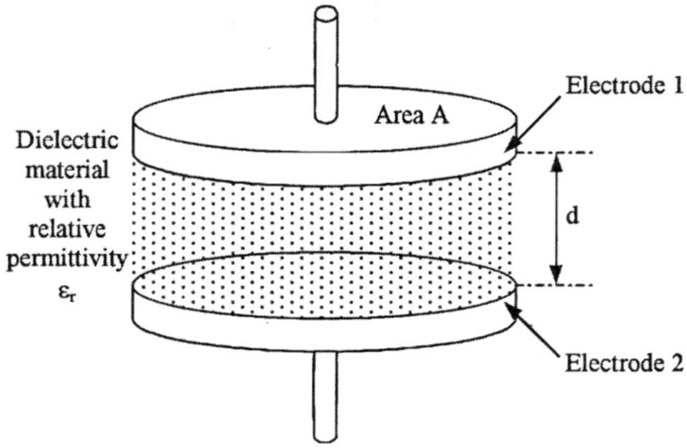

FIGURE 9.1 A typical capacitor made from two parallel plates. The capacitance between two charged bodies depends on the permittivity of the medium, the distance between the bodies, and the effective area. It can also be expressed in terms of the absolute values of the charge and the absolute values of the voltages between bodies.

TABLE 9.3 Permittivity (Dielectric Constants of Materials Used in Capacitors)

Material	Permittivity
Vacuum	1.0
Air	1.0006
Teflon	2.1
Polyethylene, etc.	2.0–3.0
Impregnated paper	4.0–6.0
Glass and mica	4.0–7.0
Ceramic (low *K*)	≤20.0
Ceramic (medium *K*)	80.0–100.0
Ceramic (high *K*)	≥1000.0

- *Low-loss capacitors* such as mica, glass, low-loss ceramic, and low-loss plastic film capacitors generally have a good capacitance stability. These capacitors are expensive and often selected in precision applications, e.g., telecommunication filters.
- *Medium-loss capacitors* have medium stability in a wide range of ac and dc applications. These are paper, plastic film, and medium-loss ceramic capacitors. Their applications include coupling, decoupling, bypass, energy storage, and some power electronic applications (e.g., motor starter, lighting, power line applications, and interference suppressions).
- *High-tolerance capacitors* such as aluminum and tantalum electrolytic capacitors deliver high capacitances. Although these capacitors are relatively larger in dimension, they are reliable and have longer service life. They are used in polarized voltage applications, radios, televisions, and other consumer goods, as well as military equipment and harsh industrial environments.

There are also specially designed capacitors (e.g., mica, glass, oil, gas, and vacuum). These capacitors are used particularly in high-voltage (35 kV) and high-current (200 A) applications.

In the majority of cases, the manufacturing process of the capacitors begins by forming one plate using metallization of one side of a flexible dielectric film. A foil such as aluminum is used as the other plate. The film/foil combination is rolled on a suitable core with alternate layers slightly extended and then heat-treated. In some cases, two-foil layers are divided by a dielectric film or paper impregnated with oil.

Generally, capacitors are two-terminal devices with one electrode as the ground terminal. However, if both terminals are separated from the common terminal, the additional capacitances between ground and electrodes might have to be taken into account. Usually, capacitance between electrodes and ground are small compared to the dominant capacitance between plates. Three-terminal capacitances exist and are manufactured in many different ways, as illustrated in Table 9.2.

As far as construction and materials and construction techniques are concerned, the capacitors can broadly be classified as: electrolytic, ceramic, paper, polymer, mica, variable capacitors, or integrated circuit capacitors.

Paper capacitors: Usually, paper capacitors are made with thin (5 to 50 μm in thickness) wood pulp. A number of sheets are used together to eliminate possible chemical and fibrous defects that may exist in each sheet. The paper sheets are placed between thin aluminum foils and convolutely wound, as shown in Figure 9.2. The moisture of the paper is removed at high-temperature vacuum drying before the capacitor is vacuum impregnated with oil, paraffin, or wax. The losses and self-inductance are sizeable and frequency dependent. The applications are usually restricted to low frequency and high voltages. When impregnated with silicone-oil, they can withstand voltages up to 300 kV.

Electrolytic capacitors: This describes any capacitor in which the dielectric layer is formed by an electrolytic method. The electrolytic capacitors in dry foil form may be similar to construction to paper film capacitors; that is, two foil layers separated by an impregnated electrolyte paper spacer are rolled together. In this case, one of the plates is formed using metallization of one side of a flexible dielectric

TABLE 9.4 Characteristics of Common Capacitors

Capacitor Types	Range (F)	Tolerance (%)	Voltage Range (V)	Temperature Range (°C)	Temperature Coefficient (ppm/°C)	Frequency Range (Hz)	Permittivity ($\varepsilon/\varepsilon_0$ C^2/Nm^2)	Dielectric Strength (C/C_0)	Dissipation Factor (%)	Insulation Resistance (MΩ/μF)	Typical Average Failure Rates (fail per 10^6 h)
Mica, glass, porcelain, and Teflon											0.0133
Mica	5p–0.01μ	5	100–600	–55/125	–50	100Hz–10GHz	7.0	1000	0.001	2.5×10^4	
Glass	5p–1000p	5	100–600	–55/125	40	100Hz–10GHz	6.6	2500	0.001	10^6	
Porcelain	100p–0.1μ	5	50–400	–55/125	120				0.1	5×10^5	
Teflon	1000p–2μ	10	50–200	–70/250	–200				0.04	5×10^6	
Ceramic											
Low-Loss	100p–1 μ	10	50–400	–55/125	±30	100Hz–10GHz	5.7	200–300	0.02	5×10^3	0.11–0.008
Disk											
Multilayer											
Plate											
High Permittivity	10p–1μ		50–30000	–55/125		1kHz–1GHz	1000–7000	100			
Disk											
Multilayer											
Plate											
Paper											
Paper	0.1μ–10μ	10	200–1600	–55/125	±800	D.C.–1MHz	4.5	500–1000	1.0	5×10^3	0.002
Metallized paper											
Plastic											
Kapton	1000p–1μ	10		–55/220	100				0.3	10^5	0.05
Polyester/Mylar	1000p–50μ	10	50–600	–55/125	400	D.C.–10GHz	2.3	1000	0.75	10^5	

Parylene	5000p–1μ	10		–55/125	±100	D.C.–10GHz		1000	0.1	10^5	
Polysulfone	1000p–1μ	5		–55/150	80				0.3	10^5	
Polycarbonate											
Axial	100p–30μ	10	50–800	–55/125	±100	D.C.–10GHz	2.8		0.2	5×10^5	
Can											
Radial											
Polypropylene											
Axial	100p–50μ	10	100–800	–55/105	–200	D.C.–10GHz			0.2	10^5	
Can											
Radial											
Polystyrene											
Axial	10p–2.7μ	10	100–600	–55/85	–100	D.C.–10GHz	10		0.05	10^6	
Can											
Radial											
Electrolytic and solid											0.04
Aluminum	0.1μ–1.6	–10/100	3–600	–40/85	2500		8–10		10	100	
Tantalum											
Axial	0.1μ–1000μ	–10/100	6–100	–55/85	800	D.C.–1kHz	25–27		4.0	20	
Can											
Radial											
TiTiN Film	10p–200p	10	6–30	–55/125	100				0.01	10^6	
Oil	0.1μ–20μ		200–10000			D.C.–1MHz	2.4	1000	0.5		
Air/Vacuum	1p–100p		2000–3600				1.0				

TABLE 9.5 Capacitor Specifications and Applications

Capacitor Types	Typical Commercial Specifications: Voltage A.C. V	Capacitance F	Tolerance %	Applications
Mica, glass, porcelain, and Teflon				High temperature, low absorption, good in RF applications, and circuit requiring long-term stability
Mica	350	2.2 p–1000 p	1	
Glass				
Porcelain				
Teflon				
Ceramic				Active filters, and high-density PCB applications, power-tuned circuits, coupling and decoupling of high-frequency circuits (PCB versions available)
Low-loss				
Disk	100	1.8 p–470 p	±2	
Multilayer	63/50	10 p–1 μ	10	
Plate	100	390 p–4700 p	10	
High permittivity				
Disk				
Multilayer				
Plate				
Paper				General-purpose motor applications
Paper				
Metallized paper				

Samples/codes

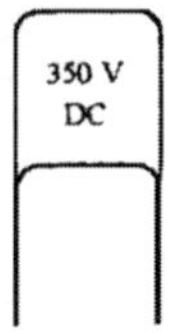

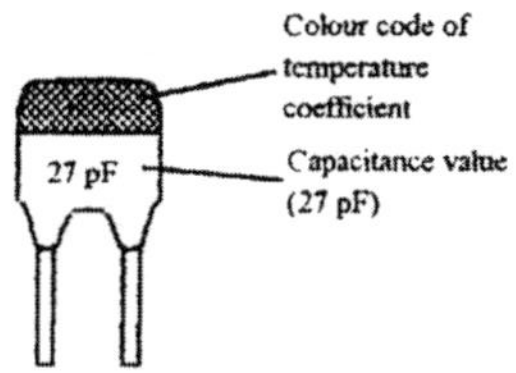

Color	α ppm/°C
Red Purple	100
Black	0
Brown	−33
Red	−75
Orange	−150
Yellow	−220
Green	−330
Blue	−470
Purple	−750

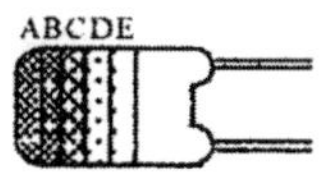

Color	Significant Figures A	Significant Figures B	Multiplier pF C	Tolerance % D	Voltage V E
Black	—	0	1	±20	
Brown	1	1	10		125
Red	2	2	10^2		160
Orange	3	3	10^3		250
Yellow	4	4	10		
Green	5	5	10^5		
Blue	6	6	10^6		
Purple	7	7	10^7		
Gray	8	8	—		
White	9	9	—	±10	

Plastic				
Kapton polyester/ mylar	63	0.1 μ–68 μ	10	Filters, timing and other high-stability applications, high quality, small low TC, tuned circuits, timing networks, stable oscillator circuits, resonance circuits and other high-performance pulse handling applications, phase shifting, pulse applications
Parylene				
Polysulfone				
Polycarbonate				
Axial	280	1 p–10 p	10	
Can	100	0.1 μ–68 μ	10	
Radial				
Polypropylene				
Axial	63	100 p–2200 p	5	
Can	400	150 p–1000 p	1	
Radial	1000	1 n–470 n	20	
Polystyrene				
Axial	450	47 p–680 p	1	
Can	1000	1 n–470 n	20	
Radial	160	10p–10000p	2.5	
Electrolytic and solid				General-purpose to high-performance applications, power supply filters, motor capacitors, switching circuits, high-voltage filter transmitters and long life applications
Aluminum	25	680 p–6800 p	20	
Tantalum				
Axial	6.3	6.8 μ–150 μ	20	
Can	35	6.8 μ–150 μ	20	
Radial	16	2.2 μ–68 μ	20	

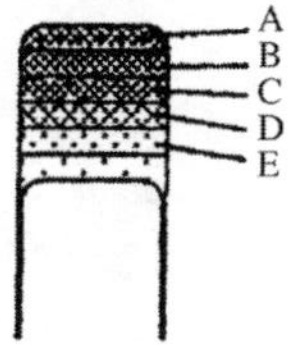

Color	Significant Figures A	Significant Figures B	Multiplier pF C	Tolerance % D	Voltage V (D.C.) E
Black	0	0	1	±20	—
Brown	1	1	10		100
Red	2	2	10^2		250
Orange	3	3	10^3		—
Yellow	4	4	10^4		400
Green	5	5	10^5		—
Blue	6	6	—		630
Purple	7	7	—		—
Gray	8	8	10^8		—
White	9	9	10^9	±10	—

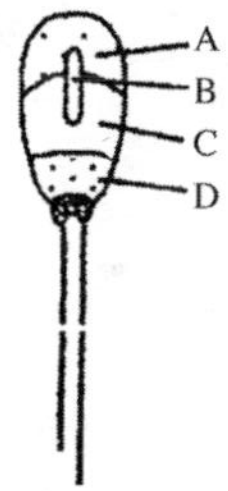

Color	Significant Figures A	Significant Figures B	Multiplier pF C	Voltage V D
Black	—	0	1	10
Brown	1	1	—	1.6
Red	2	2	—	4
Orange	3	3	—	40
Yellow	4	4	—	6.3
Green	5	5	—	16
Blue	6	6	—	—
Purple	7	7	10^{-3}	—
Gray	8	8	10^{-2}	25
White	9	9	10^{-1}	2.5

TABLE 9.6 List of Manufacturers

Bycap Inc. 5115 N. Ravenswood, Dept. T Chicago, IL 60640 Tel: (312) 561-4976 Fax: (312) 561-5095	Magnetek 902 Crescent Avenue Bridgeport, CT 06607 Tel: (800) 541-9997 Fax: (203) 335-2820
Chenelex Barr Road, P.O. Box 82 Norwich, NY 13815 Tel: (607) 344-3777 Fax: (607) 334-9076	Maxwell Laboratories Inc. 8888 Balboa Avenue San Diego, CA, 92123 Tel: (619) 576-7545 Fax: (619) 576-7545
Chicago Condenser 2900-T W. Chicago Ave. Chicago, IL 60622 Tel: (312) 227-7070 Fax: (312) 227-6646	Metuchen Capacitors Inc. 139 White Oak Lane Old Bridge, NJ 08857 Tel: (800) 679-0514 Fax: (800) 679-9959
Comet North America Inc. 11 Belden Avenue Norwalk, CT 06850 Tel: (203) 852-1231 Fax: (203) 838-3827	Murata Electronics Marketing Communications 2200 Lake Park Drive Smyrna, GA 30080 Tel: (800) 394-5592 Fax: (800) 4FAXCAT
Condenser Products 2131 Broad Street Brooksville, FL 34609 Tel: (800) 382-6874 Fax: (904) 799-0221	NWL Capacitors 204 Caroline Drive, P.O. Box 97 Snow Hill, NC 28580 Tel: (919) 747-5943 Fax: (919) 747-8979
CSI Capacitors 810 Rancheros Drive San Marcos, CA 92069-3009 Tel: (619) 747-4000 Fax: (619) 743-5094	Okaya Electric America Inc. 503 Wall Street Valparaiso, IN 46383 Tel: (219) 477-4488 Fax: (219) 477-4856
HVCO Inc. P.O. Drawer 223, 7137 Sycamore Ln. Cedarburg, WI 53012 Tel: (414) 375-0172 Fax: (414) 275-0173	

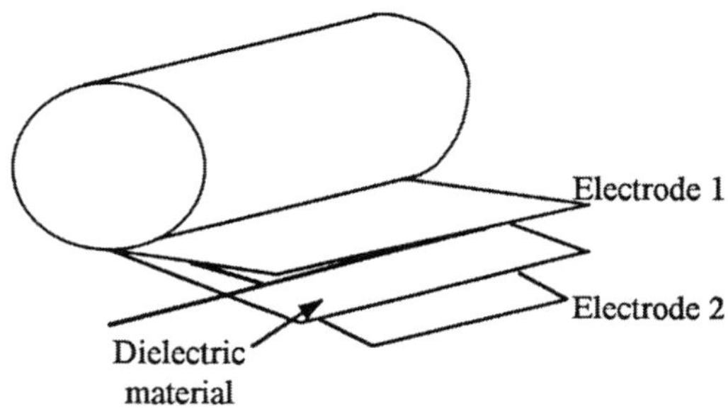

FIGURE 9.2 Construction of a typical capacitor. Dielectric material sheets are placed between electrode foils and convolutely wound. The moisture of the dielectric material is removed at high temperatures by vacuum drying before the capacitor is impregnated with oil, paraffin, or wax.

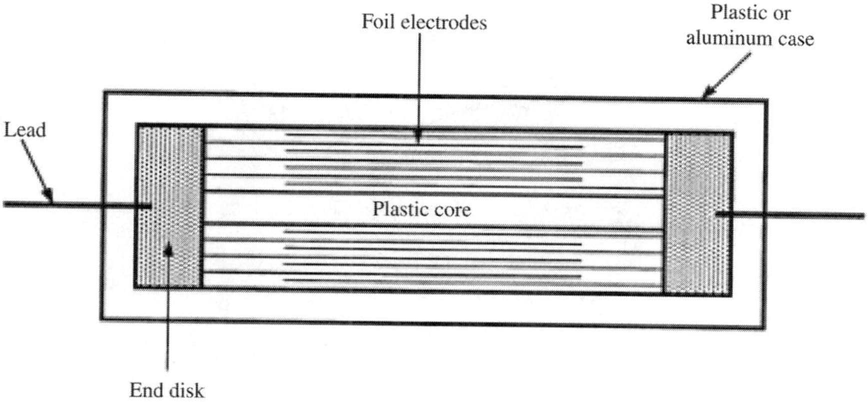

FIGURE 9.3 Construction of an electrolytic capacitor. The two foil-layer electrodes are separated by an impregnated electrolyte paper spacer and rolled together on a plastic core. Usually, a flexible metallized dielectric film is used as one of the plates and an ordinary foil is used for the other. The capacitor is then hermetically sealed in an aluminum or plastic can.

film. A foil (e.g., aluminum) is used as the plate. The capacitor is then hermetically sealed in an aluminum or plastic can, as shown in Figure 9.3. These capacitors can be divided into two main subgroups.

Tantalum electrolytic: The anode consists of sintered tantalum powder and the dielectric is Ta_2O_5, which has a high value of ε_r. A semiconductor layer MnO_2 surrounds the dielectric. The cathode made from graphite is deposited around MnO_2 before the capacitor is sealed. The form of a tantalum electrolytic capacitor includes a porous anode slug to obtain a large active surface. These capacitors are highly stable and reliable, with good temperature ranges, and are suitable for high-frequency applications.

Aluminum electrolytic capacitors: Aluminum foil is oxidized on one side as Al_2O_3. The oxide layer is the dielectric having a thickness of about 0.1 μm and a high electric field strength (7×10^5 Vmm^{-1}). A second layer acting as the cathode, made from etched Al-foil, is inserted. The two layers are separated by a spacer when the layers are rolled and mounted.

Electrolytic capacitors must be handled with caution, since in these capacitors the electrolytic is polarized. That is, the anode should always be positive with respect to the cathode. If not connected correctly, hydrogen gas will form; this damages the dielectric layer, causing a high leakage current or blow-up. These capacitors can be manufactured in values up to 1 F. They are used in not-so-critical applications such as coupling, bypass, filtering, etc. However, they are not useful at frequencies above 1 kHz or so.

Ceramic and glass capacitors: The dielectric is a ceramic material with deposited metals. They are usually rod or disk shaped. They have good temperature characteristics and are suitable in many high-frequency applications. There are many different types, such as (1): *Low K ceramic:* These capacitors are made with materials that contain a large fraction of titanium dioxide (TiO_2). The relative permittivity of these materials varies from 10 to 500, with negative temperature coefficient. The dielectric is TiO_2 + MgO + SiO_2, suitable in high-frequency applications in filters, tuned circuits, coupling and bypass circuits, etc. (2): *High K ceramic:* The dielectric contains a large fraction of barium titanate, $BaTiO_3$, mixed with $PbTiO_3$ or $PbZrO_3$ giving relative permittivity of 250 to 10,000. They have high losses, and also have high-voltage time dependence with poor stability. (3): *Miniature ceramic capacitors:* These are used in critical high-frequency applications. They are made in the ranges of 0.25 pF to 1 nF. (4): *Dielectric ceramic capacitors:* The material is a semiconducting ceramic with deposited metals on both sides. This arrangement results in two depletion layers that make up the very thin dielectric. In this way, high capacitances can be obtained. Due to thin depletion layers, only small dc voltages are allowed. They are used in small and lightweight equipment such as hearing aids.

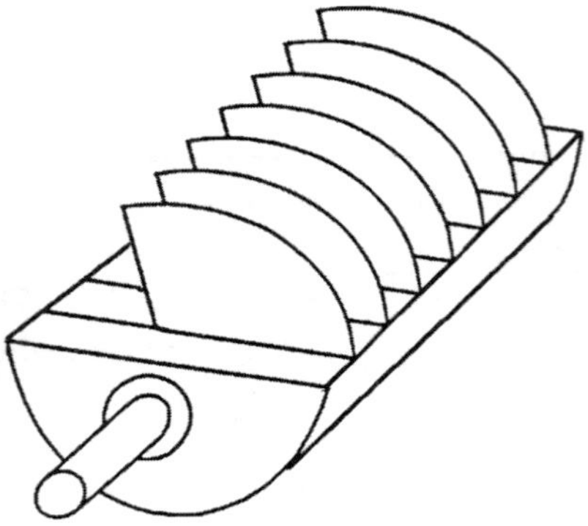

FIGURE 9.4 A variable capacitor. It consists of two assemblies of spaced plates positioned together by insulation members such that one set of plates can be rotated. The majority of variable capacitors have air as the dielectric. They are used mainly in adjustment of resonant frequency of tuned circuits in receivers and transmitters. By shaping the plates suitably, they can be made to be linear or logarithmic.

Glass capacitors are made with glass dielectric materials. The properties of glass dielectrics are similar to ceramic materials.

Polymer capacitors: Various polymers, such as polycarbonate, polystyrol, polystyrene, polyethylene, polypropylene, etc., are used as the dielectric. The construction is similar to that of paper capacitors. Polystyrene capacitors, in particular, are very stable, and are virtually frequency independent. They have low voltage ratings and are used in transistorized applications as tuning capacitors and capacitance standards.

Mica capacitors: A thin layer of mica, usually muscovite mica (≥0.003 mm) are stapled with Cu-foil or coated with a layer of deposited silver. They are then vacuum impregnated and coated with epoxy. The field strength of these capacitors is very high (10^5 V mm^{-1}) and resistivity $\rho = 10^6$ to 10^{15} Ω m. These capacitors are available in values from 1.0 pF to several microfarads for high voltage (from 100 V to 2000 V) and high-frequency applications. They have tolerances between ±20% and ±0.5%.

Variable capacitors: These capacitors usually have air as the dielectric and consist of two assemblies of spaced plates positioned together by insulation members such that one set of plates can be rotated. A typical example of variable capacitors is given in Figure 9.4. Their main use is the adjustment of resonant frequency of tuned circuits in receivers and transmitters, filters, etc. By shaping the plates, various types of capacitances can be obtained, such as: *linear capacitance*, in which capacitance changes as a linear function of rotation, and *logarithmic capacitance.*

Variable capacitors can be grouped as: precision types, general-purpose types, transmitter types, trimmer types, and special types such as phase shifters.

Precision-type variable capacitors are used in bridges, resonant circuits, and many other instrumentation systems. The capacitance swing can be from 100 pF to 5000 pF. They have excellent long-term stability with very tight tolerances.

General-purpose type variable capacitors are used as tuning capacitors in radio and other broadcasting devices. They are available in many laws such as, straight line frequency, straight line wavelength, etc. The normal capacitance swing is from 400 pF to 500 pF. In some cases, a swing of 10 pF to 600 pF are available.

Transmitter-type variable capacitors are similar to general-purpose variable capacitors, but they are specially designed for high-voltage operations. The vanes are rounded and spaced wider to avoid flashover and excessive current leakages. The swing of these capacitors can go from few picofarads up to 1000 pF. In some cases, oil filling or compressed gases are used to increase operating voltages and capacitances.

Trimmer capacitors are used for coil trimming at intermediate radio frequencies. They can be air-spaced rotary types (2 pF to 100 pF), compression types (1.5 pF to 2000 pF), ceramic-dielectric rotary types (5 pF to 100 pF), and tubular types (up to 3 pF).

Sometimes, special type variable capacitors are produced for particular applications, such as differential and phase shift capacitors in radar systems. They are used for accurate measurement of time intervals, high-speed scanning circuits, transmitters and receivers, etc.

Integrated circuit capacitors: These are capacitors adapted for use in microelectronic circuits. They include some miniature ceramic capacitors, tantalum oxide solid capacitors, and tantalum electrolyte solid capacitors. The ceramic and tantalum oxide chips are unencapsulated and are fitted with end caps for direct surface mounting onto the circuit board. The beam-leaded tantalum electrolytic chips are usually attached by pressure bonding. Typical values of these capacitors are: 1 pF to 27 nF for temperature compensating ceramic, (100 to 3000 pF) for tantalum oxide, 390 pF to 0.47 μF for general-purpose ceramic, and (0.1 to 10 μF) for tantalum electrolyte. Operating voltages range from 25 to 200 V for ceramic, 12 to 35 V for tantalum electrolyte, and 12 to 25 V for tantalum oxide.

Integrated circuit capacitors are made mostly within MOS integrated circuits as monolayer capacitors containing tantalum or other suitable deposits. The plates of the capacitors of the integrated circuits are generally formed by two heavily doped polysilicon layers formed on a thick layer of oxide. The dielectric is usually made from a thin layer of silicon oxide. These capacitors are temperature stable, with a temperature coefficient of about 20 ppm/°C. Integrated circuit capacitive sensors are achieved by incorporating a dielectric sensitive to physical variables. Usually, the metallization layer formed on top of the dielectric forms a shape to provide access to measured physical variable to the dielectric.

Voltage variable capacitors: These capacitors make use of the capacitive effect of the reversed-biased p-n junction diode. By applying different reverse bias voltages to the diode, the capacitance can be changed. Hence, the name varicap or varactor diodes is given to theses devices. Varactors are designed to provide various capacitance ranges from a few picofarads to more than 100 pF. It is also possible to make use of high-speed switching silicon diodes as voltage variable capacitors. However, they are limited by the very low maximum capacitance available. Typical applications of these varactor diodes are in the tuning circuits in radio frequency receivers. Present-day varactor diodes operate into the microwave part of the spectrum. These devices are quite efficient as frequency multipliers at power levels as great as 25 W. The efficiency of a correctly designed varactor multiplier can exceed 50% in most instances. It is also worth noting that some Zener diodes and selected silicon power-supply rectifier diodes can work effectively as varactors at frequencies as high as 144 MHz. In the case of the Zener diode, it should be operated below its reverse breakdown voltage.

9.2 Characteristics of Capacitors

Capacitors are characterized by dielectric properties, break-down voltages, temperature coefficients, insulation resistances, frequency and impedances, power dissipation and quality factors, reliability and aging, etc. Typical characteristics of common capacitors are given in Table 9.4.

Dielectric properties: Dielectrics of capacitors can be made from polar or nonpolar materials. Polar materials have dipolar characteristics; that is, they consist of molecules whose ends are oppositely charged. This polarization causes oscillations of the dipoles at certain frequencies, resulting in high losses.

In general, capacitor properties are largely determined by the dielectric properties. For example, the losses in the capacitors occur due to the current leakage and the dielectric absorption. These losses are frequency dependent, as typified by Figure 9.5.

The dielectric absorption introduces a time lag during the charging and discharging of the capacitor, thus reducing the capacitance values at high frequencies and causing unwanted time delays in pulse circuits. The leakage current, on the other hand, prevents indefinite storage of energy in the capacitor. An associated parameter to leakage currents is the leakage resistance, which is measured in megohms, but usually expressed in megohm-microfarads or ohms-farads. The leakage resistance and capacitance introduces time constants that can vary from a few days for polystyrene to several seconds in some electrolytic capacitors. It is important to mention that the leakage current does not only depend on the properties of the dielectric materials, but also depends on the construction and structure of capacitors.

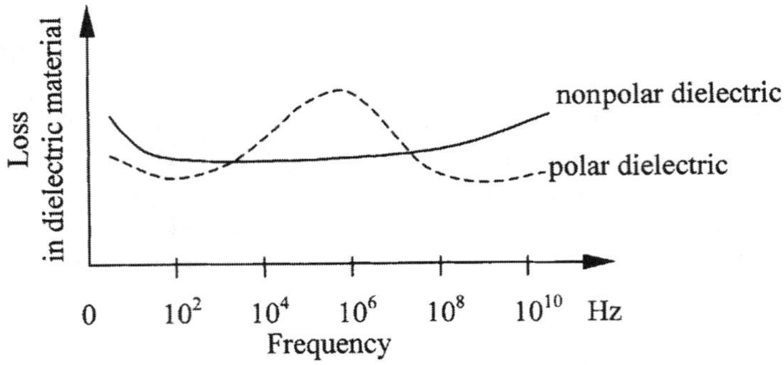

FIGURE 9.5 Frequency dependence of dielectric loss. The dielectric material of capacitors can be polar or nonpolar. In polar materials, polarization causes oscillations at certain frequencies, resulting in high losses. The dielectric losses introduce a time lag during the charging and discharging of the capacitor, thus reducing the capacitance values at high frequencies.

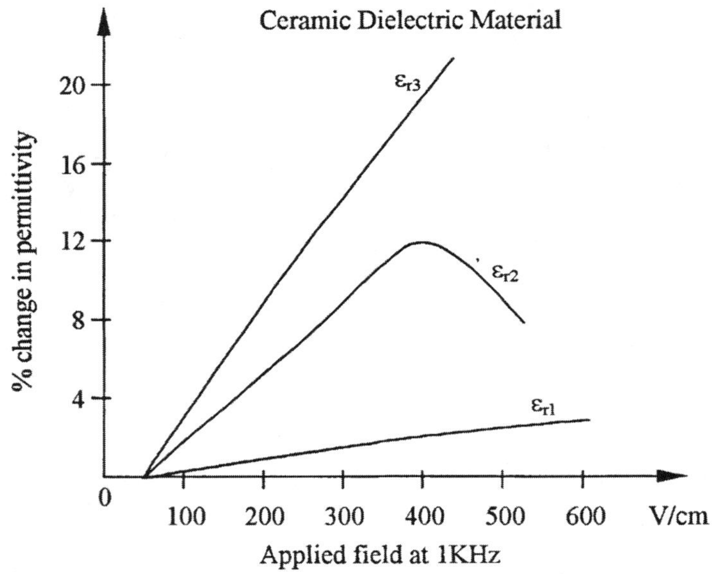

FIGURE 9.6 Changes in relative permittivity vs. field strength. The dielectric strength depends on the temperature, frequency, and applied voltage. Increases in the applied voltage cause higher changes in the dielectric strength. If the capacitor is subjected to high operating voltages, the electric field in the dielectric exceeds the breakdown value which can damage the dielectric permanently.

This is particularly true for capacitors having values less than 0.1 μF, having very thin dielectric materials between the electrodes.

Breakdown voltage: If the capacitor is subjected to high operating voltages, the electric field in the dielectric exceeds the breakdown value, which damages the dielectric permanently. The dielectric strength, which is the ability to withstand high voltages without changing properties, depends on the temperature, frequency, and applied voltage. An example of this dependence on the applied voltage is given in Figure 9.6. It is commonly known that the use of capacitors below their rated values increases the reliability and the expected lifetime. The standard voltage ratings of most capacitors are quoted by the manufacturers as 50, 100, 200, 400, and 600 V. Tantalum and electrolytic capacitors have ratings of 6, 10, 12, 15, 20, 25, 35, 50, 75, 100 V and higher.

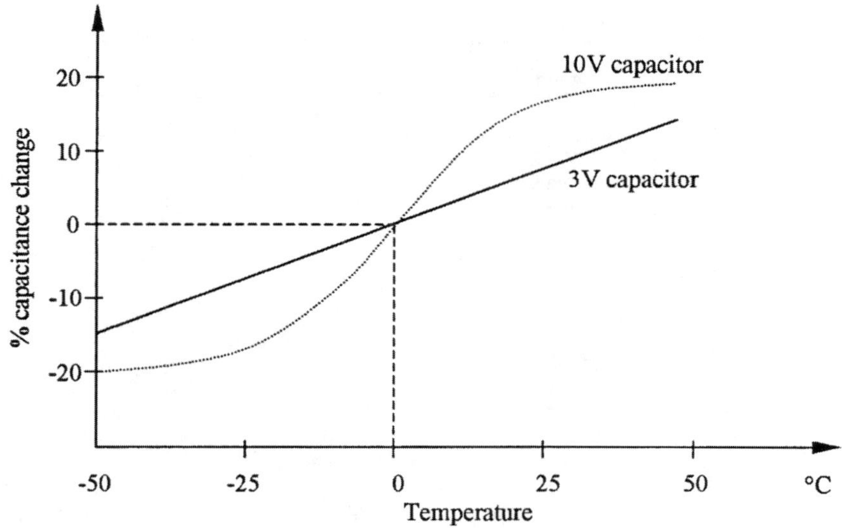

FIGURE 9.7 Temperature dependence of capacitors. The temperature characteristics of capacitors are largely dependent on the temperature properties of the dielectric materials used. The variations in capacitance due to temperature also depends on the type of capacitor and the operational voltage. The temperature coefficient of glass, teflon, mica, and polycarbonate are very small, and relatively high in ceramic capacitors.

Usually, values for surge voltages are given to indicate the ability of capacitors to withstand high transients. Typically, the surge voltages for electrolytic capacitors is 10% above the rated voltage, 50% for aluminum capacitors and about 250% for ceramic and mica capacitors.

The rated reverse voltages of electrolytic capacitors are limited to 1.5 V and, in some cases, to 15% of the rated forward voltages.

Temperature coefficient: The temperature characteristics of capacitors largely dependent on the temperature properties of the dielectric materials used, as given in Figure 9.7. The temperature coefficients of glass, teflon, mica, polycarbonate, etc. are very small, whereas in ceramic capacitors, they can be very high.

Insulation resistance: The insulation resistance of capacitors is important in many circuits. The insulation resistance is susceptible to temperature and humidity. For example, unsealed capacitors show large and rapid changes against temperature and humidity. For most capacitors, under high temperature conditions, the change in insulation resistance is an exponential function of temperature ($R_{T1} = R_{T2}e^{K(T1-T2)}$). The temperature dependence of insulation resistance of common capacitors is shown in Figure 9.8.

Frequency and impedance: Practical capacitors have increases in losses at very low and very high frequencies. At low frequencies, the circuit becomes entirely resistive and the dc leakage current becomes effective. At very high frequencies, the current passes through the capacitance and the dielectric losses become important. Approximate useable frequency ranges of capacitors are provided in Table 9.4.

An ideal capacitor should have an entirely negative reactance, but losses and inherent inductance prevents ideal operation. Depending on the construction, capacitors will resonate at certain frequencies due to unavoidable construction-based inductances. A typical impedance characteristic of a capacitor is depicted in Figure 9.9.

Power dissipation and quality factors: Ideally, a capacitor should store energy without dissipating any power. However, due to equivalent resistances, R_{eq}, some power will be dissipated. The power factor of a capacitor can be expressed as:

$$PF = \cos\theta = R_{eq} \Big/ |Z_{eq}| \tag{9.7}$$

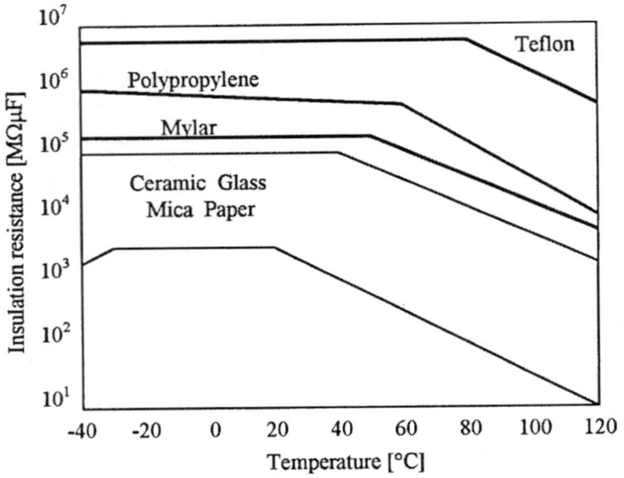

FIGURE 9.8 Temperature dependence of insulation resistance. The insulation resistance of many capacitors is not affected at low temperatures. However, under high temperature conditions, the change in insulation resistance can be approximated by an exponential relation. The insulation resistance is also susceptible to variations in humidity.

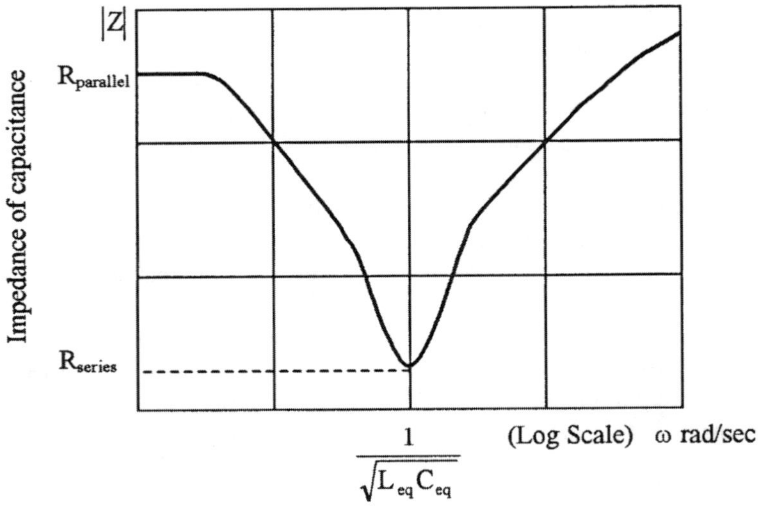

FIGURE 9.9 Frequency and impedance relation of capacitors. The losses and inherent inductance affects the ideal operation of capacitors and the capacitance impedance becomes a function of frequency. Depending on the construction, all capacitors will resonate at a certain frequency.

where θ = Phase angle
Z_{eq} = Equivalent total impedance

An important characteristic, the *dissipation factor* of capacitors, is expressed as:

$$DF = \tan\delta = R_{eq}/X_{eq} \tag{9.8}$$

where δ = Angle of loss
X_{eq} = Equivalent reactance

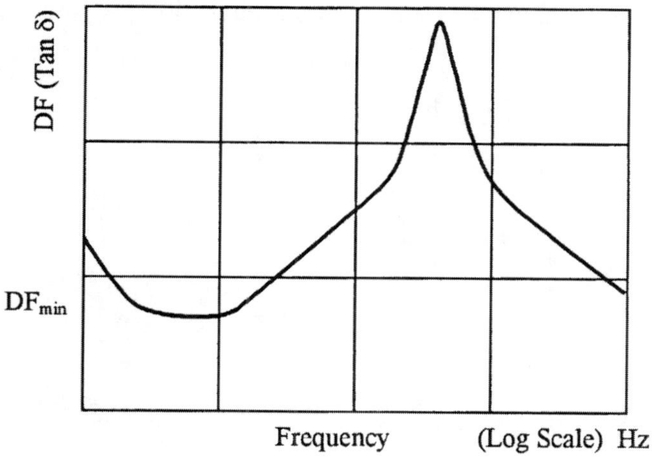

FIGURE 9.10 Power dissipation factors. In ideal operations, capacitors should store energy without dissipating power. Nevertheless, due to resistances, some power will be dissipated. The dissipation depends on frequency. The standard measurement of dissipation factor δ is determined by applying 1.0 Vrms at 1 kHz.

The dissipation factor depends on the frequency. Capacitors are designed such that this dependence is minimal. The measurement of dissipation factor δ is made at 1 kHz and 1.0 Vrms applied to the capacitor. A typical dissipation factor curve is depicted in Figure 9.10.

Selection of Capacitors and Capacitor Reliability

Capacitor Selection

Experience shows that a substantial part of component failures in electronic equipment is due to capacitors. The major cause of this can be attributed to the improper selection and inappropriate applications of capacitors. The following factors are therefore important criteria in the selection of capacitors in circuit applications: (1) the capacitance *values and tolerances* are determined by operating frequencies or by the value required for timing, energy storage, phase shifting, coupling, or decoupling; (2) the *voltages* are determined by the type and nature of the source, ac, dc, transient, surges, and ripples; (3) the *stability* is determined by operating conditions like temperature, humidity, shock, vibration, and life expectancy; (4) the *electrical properties* are determined by life expectancy, leakage current, dissipation factor, impedance, and self-resonant frequency; (5) the mechanical properties are determined by the types and construction, e.g., size, configuration, and packaging; and (6) the cost is determined by the types and physical dimensions of capacitors and the required tolerance.

Capacitor Reliability

Some of the common causes of capacitor failure are due to voltage and current overloads, high temperature and humidity, shock, vibration pressure, frequency effects, and aging. The voltage overload produces an excessive electric field in the dielectric that results in the breakdown and destruction of the dielectric. The current overload caused by rapid voltage variations results in current transients. If these currents are of sufficient amplitude and duration, the dielectric can be deformed or damaged, resulting in drastic changes in capacitance values, and thus leading to equipment malfunction. The high temperatures are mainly due to voltage and current overloads. The overheating and high temperatures accelerate the dielectric aging. This causes the plastic film to be brittle and also introduces cracks in the hermetic seals. The moisture and humidity due to severe operating environments cause corrosion, reduce the dielectric strength, and lower insulation resistances. The mechanical effects are mainly the pressure, variation, shock, and stress, which can cause mechanical damages of seals that result in electrical failures. Aging deteriorates the insulation resistance and affects the dielectric strength. The aging is usually determined by shelf-life; information about aging is supplied by the manufacturers.

TABLE 9.7 Standard Capacitors and Tolerances

Value (pF, nF, μF)\tolerance	5%	10%	20%
10	x	x	x
11	x		
12	x		
13	x		
15	x	x	x
16	x		
18	x	x	
20	x		
22	x	x	x
24	x		
27	x	x	
30	x		
33	x	x	x
36	x		
39	x		
43	x		
47	x	x	x
51	x		
56	x	x	x
62	x		
68	x	x	x
75	x		
82	x		
91	x		

Capacitor Standard Values and Tolerances

General-purpose capacitors values tend to be grouped close to each other in a bimodal distribution manner within their tolerance values. Usually, the tolerances of standard capacitors are 5%, 10%, and 20% of their values, as shown in Table 9.6. Nevertheless, tolerances of precision capacitors are much tighter — in the range of 0.25%, 0.5%, 0.625%, 1%, 2%, and 3% of the values. These capacitors are much more expensive than the standard range.

For capacitors in the small pF range, the tolerances can be given as ±1.5, ±1, ±0.5, ±0.25, and ±0.1 pF. Usually, low tolerance ranges are achieved by selecting manufactured items.

Standard capacitors are constructed from interleaved metal plates using air as the dielectric material. The area of the plates and distance between them are determined and constructed with precision. National Bureau of Standards maintains a bank of primary standard air capacitors that can be used to calibrate the secondary and working standards for laboratories and industry. Generally, smaller capacitance working standards are obtained from air capacitors, whereas larger working standards are made from solid dielectric materials. Usually, silver-mica capacitors are selected as working standards. These capacitors are very stable, have very low dissipation factors and small temperature coefficients, and have very little aging effect. They are available in decade mounting forms.

Capacitors as Circuit Components

The capacitor is used as a two-terminal element in electric circuits with the current–voltage relationship given by:

$$i(t) = C\mathrm{d}v(t)/\mathrm{d}t \tag{9.9}$$

where C is the capacitance.

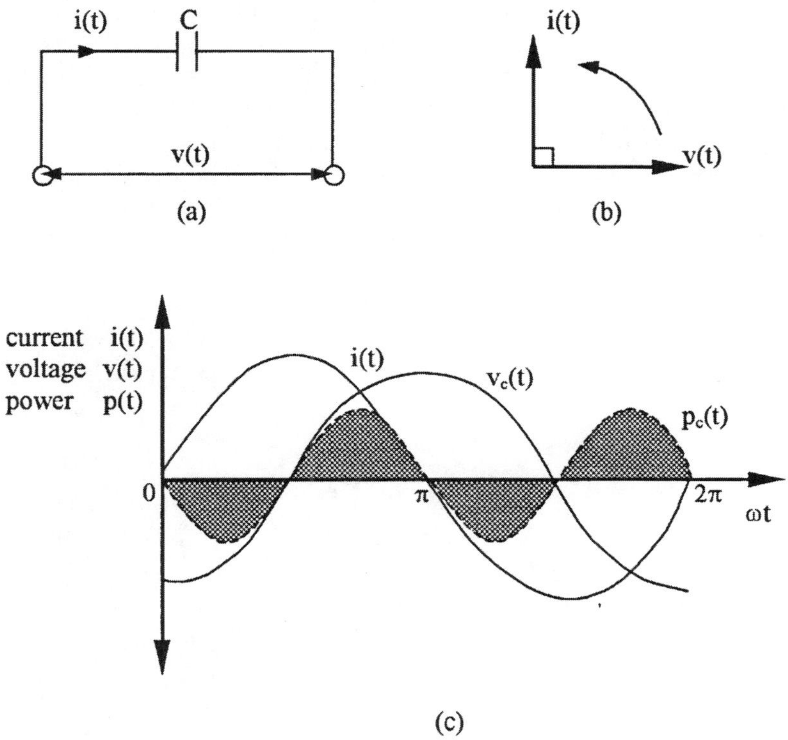

FIGURE 9.11 A capacitor as a two-terminal circuit element: (a) connection of a capacitor in electric circuits; (b) current–voltage relationship under sinusoidal operations; and (c) the power, voltage, and current relationships. The power has positive and negative values with twice the frequency of the applied voltage.

This element is represented by the circuit shape as shown in Figure 9.11(a).

From the $v(t)$, $i(t)$ relationship, the instantaneous power of this element can then be given by:

$$p(t) = v(t)i(t) \tag{9.10}$$

The stored energy in the capacitor at time t seconds can be calculated as:

$$\begin{aligned} w(t) &= \int Cv(t)\{dv(t)/dt\}dt \\ &= \frac{[Cv^2(t)]}{2} \end{aligned} \tag{9.11}$$

If the voltage across the capacitor is changing in time, the energy stored in the capacitor in the time interval t_1 to t_2 can be found using Equation 9.11 as:

$$W = (1/2)C[v^2(t_2) - v^2(t_1)] \tag{9.12}$$

Although the voltage assumed different values in time interval t_1 to t_2, if the initial and final voltages are equal (e.g., $v(t_1) = v(t_2)$), the net energy stored in the capacitor will be equal to zero. This implies that the energy stored in the capacitor is returned to the circuit during the time interval; that is, the capacitor transforms the energy without dissipation. The stored energy is associated with the electrostatic field of the capacitor and, in the absence of the electrostatic field, it will be zero.

From the voltage–current relationship in Equation 9.9, it can be seen that a capacitor as a circuit element is a passive component. That is, if the voltage is known, the current can be immediately determined by differentiation. Conversely, if the current is known, then the voltage can be determined by integration. If the voltage $v(t)$ is considered as an input variable, and the current $i(t)$ as an output variable, then the behavior of the current in a certain time interval is completely determined by the behavior of the voltage in this interval. In this case, the solution of the differential equation has the forced component only. Here, the particular solution of the voltage–current differential equation coincides with a full solution and the Laplace transform gives the relationship:

$$I(s) = sCV(s) \tag{9.13}$$

From this, the input–output relationship yields the impedance of the capacitor as:

$$Z(s) = \frac{1}{sC}$$

or

$$Y(s) = sC \tag{9.14}$$

In the stationary condition, $s \to 0$, and $Z \to \infty$, and $Y \to 0$.

In the sinusoidal condition, $s = j\omega = 2\pi f$, where f = frequency and, hence:

$$Z(j\omega) = \frac{1}{\{j\omega C\}} = \frac{-j}{\{\omega C\}} \tag{9.15}$$

and

$$Y(j\omega) = j\omega C \tag{9.16}$$

The capacitor can then be characterized under sinusoidal conditions, by a reactance of $X_C = 1/\omega C$ measured in ohms with the current leading the voltage by 90°, as shown in the phasor diagram in Figure 9.11(b).

In sinusoidal operations, the instantaneous power $p(t) = v(t)\, i(t)$ can be calculated as:

$$v(t) = V_{max} \cos\omega t = \sqrt{2} V \cos\omega t \tag{9.17}$$

Using the relationship given by Equation 9.9, the current can be written as:

$$i(t) = C\mathrm{d}v/\mathrm{d}t = -\omega C \sqrt{2}\, V \sin\omega t \tag{9.18}$$

giving:

$$p(t) = v(t)i(t) = -2\omega C V^2 \sin\omega t \cos\omega t = \frac{V^2 \sin 2\omega t}{X_C} \tag{9.19}$$

This indicates that the average power is zero because of the $\sin 2\omega t$ term, but there is a periodic storage and return of energy and the amplitude of that power is V^2/X_c. The power, voltage, and current relationship in a capacitor is given in Figure 9.11(c).

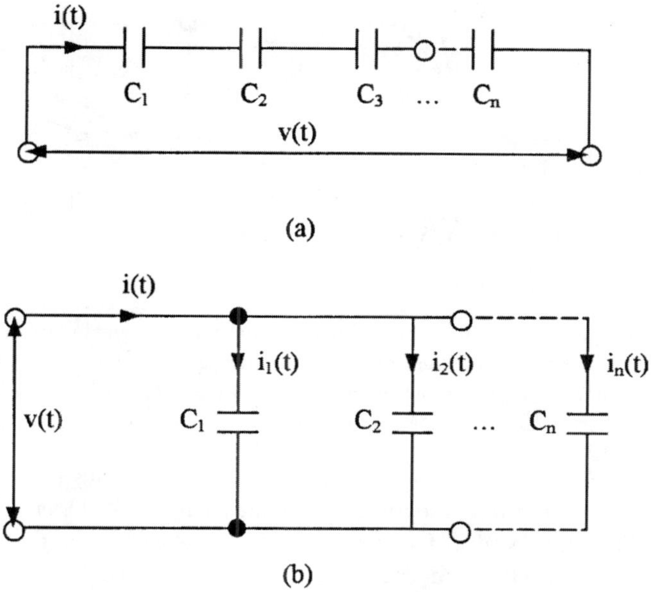

FIGURE 9.12 Series and parallel connection of capacitors: (a) series connection, and (b) parallel connection. In a series connection, the final capacitance value will always be smaller than the smallest value of the capacitor at the circuit element, whereas in parallel connection the final value is greater than the largest capacitance.

Series and Parallel Connection of Capacitors

The formulae for series and parallel connection of capacitors can be obtained from the general consideration of series and parallel connection of impedances as shown in Figures 9.12(a) and (b), respectively. For the series connection, the impedances are added such that:

$$\frac{1}{sC} = \frac{1}{sC_1} + \frac{1}{sC_2} + \cdots + \frac{1}{sC_n} \tag{9.20}$$

where C_1, C_2, ... C_n are the capacitance of the capacitors connected in series as in Figure 9.12(a). The equivalent capacitance is then given by:

$$C = \left\{ \frac{1}{C_1} + \frac{1}{C_2} + \cdots + \frac{1}{C_n} \right\}^{-1} \tag{9.21}$$

The final capacitance value will always be smaller than the smallest value.

In a similar way, the equivalent capacitance of parallel connected capacitors is

$$C = C_1 + C_2 + \ldots + C_n \tag{9.22}$$

and the final value of C is always larger than the largest capacitance in the circuit.

Distributed Capacitances in Circuits

Since capacitance is inherent whenever an electric potential exists between two conducting surfaces, its effect will be most noticeable in coils and in transmission lines at high frequencies. In the case of coils, there are small capacitances between adjacent turns, between turns that are not adjacent, between terminal leads, between turns and ground, etc. Each of the various capacitance associated with the coil stores a quantity of electrostatic energy that is determined by the capacitance involved and the fraction of the

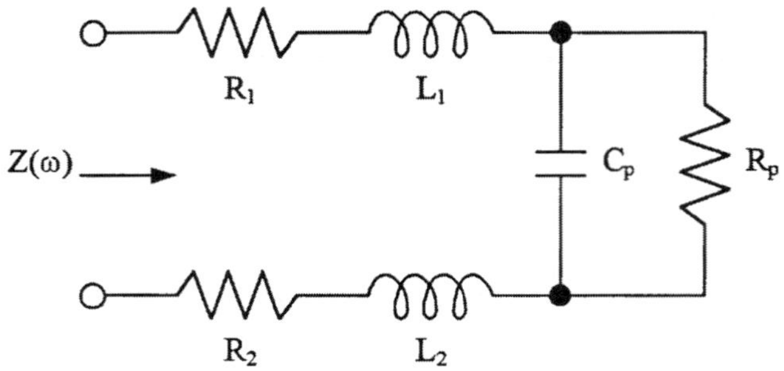

FIGURE 9.13 Capacitor equivalent circuit. A practical capacitor has resistance and inductances. Often, the electrical equivalent circuit of a capacitor can be simplified by a pure capacitance C_p and a parallel resistance R_p by neglecting resistances R_1, R_2 and inductances L_1, L_2. In low-leakage capacitors where R_p is high, the equivalent circuit can be represented by a series *RC* circuit.

total coil voltage that appears across it. The total effect is that the numerous small coil capacitances can be replaced by a single capacitor of appropriate size shunted across the coil terminals. This equivalent capacitance is called either the *distributed capacitance* or the *self-capacitance* of the coil, and it causes the coil to show parallel resonance effects under some conditions. In the case of a mismatched or unterminated transmission line, the distributed capacitance, together with the inductive effect, will create a phase difference between the voltage and current in the line. This phase difference depends on the type of termination and the electrical length of the line and, as a result, the input impedance of the line can effectively be an equivalent capacitor when its electrical length is less than a quarter wavelength for an open-circuit termination, or between a quarter wavelength and half a wavelength for a short-circuit termination.

Capacitor Equivalent Circuits

The electric equivalent circuit of a capacitor consists of a pure capacitance (C_p), plate inductances (L_1, L_2), plate resistances (R_1, R_2), and a parallel resistance R_p that represents the resistance of the dielectric or leakage resistance, as shown in Figure 9.13. The capacitors that have high leakage currents flowing through the dielectric have relatively low R_p values. Very low leakage currents are represented by extremely large R_p values. Examples of these two extremes are electrolytic capacitors that have high leakage current (low R_p), and plastic film capacitors, which have very low leakage current (high R_p). Typically, an electrolytic capacitor might easily have several microamperes of leakage current ($R_p < 1$ MΩ), while a plastic film capacitor could have a resistance greater than 100,000 MΩ.

It is usual to represent a low leakage capacitor (high R_p) by a series *RC* circuit, while those with high leakage (low R_p) are represented by parallel *RC* circuits. However, when a capacitor is measured in terms of the series *C* and *R* quantities, it is desirable to resolve them into parallel equivalent circuit quantities. This is because the (parallel) leakage resistance best represents the quality of the capacitor dielectric.

Capacitive Bridges and Measurement of Capacitance

Bridges are used to make precise measurements of unknown capacitances and associated losses in terms of some known external capacitances and resistances. Most commonly used bridges are: series-resistance-capacitance bridge, parallel-resistance-capacitance bridge, Wien bridge, and Schering bridge.

Series-Resistance-Capacitance Bridge

Figure 9.14 is a series-resistance-capacitance (*RC*) bridge, which is used for the comparison of a known capacitance with an unknown capacitance. The unknown capacitance is represented by C_x and R_x. A standard adjustable resistance R_1 is connected in series with a standard capacitor C_1. The voltage drop across R_1 balances the resistive voltage drop when the bridge is balanced. The additional resistor in series with C_x increases the total resistive component, so that small values of R_1 will not be required to achieve

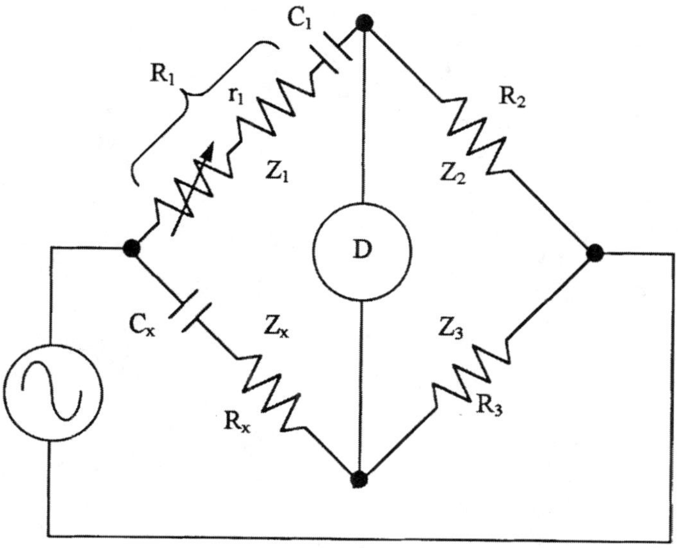

FIGURE 9.14 A series *RC* bridge. In these bridges, the unknown capacitance is compared with a known capacitance. The voltage drop across R_1 balances the resistive voltage drop in branch Z_2 when the bridge is balanced. The bridge balance is most easily achieved when capacitive branches have substantial resistive components. The resistors R_1 and either R_3 or R_4 are adjusted alternately to obtain the balance. This type of bridge is found to be most suitable for capacitors with a high-resistance dielectric, hence very low leakage currents.

balance. Generally, the bridge balance is most easily achieved when capacitive branches have substantial resistive components. To obtain balance, R_1 and either R_3 or R_4 are adjusted alternately. This type of bridge is found to be most suitable for capacitors with a high-resistance dielectric and hence very low leakage currents.

At balance:

$$Z_1 Z_3 = Z_2 Z_x \tag{9.23}$$

Substituting impedance values gives:

$$\left(R_1 - \frac{j}{\omega C_1}\right) R_3 = \left(R_x - \frac{j}{\omega C_2}\right) R_2 \tag{9.24}$$

Equating the *real* terms gives:

$$R_x = \frac{R_1 R_3}{R_2} \tag{9.25}$$

and equating *imaginary* terms gives:

$$C_x = \frac{C_1 R_2}{R_3} \tag{9.26}$$

An improved version of the series *RC* bridge is the *substitution bridge*, which is particularly useful to determine the values of capacitances at radio frequencies. In this case, a series-connected *RC* bridge is balanced by disconnecting the unknown capacitance C_x and resistance R_x, and replacing it by an adjustable standard capacitor C_s and adjustable resistor R_s. After having obtained the balance position, the unknown capacitance and resistance C_x and R_x are connected in parallel to the capacitor C_s. The capacitor C_s and

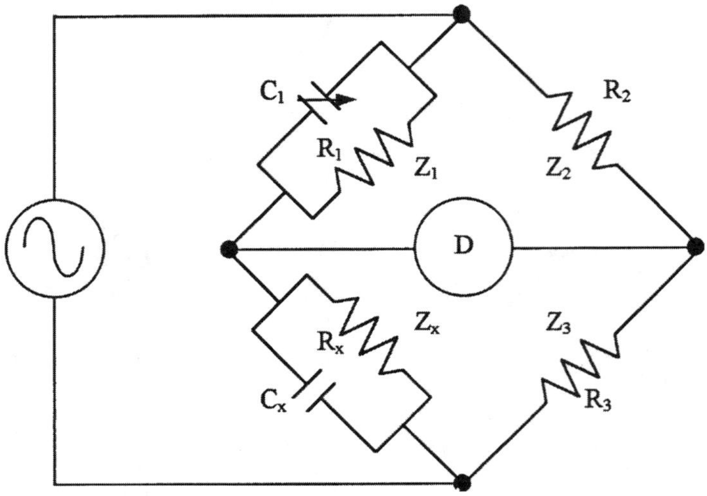

FIGURE 9.15 A parallel-resistance-capacitance bridge. The unknown capacitance is represented by its parallel equivalent circuit; C_x in parallel with R_x. The bridge balance is achieved by adjustment of R_1 and either R_3 or R_4. The parallel-resistance-capacitance bridge is found to be most suitable for capacitors with a low-resistance dielectric, hence relatively high leakage currents.

resistor R_s are adjusted again for the re-balance of the bridge. The changes in the ΔC_s and ΔR_s lead to unknown values as:

$$C_x = \Delta C_s \quad \text{and} \quad R_x = \Delta R_s (C_{s1}/C_x)^2 \tag{9.27}$$

where C_{s1} is the value of C_s in the initial balance condition.

The Parallel-Resistance-Capacitance Bridge

Figure 9.15 illustrates a parallel-resistance-capacitance bridge. In this case, the unknown capacitance is represented by its parallel equivalent circuit C_x in parallel with R_x. The Z_2 and Z_3 impedances are pure resistors with either or both being adjustable. The Z_1 is balanced by a standard capacitor C_1 in parallel with an adjustable resistor R_1. The bridge balance is achieved by adjustment of R_1 and either R_2 or R_3. The parallel-resistance-capacitance bridge is found to be most suitable for capacitors with a low-resistance dielectric, hence relatively high leakage currents. At balance:

$$\frac{1}{\left(\frac{1}{R_1} + j\omega C_1\right)} R_3 = \frac{1}{\left(\frac{1}{R_x} + j\omega C_x\right)} R_2 \tag{9.28}$$

Equating *real* terms gives:

$$R_x = \frac{R_3 R_1}{R_2} \tag{9.29}$$

and equating *imaginary* terms gives:

$$C_x = \frac{C_1 R_2}{R_3} \tag{9.30}$$

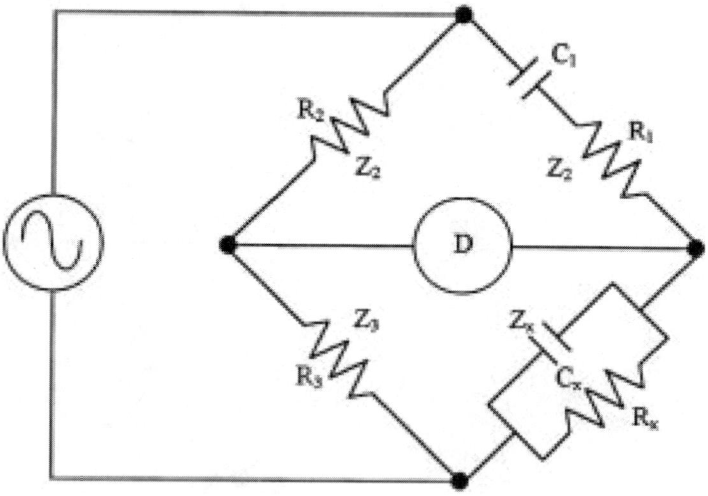

FIGURE 9.16 The Wien bridge. This bridge is used to compare two capacitors directly. It finds applications particularly in determining the frequency in *RC* oscillators. In some cases, capacitors C_1 and C_x are made equal and ganged together so that the frequency at which the null occurs varies linearly with capacitance.

The Wien Bridge

Figure 9.16 shows a Wien bridge. This is a special resistance-ratio bridge that permits two capacitances to be compared once all the resistances of the bridge are known. At balance, it can be proven that the unknown resistance and the capacitance are:

$$R_x = \frac{R_3\left(1+\omega^2 R_1^2 C_1^2\right)}{\omega^2 R_1 R_2 C_1^2} \tag{9.31}$$

and

$$C_x = \frac{C_1 R_2}{\left[R_3\left(1+\omega^2 R_1^2 C_1^2\right)\right]} \tag{9.32}$$

It can also be shown that:

$$\omega^2 = \frac{1}{R_1 C_1 R_x C_x} \tag{9.33}$$

As indicated in Equation 9.33, the Wien bridge has an important application in determining the frequency in *RC* oscillators. In frequency meters, C_1 and C_x are made equal and the two capacitors are ganged together so that the frequency at which the null occurs varies linearly with capacitances.

The Schering Bridge

Figure 9.17 illustrates the configuration of the Schering bridge. This bridge is used for measuring the capacitance, the dissipation factors, and the loss angles. The unknown capacitance is directly proportional to the known capacitance C_3. That is, the bridge equations are:

$$C_x = \frac{C_3 R_2}{R_1} \quad \text{and} \quad R_x = \frac{C_2 R_1}{C_3} \tag{9.34}$$

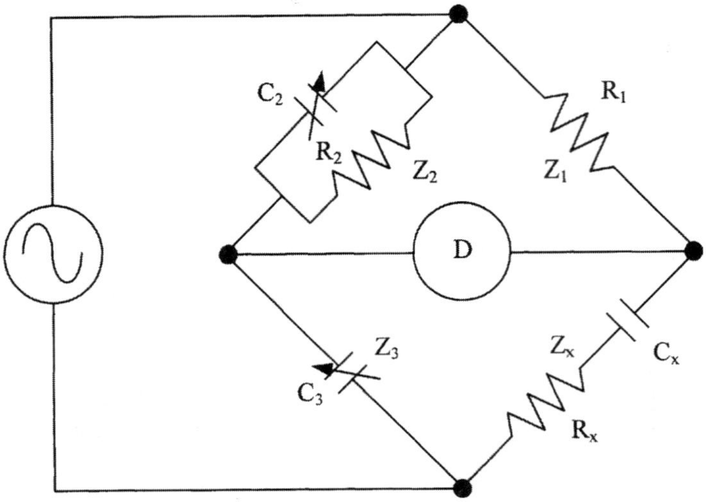

FIGURE 9.17 The Schering bridge. This bridge is particularly useful for measuring the capacitance, associated dissipation factors, and the loss angles. The unknown capacitance is directly proportional to the known capacitance C_1. The Schering bridge is frequently used as a high-voltage bridge with a high-voltage capacitor as C_1.

Usually, R_2 and R_3 are fixed, and C_2 and C_3 are made variable. Schering bridges are frequently used in highvoltage applications with high-voltage capacitor C_3. They are also used as high-frequency bridges since the use of two variable adjustment capacitors are convenient for precise balancing.

Further Information

J.C. Whitaker, *The Electronics Handbook,* Boca Raton, FL: CRC Press & IEEE Press, 1996.

D.G. Fink and D. Christiansen, *Electronics Engineers' Handbook,* New York: McGraw-Hill, 1982.

L.J. Giacoletto, *Electronics Designers' Handbook,* 2nd ed., New York: McGraw-Hill, 1977.

F.F. Mazda, *Electronics Engineers' Reference Book,* 5th ed., London: Butterworth, 1983.

10

Permittivity Measurement

Devendra K. Misra
University of Wisconsin

Dielectric materials possess relatively few free charge carriers. Most of the charge carriers are bound and cannot participate in conduction. However, these bound charges can be displaced by applying an external electric field. In such cases, the atom or molecule forms an electric dipole that maintains an electric field. Consequently, each volume element of the material behaves as an electric dipole. The dipole field tends to oppose the applied field. Dielectric materials that exhibit nonzero distribution of such bound charge separations are said to be *polarized*. The volume density of those electric dipoles is described by the volume density of polarization, $\vec{P}$. When a material is linear and isotropic in nature, the polarization density is related to applied electric field intensity, $\vec{E}$, as follows:

$$\vec{P} = \varepsilon_0 \chi_e \vec{E} \qquad (10.1)$$

where ε_0 (= 8.854×10^{-12} farad per meter) is the permittivity of free-space and χ_e is called the electric susceptibility of the material.

The electric flux density or displacement, $\vec{D}$, is defined as follows:

$$\vec{D} = \varepsilon_0 \vec{E} + \vec{P} = \varepsilon_0 \left(1 + \chi_0\right) \vec{E} = \varepsilon_0 \varepsilon_r \vec{E} = \varepsilon \vec{E} \qquad (10.2)$$

where ε is called the permittivity of the material and ε_r is its relative permittivity or dielectric constant. Electric flux density is expressed in Coulombs per square meter.

Equation 10.2 represents a relation between the electric flux density and the electric field intensity in the frequency domain. It will hold in time-domain only if the permittivity is independent of frequency. A material is called *dispersive* if its characteristics are frequency dependent. The product relation of Equation 10.2 in frequency domain will be replaced by a convolution integral for the time-domain fields.

Assuming that the fields are time-harmonic as $e^{j\omega t}$, the generalized Ampere's law can be expressed in phasor form as follows:

$$\nabla \times \vec{H} = \vec{J}^{e} + \vec{J} + j\omega \vec{D} \qquad (10.3)$$

0-8493-1733-9/04/$0.00+$1.50

where H is the magnetic field intensity in Amperes per meter and J^e is the current-source density in Amperes per square meter. J is the conduction current density in Amperes per square meter, and the last term represents the displacement current density. J^e will be zero for a source-free region.

The conduction current density is related to the electric field intensity through Ohm's law as follows.

$$\vec{J} = \sigma \vec{E} \tag{10.4}$$

where σ is the conductivity of material in siemens per meter.

From Equations 10.2 to 10.4, we have,

$$\nabla \times \vec{H} = \vec{J}^{e} + \sigma \vec{E} + j\omega\varepsilon \vec{E} \tag{10.5}$$

Conduction current represents the loss of power. There is another source of loss in dielectric materials. When a time-harmonic electric field is applied, the dipoles flip constantly back and forth. Because the charge carriers have finite mass, the field must do work to move them and they may not respond instantaneously. This means that the polarization vector will lag behind the applied electric field. This factor shows up at high frequencies. Therefore, Equation 10.5 is modified as follows:

$$\nabla \times \vec{H} = \vec{J}^{e} + \sigma \vec{E} + \omega\kappa'' \vec{E} + j\omega\varepsilon \vec{E} = \vec{J}^{e} + j\omega\left(\varepsilon - j\frac{\sigma + \omega\kappa''}{\omega}\right)\vec{E} = \vec{J}^{e} + j\omega\varepsilon^{*} \vec{E} \tag{10.6}$$

Complex relative permittivity of a material is defined as follows:

$$\varepsilon_r^{*} = \frac{\varepsilon^{*}}{\varepsilon_0} = \frac{1}{\varepsilon_0}\left(\varepsilon - j\frac{\sigma + \omega\kappa''}{\omega}\right) = \varepsilon_r' - j\varepsilon_r'' = \varepsilon_r\left(1 - j\tan\delta\right) \tag{10.7}$$

where ε_r' and ε_r'' represent real and imaginary parts of the complex relative permittivity. The imaginary part is zero for a lossless material. The term $\tan\delta$ is called the *loss-tangent*. It represents the tangent of the angle between the displacement phasor and total current, as shown in Figure 10.1. Thus, it will be close to zero for a low-loss material. Dielectric characteristics of selected substances are given in Tables 10.1 and 10.2.

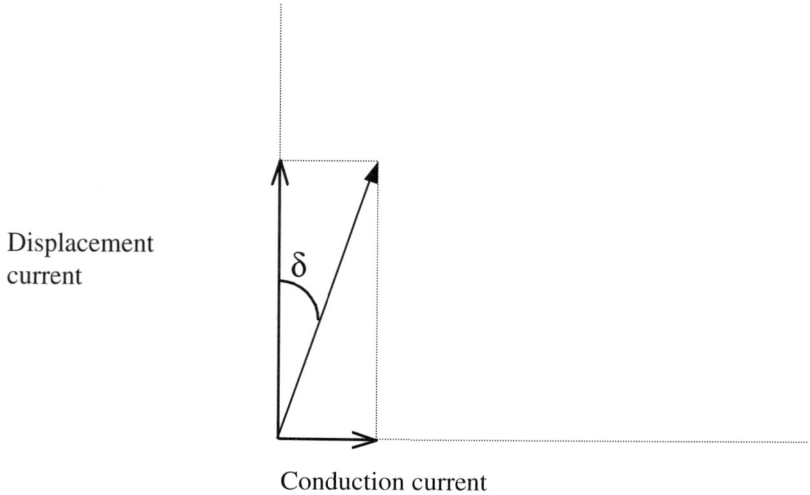

FIGURE 10.1 A phasor diagram representing displacement and loss currents.

TABLE 10.1 Complex Permittivity of Some Substances at Room Temperature

Substance	60 Hz	1 MHz	10 GHz
Nylon	3.60–j 0.06	3.14–j 0.07	2.80–j 0.03
Plexiglass	3.45–j 0.22	2.76–j 0.04	2.5–j 0.02
Polyethylene	2.26–j 0.0005	2.26–j 0.0005	2.26–j 0.0011
Polystyrene	2.55–j 0.0077	2.55–j 0.0077	2.54–j 0.0008
Styrofoam	1.03–j 0.0002	1.03–j 0.0002	1.03–j 0.0001
Teflon	2.1–j 0.01	2.1–j 0.01	2.1–j 0.0008
Glass (lead barium)	6.78–j 0.11	6.73–j 0.06	6.64–j 0.31

TABLE 10.2 Dielectric Properties of Biological Tissues at Selected Frequencies

	10 kHz		100 kHz		10 MHz		100 MHz		1 GHz	
	ε_r	σ' (S/m)	ε_r	σ' (S/m)	ε_r	σ' (S/m)	ε_r	σ' (S/m)	ε_r	σ' (S/m)
Brain (gray matter)	15×10^3	0.1	3000	0.14	300	0.3	90	0.7	60	1.2
Heart muscle	6×10^4	0.15	12×10^3	0.2	350	0.5	80	0.9	60	1.2
Kidney (cortex)	3×10^4	0.14	7000	0.2	350	0.6	85	1	60	1.5
Liver	2×10^4	0.05	7500	0.09	200	0.35	65	0.5	50	0.9
Lung (inflated)	9×10^3	0.07	2000	0.09	130	0.2	30	0.3	25	0.4
Spleen	12×10^3	0.1	3500	0.12	450	0.4	75	0.85	55	1.2
Uterus	2×10^4	0.5	2500	0.5	300	0.65	80	1	60	1.5
Skin	2×10^4	0.005	10^4	0.08	150	0.4	60	0.5	55	0.9

Source: Adapted from [9], $\sigma' = \sigma + \omega\kappa''$.

TABLE 10.3 Dielectric Dispersion Parameters for Some Liquids at Room Temperature

Substance	e_∞	ε_s	α	τ (picoseconds)
Water	5	78	0	8.0789
Methanol	5.7	33.1	0	53.0516
Ethanol	4.2	24	0	127.8545
Acetone	1.9	21.2	0	3.3423
Ethylene glycol	3	37	0.23	79.5775
Propanol	3.2	19	0	291.7841
Butanol	2.95	17.1	0.08	477.4648
Chlorobenzene	2.35	5.63	0.04	10.2920

Dispersion characteristics of a large class of materials can be represented by the following empirical equation of Cole-Cole.

$$\varepsilon_r^* = \varepsilon_\infty + \frac{\varepsilon_s - \varepsilon_\infty}{1 + (j\omega\tau)^{1-\alpha}} \tag{10.8}$$

where ε_∞ and ε_s are the relative permittivities of material at infinite and zero frequencies, respectively. ω is the signal frequency in radians per second and τ is the characteristic relaxation time in seconds. For α equal to zero, Equation 10.8 reduces to the Debye equation. Dispersion parameters for a few liquids are given in Table 10.3.

Complex permittivity of a material is determined using lumped circuits at low frequencies, and distributed circuits or free-space reflection and transmission of waves at high frequencies. Capacitance and dissipation factor of a lumped capacitor are measured using a bridge or a resonant circuit. The complex permittivity is calculated from this data. At high frequencies, the sample is placed inside a transmission line or a resonant cavity. Propagation constants of the transmission line or resonant frequency and the quality factor of the cavity resonator are used to calculate the complex permittivity. Propagation characteristics of electromagnetic waves are influenced by the complex permittivity of that medium. Therefore, a material can be characterized by monitoring the reflected and transmitted wave characteristics as well.

10.1 Measurement of Complex Permittivity at Low Frequencies [1,2]

A parallel-plate capacitor is used to determine the complex permittivity of dielectric sheets. For a separation d between the plates of area A in vacuum, the capacitance is given by

$$C_0 = 8.854\frac{A}{d}\ \text{pF} \tag{10.9}$$

where all dimensions are measured in meters. If the two plates have different areas, then the smaller one is used to determine C_0. Further, it is assumed that the field distribution is uniform and perpendicular to the plates. Obviously, the fringing fields along the edges do not satisfy this condition. As shown in Figure 10.2, the guard electrodes are used to ensure that the field distribution is close to the assumed condition. For best results, the width of the guard electrode must be at least $2d$ and the unguarded plate must extend to the outer edge of the guard electrode. Further, the gap between the guarded and guard electrodes must be as small as possible.

The radius of guarded electrode is r_1 and the inner radius of guard electrode is r_2. It is assumed that $R - r_2 \geq 2d$. The area A for this parallel plate capacitor is πr^2, where r is defined as follows:

$$r = r_1 + \Delta \tag{10.10}$$

$$\Delta = \frac{1}{2}(r_2 - r_1) - \frac{2d}{\pi}\ln\left(\cosh\frac{\pi(r_2 - r_1)}{4d}\right) = \frac{1}{2}(r_2 - r_1) - 1.4659d\ln\left(\cosh 0.7854\frac{r_2 - r_1}{d}\right) \tag{10.11}$$

Using the Debye model (i.e., $\alpha = 0$ in Equation 10.8), an equivalent circuit for a dielectric-filled parallel plate capacitor can be drawn as shown in Figure 10.3. If a step voltage V is applied to it, then the current I can be found as follows [2].

$$I = \varepsilon_\infty C_0 V\delta(t) + \frac{VC_0(\varepsilon_s - \varepsilon_\infty)}{\tau}\exp\left(-\frac{t}{\tau}\right) \tag{10.12}$$

where $\tau = RC_0(\varepsilon_s - \varepsilon_\infty)$

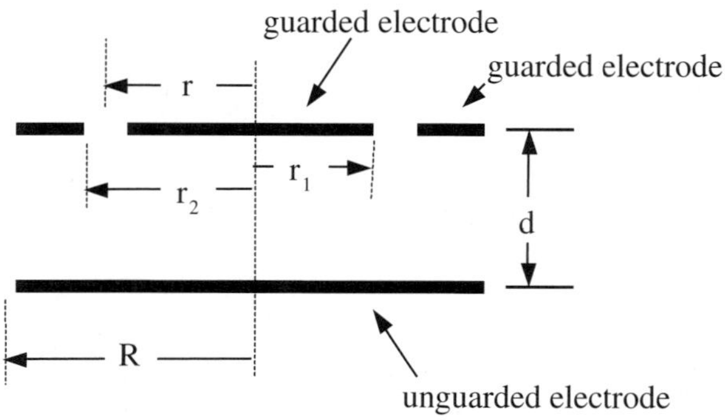

FIGURE 10.2 Geometry of a guarded capacitor.

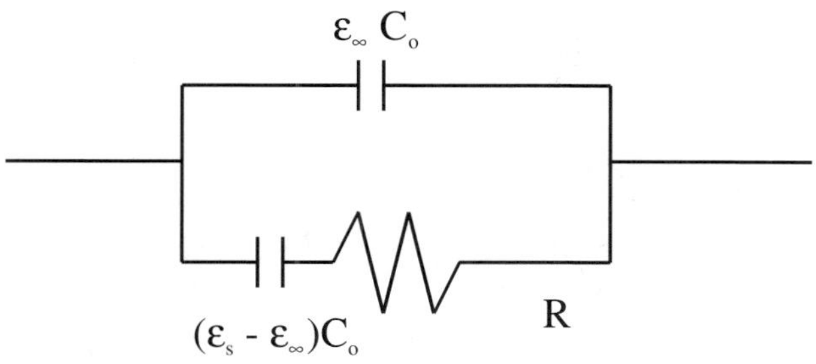

FIGURE 10.3 Equivalent circuit of a parallel-plate capacitor based on the Debye's model.

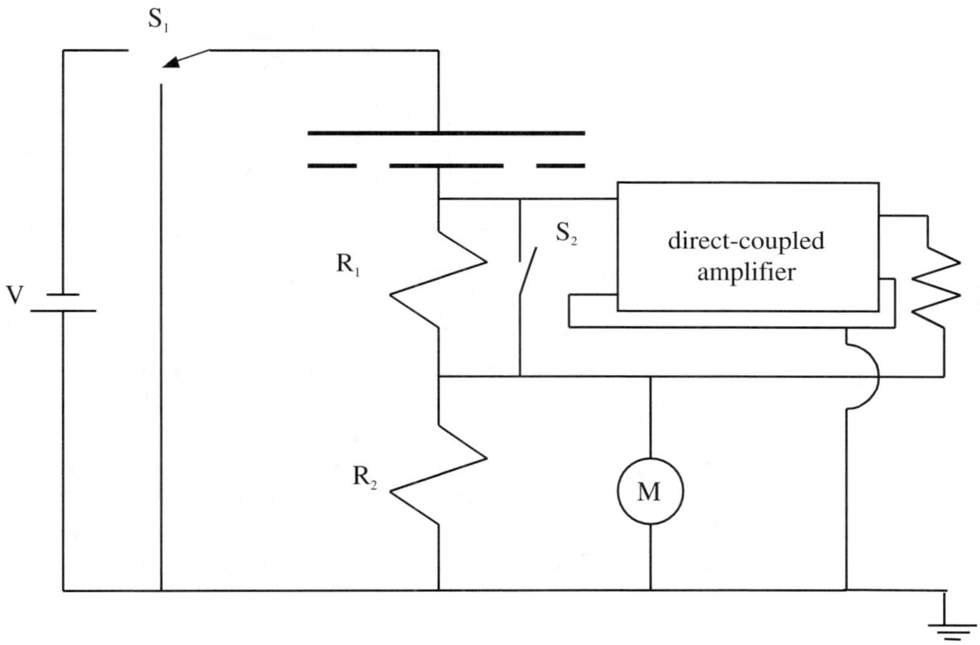

FIGURE 10.4 Circuit arrangement for the characterization of dielectric materials using a step voltage.

The first term in Equation 10.12 represents the charging current of capacitor $\varepsilon_\infty C_0$ in the upper branch. This current is not measured because it disappears instantaneously. In practice, it needs to be bypassed briefly to protect the detector from overloading or burning. The second term of Equation 10.12 represents charging current of the lower branch of an equivalent circuit. The time constant, τ, is determined following the decay characteristics of this current. Further, the resistance R can be found after extrapolating this current-time curve to $t = 0$. The discharging current characteristics are used to remove V at $t = 0$.

A typical circuit arrangement for the characterization of dielectric materials using a step voltage is shown in Figure 10.4. A standard resistor R_1 of either 10^{10} or 10^{12} Ω is connected between the guarded electrode and the load resistor R_2. A feedback circuit is used that forces the voltage drop across R_1 to be equal in magnitude but opposite in polarity to that of across R_2. It works as follows. Suppose that the node between capacitor and R_1 has a voltage V_1 with respect to the ground. It is amplified but reversed in polarity by the amplifier. Therefore, the current through R_1 will change. This process continues until the input to the amplifier is zero. The junction between R_1 and the capacitor will then be at the ground potential. Thus, the meter M measures voltage across R_2 that is negative of the voltage across R_1. Since R_1 is known, the current through it can be calculated. This current also flows through the sample. S_1 is used to switch from the charging to discharging mode while S_2 is used to provide a path for surge currents.

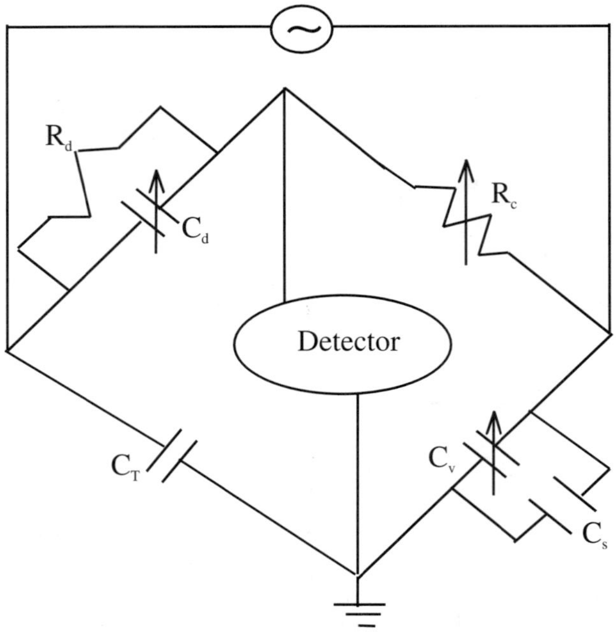

FIGURE 10.5 Schering bridge.

Capacitance and the dissipation factor of the dielectric-loaded parallel-plate capacitor are used in the medium frequency range to determine the complex permittivity of materials. A substitution method is generally employed in the Schering bridge circuit for this measurement.

In the Schering bridge shown in Figure 10.5, assume that the capacitor C_v is disconnected for the time being, and the capacitor C_s contains the dielectric sample. In case of a lossy dielectric sample, it can be modeled as an ideal capacitor C_x in series with a resistor R_x. The bridge is balanced by adjusting C_d and R_c. An analysis of this circuit under the balanced condition produces the following relations.

$$R_x = \frac{C_d R_c}{C_T} \tag{10.13}$$

and,

$$C_x = \frac{C_T R_d}{R_c} \tag{10.14}$$

Quality factor Q of a series RC circuit is defined as the tangent of its phase angle while the inverse of Q is known as the dissipation factor D. Hence,

$$Q = \frac{X_x}{R_x} = \frac{1}{\omega C_x R_x} = \frac{1}{D} \tag{10.15}$$

For a fixed R_d, the capacitor C_d can be calibrated directly in terms of the dissipation factor. Similarly, the resistor R_c can be used to determine C_x. However, an adjustable resistor limits the frequency range. A substitution method is preferred for precision measurement of C_x at higher frequencies. In this technique, a calibrated precision capacitor C_v is connected in parallel with C_s as shown in Figure 10.5 and the bridge is balanced. Assume that the settings of two capacitors at this condition are C_{d1} and C_{v1}. The capacitor C_s is then removed and the bridge is balanced again. Let the new settings of these capacitors

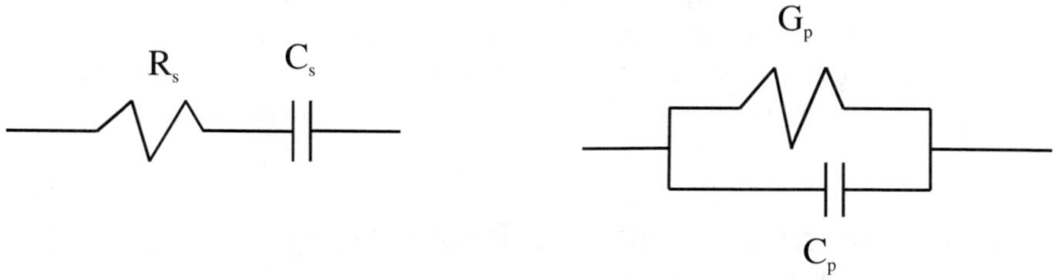

FIGURE 10.6 Series and parallel equivalent circuits of a dielectric-loaded capacitor.

be C_{d2} and C_{v2}, respectively. Equivalent circuit parameters of the dielectric-loaded capacitor C_s are then found as follows.

$$C_x = C_{v2} - C_{v1} \tag{10.16}$$

$$D_x = \frac{C_{v2}}{C_x}\delta D \tag{10.17}$$

where $\delta D = \omega R_d (C_{d1} - C_{d2})$.

Complex permittivity of the specimen is calculated from this data as follows:

$$\varepsilon_r' = \frac{C_x}{C_0} \tag{10.18}$$

and,

$$\varepsilon_r'' = \frac{C_x D_x}{C_0} \tag{10.19}$$

So far, a series *RC* circuit equivalent model is used for the dielectric-loaded capacitor. As illustrated in Figure 10.6, an equivalent parallel *RC* model can also be obtained for it. The following equations can be used to switch back and forth between these two equivalent models.

$$G_p = \frac{R_s}{R_s^2 + \dfrac{1}{\omega^2 C_s^2}} = \frac{1}{R_s}\left(\frac{1}{1+Q^2}\right) \tag{10.20}$$

$$C_p = \frac{C_s}{1+(\omega R_s C_s)^2} = \frac{C_s}{1+D^2} \tag{10.21}$$

$$R_s = \frac{G_p}{G_p^2 + \omega^2 C_p^2} = \frac{1}{G_p}\left(\frac{1}{1+Q^2}\right) \tag{10.22}$$

$$C_s = \frac{G_p^2 + \omega^2 C_p^2}{\omega^2 C_p} = C_p(1+D^2) \tag{10.23}$$

and,

$$Q = \frac{1}{D} = \frac{\omega C_p}{G_p} = \frac{1}{\omega R_s C_s} \tag{10.24}$$

Proper shielding and grounding arrangements are needed for a reliable measurement, especially at higher frequencies. Grounding and edge capacitances of the sample holder need to be taken into account for improved accuracy. Further, a guard point needs to be obtained that may require balancing in some cases. An inductive-ratio-arm capacitance bridge may be another alternative to consider for such applications [1].

10.2 Measurement of Complex Permittivity Using Distributed Circuits

Measurement techniques based on the lumped circuits are limited up to the lower end of VHF band. The characterization of materials at microwave frequencies requires the distributed circuits. A number of techniques have been developed on the basis of wave reflection and transmission characteristics inside a transmission line or in free-space. Some other methods employ a resonant cavity that is loaded with the sample. Cavity parameters are measured and the material characteristics are deduced from that. A number of these techniques are described in [3,4] that can be used for a sheet material. These techniques require cutting a piece of sample to be placed inside a transmission line or a cavity. In case of liquid or powder samples, a so-called modified infinite sample method can be used. In this technique, a waveguide termination is filled completely with the sample as shown in Figure 10.7. Since a tapered termination is embedded in the sample, the wave incident on it dissipates with negligible reflection and it looks like the sample is extending to infinity. The impedance at its input port depends on the electrical properties of filling sample. Its VSWR S and location of first minimum d from the load plane are measured using a slotted line. The complex permittivity of the sample is then calculated as follows [5].

$$\varepsilon_r' = \left(\frac{\lambda}{\lambda_c}\right)^2 + \frac{\left[1-\left(\lambda/\lambda_c\right)^2\right]\times\left[S^2\sec^4(\beta d)-\left(1-S^2\right)^2\tan^2(\beta d)\right]}{\left[1+S^2\tan^2(\beta d)\right]^2} \tag{10.25}$$

and,

$$\varepsilon_r'' = \frac{\left[1-\left(\lambda/\lambda_c\right)^2\right]\times\left[2S\left(1-S^2\right)^2\sec^4(\beta d)\tan(\beta d)\right]}{\left[1+S^2\tan^2(\beta d)\right]^2} \tag{10.26}$$

where λ is the free-space wavelength; λ_c is the cut-off wavelength for the mode of propagation in the empty guide, and β is the propagation constant in the feeding guide. It is assumed that the waveguide supports the TE_{10} mode only.

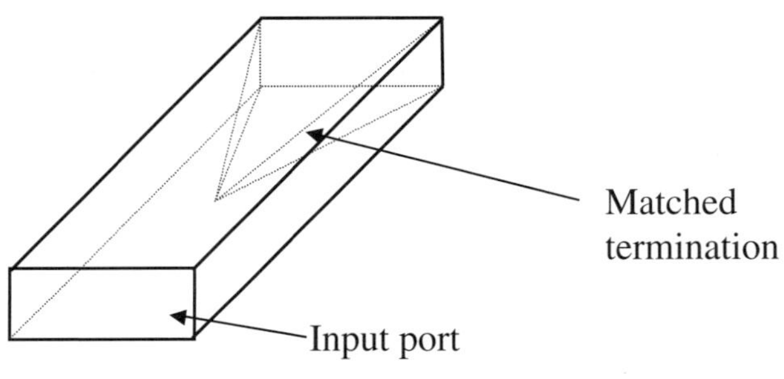

FIGURE 10.7 A waveguide termination filled with liquid or powder sample.

Resonant Cavity Method

A cavity resonator can be used to determine the complex permittivity of materials at microwave frequencies. If a cavity can be filled completely with the sample, then the following procedure can be used.

Measure the resonant frequency f_1 and the quality factor Q_1 of an empty cavity. Next, fill that cavity with the sample material and measure its new resonant frequency f_2 and quality factor Q_2. The dielectric parameters of the sample are then calculated from the following formulas [3].

$$\varepsilon_r = \left(1 + \frac{f_1 - f_2}{f_2}\right)^2 \tag{10.27}$$

and,

$$\tan\delta = \frac{1}{Q_2} - \frac{1}{Q_1}\sqrt{\frac{f_1}{f_2}} \tag{10.28}$$

On the other hand, a cavity perturbation technique will be useful for smaller samples [4]. If the sample is available in a circular cylindrical form, then it may be placed inside a TE_{101} rectangular cavity through the center of its broad face where the electric field is maximum. Its resonant frequency and quality factor with and without sample are then measured. The complex permittivity of the sample is calculated as follows.

$$\varepsilon_r' = 1 + \frac{1}{2}\frac{f_1 - f_2}{f_2}\frac{V}{v} \tag{10.29}$$

and,

$$\varepsilon_r'' = \frac{V}{4v}\frac{Q_2 - Q_1}{Q_1 Q_2} \tag{10.30}$$

where V and v are cavity and sample volumes, respectively.

Similarly, for a small spherical sample of radius r that is placed in a uniform field at the center of the rectangular cavity, the dielectric parameters are found as follows.

$$\varepsilon_r' = \frac{abd}{8\pi r^3}\frac{f_1 - f_2}{f_2} \tag{10.31}$$

and,

$$\varepsilon_r'' = \frac{abd}{16\pi r^3}\left(\frac{Q_2 - Q_1}{Q_1 Q_2}\right) \tag{10.32}$$

where a, b, and d are the width, height, and length of the rectangular cavity, respectively. For the highest accuracy with the cavity perturbation method, the shift in frequency $(f_1 - f_2)$ must be very small.

Free-Space Method for Measurement of Complex Permittivity

When a plane electromagnetic wave is incident on a dielectric interface, its reflection and transmission depend upon the contrast in the dielectric parameters. Many researchers have used it for determining the complex permittivity of dielectric materials placed in free-space. An automatic network analyzer and phase-corrected horn antennas may be used for such measurements [6]. The system is calibrated using the TRL (through, reflect, and line) technique. A time-domain gating is used to minimize error due to multiple reflections. The sample of thickness d is placed in front of a conducting plane and its reflection coefficient S_{11} is measured. A theoretical expression for this reflection coefficient is found as follows.

$$S_{11} = \frac{jZ_d \tan(\beta_d d) - 1}{jZ_d \tan(\beta_d d) + 1} \tag{10.33}$$

where,

$$Z_d = \frac{1}{\sqrt{\varepsilon_r^*}} \tag{10.34}$$

$$\beta_d = \frac{2\pi}{\lambda}\sqrt{\varepsilon_r^*} \tag{10.35}$$

and λ is the free-space wavelength of electromagnetic signal.

Equation 10.33 is solved for ε_r^* after substituting the measured S_{11}. Since it represents a nonlinear relation, an iterative numerical procedure may be used.

A Nondestructive Method for Measuring the Complex Permittivity of Materials

Most of the techniques described so far require cutting and placing part of a sample in the test fixture. Sometimes it may not be permissible to do so. Further, the dielectric parameters may change in that process. It is especially important in the case of a biological specimen to perform *in vivo* measurements. In one such technique, an open-ended coaxial line is placed in close contact with the sample and its input reflection coefficient is measured using an automatic network analyzer [7,8]. Open-ended coaxial lines are used as sensors for *in situ* measurement of electrical properties of materials because they possess several advantages over others [9–13]. Besides its usefulness over a broad frequency band, coaxial probes require almost no preparation of samples [9,13]. Further, a needle-like coaxial probe can be inserted easily into the sample for *in vivo* characterization of materials [11]. As recommended by the manufacturers, the network analyzer is calibrated initially using an open circuit, a short circuit, and a matched load. The reference plane is then moved to the measuring end of the coaxial line using a short circuit. Electrical parameters of the material are extracted subsequently via a suitable admittance model of the coaxial opening [12]. If the coaxial opening is electrically small such that its radiation fields are negligible and the sample volume is large enough to contain the aperture fields in it, then a significantly simple procedure can be adopted. This does not require the usual calibration of the network analyzer and shifting of the reference plan. It utilizes four standards (materials of known electrical properties) to calibrate the measurement setup. Typically, an open circuit, a short circuit, and two different materials of known electrical characteristics are used in this technique. Since this procedure is not based on the manufacturer's recommended calibration (does not require exact reflection coefficient data), a reflectometer arrangement with amplitude and phase measurement options can produce fairly accurate results [14]. One such arrangement is illustrated in Figure 10.8. The procedure is summarized below.

One end of the coaxial line is connected to the automatic network analyzer and the other is left open in the air. The reflection coefficient for this case is recorded as Γ_1. Next, the probe is dipped in distilled water and then in methanol, and the corresponding reflection coefficients are recorded as Γ_2 and Γ_4, respectively. With aluminum foil pressed against the opening, its reflection coefficient is recorded as Γ_3. These four measurements are used to calibrate the system. Now, a coefficient ξ is computed as follows:

$$\xi = \frac{(1+\Delta)\varepsilon_{r4}^* - \varepsilon_{r1}^* - \varepsilon_{r2}^*\Delta}{\varepsilon_{r1}^{*2} + \varepsilon_{r2}^{*2}\Delta - (1+\Delta)\varepsilon_{r4}^{*2}} \tag{10.36}$$

where ε_{r1}^*, ε_{r2}^*, and ε_{r4}^* are complex relative permittivities of air, water, and methanol at the frequency of measurement, and

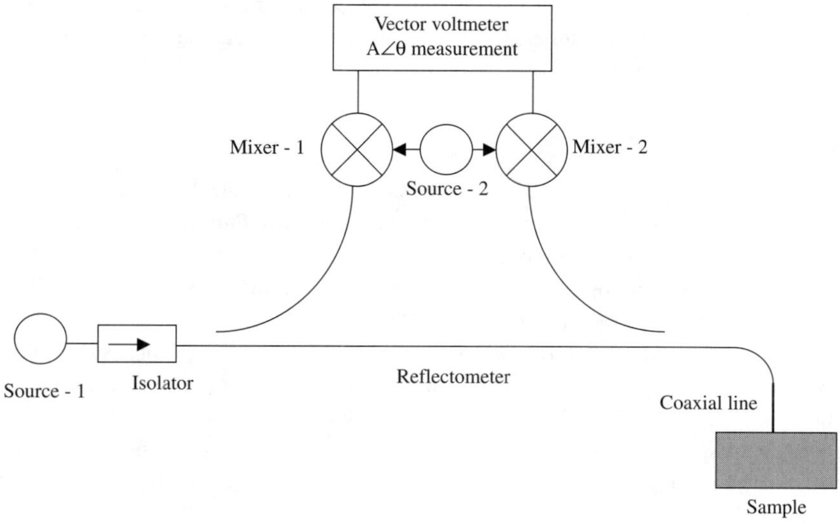

FIGURE 10.8 A reflectometer arrangement for measuring the complex permittivity of materials using a coaxial probe.

$$\Delta = \frac{(\Gamma_4 - \Gamma_1)\times(\Gamma_3 - \Gamma_2)}{(\Gamma_4 - \Gamma_2)\times(\Gamma_1 - \Gamma_3)} \tag{10.37}$$

It should be noted that the magnitudes of these reflection coefficients $|\Gamma_i|$ may occasionally exceed unity because the measurement system is not calibrated for this.

The equivalent admittance parameter (dimensionless) at the aperture is represented by

$$y = \varepsilon_r^* + \varsigma \varepsilon_r^{*2} \tag{10.38}$$

Note that y is not exactly the admittance of the aperture that is terminated by a material of complex relative permittivity ε_r^*. The parameters y_1 and y_2 are determined via this equation using ε_{r1}^*, and ε_{r2}^*, respectively.

Next, repeating the measurement with the sample material, the reflection coefficient is recorded as Γ_s. The aperture-admittance parameter with the sample is then computed as follows:

$$y_s = \frac{y_1 - \xi y_2}{1 - \xi} \tag{10.39}$$

where,

$$\xi = \frac{(\Gamma_s - \Gamma_1)\times(\Gamma_3 - \Gamma_2)}{(\Gamma_s - \Gamma_2)\times(\Gamma_1 - \Gamma_3)} \tag{10.40}$$

The complex permittivity ε_{rs}^* of the sample is determined from Equations 10.38 and 10.39 as follows:

$$\varepsilon_{rs}^* = \frac{-1 \pm \sqrt{1 + 4 y_s \varsigma}}{2\varsigma} \tag{10.41}$$

Only one of the solutions to Equation 10.41 is found to be a physically possible value because the real part of the complex permittivity must be a positive number and its imaginary part must be negative (the formulation is based on the time-harmonic representation of $e^{j\omega t}$).

References

[1] A.R. Von Hippel, *Dielectric Materials and Applications,* The M.I.T. Press, Cambridge, MA, 1961.
[2] N.E. Hill, W.E. Vaughan, A.H. Price, and M. Davies, *Dielectric Properties and Molecular Behaviour,* Van Nostrand Reinhold Co., London, 1969.
[3] H. Altschuler, Dielectric constant, in *Handbook of Microwave Measurements,* Vol. II, M. Sucher and J. Fox, Eds., Polytechnic Press, Brooklyn, NY, 1963.
[4] R. Chatterjee, *Advanced Microwave Engineering,* Ellis Horwood Ltd., Chichester, 1988.
[5] D.K. Misra, Permittivity measurement of modified infinite samples by a directional coupler and a sliding load, *IEEE Trans. Microwave Theory Tech.*, 29(1), 65–67, 1981.
[6] D.K. Ghodgaonkar, V.V. Varadan, and V.K. Varadan, A free-space method for measurement of dielectric constants and loss tangents at microwave frequencies, *IEEE Trans. Instrum. Meas.*, 38(3), 789–793, 1989.
[7] A.P. Gregory, R.N. Clarke, T.E. Hodgetts, and G.T. Symm, RF and microwave dielectric measurements upon layered materials using a reflectometric coaxial sensor, NPL Report DES 125, UK, March 1993.
[8] D. Misra, On the measurement of the complex permittivity of materials by an open-ended coaxial probe, *IEEE Microwave Guided Wave Lett.*, 5(5), 161–163, 1995.
[9] S. Gabriel, R.W. Lau, and C. Gabriel, The dielectric properties of biological tissues: II. Measurements in the frequency range 10 Hz to 20 GHz, *Phys. Med. Biol.*, 41, 2251–2269, 1996.
[10] M.F. Kabir, K.B. Khalid, W.M. Daud, and S. Aziz, Dielectric properties of rubber wood at microwave frequencies measured with an open-ended coaxial line, *Wood Fiber Sci.*, 29(4), 319–324, 1997.
[11] Y. Xu, R.G. Bosisio, A. Bonincontro, F. Pedone, and G.F. Mariutti, On the measurement of microwave permittivity of biological samples using needle-type coaxial probes, *IEEE Trans. Instrum. Meas.*, IM-42(4), 822–827, 1993.
[12] C.L. Pournaropoulos and D.K. Misra, The coaxial aperture electromagnetic sensor and its application in material characterization, *Meas. Sci. Technol.*, 8, 1191–1202, 1997.
[13] H.C.F. Martens, Reedijk, and H.B. Brom, Measurement of the complex dielectric constant down to helium temperatures. I. Reflection method from 1 MHz to 20 GHz using an open-ended coaxial line, *Rev. Scientific Instrum.*, 71(2), 473–477, 2000.
[14] S.D. Zandron, C. Pournaropoulos, and D.K. Misra, Complex permittivity measurement of materials by the open-ended coaxial probe technique, *J. Wave-Mater. Interact.*, 5 & 6(4), 329–342, 1991.

11

Electric Field Strength[1]

David A. Hill
National Institute of Standards and Technology

Motohisa Kanda
National Institute of Standards and Technology

Electric field strength is defined as the ratio of the force on a positive test charge at rest to the magnitude of the test charge in the limit as the magnitude of the test charge approaches zero. The units of electric field strength are volts per meter (V m^{-1}). Electric charges and currents are sources of electric and magnetic fields, and Maxwell's equations [1] provide the mathematical relationships between electromagnetic (EM) fields and sources.

The electric field at a point in space is a vector defined by components along three orthogonal axes. For example, in a rectangular coordinate system, the electric field $\vec{E}$ can be written as:

$$\vec{E} = \hat{x}E_x + \hat{y}E_y + \hat{z}E_z \qquad (11.1)$$

where $\hat{x}$, $\hat{y}$, and $\hat{z}$ are unit vectors and E_x, E_y, and E_z are scalar components. For electrostatic fields, the components are real scalars that are independent of time. For steady-state, time-harmonic fields, the components are complex phasors that represent magnitude and phase. The time dependence, $e^{j\omega t}$, is suppressed.

11.1 Electrostatic Fields

Electrostatic fields are present throughout the atmosphere, and there are strong electrostatic fields near high-voltage dc power lines. The commonly used electrostatic field meters generate an ac signal by periodic conductor motion (either rotation or vibration). This ac signal is proportional to the electric field strength, and field meter calibration is performed in a known electrostatic field.

Field Mills

Field mills (also called generating voltmeters) determine electric field strength by measuring modulated, capacitively induced charges or currents on metal electrodes. Two types of field mills — the shutter type

[1]Contribution of the National Institute of Standards and Technology, not subject to copyright in the United States.

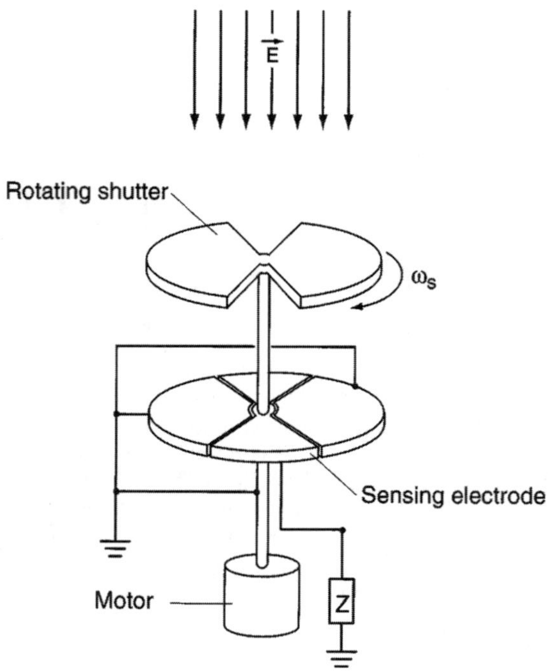

FIGURE 11.1 Shutter-type electric field mill for measurement of the polarity and magnitude of an electrostatic field.

and the cylindrical type — are described in the technical literature [2]. The shutter type is more common; a simplified version is shown in Figure 11.1. The sensing electrode is periodically exposed to and shielded from the electric field by a grounded, rotating shutter. The charge q_s induced on the sensing electrode and the current i_s between the sensing electrode and ground are both proportional to the electric field strength E normal to the electrode:

$$q_s(t) = \varepsilon_0 E\, a_s(t) \quad \text{and} \quad i_s(t) = \varepsilon_0 E \frac{da_s(t)}{dt} \tag{11.2}$$

where ε_0 is the permittivity of free space [1] and $a_s(t)$ is the effective exposed area of the sensing electrode at time t.

Thus, the field strength can be determined by measuring the induced charge or current (or voltage across the impedance Z). If the induced signal is rectified by a phase-sensitive detector (relative to the shutter motion), the dc output signal will indicate both the polarity and magnitude of the electric field [3].

Shutter-type field mills are typically operated at the ground or at a ground plane, but a cylindrical field mill can be used to measure the electric field at points removed from a ground plane. A cylindrical field mill consists of two half-cylinder sensing electrodes as shown in Figure 11.2. Charges induced on the two sensing electrodes are varied periodically by rotating the sensing electrodes about the cylinder axis at a constant angular frequency ω_c. The charge q_c induced on a half-cylinder of length L and the current i_c between the half-cylinders are given by:

$$q_c = 4\varepsilon_0 r_c L E \sin\omega_c t \quad \text{and} \quad i_c = 4\varepsilon_0 r_c L E \omega_c \cos\omega_c t \tag{11.3}$$

where r_c is the cylinder radius. Equation 11.3 is based on the two-dimensional solution for a conducting cylinder in an electric field and neglects end effects for finite L. Equation 11.3 shows that the electric field strength E can be determined from a measurement of the induced charge or current.

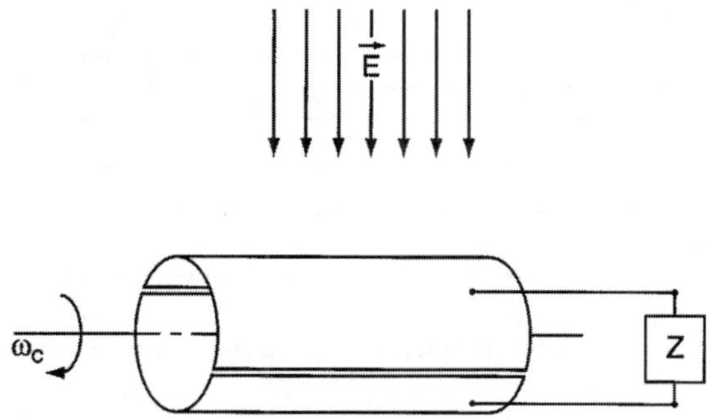

FIGURE 11.2 Cylindrical field mill for measurement of electrostatic field strength.

A third type of electric field meter uses a vibrating plate [4] to generate an ac signal that is proportional to the electric field strength. With any type of electric field strength meter, the observer should be at a sufficient distance from the measurement location to avoid perturbing the electric field.

Calibration Field

A known uniform field for meter calibration can be produced between a pair of parallel plates [2]. If a potential difference V is applied between a pair of plates with a separation d_p, the field strength away from the plate edges is V/d_p. The plate dimensions should be much larger than d_p to provide an adequate region of uniform field. Also, d_p should be much larger than the field meter dimensions so that the charge distribution on the plates is not disturbed. The parallel plates can be metal sheets or tightly stretched metal screens.

The field meter should be located in the type of environment in which it will be used. Shutter-type field mills that are intended to be located at a ground plane should be located at one of the plates. Cylindrical field mills that are not intended to be used at a ground plane should be located in the center of the region between the plates.

For simplicity, only field mills and calibration in the absence of space charge were mentioned. However, near power lines or in the upper atmosphere [5], the effects of space charge can be significant and require modifications in field mill design. A field mill for use in a space charge region and a calibration system with space charge are described in [6].

11.2 ELF and ULF Electric Fields

In this section, measurement techniques for extremely low frequency (ELF, 3 Hz to 3 kHz) and ultralow frequency (ULF, below 3 Hz) electric fields are considered. Natural ELF fields are produced by thunderstorms, and natural ULF fields are produced by micropulsations in the earth's magnetic field [7]. Geophysicists make use of these natural fields in the magnetotelluric method for remote sensing of the Earth's crust [8]. Ac power lines are dominant sources of fields at 50 Hz or 60 Hz and their harmonics.

An ac electric field strength meter [9] includes two essential parts: (1) an antenna and (2) a detector (receiver). Other possible features are a transmission line or optical link, frequency-selective circuits, amplifying and attenuating circuits, an indicating device, and a nonconducting handle. The antenna characteristics can be calculated for simple geometries or determined by calibration. For example, linear antennas are often characterized by their effective length L_{eff} [10], which determines the open-circuit voltage V_{oc} induced at the antenna terminals:

$$V_{oc} = L_{eff} E_{inc} \tag{11.4}$$

where E_{inc} is the component of the incident electric field parallel to the axis of the linear antenna. The detector could respond to the terminal voltage or current or to the power delivered to the load.

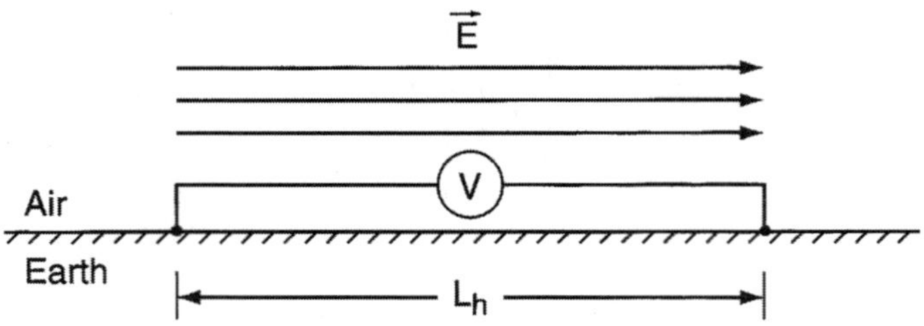

FIGURE 11.3 Grounded horizontal antenna for measurement of the horizontal component of the electric field.

Natural Horizontal Electric Field at the Earth's Surface

Magnetotelluric sounding of the Earth's crust requires measurement of the horizontal electric and magnetic fields at the Earth's surface [8]. The magnetic field is measured with a horizontal axis loop, and the electric field is measured with a horizontal wire antenna as shown in Figure 11.3. The antenna wire is insulated since it lies on the ground, but it is grounded at its end points. Nonpolarizing grounding electrodes should be used to avoid polarization potentials between the electrodes and the ground.

Since the natural electric field strength to be measured is on the order of 1 μV m^{-1}, the antenna length L_h needs to be on the order of 1 km to produce a measurable voltage. Since the effective length of a grounded antenna is equal to the physical length ($L_{eff} = L_h$), the horizontal component E_h of the electric field parallel to the antenna is equal to the open-circuit voltage divided by the antenna length:

$$E_h = V_{oc}/L_h \tag{11.5}$$

The frequencies used in typical magnetotelluric sounding range from approximately 0.1 mHz to 10 Hz. If both horizontal components of the electric field are needed, a second orthogonal antenna is required.

Free-Body Electric Field Meters

ELF electric fields in residential and industrial settings are most conveniently measured with free-body field meters [11, 12], which measure the steady-state current or charge oscillating between two halves of a conducting body in free space. (Ground reference meters [13] are also available for measuring the electric field normal to the ground or some other conducting surface.) Geometries for free-body electric field meters are shown in Figure 11.4. Commercial field meters are usually rectangular in shape, and typical dimensions are on the order of 10 cm. A large dynamic range (1 V m^{-1} to 30 kV m^{-1}) is required to cover the various field sources (ac power lines, video display terminals, mass transportation systems, etc.) of interest. (Electro-optic field meters with less sensitivity are described in [12].) A long, nonconducting handle is normally attached perpendicular to the field-meter axis for use in measurement surveys.

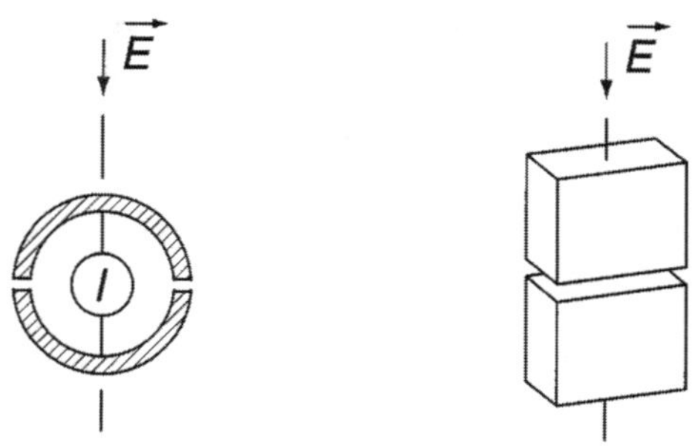

FIGURE 11.4 Electric field meters for measurement of the axial component of the electric field: (a) spherical geometry and (b) rectangular geometry.

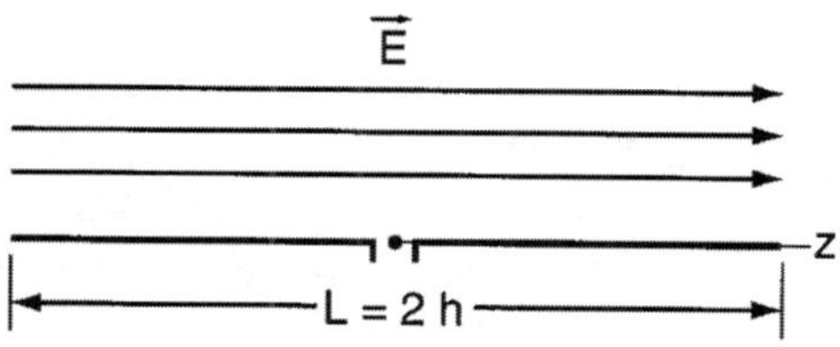

FIGURE 11.5 Dipole antenna for measurement of the axial component of the electric field.

The charge Q on half of the field meter is proportional to the incident electric field E along the meter axis:

$$Q = A\varepsilon_0 E \tag{11.6}$$

where ε_0 is the permittivity of free space [1] and A is a constant proportional to the surface area. For the spherical geometry in Figure 11.4(a), $A = 3\pi a^2$, where a is the sphere's radius. Since the current I between the two halves is equal to the time derivative of the charge, for time-harmonic fields it can be written:

$$I = j\omega A\varepsilon_0 E \tag{11.7}$$

This allows E to be determined from the measured current. For commercial field meters that are not spherical, the constant A needs to be determined by calibration. A known calibration field can be generated between a pair of parallel plates where the plates are sufficiently large compared to the separation to produce a uniform field with small edge effects. This technique produces a well-characterized field with an uncertainty less than 0.5% [11]. However, the presence of harmonic frequencies can cause less accurate meter readings in field surveys.

11.3 Radio-Frequency and Microwave Techniques

Dipole Antennas

A thin, linear dipole antenna of length L is shown in Figure 11.5. Its effective length is approximately [10]:

$$L_{\text{eff}} = \frac{\lambda}{\pi}\tan\left(\frac{\pi L}{2\lambda}\right) \tag{11.8}$$

where λ is the free-space wavelength. Resonant half-wave dipoles ($L = \lambda/2$) have an effective length of $L_{\text{eff}} = 2\,L/\pi$ and are of convenient length for frequencies from 30 to 1000 MHz. The physical length of a dipole at resonance is actually slightly shorter than $\lambda/2$ to account for the effect of a finite length-to-diameter ratio. Resonant dipoles are used as standard receiving antennas to establish a known standard field in the *standard antenna method* [9]. Commercial antennas and field meters are calibrated in such standard fields.

For $L < \lambda/2$, Equation 11.8 must be used to determine L_{eff}. For very short dipoles ($L/\lambda \ll 1$), the current distribution is approximately linear, and the effective length is approximately one half the physical length ($L_{\text{eff}} \approx L/2$). Short dipoles are frequently used as electric field probes.

Aperture Antennas

Aperture antennas are commonly used for receiving and transmitting at microwave frequencies (above 1 GHz). As receiving antennas, they are conveniently characterized by their on-axis gain g or effective area A_{eff}. Effective area is defined as the ratio of the received power P_r to the incident power density S_{inc} and can also be written in terms of the gain [9]:

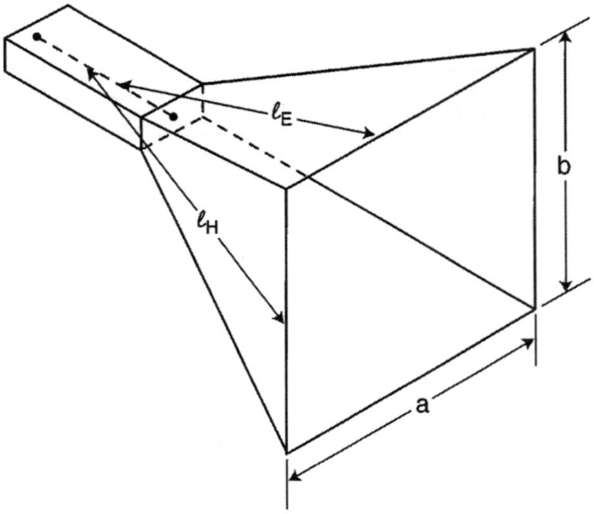

FIGURE 11.6 Pyramidal horn for measuring power density or electric field strength.

$$A_{eff} = \frac{P_r}{S_{inc}} = \frac{g\lambda^2}{4\pi} \tag{11.9}$$

Equation 11.9 applies to the case where the incident field is polarization-matched to the receiving antenna. The incident power density in a plane wave is $S_{inc} = E^2/\eta_0$, where E is the rms electric field strength and η_0 is the impedance of free space ($\approx 377\ \Omega$). Thus, the electric field strength can be determined from the received power:

$$E = \sqrt{P_r\eta_0/A_{eff}} = \lambda^{-1}\sqrt{4\pi\eta_0 P_r/g} \tag{11.10}$$

In general, the gain can be measured using the two-antenna method [9]. For a pyramidal horn antenna as shown in Figure 11.6, the gain can be calculated accurately from [14]:

$$g = \frac{32\,ab}{\pi\lambda^2} R_E R_H \tag{11.11}$$

where R_E and R_H are gain reduction factors due to the E and H plane flare of the horn. The gain reduction factors are

$$R_E = \frac{C^2(w) + S^2(w)}{w^2} \quad \text{and} \quad R_H = \frac{\pi^2\left\{\left[C(u) - C(v)\right]^2 + \left[S(u) - S(v)\right]^2\right\}}{4(u-v)^2} \tag{11.12}$$

Where:
$$w = \frac{b}{\sqrt{2\lambda l_E}} \quad \text{and} \quad \left.\begin{matrix} u \\ v \end{matrix}\right\} = \frac{\sqrt{\lambda l_H/2}}{a} \pm \frac{a}{\sqrt{2\lambda l_H}}$$

The Fresnel integrals C and S are defined as [15]:

$$C(w) = \int_0^w \cos\left(\frac{\pi}{2}t^2\right)dt \quad \text{and} \quad S(w) = \int_0^w \sin\left(\frac{\pi}{2}t^2\right)dt \tag{11.13}$$

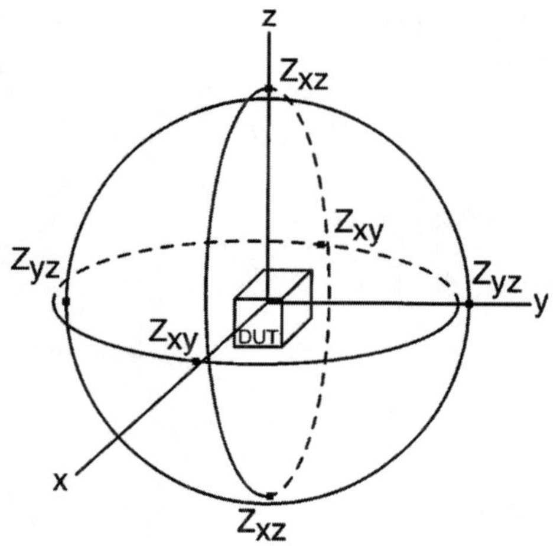

FIGURE 11.7 Geometry of the three-loop antenna system and the device under test.

Well-characterized aperture antennas are also used to generate standard fields [16] for calibrating commercial antennas and field strength meters. This method of calibration is called the *standard field method* [9]. The electric field strength E at a distance d from the transmitting antenna is:

$$E = \sqrt{\eta_0 P_{\text{del}} g/(4\pi)}/d, \tag{11.14}$$

where P_{del} is the net power delivered to the transmitting antenna and is typically measured with a directional coupler [16]. The gain g for a pyramidal horn can be calculated from Equation 11.11. The National Institute of Standards and Technology (NIST) uses rectangular open-ended waveguides from 200 MHz to 500 MHz and a series of pyramidal horns from 500 MHz to 40 GHz to generate standard fields in an anechoic chamber [16]. The uncertainty of the field strength is less than 1 dB over the entire frequency range of 200 MHz to 40 GHz.

11.4 Three-Loop Antenna System

Electronic equipment can emit unintentional electromagnetic radiation that can interfere with other electronic equipment. If the radiating source is electrically small (as is a video display terminal), then it can be characterized by equivalent electric and magnetic dipole moments. The three-loop antenna system (TLAS), shown in Figure 11.7, consists of three orthogonal loop antennas that are terminated at diametrically opposite points. The unique feature of loop antennas with double terminations [17] is that they can measure both electric and magnetic fields. For electromagnetic interference (EMI) applications, a device under test (DUT) is placed at the center of the TLAS. On the basis of six terminal measurements, the TLAS determines three equivalent electric dipole components and three equivalent magnetic dipole moments of the DUT and hence its radiation characteristics.

Here, the theory [18] is summarized for one of the three loops. The DUT in Figure 11.7 is replaced by an electric dipole moment $\vec{m}_e$ and a magnetic dipole moment $\vec{m}_m$, both located at the origin of the coordinate system. The dipole moments can be written in terms of their rectangular components:

$$\vec{m}_e = \hat{x}m_{ex} + \hat{y}m_{ey} + \hat{z}m_{ez} \quad \text{and} \quad \vec{m}_m = \hat{x}m_{mx} + \hat{y}m_{my} + \hat{z}m_{mz} \tag{11.15}$$

The loop in the xy plane has radius r_0 and has impedance loads Z_{xy} located at the intersections with the x axis ($\phi = 0, \pi$).

The solution for the current induced in the loop is based on Fourier series analysis [19]. The incident azimuthal electric field $E_\phi^i(\phi)$ tangent to the loop is:

$$E_\phi^i(\phi) = A_0 + A_1\cos\phi + B_1\sin\phi \tag{11.16}$$

Where:

$$A_0 = m_{mx}G_m$$

$$A_1 = m_{ey}G_e$$

$$B_1 = -m_{ex}G_e$$

$$G_m = \frac{\eta_0}{4\pi}\left(\frac{k^2}{r_0} - \frac{jk}{r_0^2}\right)e^{-jkr_0}$$

$$G_e = \frac{-\eta_0}{4\pi}\left(\frac{jk}{r_0} + \frac{1}{r_0^2} + \frac{1}{jkr_0^2}\right)e^{-jkr_0}$$

and $k = 2\pi/\lambda$ is the free-space wavenumber. An approximate solution [17] for the loop current $I(\phi)$ yields the following results for the load currents $I(0)$ and $I(\pi)$:

$$\begin{aligned} I(0) &= 2\pi r_0\left(\frac{m_{mz}G_mY_0}{1+2Y_0Z_{xy}} + \frac{m_{ey}G_eY_1}{1+2Y_1Z_{xy}}\right) \\ I(\pi) &= 2\pi r_0\left(\frac{m_{mz}G_mY_0}{1+2Y_0Z_{xy}} - \frac{m_{ey}G_eY_1}{1+Y_1Z_{xy}}\right) \end{aligned} \tag{11.17}$$

where Y_0 and Y_1 are the admittances for the constant and cosϕ currents [17].

One can solve Equation 11.17 for the magnetic and electric dipole components:

$$m_{mz} = \frac{I_\Sigma\left(1+2Y_0Z_{xy}\right)}{2\pi r_0G_mY_0} \quad \text{and} \quad m_{ey} = \frac{I_\Delta\left(1+2Y_1Z_{xy}\right)}{2\pi r_0G_eY_1} \tag{11.18}$$

Where:

$$I_\Sigma = \left[I(0) + I(\pi)\right]/2$$

$$I_\Delta = \left[I(0) - I(\pi)\right]/2$$

Thus, the sum current I_Σ can be used to measure the magnetic dipole moment, and the difference current I_Δ can be used to measure the electric dipole moment. The four remaining dipole components can be obtained in an analogous manner. The loop in the *xz* plane can be used to measure m_{my} and m_{ex}, and the loop in the *yz* plane can be used to measure m_{mx} and m_{ez}.

The total power P_T radiated by the source can be written in terms of the magnitudes of the six dipole components:

$$P_T = \frac{2\pi\eta_0}{3\lambda^2}\left[\left|m_{ex}\right|^2 + \left|m_{ey}\right|^2 + k^2\left(\left|m_{mx}\right|^2 + \left|m_{my}\right|^2 + \left|m_{mz}\right|^2\right)\right] \tag{11.19}$$

The expression for the power pattern is more complicated and involves the amplitudes and phases of the dipole moments [20]. The TLAS has been constructed with 1-m diameter loops and successfully tested from 3 kHz to 100 MHz [21]. It is currently being used to measure radiation from video display terminals and other inadvertent radiators.

11.5 Broadband Dipole Antennas

The EM environment continues to grow more severe and more complex as the number of radiating sources increases. Broadband antennas are used to characterize the EM environment over a wide frequency range. For electric-field measurements, electrically short dipole antennas with a high capacitive input impedance are used with a capacitive load, such as a field-effect transistor (FET). The transfer function S of frequency f is defined as the ratio of the output voltage V_L of the antenna to the incident electric field E_i [22]:

$$S(f) = \frac{V_L(f)}{E_i(f)} = \frac{h\alpha/2}{1 + C/C_a} \tag{11.20}$$

Where:

$$C_a = \frac{4\pi h}{c\eta_0(\Omega_a - 2 - \ln 4)}$$

$$\alpha = \frac{\Omega_a - 1}{\Omega_a - 2 + \ln 4}$$

$$\Omega_a = 2\ln(2h/r_a)$$

r_a = Antenna radius
C = Load capacitance
C_a = Antenna capacitance
h = Half the physical length of the dipole antenna, as shown in Figure 11.5
c = Free-space speed of light
Ω_a = Antenna thickness factor

Since the input impedance of an electrically short dipole is predominantly a capacitive reactance, a very broadband frequency response can be achieved with a high-impedance capacitive load. However, given the present state of the art, it is not possible to build a balanced, high-input impedance FET with high common-mode rejection above 400 MHz. For this reason, it is more common practice to use a high-frequency, beam-lead Schottky-barrier diode with a very small junction capacitance (less than 0.1 pF) and very high junction resistance (greater than several M3) for frequencies above 400 MHz.

The relationship between the time-dependent diode current $i_d(t)$ and voltage $v_d(t)$ is

$$i_d(t) = I_s\left[e^{\alpha_d v_d(t)} - 1\right] \tag{11.21}$$

where I_s and α_d are constants of the diode. For very small incident fields $E_i(t)$, the output detected dc voltage v_o is:

$$v_o = \frac{-b_d^2}{2\alpha_d}\left\langle \tilde{v}_i^{-2} \right\rangle \qquad v_i = E_i L_e \tag{11.22}$$

Where:

$$\tilde{v}_i(t) = v_i(t) - \langle v_i \rangle$$

$$b_d = \frac{C_a \alpha_d}{C_a + C_d}$$

C_a = Dipole capacitance
C_d = Diode capacitance
L_e = Effective length of the dipole antenna
< > Indicates time average

Thus, the dc detected voltage is frequency independent and is directly proportional to the average of $(E_i - \langle E_i \rangle)^2$.

For large incident fields $E_i(t)$, the output detected dc voltage is:

$$v_o = -\frac{b_d \tilde{V}_i}{\alpha_d} \tag{11.23}$$

where $\tilde{V}_i$ is the peak value of $v_i(t)$. Consequently, for a large incident field, v_o is also frequency independent and is directly proportional to the peak field.

Conventional dipole antennas support a standing-wave current distribution; thus, the useful frequency range of this kind of dipole is usually limited by its natural resonant frequency. In order to suppress this resonance, a resistively loaded dipole (traveling-wave dipole) has been developed. If the internal impedance per unit length $Z_l(z)$ as function of the axial coordinate z (measured from the center of the dipole) has the form:

$$Z_l(z) = \frac{60\psi}{h - |z|} \tag{11.24}$$

then the current distribution $I_z(z)$ along the dipole is that of a traveling wave. Its form is

$$I_z(z) = \frac{V_0}{60\psi(1 - j/kh)}\left[1 - \frac{|z|}{h}\right]e^{-jk|z|} \tag{11.25}$$

where $2h$ = Total physical length of the dipole
V_0 = Driving voltage

$$\psi = 2\left[\sinh^{-1}\frac{h}{a_d} - C(2ka_d, 2kh) - jS(2ka_d, 2kh)\right] + \frac{j}{kh}\left(1 - e^{-j2kh}\right) \tag{11.26}$$

$C(x,y)$ and $S(x,y)$ = Generalized cosine and sine integrals
a_d = Dipole radius

This type of resistively tapered dipole has a fairly flat frequency response from 100 kHz to 18 GHz [23].

Defining Terms

Electric field strength: The ratio of the force on a positive test charge to the magnitude of the test charge in the limit as the magnitude of the test charge approaches zero.
Electrostatic field: An electric field that does not vary with time.
Field mill: A device used to measure an electrostatic field.
Antenna: A device designed to radiate or to receive time-varying electromagnetic waves.
Microwaves: Electromagnetic waves at frequencies above 1 GHz.
Power density: The time average of the Poynting vector.
Aperture antenna: An antenna that radiates or receives electromagnetic waves through an open area.
Dipole antenna: A straight wire antenna with a center feed used for reception or radiation of electromagnetic waves.

References

1. J.A. Stratton, *Electromagnetic Theory,* New York: McGraw-Hill, 1941.
2. ANSI/IEEE Std. 1227-1990, *IEEE Guide for the Measurement of dc Electric-Field Strength and Ion Related Quantities.*
3. P.E. Secker and J.N. Chubb, Instrumentation for electrostatic measurements, *J. Electrostatics*, 16, 1–19, 1984.
4. R.E. Vosteen, Dc electrostatic voltmeters and fieldmeters, *Conference Record of Ninth Annual Meeting of the IEEE Industrial Applications Society,* October 1974.
5. P.J.L. Wildman, A device for measuring electric field in the presence of ionisation, *J. Atmos. Terr. Phys.*, 27, 416–423, 1965.
6. M. Misakian, Generation and measurement of dc electric fields with space charge, *J. Appl. Phys.*, 52, 3135–3144, 1981.
7. G.V. Keller and F.C. Frischknecht, *Electrical Methods in Geophysical Prospecting,* Oxford, U.K.: Pergamon Press, 1966.
8. A.A. Kaufman and G.V. Keller, *The Magnetotelluric Sounding Method,* Amsterdam: Elsevier, 1981.
9. IEEE Std. 291-1991, *IEEE Standard Methods for Measuring Electromagnetic Field Strength of Sinusoidal Continuous Waves,* 30 *Hz to* 30 GHz.
10. E.C. Jordan and K.G. Balmain, *Electromagnetic Waves and Radiating Systems, 2nd ed.,* Englewood Cliffs, NJ: Prentice-Hall, 1968.
11. ANSI/IEEE Std. 644-1987, *IEEE Standard Procedures for Measurement of Power Frequency Electric and Magnetic Fields from Ac Power Lines.*
12. IEEE Std. 1308-1994, *IEEE Recommended Practice for Instrumentation: Specifications for Magnetic Flux Density and Electric Field Strength Meters — 10 Hz to 3 kHz.*
13. C.J. Miller, The measurements of electric fields in live line working, *IEEE Trans. Power Apparatus Sys.*, PAS-16, 493–498, 1967.
14. E.V. Jull, *Aperture Antennas and Diffraction Theory,* Stevenage, U.K.: Peter Peregrinus, 1981.
15. M. Abramowitz and I.A. Stegun, *Handbook of Mathematical Functions,* Nat. Bur. Stand. (U.S.), Spec. Pub. AMS 55, 1968.
16. D.A. Hill, M. Kanda, E.B. Larsen, G.H. Koepke, and R.D. Orr, Generating standard reference electromagnetic fields in the NIST anechoic chamber, 0.2 to 40 GHz, Natl. Inst. Stand. Technol. Tech. Note 1335, 1990.
17. M. Kanda, An electromagnetic near-field sensor for simultaneous electric and magnetic-field measurements, *IEEE Trans. Electromag. Compat.*, EMC-26, 102–110, 1984.
18. M. Kanda and D.A. Hill, A three-loop method for determining the radiation characteristics of an electrically small source, *IEEE Trans. Electromag. Compat.*, 34, 1–3, 1992.
19. T.T. Wu, Theory of the thin circular antenna, *J. Math. Phys.*, 3, 1301–1304, 1962.
20. I. Sreenivasiah, D.C. Chang, and M.T. Ma, Emission characteristics of electrically small radiating sources from tests inside a TEM cell, *IEEE Trans. Electromag. Compat.*, EMC-23, 113–121, 1981.
21. D.R. Novotny, K.D. Masterson, and M. Kanda, An optically linked three-loop antenna system for determining the radiation characteristics of an electrically small source, *IEEE Int. EMC Symp.*, 1993, 300–305.
22. M. Kanda, Standard probes for electromagnetic field measurements, *IEEE Trans. Antennas Propagat.*, 41, 1349–1364, 1993.
23. M. Kanda and L.D. Driver, An isotropic electric-field probe with tapered resistive dipoles for broadband use, 100 kHz to 18 GHz, *IEEE Trans. Microwave Theory Techniques*, MTT-35, 124–130, 1987.

12

Magnetic Field Measurement

Steven A. Macintyre
Macintyre Electronic Design

Magnetic field strength is measured using a variety of different technologies. Each technique has unique properties that make it more suitable for particular applications. These applications can range from simply sensing the presence or change in the field to the precise measurements of a magnetic field's scalar and vector properties. A very good and exhaustive fundamental description of both mechanical and electrical means for sensing magnetic fields can be found in Lion [1]. Less detailed but more up-to-date surveys of magnetic sensor technologies can be found in [2, 3]. It is not possible to adequately describe all of these technologies in the space available. This chapter concentrates on sensors that are commonly used in magnetic field measuring instruments.

As shown in Figure 12.1, magnetic field sensors can be divided into vector component and scalar magnitude types. The vector types can be further divided into sensors that are used to measure low fields (<1 mT) and high fields (>1 mT). Instruments that measure low fields are commonly called *magnetometers.* High-field instruments are usually called *gaussmeters.*

The induction coil and fluxgate magnetometers are the most widely used vector measuring instruments. They are rugged, reliable, and relatively less expensive than the other low-field vector measuring instruments. The fiber optic magnetometer is the most recently developed low-field instrument. Although it currently has about the same sensitivity as a fluxgate magnetometer, its potential for better performance is large. The optical fiber magnetometer has not yet left the laboratory, but work on making it more rugged and field worthy is under way. The superconducting quantum interference device (SQUID) magnetometers are the most sensitive of all magnetic field measuring instruments. These sensors operate at temperatures near absolute zero and require special thermal control systems. This makes the SQUID-based magnetometer more expensive, less rugged, and less reliable.

The Hall effect device is the oldest and most common high-field vector sensor used in gaussmeters. It is especially useful for measuring extremely high fields (>1 T). The magnetoresistive sensors cover the middle ground between the low- and high-field sensors. Anisotropic magnetoresistors (AMR) are currently being used in many applications, including magnetometers. The recent discovery of the giant

0-8493-1733-9/04/$0.00+$1.50

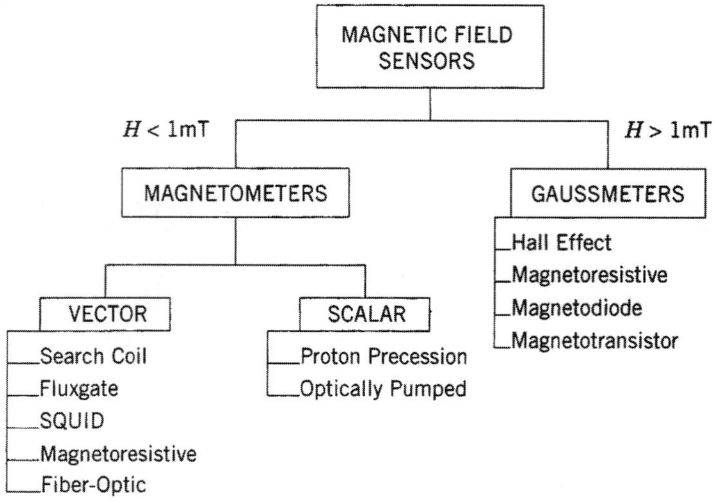

FIGURE 12.1 Magnetic field sensors are divided into two categories based on their field strengths and measurement range: magnetometers measure low fields and gaussmeters measure high fields.

TABLE 12.1 Field Strength Instrument Characteristics

Instrument	Range (mT)	Resolution (nT)	Bandwidth (Hz)	Comment
Induction coil	10^{-10} to 10^{6}	Variable	10^{-1} to 10^{6}	Cannot measure static fields
Fluxgate	10^{-4} to 0.5	0.1	Dc to 2×10^{3}	General-purpose vector magnetometer
SQUID	10^{-9} to 0.1	10^{-4}	Dc to 5	Highest sensitivity magnetometer
Hall effect	0.1 to 3×10^{4}	100	Dc to 10^{8}	Best for fields above 1T
Magnetoresistance	10^{-3} to 5	10	Dc to 10^{7}	Good for mid-range applications
Proton precession	0.02 to 0.1	0.05	Dc to 2	General-purpose scalar magnetometer
Optically pumped	0.01 to 0.1	0.005	Dc to 5	Highest resolution scalar magnetometer

magnetoresistive (GMR) effect, with its tenfold improvement in sensitivity, promises to be a good competitor for the traditional fluxgate magnetometer in medium-sensitivity applications.

The proton (nuclear) precession magnetometer is the most popular instrument for measuring the scalar magnetic field strength. Its major applications are in geological exploration and aerial mapping of the geomagnetic field. Since its operating principle is based on fundamental atomic constants, it is also used as the primary standard for calibrating magnetometers. The proton precession magnetometer has a very low sampling rate, on the order of one to three samples per second, so it cannot measure fast changes in the magnetic field. The optically pumped magnetometer operates at a higher sampling rate and is capable of higher sensitivities than the proton precession magnetometer, but it is more expensive and not as rugged and reliable.

Table 12.1 lists various magnetic field strength instruments and their characteristics.

12.1 Magnetic Field Fundamentals

An understanding of the nature of magnetic fields is necessary in order to understand the techniques used for measuring magnetic field strength. The most familiar source of a magnetic field is the bar magnet. The field it produces is shown in Figure 12.2. Magnetic field is a vector quantity; that is, it has both a magnitude and a direction. The field of a bar magnet or any other magnetized object, when measured at a distance much greater than its longest dimension, is described by Equation 12.1:

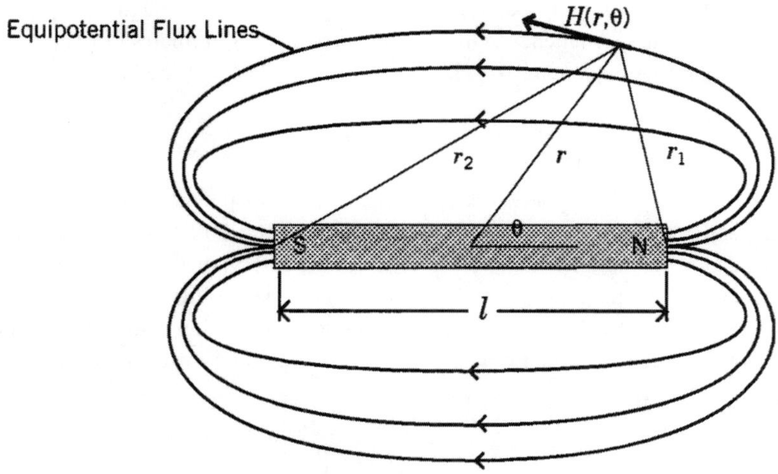

FIGURE 12.2 Magnets produce magnetic fields. A magnetic field is a vector quantity with both magnitude and direction properties.

$$\vec{H} = \frac{3(\vec{m} \times \hat{a}_r)\hat{a}_r - \vec{m}}{r^3} \tag{12.1}$$

where $\hat{a}_r$ is a unit vector along r, r is the distance between the magnetic field source and the measurement point, and $\vec{m}$ is called the magnetic dipole moment. The derivation of this equation can be found in many textbooks on electromagnetics. This is a very convenient equation for estimating the field produced by many magnetized objects.

The strength or intensity of a magnetized object depends on the density of its volume-distributed moments. This intensity is called its magnetization M, which is defined as the moments per unit volume:

$$\vec{M} = \frac{\vec{m}}{volume} \tag{12.2}$$

Like magnetic field, magnetization is a vector quantity. Magnetization is a material property that can arise from internal magnetic sources as well as be induced by an external magnetic field.

There is a third magnetic vector $\vec{B}$ called magnetic induction or flux density. In free space, magnetic field and magnetic induction are proportional to one another by a constant factor μ_0.

$$\vec{B} = \mu_0 \vec{H} \tag{12.3}$$

Things are different in matter. Equation 12.4 describes the relationship among the magnetic field, magnetic induction, and magnetization vectors in matter:

$$\vec{B} = \mu_0 \left(\vec{H} + \vec{M}\right) \tag{12.4}$$

In this case, the magnetic induction and the magnetic field vectors do not necessarily have the same direction. Some materials have anisotropic magnetic properties that make these two vectors point in different directions. The magnetization vector can consist of permanent and induced magnetization components. The permanent magnetization vector does not depend on the presence of an external field. The induced magnetization vector does depend on an external magnetic field and only exists while the inducing field is present.

Magnetic materials can be loosely classified as magnetically "soft" or magnetically "hard." In a magnetically hard material, the permanent magnetization component dominates (a magnet is an example). Magnetization in a soft magnetic material is largely induced and is described by the following equation:

$$\vec{M} = \chi \vec{H} \tag{12.5}$$

where χ is called the material's magnetic susceptibility. In an isotropic material (magnetic properties are not direction dependent), χ is a scalar quantity, and the magnetization and field vectors are proportional and aligned. In anisotropic material (magnetic properties depend on direction), χ is a tensor represented by a 3×3 matrix; therefore, the magnetization vector magnitude and direction depend on the direction and strength of the inducing field. As a result, the magnetization vector will not always align with the magnetization inducing field vectors. Equation 12.5 can be modified for magnetically "soft" material to the following:

$$\vec{B} = \mu_0 (1 + \chi) \vec{H} = \mu_0 \mu \vec{H} \tag{12.6}$$

where μ is called the relative permeability of the material.

A magnetized object with a magnetic moment $\vec{m}$ will experience torque $\vec{T}$ in the presence of a uniform magnetic field $\vec{H}$. Equation 12.7 expresses this relationship.

$$\vec{T} = \vec{m} \times \vec{H} \tag{12.7}$$

Torque is the cross-product of the magnetic moment and field vectors. The magnitude equation is:

$$T = mH \sin\theta \tag{12.8}$$

where θ is the angle between the direction of $\vec{m}$ and $\vec{H}$.

There is an intimate relationship between electric and magnetic fields. Oersted discovered that passing a current through a wire near a compass causes the compass needle to rotate. The compass was the first magnetic field strength sensor. Faraday found that he could produce an electric voltage at the terminals of a loop of wire if he moved a magnet near it. This led to the induction or search coil sensor.

Magnetic fields are produced by the flow of electric charge (i.e., electric currents). In effect, a magnetic field is a velocity-transformed electric field (through a Lorentz transformation). Current flowing through a straight wire, a loop of wire, or a solenoid will also produce a magnetic field as illustrated in Figure 12.3.

Units are always a problem when dealing with magnetic fields. The Gaussian cgs (centimeter, gram, and second) system of units was favored for many years. Since $\mu_0 = 1$ in the cgs system, magnetic field and flux density have the same numeric value in air, and their units (oerstedt for field and gauss for flux density) are often indiscriminately interchanged. This has led to great confusion. The cgs system has now been replaced by the International System of Units (SI). The SI system uses, among others, the meter (m), kilogram (kg), second (s), and ampere (A) as the fundamental units. Payne [4] gives a very good explanation of the differences between these systems of units as they relate to magnetic fields. Table 12.2 summarizes the relationships between the two systems of units.

The SI system of units is used throughout this chapter.

12.2 Low-Field Vector Magnetometers

The Induction Coil Magnetometer

The induction or search coil, which is one of the simplest magnetic field sensing devices, is based on Faraday's law. This law states that if a loop of wire is subjected to a changing magnetic flux, ϕ, through the area enclosed by the loop, then a voltage will be induced in the loop that is proportional to the rate of change of the flux:

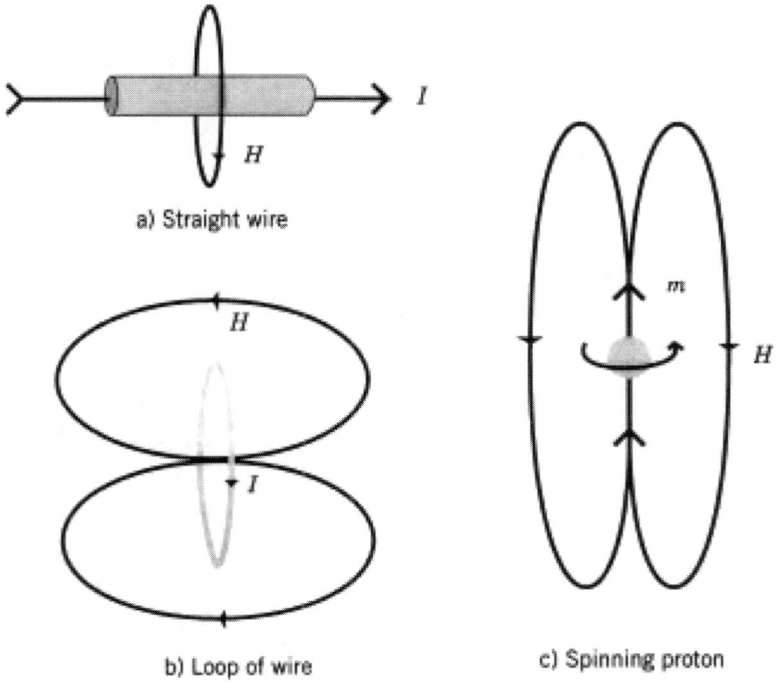

FIGURE 12.3 Magnetic fields are also produced by electric currents.

TABLE 12.2 Factors for Converting from cgs to SI Magnetic Field Units

Description	Symbol	SI unit	Gaussian cgs unit	Multiply by
Magnetic induction	B	Tesla	gauss	10^4
Magnetic field strength	H	A m^{-1}	oerstedt (oe)	$4\pi \times 10^{-3}$
Magnetization	M	A m^{-1}	emu m^3	10^{-3}
Magnetic dipole moment	m	A m^2	emu	10^3
Magnetic flux	ϕ	Weber (Wb)	maxwell	10^8
Magnetic pole strength	p	A m	emu	
Permeability of free space	μ_0	H m^{-1}	$4\pi \times 10^{-7}$	1

$$e(t) = -\frac{\mathrm{d}\phi}{\mathrm{d}t} \tag{12.9}$$

Since magnetic induction $\vec{B}$ is flux density, then a loop with cross-sectional area $\vec{A}$ will have a terminal voltage:

$$e(t) = -\frac{\mathrm{d}\left(\vec{B}\cdot\vec{A}\right)}{\mathrm{d}t} \tag{12.10}$$

for spatially uniform magnetic induction fields.

Equation 12.10 states that a temporal change in $\vec{B}$ or the mechanical orientation of $\vec{A}$ relative to $\vec{B}$ will produce a terminal voltage. If the coil remains fixed with respect to $\vec{B}$, then static fields cannot be detected; but if the loop is rotated or the magnitude of $\vec{A}$ is changed, then it is possible to measure a static field. The relationship described by Equation 12.10 is exploited in many magnetic field measuring instruments (see [1]).

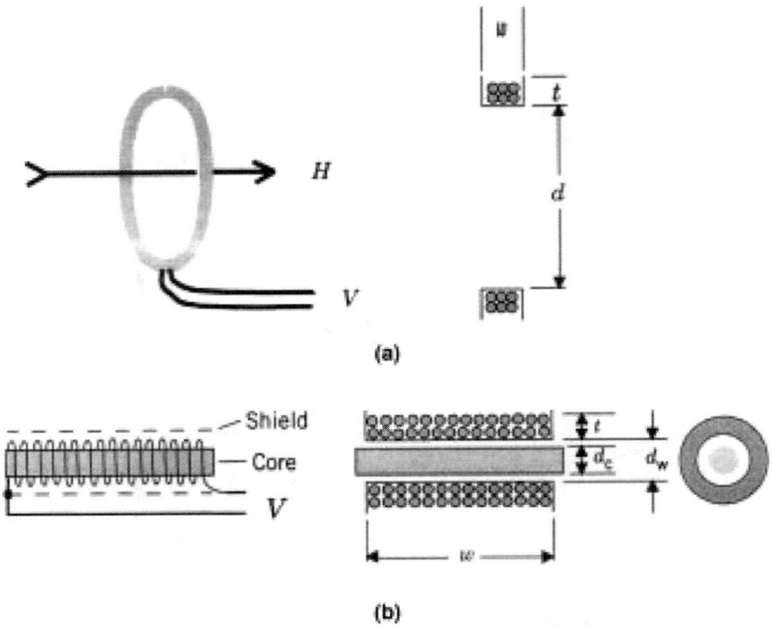

FIGURE 12.4 Induction or search coil sensors consist of a loop of wire (or a solenoid), which may or may not surround a ferromagnetic core. (a) Air core loop antenna; (b) solenoid induction coil antenna with ferromagnetic core.

Figure 12.4 shows the two most common induction coil configurations for measuring field strength: the air core loop antenna and the rod antenna. The operating principle is the same for both configurations. Substituting $\mu_0\mu_e H(t)$ for B in Equation 12.10 and, assuming the loop is stationary with respect to the field vector, the terminal voltage becomes:

$$e(t) = -\mu_0\mu_e nA\frac{dH(t)}{dt} \tag{12.11}$$

where n is the number of turns in the coil, and μ_e is the effective relative permeability of the core. The core of a rod antenna is normally constructed of magnetically "soft" material so one can assume the flux density in the core is induced by an external magnetic field and, therefore, the substitution above is valid. With an air (no) core, the effective relative permeability is one. The effective permeability of an induction coil that contains a core is usually much greater than one and is strongly dependent on the shape of the core and, to some extent, on the configuration of the winding.

Taking the Laplace transform of Equation 12.11 and dividing both sides by H, one obtains the following transfer function $T(s)$ for an induction coil antenna:

$$T(s) = -\mu_0\mu_e nAs = -Ks \left(\text{VmA}^{-1}\right) \tag{12.12}$$

where $E(s) = T(s)\ H(s)$, $E(s)$ and $H(s)$ are the Laplace transforms of $e(t)$ and $H(t)$, and s is the Laplace transform operator. Inspection of Equation 12.12 reveals that the magnitude of the coil voltage is proportional to both the magnitude and frequency of the magnetic field being measured. The coil constant or sensitivity of the loop antenna is:

$$K = \mu_0\mu_e nA \left(\text{VsmA}^{-1}\right) \tag{12.13}$$

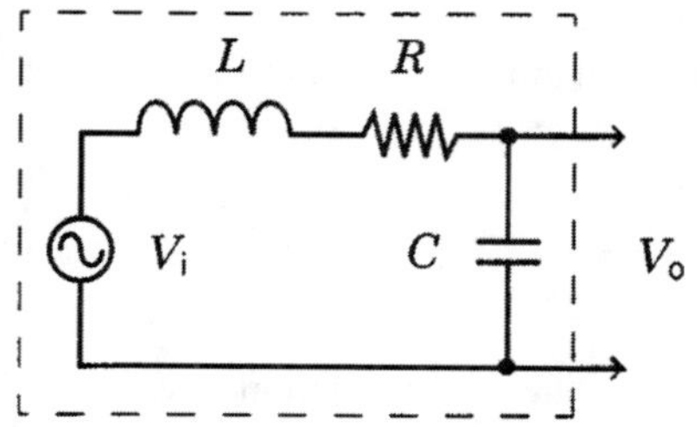

FIGURE 12.5 The induction coil equivalent circuit is a frequency-dependent voltage source in series with an inductor, resistor, and lumped capacitor.

Figure 12.5 is the equivalent circuit for an induction coil antenna. The actual voltage measured at the terminals of the loop is modified by the inductance L, resistance R, and the distributed stray and shield capacitances represented by the lumped capacitor C. These circuit parameters depend on the geometry of the core, coil, and winding.

The electrostatic shield made of nonmagnetic material shown in Figure 12.4 is an important element in the design of an induction coil. It prevents coupling of electric fields into the coil, thereby assuring that the signal seen at the coil terminals is only that due to a magnetic field. The shield should not be placed too close to the winding since it contributes to coil capacitance and noise.

The Air Core Loop Antenna

The air core loop antenna consists of a circular or rectangular loop containing one or more turns of wire and no magnetic core. The diameter of the loop is usually much greater than the dimensions of the winding cross-section. The sensitivity of a circular loop antenna with a winding inside diameter d and rectangular cross-section is approximately:

$$K = \mu_0 n \pi \frac{d^2}{4}\left[1 + 2\left(\frac{t}{d}\right) + \frac{3}{4}\left(\frac{t}{d}\right)^2\right] \tag{12.14}$$

where t is the thickness of the winding and n is the number of turns.

The resistance of the coil is:

$$R = 4n\frac{d}{d_w^2}\left(1 + \frac{t}{d}\right)\rho\Omega \tag{12.15}$$

where d_w is the diameter of the bare wire and ρ is its resistivity in Ω m (1.7×10^{-8} Ω m for copper).

The inductance of the coil is more difficult to compute since it depends heavily on the geometry of the coil. Those who are interested in computing very accurate inductance values for a wide variety of coil shapes should consult [5]. Equation 12.16 is a general expression that gives a good approximation for the inductance of a circular air core coil.

$$L = \mu_0 n^2 \pi \left(\frac{\bar{d}}{2}\right)^2 \frac{k}{w} \text{ H} \tag{12.16}$$

where w is the width of the winding, $\bar{d}$ is the average diameter, and k is Nagaoka's constant:

$$k = \frac{1}{1 + 0.45\frac{\bar{d}}{w} + 0.64\frac{t}{\bar{d}} + 0.84\frac{t}{w}} \tag{12.17}$$

The distributed capacitance of the coil contributes the most to the overall antenna capacitance. The parasitic capacitances can usually be ignored. Equation 12.18 can be used to estimate the distributed capacitance of a coil.

$$C_d = \left[\frac{\varepsilon_w \varepsilon_l}{\varepsilon_w t_l + \varepsilon_l t_w}\right] \frac{0.018544 \bar{d} w (n_l - 1)}{n_l^2} \tag{12.18}$$

where ε_w is the dielectric constant of the wire insulation, ε_l is the dielectric constant of the interlayer insulation if any, t_w is the thickness of the wire insulation, t_l is the thickness of the interlayer insulation, and n_l is the number of layers. Adding a second layer to a single-layer coil significantly increases the capacitance but, as the number of layers increases, the capacitance decreases.

The air core loop antenna is particularly useful for measuring magnetic fields with frequencies from 100 Hz to several megahertz. Because it has a linear response to magnetic field strength, it has virtually no intermodulation distortion. On the negative side, the size of the sensor can get quite large for applications that require high sensitivities at low frequencies.

The Rod Antenna

The rod antenna is a good alternative to an air core loop antenna. It is smaller in size than a loop antenna with the same sensitivity, and it can be designed to operate at lower frequencies. Unfortunately, its response to magnetic field strength can be nonlinear and the core adds noise.

Figure 12.4(b) is a typical configuration for a rod antenna. It is basically a solenoid with a magnetic core. The core can have a circular or rectangular cross-section and can be made from a ferrite, a nickel-iron alloy, an amorphous metal glass alloy, or some other material with high relative permeability. The winding can be wound directly on the core or on a form through which the core is inserted. Insulation is sometimes placed between layers of the winding to reduce distributed capacitance. An electrostatic shield is placed around the winding to attenuate any electric field coupling into the signal. The shield has a gap that runs the length of the winding. This prevents circulating currents in the shield from attenuating the magnetic field within the coil.

The most common rod antenna configuration is a core with a circular cross-section and a tightly coupled winding that runs most of the length of the core. The sensitivity of the rod antenna is computed by substituting μ_e in Equation 12.13 with the following:

$$\mu_e = 1 + \left(\frac{d_c}{d + t}\right)^2 (\bar{\mu} - 1) \tag{12.19}$$

where d_c is the core diameter and $\bar{\mu}$ is the core average effective permeability. The core effective or apparent permeability depends on its geometry and initial permeability, as well as the winding length relative to the core length. A rod becomes magnetized when a magnetic field is applied to it. In response, a magnetic field is created within the rod that opposes the externally applied field and reduces the flux density. The demagnetizing field is proportional to the magnetization and the net field H in the core is:

$$H = H' - NM \tag{12.20}$$

where H' is the applied external field, N is the demagnetizing factor, and M is the magnetization. The apparent relative permeability of a core is the ratio of the flux density B in the middle of the core to the flux density in air:

$$\frac{B}{\mu_0 H'} = \mu_a = \frac{\mu_i}{1 + N(\mu_i - 1)} \tag{12.21}$$

where μ_i is the initial relative permeability of the core material. Initial relative permeability is the slope of the B–H magnetization curve near zero applied field for a closed magnetic path.

TABLE 12.3 Demagnetizing Factors, N for Rods and Ellipsoids Magnetized Parallel to Long Axis

Dimensional ratio (length/diameter)	Rod	Prolate ellipsoid	Oblate ellipsoid
0	1.0	1.0	1.0
1	0.27	0.3333	0.3333
2	0.14	0.1735	0.2364
5	0.040	0.0558	0.1248
10	0.0172	0.0203	0.0696
20	0.00617	0.00675	0.0369
50	0.00129	0.00144	0.01472
100	0.00036	0.000430	0.00776
200	0.000090	0.000125	0.00390
500	0.000014	0.0000236	0.001576
1000	0.0000036	0.0000066	0.000784
2000	0.0000009	0.0000019	0.000392

TABLE 12.4 Magnetic Properties of Typical Core Material

Name	Composition	Manufacturer	μ_i	μ_{max}
Mild steel	0.2 C, 99 Fe		120	2000
Silicon iron	4.0 Si, 96 Fe		500	7000
CN20	Ni-Zn Ferrite	Ceramic Magnetics	800	4500
MN60	Mn-Zn Ferrite	Ceramic Magnetics	5000	10,500
"49" Alloy	48 Ni, 52 Fe	Carpenter	6500	75,000
2605S-2	Fe-based amorphous alloy	Allied-Signal	10,000	600,000
4-79 Permalloy	4 Mn, 79 Ni, 17 Fe	Magnetics	20,000	100,000
Mumetal	5 Cu, 2 Cr, 77 Ni, 16 Fe	Magnetics	20,000	100,000
HyMu "80"	4.2 Mo, 80 Ni, 15 Fe	Carpenter	50,000	200,000
2826MB	NiFe-based amorphous alloy	Allied-Signal	100,000	800,000

Note: μ_i is the slope of the magnetization curve at the origin. μ_{max} is the maximum incremental slope of the magnetization curve.

The value of N is shape dependent. As the length-to-diameter ratio m of a rod increases, N decreases and the apparent relative permeability approaches the initial permeability. Table 12.3, which is reproduced from [6], lists demagnetizing factors for a rod, prolate ellipsoid (cigar shape), and oblate ellipsoid (disk shape).

Equation 12.22 can be used to approximate the value of N for cylindrical rods with $m > 10$ and $\mu_i >$ 1000:

$$N = \frac{2.01 \times \log_{10} m - 0.46}{m^2} \tag{12.22}$$

The apparent permeability of a rod with a small m and large μ_i is almost exclusively determined by m alone. Table 12.4 lists the magnetic properties of several ferromagnetic materials that can be used to construct a core.

Bozorth [7] found that the apparent permeability of a rod is not constant throughout the length of the rod. It reaches a maximum at the center of the rod and continuously drops in value until the ends of the rod are reached. The variation in permeability can be approximated by:

$$\mu(l) = \mu_a \left[1 - F\left(\frac{l}{l_0}\right)^2 \right] \tag{12.23}$$

where l is the distance from the center of the rod to the measurement point, l_0 is the half length of the rod, and F is a constant that varies from 0.72 to 0.96. The average permeability seen by the coil is the integral of Equation 12.23 over the length of the coil:

$$\bar{\mu} = \mu_a \left[1 - F\left(\frac{l_w}{l_c}\right)^2 \right] \tag{12.24}$$

where l_w is the length of the winding and l_c is the length of the core. Equation 12.24 is substituted into Equation 12.19 to compute the rod's effective relative permeability which is used in Equation 12.13 to compute sensitivity.

The inductance of the rod antenna can be computed using the following equations:

$$L = \frac{\mu_0 \mu_e n^2 \pi (d+t)^2 l_w}{4 l_c} \tag{12.25}$$

$$\mu_e = 1 + \left(\frac{d_c}{d+t}\right)^2 \left[\mu_a f(l_w/l_c) - 1\right] \tag{12.26}$$

$$f(l_w/l_c) = 1.9088 - 0.8672(l_w/l_c) - 1.1217(l_w/l_c)^2 + 0.8263(l_w/l_c)^3 \tag{12.27}$$

The function $f(l_w/l_c)$ accounts for the variation in flux density from the middle of the winding to its ends and assumes the winding is centered about the middle of the core.

Equations 12.15 and 12.16 can be used to compute the resistance and capacitance of a rod antenna.

Signal Conditioning

To be useful, the induction coil signal must be conditioned using either a voltage or a current amplifier. Figure 12.6 illustrates the circuit configurations for both of these signal conditioning methods. The voltage amplifier can have either a single-ended or differential input and it can be tuned or untuned. The signal output of the voltage amplifier is proportional to the magnitude and frequency of the field for frequencies well below resonance. Its output will peak at the resonant frequency of the coil or at the tuning frequency. Because its output signal depends on both the frequency and strength of the magnetic field, the voltage amplifier is more suited to narrow band or tuned frequency applications.

In the current amplifier configuration, the induction coil terminals are connected to a virtual ground. As long as the product of the amplifier forward gain and the coil ohmic resistance is much greater than the feedback resistor, the output signal magnitude is independent of the frequency of the magnetic field beyond the R/L (rad s^{-1}) corner of the coil. This remains true up to the coil's resonant frequency. For this reason, the current amplifier configuration is particularly suited to broadband magnetic field strength measurements. The current amplifier configuration minimizes intermodulation distortion in induction coils with magnetic cores. The current flowing through the coil produces a magnetic field that opposes the ambient field. This keeps the net field in the core near zero and in a linear region of the B–H curve.

Current-amplifier-based induction coil magnetometers have been built that have a flat frequency response from 10 Hz to over 200 kHz. Some magnetometers designed for geophysical exploration applications have low frequency corners that extend down to 0.1 Hz. For further information on this subject, see [8, 9].

The Fluxgate Magnetometer

The fluxgate magnetometer has been and is the workhorse of magnetic field strength instruments both on Earth and in space. It is rugged, reliable, physically small, and requires very little power to operate. These characteristics, along with its ability to measure the vector components of magnetic fields over a

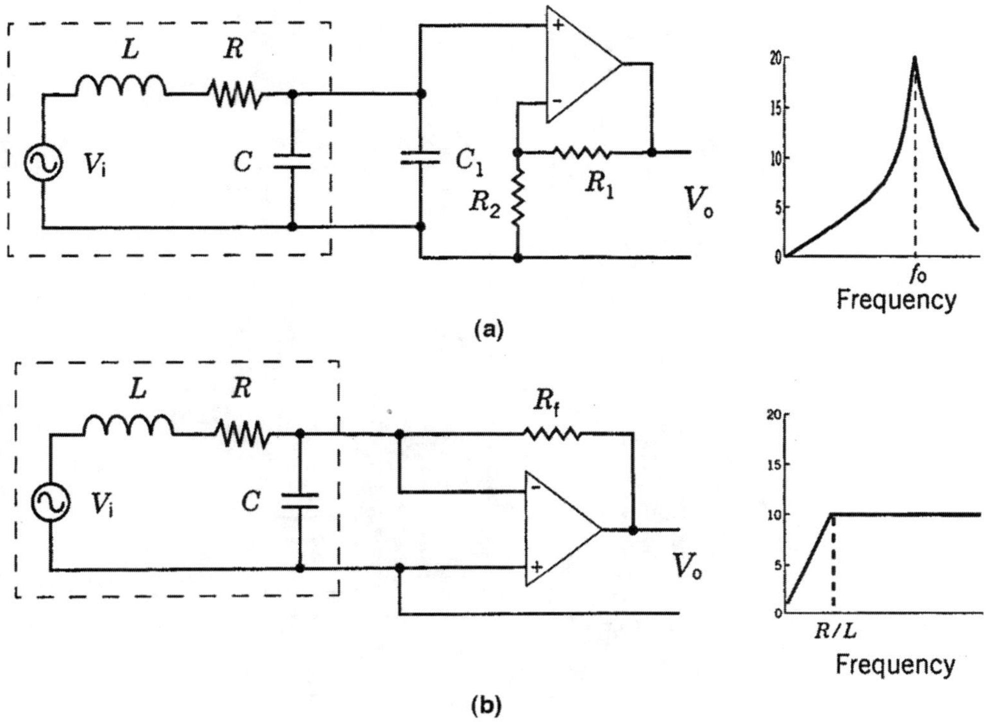

FIGURE 12.6 (a) The amplitude of a voltage-amplified induction coil signal is proportional to the frequency and strength of the field. (b) The amplitude of a current-amplified induction coil signal is only proportional to field strength beyond its L/R corner frequency.

0.1 nT to 1 mT range from dc to several kHz, make it a very versatile instrument. Geologists use them for exploration and geophysicists use them to study the geomagnetic field (about 20 μT to 75 μT on the Earth's surface). Satellite engineers use them to determine and control the attitude of spacecraft, scientists use them in their research, and the military uses them in many applications, including mine detection, vehicle detection, and target recognition. Some airport security systems use them to detect weapons.

The Fluxgate

The heart of the magnetometer is the *fluxgate*. It is the transducer that converts a magnetic field into an electric voltage. There are many different fluxgate configurations. Two of the more popular ones are shown in Figure 12.7. A very comprehensive explanation of the fluxgate principle and the different fluxgate configurations is given in [10].

The ring core fluxgate is constructed from a thin ribbon of easily saturable ferromagnetic material, such as 4-79 Permalloy wrapped around a bobbin to form a ring or toroid. As shown in Figure 12.8, an alternating current is applied through a coil that is wound about the toroid. This creates a magnetic field that circulates around the magnetic core. This magnetic field causes the flux in the ferrous material to periodically saturate first clockwise and then counterclockwise. A pick-up (signal) winding is wrapped around the outside of the toroid. While the ferrous material is between saturation extremes, it maintains an average permeability much greater than that of air. When the core is in saturation, the core permeability becomes equal to that of air. If there is no component of magnetic field along the axis of the signal winding, the flux change seen by the winding is zero. If, on the other hand, a field component is present along the signal winding axis, then each time the ferrous material goes from one saturation extreme to the other, the flux within the core will change from a low level to a high level. According to Faraday's law, a changing flux will produce a voltage at the terminals of the signal winding that is proportional to the rate of change of flux. For dc and low-frequency magnetic fields, the signal winding voltage is:

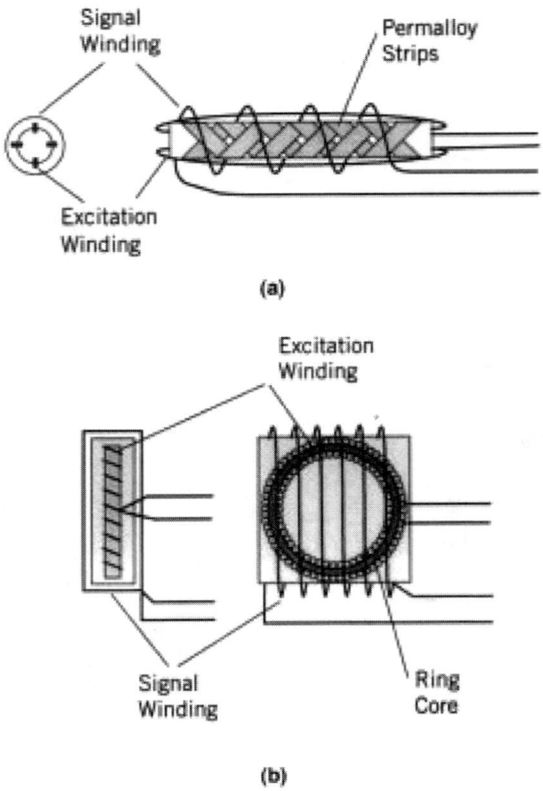

FIGURE 12.7 In Schonstedt (a) and ring core (b) fluxgate sensors, the excitation field is at right angles to the signal winding axis. This configuration minimizes coupling between the excitation field and the signal winding.

$$e(t) = nA\frac{\mathrm{d}(\mu_0\mu_e H)}{\mathrm{d}t} = nA\mu_0 H\frac{\mathrm{d}\mu_e(t)}{\mathrm{d}t} \tag{12.28}$$

where H = Component of the magnetic field being measured
n = Number of turns on the signal winding
A = Cross-sectional area of the signal winding
$\mu_e(t)$ = Effective relative permeability of the core

As the core permeability alternates from a low value to a high value, it produces a voltage pulse at the signal winding output that has an amplitude proportional to the magnitude of the external magnetic field and a phase indicating the direction of the field. The frequency of the signal is twice the excitation frequency since the saturation-to-saturation transition occurs twice each excitation period.

The discussion about effective permeability in the induction coil section applies here as well. Consult [10, 13] for comprehensive discussions about fluxgate effective permeability and signal characteristics as they relate to excitation field level, excitation waveform, and winding geometry.

Signal Conditioning

The signal from the fluxgate is an amplitude-modulated suppressed carrier signal that is synchronous with the second harmonic of the excitation signal. In a simple low-power magnetometer, this signal is converted to the base band using a synchronous demodulator, filtered, and presented as the final output. Example circuits are given in [11, 12]. The accuracy of magnetometers that use this open-loop architecture is limited by the linearity of the core's magnetization curve and is about 5% for Earth's field (60 μT) applications.

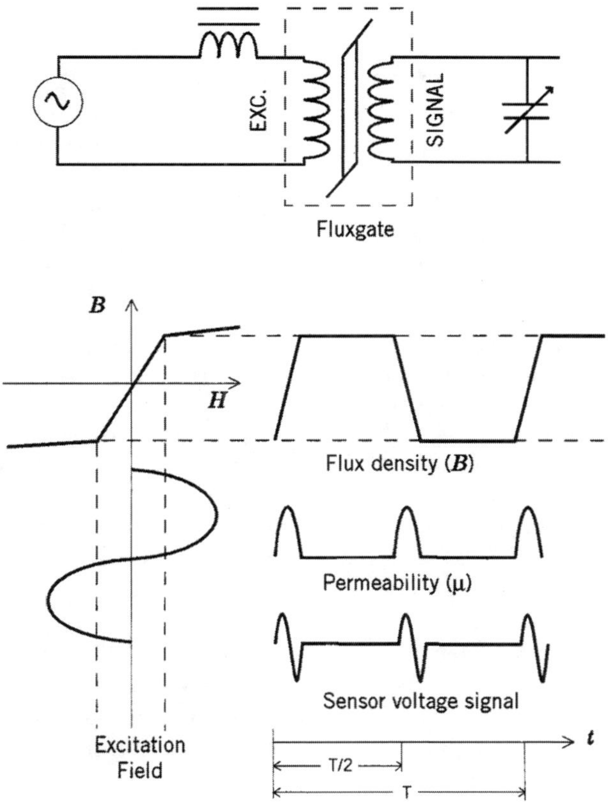

FIGURE 12.8 The excitation field of a fluxgate magnetometer alternately drives the core into positive or negative saturation, causing the core's effective permeability to switch between 1 and a large value twice each cycle.

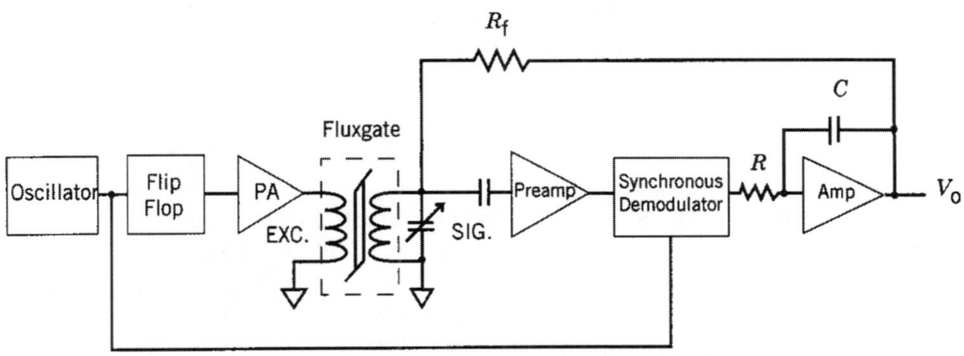

FIGURE 12.9 Typical circuit configuration for a field feedback fluxgate magnetometer. The sensor output is ac amplified, synchronously demodulated, and filtered. A magnetic field that nulls the ambient field at the sensor is produced by connecting the resistor R_f between the output and the signal winding.

More precise and stable magnetometers use magnetic field feedback rather than the open-loop structure described above. A simplified schematic of a typical second harmonic field feedback fluxgate magnetometer is shown in Figure 12.9. The circuitry to the left of the fluxgate is called the excitation circuit. It consists of an oscillator tuned to twice the excitation frequency, a flip-flop that divides the oscillator frequency by two, and a power amplifier driven by the flip-flop and, in turn, provides the excitation current to the excitation winding.

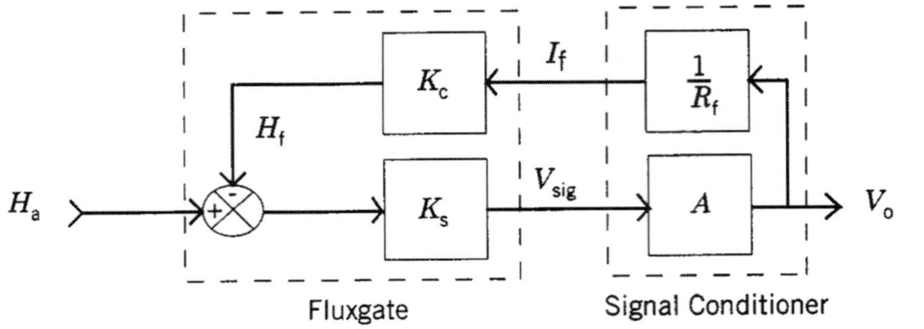

FIGURE 12.10 Block diagram of a field feedback fluxgate magnetometer. K_c is the current-to-field constant for the coil. K_s is the field-to-voltage transduction constant for the sensor. The feedback field H_f opposes the ambient field H_a, thus keeping the net sensor field very small.

The circuitry to the right of the fluxgate is called the signal channel circuit. It amplifies the output from the fluxgate signal winding, synchronously demodulates the ac signal using the oscillator signal as a reference, integrates and amplifies the base band output, and then feeds back the output through a resistor to the signal winding. The fed-back signal produces a magnetic field inside the sensor that opposes the external field. This keeps the field inside the sensor near zero and in a linear portion of the magnetization curve of the ferromagnetic core.

The flow diagram for the magnetometer is given in Figure 12.10. The external field H_a is opposed by the feedback field H_f, and the difference is converted into a voltage signal (K_s represents the transfer function from field to voltage). This signal is amplified (A), and the amplified signal is converted into a current I_f and then into the feedback field (K_c represents the transfer function from current to field). The overall transfer function for the magnetometer is:

$$\frac{V_0}{H_a} = \frac{AK_s}{1+\dfrac{K_c A K_s}{R_f}} \tag{12.29}$$

The amplifier gain is normally very high such that the second term in the denominator is much larger than one, and Equation 12.29 reduces to

$$\frac{V_0}{H_a} = \frac{R_f}{K_c} \tag{12.30}$$

Under these circumstances, the transfer function becomes almost completely determined by the ratio of R_f (the feedback resistor) to K_c (the current-to-field coil constant of the sensor winding). Both of these constants can be very well controlled. The consequence of this circuit topology is a highly stable and accurate magnetometer that is insensitive to circuit component variations with temperature or time. An accuracy of 1% over a temperature range of –80°C to 80°C is easily achievable. Accuracy and stability can be improved using a current feedback circuit, like the one described in [13], that compensates for the resistance of the signal winding or by using a separate feedback winding and a high-quality voltage-to-current converter instead of a simple feedback resistor.

The SQUID Magnetometer

Brian D. Josephson in 1962, while a graduate student at Cambridge University, predicted that superconducting current could flow between two superconductors that are separated by a thin insulation layer. The magnitude of the superconductor (critical) current through this "Josephson junction" is affected by the presence of a magnetic field and forms the basis for the SQUID magnetometer.

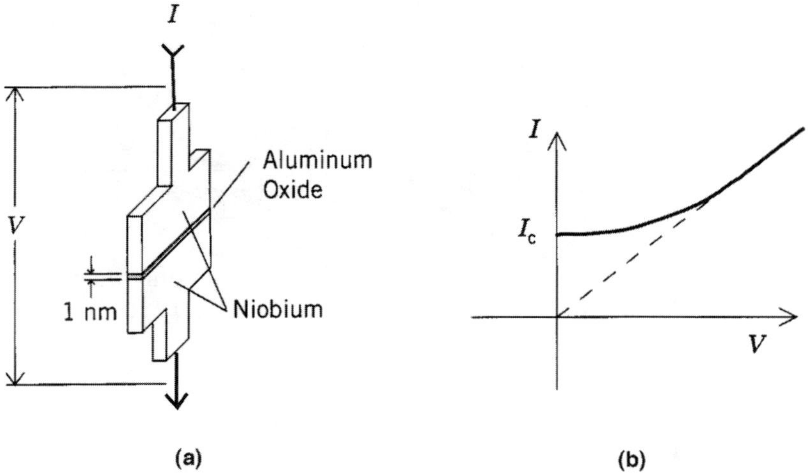

FIGURE 12.11 The Josephson junction in (a) consists of a superconductor such as niobium separated by a thin insulation layer. The voltage (V) vs. current (I) curve in (b) shows that a superconducting current flows through the junction with zero volts across the junction.

Figure 12.11 illustrates the general structure of a Josephson junction and the voltage–current (V–I) relationship. Two superconductors (e.g., niobium) are separated by a very thin insulating layer (e.g., aluminum oxide). The thickness of this layer is typically 1 nm. When the temperature of the junction is reduced to below 4.2 K (–269°C), a superconductor current will flow in the junction with 0 V across the junction. The magnitude of this current, called the critical current I_c, is a periodic function of the magnetic flux present in the junction. Its maximum magnitude occurs for flux values equal to $n\phi_0$, where ϕ_0 is one flux quantum (2 fW), and its minimum magnitude occurs for flux values equal to $(n + 1/2)\phi_0$. The period is one flux quantum. This phenomenon is called the "DC Josephson effect" and is only one of the "Josephson effects."

Magnetometers based on the Superconducting Quantum Interference Device (SQUID) are currently the most sensitive instruments available for measuring magnetic field strength. SQUID magnetometers measure the change in the magnetic field from some arbitrary field level; they do not intrinsically measure the absolute value of the field. Biomedical research is one of the more important applications of SQUID magnetometers. SQUID magnetometers and gradiometers (measure spatial variation in the magnetic field) have the high sensitivities needed to measure the weak magnetic fields generated by the body [15]. Other application areas include paleomagnetics (measuring remnant magnetism in rocks) and magnetotellurics (Earth resistivity measurements). Descriptions of these applications as well as the general theory of SQUIDs can be found in [16]. Clark [17], one of the pioneers in SQUID magnetometers, provides a good contemporary overview of SQUID technology and applications.

A DC SQUID magnetometer uses two Josephson junctions in the two legs of a toroid as shown in Figure 12.12(a). The toroid is biased with a constant current that exceeds the maximum critical current of the junctions. When the flux through the toroid is an integral multiple of ϕ_0, the voltage across the junctions is determined by the intersection of I_b and the $n\phi_0$ V–I curve (point A). As the flux increases, the critical current decreases. The V–I curve and thus the intersection point move to the right (the junction voltage increases). The critical current reaches a minimum when the flux has increased by $1/2\phi_0$ and the junction voltage is at its maximum (point B). As the flux continues to increase, the critical current increases back toward its maximum value and the junction voltage decreases. Thus, the period of the flux cycle is ϕ_0.

Signal Conditioning

Figure 12.13 is a block diagram of one implementation of a DC SQUID magnetometer that can be used for wide dynamic range field measurements. A large superconducting loop, which is exposed to the

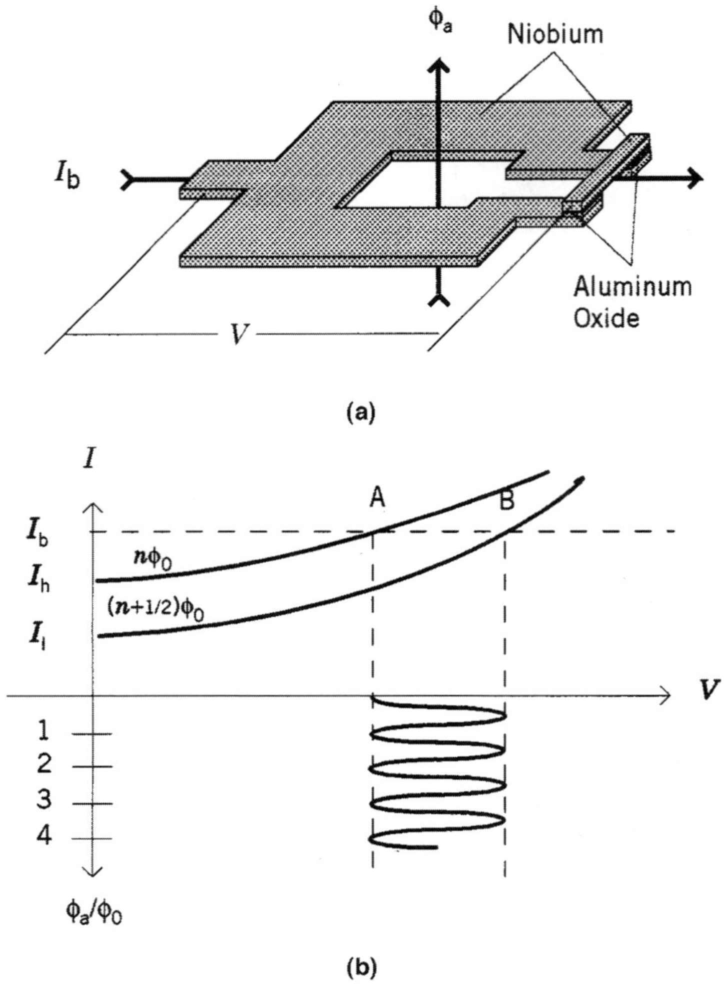

FIGURE 12.12 Use of a DC SQUID to measure magnetic flux. The DC SQUID in (a) consists of a superconductor loop and two Josephson junctions with a bias current that is greater than the maximum critical current I_h. The V–I curve in (b) illustrates how the voltage across the SQUID oscillates with a period equal to one flux quantum ϕ_0.

magnetic field being measured, is connected to a multiturn signal winding that is magnetically coupled directly to the SQUID. At cryogenic temperatures, the loop and signal winding effectively form a dc induction coil. External flux applied to the coil will generate a current in the loop that keeps the net flux within the loop constant, even for dc magnetic fields. The signal winding magnifies the flux that is applied to the SQUID.

The SQUID is magnetically biased at an optimal sensitivity point. A small ac magnetic field at 100 kHz to 500 kHz is superimposed on the bias field. The output of the SQUID is a suppressed carrier amplitude modulated signal where the amplitude indicates the change in magnetic field from the bias point, and the phase indicates the polarity of the change. The output signal is amplified and then synchronously demodulated down to the base band. The resulting dc signal is amplified and fed back through a resistor to a coil coupled to the SQUID. The current through the coil generates a magnetic field at the SQUID that opposes the applied field. This keeps the SQUID operating point very near the bias point. The scale factor of the magnetometer depends on the feedback resistor and the coil constant of the feedback winding in the same manner that it does for a field feedback fluxgate magnetometer.

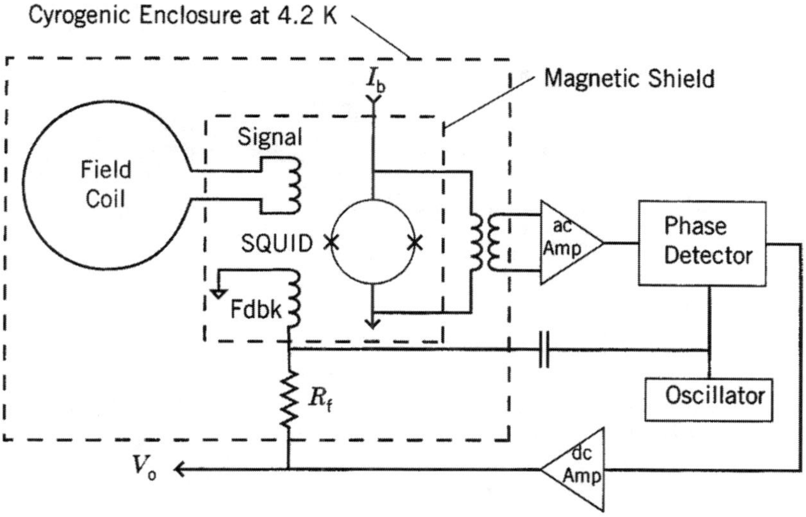

FIGURE 12.13 Wide dynamic range DC SQUID magnetometer. A magnetic field produced by connecting resistor R_f between the output and a feedback coil keeps the field in the SQUID within one flux quantum over its operating range. (Adapted from Wellstood, Heiden, and Clark, 1984.)

The pick-up loop, signal coil, SQUID, feedback coil, and feedback resistor are kept in a cryogenic temperature chamber and, except for the pick-up coil, are magnetically shielded. The rest of the circuit is operated at room temperature.

12.3 High-Field Vector Gaussmeters

The Hall Effect Gaussmeter

The Hall effect device, which is probably the most familiar and widely used sensor for measuring strong magnetic fields, is based on the discovery of the Hall effect by Edwin H. Hall in 1897. The Hall effect is a consequence of the Lorentz force law, which states that a moving charge q, when acted upon by a magnetic induction field $\vec{B}$, will experience a force $\vec{F}$ that is at right angles to the field vector and the velocity vector v of the charge as expressed by the following equation:

$$\vec{F} = -q\left(\vec{E} + \vec{v} \times \vec{B}\right) \tag{12.31}$$

The Hall effect device consists of a flat, thin rectangular conductor or semiconductor with two pairs of electrodes at right angles to one another as illustrated in Figure 12.14. An electric field E_x is applied along the x or control axis. When a magnetic field B_z is applied perpendicular to the surface of the device, the free charge, which is flowing along the x-axis as a result of E_x, will be deflected toward the y or Hall voltage axis. Since current cannot flow in the y-axis under open-loop conditions, this will cause a buildup of charge along the y-axis that will create an electric field which produces a force opposing the motion of the charge:

$$E_y = v_x B_z \tag{12.32}$$

where v_x is the average drift velocity of the electrons (or majority carriers). In a conductor that contains n free charges per unit volume having an average drift velocity of v_x, the current density is:

$$J_x = qnv_x \tag{12.33}$$

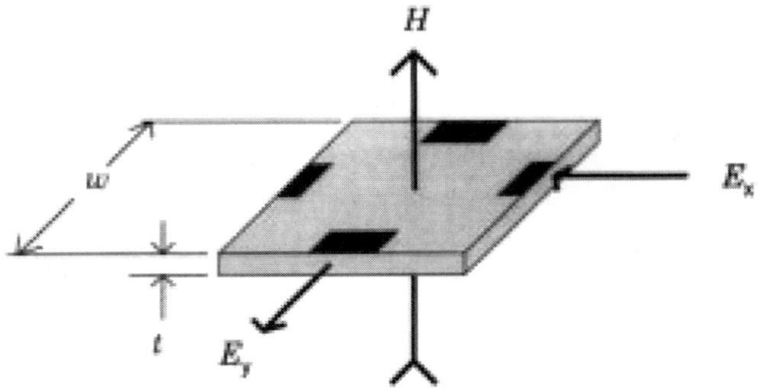

FIGURE 12.14 Hall effect sensor. A magnetic field H applied normal to the surface of the sensor, which is conducting current along the x-direction, will generate a voltage along the y-direction. E_x is the applied electric field along the x-direction, and E_y is the Hall effect electric field along the y-direction.

and

$$E_y = \frac{J_x B_z}{qn} = R_H J_x B_z \tag{12.34}$$

where R_H is called the Hall coefficient.

A semiconductor is treated in terms of the mobility μ (drift velocity/field) of the majority carrier (electron or hole) and conductivity σ. In this case,

$$E_y = \mu E_x B_z \quad \text{and} \quad E_x = \frac{J_x}{\sigma} \tag{12.35}$$

Therefore,

$$E_y = \frac{\mu}{\sigma} J_x B_z \quad \text{and} \quad R_H = \frac{\mu}{\sigma} \tag{12.36}$$

The value of R_H varies substantially from one material to another and is both temperature and field magnitude dependent. Its characteristics can be controlled to a certain extent by doping the base material with some impurities. For example, doping germanium with arsenic can reduce the temperature dependence at the expense of magnitude.

The voltage measured across the y-axis terminals is the integral of the electric field along the y-axis. If a constant control current I is flowing along the x axis, then:

$$J_x = \frac{I}{wt} \tag{12.37}$$

and the measured output voltage is:

$$e_y = \frac{R_H I B_z}{t} \tag{12.38}$$

where t is thickness (m) and w is the distance between the y-axis terminals.

Another characteristic specified by manufacturers of Hall effect devices is the magnetic sensitivity γ_b at the rated control current I_c:

$$\gamma_b = \frac{e_y}{B_z} = \frac{R_H I_c}{t} \tag{12.39}$$

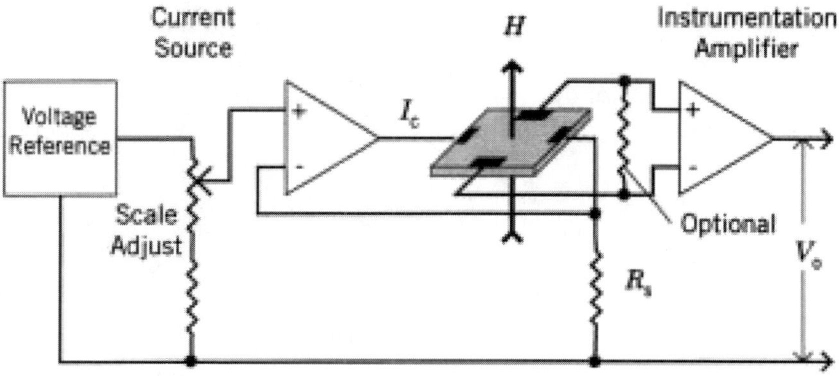

FIGURE 12.15 Example of how to construct a Hall effect gaussmeter. The operational amplifier and resistor R_s form a stable constant-current source for the Hall effect sensor. An instrumentation or differential amplifier amplifies and scales the Hall voltage. A load resistor is sometimes required across the Hall voltage output terminals.

Although conductors such as copper (Cu) can be used to make a Hall effect device, semiconductor materials, such as gallium arsenide (GaAs), indium antimonide (InSb), and indium arsenide (InAs), produce the highest and most stable Hall coefficients. InAs, because of its combined low temperature coefficient of sensitivity (<0.1%/°C), low resistance, and relatively good sensitivity, is the material favored by commercial manufacturers of Hall effect devices.

The typical control current for Hall effect devices is 100 mA, but some do operate at currents as low as 1 mA. Sensitivities range from 10 mV/T to 1.4 V/T. Linearity ranges from 1/4% to 2% over their rated operating field range. The control input and the voltage output resistance are typically in the range of 1 Ω to 3 Ω. The sensor element is usually tiny (on the order of 10 mm square by 0.5 mm thick), and a three-axis version can be housed in a very small package. These devices are most effective for measuring flux densities ranging from 50 μT to 30 T.

Signal Conditioning

A simple Hall effect gaussmeter can be constructed using the signal conditioning circuit shown in Figure 12.15. The voltage reference, operational amplifier, and sense resistor R_s form a precision constant-current source for the Hall effect device control current I_c. For best performance, the voltage reference and R_s should be very stable with temperature and time. A general-purpose operational amplifier can be used for low control currents. A power amplifier is required for control currents above 20 mA.

The Hall voltage can be conditioned and amplified by any high input impedance (>1 kΩ) differential amplifier. A precision instrumentation amplifier is a good choice because it has adequate input impedance, its gain can be determined by a stable resistor, and the amplifier zero offset trim resistor can be used to cancel the zero offset of the Hall effect device. Some devices require a load resistor across the Hall voltage terminal to achieve optimum linearity.

The zero offset and $1/f$ noise of the Hall voltage amplifier limit the performance of a Hall effect gaussmeter for low field strength measurements. Sometimes, these effects can be reduced by using an ac precision current source. The ac amplitude modulated Hall voltage can then be amplified in a more favorable frequency band and synchronously detected to extract the Hall voltage signal. If the field to be measured requires this amount of signal conditioning, it probably is better to use a fluxgate magnetometer for the application.

The Magnetoresistive Gaussmeter

The magnetoresistance effect was first reported by William Thomson (Lord Kelvin) in the middle of the 19th century. He found that a magnetic field applied to a ferromagnetic material caused its resistivity to change. The amount of change depends on the magnetization magnitude and the direction in which the

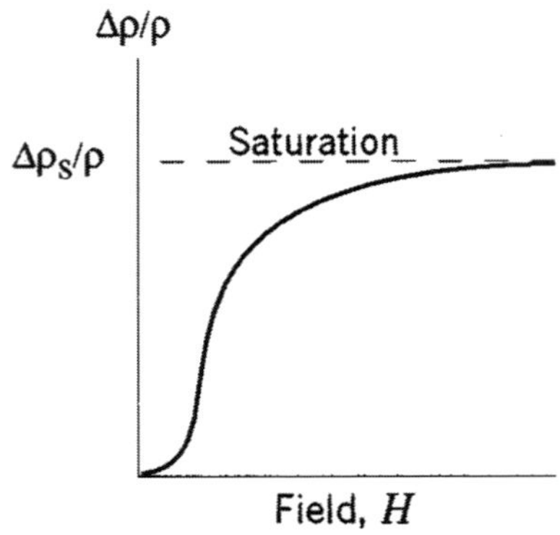

FIGURE 12.16 Change in resistivity in a ferromagnetic material. As field is applied, the resistivity changes rapidly at first. As the material approaches magnetic flux saturation, the resistivity change approaches its maximum value.

current used to measure resistivity is flowing. Nickel–iron alloys show the greatest change in resistivity (about 5% maximum). Figure 12.16 illustrates how the resistivity changes in Permalloy (a nickel–iron alloy) for a field applied parallel to the current flow. As magnetic field is increased, the change in resistivity increases and asymptotically approaches its maximum value when the material approaches saturation. Bozorth [6] points out that the shape of the curve and the magnitude of the change depend on the composition of the alloy. Permalloy with 80% Ni and 20% Fe provides a high magnetoresistance effect with near-zero magnetostriction and is a favorite material for magnetoresistors.

The change in resistivity in permalloy film [18] is also a function of the angle θ between the magnetization direction and the current direction:

$$\rho(\theta) = \rho_0 + \Delta\rho_m \cos^2\theta \tag{12.40}$$

where $\Delta\rho_m$ is the magnetoresistivity anisotropy change and ρ_0 is the resistivity for $\theta = \pi/2$.

It was mentioned earlier that magnetic materials have anisotropic magnetic properties (their magnetic properties are direction dependent). The physical shape of an object (see the discussion on demagnetizing factor above) and the conditions that exist during fabrication strongly determine its anisotropic characteristics. A thin long film of permalloy can be made to have highly uniaxial anisotropic properties if it is exposed to a magnetizing field during deposition. This characteristic is exploited in the anisotropic magnetoresistance (AMR) sensor.

The basic resistor element in an AMR is a thin rectangular shaped film as shown in Figure 12.17. One axis, called the anisotropy or easy axis, has a much higher susceptibility to magnetization than the other two. The easy axis is normally along the length of the film. Because of its thinness, the axis normal to

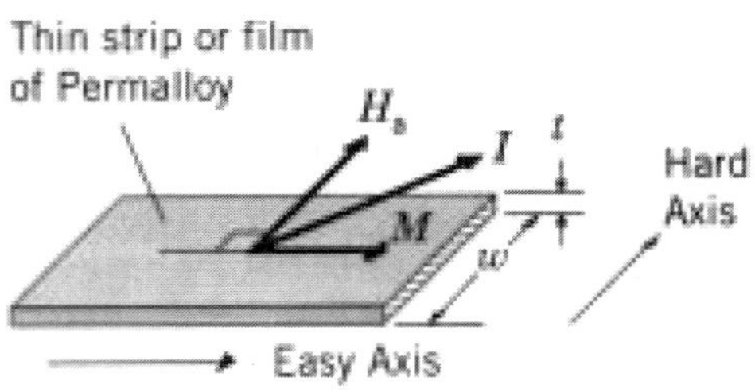

FIGURE 12.17 An AMR resistor element. During fabrication, a magnetic field is applied along the strip's length to magnetize it and establish its easy axis. Current I is passed through the film at 45° to the easy or anisotropic axis. A magnetic field H_a applied at right angles to the magnetization vector M causes the magnetization vector to rotate and the magnetoresistance to change.

the film has virtually no magnetic susceptibility. The axis transverse to the easy axis (across the width) has very little susceptibility as well.

A bias field H_b is used to saturate the magnetization along the easy axis and establish the magnetization direction for zero external field. For a simplified analysis, the film can be modeled as a single domain. The effect of an external field in the plane of the film and normal to the anisotropy axis is to rotate the magnetization vector and, according to Equation 12.40, change the resistivity. Kwiatkowski and Tumanski [19] stated that the change in resistance of the film can be approximated by Equation 12.41:

$$\Delta R \approx R_s \frac{\Delta\rho_m}{\rho}\left(h_a^2 \cos 2\theta + h_a \sqrt{1-h_a^2}\sin 2\theta - \frac{1}{2}\cos 2\theta\right) \quad (12.41)$$

where h_a is the normalized externally applied field (i.e., $h_a = H_a/H_k$), R_s is the nominal resistance, and $\Delta\rho_m/\rho$ is the maximum resistivity change. H_k is the anisotropy field. Optimum linear performance is obtained when $\theta = \pi/4$ and Equation 12.41 reduces to:

$$\Delta R \approx R_s \frac{\Delta\rho_m}{\rho}\frac{1}{H_k + H_b} H_a \quad (12.42)$$

The anisotropy field is given by:

$$H_k = \sqrt{H_{k0}^2 + (NM_s)^2} \quad (12.43)$$

where H_{k0} is the film anisotropy field, N is the demagnetizing factor ($\approx$ thickness(t)/width(w)) and M_s is the saturation magnetization.

An AMR is constructed using long thin film segments of deposited permalloy. During deposition, a magnetic field is applied along the length of the film to establish its easy axis of magnetization. The shape of the film also favors the length as an easy axis. As shown in Figure 12.18, a series of these permalloy films is connected together to form the magnetoresistor. The current is forced to flow at a 45° angle to the easy axis by depositing thin strips of highly conductive material (e.g., gold) across the permalloy film. The level of magnetization of the film is controlled by a bias field that is created through the deposition

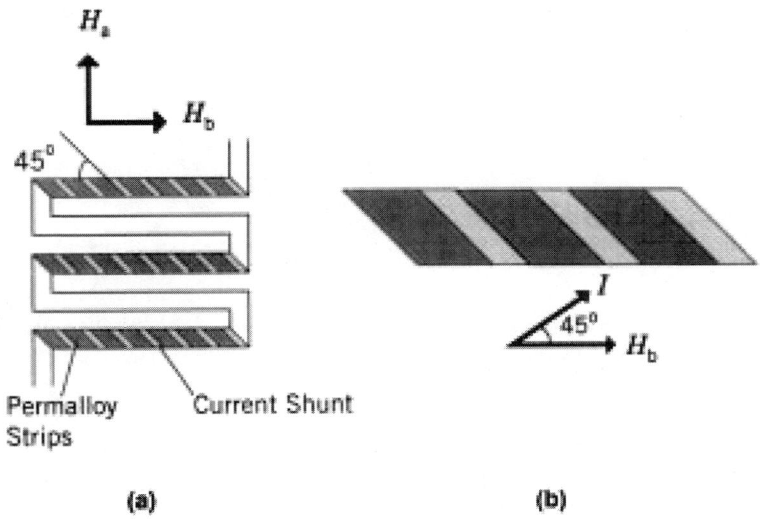

FIGURE 12.18 Magnetoresistor construction. (a) A typical AMR element consists of multiple strips of permalloy connected together in a serpentine pattern. Current shunts force the current to flow through the permalloy at 45° to the easy axis. (b) A close-up view.

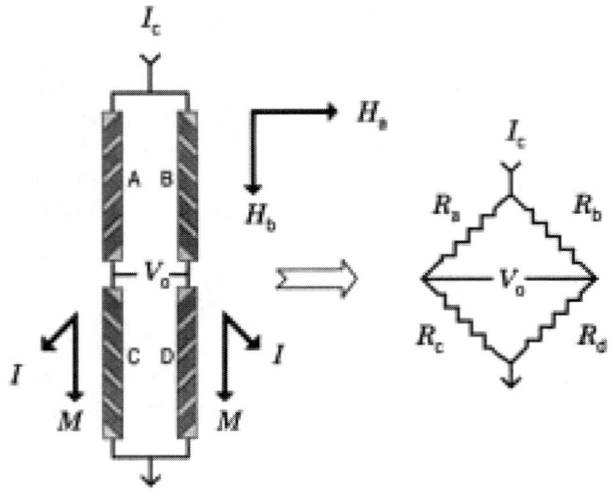

FIGURE 12.19 AMR bridge sensor. In an AMR bridge, the current shunts of resistors A and D are the same and reversed from B and C. Thus, the resistors on diagonal legs of the bridge have the same response to an applied field and opposite that of the other diagonal pair. Bridge leg resistance varies from 1 kΩ to 100 kΩ.

of a thin layer of cobalt over the resistors, which is then magnetized parallel to the easy axis of the permalloy.

A typical AMR sensor suitable for a gaussmeter or magnetometer consists of four AMRs connected in a Wheatstone bridge as shown in Figure 12.19. The transfer function polarity of the A and D resistors is made to be opposite that of the B and C resistors by rotating the current shunt 90°. This complementary arrangement enhances the output voltage signal for a given field by a factor of four over a single resistor. Kwiatkowski and Tumanski [19] showed that the transfer function for the bridge configuration is described by:

$$v = IR_s \frac{\Delta\rho_m}{\rho} \cos 2\Delta\varepsilon h_a \sqrt{1-h_a^2} \tag{12.44}$$

Where:

$$\cos 2\Delta\varepsilon = \frac{H_{k0}^2 + H_k^2 - (NM_s)^2}{2H_{k0}H_k} \tag{12.45}$$

$$h_a = \frac{H_a}{H_k + H_b} \tag{12.46}$$

For best linearity, $H_a < 0.1\, H_k$. The linearity of the bridge can be controlled during fabrication by adjusting the l/w ratio and H_{k0}. The bias field can also be used to optimize linearity and establish the measurement field range. Some transfer functions for a typical AMR bridge [20] are shown in Figure 12.20. A more comprehensive discussion of AMR theory can be found in [21–23].

Signal Conditioning

Conventional Wheatstone bridge signal conditioning circuits can be used to process the AMR bridge. The bridge sensitivity and zero offset are proportional to the bridge voltage, so it is important to use a well-regulated supply with low noise and good temperature stability.

As Equation 12.44 shows, the polarity of the transfer function is determined by the polarity of $(H_k + H_b)$. If the sensor is exposed to an external field that is strong enough to reverse this field, then the transfer function polarity will reverse. To overcome this ambiguity, the polarity should be established

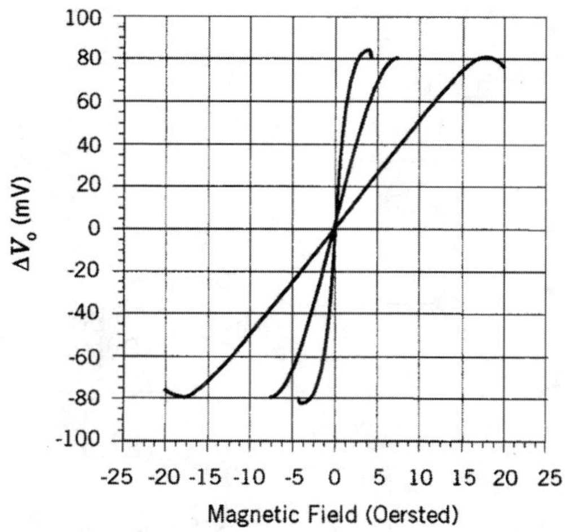

FIGURE 12.20 Typical AMR bridge sensor transfer functions. The sensitivity of an AMR bridge can be adjusted by changing its bias field. Increases in sensitivity are accompanied by corresponding decreases in range.

prior to making a measurement. This can be accomplished by momentarily applying a strong magnetic field along the easy axis of the AMR bridge. Some commercial AMR bridges come with a built-in method for performing this action.

Figure 12.21 is a block diagram for a signal conditioner that takes advantage of the bias field polarity flipping property to eliminate zero offset errors and low frequency $1/f$ noise. A square wave oscillator is used to alternately change the direction of the bias field and thus the polarity of the transfer function. The duration of the current used to set the bias field direction should be short in order to minimize power consumption. The amplitude of the ac signal from the bridge is proportional to the field magnitude, and its phase relative to the oscillator gives the field direction. This signal can be amplified and then phase-detected to extract the field-related voltage. Optionally, the output signal can be fed back through a coil that produces a magnetic field opposing the field being measured. This feedback arrangement makes the AMR bridge a null detector and minimizes the influence of changes in its transfer function on overall performance. Of course, the added circuitry increases the size, cost, and complexity of the instrument.

12.4 Scalar Magnetometers

Scalar magnetometers measure the magnitude of the magnetic field vector by exploiting the atomic and nuclear properties of matter. The two most widely used scalar magnetometers are the proton precession and the optically pumped magnetometer. When operated under the right conditions, these instruments have extremely high resolution and accuracy and are relatively insensitive to orientation. They both have several common operating limitations. The instruments require the magnetic field to be uniform throughout the sensing element volume. They have a limited magnetic field magnitude measurement range: typically 20 μT to 100 μT. And they have limitations with respect to the orientation of the magnetic field vector relative to the sensor element.

The proton precession magnetometer uses a strong magnetic field to polarize the protons in a hydrocarbon and then detects the precession frequency of the protons while they decay to the nonpolarized state after the polarizing field is turned off. The precession frequency is proportional to the magnitude of any ambient magnetic field that is present after the polarizing field is removed. This sampling of the magnetic field strength through the polarize-listen sequence makes the proton precession magnetometer response very slow. Maximum rates of only a few samples per second are typical. Because of its dependence on atomic constants, the proton precession magnetometer is the primary standard for calibrating systems used to generate magnetic fields and calibrate magnetometers.

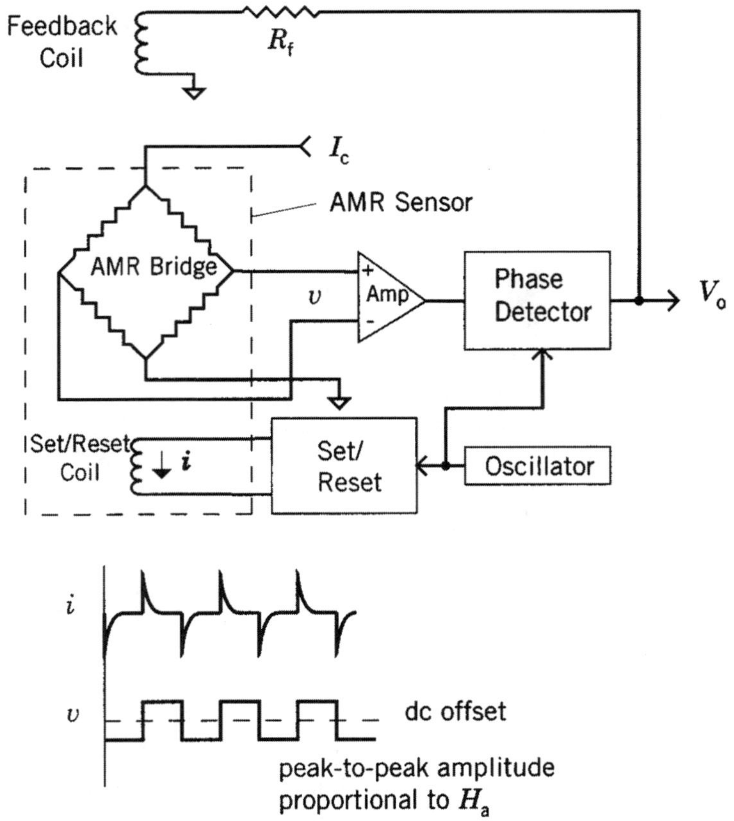

FIGURE 12.21 Example AMR gaussmeter. The magnetization direction can be alternately flipped to eliminate zero offset. The resulting ac signal can then be amplified and synchronously phase-detected to recover the field-related signal. Optionally, the range and stability of the AMR gaussmeter can be increased by connecting the output voltage through a resistor to a feedback coil that produces a field that nulls the applied field.

The optically pumped magnetometer is based on the Zeeman effect. Zeeman discovered that applying a field to atoms, which are emitting or absorbing light, will cause the spectral lines of the atoms to split into a set of new spectral lines that are much closer together than the normal lines. The energy-related frequency interval between these hyperfine lines is proportional to the magnitude of the applied field. These energy levels represent the only possible energy states that an atom can possess. The optically pumped magnetometer exploits this characteristic by optically stimulating atoms to produce an over-populated energy state in one of the hyperfine spectral lines and then causing the energy state to depopulate using an RF magnetic field. The RF frequency required to depopulate the energy state is equal to the spectral difference of the hyperfine lines produced by a magnetic field and, therefore, is proportional to the magnetic field strength. The optically pumped magnetometer can be used to sample the magnetic field at a much higher rate than the proton precession magnetometer and generally can achieve a higher resolution. The sample rate and instrument resolution are interdependent.

The Proton Precession Magnetometer

The proton precession magnetometer works on the principle that a spinning nucleus, which has both angular momentum $\vec{L}$ and a magnetic moment $\vec{\mu}_p$, will precess about a magnetic field like a gyroscope, as shown in Figure 12.22. The precession frequency ω_p is proportional to the applied field. When the magnetic field $\vec{H}_a$ is applied to the nucleus, it will produce a torque:

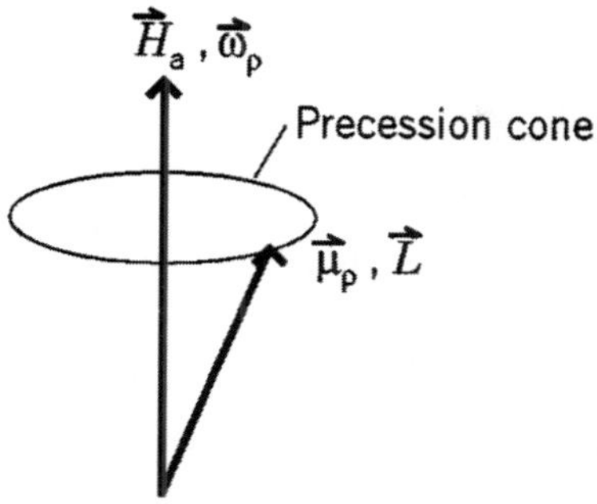

FIGURE 12.22 Nuclear precession. A spinning proton with angular momentum L and magnetic moment μ_p, when subjected to a magnetic field H_a, will precess about the field at an angular rate ω_p equal to $\mu_p H_a / L$.

$$\vec{T} = \vec{\mu}_p \times \vec{H}_a \tag{12.47}$$

on the nucleus. Because the nucleus has angular momentum, this torque will cause the nucleus to precess about the direction of the field. At equilibrium, the relationship between the torque, precession rate, and angular momentum is:

$$\mu_p \times \vec{H}_a = \vec{\omega}_p \times \vec{L} \tag{12.48}$$

Solving for the magnitude of the (Larmor) precession frequency, one finds that:

$$\omega_p = \left(\frac{\mu_p}{L}\right) H_a = \gamma H_a \tag{12.49}$$

where γ is called the gyromagnetic ratio and equals $(2.6751526 \pm 0.0000008) \times 10^{-8}\ T^{-1}\ s^{-1}$.

Figure 12.23 is a block diagram of a proton precession magnetometer. The sensor is a container of hydrocarbon rich in free hydrogen nuclei. A solenoid wrapped around the container is used to both polarize the nuclei and detect the precession caused by the ambient field. Before the polarizing field is applied, the magnetic moments of the nuclei are randomly oriented, and the net magnetization is zero. Application of the polarizing field (typically 3 mT to 10 mT) causes the nuclei to precess about the field. The precession axis can be parallel or antiparallel (nuclear magnetic moment pointing in the direction of the field) to the applied field. From a quantum mechanical standpoint, the antiparallel state is a lower energy level than the parallel state. In the absence of thermal agitation, which causes collisions between atoms, the fluid would remain unmagnetized. When a collision occurs, the parallel precession-axis nuclei lose energy and switch to the antiparallel state. After a short time, there are more nuclei with magnetic moments pointing in the direction of the field than away from it, and the fluid reaches an equilibrium magnetization M_0. The equation that relates magnetization buildup to time is:

$$M(t) = M_0\left(1 - e^{-t/\tau_e}\right) \tag{12.50}$$

where τ_e is the spin-lattice relaxation time.

The equilibrium magnetization is based on thermodynamic considerations. From Boltzmann statistics for a system with spins of 1/2:

$$\frac{n_p}{n_a} = e^{2\mu_p H_a / kT} \tag{12.51}$$

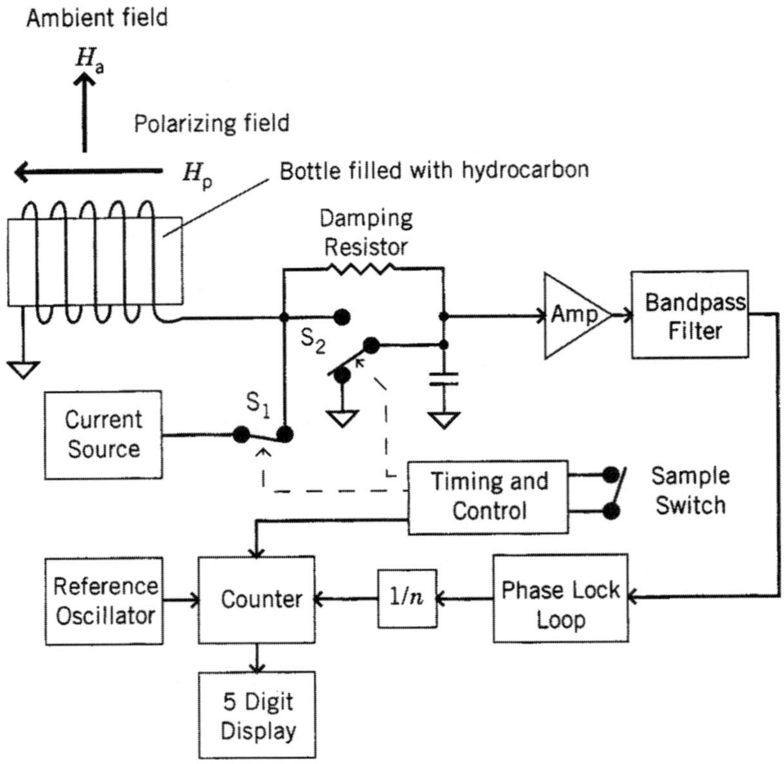

FIGURE 12.23 Typical proton precession magnetometer. A polarizing field is applied to the hydrocarbon when S1 is closed. The amplifier input is shorted to prevent switching transients from overdriving it. After a few seconds, S1 is opened and the coil is connected to the signal processor to measure the Larmor frequency.

where n_p is the number of precession spin axes parallel to H_a, n_a is the number of precession spin axes antiparallel to H_a, k is Boltzmann's constant, and T is temperature (kelvin). If n is the number of magnetic moments per unit volume, then:

$$n = n_p + n_a = n_a\left(1 + e^{2\mu_p H_a/kT}\right) \tag{12.52}$$

and

$$M_0 = \left(n_p - n_a\right)\mu_p \approx \frac{n\mu_p^2 H_a}{kT} \tag{12.53}$$

Once the fluid has reached equilibrium magnetization, the field is removed and the nuclei are allowed to precess about the local ambient field until they become randomized again. This process of excitation–relaxation can take as long as several seconds.

The hydrocarbon spin-lattice relaxation time can be adjusted by mixing paramagnetic salts, such as ferrous nitrate, into the solution. The trade-off in reduced relaxation time is increased signal-to-noise and resolution. Benzene is a good general-purpose hydrocarbon that can be used in a proton precession magnetometer.

Signal Conditioning

The block diagram shown in Figure 12.23 is an example of the signal conditioning required for a proton precession magnetometer. The coil surrounding the bottle containing the hydrocarbon serves two purposes.

At the beginning of a measurement, the current source is connected to the coil to generate the magnetic field that polarizes the fluid. This field is on the order of 10 mT. After a few seconds, the current source is disconnected and the coil, which now has a decaying nuclear precession signal at its output, is connected to the input of the amplifier. The signal is amplified, filtered, and then the period of the Larmor frequency is measured, averaged, scaled, and presented to the user in magnetic field units on a digital display.

The scale factor of the proton precession magnetometer is based on the gyromagnetic ratio, which is 0.042579 Hz nT^{-1}. High resolution, up to 0.01 nT, is achieved by measuring the period of the signal rather than the frequency. The signal frequency can be divided down and used to gate a high-frequency oscillator that is driving a counter.

The sampling of the field is controlled manually in many commercially available proton precession magnetometers. Some magnetometers have an internally controlled sample rate. The sample rate and resolution are inversely related to one another. A higher sample rate produces a poorer resolution.

The Optically Pumped Magnetometer

As explained earlier, the optically pumped magnetometer is based on the Zeeman effect. This effect is most pronounced in alkaline vapors (rubidium, lithium, cesium, sodium, and potassium). Figure 12.24 is the hyperfine spectral structure for the valence electrons of rubidium (Rb) 85, which is commonly used in these types of magnetometers. The energy-related frequency interval between these hyperfine lines is proportional to the applied field. The magnetic quantum number m is related to the angular momentum number and specifies the possible component magnitude of the magnetic moment along the applied field. The optically pumped magnetometer takes advantage of this characteristic.

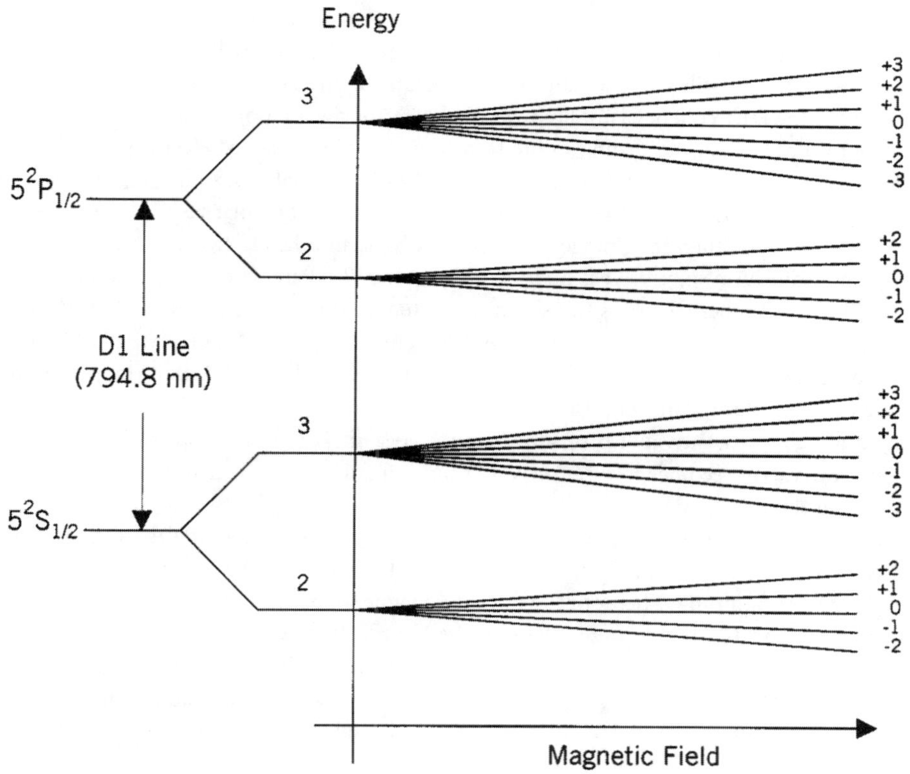

FIGURE 12.24 Rb-85 energy diagram. When a magnetic field is applied, the energy levels split into Zeeman sublevels that diverge as the field increases. Quantum mechanical factors determine the number of sublevels at each primary energy level.

TABLE 12.5 The Allowable Change in *m* When Jumping from One Energy Level to Another Depends on the Polarization of the Light Causing the Transition

Polarization	*m*
Left circular	–1
Parallel	0
Right circular	1

TABLE 12.6 The Change in Frequency for a Change in Field Is Much Higher in Optically Pumped Magnetometers Than in Proton Precession Magnetometers

Alkali	Scale factor ($Hz\ nT^{-1}$)
Rb-85	4.66737
Rb-87	~7
Cesium	3.4986

Transitions occur between levels of different *m* values and obey the rule that the change in *m* can only have the values 0, 1, and –1. Table 12.5 lists the relationship between the polarization of the light stimulating the transition and the allowable change in *m*.

When not optically excited, the energy states of the valence electrons will be distributed according to Boltzmann statistics and will be in a state of equilibrium. If the electrons are excited with circularly polarized light at the D1 frequency (794.8 nm wavelength), they will absorb photons and transition from the $^2S_{1/2}$ state to the $^2P_{1/2}$ state according to the transition rules. The excited electrons will then fall back in a random fashion to the lower states, being distributed with an equal probability among all the *m* states.

But the rules state that the change in *m* can only be 1 or –1 for polarized light. If one uses right circularly polarized light, then the change in *m* can only be 1, and the electrons in the $m = 3$ level of the $^2S_{1/2}$ state cannot transition since there is no $m = 4$ level at the $^2P_{1/2}$ state. Therefore, these electrons remain in the $m = 3$ state. All other electrons transition to the higher state and then fall back to the lower state with equal probability of arriving at any of the *m* levels, including $m = 3$. Thus, the $m = 3$ level fills up, and the other levels empty until all the electrons are in the $m = 3$ level, and no more transitions to the higher state can take place. Pumping stops.

When pumping begins, the vapor is opaque. As time goes on, less electrons are available for absorbing photons, and the vapor becomes more transparent until, finally, pumping action stops and the vapor is completely transparent.

If a small RF magnetic field at the Larmor frequency is applied at right angles to the magnetic field being measured, the electrons in the $m = 3$ state will be depumped to the other *m* levels, making them available for further pumping. The optically pumped magnetometer exploits this situation in a positive feedback arrangement to produce an oscillator at the Larmor frequency.

The scale factors for optically pumped magnetometers are significantly higher than for the proton precession magnetometer. Table 12.6 lists these scale factors for a number of alkali vapors.

As a result, the sample rate and resolution can be much higher. A resolution of 0.005 nT is possible. Sampling rates can be as high as 15 samples per second.

Signal Conditioning

Descriptions of several optically pumped magnetometers and their operating principles can be found in [24–26]. There are a number of different signal conditioning arrangements that can be used to derive a

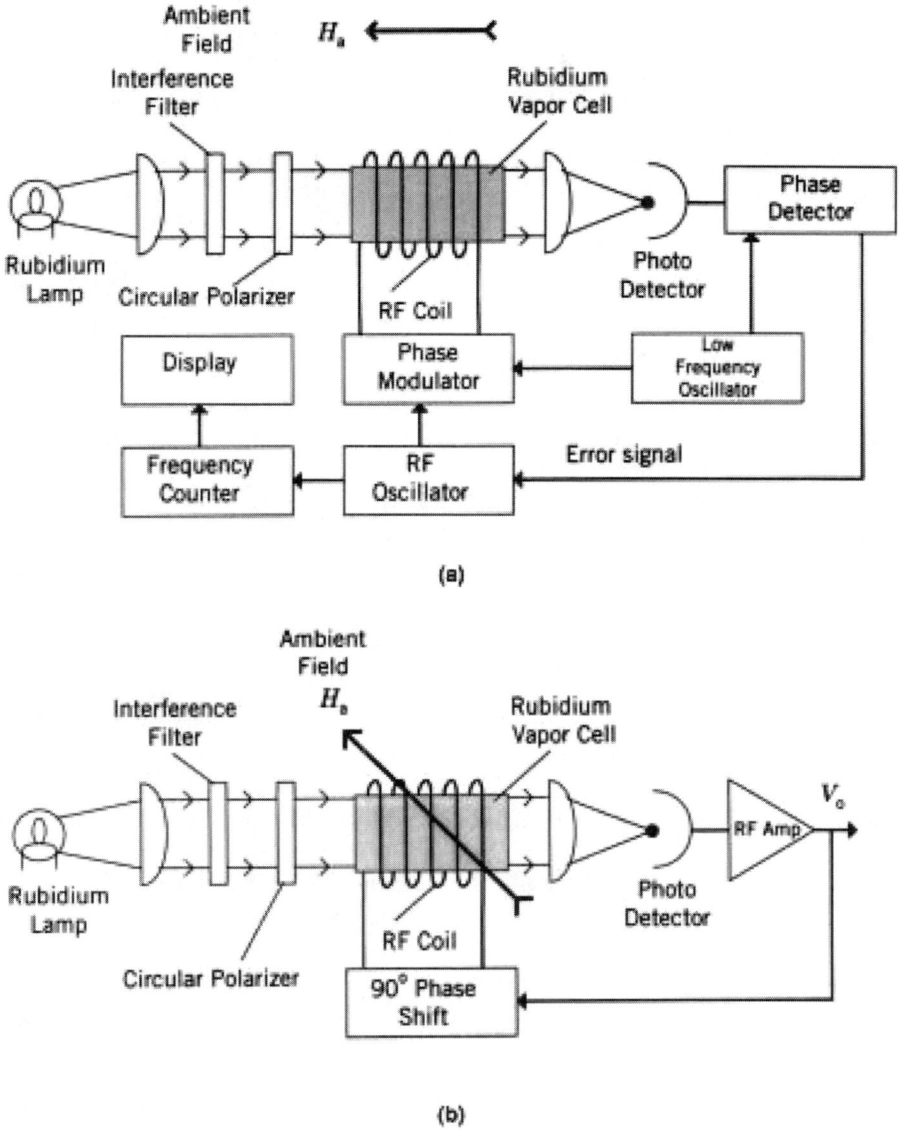

FIGURE 12.25 Two examples of optically pumped scalar magnetometers. The servoed magnetometer: (a) slightly modulates the RF field at a low frequency, causing the vapor transmissivity to modulate. A phase detector provides an error signal that is used to lock the RF oscillator to the Larmor frequency. (b) A self-oscillating magnetometer: the transmissivity of the vapor, at right angles to the applied field, is made to oscillate at the Larmor frequency by phase-shifting the detected light modulation and feeding it back to the RF field generator. (Adapted from Hartmann, 1972.)

useful readout of the measured fields. Two of the more common methods are described in [26] and are shown in Figure 12.25.

In the servoed type shown in Figure 12.25(a), the magnetic field being measured and the RF field are coaxial. The frequency of the RF oscillator is modulated with a fixed low-frequency oscillator. This causes the RF frequency to sweep through the Larmor frequency. If the sweeped RF oscillator is not centered about the Larmor frequency, the photocell output signal will contain a fundamental component of the RF modulation frequency. The phase of the signal relative to the modulator oscillator determines whether the central RF frequency is above or below the Larmor frequency. The photocell output is phase-detected to produce an error voltage that is used to drive the RF frequency toward the Larmor frequency. The RF frequency can be measured to determine the magnetic field. If a linear voltage controlled oscillator is used as the RF oscillator, its control voltage can also be used as an output since it is a measure of the Larmor frequency.

The auto-oscillating type shown in Figure 12.25(b) is based on the transmission of a polarized beam that is at right angles to the field being measured. The intensity of this cross-beam will be modulated at the Larmor frequency. The photocell signal will be shifted by 90° relative to the RF field. By amplifying the photocell signal, shifting it 90° and feeding it back to drive the RF field coil, an oscillator is created at the Larmor frequency. In practice, only one light source is used, and the field being measured is set at an angle of 45°.

Defining Terms

Anisotropic: The material property depends on direction.

Gaussmeter: An instrument used to measure magnetic fields greater than 1 mT.

Induced magnetization: The object's magnetization is induced by an external magnetic field and disappears when the inducing field is removed.

Initial permeability: The slope at the origin of the magnetization curve.

Isotropic: The material property is the same in all directions.

Magnetic dipole moment: A vector quantity that describes the strength and direction of a magnetic field source, such as a small current loop or spinning atomic nucleus.

Magnetically "hard" material: The material has a significant residual (permanent) magnetization after an external magnetic field is removed.

Magnetically "soft" material: The material's magnetization is induced by an external magnetic field and the material has no significant residual (permanent) magnetization after the field is removed.

Magnetization curve: A plot of flux density B vs. magnetic field H for an initially unmagnetized ferromagnetic material.

Magnetization: A vector quantity describing the average density and direction of magnetic dipole moments.

Magnetometer: An instrument used to measure magnetic fields with magnitudes up to 1 mT.

Magnetoresistance: The change in the electrical resistivity of a material in response to an applied magnetic field.

Maximum permeability: The maximum slope of the line drawn from the origin of the magnetization curve to a point on the magnetization curve.

Permanent magnetization: The source of an object's magnetization is internal and does not depend on the presence of an external field.

Permeability: A function that describes the relationship between an applied magnetic field and the resulting flux density.

Relative permeability: The permeability of a material normalized (divided) by the permeability of a vacuum.

Scalar magnetometer: A magnetometer that measures the magnitude of a magnetic field vector.

Vector magnetometer: A magnetometer that measures one or more of the individual components of a magnetic field vector.

References

1. K.S. Lion, *Instrumentation in Scientific Research. Electrical Input Transducers*, New York: McGraw-Hill, 1959.
2. H.R. Everett, *Sensors for Mobile Robots: Theory and Application*, Wellesley, MA: A.K. Peters, 1995.
3. J.E. Lenz, A review of magnetic sensors, *Proc. of IEEE*, 78, 973–989, 1990.
4. M.A. Payne, SI and Gaussian cgs units, conversions and equations for use in geomagnetism, *Phys. of Earth and Planitary Interiors*, 26, P10–P16, 1981.

5. F.W. Grover, *Induction Calculations. Working Formulas and Tables*, New York: Dover Publications, 1973.
6. R.M. Bozorth, *Ferromagnetism*, New York: D. Van Nostrand, 1951.
7. R.M. Bozorth and D.M. Chapin, Demagnetization factors of rods, *J. Appl. Phys.*, 13, 320–326, 1942.
8. S. A. Macintyre, A portable low noise low current three-axis search coil magnetometer, *IEEE Trans. Magnetics*, MAG-16, 761–763, 1980.
9. J.P. Hauser, A 20-Hz to 200-kHz magnetic flux probe for EMI surveys, *IEEE Trans. Electromagnetic Compatibility*, 32, 67–69, 1990.
10. F. Primdahl, The fluxgate magnetometer, *J. Phys. E: Sci. Instrum.*, 1, 242–253, 1979.
11. C.J. Pellerin and M.H. Acuna, A miniature two-axis fluxgate magnetometer, *NASA Technical Note*, TN D-5325, NASA, 1970.
12. S.V. Marshall, A gamma-level portable ring-core magnetometer, *IEEE Trans. Magnetics*, MAG-7, 183–185, 1971.
13. W.A. Geyger, *Nonlinear-Magnetic Control Devices*, New York: McGraw-Hill, 1964.
14. M. Acuna, C. Scearce, J. Seek, and J. Schelfiele, The MAGSAT vector magnetometer — a precise fluxgate magnetometer for the measurement of the geomagnetic field, *NASA Technical Report.*
15. D. Cohen, Measurements of the magnetic field produced by the human heart, brain and lung, *IEEE Trans. Magnetics*, MAG-11, 694–700, 1975.
16. C.M. Falco and I.K. Schuller, SQUIDs and their sensitivity for geophysical applications, *SQUID Applications to Geophysics*, The Society of Exploration Geophysicists, 13–18, 1981.
17. J. Clark, SQUIDs, *Sci. Am.*, 46–53, August 1994.
18. T.H. Casselman and S.A. Hanka, Calculation of the performance of a magnetoresistive permalloy magnetic field sensor, *IEEE Trans. Magnetics*, MAG-16, 461–464, 1980.
19. W. Kwaitkawski and S. Tumanski, The permalloy magnetoresistive sensors-properties and applications, *J. Phys. E: Sci. Instrum.*, 19, 502–515, 1986.
20. Permalloy Magnetic Sensors, Honeywell Technical Note.
21. U. Dibbern and A. Petersen, The magnetoresistor sensor — a sensitive device for detecting magnetic field variations, *Electronic Components and Applications*, 5(3), 148–153, 1983.
22. S. Tumanski and M.M. Stabrowski, Optimization of the performance of a thin film magnetoresistive sensor, *IEEE Trans. Magnetics*, MAG-20, 963–965, 1984.
23. L.W. Parson and Z.M. Wiatr, Rubidium vapor magnetometer, *J. Sci. Instrum.*, 39, 292–299, 1962.
24. W.H. Farthing and W.C. Folz, Rubidium vapor magnetometer for near Earth orbiting spacecraft, *Rev. Sci. Instrum.*, 38, 1023–1030, 1967.
25. F. Hartmann, Resonance magnetometers, *IEEE Trans. Magnetics*, MAG-8, 66–75, 1972.
26. F. Wellstood, C. Heiden, and J. Clark, Integrated DC SQUID magnetometer with high slew rate, *Rev. Sci. Instrum.*, 66, 952–957, 1984.

13

Permeability and Hysteresis Measurement

Jeff P. Anderson
LTV Steel Corporation

Richard J. Blotzer

Magnetic fields are typically conceptualized with so-called "flux lines" or "lines of force." When such flux lines encounter any sort of matter, an interaction takes place in which the number of flux lines is either increased or decreased. The original magnetic field therefore becomes amplified or diminished in the body of matter as a result of the interaction. This is true whether the matter is a typical "magnetic" material, such as iron or nickel, or a so-called "nonmagnetic" material, such as copper or air.

The *magnetic permeability* of a substance is a numerical description of the extent to which that substance interacts with an applied magnetic field. In other words, permeability refers to the degree to which a substance can be magnetized.

Different substances possess varying degrees of magnetization. The aforementioned examples of strongly magnetic materials have the ability to strengthen an applied magnetic field by a factor of several thousand. Such highly magnetizable materials are called *ferromagnetic.* Certain other substances, such as Al, only marginally increase an applied magnetic field. Such weakly magnetizable materials are called *paramagnetic.* Still other substances, such as Cu and the rare gases, slightly weaken an applied magnetic field. Such "negatively magnetizable" substances are called *diamagnetic.*

In common parlance, diamagnetic and paramagnetic substances are often called *nonmagnetic.* However, as detailed below, all substances are magnetic to some extent. Only empty space is truly nonmagnetic.

The term *hysteresis* has been used to describe many instances where an effect lags behind the cause. However, Ewing was apparently the first to use the term in science [1] when he applied it to the particular magnetic phenomenon displayed by ferromagnetic materials. Magnetic hysteresis occurs during the cyclical magnetization of a ferromagnet. The magnetization path created while increasing an externally applied field is not retraced on subsequent decrease (and even reversal) of the field. Some magnetization, known as *remanence,* remains in the ferromagnet after the external field has been removed. This remanence, if appreciable, allows for the permanent magnetization observed in common bar magnets.

0-8493-1733-9/04/$0.00+$1.50

TABLE 13.1 Conversion Factors Between the mks and cgs Systems for Important Quantities in Magnetism

Quantity	mks	cgs
H, applied field	A/m	$= 4\pi \times 10^{-3}$ Oe
B, flux density	Wb/m^2	$= 10^4$ G
M, magnetization	Wb/m^2	$= 10^4/4\pi$ emu/cm^3
κ, susceptibility	Wb/(A·m)	$= 16\pi^2 \times 10^{-7}$ emu/Oe·cm^3

13.1 Definition of Permeability

Let an externally applied field be described by the vector quantity **H.** This field may be produced by a solenoid or an electromagnet. Regardless of its source, **H** has units of ampere turns per meter (A m^{-1}). On passing through a body of interest, **H** magnetizes the body to a degree, **M,** formally defined as the magnetic moment per unit volume. The units of **M** are usually webers per square meter. A secondary coil (and associated electronics) is typically used to measure the combined effects of the applied field and the body's magnetization. This sum total flux-per-unit-area (flux density) is known as the induction, **B,** which typically has units of Wb/m^2, commonly referred to as a Tesla (T). Because **H, M,** and **B** are usually parallel to one another, the vector notation can be dropped, so that:

$$B = \mu_0 H + M \tag{13.1}$$

where μ_0 is the permeability of free space ($4\pi \times 10^{-7}$ Wb/A^{-1} m^{-1}).

The absolute permeability, μ, of a magnetized body is defined as the induction achieved for a given strength of applied field, or:

$$\mu = \frac{B}{H} \tag{13.2}$$

Often, the absolute permeability is normalized by μ_0 to result in the relative permeability, μ_r ($=\mu/\mu_0$). This relative permeability is numerically equal and physically equivalent to the cgs version of permeability. This, unfortunately, is still in common usage, and often expressed in units of gauss per oersted (G Oe^{-1}), although the cgs permeability is actually dimensionless. Conversion factors between the mks and cgs systems are listed in Table 13.1 for the important quantities encountered.

13.2 Types of Material Magnetization

All substances fall into one of three magnetic groups: diamagnetic, paramagnetic, or ferromagnetic. Two important subclasses, antiferromagnetic and ferrimagnetic, will not be included here. The interested reader can find numerous discussions of these subclasses; for example, see [1].

Diamagnetism

Diamagnetic and paramagnetic (see next section) substances are usually characterized by their magnetic susceptibility rather than permeability. Susceptibility is derived by combining Equations 13.1 and 13.2, viz.

$$\mu_r = 1 + \frac{M}{\mu_0 H} = 1 + \frac{\kappa}{\mu_0} \tag{13.3}$$

where κ is the susceptibility with units of Wb A^{-1} m^{-1}. This so-called *volume susceptibility* is often converted to a mass susceptibility (χ) or a molar susceptibility (χ_M). Values for the latter are readily

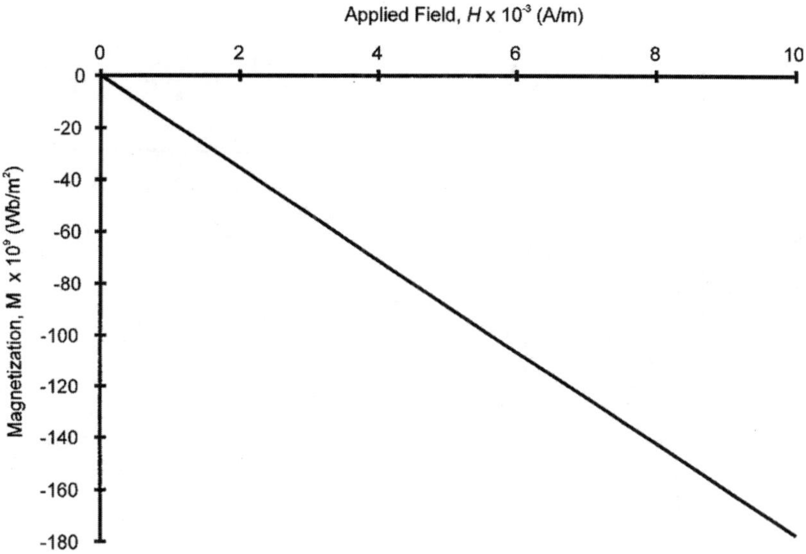

FIGURE 13.1 For diamagnetic substances, magnetization M is small and opposite the applied field H as in this schematic example for graphite ($\kappa = -1.78 \times 10^{-11}$ Wb A^{-1} m^{-1}).

available for many pure substances and compounds [2]. Susceptibility is also often called "intrinsic permeability" [3].

In any atom, the orbiting and spinning electrons behave like tiny current loops. As with any charge in motion, a magnetic moment is associated with each electron. The strength of the moment is typically expressed in units of Bohr magnetons.

Diamagnetism represents the special case in which the moments contributed by all electrons cancel. The atom as a whole possesses a net zero magnetic moment. An applied field, however, can induce a moment in the diamagnetic material, and the induced moment opposes the applied field. The magnetization, **M,** in Equation 13.3 is therefore antiparallel to the applied field, **H,** and the susceptibility, κ, is negative. For diamagnetic materials, $\mu < 1$. Figure 13.1 shows a schematic M vs. H curve for graphite with $\kappa = -1.78 \times 10^{-11}$ Wb A^{-1} m^{-1}. Note that κ is a constant up to very high applied field values.

Paramagnetism

In a paramagnetic substance, the individual electronic moments do not cancel and the atom possesses a net nonzero moment. In an applied field, the weak diamagnetic response is dominated by the atom's tendency to align its moment parallel with the applied field's direction. Paramagnetic materials have relatively small positive values for κ, and $\mu > 1$.

Thermal energy retards a paramagnet's ability to align with an applied field. Over a considerable range of applied field and temperature, the paramagnetic susceptibility is constant. However, with very high applied fields and low temperatures, a paramagnetic material can be made to approach saturation — the condition of complete alignment with the field. This is illustrated in Figure 13.2 for potassium chromium alum, a paramagnetic salt. Even at a temperature as low as 1.30 K, an applied field in excess of about 3.8×10^6 A m^{-1} is necessary to approach saturation. [Note in Figure 13.2, that 1 Bohr magneton = 9.27×10^{-24} J T^{-1}.]

Ferromagnetism

Ferromagnetic substances are actually a subclass of paramagnetic substances. In both cases, the individual electronic moments do not cancel, and the atom has a net nonzero magnetic moment that tends to align itself parallel to an applied field. However, a ferromagnet is much less affected by the randomizing action

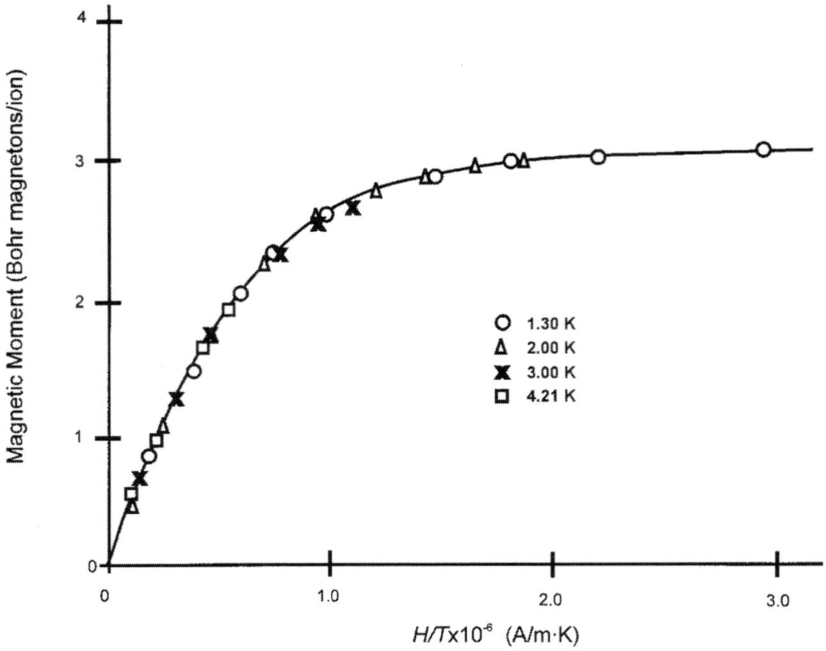

FIGURE 13.2 For paramagnetic substances, the susceptibility is constant over a wide range of applied field and temperature. However, at very high *H* and low *T*, saturation can be approached, as in this example for potassium chromium alum. (After W. E. Henry, *Phys. Rev.*, 88, 559-562, 1952.)

of thermal energy compared to a paramagnet. This is because the individual atomic moments of a ferromagnet are coupled in rigid parallelism, even in the absence of an applied field.

With no applied field, a demagnetized ferromagnet is comprised of several magnetic domains. Within each domain, the individual atomic moments are parallel to one another, or coupled, and the domain has a net nonzero magnetization. However, the direction of this magnetization is generally opposed by a neighboring domain. The vector sum of all magnetizations among the domains is zero. This condition is called the *state of spontaneous magnetization.*

With an increasing applied field, domains with favorable magnetization directions, relative to the applied field direction, grow at the expense of the less favorably oriented domains. This process is schematically illustrated in Figure 13.3. The exchange forces responsible for the ferromagnetic coupling are explained by Heisenberg's quantum mechanical model [4]. Above a critical temperature known as the Curie point, the exchange forces disappear and the formerly ferromagnetic material behaves exactly like a paramagnet.

During magnetization, ferromagnets show very different characteristics from diamagnets and paramagnets. Figure 13.4 is a so-called *B–H* curve for a typical soft ferromagnet. Note that *B* is no longer linear with *H* except in the very low- and very high-field regions. Because of this, the permeability μ for a ferromagnet must always be specified at a certain applied field, *H*, or, more commonly, a certain achieved induction, *B*. Note that μ is the slope of the line connecting a point of interest on the *B–H* curve to the origin. It is not the slope of the curve itself, although this value, dB/dH, is called the *differential permeability* [3].

Another ferromagnetic characteristic evident in Figure 13.4 is *saturation*. Once the applied field has exceeded a certain (and relatively low) value, the slope of the magnetization curve assumes a constant value of unity. At this point, *M* in Equation 13.1 has reached a maximum value, and the ferromagnet is said to be saturated. For all practical purposes, all magnetic moments in the ferromagnet are aligned with the applied field at saturation. This maximum magnetization is often called the saturation induction, B_s [5]. Note that B_s is an intrinsic property — it does not include the applied field in its value.

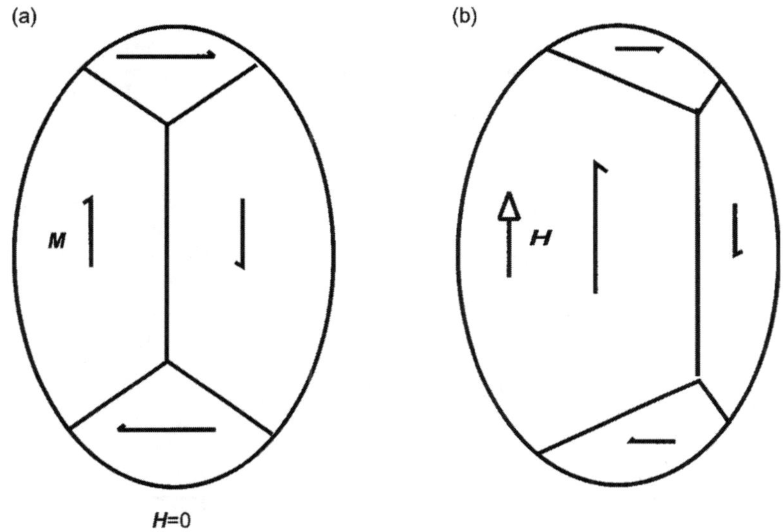

FIGURE 13.3 With no applied field (a) a ferromagnet assumes spontaneous magnetization. With an applied field (b) domains favorably oriented with ***H*** grow at the expense of other domains.

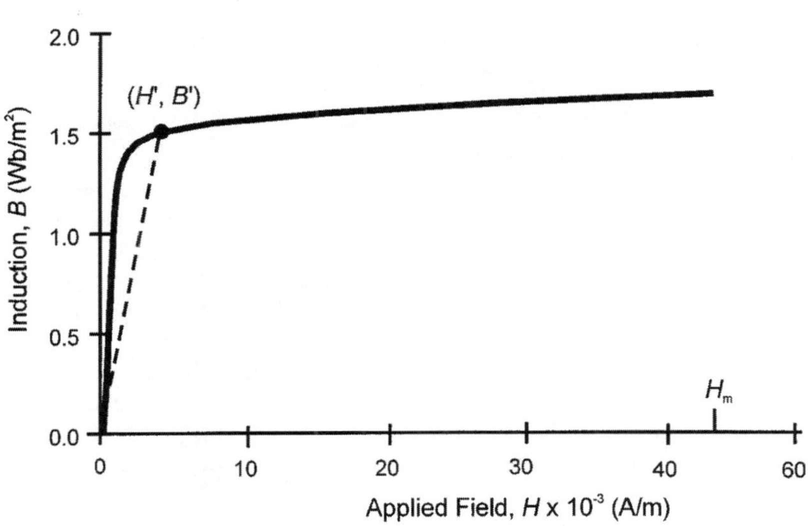

FIGURE 13.4 Magnetization (*B–H*) curve for a typical soft ferromagnet. Permeability at point (H', B') is the slope of the dashed line.

13.3 Definition of Hysteresis

If H is decreased from H_M in Figure 13.4, B does not follow the original magnetization path in reverse. Even if H is repeatedly cycled from H_M to $-H_M$, B will follow a path on increasing H that is different from decreasing H. The cyclical B–H relationship for a typical soft ferromagnet is displayed by the hysteresis loops in Figure 13.5. Two loops are included: a minor loop inside a major loop generated by independent measurements. The two differ in the value of maximum applied field: $H_{M'}$ for the minor loop was below saturation while H_{M2} for the major loop was near saturation. Both loops are symmetrical about the origin as a point of inversion since in each case $H_M = |-H_M|$.

Notice for the minor loop that when the applied field is reduced from $H_{M'}$ to 0, the induction does not also decrease to zero. Instead, the induction assumes the value B_r, known as the *residual induction.*

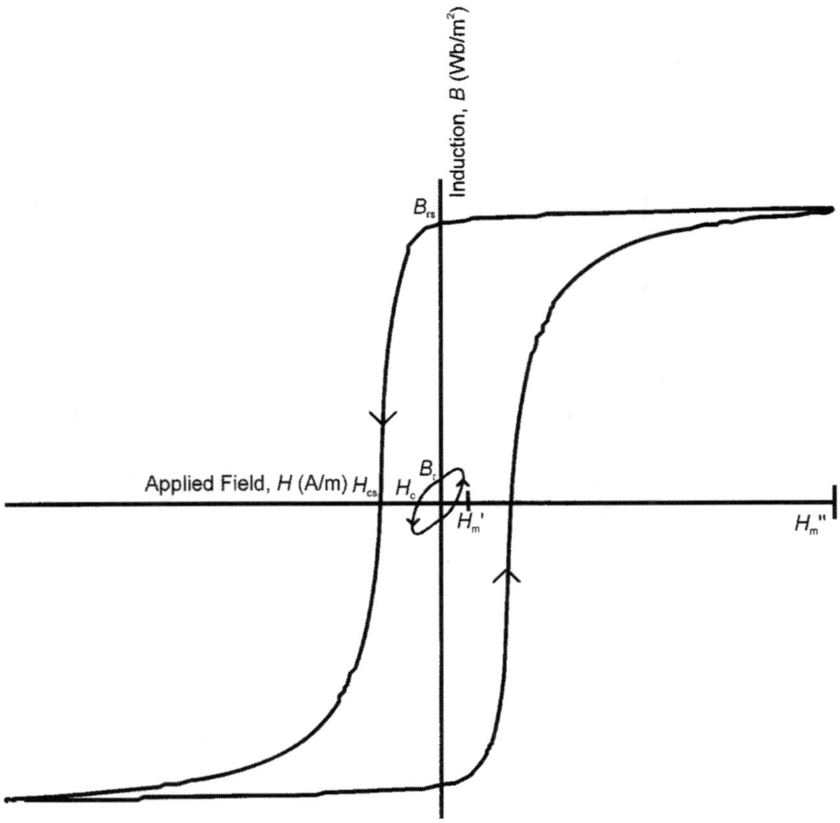

FIGURE 13.5 Major and minor dc hysteresis loops for a typical soft ferromagnet. Labeled points of interest are described in the text.

If the peak applied field exceeds the point of saturation, as for the major loop in Figure 13.5, B_r assumes a maximum value known as the *retentivity*, B_{rs}.

Note that B_r and B_{rs} are short-lived quantities observable only during cyclical magnetization conditions. When the applied field is removed, B_r rapidly decays to a value B_d, known as the *remanent induction*. B_d is a measure of the permanent magnetization of the ferromagnet. If the maximum applied field was in excess of saturation, B_{rs} rapidly decays to a maximum value of permanent magnetization, or *remanence*, B_{dm}.

The minor loop in Figure 13.5 shows that in order to reduce the induction B to zero, a reverse applied field, H_c, is needed. This is known as the *coercive force*. If the maximum applied field was in excess of saturation, the coercive force assumes a maximum value, H_{cs}, known as the *coercivity*. Note that H_c and H_{cs} are usually expressed as positive quantities, although they are negative fields relative to $H_{M'}$ and H_{M2}.

The hysteresis loops in Figure 13.5 are known as *dc loops*. Typical sweep times for such loops range from 10 s to 120 s. At faster sweep times, the coercivity will show a frequency dependence, as shown experimentally in Figure 13.6. For soft magnetic materials, this dependence can be influenced by the metallurgical condition of the ferromagnet [6].

13.4 Core Loss

During ac magnetization, some of the input energy is converted to heat in ferromagnetic materials. This heat energy is called *core loss* and is classically comprised of three parts. The first, *hysteresis loss*, P_h, is proportional to the ac frequency, f, and the area of the (slow-sweep) dc hysteresis loop:

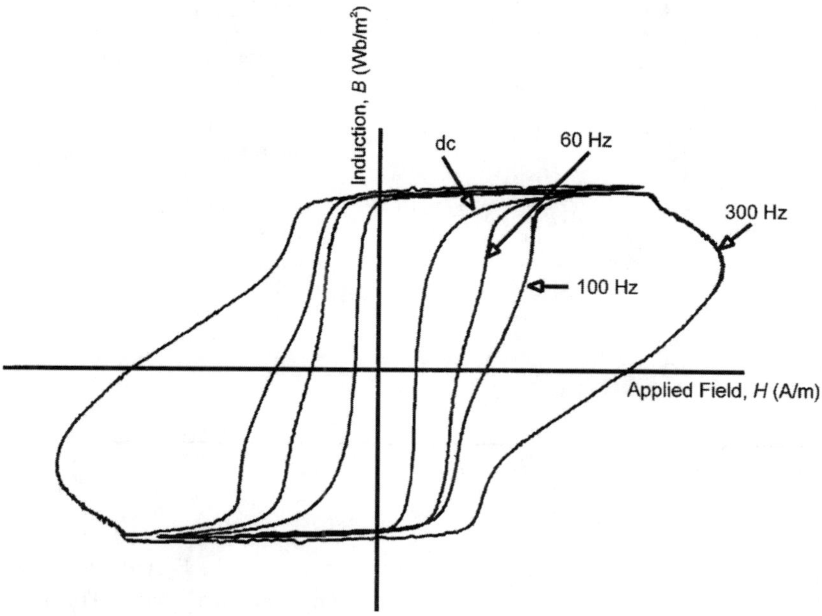

FIGURE 13.6 With increasing test frequency, coercivity for a soft ferromagnet also increases.

$$P_h = kf \int BdH \tag{13.4}$$

The second part is the *loss due to eddy current formation*, P_e. In magnetic testing of flat-rolled strips (e.g., the Epstein test; see next section), this core loss component is classically expressed as

$$P_e = \frac{(\pi Bft)^2}{6d\rho} \tag{13.5}$$

where B = Peak induction
t = Strip thickness
d = Material density
ρ = Material resistivity

The sum total $P_h + P_e$ almost never equals the observed total core loss, P_t. The discrepancy chiefly originates from the assumptions made in the derivation of Equation 13.5. To account for the additional observed loss, an anomalous loss term, P_a, has often been included, so that

$$P_t = P_h + P_e + P_a \tag{13.6}$$

13.5 Measurement Methods

Reference [3] is a good source for the various accepted test methods for permeability and hysteresis in diamagnetic, paramagnetic, and ferromagnetic materials. Unfortunately, only a few of the instruments described there are available commercially. Examples of these are listed in Table 13.2.

The instruments in Table 13.2 include hysteresigraphs (LDJ models 3500, 5600, and 5500) and vibrating sample magnetometers (LDJ and Lakeshore Cryotronics VSM models). Also included are two Donart models of Epstein testers. The Epstein test is commonly used to characterize flat-rolled soft ferromagnets

TABLE 13.2 Commercially Available Instruments for Measurement of Permeability and Hysteresis

Manufacturer	Model	Power	Material Type[a]	Ferromagnetic Type	Hysteresis Loop?	Core Loss?	Cost ($U.S.)
LDJ	3500/5600	Ac/dc	F	Soft & hard	Y	Y	30–90k
Troy, MI	5500	Dc	F	Soft & hard	Y	N	30–90k
(810) 528-2202	VSM	Dc	D, P, F	Soft & hard	Y	N	50–110k
Lakeshore Cryotronics	VSM	Dc	D, P, F	Soft & hard	Y	N	45–120k
Westerville, OH	Susceptometer	Ac	F	Soft & hard	N	N	50–110k
(614) 891-2243	Magnetometer	Dc	F	Soft & hard	N	N	50–110k
Donart Electronics	3401	Dc	F	Soft	Y	N	20k+
Pittsburgh, PA	MS-2	Ac	F	Soft	N	Y	20k+
(412) 796-5941							
Soken/Magnetech	DAC-BHW-2	Ac	F	Soft	Y	Y	38k+
Racine, WI							
(501) 922-6899							

[a] D — diamagnetic, P — paramagnetic, F — ferromagnetic.

such as silicon electrical steels in sheet form. A recent alternative to the Epstein test is the single-sheet test method. The Soken instrument in Table 13.2 is an example of such a tester. This method requires much less sample volume than the Epstein test. It can also accommodate irregular sample geometries. However, the Soken instrument is not yet accepted by the American Society for Testing and Materials (ASTM) for reasons explained in the next section.

Note that all instruments in Table 13.2 can measure permeability (or susceptibility), but not all can provide hysteresis loop measurements. Diamagnetic and paramagnetic materials generally require VSM instruments unless one is prepared to construct their own specialty apparatus [3]. All instruments in Table 13.2 can measure ferromagnetic materials, although only a few can accommodate hard (i.e., permanently magnetizable) ferromagnets.

The price ranges in Table 13.2 account for such options as temperature controls, specialized test software, high-frequency capabilities, etc.

13.6 Validity of Measurements

For a ferromagnet under sinusoidal ac magnetization, the induction will show a waveform distortion (i.e., B is nonsinusoidal) once H_m exceeds the knee of the B–H curve in Figure 13.4. Brailsford [7] has discussed such waveform distortion in detail. With one exception, all ac instruments in Table 13.2 determine H from its sinusoidal waveform and B from its distorted waveform.

The single exception is the Soken instrument. Here, feedback amplification is employed to deliberately distort the H waveform in a way necessary to force a sinusoidal B waveform. In general, this will result in a smaller measured value for permeability compared to the case where feedback amplification is not used. Some suggest this to be the more precise method for permeability measurement, but the use of feedback amplification has prevented instruments such as the Soken from achieving ASTM acceptance to date.

Defining Terms

Permeability: The extent to which a material can be magnetized.

Hysteresis: A ferromagnetic phenomenon in which the magnetic induction B is out of phase with the magnetic driving force H.

References

1. B.D. Cullity, *Introduction to Magnetic Materials*, Reading, MA: Addison-Wesley, 1972.
2. D.R. Lide (Ed.), *CRC Handbook of Chemistry and Physics*, Boca Raton, FL: CRC Press, 1992-3.
3. Anonymous, 1995 *Annual Book of ASTM Standards*, Philadelphia, PA: ASTM, 3.04, 1995.
4. C. Kittel, *Introduction to Solid State Physics, 5th ed.*, New York: John Wiley & Sons, 1976.
5. Page 11 of Ref. [3].
6. R. Thomas and G. Morgan, *Proc. Eleventh Ann. Conf. on Properties and Applications of Magnetic Materials,* Chicago, 1992.
7. F. Brailsford, *Magnetic Materials, 3rd ed.*, London: Metuen and New York: Wiley, 1960.

14

Inductance Measurement

Michał Szyper
University of Mining and Metallurgy

Inductance is an electrical parameter that characterizes electric circuit elements (two- or four-terminal networks) that become magnetic field sources when current flows through them. They are called inductors, although inductance is not a unique property of them. Electric current i (A) and magnetic flux Φ (Wb) are interdependent; that is, they are coupled. Inductance is a measurable parameter; therefore, it has a physical dimension, a measurement unit (the henry), as well as reference standards. Inductance is a property of all electrical conductors. It is found as *self-inductance L* of a single conductor and as *mutual inductance M* in the case of two or more conductors. Inductors can vary in construction, but windings in the form of coils are the most frequent. In this case, they have characteristic geometrical dimensions: surface A, length l, and N number of turns of windings. The path of magnetic flux can be specially shaped by magnetic cores. Figure 14.1 shows examples of different inductive elements of electric circuits.

Figures 14.1a and b present windings with self-inductance made as a coreless coil (a), and wound on a ferromagnetic core that is the concentrator of the magnetic field (b). A transformer loaded by impedance Z_L and an electromagnet loaded by impedance of eddy currents in the metal board, shown in Figures 14.1c and d, will have not only self-inductance of windings, but also mutual inductance between windings (c) or between winding and eddy currents (d). Self-inductances and mutual inductances can be detected in busbars of electric power stations as shown in Figure 14.1e, and also on tracks on printed circuit boards as in Figure 14.1f. Couplings between currents and electromagnetic fields can be made deliberately but can also be harmful, e.g., due to energy losses caused by eddy currents, or due to electromagnetic disturbances induced in the tracks of printed circuit boards. Inductors made as windings on ferromagnetic cores have numerous other applications.

The presence of a ferromagnetic core changes the shape of a magnetic field and increases inductance; but in the case of closed cores, it also causes nonlinear relations between inductance and current as well as current frequency. Closed magnetic cores then have a nonlinear and ambiguous magnetization characteristic $\Phi(i)$ because of magnetic saturation and hysteresis. Inductances of open magnetic cores with an air gap are mainly dependent on the length of the magnetic path in the air. Real coils made of metal (e.g., copper) wire also show resistance R. However, if resistance is measured at the coil terminals with ac current, it depends not only on the cross section, length, and resistivity of the wire, but also on losses of active power in the magnetic core. These losses depend on both the current value and frequency.

0-8493-1733-9/04/$0.00+$1.50

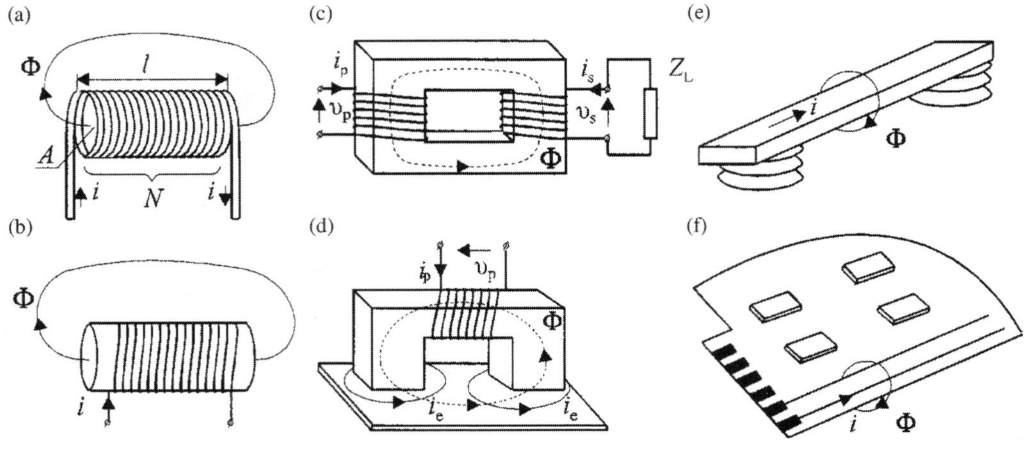

FIGURE 14.1 Examples of different inductive elements: coreless coil (a), coil with ferromagnetic concentrating core (b), transformer (c), electromagnet (d), element of electrical power station bus-bars (e), and printed circuit board with conductive tracks (f).

Resistances of inductive elements can also depend on the skin effect. This phenomenon consists of the flow of electric current through a layer of the conductor, near its outer surface, as the result of the effects of the conductor's own magnetic field generated by the current flowing inside the conductor. Notice that the coils have also interturns and stray (to Earth) capacitances *C*. From its terminals, the inductive elements can then be described by impedances (two-terminal networks) or transmittances (four-terminal networks), values that are experimentally evaluated by current and voltage measurements, or by comparing them with reference impedances. The measured values and the equivalent circuit models of inductive elements are used for evaluation of model parameters: self-inductances, mutual inductances, resistances, and capacitances.

14.1 Definitions of Inductance

Self-inductance is defined as the relation between current *i* flowing through the coil and voltage *v* measured at its terminals [1].

$$v = L\frac{\mathrm{d}i}{\mathrm{d}t} \tag{14.1}$$

Using Equation 14.1, the unit of inductance, i.e., the henry, can be defined as follows:

> One henry (1 H) is the inductance of a circuit in which an electromotive force of one volt (1 V) is induced, when the current in the circuit changes uniformly by one ampere (1 A) per second (1 s).

The definition of the unit implies the use of uniform-ramp current excitation (the derivative is constant); in practice, however, mainly sinusoidal current excitation is used in inductance measurement.

Mutual inductance *M* of windings coupled by magnetic flux **Φ** is a parameter that depends on the coupling coefficient. The coupling coefficient is defined as *perfect coupling* in the case in which the total flux of one winding links the second one; *partial coupling* is the case in which only a fraction of flux links the second one; and *zero coupling* is the case in which no part of the flux of one winding links the second one. Assuming that, as in Figure 14.1c, the second N_s winding is not loaded (i.e., $i_s = 0$) and the flux **Φ** is a part of the total flux produced by current i_p, then voltage v_s is described by:

$$v_s = N_s\frac{\mathrm{d}\Phi}{\mathrm{d}t} = \pm k\sqrt{L_p L_s}\frac{\mathrm{d}i_p}{\mathrm{d}t} = \pm M\frac{\mathrm{d}i_p}{\mathrm{d}t} \tag{14.2}$$

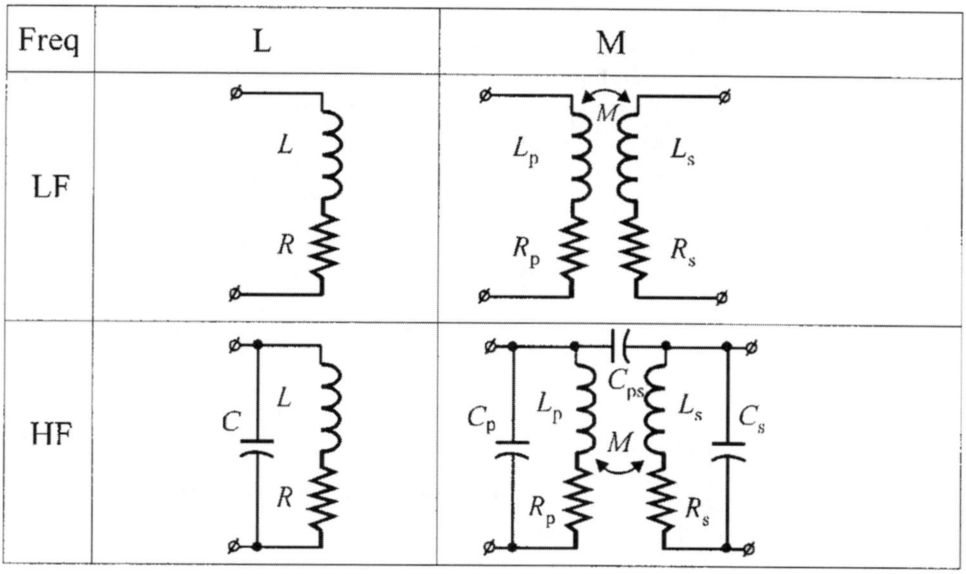

FIGURE 14.2 Basic equivalent circuits of inductive elements with self-inductance *L*, mutual inductance *M* for low (LF) and high (HF) frequencies.

where $\pm k$ is the coupling coefficient of the primary (p) and secondary (s) windings. Its sign depends on the direction of the windings. Because of the similarity between Equations 14.1 and 14.2, mutual inductance is defined similarly to self-inductance. The definitions of self-inductance and mutual inductance described above are correct when *L* and *M* can be assumed constant, i.e., not depending on current, frequency, or time.

14.2 Equivalent Circuits and Inductive Element Models

Equivalent circuits of inductive elements and their mathematical models are built on the basis of analysis of energy processes in the elements. They are as follows: energy storage in the parts of the circuits represented by lumped inductances (*L*) and capacitances (*C*), and dissipation of energy in the parts of the circuits represented by lumped resistances (*R*). Essentially, the above-mentioned energy processes are never lumped, so the equivalent circuits and mathematical models of inductive elements only approximate reality. Choosing an equivalent circuit (model), one can influence the quality of the approximation. In Figure 14.2, the basic models of typical coreless inductive elements of electric circuits are presented, using the type of inductance (self-inductance *L*, mutual inductance *M*) and frequency band (low LF, high HF) as criteria. Models of inductive elements with ferromagnetic cores and problems concerning model parameter evaluation are beyond the scope of this chapter. The equivalent circuits of coreless inductive elements at low frequencies contain only inductances and series resistances. At high frequencies, parallel and coupling equivalent capacitances are included. Calculations of *LCR* values of complicated equivalent circuits (with many *LCR* elements) are very tedious, so often for that purpose a special dedicated processor with appropriate software is provided in the measuring device.

In metrology, complex notation [1] is frequently used to describe linear models of inductive elements. In complex notation, the impedances and transmittances are as follows:

$$Z = R_z + jX_z = Z_m \exp(j\phi_z),\quad Z_m = \sqrt{R_z^2 + X_z^2},\quad \phi_z = \arctan\frac{X_z}{R_z} \tag{14.3}$$

$$T = R_T + jX_T = T_m \exp(j\phi_T),\quad T_m = \sqrt{R_T^2 + X_T^2},\quad \phi_T = \arctan\frac{X_T}{R_T} \tag{14.4}$$

The components of impedances and transmittances can be described as algebraic functions of the *LCR* elements determined for the equivalent circuit. In each case, the forms of the functions depend on the assumed equivalent circuit. By measuring the real and imaginary components or modules and phase angles of impedances or transmittances and comparing them to corresponding quantities determined for the equivalent circuit, and then solving the obtained algebraic equations, the *LCR* values of the equivalent circuit can be determined.

An example for the *LF* equivalent circuit in Figure 14.2 can be obtained:

$$Z = R + j\omega L = R_z + jX_z \rightarrow R = R_z,\ \omega L = X_z \tag{14.5}$$

$$T = j\omega M = R_T + jX_T \rightarrow R_T = 0,\ \omega M = X_T \tag{14.6}$$

For the HF equivalent circuits shown in Figure 14.2, the models are more complex. More particularly, models of circuits with mutual inductance can represent:

- Ideal transformers, with only their self-inductances and mutual inductances of the coils
- Perfect transformers, without losses in the core
- Real transformers, having a specific inductance, resistance, and capacity of the coils, and also a certain level of losses in the core

The quality factor Q and dissipation factor D are defined for the equivalent circuits of inductive elements. In the case of a circuit with inductance and resistance only, they can be defined as follows:

$$Q = \frac{1}{D} = \frac{\omega L}{R} = \tau\omega \tag{14.7}$$

where parameter τ is a time constant.

The presented models with lumped inductance are not always a sufficiently good approximation of the real properties of electric circuits. This particularly applies to circuits made with geometrically large wires (i.e., of a significant length, surface area, or volume). In such cases, the models applied use an adequately distributed inductance (linearly, over the surface, or throughout the volume). Inductances determined to be the coefficients of such models depend on the geometrical dimensions and, in the case of surface conductivity or conduction by the surface or volume, by the frequencies of the currents flowing in the conductor lines.

A complex formulation is used to represent and analyze circuits with lumped and linear inductances. Sometimes the same thing can be done using simple linear differential equations or the corresponding integral operators.

The analytical methodology for such circuits is described in [1]. In the case of nonlinear inductances, the most frequently used method is the linearized equivalent circuit, also represented in a simplified form using a complex formulation. Circuits with distributed inductance are represented by partial differential equations.

14.3 Measurement Methods

Impedance (or transmittance) measurement methods for inductors are divided into three basic groups:

1. Current and voltage methods based on impedance/transmittance determination.
2. Bridge and differential methods based on comparison of the voltages and currents of the measured and reference impedances until a state of balance is reached.
3. Resonance methods based on physical connection of the measured inductor and a capacitor to create a resonant system.

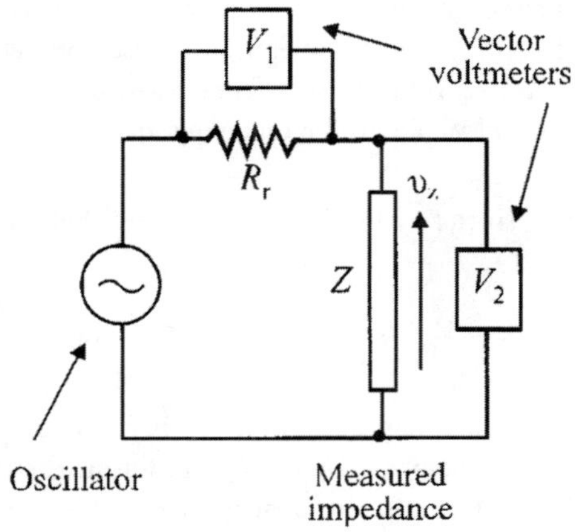

FIGURE 14.3 Circuit diagram for impedance measurement by current and voltage method.

Current–Voltage Methods

Current–voltage measurement methods are used for all types of inductors. A current–voltage method using vector voltmeters is shown in Figure 14.3. It is based on evaluation of the modules and phase angles of impedances (in the case of self-inductance) or transmittances (in the case of mutual inductance) using Equations 14.8 and 14.9.

$$Z = \frac{v_2}{i} = R_r \frac{v_2}{v_1} = R_r \frac{V_{m2} \exp(j\phi_2)}{V_{m1} \exp(j\phi_1)} = R_r \frac{V_{m2}}{V_{m1}} \exp j(\phi_2 - \phi_1) = Z_m \exp(j\phi_z) \tag{14.8}$$

$$T = \frac{v_s}{i_p} = R_r \frac{v_2}{v_{pr}} = R_r \frac{V_{ms} \exp(j\phi_s)}{V_{mpr} \exp(j\phi_{pr})} = R_r \frac{V_{ms}}{V_{mpr}} \exp j(\phi_s - \phi_{pr}) = T_m \exp(j\phi_T) \tag{14.9}$$

where R_r = Sampling resistor used for measurement of current
v_1 = Voltage proportional to the current
v_2 = Voltage across the measured impedance
v_{pr} = Voltage proportional to primary current
v_s = Voltage of the secondary winding of the circuit with mutual inductance

A block diagram illustrating the principle of the vector voltmeter is shown in Figure 14.4a. The system consists of the multiplier or gated synchronous phase-sensitive detector (PSD) [3, 4] of the measured

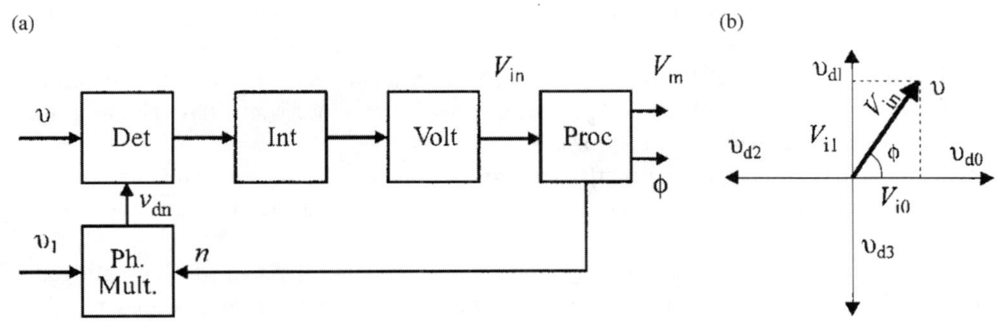

FIGURE 14.4 Block diagram (a) and phasor diagram (b) illustrating the principle of operation of a vector voltmeter. Block abbreviations: "Det" — phase-sensitive amplitude detector, "Int" — integrator, "Volt" — voltmeter, "Proc" — processor, "Ph. mult" — controlled phase multiplexer.

voltage v with the switching system of the phase of the reference voltage v_{dn}, integrator, digital voltmeter, and processor. The principle of vector voltmeter operation is based on determination of the magnitude V_m and phase angle ϕ of the measured voltage v in reference to voltage v_1, which is proportional to the current i. Assume that voltages v and v_{dn} are in the following forms:

$$v = V_m \sin(\omega t + \phi) = V_m(\sin\omega t \cos\phi + \cos\omega t \sin\phi) \tag{14.10}$$

$$v_{dn} = V_{md} \sin\left(\omega t + n\frac{\pi}{2}\right), \quad n = 0,1,2,3 \tag{14.11}$$

Phase angle n $\pi/2$ of voltage v_{dn} can take values from the set $\{0, \pi/2, \pi, 3/2\ \pi\}$ by choosing the number n that gives the possibility of detecting the phase angle ϕ in all four quadrants of the Cartesian coordinate system, as is shown in Figure 14.4b. A multiplying synchronous phase detector multiplies voltages v and v_{dn} and bilinearly, and the integrator averages the multiplication result during time T_i.

$$V_{in} = \frac{1}{T_i}\int_0^{T_i} v v_{dn}\, dt \tag{14.12}$$

Averaging time $T_i = k\ T$, $k = 1,2,3\ldots$ is a multiple of the period T of the measured voltage. From Equations 14.10 through 14.12, an example for $0 \le \phi \le \pi/2$ (e.g., $n = 0$ and $n = 1$), a pair of numbers is obtained:

$$V_{i0} = 0.5\ V_m\ V_{md}\ \cos\phi,\quad V_{i1}\ 0.\ V_m\ V_m\ \sin\phi \tag{14.13}$$

which are the values of the Cartesian coordinates of the measured voltage v. The module and phase angle of voltage are calculated from:

$$V_m = \frac{2}{V_{md}}\sqrt{V_{i0}^2 + V_{i1}^2},\quad \phi = \arctan\frac{V_{i1}}{V_{i0}} \tag{14.14}$$

Both coordinates of the measured voltage v can be calculated in a similar way in the remaining quadrants of the Cartesian coordinate system. A vector voltmeter determines the measured voltage as vector (phasor) by measurement of its magnitude and angle as shown in Figure 14.4.

The current and voltage method of impedance or transmittance measurement is based on measurement of voltages $v_1 = iR_r$ and v_2 or $v_{pr} = iR_r$ and v_s, and the use of Equations 14.8 or 14.9. Calculation of the voltage measurement results and control of number n is performed by a processor. References [11–13] contain examples of PSD and vector voltmeter applications. Errors of module and phase angle measurement of impedance when using vector voltmeters are described in [11] as being within the range of 1 to 10% and between 10^{-5} rad and 10^{-3} rad, respectively, for a frequency equal to 1.8 GHz. Publications [9] and [10] contain descriptions of applications of comparative methods which have had an important influence on the development of the methods.

Another method of comparative current/voltage type is a modernized version of the "three-voltmeter" method [7]. A diagram of a measurement system illustrating the principle of the method is shown in Figure 14.5a and b.

The method is based on the properties of an operational amplifier (OA), in which output voltage v_2 is proportional to input voltage v_1 and to the ratio of the reference resistance R_r to measured impedance Z. The phasor difference v_3 of voltages v_1 and v_2 can be obtained using the differential amplifier (DA). The three voltages (as in Figure 14.5b) can be used for the module Z_m and phase φ calculation using relations:

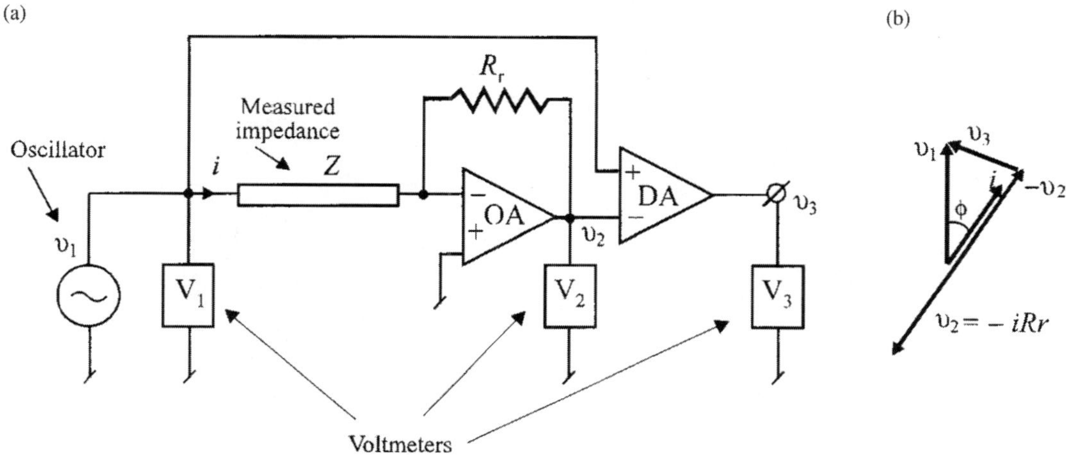

FIGURE 14.5 Block diagram (a) and phasor diagram (b) of the "three-voltmeter" method. Operational and differential amplifiers are represented by blocks "OA" and "DA."

$$v_2 = -iR_r = -\frac{R_r}{Z}v_1 \rightarrow Z_m = \frac{V_1}{V_2}R_r \tag{14.15}$$

$$v_3 = v_1 - v_2 \rightarrow \phi = \arccos\frac{V_1^2 + V_2^2 - V_3^2}{2V_1V_2} \tag{14.16}$$

where V_1, V_2, V_3 are the results of rms voltage measurements in the circuit. The advantage of the method lies in limiting the influence of stray capacitances as a result of attaching one of the terminals of the measured impedance to a point of "virtual ground." However, to obtain small measurement errors, especially at high frequencies, amplifiers with very good dynamic properties must be used.

Joint errors of inductance measurement, obtained by current and voltage methods, depend on the following factors: voltmeter errors, errors in calculating resistance R_r, system factors (residual and leakage inductances, resistances, and capacitances), and the quality of approximation of the measured impedances by the equivalent circuit.

Bridge Methods

There are a variety of bridge methods for measuring inductances. Bridge principles of operation and their circuit diagrams are described in [2–5] and [7]. The most common ac bridges for inductance measurements and the formulae for calculating measurement results are shown in Figure 14.6.

The procedure referred to as *bridge balancing* is based on a proper selection of the reference values of the bridge so as to reduce the differential voltage to zero (as referred to the output signal of the balance indicator). It can be done manually or automatically.

The condition of the balanced bridge $v_0 = 0$ leads to the following relation between the impedances of the bridge branches; one of them (e.g., Z_1) is the measured impedance:

$$Z_1Z_3 = Z_2Z_4 \rightarrow (R_1 + jX_1)(R_3 + jX_3) = (R_2 + jX_2)(R_4 + jX_4) \tag{14.17}$$

Putting Equation 14.17 into complex form and using expressions for the impedances of each branch, two algebraic equations are obtained by comparing the real and imaginary components. They are used to determine the values of the equivalent circuit elements of the measured impedance. In the most simple case, they are L and R elements connected in series. More complicated equivalent circuits need more

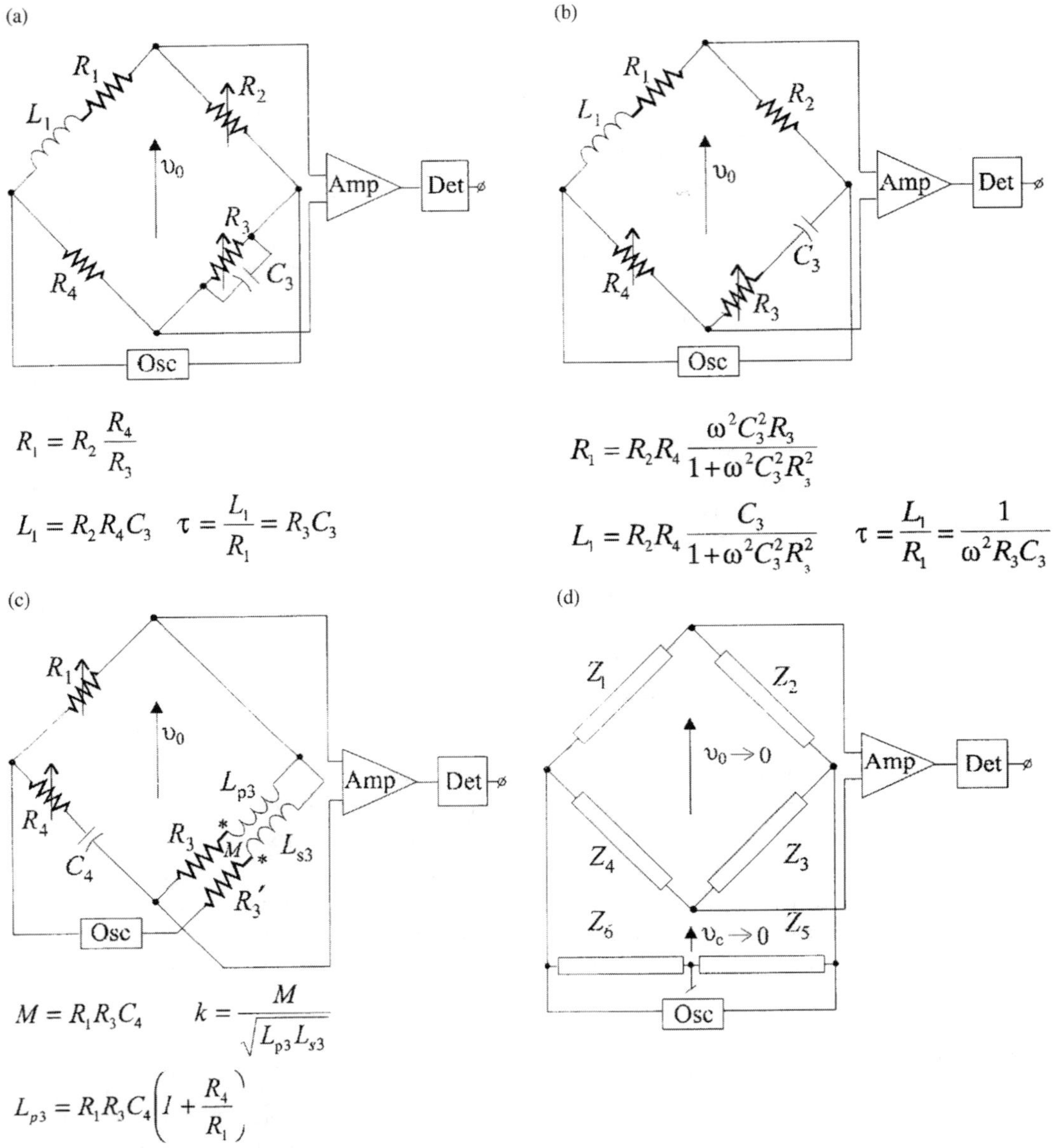

FIGURE 14.6 Bridge circuits used for inductance measurements: Maxwell-Wien bridge (a), Hay bridge (b), Carey-Foster bridge (c), and ac bridge with Wagner branch (d). Block abbreviations: "Osc" — oscillator, "Amp" — amplifier, "Det" — amplitude detector.

equations to determine the equivalent circuit parameters. Additional equations can be obtained from measurements made at different frequencies.

In self-balancing bridges, vector voltmeters preceded by an amplifier of the out-of-balance voltage v_0 are used as "zero" detectors. The detector output is coupled with variable standard bridge components.

The Maxwell-Wien bridge shown in Figure 14.6a is one of the most popular ac bridges. Its range of measurement values is large and the relative error of measurement is about 0.1% of the measured value. It is used in the wide-frequency band 20 Hz to 1 MHz. The bridge is balanced by varying the R_2 and R_3 resistors or by varying R_3 and capacitor C_3. Some difficulties can be expected when balancing a bridge with inductors with high time constants.

The Hay bridge presented in Figure 14.6b is also used for measurement of inductors, particularly those with high time constants. The balance conditions of the bridge depend on the frequency value, so the frequency should be kept constant during the measurements, and the bridge supply voltage should be free from higher harmonic distortions. The dependence of bridge balance conditions on frequency also limits the measurement ranges. The bridge is balanced by varying R_3 and R_4 resistors and by switching capacitor C_3.

Mutual inductance M of two windings with self-inductances L_p and L_s can be determined by Maxwell-Wien or Hay bridges. For this, two inductance measurements have to be made for two possible combinations of the series connection of both coupled windings: one of them for the corresponding directions of the windings, and one for the opposite directions. Two values of inductances L_1 and L_2 are obtained as the result of the measurements:

$$L_1 = L_p + L_s + 2\,M, \quad L_2 = L_p + L_s - 2\,M \tag{14.18}$$

Mutual inductance is calculated from:

$$M = 0.25\left(L_1 - L_2\right) \tag{14.19}$$

The Carey-Foster bridge described in Figure 14.6c is used for mutual inductance measurement. The self-inductances of the primary and secondary windings can be determined by two consecutive measurements. The expressions presented in Figure 14.6c yield the magnetic coupling coefficient k. The bridge can be used in a wide frequency range. The bridge can be balanced by varying R_1 and R_4 resistances and switching the remaining elements.

For correct measurements when using ac bridges, it is essential to minimize the influence of harmful couplings among the bridge elements and connection wires, between each other and the environment. Elimination of parallel (common) and series (normal) voltage distortions is necessary for high measurement resolution. These couplings are produced by the capacitances, inductances, and parasitic resistances of bridge elements to the environment and among themselves. Because of their appearance, they are called stray couplings. Series voltage distortions are induced in the bridge circuit by varying common electromagnetic fields. Parallel voltage distortions are caused by potential differences between the reference point of the supply voltage and the points of the out-of-balance voltage detector system.

Magnetic shields applied to connection wires and bridge-balancing elements are the basic means of minimizing the influence of parasitic couplings and voltage distortions [8]. All the shields should be connected as a "star" connection; that is, at one point, and connected to one reference ("ground") point of the system. For these reasons, amplifiers of out-of-balance voltage with symmetric inputs are frequently used in ac bridges, as they reject parallel voltage distortions well.

When each of the four nodes of the bridge has different stray impedances to the reference ground, an additional circuit called a Wagner branch is used (see Figure 14.6d). By varying impedances Z_5 and Z_6 in the Wagner branch, voltage v_c can be reduced to zero; by varying the other impedances, voltage v_0 can also be reduced to zero. In this way, the bridge becomes symmetrical in relation to the reference ground point and the influence of the stray impedances is minimized.

The joint error of the inductance measurement results (when using bridge methods) depends on the following factors: the accuracy of the standards used as the bridge elements, mainly standard resistors and capacitors; errors of determining the frequency of the bridge supplying voltage (if it appears in the expressions for the measured values); errors of the resolution of the zero detection systems (errors of state of balance); errors caused by the influence of residual and stray inductances; resistances and capacitances of the bridge elements and wiring; and the quality of approximation of the measured impedances in the equivalent circuit.

The errors of equivalent resistance measurements of inductive elements using bridge methods are higher than the errors of inductance measurements. The number of various existing ac bridge systems

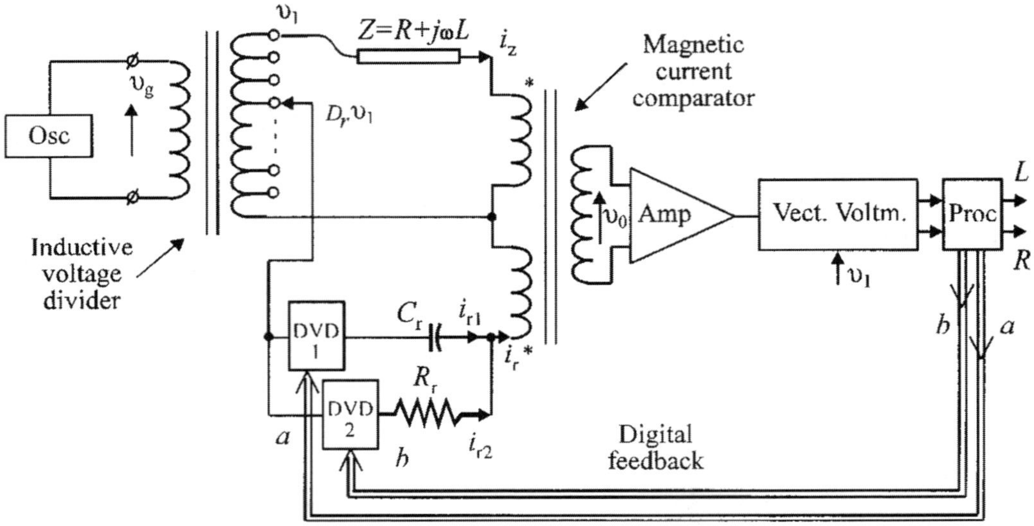

FIGURE 14.7 Scheme of the differential method. Block abbreviations: "Osc", "Amp" — as in Figure 14.6, "Vect. voltm" — vector voltmeter, "Proc" — processor, "DVD" — digital voltage divider.

is very high. Often the bridge system is built as a universal system that can be configured for different applications by switching elements. One of the designs of such a system is described in [3]. Reference [13] describes the design of an automatic bridge that contains, as a balancing element, a multiplying digital-to-analog converter (DAC), controlled by a microcontroller.

Differential Methods

Differential methods [7] can be used to build fully automatic digital impedance meters (of real and imaginary components or module and phase components) that can also measure inductive impedances. Differential methods of measurement are characterized by small errors, high resolution, and a wide frequency band, and often utilize precise digital control and digital processing of the measurement results. The principle of differential methods is presented through the example of a measuring system with an inductive voltage divider and a magnetic current comparator (Figure 14.7). An inductive voltage divider (IVD) is a precise voltage transformer with several secondary winding taps that can be used to vary the secondary voltage in precisely known steps [7]. By combining several IVDs in parallel, it is possible to obtain a precise voltage division, usually in the decade system. The primary winding is supplied from a sinusoidal voltage source. A magnetic current comparator (MCC) is a precise differential transformer with two primary windings and a single secondary winding. The primary windings are connected in a differential way; that is, the magnetic fluxes produced by the currents in these windings subtract. The output voltage of the secondary winding depends on the current difference in the primary windings. MCCs are characterized by very high resolution and low error but are expensive. In systems in common use, precise control of voltages (IVD) is provided by digitally controlled (sign and values) digital voltage dividers (DVD), and the magnetic comparator is replaced by a differential amplifier.

The algorithms that enable calculation of L and R element values of the series equivalent circuit of an inductive impedance Z in a digital processor result from the mathematical model described in Figure 14.7. When the system is in a state of equilibrium, the following relations occur:

$$v_0 = 0 \rightarrow i_z - i_r = 0 \tag{14.20}$$

$$i_z = v_1 \frac{1}{R + j\omega L}, \quad i_r = D_r v_1 \left(\frac{b}{R_r} - a\, j\omega C_r \right) \tag{14.21}$$

where $0 < D_r \leq 1$ is the coefficient of v_1 voltage division and the values a and b are equivalent to the binary signals used by the processor to control DVD. Multiple digital-to-analog converters (DACs) are used as digitally controlled voltage dividers. They multiply voltage $D_r v_1$ by negative numbers ($-a$), which is needed in the case of using a standard capacitor C_r for measurements of inductance. After substituting Equation 14.21 into Equation 14.20 and equating the real and imaginary parts, the following formulae are obtained:

$$L = \frac{a R_r^2 C_r}{D_r \left(b^2 + a^2 \omega^2 R_r^2 C_r^2\right)} \tag{14.22}$$

$$R = \frac{b R_r}{D_r \left(b^2 + a^2 \omega^2 R_r^2 C_r^2\right)} \tag{14.23}$$

The lengths N of the code words $\{an\} \ll a$ and $\{bn\} \ll b$, $n = 1, 2, \ldots, N$ determine the range and resolution of the measurement system; that is, the highest measurable inductance and resistance values and the lowest detectable values. The range can be chosen automatically by changing the division coefficients D_r. Achieved accuracy is better than 0.1% in a very large range of impedances and in a sufficiently large frequency range. Measurements can be periodically repeated and their results can be stored and processed.

Resonance Methods

Resonance methods are a third type of inductance measurement method. They are based on application of a series or parallel resonance LC circuits as elements of either a bridge circuit or a two-port (four-terminal) "T"-type network. Examples of both circuit applications are presented in Figure 14.8.

In the bridge circuit shown in Figure 14.8a, which contains a series resonance circuit, the resonance state is obtained by varying the capacitor C_r, and then the bridge is balanced ($v_0 = 0$) using the bridge resistors. From the resonance and balance conditions, the following expressions are obtained:

$$L = \frac{1}{\omega^2 C_r}, \quad R = R_2 \frac{R_4}{R_3} \tag{14.24}$$

To calculate the values of the LR elements of the series equivalent circuit of the measured impedance, it is necessary to measure (or know) the angular frequency ω of the supply voltage. The frequency band is limited by the influence of unknown interturn capacitance value. In the "shunted T" network presented in Figure 14.8b, the state of balance (i.e., the minimal voltage v_0 value) is achieved by tuning the multiloop

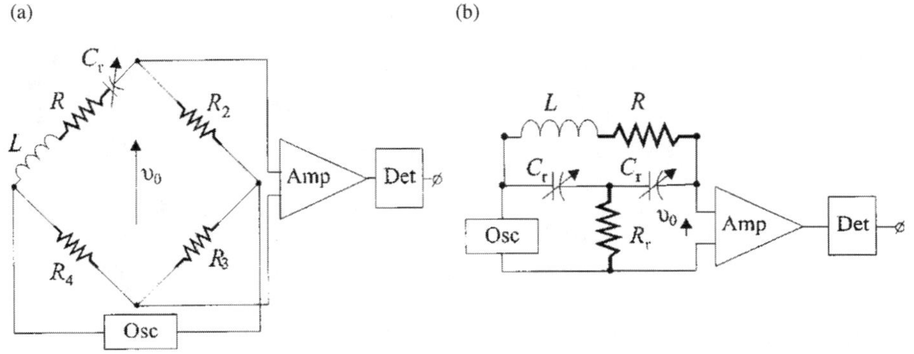

FIGURE 14.8 Circuits diagrams applied in resonance methods: bridge circuit with series resonance circuit (a), and two-port "shunted T" type with parallel resonance circuit (b). Block abbreviations as in Figure 14.6.

TABLE 14.1 Basic Features of Selected Types of *LCR* Meters

Manufacturer, Model (Designation)	Measurement Range of Inductance	Frequency	Basic Accuracy	Price
Leader LCR 740 (LCR bridge)	0.1 μH–1100 H	int. 1 kHz ext. 50 Hz–40 kHz	0.5%	$545
Electro Scientific Industries 253 (Digital impedance meter)	200 μH–200 H	1 kHz	(3.5 digit)	$995
Stanford RS SR 715 (LCR meter)	0.1 nH–100 kH	100 Hz–10 kHz	0.2%	$1295 ($1425)
Wayne Kerr 4250	0.01 nH–10 kH	120 Hz–100 kHz	0.1%	$3500
General Radio 1689 (Precision LCR meter)	0.00001 mH–99.999 H	12 Hz–100 kHz	0.02%	$4120
Hewlett-Packard 4284A (Precision LCR meter)	0.01 nH–99.9999 kH	20 Hz–1 MHz	0.05%	$9500

LCR circuit to parallel resonance. The circuit analysis [7] is based on the "star-delta" transformation of the $C_r R_r C_r$ element loop and leads to the relations for L and R values:

$$L = \frac{2}{\omega^2 C_r}, \quad R = \frac{1}{\omega^2 C_r^2 R_r} \tag{14.25}$$

According to reference [7], a "double T" network can be used for inductive impedance measurements at high frequencies (up to 100 MHz).

14.4 Instrumentation

Instruments commonly used for inductance measurements are built as universal and multifunctional devices. They enable automatic (triggered or cyclic) measurements of other parameters of the inductive elements: capacitance, resistance, quality, and dissipation factors. Equipped with interfaces, they can also work in digital measuring systems. Table 14.1 contains a review of the basic features of the instruments called *LCR* meters, specified on the basis of data accessible from the Internet [15]. Proper design of *LCR* meters limits the influence of the factors causing errors of measurement (i.e., stray couplings and distortions). The results of inductance measurements also depend on how the measurements are performed, including how the measured inductor is connected to the meter. Care should be taken to limit such influences as inductive and capacitive couplings of the inductive element to the environment, sources of distortion, and the choice of operating frequency.

References

1. R.C. Dorf, *Introduction to Electric Circuits*, New York: John Wiley & Sons, 1989.
2. J.P. Bentley, *Principles of Measurement Systems, 2nd ed.*, Harlow: Longman Group U.K., New York: John Wiley & Sons, 1988.
3. A.D. Helfrick and W.D. Cooper, *Modern Electronic Instrumentation and Measurement Techniques*, Englewood Cliffs, NJ: Prentice-Hall, 1990.
4. L.D. Jones and A.F. Chin, *Electronic Instruments and Measurements, 2nd ed.*, Englewood Cliffs, NJ: Prentice-Hall, 1991.
5. M.U. Reissland, *Electrical Measurement*, New York: John Wiley & Sons, 1989.
6. P.H. Sydenham (Ed.), *Handbook of Measurement Science, Vol.* 1, *Theoretical Fundamentals*, New York: John Wiley & Sons, 1982.

7. P.H. Sydenham (Ed.), *Handbook of Measurement Science, Vol. 2, Practical Fundamentals,* New York: John Wiley & Sons, 1983.
8. R.L. Bonebreak, *Practical Techniques of Electronic Circuit Design, 2nd ed.*, New York: John Wiley & Sons, 1987.
9. S. Hashimoto and T. Tamamura, An automatic wide-range digital LCR meter, *Hewlett-Packard J.*, 8, 9–15, 1976.
10. T. Wakasugi, T. Kyo, and T. Tamamura, New multi-frequency LCZ meters offer higher-speed impedance measurements, *Hewlett-Packard J.*, 7, 32–38, 1983.
11. T. Yonekura, High-frequency impedance analyzer, *Hewlett-Packard J.*, 5, 67–74, 1994.
12. M.A. Atmanand, V.J. Kumar, and V.G.K. Murti, A novel method of measurement of *L* and *C, IEEE Trans. Instrum. Meas.*, 4, 898–903, 1995.
13. M.A. Atmanand, V.J. Kumar, and V.G.K. Murti, A microcontroller- based quasi-balanced bridge for the measurement of *L, C* and *R, IEEE Trans. Instrum. Meas.,* 3, 757–761, 1996.
14. J. Gajda and M. Szyper, Electromagnetic transient states in the micro resistance measurements, *Systems Analysis Modeling Simulation,* Amsterdam: Overseas Publishers Association, 22, 47–52, 1996.
15. Information from *Internet* on *LCR* meters.

15

Immittance Measurement

Achim Dreher

German Aerospace Center

Electronic circuits consist of numerous elements that can be lumped, distributed, or a combination of both. The components are regarded as *lumped* if their size is much smaller than the signal wavelength. This condition holds for resistors, inductors, capacitors, transformers, diodes, transistors, or similar devices operating in printed circuits at frequencies up to a few hundred megahertz or even higher in small integrated circuits. In the microwave or millimeter-wave region, the elements and their connecting transmission lines must be considered as *distributed.* While in lumped circuits a change of voltage or current at one single point immediately affects these quantities at all other points, in distributed circuits the propagation properties now have to be taken into account. The same holds for long connecting cables even at lower frequencies.

To describe the effect of any element within an electronic circuit or of the connection of different circuits, the *immittance* is used as a characteristic quantity. It simply provides a relation of sinusoidal voltage and current at the terminals of the element as a function of frequency. The immittance therefore also characterizes arbitrarily complicated networks considered as one port. This is useful, since in practice the single elements are interconnected to networks. On the other hand, the elements themselves are not ideal. A resistor, for example, made of wound resistive wire, has parasitic components such as capacitance and inductance of winding and terminals. It must be represented by an equivalent circuit forming a complex network [1].

The word "immittance" was proposed by Bode [2] and is a combination of the words "impedance" and the reverse quantity called "admittance." These terms do not only occur in electrodynamics but wherever wave propagation takes place — in acoustics as well as in elasticity. The emphasis of this chapter is on lumped networks and guided electromagnetic waves. Readers interested in more general propagation and scattering phenomena are referred to [3].

0-8493-1733-9/04/$0.00+$1.50

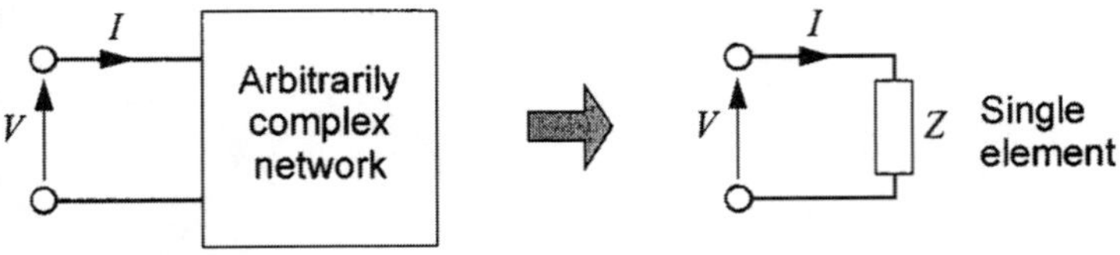

FIGURE 15.1 An arbitrarily complex network can be replaced by its impedance for a given frequency without changing the electrical properties at the terminal.

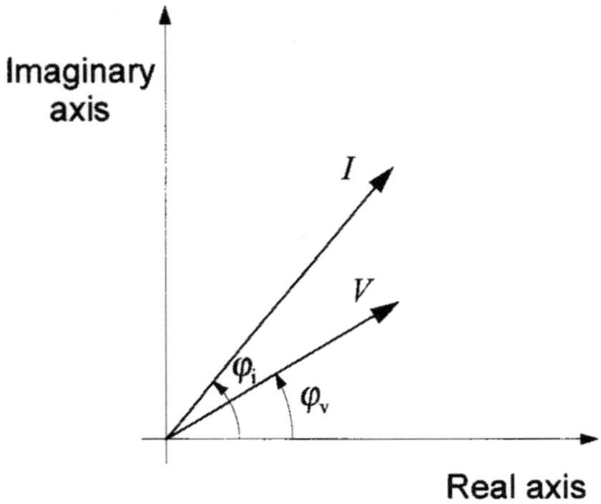

FIGURE 15.2 Voltage and current phasors in the complex plane.

15.1 Definitions

Assume a stable linear and time-invariant (LTI) network with only one port. Linearity and time independence are generally met for combinations of passive elements but also for active devices with small-signal driving under constant physical conditions (temperature, humidity, dimensions, etc.). In the steady state, a voltage $v(t) = V_m \cos(\omega t + \varphi_v)$ with amplitude V_m varying harmonically with the angular frequency $\omega = 2\pi f$ which is applied to the terminal then only produces voltages and currents of the same frequency within the network (Figure 15.1). Using complex notation:

$$v(t) = \mathrm{Re}\left\{V e^{j\omega t}\right\} \quad \text{with} \quad V = V_m e^{j\varphi_v} \tag{15.1}$$

the current flowing into the network is given by:

$$i(t) = I_m \cos(\omega t + \varphi_i) = \mathrm{Re}\left\{I e^{j\omega t}\right\} \quad \text{with} \quad I = I_m e^{j\varphi_i} \tag{15.2}$$

The phasors V and I are time independent and can be represented in the complex plane (Figure 15.2). Relating voltage and current at the terminal, the network is uniquely described by means of a complex frequency-dependent quantity, the impedance Z:

$$Z = \frac{V}{I} = \frac{V_m}{I_m} e^{j(\varphi_v - \varphi_i)} = |Z| e^{j\varphi_z} \tag{15.3}$$

For a given frequency, an arbitrarily complex network within a circuit thus can be replaced by a single element without changing the electrical properties at the terminals. Sometimes it is more convenient to use the inverse of Z, the admittance Y:

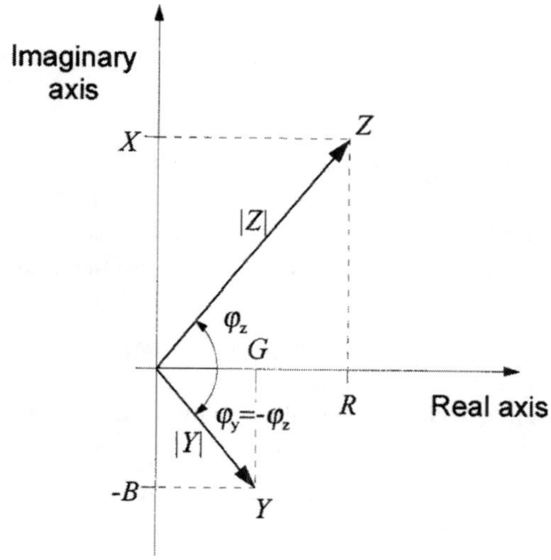

FIGURE 15.3 Representation of impedance and admittance in the complex plane showing the relations between rectangular and polar coordinates. Note that the units are different for each vector.

$$Y = \frac{1}{Z} = \frac{I}{V} = |Y| e^{j\varphi_y} \quad \text{with} \quad \varphi_y = \varphi_i - \varphi_v = -\varphi_z \tag{15.4}$$

Both quantities are combined to form the word "immittance." Figure 15.3 shows their representation in the complex plane. Equations 15.3 and 15.4 give the definition in polar coordinates. In data sheets, they are often written as:

$$|Z| \angle \varphi_z, \; |Y| \angle \varphi_y \tag{15.5}$$

Using Euler's identity $e^{j\varphi} = \cos\varphi + j\sin\varphi$, one obtains in rectangular coordinates:

$$\begin{aligned} Z &= |Z|\cos\varphi_z + j|Z|\sin\varphi_z = R + jX \\ Y &= |Y|\cos\varphi_y + j|Y|\sin\varphi_y = G + jB \end{aligned} \tag{15.6}$$

From Figure 15.3, the following relations between rectangular and polar coordinate representation can be deduced immediately:

$$\begin{aligned} &R = |Z|\cos\varphi_z \quad |Z| = \sqrt{R^2 + X^2} \qquad G = |Y|\cos\varphi_y \quad |Y| = \sqrt{G^2 + B^2} \\ &X = |Z|\sin\varphi_z \quad \varphi_z = \tan^{-1}\left(\frac{X}{R}\right) \qquad B = |Y|\sin\varphi_y \quad \varphi_y = \tan^{-1}\left(\frac{B}{G}\right) \end{aligned} \tag{15.7}$$

The real parts are the resistance R and the conductance G. They indicate the losses within the network. The imaginary parts, which are termed reactance X and susceptance B, respectively, are a measure of the reactive energy stored in the network during one period. In general, all these quantities are frequency dependent.

Note that the correct sign of the imaginary parts must be used: the angle φ is in the range of $-180° < \varphi \leq 180°$ and $\varphi < 0$ always corresponds to $X, B < 0$.

For elements with low losses, the loss angle δ or loss factor D are often given instead of the phases φ_z and φ_y. They are always positive small quantities and tend to 0 for a lossless device

$$D = \tan\delta = \tan\left(\frac{\pi}{2} - |\varphi_z|\right) = \tan\left(\frac{\pi}{2} - |\varphi_y|\right) = \left|\frac{R}{X}\right| = \left|\frac{G}{B}\right| \quad (15.8)$$

The inverse quantity is the quality factor $Q = 1/D$. It involves a ratio of stored electric energy to power dissipated. A high Q indicates a nearly pure reactive component.

In high-power electronics, it is necessary to reduce losses on transmission lines and therefore avoid currents associated with reactive load components. To obtain a criterion for the application and efficiency of compensation techniques, a power factor is defined. From complex power representation:

$$P = VI^* = |P|(\cos\varphi + j\sin\varphi) \quad (15.9)$$

(the asterisk indicates the conjugate complex number) follows from Equations 15.3 and 15.4.

$$P = |I|^2 Z = |I|^2 |Z|(\cos\varphi_z + j\sin\varphi_z) = |V|^2 Y^* = |V|^2 |Y|(\cos\varphi_y - j\sin\varphi_y) \quad (15.10)$$

and since the effective power is given by the real part of P:

$$P_{\text{eff}} = \text{Re}\{P\} = |P|\cos\varphi \quad (15.11)$$

the power factor is:

$$\cos\varphi = \cos\varphi_z = \cos\varphi_y \quad (15.12)$$

In general, rms values are used for the phasors. Otherwise, a factor 1/2 has to be taken into account in Equations 15.9 and 15.10, since $|P| = \frac{1}{2}V_m I_m$ for sinusoidal quantities.

It can also be seen from Equations 15.9 and 15.10 that the immittances are directly related to the apparent power:

$$|P| = |V||I| = |I|^2 |Z| = |V|^2 |Y| \quad (15.13)$$

15.2 Ideal Lumped Components

The immittances of the fundamental passive circuit elements are derived from their instantaneous voltage current relations using Equations 15.1 through 15.4 and the differentiation rules.

Resistances

From Equation 15.14:

$$v(t) = Ri(t) \quad (15.14)$$

it follows $V = RI$ and thus $Z = R$ or $Y = G$. The immittance of a resistance is real and identical to its dc resistance or conductance.

Inductances

Voltage and current are related via the differential equation:

$$v(t) = L\frac{di(t)}{dt} \quad (15.15)$$

with inductance L, from which follows that $V = j\omega LI$ and

$$Z = j\omega L = jX_L, \quad Y = \frac{1}{j\omega L} = -j\frac{1}{X_L} = -jB_L \tag{15.16}$$

Capacitances

From Equation 15.17:

$$i(t) = C\frac{dv(t)}{dt} \tag{15.17}$$

with capacitance C, it follows that $I = j\omega CV$ and

$$Y = j\omega C = jB_C, \quad Z = \frac{1}{j\omega C} = -j\frac{1}{B_C} = -jX_C \tag{15.18}$$

The immittance of ideal inductors and capacitors is purely imaginary with different signs according to the phase shift of ±90° between voltage and current. A general element or network is therefore called inductive or capacitive at a given frequency corresponding to the sign of the imaginary part of its impedance. Note, however, that the frequency dependence can be much more complicated than for these ideal elements and the impedance can even change several times between capacitive and inductive characteristic.

15.3 Distributed Elements

At high frequencies, the size of the elements may no longer be small compared to the signal wavelength. Propagation effects must then be taken into account and the components can no longer be described by means of simple lumped equivalent circuits. If at all possible, they are replaced by transmission line circuits, which are easier to characterize; they realize the required electrical properties more exactly within a defined frequency range.

Transmission Lines

Assuming a simplifying transverse electromagnetic wave (TEM mode) with no field components in the propagation direction, voltages and currents can be uniquely defined and are given as solutions of the corresponding wave equations [4]:

$$\frac{d^2V}{dz^2} - \gamma^2 V = 0, \quad \frac{d^2I}{dz^2} - \gamma^2 I = 0 \tag{15.19}$$

They vary along the line in the z-direction according to:

$$V(z) = V_0^+ e^{-\gamma z} + V_0^- e^{\gamma z}, \quad I(z) = I_0^+ e^{-\gamma z} + I_0^- e^{\gamma z} \tag{15.20}$$

These solutions are sums of forward ($e^{-\gamma z}$) and backward ($e^{\gamma z}$) traveling waves with amplitudes V_0^+, I_0^+ and V_0^-, I_0^- and a propagation constant:

$$\gamma = \sqrt{(R' + j\omega L')(G' + j\omega C')} \tag{15.21}$$

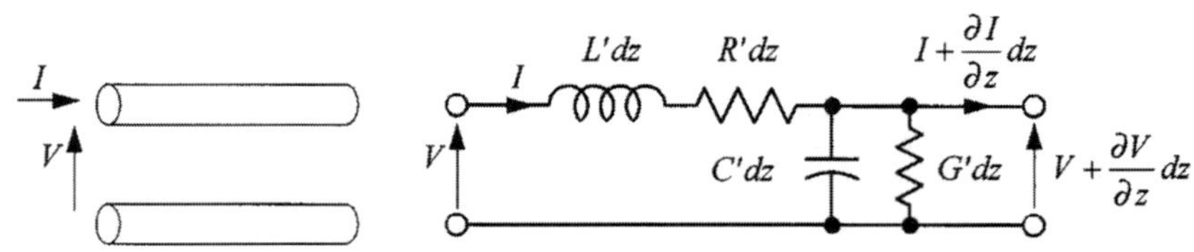

FIGURE 15.4 Equivalent circuit of a differential length of transmission line. The wave equations can be obtained by simply applying Kirchhoff's laws to voltages and currents.

The equivalent circuit of the transmission line is shown in Figure 15.4. The energy storage in the electric field is accounted for by the distributed shunt capacitance C' per unit length, while the effect of the magnetic field is represented by the series inductance L' per unit length. The series resistance R' per unit length and the shunt conductance G' per unit length represent the power losses in the conductors and in the dielectric, respectively. The amplitudes of voltage and current are related by means of the characteristic impedance Z_0:

$$Z_0 = \frac{V^+}{I^+} = -\frac{V^-}{I^-} = \sqrt{\frac{R' + j\omega L'}{G' + j\omega C'}} \tag{15.22}$$

Of special interest for the use within a network is the input impedance Z_{in} of the transmission line. It depends also on the termination Z_L at the other end of the line. For a transmission line of length l, it is given by:

$$Z_{in} = Z_0 \frac{Z_L + Z_0 \tanh \gamma l}{Z_0 + Z_L \tanh \gamma l} \tag{15.23}$$

that is, a transmission line transforms the impedance Z_L into Z_{in} at the input.

A quantity more suitable to wave propagation and measurement at high frequencies is the reflection coefficient Γ. It is defined by the relation of the voltages associated with forward and backward traveling waves. At the end of the line, using $V(l) = Z_L I(l)$, one finds:

$$\Gamma = \frac{V_0^- e^{\gamma l}}{V_0^+ e^{-\gamma l}} = \frac{Z_L - Z_0}{Z_L + Z_0} \tag{15.24}$$

For devices that support quasi or strong non-TEM waves like microstrip lines, hollow waveguides, dielectric and optical waveguides, a voltage cannot be uniquely defined. That is why several definitions of the characteristic impedance Z_0 exist [5].

15.4 Interconnections and Graphical Representations

Since Kirchhoff 's laws for voltages and currents also holds for complex quantities, the rules for series and parallel connections of resistances and susceptances in the dc case apply as well for immittances.

Series connection:

$$Z = \sum_i Z_i \quad \frac{1}{Y} = \sum_i \frac{1}{Y_i} \tag{15.25}$$

Parallel connection:

$$Y = \sum_i Y_i \quad \frac{1}{Z} = \sum_i \frac{1}{Z_i} \tag{15.26}$$

As an example, consider a simplified equivalent circuit of a resistor with the nominal value R_0 (Figure 15.5). Gradually using the rules for series and parallel connection and the impedances for

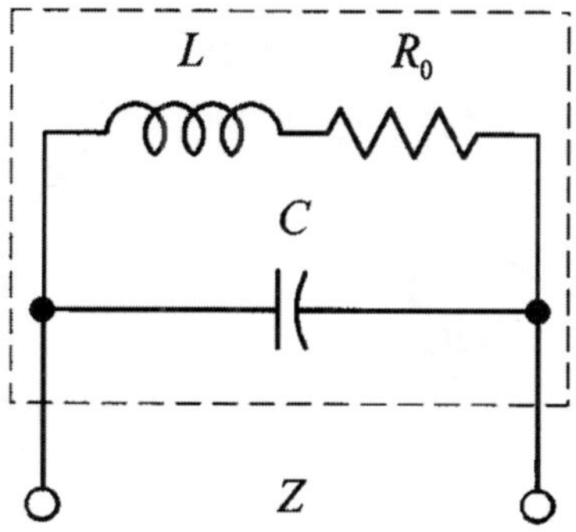

FIGURE 15.5 The simple equivalent circuit of a wire-wound resistor with nominal value R_0, inductance of the winding L, and capacitance of winding and terminal C. It is valid for a wide frequency range.

inductances (Equation 15.16) and capacitances (Equation 15.18), the impedance of the real resistor with parasitic elements as given leads to:

$$Z = \frac{R_0 + j\omega L}{1 - \omega^2 LC + j\omega R_0} \tag{15.27}$$

The magnitude and phase of Z/R_0 as a function of ω/ω_0 are shown in Figure 15.6 with $\omega_0 = 1/\sqrt{LC}$ as the resonant frequency defined by the parasitic elements, which might be caused by the windings of a wire-wound resistor. The network is inductive for low ($\varphi_z > 0$) and capacitive for high frequencies. An alternative representation is to plot real and imaginary parts in the impedance plane with the frequency as parameter as indicated by the labels (Figure 15.7). This version, called the *locus*, is very suitable to see immittance changes caused by parameters like frequency or adjustable elements within the network. Note that both real and imaginary parts are parameter dependent and vary with frequency.

In high-frequency applications, one obtains the impedance more easily from the reflection coefficient. Rewriting Equation 15.24 in the form:

$$\Gamma = \frac{\overline{Z}_L - 1}{\overline{Z}_L + 1} \quad \text{with} \quad \overline{Z}_L = \frac{Z_L}{Z_0} \tag{15.28}$$

defines a transformation of which the graphical representation has been called the Smith chart (Figure 15.8). It can be regarded as two coordinate systems lying one on top of the other. The reflection coefficient is given in polar coordinates around the center, the circles give the real and imaginary part of the associated impedance. The Smith chart is very useful for solving transmission line and waveguide impedance matching problems [6].

15.5 Measurement Techniques

Since immittances are complex quantities, one must determine two parameters: magnitude and phase or real and imaginary part, described as vector measurements. There exist several techniques depending on frequency range and required accuracy [7].

Current–Voltage Methods

A simple way to measure immittances follows directly from the defining Equation 15.3. Applying a well-known sinusoidal voltage to the terminal and measuring magnitude and phase of the current gives the desired quantity (Figure 15.1). However, the internal impedance Z_A of the ammeter should be known exactly and the unknown impedance is then given by:

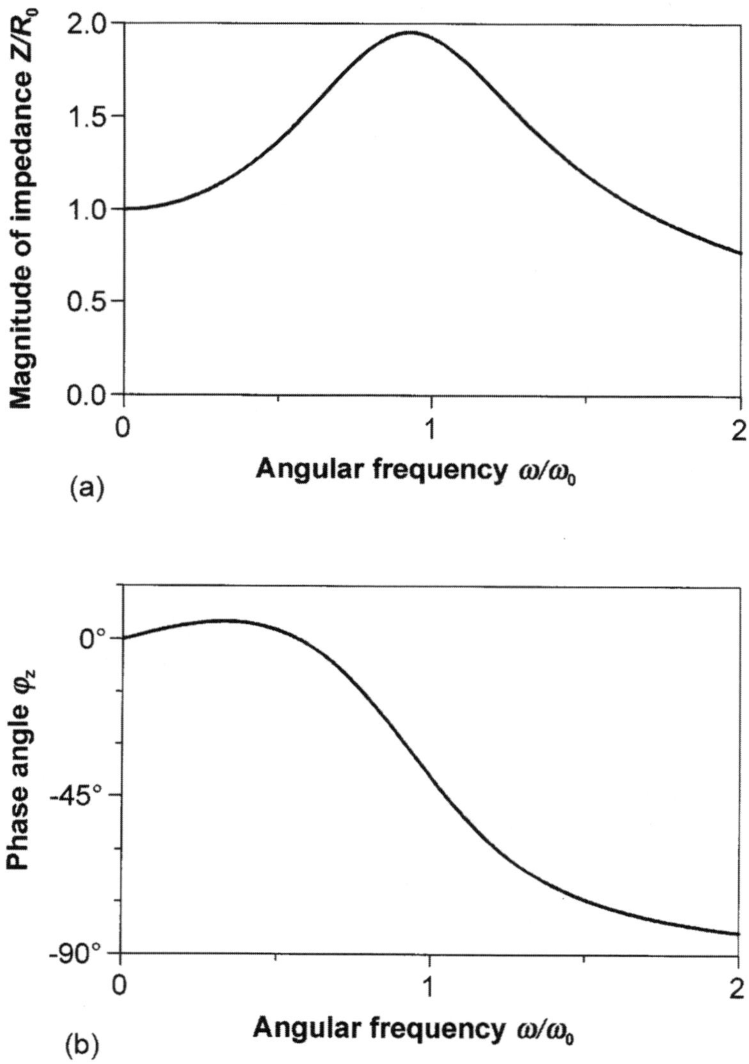

FIGURE 15.6 Normalized magnitude (a) and phase (b) of the impedance of a wire-wound resistor varying with frequency. ω_0 is the resonant frequency defined by the parasitic elements.

$$Z = \frac{V}{I} - Z_A \tag{15.29}$$

In practical applications, impedances below 1000 Ω are measured by passing a predetermined current through the unknown device and measuring the voltage across it. Phase angle information is obtained by comparing the relative phase between voltage and current by means of a phase detector [8].

A variant on this method using only the better practicable voltage measurements is shown in Figure 15.9. The accurately known resistor R must be small compared to Z_x and to the internal resistance of V_2. One finds that:

$$Z_x = \left(\frac{V_1}{V_2} - 1\right)R, \text{ or } Z_x \approx \frac{V_1}{V_2}R \text{ if } R \ll |Z_x| \tag{15.30}$$

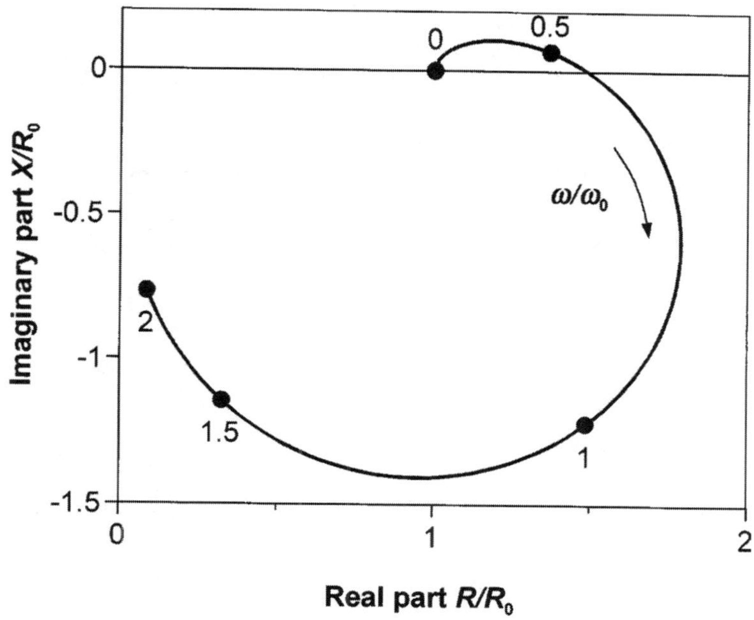

FIGURE 15.7 Normalized impedance of a wire-wound resistor in the complex plane. The arrow indicates the direction of increasing frequency.

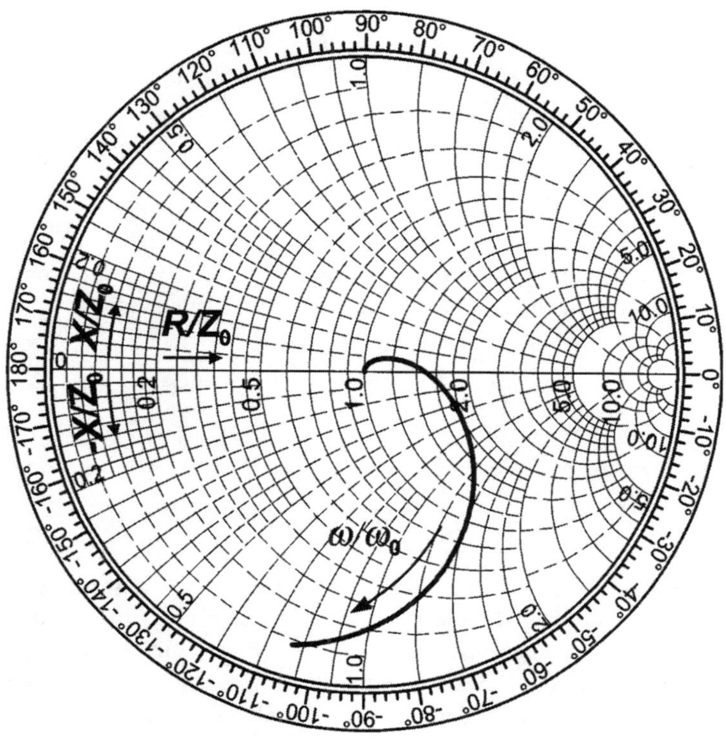

FIGURE 15.8 Smith chart representation of the impedance of a wire-wound resistor.

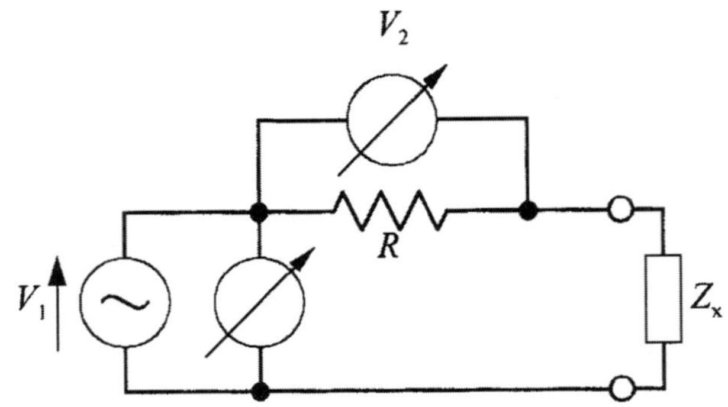

FIGURE 15.9 Determination of an impedance Z_x by phase-sensitive voltage measurements, only using a well-known resistor R.

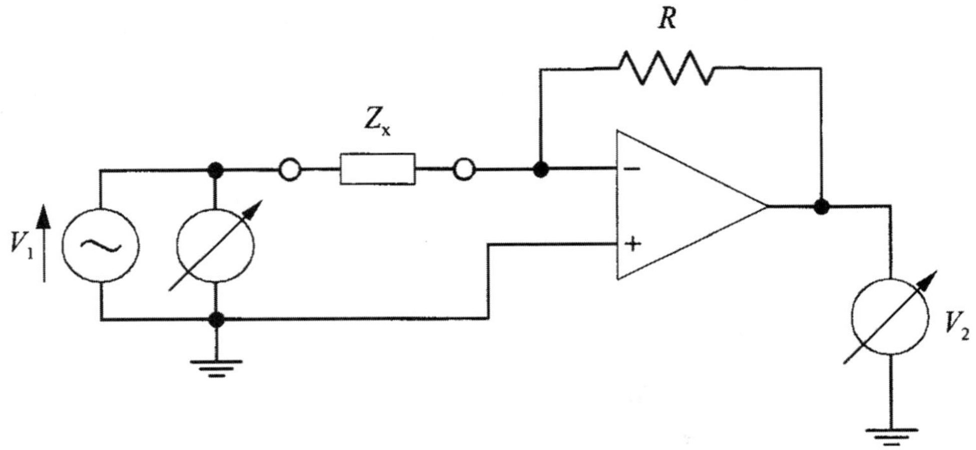

FIGURE 15.10 Impedance measurement with an inverting operational amplifier circuit. Its advantages are high input and low output resistance.

The measurement can be enhanced using an operational amplifier with high input and low output resistance in an inverting circuit (Figure 15.10). The unknown is then given by

$$Z_x = -\frac{V_1}{V_2}R \tag{15.31}$$

Practical implementations use operational amplifiers as part of an autobalancing bridge; see [7, 8].

Bridge Methods

Alternating current bridges are low-cost standard laboratory devices to measure impedances over a wide frequency range from dc up to 300 MHz with very high precision (Figure 15.11). A comprehensive survey is given in [1]. Their main advantage is that only a zero indicator in the diagonal branch is necessary. For this reason, the internal impedance does not influence the accuracy and the null point can be detected with a high-sensitivity ac galvanometer as well as with headphones in the audio frequency range.

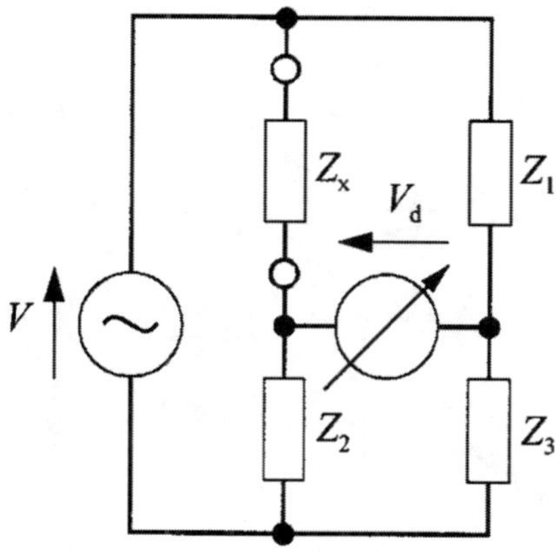

FIGURE 15.11 Impedance measurement by bridge methods. The bridge is balanced when the voltage V_d across the diagonal branch is adjusted to zero by tuning Z_1, Z_2, or Z_3.

If the bridge is balanced, the unknown immittance is given by:

$$Z_x = \frac{Z_1}{Z_3} Z_2 \text{ or } Y_x = \frac{Z_3}{Z_1} Y_2 \tag{15.32}$$

Since the quantities are complex, Equation 15.32 involves the adjustment of two parameters: magnitude and phase:

$$|Z_x| = \left|\frac{Z_1}{Z_3}\right| |Z_2|, \quad \varphi_x = \varphi_1 - \varphi_3 + \varphi_2 \tag{15.33}$$

or real and imaginary parts, respectively.

An important property of an impedance bridge is the sensitivity ε:

$$\varepsilon = \left|\frac{\partial V_d}{\partial Z_x}\right| = V \frac{Z_2}{(Z_2 + Z_x)^2} \tag{15.34}$$

or (independent of Z_x)

$$\varepsilon = V \frac{Z_3^2}{Z_2(Z_1 + Z_3)^2} \tag{15.35}$$

in the vicinity of zero crossing when the bridge is balanced.

The precision of the measurement not only depends on the exact zero adjustment, which can be enhanced by choosing the elements and the voltage according to Equation 15.35 to obtain a high sensitivity, but also on the realization of $Z_1 \ldots Z_3$. Mostly, these are connections of resistors and capacitors. Inductors are avoided because they always have a resistive component and it is difficult and expensive to manufacture inductors with exactly defined and reproducible electrical properties. There exist various types of bridges depending on how the elements are designed and interconnected. To choose the correct configuration, it must be known whether the unknown impedance is capacitive or inductive; otherwise, a zero adjustment is not always possible since the balancing condition cannot be fulfilled. Bridges are

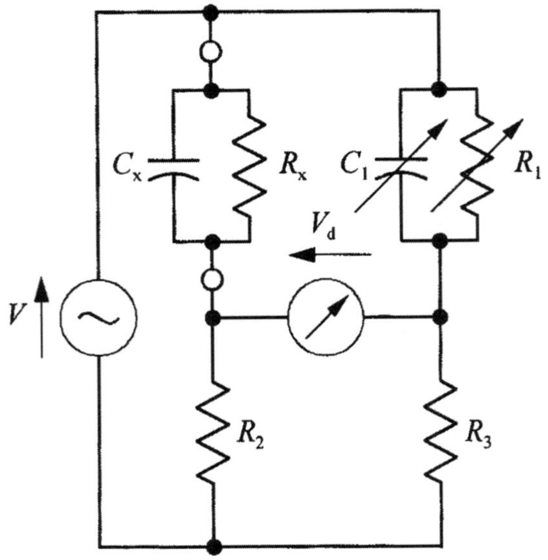

FIGURE 15.12 Wheatstone bridge for the capacitance and dissipation factor measurement of capacitors. The balancing condition is frequency independent. The resistor R_1 and the capacitor C_1 must be tuned successively until the bridge is balanced.

therefore principally used to measure capacitances and inductances as well as loss and quality factors of capacitors and coils. Since magnitude and phase conditions must be matched simultaneously, two elements must be tuned. To obtain a wide measurement range, the variable elements are designed as combinations of switchable and tunable capacitors and resistors. The sensitivity of the zero indicator can be changed for global search and final adjustment. Unfortunately, magnitude and phase cannot be adjusted independently of each other. If the balancing is performed by hand, a suitable strategy is to search the minimum voltage by tuning each element successively.

Frequently used bridges are the Wheatstone bridge (Figure 15.12) for the measurement of lossy capacitances, and the Hay bridge (Figure 15.13) to determine inductivity and quality factor of coils. Because of its symmetrical structure, the balancing condition for the Wheatstone bridge is simply:

$$R_x = \alpha R_1, \quad C_x = \alpha C_1, \quad \alpha = \frac{R_3}{R_2} \tag{15.36}$$

which is independent of frequency.

The measurement of a coil with the Hay bridge requires that:

$$R_x + j\omega L_x = \frac{j\omega C_3 R_1 R_2}{1 + j\omega R_3 C_3} = \frac{\omega^2 C_3^2 R_1 R_2 R_3}{1 + \left(\omega R_3 C_3\right)^2} + j\omega \frac{C_3 R_1 R_2}{1 + \left(\omega R_3 C_3\right)^2} \tag{15.37}$$

from which the quality factor is obtained as:

$$Q = \frac{\omega L_x}{R_x} = \frac{1}{\omega R_3 C_2} \tag{15.38}$$

The inductance of high-Q coils can be determined frequency independent since

$$L_x \approx R_1 R_2 C_3 \text{ if } \left(\omega R_3 C_3\right)^2 \ll 1 \tag{15.39}$$

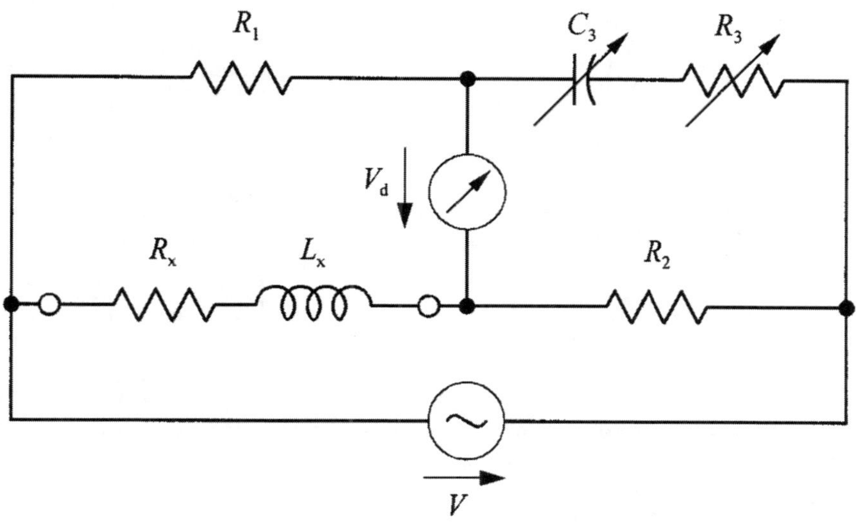

FIGURE 15.13 Hay bridge for the measurement of the inductance and the quality factor of coils. If Q is sufficiently high, the inductance can be determined nearly frequency independent.

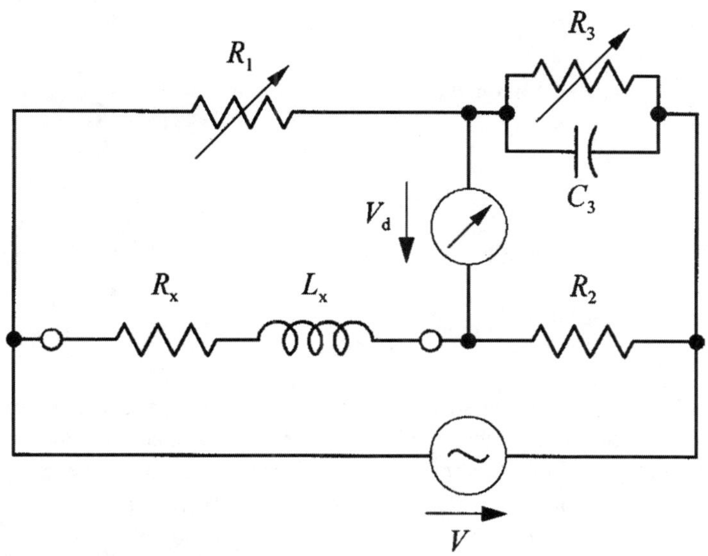

FIGURE 15.14 Maxwell bridge with simple and frequency-independent balancing conditions. Despite these advantages, it is not recommended for high-Q coils because of a very large R_1 value.

A very interesting alternative is the Maxwell bridge (Figure 15.14), since it requires only resistors as variable elements, which can be manufactured with high precision. The balancing is frequency independent and leads to:

$$R_x = \frac{R_1 R_2}{R_3}, \quad L_x = R_1 R_2 C_3, \quad Q = \omega C R_1 \tag{15.40}$$

Nevertheless, the Hay bridge is preferred for high-Q coils, because a very large value of R_1 is required for the Maxwell bridge leading to a disadvantageous balancing [9].

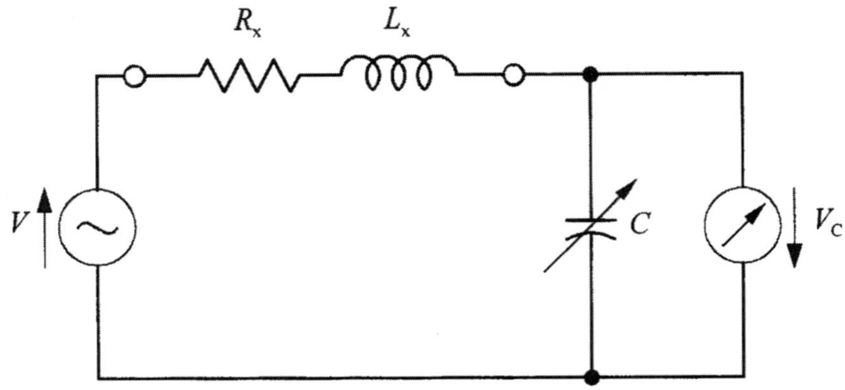

FIGURE 15.15 Coil as part of a resonance circuit to determine inductance and quality factor. The capacitor C is tuned to maximum voltage V_C.

Resonant Method

Using the coil as part of a resonance circuit as in Figure 15.15 and tuning C to maximum voltage, the quality factor can be measured directly as:

$$Q = \left| \frac{V_{c,\,max}}{V} \right| = \frac{1}{\omega R_x C} \tag{15.41}$$

The unknown inductance is then obtained from the test frequency by means of the resonance condition:

$$L_x = \frac{1}{\omega^2 C} \tag{15.42}$$

If a capacitor with sufficiently low losses is used, Q values as high as 1000 can be measured.

Network Analysis Methods

Frequency Domain

In the case of distributed elements, measurements of currents and voltages depend on the position and are often not directly applicable to high-frequency devices like waveguides or microstrip lines. For that reason the determination of immittances is derived from measuring the reflection coefficient. Equation 15.23 shows the importance of defining a proper measurement plane. This is the cross-section of the line or waveguide perpendicular to the direction of propagation at a definite length l_0, where the reflection coefficient has to be measured. It can then be transformed along the line toward load or source using this relation or the Smith chart. Exact microwave measurements are very sophisticated and need a lot of practical experience. Further details can be found in the literature [5, 10–12].

Automated and precise immittance measurements over a wide frequency range are best carried out with a vector network analyzer [11]. Unfortunately, this is also the most expensive method. The principle of measurement is shown in Figure 15.16. A power divider splits the incident signal into a transmitted and a reference part. The directional bridge or coupler separates forward and backward traveling waves, and the reflected signal now appears in the branch with the phase-sensitive voltmeter V_2. Using a bridge with impedances matched to the line ($Z_1 = Z_2 = Z_3 = Z_0$), the voltage in the diagonal branch is given by (Figure 15.11):

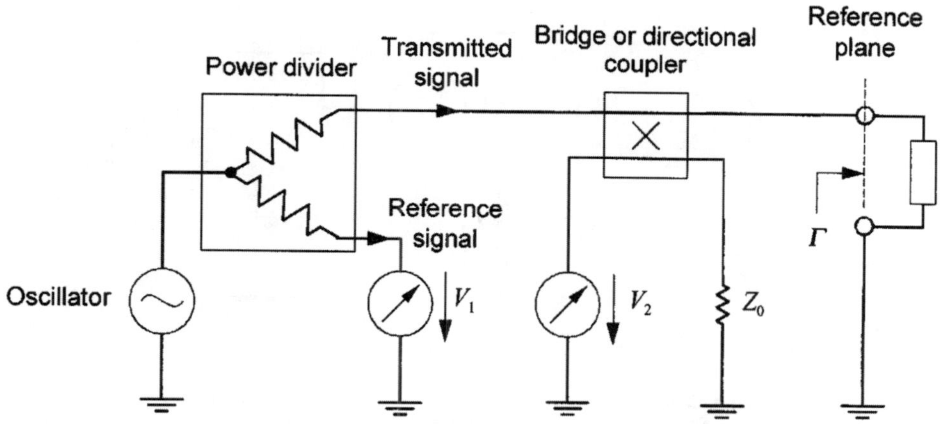

FIGURE 15.16 Schematic of network analyzer measurements. The voltage ratio V_2/V_1 of reflected wave and reference signal is proportional to the reflection coefficient Γ. The impedance Z_x can then be computed.

$$V_d = \frac{Z_x - Z_0}{2(Z_x + Z_0)} V = V_2 \tag{15.43}$$

and thus the reflection coefficient:

$$\Gamma = \frac{Z_x - Z_0}{Z_x + Z_0} = \alpha \frac{V_2}{V_1} \tag{15.44}$$

is directly proportional to the voltage ratio.

Network analyzers use an automatic error correction to eliminate the effect of internal and external couplers and junctions. Because of that, a calibration procedure with standard terminations is necessary. These terminations must be manufactured very precisely, since they define the measurement plane and determine the overall measurement error.

Time Domain

It is often necessary to locate an impedance step, whether to find out the distance of a cable defect or to track down the origin of reflections within a connection. To this end, high-performance vector network analyzers have a Fourier transform procedure. But there also exist cheaper time domain reflectometers (TDR) [13, 14]. They use an incident step or impulse signal (Figure 15.17) and the reflected signal is separated by means of a directional coupler and displayed on a CRT in the time domain. From the shape of the signal, the impedance step can be localized by means of the time delay:

$$l = \frac{1}{2} v_g t \tag{15.45}$$

with v_g as signal or group velocity on the line varying from step to step. Characteristic and magnitude of the impedance can only be estimated, since phase information is usually not available. TDR measurements are restricted to the localization of impedance steps and not to be recommended for exact measurements. Moreover, additional pulse deformations occur in dispersive waveguides.

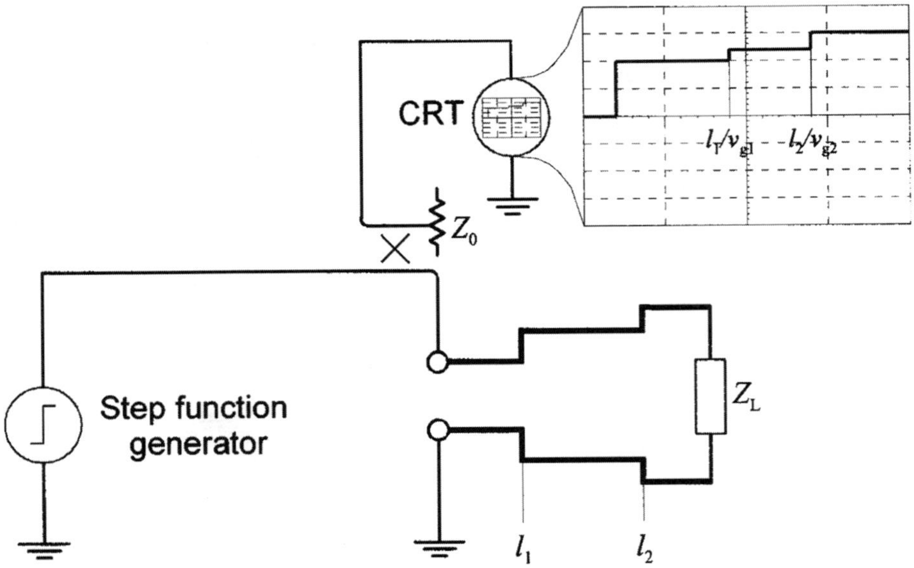

FIGURE 15.17 Detection and measurement of impedance steps on a line or waveguide with a time domain reflectometer (TDR). Since phase information is usually not available, the characteristics and magnitudes of the impedances can only be estimated. Notice that the group or signal velocity v_g varies from step to step.

15.6 Instrumentation and Manufacturers

A broad range of instrumentation for measuring immittance is available. Some of these instruments are included in Table 15.1. Table 15.2 provides the names and addresses of some companies that produce immittance-measuring instrumentation.

TABLE 15.1 Instruments for Immittance Measurements

Manufacturer	Model Number	Description
Agilent Technologies	E5100	Network analyzers 10 kHz–300 MHz
Agilent Technologies	ENA Series	Network analyzers 300 kHz–8.5 GHz
Agilent Technologies	PNA Series	Network analyzers 300 kHz–67 GHz
Agilent Technologies	8510C	Network analyzer systems 45 MHz–110 GHz
Agilent Technologies	4263B	LCR meter 100 Hz–100 kHz
Agilent Technologies	4284A	LCR meter 20 Hz–1 MHz
Agilent Technologies	4285A	LCR meter 75 kHz–30 MHz
Anritsu	ME7808A	Vector network analyzer 40 MHz–110 GHz
Fluke	PM 6303A	Automatic RCL meter
Fluke	PM 6304	Automatic RCL meter
Keithley	3321	LCZ meter, 4 test frequencies to 100 kHz
Keithley	3322	LCZ meter, 11 test frequencies to 100 kHz
Keithley	3330	LCZ meter 40 Hz–100 kHz (201 test frequencies)
Quadtech	1710	LCR meter 20 Hz–200 kHz
Quadtech	7000	LCR meter Series 10 Hz–2 MHz
Quadtech	1910	Inductance analyzer 20 Hz–1 MHz
Rohde & Schwarz	ZVR	Network analyzer
SST	SR715/720	LCR meter
TTi	LCR400	Precision LCR bridge
Voltech	ATi	Transformer tester

TABLE 15.2 Companies Producing Immittance Measurement Equipment

Agilent Headquarters 395 Page Mill Rd. P.O. Box 10395 Palo Alto, CA 94303 Tel: (877) 4-Agilent www.agilent.com	Rohde & Schwarz, Inc. 7150-K Riverwood Drive Columbia, MD 21046 Tel: (410) 910-7800 www.rsd.de/www/dev_center.nsf/USA
Anritsu Co. 1155 East Collins Blvd. Richardson, TX 75081 Tel: (800) ANRITSU (267-4878) www.global.anritsu.com	SRS Stanford Research Systems 1290-D Reamwood Ave. Sunnyvale, CA 94089 Tel: (408) 744-9040 www.thinksrs.com
Fluke Corporation 6929 Seaway Boulevard P.O. Box 9090 Everett, WA 98206 Tel: (800) 44-FLUKE www.fluke.com	TTi (Thurlby Thandar Instruments Ltd.) Glebe Road Huntingdon Cambs. PE29 7DR U.K. Tel: +44-1480-412451 www.tti-test.com
Keithley Instruments, Inc. 28775 Aurora Road Cleveland, OH 44139 Tel: (800) 552-1115 www.keithley.com	Voltech Instruments, Inc. 11637 Kelly Road Suite 306 Fort Myers, FL 33908-2544 Tel: (239) 437-0494 www.voltech.com
QuadTech, Inc. 5 Clock Tower Place Suite 210 East Maynard, MA 01754 Tel: (800) 253-1230 www.quadtechinc.com	

Defining Terms

Admittance (*Y*): The reciprocal of impedance.

Immittance: A response function for which one variable is a voltage and the other a current. Immittance is a general term for both impedance and admittance, used where the distinction is irrelevant.

Impedance (*Z*): The ratio of the phasor equivalent of a steady-state sine-wave voltage to the phasor equivalent of a steady-state sine-wave current. The real part is the *resistance*, the imaginary part is the *reactance*.

Phasor: A complex number, associated with sinusoidally varying electrical quantities, such that the absolute value (modulus) of the complex number corresponds to either the peak amplitude or root-mean-square (rms) value of the quantity, and the phase (argument) to the phase angle at zero time. The term "phasor" can also be applied to impedance and related complex quantities that are not time dependent.

Reflection coefficient: At a given frequency, at a given point, and for a given mode of propagation, the ratio of voltage, current, or power of the reflected wave to the corresponding quantity of the incident wave.

References

1. B.M. Oliver and J.M. Cage, *Electronic Measurements and Instrumentation,* New York: McGraw-Hill, 1971.
2. H.W. Bode, *Network Analysis and Feedback Amplifier Design,* Princeton: Van Nostrand, 1959.

3. A.T. de Hoop, *Handbook of Radiation and Scattering of Waves,* London: Academic Press, 1995.
4. R.E. Collin, *Foundations for Microwave Engineering,* New York: McGraw-Hill, 1992.
5. P.I. Somlo and J.D. Hunter, *Microwave Impedance Measurement,* London: Peter Peregrinus, 1985.
6. R.L. Thomas, *A Practical Introduction to Impedance Matching,* Dedham, MA: Artech House, 1976.
7. M. Honda, *The Impedance Measurement Handbook,* Yokogawa: Hewlett-Packard, 1989.
8. Anonymous, *4800A Vector Impedance Meter Operating and Service Manual,* Rockaway: Hewlett-Packard, 1967.
9. C.R. Paul, S.A. Nasar, and L.E. Unnewehr, *Introduction to Electrical Engineering,* New York: McGraw-Hill, 1992.
10. T.S. Laverghetta, *Modern Microwave Measurements and Techniques,* Norwood, MA: Artech House, 1988.
11. A.E. Bailey (Ed.), *Microwave Measurement,* London: Peter Peregrinus, 1985.
12. T.S. Laverghetta, *Handbook of Microwave Testing,* Dedham, MA: Artech House, 1981.
13. E.K. Miller (Ed.), *Time-Domain Measurements in Electromagnetics,* New York: Van Nostrand Reinhold, 1986.
14. Anonymous, *TDR Fundamentals for Use with HP* 54120*T Digitizing Oscilloscope and TDR,* Application notes AN 62, Palo Alto, CA: Hewlett-Packard, 1988.

Further Information

L.S. Bobrow, *Fundamentals of Electrical Engineering,* New York: Oxford University Press, 1996.

W.H. Roadstrum and D.H. Wolaver, *Electrical Engineering for all Engineers,* New York: John Wiley & Sons, 1994.

A.S. Morris, *Principles of Measurement and Instrumentation,* London: Prentice Hall International (U.K.), 1993.

G.H. Bryant, *Principles of Microwave Measurement,* London: Peter Peregrinus, 1988.

Anonymous, *Low Level Measurements,* Cleveland, OH: Keithley Instruments, 1984.

16

Q Factor Measurement

Albert D. Helfrick
Embry-Riddle Aeronautical University

Q factor is a method of characterizing the rate of dissipation of energy from an oscillating system. *Q* is defined as 2π times the energy stored in a resonant system divided by the energy dissipated per cycle. The term system used in this context refers to any type of resonance: mechanical, electrical, nuclear, etc. For the purposes of this Handbook, *Q* will be that of an electric circuit. Also, for the discussion of *Q*, very low values of *Q*, typically less than 10, will not be considered as these low values of *Q* produce highly damped oscillations that are more exponential than oscillatory and the concept of *Q* does not fit.

A common interpretation of *Q* is quality factor that explains the use of the letter, *Q*, but this is misleading. A component that has a high *Q* is not always beneficial and may not be of high quality. In many circuits, a component requires a specific value of *Q* rather than "higher is better." In other cases, a high *Q* is an advantage.

The *Q* factors encountered in common circuit components range from a low of about 50 for many inductors to nearly 1 million found in high-quality quartz crystals. *Q* can be applied to a resonant circuit or to capacitors and inductors. When *Q* is applied to a component, such as an inductor, the *Q* would be that obtained if the inductor were used in a resonant circuit with a capacitor that dissipates no energy. In this case, the value of *Q* depends on the frequency.

For most *LC* resonant circuits, the losses in the inductor dominate and the *Q* of the inductor is, essentially, the *Q* of the circuit. It is easy to make very low-loss capacitors even in the UHF region. On the other hand, varactor diodes have considerably more loss than fixed capacitors, and a varactor can play a more significant role in setting the circuit *Q*.

16.1 Basic Calculation of *Q*

Figure 16.1 shows a simple resonant circuit. The capacitor stores energy in the electric field and the inductor in the magnetic field. The circuit oscillates with the energy transferring between the two elements. For the ideal elements shown, this continues forever. Since the definition of *Q* has the energy lost per cycle in the denominator — which is zero — the result is an infinite *Q*.

In practice, circuit elements are not perfect and the energy initially contained within this circuit would be lost by the circuit and the oscillations would decrease in amplitude as the energy diminished. Energy loss in a circuit is represented by that in a resistor which can be included in one of two ways. The first way is shown in Figure 16.2(a), where the resistor is in series with the capacitor and inductor. A second representation is a parallel resistor as shown in Figure 16.2(b).

0-8493-1733-9/04/$0.00+$1.50

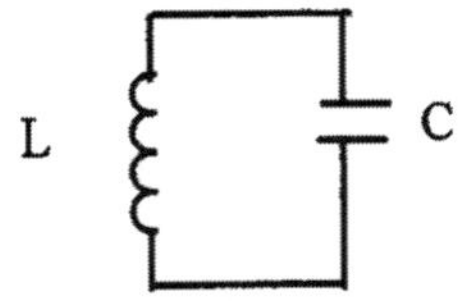

FIGURE 16.1 A simple resonant circuit with no loss elements.

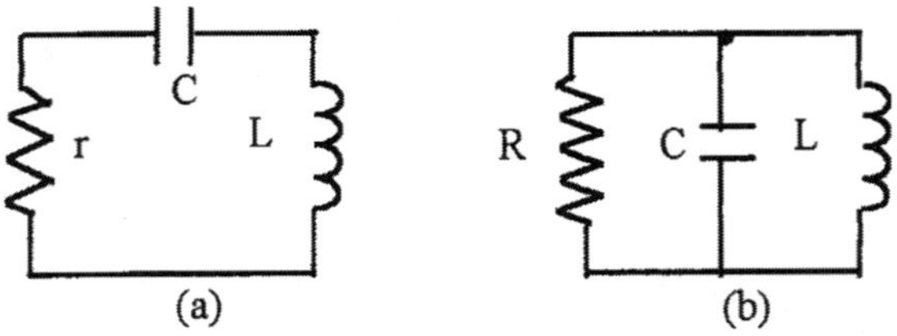

FIGURE 16.2 (a) A simple series resonant circuit with the equivalent resistance. (b) A parallel resonant circuit with a parallel equivalent resistance. For the same Q circuit, the values of the resistors are not the same.

To derive the relationship between the circuit element values and the Q of the circuit, either the current or voltage of the circuit can be used in the equations. Current is the obvious choice for a series circuit, while voltage is the common thread for the parallel circuit. Assume that the amplitude of the current through the series circuit of Figure 16.2(a) is given by:

$$i(t) = I(t)\cos(2\pi f_0 t) \tag{16.1}$$

where $f_0 = \text{Resonant frequency} = f_0 = \dfrac{1}{2\pi\sqrt{LC}}$ (16.2)

and $I(t)$ is the amplitude, which is decreasing in some unspecified fashion. The circuit's peak current occurs when the cosine function is equal to 1 and all of the energy is contained in the inductor and is equal to (1/2) $LI^2(t)$.

Assume that a relatively high Q is present in this circuit and that $I(t)$ changes by only a slight amount during the time of one cycle. During this cycle, the peak current is $I(t)$, and the rms value of the current is (0.707) $I(t)$. Therefore, the energy dissipated in one cycle is (0.5) $I^2(t)r/f_0$. Substituting these values in the definition of Q yields:

$$Q = 2\pi \frac{\frac{1}{2} LI^2(t)}{\frac{1}{2} \frac{rI^2(t)}{f_0}} = \frac{2\pi f_0 L}{r} = \frac{X_L}{r} \tag{16.3}$$

where X_L is the reactance of the inductor. The same procedure can be used with the parallel resonant circuit of Figure 16.2(b) using voltage equations to obtain the relationship between a parallel resistance and Q, which is:

$$Q = \frac{R}{X_L} \tag{16.4}$$

It is very important to understand the nature of the circuit resistance in Figure 16.2(a) and (b). This resistance represents all of the losses in the resonant circuit. These losses are from a variety of sources,

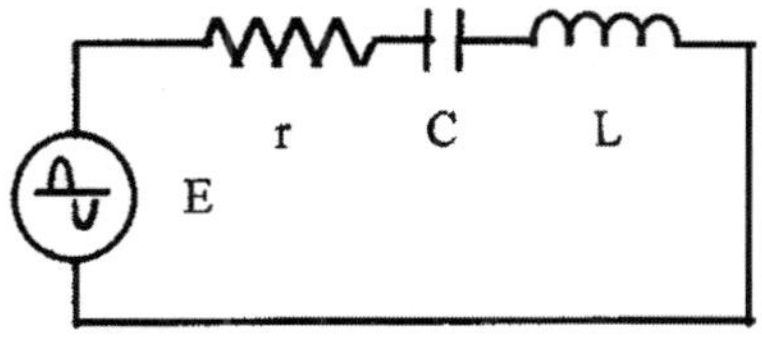

FIGURE 16.3 A series resonant circuit showing a driving source.

such as the resistance of the wire to make an inductor or the leakage current of a capacitor. It can also represent the deliberate addition of resistors to set the *Q* of the circuit to a specific value. The resistance shown in the circuits of Figure 16.2 represents the equivalent resistance of all of the energy losses. This resistance cannot be measured with an ohmeter as the value of the equivalent resistor is a function of frequency and other variables such as signal level. Some of the loss in a resonant circuit is due to radiation, which is a function of frequency. The resistance of conductors is mostly due to skin effect, which increases with increasing frequency. The losses in the ferromagnetic materials used for making some inductors are nonlinear; thus, the equivalent resistance is not only a function of frequency but of signal level.

Most resonant circuits are not stand-alone circuits as shown in Figure 16.2, but are a part of other circuits where there are sources and loads. These additional resistances further remove energy from the resonant circuit. The *Q* of a resonant circuit when there are sources and loads is called the *loaded Q*. In most applications of resonant circuits, the *Q* of the resonance is set by the external loads rather than the capacitor and inductor that form the resonance.

16.2 Bandwidth and *Q*

The *bandwidth* of a resonant circuit is a measure of how well a resonant circuit responds to driving signals of a frequency near the resonant frequency and is a function of *Q*. The relationship between the 3 dB bandwidth and *Q* will be derived.

Applying a driving signal to a resonant circuit can overcome the losses of the circuit and cause the resonant circuit to oscillate indefinitely. As an example of this, consider the voltage generator in the series resonant circuit shown in Figure 16.3.

When the frequency of the voltage source is equal to the resonant frequency of the circuit, the equivalent impedance of the series resonant circuit is the resistance of the circuit and the current in the circuit, simply *E*/*r*.

At frequencies higher or lower than the resonant frequency, the impedance of the circuit is greater because the net reactance is not zero and the circuit current will be less than at resonance.

At what frequency will the circuit current be 3 dB less than at resonance? This frequency is where the circuit impedance is 1.414 that of the impedance at resonance. This is the frequency where the reactive part of the impedance is equal to the real part. This situation occurs at two frequencies. Below the resonant frequency, the net reactance is capacitive and is equal to *r*, while at a second frequency above the resonant frequency, the reactance is inductive and equal to *r*. This can be represented by two equations for the two frequencies:

For $f_1 > f_0$,

$$\left|X_L - X_C\right| = 2\pi f_1 L - \frac{1}{2\pi f_1 C} = \frac{\left(\frac{f_1}{f_0}\right)^2 - 1}{2\pi f_1 C} = r \tag{16.5}$$

$$\left(\frac{f_1}{f_0}\right)^2 - \frac{1}{Q}\left(\frac{f_1}{f_0}\right) - 1 = 0 \qquad \left(\frac{f_1}{f_0}\right) = \frac{1}{2Q} \pm \sqrt{\frac{1}{4Q^2} + 1}$$

For $f_2 < f_0$,

$$|X_L - X_C| = \frac{1}{2\pi f_2 C} - 2\pi f_2 L = \frac{-\left(\frac{f_2}{f_0}\right)^2 - 1}{2\pi f_2 C} = r \quad (16.6)$$

$$\left(\frac{f_2}{f_0}\right)^2 + \frac{1}{Q}\left(\frac{f_2}{f_0}\right) - 1 = 0 \qquad \left(\frac{f_2}{f_0}\right) = \frac{1}{2Q} \pm \sqrt{\frac{1}{4Q^2} + 1}$$

$$f_1 - f_2 = \frac{f_0}{Q} = \text{bandwidth}$$

Measuring *Q*

There are a number of methods of measuring *Q* using a variety of bridges, several of which are described in [1]. One method of measuring a capacitive or inductive *Q* is to place the component in a resonant circuit. When the *Q* to be measured of a device that is, in itself, a resonant circuit such as quartz crystal, similar techniques are used except the device's own resonance is used for the measurement. Circuit voltages or currents are measured at the resonance frequency and the *Q* is determined.

16.3 The *Q*-Meter

One simple and very popular method of measuring *Q* is with a device called, appropriately, the *Q*-meter. Consider the resonant circuit in Figure 16.4 for measuring the *Q* of inductors. This circuit has a very low-loss capacitor of known value and a constant voltage source.

The usual components measured by the *Q*-meter are inductors. It was previously mentioned that inductors are the weak link in resonant circuits, and the *Q* of a circuit is usually set by the inductor. The *Q*-meter can measure capacitance and capacitor *Q*. In this theoretical circuit, the circuit resistance is the equivalent series resistance of the inductor under test. This is due to the fact the variable capacitor is assumed to be lossless, the generator has zero resistance, and the voltmeter does not appreciably load the circuit. In a real circuit, it is not possible to achieve this situation, but the instrument is designed to approach this goal.

To measure the *Q* of an inductor using the *Q*-meter, the generator is set to the desired frequency while the variable capacitor tunes the circuit to resonance as indicated by the peak reading of the voltmeter.

At resonance, the impedance of the circuit is simply the equivalent series resistance of the inductor. This sets the current of the circuit as:

$$I = E/R_X \quad (16.7)$$

where E is the generator voltage and R_X is the equivalent resistance of the inductor.

Because the circuit is at resonance, the voltages of the two reactances are equal and of opposite phase. Those voltages are:

$$V = IX_C \quad \text{or} \quad V = IX_L \quad (16.8)$$

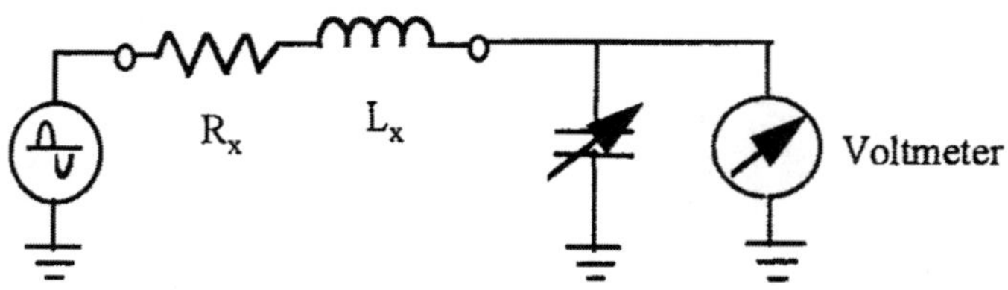

FIGURE 16.4 The basic circuit of a *Q*-meter showing the signal source, the inductor under test, and the voltmeter.

where X_C is the capacitive reactance and X_L is the inductive reactance, which are numerically equal at resonance.

Substituting the relationship of the circuit current, the result is:

$$V = EX_L/R_X = EX_C/R_X = EQ \tag{16.9}$$

Therefore, the voltage across the reactances is equal to *Q* times the applied voltage. If, as an example, the voltage source were 1 V, the voltmeter would read *Q* directly. *Q* values of several hundred are common and, therefore, voltages of several hundred volts could be measured. Modern circuits do not typically encounter voltages of this magnitude, and many components cannot withstand this potential. Therefore, most *Q*-meters use a much smaller source voltage, typically 20 mV.

If the frequency of the source and the circuit capacitance are known, it is possible to calculate the inductance of the unknown.

16.4 Other *Q* Measuring Techniques

There are very few *Q*-meters being manufactured today, although there are thousands of old *Q*-meters still in use. Because the *Q*-meter was the only accepted method of measuring *Q* for so many years, it will take decades before alternative methodologies overtake the *Q*-meter.

Measuring Q without a Q-meter involves a variety of *RLC* measuring instruments that measure the vector impedance and calculate Q. The calculated Q value is not as valid as that determined with a Q-meter unless the *RLC* meter is capable of measuring Q at the desired frequency. Only the more sophisticated, and expensive, *RLC* meters allow the use of any test frequency. Despite the new sophisticated *RLC* measuring instruments, there is an adapter for one model *RLC* instrument that allows the classic Q-meter-style measurement to be made.

Because vector impedance measurement is covered elsewhere in this Handbook, the remainder of this section will be devoted to measurements using the *Q*-meter.

16.5 Measuring Parameters Other than *Q*

In addition to *Q*, the *Q*-meter can be used to measure inductance, the *Q* or dissipation factor of a capacitor, and the distributed capacitance, C_d, of an inductor.

If an inductor with capacitance C_d is placed in the *Q*-meter circuit, the total circuit capacitance includes both the *Q*-meter's capacitance plus the additional C_d. Therefore, when resonance is achieved, the actual resonating capacitance is more than what is indicated on the *Q*-meter capacitor's dial. If C_d is not included in the calculation of inductance, the resulting value would be too large.

In many applications, the actual inductance is not the important parameter to be measured by the *Q*-meter. The actual parameter being determined is "what capacitance is required to resonate the particular inductor at a specific frequency," regardless of C_d.

In other applications, such as inductors that are to be used in wide-range oscillators, where C_d will limit the tuning range, C_d is an important parameter.

The *Q*-meter can be used to determine C_d, which will also allow for an accurate calculation of inductance. Determining C_d is a matter of resonating the inductor under test at more than one frequency.

To understand how the two-frequency measurement will allow the determination of C_d, assume an inductor is resonated at a frequency f_1. The relationship between the applied frequency, f_1, and the capacitor of the *Q*-meter to obtain resonance is:

$$f_1 = \frac{1}{2\pi\sqrt{L(C_1 + C_d)}} \tag{16.10}$$

where C_1 is the capacitance set on the *Q*-meter.

Resonating the same inductor at a second, higher, frequency, f_2, requires a Q-meter capacitance of C_2 such that:

$$f_2 = \frac{1}{2\pi\sqrt{L(C_2 + C_d)}} \tag{16.11}$$

This implies that C_2 is a smaller capacitance than C_1. Using these two equations and solving for C_d, the following result is obtained.

$$C_d = \frac{C_2 f_2^2 - C_1 f_1^2}{f_1^2 - f_2^2} \tag{16.12}$$

A convenient relationship between f_1 and f_2 is to set $f_2 = 1.414\ f_1$. With frequencies thus related, the distributed capacitance is:

$$C_d = C_1 - 2C_2 \tag{16.13}$$

C_d causes errors in the measurement of Q because of current through the C_d. The Q measured by the Q-meter is called the "effective Q." Since large inductors with significant C_d are no longer in common use because of the use of active filters, the distinction between effective Q and real Q is seldom considered. For additional information about effective Q and distributed capacitance, see [2].

To measure capacitors on the Q-meter, a relatively high Q inductor is connected to the inductance terminals on the Q-meter and resonated. The capacitor to be measured is connected to the capacitor terminals, which increases the circuit capacitance. The Q-meter variable capacitor is adjusted to regain resonance, which requires that the capacitance be reduced by an amount equal to the unknown capacitor.

The Q of the capacitor can be measured. The addition of the capacitor reduces the circuit Q because of the additional loss introduced by the capacitor. In the description of the Q-meter, the Q-meter's variable capacitor is assumed to have no loss.

Measuring the Q of a capacitor using the Q-meter is seldom done. This is because most capacitors have very high Q values. There are special cases, such as measuring the Q of a transducer or a varactor diode, where low Q values are encountered.

The unknown capacitor is connected to the CAP terminals of the Q-meter, which are simply in parallel with the Q-meter's variable capacitor. The variable capacitor in the Q-meter is set to the minimum capacitance and a suitable inductor is placed across the IND (inductor) terminals. The Q-meter is resonated using the frequency control rather than the variable capacitance.

For best results, the Q of the inductor must be considerably greater than that of the unknown capacitor, and the unknown capacitance must be considerably greater than the internal capacitance. If these criteria are met, the Q of the unknown capacitor can be read from the Q-meter. If these criteria are compromised, corrections can be made but the equations become complex and the accuracy degrades.

Most Q-meters provide measurement ranges to about 500 or to 1000. This is sufficient for measuring inductors, which was the main purpose of the Q-meter. For measuring high-Q devices such as ceramic resonators with Q values greater than 1000 or for quartz resonators with Q values extending into the hundreds of thousands, the Q-meter technique is insufficient. A high-Q circuit implies very little energy loss, and the energy that must be removed to make a measurement must be very small if Q is to be measured accurately.

Defining Terms

Bandwidth: A measurement of the amount of frequency spectrum occupied by a signal or the equivalent spectrum covered by a circuit that passes a finite band of frequencies. There are a number of methods of defining bandwidth, depending on the nature of the spectrum. Relative to resonant circuits, the bandwidth is measured between the –3 dB points of the passband.

Distributed capacitance: The amount of capacitance added to an inductor typically from the capacitance due to adjacent wires in a solenoid-type inductor. The distributed capacitance is given as a single capacitance figure for a specific inductor and can be defined as the equivalent capacitance across the entire coil. This would also allow the inductor to have a self-resonant frequency where the inductor resonates with the distributed capacitance with no external capacitance.

Effective inductance: Due to distributed capacitance, less capacitance than that calculated from an inductance value is required to resonate a circuit. If the actual capacitance required to resonate a circuit is used to calculate an inductance value, the resulting inductance value will be higher than the theoretical inductance value. This higher value is called the "effective inductance." The actual inductor cannot be considered as a pure inductance of a value equal to the effective inductance because the real inductor has a resonant frequency that a pure inductance does not.

Q: A measurement of the rate of energy loss in a resonant circuit.

Q-Meter: An instrument for measuring *Q* factor by resonating the circuit and measuring the voltage across the reactances.

References

1. *Reference Data for Engineers: Radio Electronics, Computer and Communications, 8th ed.,* Sams, Englewood Cliffs, NJ: Prentice-Hall Computer Publishing, 1993.
2. A. Helfrick and W. Cooper, *Modern Instrumentation and Measurement Techniques,* Englewood Cliffs, NJ: Prentice-Hall, 1990.
3, D. Kajfez, Q factor, thesis, Vector Fields, 1994.

17

Distortion Measurement

Michael F. Toner
Northern Telecom Ltd.

Gordon W. Roberts
McGill University

A sine-wave signal will have only a single-frequency component in its spectrum; that is, the frequency of the tone. However, if the sine wave is transmitted through a system (such as an amplifier) having some nonlinearity, then the signal emerging from the output of the system will no longer be a pure sine wave. That is, the output signal will be a distorted representation of the input signal. Since only a pure sine wave can have a single component in its frequency spectrum, this situation implies that the output must have other frequencies in its spectral composition. In the case of *harmonic distortion,* the frequency spectrum of the distorted signal will consist of the fundamental (which is the same frequency as the input sine wave) plus harmonic frequency components that are at integer multiples of the fundamental frequency. Taken together, these will form a Fourier representation of the distorted output signal. This phenomenon can be described mathematically. Refer to Figure 17.1, which depicts a sine-wave input signal $x(t)$ at frequency f_1 applied to the input of a system $A(x)$, which has an output $y(t)$. Assume that system $A(x)$ has some nonlinearity. If the nonlinearity is severe enough, then the output $y(t)$ might have excessive harmonic distortion such that its shape no longer resembles the input sine wave. Consider the example where the system $A(x)$ is an audio amplifier and $x(t)$ is a voice signal. Severe distortion can result in a situation where the output signal $y(t)$ does not represent intelligible speech. The *total harmonic distortion* (THD) is a figure of merit that is indicative of the quality with which the system $A(x)$ can reproduce an input signal $x(t)$. The output signal $y(t)$ can be expressed as:

$$y(t) = a_0 + \sum_{\mathrm{k}=1}^{N} a_{\mathrm{k}} \cos\left(2\pi k f_1 t + \theta_{\mathrm{k}}\right) \qquad (17.1)$$

where the a_k, $k = 0, 1, \ldots, N$ are the magnitudes of the Fourier coefficients, and θ_k, $k = 0, 1, \ldots, N$ are the corresponding phases. The THD is defined as the percentage ratio of the rms voltage of all harmonics components above the fundamental frequency to the rms voltage of the fundamental. Mathematically, the definition is written:

$$\mathrm{THD} = \frac{\sqrt{\sum_{\mathrm{k}=2}^{N} a_{\mathrm{k}}^2}}{a_1} \times 100\% \qquad (17.2)$$

0-8493-1733-9/04/$0.00+$1.50

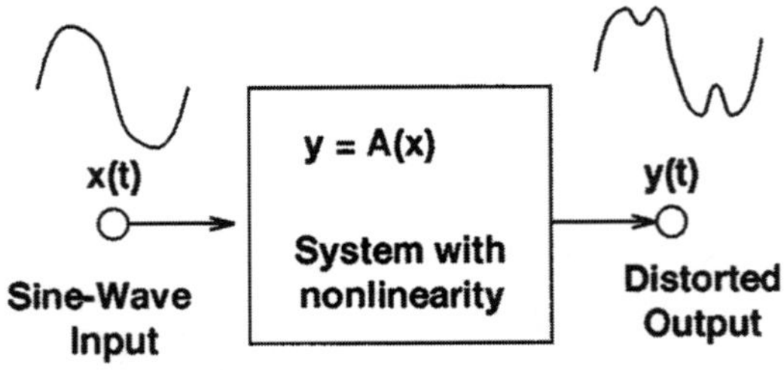

FIGURE 17.1 Any system with a nonlinearity gives rise to distortion.

If the system has good linearity (which implies low distortion), then the THD will be a smaller number than that for a system having poorer linearity (higher distortion). To provide the reader with some feeling for the order of magnitude of a realistic THD, a reasonable audio amplifier for an intercom system might have a THD of about 2% or less, while a high-quality sound system might have a THD of 0.01% or less. For the THD to be meaningful, the bandwidth of the system must be such that the fundamental and the harmonics will lie within the passband. Therefore, the THD is usually used in relation to low-pass systems, or bandpass systems with a wide bandwidth. For example, an audio amplifier might have a 20 Hz to 20 kHz bandwidth, which means that a 1 kHz input sine wave could give rise to distortion up to the 20th harmonic (i.e., 20 kHz), which can lie within the passband of the amplifier. On the other hand, a sine wave applied to the input of a narrow-band system such as a radio frequency amplifier will give rise to harmonic frequencies that are outside the bandwidth of the amplifier. These kinds of narrow-band systems are best measured using *intermodulation distortion*, which is treated elsewhere in this Handbook. For the rest of the discussion at hand, consider the example system illustrated in Figure 17.1 which shows an amplifier system $A(x)$ that is intended to be linear but has some undesired nonlinearities. Obviously, if a linear amplifier is the design objective, then the THD should be minimized.

17.1 Mathematical Background

Let $y = A(x)$ represent the input-output transfer characteristic of the system $A(x)$ in Figure 17.1 containing the nonlinearity. Expanding into a power series yields

$$A(x) = \sum_{k=0}^{\infty} c_k x^k = c_0 + c_1 x + c_2 x^2 + c_3 x^3 + \ldots \tag{17.3}$$

Let the input to the system be $x = \cos(2\pi f_0 t)$. Then the output will be

$$y = A(x) = c_0 + c_1 A_0 \cos(2\pi f_0 t) + c_2 A_0^2 \cos^2(2\pi f_0 t) + c_3 A_0^3 \cos^3(2\pi f_0 t) + \ldots \tag{17.4}$$

This can be simplified using the trigonometric relationships:

$$\cos^2(\theta) = \frac{1}{2} - \frac{1}{2}\cos(2\theta) \tag{17.5}$$

$$\cos^3(\theta) = \frac{3}{4}\cos(\theta) + \frac{1}{4}\cos(3\theta) \tag{17.6}$$

$$\cos^4(\theta) = \frac{1}{8} - \frac{1}{2}\cos(2\theta) + \frac{1}{8}\cos(4\theta) \tag{17.7}$$

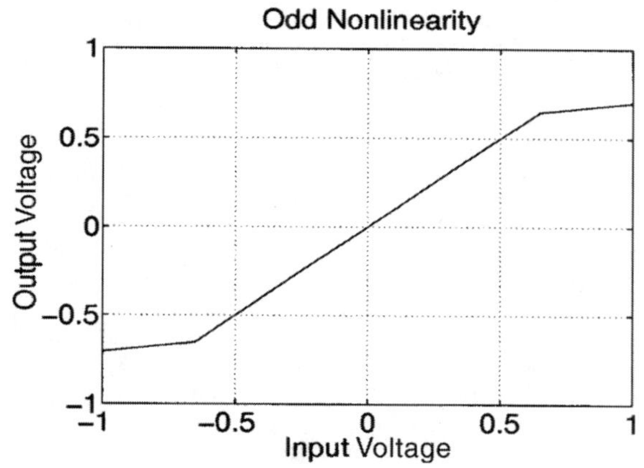

FIGURE 17.2 An odd nonlinearity with $f(-x) = -f(x)$.

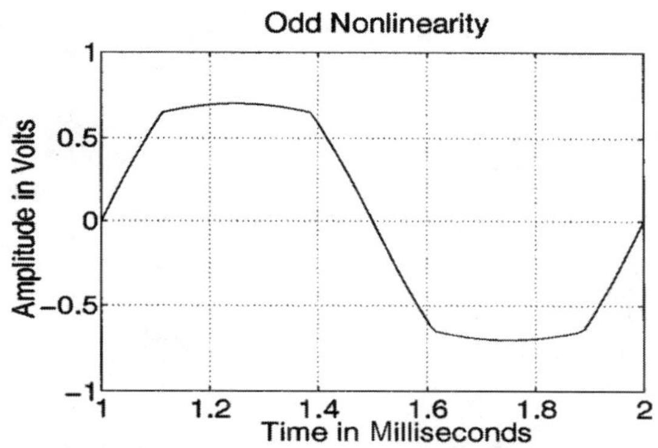

FIGURE 17.3 Distortion due to symmetrical two-sided clipping.

$$\cos^5(\theta) = \frac{5}{8}\cos(\theta) + \frac{5}{16}\cos(3\theta) + \frac{1}{16}\cos(5\theta) \tag{17.8}$$

and so on. Performing the appropriate substitutions and collecting terms results in an expression for the distorted signal $y(t)$ that is of the form shown in Equation 17.1. The THD can then be computed from Equation 17.2.

Closer inspection of Equations 17.6 and 17.8 reveal that a cosine wave raised to an odd power gives rise to only the fundamental and odd harmonics, with the highest harmonic corresponding to the highest power. A similar phenomenon is observed for a cosine raised to even powers; however, the result is only a dc component and even harmonics without any fundamental component. In fact, any nonlinear system that possesses an odd input-output transfer characteristic $A(x)$ (i.e., the function $A(x)$ is such that $-A(x) = A(-x)$) will give rise to odd harmonics only. Consider Figure 17.2, which illustrates an example of two-sided symmetrical clipping. It is an odd function. The application of a sine wave to its input will result in a waveform similar to that shown in Figure 17.3, which has only odd harmonics as shown in Figure 17.4. The majority of physical systems are neither odd nor even. (An even function is one that has the property $A(x) = A(-x)$; for example, a full-wave rectifier.) Consider the enhancement NMOS transistor illustrated in Figure 17.5, which has the square-law characteristic shown. Assume that the voltage V_{GS} consists of a dc bias plus a sine wave such that V_{GS} is always more positive than V_T (the threshold voltage). Then the current flowing in the drain of this NMOS transistor could have the appearance shown in Figure 17.6. It is observed that the drain current is distorted, since the positive-going side has a greater swing than the negative-going side. The equation for the drain current can be

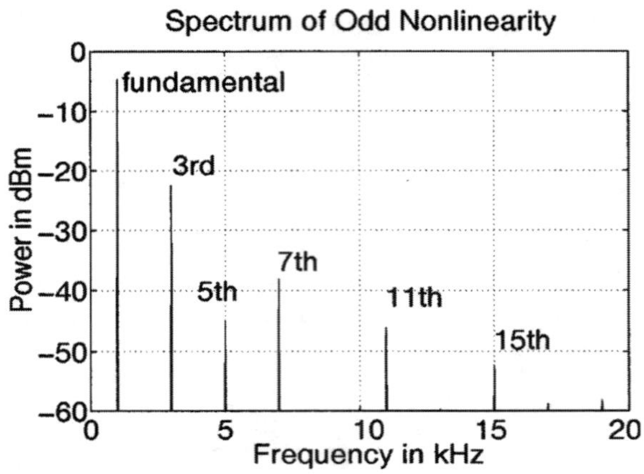

FIGURE 17.4 Frequency spectrum of signal distorted by symmetrical two-sided clipping.

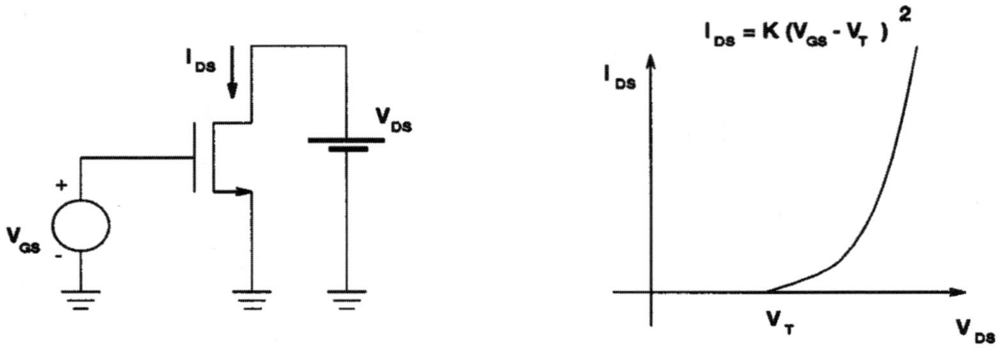

FIGURE 17.5 NMOS enhancement transistor is actually a nonlinear device. It is neither odd nor even in the strict sense.

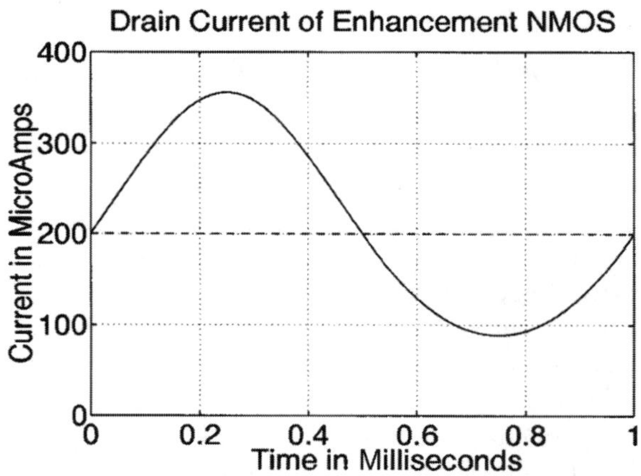

FIGURE 17.6 Showing how the drain current of the enhancement NMOS device is distorted.

derived mathematically as follows. A MOS transistor operating in its saturation region can be approximated as a square-law device:

$$I_{\mathrm{DS}} = \frac{\mu C_{\mathrm{ox}}}{2} \frac{W}{L} \left(V_{\mathrm{GS}} - V_{\mathrm{T}}\right)^2 \tag{17.9}$$

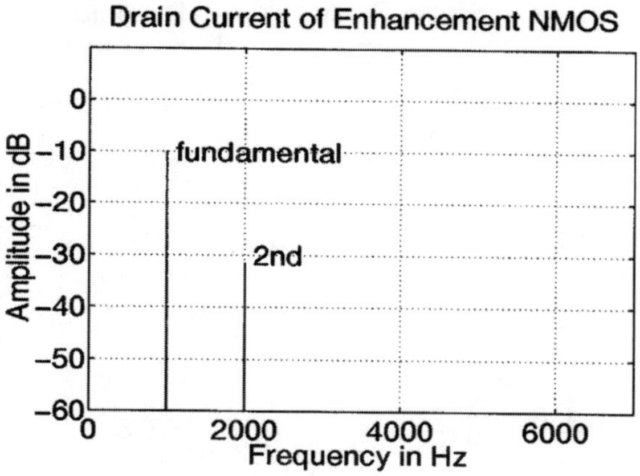

FIGURE 17.7 The drain current contains a dc bias, the fundamental, and the second harmonic only, for an ideal device.

If the gate of the *n*-channel enhancement MOSFET is driven by a voltage source consisting of a sine-wave generator in series with a dc bias, i.e.:

$$V_{GS} = V_B + A_0 \sin(2\pi f_0 t) \tag{17.10}$$

then the current in the drain can be written as:

$$I_{DS} = \frac{\mu C_{ox}}{2} \frac{W}{L} \left\{\left[V_B + A_0 \sin(2\pi f_0 t)\right] - V_T\right\}^2 \tag{17.11}$$

Expanding and using the trigonometric relationship:

$$\sin^2(\theta) = \frac{1}{2} - \frac{1}{2}\sin\left(2\theta + \frac{\pi}{2}\right) \tag{17.12}$$

Equation 17.11 can be rewritten as:

$$I_{DS} = \frac{\mu C_{ox}}{2} \frac{W}{L}\left[(V_B - V_T)^2 + \frac{A_0^2}{2} + 2(V_B - V_T)A_0 \sin(2\pi f_0 t) - \frac{A_0^2}{2}\sin(4\pi f_0 t) + \frac{\pi}{2}\right] \tag{17.13}$$

which clearly shows the dc bias, the fundamental, and the second harmonic that are visible in the spectrum of the drain current I_{DS} in Figure 17.7. There is one odd harmonic (i.e., the fundamental) and two even harmonics (strictly counting the dc component and the second harmonic). This particular transfer characteristic is neither odd nor even. Finally, for an ideal square-law characteristic, the second harmonic is the highest frequency component generated in response to a sine-wave input. Another example of a transfer characteristic that is neither even nor odd is single-sided clipping as shown in Figure 17.8, which gives rise to the distortion of Figure 17.9. One last example of an odd input-output transfer characteristic is symmetrical cross-over distortion as depicted in Figure 17.10. The distorted output in response to a 1 kHz sine-wave input is shown in Figure 17.11. The spectrum of the output signal is shown in Figure 17.12. Note that only odd harmonics have been generated.

To round out the discussion, consider a mathematical example wherein the harmonics are derived algebraically. Consider an input-output transfer function $f(x) = c_1 x + c_3 x^3 + c_5 x^5$ that has only odd powers of x. If the input is a cosine $x = A_0 \cos(2\pi f_0 t)$, then the output will be of the form:

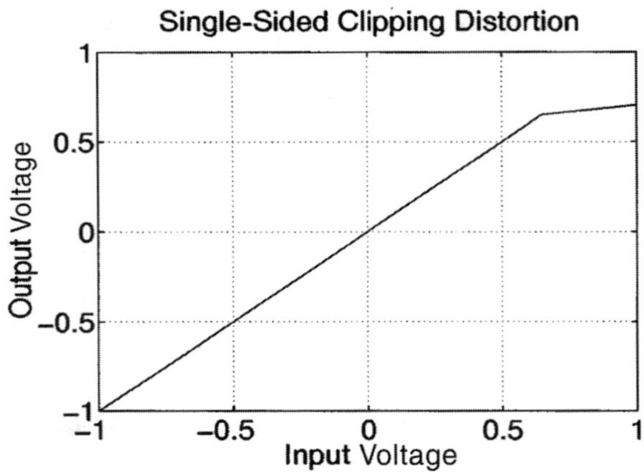

FIGURE 17.8 Single-sided clipping is neither even nor odd.

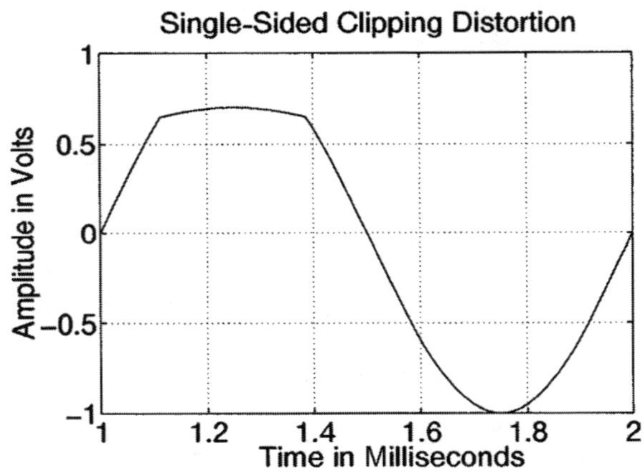

FIGURE 17.9 Distortion due to single-sided clipping.

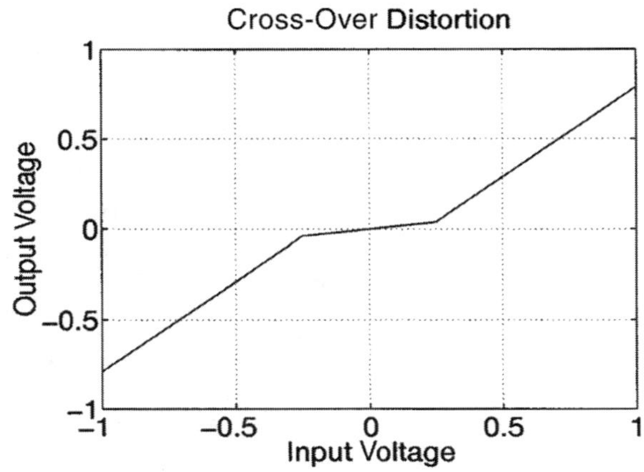

FIGURE 17.10 Symmetrical cross-over distortion is odd.

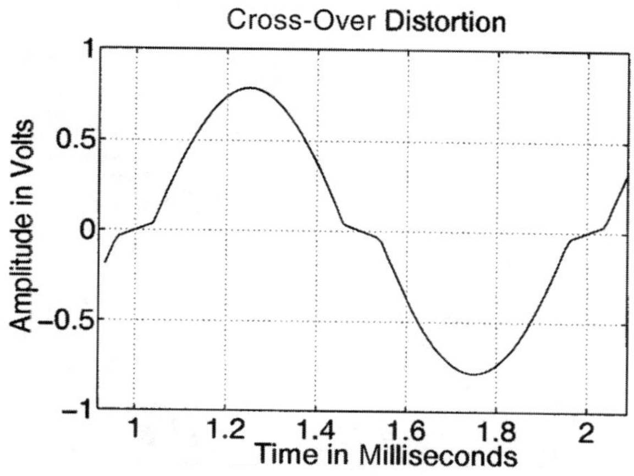

FIGURE 17.11 An example of cross-over distortion.

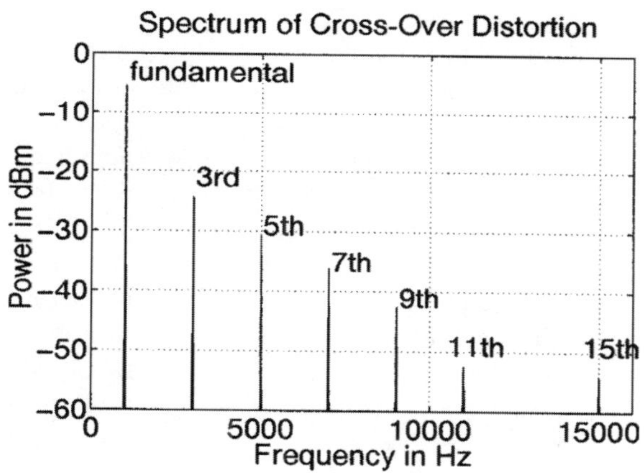

FIGURE 17.12 Symmetrical cross-over distortion gives rise to odd harmonics.

$$y(t) = f(x) = c_1A_0\cos(2\pi f_0 t) + c_3A_0^3\cos^3(2\pi f_0 t) + c_5A_0^5\cos^5(2\pi f_0 t) \tag{17.14}$$

This can be simplified using the trigonometric relationships given in Equations 17.5 through 17.8 with the following result:

$$y(t) = \left(c_1A_0 + \frac{3c_3A_0^3}{4} + \frac{5c_5A_0^5}{8}\right)\cos(2\pi f_0 t) + \left(\frac{c_3A_0^3}{4} + \frac{5c_5A_0^5}{16}\right)\cos(2\pi 3 f_0 t) + c_5A_0^5\cos^5(2\pi 5 f_0 t) \tag{17.15}$$

Clearly, only the fundamental plus the third and fifth harmonics are present. Should the exercise be repeated for an input-output transfer function consisting of only even powers of x, then only a dc offset plus even harmonics (not the fundamental) would be present in the output.

17.2 Intercept Points (IP)

It is often desirable to visualize how the various harmonics increase or decrease as the amplitude of the input sine wave $x(t)$ is changed. Consider the example of a signal applied to a nonlinear system $A(x)$ having single-sided clipping distortion as shown in Figure 17.8. The clipping becomes more severe as the

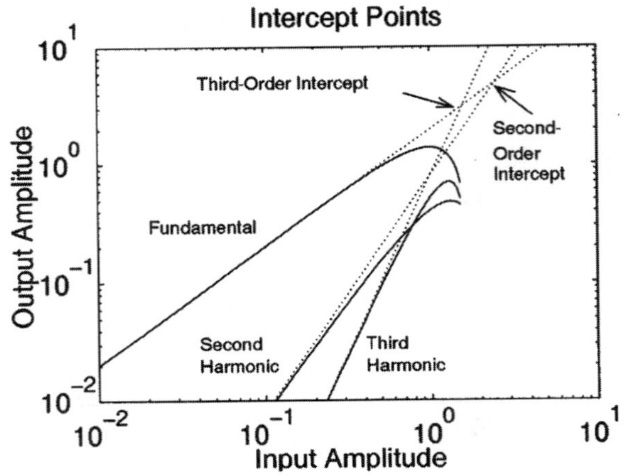

FIGURE 17.13 An example showing the second- and third-order intercept points for a hypothetical system. Both axes are plotted on a logarithmic scale.

amplitude of the input signal $x(t)$ increases in amplitude, so the distortion of the output signal $y(t)$ becomes worse. The *intercept point* (IP) is used to provide a figure of merit to quantify this phenomenon. Consider Figure 17.13, which shows an example of the power levels of the first three harmonics of the distorted output $y(t)$ of a hypothetical system $A(x)$ in response to a sine-wave input $x(t)$. It is convenient to plot both axes on a log scale. It can be seen that the power in the harmonics increases more quickly than the power in the fundamental. This is consistent with the observation of how clipping becomes worse as the amplitude increases. It is also consistent with the observation that, in the equations above, the higher harmonics will rapidly become more prominent because they are proportional to higher exponential powers of the input signal amplitude. The intercept point for a particular harmonic is the power level where the extrapolated line for that harmonic intersects with the extrapolated line for the fundamental. The second-order intercept is often abbreviated IP2, the third-order intercept abbreviated IP3, etc.

17.3 Measurement of the THD

Classical Method

The traditional method of measuring THD is shown in Figure 17.14. A sine-wave test stimulus $x(t)$ is applied to the input of the system $A(x)$ under test. The system output $y(t)$ is fed through a bandpass filter tuned to the frequency of the input stimulus to extract the signal. Its power p_1 can be measured with a power meter. The bandpass filter is then tuned to each of the desired harmonics in turn and the measurement is repeated to determine the required p_i. The THD is then calculated from:

$$\mathrm{THD} = \sqrt{\frac{\sum_{k=2}^{N} p_k}{p_1}} \times 100\% \tag{17.16}$$

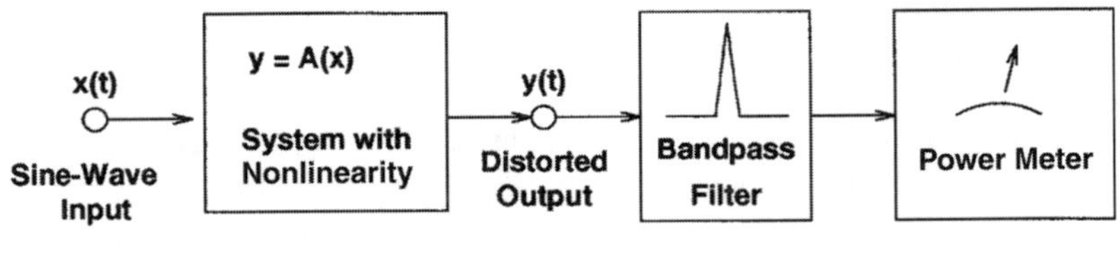

FIGURE 17.14 Illustrating the classical method of measuring THD.

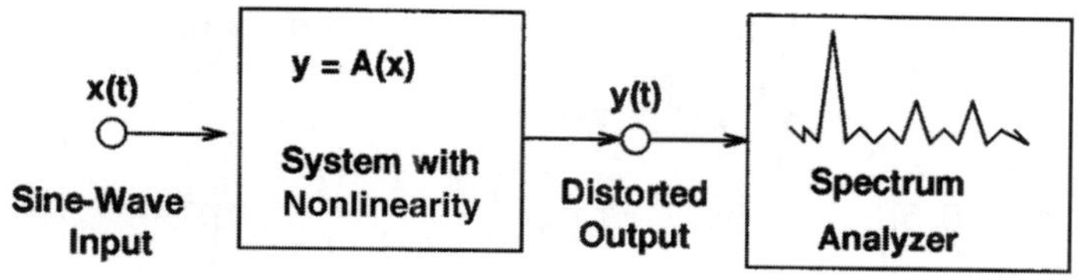

FIGURE 17.15 Illustrating measurement of THD using a spectrum analyzer.

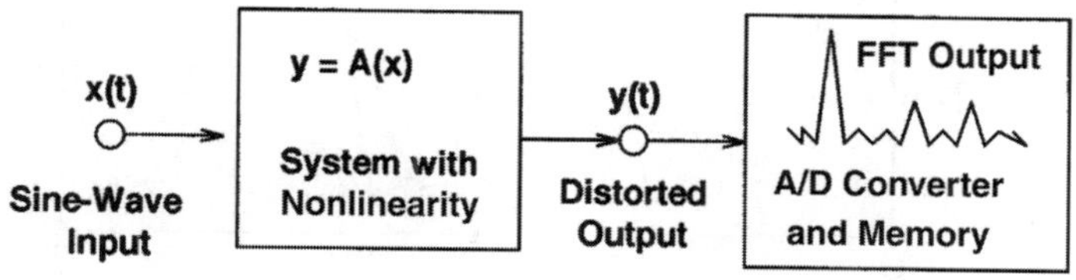

FIGURE 17.16 Illustrating the measurement of THD using FFT.

In the case of an ordinary audio amplifier, nearly all of the power in the distorted output signal is contained in the first 10 or 11 harmonics. However, in more specialized applications, a much larger number of harmonics might need to be considered.

Spectrum Analyzer Method

THD measurements are often made with a spectrum analyzer using the setup shown in Figure 17.15. The readings for the power levels of each of the desired harmonic components in the frequency spectrum of the distorted signal $y(t)$ are collected from the spectrum analyzer, usually in units of dB. They are converted to linear units by means of the relationship:

$$a_i = 10^{r_i/20} \tag{17.17}$$

where r_i is the reading for the ith component in dB. The THD is then computed from Equation 17.2. The spectrum analyzer method can be considered as an extension of the classical method described above, except that the spectrum analyzer itself is replacing both the bandpass filter and the power meter.

DSP Method

Digital signal processing (DSP) techniques have recently become popular for use in THD measurement. In this method, the distorted output $y(t)$ is digitized by a precision A/D converter and the samples are stored in the computer's memory as shown in Figure 17.16. One assumes that the samples have been collected with a uniform sample period T_s and that appropriate precautions have been taken with regard to the Nyquist criterion and aliasing. Let $y(n)$ refer to the nth stored sample. A fast Fourier transform (FFT) is executed on the stored data using the relationship:

$$Y(k) = \sum_{n=0}^{N-1} y(n) e^{-j(2\pi/N)kn} \tag{17.18}$$

where N is the number of samples that have been captured. The frequency of the input test stimulus is chosen such that the sampling is coherent. Coherency in this context means that if N samples have been captured, then the input test stimulus is made to be a harmonic of the primitive frequency f_P, which is defined as:

$$f_p = \frac{f_s}{N} = \frac{1}{T_s N} \tag{17.19}$$

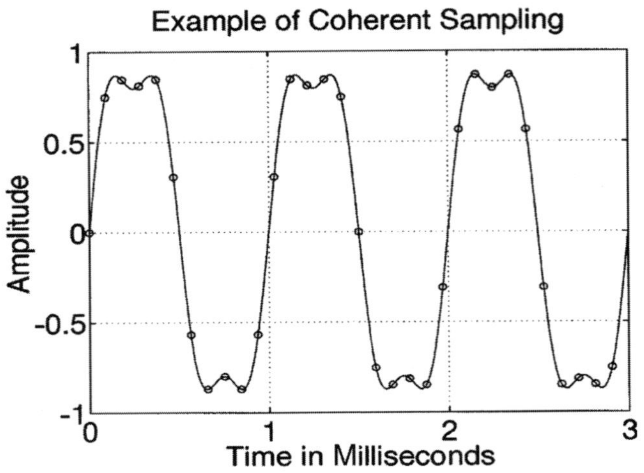

FIGURE 17.17 With coherent sampling, each sample occurs on a unique point of the signal.

One can view the primitive frequency f_P as the frequency of a sinusoidal signal whose period is exactly equal to the time interval formed by the N-points. Thus, the frequency of the test stimulus can be written as:

$$f_0 = M \times f_p = M \times \frac{f_s}{N} = \frac{M}{N} \times f_s \tag{17.20}$$

where M and N are integers. To maximize the information content collected by a set of N-points, M and N are selected so that they have no common factors, i.e., relatively prime. This ensures that every sample is taken at a different point on the periodic waveform. An example is provided in Figure 17.17, where $M = 3$ and $N = 32$. The FFT is executed on the distorted signal as per Equation 17.18, and then the THD is computed from:

$$\text{THD} = \frac{\sqrt{\sum_{k=2}^{N} \left|Y(k \times M)\right|^2}}{\left|Y(M)\right|} \times 100\% \tag{17.21}$$

17.4 Conclusions

The *total harmonic distortion* (THD) is a figure of merit for the quality of the transmission of a signal through a system having some nonlinearity. Its causes and some methods of measuring it have been discussed. Some simple mathematical examples have been presented. However, in real-world systems, it is generally quite difficult to extract all of the parameters c_k in the transfer characteristic. The examples were intended merely to assist the reader's understanding of the relationship between even and odd functions and the harmonics that arise in response to them.

Defining Terms

Total harmonic distortion (THD): A numerical figure of merit of the quality of transmission of a signal, defined as the ratio of the power in all the harmonics to the power in the fundamental.

Fundamental: The lowest frequency component of a signal other than zero frequency.

Harmonic: Any frequency component of a signal that is an integer multiple of the fundamental frequency.

Distortion: The effect of corrupting a signal with undesired frequency components.

Nonlinearity: The deviation from the ideal of the transfer function, resulting in such effects as clipping or saturation of the signal.

Further Information

D.O. Pederson and K. Mayaram, *Analog Integrated Circuits for Communications,* New York: Kluwer Academic Press, 1991.

M. Mahoney, *DSP-Based Testing of Analog and Mixed-Signal Circuits,* Los Almos, CA: IEEE Computer Society Press, 1987.

18

Noise Measurement

W. Marshall Leach, Jr.
Georgia Institute of Technology

This chapter describes the principal sources of electric noise and discusses methods for the measurement of noise. The notations for voltages and currents correspond to the following conventions: dc quantities are indicated by an upper-case letter with upper-case subscripts, e.g., V_{BE}. Instantaneous small-signal ac quantities are indicated by a lower-case letter with lower-case subscripts, e.g., v_n. The mean-square value of a variable is denoted by a bar over the square of the variable, e.g., $\overline{v_n^2}$, where the bar indicates an arithmetic average of an ensemble of functions. The root-mean-square or rms value is the square root of the mean-square value. Phasors are indicated by an upper-case letter and lower case subscripts, e.g., V_n. Circuit symbols for independent sources are circular, symbols for controlled sources are diamond shaped, and symbols for noise sources are square. In the evaluation of noise equations, Boltzmann's

0-8493-1733-9/04/$0.00+$1.50

constant is $k = 1.38 \times 10^{-23}$ J K^{-1} and the electronic charge is $q = 1.60 \times 10^{-19}$ C. The standard temperature is denoted by T_0 and is taken to be $T_0 = 290$ K. For this value, $4kT_0 = 1.60 \times 10^{-20}$ J and the thermal voltage is $V_T = kT_0/q = 0.025$ V.

18.1 Thermal Noise

Thermal noise or *Johnson noise* is generated by the random collision of charge carriers with a lattice under conditions of thermal equilibrium [1–7]. Thermal noise in a resistor can be modeled by a series voltage source or a parallel current source having the mean-square values

$$\overline{v_t^2} = 4kTR\Delta f \tag{18.1}$$

$$\overline{i_t^2} = \frac{4kT\Delta f}{R} \tag{18.2}$$

where R is the resistance and Δf is the bandwidth in hertz (Hz) over which the noise is measured. The equation for $\overline{v_t^2}$ is commonly referred to as the *Nyquist formula.* Thermal noise in resistors is independent of the resistor composition.

The *crest factor* for thermal noise is the ratio of the peak value to the rms value. A common definition for the peak value is the level that is exceeded 0.01% of the time. The amplitude distribution of thermal noise is modeled by a gaussian or normal probability density function. The probability that the instantaneous value exceeds 4 times the rms value is approximately 0.01%. Thus, the crest factor is approximately 4.

18.2 Spectral Density

The *spectral density* of a noise signal is defined as the mean-square value per unit bandwidth. For the thermal noise generated by a resistor, the voltage and current spectral densities, respectively, are given by:

$$S_v(f) = 4kTR \tag{18.3}$$

$$S_i(f) = \frac{4kT}{R} \tag{18.4}$$

Because these are independent of frequency, thermal noise is said to have a uniform or flat distribution. Such noise is also called *white noise.* It is called this by analogy to white light, which also has a flat spectral density in the optical band.

18.3 Fluctuation Dissipation Theorem

Consider any system in thermal equilibrium with its surroundings. If there is a mechanism for energy in a particular mode to leak out of that mode to the surroundings in the form of heat, then energy can leak back into that mode from the surrounding heat by the same mechanism. The fluctuation dissipation theorem of quantum mechanics states that the average energy flow in each direction is the same. Otherwise, the system would not be in equilibrium.

Mathematically, the fluctuation dissipation theorem states, in general, that the generalized mean-square force $\overline{\Im^2}$ acting on a system in the frequency band from f_1 to f_2 is given by:

$$\overline{\Im^2} = 4kT\int_{f_1}^{f_2} \mathrm{Re}\left[Z(f)\right] df \tag{18.5}$$

where Re $[Z(f)]$ is the real part of the system impedance $Z(f)$ and f is the frequency in hertz (Hz). For a mechanical system, the generalized force is the mechanical force on the system and the impedance is force divided by velocity. For an electric system, the generalized force is the voltage and the impedance is the ratio of voltage to current.

Equation 18.1 is a statement of the fluctuation dissipation theorem for a resistor. The theorem can be used to calculate the mean-square thermal noise voltage generated by any two-terminal network containing resistors, capacitors, and inductors. Let $Z(f)$ be the complex impedance of the network. The mean-square open-circuit thermal noise voltage is given by:

$$\overline{v_t^2} = 4kT\int_{f_1}^{f_2} \mathrm{Re}\left[Z(f)\right]df \simeq 4kT\,\mathrm{Re}\left[Z(f)\right]\Delta f \tag{18.6}$$

where $\Delta f = f_2 - f_1$ and the approximation holds if Re $[Z(f)]$ is approximately constant over the band.

18.4 Equivalent Noise Resistance and Conductance

A mean-square noise voltage can be represented in terms of an *equivalent noise resistance* [8]. Let $\overline{v_n^2}$ be the mean-square noise voltage in the band Δf. The noise resistance R_n is defined as the value of a resistor at the standard temperature $T_0 = 290$ K that generates the same noise. It is given by:

$$R_n = \frac{\overline{v_n^2}}{4kT_0\Delta f} \tag{18.7}$$

A mean-square noise current can be represented in terms of an *equivalent noise conductance.* Let $\overline{i_n^2}$ be the mean-square noise current in the band Δf. The noise conductance G_n is defined as the value of a conductance at the standard temperature that generates the same noise. It is given by:

$$G_n = \frac{\overline{i_n^2}}{4kT_0\Delta f} \tag{18.8}$$

18.5 Shot Noise

Shot noise is caused by the random emission of electrons and by the random passage of charge carriers across potential barriers [1–7]. The shot noise generated in a device is modeled by a parallel noise current source. The mean-square shot noise current in the frequency band Δf is given by:

$$\overline{i_{sh}^2} = 2qI\Delta f \tag{18.9}$$

where I is the dc current through the device. This equation is commonly referred to as the *Schottky formula.* Like thermal noise, shot noise is white noise and has a crest formula of approximately 4.

18.6 Flicker Noise

The imperfect contact between two conducting materials causes the conductivity to fluctuate in the presence of a dc current [1–7]. This phenomenon generates what is called *flicker noise* or *contact noise.* It is modeled by a noise current source in parallel with the device. The mean-square flicker noise current in the frequency band Δf is given by:

$$\overline{i_{\mathrm{f}}^2} = \frac{K_f I^m \Delta f}{f^n} \tag{18.10}$$

where K_{f} is the flicker noise coefficient, I is the dc current, m is the flicker noise exponent, and $n \simeq 1$. Other names for flicker noise are *1/f noise* (read "one-over-f-noise"), *low-frequency noise,* and *pink noise.* The latter comes from the optical analog of pink light, which has a spectral density that increases at lower frequencies.

18.7 Excess Noise

In resistors, flicker noise is caused by the variable contact between particles of the resistive material and is called *excess noise.* Metal film resistors generate the least excess noise, carbon composition resistors generate the most, with carbon film resistors lying between the two. In modeling excess noise, the flicker noise exponent has the value $m = 2$. The mean-square excess noise current is often written as a function of the *noise index NI* as follows:

$$\overline{i_{\mathrm{ex}}^2} = \frac{10^{NI/10}}{10^{12} \ln 10} \times \frac{I^2 \Delta f}{f} \tag{18.11}$$

where I is the dc current through the resistor. The noise index is defined as the number of μA of excess noise current in each decade of frequency per A of dc current through the resistor. An equivalent definition is the number of μV of excess noise voltage in each decade of frequency per volt of dc drop across the resistor. In this case, the mean-square excess noise voltage generated by the resistor is given by:

$$\overline{v_{\mathrm{ex}}^2} = \frac{10^{NI/10}}{10^{12} \ln 10} \times \frac{V^2 \Delta f}{f} \tag{18.12}$$

where $V = IR$ is the dc voltage across the resistor.

18.8 Burst Noise

Burst noise or *popcorn noise* is caused by a metallic impurity in a *pn* junction [4]. When amplified and reproduced by a loudspeaker, it sounds like corn popping. When viewed on an oscilloscope, it appears as fixed amplitude pulses of randomly varying width and repetition rate. The rate can vary from less than one pulse per minute to several hundred pulses per second. Typically, the amplitude of burst noise is 2 to 100 times that of the background thermal noise.

18.9 Partition Noise

Partition noise occurs when the charge carriers in a current have the possibility of dividing between two or more paths. The noise is generated in the resulting components of the current by the statistical process of partition [9]. Partition noise occurs in BJTs where the current flowing from the emitter into the base can take one of two paths. The recombination of injected carriers in the base region corresponds to the current flow in one path. This current flows in the external base lead. The current carried to the collector corresponds to the current flow in the second path. Because the emitter current exhibits full shot noise, the base and collector currents also exhibit full shot noise. However, the base and collector noise currents are correlated because they have equal and opposite partition components. Partition noise in the BJT can be accounted for if all shot noise is referred to two uncorrelated shot noise current sources, one from base to emitter and the other from collector to emitter [10]. This noise model of the BJT is described here.

18.10 Generation–Recombination Noise

Generation–recombination noise in a semiconductor is generated by the random fluctuation of free carrier densities caused by spontaneous fluctuations in the generation, recombination, and trapping rates [7]. In BJTs, it occurs in the base region at low temperatures. The generation–recombination gives rise to fluctuations in the base spreading resistance which are converted into a noise voltage due to the flow of a base current. In junction FETs, it occurs in the channel at low temperatures. Generation–recombination causes fluctuations of the carrier density in the channel, which gives rise to a noise voltage when a drain current flows. In silicon junction FETs, the effect occurs below 100 K. In germanium junction FETs, it occurs at lower temperatures. The effect does not occur in MOS FETs.

18.11 Noise Bandwidth

The *noise bandwidth* of a filter is the bandwidth of an ideal filter having a constant passband gain which passes the same rms noise voltage, where the input signal is white noise [1–7]. The filter and the ideal filter are assumed to have the same gains. Let $A_v(f)$ be the complex voltage gain transfer function of a filter, where f is the frequency in Hz. Its noise bandwidth B_n is given by:

$$B_n = \frac{1}{A_{vo}^2}\int_0^{\infty} |A_v(f)|^2 \, df \tag{18.13}$$

where A_{vo} is the maximum value of $|A_v(f)|$. For a white noise input voltage with the spectral density $S_v(f)$, the mean-square noise voltage at the filter output is $\overline{v_{no}^2} = A_{vo}^2 S_v(f) B_n$.

Two classes of low-pass filters are commonly used in making noise measurements. The first has n real poles, all with the same frequency, having the magnitude-squared transfer function given by:

$$|A_v(f)|^2 = \frac{A_{vo}^2}{\left[1+(f/f_0)^2\right]^n} \tag{18.14}$$

where f_0 is the pole frequency. The second is an n-pole Butterworth filter having the magnitude-squared transfer function given by:

$$|A_v(f)|^2 = \frac{A_{vo}^2}{1+(f/f_3)^{2n}} \tag{18.15}$$

where f_3 is the –3 dB frequency. Table 18.1 gives the noise bandwidths as a function of the number of poles n for $1 \le n \le 5$. For the real-pole filter, B_n is given as a function of both f_0 and f_3. For the Butterworth filter, B_n is given as a function of f_3.

TABLE 18.1 Noise Bandwidth B_n of Low-Pass Filters

Number of poles	Slope dB/dec	Real pole B_n		Butterworth B_n
1	20	$1.571\, f_0$	$1.571\, f_3$	$1.571\, f_3$
2	40	$0.785\, f_0$	$1.220\, f_3$	$1.111\, f_3$
3	60	$0.589\, f_0$	$1.155\, f_3$	$1.042\, f_3$
4	80	$0.491\, f_0$	$1.129\, f_3$	$1.026\, f_3$
5	100	$0.420\, f_0$	$1.114\, f_3$	$1.017\, f_3$

18.12 Noise Bandwidth Measurement

The noise bandwidth of a filter can be measured with a white noise source with a known voltage spectral density $S_v(f)$. Let $\overline{v_n^2}$ be the mean-square noise output voltage from the filter when it is driven by the noise source. The noise bandwidth is given by:

$$B_n = \frac{\overline{v_o^2}}{A_{vo}^2 S_v(f)} \tag{18.16}$$

If the spectral density of the source is not known, the noise bandwidth can be determined if another filter with a known noise bandwidth is available. With both filters driven simultaneously, let $\overline{v_{o1}^2}$ and $\overline{v_{o2}^2}$ be the two mean-square noise output voltages, B_{n1} and B_{n2} the two noise bandwidths, and A_{vo1} and A_{vo2} the two maximum gain magnitudes. The noise bandwidth B_{n2} is given by:

$$B_{n2} = B_{n1} \frac{\overline{v_{o2}^2}}{\overline{v_{o1}^2}} \left(\frac{A_{vo1}}{A_{vo2}} \right)^2 \tag{18.17}$$

The white noise source should have an output impedance that is low enough so that the loading effect of the filters does not change the spectral density of the source.

18.13 Spot Noise

Spot noise is the rms noise in a band divided by the square root of the noise bandwidth. For a noise voltage, it has the units $V/\sqrt{Hz}$, which is read "volts per root Hz." For a noise current, the units are $A/\sqrt{Hz}$. For white noise, the spot noise in any band is equal to the square root of the spectral density. Spot noise measurements are usually made with a bandpass filter having a bandwidth that is small enough so that the input spectral density is approximately constant over the filter bandwidth. The spot noise voltage at a filter output is given by $\sqrt{(\overline{v_{no}^2}/B_n)}$, where $\overline{v_{no}^2}$ is the mean-square noise output voltage and B_n is the filter noise bandwidth. The spot noise voltage at the filter input is obtained by dividing the output voltage by A_{vo}, where A_{vo} is the maximum value of $|A_v(f)|$.

A filter that is often used for spot noise measurements is a second-order bandpass filter. The noise bandwidth is given by $B_n = \pi B_3/2$, where B_3 is the –3 dB bandwidth. A single-pole high-pass filter having a pole frequency f_1 cascaded with a single-pole low-pass filter having a pole frequency f_2 is a special case of bandpass filter having two real poles. Its noise bandwidth is given by $B_n = \pi(f_1 + f_2)/2$. The –3 dB bandwidth in this case is $f_1 + f_2$, not $f_2 - f_1$.

18.14 Addition of Noise Voltages

Consider the instantaneous voltage $v = v_n + i_n R$, where v_n is a noise voltage and i_n is a noise current. The mean-square voltage is calculated as follows:

$$\overline{v^2} = \overline{(v_n + i_n R)^2} = \overline{v_n^2} + 2\rho\sqrt{\overline{v_n^2}}\sqrt{\overline{i_n^2}}\,R + \overline{i_n^2}R^2 \tag{18.18}$$

where ρ is the *correlation coefficient* defined by:

$$\rho = \frac{\overline{v_n i_n}}{\sqrt{\overline{v_n^2}}\sqrt{\overline{v_n^2}}} \tag{18.19}$$

For the case $\rho = 0$, the sources are said to be uncorrelated or independent. It can be shown that $-1 \le \rho \le 1$.

In ac circuit analysis, noise signals are often represented by phasors. The square magnitude of the phasor represents the mean-square value at the frequency of analysis. Consider the phasor voltage $V = V_n + I_n Z$, where V_n is a noise phasor voltage, I_n is a noise phasor current, and $Z = R + jX$ is a complex impedance. The mean-square voltage is given by:

$$\overline{v^2} = \overline{\left(V_n + I_n R\right)\left(V_n^* + I_n^* Z^*\right)}$$
$$= \overline{v_n^2} + 2\sqrt{\overline{v_n^2}}\sqrt{\overline{i_n^2}}\,\mathrm{Re}\left(\gamma Z^*\right) + \overline{i_n^2}\left|Z\right|^2 \tag{18.20}$$

where the * denotes the complex conjugate and γ is the *complex correlation coefficient* defined by:

$$\gamma = \gamma_r + j\gamma_i = \frac{\overline{V_n I_n^*}}{\sqrt{\overline{v_n^2}}\sqrt{\overline{i_n^2}}} \tag{18.21}$$

It can be shown that $\left|\gamma\right| \le 1$.

Noise equations derived by phasor analysis can be converted easily into equations for real signals. However, the procedure generally cannot be done in reverse. For this reason, noise formulas derived by phasor analysis are the more general form.

18.15 Correlation Impedance and Admittance

The *correlation impedance* Z_γ and *correlation admittance* Y_γ between a noise phasor voltage V_n and a noise phasor current I_n are defined by [8]:

$$Z_\gamma = R_\gamma + jX_\gamma = \gamma\sqrt{\frac{\overline{v_n^2}}{\overline{i_n^2}}} \tag{18.22}$$

$$Y_\gamma = G_\gamma + jB_\gamma = \gamma^*\sqrt{\frac{\overline{i_n^2}}{\overline{v_n^2}}} \tag{18.23}$$

where $\overline{v_n^2}$ is the mean-square value of V_n, $\overline{i_n^2}$ is the mean-square value of I_n, and γ is the complex correlation coefficient between V_n and I_n. With these definitions, it follows that $\overline{V_n I_n^*} = \overline{i_n^2}\, Z_\gamma = \overline{v_n^2}\, Y_\gamma^*$.

18.16 The $v_n - i_n$ Amplifier Noise Model

The noise generated by an amplifier can be modeled by referring all internal noise sources to the input [1–7], [11]. In order for the noise sources to be independent of the source impedance, two sources are required — a series voltage source and a shunt current source. In general, the sources are correlated.

Figure 18.1 shows the amplifier noise model, where V_s is the source voltage, $Z_s = R_s + jX_s$ is the source impedance, V_{ts} is the thermal noise voltage generated by R_s, $A_v = V_o/V_i$ is the complex voltage gain, and Z_i is the input impedance. The output voltage is given by:

$$V_o = \frac{A_v Z_i}{Z_s + Z_i}\left(V_s + V_{ts} + V_n + I_n Z_s\right) \tag{18.24}$$

The *equivalent noise input voltage* is the voltage in series with V_s that generates the same noise voltage at the output as all noise sources in the circuit. It is given by $V_{ni} = V_{ts} + V_n + I_n Z_s$. The mean-square value is:

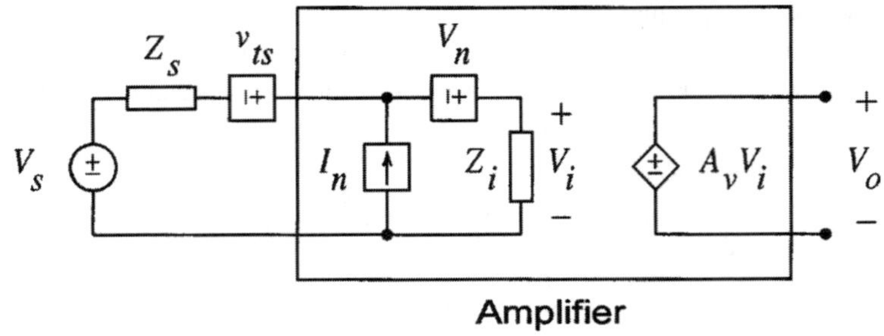

FIGURE 18.1 Amplifier $v_n - i_n$ noise model.

$$\overline{v_{ni}^2} = 4kTR_sB_n + \overline{v_n^2} + 2\sqrt{\overline{v_n^2}}\sqrt{\overline{i_n^2}}\,\mathrm{Re}\left(\gamma Z_s^*\right) + \overline{i_n^2}\left|Z_s\right|^2 \tag{18.25}$$

where B_n is the amplifier noise bandwidth and γ is the complex correlation between V_n and I_n. For $|Z_s|$ very small, $\overline{v_{ni}^2} \simeq \overline{v_n^2}$ and γ is not important. Similarly, for $|Z_s|$ very large, $\overline{v_{ni}^2} \simeq \overline{i_n^2}\,|Z_s|^2$ and γ is again not important.

When the source is represented by a Norton equivalent consisting of a source current i_s in parallel with a source admittance $Y_s = G_s + jB_s$, the noise referred to the input must be represented by an *equivalent noise input current*. The mean-square value is given by:

$$\overline{i_{ni}^2} = 4kTG_sB_n + \overline{v_n^2}\left|Y_2\right|^2 + 2\sqrt{\overline{v_n^2}}\sqrt{\overline{i_n^2}}\,\mathrm{Re}\left(\gamma Y_s\right) + \overline{i_n^2} \tag{18.26}$$

18.17 Measuring $\overline{v_{ni}^2}$, $\overline{v_n^2}$, and $\overline{i_n^2}$

For a given Z_s, the mean-square equivalent noise input voltage can be measured by setting $V_s = 0$ and measuring the mean-square noise output voltage $\overline{v_{no}^2}$. It follows that $\overline{v_{ni}^2}$ is given by:

$$\overline{v_{ni}^2} = \frac{\overline{v_{no}^2}}{\left|A_v\right|^2} \times \left|1 + \frac{Z_s}{Z_i}\right|^2 \tag{18.27}$$

To measure $\overline{v_n^2}$ and $\overline{i_n^2}$, $\overline{v_{no}^2}$ is measured with $Z_s = 0$ and with Z_s replaced by a large-value resistor. It follows that $\overline{v_n^2}$ and $\overline{i_n^2}$ are then given by:

$$\overline{v_n^2} = \frac{\overline{v_{no}^2}}{\left|A_v\right|^2} \quad \text{for } Z_s = 0 \tag{18.28}$$

$$\overline{i_n^2} = \left|\frac{1}{R_s} + \frac{1}{Z_i}\right|^2 \frac{\overline{v_{no}^2}}{\left|A_v\right|^2} \quad \text{for } Z_s = R_s \text{ and } R_s \text{ large} \tag{18.29}$$

18.18 Noise Temperature

The internal noise generated by an amplifier can be expressed as an equivalent *input-termination noise temperature* [12]. This is the temperature of the source resistance that generates a thermal noise voltage equal to the internal noise generated in the amplifier when referred to the input. The noise temperature T_n is given by:

$$T_n = \frac{\overline{v_{ni}^2}}{4kR_sB_n} - T \tag{18.30}$$

where $\overline{v_{ni}^2}$ is the mean-square equivalent input noise voltage in the band B_n, R_s is the real part of the source output impedance, and T is the temperature of R_s.

18.19 Noise Reduction with a Transformer

Let a transformer be connected between the source and the amplifier in Figure 18.1. Let n be the transformer turns ratio, R_1 the primary resistance, and R_2 the secondary resistance. The equivalent noise input voltage in series with the source voltage V_s has the mean-square value:

$$\begin{aligned}\overline{v_{ni}^2} &= 4kT\left(R_s + R_1 + \frac{R_2}{n^2}\right)\Delta f + \frac{\overline{v_n^2}}{n^2} \\ &+ 2\sqrt{\overline{v_n^2}}\sqrt{\overline{i_n^2}}\,\mathrm{Re}\left[\gamma\left(Z_s^* + R_1 + \frac{R_2}{n^2}\right)\right] + n^2\overline{i_n^2}\left|Z_s + R_1 + \frac{R_2}{n^2}\right|^2\end{aligned} \tag{18.31}$$

In general, $R_2/R_1 \propto n$, which makes it difficult to specify the value of n that minimizes $\overline{v_{ni}^2}$. For $R_1 + R_2/n^2 \ll |Z_s|$, it is minimized when:

$$n^2 = \frac{1}{|Z_s|}\sqrt{\frac{\overline{v_n^2}}{\overline{i_n^2}}} \tag{18.32}$$

18.20 The Signal-to-Noise Ratio

The *signal-to-noise ratio* of an amplifier is defined by:

$$\mathrm{SNR} = \frac{\overline{v_{so}^2}}{\overline{v_{no}^2}} \tag{18.33}$$

where $\overline{v_{so}^2}$ is the mean-square signal output voltage and $\overline{v_{no}^2}$ is the mean-square noise output voltage. The SNR is often expressed in dB with the equation $\mathrm{SNR}=10\log(\overline{v_{so}^2}/\overline{v_{no}^2})$. In measuring the SNR, a filter should be used to limit the bandwidth of the output noise to the signal bandwidth of interest. An alternative definition of the SNR that is useful in making calculations is:

$$\mathrm{SNR} = \frac{\overline{v_s^2}}{\overline{v_{ni}^2}} \tag{18.34}$$

where $\overline{v_s^2}$ is the mean-square signal input voltage and $\overline{v_{ni}^2}$ is the mean-square equivalent noise input voltage.

18.21 Noise Factor and Noise Figure

The *noise factor* F of an amplifier is defined by [1–8]:

$$F = \frac{\overline{v_{no}^2}}{\overline{v_{nos}^2}} \tag{18.35}$$

where $\overline{v_{no}^2}$ is the mean-square noise output voltage with the source voltage zeroed and $\overline{v_{nos}^2}$ is the mean-square noise output voltage considering the only source of noise to be the thermal noise generated by the source resistance R_s. The *noise figure* is the noise factor expressed in dB and is given by:

$$\mathrm{NF} = 10 \log F \tag{18.36}$$

For the $v_n - i_n$ amplifier noise model, the noise factor is given by:

$$F = \frac{\overline{v_{ni}^2}}{4kT R_s B_n} = 1 + \frac{\overline{v_n^2} 2\sqrt{\overline{v_n^2}}\sqrt{\overline{i_n^2}}\,\mathrm{Re}\left(\gamma Z_s^*\right) + \overline{i_n^2}\left|Z_s\right|^2}{4kT R_s B_n} \tag{18.37}$$

where B_n is the amplifier noise bandwidth. The value of Z_s that minimizes the noise figure is called the *optimum source impedance* and is given by:

$$Z_{so} = R_{so} + jX_{so} = \left[\sqrt{1-\gamma_i^2} - j\gamma_i\right]\sqrt{\frac{\overline{v_n^2}}{\overline{i_n^2}}} \tag{18.38}$$

where $\gamma_i = \mathrm{Im}\ (\gamma)$. The corresponding value of F is denoted by F_0 and is given by:

$$F_0 = 1 + \frac{\sqrt{\overline{v_n^2}}\sqrt{\overline{i_n^2}}}{2kTB_n}\left(\gamma_r + j\sqrt{1-\gamma_i^2}\right) \tag{18.39}$$

It follows that F can be expressed in terms of F_0 as follows:

$$F = F_0 + \frac{G_n}{R_{ns}}\left[\left(R_s - R_{so}\right)^2 + \left(X_s - X_{so}\right)^2\right] \tag{18.40}$$

where G_n is the noise conductance of I_n and R_{ns} is the noise resistance of the source. These are given by:

$$G_n = \frac{\overline{i_n^2}}{4kT_0B_n} \tag{18.41}$$

$$R_{ns} = \frac{\overline{v_{ts}^2}}{4kT_0B_n} = \frac{TR_s}{T_0} \tag{18.42}$$

When the source is represented by a Norton equivalent consisting of a source current i_s in parallel with a source admittance $Y_s = G_s + jB_s$, the *optimum source admittance* is given by:

$$Y_{so} = G_{so} + jB_{so} = \left[\sqrt{1-\gamma_i^2} + j\gamma_i\right]\sqrt{\frac{\overline{i_n^2}}{\overline{v_n^2}}} \tag{18.43}$$

which is the reciprocal of Z_{so}. The noise factor can be written as:

$$F = F_0 + \frac{R_n}{G_{ns}}\left[\left(G_s - G_{so}\right)^2 + \left(B_s - B_{so}\right)^2\right] \tag{18.44}$$

where R_n is the noise resistance of V_n and G_{ns} is the noise conductance of the source. These are given by:

$$R_n = \frac{\overline{v_n^2}}{4kT_0B_n} \tag{18.45}$$

$$G_{ns} = \frac{\overline{i_{ts}^2}}{4kT_0B_n} = \frac{TG_s}{T_0} \tag{18.46}$$

18.22 Noise Factor Measurement

The noise factor can be measured with a calibrated white noise source driving the amplifier. The source output impedance must equal the value of Z_s for which F is to be measured. The source temperature must be the standard temperature T_0. First, measure the amplifier noise output voltage over the band of interest with the source voltage set to zero. For the amplifier model of Figure 18.1, the mean-square value of this voltage is given by:

$$\overline{v_{no1}^2} = \left|\frac{A_{vo}Z_i}{Z_s + Z_i}\right|^2 \left[4kT_0R_sB_n + \overline{v_n^2} + 2\sqrt{\overline{v_n^2}}\sqrt{\overline{i_n^2}}\,\mathrm{Re}\left(\gamma Z_s^*\right) + \overline{i_n^2}\left|Z_s\right|^2\right] \tag{18.47}$$

The source noise voltage is then increased until the output voltage increases by a factor r. The new mean square output voltage can be written as:

$$r^2\overline{v_{no1}^2} = \left|\frac{A_{vo}Z_i}{Z_s + Z_i}\right|^2 \left[\left(S_v(f) + 4kT_0R_s\right)B_n + \overline{v_n^2} + 2\sqrt{\overline{v_n^2}}\sqrt{\overline{i_n^2}}\,\mathrm{Re}\left(\gamma Z_s^*\right) + \overline{i_n^2}\left|Z_s\right|^2\right] \tag{18.48}$$

where $S_v(f)$ is the open-circuit voltage spectral density of the white noise source.

The above two equations can be solved for F to obtain:

$$F = \frac{S_s(f)}{\left(r^2 - 1\right)4kT_0R_s} \tag{18.49}$$

A common value for r is $\sqrt{2}$. The gain and noise bandwidth of the amplifier are not needed for the calculation. If a resistive voltage divider is used between the noise source and the amplifier to attenuate the input signal, the source spectral density $S_s(f)$ is calculated or measured at the attenuator output with it disconnected from the amplifier input.

If the noise bandwidth of the amplifier is known, its noise factor can be determined by measuring the mean-square noise output voltage $\overline{v_{no}^2}$ with the source voltage set to zero. The noise factor is given by:

$$F = \left|1 + \frac{Z_s}{Z_i}\right|^2 \frac{\overline{v_{no}^2}}{4kT_0R_sB_nA_{vo}^2} \tag{18.50}$$

This expression is often used with $B_n = \pi B_3/2$, where B_3 is the –3 dB bandwidth. Unless the amplifier has a first-order low-pass or a second-order bandpass frequency response characteristic, this is only an approximation.

18.23 The Junction Diode Noise Model

When forward biased, a diode generates both shot noise and flicker noise [1–7]. The noise is modeled by a parallel current source having the mean-square value:

$$\overline{i_n^2} = 2qI\Delta f + \frac{K_f I^m \Delta f}{f} \tag{18.51}$$

where I is the dc diode current. A plot of $\overline{i_n^2}$ vs. f for a constant Δf exhibits a slope of –10 dB/decade for low frequencies and a slope of zero for higher frequencies.

Diodes are often used as noise sources in circuits. Specially processed zener diodes are marketed as solid-state noise diodes. The noise mechanism in these is called *avalanche noise,* which is associated with the diode reverse breakdown current. For a given breakdown current, avalanche noise is much greater than shot noise in the same current.

18.24 The BJT Noise Model

Figure 18.2(a) shows the BJT noise model [1–7]. The base spreading resistance r_x is modeled as an external resistor; v_{tx} is the thermal noise generated by r_x; i_{shb}, and i_{fb}, respectively, are the shot noise and flicker noise in the base bias current I_B; and i_{shc} is the shot noise in the collector bias current I_C. The sources have the mean-square values of:

$$\overline{v_{tx}^2} = 4kTr_x\Delta f \tag{18.52}$$

$$\overline{i_{shb}^2} = 2qI_B\Delta f \tag{18.53}$$

$$\overline{i_{fb}^2} = \frac{K_f I_B^m \Delta f}{f} \tag{18.54}$$

$$\overline{i_{shc}^2} = 2qI_C\Delta f \tag{18.55}$$

Let the resistances to signal ground seen looking out of the base and the emitter, respectively, be denoted by R_1 and R_2. The mean-square equivalent noise input voltages in series with the base or the emitter that generates the same collector noise current is given by:

$$\begin{aligned}\overline{v_{ni}^2} &= 4kT(R_1 + r_x + R_2)\Delta f + \left(2qI_B\Delta f + \frac{K_f I_B \Delta f}{f}\right)(R_1 + r_x + R_2)^2 \\ &+ 2qI_C\Delta f\left(\frac{R_1 + r_x + R_2}{\beta} + \frac{V_T}{I_C}\right)^2\end{aligned} \tag{18.56}$$

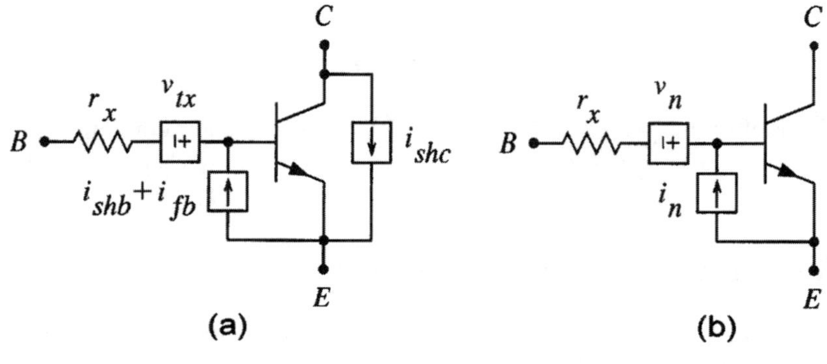

FIGURE 18.2 (a) BJT noise model. (b) BJT $v_n - i_n$ noise model.

At frequencies where flicker noise can be neglected, the value of I_C that minimizes $\overline{v_{ni}^2}$ is called the *optimum collector current.* It is given by:

$$I_{C_{opt}} = \frac{V_T}{R_1 + r_x + R_2} \times \frac{\beta}{\sqrt{1+\beta}} \quad (18.57)$$

The corresponding value of $\overline{v_{ni}^2}$ is given by:

$$\overline{v_{ni_{min}}^2} = 4kT\left(R_1 + r_x + R_2\right)\Delta f \times \frac{\sqrt{1+\beta}}{\sqrt{1+\beta} - 1} \quad (18.58)$$

If N identical BJTs that are identically biased are connected in parallel, the equivalent noise input voltage is given by Equation 18.56 with r_x replaced with r_x/N, I_B replaced with NI_B, and I_C replaced with NI_C. In this case, R_1 and R_2, respectively, are the resistances to signal ground seen looking out of the parallel connected bases and the parallel connected emitters. For N fixed, the value of I_C that minimizes $\overline{v_{no}^2}$ is given by Equation 18.57 with R_1 replaced with NR_1 and R_2 replaced with NR_2. The corresponding value of $\overline{v_{ni_{min}}^2}$ is given by Equation 18.58 with r_x replaced with r_x/N. It follows that parallel connection of BJTs can be used to reduce the thermal noise of r_x, provided the devices are optimally biased.

The BJT $v_n - i_n$ noise model is given in Figure 18.2(b), where r_x is a noiseless resistor, for its thermal noise is included in v_n. The mean-square values of v_n and i_n and the correlation coefficient are given by:

$$\overline{v_n^2} = 4kTr_x\Delta f + 2kT\frac{V_T}{I_C}\Delta f \quad (18.59)$$

$$\overline{i_n^2} = 2qI_B\Delta f + \frac{K_f I_B^m \Delta f}{f} + \frac{2qI_C\Delta f}{\beta^2} \quad (18.60)$$

$$\rho = \frac{2kT\Delta f}{\beta\sqrt{\overline{v_n^2}}\sqrt{\overline{i_n^2}}} \quad (18.61)$$

where $\beta = I_C/I_B$ is the current gain. An alternative model puts r_x inside the BJT. For this model, the expressions for i_n and ρ are more complicated than the ones given here.

The $v_n - i_n$ noise model of Figure 18.1 does not have a noiseless resistor in series with its input. Before formulae that are derived for this model are applied to the BJT model of Figure 18.2(b), the formulae must be modified to account for the noiseless r_x. For the common-emitter amplifier, for example, the source resistance R_s would be replaced in the expression for $\overline{v_{ni}^2}$ by $R_s + r_x$ in all occurrences except in terms that represent the thermal noise of R_s.

The value of the base spreading resistance r_x depends on the method used to measure it. For noise calculations, r_x should be measured with a noise technique. A test circuit for measuring r_x is shown in Figure 18.3. The emitter bias current I_E and the collector bias voltage V_C are given by:

$$I_E = \frac{-V_{BE} - V_{EE}}{R_E} \quad (18.62)$$

$$V_C = V_{CC} - \alpha I_E R_C \quad (18.63)$$

where $\alpha = \beta/(1 + \beta)$. Capacitors C_1 and C_2 should satisfy $C_1 \gg I_E/(2\pi f V_T)$ and $C_2 \gg 1/(2\pi f R_C)$, where f is the lowest frequency of interest. To minimize the noise contributed by R_C, R_F, and the op-amp, a low-noise op-amp should be used and R_F should be much larger than R_C. The power supply rails must be properly decoupled to minimize power supply noise.

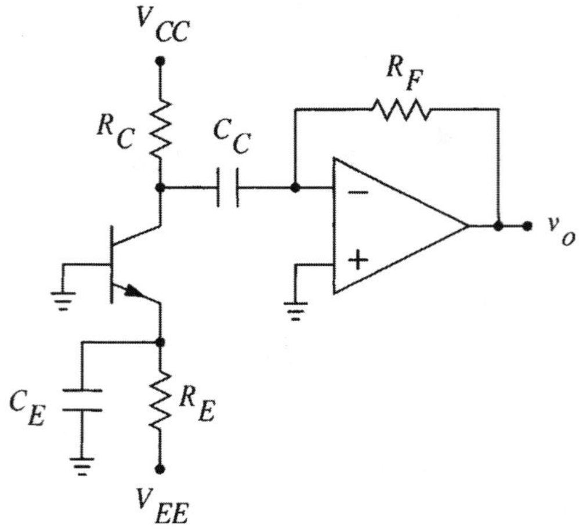

FIGURE 18.3 Test circuit for measuring r_x of a BJT.

To prevent flicker noise from affecting the data, the op-amp output voltage must be measured over a noise bandwidth where the spectral density is white. Denote the mean-square op-amp output voltage over the band B_n with the BJT in the circuit by $\overline{v_{no1}^2}$. Denote the mean-square voltage over the band B_n with the BJT removed by $\overline{v_{no2}^2}$. The base spreading resistance r_x can be obtained by solving:

$$r_x^2\left[\frac{A}{\beta^2} - 2qI_B B_n\right] + r_x\left[\frac{2AV_T}{\beta I_C} - 4kT B_n\right] + \frac{AV_T^2}{I_C^2} = 0 \tag{18.64}$$

Where:

$$A = \frac{\overline{v_{no1}^2} - \overline{v_{no2}^2}}{R_F^2} - 2qI_C B_n \tag{18.65}$$

The test circuit of Figure 18.3 can be used to measure the flicker noise coefficient K_f and the flicker noise exponent m. The plot of $(\overline{v_{no1}^2} - \overline{v_{no2}^2})$ vs. frequency for a constant noise bandwidth Δf must be obtained, e.g., with a signal analyzer. In the white noise range, the slope of the plot is zero. In the flicker noise range, the slope is –10 dB per decade. The lower frequency at which $(\overline{v_{no1}^2} - \overline{v_{no2}^2})$ is 3 dB greater than its value in the white noise range is the flicker noise corner frequency f_f. It can be shown that:

$$K_f I_B^m = \frac{\left(\overline{v_{no1}^2} - \overline{v_{no2}^2}\right) f_f}{2R_F^2 \Delta f} \times \left(\frac{r_x}{\beta} + \frac{V_T}{I_C}\right)^2 \tag{18.66}$$

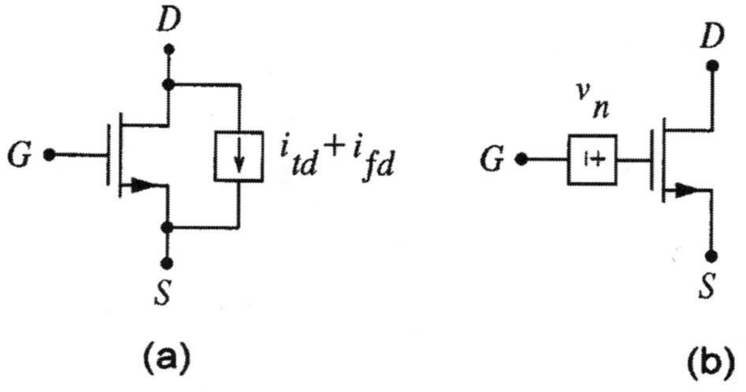

FIGURE 18.4 (a) FET noise model. (b) FET v_n noise model.

By repeating the measurements for at least two values of I_C, this equation can be used to solve for both K_f and m. Unless I_C is large, the r_x/β term can usually be neglected compared to the V_T/I_C term. The value of the flicker noise exponent is usually in the range $1 < m < 3$, but is often taken as unity. If it is assumed that $m = 1$, the value of K_f can be calculated by making the measurements with only one value of I_C.

18.25 The FET Noise Model

Figure 18.4(a) shows the MOSFET noise equivalent circuit, where i_{td} is the channel thermal noise current and i_{fd} is the channel flicker noise current [1–7]. The mean-square values of these currents are given by:

$$\overline{i_{td}^2} = \frac{8kT\Delta f}{3g_m} \tag{18.67}$$

$$\overline{i_{fd}^2} = \frac{K_f \Delta f}{4KfL^2 C_{ox}} \tag{18.68}$$

where K is the transconductance parameter, $g_m = 2\sqrt{KI_D}$ is the transconductance, L is the effective length of the channel, and C_{ox} is the gate oxide capacitance per unit area.

Let the resistances to signal ground seen looking out of the gate and the source, respectively, be denoted by R_1 and R_2. The mean-square equivalent noise input voltage in series with either the gate or the source that generates the same drain noise current is given by:

$$\overline{v_{ni}^2} = 4kT(R_1 + R_2)\Delta f + \frac{4kT\Delta f}{3\sqrt{KI_D}} + \frac{K_f \Delta f}{4KL^2 C_{ox} f} \tag{18.69}$$

where it is assumed that the MOSFET bulk is connected to its source in the ac circuit. If N identical MOSFETs that are identically biased are connected in parallel, the equivalent noise input voltage is given by Equation 18.69 with the exception that the second and third terms are divided by N.

The noise sources in Figure 18.4(a) can be reflected into a single source in series with the gate. The circuit is shown in Figure 18.4(b). The mean-square value of v_n is given by:

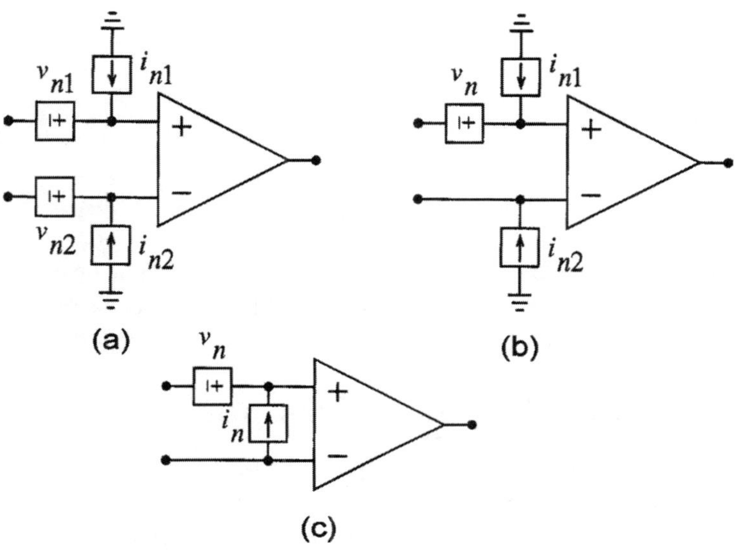

FIGURE 18.5 Op-amp noise models.

$$\overline{v_n^2} = \frac{8kT\Delta f}{3g_m} + \frac{K_f \Delta f}{4KfL^2C_{ox}} \quad (18.70)$$

The FET flicker noise coefficient K_f can be measured by replacing the BJT in Figure 18.3 with the FET. On a plot of $(\overline{v_{no1}^2} - \overline{v_{no2}^2})$ as a function of frequency for a constant noise bandwidth, the flicker noise corner frequency f_f is the lower frequency at which $(\overline{v_{no1}^2} - \overline{v_{no2}^2})$ is up 3 dB above the white noise level. A signal analyzer can be used to display this plot. The flicker noise coefficient is given by:

$$K_f = \frac{32kT\,K\,f_f L^2 C_{ox}}{3g_m} \quad (18.71)$$

The MOSFET circuits and equations also apply to the junction FET with the exception that the L_2 and C_{ox} terms are omitted from the formulae. This assumes that the junction FET gate-to-channel junction is reverse biased, which is the usual case. Otherwise, shot noise in the gate current must be modeled.

18.26 Operational Amplifier Noise Models

Variations of the $v_n - i_n$ amplifier noise model are used in specifying op-amp noise performance. The three most common models are given in Figure 18.5. In Figures 18.5(b) and (c), v_n can be placed in series with either input [4, 6]. In general, the sources in each model are correlated. In making calculations that use specified op-amp noise data, it is important to use the noise model for which the data apply.

18.27 Photodiode Detector Noise Model

Figure 18.6(a) shows the circuit symbol of a photodiode detector [4]. When reverse biased by a dc source, an incident light signal causes a signal current to flow in the diode. The diode small-signal noise model is shown in Figure 18.6(b), where i_s is the signal current (proportional to the incident light intensity), i_n is the diode noise current, r_d is the small-signal resistance of the reverse-biased junction, c_d is the small-signal junction capacitance, r_c is the cell resistance (typically <50 Ω), and v_{tc} is the thermal noise voltage generated by r_c. The noise current i_n consists of three components: shot noise i_{sh}, flicker noise i_f, and carrier generation–recombination noise i_{gr}. The first three have the mean-square values:

$$\overline{v_{tc}^2} = 4kTr_c\Delta f \quad (18.72)$$

$$\overline{i_{sh}^2} = 2qI_D\Delta f \quad (18.73)$$

$$\overline{i_f^2} = \frac{K_f I_D^m \Delta f}{f} \quad (18.74)$$

where I_D is the reverse-biased diode current. The carrier generation–recombination noise has a white spectral density up to a frequency determined by the carrier lifetime. Because the detector has a large output resistance, it should be used with amplifiers that exhibit a low input current noise.

18.28 Piezoelectric Transducer Noise Model

Figure 18.7(a) shows the circuit symbol of a piezoelectric transducer [4]. This transducer generates an electric voltage when a mechanical force is applied between two of its surfaces. An approximate equivalent circuit that is valid for frequencies near the transducer mechanical resonance is shown in Figure 18.7(b). In this circuit, C_e represents the transducer electric capacitance, while C_s, L_s, and R_s are chosen to have

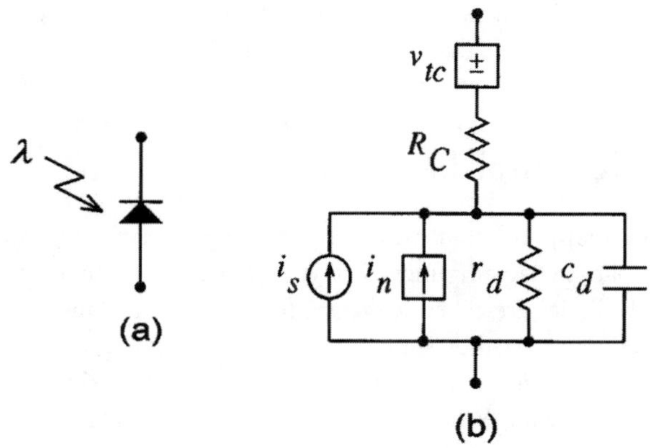

FIGURE 18.6 (a) Photodiode symbol. (b) Small-signal noise model of photodiode.

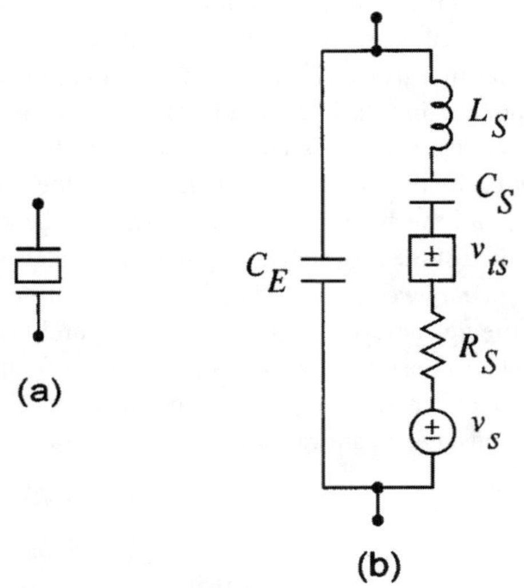

FIGURE 18.7 (a) Piezoelectric transducer symbol. (b) Noise model of piezoelectric transducer.

a resonant frequency and quality factor numerically equal to those of the transducer mechanical resonance. The source v_s represents the signal voltage, which is proportional to the applied force. The source v_{ts} represents the thermal noise generated by R_s. It has a mean-square value of:

$$\overline{v_{ts}^2} = 4kTR_s\Delta f \tag{18.75}$$

This noise component is negligible in most applications.

The piezoelectric transducer has two resonant frequencies: a short-circuit resonant frequency f_{sc} and an open-circuit resonant frequency f_{oc} given by:

$$f_{sc} = \frac{1}{2\pi\sqrt{L_s C_s}} \tag{18.76}$$

$$f_{oc} = \frac{1}{2\pi\sqrt{L_1 C_1}} \tag{18.77}$$

where $C_1 = C_sC_e/(C_s + C_e)$. It is normally operated at the open-circuit resonant frequency where the transducer output impedance is very high. For this reason, it is should be used with amplifiers that exhibit a low input current noise.

18.29 Parametric Amplifiers

A *parametric amplifier* is an amplifier that uses a time varying reactance to produce amplification [13]. In low-noise microwave parametric amplifiers, a reverse biased *pn* junction diode is used to realize a variable capacitance. Such diodes are called *varactors,* for variable reactance. The depletion capacitance of the reverse-biased junction is varied by simultaneously applying a signal current and a pump current at different frequencies. The nonlinear capacitance causes frequency mixing to occur between the signal frequency and the pump frequency. When the power generated by the frequency mixing exceeds the signal input power, the diode appears to have a negative resistance and signal amplification occurs. The only noise that is generated is the thermal noise of the effective series resistance of the diode, which is very small.

Figure 18.8 shows a block diagram of a typical parametric amplifier. The varactor diode is placed in a resonant cavity. A circulator is used to isolate the diode from the input and output circuits. A pump signal is applied to the diode to cause its capacitance to vary at the pump frequency. The filter isolates the pump signal from the output circuit. The idler circuit is a resonant cavity that is coupled to the diode cavity to reduce the phase sensitivity. Let the signal frequency be f_s, the pump frequency be f_p, and the resonant frequency of the idler cavity be f_i. In cases where $f_p = f_s + f_i$, the varying capacitance of the diode looks like a negative resistance and the signal is amplified. If $f_i = f_s$, the amplifier is called a *degenerate* amplifier. This is the simplest form of the parametric amplifier and it requires the lowest pump frequency and power to operate. For the *nondegenerate* amplifier, $f_p > 2f_s$. In both cases, the input and the output are at the same frequency. In the *up-converter* amplifier, $f_p = f_i - f_s$ and $f_p > 2f_s$. In this case, the varying capacitance of the diode looks like a positive resistance and the signal frequency output is not amplified. However, there is an output at the idler frequency that is amplified. Thus, the output frequency is higher than the input frequency. The conversion gain can be as high as the ratio of the output frequency to the input frequency.

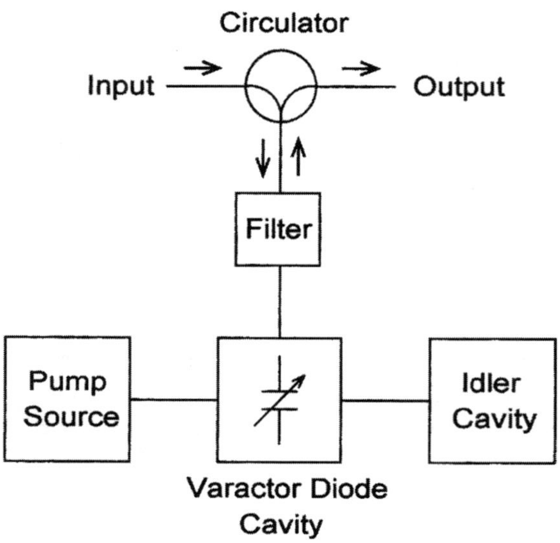

FIGURE 18.8 Block diagram of a typical parametric amplifier.

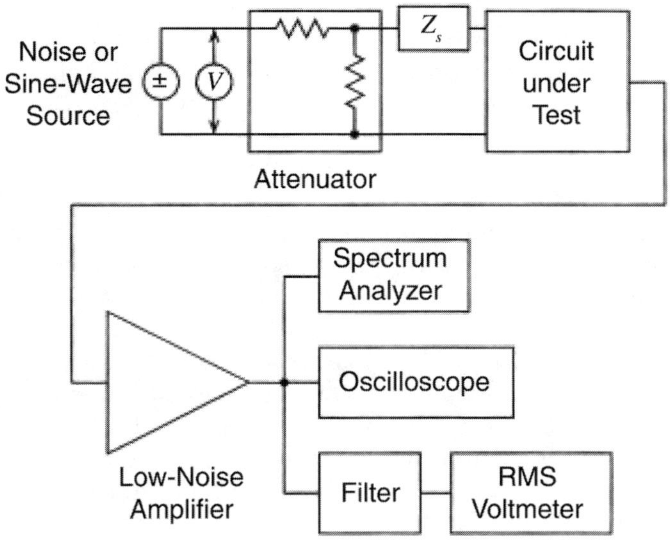

FIGURE 18.9 Noise measuring setup.

18.30 Measuring Noise

A typical setup for measuring noise is shown in Figure 18.9. To prevent measurement errors caused by signals coupling in through the ground and power supply leads, the circuit under test and the test set must be properly grounded and good power supply decoupling must be used [6]. For measurement schemes requiring a sine-wave source, an internally shielded oscillator is preferred over a function generator. This is because function generators can introduce errors caused by radiated signals and signals coupled through the ground system.

When making measurements on a high-gain circuit, the input signal must often be attenuated. Attenuators that are built into sources might not be adequately shielded, so that errors can be introduced by radiated signals. These problems can be minimized if a shielded external attenuator is used between the source and the circuit under test. Such an attenuator is illustrated in Figure 18.9. When a high attenuation is required, a multi-stage attenuator is preferred. For proper frequency response, the attenuator might require frequency compensation [4]. Unless the load impedance on the attenuator is large compared to its output impedance, both the attenuation and the frequency compensation can be a function of the load impedance.

Figure 18.9 shows a source impedance Z_s in series with the input to the circuit under test. This impedance is in series with the output impedance of the attenuator. It must be chosen so that the circuit under test has the desired source impedance termination for the noise measurements.

Because noise signals are small, a low-noise amplifier is often required to boost the noise level sufficiently so that it can be measured. Such an amplifier is shown in Figure 18.9. The noise generated by the amplifier will add to the measured noise. To correct for this, first measure the mean-square noise voltage with the amplifier input terminated in the output impedance of the circuit under test. Then subtract this from the measured mean-square noise voltage with the circuit under test driving the amplifier. The difference is the mean-square noise due to the circuit. Ideally, the amplifier should have no effect on the measured noise.

The noise voltage over a band can be measured with either a spectrum analyzer or with a filter having a known noise bandwidth and a voltmeter. The noise can be referred to the input of the circuit under test by dividing by the total gain between its input and the measuring device. The measuring voltmeter should have a bandwidth that is at least 10 times the noise bandwidth of the filter. The *voltmeter crest factor* is the ratio of the peak input voltage to the full-scale rms meter reading at which the internal meter circuits overload. For a sine-wave signal, the minimum voltmeter crest factor is $\sqrt{2}$. For noise measurements, a higher crest factor is required. For gaussian noise, a crest factor of 3 gives an error less than 1.5%. A crest factor of 4 gives an error less than 0.5%. To avoid overload on noise peaks caused by an inadequate crest factor, measurements should be made on the lower one-third to one-half of the voltmeter scale.

A true rms voltmeter is preferred over one that responds to the average rectified value of the input voltage but is calibrated to read rms. When the latter type of voltmeter is used to measure noise, the reading will be low. For gaussian noise, the reading can be corrected by multiplying the measured voltage by 1.13. Noise voltages measured with a spectrum analyzer must also be corrected by the same factor if the spectrum analyzer responds to the average rectified value of the input voltage but is calibrated to read rms.

Noise measurements with a spectrum analyzer require a knowledge of the noise bandwidth of the instrument. For a conventional analyzer, the bandwidth is proportional to frequency. When white noise is analyzed, the display exhibits a slope of +10 dB per decade. However, the measured voltage level at any frequency divided by the square root of the noise bandwidth of the analyzer is a constant equal to the spot-noise value of the input voltage at that frequency. Bandpass filters that have a bandwidth proportional to the center frequency are called *constant-Q filters.* For a second-order constant-Q filter, the noise bandwidth is given by $B_n = \pi f_0/2Q$, where f_0 is the center frequency and Q is the quality factor. The latter is given by $Q = f_0 / B_3$, where B_3 is the –3 dB bandwidth. These equations are often used to estimate the noise bandwidth of bandpass filters that are not second order.

A second type of spectrum analyzer is called a *signal analyzer.* Such an instrument uses digital signal processing techniques to calculate the spectrum of the input signal as a discrete Fourier transform. The noise bandwidth of these instruments is a constant so that the display exhibits a slope of zero when white noise is the input signal.

Fairly accurate rms noise measurements can be made with an oscilloscope. A filter should be used to limit the noise bandwidth at its input. Although the procedure is subjective, the rms voltage can be estimated by dividing the observed peak-to-peak voltage by 6 [4]. One of the advantages of using the oscilloscope is that nonrandom noise that can affect the measurements can be identified, e.g., a 60 Hz hum signal.

Another oscilloscope method is to display the noise simultaneously on both inputs of a dual-channel oscilloscope that is set in the dual-sweep mode. The two channels must be identically calibrated and the sweep rate must be set low enough so that the displayed traces appear as bands. The vertical offset between the two bands is adjusted until the dark area between them just disappears. The rms noise voltage is then measured by grounding the two inputs and reading the vertical offset between the traces.

Defining Terms

Burst noise: Noise caused by a metallic impurity in a *pn* junction that sounds like corn popping when amplified and reproduced by a loudspeaker. Also called *popcorn noise.*

Crest factor: The ratio of the peak value to the rms value.

Equivalent noise input current: The noise current in parallel with an amplifier input that generates the same noise voltage at its output as all noise sources in the amplifier.

Equivalent noise input voltage: The noise voltage in series with an amplifier input that generates the same noise voltage at its output as all noise sources in the amplifier.

Equivalent noise resistance (conductance): The value of a resistor (conductance) at the standard temperature $T_0 = 290$ K that generates the same mean-square noise voltage (current) as a source.

Excess noise: Flicker noise in resistors.

Flicker noise: Noise generated by the imperfect contact between two conducting materials causing the conductivity to fluctuate in the presence of a dc current. Also called *contact noise, 1/f noise,* and *pink noise.*

Generation–recombination noise: Noise generated in a semiconductor by the random fluctuation of free carrier densities caused by spontaneous fluctuations in the generation, recombination, and trapping rates.

Noise bandwidth: The bandwidth of an ideal filter having a constant passband gain that passes the same rms noise voltage as a filter, where the input signal is white noise.

Noise factor: The ratio of the mean-square noise voltage at an amplifier output to the mean-square noise voltage at the amplifier output considering the thermal noise of the input termination to be the only source of noise.

Noise figure: The noise factor expressed in dB.

Noise index: The number of μA of excess noise current in each decade of frequency per A of dc current through a resistor. Also, the number of μV of excess noise voltage in each decade of frequency per V of dc voltage across a resistor.

Noise temperature: The internal noise generated by an amplifier expressed as an equivalent input-termination noise temperature.

Nyquist formula: Expression for the mean-square thermal noise voltage generated by a resistor.

Optimum source impedance (admittance): The complex source impedance (admittance) that minimizes the noise factor.

Parametric amplifier: An amplifier that uses a time-varying reactance to produce amplification.

Partition noise: Noise generated by the statistical process of partition when the charge carriers in a current have the possibility of dividing between two or more paths.

Shot noise: Noise caused by the random emission of electrons and by the random passage of charge carriers across potential barriers.

Schottky formula: Expression for the mean-square shot noise current.

Signal-to-noise ratio: The ratio of the mean-square signal voltage to the mean-square noise voltage at an amplifier output.

Spectral density: The mean-square value per unit bandwidth of a noise signal.

Spot noise: The rms noise in a band, divided by the square root of the noise bandwidth.

Thermal noise: Noise generated by the random collision of charge carriers with a lattice under conditions of thermal equilibrium. Also called *Johnson noise.*

Varactor diode: A diode used as a variable capacitance.

Voltmeter crest factor: The ratio of the peak input voltage to the full-scale rms meter reading at which the internal meter circuits overload.

White noise: Noise that has a spectral density that is flat, i.e., not a function of frequency.

References

1. P.R. Gray and R.G. Meyer, *Analysis and Design of Analog Integrated Circuits,* New York: John Wiley & Sons, 1993.
2. P. Horowitz and W. Hill, *The Art of Electronics,* 2nd ed., New York: Cambridge Press, 1983.
3. W.M. Leach, Jr., Fundamentals of low-noise analog circuit design, *Proc. IEEE,* 82(10), 1515–1538, 1994.
4. C.D. Motchenbacher and J.A. Connelly, *Low Noise Electronic System Design,* New York: Wiley, 1993.
5. Y. Netzer, The design of low-noise amplifiers, *Proc. IEEE,* 69, 728-741, 1981.
6. H.W. Ott, *Noise Reduction Techniques in Electronic Systems,* 2nd ed., New York: Wiley, 1988.
7. A. Van der Ziel, Noise in solid-state devices and lasers, *Proc. IEEE,* 58, 1178-1206, 1970.
8. H.A. Haus et al., Representation of noise in linear twoports, *Proc. IRE,* 48, 69–74, 1960.
9. R.E. Burgess, Electronic fluctuations in semiconductors, *Br. J. Appl. Phys.,* 6, 185–190, 1955.
10. H. Fukui, The noise performance of microwave transistors, *IEEE Trans. Electron Devices,* ED-13, 329–341, 1966.
11. H. Rothe and W. Dahlke, Theory of noisy fourpoles, *Proc. IRE,* 44, 811–818, 1956.

12. H.A. Haus et al., IRE standards on methods of measuring noise in linear twoports, 1959, *Proc. IRE,* 48, 60–68, 1960.
13. A.L. Lance, *Introduction to Microwave Theory and Measurements,* New York: McGraw-Hill, 1964.

Further Information

H. Fukui, *Low-Noise Microwave Transistors & Amplifiers,* New York: IEEE Press, 1981.

M.S. Gupta, Ed., *Electrical Noise: Fundamentals & Sources,* New York: IEEE Press, 1977.

A. Van der Ziel, *Noise: Sources, Characterization, Measurements,* Englewood Cliffs, NJ: Prentice-Hall, 1970.

J.C. Bertails, Low frequency noise considerations for MOS amplifier design, *IEEE J. Solid-State Circuits,* SC-14, 773–776, 1979.

C.A. Liechti, Microwave field-effect transistors — 1976, *IEEE Trans. Microwave Theory and Tech.,* MTT-24, 279–300, 1976.

M. Steyaert, Z.Y. Chang, and W. Sansen, Low-noise monolithic amplifier design: bipolar vs. CMOS, *Analog Integrated Circuits and Signal Processing,* 1(1), 9–19, 1991.

19

Microwave Measurement

A. Dehé
Institut für Hochfrequenztechnik, Technische Universität Darmstadt

K. Beilenhoff
Institut für Hochfrequenztechnik, Technische Universität Darmstadt

K. Fricke
Institut für Hochfrequenztechnik, Technische Universität Darmstadt

H. Klingbeil
Institut für Hochfrequenztechnik, Technische Universität Darmstadt

V. Krozer
Institut für Hochfrequenztechnik, Technische Universität Darmstadt

H. L. Hartnagel
Institut für Hochfrequenztechnik, Technische Universität Darmstadt

Microwave measurements cover the frequency range from 0.5 GHz to about 20 GHz. Frequencies from 30 GHz to 300 GHz are often referred to as mm-waves. In the following, the most important measurements are described.

19.1 Power Measurement

Exact microwave power measurement is in demand for development, fabrication, and installation of modern telecommunication networks. It is essential for attenuation, insertion, and return loss measurements, as well as for noise measurement and six-port network analysis. This chapter gives a brief overview about the basics of microwave power measurement. Detailed information about this subject can be found in [1–4]. Power detectors usually consist of a sensor that transfers the microwave signal into a dc or low-frequency signal and a power meter to read out the measured power levels. The sensor includes a load impedance for the microwave source (Figure 19.1).

First, several power definitions need to be clarified:

- *Conjugate available power* (P_{CA}): The maximum available power a signal generator can transmit. This power is delivered by the generator if the load impedance is equal to the complex conjugate of the generator's source impedance. Since measurement techniques often require different lengths of waveguides or coaxial lines, the conjugate available power can be achieved only by tuning.
- Z_0 *Available power* (P_{Z0}): The power that is transferred via a coaxial line of characteristic impedance Z_0 into a load impedance Z_L equal to Z_0, while the generator has the impedance Z_G. Consequently, the available power level is reduced due to generator mismatch: $P_{Z0} = P_{CA}\,(1 - |\Gamma_G|^2)$, where Γ_G is

0-8493-1733-9/04/$0.00+$1.50

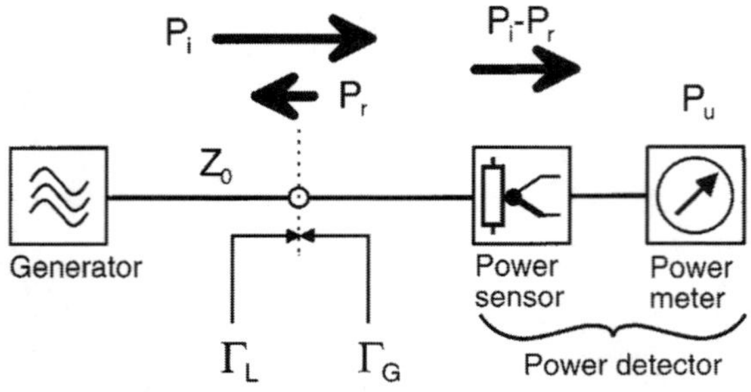

FIGURE 19.1 Setup of a power measurement in load configuration.

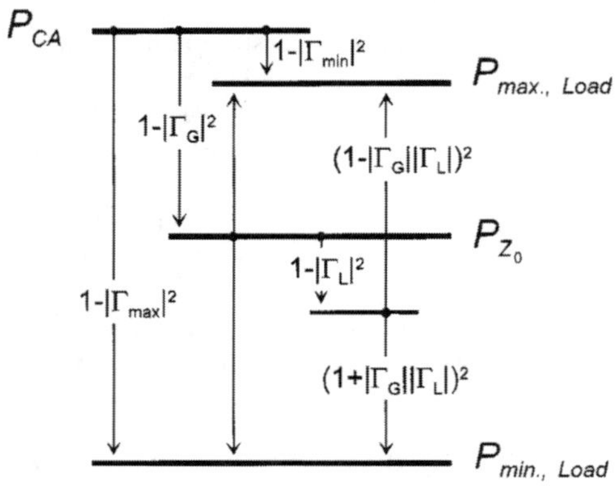

FIGURE 19.2 Relationships between conjugate power P_{CA}, Z_0 available power P_{Z0}, and the power levels available in the load of the power sensor.

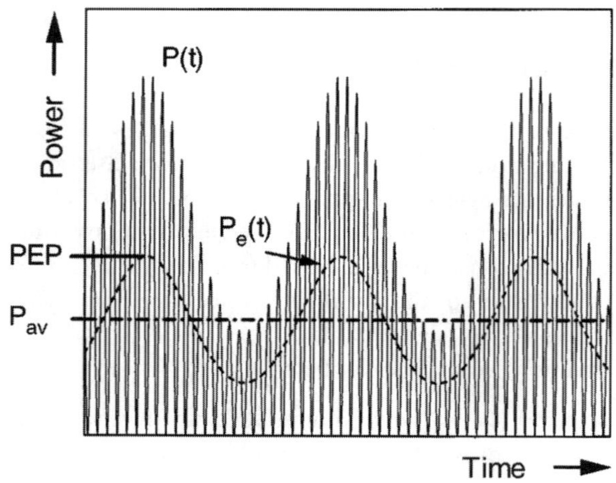

FIGURE 19.3 Power definitions for modulated signals.

the reflection coefficient of the generator. Figure 19.2 shows the relationships between P_{CA}, P_{Z0}, and the maximum and minimum power levels that can be measured in the load of a power sensor.

- *Average power* (P_{av}): The result of an averaging over many periods of the lowest modulation frequency of a modulated signal $P(t)$ (Figure 19.3).

- *Envelope power* ($P_e(t)$): The power averaged over the period of the carrier frequency.
- *Peak envelope power* (PEP): The maximum of $P_e(t)$.
- *Pulse power* (P_P): Defined for pulsed signals. If the pulse width τ and the repetition frequency $1/T$ is known, P_P is given by the measured average power:

$$P_P = P_{av} \frac{T}{\tau}$$

Measurement Errors and Uncertainties

Measurement errors occur due to mismatch as well as inside the power sensor and in the power meter. After correcting for these errors, the measurement uncertainties remain. Typically 75% of the total uncertainty belongs to mismatch and the smallest part is due to the power meter. The uncertainties and errors can be power dependent, frequency dependent, or independent of both. Of course, the *total uncertainty* must be calculated from the different uncertainties u_i as the root-sum-of-the-squares: rss = $\sqrt{\Sigma u_i^2}$, provided that the errors are all independent. A pessimistic error definition is the *worst-case uncertainty,* which simply sums up all extreme values of the independent errors.

Mismatch errors occur due to the fact that neither the generator (G), the line, nor the load (L) exhibit exactly the characteristic impedance Z_0. Using the modulus of the reflection coefficient $|\Gamma|$, the available power of the generator P_G can be expressed as [3]:

$$P_G = P_L \frac{|1 - \Gamma_G \Gamma_L|^2}{\left(1 - |\Gamma_G|^2\right)\left(1 - |\Gamma_L|^2\right)} \tag{19.1}$$

The reflection coefficients can also be expressed in terms of the voltage standing wave ratio (VSWR):

$$|\Gamma| = \frac{\text{VSWR} - 1}{\text{VSWR} + 1} = \frac{Z_L - Z_0}{Z_L + Z_0} \tag{19.2}$$

As mentioned, the knowledge of Z_0 available power P_{Z0} is sufficient in most applications. Then the ratio between the Z_0 available power and the absorbed power is given by:

$$\frac{P_{Z_0}}{P_i - P_r} = \frac{|1 - \Gamma_G \Gamma_L|^2}{1 - |\Gamma_L|^2} \tag{19.3}$$

where the subscripts "i" and "r" denote the incident and reflected power, respectively. Expressing Equation 19.3 in dB gives the Z_0 *mismatch loss* while the term $-10 \log(1 - |\Gamma_L|^2)$ is called the *mismatch loss,* which accounts for reflected power from the load only.

Since the reflection coefficients are seldomly known completely but only their magnitudes are known, a *mismatch uncertainty* is defined as:

$$M_u = \left\{\left(1 \pm |\Gamma_G||\Gamma_L|^2\right) - 1\right\} 100\% \tag{19.4}$$

Conversion errors are due to the specific characteristics of the individual power sensor. The conversion efficiency is frequently dependent and a *calibration factor* (CF) is defined for each power detector:

$$\text{CF} = \frac{P_u}{P_i} \tag{19.5}$$

where P_u is the uncorrected power and P_i is the actual incident power (Figure 19.1). Sometimes also the *effective efficiency* η_e is used:

$$\eta_e = \frac{P_u}{P_i - P_r} \tag{19.6}$$

Both quantities are related via the reflection coefficient of the load of the power sensor:

$$\text{CF} = \eta_e \left(1 - |\Gamma_L|^2\right) \tag{19.7}$$

The calibration factor is used to correct for efficiency loss and it also accounts for the mismatch loss. Still a remaining *calibration factor uncertainty* has to be taken into account. It is specified for each sensor. The calibration data are usually traceable to a national bureau of standards. Power sensors under test can be compared to the standards using high directivity couplers or power splitters [2].

The next described errors are due to the electronics inside the power meter.

Some power meters exhibit an internal reference oscillator to verify and adjust for the sensitivity of the diode or thermocouple sensor. The *reference power uncertainty* is specified by the manufacturer. Since this reference has its own reflection coefficient, it is related to a *reference oscillator mismatch uncertainty.*

Instrumentation uncertainty depends on the circuit limitations of the power meter and is specified by the manufacturer.

The *±1 Count Error* is for digital output and can be expressed by the relative power reading of the last significant digit.

Autozero can be used on all measurement ranges. Zero-setting immediately prior to the measurement can reduce *drift errors* when measuring in the lower ranges. Still, *zero set errors* remain due to noise during the zero-setting operation. This error can be very serious for measurement of low power levels [5]. *Zero carryover* is caused by ADC quantization errors in the zero readings for all measurement ranges except the most sensitive one. ADC quantization also causes a *power quantization error.* If very low power levels have to be measured, averaging can reduce random noise at the expense of measurement speed.

Power Sensors

A large variety of power sensors is available for the frequency range from dc up to 110 GHz and for minimum detectable power levels as low as 100 pW. The sensors differ in measurement principle and hence the correct choice depends on the application. A detailed description of power sensors is given in [2, 3]. Thermal sensors transform the absorbed microwave power into heat that is measured with temperature-sensitive elements:

- *Calorimeters* measure the heat absorbed in a fluid (e.g., water) of well-known specific heat. Applying the substitution principle, their precision can be enhanced. Because of their high stability, they are used in the National Institute of Standards. The manufacturer usually references the sensors to these standards.
- *Bolometers* and *thermistors* [3] make use of the temperature-dependent resistivity change of a resistive load, which is highly nonlinear. Hence, dc power substitution is used to keep the load at constant temperature. The substituted dc power is the measurement. Self-balancing bridges are used for this purpose but need careful ambient temperature compensation. The effective efficiency of bolometers is known very exactly. However, they have only relatively small dynamic range (typical 10 μW to 10 mW). Currently, liquid nitrogen-cooled, high-temperature superconducting bolometers with extremely high sensitivity of several thousands volts per watt are used [6]. A comprehensive overview of bolometers is given in [7].

- *Thermocouple sensors* are based on microwave power conversion into heat via a matched load impedance. Its temperature is controlled by thermocouples utilizing the thermoelectric effect in thin films or semiconductors [8]. One has to distinguish between realizations where the thermocouple itself represents the load impedance [3] and galvanically decoupled thermocouples [9]. The main advantages of the latter sensors are the better flatness of the response from dc to microwave and a lower VSWR. The design of such sensors is simple and a silicon micromachining technique can be applied to enhance the sensitivity [10]. The lowest measurable power levels are 1 μW.

Thermal sensors are well-suited to absolute power measurement, especially with respect to their high linearity. Their drawback is the relatively long response time (>10 ms), which limits these sensors to the measurement of average power. The high speed of Schottky diodes predestines them for the measurement of peak power, envelope power, and peak envelope power.

- *Diode sensors* consist of Schottky barrier diodes that are zero-biased and work in the square-law part of their *I–V* characteristics. For low power levels (<–20 dBm), these devices are very linear in response and measure the average power correctly. Still, diode sensors exhibit nonquadratic contributions [11], which can be important to account for when accurate measurement is required. Minimum detectable power is –70 dBm where the signal level is of the order of 50 nV and requires sophisticated amplification. The diode sensor is part of the sensors described below.
- *Peak power sensors* are specially designed for peak power measurements and account for measurement errors due to waveform and power level, although any diode detector can be used for this purpose if the peak voltages are ≤1 V.
- *Peak envelope analyzers* are designed to detect the envelope power. This is not possible with a simple diode sensor because the electronic setup of the diode must be different.
- *Feedthrough power sensors* are used to measure microwave power in transmission configuration. They have minor losses of approximately 0.5 dB and a limited bandwidth of typically 0.1 GHz to 1 GHz. The limiting device in these systems is a directional coupler with power splitters. Such a measurement can also be implemented with discrete elements. The characteristic figure of merit of these devices is the *directivity* (a_D), relating the read incident power P_i to the read reflected power P_r in case of reflection free load ($\Gamma_L = 0$):

$$a_D = 10\log\frac{P_i}{P_r}[\text{dB}] \tag{19.8}$$

The directivity should be as high as possible.

The above power sensors are discrete devices. The maturity in microwave monolithic integrated circuit (MMIC) design and fabrication allows the integrated realization of diode power sensors; for example, in an integrated six-port reflectometer [12]. Activities to fabricate thermal power sensors integrable to MMIC-typical processes [13] can presently be implemented on commercial processes, such as the Philips Lemeill HEMT processes with additional postmicromachining [14].

Commercially Available Power Measurement Systems

A collection of different power sensors and corresponding measurement units is shown in Table 19.1. All power meters have GPIB interfaces for easy use in automated measurement.

19.2 Frequency Measurement

Frequency measurement in the microwave regime is usually part of a more complex measurement procedure, for example, determining the scattering parameters and filter characteristics of a DUT. If one

TABLE 19.1 Available Commercial Power Sensors and Power Meters

Supplier	Power sensor *Power meter	Frequency range	Dynamic range	VSWR	Remark	~Price $U.S.
		Power Sensors in Matched Load Configuration and Related Power Meter				
Hewlett-Packard	HP8478B	10 MHz–18 GHz		1.1–1.75	Thermistor	
Rohde & Schwarz	NRV-Z52	Dc–26.5 GHz	1 μW–100 mW	1.11–1.22	Thermocouple, up to 30 W available	1790
Marconi	6913/6914S	10 MHz–26.5/46 GHz	1 μW–100 mW	1.1–1.4/3.6	Thermocouple, up to 3 W available	3000/3900
Boonton	51100(9E)	10 MHz–18 GHz	10 μW–100 μW	1.18–1.28	Thermocouple	
Hewlett-Packard	HP8485A	50 MHz–26.5 GHz	1 μW–100 mW	1.10–1.25	Thermocouple	
	HP8487A	50 MHz–50 GHz	1 μW–100 mW	1.10–1.50	Thermocouple	
	HPR/Q/W8486A	26.5–40/33–50/75–110 GHz	1 μW–100 mW	1.4/1.5/1.08	Rectangular waveguide, thermocouple	
Rohde & Schwarz	NRV-Z6	50 MHz–26.5 GHz	1 nW–20 mW	1.2–1.4	Diode	1980
Marconi	6923/6924S	10 MHz–26.5/46 GHz	0.1 nW/0.1 μW–10 μW	1.12–1.5/3.6	Diode	4300/5000
Hewlett-Packard	HP8487D	50 MHz–50 GHz	0.1 nW–10 μW	1.15–1.89	Diode	
Rohde & Schwarz	*NRVS/D	Dc–26.5 GHz	0.4 nW–30 W		One/two channel	2710/5290
Boonton	*4230A	10 kHz–100 GHz	0.1 nW–25 W			
Marconi	*6960B	30 kHz–46 GHz	0.1 nW–30 W			6900
	*6970	30 kHz–46 GHz	0.1 nW–30 W		Hand portable	3800
Hewlett-Packard	*HP437B	100 kHz–110 GHz	0.07 nW–25 W			
			Power Analyzer			
Hewlett-Packard	HP84812/13/14A	500 MHz–18/26.5/40 GHz	0.6 μW–100 mW	1.25–1.60	Resolution 100 ps	
	*HP8991A	500 MHz–40 GHz	0.5 μW–100 mW		Rise/fall time 5 ns	
		Peak Power Sensors and Related Power Meter				
Rohde & Schwarz	NRV-Z31/33	0.03–6 GHz	1 μW–20 mW/1 mW–20 W	1.05–1.33	With NRVS	2020/2500
Hewlett-Packard	HP84812/3/4A	500 MHz–18/26.5/40 GHz	1 μW–100 mW	1.25/1.35/1.50		
Boonton	56340	500 MHz–40 GHz		1.25–2.00	Dual diode risetime <15 ns	
	*HP8990A	500 MHz–40 GHz				
Boonton	4500A	1 MHz–40 GHz	0.1 μW–100 mW			
		Feedthrough Power Sensor and Related Power Meter				
Rohde & Schwarz	NAS-Z7	1.71–1.99 GHz	0.01–30 W	<1.15	a_D > 26 dB, GSM, DCS1800/1900	1250
	*NAS	0.001–1.99 GHz	10 mW–1200 W			1120

TABLE 19.2 Digital Microwave Frequency Counter

Supplier	Counter	Frequency range	Resolution	Sensitivity	Remark	~Price $US
Hewlett-Packard	HP5351B	26.5 GHz	1 Hz	–40 dBm		7,500
Hewlett-Packard	HP5352B	40(46) GHz	1 Hz	–30 dBm		11,800

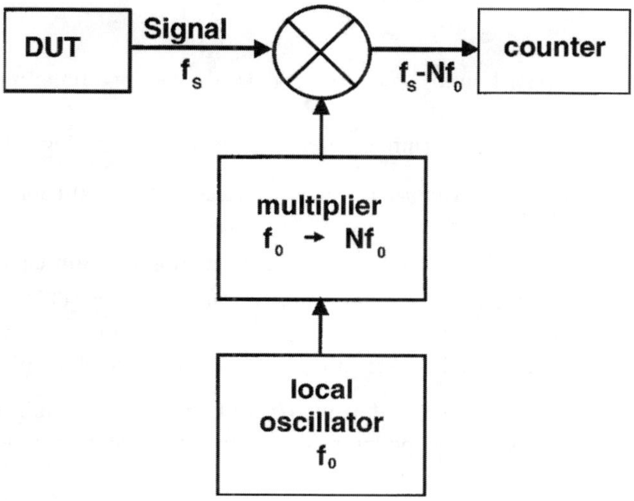

FIGURE 19.4 Block diagram of a digital frequency counter.

is only interested in frequency or a higher accuracy is required, one must use direct frequency measurement systems.

Two different techniques can be distinguished. The rather old-fashioned way is to use mechanically tunable resonators, the so-called *wave meters.* These are not explained in detail here. *Digital frequency counters* are an alternative and are now the state of the art (Table 19.2).

The digital frequency counter measurement system is based on the principle of counting the zero crossovers of a continuous sinusoidal signal. At low frequencies, this method can be used directly; whereas in the microwave region, direct digital counters are not available because of their limited bandwidth. Thus, a modified measurement system must be used.

The standard digital frequency counter usually consists of a mixer, a local oscillator (frequency f_0) in the lower frequency regime, several multipliers, and the digital counter. The principal measurement technique is shown in Figure 19.4.

An extremely stable local oscillator (quartz oscillator) is used to provide the reference signal used in the measurement system. This signal is multiplied by a factor of N and mixed with the RF signal of the DUT. The IF in the low-frequency range can be easily counted and, thus, the frequency of the signal f_s can be calculated according to the following equation.

$$f_s = f_{IF} + Nf_0 \tag{19.9}$$

For this method, an extremely stable low-frequency oscillator (often temperature controlled) must be provided and, in order to allow a sufficient bandwidth, a high number of multipliers must be implemented in this system.

A pulsed oscillator with extremely short risetime can circumvent this problem. In the frequency domain, this signal is given by spectral lines at $f = if_0$, where f_0 denotes the fundamental frequency of the pulses. Using a bandpass filter, a single frequency can be separated and transferred to the mixer.

19.3 Spectrum Analysis

The expression spectrum analysis subsumes the measurements that are performed to obtain the Fourier transformation $S(f)$ of a given signal $s(t)$. The Fourier transformation of $s(t)$ in the frequency domain is defined by the equation:

$$S(f) = \int_{-\infty}^{+\infty} s(t) e^{-j2\pi ft} \mathrm{d}t \tag{19.10}$$

In practice, the lower and upper bounds of the integral are limited by a finite measurement time that must be fixed by the user.

For the analysis of an unknown spectrum, different methods can be distinguished:

- *Wave analyzers and selective voltmeters:* These devices utilize a tunable filter for frequency-selective measurements.
- *Spectrum analyzers* rely on the principle of heterodyne mixing with subsequent bandpass filtering.
- Calculation of the spectrum using a *fast Fourier transformation* (FET). This method can be employed only for lower frequencies, since a digital-to-analog converter is needed. For microwave frequencies, the calculation of the spectrum using the FFT is, therefore, difficult to realize.

Since the spectrum analyzer is most suitable for microwave frequencies, it will be described in detail in the following. Brief introductions into the spectrum analyzer measurement techniques are given in [15, 16].

Spectrum Analyzer

The spectrum analyzer is most suitable for the analysis of microwave signals. It is a general-purpose instrument for measurements in the frequency domain and provides the user with the amplitude, power, or noise density of a signal depicted vs. the frequency. The frequency scale is in most cases linear; the vertical axes can be either linear or logarithmic. Spectrum analyzers are available from a few hertz up to more than 100 GHz. They give a quick overview of the spectral power distribution of a signal. Spectrum analyzers have a large dynamic range, a resolution bandwidth of a few hertz, and a reasonable frequency resolution.

The spectrum analyzer is suitable for the following measurements:

- *Measurement of absolute and relative frequency:* Frequency drift (unstabilized oscillators), spectral purity, and frequency of harmonics.
- *Absolute and relative amplitude:* Gain of frequency multipliers, harmonics of periodic signals, intermodulation (IM) distortion, and harmonic distortion.
- *Scalar network analysis (if equipped with a tracking generator):* Frequency response of amplifiers and filters.
- *Electromagnetic interference (EMI) measurements:* Broadband spectra.
- *Measurements of modulated signals:* AM, FM, or PM.
- *Noise:* Many spectrum analyzers can be used for noise measurements of active devices.
- *Phase noise:* Phase noise of oscillators can be analyzed with spectrum analyzers [17].

Spectrum Analyzer Setup

The spectrum analyzer is basically an electronically tunable filter that allows the measurement of the amplitude, power, or noise at a certain frequency. Using the example shown in Figure 19.5, the principle of operation can be explained as follows.

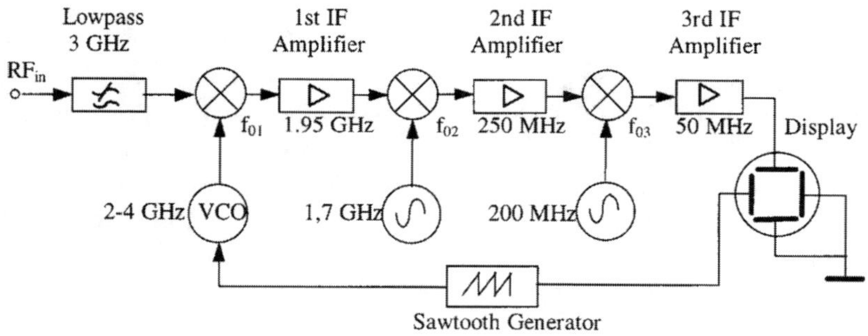

FIGURE 19.5 Simplified setup of a spectrum analyzer.

The tunable filter used to separate the frequencies to be measured is realized using a chain of mixers and IF amplifiers. In this case, three mixers convert a given input signal frequency f_s to the IF passband of the last IF amplifier. At least one of the oscillators is tunable (VCO) in order to scan the input frequency f_s. Sometimes, more than one tunable oscillator is used.

The first mixer and the following IF amplifier with a bandpass center frequency of 1.95 GHz in the given example selects the input frequency to be analyzed according to:

$$f_s = f_{o1} - 1.95 \text{ GHz} \tag{19.11}$$

The input frequency for the spectrum analyzer shown in Figure 19.4 can be due to the scan of the frequency of the tunable oscillator f_{o1}, between 50 MHz and 2.05 GHz. However, the image frequency f_{is} will also be mixed to the IF:

$$f_{is} = f_{o1} + 1.95 \text{ GHz} \tag{19.12}$$

Since f_{is} is in the range of 3.95 GHz to 5.95 GHz, a bandpass filter with a cut-off frequency of 3 GHz is used at the input to reject the image frequency.

Because it is difficult to realize a narrow bandpass at 1.95 GHz, the signal is converted to a lower frequency in the megahertz range. At these frequencies, stable quartz filters with high-quality factors can be employed. In principle, the RF frequency could be mixed down to the last IF section in one step. However, in such an arrangement, it would be difficult to suppress the image frequencies. In any case, the image frequency of each stage should be rejected by the preceding IF amplifier as shown in Figure 19.6 for the second stage of the given example.

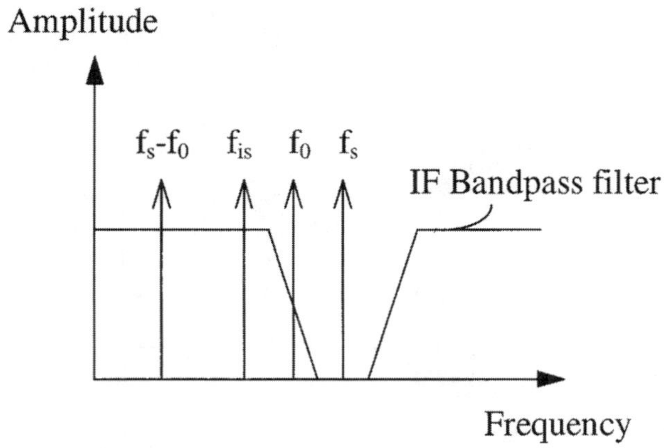

FIGURE 19.6 Example for the blocking of the image frequency by the IF filter in an arrangement according to Figure 19.5.

The last IF amplifier is very important for the performance of the complete system. In nearly all spectrum analyzers, its bandwidth — the so-called resolution bandwidth — can be adjusted in steps. For separation of closely spaced spectral components, the bandwidth should be very small. Most spectrum analyzers offer a minimum bandwidth of a few hertz. On the other hand, larger bandwidths are needed, since for a narrow IF filter only a slow scan speed can be allowed (see below) and, therefore, the measurement over several frequency decades would result in a large sweep time. The shape of the IF filter is important for the capability of the spectrum analyzer to separate close spectral components. The performance of the IF filter is described by the shape factor. It is defined by the ratio of the 60 dB to the 3 dB bandwidth of the IF filter. The IF filter can be switched between linear and logarithmic amplification. This is performed numerically in most cases.

A sawtooth generator produces the control voltage for the voltage-controlled oscillator (VCO) and the *x*-deflection voltage of the screen. A significant error is introduced in the frequency scale by a not-ideal voltage-to-frequency characteristic of the VCO. Therefore, in many cases, synthesizers with a quartzstabilized phase-locked loop [18] are used.

The detector has to be sensitive either to the amplitude, the power, or the noise ($\mathrm{mV}\sqrt{\mathrm{Hz}^{-1}}$). In a modern spectrum analyzer, digital signal processing is used for this purpose.

It is important to note that there are restrictions on the minimum sweep time T_s. In order to avoid settling errors of the narrow band pass IF amplifier with bandwidth *B*, the scan time should be for a frequency span *S* larger than [19, 20]:

$$T_s > 20S/B^2 \tag{19.13}$$

Most of the spectrum analyzers control the scan time automatically according to this equation.

Harmonic Mixing

For higher frequencies, the harmonic mixing technique is widely used. If the first 3-GHz low-pass filter in Figure 19.5 is omitted and the VCO produces harmonics, a larger number of input frequencies are converted to the passband of the spectrum analyzer. For the example shown, the possible input signals are depicted in Figure 19.7 vs. the VCO frequency according to:

$$f_s = nf_{01} \pm 1.95\ \mathrm{GHz} \tag{19.14}$$

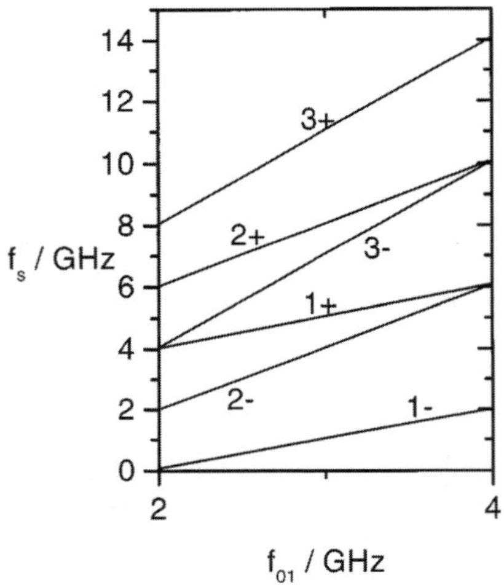

FIGURE 19.7 Measured RF frequency f_s vs. the frequency of the VCO (voltage-controlled oscillator) f_{01}. Parameters are *n* and the plus or the minus sign in Equation 19.4.

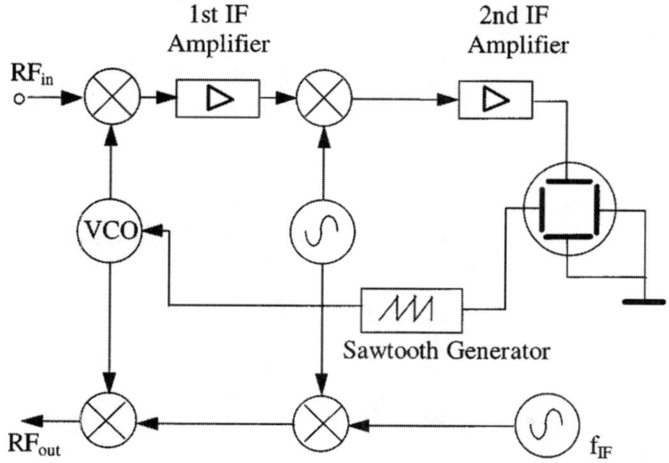

FIGURE 19.8 Principle of a tracking generator, which delivers at the port RF_{out} a signal that is precisely in the passband of the spectrum analyzer.

The notation of the numbers in the figure are the harmonic number n and the plus or the minus sign in the above equation. For a frequency f_{o1} = 3 GHz of the VCO, the following frequencies will appear at the same frequency location on the screen: 1.05 GHz (1–), 4.95 GHz (1+), 4.05 GHz (2–), 7.95 GHz (2+), 7.05 GHz (3–), 10.95 GHz (3+). A tracking preselection filter (see below), which is scanned with the VCO, can select one of these harmonics.

For further extension of the frequency in the upper mm-wave range, external mixers are used. With such an arrangement, frequencies higher than 500 GHz can be measured. Additionally, equipment for mixing of signals in the optical range is offered by some companies.

Tracking Preselection

For small input signals, the spectrum analyzer can be considered a linear device. However, if the input level increases, harmonics and intermodulation products are generated due to the nonlinearities of a mixer. These products result in spurious signals on the screen of the spectrum analyzer. In addition, image frequencies will appear on the screen, as demonstrated above. To avoid these spurious responses, a tracking preselection filter is employed at the input of the spectrum analyzer. A tracking preselection filter is an electronically tuned bandpass filter usually realized using a YIG filter.

Tracking Generator

Spectrum analyzers are often equipped with a tracking generator. The principle is shown in Figure 19.8. A local oscillator, with a frequency exactly on the center frequency of the IF amplifier, is mixed by an identical setup as in the analyzer path. Using the same local oscillator as in the analyzer path ensures that the frequency of the tracking generator follows precisely the center frequency of the swept window of the analyzed frequency band.

The tracking generator can be used for network analysis. If the tracking frequency is used as an input signal of a two-port, the amplitude of the output can be measured with the spectrum analyzer. Such a network analyzer has the advantage of being sensitive only in a very narrow band. Thus, third-order intermodulation products and noise are suppressed. However, only scalar measurements can be performed.

Commercially Available Spectrum Analyzers

A number of commercially available, general-purpose spectrum analyzers for the microwave and mm-wave range are listed in Table 19.3. Only a small number of spectrum analyzers available on the market

TABLE 19.3 Commercially Available Spectrum Analyzers

Company/ model	Frequency range	Min. res. bandw.	Amplitude accuracy	Remarks	Price $U.S.
Anritsu					
MS2602A	100 Hz–8.5 GHz		1.1 dB		30,600
Avantek					
3365	100 Hz–8 GHz	10 Hz		Portable, tracking	58,000
3371	100 Hz–26.5 GHz	10 Hz		Portable, tracking	66,000
R3272	9 kHz–26.5 GHz	300 Hz	1 dB	External mixer 325 GHz	38,000
Hewlett-Packard					
HP4196A	2 Hz–1.8 GHz	1 Hz	1 dB		
HP8590L	9 kHz–1.8 GHz	1 kHz	1.7 dB		9,080
HP8560E	30 Hz–2.9 GHz	1 Hz	1.85 dB	Portable	27,530
HP8596E	9 kHz–12.8 GHz	30 Hz	2.7 dB		25,090
HP8593E	9 kHz–22 GHz	30 Hz	2.7 dB		27,435
HP8564E	9 kHz–40 GHz	1 Hz	3 dB		50,890
HP8565E	9 kHz–50 GHz	1 Hz	3 dB		67,245
Marconi					
2370	30 Hz–1.25 GHz	5 Hz	5 Hz	With frequency extender	
2383	30 Hz–4.2 GHz	3 Hz	1 dB	Tracking	
Rohde & Schwarz					
FSEA30	20 Hz–3.5 GHz	1 Hz	1 dB		43,000
FSEB30	20 Hz–7 GHz	1 Hz	1 dB		52,000
FSEM30	20 Hz–26.5 GHz	1 Hz	1 dB	External mixer 110 GHz	64,000
Tektronix					
2714	9 kHz–1.8 GHz	300 Hz	2 dB	AM/FM demodulation 50 Ω/75 Ω	
2784	100 Hz–40 GHz	3 Hz	1.5 dB	Counter 1.2 THz, external mixer 325 GHz	

are presented. Most of the companies offer special equipment for production quality control. These spectrum analyzers can be computer controlled for fixed measurement routines. On request, spectrum analyzers with special options like fixed frequency operation, multichannel operation, support of external mixers, integrated frequency counters, and digital storage are available.

Typical specifications of microwave spectrum analyzers include:

- The frequency span is several gigahertz.
- With external mixers, the upper frequency limit can be extended to more than 100 GHz.
- The frequency accuracy is between 10^{-5} and 10^{-7}.
- The resolution bandwidth (i.e., the effective bandwidth of the narrow IF filter) can be adjusted in steps between 1 Hz and a few megahertz.
- The resolution bandwidth shape factor is typically 10:1.
- The amplitude accuracy is about 1 dB.
- The noise floor is at about –140 dBm.

19.4 Cavity Modes and Cavity *Q*

Cavities are used in a variety of applications. For example, they can be used to construct filters and they serve as those elements in microwave generators (e.g., klystrons) that determine the operating frequency. Cavities can also be applied in order to measure the frequency or the wavelength of microwaves (wavemeter). The most important parameters of a cavity are its resonant frequency and its q-factor. The latter determines the sharpness of the resonance, or, in case of filters, the bandwidth of the passband.

A cavity can be defined as a volume that is almost completely surrounded by a metallic surface. At one or two positions, coupling elements are applied to the metal surface in order to connect the cavity to other circuit elements. In the case of one coupling element, one speaks of a single-ended cavity; whereas a transmission cavity has two coupling elements. The coupling elements are usually small compared to the dimensions of the cavity. The inside of the cavity is filled with dielectric material (e.g., air: $\varepsilon_r = 1$).

Cavities are used as resonators for microwave applications. Therefore, they are comparable to low-frequency resonant circuits consisting of an inductance L and a capacitance C. In the low-frequency range, lumped elements (inductors and capacitors) are used, which are small in comparison with the wavelength. In contrast, cavities are distributed elements with dimensions comparable to the wavelength. This results in comparatively small losses. Since cavities are distributed elements, it is in general no longer possible to determine L and C directly. Instead of L and C, the resonant frequency $f_0 = \omega_0 /2\pi$ is the most important property of a cavity.

Neither lumped elements nor cavities are completely lossless. One must take into account the finite conductivities of the materials, resulting in a resistance R when analyzing resonant circuits. For cavities, however, R cannot be determined directly due to the same reasons that hold for L and C. Therefore, a different parameter, the q-factor Q plays a similar role for cavities. Q is proportional to the stored electric and magnetic energy W, divided by the power loss P:

$$Q = \frac{\omega_0 W}{P} \tag{19.15}$$

Although cavities are distributed elements, one is able to show that their behavior near resonance can be described by a simple parallel resonant circuit consisting of R, L, and C if the reference plane is chosen appropriately. Therefore, the basic properties of such a parallel resonant circuit are analyzed in the following.

Assume that the losses are small ($Q \gg 1$), which is desirable in practice. Therefore, the resistance R of the parallel resonant circuit is comparatively large. In this case, the resonant frequency does not depend on R in a first-order approximation:

$$\omega_0 \approx 1/\sqrt{LC} \tag{19.16}$$

If Equation 19.15 is applied to the analyzed parallel resonant circuit, one obtains:

$$Q \approx \frac{R}{\omega_0 L} \approx R\sqrt{\frac{C}{L}} \tag{19.17}$$

One can characterize the width of a resonance curve by those angular frequencies $\omega_1 = 2\pi f_1$ and $\omega_2 = 2\pi f_2$ where the power has decreased to one half of its maximum value (–3 dB). The impedance has then decreased to $1/\sqrt{2} \approx 70.7\%$ of its maximum value. This enables calculation of the angular frequencies ω_1 and w_2:

$$\omega_1 \approx \omega_0 - \frac{1}{2RC}, \quad \omega_2 \approx \omega_0 + \frac{1}{2RC} \tag{19.18}$$

Using these relations, one can easily derive the following equation, which is equivalent to Equation 19.17.

$$Q \approx \frac{\omega_0}{\omega_2 - \omega_1} \approx \frac{f_0}{f_2 - f_1} \tag{19.19}$$

This expression shows that Q is a symbol for the sharpness of resonance. Furthermore, it leads to the first principle as to how the q-factor can be measured. This principle, which is often referred to as the bandpass method, is based on the measurement of a resonance curve. For example, one can measure the reflection coefficient S_{11} with a network analyzer. From this curve, the 3 dB-bandwidth $\Delta f = f_2 - f_1$ and the resonant frequency $f_0 = (f_1 + f_2)/2$ can be easily determined. The application of Equation 19.19 yields the desired Q.

A second principle used to measure the q-factor is based on the transient behavior of the cavity after excitation with a pulse. In this case, the energy decays exponentially, and the time constant of the decay is proportional to Q^{-1}. This can again be shown by analyzing the equivalent parallel resonant circuit. Up until now, all signals were time-harmonic, which enabled a solution by complex quantities. This time, the corresponding differential equation must be solved. One can easily show that the amplitude of the voltage is proportional to $e^{-\omega_0 t/2Q}$, which means a $e^{-\omega_0 t/2Q}$ dependency of the energy exists. Measuring the decay time constant, therefore, enables one to determine the q-factor.

The Q defined here is the unloaded Q of the cavity; that is, no further resistance is connected to the cavity. Sometimes it is desirable to determine the loaded Q (Q_L), which takes into account such resistances. Detailed information about the loaded Q can be found in [21–23]. Although the q-factor is not influenced by the coupling structure if it is lossless, the coupling structure is of great importance. Further information about coupling parameters is presented in [21–23]. Detailed descriptions of measurement methods based on the above mentioned principles can also be found in [21, 22].

Up to now, all analyses were based on the equivalent resonant circuit. There are many other effects that can be explained by this analogy. For example, the energy in both the cavity and in its equivalent circuit oscillates between the magnetic and electric fields. Some properties of cavities, however, can only be seen by examining the electric and magnetic fields themselves. These are governed by Maxwell's equations.

A solution of Maxwell's equations (which can be very complicated for a given cavity) shows that all cavities have an infinite number of resonant frequencies. This could not be seen by analyzing the equivalent parallel resonant circuit. A description of the cavity by a discrete parallel resonant circuit is only valid in the vicinity of *one* of these resonant frequencies.

Furthermore, different modes can exist that have the same resonant frequency. This phenomenon is called *degeneration.* If the resonator will operate at such a frequency, where degenerate modes exist, one has to take care that the companion mode is suppressed. This can be accomplished by an appropriate choice of the positions of the coupling elements.

Even if the desired mode does not have any companion mode, one should take care that no other mode exists in the operating range of the cavity (this can be accomplished with a mode chart [22, 23]); otherwise, these modes must be damped sufficiently in order to be sure that an observed resonance corresponds to the desired mode.

19.5 Scattering Parameter Measurements

Scattering parameters describe multiple port structures in terms of wave variables. The introduction of scattering parameters (S-parameters) arises naturally in microwave circuits and systems, due to the lack of a unique definition for currents and voltages at these frequencies. Most circuits and systems at high frequencies are efficiently described in terms of S-parameters.

This section describes the fundamentals and properties of S-parameters, together with network analysis based on S-parameter calculations and measurements. Measurement procedures are outlined and the most frequently used systems for S-parameter measurement are described. Finally, information on hardware required for the experimental determination of S-parameters is provided, together with the corresponding suppliers.

Introduction and Fundamentals

At high frequencies, the wave variables are a natural extension of the voltages and currents at port terminals. In electric systems where the voltages and currents cannot be uniquely defined, the power flow in a waveguide can be described via wave variables. Whenever a TEM mode of wave propagation cannot be assumed, the currents and voltages are dependent on the integration path. This situation is encountered in all rectangular, circular, and passive waveguide structures, even in the case of lossless wave propagation. It is also true for all guiding structures if losses are to be considered along the path of wave propagation [24–27]. For the case of wave propagation along a transmission line, the wave variables a and b are defined as follows:

$$\begin{aligned} a(z) &= \frac{1}{2}\left(\frac{U(z)}{\sqrt{Z_0}} + I(z)\sqrt{Z_0}\right) \\ &= \frac{U_+}{\sqrt{Z_0}} = I_+\sqrt{Z_0} \\ b(z) &= \frac{1}{2}\left(\frac{U(z)}{\sqrt{Z_0}} - I(z)\sqrt{Z_0}\right) \\ &= \frac{U_-}{\sqrt{Z_0}} = I_-\sqrt{Z_0} \end{aligned} \qquad (19.20)$$

The propagation is along the z-direction. The characteristic impedance of the transmission line is Z_0, and $U(z)$ and $I(z)$ are the voltage and current, respectively, at location z along the line. The variables $a(z)$ and $b(z)$ are the complex amplitudes of the modes on the line. The voltage U_+ and U_- and the currents I_+ and I_- denote the voltage and current amplitudes, respectively, in forward and reverse direction. The wave variables are related to the power in the following form:

$$\begin{aligned} P_+ &= \frac{1}{2}\frac{|U_+|^2}{Z_0} = \frac{1}{2}|I_+|^2 Z_0 = \frac{1}{2}|a(z)|^2 \\ P_- &= \frac{1}{2}\frac{|U_-|^2}{Z_0} = \frac{1}{2}|I_-|^2 Z_0 = \frac{1}{2}|b(z)|^2 \end{aligned} \qquad (19.21)$$

In Equation 19.21, it is assumed that the system is excited by a pure sinusoid and that the characteristic impedance is purely real. The wave variables have the dimensions of $\sqrt{W}$.

Strictly speaking, wave variables and S-parameters can only be applied to linear networks. This is important because many publications are devoted to so-called large-signal S-parameter measurements. The interpretation of such results is not simple and great care must be employed in the correct determination of the characteristic impedance of the system [26]. In the case of analysis of large-signal or nonlinear circuits, two methods exist to introduce the wave variables:

- Volterra series representation [28]
- Harmonic-balance method [26]

Both methods transform the nonlinear circuit into a number of linear circuits at different frequencies, and then change the terminal voltages and currents into wave variables. This situation is sketched in Figure 19.9. Particular attention must be paid to the definition of the characteristic impedance, which can vary between different frequencies. An in-depth treatment of wave variables can be found in [26].

A further utilization of wave variables can be found in the noise analysis of microwave circuits. According to Figure 19.10, a noisy multiport can be analyzed by an associated noiseless two-port with the according noise sources c_i at the corresponding ports of the circuit.

Calculations and Analysis with S-Parameters

The characterization of multiports with S-parameters requires embedding of the multiport into a system consisting of a signal source with a characteristic impedance and appropriate terminations of all ports. This situation is shown in Figure 19.11. The outgoing wave parameters b are reflections at the corresponding ports. The wave variables are related to the scattering parameters of a two-port in the following manner:

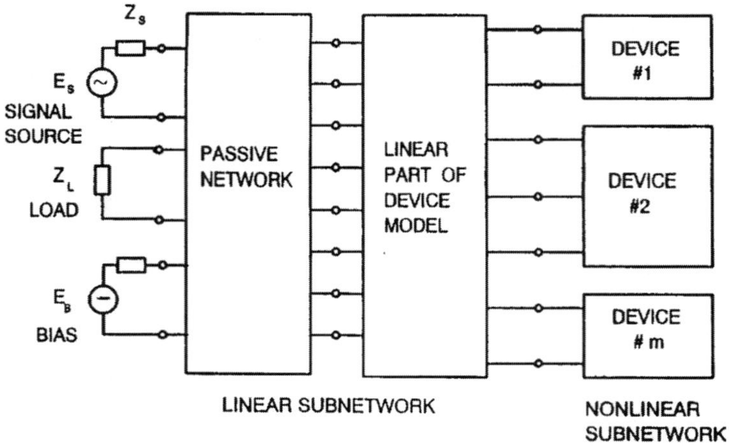

FIGURE 19.9 Schematic illustration of application of wave variables to nonlinear circuits.

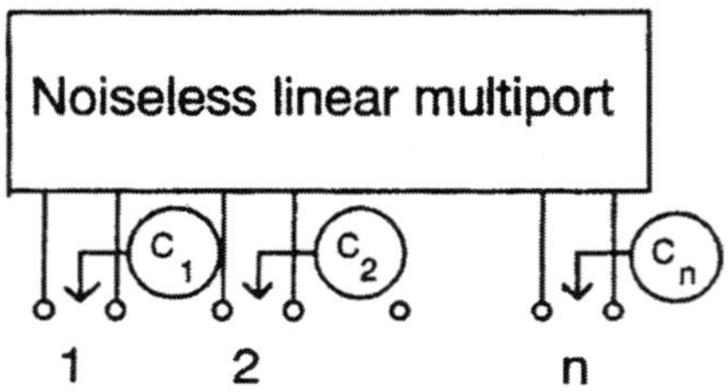

FIGURE 19.10 Schematic illustration of application of wave variables to noisy circuits.

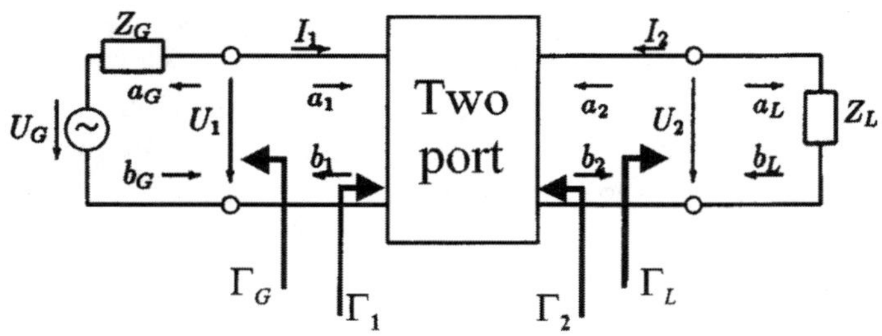

FIGURE 19.11 Two-port network indicating the wave variables and the scattering parameters. The subscripts G and L indicate the generator and the load, respectively.

$$\begin{pmatrix} b_1 \\ b_2 \end{pmatrix} = (S)\begin{pmatrix} a_1 \\ a_2 \end{pmatrix} = \begin{pmatrix} S_{11} & S_{12} \\ S_{21} & S_{22} \end{pmatrix}\begin{pmatrix} a_1 \\ a_2 \end{pmatrix} \tag{19.22}$$

For the determination of the individual scattering matrix elements, all ports of the network must be terminated in their characteristic impedance. The impedances at the corresponding ports need not be equal for all ports. The S-parameters are, in general, complex and are defined with respect to reference planes. These reference planes can be the network terminals, but could also be shifted to other locations in the circuit if this is desirable. The scattering matrix can be transformed into all circuit representations. Table 19.4 indicates the conversion formulae between the S-parameters and $Z\,Y\,h$ parameters for arbitrary characteristic impedances. Additional conversions to $ABCD$ and T parameters can be found in [29].

TABLE 19.4 Equations for the Conversion Between Different Two-Port Parameters

$S_{11}=\dfrac{\left(X_{11}-Z_{01}^{*}\right)\left(X_{22}+Z_{02}\right)-X_{12}X_{21}}{\left(X_{11}-Z_{01}\right)\left(X_{22}+Z_{02}\right)-X_{12}X_{21}}$	$X_{11}=\dfrac{\left(Z_{01}^{*}+S_{11}Z_{01}\right)\left(1-S_{22}\right)+S_{12}S_{21}Z_{01}}{\left(1-S_{11}\right)\left(1-S_{22}\right)-S_{12}S_{21}}$
$S_{12}=\dfrac{2X_{12}\sqrt{R_{01}R_{02}}}{\left(X_{11}+Z_{01}\right)\left(X_{22}+Z_{02}\right)-X_{12}X_{21}}$	$X_{12}=\dfrac{2S_{12}\sqrt{R_{01}R_{02}}}{\left(1-S_{11}\right)\left(1-S_{22}\right)-S_{12}S_{21}}$
$S_{21}=\dfrac{2X_{21}\sqrt{R_{01}R_{02}}}{\left(X_{11}+Z_{01}\right)\left(X_{22}+Z_{02}\right)-X_{12}X_{21}}$	$X_{21}=\dfrac{2S_{21}\sqrt{R_{01}R_{02}}}{\left(1-S_{11}\right)\left(1-S_{22}\right)-S_{12}S_{21}}$
$S_{22}=\dfrac{\left(X_{11}+Z_{01}\right)\left(X_{22}-Z_{02}^{*}\right)-X_{12}X_{21}}{\left(X_{11}+Z_{01}\right)\left(X_{22}+Z_{02}\right)-X_{12}X_{21}}$	$X_{22}=\dfrac{\left(1-S_{11}\right)\left(Z_{02}^{*}+S_{22}Z_{02}\right)-S_{12}S_{21}Z_{02}}{\left(1-S_{11}\right)\left(1-S_{22}\right)-S_{12}S_{21}}$
$S_{11}=\dfrac{\left(1-Y_{11}Z_{01}^{*}\right)\left(1+Y_{22}Z_{02}\right)+Y_{12}Y_{21}Z_{01}^{*}Z_{02}}{\left(1+Y_{11}Z_{01}\right)\left(1+Y_{22}Z_{02}\right)-Y_{12}Y_{21}Z_{01}Z_{02}}$	$Y_{11}=\dfrac{\left(1-S_{11}\right)\left(Z_{02}^{*}+S_{22}Z_{02}\right)+S_{12}S_{21}Z_{02}}{\left(Z_{01}^{*}+S_{11}Z_{01}\right)\left(Z_{02}^{*}+S_{22}Z_{02}\right)-S_{12}S_{21}Z_{01}Z_{02}}$
$S_{12}=\dfrac{-2Y_{12}\sqrt{R_{01}R_{02}}}{\left(1+Y_{11}Z_{01}\right)\left(1+Y_{22}Z_{02}\right)-Y_{12}Y_{21}Z_{01}Z_{02}}$	$Y_{12}=\dfrac{-2S_{12}\sqrt{R_{01}R_{02}}}{\left(Z_{01}^{*}+S_{11}Z_{01}\right)\left(Z_{02}^{*}+S_{22}Z_{02}\right)-S_{12}S_{21}Z_{01}Z_{02}}$
$S_{21}=\dfrac{-2Y_{21}\sqrt{R_{01}R_{02}}}{\left(1+Y_{11}Z_{01}\right)\left(1+Y_{22}Z_{02}\right)-Y_{12}Y_{21}Z_{01}Z_{02}}$	$Y_{21}=\dfrac{-2S_{21}\sqrt{R_{01}R_{02}}}{\left(Z_{01}^{*}+S_{11}Z_{01}\right)\left(Z_{02}^{*}+S_{22}Z_{02}\right)-S_{12}S_{21}Z_{01}Z_{02}}$
$S_{22}=\dfrac{\left(1+Y_{11}Y_{11}\right)\left(1-Y_{22}Z_{02}^{*}\right)+Y_{12}Y_{21}Z_{01}Z_{02}^{*}}{\left(1+Y_{11}Z_{01}\right)\left(1+Y_{22}Z_{02}\right)-Y_{12}Y_{21}Z_{01}Z_{02}}$	$Y_{22}=\dfrac{\left(Z_{01}^{*}+S_{11}Z_{01}\right)\left(1-S_{22}\right)-S_{12}S_{21}Z_{01}}{\left(Z_{01}^{*}+S_{11}Z_{01}\right)\left(Z_{02}^{*}+S_{22}Z_{02}\right)-S_{12}S_{21}Z_{01}Z_{02}}$
$S_{11}=\dfrac{\left(h_{11}-Z_{01}^{*}\right)\left(1+h_{22}Z_{02}\right)-h_{12}h_{21}Z_{02}}{\left(h_{11}+Z_{01}\right)\left(1+h_{22}Z_{02}\right)-h_{12}h_{21}Z_{02}}$	$h_{11}=\dfrac{\left(Z_{01}^{*}+S_{11}Z_{01}\right)\left(Z_{02}^{*}+S_{22}Z_{02}\right)+S_{12}S_{21}Z_{01}Z_{02}}{\left(1-S_{11}\right)\left(Z_{02}^{*}+S_{22}Z_{02}\right)-S_{12}S_{21}Z_{02}}$
$S_{12}=\dfrac{2h_{12}\sqrt{R_{01}R_{02}}}{\left(h_{11}+Z_{01}\right)\left(1+h_{22}Z_{02}\right)-h_{12}h_{21}Z_{02}}$	$h_{12}=\dfrac{2S_{12}\sqrt{R_{01}R_{02}}}{\left(1-S_{11}\right)\left(Z_{02}^{*}+S_{22}Z_{02}\right)-S_{12}S_{21}Z_{02}}$
$S_{12}=\dfrac{2h_{21}\sqrt{R_{01}R_{02}}}{\left(h_{11}+Z_{01}\right)\left(1+h_{22}Z_{02}\right)-h_{12}h_{21}Z_{02}}$	$h_{21}=\dfrac{2S_{21}\sqrt{R_{01}R_{02}}}{\left(1-S_{11}\right)\left(Z_{02}^{*}+S_{22}Z_{02}\right)-S_{12}S_{21}Z_{02}}$
$S_{22}=\dfrac{\left(h_{11}+Z_{01}\right)\left(h_{22}-Z_{02}\right)-h_{12}h_{21}Z_{02}^{*}}{\left(h_{11}+Z_{01}\right)\left(h_{22}Z_{02}\right)-h_{12}h_{21}Z_{02}}$	$h_{22}=\dfrac{\left(1-S_{11}\right)\left(1-S_{22}\right)-S_{12}S_{21}}{\left(1-S_{11}\right)\left(Z_{02}^{*}\right)-S_{12}S_{21}Z_{02}}$

The scattering parameters in the case of a noisy two-port as indicated schematically in Figure 19.10 are defined as [26]:

$$\begin{pmatrix} b_1 \\ b_2 \end{pmatrix} = \begin{pmatrix} S_{11} & S_{12} \\ S_{21} & S_{22} \end{pmatrix} \begin{pmatrix} a_1 \\ a_2 \end{pmatrix} + \begin{pmatrix} c_1 \\ c_2 \end{pmatrix} \tag{19.23}$$

The noise wave sources c_1 and c_2 represent the noise generated in the circuit and are therefore complex variables varying with time. They are characterized by a correlation matrix C_s as follows:

$$C_s = \overline{\begin{pmatrix} c_1 \\ c_2 \end{pmatrix} \begin{pmatrix} c_1 \\ c_2 \end{pmatrix}^{H}} = \begin{pmatrix} \overline{\left|c_1\right|^2} & \overline{c_1 c_2^*} \\ \overline{c_2 c_1^*} & \overline{\left|c_2\right|^2} \end{pmatrix} \tag{19.24}$$

where the bar indicates times averaging, $(\cdot)_H$ denotes the Hermitian conjugate, and * stands for the complex conjugate.

For the calculation of the cascade connection of two networks, it is desirable to convert the S-parameters to T parameters defined in the following way:

$$\begin{pmatrix} b_1 \\ a_1 \end{pmatrix} = \begin{pmatrix} T_{11} & T_{12} \\ T_{21} & T_{22} \end{pmatrix} \cdot \begin{pmatrix} a_2 \\ b_2 \end{pmatrix} \tag{19.25}$$

The conversion between S-parameters and T parameters is given below.

$$T_{11} = S_{12} - \frac{S_{11}S_{22}}{S_{21}} \tag{19.26}$$

$$T_{12} = \frac{S_{11}}{S_{21}} \tag{19.27}$$

$$T_{21} = -\frac{S_{22}}{S_{21}} \tag{19.28}$$

$$T_{22} = \frac{1}{S_{21}} \tag{19.29}$$

It should be emphasized that different definitions of the T parameters exist in the literature [2, 24–27, 30]. Power gain, mismatch, insertion loss, etc. can be efficiently described with the help of scattering parameters [2, 30].

Measurement of S-Parameters

Network analyzers are generally used for the measurement of S-parameters. A schematic configuration of a network analyzer is indicated in Figure 19.12. The network analyzer consists of two structures to separate the signals and a heterodyne receiver. According to the definitions, the measurement is performed in two steps as indicated in Figure 19.13. Different error models are employed for the calibration of the network analyzer. The most simple is the one-port error model, which consists of contributions due to directivity, source mismatch, and frequency response. A flow diagram is illustrated in Figure 19.14.

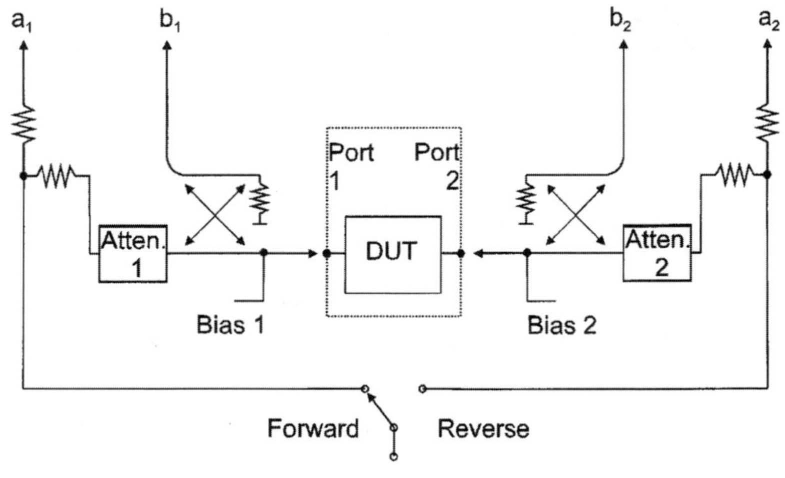

FIGURE 19.12 Schematic illustration of a network analyzer configuration.

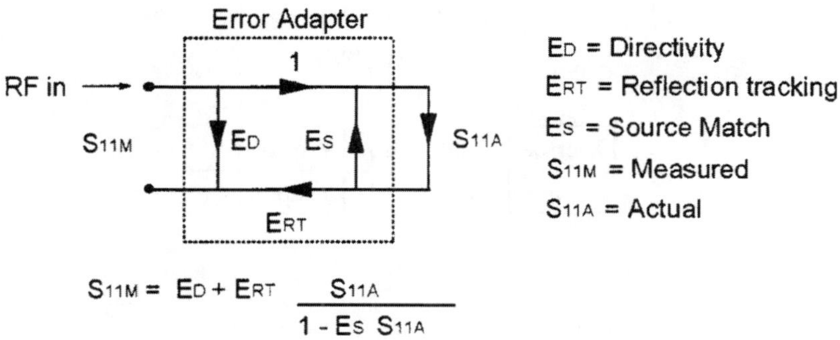

FIGURE 19.13 Flow diagram of the measurement procedure.

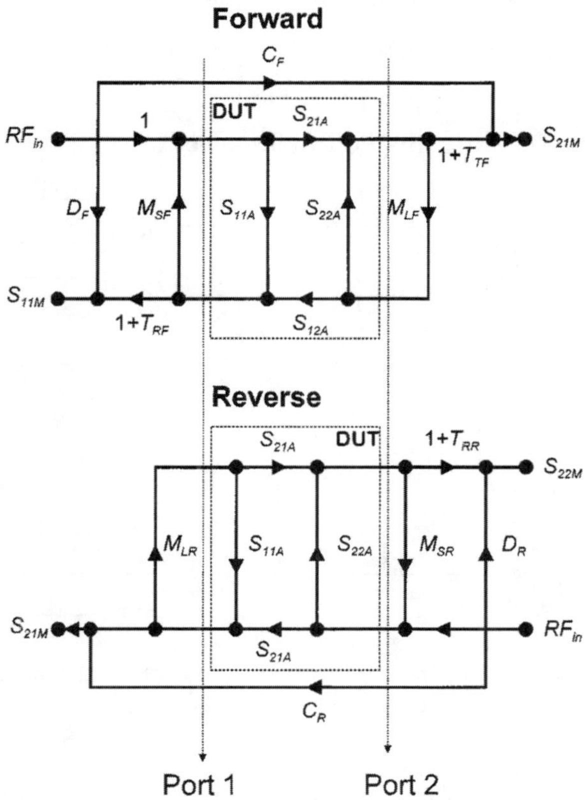

FIGURE 19.14 Flow diagram of the two-port error model.

For the characterization of active and passive two-ports, a more sophisticated error model is required. The signal flow graphs of the full two-port model is drawn in Figure 19.15. To determine the *S*-parameters of the device under test, a de-embedding procedure is required [31–43]. The two-port error model is then divided into two error adapters and the actual device under test (DUT), as depicted in Figure 19.16. The example shown makes reference to the so-called "TRL" calibration procedure. This name abbreviates the three calibration standards utilized in this method: a *through* standard with zero length, a *reflecting* standard, and a *line* standard. This method cannot be applied to on-wafer measurements at low frequencies, due to the excessive line length required for a broadband measurement. Other error correction methods are summarized in Table 19.5 [44]. In addition to the measurements given in the table, a known reference impedance and port 1 to port 2 connection are required. Furthermore, at higher frequencies (above ≈15 GHz), a calibration measurement of the isolation should be performed. For this purpose,

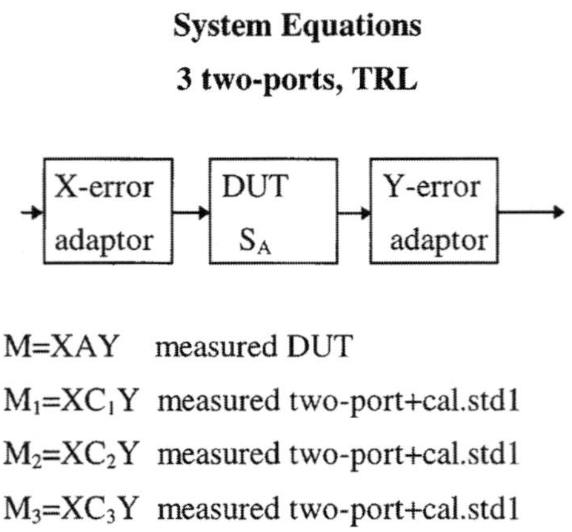

FIGURE 19.15 De-embedding structure for the calibration of a network analyzer.

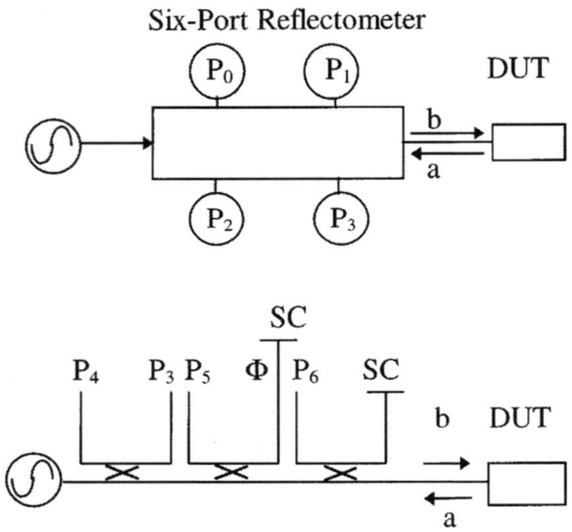

FIGURE 19.16 A network analyzer based on a six-port reflectometer and a possible realization using three couplers and four power detectors, two short-circuits (SC), and one phase shifter (Φ).

TABLE 19.5 Summary of Different Calibration Methods

TOSL	Through standard (T) with known length; Fulfills 4 conditions	3 known reflections (OSL) on port 1; Fulfills 3 conditions	3 known reflections (OSL) on port 2; Fulfills 3 conditions
TRL & LRL	Through or line standard (T) or (L) with known length; Fulfills 4 conditions	Unknown equal reflection standard (R) on port 1 and port 2; Fulfills 1 condition	Line (L) with known S_{11} and S_{22}; Fulfills 2 conditions
TRM & LRM	Through or line standard (T) or (L) with known length; Fulfills 4 conditions	Unknown equal reflection standard (R) on port 1 and port 2; Fulfills 1 condition	Known match (M) on port 1 and port 2; Fulfills 2 conditions
TXYZ & LXYZ	Through or line standard (T) or (L) with known length; Fulfills 4 conditions	3 known reflection standards (XYZ) on port 1 or port 2; Fulfills 3 conditions	
UXYZ	Unknown line standard (U) with $S_{11} = S_{21}$; Fulfills 1 condition	3 known reflection standards (XYZ) on port 1; Fulfills 3 conditions	3 known reflection standards (XYZ) on port 2; Fulfills 3 conditions

TABLE 19.6 Companies Supplying Network Analyzers for S-Parameter Measurements

Company	Frequency min	Frequency max	Method	Cal. Methods
Hewlett-Packard HP 8510	45 MHz	110 GHz	Heterodyne	SOLT, TLR, LRL, LRM, TRM
Wiltron	45 MHz	110 GHz	Heterodyne	SOLT, TLR, LRL, LRM, TRM
Rhode & Schwarz	10 Hz	4 GHz	Heterodyne	SOLT, TLR, LTL, LRM, TRM
AB Millimeterique	2 GHz	800 GHz	Heterodyne	TLR, LRL, proprietary

both ports are terminated by their characteristic impedances and a transmission measurement is performed. This measurement determines the values of C_F and C_R in the calibration flow diagram (see Figure 19.14).

Another possibility for performing vector network measurement is based on multiport reflectometers [45]. The advantage of such systems is the reduced complexity of the network analyzer receiver. A possible realization of a reflectometer structure is the so-called *six-port reflectometer.* The reflectometer consists of three couplers and four power sensors. No frequency conversion is required, which simplifies the test equipment.

Commercially Available Network Analyzers

Table 19.6 shows some of the current suppliers for network analysis. The frequency range is 10 Hz up to 800 GHz.

References

1. A. Fantom, *Radio Frequency and Microwave Power Measurements,* London: Peter Peregrinus, 1990.
2. A.L. Lance, *Microwave Measurements* in *Handbook of Microwave Optical Components,* Vol. 1, K. Chang (Ed.), New York: John Wiley & Sons, 1989.
3. J. Minck, Fundamentals of RF and microwave power measurements, Hewlett-Packard Application Note 64-1A, 1997.
4. G.H. Bryant, *Principles of Microwave Measurements,* London: Peter Peregrinus, 1988.
5. *Microwave Powermate,* Marconi Instruments Ltd., 1989.
6. D. Janik, H. Wolf, and R. Schneider, High-Tc edge bolometer for detecting guided millimeter waves, *IEEE Trans. Appl. Superconduct.,* 3, 2148–2151, 1993.
7. P.L. Richards, Bolometers for infrared and millimeter waves, *J. Appl. Phys.,* 76(1), 1–24, 1994.
8. D.M. Rowe, *Handbook of Thermoelectrics,* Boca Raton, FL: CRC Press, 1995.
9. P. Kopystynski, E. Obermayer, H. Delfs, W. Hohenester, and A. Löser, Silicon power sensor from dc to microwave, in *Micro Systems Technologies 90,* Berlin: Springer, 1990, 605–610.
10. G.C.M. Meijer and A.W. Herwaarden, *Thermal Sources,* Bristol: IOP Ltd., 1994.
11. T. Närhi, Nonlinearity characterisation of microwave detectors for radiometer applications, *Electron. Lett.,* 32, 224–225, 1996.
12. F. Wiedmann, B. Huyart, E. Bergeault, and L. Jallet, New structure for a six-port reflectometer in monolithic microwave integrated-circuit technology, *IEEE Trans. Instrument. Measure.,* 46(2), 527–530, 1997.
13. A. Dehé, V. Krozer, B. Chen, and H. L. Hartnagel, High-sensitivity microwave power sensor for GaAs-MMIC implementation, *Electron. Lett.,* 32(23), 2149–2150, 1996.
14. Circuits Multi Projects, *Information CMP — 42.* Grenoble, France, December 1996.
15. Hints for making better spectrum analyzer measurements, Hewlett-Packard Application Note, No. 1286-1, 1997.
16. C. Brown, Spectrum Analysis Basics, *Hewlett-Packard 1997 Back to Basics Seminar,* 1997.

17. A. Kiiss, Microwave instrumentation and measurements, in *Handbook of Microwave Technology,* T.K. Ishii (Ed.), San Diego: Academic Press, 1995, Vol. 2, 562.
18. H. Brand, Spectrum analyzers: precision test instruments, *Microwave J.,* 37(3), 98, 1994.
19. Spectrum analysis: Spectrum analysis basics, Hewlett-Packard Application Note 150, November 1989.
20. T.K. Ishii, Spectrum Analysis: Amplitude and Frequency Modulation, Application Note 150-1, January 1989.
21. M. Sucher and J. Fox, *Handbook of Microwave Measurements,* Vol. II, Polytechnic Press of the Polytechnic Institute of Brooklyn, 1963, chap. VIII.
22. E.L. Ginzton, *Microwave Measurements,* New York: McGraw-Hill, 1957, chap. 7 and 8.
23. F.E. Terman and J.M. Pettit, *Electronic Measurements,* New York: McGraw-Hill, 1952, chapter 4–15.
24. C.A. Lee and G.C. Dalman, *Microwave Devices, Circuits and Their Interaction,* New York: John Wiley & Sons, 1994.
25. G.D. Vendelin, A.M. Pavio, and U.L. Rohde, *Microwave Circuit Design Using Linear and Nonlinear Techniques,* New York: John Wiley & Sons, 1990.
26. J. Dobrowolski, *Microwave Circuit Analysis with Wave Variables,* Norwood, MA: Artech House, 1991.
27. G. Gonzalez, *Microwave Transistor Amplifiers,* Englewood Cliffs, NJ: Prentice-Hall, 1984.
28. D. Weiner and G. Naditch, A scattering variable approach to the Volterro analysis of nonlinear systems, *IEEE Trans. Microwave Theory & Technique,* MTT-24(7), 422–433, 1976.
29. D.A. Frickey, Conversion between S, Z, Y, H, ABSD, and T parameters which are valid for complex source and load impedances, *IEEE Trans. Microwave Theory & Technique,* MTT-42, 205-211, 1994.
30. T.K. Ishii, *Handbook of Microwave Technology,* San Diego, CA: Academic Press, 1995.
31. R.A. Hackborn, An automatic network analyzer system, *Microwave J.,* 45–52, 1968.
32. S. Rehnmark, On the calibration process of automatic network analyzer systems, *IEEE Trans. Microwave Theory & Technique,* MTT-22(4), 457–458, 1974.
33. J. Fitzpatrick, Error models of vector systems measurements, *Microwave J.,* 21(5), 63–66, 1978.
34. D. Rytting, An analysis of vector measurement accuracy enhancement techniques, *RF & Microwave Measurement Symp. and Exhibition,* Hewlett Packard, 1982.
35. N.R. Franzen and R.A. Speciale, A new procedure for system calibration and error removal in automated S-parameter measurements, *5th European Microwave Conf.,* 1975, 69–73.
36. R.A. Soares and C.A. Hoer, A unified mathematical approach to two-port calibration techniques and some applications, *IEEE Trans. Microwave Theory & Techniques,* MTT-37(11), 1669–1674, 1989.
37. R.A. Soares, *GaAs MESFET Circuit Design,* Norwood, MA: Artech House, 1988.
38. Cascade Microtech, *Microwave Wafer Probe Calibration Standards: HP8510 Network Analyzer Input,* Cascade Microtech Instruction Manual, 1990.
39. Understanding the fundamental principles of vector network analysis, Hewlett-Packard Application Note 1287-1, 1997.
40. B. Donecker, Determining the measurement accuracy of the HP8510 microwave network analyzer, *RF & Microwave Measurement Symp. and Exhibition,* Hewlett-Packard, March 1985.
41. Exploring the architectures of network analyzers, Hewlett-Packard Application Note 1287-2, 1997.
42. Applying error correction to network analyzer measurements, Hewlett-Packard Application Note 1287-3, 1997.
43. D. Ballo, Network analyzer basics, *Hewlett-Packard 1997 Back to Basics Seminar,* 1997.
44. H.J. Eul and B. Schiek, Thru-Match-Reflect: one result of a rigorous theory for de-embedding and network analyzer calibration, *18th European Microwave Conf.,* 1988, 909–914.
45. G.F. Engen and C.A. Hoer, Thru-Reflect-Line: an improved technique for calibrating the dual 6-port automatic network analyzer, *IEEE Trans. Microwave Theory & Technique,* MTT-27(12), 993–998, 1979.

II

Signal Processing

20

Amplifiers and Signal Conditioners

Ramón Pallás-Areny
Universitat Politechnica de Catalyna

20.1 Introduction

Signals from sensors do not usually have suitable characteristics for display, recording, transmission, or further processing. For example, they may lack the amplitude, power, level, or bandwidth required, or they may carry superimposed interference that masks the desired information.

Signal conditioners, including amplifiers, adapt sensor signals to the requirements of the receiver (circuit or equipment) to which they are to be connected. The functions to be performed by the signal conditioner derive from the nature of both the signal and the receiver. Commonly, the receiver requires a single-ended, low-frequency (dc) voltage with low output impedance and amplitude range close to its power-supply voltage(s). A typical receiver here is an analog-to-digital converter (ADC). Signals from sensors can be analog or digital. Digital signals come from position encoders, switches, or oscillator-based sensors connected to frequency counters. The amplitude for digital signals must be compatible with logic levels for the digital receiver, and their edges must be fast enough to prevent any false triggering. Large voltages can be attenuated by a voltage divider and slow edges can be accelerated by a Schmitt trigger.

Analog sensors are either self-generating or modulating. *Self-generating sensors* yield a voltage (thermocouples, photovoltaic, and electrochemical sensors) or current (piezo- and pyroelectric sensors) whose bandwidth equals that of the measurand. Modulating sensors yield a variation in resistance, capacitance, self-inductance or mutual inductance, or other electrical quantities. *Modulating sensors* need to be excited

0-8493-1733-9/04/$0.00+$1.50

or biased (semiconductor junction-based sensors) in order to provide an output voltage or current. Impedance variation-based sensors are normally placed in voltage dividers, or in Wheatstone bridges (resistive sensors) or ac bridges (resistive and reactance-variation sensors). The bandwidth for signals from modulating sensors equals that of the measured in dc-excited or biased sensors, and is twice that of the measurand in ac-excited sensors (sidebands about the carrier frequency) (see Chapter 21). Capacitive and inductive sensors require an ac excitation, whose frequency must be at least ten times higher than the maximal frequency variation of the measurand. Pallás-Areny and Webster [1] give the equivalent circuit for different sensors and analyze their interface.

Current signals can be converted into voltage signals by inserting a series resistor into the circuit. Graeme [2] analyzes current-to-voltage converters for photodiodes, applicable to other sources. Henceforth, we will refer to voltage signals to analyze transformations to be performed by signal conditioners.

20.2 Dynamic Range

The **dynamic range** for a measurand is the quotient between the measurement range and the desired resolution. Any stage for processing the signal from a sensor must have a dynamic range equal to or larger than that of the measurand. For example, to measure a temperature from 0 to 100°C with 0.1°C resolution, we need a dynamic range of at least (100 – 0)/0.1 = 1000 (60 dB). Hence, a 10-bit ADC should be appropriate to digitize the signal because $2^{10} = 1024$. Let us assume we have a 10-bit ADC whose input range is 0 to 10 V; its resolution will be 10 V/1024 = 9.8 mV. If the sensor sensitivity is 10 mV/°C and we connect it to the ADC, the 9.8 mV resolution for the ADC will result in a 9.8 mV/(10 mV/°C) = 0.98°C resolution! In spite of having the suitable dynamic range, we do not achieve the desired resolution in temperature because the output range of our sensor (0 to 1 V) does not match the input range for the ADC (0 to 10 V).

The basic function of voltage amplifiers is to amplify the input signal so that its output extends across the input range of the subsequent stage. In the above example, an amplifier with a gain of 10 would match the sensor output range to the ADC input range. In addition, the output of the amplifier should depend only on the input signal, and the signal source should not be disturbed when connecting the amplifier. These requirements can be fulfilled by choosing the appropriate amplifier depending on the characteristics of the input signal.

20.3 Signal Classification

Signals can be classified according to their amplitude level, the relationship between their source terminals and ground, their bandwidth, and the value of their output impedance. Signals lower than around 100 mV are considered to be low level and need amplification. Larger signals may also need amplification depending on the input range of the receiver.

Single-Ended and Differential Signals

A *single-ended signal* source has one of its two output terminals at a constant voltage. For example, Figure 20.1a shows a voltage divider whose terminal L remains at the power-supply reference voltage regardless of the sensor resistance, as shown in Figure 20.1b. If terminal L is at ground potential (grounded power supply in Figure 20.1a), then the signal is single ended and grounded. If terminal L is isolated from ground (for example, if the power supply is a battery), then the signal is single ended and floating. If terminal L is at a constant voltage with respect to ground, then the signal is single ended and driven off ground. The voltage at terminal H will be the sum of the signal plus the off-ground voltage. Therefore, the off-ground voltage is common to H and L; hence, it is called the **common-mode voltage.** For example, a thermocouple bonded to a power transistor provides a signal whose amplitude depends on the temperature of the transistor case, riding on a common-mode voltage equal to the case voltage.

A *differential signal* source has two output terminals whose voltages change simultaneously by the same magnitude but in opposite directions. The Wheatstone bridge in Figure 20.1c provides a differential

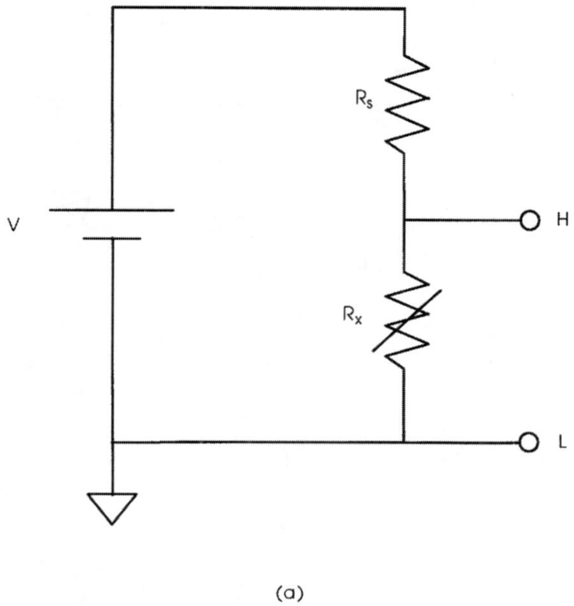

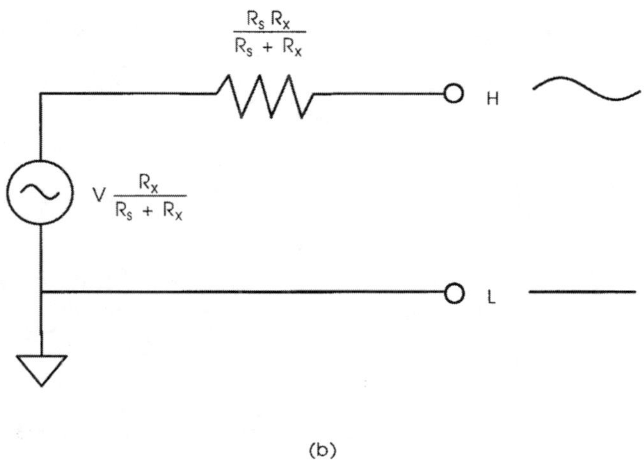

FIGURE 20.1 Classes of signals according to their source terminals. A voltage divider (a) provides a single-ended signal (b) where terminal L is at a constant voltage. A Wheatstone bridge with four sensors (c) provides a balanced differential signal which is the difference between two voltages v_H and v_L having the same amplitude but opposite signs and riding on a common-mode voltage V_c. For differential signals much smaller than the common-mode voltage, the equivalent circuit in (e) is used. If the reference point is grounded, the signal (single-ended or differential) will be grounded; if the reference point is floating, the signal will also be floating.

signal. Its equivalent circuit (Figure 20.1d) shows that there is a differential voltage ($v_d = v_H - v_L$) proportional to x and a common-mode voltage ($V_c = V/2$) that does not carry any information about x. Further, the two output impedances are balanced. We thus have a balanced differential signal with a superimposed common-mode voltage. Were the output impedances different, the signal would be unbalanced. If the bridge power supply is grounded, then the differential signal will be grounded; otherwise, it will be floating. When the differential signal is very small as compared with the common-mode voltage, in order to simplify circuit analysis it is common to use the equivalent circuit in Figure 20.1e. Some differential signals (grounded or floating) do not bear any common-mode voltage.

Signal conditioning must ensure the compatibility between sensor signals and receivers, which will depend on the relationship between input terminals and ground. For example, a differential and grounded

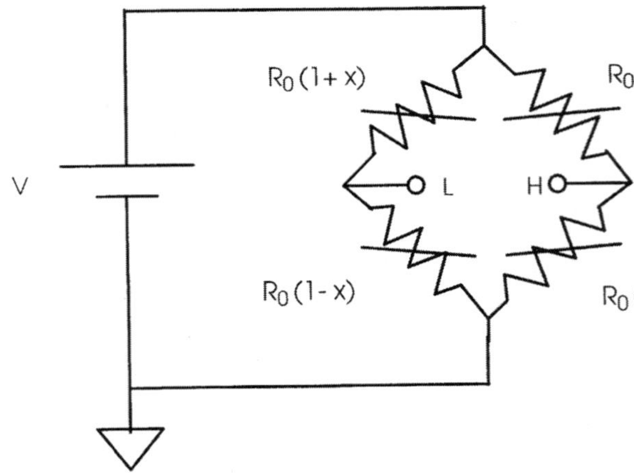

(c)

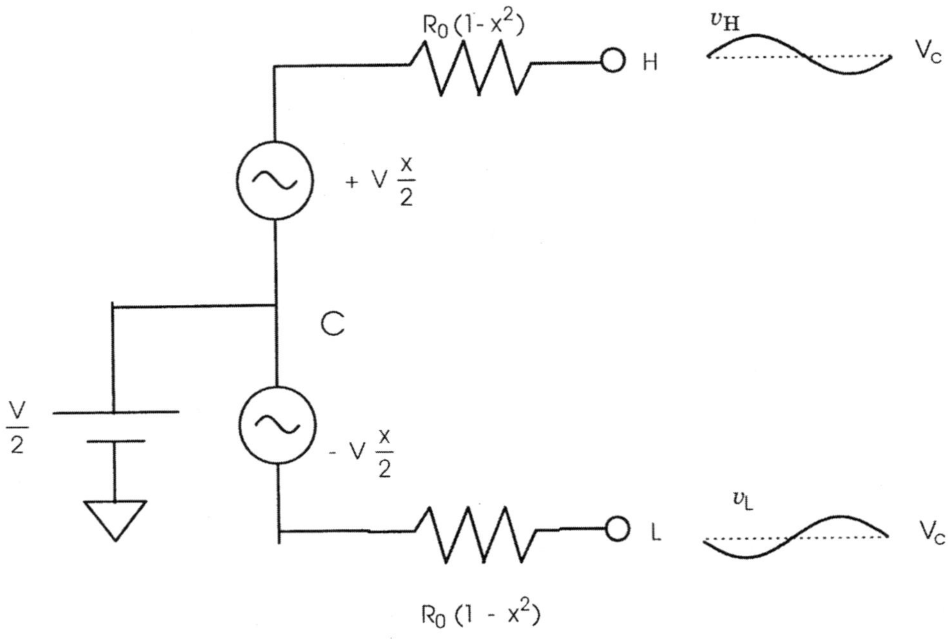

(d)

FIGURE 20.1 (continued)

signal is incompatible with an amplifier having a grounded input terminal. Hence, amplifiers must also be described according to their input topology.

Narrowband and Broadband Signals

A *narrowband signal* has a very small frequency range relative to its central frequency. Narrowband signals can be dc, or static, resulting in very low frequencies, such as those from a thermocouple or a weighing scale, or ac, such as those from an ac-driven modulating sensor, in which case the exciting frequency (carrier) becomes the central frequency (see Chapter 21).

Broadband signals, such as those from sound and vibration sensors, have a large frequency range relative to their central frequency. Therefore, the value of the central frequency is crucial; a signal ranging

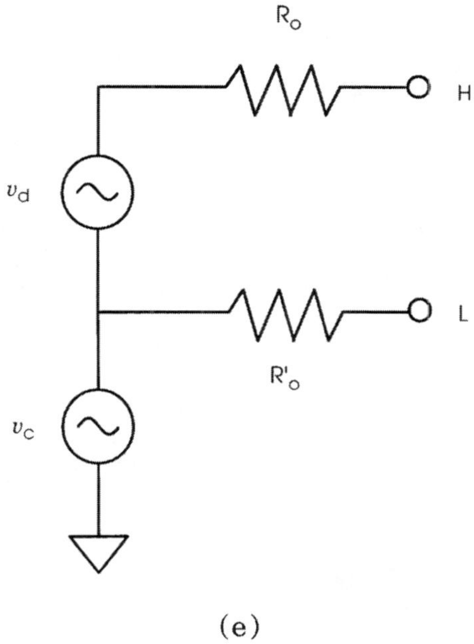

FIGURE 20.1 (continued)

from 1 Hz to 10 kHz is a broadband instrumentation signal, but two 10 kHz sidebands around 1 MHz are considered to be a narrowband signal. Signal conditioning of ac narrowband signals is easier because the conditioner performance only needs to be guaranteed with regard to the carrier frequency.

Low- and High-Output-Impedance Signals

The output impedance of signals determines the requirements of the input impedance of the signal conditioner. Figure 20.2a shows a voltage signal connected to a device whose input impedance is Z_d. The voltage detected will be

$$v_d = v_s \frac{Z_d}{Z_d + Z_s} \tag{20.1}$$

Therefore, the voltage detected will equal the signal voltage only when $Z_d \gg Z_s$; otherwise $v_d \neq v_s$ and there will be a *loading effect.* Furthermore, it may happen that a low Z_d disturbs the sensor, changing the value of v_s and rendering the measurement useless or, worse still, damaging the sensor.

At low frequencies, it is relatively easy to achieve large input impedances even for high-output-impedance signals, such as those from piezoelectric sensors. At high frequencies, however, stray input capacitances make it more difficult. For narrowband signals this is not a problem because the value for Z_s and Z_d will be almost constant and any attenuation because of a loading effect can be taken into account later. However, if the impedance seen by broadband signals is frequency dependent, then each frequency signal undergoes different attenuations which are impossible to compensate for.

Signals with very high output impedance are better modeled as current sources, Figure 20.2b. The current through the detector will be

$$i_d = i_s \frac{Z_s}{Z_d + Z_s} \tag{20.2}$$

In order for $i_d = i_s$, it is required that $Z_d \ll Z_s$ which is easier to achieve than $Z_d \gg Z_s$. If Z_d is not low enough, then there is a *shunting effect.*

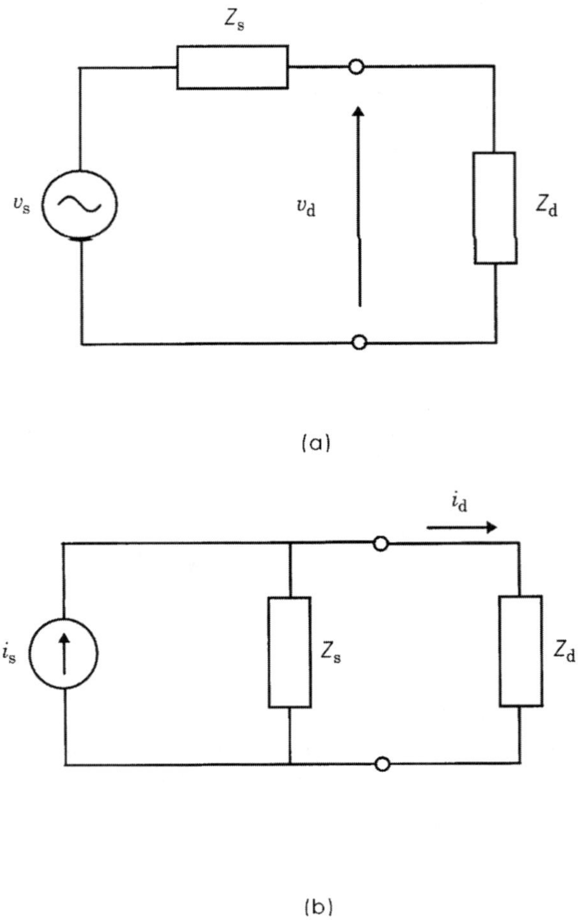

FIGURE 20.2 Equivalent circuit for a voltage signal connected to a voltage detector (a) and for a current signal connected to a current detector (b). We require $Z_d \gg Z_o$ in (a) to prevent any loading effect, and $Z_d \ll Z_s$ in (b) to prevent any shunting effect.

20.4 General Amplifier Parameters

A *voltage amplifier* produces an output voltage which is a proportional reproduction of the voltage difference at its input terminals, regardless of any common-mode voltage and without loading the voltage source. Figure 20.3a shows the equivalent circuit for a general (differential) amplifier. If one input terminal is connected to one output terminal as in Figure 20.3b, the amplifier is single ended; if this common terminal is grounded, the amplifier is single ended and grounded; if the common terminal is isolated from ground, the amplifier is single ended and floating. In any case, the output power comes from the power supply, and the input signal only controls the shape of the output signal, whose amplitude is determined by the *amplifier gain,* defined as

$$G = \frac{v_o}{v_d} \tag{20.3}$$

The ideal amplifier would have any required gain for all signal frequencies. A practical amplifier has a gain that rolls off at high frequency because of parasitic capacitances. In order to reduce noise and reject

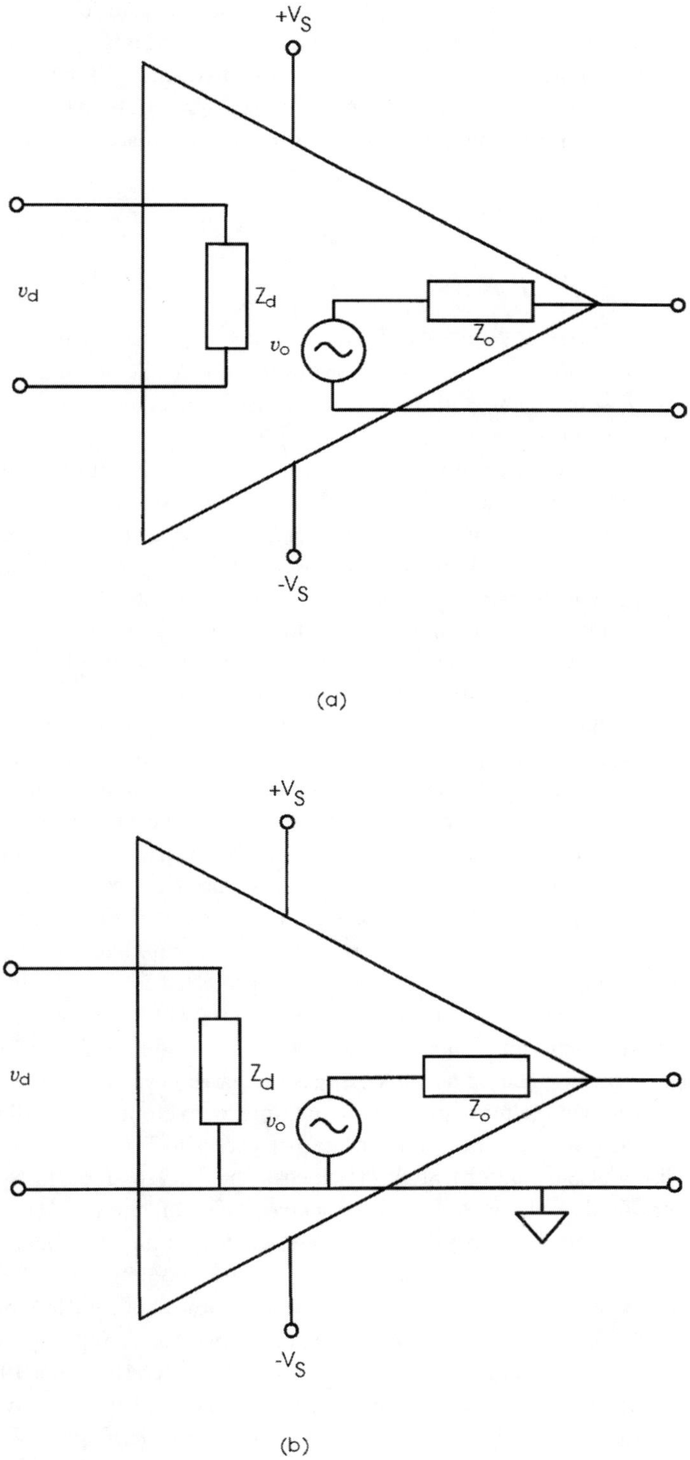

FIGURE 20.3 General amplifier, differential (a) or single ended (b). The input voltage controls the amplitude of the output voltage, whose power comes from the power supply.

interference, it is common to add reactive components to reduce the gain for out-of-band frequencies further. If the gain decreases by n times 10 when the frequency increases by 10, we say that the gain (downward) slope is 20 n dB/decade. The corner (or –3 dB) frequency f_0 for the amplifier is that for which the gain is 70% of that in the bandpass. (*Note:* 20 log 0.7 = –3 dB). The *gain error* at f_0 is then 30%, which is too large for many applications. If a maximal error ε is accepted at a given frequency f, then the corner frequency for the amplifier should be

$$f_0 = \frac{f(1-\varepsilon)}{\sqrt{2\varepsilon - \varepsilon^2}} \approx \frac{f}{\sqrt{2\varepsilon}} \tag{20.4}$$

For example, $\varepsilon = 0.01$ requires $f_0 = 7f$, $\varepsilon = 0.001$ requires $f_0 = 22.4f$. A broadband signal with frequency components larger than f would undergo amplitude distortion. A narrowband signal centered on a frequency larger than f would be amplified by a gain lower than expected, but if the actual gain is measured, the gain error can later be corrected.

Whenever the gain decreases, the output signal is delayed with respect to the output. In the above amplifier, an input sine wave of frequency f_0 will result in an output sine wave delayed by 45° (and with relative attenuation 30% as compared with a sine wave of frequency $f \gg f_0$). Complex waveforms having frequency components close to f_0 would undergo shape (or phase) distortion. In order for a waveform to be faithfully reproduced at the output, the phase delay should be either zero or proportional to the frequency (linear phase shift). This last requirement is difficult to meet. Hence, for broadband signals it is common to design amplifiers whose bandwidth is larger than the maximal input frequency. Narrowband signals undergo a delay which can be measured and corrected.

An ideal amplifier would have infinite *input impedance.* Then no input current would flow when connecting the signal, Figure 20.2a, and no energy would be taken from the signal source, which would remain undisturbed. A practical amplifier, however, will have a finite, yet large, input impedance at low frequencies, decreasing at larger frequencies because of stray input capacitances. If sensors are connected to conditioners by coaxial cables with grounded shields, then the capacitance to ground can be very large (from 70 to 100 pF/m depending on the cable diameter). This capacitance can be reduced by using driven shields (or guards) (see Chapter 29). If twisted pairs are used instead, the capacitance between wires is only about 5 to 8 pF/m, but there is an increased risk of capacitive interference.

Signal conditioners connected to remote sensors must be protected by *limiting* both *voltage* and *input currents.* Current can be limited by inserting a power resistor (100 Ω to 1 kΩ, 1 W for example), a PTC resistor, or a fuse between each signal source lead and conditioner input. Input voltages can be limited by connecting diodes, zeners, metal-oxide varistors, gas-discharge devices, or other surge-suppression nonlinear devices, from each input line to dc power-supply lines or to ground, depending on the particular protecting device. Some commercial voltage limiters are Thyzorb® and Transzorb® (General Semiconductor), Transil® and Trisil® (SGS-Thomson), SIOV® (Siemens), and TL7726 (Texas Instruments).

The ideal amplifier would also have zero *output impedance.* This would imply no loading effect because of a possible finite input impedance for the following stage, low output noise, and unlimited output power. Practical amplifiers can indeed have a low output impedance and low noise, but their output power is very limited. Common signal amplifiers provide at best about 40 mA output current and sometimes only 10 mA. The power gain, however, is quite noticeable, as input currents can be in the picoampere range (10^{-12} A) and input voltages in the millivolt range (10^{-3} V); a 10 V, 10 mA output would mean a power gain of 10^{14}! Yet the output power available is very small (100 mW). Power amplifiers are quite the opposite; they have a relatively small power gain but provide a high-power output. For both signal and power amplifiers, output power comes from the power supply, not from the input signal.

Some sensor signals do not require amplification but only *impedance transformation,* for example, to match their output impedance to that of a transmission line. Amplifiers for impedance transformation (or matching) and $G = 1$ are called buffers.

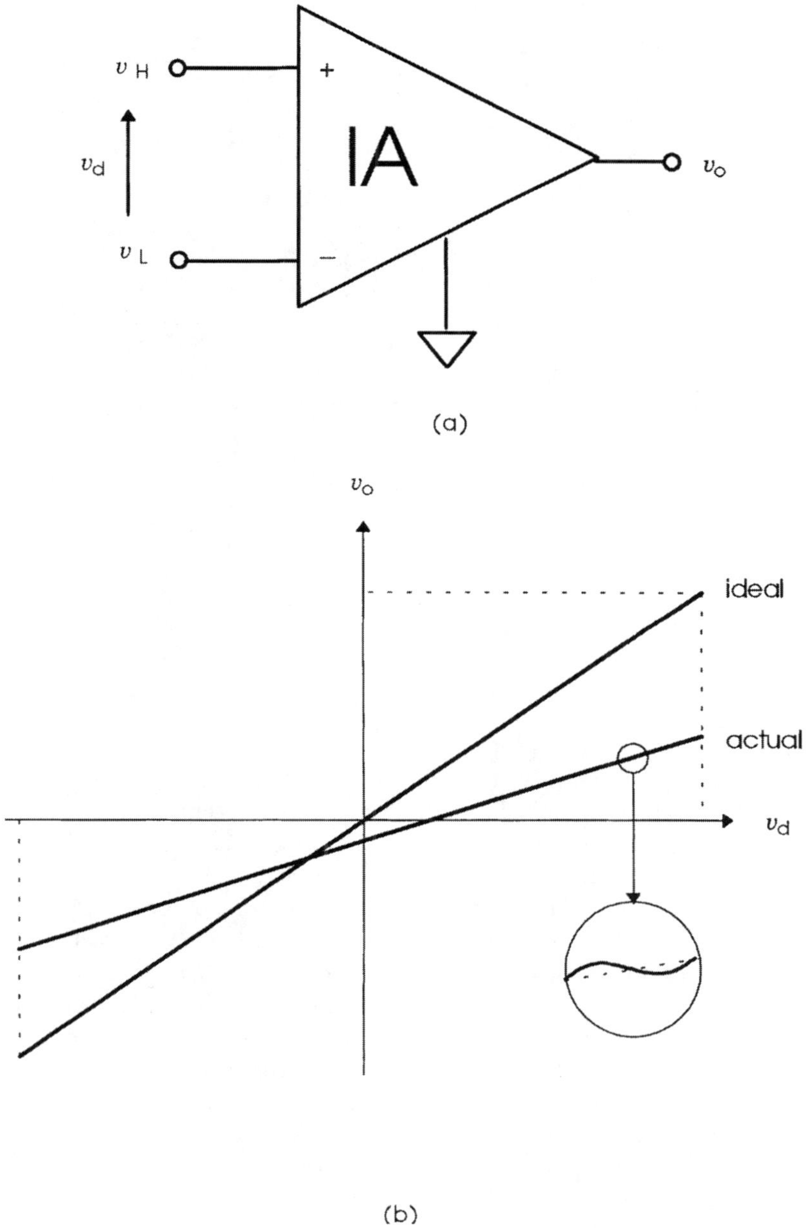

FIGURE 20.4 Instrumentation amplifier. (a) Symbol. (b) Ideal and actual input/output relationship. The ideal response is a straight line through the point (0,0) and slope G.

20.5 Instrumentation Amplifiers

For instrumentation signals, the so-called **instrumentation amplifier** (IA) offers performance closest to the ideal amplifier, at a moderate cost (from \$1.50 up). Figure 20.4a shows the symbol for the IA and Figure 20.4b its input/output relationship; ideally this is a straight line with slope G and passing through the point (0,0), but actually it is an off-zero, seemingly straight line, whose slope is somewhat different from G. The output voltage is

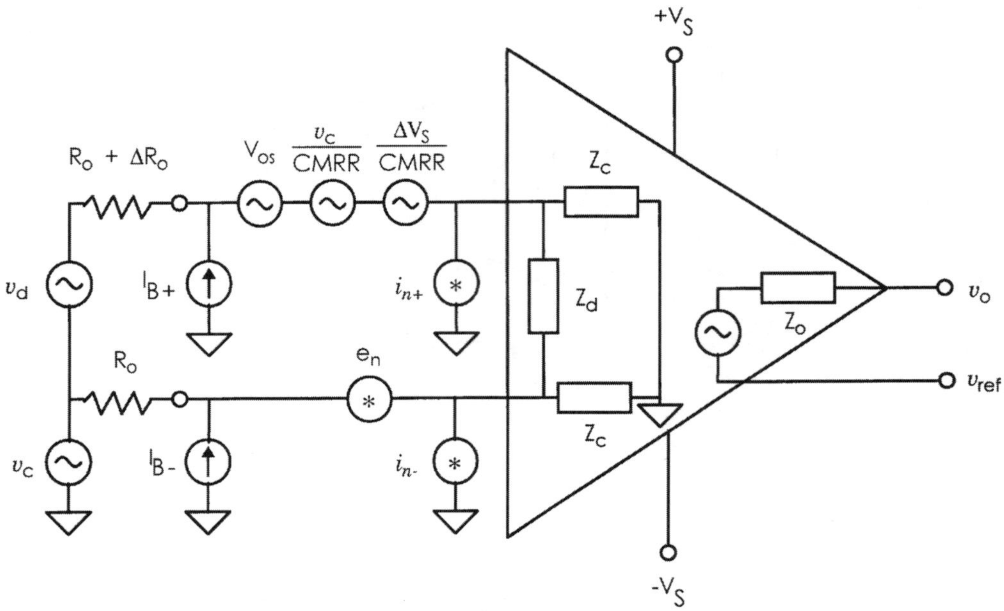

FIGURE 20.5 A model for a practical instrumentation amplifier including major error sources.

$$v_o = v_a + (v_{os} + v_b + v_r + v_n)G + v_{ref} \tag{20.5}$$

where v_a depends on the input voltage v_d, the second term includes offset, drift, noise, and interference-rejection errors, G is the designed gain, and v_{ref} is the reference voltage, commonly 0 V (but not necessarily, thus allowing output level shifting). Equation 20.5 describes a worst-case situation where absolute values for error sources are added. In practice, some cancellation between different error sources may happen.

Figure 20.5 shows a circuit model for *error analysis* when a practical IA is connected to a signal source (assumed to be differential for completeness). Impedance from each input terminal to ground (Z_c) and between input terminals (Z_d) are all finite. Furthermore, if the input terminals are both connected to ground, v_o is not zero and depends on G; this is modeled by V_{os}. If the input terminals are grounded through resistors, then v_o also depends on the value of these resistors; this is modeled by current sources I_{B+} and I_{B-}, which represent input bias or leakage currents. These currents need a return path, and therefore a third lead connecting the signal source to the amplifier, or a common ground, is required. Neither V_{os} nor I_{B+} nor I_{B-} is constant; rather, they change with temperature and time: slow changes (<0.01 Hz) are called drift and fast changes are described as noise (hence the noise sources e_n, i_{n+}, and i_{n-} in Figure 20.5). Common specifications for IAs are defined in Reference 3.

If a voltage v_c is simultaneously applied to both inputs, then v_o depends on v_c and its frequency. The *common-mode gain* is

$$G_c(f) = \frac{V_o(v_d = 0)}{V_c} \tag{20.6}$$

In order to describe the output voltage due to v_c as an input error voltage, we must divide the corresponding $v_o(v_c)$ by G (the normal- or differential-mode gain, $G = G_d$). The **common-mode rejection ratio** (CMRR) is defined as

$$\text{CMRR} = \frac{G_d(f)}{G_c(f)} \tag{20.7}$$

and is usually expressed in decibels ($\{\text{CMRR}\}_{\text{dB}}$ = 20 log CMRR). The input error voltage will be

$$\frac{v_o(v_c)}{G_d} = \frac{G_c v_c}{G_d} = \frac{v_c}{\text{CMRR}} \tag{20.8}$$

In the above analysis we have assumed $Z_c \ll R_o$; otherwise, if there were any unbalance (such as that for the source impedance in Figure 20.5), v_c at the voltage source would result in a differential-mode voltage at the amplifier input,

$$\begin{aligned} v_d(v_c) &= v_c\left(\frac{R_o + \Delta R_o}{Z_c + R_o + \Delta R_o} - \frac{R_o}{Z_c + R_o}\right) \\ &= v_c \frac{Z_c \Delta R_o}{(Z_c + R_o + \Delta R_o)(Z_c + R_o)} \approx v_c \frac{\Delta R_o}{Z_c} \end{aligned} \tag{20.9}$$

which would be amplified by G_d. Then, the effective common-mode rejection ratio would be

$$\frac{1}{\text{CMRR}_e} = \frac{\Delta R_o}{Z_c} + \frac{1}{\text{CMRR}} \tag{20.10}$$

where the CMRR is that of the IA alone, expressed as a fraction, not in decibels. Stray capacitances from input terminals to ground will decrease Z_c, therefore reducing CMRR_e.

The ideal amplifier is unaffected by power supply fluctuations. The practical amplifier shows output fluctuations when supply voltages change. For slow changes, the equivalent input error can be expressed as a change in input offset voltages in terms of the *power supply rejection ratio* (PSRR),

$$\text{PSRR} = \frac{\Delta V_{os}}{\Delta V_s} \tag{20.11}$$

The terms in Equation 20.5 can be detailed as follows. Because of gain errors we have

$$v_a = v_d\left(G + e_G + \frac{\Delta G}{\Delta T} \times \Delta T + e_{\text{NLG}}\right) \tag{20.12}$$

where G is the differential gain designed, e_G its absolute error, $\Delta G/\Delta T$ its thermal drift, ΔT the difference between the actual temperature and that at which the gain G is specified, and e_{NLG} is the nonlinearity gain error, which describes the extent to which the input/output relationship deviates from a straight line (insert in Figure 20.4b). The actual temperature T_J is calculated by adding to the current ambient temperature T_A the temperature rise produced by the power P_D dissipated in the device. This rise depends on the thermal resistance θ_{JA} for the case

$$T_J = T_A + P_D \times \theta_{JA} \tag{20.13}$$

where P_D can be calculated from the respective voltage and current supplies

$$P_D = |V_{S+}||I_{S+}| + |V_{S-}||I_{S-}| \tag{20.14}$$

The terms for the equivalent *input offset error* will be

$$v_{os} = V_{os}(T_a) + \frac{\Delta V_{os}}{\Delta T} \times (T_J - T_a) \tag{20.15}$$

$$v_{\rm b} = \left(I_{\rm B+} - I_{\rm B-}\right)R_{\rm o} + I_{\rm B+}\Delta R_{\rm o} = I_{\rm os}R_{\rm o} + I_{\rm B}\Delta R_{\rm o} \tag{20.16}$$

where $T_{\rm a}$ is the ambient temperature in data sheets, $I_{\rm os} = I_{\rm B+} - I_{\rm B-}$ is the offset current, $I_{\rm B} = (I_{\rm B+} + I_{\rm B-})/2$, and all input currents must be calculated at the actual temperature,

$$I = I(T_{\rm a}) + \frac{\Delta I}{\Delta T} \times \left(T_{\rm J} - T_{\rm a}\right) \tag{20.17}$$

Error contributions from finite *interference rejection* are

$$v_{\rm r} = \frac{v_{\rm c}}{\rm CMRR_e} + \frac{\Delta V_{\rm s}}{\rm PSRR} \tag{20.18}$$

where the $\rm CMRR_e$ must be that at the frequency for $v_{\rm c}$, and the PSRR must be that for the frequency of the ripple $\Delta V_{\rm s}$. It is assumed that both frequencies fall inside the bandpass for the signal of interest $v_{\rm d}$.

The equivalent *input voltage noise* is

$$v_{\rm n} = \sqrt{e_{\rm n}^2 B_{\rm e} + i_{\rm n}^2 - R_{\rm o}^2 B_{\rm i+} + i_{\rm n}^2 - R_{\rm o}^2 B_{\rm i-}} \tag{20.19}$$

where $e_{\rm n}^2$ is the voltage noise power spectral density of the IA, $i_{\rm n+}^2$ and $i_{\rm n-}^2$ are the current noise power spectral densities for each input of the IA, and $B_{\rm e}$, $B_{\rm i+}$, and $B_{\rm i-}$ are the respective noise equivalent bandwidths of each noise source. In Figure 20.5, the transfer function for each noise source is the same as that of the signal $v_{\rm d}$. If the signal bandwidth is determined as $f_{\rm h} - f_{\rm l}$ by sharp filters, then

$$B_{\rm e} = f_{\rm h} - f_{\rm l} + f_{\rm ce} \ln\frac{f_{\rm h}}{f_{\rm l}} \tag{20.20}$$

$$B_{\rm i+} = B_{\rm i-} = f_{\rm h} - f_{\rm l} + f_{\rm ci} \ln\frac{f_{\rm h}}{f_{\rm l}} \tag{20.21}$$

where $f_{\rm ce}$ and $f_{\rm ci}$ are, respectively, the frequencies where the value of voltage and current noise spectral densities is twice their value at high frequency, also known as corner or 3 dB frequencies.

Another noise specification method states the peak-to-peak noise at a given low-frequency band ($f_{\rm A}$ to $f_{\rm B}$), usually 0.1 to 10 Hz, and the noise spectral density at a frequency at which it is already constant, normally 1 or 10 kHz. In these cases, if the contribution from noise currents is negligible, the equivalent input voltage noise can be calculated from

$$v_{\rm n} = \sqrt{v_{\rm nL}^2 + v_{\rm nH}^2} \tag{20.22}$$

where $v_{\rm nL}$ and $v_{\rm nH}$ are, respectively, the voltage noise in the low-frequency and high-frequency bands expressed in the same units (peak-to-peak or rms voltages). To convert rms voltages into peak-to-peak values, multiply by 6.6. If the signal bandwidth is from $f_{\rm l}$ to $f_{\rm h}$, and $f_{\rm l} = f_{\rm A}$ and $f_{\rm h} > f_{\rm B}$, then Equation 20.22 can be written

$$v_{\rm n} = \sqrt{v_{\rm nL}^2 + (6.6e_{\rm n})^2 (f_{\rm h} - f_{\rm B})} \tag{20.23}$$

where $v_{\rm nL}$ is the peak-to-peak value and $e_{\rm n}$ is the rms voltage noise as specified in data books. Equation 20.23 results in a peak-to-peak calculated noise that is lower than the real noise, because noise spectral density is not constant from $f_{\rm B}$ up. However, it is a simple approach providing useful results.

For signal sources with high output resistors, thermal and excess noise from resistors (see Chapter 18) must be included. For first- and second-order filters, noise bandwidth is slightly larger than signal bandwidth. Motchenbacher and Connelly [4] show how to calculate noise bandwidth, resistor noise, and noise transfer functions when different from signal transfer functions.

Low-noise design always seeks the minimal bandwidth required for the signal. When amplifying low-frequency signals, if a large capacitor C_i is connected across the input terminals in Figure 20.5, then noise and interference having a frequency larger than $f_0 = 1/2\pi(2R_o)C_i$ ($f_0 \ll f_s$) will be attenuated.

Another possible source of error for any IA, not included in Equation 20.5, is the *slew rate limit* of its output stage. Because of the limited current available, the voltage at the output terminal cannot change faster than a specified value SR. Then, if the maximal amplitude A of an output sine wave of frequency f exceeds

$$A = \frac{SR}{2\pi f} \tag{20.24}$$

there will be a waveform distortion.

Table 20.1 lists some basic specifications for IC instrumentation amplifiers whose gain G can be set by an external resistor or a single connection.

Instrumentation Amplifiers Built from Discrete Parts

Instrumentation amplifiers can be built from discrete parts by using operational amplifiers (op amps) and a few resistors. An *op amp* is basically a differential voltage amplifier whose gain A_d is very large (from 10^5 to 10^7) at dc and rolls off (20 dB/decade) from frequencies of about 1 to 100 Hz, becoming 1 at frequencies from 1 to 10 MHz for common models (Figure 20.6a), and whose input impedances are so high (up to $10^{12}\ \Omega \parallel 1$ pF) that input currents are almost negligible. Op amps can also be modeled by the circuit in Figure 20.5, and their symbol is that in Figure 20.4a, deleting IA. However, because of their large gain, op amps cannot be used directly as amplifiers; a mere 1 mV dc input voltage would saturate any op amp output. Furthermore, op amp gain changes from unit to unit, even for the same model, and for a given unit it changes with time, temperature, and supply voltages. Nevertheless, by providing external feedback, op amps are very flexible and far cheaper than IAs. But when the cost for external components and their connections, and overall reliability are also considered, the optimal solution depends on the situation.

Figure 20.6b shows an amplifier built from an op amp with external feedback. If input currents are neglected, the current through R_2 will flow through R_1 and we have

$$v_d = v_s - v_o \frac{R_1}{R_1 + R_2} \tag{20.25}$$

$$v_o = A_d v_d \tag{20.26}$$

Therefore,

$$\frac{v_o}{v_s} = \frac{A_d\left(1 + \frac{R_2}{R_1}\right)}{A_d + 1 + \frac{R_2}{R_1}} = \frac{G_i}{1 + \frac{G_i}{A_d}} \tag{20.27}$$

where $G_i = 1 + R_2/R_1$ is the ideal gain for the amplifier. If G_i/A_d is small enough (G_i small, A_d large), the gain does not depend on A_d but only on external components. At high frequencies, however, A_d becomes smaller and, from Equation 20.27, $v_o < G_i v_s$ so that the bandwidth for the amplifier will reduce for large gains. Franco [5] analyzes different op amp circuits useful for signal conditioning.

TABLE 20.1 Basic Specifications for Some Instrumentation Amplifiers

	AD624A	AMP02F	INA110KP	LT1101AC	Units
Gain range	1–1000	1–1000	1–500	10,100	V/V
Gain error, e_G					
$G = 1$	±0.05	0.05	±0.02	n.a.	%
$G = 10$	n.s.	0.40	±0.05	±0.04	%
$G = 100$	±0.25	0.50	±0.10	±0.04	%
$G = 1000$	±1.0	0.70	n.a.	n.a.	%
Gain nonlinearity error e_{NLG}[a]					
$G = 1$	±0.005	0.006	±0.005	n.a.	%
$G = 10$	n.s.	0.006	±0.005	±0.0008	%
$G = 100$	±0.005	0.006	±0.01	±0.0008	
$G = 1000$	±0.005	0.006	n.a.	n.a.	%
Gain drift $\Delta G/\Delta T$					
$G = 1$	5	50	±10	n.a.	µV/V/°C
$G = 10$	n.s.	50	±10	5	µV/V/°C
$G = 100$	10	50	±20	5	µV/V/°C
$G = 1000$	25	50	n.a.	n.a.	µV/V/°C
V_{os}	200 + 5/G	200	±(1000 + 5000/G)	160	µV
$\Delta v_{os}/\Delta T$	2 + 50/G	4	±(2 + 50/G)	2	µV/°C
I_B	±50	20	0.05	10	nA
$\Delta I_B/\Delta T$	±50 typ	250 typ	b	30	pA/°C
I_{os}	±35	10	0.025	0.90	nA
$\Delta I_{os}/\Delta T$	±20 typ	15 typ	n.s.	7.0	pA/°C
Z_d	1 ‖ 10 typ	10 typ	5000 ‖ 6 typ	12	GΩ
Z_c	1 ‖ 10 typ	16.5 typ	2000 ‖ 1 typ	7	GΩ
CMRR at dc					
$G = 1$	70 min	80 min	70 min	n.a.	dB
$G = 10$	n.s.	100 min	87 min	82	dB
$G = 100$	100 min	115 min	100 min	98	dB
$G = 1000$	110 min	115 min	n.a.	n.a.	dB
PSRR at dc					
$G = 1$	70 min	80 min	c	n.a.	dB
$G = 10$	n.s.	100 min	c	100	dB
$G = 100$	95 min	115 min	c	100	dB
$G = 1000$	100 min	115 min	n.a.	n.a.	dB
Bandwidth (–3 dB) (typ)					
$G = 1$	1000	1200	2500	n.a.	kHz
$G = 10$	n.s.	300	2500	37	kHz
$G = 100$	150	200	470	3.5	kHz
$G = 1000$	25	200	n.a.	n.a.	kHz
Slew rate (typ)	5.0	6	17	0.1	V/µs
Settling time to 0.01%					
$G = 1$	15 typ	10 typ	12.5	n.a.	µs
$G = 10$	15 typ	10 typ	7.5	n.a.	µs
$G = 100$	15 typ	10 typ	7.5	n.a.	µs
$G = 1000$	75 typ	10 typ	n.a.	n.a.	µs
e_n (typ)					
$G = 1$	4	120	66	n.a.	nV/$\sqrt{Hz}$
$G = 10$	4	18	12	43	nV/$\sqrt{Hz}$
$G = 100$	4	10	10	43	nV/$\sqrt{Hz}$
$G = 1000$	4	9	n.a.	n.a.	nV/$\sqrt{Hz}$
v_n 0.1 to 10 Hz (typ)					
$G = 1$	10	10	1	0.9	µVp-p
$G = 10$	n.s.	1.2	1	0.9	µVp-p
$G = 100$	0.3	0.5	1	0.9	µVp-p
$G = 1000$	0.2	0.4	1	0.9	µVp-p

TABLE 20.1 (continued) Basic Specifications for Some Instrumentation Amplifiers

	AD624A	AMP02F	INA110KP	LT1101AC	Units
i_n 0.1 to 10 Hz (typ)	60	n.s.	n.s.	2.3	pAp-p
i_n (typ)	n.s.	400	1.8	20	$fA/\sqrt{Hz}$

Note: All parameter values are maximum, unless otherwise stated (typ = typical; min = minimum; n.a. = not applicable; n.s. = not specified). Measurement conditions are similar; consult manufacturers' data books for further detail.

[a] For the INA110, the gain nonlinearity error is specified as percentage of the full-scale output.

[b] Input current drift for the INA110KP approximately doubles for every 10°C increase, from 25°C (10 pA-typ) to 125°C (10 nA-typ).

[c] The PSRR for the INA110 is specified as an input offset ±(10 + 180/*G*) μV/V maximum.

Figure 20.7 shows an IA built from three op amps. The input stage is fully differential and the output stage is a *difference amplifier* converting a differential voltage into a single-ended output voltage. Difference amplifiers (op amp and matched resistors) are available in IC form: AMP 03 (Analog Devices) and INA 105/6 and INA 117 (Burr-Brown). The gain equation for the complete IA is

$$G = \left(1 + 2\frac{R_2}{R_1}\right)\frac{R_4}{R_3} \tag{20.28}$$

Pallás-Areny and Webster [6] have analyzed matching conditions in order to achieve a high CMRR. Resistors R_2 do not need to be matched. Resistors R_3 and R_4 need to be closely matched. A potentiometer connected to the v_{ref} terminal makes it possible to trim the CMRR at low frequencies.

The *three-op-amp IA* has a symmetrical structure making it easy to design and test. IAs based on an IC difference amplifier do not need any user trim for high CMRR. The circuit in Figure 20.8 is an IA that lacks these advantages but uses only two op amps. Its gain equation is

$$G = 1 + \frac{R_2}{R_1} + \frac{2R_2}{R_G} \tag{20.29}$$

R_1 and R_2 must be matched and R_G should be comparable to R_2.

Another approach to build an IA is by the *switched-capacitor* technique (Figure 20.9). Switches SW1 and SW2 close together and charge C_S (1 μF) to the voltage difference $v_H - v_L$; next, SW1 and SW2 open and SW3 and SW4 close, so that C_H (0.1 to 1 μF) also charges to $v_H - v_L$. Then SW1 and SW2 close again, SW3 and SW4 open, and so on. While C_S is being charged C_H holds the previous voltage difference. Therefore, the maximal frequency for the input signal must be at least ten times lower than the switching frequency. This circuit has a high CMRR because the charge at C_S is almost insensitive to the input common-mode voltage. Furthermore, it converts the differential signal to a single-ended voltage. The LTC 1043 (Linear Technology) includes two sets of four switches to implement this circuit.

Composite Instrumentation Amplifiers

Instrumentation amplifiers have a very limited bandwidth. They achieve a gain of 10 at 2.5 MHz, at best. Moreover, their inputs must be either dc-coupled or, if ac-coupled with input series capacitors, there must be a path for bias currents; if that path is a resistor from each input to ground, then the common-mode input impedance Z_c decreases and noise may increase.

A larger bandwidth for a given gain can be obtained by cascade connection of two or more amplifiers. However, if the additional gain is provided by a single-ended amplifier after the IA, then the overall CMRR is that of the IA, which is small at high frequencies. The circuit in Figure 20.10a is a broadband IA with large CMRR because the CMRR for the second stage is multiplied by the differential gain for the first stage, which can be very high if implemented by broadband op amps. The overall gain is

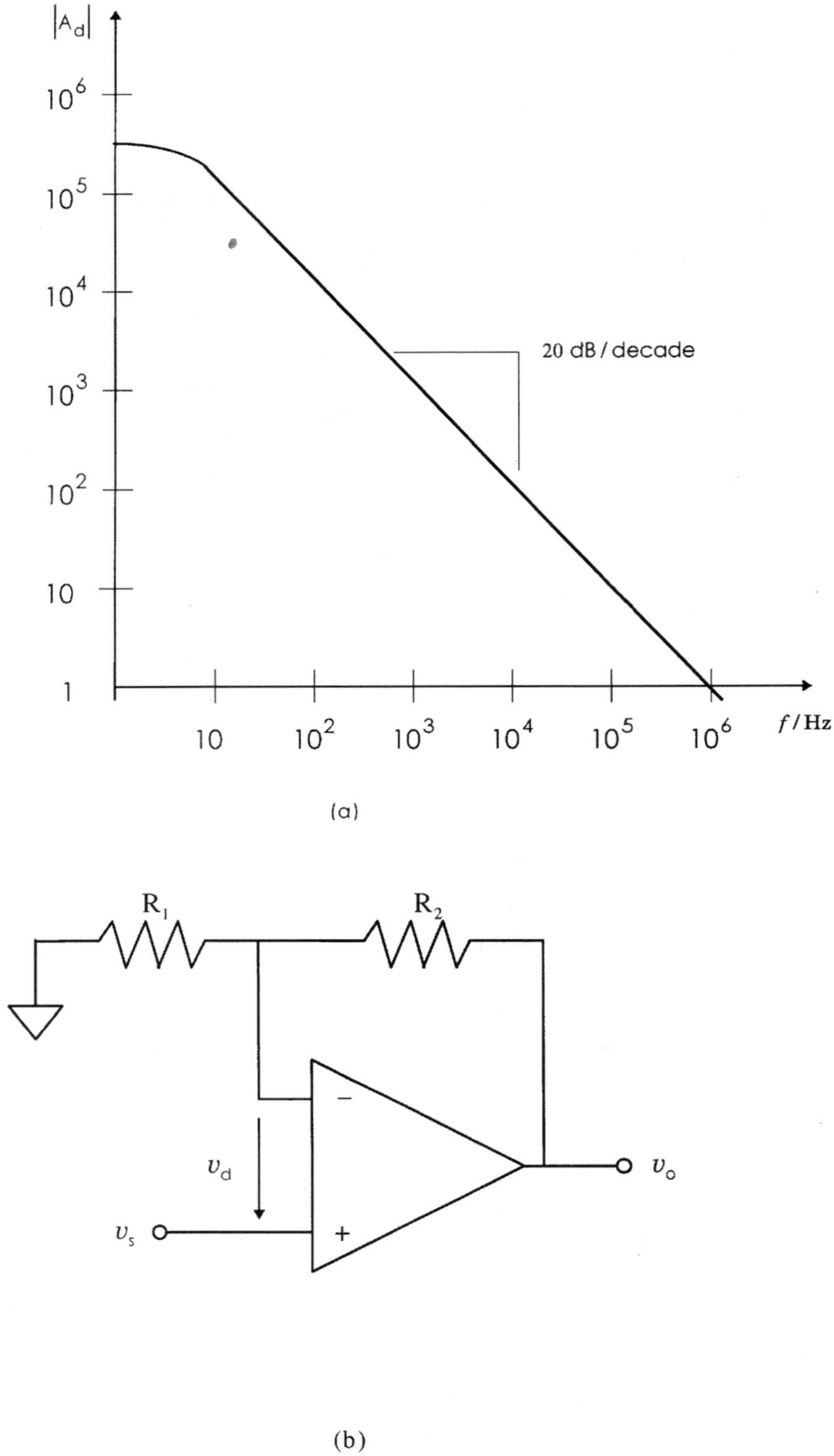

FIGURE 20.6 (a) Open loop gain for an op amp. (b) Amplifier based on an op amp with external feedback.

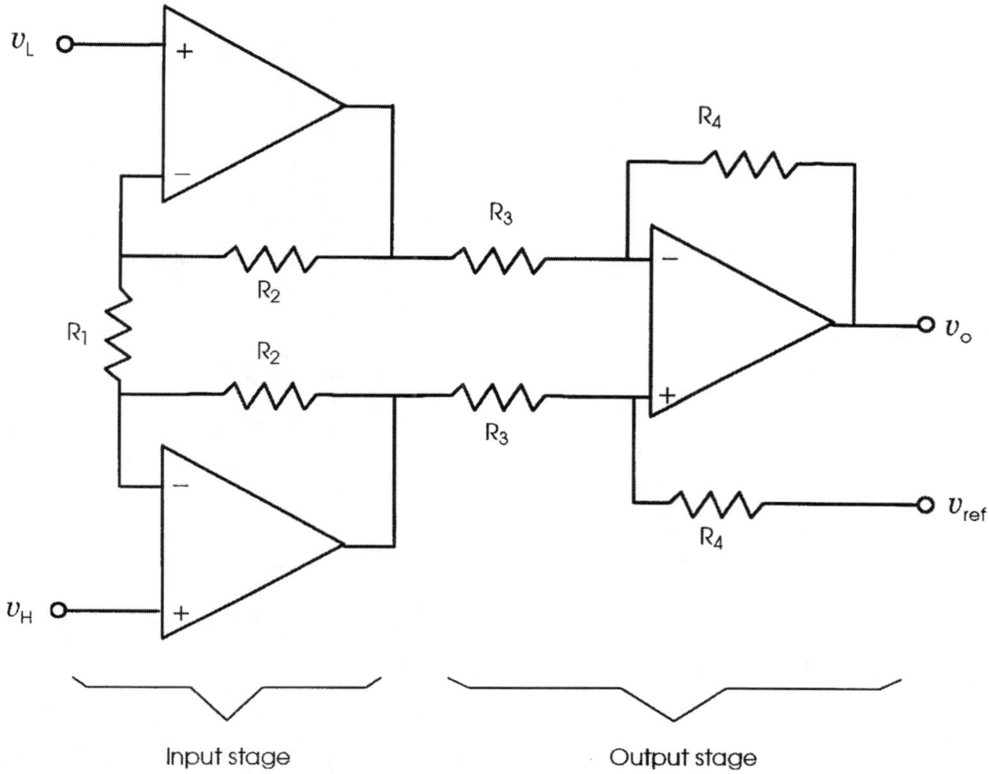

FIGURE 20.7 Instrumentation amplifier built from three op amps. R_3 and R_4 must be matched.

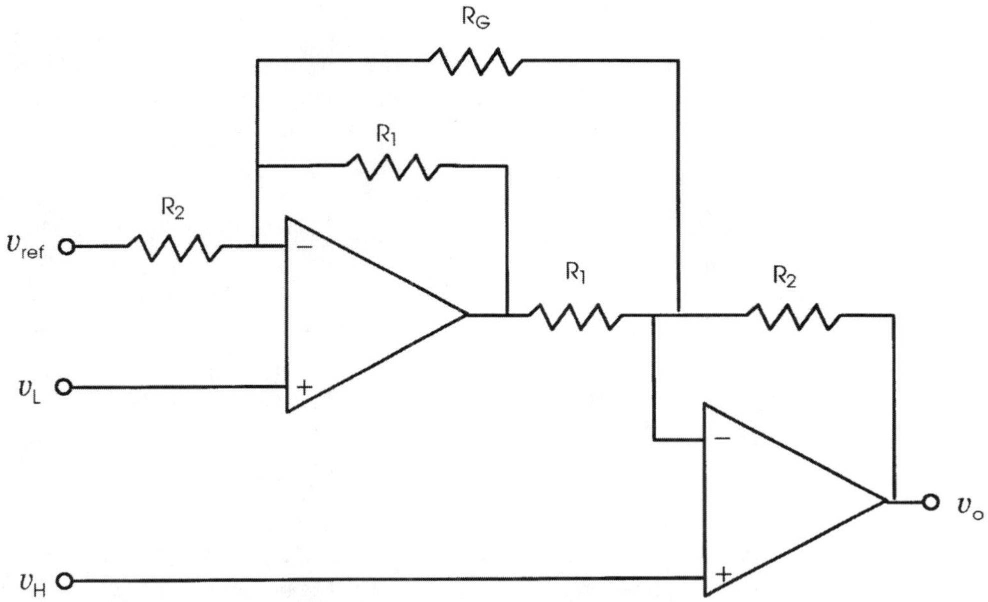

FIGURE 20.8 Instrumentation amplifier built from two op amps. R_1 and R_2 must be matched.

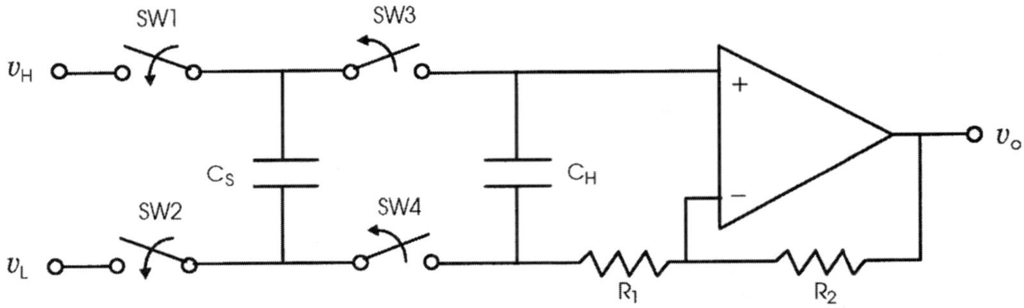

FIGURE 20.9 Instrumentation amplifier based on the switched-capacitor technique. First switches SW1 and SW2 close while SW3 and SW4 are open, and *CS* charges to $v_H - v_L$. Then SW1 and SW2 open and SW3 and SW4 close, charging C_H to $v_H - v_L$.

$$G = G_1 G_2 = \left(1 + \frac{2R_b}{R_a}\right) G_{IA} \tag{20.30}$$

An *IA* can be *ac-coupled* by feeding back its dc output to the reference terminal as shown in Figure 20.10b. The high-pass corner frequency is $f_0 = 1/(2\pi R_0 C_0)$.

20.6 Single-Ended Signal Conditioners

Floating signals (single ended or differential) can be connected to amplifiers with single-ended grounded input. Grounded single-ended can be connected to *single-ended amplifiers,* provided the difference in ground potentials from signal to amplifier is not too large. Figure 20.11a shows a simple single-ended amplifier based on an IA. However, op amps are better suited than IAs for single-ended amplifiers and signal conditioners performing additional functions.

Figure 20.11b shows an *inverting amplifier* whose gain is $G = -R_2/R_1$, and whose input impedance is R_1. The capacitor on the dashed line (10 pF or larger) prevents gain peaking and oscillation. If a capacitor C replaces R_2, input signals are integrated and inverted. If C replaces R_1 instead, input signals are differentiated and inverted. The circuit in Figure 20.11c has $G = 1$ for dc and signals of low frequency relative to $f_1 = 1/(2\pi R_1 C_1)$ (offset and drift included) and $G = 1 + R_2/R_1$ for high-frequency signals. The circuit in Figure 20.11d calculates the average for n voltages. The difference between two voltages can be obtained from the difference amplifier in Figure 20.7 (output stage).

Op amps must be carefully selected according to the application. For dc circuits, chopper op amps offer the best performance. For low-impedance signals, op amps with bipolar input transistors are better. For high-impedance signals, op amps with FET input transistors offer lower input currents, but they have larger drift and voltage noise. Table 20.2 lists some parameters for several different op amps. Some manufacturers provide selection guides on floppy disk which suggest the most appropriate model for a set of user-defined values for some parameters.

20.7 Carrier Amplifiers

A **carrier amplifier** is a conditioner for extremely narrowband ac signals from ac-driven sensors. A carrier amplifier is made of a sine wave oscillator, to excite the sensor bridge, an ac voltage amplifier for the bridge output, a synchronous demodulator (see Chapter 24), and a low-pass filter (Figure 20.12). The NE5520/1 (Philips) are carrier amplifiers in IC form intended for (but not limited to) LVDTs driven at a frequency from 1 to 20 kHz.

Carrier amplifiers make it possible to recover the amplitude and phase of the modulating signal after amplifying the output modulated waveform from the bridge. This is useful first because ac amplifiers

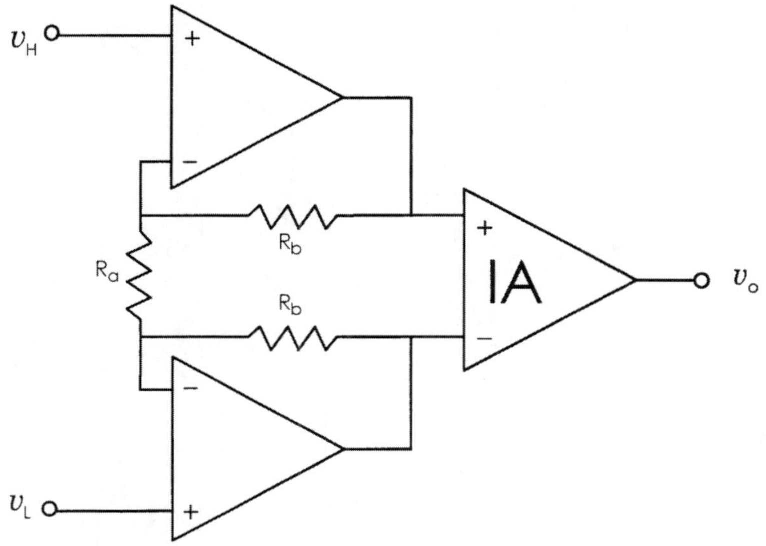

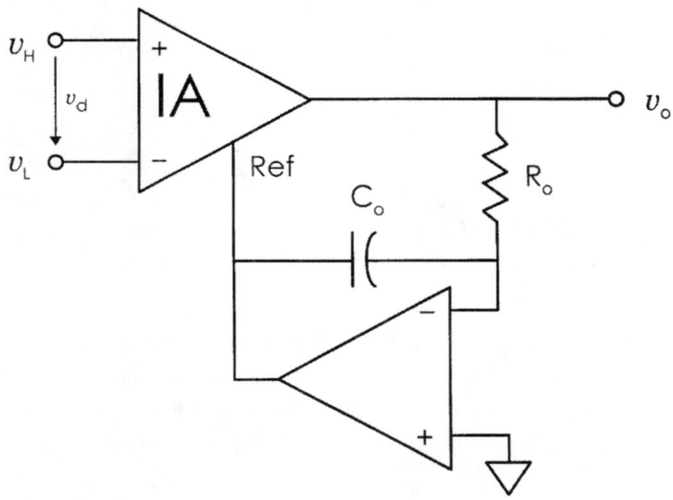

FIGURE 20.10 Composite instrumentation amplifiers. (a) Broadband IA with large CMRR; (b) ac-coupled IA.

are not affected by offset, drift, or low-frequency noise, and therefore the bridge output can easily be amplified. Second, the *phase-sensitive demodulator* yields not only the amplitude but also the sign of the measurand. If the measurement range includes positive and negative values for the measurand, phase detection is essential.

A further advantage of carrier amplifiers is their extremely narrow frequency response, determined by the output low-pass filter. In the demodulator, the product of the modulated carrier of frequency f_c by the reference signal, also of frequency f_c, results in a baseband component and components at nf_c ($n \geq 2$). The output low-pass filter rejects components other than the baseband. If the corner frequency for this

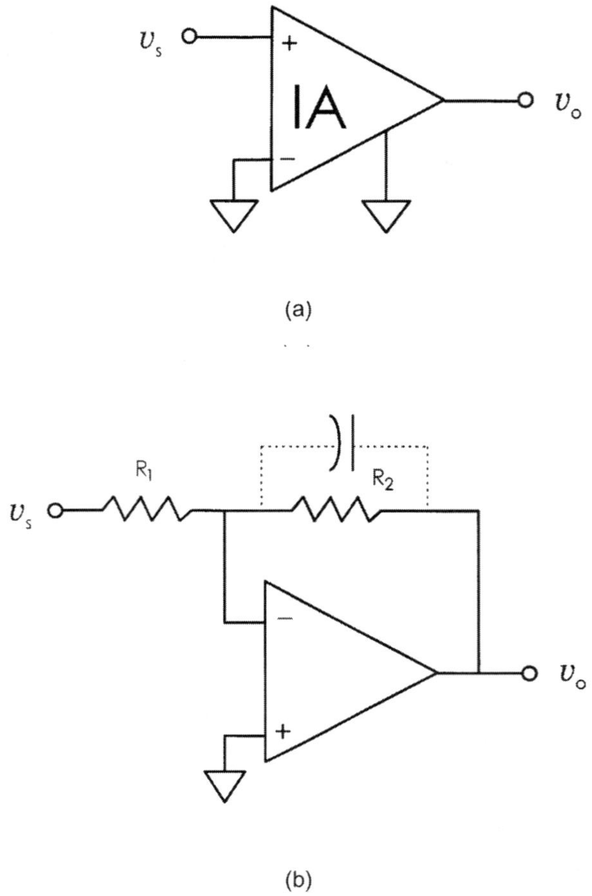

FIGURE 20.11 Single-ended amplifiers and signal conditioners. (a) Noninverting amplifier based on an IA; (b) inverting amplifier based on an op amp; (c) ac amplifier; (d) voltage averager.

filter is f_0, then the passband for the system is $f_c \pm f_o$. Therefore, any interference of frequency f_i added to the modulated signal will be rejected if falling outside that passband. The ability to discriminate signals of interest from those added to them is described by the *series* (or *normal*) *mode rejection ratio* (SMRR), and is usually expressed in decibels. In the present case, using a first-order low-pass filter we have

$$\text{SMRR} = 20 \log \frac{v_o(f_c)}{v_o(f_i)} = 20 \log \frac{\sqrt{1+(f_c - f_i)^2}}{f_0} \approx 20 \log \frac{|f_c - f_i|}{f_0} \qquad (20.31)$$

A power-line interference superimposed on a 10 kHz carrier will undergo an 80-dB attenuation if the output low-pass filter has $f_0 = 1$ Hz. The same interference superimposed on the baseband signal would be attenuated by only 35 dB.

Carrier amplifiers can be built from a precision sine wave oscillator — AD2S99 (Analog Devices), 4423 (Burr-Brown), SWR300 (Thaler) — or a discrete-part oscillator, and a demodulator (plus the output filter). Some IC demodulators are based on switched amplifiers (AD630, OPA676). The floating capacitor in Figure 20.9 behaves as a synchronous demodulator if the switch clock is synchronous with the carrier, and its duty cycle is small (less than 10%), so that switches SW1 and SW2 sample the incoming modulated waveform for a very short time [7].

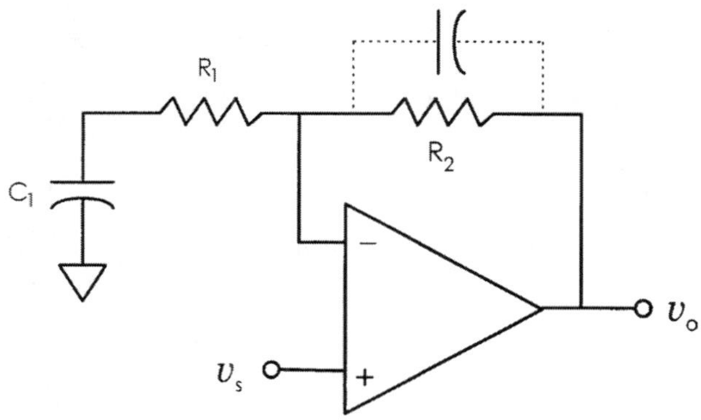

(c)

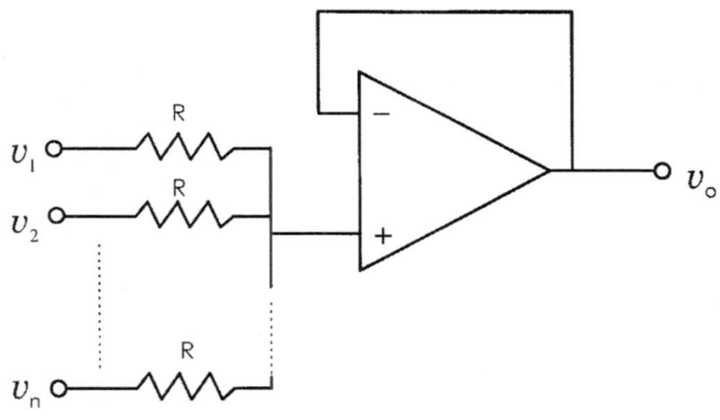

(d)

FIGURE 20.11 (continued)

20.8 Lock-In Amplifiers

A *lock-in amplifier* is based on the same principle as a carrier amplifier, but instead of driving the sensor, here the carrier signal drives the experiment, so that the measurand is frequency translated. Lock-in amplifiers are manufactured as equipment intended for recovering signals immersed in high (asynchronous) noise. These amplifiers provide a range of driving frequencies and bandwidths for the output filter. Some models are vectorial because they make it possible to recover the in-phase and quadrature (90° out-of-phase) components of the incoming signal, by using two demodulators whose reference signals are delayed by 90°. Still other models use bandpass filters for the modulated signal and two demodulating stages. Meade [8] analyzes the fundamentals, specifications, and applications of some commercial lock-in amplifiers.

20.9 Isolation Amplifiers

The maximal common-mode voltage withstood by common amplifiers is smaller than their supply voltage range and seldom exceeds 10 V. Exceptions are the INA 117 (Burr-Brown) and similar difference amplifiers whose common-mode range is up to ±200 V, and the IA in Figure 20.9 when implemented by high-voltage

TABLE 20.2 Basic Specifications for Operational Amplifiers of Different Technologies

	V_{os}, μV	$(\Delta v_{os}/\Delta T)_{av}$, μV/°C	I_B, pA	$\Delta I_B/\Delta T$, pA/°C	I_{os}, pA	$BW_{typ}(G = 1)$, MHz	e_n(1 kHz), $nV/\sqrt{Hz}$	f_{ce}, Hz	$v_{n(p\text{-}p)}$, μV	i_n(1 kHz), $fA/\sqrt{Hz}$
					Bipolar					
μA741	6000	15	500000	500	200000	1.5	20	200	—	550
LM358A	3000	20	100000	—	±30000	1	—	—	—	—
LT1028	80	0.8	180000	—	100000	75	0.9	3.5	0.035	1000
OP07	75	1.3	3000	50	2800	0.6	9.6	10	0.35	170
OP27C	100	1.8	80000	—	75000	8	3.2	2.7	0.09	400
OP77A	25	0.3	2000	25	1500	0.6	9.6	10	0.35	170
OP177A	10	0.1	1500	25	1000	0.6	—	—	0.8	—
TLE2021C	600	2	70000	80	3000	1.2	30	—	0.47	90
TLE2027C	100	1	90000	—	90000	13	2.5	—	0.05	400
					FET Input					
AD549K	250	5	0.1	b	0.03 typ	1	35	—	4	0.16
LF356A	2000	5	50	b	10	4.5	12	—	—	10
OPA111B	250	1	1	b	0.75	2	7	200	1.2	0.4
OPA128J	1000	20	0.3	b	65	1	27	—	4	0.22
TL071C	10000	18	200	b	100	3	18	300	4	10
TLE2061C	3000	6	4 typ	b	2 tip	2	60	20	1.2	1
					CMOS					
ICL7611A	2	10 typ	50	b	30	0.044	100	800	—	10
LMC660C	6000	1.3 typ	20	b	20	1.4	22	—	—	0.2
LMC6001A	350	10	0.025	b	0.005	1.3	22	—	—	0.13
TLC271CP	10000	2 typ	0.7 typ	c	0.1 typ	2.2	25	100	—	n.s.
TLC2201C	500	0.1 typ	1 typ	d	0.5 typ	1.8	8	—	0.7	0.6
					BiMOS					
CA3140	15000	8	50	b	30	4.5	40	—	—	—
					CMOS Chopper					
LTC1052	5	0.05	30	e	30	1.2	—	—	1.5	0.6
LTC1150C	5	0.05	100	f	200	2.5	—	—	1.8	1.8
MAX430C	10	0.05	100	g	200	0.5	—	—	1.1	10
TLC2652AC	1	0.03	4 typ	d	2 typ	1.9	23	—	2.8	4
TLC2654C	20	0.3	50 typ	0.65	30 typ	1.9	13	—	1.5	4
TSC911A	15	0.15	70	—	20	1.5	—	—	11	—

Specified values are maximal unless otherwise stated and those for noise, which are typical (typ = typical, av = average; nonspecified parameters are indicated by a dash).

[a] Values estimated from graphs.

[b] I_B doubles every 10°C.

[c] I_B doubles every 7.25°C.

[d] I_B is almost constant up to 85°C.

[e] I_B is almost constant up to 75°C.

[f] I_{B+} and I_{B-} show a different behavior with temperature.

[g] I_B doubles every 10°C above about 65°C.

switches (relays, optorelays). Signals with large off-ground voltages, or differences in ground potentials exceeding the input common-mode range, result in permanent amplifier damage or destruction, and a safety risk, in spite of an exceptional CMRR: a 100 V common-mode 60 Hz voltage at the input of a common IA having a 120 dB CMRR at power-line frequency does not result in a 100 V/10^6 = 100 μV output, but a burned-out IA.

Figure 20.13a shows a signal source grounded at a point far from the amplifier ground. The difference in voltage between grounds v_i not only thwarts signal measurements but can destroy the amplifier. The

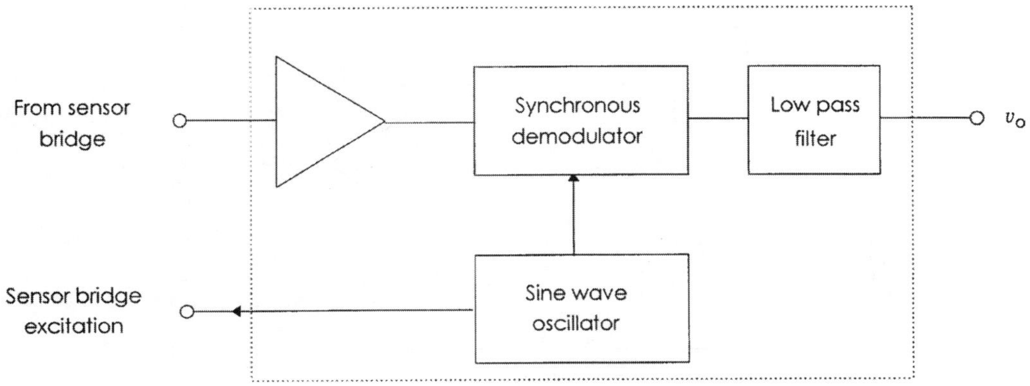

FIGURE 20.12 Elements for a carrier amplifier.

solution is to prevent this voltage from forcing any large current through the circuit and at the same time to provide an information link between the source and the amplifier. Figure 20.13b shows a solution: the signal source and the amplifier have separated (isolated) power supplies and the signal is coupled to the amplifier through a transformer acting as an isolation barrier for v_i. Other possible barriers are optocouplers (IL300-Siemens) and series capacitors (LTC1145-Linear Technology). Those barriers impose a large series impedance (isolation impedance, Z_i) but do not usually have a good low-frequency response, hence the need to modulate and then demodulate the signal to transfer through it. The subsystem made of the modulator and demodulator, plus sometimes an input and an output amplifier and a dc–dc converter for the separate power supply, is called an **isolation amplifier.** The ability to reject the voltage difference across the barrier (isolation-mode voltage, v_i) is described by the **isolation mode rejection ratio** (IMRR), expressed in decibels,

$$\text{IMRR} = 20 \log \frac{\text{OUTPUT Voltage}}{\text{ISOLATION-MODE Voltage}} \tag{20.32}$$

Ground isolation also protects people and equipment from contact with high voltage because Z_i limits the maximal current. Some commercial isolation amplifiers are the AD202, AD204, and AD210 (Analog Devices) and the ISOxxx series (Burr-Brown).

Table 20.3 summarizes the compatibility between signal sources and amplifiers. When grounded, amplifiers and signals are assumed to be grounded at different physical points.

20.10 Nonlinear Signal-Processing Techniques

Limiting and Clipping

Clippers or *amplitude limiters* are circuits whose output voltage has an excursion range restricted to values lower than saturation voltages. Limiting is a useful signal processing technique for signals having the information encoded in parameters other than the amplitude. For example, amplitude limiting is convenient before homodyne phase demodulators. *Limiting* can also match output signals levels to those required for TTL circuits (0 to 5 V). Limiting avoids amplifier saturation for large input signal excursions, which would result in a long recovery time before returning to linear operation.

Limiting can be achieved by op amps with diodes and zeners in a feedback loop. Figure 20.14a shows a positive voltage clipper. When v_s is positive, v_o is negative and R_2/R_1 times larger; the diode is reverse biased and the additional feedback loop does not affect the amplifier operation. When v_s is negative and large enough, v_o forward biases the diode (voltage drop V_f) and the zener clamps at V_z, the output amplitude thus being limited to $v_o = V_f + V_z$ until $|v_s| < (V_f + V_z)R_1/R_2$ (Figure 20.14b). The circuit then acts again as an inverting amplifier until v_s reaches a large negative value.

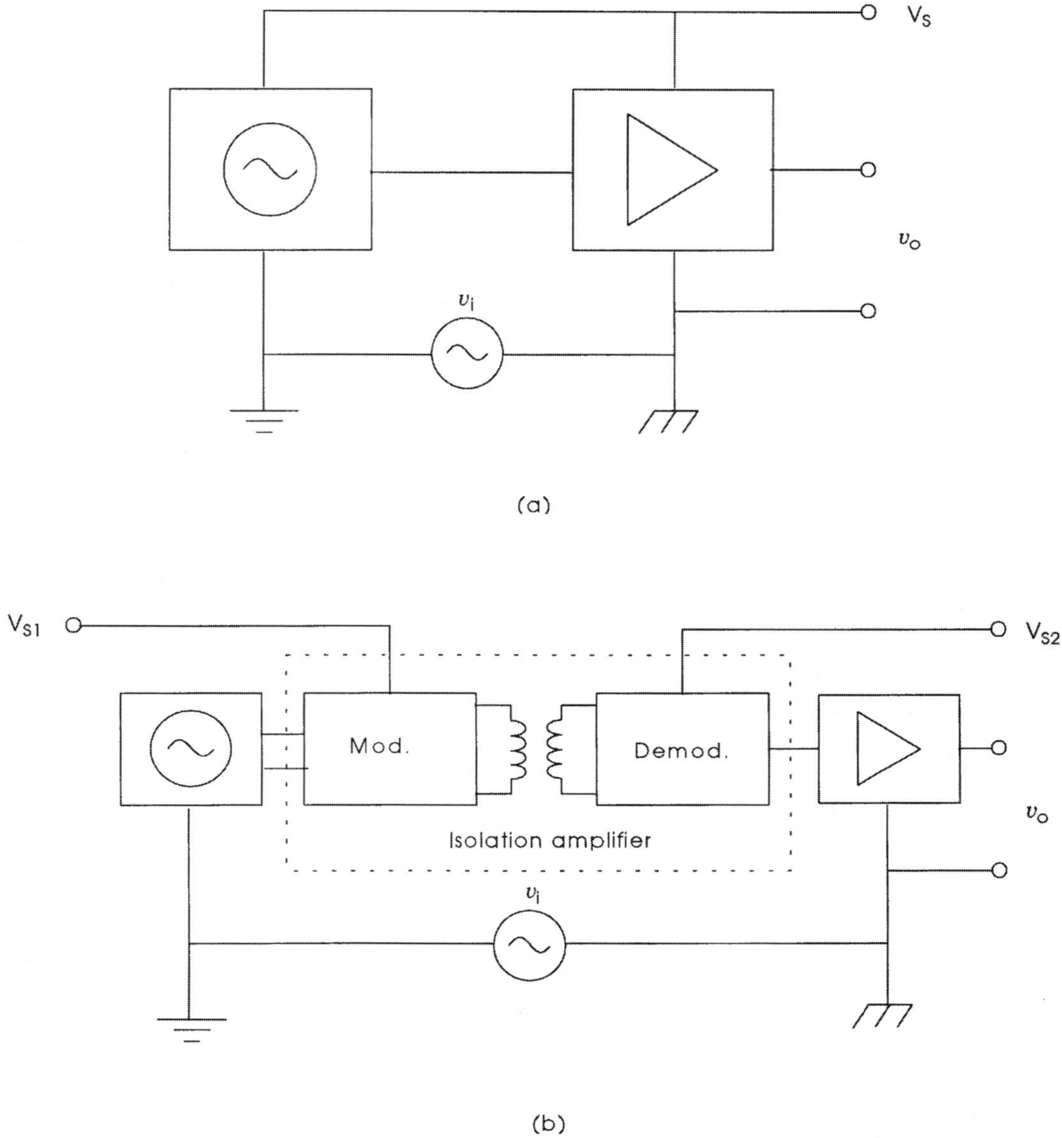

FIGURE 20.13 (a) A large difference in ground potentials damages amplifiers. (b) An isolation amplifier prevents large currents caused by this difference from flowing through the circuit.

A negative voltage clipper can be designed by reversing the polarity of the diode and zener. To limit the voltage in both directions, the diode may be substituted by another zener diode. The output is then limited to $|v_o| < V_{z1} + V_{f2}$ for negative inputs to $|v_o| < V_{f1} + V_{z2}$ for positive inputs. If $V_{z1} = V_{z2}$, then the voltage limits are symmetrical. Jung [9] gives component values for several precision limiters.

Logarithmic Amplification

The dynamic range for common linear amplifiers is from 60 to 80 dB. Sensors such as photodetectors, ionizing radiation detectors, and ultrasound receivers can provide signals with an amplitude range wider than 120 dB. The only way to encompass this wide amplitude range within a narrower range is by amplitude compression. A logarithmic law compresses signals by offering equal-output amplitude changes in response to a given ratio of input amplitude increase. For example, a scaling of 1 V/decade means that the output would change by 1 V when the input changes from 10 to 100 mV, or from 100 mV

TABLE 20.3 Compatibility between Signal Sources and Conditioners

Conditioner input \ Signal source	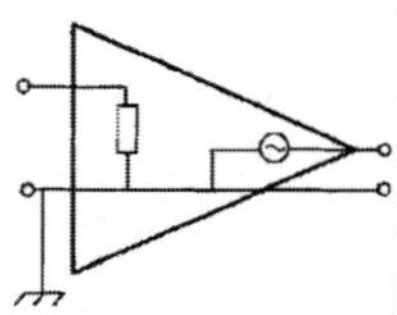	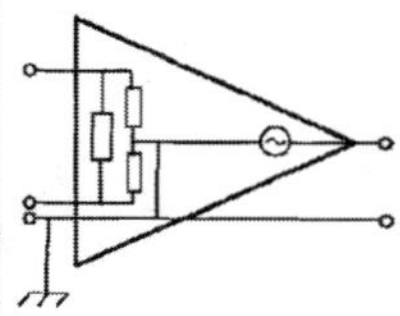	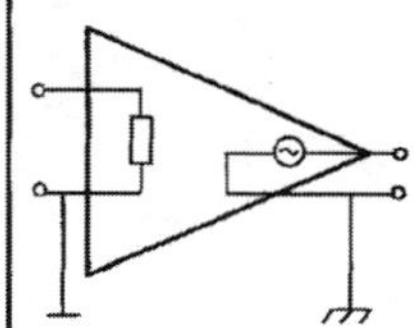	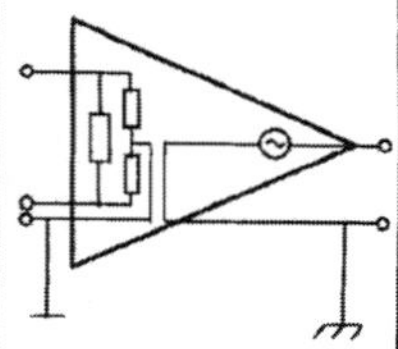
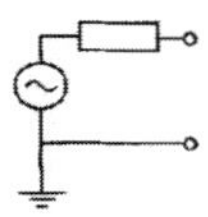	Incompatible unless grounds are very close	Compatible if CMRR is large	Compatible	Compatible
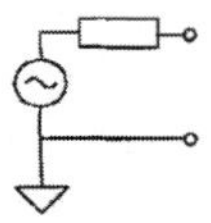	Compatible	Compatible	Compatible	Compatible
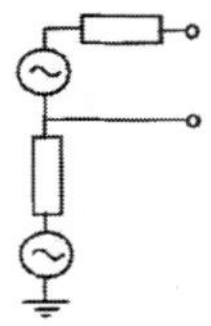	Incompatible unless grounds are very close	Compatible if CMRR is large	Compatible for large Z_i	Compatible
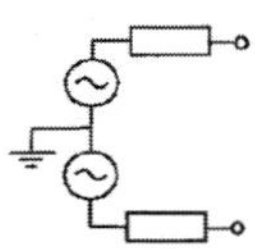	Incompatible	Compatible	Compatible for large Z_i	Compatible
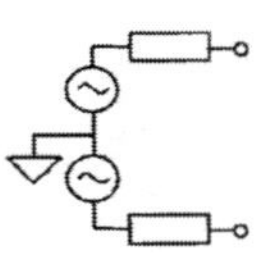	Compatible	Compatible	Compatible	Compatible
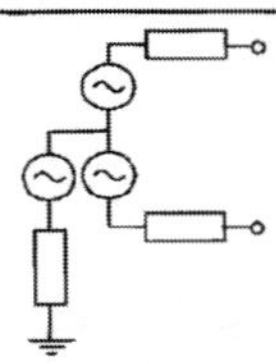	Incompatible	Compatible if CMRR is large	Compatible for large Z_i	Compatible

Note: When grounded, signals sources and amplifiers are assumed to be grounded at different points. Isolation impedance is assumed to be very high for floating signal sources but finite (Z_i) for conditioners.

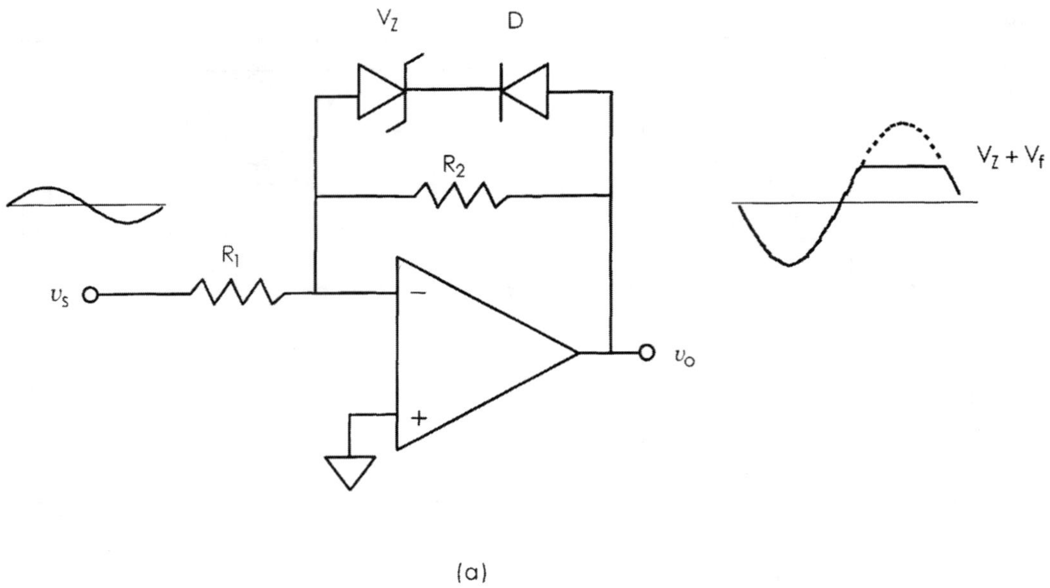

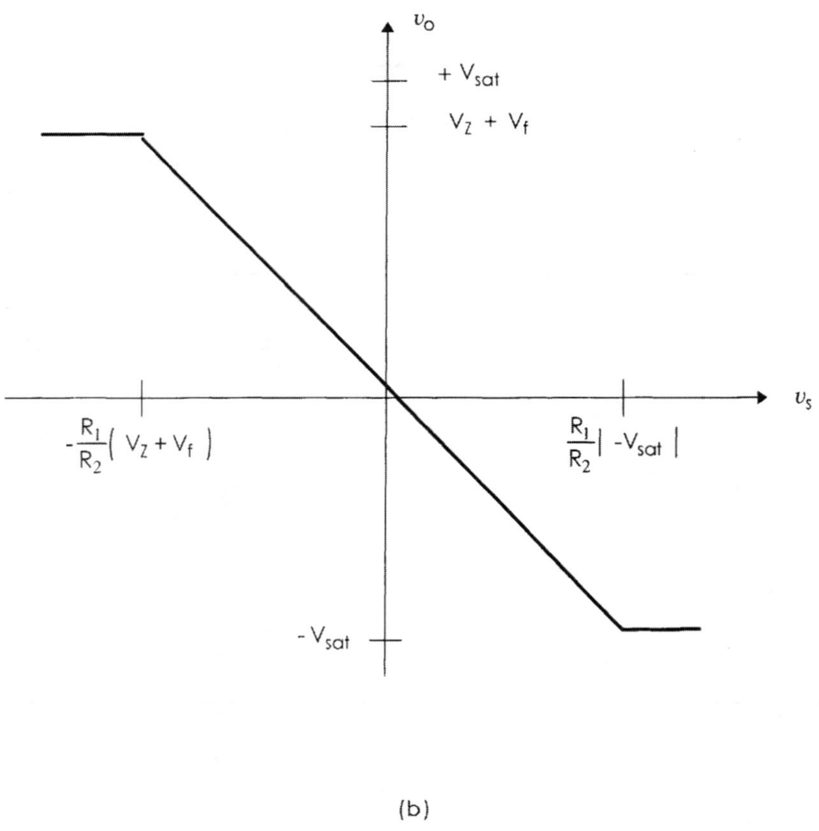

FIGURE 20.14 Voltage limiter. (a) Circuit based on op amp and diode network feedback. (b) Input/output relationship.

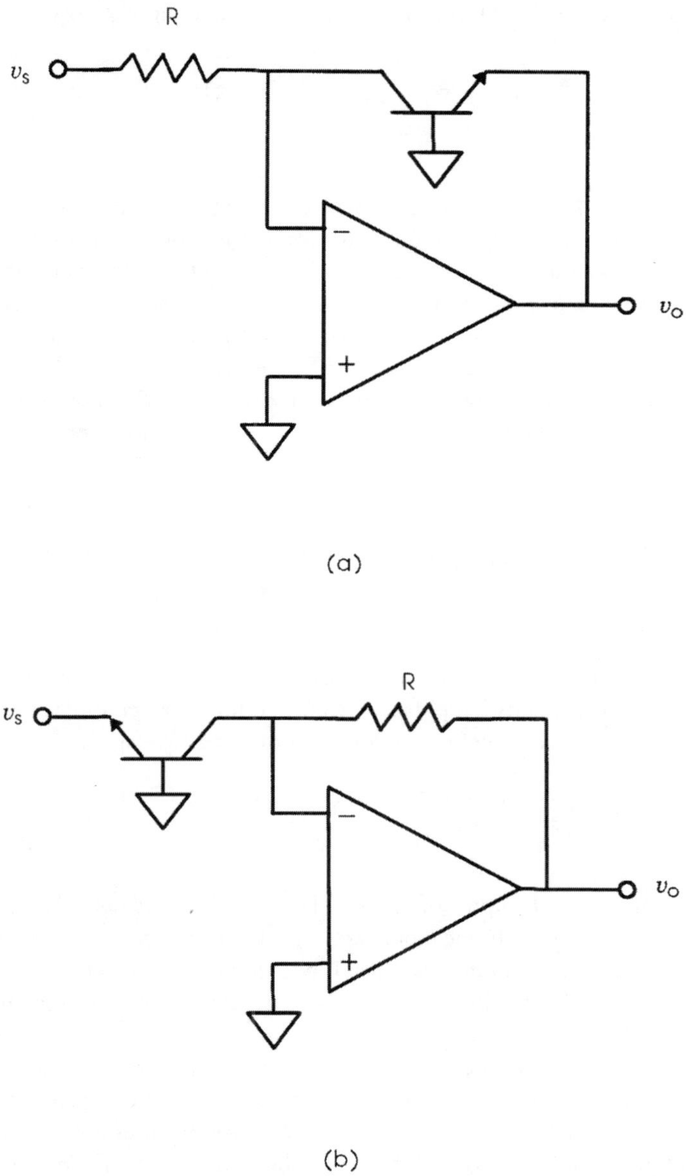

FIGURE 20.15 Basic circuit for logarithmic (a) and antilog or exponential (b) conversion using the transdiode technique. Practical converters include additional components for error reduction and protection.

to 1 V. Therefore, *logarithmic amplifiers* do not necessarily amplify (enlarge) input signals. They are rather converters providing a voltage or current proportional to the ratio of the input voltage, or current, to a reference voltage, or current.

Logarithmic conversion can be obtained by connecting a bipolar transistor as a feedback element of an op amp, Figure 20.15a. The collector current i_C and the base-emitter voltage have an exponential relationship. From the Ebers–Moll model for a transistor, if $v_{CB} = 0$, then

$$i_C = I_S\left(e^{v_{BE}/v_T} - 1\right) \tag{20.33}$$

where $v_T = kT/q = 25$ mV at room temperature, and I_S is the saturation current for the transistor. In Figure 20.15a the input voltage is converted into an input current and the op amp forces the collector current of the transistor to equal the input current, while maintaining $v_{CB} \approx 0$ V. Hence, provided $i_C \gg I_S$, for $v_s > 0$,

$$v_o = \frac{v_T}{\log e} \log \frac{v_s}{RI_S} \tag{20.34}$$

The basic circuit in Figure 20.15a must be modified in order to provide temperature stability, phase compensation, and scale factor correction; reduce bulk resistance error; protect the base-emitter junction; accept negative input voltages and other improvements. Wong and Ott [10] and Peyton and Walsh [11] describe some common circuit techniques to implement these and additional functions. The LOG100 (Burr-Brown) is a logarithmic converter using this so-called transdiode technique. The AD640 (Analog Devices) and TL441 (Texas Instruments) use different techniques.

Figure 20.15b shows a basic *antilog* or *exponential converter* for negative input voltages. The transistor and the resistor have interchanged positions with respect to Figure 20.15b. For $v_s < 0$,

$$v_o = I_S R e^{v_s/v_T} \tag{20.35}$$

Positive voltages require an input *pnp* transistor instead.

Multiplication and Division

Analog multiplication is useful not only for analog computation but also for modulation and demodulation, for voltage-controlled circuits (amplifiers, filters), and for linearization [12]. An *analog multiplier* (Figure 20.16a) has two input ports and one output port offering a voltage

$$v_o = \frac{v_x v_y}{V_m} \tag{20.36}$$

where V_m is a constant voltage. If inputs of either polarity are accepted, and their signs preserved, the device is a *four-quadrant multiplier.* If one input is restricted to have a defined polarity but the other can change sign, the device is a *two-quadrant multiplier.* If both inputs are restricted to only one polarity, the device is a *one-quadrant multiplier.* By connecting both inputs together, we obtain a voltage squarer (Figure 20.16b).

Wong and Ott[10] describe several multiplication techniques. At low frequencies, one-quadrant multipliers can be built by the log–antilog technique, based on the mathematical relationships $\log A + \log B = \log AB$ and then antilog $(\log AB) = AB$. The AD538 (Analog Devices) uses this technique. Currently, the most common multipliers use the transconductance method, which provides four-quadrant multiplication and differential ports. The AD534, AD633, AD734, AD834/5 (Analog Devices), and the MPY100 and MPY600 (Burr-Brown), are *transconductance multipliers.* A digital-to-analog converter can be considered a multiplier accepting a digital input and an analog input (the reference voltage). A multiplier can be converted into a *divider* by using the method in Figure 20.16c. Input v_x must be positive in order for the op amp feedback to be negative. Then

$$v_o = -V_m \frac{R_2}{R_1} \frac{v_z}{v_x} \tag{20.37}$$

The log–antilog technique can also be applied to dividing two voltages by first subtracting their logarithms and then taking the antilog. The DIV100 (Burr-Brown) uses this technique. An analog-to-digital converter can be considered a divider with digital output and one dc input (the reference voltage). A multiplier can also be converted into a square rooter as shown in Figure 20.16d. The diode is required to prevent circuit latch-up [10]. The input voltage must be negative.

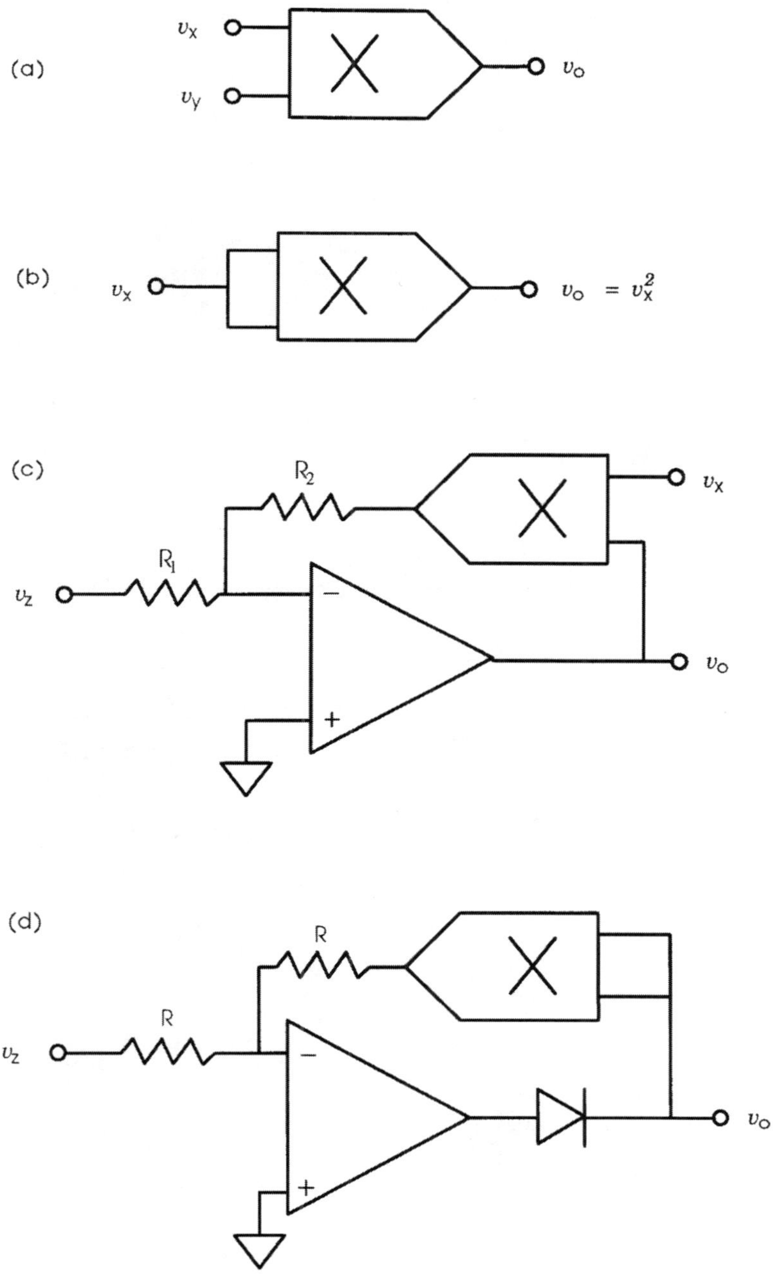

FIGURE 20.16 (a) Symbol for an analog multiplier. (b) Voltage squarer from an analog multiplier. (c) Two-quadrant analog divider from a multiplier and op amp feedback. (d) Square rooter from a multiplier and op amp feedback.

20.11 Analog Linearization

Nonlinearity in instrumentation can result from the measurement principle, from the sensor, or from sensor conditioning. In pressure-drop flowmeters, for example, the drop in pressure measured is proportional to the square of the flow velocity; hence, flow velocity can be obtained by taking the square root of the pressure signal. The circuit in Figure 20.16d can perform this calculation. Many sensors are linear only in a restricted measurand range; other are essentially nonlinear (NTC, LDR); still others are

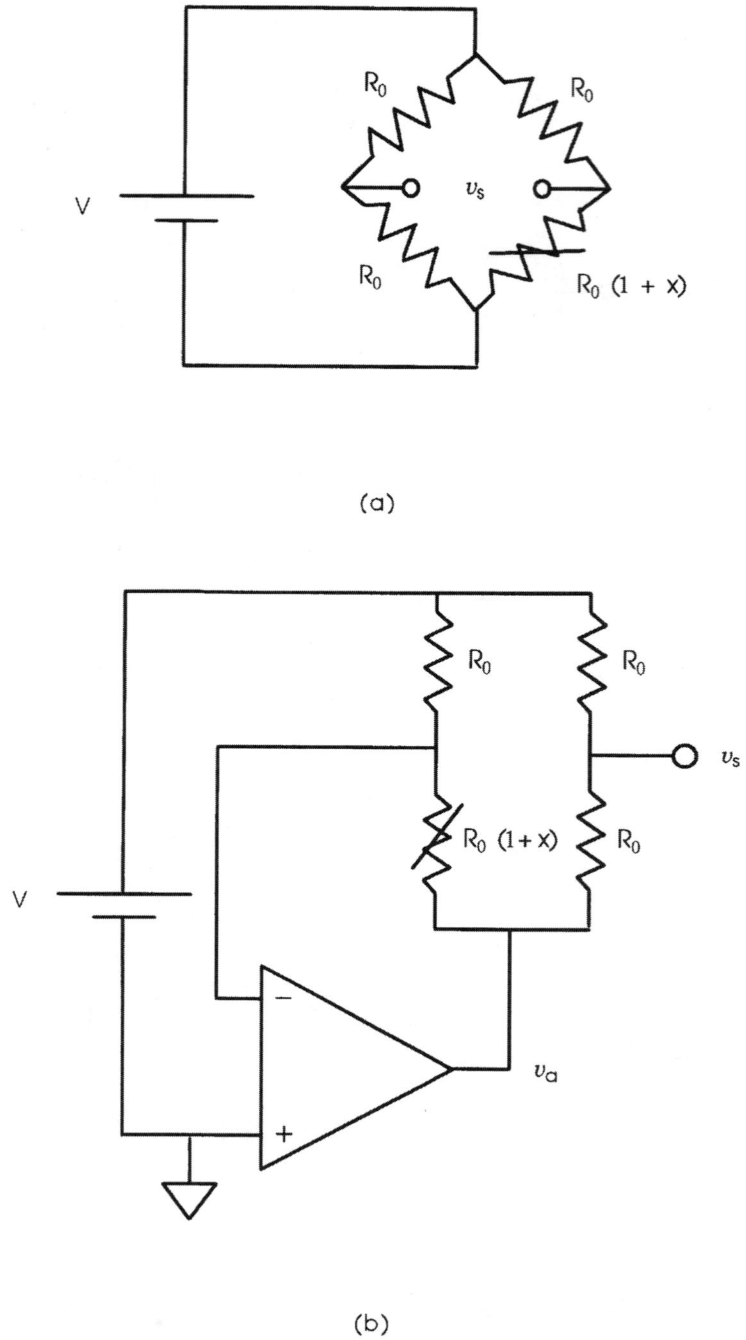

FIGURE 20.17 (a) A Wheatstone bridge supplied at a constant voltage and including a single sensor provides a nonlinear output voltage. (b) By adding an op amp which forces a constant current through the sensor, the output voltage is linearized.

linear in some ranges but nonlinear in other ranges of interest (thermocouples). Linearization techniques for particular sensors are described in the respective chapters.

Nonlinearity attributable to sensor conditioning is common, for example, when resistive (linear) sensors are placed in voltage dividers or bridges. The Wheatstone bridge in Figure 20.17a, for example, includes a linear sensor but yields a nonlinear output voltage,

TABLE 20.4 Special-Purpose Integrated Circuit Signal Conditioners

Model	Function	Manufacturer
4341	rms-to-dc converter	Burr-Brown
ACF2101	Low-noise switched integrator	Burr-Brown
AD1B60	Intelligent digitizing signal conditioner	Analog Devices
AD2S93	LVDT-to-digital converter (ac bridge conditioner)	Analog Devices
AD594	Thermocouple amplifier with cold junction compensation	Analog Devices
AD596/7	Thermocouple conditioner and set-point controllers	Analog Devices
AD598	LVDT signal conditioner	Analog Devices
AD636	rms-to-dc (rms-to-dc converter)	Analog Devices
AD670	Signal conditioning ADC	Analog Devices
AD698	LVDT signal conditioner	Analog Devices
AD7710	Signal conditioning ADC with RTD excitation currents	Analog Devices
AD7711	Signal conditioning ADC with RTD excitation currents	Analog Devices
IMP50E10	Electrically programmable analog circuit	IMP
LM903	Fluid level detector	National Semiconductor
LM1042	Fluid level detector	National Semiconductor
LM1819	Air-core meter driver	National Semiconductor
LM1830	Fluid detector	National Semiconductor
LT1025	Thermocouple cold junction compensator	Linear Technology
LT1088	Wideband rms-to-dc converter building block	Linear Technology
LTK001	Thermocouple cold junction compensator and matched amplifier	Linear Technology
TLE2425	Precision virtual ground	Texas Instruments

$$v_s = V\left(\frac{1+x}{2+x} - \frac{1}{2}\right) = \frac{Vx}{2(2+x)} \tag{20.38}$$

The nonlinearity arises from the dependence of the current through the sensor on its resistance, because the bridge is supplied at a constant voltage. The circuit in Figure 20.17b provides a solution based on one op amp which forces a constant current V/R_0 through the sensor. The bridge output voltage is

$$v_s = \frac{V + v_a}{2} = V\frac{x}{2} \tag{20.39}$$

In addition, v_s is single ended. The op amp must have a good dc performance.

20.12 Special-Purpose Signal Conditioners

Table 20.4 lists some signal conditioners in IC form intended for specific sensors and describes their respective functions. The decision whether to design a signal conditioner from parts or use a model from Table 20.4 is a matter of cost, reliability, and availability. Signal conditioners are also available as subsystems (plug-in cards and modules), for example, series MB from Keithley Metrabyte, SCM from Burr-Brown, and 3B, 5B, 6B, and 7B from Analog Devices.

Defining Terms

Carrier amplifier: Voltage amplifier for narrowband ac signals, that includes in addition a sine wave oscillator, a synchronous demodulator, and a low-pass filter.

Common-mode rejection ratio (CMRR): The gain for a differential voltage divided by the gain for a common-mode voltage in a differential amplifier. It is usually expressed in decibels.

Common-mode voltage: The average of the voltages at the input terminals of a differential amplifier.

Differential amplifier: Circuit or device that amplifies the difference in voltage between two terminals, none of which is grounded.

Dynamic range: The measurement range for a quantity divided by the desired resolution.

Instrumentation amplifier: Differential amplifier with large input impedance and low offset and gain errors.

Isolation amplifier: Voltage amplifier whose ground terminal for input voltages is independent from the ground terminal for the output voltage (i.e., there is a large impedance between both ground terminals).

Isolation Mode Rejection Ratio (IMRR): The amplitude of the output voltage of an isolation amplifier divided by the voltage across the isolation impedance yielding that voltage.

Signal conditioner: Circuit or device that adapts a sensor signal to an ensuing circuit, such as an analog-to-digital converter.

Voltage buffer: Voltage amplifier whose gain is 1, or close to 1, and whose input impedance is very large while its output impedance is very small.

References

1. R. Pallás-Areny and J.G. Webster, *Sensors and Signal Conditioning*, 2nd ed., New York: John Wiley & Sons, 2001.
2. J. Graeme, *Photodiode Amplifiers, Op Amp Solutions*, New York: McGraw-Hill, 1996.
3. C. Kitchin and L. Counts, *Instrumentation Amplifier Application Guide*, 2nd ed., Application Note, Norwood, MA: Analog Devices, 1992.
4. C.D. Motchenbacher and J.A. Connelly, *Low-Noise Electronic System Design*, New York: John Wiley & Sons, 1993.
5. S. Franco, *Design with Operational Amplifiers and Analog Integrated Circuits*, 2nd ed., New York: McGraw-Hill, 1998.
6. R. Pallás-Areny and J.G. Webster, Common mode rejection ratio in differential amplifiers, *IEEE Trans. Instrum. Meas.*, 40, 669–676, 1991.
7. R. Pallás-Areny and O. Casas, A novel differential synchronous demodulator for ac signals, *IEEE Trans. Instrum. Meas.*, 45, 413–416, 1996.
8. M.L. Meade, *Lock-in Amplifiers: Principles and Applications*, London: Peter Peregrinus, 1984.
9. W.G. Jung, *IC Op Amp Cookbook*, 3rd ed., Indianapolis, IN: Howard W. Sams, 1986.
10. Y.J. Wong and W.E. Ott, *Function Circuits Design and Application*, New York: McGraw-Hill, 1976.
11. A.J. Peyton and V. Walsh, *Analog Electronics with Op Amps*, Cambridge, U.K.: Cambridge University Press, 1993.
12. D.H. Sheingold, Ed., *Multiplier Application Guide*, Norwood, MA: Analog Devices, 1978.

Further Information

B.W.G. Newby, *Electronic Signal Conditioning*, Oxford, U.K.: Butterworth-Heinemann, 1994, is a book for those in the first year of an engineering degree. It covers analog and digital techniques at beginners' level, proposes simple exercises, and provides clear explanations supported by a minimum of equations.

P. Horowitz and W. Hill, *The Art of Electronics*, 2nd ed., Cambridge, U.K.: Cambridge University Press, 1989. This is a highly recommended book for anyone interested in building electronic circuits without worrying about internal details for active components.

M.N. Horenstein, *Microelectronic Circuits and Devices*, 2nd ed., Englewood Cliffs, NJ: Prentice-Hall, 1996, is an introductory electronics textbook for electrical or computer engineering students. It provides many examples and proposes many more problems, for some of which solutions are offered.

J. Dostál, *Operational Amplifiers*, 2nd ed., Oxford, U.K.: Butterworth-Heinemann, 1993, provides a good combination of theory and practical design ideas. It includes complete tables which summarize errors and equivalent circuits for many op amp applications.

T.H. Wilmshurst, *Signal Recovery from Noise in Electronic Instrumentation,* 2nd ed., Bristol, U.K.: Adam Hilger, 1990, describes various techniques for reducing noise and interference in instrumentation. No references are provided and some demonstrations are rather short, but it provides insight into very interesting topics.

Manufacturers' data books provide a wealth of information, albeit nonuniformly. Application notes for special components should be consulted before undertaking any serious project. In addition, application notes provide handy solutions to difficult problems and often inspire good designs. Most manufacturers offer such literature free of charge. The following have shown to be particularly useful and easy to obtain: *1993 Applications Reference Manual,* Analog Devices; *1994 IC Applications Handbook,* Burr-Brown; *1990 Linear Applications Handbook* and *1993 Linear Applications Handbook,* Vol. II, Linear Technology; *1994 Linear Application Handbook,* National Semiconductor; *Linear and Interface Circuit Applications,* Vols. 1, 2, and 3, Texas Instruments.

R. Pallás-Areny and J.G. Webster, *Analog Signal Processing,* New York: John Wiley & Sons, 1999, offers a design-oriented approach to processing instrumentation signals using standard analog integrated circuits, that relies on signal classification, analog domain conversions, error analysis, interference rejection and noise reduction, and highlights differential circuits.

21

Modulation

David M. Beams
University of Texas at Tyler

21.1 Introduction

It is often the case in instrumentation and communication systems that an information-bearing signal may not be in an optimal form for direct use. In such cases, the information-bearing signal may be used to alter some characteristic of a second signal more suited to the application. This process of altering one signal by means of another is known as **modulation**; the original information is called the *baseband signal,* and the signal modulated by the baseband signal is termed the *carrier* (because it "carries" the information). Recovery of the original information requires a suitable demodulation process to reverse the modulation process.

A prominent use of modulation techniques is found in radio communication. The extremely long wavelengths of electromagnetic waves at frequencies found in a typical audio signal make direct transmission impractical, because of constraints on realistic antenna size and bandwidth. Successful radio communication is made possible by using the original audio (baseband) signal to modulate a carrier signal of a much higher frequency and transmitting the modulated carrier by means of antennas of feasible size. Another example is found in the use of modems to transmit digital data by the telephone network. Digital data are not directly compatible with analog local subscriber connections, but these data may be used to modulate audible signals which may be carried over local telephone lines. Instrumentation systems use modulation techniques for telemetry (where the distances may be on the order of centimeters for implanted medical devices to hundreds of millions of kilometers for deep-space probes), for processing signals in ways for which the original signals are unsuited (such as magnetic recording of low-frequency and dc signals), and for specialized amplification purposes (carrier and lock-in amplifiers).

Techniques that modulate the amplitude of the carrier are full-carrier **amplitude modulation** (AM), reduced- or suppressed-carrier double-sideband amplitude modulation (DSB), single-sideband suppressed-carrier modulation (SSB), vestigial-sideband modulation (VSB), and on–off keying (OOK). Techniques that modulate the frequency or phase angle of the carrier include **frequency modulation** (FM), phase modulation (PM), frequency-shift keying (FSK), and phase-shift keying (PSK). Simultaneous variation

0-8493-1733-9/04/$0.00+$1.50

of amplitude and phase are applied in quadrature amplitude modulation (QAM). Each technique has its own particular uses. Full-carrier AM is used in radio broadcasting; VSB is used in television broadcasting. DSB appears in instrumentation systems utilizing carrier amplifiers and modulating sensors, while SSB finds use in certain high-frequency radio communications. FM is used in broadcasting (radio and television audio) and point-to-point mobile communications. OOK is commonly used to transmit digital data in optical fiber links. FSK, PSK, and QAM are found in digital communications; analog QAM carries the chrominance (color) information in color television broadcasting. The emphasis of this particular chapter will be instrumentation systems; those interested principally in communications applications could begin by consulting References 1 through 4.

21.2 Generalized Modulation

We begin by making two assumptions: (1) the highest frequency present in the baseband signal is considerably less than the carrier frequency and (2) the results derived in the following chapter pertain to sinusoidal carriers but may be extended to other periodic carrier signals (such as square waves and triangle waves). Equation 21.1 gives a general expression for a modulated sinusoidal carrier signal of radian frequency ω_c:

$$f_s(t) = A_c(t)\cos\left[\omega_c t + \phi(t)\right] \tag{21.1}$$

Information may be carried by $f_s(t)$ by modulation of its amplitude $A_c(t)$, its phase angle $\phi(t)$, or, in some cases, both (note that frequency modulation is a form of phase **angle modulation**). Equation 21.1 may be recast in an equivalent form:

$$f_s(t) = f_i(t)\cos(\omega_c t) - f_q(t)\sin(\omega_c t) \tag{21.2}$$

where:

$$f_i(t) = A_c(t)\cos\left[\phi(t)\right]$$

$$f_q(t) = A_c(t)\sin\left[\phi(t)\right]$$

Equation 21.2 gives $f_s(t)$ as the sum of a cosinusoidal carrier term with time-varying amplitude $f_i(t)$ and a sinusoidal (quadrature) carrier term with time-varying amplitude $f_q(t)$ and is thus known as the carrier-quadrature description of $f_s(t)$. The terms $f_i(t)$ and $f_q(t)$ are known, respectively, as the in-phase and quadrature components of $f_s(t)$. Carlson [1] gives the Fourier transform of a signal in carrier-quadrature form:

$$F_s(\omega) = \frac{1}{2}\left[F_i(\omega-\omega_c) + F_i(\omega+\omega_c)\right] + \frac{j}{2}\left[F_q(\omega-\omega_c) + F_q(\omega+\omega_c)\right] \tag{21.3}$$

where $F_i(\omega)$: $f_i(t)$ and $F_q(\omega)$: $f_q(t)$ are Fourier transform pairs. Notice that the spectra of both $F_i(\omega)$ and $F_q(\omega)$ are both translated by $\pm\omega_c$. Modulation of the carrier in any sense causes energy to appear at frequencies (known as *sidebands*) other than the carrier frequency. Sidebands will be symmetrically distributed relative to the carrier in all but the specialized cases of VSB and SSB.

21.3 Amplitude Modulation

AM that appears in instrumentation systems takes the form of double-sideband AM on which we will focus some degree of attention. VSB and SSB are encountered in communications systems but not in instrumentation systems; interested readers may refer to References 1 through 3.

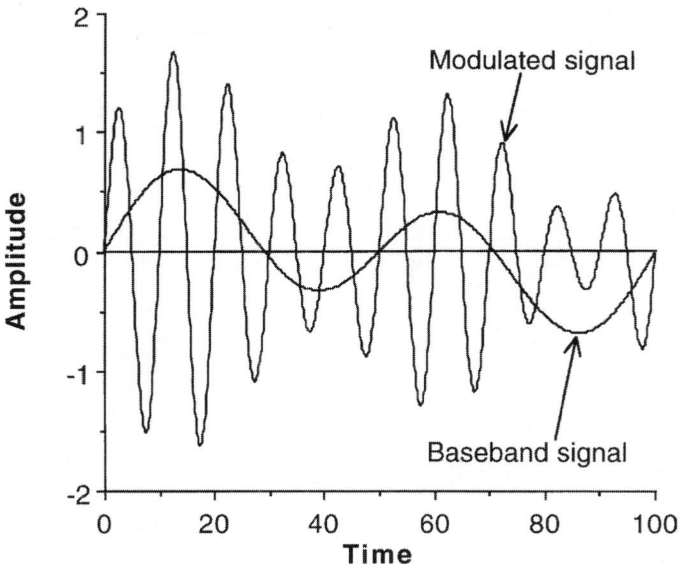

FIGURE 21.1 Time-domain representation of a baseband and the resulting full-carrier AM signal. The time scale is arbitrary.

Double-Sideband Amplitude Modulation

AM applied to a sinusoidal carrier is described by

$$f_s(t) = A_c\left[k + \mu f_m(t)\right]\cos(\omega_c t) \tag{21.4}$$

where A_c is the amplitude of the unmodulated carrier, k is the proportion of carrier present in the modulated signal, μ is the modulation index, $f_m(t)$ is the modulating baseband signal (presumed to be a real bandpass signal), and ω_c is the carrier radian frequency. The modulation index relates the change in amplitude of the modulated signal to the amplitude of the baseband signal. The value of k ranges from 1 in full-carrier AM to 0 in suppressed-carrier double-sideband modulation. The peak value of the modulated signal is $k + \mu f_m(t)$ which may take on any positive value consistent with the dynamic range of the modulator and demodulating system; note that phase reversal of the carrier occurs if $k + \mu f_m(t)$ becomes negative. Figure 21.1 represents a full-carrier AM signal and its baseband signal. Recasting Equation 21.4 in carrier-quadrature form gives $f_i = A_c[k + \mu f_m(t)]$ and $f_q = 0$. The Fourier transform of this signal is

$$F_s(\omega) = \frac{A_c}{2}\left\{k\delta(\omega - \omega_c) + k\delta(\omega + \omega_c) + \mu\left[F_m(\omega - \omega_c) + F_m(\omega + \omega_c)\right]\right\} \tag{21.5}$$

where $\delta(\omega - \omega_c)$ and $\delta(\omega + \omega_c)$ are unit impulses at $+\omega_c$ and $-\omega_c$, respectively, and represent the carrier component of $F_S(\omega)$. The frequency domain representation of $F_S(\omega)$ also contains symmetric sidebands about the carrier with the upper sideband arising from the positive-frequency component of $F_m(\omega)$ and the lower sideband from the negative-frequency component. A double-sideband AM signal thus has a bandwidth twice as large as that of the baseband signal.

Figure 21.2 shows a time-domain representation of a suppressed-carried DSB signal with the same baseband modulation as in Figure 21.1. The information in a full-carrier AM signal is found in the

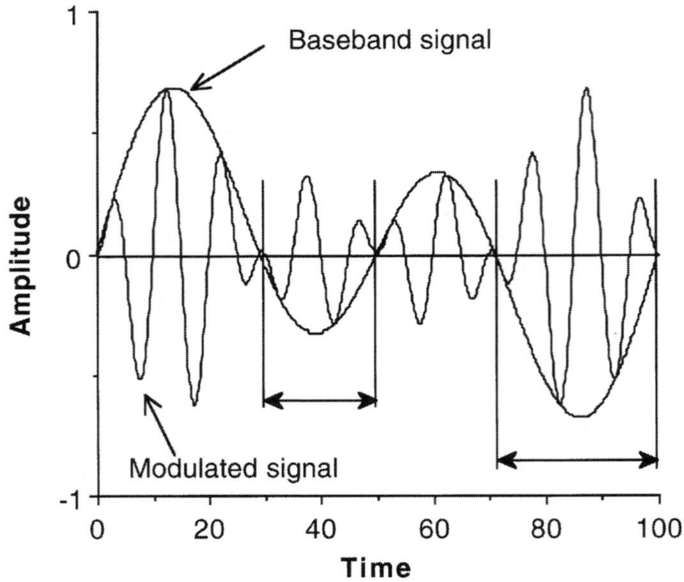

FIGURE 21.2 Time-domain representation of a DSB suppressed-carrier signal. Regions demarcated by double arrows indicate phase inversion of the modulated signal relative to the unmodulated carrier. This is in contrast to full-carrier AM in which the modulated signal is always in phase with the carrier.

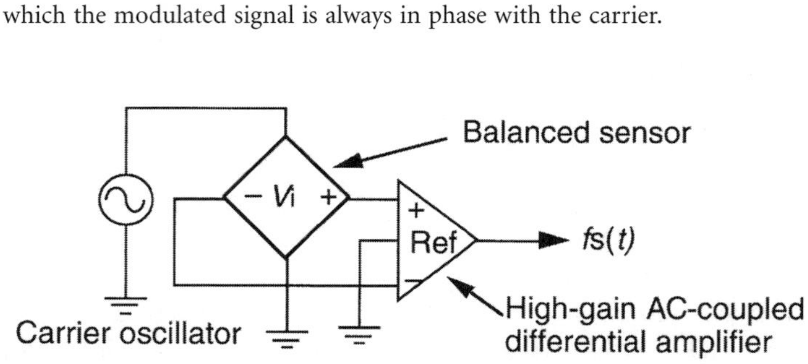

FIGURE 21.3 A balanced sensor with ac excitation and a differential-amplifier output stage. Typical sensors which might be found in this role are resistive Wheatstone bridges, differential-capacitance pressure sensors or accelerometers, or linear-variable differential transformers (LVDTs). Variation in the sensor measurand produces a DSB suppressed-carrier signal at the amplifier output.

time-varying amplitude of the modulated signal, but the information carried by the suppressed-carrier DSB signal is found both in the amplitude and instantaneous phase of the modulated signal (note that the phase of the DSB signal is inverted relative to the carrier when the baseband signal is negative and in phase when the baseband signal is positive). AM is a linear modulation technique; the sum of multiple AM signals produced from a common carrier by different baseband signals is the same as one AM signal produced by the sum of the baseband signals.

Generation of Double-Sideband AM Signals

AM in radio transmitters is frequently performed by applying the modulating waveform to the supply voltage to a nonlinear radiofrequency power amplifier with a resonant-circuit load as described by Carlson [1]. Low-level AM may be achieved by direct multiplication of the carrier signal by $[k + \mu f_m(t)]$.

AM signals often arise in instrumentation systems as the result of the use of an ac drive signal to a modulating sensor. Figure 21.3 shows an example in which a balanced sensor is excited by a sinusoidal carrier. The output voltage of the differential amplifier will be zero when the sensor is balanced; a nonzero

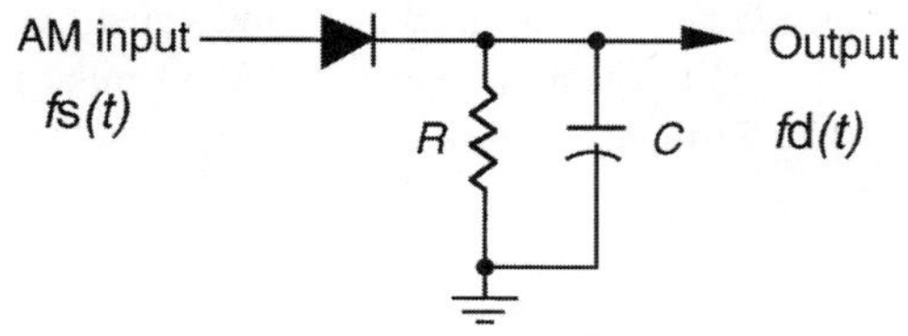

FIGURE 21.4 Envelope detector for full-carrier AM signals.

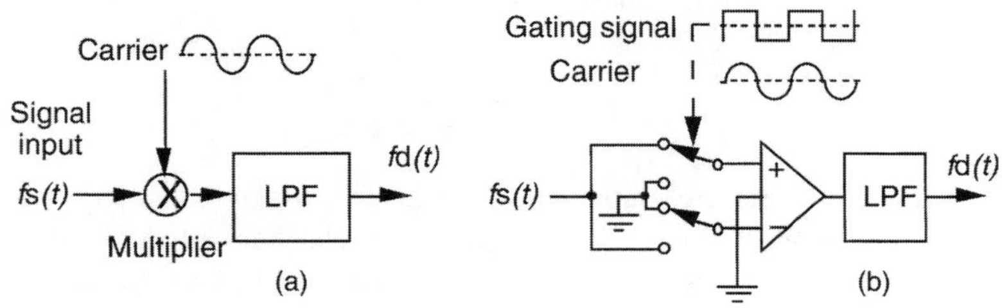

FIGURE 21.5 Multiplying (a) and switching (b) synchronous demodulators. The blocks marked **LPF** represent low-pass filters.

output voltage appears when the sensor is unbalanced. The magnitude of the voltage indicates the degree of imbalance in the sensor, and the phase of the output voltage relative to the carrier determines the direction of imbalance. The suppressed-carrier DSB signal of Figure 21.2 would be seen at the amplifier output if we substitute the sensor measurand for the baseband signal of Figure 21.2. The technique of applying ac excitation to a balanced sensor may be required for inductive or capacitive sensors; it may also be desirable in the case of resistive sensors (such as strain-gage bridges) requiring high-gain amplifiers. In these circumstances, amplification of an ac signal minimizes both $1/f$ noise and dc offset problems associated with high-gain dc-coupled amplifiers.

Envelope Demodulation of Double-Sideband AM Signals

Full-carrier AM signals are readily demodulated by the simple envelope detector shown in Figure 21.4. The components of the *RC* low-pass filter are chosen such that $\omega_m \ll (1/RC) \ll \omega_c$. Envelope detection, however, cannot discriminate phase and is thus unsuitable for demodulation of signals in which phase reversal of the carrier occurs (such as reduced-carrier or suppressed-carrier signals). Synchronous demodulation is required for such signals.

Synchronous Demodulation of Double-Sideband AM Signals

Figure 21.5 shows two methods of synchronous demodulation. In Figure 21.5(a), the modulated signal is multiplied by $\cos(\omega_c t)$; in Figure 21.5(b), the modulated signal is gated by a square wave synchronous with $\cos(\omega_c t)$. Consider the multiplying circuit of Figure 21.5(a); the Fourier transform $F_d(\omega)$ of $f_d(t) = f_s(t)\cos(\omega_c t)$ is given by

$$F_d(\omega) = \frac{(A_c k)}{4}\left[\delta(\omega - 2\omega_c) + \delta(\omega + 2\omega_c) + 2\delta(\omega)\right] + \frac{(\mu A_c)}{4}\left[F_m(\omega - 2\omega_c) + F_m(\omega - 2\omega_c) + 2F_m(\omega)\right] \quad (21.6)$$

The spectral components translated by $\pm 2\omega_c$ may be removed by low-pass filtering; the result, translated into the time domain, is

$$f_d(t) = \frac{A_c k}{2} + \frac{\mu A_c}{2} f_m(t) \quad (21.7)$$

We thus have a dc component of $A_c k/2$ and the original baseband signal $f_m(t)$ multiplied by a scale factor of $\mu A_c/2$. The gating circuit of Figure 21.5(b) may be analyzed in a similar manner; the action of gating is equivalent to multiplying $f_s(t)$ by a square wave with levels of ±1. The Fourier series representation of such a square wave is given by

$$f_g(t) = \frac{4}{\pi}\sum_{n=0}^{\infty}\frac{(-1)^n}{(2n+1)}\cos\left[(2n+1)\omega_c t\right] \tag{21.8}$$

Low-pass filtering of the product of Equations 21.4 and 21.8 gives

$$f_d(t) = \frac{2A_c k}{\pi} + \frac{2\mu A_c}{\pi} f_m(t) \tag{21.9}$$

The baseband signal is again recovered, although the scale factor is somewhat larger than that of the multiplying case. Demodulators like those of Figure 21.5(a) are called multiplying demodulators; circuits Figure 21.5(b) are known as switching demodulators. Nowicki [5] and Meade [6] discuss and compare both types of synchronous demodulators in greater detail.

We have so far made the implicit assumption that the demodulating signal is perfectly synchronized with the carrier of the modulated signal. Let us now consider the case where there exists a phase shift between these two signals. Assume that a signal expressed in carrier-quadrature form is multiplied by a demodulating signal $\cos(\omega_c t + \theta)$. The result, after suitable low-pass filtering, is

$$f_d(t) = \frac{f_i(t)}{2}\cos(\theta) + \frac{f_q(t)}{2}\sin(\theta) \tag{21.10}$$

Equation 21.10 is an important result; we see that both the level and polarity of the demodulated signal are functions of the synchronization between the demodulating signal and the modulated carrier. Synchronous demodulation is thus often called phase-sensitive demodulation. It was previously mentioned that a DSB AM signal has no quadrature component $f_q(t)$; a phase shift of an odd integral multiple of $\pi/2$ radians between the carrier and the demodulating signal will produce a synchronous demodulator output of zero. Use of a square wave gating signal in place of the sinusoid would produce the same result as Equation 21.10 except that the recovered signals would be multiplied by $2/\pi$ instead of 1/2.

Figure 21.6 shows the effect of phase between the gating signal and the incoming signal in a switching phase-sensitive demodulator. A sinusoid with a constant amplitude of 1 is gated by a square wave which has levels of ±1; the amplifier outputs before low-pass filtering are shown for phase offsets of 0, $\pi/4$, and $\pi/2$ radians. The dc levels which would be recovered by low-pass filtering are also shown. Note that the demodulator output with phase offset of zero takes the form of a full-wave rectified sine which has only positive excursions. As the phase offset between the gating signal and the sinusoid increases, however, the dc component decreases as the gated signal shows increasing negative excursions and decreasing positive excursions. An offset of $\pi/2$ radians produces an output whose positive and negative excursions are symmetric and thus has no dc component. Synchronization of the demodulating signal with the modulated carrier is thus crucial for accurate demodulation. Synchronization of the demodulating signal with the modulated carrier is straightforward in instrumentation applications in which the carrier signal may be provided directly to the demodulator, such as the electrical impedance tomography applications discussed by Webster [7]. This becomes more difficult if the carrier cannot be directly provided but must be inferred or recovered from the incoming signal. Hershberger [8] employs a phase-locked loop to perform the carrier synchronization in a synchronous detector for radio receiver applications. Modulated signals which contain both in-phase and quadrature components may be demodulated by an I–Q demodulator which is comprised of two parallel synchronous demodulators (one driven by $\cos(\omega_c t)$ and the other by $\sin(\omega_c t)$); Breed [9] notes that any form of modulation may in principle be demodulated

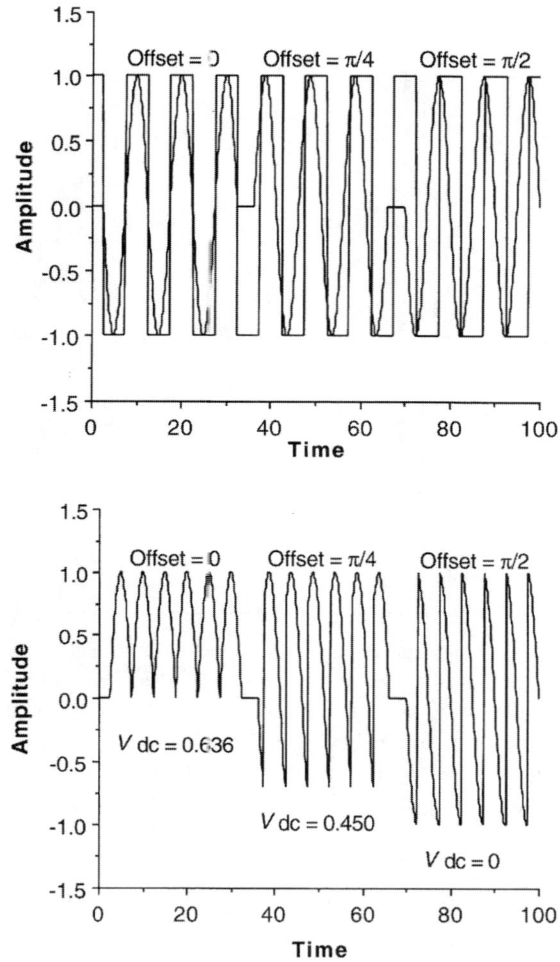

FIGURE 21.6 Gating of a constant-amplitude sinusoidal by a square wave in a switching phase-sensitive demodulator with various phase offsets between the carrier and the square wave. The upper trace shows the carrier and the square wave; the lower trace shows the signals which appear at the output of the differential amplifier in Figure 21.5(b). The dc output voltages of the low-pass filter are indicated for each value of phase offset.

by this method although less expensive techniques are suitable for many applications (such as the envelope detector for full-carrier AM). A common instrumentation application for simultaneous I–Q demodulation is the electrical impedance bridge (often called an *LCR* bridge).

Synchronous demodulators are valuable in lock-in amplifier applications for the recovery of signals otherwise masked by noncoherent noise. Assume that the low-pass filter of a synchronous demodulator has a bandwidth of Ω; only those components of the incoming signal which lie within $\pm\Omega$ of ω_c will appear at the low-pass filter output. A demodulator with extremely high selectivity may be built by use of a narrow low-pass filter, providing remarkable improvement in output signal-to-noise ratio. The use of an AD630 switching demodulator to recover a signal from broadband noise whose rms value is 100 dB greater than that of the modulated signal is shown in Reference 10. Synchronous demodulation may also be of benefit in applications in which the input signal-to-noise ratio is not so extreme; components of noise which are in quadrature with respect to the carrier produce no output from a synchronous demodulator, whereas an envelope detector responds to the instantaneous sum of the signal and all noise components.

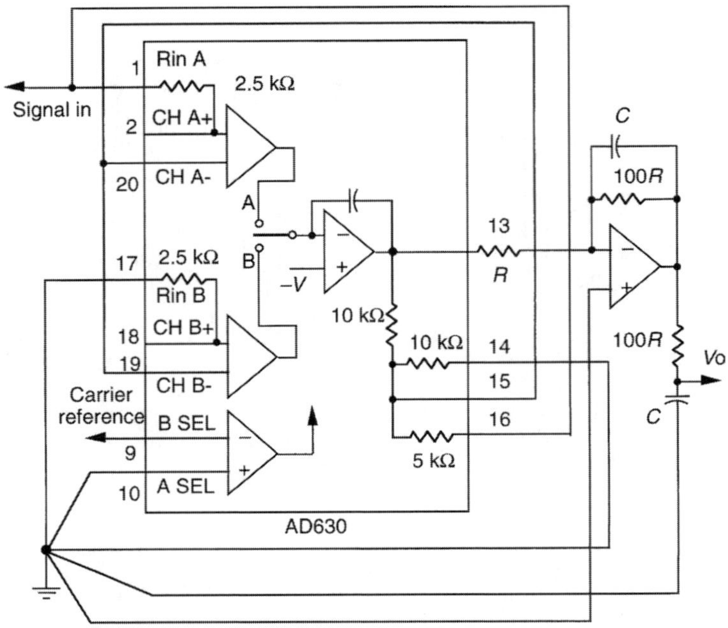

FIGURE 21.7 Lock-in amplifier circuit utilizing the Analog Devices AD630 switching balanced demodulator. This particular device is attractive for the small number of external components required and its high performance. The two-pole low-pass filter shown has a dc gain of 100 and a corner frequency of 0.00644/*RC*. The principal drawback of this filter is its dc output impedance which is 100*R*. This limitation could be avoided by replacing this filter by a pair of single-pole active filters or a single-amplifier two-pole filter (such as the Sallen-Key type of network). Resistances depicted within the outline of the AD630 are internal to the device itself; numbers at the periphery of the device are pin numbers. Pins labeled CH A+, CH A–, CH B+, and CH B– indicate the sense of the A and B channels of the device. Pins Rin A and Rin B are electrically connected to the CH A+ and CH B+ pins, respectively, through internal 2.5 kΩ resistances.

Examples

Figure 21.7 shows a lock-in amplifier application of the AD630; this particular integrated circuit has on-chip precision resistances which allow the device to be used with a minimal number of external components. Webster [11] shows the design of a diode-ring phase-sensitive demodulator. High-speed CMOS switches (such as the 74HC4053 and ADG201HS) may also be used in switching phase-sensitive demodulators. An example of this is shown in Figure 21.8 below in which presents a method for measurement of both amplitude and phase of a sinusoidal ac voltage. This method requires that gating signals synchronous with cos(ωt) and sin(ωt) be supplied to a pair of independent single-pole, single-throw switches. Assuming that the input signal V_{in} may be represented as $A\cos(\omega t+\varphi)$, the dc output voltages from the two *RC* low-pass filters are given by:

$$V_{OI} = \frac{A}{\pi}\cos(\phi) \tag{21.11a}$$

$$V_{OQ} = \frac{-A}{\pi}\sin(\phi) \tag{21.11b}$$

where the phase angle φ is measured relative to cos(ωt). Note that switching the input of the low-pass filter between the signal and ground produces dc output voltages half as large as those produced by the technique of Figure 21.5(b). The amplitude *A* and phase φ of the signal may be readily computed from V_{OI} and V_{OQ}:

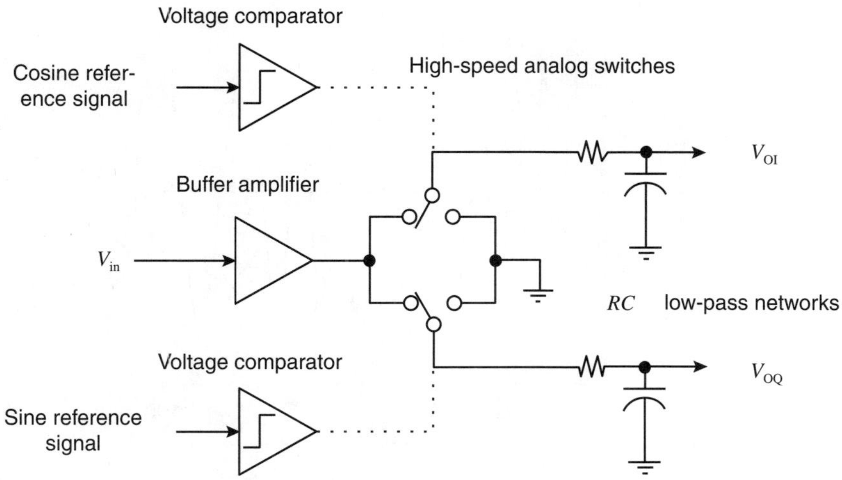

FIGURE 21.8 Use of phase-sensitive detection to measure amplitude and phase of a sinusoidal signal. Phase is measured with respect to cos(ωt).

$$A = \pi\sqrt{V_{OI}^2 + V_{OQ}^2} \tag{21.12a}$$

$$\phi = \tan^{-1}\left(\frac{-V_{OQ}}{V_{OI}}\right) \tag{21.12b}$$

Beams [12] applies this technique with a virtual-instrument program to the automated measurement of frequency response of linear networks.

21.4 Angle (Frequency and Phase) Modulation

Recall from Equation 21.1 that we may modulate a carrier by varying its phase angle in accordance with the baseband signal. Consider a signal of the form:

$$f_s(t) = A_c \cos\left[\omega_c t + \Delta\phi x_m(t)\right] \tag{21.13}$$

The instantaneous phase angle is $\omega_c t + \Delta\phi x_m(t)$; the phase relative to the unmodulated carrier is $\Delta\phi x_m(t)$. The carrier is thus phase-modulated by $x_m(t)$. The instantaneous frequency is $\omega_c + \Delta\phi[dx_m(t)/dt]$; if $x_m(t)$ is the integral of baseband signal $f_m(t)$, the instantaneous frequency becomes $\omega_c + \Delta\phi f_m(t)$. The frequency deviation $\Delta\phi f_m(t)$ is proportional to the baseband signal; the carrier is frequency modulated by $f_m(t)$. We may write the general expression for an FM signal as

$$f_s(t) = A_c \cos\left[\omega_c t + \Delta\omega \int_{-\infty}^{t} f_m(\tau)d\tau\right] \tag{21.14}$$

(The change in notation from $\Delta\phi$ to $\Delta\omega$ is intended to emphasize that the signal represented by Equation 21.14 is frequency modulated.) Figure 21.9 shows a time-domain representation of an FM signal. Note the signal has constant amplitude; this is also true of a phase-modulated signal.

Consider frequency modulation with a baseband signal $f_m = A_m \cos(\omega_m t)$; substitution into Equation 21.14 and performing the integration gives

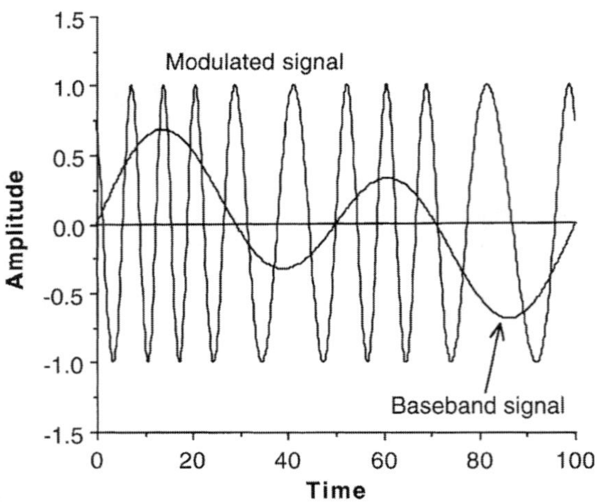

FIGURE 21.9 Time-domain representation of an FM signal and its baseband signal. The time scale is arbitrary.

$$f_s(t) = A_c \cos\left[\omega_c t + \beta \sin(\omega_m t)\right] \tag{21.15}$$

in which β (called the modulation index) has replaced $\Delta\omega A_m/\omega_m$. The carrier-quadrature form of Equation 21.15 is

$$f_s(t) = A_c\left\{\cos\left[\beta \sin(\omega_m t)\right]\cos(\omega_c t) - \sin\left[\beta \sin(\omega_m t)\right]\sin(\omega_c t)\right\} \tag{21.16}$$

We note from Equation 21.16 that, unlike AM, the FM signal contains both in-phase and quadrature components and that the amplitudes of both are nonlinear functions of the modulation index. The terms $\cos[\beta\sin(\omega_m t)]$ and $\sin[\beta\sin(\omega_m t)]$ are generally expanded in terms of Bessel functions (see Schwartz [3] for detailed analysis). The Bessel function expression of an FM signal is

$$f_s(t) = A_c \sum_{n=-\infty}^{\infty} J_n(\beta)\cos(\omega_c + n\omega_m)t \tag{21.17}$$

where J_n represents a Bessel function of the first kind of order n. Beyer [13] provides tables of $J_0(\beta)$ and $J_1(\beta)$ and gives formulae for computing higher-order Bessel functions. We also note that $J_{-n}(\beta) = (-1)^n J_n(\beta)$. The approximations $J_0(\beta) = 1$ and $J_1(\beta) = (\beta/2)$ are valid for low values of the modulation index ($\beta < 0.2$); the higher-order Bessel functions are negligible under these circumstances. A carrier with sinusoidal frequency modulation with a low modulation index will show sidebands spaced at $\pm\omega_m$ about the carrier; such narrowband FM would be indistinguishable from full-carrier AM on a spectrum analyzer display (which shows the amplitudes of spectral components but not their phase relationships). As modulation index increases, however, new sideband pairs appear at $\pm 2\omega_m$, $\pm 3\omega_m$, $\pm 4\omega_m$, etc. as the higher-order Bessel functions become significant. The amplitude of the carrier component of $f_s(t)$ varies with $J_0(\beta)$; the carrier component disappears entirely for certain values of modulation index. These characteristics are unlike AM in which the carrier component of the modulated signal is constant and in which only one sideband pair is produced for each spectral component of the baseband signal. FM is an inherently nonlinear modulation process; the sum of multiple FM signals derived from a single carrier with individual baseband signals does not give the same result as frequency modulation of the carrier by the sum of the baseband signals. The spectrum of a phase-modulated signal is similar to that of an FM signal, but the modulation index of a phase-modulated signal does not vary with ω_m. Figure 21.9 shows a time-domain representation of an FM signal and the original baseband signal.

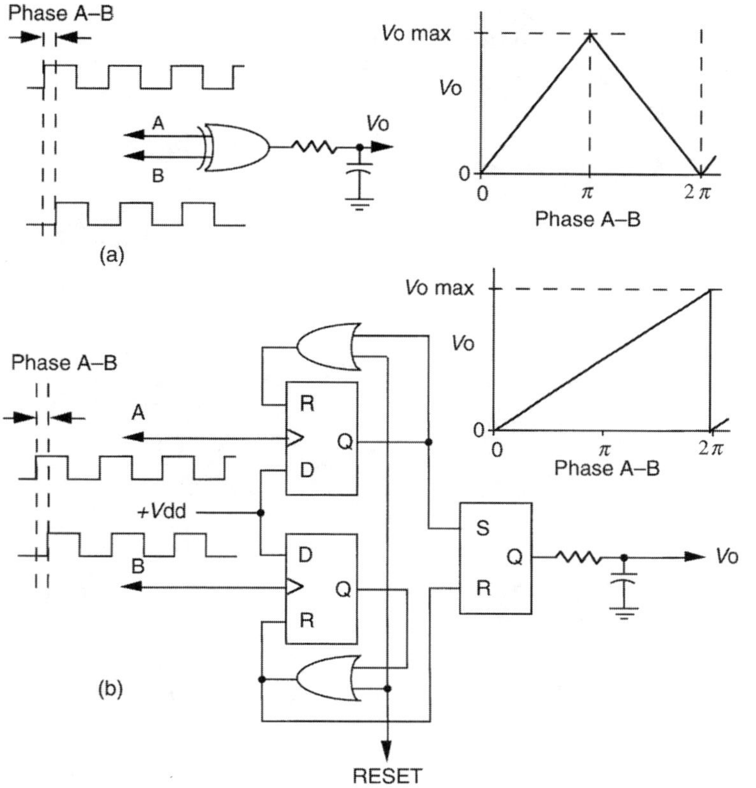

FIGURE 21.10 Examples of digital phase detectors. The exclusive-OR gate in (a) requires input signals with 50% duty cycle and produces an output voltage proportional to phase shift over the range of 0 to π radians (0° to 180°). Phase shifts between π and 2π radians produce an output voltage negatively proportional to phase. The edge-triggered RS flip-flop circuit in (b) functions with signals of arbitrary duty cycle and has a monotonic relationship between phase and output voltage over the full range of 0 to 2π radians. The circuit may be initialized by means of the RESET line. D-type flip-flops with RS capability (such as the CD4013) may be used in this circuit.

Generation of Phase- and Frequency-Modulated Signals

FM signals may arise directly in instrumentation systems such as turbine-type flowmeters or Doppler velocity sensors. Direct FM signals may be generated by applying the baseband signal to a voltage-controlled oscillator (VCO); Sherwin and Regan [14] demonstrate this technique in a system which generates direct FM using the LM566 VCO to transmit analog information over 60 Hz power lines. In radiofrequency applications, the oscillator frequency may be varied by application of the baseband signal to a voltage-variable reactance (such as a varactor diode). Indirect FM may be generated by phase modulation of the carrier by the integrated baseband signal as in Equation 21.14; DeFrance [15] gives an example of a phase modulator circuit.

Demodulation of Phase- and Frequency-Modulated Signals

PM signals may be demodulated by the synchronous demodulator circuits previously described; they are, however, sensitive to the signal amplitude as well as phase and require a limiter circuit to produce an output proportional to phase alone. Figure 21.10 shows simple digital phase demodulators. Figure 21.11 shows three of the more common methods of FM demodulation. Figure 21.11(a) shows a quadrature detector of the type commonly used in integrated-circuit FM receivers. A limiter circuit suppresses noise-induced amplitude variations in the modulated signal; the limiter output then provides a reference signal to a synchronous (phase-sensitive) demodulator. The limiter output voltage is also coupled (via a small

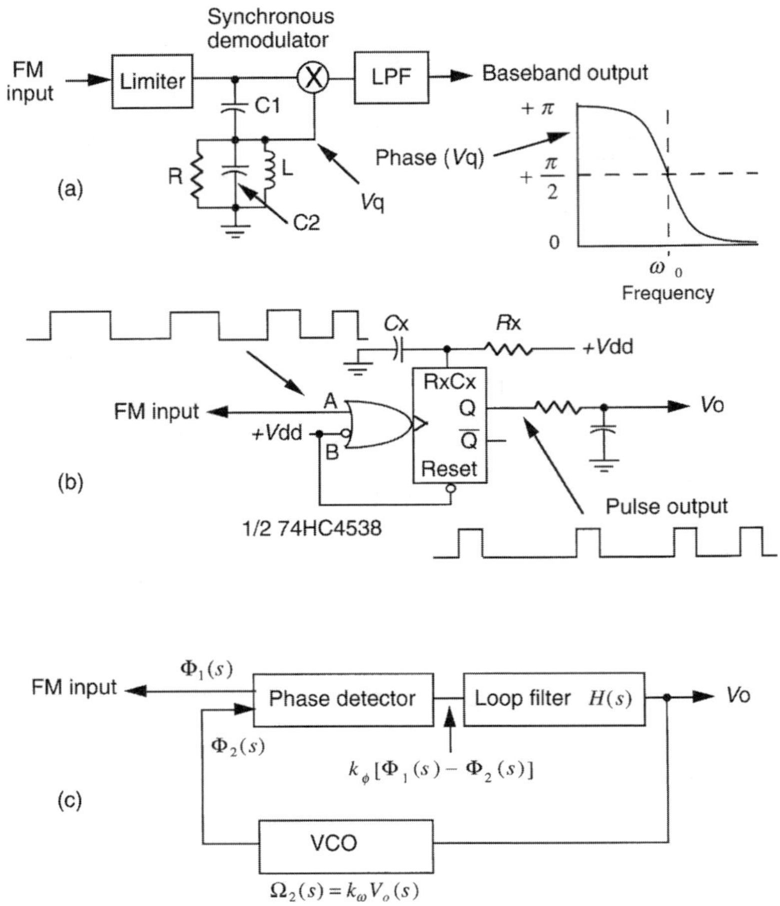

FIGURE 21.11 FM demodulators. A quadrature detector circuit is shown in (a). The phase shift of V_q (the voltage across the quadrature network consisting of L, $C2$, and R) relative to the output voltage of the limiter varies linearly with frequency for frequencies close to ω_0. The synchronous (phase-sensitive) demodulator produces an output proportional to the phase shift. A frequency-to-voltage converter is shown in (b). A monostable multivibrator (one-shot) produces a pulse of fixed width and amplitude with each cycle of the modulated signal. The average voltage of these pulses is proportional to the modulated signal frequency. The 74HC4538 contains two independent edgetriggered monostable multivibrators which may be triggered by a rising edge (as shown) or a falling edge (by applying the clock pulse to the B input and connecting the A input to V_{dd}). A phase-locked loop is shown in (c). The operation of the phase-locked loop is described in the text.

capacitor $C1$) to a quadrature network consisting of L, R, and $C2$. The phase of the voltage across the quadrature network relative to the limiter output is given by

$$\phi\left[V_q(\omega)\right] = \tan^{-1}\frac{-\omega L/R}{\left[1-\left(\omega/\omega_0\right)^2\right]} \tag{21.18a}$$

where

$$\omega_0 = \frac{1}{\sqrt{L\left(C_1 + C_2\right)}} \tag{21.18b}$$

The variation in phase is nearly linear for frequencies close to ω_0. The phase-sensitive demodulator recovers the baseband signal from the phase shift between the quadrature-network voltage and the reference signal. The quadrature network also causes the amplitude of V_q to vary with frequency, but the variation in phase with respect to frequency predominates over amplitude variation in the vicinity of ω_0. There is also generally enough voltage across the quadrature network to force the phase-sensitive demodulator into a nonlinear regime in which its output voltage is relatively insensitive to the amplitude of the voltage across the quadrature network. Figure 21.11(b) shows a frequency-to-voltage converter which consists of a monostable (one-shot) triggered on each zero crossing of the modulated signal. The pulse output of the one-shot is integrated by the low-pass filter to recreate the original baseband signal. Figure 21.11(c) shows a phase-locked loop. The phase comparator circuit produces an output voltage proportional to the phase difference between the input signal and the output of a VCO; that voltage is filtered and used to drive the VCO. Assume that $\phi_1(t)$ and $\omega_1(t)$ represent the phase and frequency of the input signal as functions of time with corresponding Laplace transforms $\Phi_1(s)$ and $\Omega_1(s)$. The VCO output phase and frequency are represented by $\phi_2(t)$ and $\omega_2(t)$ with corresponding Laplace transforms $\Phi_2(s)$ and $\Omega_2(s)$. The phase detector produces an output voltage given by

$$v_\phi(t) = k_\phi\left[\phi_1(t) - \phi_2(t)\right] \tag{21.19}$$

The corresponding Laplace transform expression is

$$V_\phi(s) = k_\phi\left[\Phi_1(s) - \Phi_2(s)\right] \tag{21.20}$$

This voltage is filtered by the loop filter and is applied to the VCO which produces a frequency proportional to the control voltage. The Laplace transform of the VCO frequency is

$$\Omega_2(s) = k_\omega k_\phi H(s)\left[\Phi_1(s) - \Phi_2(s)\right] \tag{21.21}$$

Since $\Phi_1(s) = \Omega_1(s)/s$ and $\Phi_2(s) = \Omega_2(s)/s$, we may write

$$\Omega_2(s) = \left[\frac{k_\omega k_\phi H(s)}{s + k_\omega k_\phi H(s)}\right]\Omega_1(s) \tag{21.22}$$

Assume that the input frequency is a step function of height ω_1; this gives $\Omega_1(s) = \omega_1/s$. Inserting this into Equation 21.22 gives

$$\Omega_2(s) = \left[\frac{k_\omega k_\phi H(s)}{s + k_\omega k_\phi H(s)}\right]\frac{\omega_1}{s} \tag{21.23}$$

Application of the final value theorem of the Laplace transform to Equation 21.24 gives ω_1 as the asymptotic limit of $\omega_2(t)$; the output frequency of the VCO matches the input frequency. The VCO input voltage is ω_1/k_ω and is thus proportional to the input frequency. If the input frequency is varied by some baseband signal, the VCO input voltage follows that baseband signal. The phase-locked loop thus may serve as an FM detector; Taub and Schilling [2] give an extensive analysis. If we assume the simplest loop filter transfer function $H(s) = 1$, the bracketed term of Equation 21.21 takes the form of a single-pole low-pass filter with corner frequency $\omega_{-3\text{dB}} = k_\omega k_\phi$. The VCO output frequency tracks slowly varying input frequencies quite well, but the response to rapidly varying input frequencies is limited by the low-pass behavior of the loop. Baseband spectral components which lie below the loop corner frequency are recovered without distortion, but baseband spectral components beyond the loop corner frequency are

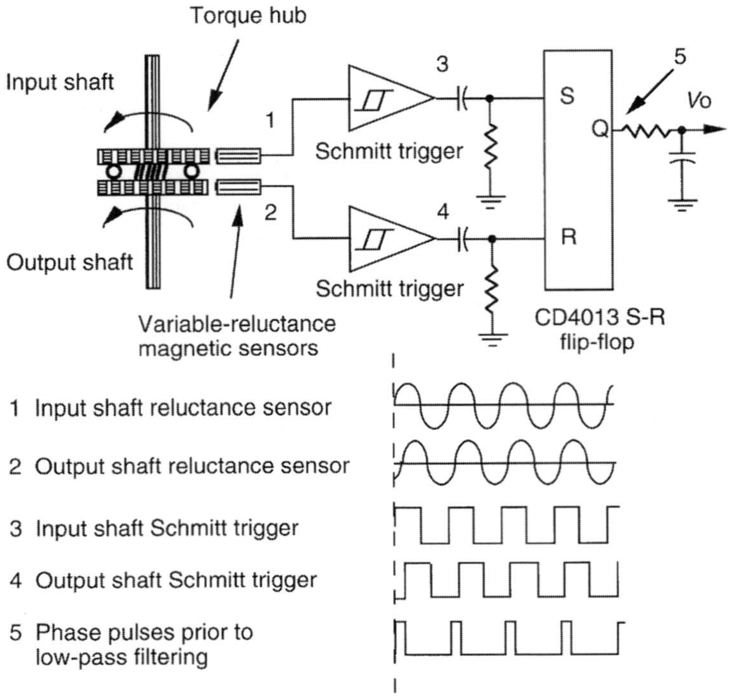

FIGURE 21.12 Use of phase-sensitive demodulation to measure torque in a rotating shaft as described by Sokol et al. [16]. Typical waveforms and the points in the circuit where they are found are indicated. The *RC* coupling to the S and R terminals of the CD4013 flip-flop provides for edge triggering.

attenuated. A single-pole *RC* low-pass filter is often used for the loop filter; the transfer function of the complete phase-locked loop with such a filter is:

$$\frac{\Omega_2(s)}{\Omega_1(s)} = \left[\frac{k_\omega k_\phi p_1}{s^2 + sp_1 + k_\omega k_\phi p_1}\right] \tag{21.24}$$

where $p_1 = 1/RC$. Note that the loop transfer function is that of a second-order low-pass filter and thus may exhibit dynamics such as resonance and underdamped transient response depending on the values of the parameters k_ω, k_ϕ, and p_1.

Examples

Sokol et al. [16] describe the use of phase-sensing techniques to a shaft torque-sensing application in Figure 21.12. Two identical gears are coupled by springs in a device known as a torque hub. The teeth of the gears induce signals in variable-reluctance magnetic sensors as the shaft rotates, and the Schmitt trigger circuits transform the reluctance sensor outputs (approximately sinusoidal signals with amplitudes proportional to shaft rotational velocity) into square waves of constant amplitude. The springs compress with increasing torque, shifting the phase of the magnetic pickup signals (and hence the square waves) relative to each other. The relative phase of the square waves is translated to a dc voltage by a flip-flop phase detector and *RC* low-pass filter. The suppression of amplitude variation by the Schmitt trigger circuits causes no loss of information; the torque measurement is conveyed by phase alone.

Cohen et al. [17] in Figure 21.13 describe the use of a direct FM technique to perform noninvasive monitoring of human ventilation. Elastic belts with integral serpentine inductors encircle the chest and abdomen; these inductors are parts of resonant tank circuits of free-running radiofrequency oscillators. Ventilation causes the inductor cross-sectional areas to vary, changing their inductances and thus varying

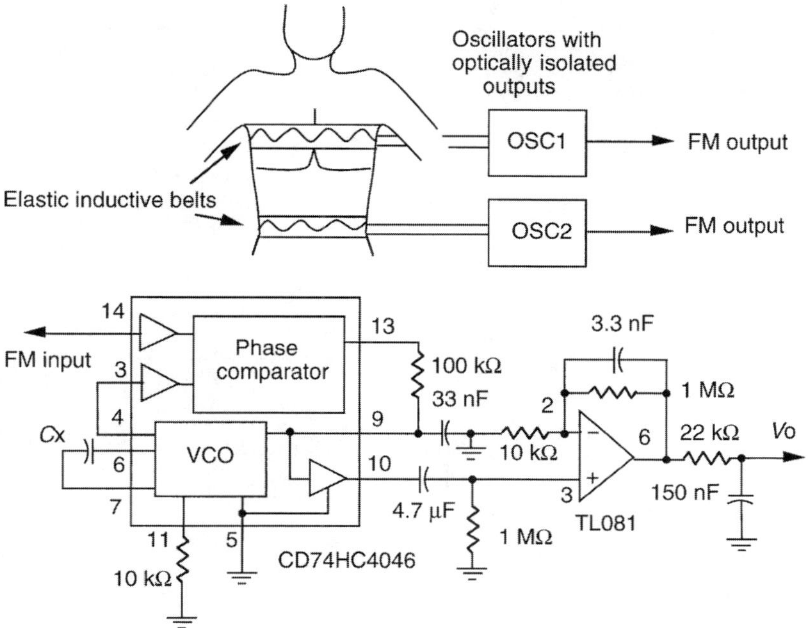

FIGURE 21.13 Use of FM techniques in measurement of human ventilation (adapted from Cohen et al. [17]). An independent oscillator was used for each inductive belt, and the oscillator outputs were optically coupled to the demodulator circuits for the purpose of electrical isolation of the subject from equipment operated from the 60 Hz ac line. One of two demodulators is shown; the value of C_x was either 470 pF for use at a nominal frequency of 850 kHz or 220 pF for use at a nominal frequency of 1.5 MHz. Numbers adjacent to the CD74HC4046 and TL081 devices are pin numbers.

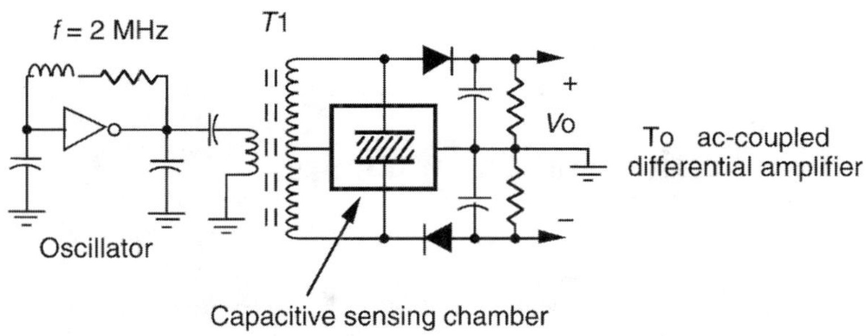

FIGURE 21.14 FM demodulator technique applied to measurement of the flow of small seeds in an agricultural machine application. The relative permittivity of the seeds causes the resonant frequency of the discriminator circuit to shift and to produce a change in the output voltage. The claimed sensitivity of this circuit is on the order of tens of femtofarads. The cross-hatched region of the sensing chamber represents the active area through which the seeds flow.

the frequencies of the oscillators. Figure 21.13 also shows one of two phase-locked loop demodulator circuits which are identical except for the value of one capacitor in the VCO circuit. The phase-locked loop utilizes a single-pole *RC* low-pass loop filter.

Bachman [18,19] in Figure 21.14 describes a novel application of FM techniques to provide a sensing function which would otherwise be difficult to achieve (measurement of deviations in mass flow of fine seeds in an agricultural implement). A capacitive sensing cell was constructed and made part of a discriminator circuit driven at a constant frequency; the flow of seeds through the sensing capacitor shifts the resonant frequency of the discriminator and produces a change in the discriminator output voltage. An ac-coupled differential amplifier provides an output proportional to changes in the mass flow.

TABLE 21.1 Integrated Circuits Used with Modulation Techniques

Designation	Function	Manufacturer(s)	Approximate Price, $
AD630JN	Balanced modulator/demodulator	Analog Devices	10.19
74HC4053	High-speed analog switch	Fairchild, TI	0.67
ADG201HSJN	High-speed analog switch	Analog Devices	3.10
AD532JH	Four-quadrant multiplier	Analog Devices	15.91
AD633JN	Four-quadrant multiplier	Analog Devices	3.73
74HC4046	Phase-locked loop	Fairchild, TI	2.00
AD650JN	Voltage-to-frequency, frequency-to-voltage converter	Analog Devices	9.15
AD652JP	Voltage–to-frequency converter	Analog Devices	10.46
AD654JN	Voltage-to-frequency converter	Analog Devices	4.15
74HC4538	Dual retriggerable monostable (one-shot)	Fairchild, TI	0.72

TABLE 21.2 Companies That Make Integrated Circuits for Modulating

Analog Devices, Inc.
One Technology Way
Box 9106
Norwood, MA 02062
Phone: (617) 329-4700
www.analog.com

Fairchild Semiconductor Corp.
82 Running Hill Road
South Portland, ME 04106
Phone: (207) 775-8100
www.fairchildsemi.com

Texas Instruments, Inc.
12500 TI Blvd.
Dallas, TX 75243-4136
www.ti.com

21.5 Instrumentation and Components

Integrated Circuits

Table 21.1 lists certain integrated circuits which may be used in application of the modulation techniques covered in this chapter. This is not an exhaustive list of all useful types nor of all manufacturers. Most integrated circuits are available in a number of packages and performance grades; the prices given are indicative of the pricing of the least expensive versions purchased in small quantities. Table 21.2 provides contact information for these vendors.

Instrumentation

Oscilloscopes and spectrum analyzers are frequently employed in analysis of modulated signals or systems utilizing modulation techniques; both types of instruments are available from a number of manufacturers (e.g., Agilent Technologies, Tecktronix). Certain types of specialized instrumentation are also available, such as scalar and vector modulation analyzers, signal sources with digital or analog modulation capability, and instruments for analysis of color-television signals (vectorscopes). Representative manufacturers include Agilent Technologies, Tektronix, and Rohde & Schwarz. Additional information is available from their respective on-line sources.

Defining Terms

Modulation: The process of encoding the source information onto a bandpass signal with a carrier frequency f_c.

Amplitude modulation (AM): Continuous wave modulation, where the amplitude of the carrier varies linearly with the amplitude of the modulating signal.

Angle modulation: Continuous wave modulation, where the angle of the carrier varies linearly with the amplitude of the modulating signal.

Frequency modulation (FM): Continuous wave modulation, where the frequency of the carrier varies linearly with the amplitude of the modulating signal.

References

1. A.B. Carlson, *Communication Systems,* 3rd ed., New York: McGraw-Hill, 1986.
2. H. Taub and D.L. Schilling, *Principles of Communication Systems,* New York: McGraw-Hill, 1971.
3. M. Schwartz, *Information Transmission, Modulation, and Noise,* 4th ed., New York: McGraw-Hill, 1990.
4. W. Tomasi, *Advanced Electronic Communications Systems,* 2nd ed., Englewood Cliffs, NJ: Prentice-Hall, 1992.
5. D.J. Nowicki, Voltage measurement and signal demodulation, in J. G. Webster, Ed., *Electrical Impedance Tomography,* Bristol, UK: Adam Hilger, 1990.
6. M.L. Meade, *Lock-in Amplifiers: Principles and Applications,* London: Peregrinus, 1984.
7. J.G. Webster, Ed., *Electrical Impedance Tomography,* Bristol, UK: Adam Hilger, 1990.
8. D.L. Hershberger, Build a synchronous detector for AM radio, *Pop. Electron.,* 20 (4), 61, 66–71, 1982.
9. G.A. Breed, Receiver basics—part 2. Fundamental receiver architectures, *RF Design,* 17 (3), 84, 86, 88–89, 1994.
10. Anonymous, AD630 balanced modulator/demodulator, in *Analog Devices* 1992 *Special Linear Reference Manual,* pp. 2.35–2.41, 1992.
11. J.G. Webster, Ed., *Medical Instrumentation: Application and Design.,* 3rd ed., New York: John Wiley & Sons, 1998.
12. D.M. Beams. Project TUNA — the development of a LabVIEW virtual instrument as a class project in a junior-level electronics course, *Comput. Educ. J.* 11, 1, 28–33, 2001. Additional information available at http://www.eng.uttyl.edu/usr/dbeams/home/Tuna/Project_TUNA.htm.
13. W.H. Beyer, *CRC Standard Mathematical Tables,* 28th ed., Boca Raton, FL: CRC Press, 1987.
14. J. Sherwin and T. Regan, FM remote speaker system, National Semiconductor Application Note AN-146, in *National Semiconductor Corp.* 1991 *Linear Applications Databook,* 1991.
15. J.J. DeFrance, *Communication Electronics Circuits,* 2nd ed., San Francisco, CA: Rinehart Press, 1972.
16. D.G. Sokol, R.B. Whitaker, J.J. Lord, and D.M. Beams, *Combine data center,* U.S. Patent No. 4,376,298, 1983.
17. K.P. Cohen, D. Panescu, J.H. Booske, J.G. Webster, and W.J. Tompkins, Design of an inductive plethysmograph for ventilation measurement, *Physiol. Meas.,* 15, 217–229, 1994.
18. W.J. Bachman, *Capacitive-type seed sensor for a planter monitor,* U.S. Patent No. 4,782,282, 1988.
19. W.J. Bachman, Private communication, 1995.

22 Filters

Rahman Jamal
National Instruments Germany

Robert Steer
Frequency Devices

22.1 Introduction

In its broadest sense, a filter can be defined as a signal processing system whose output signal, usually called the *response,* differs from the input signal, called the *excitation,* such that the output signal has some prescribed properties. In more practical terms an electric filter is a device designed to suppress, pass, or separate a group of signals from a mixture of signals according to the specifications in a particular application. The application areas of filtering are manifold, for example to band-limit signals before sampling to reduce aliasing, to eliminate unwanted noise in communication systems, to resolve signals into their frequency components, to convert discrete-time signals into continuous-time signals, to demodulate signals, etc. Filters are generally classified into three broad classes: *continuous-time, sampled-data,* and *discrete-time* filters depending on the type of signal being processed by the filter. Therefore, the concept of signals are fundamental in the design of filters.

A *signal* is a function of one or more independent variables such as time, space, temperature, etc. that carries information. The independent variables of a signal can either be continuous or discrete. Assuming that the signal is a function of time, in the first case the signal is called continuous-time and in the second, discrete-time. A continuous-time signal is defined at every instant of time over a given interval, whereas a discrete-time signal is defined only at discrete-time instances. Similarly, the *values* of a signal can also be classified in either continuous or discrete.

In real-world signals, often referred to as analog signals, both amplitude and time are continuous. These types of signals cannot be processed by digital machines unless they have been converted into

0-8493-1733-9/04/$0.00+$1.50

discrete-time signals. By contrast, a digital signal is characterized by discrete signal values, that are defined only at discrete points in time. Digital signal values are represented by a finite number of digits, which are usually binary coded. The relationship between a continuous-time signal and the corresponding discrete-time signal can be expressed in the following form:

$$x(kT) = x(t)_{t=kT}, \quad k = 0, 1, 2, \ldots, \tag{22.1}$$

where T is called the sampling period.

Filters can be classified on the basis of the input, output, and internal operating signals. A continuous data filter is used to process continuous-time or analog signals, whereas a digital filter processes digital signals. Continuous data filters are further divided into *passive* or *active* filters, depending on the type of elements used in their implementation. Perhaps the earliest type of filters known in the engineering community are *LC* filters, which can be designed by using discrete components like inductors and capacitors, or crystal and mechanical filters that can be implemented using *LC* equivalent circuits. Since no external power is required to operate these filters, they are often referred to as *passive* filters. In contrast, *active* filters are based on active devices, primarily *RC* elements, and amplifiers. In a sampled data filter, on the other hand, the signal is sampled and processed at discrete instants of time. Depending on the type of signal processed by such a filter, one may distinguish between an *analog sampled data* filter and a *digital* filter. In an analog sampled data filter the sampled signal can principally take any value, whereas in a digital filter the sampled signal is a digital signal, the definition of which was given earlier. Examples of analog sampled data filters are switched capacitor (SC) filters and charge-transfer device (CTD) filters made of capacitors, switches, and operational amplifiers.

22.2 Filter Classification

Filters are commonly classified according to the filter function they perform. The basic functions are: low-pass, high-pass, bandpass, and bandstop. If a filter passes frequencies from zero to its cutoff frequency Ω_c and stops all frequencies higher than the cutoff frequencies, then this filter type is called an ideal **lowpass filter.** In contrast, an ideal **high-pass filter** stops all frequencies below its cutoff frequency and passes all frequencies above it. Frequencies extending from Ω_1 to Ω_2 are passed by an ideal **bandpass filter,** while all other frequencies are stopped. An ideal bandstop filter stops frequencies from Ω_1 to Ω_2 and passes all other frequencies. Figure 22.1 depicts the magnitude functions of the four basic **ideal filter** types.

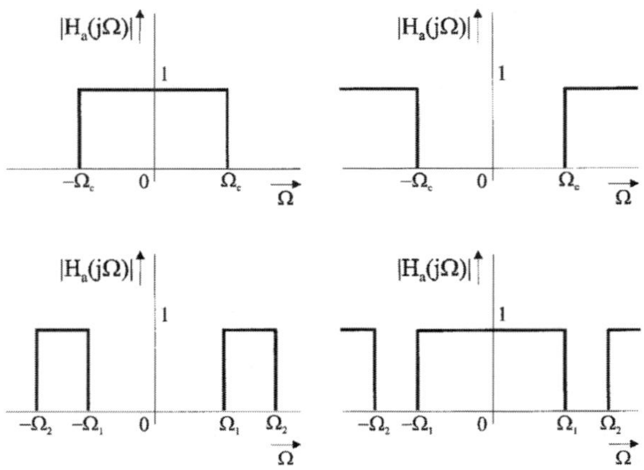

FIGURE 22.1 The magnitude function of an ideal filter is 1 in the passband and 0 in the stopband as shown for (a) low-pass, (b) high-pass, (c) bandpass, and (d) stopband filters.

So far we have discussed ideal filter characteristics having rectangular magnitude responses. These characteristics, however, are physically not realizable. As a consequence, the ideal response can only be approximated by some nonideal realizable system. Several classical approximation schemes have been developed, each of which satisfies a different criterion of optimization. This should be taken into account when comparing the performance of these filter characteristics.

22.3 The Filter Approximation Problem

Generally the input and output variables of a linear, time-invariant, causal filter can be characterized either in the time-domain through the convolution integral given by

$$y(t) = \int_0^t h_a(t-\tau)x(t)d\tau \tag{22.2}$$

or, equivalently, in the frequency-domain through the transfer function

$$H_a(s) = \frac{Y(s)}{X(s)} = \frac{\sum_{i=0}^{N} b_i s^i}{\sum_{i=0}^{N} a_i s^i} \Leftrightarrow H_a(s) = \frac{b_N}{a_N} \prod_{i=1}^{N} \left(\frac{s - s_{0i}}{s - s_{\infty i}} \right) \tag{22.3}$$

where $H_a(s)$ is the Laplace transform of the impulse response $h_a(t)$ and $X(s)$, $Y(s)$ are the Laplace transforms of the input signal $x(t)$ and the output or the filtered signal $y(t)$. $X(s)$ and $Y(s)$ are polynomials in $s = \sigma + j\Omega$ and the overall transfer function $H_a(s)$ is a real rational function of s with real coefficients. The zeroes of the polynomial $X(s)$ given by $s = s_{\infty i}$ are called the poles of $H_a(s)$ and are commonly referred to as the *natural frequencies* of the filter. The zeros of $Y(s)$ given by $s = s_{0i}$ which are equivalent to the zeroes of $H_a(s)$ are called the *transmission zeros* of the filter. Clearly, at these frequencies the filter output is zero for any finite input. Stability restricts the poles of $H_a(s)$ to lie in the left half of the s-plane excluding the $j\Omega$–axis, that is $\mathrm{Re}\{s_{\infty i}\} < 0$. For a stable transfer function $H_a(s)$ reduces to $H_a(j\Omega)$ on the $j\Omega$–axis, which is the continuous-time Fourier transform of the impulse response $h_a(t)$ and can be expressed in the following form:

$$H_a(j\Omega) = |H_a(j\Omega)| d^{j\theta(\Omega)} \tag{22.4}$$

where $|H_a(j\Omega)|$ is called the magnitude function and $\theta(\Omega) = \arg H_a(j\Omega)$ is the phase function. The gain magnitude of the filter expressed in decibels (dB) is defined by

$$\alpha(\Omega) = 20 \log |H_a(j\Omega)| = 10 \log |H_a(j\Omega)|^2 \tag{22.5}$$

Note that a filter specification is often given in terms of its attenuation, which is the negative of the gain function also given in decibels. While the specifications for a desired filter behavior are commonly given in terms of the loss response $\alpha(\Omega)$, the solution of the filter approximation problem is always carried out with the help of the characteristic function $C(j\Omega)$ giving

$$\alpha(\Omega) = 10 \log \left[1 + |C(j\Omega)|^2 \right] \tag{22.6}$$

Note that $\alpha(\Omega)$ is not a rational function, but $C(j\Omega)$ can be a polynomial or a rational function and approximation with polynomial or rational functions is relatively convenient. It can also be shown that

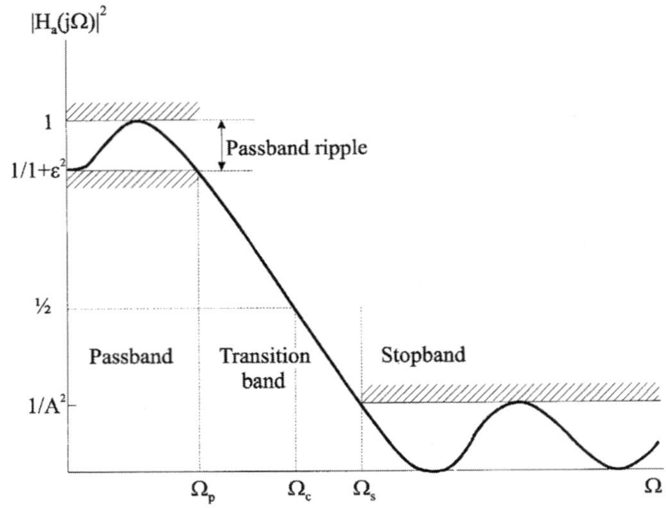

FIGURE 22.2 The squared magnitude function of an analog filter can have ripple in the passband and in the stopband.

frequency-dependent properties of $|C(j\Omega)|$ are in many ways identical to those of $\alpha(\Omega)$. The approximation problem consists of determining a desired response $|H_a(j\Omega)|$ such that the typical specifications depicted in Figure 22.2 are met. This so-called tolerance scheme is characterized by the following parameters:

- Ω_p Passband cutoff frequency (rad/s)
- Ω_s Stopband cutoff frequency (rad/s)
- Ω_c –3 dB cutoff frequency (rad/s)
- ε Permissible error in passband given by $\varepsilon = (10^{r/10} - 1)^{1/2}$, where r is the maximum acceptable attenuation in dB; note that $10 \log 1/(1 + \varepsilon^2)^{1/2} = -r$
- $1/A$ Permissible maximum magnitude in the stopband, i.e., $A = 10^{\alpha/20}$, where α is the minimum acceptable attenuation in dB; note that $20 \log (1/A) = -\alpha$.

The **passband** of a low-pass filter is the region in the interval $[0,\Omega_p]$ where the desired characteristics of a given signal are preserved. In contrast, the **stopband** of a low-pass filter (the region $[\Omega_s,\infty]$) rejects signal components. The **transition** band is the **region** between $(\Omega_x - \Omega_p)$, which would be 0 for an ideal filter. Usually, the amplitudes of the permissible ripples for the magnitude response are given in decibels.

The following sections review four different classical approximations: Butterworth, Chebyshev Type I, elliptic, and Bessel.

Butterworth Filters

The frequency response of an Nth-order Butterworth low-pass filter is defined by the squared magnitude function

$$|H_a(j\Omega)|^2 = \frac{1}{1+(\Omega/\Omega_c)^{2N}} \tag{22.7}$$

It is evident from the Equation 22.7 that the Butterworth approximation has only poles, i.e., no finite zeros and yields a maximally flat response around zero and infinity. Therefore, this approximation is also called maximally flat magnitude (MFM). In addition, it exhibits a smooth response at all frequencies and a monotonic decrease from the specified cutoff frequencies.

Equation 22.7 can be extended to the complex s-domain, resulting in

$$H_a(s)H_a(-s)=\frac{1}{1+\left(s/j\Omega_c\right)^{2N}} \tag{22.8}$$

The poles of this function are given by the roots of the denominator

$$s_k=\Omega_c e^{j\pi[1/2+(2k+1)/2N]},\quad k=0,1,\ldots,2N-1 \tag{22.9}$$

Note that for any N, these poles lie on the unit circle of radius Ω_c in the s-plane. To guarantee stability, the poles that lie in the left half-plane are identified with $H_a(s)$. As an example, we will determine the transfer function corresponding to a third-order Butterworth filter, i.e., $N = 3$.

$$H_a(s)H_a(-s)=\frac{1}{1+\left(-s^2\right)^3}=\frac{1}{1-s^6} \tag{22.10}$$

The roots of denominator of Equation 22.10 are given by

$$s_k=\Omega_c e^{j\pi[1/2+(2k+1)/6]},\quad k=0,1,2,3,4,5 \tag{22.11}$$

Therefore, we obtain

$$\begin{aligned}
s_0&=\Omega_c e^{j\pi 2/3}=-1/2+j\sqrt{2}/2\\
s_1&=\Omega_c e^{j\pi}=-1\\
s_2&=\Omega_c e^{j\pi 4/3}=-1/2-j\sqrt{3}/2\\
s_3&=\Omega_c e^{j\pi 5/3}=1/2-j\sqrt{3}/2\\
s_4&=\Omega_c e^{j2\pi}=1\\
s_5&=\Omega_c e^{j\pi/3}=1/2+j\sqrt{3}/2
\end{aligned} \tag{22.12}$$

The corresponding transfer function is obtained by identifying the left half-plane poles with $H_a(s)$. Note that for the sake of simplicity we have chosen $\Omega_c = 1$.

$$H_a(s)=\frac{1}{(s+1)\left(s+1/2-j\sqrt{3}/2\right)\left(s+1/2+j\sqrt{3}/2\right)}=\frac{1}{1+2s+2s^2+s^3} \tag{22.13}$$

Table 22.1 gives the Butterworth denominator polynomials up $N = 5$.

TABLE 22.1 Butterworth Denominator Polynomials

Order(N)	Butterworth Denominator Polynomials of $H(s)$
1	$s + 1$
2	$s^2 + \sqrt{2}s + 1$
3	$s^3 + 2s^2 + 2s + 1$
4	$s^4 + 2.6131s^3 + 3.4142s^2 + 2.6131s + 1$
5	$s^5 + 3.2361s^4 + 5.2361s^3 + 5.2361s^2 + 3.2361s + 1$

TABLE 22.2 Butterworth and Bessel Poles

	Butterworth Poles				Bessel Poles (–3 dB)			
	Re	Im(±*j*)			Re	Im(±*j*)		
N	*a*	*b*	Ω	*Q*	*a*	*b*	Ω	*Q*
1	–1.000	0.000	1.000	—	–1.000	0.000	1.000	—
2	–0.707	0.707	1.000	0.707	–1.102	0.636	1.272	0.577
3	–1.000	0.000	1.000	—	–1.323	0.000	1.323	—
	–0.500	0.866	1.000	1.000	–1.047	0.999	1.448	0.691
4	–0.924	0.383	1.000	0.541	–1.370	0.410	1.430	0.522
	–0.383	0.924	1.000	1.307	–0.995	1.257	1.603	0.805
5	–1.000	0.000	1.000	—	–1.502	0.000	1.502	—
	–0.809	0.588	1.000	0.618	–1.381	0.718	1.556	0.564
	–0.309	0.951	1.000	1.618	–0.958	1.471	1.755	0.916
6	–0.966	0.259	1.000	0.518	–1.571	0.321	1.604	0.510
	–0.707	0.707	1.000	0.707	–1.382	0.971	1.689	0.611
	–0.259	0.966	1.000	1.932	–0.931	1.662	1.905	1.023
7	–1.000	0.000	1.000	—	–1.684	0.000	1.684	—
	–0.901	0.434	1.000	0.555	–1.612	0.589	1.716	0.532
	–0.623	0.782	1.000	0.802	–1.379	1.192	1.822	0.661
	–0.223	0.975	1.000	2.247	–0.910	1.836	2.049	1.126
8	–0.981	0.195	1.000	0.510	–1.757	0.273	1.778	0.506
	–0.831	0.556	1.000	0.601	–1.637	0.823	1.832	0.560
	–0.556	0.831	1.000	0.900	–1.374	1.388	1.953	0.711
	–0.195	0.981	1.000	2.563	–0.893	1.998	2.189	1.226

Table 22.2 gives the Butterworth poles in real and imaginary components and in frequency and Q.

In the next example, the order N of a low-pass Butterworth filter is to be determined whose cutoff frequency (–3 dB) is $\Omega_c = 2$ kHz and stopband attenuation is greater than 40 dB at $\Omega_s = 6$ kHz. Thus the desired filter specification is

$$20 \log\left|H_a(j\Omega)\right| \le -40, \quad \Omega \ge \Omega_s \tag{22.14}$$

or equivalently,

$$\left|H_a(j\Omega)\right| \le 0.01, \quad \Omega \ge \Omega_s \tag{22.15}$$

It follows from Equation 22.7

$$\frac{1}{1+\left(\Omega_s/\Omega_c\right)^{2N}} = (0.01)^2 \tag{22.16}$$

Solving the above equation for N gives $N = 4.19$. Since N must be an integer, a fifth-order filter is required for this specification.

Chebyshev Filters or Chebyshev I Filters

The frequency response of an Nth-order Chebyshev low-pass filter is specified by the squared-magnitude frequency response function

$$\left|H_a(j\Omega)\right|^2 = \frac{1}{1+\varepsilon^2 T_N^2\left(\Omega/\Omega_p\right)} \tag{22.17}$$

where $T_N(x)$ is the Nth-order Chebyshev polynomial and ε is a real constant less than 1 which determines the ripple of the filter. Specifically, for nonnegative integers N, the Nth-order Chebyshev polynomial is given by

$$T_N(x) = \begin{cases} \cos\left(N \cos^{-1} x\right), & |x| \le 1 \\ \cosh\left(N \cosh^{-1} x\right), & |x| \ge 1 \end{cases} \tag{22.18}$$

High-order Chebyshev polynomials can be derived from the recursion relation

$$T_{N+1}(x) = 2xT_N(x) - T_{N-1}(x) \tag{22.19}$$

where $T_0(x) = 1$ and $T_1(x) = x$.

The Chebyshev approximation gives an **equiripple** characteristic in the passband and is maximally flat near infinity in the stopband. Each of the Chebyshev polynomials has real zeros that lie within the interval (–1,1) and the function values for $x \in [-1,1]$ do not exceed +1 and –1.

The pole locations for Chebyshev filter can be determined by generating the appropriate Chebyshev polynomials, inserting them into Equation 22.17, factoring, and then selecting only the left half plane roots. Alternatively, the pole locations P_k of an *N*th-order Chebyshev filter can be computed from the relation, for $k = 1 \rightarrow N$

$$P_k = -\sin \Theta_k \sinh \beta + j \cos \Theta_k \cosh \beta \tag{22.20}$$

where $\Theta_k = (2k - 1)\pi/2N$ and $\beta = \sinh^{-1}(1/\varepsilon)$.

Note: P_{N-k+1} and P_k are complex conjugates and when N is odd there is one real pole at

$$P_{N+1} = -2 \sinh \beta$$

For the Chebyshev polynomials, Ω_p is the last frequency where the amplitude response passes through the value of ripple at the edge of the passband. For odd N polynomials, where the ripple of the Chebyshev polynomial is negative going, it is the $[-1/(1 + \varepsilon_2)]_{(1/2)}$ frequency and for even N, where the ripple is positive going, is the 0 dB frequency.

The Chebyshev filter is completely specified by the three parameters ε, Ω_p, and N. In a practical design application, ε is given by the permissible passband ripple and Ω_p is specified by the desired passband cutoff frequency. The order of the filter, i.e., N, is then chosen such that the stopband specifications are satisfied.

Elliptic or Cauer Filters

The frequency response of an *N*th-order elliptic low-pass filter can be expressed by

$$\left|H_a(j\Omega)\right|^2 = \frac{1}{1 + \varepsilon^2 F_N^2\left(\Omega/\Omega_p\right)} \tag{22.21}$$

where $F_N(\cdot)$ is called the Jacobian elliptic function. The elliptic approximation yields an equiripple passband and an equiripple stopband. Compared with the same-order Butterworth or Chebyshev filters, the elliptic design provides the sharpest transition between the passband and the stopband. The theory of elliptic filters, initially developed by Cauer, is involved, therefore for an extensive treatment refer to Reference 1.

Elliptic filters are completely specified by the parameters ε, α, Ω_p, Ω_s, and N

where ε = passband ripple
a = stopband floor
Ω_p = the frequency at the edge of the passband (for a designated passband ripple)
Ω_s = the frequency at the edge of the stopband (for a designated stopband floor)
N = the order of the polynomial

In a practical design exercise, the desired passband ripple, stopband floor, and Ω_s are selected and N is determined and rounded up to the nearest integer value. The appropriate Jacobian elliptic function must be selected and $H_a(j\Omega)$ must be calculated and factored to extract only the left plane poles. For some synthesis techniques, the roots must expanded into polynomial form.

This process is a formidable task. While some filter manufacturers have written their own computer programs to carry out these calculations, they are not readily available. However, the majority of applications can be accommodated by use of published tables of the pole/zero configurations of low-pass elliptic transfer functions. An extensive set of such tables for a common selection of passband ripples, stopband floors, and shape factors is available in Reference 2.

Bessel Filters

The primary objectives of the preceding three approximations were to achieve specific loss characteristics. The phase characteristics of these filters, however, are nonlinear. The Bessel filter is optimized to reduce nonlinear phase distortion, i.e., a maximally flat delay. The transfer function of a Bessel filter is given by

$$H_a(s) = \frac{B_0}{B_N(s)} = \frac{B_0}{\sum_{k=0}^{N} B_k s^k}, \quad B_k = \frac{(2N-k)!}{2^{N-k} k!(N-k)!} \quad k = 0, 1, \ldots, N \tag{22.22}$$

where $B_N(s)$ is the Nth-order Bessel polynomial. The overall squared-magnitude frequency response function is given by

$$\left|H_a(j\Omega)\right|^2 = 1 - \frac{\Omega^2}{2N-1} + \frac{2(N-1)\Omega^4}{(2N-1)^2(2N-3)} + \cdots \tag{22.23}$$

To illustrate Equation 22.22 the Bessel transfer function for $N = 4$ is given below:

$$H_a(s) = \frac{105}{105 + 105s + 45s^2 + 10s^3 + s^4} \tag{22.24}$$

Table 22.2 lists the factored pole frequencies as real and imaginary parts and as frequency and Q for Bessel transfer functions that have been normalized to $\Omega_c = -3$ dB.

22.4 Design Examples for Passive and Active Filters

Passive *R, L, C* Filter Design

The simplest and most commonly used passive filter is the simple, first-order ($N = 1$) *R–C* filter shown in Figure 22.3. Its transfer function is that of a first-order Butterworth low-pass filter. The transfer function and –3 dB Ω_c are

$$H_a(s) = \frac{1}{RCs+1} \quad \text{where} \quad \Omega_c = \frac{1}{RC} \tag{22.25}$$

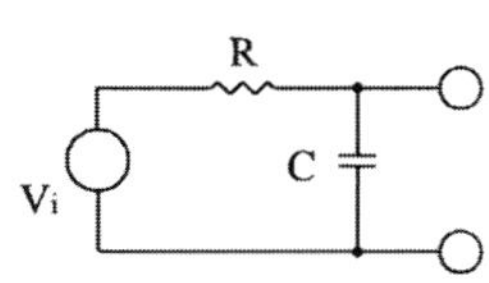

FIGURE 22.3 A passive first-order *RC* filter can serve as an antialiasing filter or to minimize high-frequency noise.

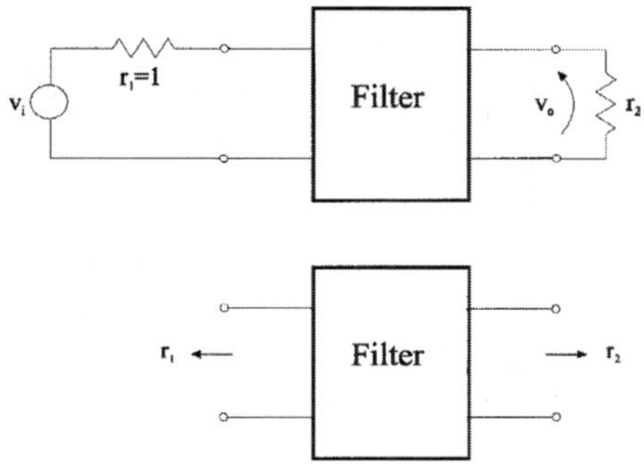

FIGURE 22.4 A passive filter can have the symbolic representation of a doubly terminated filter.

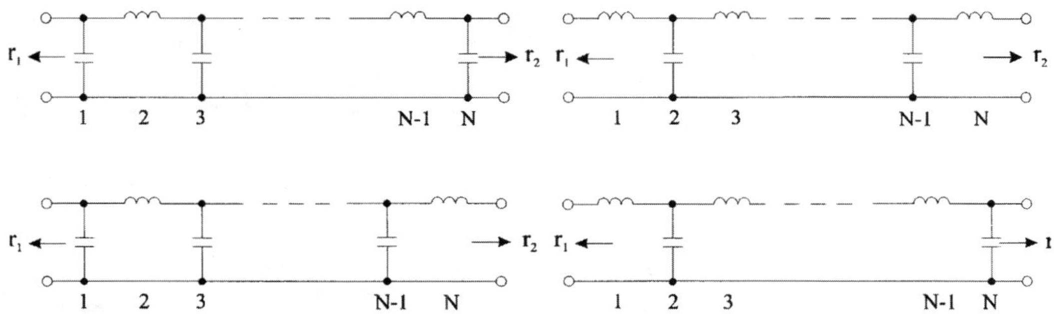

FIGURE 22.5 Even and odd N passive all-pole filter networks can be realized by several circuit configurations (N odd, above; N even, below).

While this is the simplest possible filter implementation, both source and load impedance change the dc gain and/or corner frequency and its rolloff rate is only first order, or –6 dB/octave.

To realize higher-order transfer functions, passive filters use R, L, C elements usually configured in a ladder network. The design process is generally carried out in terms of a doubly terminated two-port network with source and load resistors R_1 and R_2 as shown in Figure 22.4. Its symbolic representation is given below.

The source and load resistors are normalized in regard to a reference resistance $R_B = R_1$, i.e.,

$$r_i = \frac{R_1}{R_B} = 1, \quad r_2 = \frac{R_2}{R_B} = \frac{R_2}{R_1} \tag{22.26}$$

The values of L and C are also normalized in respect to a reference frequency to simplify calculations. Their values can be easily scaled to any desired set of actual elements.

$$l_v = \frac{\Omega_B L_v}{R_B}, \quad c_v = \Omega_B C_v R_B \tag{22.27}$$

Low-pass filters, whose magnitude-squared functions have no finite zero, i.e., whose characteristic functions $C(j\Omega)$ are polynomials, can be realized by lossless ladder networks consisting of inductors as the series elements and capacitors as the shunt elements. These types of approximations, also referred to as *all-pole approximations,* include the previously discussed Butterworth, Chebyshev Type I, and Bessel filters. Figure 22.5 shows four possible ladder structures for even and odd N, where N is the filter order.

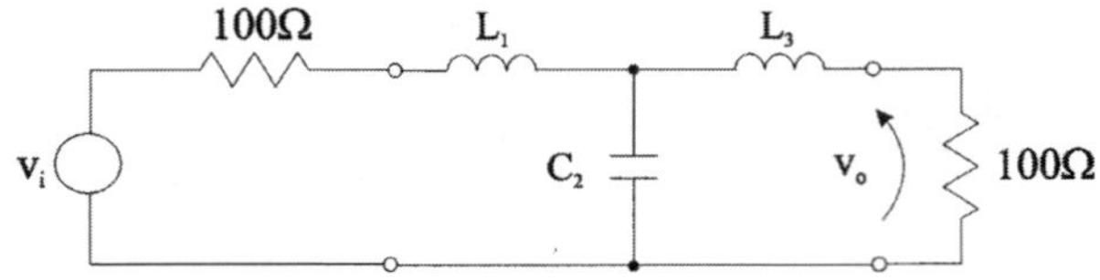

FIGURE 22.6 A third-order passive all-pole filter can be realized by a doubly terminated third-order circuit.

TABLE 22.3 Element Values for Low-Pass Filter Circuits

		N = 2, Element Number		N = 3, Element Number		
Filter Type	r_2	1	2	1	2	3
Butterworth	∞	1.4142	0.7071	1.5000	1.3333	0.5000
	1	1.4142	1.4142	1.0000	2.0000	1.0000
Chebyshev type I	∞	0.7159	0.4215	1.0895	1.0864	0.5158
0.1-dB ripple	1	—	—	1.0316	1.1474	1.0316
Chebyshev type I	∞	0.9403	0.7014	1.3465	1.3001	1.5963
0.5 dB ripple	1	—	—	1.5963	1.0967	1.5963
Bessel	∞	1.0000	0.3333	0.8333	0.4800	0.1667
	1	1.5774	0.4227	1.2550	0.5528	0.1922

In the case of doubly terminated Butterworth filters, the normalized values are precisely given by

$$a_\nu = 2\sin\left(\frac{(2\nu-1)\pi}{2N}\right), \quad \nu = 1, \ldots, N \tag{22.28}$$

where a_ν is the normalized L or C element value. As an example we will derive two possible circuits for a doubly terminated Butterworth low-pass of order 3 with $R_B = 100\ \Omega$ and a cutoff frequency $\Omega_c = \Omega_B = 10$ kHz. The element values from Equation 22.28 are

$$\begin{aligned} l_1 &= 2\sin\left(\frac{(2-1)\pi}{6}\right) = 1 \Rightarrow L_1 = \frac{R_B}{\Omega_c} = 1.59\text{ mH} \\ c_2 &= 2\sin\left(\frac{(4-1)\pi}{6}\right) = 2 \Rightarrow C_2 = \frac{2}{\Omega_c R_B} = 3.183\text{ nF} \\ l_3 &= 2\sin\left(\frac{(6-1)\pi}{6}\right) = 1 \Rightarrow L_3 = \frac{R_B}{\Omega_c} = 1.59\text{ mH} \end{aligned} \tag{22.29}$$

A possible realization is shown in Figure 22.6.

Table 22.3 gives normalized element values for the various all-pole filter approximations discussed in the previous section up to order 3 and is based on the following normalization:

1. $r_1 = 1$;
2. All the cutoff frequencies (end of the ripple band for the Chebyshev approximation) are Ω_c = 1 rad/s;
3. r_2 is either 1 or ∞, so that both singly and doubly terminated filters are included.

The element values in Table 22.3 are numbered from the source end in the same manner as in Figure 22.4. In addition, empty spaces indicate unrealizable networks. In the case of the Chebyshev filter, the amount of ripple can be specified as desired, so that in the table only a selective sample can be given. Extensive tables of prototype element values for many types of filters can be found in Reference 4.

The example given above, of a Butterworth filter of order 3, can also be verified using Table 22.3. The steps necessary to convert the normalized element values in the table into actual filter values are the same as previously illustrated.

In contrast to all-pole approximations, the characteristic function of an elliptic filter function is a rational function. The resulting filter will again be a ladder network but the series elements may be parallel combinations of capacitance and inductance and the shunt elements may be series combinations of capacitance and inductance.

Figure 22.5 illustrates the general circuits for even and odd N, respectively. As in the case of all-pole approximations, tabulations of element values for normalized low-pass filters based on elliptic approximations are also possible. Since these tables are quite involved the reader is referred to Reference 4.

Active Filter Design

Active filters are widely used and commercially available with cutoff frequencies from millihertz to megahertz. The characteristics that make them the implementation of choice for several applications are small size for low-frequency filters because they do not use inductors; precision realization of theoretical transfer functions by use of precision resistors and capacitors; high input impedance that is easy to drive and for many circuit configurations the source impedance does not effect the transfer function; low-output impedance that can drive loads without effecting the transfer function and can drive the transient, switched-capacitive, loads of the input stages of A/D converters and low (N+THD) performance for pre-A/D antialiasing applications (as low as –100 dBc).

Active filters use R, C, A (operational amplifier) circuits to implement polynomial transfer functions. They are most often configured by cascading an appropriate number of first- and second-order sections.

The simplest first-order ($N = 1$) active filter is the first-order passive filter of Figure 22.3 with the addition of a unity gain follower amplifier. Its cutoff frequency (Ω_c) is the same as that given in Equation 22.25. Its advantage over its passive counterpart is that its operational amplifier can drive whatever load that it can tolerate without interfering with the transfer function of the filter.

The vast majority of higher-order filters have poles that are not located on the negative real axis in the s-plane and therefore are in complex conjugate pairs that combine to create second-order pole pairs of the form:

$$H(s) = s^2 + \frac{\omega_p}{Q}s + \omega_p^2 \Leftrightarrow s^2 + 2as + a^2 + b^2 \tag{22.30}$$

where $p_1, p_2 = a \pm jb$

$$\omega_p^2 = a^2 + b^2$$

$$Q = \frac{\omega_p}{2a} = \frac{\sqrt{(a^2 + b^2)}}{2a}$$

The most commonly used two-pole active filter circuits are the *Sallen and Key* low-pass resonator, the *multiple feedback* bandpass, and the *state variable* implementation as shown in Figure 22.7a, b, and c. In the analyses that follow, the more commonly used circuits are used in their simplest form. A more comprehensive treatment of these and numerous other circuits can be found in Reference 20.

The Sallen and Key circuit of Figure 22.7a is used primarily for its simplicity. Its component count is the minimum possible for a two-pole active filter. It cannot generate stopband zeros and therefore is limited in its use to monotonic roll-off transfer functions such as Butterworth and Bessel filters. Other limitations are that the phase shift of the amplifier reduces the Q of the section and the capacitor ratio becomes large for high-Q circuits. The amplifier is used in a follower configuration and therefore is subjected to a large common mode input signal swing which is not the best condition for low distortion performance. It is recommended to use this circuit for a section $Q < 10$ and to use an amplifier whose gain bandwidth product is greater than 100 f_p.

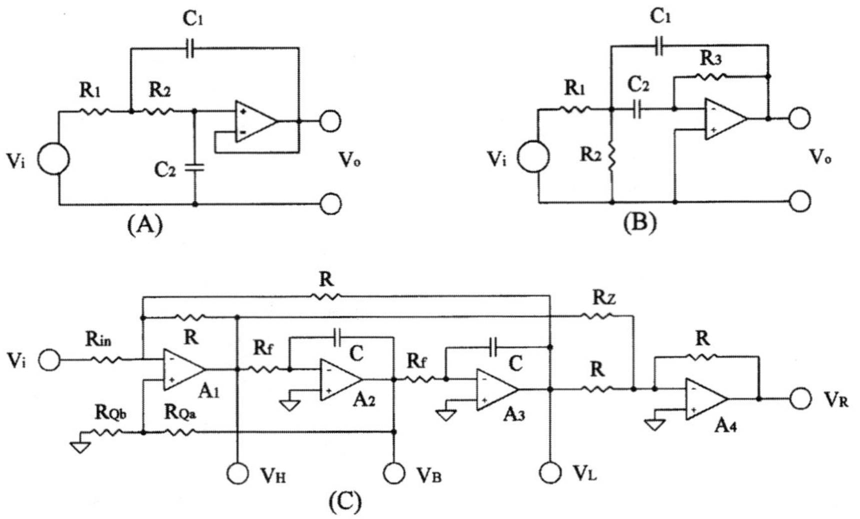

FIGURE 22.7 Second-order active filters can be realized by common filter circuits: (A) Sallen and Key low-pass, (B) multiple feedback bandpass, (C) state variable.

The transfer function and design equations for the Sallen and Key circuit of Figure 22.7a are

$$H(s) = \frac{\dfrac{1}{R_1R_2C_1C_2}}{s^2 + \dfrac{1}{R_1C_2}s + \dfrac{1}{R_1R_2C_1C_2}} = \frac{\omega_p^2}{s^2 + \dfrac{\omega_p}{Q}s + \omega_p^2} \tag{22.31}$$

from which obtains

$$\omega^2 = \frac{1}{R_1R_2C_1C_2}, \quad Q = \omega_p R_1C_2 = \sqrt{\frac{R_1C_2}{R_2C_1}} \tag{22.32}$$

$$R_1, R_2 = \frac{1}{4\pi f_p QC_2}\left[1 \pm \frac{4Q^2C_2}{C_1}\right] \tag{22.33}$$

which has valid solutions for

$$\frac{C_1}{C_2} \geq 4Q^2 \tag{22.34}$$

In the special case where

$$\begin{aligned} &R_1 = R_2 = R, \text{ then} \\ &C = 1/2\pi Rf_p, \;\; C_1 = 2QC, \text{ and } \; C_2 = C/2Q \end{aligned} \tag{22.35}$$

The design sequence for Sallen and Key low-pass of Figure 22.7a is as follows:

For a required f_p and Q, select C_1, C_2 to satisfy Equation 22.34. Compute R_1, R_2 from Equation 22.33 (or Equation 22.35 if R_1 is chosen to equal R_2) and scale the values of C_1 and C_2 and R_1 and R_2 to desired impedance levels.

As an example, a three-pole low-pass active filter is shown in Figure 22.8. It is realized with a buffered single-pole *RC* low-pass filter section in cascade with a two-pole Sallen and Key section.

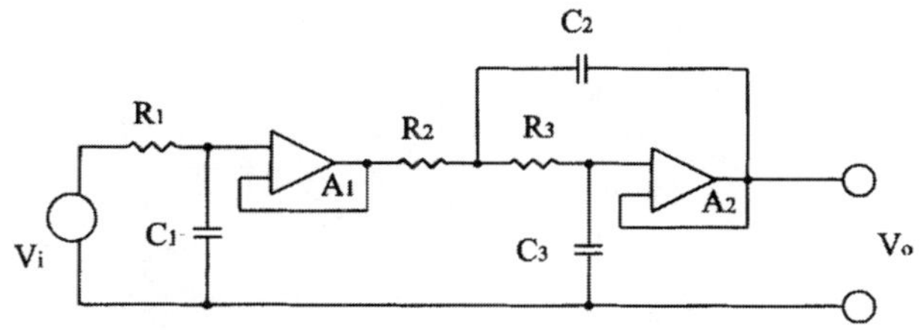

FIGURE 22.8 A three-pole Butterworth active can be configured with a buffered first-order *RC* in cascade with a two-pole Sallen and Key resonator.

To construct a three-pole Butterworth filter, the pole locations are found in Table 22.2 and the element values in the sections are calculated from Equation 22.25 for the single real pole and in accordance with the Sallen and Key design sequence listed above for the complex pole pair.

From Table 22.2, the normalized pole locations are

$$f_{p1} = 1.000, \qquad f_{p2} = 1.000, \qquad \text{and} \quad Q_{p2} = 1.000$$

For a cutoff frequency of 10 kHz and if it is desired to have an impedance level of 10 kΩ, then the capacitor values are computed as follows:

For $R_1 = 10$ kΩ:

$$\text{from Equation 22.25,}\ C_1 = \frac{1}{2\pi R_1 r_{p1}} = \frac{1}{2\pi(10{,}000)(10{,}000)} = \frac{10^{-6}}{200\pi} = 0.00159\ \mu\text{F}$$

For $R_2 = R_3 = R = 10$ kΩ:

$$\text{from Equation 22.35,}\ C = \frac{1}{2\pi R f_{p2}} = \frac{1}{2\pi(10{,}000)(10{,}000)} = \frac{10^{-6}}{200\pi} = 0.00159\ \mu\text{F}$$

from which

$$C_2 = 2QC = 2(0.00159)\ \mu\text{F} = 0.00318\ \mu\text{F}$$

$$C_3 = C/2Q = 0.5(0.00159)\ \mu\text{F} = 0.000795\ \mu\text{F}$$

The *multiple feedback* circuit of Figure 22.7b is a minimum component count, two-pole (or one-pole pair), bandpass filter circuit with user definable gain. It cannot generate stopband zeros and therefore is limited in its use to monotonic roll-off transfer functions. Phase shift of its amplifier reduces the Q of the section and shifts the f_p. It is recommended to use an amplifier whose open loop gain at f_p is $> 100Q^2H_p$.

The design equations for the *multiple feedback* circuit of Figure 22.4b are

$$H(s) = \frac{\dfrac{s}{R_1C_1}}{s^2 + \dfrac{s}{R_3}\left(\dfrac{1}{C_1} + \dfrac{1}{C_2}\right) + \dfrac{(R_1+R_2)}{R_1R_2R_3C_1C_2}} = -\frac{\dfrac{s\omega_p H_p}{Q}}{s^2 + \dfrac{s\omega_p}{Q} + \omega_p^2} \tag{22.36}$$

when $s = j\omega_p$, the gain H_p is

$$H_p = \frac{R_3C_2}{R_1(C_1+C_2)} \tag{22.37}$$

From Equation 22.36 and 22.37 for a required set of ω_p, Q, and H_p:

$$R_1 = \frac{Q}{C_1 H_p \omega_p}, \quad R_2 = \frac{Q}{\omega_p}\left(\frac{1}{Q^2(C_1+C_2) - H_p C_1}\right), \quad R_3 = \frac{R_1 H_p (C_1+C_2)}{C_2} \tag{22.38}$$

For R_2 to be realizable,

$$Q^2(C_1+C_2) \geq H_p C_1 \tag{22.39}$$

The design sequence for a *multiple feedback* bandpass filter is as follows:

Select C_1 and C_2 to satisfy Equation 22.39 for the H_p and Q required. Compute R_1, R_2, and R_3. Scale R_1, R_2, R_3, C_1, and C_2 as required to meet desired impedance levels.

Note that it is common to use $C_1 = C_2 = C$ for applications where $H_p = 1$ and $Q > 0.707$.

The *state variable* circuit of Figure 22.7c is the most widely used active filter circuit. It is the basic building block of programmable active filters and of switched capacitor designs. While it uses three or four amplifiers and numerous other circuit elements to realize a two-pole filter section, it has many desirable features. From a single input it provides low-pass (V_L), high-pass (V_H), and bandpass (V_B) outputs and by summation into an additional amplifier (A_4) (or the input stage of the next section) a band reject (V_R) or stopband zero can be created. Its two integrator resistors connect to the virtual ground of their amplifiers (A_2, A_3) and therefore have no signal swing on them. Therefore, programming resistors can be switched to these summing junctions using electronic switches. The sensitivity of the circuit to the gain and phase performance of its amplifiers is more than an order of magnitude less than single amplifier designs. The open-loop gain at f_p does not have to be multiplied by either the desired Q or the gain at dc or f_p. Second-order sections with Q up to 100 and f_p up to 1 MHz can be built with this circuit.

There are several possible variations of this circuit that improve its performance at particular outputs. The input can be brought into several places to create or eliminate phase inversions; the damping feedback can be implemented in several ways other than the R_{Qa} and R_{Qb} that are shown in Figure 22.7c and the f_p and Q of the section can be adjusted independently from one another. dc offset adjustment components can be added to allow the offset at any one output to be trimmed to zero.

For simplicity of presentation, Figure 22.7c makes several of the resistors equal and identifies others with subscripts that relate to their function in the circuit. Specifically, the feedback amplifier A_1, that generates the V_H output has equal feedback and input resistor from the V_L feedback signal to create unity gains from that input. Similarly, the "zero summing" amplifier, A_4 has equal resistors for its feedback and input from V_L to make the dc gain at the V_R output the same as that at V_L. More general configurations with all elements included in the equation of the transfer function are available in numerous reference texts including Reference 20.

The *state variable* circuit, as configured in Figure 22.7c, has four outputs. Their transfer functions are

$$V_L(s) = -\frac{R}{R_i (R_f C)^2}\left(\frac{1}{D(s)}\right) \tag{22.40a}$$

$$V_B(s) = \frac{R}{R_i}\left(\frac{\frac{s}{(R_f C)}}{D(s)}\right) \tag{22.40b}$$

$$V_{\mathrm{H}}(s) = -\frac{R}{R_{\mathrm{i}}}\left(\frac{s^2}{D(s)}\right) \tag{22.40c}$$

$$V_{\mathrm{R}}(s) = \frac{R}{R_{\mathrm{i}}(R_{\mathrm{f}}C)^2}\left(\frac{\left(\frac{R_{\mathrm{z}}}{R}\right)s^2 + 1}{D(s)}\right) \tag{22.40d}$$

where

$$D(s) = s^2 + \frac{a}{R_{\mathrm{f}}C}s + \frac{1}{(R_{\mathrm{f}}C)^2} = s^2 + \frac{\omega_{\mathrm{p}}}{Q}s + \omega_{\mathrm{p}}^2 \quad a = \frac{R_{\mathrm{Qb}}}{(R_{\mathrm{Qa}} + R_{\mathrm{Qb}})}\left(2 + \frac{R}{R_{\mathrm{i}}}\right) \tag{22.41}$$

Note that the dc gain at the low-pass output is

$$V_{\mathrm{L}}(0) = -\frac{R}{R_{\mathrm{i}}} \tag{22.42a}$$

from which obtains

$$\omega_{\mathrm{p}} = \frac{1}{R_{\mathrm{f}}C} \quad \text{and} \quad \frac{1}{Q} = \frac{R_{\mathrm{Qb}}}{(R_{\mathrm{Qa}} + R_{\mathrm{Qb}})}\left(2 + \frac{R}{R_{\mathrm{i}}}\right) \tag{22.42b}$$

The design sequence for the state variable filter of Figure 22.7c is

Select the values of R_{f} and C to set the frequency ω_{p}, the values of R_{i} for the desired dc gain and R_{Qa} and R_{Qb} for the desired Q and dc gain.

22.5 Discrete-Time Filters

A digital filter is a circuit or a computer program that computes a discrete output sequence from a discrete input sequence. Digital filters belong to the class of discrete-time LTI (linear time invariant) systems, which are characterized by the properties of causality, recursibility, and stability, and may be characterized in the time domain by their impulse response and in the transform domain by their transfer function. The most general case of a discrete-time LTI system with the input sequence denoted by $x(kT)$ and the resulting output sequence $y(kT)$ can be described by a set *of linear difference equations with constant coefficients*.

$$y(kT) = \sum_{\mu=0}^{N} b_{\mu}x(kT - \mu T) - \sum_{\mu=1}^{N} a_{\mu}y(kT - \mu T) \tag{22.43}$$

where $a_0 = 1$. An equivalent relation between the input and output variables can be given through the convolution sum in terms of the impulse response sequence $h(kT)$:

$$y(kT) = \sum_{\mu=0}^{N} h(kT)x(kT - \mu T) \tag{22.44}$$

The corresponding transfer function is given by

$$H(z)=\frac{Y(z)}{X(z)}=\frac{\sum_{\mu=0}^{N} b_\mu z^{-\mu}}{1+\sum_{\mu=1}^{N} a_\mu z^{-\nu}} \Leftrightarrow H(z)=b_0\prod_{\mu=1}^{N}\left(\frac{z-z_{0\mu}}{z-z_{\infty\mu}}\right) \tag{22.45}$$

where $H(z)$ is the z-transform of the impulse response $h(kT)$ and $X(z)$, $Y(z)$ are the z-transform of the input signal $x(kT)$ and the output or the filtered signal $y(kT)$. As can be seen from Equation 22.44, if for at least one μ, $a_\mu \neq 0$, the corresponding system is recursive; its impulse response is of infinite duration — **infinite impulse response (IIR) filter**. If $a_\mu = 0$, the corresponding system is nonrecursive — **finite impulse response (FIR) filter**; its impulse response is of finite duration and the transfer function $H(z)$ is a polynomial in z^{-1}. The zeros of the polynomial $X(z)$ given by $z = z_{\infty i}$ are called the poles of $H(z)$ and are commonly referred to as the *natural frequencies* of the filter. The condition for the stability of the filter is expressed by the constraint that all the poles of $H(z)$ should lie inside the unit circle, that is $|z_{\infty i}| < 1$. The zeros of $Y(z)$ given by $z = z_{0t}$ which are equivalent to the zeros of $H(z)$ are called the *transmission zeros* of the filter. Clearly, at these frequencies the output of the filter is zero for any finite input.

On the unit circle, the transfer function frequency $H(z)$ reduces to the frequency response function $H(e^{j\omega T})$, the discrete-time Fourier transform of $h(kT)$, which in general is complex and can be expressed in terms of magnitude and phase

$$H\left(e^{j\omega T}\right)=\left|H\left(e^{j\omega T}\right)\right|e^{j\theta(\omega)} \tag{22.46}$$

The gain function of the filter is given as

$$\alpha(\Omega)=20\log_{10}\left|H\left(e^{j\omega T}\right)\right| \tag{22.47}$$

It is also common practice to call the negative of the gain function the attenuation. Note that the attenuation is a positive number when the magnitude response is less than 1.

Figure 22.9 gives a block diagram realizing the difference equation of the filter, which is commonly referred to as the *direct-form I* realization. Notice that the element values for the multipliers are obtained directly from the numerator and denominator coefficients of the transfer function. By rearranging the structure in regard to the number of delays, one can obtain the canonic structure called *direct-form II* shown in Figure 22.10, which requires the minimum number of delays.

Physically, the input numbers are samples of a continuous signal and real-time digital filtering involves the computation of the iteration of Equation 22.43 for each incoming new input sample. Design of a filter consists of determining the constants a_μ and b_μ that satisfies a given filtering requirement. If the filtering is performed in real time, then the right side of Equation 22.46 must be computed in less than the sampling interval T.

22.6 Digital Filter Design Process

The digital filter design procedure consists of the following basic steps:

1. Determine the desired response. The desired response is normally specified in the frequency domain in terms of the desired magnitude response and/or the desired phase response.
2. Select a class of filters (e.g., linear-phase FIR filters or IIR filters) to approximate the desired response.
3. Select the best member in the filter class.

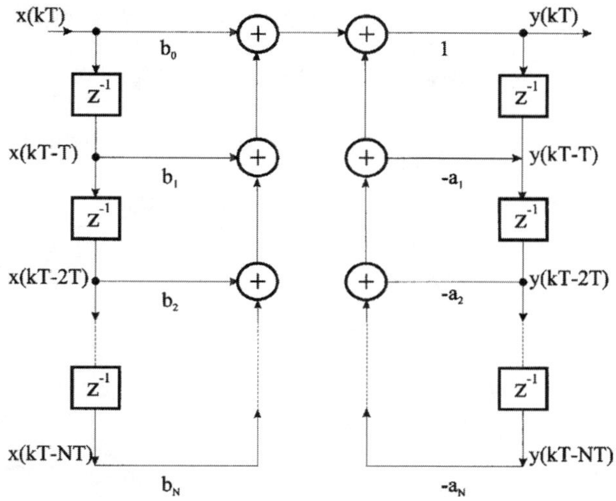

FIGURE 22.9 The difference equation of a digital filter can be realized by a direct-form I implementation that uses separate delay paths for the *X* and *Y* summations.

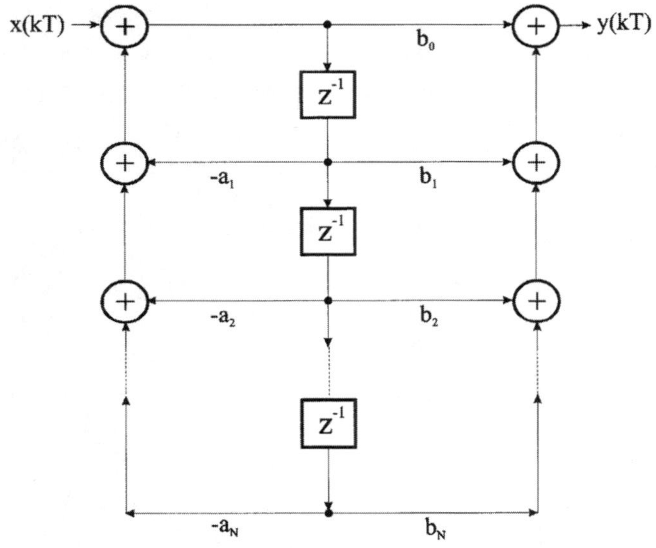

FIGURE 22.10 A direct-form II implementation of the difference equations minimizes the number of delay elements.

4. Implement the best filter using a general-purpose computer, a DSP, or a custom hardware chip.
5. Analyze the filter performance to determine whether the filter satisfies all the given criteria.

22.7 FIR Filter Design

In many digital signal-processing applications, FIR filters are generally preferred over their IIR counterparts, because they offer a number of advantages compared with their IIR equivalents. Some of the good properties of FIR filters are a direct consequence of their nonrecursive structure. First, FIR filters are inherently stable and free of limit cycle oscillations under finite-word length conditions. In addition, they exhibit a very low sensitivity to variations in the filter coefficients. Second, the design of FIR filters with exactly *linear phase* (constant group delay) vs. frequency behavior can be accomplished easily. This property is useful in many application areas, such as speech processing, phase delay equalization, image processing, etc.

Finally, there exists a number of efficient algorithms for designing optimum FIR filters with arbitrary specifications. The main disadvantage of FIR filters over IIR filters is that FIR filter designs generally require, particularly in applications requiring narrow transition bands, considerably more computation to implement.

An FIR filter of order N is described by a difference equation of the form

$$y(kT) = \sum_{\mu=0}^{N} b_\mu x(kT - \mu T) \tag{22.48}$$

and the corresponding transfer function is

$$H(z) = \frac{Y(z)}{X(z)} = \sum_{\mu=0}^{N} b_\mu z^{-\mu} \tag{22.49}$$

The objective of FIR filter design is to determine $N \pm 1$ coefficients given by

$$h(0), h(1), \ldots, h(N) \tag{22.50}$$

so that the transfer function $H(e^{j\omega T})$ approximates a desired frequency characteristic. Note that because Equation 22.47 is also in the form of a convolution summation, the impulse response of an FIR filter is given by

$$h(kT) = \begin{cases} b_\mu, & k = 0, 1, \ldots, N \\ 0 & \text{otherwise} \end{cases} \tag{22.51}$$

Two equivalent structures for FIR filters are given in Figure 22.11.

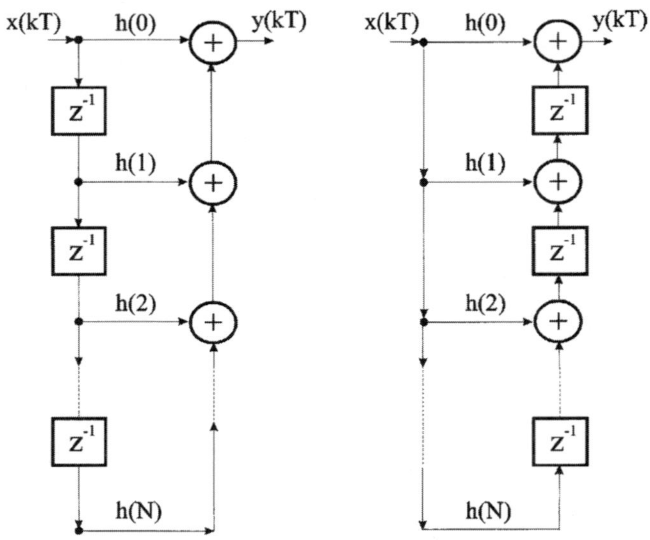

FIGURE 22.11 The sequence of the delays and summations can be varied to produce alternative direct-form implementations.

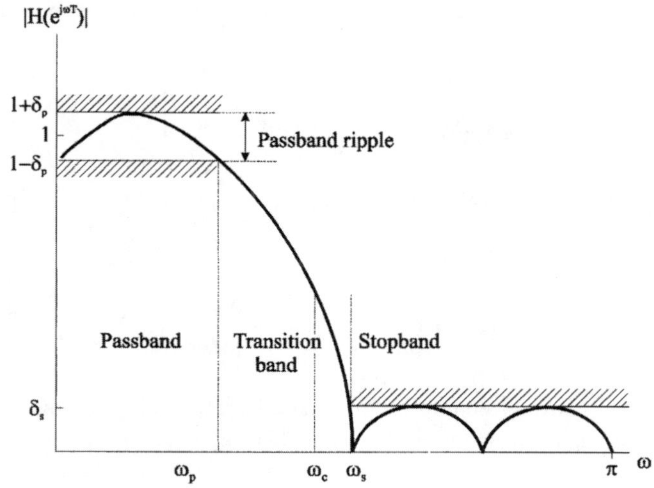

FIGURE 22.12 Tolerance limits must be defined for an FIR low-pass filter magnitude response.

The accuracy of an FIR approximation is described by the following parameters:

δ_p passband ripple
δ_s stopband attenuation
$\Delta\omega$ transition bandwidth

These quantities are depicted in Figure 22.12 for a prototype low-pass filter. δ_p and δ_s characterize the permissible errors in the passband and in the stopband, respectively. Usually, the passband ripple and stopband attenuation are given in decibels, in which case their values are related to the parameters δ_p and δ_s by

$$\text{Passband ripple (dB): } A_p = -20\log_{10}(1-\delta_p) \tag{22.52}$$

$$\text{Stopband ripple (dB): } A_s = 020\log_{10}(\delta_s) \tag{22.53}$$

Note that due to the symmetry and periodicity of the magnitude response of $|H(e^{j\omega T})|$, it is sufficient to give the filter specifications in the interval $0 \le \omega \le \pi$.

Windowed FIR Filters

Several design techniques can be employed to synthesize linear-phase FIR filters. The simplest implementation is based on *windowing*, which commonly begins by specifying the ideal frequency response and expanding it in a Fourier series and then truncating and smoothing the ideal impulse response by means of a window function. The truncation results in large ripples before and after the discontinuity of the ideal frequency response known as the Gibbs phenomena, which can be reduced by using a window function that tapers smoothly at both ends. Filters designed in this way possess equal passband ripple and stopband attenuation, i.e.,

$$\delta_p = \delta_s = \delta \tag{22.54}$$

To illustrate this method, let us define an ideal desired frequency response that can be expanded in a Fourier series

$$H_d(e^{j\omega T}) = \sum_{k=-\infty}^{\infty} h_d(kT)e^{-jk\omega T} \tag{22.55}$$

where $h_d(kT)$ is the corresponding impulse response sequence, which can be expressed in terms of $H_d(e^{j\omega T})$ as

$$h_d(kT) = \frac{1}{2\pi}\int_{-\pi}^{\pi} H_d\left(e^{j\omega T}\right) e^{jk\omega T}\, d\omega \tag{22.56}$$

The impulse response of the desired filter is then found by weighting this ideal impulse response with a window $w(kT)$ such that

$$h(kT) = \begin{cases} w(kT)h_d(kT), & 0 \le k \le N \\ 0, & \text{otherwise} \end{cases} \tag{22.57}$$

Note that for $w(kT)$ in the above-given interval we obtain the rectangular window. Some commonly used windows are Bartlett (triangular), Hanning, Hamming, Blackmann, etc., the definitions of which can be found in Reference 15.

As an example of this design method, consider a low-pass filter with a cutoff frequency of ω_c and a desired frequency of the form

$$H_d\left(e^{j\omega T}\right) = \begin{cases} e^{-j\omega NT/2}, & |\omega| \le \omega_c \\ 0, & \omega_c < |\omega| \le \pi, \end{cases} \tag{22.58}$$

Using Equation 22.56 we obtain the corresponding ideal impulse response

$$h_d(kT) = \frac{1}{2\pi}\int_{-\omega_c}^{\omega_c} e^{-j\omega TN/2} e^{jk\omega T}\, d\omega = \frac{\sin\left[\omega_c\left(kT - TN/2\right)\right]}{\pi\left(kT - TN/2\right)} \tag{22.59}$$

Choosing $N = 4$, $\omega_c = 0.6\pi$, and a Hamming window defined by

$$w(kT) = \begin{cases} 0.54 - 0.46\cos\left(2\pi kT/N\right), & 0 \le k \le N \\ 0, & \text{otherwise} \end{cases} \tag{22.60}$$

we obtain the following impulse response coefficients:

$$\begin{aligned} h(0) &= -0.00748 \\ h(1) &= 0.12044 \\ h(2) &= -0.54729 \\ h(3) &= 0.27614 \\ h(4) &= -0.03722 \end{aligned} \tag{22.61}$$

Optimum FIR Filters

As mentioned earlier, one of the principal advantages of FIR filters over their IIR counterparts is the availability of excellent design methods for optimizing arbitrary filter specifications. Generally, the design criterion for the optimum solution of an FIR filter design problem can be characterized as follows. The maximum error between the approximating response and the given desired response has to be minimized, i.e.,

$$E\left(e^{j\omega T}\right) = W_{\mathrm{d}}\left(e^{j\omega T}\right)\left\|H_{\mathrm{d}}\left(e^{j\omega T}\right)\right| - \left|H\left(e^{j\omega T}\right)\right\| \tag{22.62}$$

where $E(e^{j\omega T})$ is the weighted error function on a close range X of $[0,\pi]$ and $W_{\mathrm{d}}(e^{j\omega T})$ a weighting function, which emphasizes the approximation error parameters in the design process. If the maximum absolute value of this function is less than or equal to ε on X, i.e.,

$$\varepsilon = \max_{\omega \in X}\left|E\left(e^{j\omega T}\right)\right| \tag{22.63}$$

the desired response is guaranteed to meet the given criteria. Thus, this optimization condition implies that the best approximation must have an equiripple error function. The most frequently used method for designing optimum magnitude FIR filters is the Parks–McClellan algorithm. This method essentially reduces the filter design problem into a problem in polynomial approximation in the Chebyshev approximation sense as discussed above. The maximum error between the approximation and the desired magnitude response is minimized. It offers more control over the approximation errors in different frequency bands than is possible with the window method. Using the Parks–McClellan algorithm to design FIR filters is computationally expensive. This method, however, produces optimum FIR filters by applying time-consuming iterative techniques. A FORTRAN program for the Parks–McClellan algorithm can be found in the IEEE publication Programs for DSP in Reference 12. As an example of an equiripple filter design using the Parks–McClellan algorithm, a sixth-order low-pass filter with a passband $0 \le \omega \le 0.6\pi$, a stopband $0.8\pi \le \omega \le \pi$, and equal weighting for each band was designed by means of this program.

The resulting impulse response coefficients are

$$\begin{aligned} h(0) &= h(6) = -0.00596 \\ h(1) &= h(5) = -0.18459 \\ h(2) &= h(4) = 0.25596 \\ h(3) &= 0.70055 \end{aligned} \tag{22.64}$$

Design of Narrowband FIR Filters

When using conventional techniques to design FIR filters with especially narrow bandwidths, the resulting filter lengths may be very high. FIR filters with long filter lengths often require lengthy design and implementation times, and are more susceptible to numerical inaccuracy. In some cases, conventional filter design techniques, such as the Parks–McClellan algorithm, may fail the design altogether. A very efficient algorithm called the interpolated finite impulse response (IFIR) filter design technique can be employed to design narrowband FIR filters. Using this technique produces narrowband filters that require far fewer coefficients than those filters designed by the direct application of the Parks–McClellan algorithm. For more information on IFIR filter design, see Reference 7.

22.8 IIR Filter Design

The main advantage of IIR filters over FIR filters is that IIR filters can generally approximate a filter design specification using a lower-order filter than that required by an FIR design to perform similar filtering operations. As a consequence, IIR filters execute much faster and do not require extra memory, because they execute in place. A disadvantage of IIR filters, however, is that they have a nonlinear phase response. The two most common techniques used for designing IIR filters will be discussed in this section. The first approach involves the transformation of an analog prototype filter. The second method is an optimization-based approach allowing the approximation of an arbitrary frequency response.

The transformation approach is quite popular because the approximation problem can be reduced to the design of classical analog filters, the theory of which is well established, and many closed-form design methods exist. Note that this is not true for FIR filters, for which the approximation problems are of an entirely different nature. The derivation of a transfer function for a desired filter specification requires the following three basic steps:

1. Given a set of specifications for a digital filter, the first step is to map the specifications into those for an equivalent analog filter.
2. The next step involves the derivation of a corresponding analog transfer function for the analog prototype.
3. The final step is to translate the transfer function of the analog prototype into a corresponding digital filter transfer function.

Once the corresponding analog transfer function for the analog prototype is derived, it must be transformed using a transformation that maps $H_a(s)$ into $H(z)$. The simplest and most appropriate choice for s is the well-known bilinear transform of the z-variable

$$s = \frac{2\left(1-z^{-1}\right)}{T_d\left(1+z^{-1}\right)} \Leftrightarrow z = \frac{1+\left(T_d/2\right)s}{1-\left(T_d/2\right)s} \tag{22.65}$$

which maps a stable analog filter in the s-plane into a stable digital filter in the z-plane. Substituting s with the right-hand side of Equation 22.63 in $H_a(s)$ results in

$$H(z) = H_a\left(\frac{2\left(1-z^{-1}\right)}{T_d\left(1+z^{-1}\right)}\right) \Rightarrow H\left(e^{j\omega T}\right)\Big|_{z=e^{j\omega T}} = H_a\left(\frac{2j}{T_d}\tan\left(\frac{\omega T}{2}\right)\right) \tag{22.66}$$

As it can be seen from Equation 22.66, the analog frequency domain (imaginary axis) maps onto the digital frequency domain (unit circle) nonlinearly. This phenomena is called frequency warping and must be compensated in a practical implementation. For low frequencies Ω and ω are approximately equal. We obtain the following relation between the analog frequency Ω and the digital frequency ω

$$\Omega = \frac{2}{T_d}\tan\left(\omega T/2\right) \tag{22.67}$$

$$\omega = \frac{2}{T}\arctan\left(\Omega T_d/2\right) \tag{22.68}$$

The overall bilinear transformation procedure is as follows:

1. Convert the critical digital frequencies (e.g., ω_p and ω_s for low-pass filters) to the corresponding analog frequencies in the s-domain using the relationship given by Equation 22.67.
2. Derive the appropriate continuous prototype transfer function $H_a(s)$ that has the properties of the digital filter at the critical frequencies.
3. Apply the bilinear transform to $H_a(s)$ to obtain $H(z)$ which is the required digital filter transfer function.

To illustrate the three-step IIR design procedure using the bilinear transform, consider the design of a second-order Butterworth low-pass filter with a cutoff frequency of $\omega_c = 0.3\pi$. The sampling rate of the digital filter is to be $f_s = 10$ Hz, giving $T = 0.1$ s. First, we map the cutoff frequency to the analog frequency

$$\Omega_c = \frac{2}{0.1}\tan\left(0.15\pi\right) = 10.19 \text{ rad/s} \tag{22.69}$$

The poles of the analog Butterworth filter transfer function $H_a(s)$ are found using Equation 22.11. As explained earlier, these poles lie equally spaced in the s-plane on a circle of radius Ω_c.

$$H_a(s) = \frac{1}{s^2 + \sqrt{2}\Omega_c s + \Omega_c^2} \tag{22.70}$$

Application of the bilinear transformation

$$s = \frac{2(1 - z^{-1})}{0.1(1 + z^{-1})} \tag{22.71}$$

gives the digital transfer function

$$H(z) = \frac{0.00002 + 0.00004z^{-1} + 0.00002z^{-2}}{1 - 1.98754z^{-1} + 0.98762z^{-2}} \tag{22.72}$$

The above computations were carried out using Reference 9, which greatly automates the design procedure.

Design of Arbitrary IIR Filters

The IIR filter design approach discussed in the previous section is primarily suitable for frequencyselective filters based on closed-form formulas. In general, however, if a design other than standard lowpass, high-pass, bandpass, and stopband is required, or if the frequency responses of arbitrary specifications are to be matched, in such cases it is often necessary to employ algorithmic methods implemented on computers. In fact, for nonstandard response characteristics, algorithmic procedures may be the only possible design approach. Depending on the error criterion used, the algorithmic approach attempts to minimize the approximation error between the desired frequency response $H_d(e^{j\omega T})$ and $H(e^{j\omega T})$ or between the time-domain response $h_d(kT)$ and $h(kT)$. Computer software is available for conveniently implementing IIR filters approximating arbitrary frequency response functions [8,9].

Cascade-Form IIR Filter Structures

Recall that theoretically there exist an infinite number of structures to implement a digital filter. Filters realized using the structure defined by Equation 22.44 directly are referred to as direct-form IIR filters. The direct-form structure, however, is not employed in practice except when the filter order $N \leq 2$, because they are known to be sensitive to errors introduced by coefficient quantization and by finite-arithmetic conditions. Additionally, they produce large round-off noise, particularly for poles closed to the unit circle.

Two less-sensitive structures can be obtained by partial fraction expansion or by factoring the right-hand side of Equation 22.46 in terms of real rational functions of order 1 and 2. The first method leads to *parallel connections* and the second one to *cascade connections* of corresponding lower-order sections, which are used as building blocks to realize higher-order transfer functions. In practice, the cascade form is by far the preferred structure, since it gives the freedom to choose the pairing of numerators and denominators and the ordering of the resulting structure. Figure 22.13 shows a cascade-form implementation, whose overall transfer function is given by

$$H(z) = \prod_{k=1}^{M} H_k(z) \tag{22.73}$$

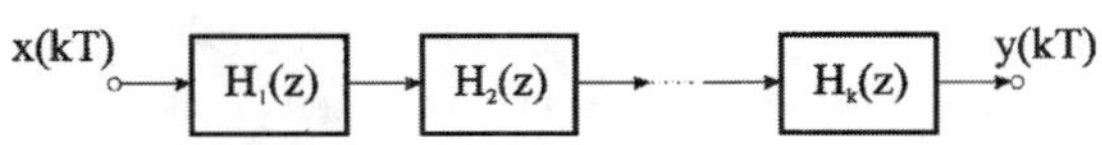

FIGURE 22.13 An IIR filter can be implemented by a cascade of individual transfer functions.

where the transfer function of the kth building block is

$$H_k(z) = \frac{b_{0k} + b_{1k}z^{-1} + b_{2k}z^{-2}}{1 + a_{1k}z^{-2} + a_{2k}z^{-2}} \tag{22.74}$$

Note this form is achieved by factoring Equation 22.45 into second-order sections.

There are, of course, many other realization possibilities for IIR filters, such as state-space structures [9], lattice structures [10], and wave structures. The last is introduced in the next section.

22.9 Wave Digital Filters

It was shown earlier that for recursive digital filters the approximation problem can be reduced to classical design problems by making use of the bilinear transform. For wave digital filters (WDFs) this is carried one step farther in that the structures are obtained directly from classical circuits. Thus, to every WDF there corresponds an *LCR* reference filter from which it is derived. This relationship accounts for their excellent properties concerning coefficient sensitivity, dynamic range, and all aspects of stability under finite-arithmetic conditions. The synthesis of WDFs is based on the wave network characterization; therefore, the resulting structures are referred to as wave digital filters. To illustrate the basic idea behind the theory of WDFs, consider an inductor L, which is electrically described by $V(s) = sLI(s)$. In the next step we define wave variables $A_1(s)$ and $B_1(s)$ as

$$\begin{aligned} A_1(s) &= V(s) + RI(s) \\ B_1(s) &= V(s) - RI(s) \end{aligned} \tag{22.75}$$

where R is called the port resistance. Substituting $V(s) = sLI(s)$ in the above relation and replacing s in $A_1(s)$ and $B_1(s)$ with the bilinear transform given by Equation 22.65, we obtain

$$B(z) = \frac{(1 - z^{-1})L - (1 + z^{-1})R}{(1 - z^{-1})L + (1 + z^{-1})R} A(z) \tag{22.76}$$

Letting $R = L$, the above relation reduces to

$$B(z) = -z^{-1} A(z) \tag{22.77}$$

Thus an inductor translates into a unit delay in cascade with an inverter in the digital domain. Similarly, it is easily verified that a capacitance can be simulated by a unit delay and a resistor by a digital sink. Figure 22.14 shows the digital realizations of impedances and other useful one-port circuit elements.

To establish an equivalence with classical circuits fully, the interconnections are also simulated by so-called wave adaptors. The most important of these interconnections are series and parallel connections, which are simulated by series and parallel adaptors, respectively. For most filters of interest, only two- and three-port adaptors are employed. For a complete design example consider Figure 22.15.

For a given *LC* filter, one can readily derive a corresponding WDF by using the following procedure. First, the various interconnections in the *LC* filter are identified as shown in Figure 22.15. In the next step the electrical elements in the *LC* filter are replaced by its digital realization using Figure 22.15. Finally, the interconnections are substituted using adaptors. Further discussions and numerical examples dealing with WDFs can be found in References 3, 13, and 14.

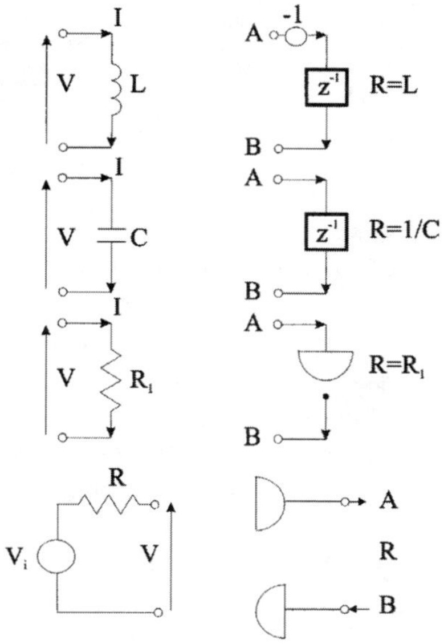

FIGURE 22.14 Digital filter implementations use functional equivalents to one-port linear filter elements.

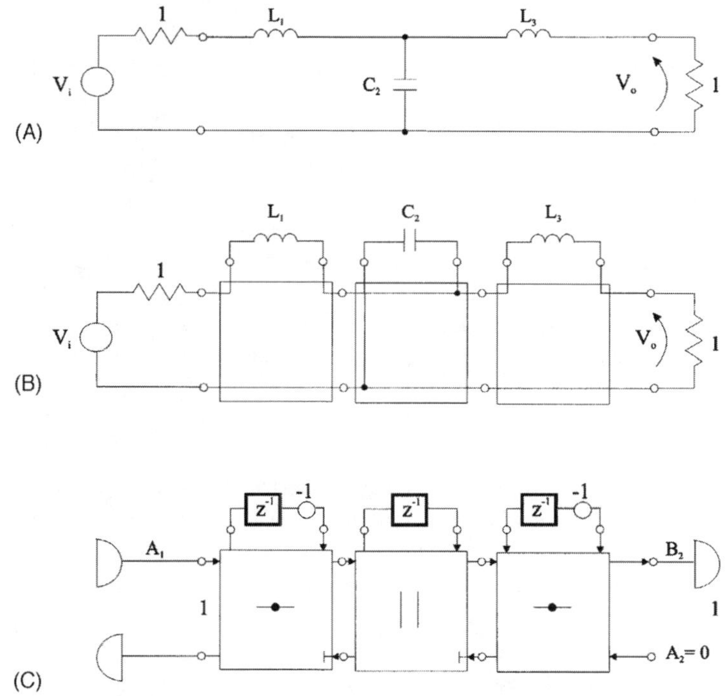

FIGURE 22.15 Digital wave filters establish equivalence with classical filter circuits by use of wave adapter substitutions: (A) *LC* reference low-pass; (B) identification of wire interconnections; (C) corresponding wave digital filter.

22.10 Antialiasing and Smoothing Filters

In this section two practical application areas of filters in the analog conditioning stage of a data acquisition system are discussed. A block diagram of a typical data acquisition system is shown in Figure 22.16, consisting of an **antialiasing filter** before the analog-to-digital converter (ADC) and a **smoothing filter** after the digital-to-analog converter (DAC).

FIGURE 22.16 A data acquisition system with continuous time inputs and outputs uses antialias prefiltering, an A/D converter, digital signal processing, a D/A converter, and an output smoothing filter.

For a complete discrete reconstruction of a time-continuous, band-limited input signal having the spectrum $0 \le f \le f_{max}$, the sampling frequency must be, according to the well-known Shannon's sampling theorem, at least twice the highest frequency in the time signal. In our case, in order to be able to represent frequencies up to f_{max}, the sampling frequency $f_s = 1/T > 2f_{max}$. The necessary band limiting to $f \le f_{max}$ of the input time-continuous signal is performed by a low-pass filter, which suppresses higher spectral components greater than f_{max}. Violation of this theorem results in alias frequencies. As a result, frequency components above $f_x/2$, the so-called Nyquist frequency, appear as frequency components below $f_x/2$. Aliasing is commonly addressed by using antialiasing filters to attenuate the frequency components at and above the Nyquist frequency to a level below the dynamic range of an ADC before the signal is digitized. Ideally, a low-pass filter with a response defined by

$$H(j\Omega) = \begin{cases} 1, & |\Omega| \le \pi/T \\ 0, & |\Omega| \le \pi/T \end{cases} \tag{22.78}$$

is desired to accomplish this task. In practice, a variety of techniques based on the principles of continuous-time analog low-pass filter design can be employed to approximate this "brick-wall" type of characteristic. Antialiasing filters typically exhibit attenuation slopes in the range from 45 to 120 dB/octave and stopband rejection from 75 to 100 dB. Among the types of filters more commonly used for antialias purposes are the Cauer elliptic, Bessel, and Butterworth. The optimum type of filter depends on which kinds of imperfections, e.g., gain error, phase nonlinearity, passband and stopband ripple, etc., are most likely to be tolerated in a particular application. For example, Butterworth filters exhibit very flat frequency response in the passband, while Chebyshev filters provide steeper attenuation at the expense of some passband ripple. The Bessel filter provides a linear phase response over the entire passband but less attenuation in the stopband. The Cauer elliptic filter, with its extremely sharp roll-off, is especially useful as an antialiasing filter for multichannel digitizing data acquisition systems. However, the large-phase nonlinearity makes it more appropriate for applications involving analysis of the frequency content of signals as opposed to phase content or waveform shape.

Many considerations discussed above also apply to smoothing filters. Due to the sampling process, the frequency response after the digital-to-analog conversion becomes periodic with a period equal to the sampling frequency. The quantitization steps that are created in the DAC reconstruction of the output waveform and are harmonically related to the sampling frequency must be suppressed through a low-pass filter having the frequency response of Equation 22.78, also referred to as a smoothing or reconstruction filter. While an antialiasing filter on the input avoids unwanted errors that would result from undersampling the input, a smoothing filter at the output reconstructs a continuous-time output from the discrete-time signal applied to its input.

Consideration must be given to how much antialiasing protection is needed for a given application. It is generally desirable to reduce all aliasable frequency components (at frequencies greater than half of the sampling frequency) to less than the LSB of the ADC being used. If it is possible that the aliasable input can have an amplitude as large as the full input signal range of the ADC, then it is necessary to attenuate it by the full 2^N range of the converter. Since each bit of an ADC represents a factor of 2 from the ones adjacent to it, and 20 log(2) = 6 dB, the minimum attenuation required to reduce a full-scale input to less than a LSB is

$$\alpha < -20N(6\text{ dB}) \tag{22.79}$$

where N is the number of bits of the ADC.

The amount of attenuation required can be reduced considerably if there is knowledge of the input frequency spectrum. For example, some sensors, for reasons of their electrical or mechanical frequency response, might not be able to produce a full-scale signal at or above the Nyquist frequency of the system and therefore "full-scale" protection is not required. In many applications, even for 16-bit converters that, in the worst case, would require 96 dB of antialias protection, 50 to 60 dB is adequate.

Additional considerations in antialias protection of the system are the noise and distortion that are introduced by the filter that is supposed to be eliminating aliasable inputs. It is possible to have a perfectly clean input signal which, when it is passed through a prefilter, gains noise and harmonic distortion components in the frequency range and of sufficient amplitude to be within a few LSBs of the ADC. The ADC cannot distinguish between an actual signal that is present in the input data and a noise or distortion component that is generated by the prefilter. It is necessary that both noise and distortion components in the output of the antialias filter must also be kept within an LBS of the ADC to ensure system accuracy.

22.11 Switched-Capacitor Filters

Switched-capacitor (SC) filters, also generally referred to as analog sampled data filters, provide an alternative to conventional active-*RC* filters and are commonly used in the implementation of adjustable antialiasing filters. SC filters comprise switches, capacitors, and op amps. Essentially, an SC replaces the resistor in the more traditional analog filter designs. Because the impedance of the SC is a function of the switching frequency, one can vary the cutoff frequency of the SC filter by varying the frequency of the clock signal controlling the switching. The main advantage of SC filters is that they can be implemented in digital circuit process technology, since the equivalent of large resistors can be simulated by capacitors having small capacitance values.

When using SC filters, one must also be aware that they are in themselves a sampling device that requires antialias protection on the input and filtering on their outputs to remove clock feedthrough. However, since clock frequencies are typically 50 to 100 times f_c of the filter, a simple first or second *RC* filter on their inputs and outputs will reduce aliases and noise sufficiently to permit their use with 12- to 14-bit ADCs. One need also to consider that they typically have dc offset errors that are large, vary with time, temperature, and programming or clock frequency. Interested readers may refer to References 5 and 14.

22.12 Adaptive Filters

Adaptive filtering is employed when it is necessary to realize or simulate a system whose properties vary with time. As the input characteristics of the filter change with time, the filter coefficients are varied with time as a function of the filter input. Some typical applications of adaptive filtering include spectral estimation of speech, adaptive equalization, echo cancellation, and adaptive control, to name just a few. Depending on the application, the variations in the coefficients are carried out according to an optimization criterion and the adaptation is performed at a rate up to the sampling rate of the system. The self-adjustment capability of adaptive filter algorithms is very valuable when the application environment cannot be precisely described. Some of the most widely used adaptive algorithms are LMS (least-mean square), RLS (recursive least-squares), and frequency domain, also known as block algorithm. The fundamental concept of an adaptive filter is depicted in Figure 22.17.

An adaptive filter is characterized by the filter input $x(kT)$ and the desired response $d(kT)$. The error sequence $\varepsilon(kT)$ formed by

$$\varepsilon(kT) = \sum_{\mu=0}^{N-1} w_\mu(kT)\, x(kT - \mu T) \tag{22.80}$$

and $x(kT),\ldots,x(kT - T(N-1))$ serve as inputs to an adaptive algorithm that recursively determines the coefficients $w_0(kT + T),\ldots,w_{N-1}(kT + T)$. A number of adaptive algorithms and structures can be found

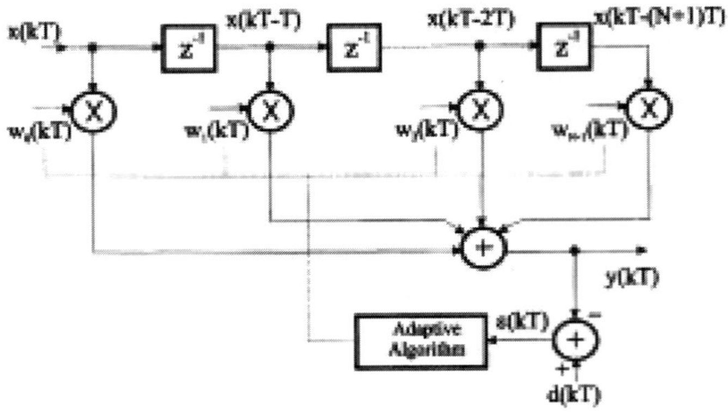

FIGURE 22.17 An adaptive filter uses an adaptive algorithm to change the performance of a digital filter in response to defined conditions.

in the literature that satisfy different optimization criteria in different application areas. For more detailed developments refer to References 1, 15, and 16.

Defining Terms

Antialiasing filter: Antialiasing filters remove any frequency elements above the Nyquist frequency. They are employed before the sampling operation is conducted to prevent aliasing in the sampled version of the continuous-time signal.

Bandpass filter: A filter whose passband extends from a lower cutoff frequency to an upper cutoff frequency. All frequencies outside this range are stopped.

Equiripple: Characteristic of a frequency response function whose magnitude exhibits equal maxima and minima in the passband.

Finite impulse response (FIR) filter: A filter whose response to a unit impulse function is of finite length, i.e., identically zero outside a finite interval.

High-pass filter: A filter that passes all frequencies above its cutoff frequency and stops all frequencies below it.

Ideal filter: An ideal filter passes all frequencies within its passband with no attenuation and rejects all frequencies in its stopband with infinite attenuation. There are five basic types of ideal filters: low-pass, high-pass, bandpass, stopband, and all-pass.

Infinite impulse response (IIR) filter: A filter whose response to a unit impulse function is of infinite length, i.e., nonzero for infinite number of samples.

Low-pass filter: A filter that attenuates the power of any signals with frequencies above its defined cutoff frequency.

Passband: The range of frequencies of a filter up to the cutoff frequency.

Stopband: The range of frequencies of a filter above the cutoff frequency.

Transition region: The range of frequencies of a filter between a passband and a stopband.

References

1. S. Mitra and J. Kaiser, *Handbook for Digital Signal Processing*, New York: John Wiley & Sons, 1993.
2. E. Christian and E. Eisenmann, *Filter Design Tables and Graphs*, New York: John Wiley & Sons, 1966.
3. A. Antoniou, *Digital Filter: Analysis and Design*, New York: McGraw-Hill, 1979.
4. R. Saal, *Handbuch zum Filterentwurf [Handbook of filter design]*, Frankfurt: Allgemeine Elektricitäts-Gesellschaft AEG-Telefunken, 1979.
5. G. Temes and S. Mitra, *Modern Filter Theory and Design*, New York: John Wiley & Sons, 1973.
6. E. Cunningham, *Digital Filtering*, New York: John Wiley & Sons, 1995.

7. P.P. Vaidyanathan, *Multirate Systems and Filter Banks*, Englewood Cliffs, NJ: Prentice-Hall, 1993.
8. M. Cerna and R. Jamal, The design of digital filters using graphical programming techniques, *Proc. MessComp*, 232–238, 1995.
9. *LabVIEW Digital Filter Design Toolkit*, Austin, TX: National Instruments, 1996.
10. R.A. Roberts and C.T. Multis, *Digital Signal Processing*, Reading, MA: Addison-Wesley, 1987.
11. A.H. Gray, Jr. and J.D. Markel, Digital lattice and ladder filter synthesis, *IEEE Trans. Acoust. Speech Signal Process.*, ASSP-23: 268–277, 1975.
12. DSP Committee, IEEE ASSP, Eds., *Programs for Digital Signal Processing*, New York: IEEE Press, 1979.
13. F. Taylor, *Digital Filter Design Handbook*, New York: Marcel Dekker, 1983.
14. A. Fettweis, *Wave digital filters: theory and practice, Proc. IEEE*, 74, 270–327, 1986.
15. W.K. Chen, *The Circuits and Filters Handbook*, Boca Raton, FL: CRC Press, 1995.
16. M.L. Honig and D. Messerschmitt, *Adaptive Filters — Structures, Algorithms, and Applications*, Boston: Kluwer Academic Publishers, 1984.
17. M. Bellanger, *Digital Processing of Signals — Theory and Practice*, New York: John Wiley & Sons, 1988.
18. R.C. Dorf, Ed., *The Electrical Engineering Handbook*, Boca Raton, FL: CRC Press, 1993.
19. A. Zverev, *Handbook of Filter Synthesis*, New York: John Wiley & Sons, 1967.
20. C. Lindquist, *Active Network Design*, Long Beach, CA: Steward & Sons, 1977.

Ronney B. Panerai
University of Leicester

A. Roberto Ambrosini
Institute of Radioastronomy

C. Bortolotti
Institute of Radioastronomy

N. D'Amico
Institute of Radioastronomy

G. Grueff
Institute of Radioastronomy

S. Mariotti
Institute of Radioastronomy

S. Montebugnoli
Institute of Radioastronomy

A. Orfei
Institute of Radioastronomy

G. Tomassetti
University of L'Aquila

23

Spectrum Analysis and Correlation

23.1 FFT Spectrum Analysis and Correlation

Ronney B. Panerai

Most sensors and instruments described in previous sections of this handbook can produce continuous measurements in time or sequential measurements at fixed or variable time intervals, as represented in Figure 23.1. The temporal patterns resulting from such measurements are usually referred to as *signals.* Signals can either be *continuous* or *discrete in time* (Figure 23.1). The main objective of *spectral analysis* is to provide an estimate of the distribution of signal power at different frequencies. Spectral analysis and correlation techniques are an aid to the interpretation of signals and to the systems that generate them. These methods are now widely used for the analysis and interpretation of measurements performed in medicine, geophysics, vibration analysis, communications, and several other areas.

Although the original concept of a **signal** involves measurements as a function of time (Figure 23.1), this term has been generalized to include measurements along other dimensions, e.g., distance. In addition, signals can have multiple dimensions — the instantaneous velocity of an airplane can be regarded as a four-dimensional signal since it depends on time and three spatial coordinates.

With the growing availability of signal-processing computer packages and dedicated instruments, most readers will perform spectral analysis and correlation at the "touch of a button," visualizing results on a screen or as a computer plot. These "black-box" systems are useful for saving time and money, but users should be aware of the limitations of the fundamental techniques and circumstances in which inappropriate use can lead to misleading results. This chapter presents the basic concepts of spectral analysis and correlation based on the **fast Fourier transform (FFT)** approach. FFT algorithms allow the most efficient computer implementation of methods to perform spectral analysis and correlation and have become the most popular option. Nevertheless, other approaches, such as parametric techniques, wavelet transforms, and time-frequency analysis are also available. These will be briefly discussed and the interested reader will be directed to the pertinent literature for applications that might benefit from alternative approaches.

0-8493-1733-9/04/$0.00+$1.50

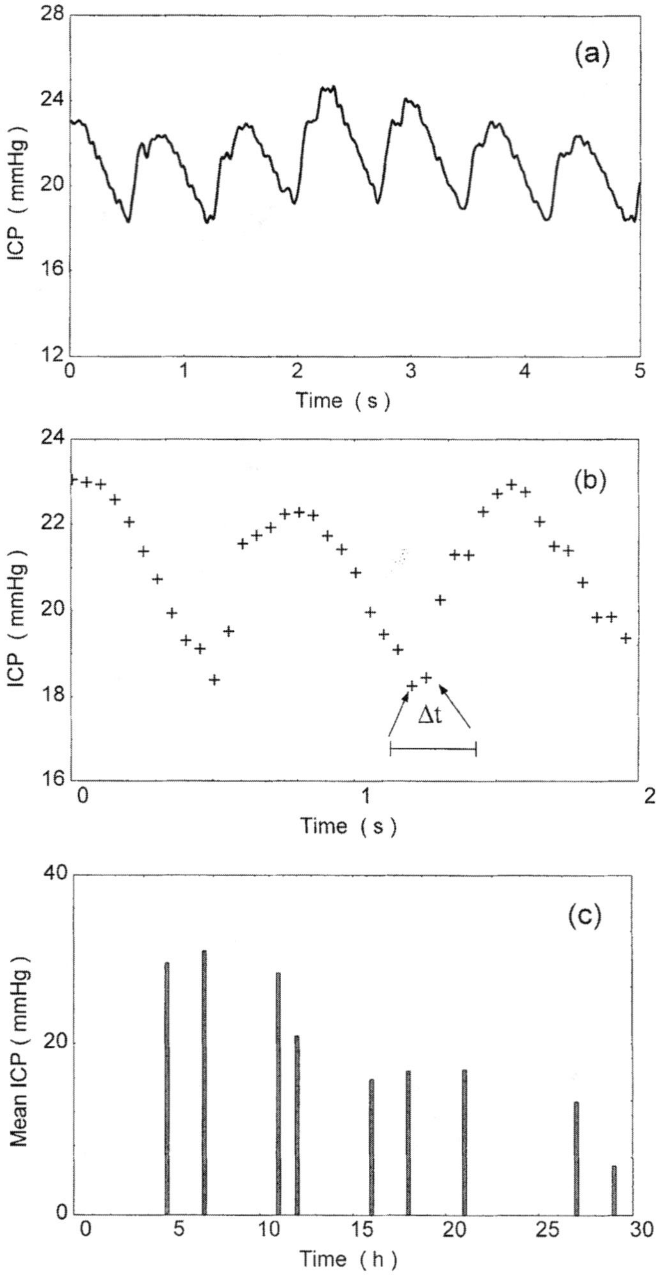

FIGURE 23.1 Examples of continuous and discrete-time signals. (a) Continuous recording of intracranial pressure in a head-injured patient. (b) Intracranial pressure measurements obtained at regular intervals of 50 ms. (c) Non-uniformly spaced measurements of mean intracranial pressure over a period of 30 h following surgery.

Fundamental Concepts

Spectral Analysis

Practical applications of spectral and correlation analysis are performed on discrete-time signals (Figure 23.1). These are obtained either from a sequence of discrete measurements or from the transformation of a continuous signal (Figure 23.1) to digital format using an **analog-to-digital converter (ADC).** When the latter is adopted to allow computer analysis of an originally continuous signal, two main characteristics of the ADC need to be considered. The first is the number of bits available to represent each sample, as this will determine the resolution and accuracy of the sampled signal. The second important consideration is the *sampling interval* Δt (Figure 23.1). From the *Nyquist theorem,*[1] the maximum value of Δt must be such that the *sampling frequency* $f_s = 1/\Delta t$ is at least twice the highest frequency of interest in the original signal. If this rule is not followed, spectral and correlation estimations might be considerably distorted by a phenomenon called *aliasing.*[2] Low-pass filtering before ADC is always recommended to limit the bandwidth of the continuous signal to allow the correct choice of f_s or Δt. In practice, the sampling frequency is usually much higher than the minimum required by the Nyquist theorem to provide a better visual representation of the sampled data.

Let x_n represent a discrete-time signal with samples at $n = 0, 1, 2, \ldots, N-1$. The Fourier theorem[1,2] states that it is possible to decompose x_n as a sum of cosine and sine waveforms of different frequencies using an appropriate combination of amplitude coefficients. Therefore,

$$x_n = a_0 + \sum_{k=1}^{N-1} a_k \cos\left(\frac{2\pi kn}{N}\right) + \sum_{k=1}^{N-1} b_k \sin\left(\frac{2\pi kn}{N}\right) \tag{23.1}$$

where $k = 1, 2, \ldots, N-1$ determines the frequency of each cosine and sine waveforms as $f_k = k/N\Delta t$. The corresponding coefficients are calculated from

$$a_0 = \frac{1}{N}\sum_{n=0}^{N-1} x_n \tag{23.2a}$$

$$a_k = \frac{1}{N}\sum_{n=0}^{N-1} x_n \cos\left(\frac{2\pi kn}{N}\right) \tag{23.2b}$$

$$b_k = \frac{1}{N}\sum_{n=0}^{N-1} x_n \sin\left(\frac{2\pi kn}{N}\right) \tag{23.2c}$$

Note that Equation 23.2a represents the mean value of x_n and that the argument $2\pi kn/N$ is the same for the *direct* (Equation 23.2) and *inverse* (Equation 23.1) **discrete Fourier transforms (DFT).**

From Euler's formula,[3] it is possible to combine the cosine and sine terms to express the DFT in exponential form:

$$e^{j\theta} = \cos\theta + j\sin\theta \tag{23.3}$$

leading to

$$x_n = \sum_{k=0}^{N-1} c_k e^{j(2\pi kn/N)} \tag{23.4}$$

with

$$c_k = \frac{1}{N}\sum_{n=0}^{N-1} x_n e^{-j(2\pi kn/N)} \tag{23.5}$$

where c_k is now a complex value related to the original cosine and sine coefficients by

$$c_0 = a_0 \tag{23.6a}$$

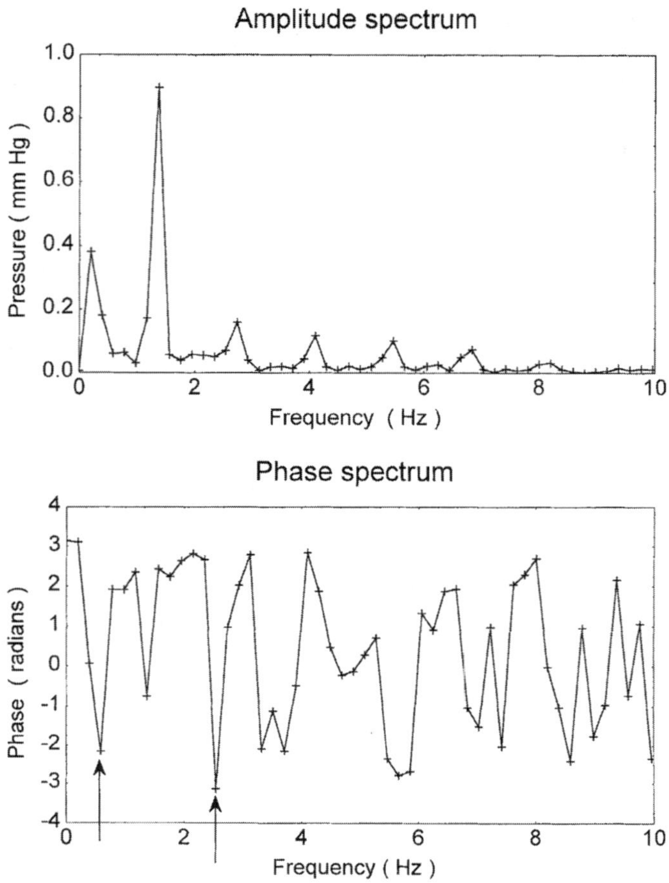

FIGURE 23.2 Amplitude and phase spectra of the intracranial pressure signal represented in Figure 23.1a after analog-to-digital conversion with a sampling interval of 20 ms. The main peak in the amplitude spectrum corresponds to the frequency of the cardiac cycle in Figure 23.1a. Wraparound of the phase spectrum is apparent in the third and 13th harmonics (arrows). Both spectra have been plotted to 10 Hz only.

$$c_k = a_k - jb_k \quad k = 1, 2, \ldots, N-1 \tag{23.6b}$$

A graphic representation of the a_k, b_k, or c_k coefficients for each value of k (or f_k) constitutes the frequency spectrum of x_n, expressing the relative contribution of different sinusoidal frequencies to the composition of x_n (Equation 23.4). Since c_k is complex (Equation 23.6b), a more meaningful physical interpretation of the spectrum is obtained with the *amplitude* and *phase* spectra, defined as

$$A_k = \left(a_k^2 + b_k^2\right)^{1/2} = \left|c_k\right| \tag{23.7a}$$

$$\theta_k = \tan^{-1}\left(-\frac{b_k}{a_k}\right) \tag{23.7b}$$

Figure 23.2 shows the amplitude (or magnitude) and phase spectra for the signal in Figure 23.1a, sampled at intervals $\Delta t = 20$ ms. The signal was low-pass-filtered at 20 Hz before ADC. The total duration is given by $T = N\Delta t = 5$ s, corresponding to $N = 250$ samples. Before calculating the spectral coefficients, the mean value of the complete record was removed (dc term) and any linear trends were removed by fitting a straight line to the data (detrending). As will be discussed below, it is also important to apply a window to the data, to minimize the phenomenon of leakage. For $k > N/2$ both spectra present symmetrical values. This can be easily demonstrated from the fact that cosine (and sine^2) functions have even symmetry while sine has odd symmetry. From Equation 23.7 it follows that A_k and θ_k have *even* and *odd* symmetry, respectively.[4] Consequently, only half the spectral components ($k \le N/2$) are required to give a complete description of x_n in the frequency domain.

The amplitude spectra indicates the combined amplitude of the cosine and sine terms to reconstruct x_n; the phase spectra reflects the relative phase differences (or time delays) between the sinusoidal waveforms to generate the temporal pattern of x_n. The amplitude spectra also reflects the signal power at different frequencies. For simplicity, the power spectrum can be defined as

$$P_k = A_k^2 = |c_k|^2 \tag{23.8}$$

Direct implementation of Equation 23.8, however, leads to spectral power estimates which are biased and inconsistent. More appropriate procedures for estimating the **power spectrum** (or *power density spectrum*) will be discussed later.

Parseval's theorem[5] demonstrates that the total signal energy can be computed either in time or frequency domain:

$$\frac{1}{N}\sum_{n=0}^{N-1} x_n^2 = \sum_{k=0}^{N-1} P_k \tag{23.9}$$

If x_n has zero mean, the left-hand side of Equation 23.9 is the biased estimator of signal variance.[6] Although most applications of spectral analysis concentrate on the characteristics of the amplitude or power spectra, it is important to bear in mind that the phase spectrum is also responsible for the temporal pattern of x_n. As an example, while both the Dirac impulse function and white noise have a flat, constant amplitude (or power) spectra,[6] it is the difference in the phase spectra which accounts for the different morphologies in the time domain.

Interpretation of the amplitude and phase spectra of both theoretical functions and sampled data is facilitated by taking into account several properties of the DFT (Equations 23.4 and 23.5), namely, *symmetry, linearity, shifting, duality,* and *convolution.*[7] To these, a very important property of Equations 23.1 and 23.4 must be added. Since cosine and sine functions are periodic, and exist for $-\infty < t < \infty$, Equations 23.1 and 23.4 will reconstruct x_n not only in the interval of interest ($0 \le t \le T$) but also at all other multiple intervals $pT \le t \le (p + 1)T$ ($p = 0, \pm 1, \pm 2, \ldots$). As a consequence, spectral estimations obtained with the DFT inherently assume that x_n is *periodic* with period $T = N/\Delta t$. As discussed in the following sections, this property needs to be taken into account when performing spectral analysis with the DFT and FFT.

Correlation Analysis

The basic concept of the correlation coefficient, as a measure of the strength of linear relationship between two variables[6] can be extended to signal analysis with the definition of the **cross-correlation function (CCF)** as:[5]

$$r_{xy}(p) = \frac{1}{N}\sum_{n=0}^{N-1} x_n y_{n-p} \qquad p = 0, \pm 1, \pm 2, \ldots \tag{23.10}$$

where x_n and y_n are zero-mean, discrete-time signals defined in the interval $n = 0, 1, 2, \ldots, N-1$. For each value of p, the cross correlation is computed by shifting y_n by $p\Delta t$ and calculating the average product in Equation 23.10. If x_n and y_n are unrelated, the sum of positive and negative products will tend to zero. Conversely, if y_n tends to follow x_n, but with a time delay D, $r_{xy}(p)$ will show a peak at $p = D/\Delta t$. This property of the CCF is illustration in Figure 23.3. As noted by Bergland,[8] cross correlation can be viewed as "one signal searching to find itself in another signal."

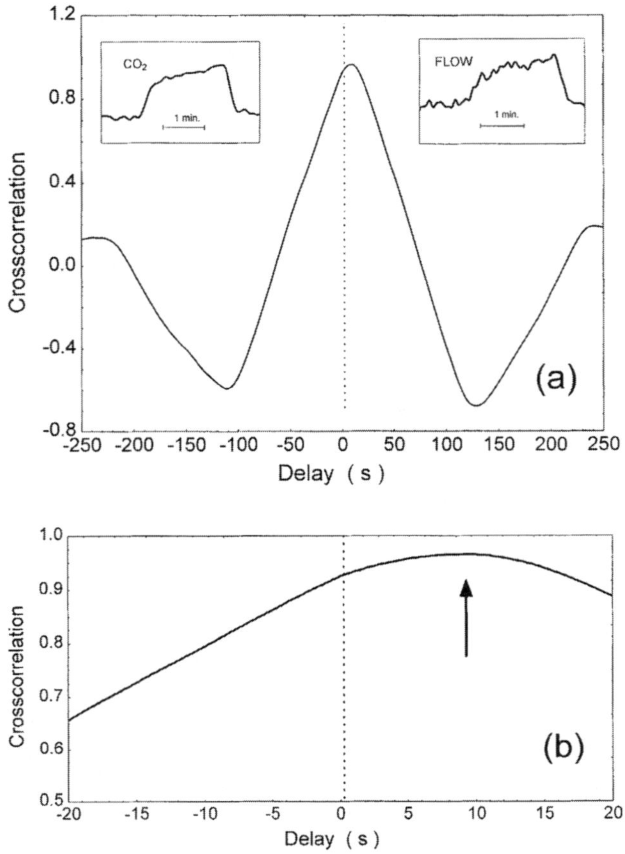

FIGURE 23.3 CCF between changes in arterial CO_2 and blood flow to the brain. Arterial CO_2 was estimated from end-tidal measurements and cerebral blood flow with Doppler ultrasound in the middle cerebral artery. (a) CCF and original signals (inserts). The cross-correlation value of approximately 1.0, observed at time delays near zero, reflects the similar temporal patterns between the two measurements. The negative cross correlations are obtained when either signal is shifted by approximately the duration of the plateau phase, which lasts 2 min. (b) Enlarging the scale around delay = 0 shows that the peak cross correlation occurs at 10 s, reflecting the time it takes for the flow to respond to the CO_2 change. (Data kindly provided by Dr. Joanne Dumville, Mr. A. Ross Naylor, and Prof. David H. Evans, University of Leicester, U.K.)

For $y_n = x_n$, $r_{xy}(p)$ becomes the **autocorrelation function (ACF):**

$$r_{xx}(p) = \frac{1}{N}\sum_{n=0}^{N-1} x_n x_{n-p} \tag{23.11}$$

and it is intuitive that the maximum value of $r_{xx}(p)$ occurs for $p = 0$ with

$$r_{xx}(0) = \frac{1}{N}\sum_{n=0}^{N-1} x_n^2 \tag{23.12}$$

which represents the signal variance or total energy. Therefore, for signals with unit standard deviation, the autocorrelation peak is equal to 1.

The Wiener–Khintchine theorem[9] demonstrates that the autocorrelation function and the power spectrum constitute a Fourier transform pair, that is,

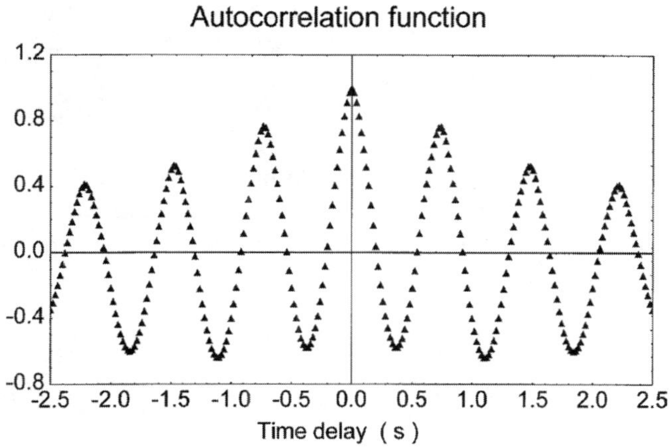

FIGURE 23.4 ACF of the discrete-time version of the signal in Figure 23.1a. The periodicity of the ACF reflects the quasi-periodic pattern of the intracranial pressure signal (Figure 23.1a).

$$S_k = \sum_{p=0}^{N-1} r_{xx}(p) e^{-j(2\pi kp/N)} \tag{23.13}$$

where S_k is usually called the autospectra of x_n.[6] Equation 23.13 indicates that it is possible to estimate the power spectra from a previous estimate of the autocorrelation function. As a transform pair, the autocorrelation function can also be derived from the autospectra by substituting S_k for c_k in Equation 23.4.

From Equation 23.11 it is clear that $r_{xx}(p)$ has even symmetry, that is, $r_{xx}(+p) = r_{xx}(-p)$. This property is apparent in Figure 23.4, which shows the estimated autocorrelation function for the signal in Figure 23.1a. Another characteristic of ACF, which can be visualized in Figure 23.4, is the occurrence of secondary peaks reflecting the presence of an oscillatory component in x_n (Figure 23.1a).

Fast Fourier Transform

The FFT is not a single algorithm but rather a large family of algorithms which can increase the computational efficiency of the DFT. The main ideas behind the formulation of FFT algorithms are discussed below. A detailed description of the different algorithms that have been proposed is beyond the scope of this introduction; this can be found in References 5 through 7 and 10 through 14.

For both software and hardware implementations of Equations 23.4 and 23.5, the computational efficiency is usually expressed by the number of complex multiplications and additions required or, simply, by the *number of operations*.[10] Straight implementation of either Equation 23.4 or 23.5 leads to N^2 operations. Typically, FFT algorithms can reduce this number to $N \log_2 N$. For $N = 1024$ the FFT algorithm is 100 times faster than the direct implementation of Equation 23.4 or 23.5.

The essence of all FFT algorithms is the periodicity and symmetry of the exponential term in Equations 23.4 and 23.5, and the possibility of breaking down a transform into a sum of smaller transforms for subsets of data. Since n and k are both integers, the exponential term is periodic with period N. This is commonly represented by

$$W_N = e^{-j(2\pi/N)} \tag{23.14}$$

and Equation 23.5 can be written as

$$c_k = \frac{1}{N} \sum_{n=0}^{N-1} x_n W_N^{kn} \quad k = 0, 1, 2, \ldots, N-1 \tag{23.15}$$

In many applications the terms W_N^{kn} are called **twiddle factors.** Assuming $N = 8$, calculation of the DFT with Equation 23.15 will require 64 values of W_8^{kn}. Apart from the minus sign, a simple calculation can show that there are only four different values of this coefficient, respectively: 1, j, $(1 + j)/\sqrt{2}$, and $(1 - j)/\sqrt{2}$.[4] Consequently, only these four complex factors need to be computed, representing a significant savings in number of operations.

Most FFT algorithms are based on the principle of **decimation-in-time,** involving the decomposition of the original time (or frequency) sequence into smaller subsequences. To understand how this decomposition can reduce the number of operations, assume that N is even. In this case it is possible to show that Equation 23.15 can be written as:[4,5,7,11]

$$c_k = \frac{1}{N}\sum_{r=0}^{(N/2)-1} x_r^e \cdot W_{N/2}^{kr} + \frac{1}{N}\sum_{r=0}^{(N/2)-1} x_r^o \cdot W_{N/2}^{kr} \tag{23.16}$$

where x_r^e and x_r^o represent the even- and odd-order samples of x_n, respectively. Comparing Equations 23.15 and 23.16, it is clear that the latter represents two DFTs with dimension $N/2$, involving $2(N/2)^2$ operations rather than the N^2 operations required by Equation 23.15. This process of decimation-in-time can be carried out further to improve computational performance. In the general case, N can be decomposed into q factors:

$$N = \prod_{i=1}^{q} r_i = r_1 r_2 \ldots r_q \tag{23.17}$$

The number of operations required is then:[6]

$$\text{number of operations} = N\sum_{i=1}^{q} r_i \tag{23.18}$$

In the original algorithm of Cooley and Tukey,[10] $r_i = 2$ and $N = 2^q$. In this case the theoretical number of operations required would be $2Nq = 2N \log_2 N$. As pointed out in Reference 6, further improvements in efficiency are possible because of the symmetry of the twiddle factors. The efficiency gain of most FFT algorithms using radix-2, i.e., $N = 2^q$ is

$$\text{efficiency gain} = \frac{N^2}{N \log_2 N} = \frac{N}{\log_2 N} \tag{23.19}$$

For $N = 1024$, $q = 10$ and the efficiency gain is approximately 100. Specific applications might benefit from other decompositions of the original sequence. Cases of particular interest are radix-4 and radix-8 FFTs.[14] However, as shown by Rabiner and Gold,[11] (p. 585), it is not possible to generalize the superiority of radix-8 over radix-4 algorithms.

In general, most FFT algorithms accept complex x_n sequences in Equation 23.5. By limiting x_n to the most common situation of real-valued signals, it is possible to obtain more efficient algorithms as demonstrated by Sorensen et al.[15] Uniyal[16] performed a comparison of different algorithms for real-valued sequences showing that performance is architecture dependent. For machines with a powerful floating point processor, the best results were obtained with Brunn's algorithm.[17]

The application of FFT algorithms for spectral and correlation analysis is discussed in the following sections.

FFT Spectral Analysis

For some deterministic signals, x_n can be expressed by a mathematical function and the amplitude and phase spectra can be calculated as an exact solution of Equations 23.5 and 23.7. The same is true for the power spectra (Equations 23.8 and 23.13). Examples of this exercise can be found in many textbooks.[1,2,4,5,7]

In most practical applications, there is a need to perform spectral analysis of experimental measurements, corresponding to signals which, in general, cannot be described by simple mathematical functions. In this case the spectra has to be estimated by a numerical solution of Equations 23.5 through 23.8, which can be efficiently implemented on a digital computer with an FFT algorithm. For estimation of the power spectrum, this approach is often classified as *nonparametric,* as opposed to other alternatives which are based on parametric modeling of the data such as *autoregressive* methods.[18] Considerable distortions can result from applications of the FFT unless attention is paid to the following characteristics and properties of the measured signal and the DFT/FFT.

Limited Observation of Signal in Time

Limited observation of a signal x_n in time can be seen as the multiplication of the original signal x_n^∞ by a rectangular window of duration $T = N\Delta t$ as exemplified for a single sinusoid in Figure 23.5. The DFT assumes that x_n is periodic, with period T, as mentioned previously. Instead of a single harmonic at the frequency of the original sinusoid, the power spectrum estimated with the FFT will have power at other harmonics as indicated by the spectrum in Figure 23.5c. The spectral power, which should have been concentrated on a single harmonic (Figure 23.5c, dashed line), has "leaked" to neighboring harmonics and for this reason this phenomenon is usually called *leakage.* The morphology of the distorted spectrum of Figure 23.5c can be explained by the fact that the Fourier transform of a rectangular window function (Figure 23.5b) is given by a *sinc* function (sin x/x) which presents decreasing side lobes.[1,2] Multiplication in time corresponds to the convolution operation in the frequency domain.[1,2] In the general case of signals comprising several harmonics, the *sinc* functions will superimpose and the resulting spectrum is then a distorted version of the "true" spectrum. As the individual *sinc* functions superimpose to produce the complete spectrum, a *picket-fence* effect is also generated.[8] This means that spectral leakage not only adds spurious power to neighboring harmonics but also restricts the frequency resolution of the main spectral peaks. The effects of spectral leakage can be reduced by (1) increasing the period of observation and (2) multiplying the original signal x_n by a window function with a smooth transition as represented by the dashed line window in Figure 23.5b. The Fourier transform of a window function with tapered ends has smaller side lobes, thus reducing the undesirable effects leakage. A large number of tapering windows have been proposed, as reviewed by Harris.[19] As an example, the four-term Blackman–Harris window, defined as

$$w_n = a_0 - a_1 \cos\left(\frac{2\pi n}{N}\right) + a_2 \cos\left(\frac{4\pi n}{N}\right) - a_3 \cos\left(\frac{6\pi n}{N}\right) \qquad n = 0, 1, 2, \ldots N-1 \qquad (23.20)$$

produces side lobe levels of –92 dB if the a_i coefficients are chosen as $a_0 = 0.35875$, $a_1 = 0.48829$, $a_2 = 0.14128$, $a_3 = 0.01168$.[19] Windows also play an important role in the sampling properties of power spectral estimates, as will be discussed later. Windowing attenuates the contribution of signal samples at the beginning and end of the signal and, therefore, reduces its effective signal duration. This effect is reflected by the equivalent noise bandwidth (ENBW) defined as[19]

$$\text{ENBW} = \frac{\sum_{n=0}^{N-1} w_n^2}{\left[\sum_{n=0}^{N-1} w_n\right]^2} \qquad (23.21)$$

For a rectangular window ENBW = 1.0 and for the Blackman–Harris window (Equation 23.20) the corresponding value is 2.0. The majority of other window shapes have intermediate values of ENBW.[19]

Effects of "Zero-Padding"

Most FFT algorithms operate with $N = 2^q$ samples; the choice of q is many times critical. Since frequency resolution is inversely proportional to N, in many circumstances a value of q leading to $2^q > N$ is preferable

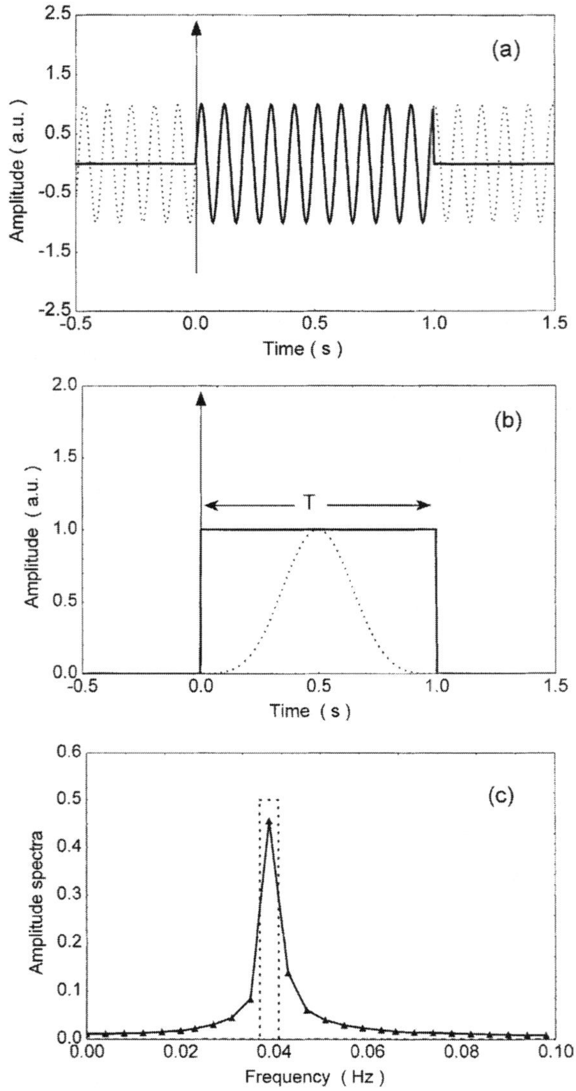

FIGURE 23.5 Effect of limited observation time T on the amplitude spectra of a sinusoidal component. (a) Observation of a single harmonic (dashed line) for a limited period of time T is equivalent to the multiplication for the rectangular function represented in (b). The Blackman–Harris window is also represented in (b) (dashed line). (c) Truncating a single harmonic produces spectral estimates smeared by *leakage* (solid line) as compared with the theoretical result (dashed line) with width equal to the frequency resolution ($f_r \approx 0.004$ Hz).

to the option of limiting the signal to $N' = 2^{q-1}$ samples with $N' < N$. The most common and simple way of extending a signal to comply with the 2^q condition is by **zero-padding.** For signals with zero mean and with first and last values around zero, this can be accomplished by complementing the signal with Q zeros to achieve the condition $N + Q = 2^q$. For signals with end points different from zero, these values can be used for padding. If initial and final values differ significantly, a linear interpolation from the last to the first point is also a practical option. However, with the application of windowing, most signals will have similar initial and final points and these can be used for zero-padding. As discussed in the next section, zero-padding has important applications for the estimation of correlation functions via FFT. For spectral analysis, it is relatively simple to demonstrate that adding Q zeros corresponds to oversampling the N point original spectrum with a new frequency resolution which is $(N + Q)/N$ times greater than the original resolution. Consequently, although zero-padding does not introduce major distortions, it produces the false illusion of higher resolution than warranted by the available N measured signal samples.

Phase Spectrum Estimation

The use of Equation 23.7b to estimate the phase spectrum is fraught with a different kind of problem, resulting from the indetermination of the tan^{-1} function to discriminate between phase angles with absolute values greater than π. This problem is illustrated in Figure 23.2b showing that phase angles decrease continuously until reaching $-\pi$ and then "jump" to continue decreasing from the $+\pi$ value. This feature of the phase spectrum is called **wraparound.** Methods to "unwrap" the phase spectrum have been proposed,[20] but a general satisfactory solution to this problem is not available. In some cases the shifting property of the DFT[5,7] can be used to "rotate" the original signal in time, thus minimizing the slope of the phase spectrum and, consequently, the occurrence of wraparound.

Sampling Properties of Spectral Estimators

The most straightforward approach to computing the power spectrum is to use Equation 23.8. This method is known as the **periodogram.**[5] Application is limited to signals which are **stationary,** meaning stable statistical properties (such as the mean and the variance) along time. For measurements performed on *nonstationary* systems, such as speech or systems with time-varying parameters, other methods of spectral estimation are available and will be mentioned later. It is possible to demonstrate that when the period of observation T tends to infinity, Equation 23.8 gives an unbiased estimate of the power spectrum. In practice, due to finite values of T, the phenomenon of spectral leakage described above will lead to power spectral estimates which are *biased.*

The second inherent problem with the periodogram is the *variance* of the resulting spectral estimates. Assuming x_n to follow a Gaussian distribution, it follows that a_k and b_k will also be Gaussian because Equation 23.2 represents a linear transformation. Since Equations 23.7a and 23.8 involve the sum of two squared Gaussian variates, P_k will follow a χ^2 distribution with two degrees of freedom.[6] In this case the mean and the standard deviation of the power spectral estimate will be the same, *independently of the frequency considered.* As a consequence, power spectral estimates obtained from Equation 23.8, using a simple sample x_n, should be regarded as highly unreliable. In addition, the variance or standard deviation of this χ^2 distribution does not decrease with increases in sample duration N. This indicates that the periodogram (Equation 23.8) is an *inconsistent* estimator of the power spectrum.

For a χ^2 distribution with m degrees of freedom, the coefficient of variation is given by

$$CV\left[\chi_m^2\right] = \frac{\sqrt{2m}}{m} = \sqrt{\frac{2}{m}} \tag{23.22}$$

showing that it is possible to improve the reliability of power spectral estimates by increasing m. This can be achieved by replacing Equation 23.8 by[21]

$$\hat{P}_k = \frac{1}{L}\sum_{l=1}^{L} c_{k,l}^2 \qquad k = 0, 1, 2, \ldots N-1 \tag{23.23}$$

with L representing a number of separate samples x_n each with length $T = N\Delta t$. If only one record of x_n can be obtained under stationary conditions, it is possible to break down this record into L segments to obtain an improved estimate of the power spectrum with variance reduced by a factor of L. However, the spectral resolution, given by $f_r = 1/T$, will be reduced by the same factor L, thus indicating an inescapable compromise between resolution and variance.

A *modified periodogram* was introduced by Welch[21] consisting of the multiplication of x_n by a triangular, or other window shape, before computing the individual spectral samples with Equation 23.5. The application of a window justifies overlapping adjacent segments of data by as much as 50%. For a signal with a total duration of N samples, the combination of overlapping with segmentation (Equation 23.23) can lead to a further reduction of the spectral variance by a factor of 11/18.

Averaging L spectral samples as indicated by Equation 23.23 represents one approach to improve spectral estimation by means of *smoothing*. A similar effect can be obtained with the **correlogram**. Equation 23.13 indicates that it is possible to estimate the power spectrum from the autocorrelation function. Limiting the number of shifts of the autocorrelation function to $p \ll N$ is equivalent to smoothing the original spectrum by convolution with the Fourier transform of a Bartlett (triangular) window.[22] As discussed in the next section, the autocorrelation function can also be computed more efficiently with the FFT and it can be shown that in this case it involves a smaller number of numerical operations than the Welch method based on the periodogram.[5]

FFT Correlation Analysis

Before considering the application of FFT algorithms to compute auto- and cross-correlation functions, it is important to discuss their sampling properties using Equations 23.10 and 23.11 as estimators. Assuming that variables x_n and y_n are not defined outside the interval $0 \le n \le N-1$, it follows from Equation 23.10 that as p increases and the two functions "slide" past each other, the effective number of summed products is $N-|p|$ rather than N as implied by Equations 23.10 and 23.11. For this reason these equations are often rewritten as

$$r_{xy}(p) = \frac{1}{N-|p|}\sum_{n=0}^{N-|p|-1} x_n y_{n-1} \qquad p = 0, \pm 1, \pm 2, \ldots \tag{23.24}$$

The main justification for this modification, however, is that Equations 23.10 and 23.11 lead to biased estimations of correlation functions while Equation 23.24 is *unbiased.*

Equation 23.24 normally assumes that x_n and y_n are standardized variables with zero mean and unit variance. If the mean values are different from zero, Equation 23.10 and 23.11 will produce distorted estimates with a "pyramid effect" due to the presence of the dc term. However, this effect is compensated for in Equation 23.24 and in this case the effect of the mean value is to add a constant term:

$$r_{xy}(p) = r'_{xy}(p) - m_x m_y \tag{23.25}$$

where $r'_{xy}(p)$ is the cross correlation of variables with mean values m_x and m_y, respectively.

Similarly to the DFT, Equations 23.10 and 23.11 involve N^2 operations and Equation 23.24 slightly less. Since the autocorrelation and the power spectra constitute a Fourier transform pair (Equation 23.13), the computation of correlation functions can also be sped up by means of an FFT algorithm. For the sake of generality, the **cross spectrum** of x_n and y_n can be defined as[6]

$$C_{xy}(f_k) = X(f_k)Y^*(f_k) \tag{23.26}$$

with $X(f_k)$ and $Y(f_k)$ representing the Fourier transforms of x_n and y_n, respectively. The generalized Wiener–Khintchine theorem then gives the cross-correlation function as

$$r_{xy}(p) = \sum_{k=0}^{N-1} C_{xy}(f_k)e^{-j(2\pi kp/N)} \tag{23.27}$$

Therefore, "fast" correlation functions can be computed[23] using the forward FFT to calculate $X(f_k)$ and $Y(f_k)$ and then the inverse FFT to obtain $r_{xy}(p)$ with Equation 23.27. Obviously, when autocorrelation functions are being computed with this method, only one transform is necessary to obtain the autospectra (Equation 23.13) instead of the cross spectra (Equation 23.26).

When correlation functions are computed with the FFT, it is critical to pay attention again to the periodicity of the transformed variables as an intrinsic property of the DFT. When the two functions in either Equation 23.10 or 23.11 are displaced by p samples, for periodic functions there will be nonzero products outside the range $0 \le n \le N - 1$, thus leading to significant errors in the estimated auto- or cross-correlation functions. In this case the resulting estimates are called *circular* correlations.[6] This error can be avoided by zero-padding the original signals from $n = N$ to $n = 2N - 1$ and computing the FFTs with $2N$ points. The resulting correlation functions will be noncircular and, in the range $0 \le p \le N - 1$, will agree with correlations computed with the original Equations 23.10 or 23.11. Finally, to remove bias the results of Equation 23.27 should also be multiplied by $N/(N - |p|)$ to agree with Equation 23.24.

Further Information

Software for FFT special analysis is available from multiple sources. Off-the-shelf software ranges from specialized packages for digital signal processing, such as DADiSP, to statistical packages which include FFT analysis of time series. Mathematical and engineering packages such as MATLAB also include routines for FFT spectral and correlation analysis. For a review of available options see Reference 24. For readers who want to implement their own software, FFT routines can be found in Reference 4, 13 through 15, and 18. Additional references are 25 through 27.

Hardware implementations of FFT algorithms are common in areas requiring real-time spectral analysis as in the case of blood flow velocity measurement with Doppler ultrasound. For a review of hardware implementations see References 11 and 28. Developments in this area follow the pace of change in VLSI technology.[29]

One of the limitations of the FFT is the fact that frequency resolution is the inverse of the signal observation time. Improved resolution can be obtained with *parametric* methods of spectral analysis and their application is particularly relevant when only short segments of data are available or when it is necessary to discriminate between frequency harmonics which are closely spaced in the spectrum. Broadly speaking, parametric methods assume that the data follow spectral densities with a known pole-zero structure of variable complexity, characterized by a given *model order.* All-zero models correspond to the *moving average* structure while the all-pole version represents the *autoregressive* model. The general case is the autoregressive-moving average model (ARMA). For a comprehensive review of these methods see Reference 30; further information and software implementations can be found in References 18 and 27.

Nonstationary signals present a particular problem. In cases where the signal statistical properties change relatively slowly with time, it is possible to select short segments of quasi-stationary data and to use the DFT or parametric methods to estimate the spectra as mentioned previously. However, when these changes in systems parameters or statistical moments are fast in relation to the phenomenon under observation (e.g., speech or seismic data), this approach is not feasible because of the poor frequency resolution resulting from short observation times. Methods proposed to cope with signal nonstationarity often depend on the underlying cause of nonstationary behavior.[9,31] More general methods, known as *time-frequency distributions,* are now favored by most investigators.[32] The Wigner–Ville and Choi–Williams transforms are some of the more widely used of these time-frequency distributions. In each case the signal is described by a simultaneous function of time *and* frequency and hence is graphically represented by a three-dimensional plot having time and frequency as dependent variables.

A different approach to the analysis of nonstationary data is the application of *wavelets.*[33] This alternative also has advantages in the representation of fast transients and in applications requiring data compression and pattern classification. Similarly to the sine and cosine functions, which are the basis of Fourier analysis, wavelets are orthogonal functions which can be used to decompose and reconstruct signals using a finite set of coefficients obtained by a *wavelet transform* (WT). The main difference between wavelets and sinusoids, however, is that the former are limited in time. In addition, the complete orthogonal set of wavelets can be obtained simply by expansion (or compression) and scaling of a single function, known as the *mother wavelet.* Because of their limited time duration wavelets can provide a much more synthetic decomposition of fast transients, or sharp edges in image analysis, than it is possible to obtain with the DFT. Their property of expansion/contraction of a single mother wavelet can also overcome a major limitation of the DFT, that is, to allow good frequency resolution at both low and high frequencies. For applications of the WT and commercially available software see References 34 and 35.

Defining Terms

Analog-to-digital conversion: The process of converting a continuous signal to a discrete time sequence of values usually sampled at uniform time intervals.

Autocorrelation function (ACF): A measure of longitudinal variability of a signal which can express the statistical dependence of serial samples.

Correlogram: Numerical calculation and graphical representation of the ACF or CCF.

Cross-correlation function (CCF): A measure of similarity between signals in the time domain which also allows the identification of time delays between transients.

Cross-spectrum: The complex product of the power spectra of two different signals.

Decimation-in-time: The process of breaking down a time series into subsequences to allow more efficient implementations of the FFT.

Discrete Fourier transform (DFT): The usual method to obtain the Fourier series of a discrete time signal.

Fast Fourier transform (FFT): Algorithm for the efficient computation of the DFT.

Periodogram: A family of methods to estimate the power spectrum using the DFT.

Power spectrum: The distribution of signal power as a function of frequency.

Signal: Continuous or discrete representation of a variable or measurement as a function of time or other dimension.

Stationarity: Property of signals which have statistical moments invariant with time.

Twiddle factors: Exponential term in the DFT whose periodicity allows repeated use and hence considerable savings of computation time in the FFT.

Wraparound: Overflow of phase spectral estimations above $|\pi|$ due to the uncertainty of the $\tan^{-1}$ function.

Zero-padding: Extension of a signal with zeros, constant values, or other extrapolating functions.

References

1. B.P. Lathi, *Communication Systems,* New York: John Wiley & Sons, 1968.
2. R.N. Bracewell, *The Fourier Transform and Its Applications,* Englewood Cliffs, NJ: Prentice-Hall, 1988.
3. E. Kreyszig, *Advanced Engineering Mathematics,* New York: John Wiley & Sons, 1962.
4. P.A. Lynn and W. Fuerst, *Introductory Digital Signal Processing with Computer Applications,* Chichester: John Wiley & Sons, 1989.
5. J.G. Proakis and D.G. Manolakis, *Digital Signal Processing: Principles, Algorithms, and Applications,* 2nd ed., New York: Macmillan, 1992.
6. J.S. Bendat and A.G. Piersol, *Random Data: Analysis and Measurement Procedures,* 2nd ed., New York: John Wiley & Sons, 1986.
7. A.V. Oppenheim and R.W. Schafer, *Discrete-Time Signal Processing,* Englewood Cliffs, NJ: Prentice-Hall, 1989.
8. G.D. Bergland, A guided tour of the fast Fourier transform, *IEEE Spectrum,* 6: 41–52, 1969.
9. M.B. Priestley, *Spectral Analysis and Time Series,* London: Academic Press, 1981.
10. J.W. Cooley and J.W. Tukey, An algorithm for the machine computation of complex Fourier series, *Math. Computation,* 19: 297–301, 1965.
11. L.R. Rabiner and B. Gold, *Theory and Application of Digital Signal Processing,* Englewood Cliffs, NJ: Prentice-Hall, 1975.
12. E.O. Brigham, *The Fast Fourier Transform and Its Applications,* 2nd ed., Englewood Cliffs, NJ: Prentice-Hall, 1988.
13. Digital Signal Processing Committee, *Programs for Digital Signal Processing,* New York: IEEE Press, 1979.

14. R.C. Singleton, An algorithm for computing the mixed radix fast Fourier transform, *IEEE Trans. Audio Electroacoust.,* 17: 93–103, 1969.
15. H.V. Sorensen, D.L. Jones, M.T. Heideman, and C.S. Burrus, Real-valued fast Fourier transform algorithms, *IEEE Trans. Acoust. Speech Signal Proc.,* 35: 849–863, 1987.
16. P.R. Uniyal, Transforming real-valued sequences: fast Fourier versus fast Hartley transform algorithms, *IEEE Trans. Signal Proc.,* 41: 3249–3254, 1994.
17. G. Brunn, z-Transform DFT filters and FFT, *IEEE Trans. Acoust. Speech Signal Proc.,* 26: 56–63, 1978.
18. S.M. Kay, *Modern Spectral Estimation,* Englewood Cliffs, NJ: Prentice-Hall, 1988.
19. F.J. Harris, On the use of windows for harmonic analysis with the discrete Fourier transform, *Proc. IEEE,* 66: 51–83, 1978.
20. J.M. Tribolet, A new phase unwrapping algorithm, *IEEE Trans. Acoust. Speech Signal Proc.,* 25: 170–177, 1977.
21. P.D. Welch, The use of fast Fourier transform for the estimation of power spectra: a method based on time averaging over short, modified periodograms, *IEEE Trans. Audio Electroacoust.,* 15: 70–73, 1967.
22. R.B. Blackman and J.W. Tukey, *The Measurement of Power Spectra,* New York: Dover Publications, 1958.
23. T.G. Stockham, Jr., High-speed convolution and correlation, *1966 Spring Joint Computer Conference, AFIPS Proc.,* 28: 229–233, 1966.
24. R. Braham, Math & visualization: new tools, new frontiers, *IEEE Spectrum,* 32(11): 19–36, 1995.
25. W.H. Press, B.P. Flannery, S.A. Teukolsky, and W.T. Vetterling, *Numerical Recipes: The Art of Scientific Computing,* Cambridge, UK: Cambridge University Press, 1986.
26. D.M. Monro, Complex discrete fast Fourier transform, *Appl. Stat.,* 24: 153–160, 1975.
27. S.L. Marple, Jr., *Digital Spectral Analysis with Applications,* Englewood Cliffs, NJ: Prentice-Hall, 1987.
28. G.D. Bergland, Fast Fourier transform hardware implementations — an overview, *IEEE Trans. Audio Electroacoust.,* 17: 104–108, 1969.
29. E. Bidet, D. Castelain, C. Joanblanq, and P. Senn, A fast single-chip implementation of 8192-complex point FFT, *IEEE J. Solid-State Circuits,* 30: 300–305, 1995.
30. S.M. Kay and S.L. Marple, Jr., Spectrum analysis — a modern perspective, *Proc. IEEE,* 69: 1380–1419, 1981.
31. J. Leuridan, H.V. Van der Auweraer, and H. Vold, The analysis of nonstationary dynamic signals, *Sound Vibration,* 28: 14–26, 1994.
32. L. Cohen, Time-frequency distributions — a review, *Proc. IEEE,* 77: 941–981, 1989.
33. I. Daubechies, Orthonormal bases of compactly supported wavelets, *Commun. Pure Appl. Math.,* 41: 909–996, 1988.
34. A. Aldroubi and M. Unser, Eds., *Wavelets in Medicine and Biology,* Boca Raton, FL: CRC Press, 1996.
35. A. Bruce, D. Donoho, and H.Y. Gao, Wavelet analysis, *IEEE Spectrum,* 33: 26–35, 1996.

23.2 RF/Microwave Spectrum Analysis[1]

A. Ambrosini, C. Bortolotti, N. D'Amico, G. Grueff, S. Mariotti, S. Montebugnoli, A. Orfei, and G. Tomassetti

A *signal* is usually defined by a time-varying function carrying some sort of information. Such a function most often represents a time-changing electric or magnetic field, whose propagation can be in free space or in dielectric materials constrained by conductors (waveguides, coaxial cables, etc.). A signal is said to be periodic if it repeats itself exactly after a given time T called the period. The inverse of the period T, measured in seconds, is the frequency f measured in hertz (Hz).

[1]All figures have been reproduced courtesy of Hewlett Packard, Rohde Schwarz, Hameg, Tektronix companies, and IEEE *Microwave Measurements.*

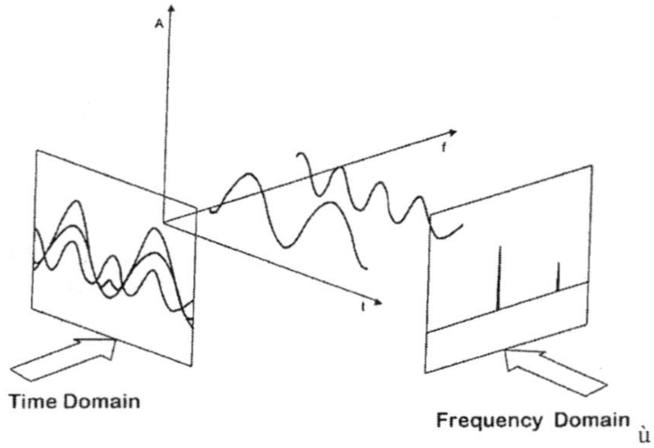

FIGURE 23.6 How the same signal can be displayed.

A periodic signal can always be represented in terms of a sum of several (possibly infinite) sinusoidal signals, with suitable amplitude and phase, and having frequencies that are integer multiples of the signal frequency. Assuming an electric signal, the square of the amplitudes of such sinusoidal signals represent the power in each sinusoid, and is said to be the power spectrum of the signal. These concepts can be generalized to a nonperiodic signal; in this case, its representation (spectrum) will include a continuous interval of frequencies, instead of a discrete distribution of integer multiples of the fundamental frequency. The representation of a signal in terms of its sinusoidal components is called Fourier analysis. The (complex) function describing the distribution of amplitudes and phases of the sinusoids composing a signal is called its Fourier transform (FT). The Fourier analysis can be readily generalized to functions of two or more variables; for instance, the FT of a function of two (spatial) variables is the starting point of many techniques of image processing. A time-dependent electrical signal can be analyzed directly as a function of time with an *oscilloscope* which is said to operate in the *time domain*. The time evolution of the signal is then displayed and evaluated on the vertical and horizontal scales of the screen.

The *spectrum analyzer* is said to operate in the *frequency domain* because it allows one to measure the harmonic content of an electric signal, that is, the power of each of its spectral components. In this case the vertical and horizontal scales read powers and frequencies. The two domains are mathematically well defined and, through the FT algorithm, it is not too difficult to switch from one response to the other. Their graphical, easily perceivable representation is shown in Figure 23.6 where the two responses are shown lying on orthogonal planes. It is trivial to say that the easiest way to make a Fourier analysis of a time-dependent signal is to have it displayed on a spectrum analyzer. Many physical processes produce (electric) signals whose nature is not deterministic, but rather stochastic, or random (noise). Such signals can also be analyzed in terms of FT, although in a statistical sense only.

A time signal is said to be band-limited if its FT is nonzero only in a finite interval of frequencies, say $(F_{max} - F_{min}) = B$. Usually, this is the case and an average frequency F_0 can be defined. Although the definition is somewhat arbitrary, a (band-limited) signal is referred to as RF (radio frequency) if F_0 is in the range 100 kHz to 1 GHz and as a microwave signal in the range 1 to 1000 GHz. The distinction is not fundamental theoretically, but it has very strong practical implications in instrumentation and spectral measuring techniques. A band-limited signal can be described further as narrowband, if $B/F_0 \ll 1$, or wideband otherwise.

The first step in performing a spectral analysis of a narrowband signal is generally the so-called heterodyne downconversion: it consists in the mixing ("beating") of the signal with a pure sinusoidal signal of frequency F_L, called local oscillator (LO). In principle, mixing two signals of frequency F_0 and F_L in any nonlinear device will result in a signal output containing the original frequencies as well as the difference $(F_0 - F_L)$ and the sum $(F_0 + F_L)$ frequencies, and all their harmonic (multiple) frequencies. In the practical case, a purely quadratic mixer is used, with an LO frequency $F_L < F_0$; the output will include the frequencies $(F_0 - F_L)$, $2F_L$, $2F_0$, and $(F_0 + F_L)$, and the first term (called the intermediate frequency or IF) will be easily separated from the others, which have a much higher frequency. The bandwidth of the IF signal will be the same as the original bandwidth B; however, to preserve the original information fully in the IF signal, stringent limits must be imposed on the LO signal, because any deviation from a pure sinusoidal law will show up in the IF signal as added phase and amplitude noise, corrupting the original spectral content. The process of downconverting a (band-limited) signal is generally necessary to perform spectral analysis in the very high frequency (microwave) region, to convert the signal to a frequency range more easily handled technically.

When the heterodyne process is applied to a wideband signal (or whenever $F_L > F_{min}$) "negative" frequencies will appear in the IF signal. This process is called *double sideband* mixing, because a given IF bandwidth B (i.e., $(F_L + B/2)$ will include two separate bands of the original signal, centered at F_L + IF ("upper" sideband) and F_L – IF ("lower" sideband). This form of mixing is obviously undesirable in spectrum analysis, and input filters are generally necessary to split a wideband signal in several narrow-band signals before downconversion. Alternatively, special mixers can be used that can deliver the upper and lower sidebands to separate IF channels. A band-limited signal in the frequency interval $(F_{max} - F_{min}) = B$ is said to be converted to baseband when the LO is placed at $F_L = F_{min}$, so that the band is converted to the interval (B–0). No further lowering of frequency is then possible, unless the signal is split into separate frequency bands by means of filters.

After downconversion, the techniques employed to perform power spectrum analysis vary considerably depending on the frequencies involved. At lower frequencies, it is possible to employ analog-to-digital converters (ADC) to get a discrete numerical representation of the analog signal, and the spectral analysis is then performed numerically, either by direct computation of the FT (generally via the fast Fourier transform, FFT, algorithm) or by computation of the signal autocorrelation function, which is directly related to the square modulus of the FT via the Wiener–Khinchin theorem. Considering that the ADC must sample the signal at least at the Nyquist rate (i.e., at twice the highest frequency present) and with adequate digital resolution, this process is feasible and practical only for frequencies (bandwidths) less than a few megahertz. Also, the possibility of a real-time analysis with high spectral resolution may be limited by the availability of very fast digital electronics and special-purpose computers. The digital approach is the only one that can provide extremely high spectral resolution, up to several hundred thousand channels. For high frequencies, several analog techniques are employed.

A Practical Approach to Spectrum Analysis [1]

Spectrum analysis is normally done in order to verify the harmonic content of oscillators, transmitters, frequency multipliers, etc. or the spurious components of amplifiers and mixer. Other specialized applications are possible, such as the monitoring of radio frequency interference (RFI), electromagnetic interference (EMI), and electromagnetic compatibility (EMC). These applications, as a rule, require an antenna connection and a low-noise, external amplifier. Which are then the specifications to look for in a good spectrum analyzer? We would suggest:

1. It should display selectable, very wide bands of the EM radio spectrum with power and frequency readable with good accuracy.
2. Its selectivity should range, in discrete steps, from few hertz to megahertz so that sidebands of a selected signal can be spotted and shown with the necessary details.
3. It should possess a very wide dynamic range, so that signals differing in amplitude six to eight orders of magnitude can be observed at the same time on the display.
4. Its sensitivity must be compatible with the measurements to be taken. As already mentioned, specialized applications may require external wideband, low-noise amplifiers and an antenna connection.
5. Stability and reliability are major requests but they are met most of the time.

Occasionally a battery-operated option for portable field applications may be necessary. A block diagram of a commercial spectrum analyzer is shown in Figure 23.7.

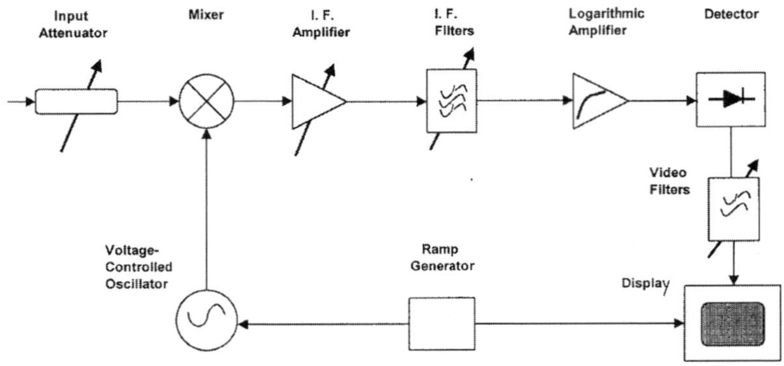

FIGURE 23.7 Block diagram of a commercial spectrum analyzer.

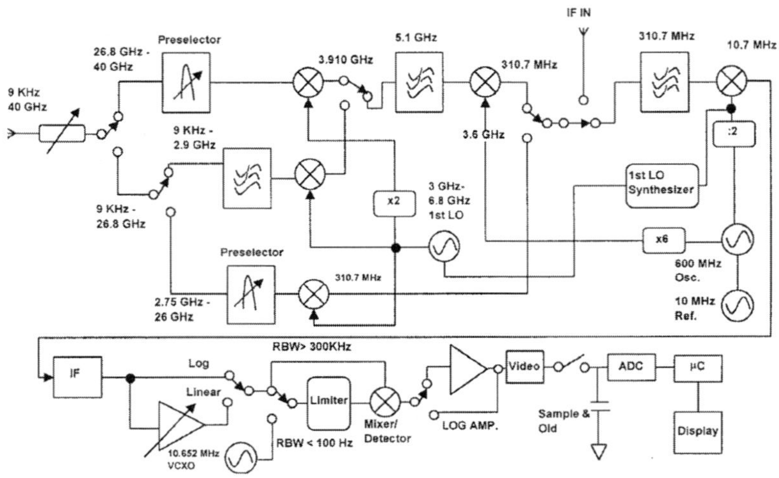

FIGURE 23.8 Standard block diagram of a modern spectrum analyzer.

Referring to Figure 23.7 we can say that we are confronted with a radio-receiver-like superhet with a wideband input circuit. The horizontal scale of the instrument is driven by a ramp generator which is also applied to the voltage-controlled LO [2].

A problem arises when dealing with a broadband mixing configuration like the one shown above, namely, avoiding receiving the image band.

The problem is successfully tackled here by upconverting the input band to a high-valued IF. An easily designed input low-pass filter, not shown in the block diagram for simplicity, will now provide the necessary rejection of the unwanted image band.

Nowadays, with the introduction of YIG bandpass filter preselectors, tunable over very wide input bands, upconversion is not always necessary. Traces of unwanted signals may, however, show up on the display although at very low level (less than –80 dBc) on good analyzers.

A block diagram of a commercial spectrum analyzer exploiting both the mentioned principles is shown in Figure 23.8. This instrument includes a very important feature which greatly improves its performance: the LO frequency is no longer coming from a free-running source but rather from a synthesized unit referenced to a very stable quartz oscillator. The improved quality of the LO both in terms of its own noise and frequency stability, optimizes several specifications of the instrument, such as frequency determining accuracy, finer resolution on display, and reduced noise in general.

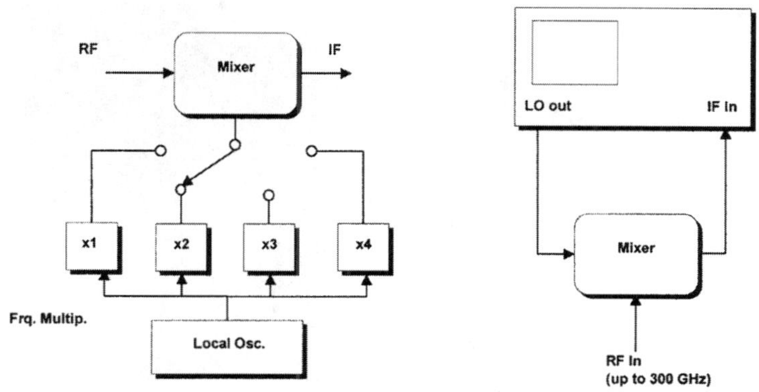

FIGURE 23.9 Encreasing the input bandwidth characteristics.

Further, a stable LO generates stable harmonics which can then be used to widen the input-selected bands up to the millimeter region. As already stated, this option requires external devices, e.g., a mixer-amplifier as shown in Figure 23.9a and b.

The power reference on the screen is the top horizontal line of the reticle. Due to the very wide dynamic range foreseen, the use of a log scale (e.g., 10 dB/square) seems appropriate. Conventionally, 1 mW is taken as the zero reference level: accordingly, dBm are used throughout.

The noise power level present on the display without an input signal connected (noise floor) is due to the input random noise multiplied by the IF amplifier gain. Such a noise is always present and varies with input frequency, IF selectivity, and analyzer sensitivity (in terms of noise figure).

The "on display dynamic range" of the analyzer is the difference between the maximum compression-free level of the input signal and the noise floor. As a guideline, the dynamic range of a good instrument could be of the order of 70 to 90 dB.

An input attenuator, always available on the front panel, allows one to apply more power to the analyzer while avoiding saturation and nonlinear readings. The only drawback is the obvious sensitivity loss. One should not expect a spectrum analyzer to give absolute power level readings to be better than a couple of dB.

For the accurate measurement of power levels, the suggestion is to use a power meter. An erratic signal pattern on display and a fancy level indication may be caused by the wrong setting of the "scan time" knob. It must be realized that high-resolution observation of a wide input band requires the proper scanning time. An incorrect parameter setting yields wrong readings but usually an optical alarm is automatically switched on to warn the operator.

The knowledge of the noise floor level allows a good valuation of the noise temperature, T_n (and therefore of the sensitivity), of the analyzer, a useful parameter on many occasions. The relations involved are as follows.

The Nyquist relation states that

$$P = kT_nB$$

where P = noise floor power level read on the display (W)
k = Boltzmann constant = 1.38×10^{-23} (J/K)
B = passband of the selected IF (Hz)

therefore,

$$T_n = P/(kB)$$

Usually engineers prefer to quote the noise figure of receivers. By definition we can write

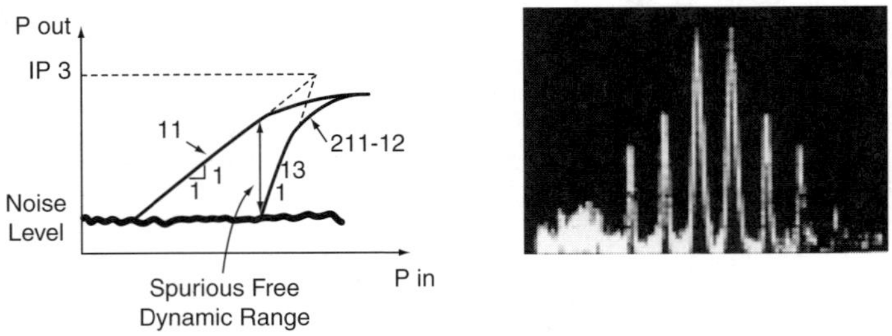

FIGURE 23.10 (a) Spurious free dynamic range. (b) Higher-order spurious.

$$N = \left(T_0/T_0\right) + 1$$

where N = noise factor
T_0 = 290 K
F (noise figure) = 10 log N

A typical F for a good spectrum analyzer is of the order of 30 dB.

It must be said, however, that the "ultimate sensitivity" of the spectrum analyzer will depend not only on its noise figure but also on the setting of other parameters like the video filter, the IF bandwidth, the insertion of averaging functions, the scan speed, the detector used, etc.

As a rough estimate a noise floor level of –130/–140 dBm is very frequently met by a good instrument.

Another criterion to select a spectrum analyzer is a good "IMD dynamic range," that is, the tendency to create spurious signals by intermodulation due to saturation.

This figure is generally quoted by the manufacturers, but it is also easily checked by the operator by injecting two equal amplitude sinusoidal signals at the input socket of the analyzer. The frequency separation between the two should be at least a couple of "resolution bandwidths," i.e., the selected IF bandwidth. As the input levels increase, spurious lines appear at the sum and difference frequencies and spacing of the input signals.

The range in decibels between the nonoverloaded input signals on display and the barely noticeable spurious lines is known as the "spurious free dynamic range," shown graphically in Figure 23.10a, where the third-order "intercept point" is also graphically determined. If input power is increased, higher-order spurious signals appear, as shown in Figure 23.10b. The input connector of most spectrum analyzers is of the 50 Ω coaxial type. Past instruments invariably used N-type connectors because of their good mechanical and electrical behavior up to quite a few gigahertz. Today SMA or K connectors are preferred.

External millimeter wave amplifiers and converters use waveguide input terminations. As is discussed in the next section, multipurpose analyzers are available where power meter, frequency counter, tracking generator, etc. can all be housed in the same cabinet. The economic and practical convenience of these units must be weighed on a case-by-case basis.

Finally, we mention that spectrum analyzers are available equipped with AM and FM detectors to facilitate their use in the RFI monitoring applications.

What Is the Right Spectrum Analyzer for My Purpose?

Several manufacturers offer a large number of spectrum analyzer models; the choice may be made on the basis of application field (i.e., CATV, mobile telephony, service, surveillance, R&D, etc.), performance (resolution bandwidth, frequency range, accuracy, battery operation etc.), or cost.

In addition, it is important to know that most spectrum analyzers need some accessories generally not furnished as a standard: for example, a connectorized, coaxial, microwave cable is always required; a directional coupler, or power divider, or handheld sampler antenna may be very useful to pick up the signals; and a personal computer is useful to collect, store, reduce, and analyze the data.

There are four main families of RF and microwave spectrum analyzers.

Family 1

The bench instruments are top performance, but also large, heavy, and the most expensive class, intended for metrology, certification, factory reference, and for radio surveillance done by government and military institutions.

The frequency ranges span from a few tens of hertz up to RF (i.e., 2.9 GHz), up to microwave region (i.e., 26.5 GHz), or up to near millimeter wavelength (i.e., 40 GHz). This class of instruments includes lower noise figures, approximately 20 dB, and may be decreased down to 10 to 15 dB with an integrated preamplifier. The synthesized local oscillator has a good phase noise (typically 10 dB better than other synthesized spectrum analyzers) for precise, accurate, and stable measurement. Also this class of instruments, by sharing the display unit, can be integrated with plug-in instruments like a power meter (for more accurate power measurements) or a tracking generator (for network analysis and mixer testing).

The interface to a computer (and a printer) such IEEE-488 or RS-232 is standard; it allows remote control and data readings; this class of spectrum analyzer often has a powerful microprocessor, RAM, and disks for storing data and performing statistical and mathematical analysis.

The best known families are the Hewlett-Packard series, 71xxxx [3] and the Rhode & Schwarz series FSxx. [4]. Indicative prices are between $50,000 and $90,000.

Family 2

Less expensive bench instruments, the workhorse class of spectrum analyzers, portable and lightweight, are associated with a synthesized local oscillator, that includes a frequency range from a few kilohertz up to RF region (i.e., 2.9 GHz), microwave region (i.e., 26.5 GHz), or near millimeter wavelengths (i.e., 40 to 50 GHz). A typical noise figure of 30 dB is good enough to ensure most measurements. A large number of filters down to few hertz of resolution are offered; digital filters are preferable to analog ones, because they give a faster refresh rate of the trace on the display. This kind of spectrum analyzer nearly always has the capability to extend the frequency range up to millimeter and submillimeter wavelengths with an external mixer. One of the most important features for a spectrum analyzer in this class is the quality of the local oscillator; it should be synthesized (PLL) to achieve stability, precision, accuracy, and low phase noise. Demodulation is also an important feature to listen to AM, FM on the loudspeaker and to display TV pictures or complex modulations onto the screen, which is often required by people working on surveillance, TV, and mobile telephone. The interface to a computer such as IEEE-488 or RS232 is standard in a large number of spectrum analyzers, and allows the remote control and data reading, storing, and manipulation.

This kind of instrument may integrate a tracking generator, a frequency counter, and other instruments that can transform the spectrum analyzer into a compact, full-featured RF and microwave laboratory.

The most popular families are the Hewlett-Packard series 856xx [3, 5], Rhode & Schwarz series FSExxx [4], Anritsu series MS26x3 [6], IFR mod. AN930 [7], and Marconi Instruments series 239x [9]. The Tektronix production should be taken in account. Prices typically span from $30,000 to $60,000.

Family 3

The entry level, a more economical class of spectrum analyzer, is intended for field use or for one specific application. If your need is mainly EMI/EMC, CATV, mobile telephone, or surveillance, perhaps you do not need the extreme stability of a synthesized local oscillator, and a frequency range up to 2 GHz may be enough; however, if you need some special functions such as "quasi-peak detector" or "occupied bandwidth measurement," two functions that are a combination of a mathematical treatment with some legislative aspects, these are easily measured with a spectrum analyzer including those functions. As the normatives can change, the capability to easily upgrade the measurement software is important; some models come with a plug-in memory card, some others with 3.5″ disks.

A large number of spectrum analyzer models are tailored to meet the specific needs of a customer. This is the case with the HP series 859x [3], Tektronix series 271x [10], IFR series A-xxxx [8], Anritsu MS2651 [6], and Advantest series U4x4x [4]. Costs typically are around $10, 000 to $20,000.

Family 4

The most economical class of spectrum analyzer, with prices around \$2,000 to \$6000, includes instruments that perform only the basic functions with a limited frequency range and filter availability and without digital capability. They are intended for service, for general-purpose measurements (i.e., IP_3, harmonic distortion) or for precertification in EMI/EMC measurements. One of the most popular series is the Hameg series HM50xx [11].

In this class are some special spectrum analyzers that come on a personal computer (PC) board. Such spectrum analyzers, generally cheap (typically \$3,000 to \$5,000), with frequency range up to 2 GHz, may include PLL local oscillators, tracking generators, and other advanced characteristics. The input is through a coaxial connector on the board, the output and the control is done by a virtual instrument running on the PC. One model is made by DKD Instruments [12].

Other unusual RF spectrum analyzers working in conjunction with a PC and worth noting are the instruments for EMI/EMC measurements and reduction in power lines and power cords. For this type of instrument, the core is not the hardware but the software that performs the measurement according to international standards and may guide the engineer to meet the required compatibility. An example is given by Seaward Electronic Sceptre [13].

Advanced Applications

New technological approaches and the use of spectrum analysis concepts in radioastronomy constitute some advanced spectrum analysis applications. Autocorrelators, with a typical frequency resolution of ~5/25 kHz, have been extensively used in radioastronomy. Their performance is well documented; the autocorrelation function is computed online and recorded. Later the FFT of the function is computed off line in order to get the power spectrum. Recently, the Tektronix 3054 Fourier Analyzer, based on a bank of programmable filters, was introduced as an alternative approach. The state of the art in integrated digital signal processors (DSPs) allows an alternative approach to spectrum analysis. By paralleling several of these DSPs, one is able to compute online the FFT directly on a very wide input bandwidth (several tens of megahertz).

By using this technique, high time and frequency resolution can be achieved.

A system based on the Sharp LH9124-LH9320 chip set is described in Figure 23.11. It is based on VME boards: one or two 10-bit, 40-MS/s ADCs and two boards in charge to compute the FFT of the incoming streams of data, in real time [14]. A following block computes the power and averages on the board up to 64 K spectra before storing the result on disk or tape. The FFT boards are powered by one of the fastest state-of-the-art DSPs (Sharp LH9124). The overall system is controlled by an embedded FORCE 3 Sparcstation. The LH 9124 DSP works with 24+24 bits (with 6 exponent bit) in block floating point. The system architecture allows expansion of the input bandwidth and the number of channels by paralleling more DSP boards. All the computing core is housed in a VME crate and is able to produce single-sided spectra from 1024 frequency bins to 131072 bins at an input bandwidth of 12 MHz without losing data or with 56% of time efficiency at 20 MHz. Single- or double-channel operation mode is provided. In a single-channel mode the main features of the system are reported as

Input bandwidth	0.5–20 MHz
Time efficiency	100% at 12 MHz (56% at 20 MHz)
FFT size	1K, 2K, 256K (points)
Avgs out format	<256 averages → integer 24 bits
	>256 averages → float 32 bits
Windows	Hanning, Hamming, Kaiser Bessel

This spectrometer was developed (1993) as a cost-effective system for both the NASA-SETI (Search for Extraterrestrial Intelligence) program [15,16] and for radioastronomical spectroscopy [17] at the CNR Institute of Radio Astronomy of Bologna. The digital spectrometer was first used to investigate the effects of the Jupiter/SL9 comet impacts (July 1994) [18,19]. In this application, the high time resolution

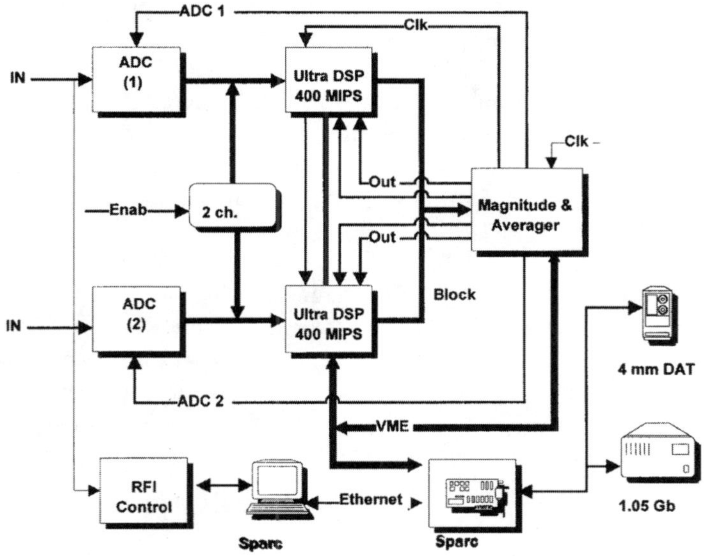

FIGURE 23.11 Block diagram of spectrum analyzer for radioastronomy.

of the spectrometer (a 16K points FFT every 1.3 ms) was exploited to compensate for the fast planet rotational velocity Doppler shift.

The system has been successfully used at the 32 m dish radiotelescope near Bologna in many line observations with unique results. Note that the use of such a high time and resolution system in radio astronomy may help to observe the molecular line in a very precise and unusual way. The whole pattern (a couple of megahertz wide) of a NH_3 molecule line coming from the sky was obtained in flash mode with a frequency resolution high enough to distinguish the different components. The same machine can be used for high-time-resolution observations of pulsar and millisecond pulsar. In those cases, the possibility of performing the FFT of the RF signal online allows coherent dedispersion of the pulses. This new technological approach in computing the FFT may be successfully addressed to many different fields, such as image processing, medical diagnostic systems, radio surveillance, etc.

References

1. A.E. Bailey, Ed., *Microwave Measurements,* 2nd ed., Peter Peregrins on behalf of IEEE, London, 1989.
2. Hewlett-Packard, *A.N 63* and *A.N 243,* Hewlett-Packard Company.
3. Hewlett-Packard, *Test & Measurement Catalog 1997,* pp. 224–257, Hewlett-Packard Company.
4. Rohde & Schwarz, *Catalog 96/97 Test & Measurement Products* pp. 109–145, Rohde & Schwarz Company.
5. Hewlett-Packard, *HP 8560E, HP8561E and HP8563E Portable Spectrum Analizers Technical Data,* Hewlett-Packard Company, pp. 5091–3274E.
6. Anritsu-Wiltron, *Electronic Measuring Instruments,* pp. 67–100, Anritsu-Wiltron Company.
7. IFR, *The AN 930 Microwave Spectrum Analyzer,* IFR System, Inc.
8. IFR, *The A-8000 & A-7500 Spectrum Analyzer,* IFR System, Inc.
9. Marconi Instruments, *Test and Measurement 96/97 Edition Instrument System*, pp. 4-1–4-5, Marconi Instruments Company.
10. Tektronix, *Measurement Products Catalog 1997/1998,* pp. 145–171, Tektronix Company.
11. Hameg Instruments, *Catalog 1994,* Hameg GmbH.
12. DKD Instruments, *Model 1800 Data Sheet,* DKD Instruments.
13. C. Galli, Analizzatori di spettro: rassegna commerciale, *Sel. Elettron.,* 15, 1996 [in Italian].

14. S. Montebugnoli et al., "A new 6 MHz 128000 channels spectrum analyzer," IAA-95-IAA.9.1.10. 46th International Astronautical Congress, October, Oslo, Norway, 1995.
15. S. Montebugnoli, C. Bortolotti, S. Buttaccio, A. Cattani, N. D'Amico, G. Grueff, A. Maccaferri, G. Maccaferri, A. Orfei, M. Roma, G. Tuccari, and M. Tugnoli, "Upgrade of the mini spectrum analyzer" in *5th Proceedings of the International Conference on Bioastronomy,* Capri, Vol. I, July 1996.
16. A. Orfei, S. Montebugnoli, C. Miani, J. Monari, and G. Tuccari, "The SETI facilities at the Medicina/Noto sites," (preprint) IAA-96-IAA.9.1.11, 47th International Astronautical Congress, October, Beijing, China, 1996.
17. S. Montebugnoli, C. Bortolotti, S. Buttaccio, A. Cattani, N. D'Amico, G. Grueff, A. Maccaferri, G. Maccaferri, A. Orfei, M. Roma, G. Tuccari, and M. Tugnoli, "A new high-resolution digital spectrometer for radioastronomy applications," *Rev. Sci. Instrum.* 67, 365–370, 1996.
18. C.B. Cosmovici et al., "Detection of the 22 GHz line of water during and after the SL-9/Jupiter event," European SL-9/Jupiter Workshop–February, ESO Headquarters, Munchen, Germany, 1995.
19. C.B. Cosmovici, S. Montebugnoli, A. Orfei, S. Pogrebenko, and P. Colom, "First evidence of planetary water MASER emission induced by the comet/Jupiter catastrophic impact," *Planetary Space Sci.,* 44, 735–773, 1996.

24

Applied Intelligence Processing

Peter H. Sydenham
University of South Australia

Rodney Pratt
University of South Australia

24.1 Introduction

For a variety of reasons, the signal that comes from the output of a sensor will usually need to be processed, as has been explained elsewhere. This may be needed for a single sensor alone, as found in the conversion from an electronic resistance temperature sensor into the equivalent digital display number.

Alternatively and increasingly so, as is now discussed, the outputs from several sensors need to be combined into a single signal to form a mapping of those sources from "many to one." An example of this would be when a set of sensors measuring many different variables is used to determine if a machine tool is correctly making a component.

There are now many ways that can be used to combine signals, each having its own features that make it the best to use in a given circumstance. Signal-processing methods range from the well proven and mostly used method using processing based on mathematical relationships that are very precise — these are explained elsewhere. These are generally executed using digital computation and are often referred to as digital signal processing (DSP) methods. Alternatively, it has now been shown conclusively that less quantitative methods can be also used to great effect despite their lack of complete formal mathematical formulation. These are here called the (applied intelligence) methods, a convention developed to distinguish man-made systems from the very broad, and oversold, use of the term *artificial intelligence.*

Many seemingly different methods exist in the latter group but, as will be shown here, they are all part of a continuum of ways that range from use of subjective to exactly objective procedures. These are not well explained as a group because they are presented in the literature as different methods used in isolation of each other. This account shows how they all tie together, thus making it easier to decide which is appropriate in a given application. They are particularly useful for handling highly nonlinear situations where algorithms cannot be realized.

While we appear to prefer the total objectivity of a mathematically formulated method of signal processing, it is now well proven that the AI methods often are better choices to use in terms of better

0-8493-1733-9/04/$0.00+$1.50

speed of performance and often lower cost of processing. Often they are the only solution since the algorithmic approach cannot be deployed because of the lack of a suitable mathematical formulation or powerful enough processor to run the algorithm.

Signal processing in the modern instrument, therefore, will often make use of many different methods. This account is an introduction to the characteristics of the various forms and is written to assist selection. Space limitations prevent presentation of each kind in detail.

24.2 Overview of Algorithmic Methods

Traditionally the most popular method used to develop mapping models is that of mathematical modeling. The mathematical model is usually what is sought, as it provides the highest level of understanding about the subject and the most precise representation of the behavior. The major disadvantage of mathematical models is that they can quickly become so complex that implementation of these models in measurement systems is often impractical.

In this class, the single, or set of multiple, input signal(s) to the data processor is converted to the output form using tightly formulated mathematical description. This relationship is called the algorithm. Strict relationships hold; the relationship is said to be *formal*, meaning that for any given input the output will always be the same. The algorithm supports only one interpretation.

This method of signal processing is the most highly developed method and is certainly one to aim for because it is devoid of ambiguity. All will agree on how it will respond. It carries a comforting level of understanding and, thus, acceptance.

Algorithmic methods can be very accurate, traceable, and can be calibrated with relative ease and agreement. They are the basis of many instrumentation systems. The origin of their use in instrumentation goes back to the early days of computing using, first, mechanical computational machines (late 1800s to around 1930) and then analog electric devices (early 1900s to 1960s), all of which were mostly replaced by the use of digital computers commencing around 1950. All of these algorithmic methods of processing can be simplistically regarded as embodiments of a mathematical equation inside a suitable technological machine.

As the demanded complexity and performance requirements grew over time, so did the demands on the detail of the algorithm and the means to model it inside a computational machine. Mathematical description eventually reaches limits of definition as the models push the boundaries of mathematical methods and human development. Too often, this arises before adequate detail is able to be built into the model. The algorithm is then an inadequate model of the need.

As the algorithm increases in complexity, the processing power needed must be increased to maintain both fidelity and speed of processing. Despite great advances being made in algorithm development and in computer power, the algorithmic methodology eventually encountered mathematical and technological barriers in many fields. The method is seen to not always be the best to use because of lack of an adequate algorithm or the high cost of computing.

In instrumentation, another factor also arises. Fast, detailed processing brings with it the need for increasing electrical bandwidth requirements in signal transmission. This increases implementation costs and also eventually reaches technological constraints.

24.3 Overview of Applied Intelligence Methods

Fortunately, the solutions that may overcome these limiting constraints in many circumstances were developing in other fields under the general name of artificial intelligence (now called applied intelligence in engineering), as new forms of mathematics and in other fields, such as decision theory.

Principally, a key limitation of the algorithmic method is that its unforgiving level of formalism carries with it a depth of processing exactitude that is often not warranted.

Other methods have emerged that allow vaguely subjective, as opposed to tightly objective, processing to be applied to good effect.

These AI methods have gradually gained acceptance to the degree that many are now routinely used and are supported by dedicated applications software and electronic integrated circuitry.

At first, these many alternatives were seen to be isolated methods. Gradually, the literature has shown trends to merge them in pairs. Their use in a more widely mixed form is still limited. This account seeks to give a comprehensive appreciation of the commonly met AI processing methods by placing them into relative perspective.

It is interesting to contemplate that natural world computing in animals does not appear to make much use of algorithmic methods, but does make extensive use of the methods presented here in the AI class.

The paradigm invoked here is that experience has shown that informal methods based on knowledge-based systems (KBS) can produce mappings of many inputs to one by use of less than completely formal description.

The AI methods can yield surprisingly efficient solutions to previously unsolved needs. They often can outperform algorithmic methods or carry out a similar task with far less computing power. They are all associated with multiple input processing and can be applied to forming decisions from data supplied by sensors. Each situation has to be judged on the balance between use of computing effort and effective processing.

On the downside, they lack formality and thus may be very hard to calibrate and authenticate. They, not having adequate scientific foundation and a solid formal base of operation, are not easily accepted as "sound." They are often hard to comprehend by a second party, for their description is not always adequately documented or done to any agreed convention. As their principles vary widely, they must be well understood before application is developed.

For all of these negative factors, they often are able to provide "more performance for less cost" and thus will be increasingly adopted.

Their rising level of use should not suggest the algorithmic methods will become obsolete, but more that the instrument designer now has a much larger set of processing tools available.

24.4 Mapping, in General

The high-level purpose of most signal processing is to yield knowledge of a situation so that decisions can be made.

For example, consider a health-monitoring system installed on an aircraft engine. A set of sensors of different measurand types and locations is installed at various critical points on the engine — temperatures, pressures, flow rates, metal content of the lubricating oil, and more. The data from these are collected and transmitted to a central processor using a common digital bus. The many inputs then need to be combined in some way to decide such conditions as emergency states, when to change the oil, and engine efficiency. This combination is a "mapping of many to a few."

These are not always simple mappings, for there is no adequate algorithm available to give a mathematical description for such things as degradation of oil condition. However, human intuition can be used quite effectively to obtain answers — the human mind is very capable of carrying out such mapping functions. This form of natural processing makes use of what are technically called "heuristics" — but more commonly known as "rules of thumb."

Consider the question, "How could we decide, using an automated measurement system, when loaves being made in a bakery are satisfactory to sell?" As the way to decide this almost all people asked would suggest that the weight, size, crustiness, appearance, and softness inside would be the parameters that must all be satisfactory (that is, lie within a small range of values for each) to be declared suitable. Weight and size are easily measured; the others are not for they are really heuristics, as is the choice of the set of parameters.

The thought process implemented here is that the designer starts with a desire to know something about a situation. Consider how we could automatically monitor the "risk of working in a hazardous place" in order to give an alarm at set levels. In this kind of situation a study of the problem will lead to identification of key parameters. These parameters can each be assigned safety functions that express how each parameter varies with system changes. With this framework it is then possible to set up a signal-processing system that continuously calculates the risk level. This form of solution is based on ideas embodied in the wide field of decision-making theory.

The heart of application of AI methods of signal processing in instrumentation lies with appreciation of decision-theory methods.

A range of multivariable mappings methods using AI ideas have emerged. Those well established in instrumentation are:

- Representational measurement theory and ways sets are mapped into other sets (whose usefulness is still emerging).
- Rule and frame representation of heuristic knowledge and ways they are used to form expert systems and other KBSs.
- Binary Boolean trees as crisp logical mappings; which is the foundation of the fuzzy logic method based on fuzzy set theory.

Another class of AI methodology that does not fit the same sequence, yet includes powerful methods, is one that assists optimization of the mapping setup. There are two main methods in use:

- Genetic algorithm and its use to optimize fuzzy logic and other multisensor setups, and the
- Artificial neural net, a mapping method that learns by itself, from experience, how to achieve an optimal mapping in a given situation that it has been taught to work in.

24.5 Basics of Decision Theory

Rules about a Decision-Making Process

Before entering into the detail of the AI signal-processing methods, it is necessary to develop a foundation about the ways in which decisions can be made by computers using sensed information. It is not that well appreciated, but setting up a mapping-type signal-processing situation is actually implementing a decision-making information system. General appreciation of decision making can be found in the introductory work of Kaufmann [3] .

Unlike the human brain decision maker which carries out smart thinking with ease to build a machine counterpart, an engineered object needs effective externalization of the process involved. This begins by developing appreciation of the basic rules that always apply about a decision-making situation. These have been summarized by Baker et al. [1] and condense to:

1. There must be a clearly expressed criterion for making a judgment of the options available, which must be such that others will understand how the judgment was made.
2. All decisions will involve choosing alternative strategies to arrive at the best one to use. This will involve assigning score numbers to the selected parameters and deciding how to process the set of numbers.
3. A decision is made by a choice of competing alternatives in the face of given restraints. Decisions can only rarely be made in the general sense but will be made for a given set of circumstances. The complexity of a problem rises rapidly as the number of parameters rises.
4. The process used attempts to achieve some payoff as a value added or lost. It aims to rank the various mapping alternatives to advise the apparently best to use. Note that once a decision-making mapping is built, the information about alternatives is no longer available as it will only usually embed one set of parameters as a single process.
5. A decision matrix carrying the competing parameters results. A method of combining the matrix constituents is needed, and, again, there is no singularly definitive, absolutely correct way to process the matrix.

In setting up a signal-processing mapping, these rules will need to be addressed. They will be embedded in the software of the hardware processor as its operational strategy. Considerable creativity is needed by the designer of the processor, for much of the setup of decision-making methods requires subjective human interpretation in several steps of the process. Decision making is really only needed when there is no exact and obvious answer. The devices built to mimic the human process will never be perfect. There will be much debate about which are the best methods and parameters to use. Engineers must live with this situation and make machines that will make good decisions, that are as close to perfect as possible.

Extracting Parameters

The first step in setting up a decision mapping is to understand the need. That means researching it by observation and from literature on the topic. This sets up the specific knowledge base to allow one to progress to the next step.

Then comes the need to define the key parameters of the situation. There is no organized way to develop these. They arise from inventive and innovative thought processes that seem to be based on prior learning.

To streamline this intuitive step, it is useful to apply some ordered processes that assist in externalizing appropriate parameters. Three methods are now briefly described.

Slip Writing

A group of people familiar with the problem area are read a brief statement of the problem by a person independent from the problem. An example could be "What do you think are the main parameters that a person uses to decide if a loaf of bread is fresh?"

Without much time to reflect the group is then asked to write down the key parameters that come to mind immediately as they work without talking about the problem as a group. They write down each parameter on a separate piece of paper, doing this as fast as ideas come to them. This only happens for a few minutes. The slips of paper are then collected and classified. The whole process takes around 10 min and is known as slip writing.

It will usually be found that there is common agreement about the majority of parameters with some quite unexpected ones also arising.

Slip writing is a good way to find consensus. It probes the mind well and can bring out appreciation of factors that open discussion might easily inhibit. It is important in this method to decouple the thoughts of each person during the process; otherwise the real parameters may not be externalized because some people may exert influence on others of the group.

Brainstorming and Think Tanks

If participants are shown what others are thinking and are encouraged to debate issues, it is possible to gain consensus and also allow group participants the opportunity to help each other be innovative at the same time. This method works best when the participants are prepared to go into open discussion. Several similar processes are those known as brainstorming or carrying out a think-tank session.

Here a problem in need of solution is written down as a well-prepared statement by the session organizer. A team of experts, each covering the expected aspects of the problem area, are selected and sent the statement along with any supporting exhibits. Each person considers, over a few days, how he or she might contribute a solution.

The group is then assembled. The problem is first reviewed by the session leader and each person is then asked for ideas. As ideas are externalized they are recorded in very brief form — large sheets of butcher paper are suitable. These sheets must be readable by all in the group and be prepared instantly to keep up with the thoughts of the group.

It will be found that innovative ideas will arise as candidate solutions are put up and seen by others in the group. This method encourages group-driven inventiveness.

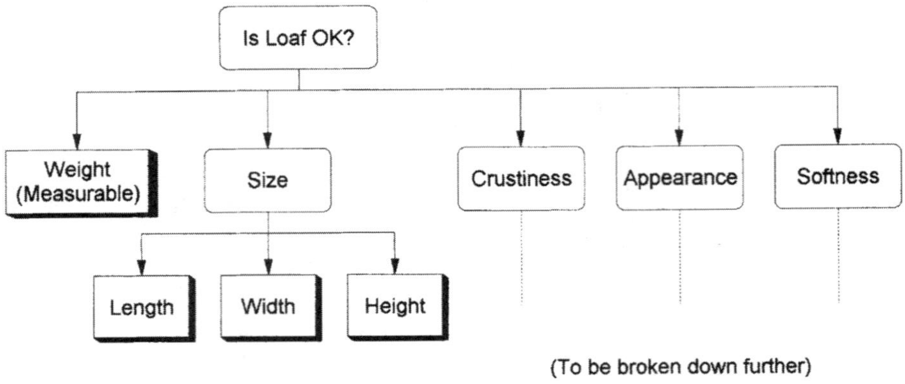

FIGURE 24.1 Knowledge trees allow facts and their relationships to be captured in pictorial form.

Gradually the group will settle on a few solutions that it feels have a good chance of succeeding. This list is then ordered in priority of likelihood of success, The session leader then writes up the outcomes, ready for further investigation.

Knowledge Trees

The final method to be described here for developing parameters of a decision, called knowledge trees, has the merit of ordering the relative place of parameters as well as encouraging inventiveness of solutions. It also provides a mapping structure. This procedure is based on the age-old realization that we think problems through by breaking them down into ever smaller subproblems until we feel able to solve them. Overall solution is then very much a matter (but not entirely so in practice) of implementing the solution of all subproblems and combining them by a process called integration.

The need is first written down. For example, "How would we measure the quality of loaves of bread?" or in a shorter form "Is the loaf OK?"

This forms the top-level parameter of the tree given as Figure 24.1. Consideration of the situation at hand then leads to realization of the collection of parameters relevant to get this answer. These might be weight, size, crustiness, appearance, and softness. They may have been externalized by a group process, such as slip writing, or created by the individual.

Each branch on the tree is then visited to see how that parameter might be measured. Only one of these can be measured as it stands — it is reasonable to assume that weight can be measured with scales.

Size is not so easy to measure as it is expressed for there is inadequate definition. More thought will yield another level to the tree — length, width, and height — for this parameter. As linear measurements, these can also be measured with ease.

When building the branching downward, a thought process has decided how the parameters map upward and downward. Size dictates the mapping of three parameters into one, so there must also be a defined mapping model for that mapping.

Note also that to branch downward, the thought process used has actually been driven by some heuristics. Each branching has been driven by rules of some kind — but more on that in the rule-based decision-making method covered below.

It is also easy to see why the other parameters could be measured with automated instrumentation. Softness could be assessed in terms of the squeeze factor, which is actually measurable as the compliance of the loaf at the center point of a side of the loaf. Appearance would map down the tree into color, texture, and graininess of the image. It is left to the reader to think up and draw a complete tree.

When all branch ends have been reticulated down to the point where they can be measured, the system mapping can be implemented with a human-made sensing system. The parameters are externalized and the mapping process is largely decided.

Use of tree-based thinking is a simple, yet powerful, way of keeping track of decision making for a complex situation. The recorded form also allows others to see the thought process used.

Size	Crustiness	Appearance	Softness	Parameters
$\frac{\Rightarrow Si}{W}$	$\frac{C}{\Rightarrow W}$	$\frac{A}{\Rightarrow W}$	$\frac{\Rightarrow So}{W}$	**Weight**
	$\frac{\Rightarrow C}{Si}$	$\frac{A}{\Rightarrow Si}$	$\frac{\Rightarrow So}{Si}$	**Size**
		$\frac{\Rightarrow A}{C}$	$\frac{So}{\Rightarrow C}$	**Crustiness**
			$\frac{\Rightarrow So}{A}$	**Appearance**

Scores (number of times preferred)

Si	C	A	So	W
2	2	1	3	2

FIGURE 24.2 Triangle of pairs assessment is a simple way to decide which choice to make. This table gives the workings for the grading of bread as in the example of Figure 24.1

Two Examples of Decision-Assistance Methods

Triangle of Pairs

Having now seen how to externalize parameters and how they might be interrelated using trees, we can move on to investigate how to set up a suitable decision-making process.

Knowing the parameters is not enough yet to design a multisensor mapping processor. The relative importance of the parameters is also a key factor. Furthermore, the tree is not the only way to combine sensor signals.

Two, of many, examples of decision-assistance methods are now outlined to illustrate these points.

The first, the triangle of pairs (TOP) method, allows parameters to be ranked against others on the binary-only basis of which is preferred of each two compared. In the bread example, compare some of the parameters for their relative importance. Crustiness is preferred to weight. Softness is preferred to size and so on until all pairs have been considered. If all combinations are carried through and recorded as a matrix, a triangle of pairs results, as in Figure 24.2.

Having formed a matrix of competing parameters, the next step is to decide how the matrix can be processed. This is where much debate can occur. For the TOP method, however, the binary nature of the choices means a simple count of first preferences gives the ordered preference of parameters — at least as that person assessed it!

We will see later how the idea is extended by giving parameters a varying degree of "preference" rather that the simple binary choice allowed here.

Utility Analysis

A more fully developed method for making decisions in complex and subjective situations is one called utility analysis (UA). This is a process that can be applied to find the usefulness of a design, piece of equipment, or any similar situation where one can externalize and prioritize a set of measurable parameters. Although mostly applied to decision making as a paper study, the process is amenable to the creation of a multisensor mapping processor.

Appreciation of this process is easily obtained by working through an example. Consider the need to set up a method for grading the quality of bread made in an automated bakery.

The first step is to decide the parameters that will affect the choice finally made. We select weight, size, and appearance as the key parameters to illustrate the method. (More parameters might be needed in a real situation.) We also decide that these are important according to the relative weighting ratios of 1.0:0.2:0.8.

Next, utility curves must be set up, one for each parameter. These show how the usefulness of the parameter changes as the parameter ranges.

The simplest way to obtain these graphs is to use one's own intuition, but a better way is to make use of some form of consensus-forming procedure as discussed above. Figure 24.3 shows what these three functions might look like. As a guide to their construction, if the weight of the loaf is too low, it fails to comply with legal requirements and thus has zero utility as a product below the allowed uncertainty of weight. The size factor depends on the type of bread. Here it is assumed it is for sandwich making, in which case the size can be too small for sliced meats or too big for a toaster. Appearance can only be measured by mapping downward to a set of measurands; it is convenient to plot the function in more vaguely defined terms in this method.

Note that the weighting ratios have already been incorporated into the utility charts by setting their best values at the correct percentage.

It is necessary to reinforce the fact that the set of graphs is needed for the parameters, not for each case to be considered. With the charts set up, the selection process can begin.

A matrix is now created, as also shown in Figure 24.3. The actual measured weight value (960 g) for loaf 1 (under automatic inspection) is compared with the graph of weight to yield a utility of 0.4. The size is done likewise, using the size graph to get 0.2, and the appearance sensor set tells us it is between poor and good to give us 0.4. Each loaf is subjected to the same process. The other two loaves have different sets of scores.

The combined usefulness of a given loaf is now to be decided by processing the set of numbers for that loaf. Here is where some difficulty arises, because there are many ways to combine the three scores for each loaf. One way often used is simply to sum the values, as is done in the example. Note that this can be satisfactory unless a zero or other unacceptable value arises, in which case more mapping processing is needed. Assume that an acceptable combined score has been determined to be 1.8 or higher (with the best, due to the weightings, 2.0).

Having carried out the processing of the matrix, we find that the first loaf is not acceptable and the other two are equally acceptable.

What we have done is to form an automatic measuring system that can make assisted and graded decisions. There is certainly a degree of human subjectivity inherent in the method, but by recording what is taking place the task can be automated and it can also be evaluated in the light of experience, correcting choices of parameters and weightings.

Other decision-making methods exist: they use elements of the above concepts in a variety of assemblages.

The above has been a very rapid introduction to decision-making methods. These are rarely taught as the foundation of the now popularized AI signal-processing methods given next. They are, however, the conceptual basis of the AI processes.

24.6 Principal AI Methods

Measurement Theory

Sensors observe a finite number of variables associated with real-world events. Measurement systems are required to convert the information from these sensors into knowledge. This process often involves what is known as mapping.

Depending on the loss or preservation of information, the mapping process can be classified as either

1. transformation, or
2. abstraction.

If a rule exists that assigns every element of x in the set X (written $x \in X$) to precisely one element $y \in Y$, we say that a function exists that maps the set X to the set Y, and this is represented as Figure 24.4 and symbolically by

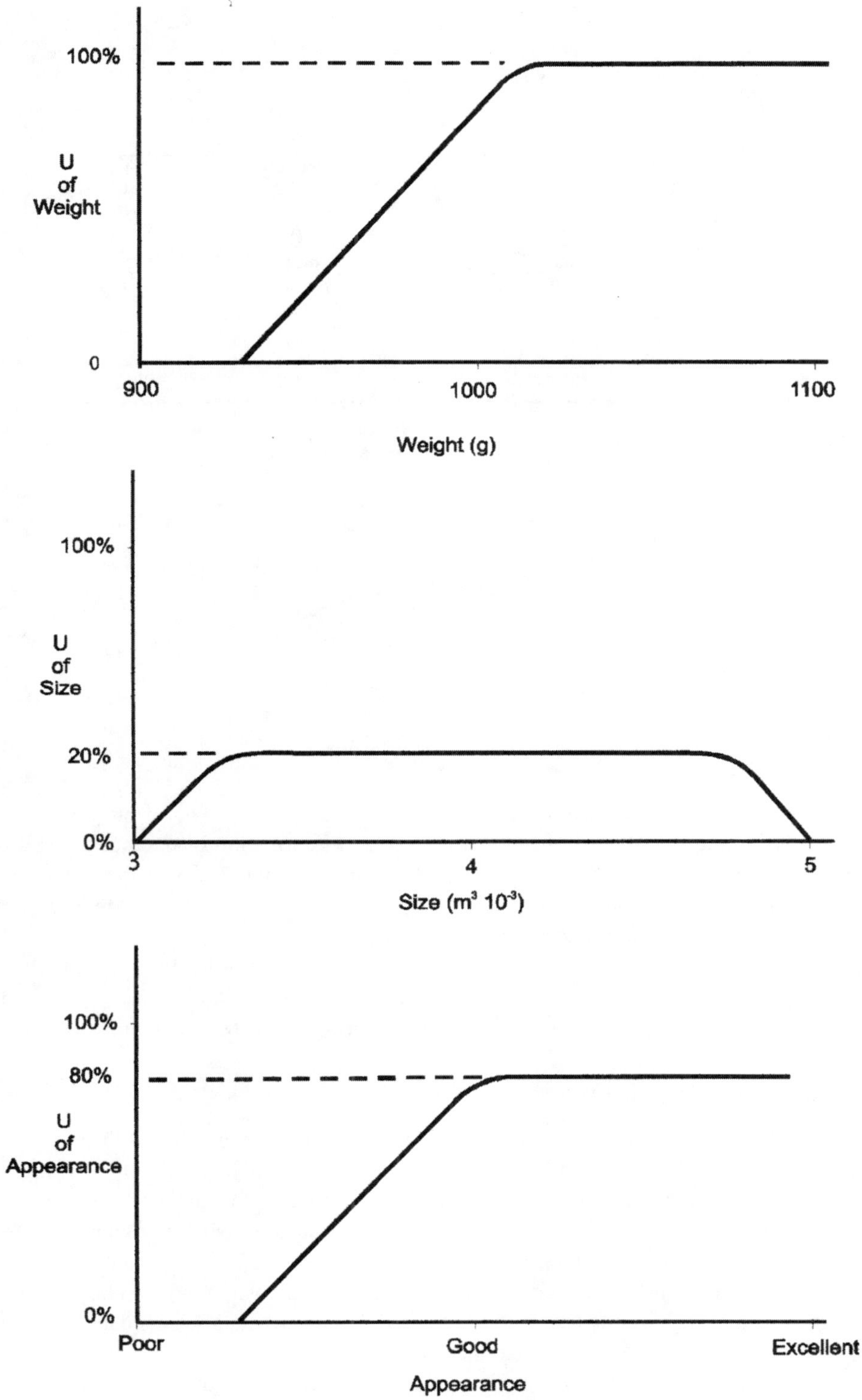

FIGURE 24.3 Utility analysis is a more detailed way to automate loaf inspection with a set of sensors. Here are shown three of the utility functions for the example in Figure 24.1 along with the scoring matrix for loaves passing the inspection point.

Parameter / Object Under Assessment	Weight (g)	Size ($m^3\ 10^{-3}$)	Appearance (grade)	Score (Σ)
Loaf 1	(960) 0.4	(4) 0.2	(poor/good) 0.4	1.0
Loaf 2	(1000) 1.0	(3.5) 0.2	(good/excellent) 0.8	2.0
Loaf 3	(1100) 1.0	(4.9) 0.1	(excellent) 0.8	1.9
•	•	•	•	•
•	•	•	•	•
Loaf 'n'	etc	etc	etc	etc

FIGURE 24.3 (continued)

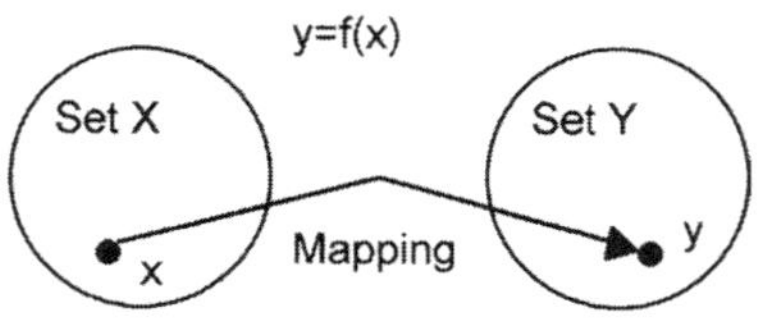

FIGURE 24.4 Pictorial representation of the set theoretical mapping process.

$$y = f(x)(x \in X)$$

A transformation defines a one-to-one mapping on the set of all elements of x into y, in which for all $x \in X$ there exists a unique value in Y; therefore the inverse transformation is guaranteed, i.e., $x = f^{-1}(y)$.

An abstraction differs from a transformation in that it can map a number of elements $x_i \in X$ into the same y, or a many-to-one mapping; hence, no inverse operator can exist such that the inverse image of X can be uniquely retrieved from Y. Therefore, abstraction is usually characterized by a loss of information.

It is possible to define the set theoretical relationship as a formal mathematical model and then embed that in the signal processor. Although this is a clearly useful method of formalizing mappings, this approach is not commonly used at the time of writing but can be expected to find more favor in the future.

Rule- and Frame-Based Systems

In early stages of problem solving, we seem naturally to look to rules of thumb to get the solution started. Even the algorithmic methods start with this form of thinking, for one has to decide what the elements of the decision are and how they might be assembled into a strategy for implementation in the signal processor. As the rules are externalized one's thought patterns also usually build knowledge trees that show the relationship between the rules.

Consideration of the knowledge supporting the structure of a tree will reveal that the decision needed about which way to branch as one moves through a tree is actually the implementation of a rule that has relevance at that junction. Rules link parameters together.

Figure 24.5 is a tree giving some of the design options for improving the performance of an audio speaker system. This tree has been built by applying a designer's knowledge of speaker system design. The heuristic rule set for the top branch is stated as

IF speaker output (W) increases
AND distortion (D) is reduced

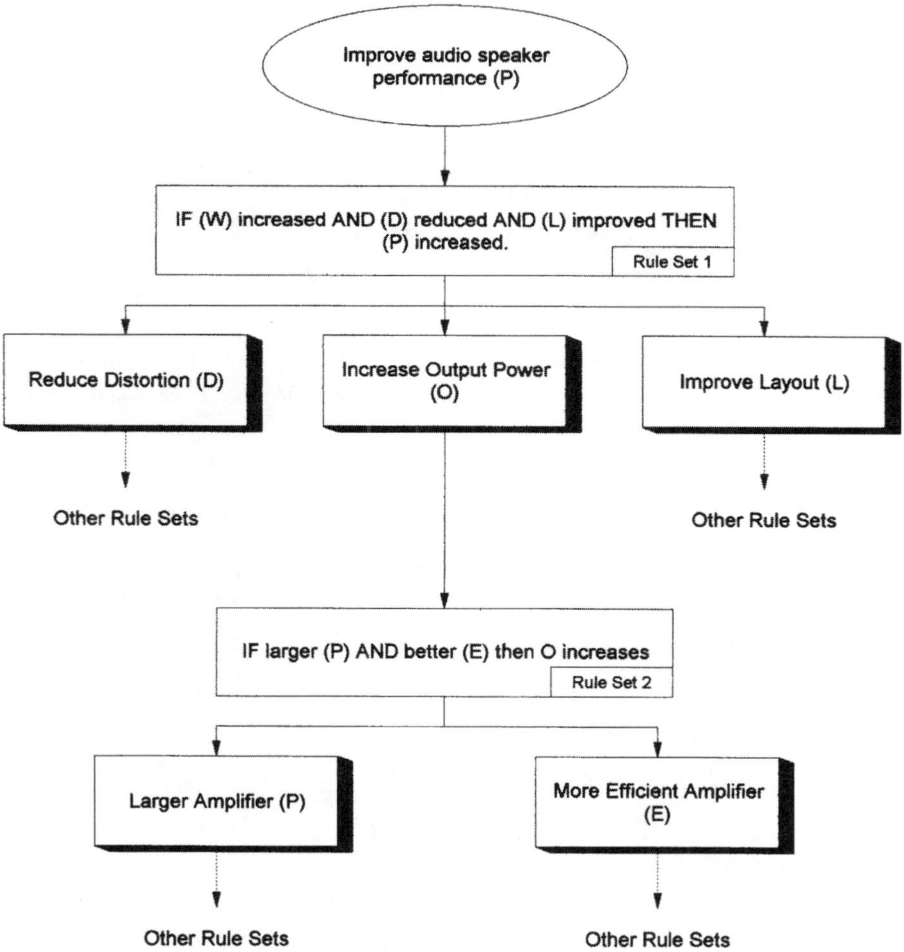

FIGURE 24.5 Part of a design knowledge tree for improving the performance of an audio speaker system.

AND positioning (L) improved
THEN audio output (O) is improved

No part of the rule set could be first realized using formal mathematical or algorithmic thinking. Intuition, leap, common sense, and other terms that describe the human intelligence process must be applied to commence a solution.

At some stage a rule may become describable by an algorithm — when that occurs a formal mathematical expression can be used to embed relationships. However, this is often not the case and so methods have been developed in computers to process heuristics.

The full set of AI techniques were originally all rolled into what became known as KBSs but this term is so overused that it has lost specific meaning.

Among the first AI processing methods were special computing ways to process the logic of a set of rules. These became known as expert systems (ES). In the example above, the rule tree is very sparse; a practical system for decision making is likely to have from 100 to several thousand rules.

The rules are considered by an inference engine (a software program) that is able to carry out Boolean logical operations of AND, OR, etc. to yield the outcome appropriate to the set of rules relevant to the problem at hand.

Trees can be traversed from the top down (downward chaining) or from the bottom up (upward chaining) and modern ES software applications carry out these operations with great sophistication.

KBS methods generally suffer from the feature that they seem to give answers all too easily, for they use only a few of the many rules available to come to a solution in a given situation. To help users feel more confident in their application, features are often offered that include plotting of the chaining used to get the solution or stating the rule set used.

Rule-based software applications are sold as empty shells. The user fills the shells with the rule set to make the application specific. These applications are now commonly used. They are relatively easy to use without the need for a competent computer programmer.

Rules are a primitive way to express knowledge. A better form of representation is the frame. This has the ability to hold more knowledge than a single rule and is a small database about a limited area of the system of interest. Frames are like objects in object-oriented programming. Advanced ES shells operate with frames.

ES shells are now very capable entities for decision making. They are a significant tool in the instrument signal processor's toolbox. Space restricts more explanation but enough has been stated here to allow further development of AI methods.

Some basic characteristics about rule- and frame-based processing are as follows (these apply variously):

- The software program is often self-explanatory as the rules can be read as (almost) normal language.
- They need considerable computer power as the rule number increases.
- They are relatively slow to yield a solution but are best used for cases where slow outcomes are applicable.
- They need special software.
- They are not that well known, so their application may be slow to find favor.

ESs soon run out of usefulness if the problem becomes complex. The computer search becomes too slow because the number of rules needed rapidly rises with problem complexity. The approach used today for large-problem-solving systems is to build an ES for each facet of the problem. These small-problem AI units are called agents. A set of agents is then combined using a conceptual software-based blackboard that calls on the agents to investigate a problem put to the system.

The chaining operation basically only carries out simple Boolean algebra operations using system parameters represented by a description called a rule. The system has no understanding of the wording of the rule. Thus, it is only processing as though the tree branching is either one way or the other. In its simplest form, it contains no concept of making that branching decision with a graded concept of which way to go.

Real life is full of unclear logical operations. The outcome of a decision will be clearly this way or that, but just where the change arises is problematic. The changeover point is unclear because the variables of the rule are fuzzy. It is desirable to process rules with regard of their likelihood of relevance depending on the state and selection of other rules. As is explained below, the fuzzy logic method of carrying out a mapping is another way that allows the rule to have more variability than the two-state exactness of binary logic.

Fuzzy Logic

This explanation is not intended to be a rigorous mathematical examination of fuzzy sets and fuzzy logic but rather explain, through example, the application of fuzzy techniques in measurement systems, specifically in respect to mapping models. For more detail, see Mauris et al. [4].

The simple example used here is a measurement system, Figure 24.6, that maps two input sensors (temperature and humidity) into one output value (comfort index). Clearly there is no formally accepted definition of comfort index as it is a subjective assessment. One of the advantages of fuzzy sets is that they are usually intended to model people's cognitive states.

In the mid 1960s Professor Lofti Zadeh recognized the deficiencies of Boolean logic, in that its TRUE/FALSE nature did not deal well with the shades of gray that exist in real life situations.

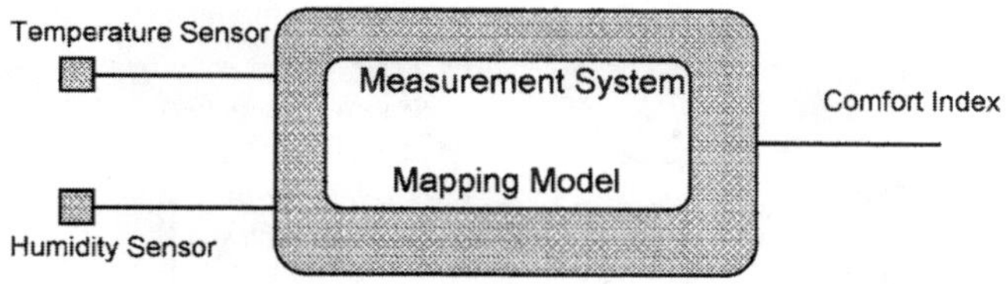

FIGURE 24.6 A simple comfort index measurement system uses temperature and humidity variables.

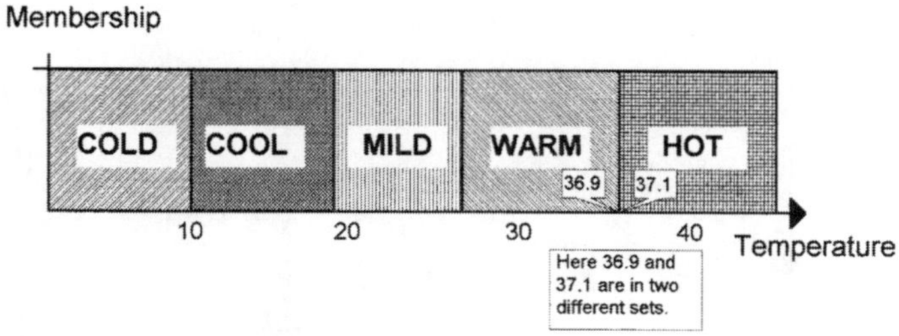

FIGURE 24.7 Conventional crisp set for the measurement system of Figure 24.6.

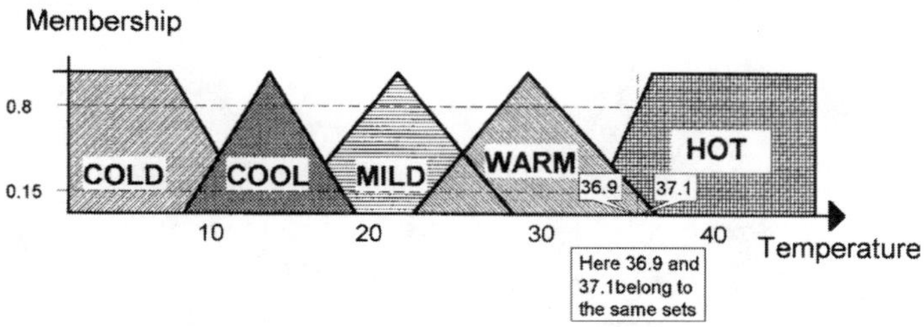

FIGURE 24.8 Fuzzy set representation temperature regimes in the comfort controller example.

Boolean logic uses classical set theory where an element is either viewed as entirely true or completely false $A = \{0,1\}$. These are often referred to as a crisp sets, Figure 24.7. In a crisp set the transition between sets is instantaneous, i.e., 36.9°C is considered warm whereas 37.1°C is considered hot. Hence, small changes in the input values can result in significant changes in the model output. Clearly, the real world is not like this.

Fuzzy logic uses a multivalued set where degrees of membership are represented by a number between 0 and 1 $A = [0,1]$ $\mu_{A:}$ $U \rightarrow [0,1]$, where μ_A is the membership function. With fuzzy logic the transition between sets is gradual and small changes in input values result in a more graceful change in the model output, Figure 24.8.

Fuzzy Expert Systems

A fuzzy expert system, Figure 24.9, combines fuzzy membership functions and rules, in place of the often all-too-crisp Boolean logic, to reason about data. The fuzzy expert system is usually composed of three processing sections:

Step 1. Fuzzification
Step 2. Rule evaluation
Step 3. Defuzzification

Step 1 — Fuzzification.
In fuzzification crisp inputs from input sensors are converted into fuzzy inputs using the membership functions in the knowledge base. A fuzzy input value is generated for each linguistic label of each input. For example, in Figure 24.8, for an input temperature of 37°C the fuzzy input is COLD(0.0), COOL(0.0),

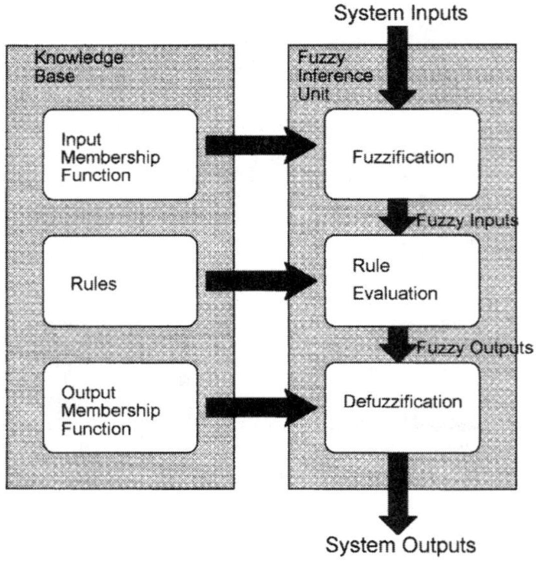

FIGURE 24.9 A fuzzy inference system combines rules in a way that allows them to be fuzzy in nature, that is, not crisp.

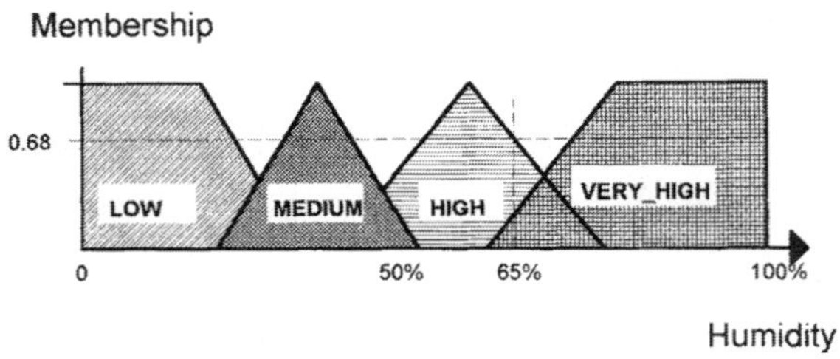

FIGURE 24.10 Humidity membership function for the controller example.

MILD(0.0), WARM(0.15), HOT(0.80). A similar set of fuzzy inputs are generated for the humidity sensor, Figure 24.10.

Step 2 — Rule Evaluation.
Rules provide a link between fuzzy inputs and fuzzy outputs. Rules are usually expressed in the form of IF … AND/OR … THEN … statements.

For example,

'IF' the TEMPERATURE is HOT 'AND' the HUMIDITY is HIGH 'THEN' it is UNCOMFORTABLE.
'IF' the TEMPERATURE is MILD 'AND' the HUMIDITY is MEDIUM 'THEN' it is COMFORTABLE.
'IF' the TEMPERATURE is WARM 'AND' the HUMIDITY is LOW 'THEN' it is UNCOMFORTABLE.

Rules can also be expressed in the form of a table or matrix, called the Fuzzy Associative Matrix. This matrix, Table 24.1, provides a complete description of the system performance for all combinations of inputs.

The function of the rule evaluation step is to evaluate the relative strengths or truth of each of the rules in order to determine which rules dominate. In this example the rules contain AND relationships and therefore the overall rule strength must be the minimum (MIN) value of the two strengths of the input values.

TABLE 24.1 The Fuzzy Associative Matrix Links Inputs and Outputs

Hum/Temp	Cold	Cool	Mild	Warm	Hot
Low	Uncomfortable	Uncomfortable	Uncomfortable	Uncomfortable	Uncomfortable
Medium	Uncomfortable	Acceptable	Comfortable	Acceptable	Uncomfortable
High	Uncomfortable	Acceptable	Acceptable	Acceptable	Uncomfortable
Very high	Uncomfortable	Uncomfortable	Uncomfortable	Uncomfortable	Uncomfortable

For example, at a temperature of 37°C and a humidity of 65% the rule strengths of the three example rules are:

'IF' the TEMPERATURE is HOT(0.8) 'AND' HUMIDITY is HIGH(0.68) 'THEN' it is UNCOMFORTABLE (Rule Strength = MIN(0.8,0.68) = 0.68).

'IF' the TEMPERATURE is MILD(0.0) 'AND' the HUMIDITY is MEDIUM(0.0) 'THEN' it is COMFORTABLE (Rule Strength = MIN(0.0,0.0) = 0.0).

'IF' the TEMPERATURE is WARM 'AND' the HUMIDITY is LOW 'THEN' it is UNCOMFORTABLE (Rule Strength = MIN(0.15,0.0) = 0.0).

All rules must be evaluated (in this case all 20 in the matrix) to determine each rule strength. If two rule strengths exist for one fuzzy output label, then the maximum (MAX) rule strength is used because this represents the rule which is most true; that is, in the previous example the fuzzy output for UNCOMFORTABLE is the MAX(0.68,0.0) = 0.68.

Step 3 — Defuzzification.
Now that rule strengths exist for all the output fuzzy labels, a crisp output value can be determined from the fuzzy output values. The most common method used to defuzzify the fuzzy output value is the center of gravity (COG) method. The fuzzy rule strengths determined from rule evaluation are used to truncate the top of the output membership functions. Given this area curve, the COG or balance point can then be calculated. For example, in Figure 24.11, for fuzzy output values of

UNCOMFORTABLE = 0.68
ACCEPTABLE = 0.2
COMFORTABLE = 0

the COG or crisp "Comfort Index" evaluates to 25.

Fuzzy logic signal processing is now a well-developed method. It is supported by copious teaching material including those on the Internet and on CD ROMs provided by manufacturers of this form of special integrated circuit chip sets and setup software. Possibly its best known application has been in clothes washing machines where the wash parameters are set to suit the load. Although there are still limitations, this method can be considered to be a mature procedure.

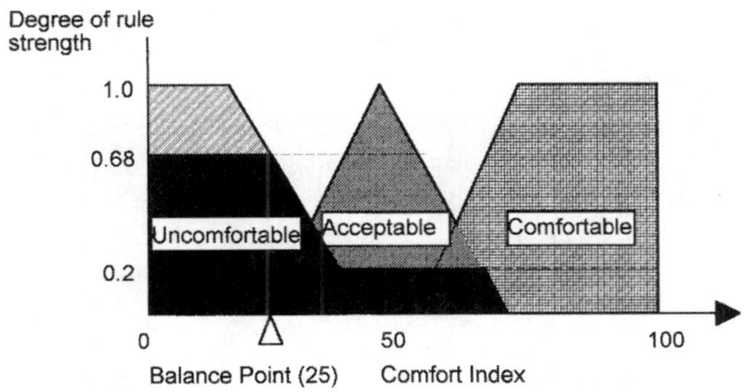

FIGURE 24.11 Output membership functions for controller. Each type of shaded area represents the three states of comfort.

Genetic Algorithms

The methods discussed thus far all have required the user to advise the system about the parameters to use. That is, they need to be taught efficient mappings.

In sharp contrast to these types of systems, there also exist other AI methods that possess the ability to learn, by themselves, what are the more optimal mapping configurations. Two main techniques are used — genetic algorithms (GAs) and artificial neural networks (ANNs). These self-learning processes both work well in certain situations and are now commonly used in signal processing. On the face of it, both seem to possess magical properties because they defy logical thought processes and a clear understanding of how they actually operate.

We begin with an overview of GAs. These make use of the basic principles by which natural selection found in living things, that is, from genetics, is able to improve gradually the fitness of species. That the genetic principles found in nature can be used in human-made applications is accredited to pioneering work of John Holland in the mid 1970s. Today, it is a large field in algorithm optimization research; see Tang et al. [7]. A very simplistic description now follows to give some insight.

The concept starts with the selection of a set of features, Figure 24.12 (these can take a wide range of forms and are not just measurement parameters), that represent the essential features of a system of interest. As examples, the DNA molecule carries the code of the characteristics of living beings and a computer string can carry a coded message that represents the features of the behavior of some human-devised system.

Various types of events (crossover, mutation, inversion are commonly encountered methods) can slightly alter the code of any particular string. When this happens, the new string then represents another closely similar, but different system having new properties. Consider, next, that a number of slightly different code strings have been formed.

When a change takes place in the code of a string, it is assessed against the other strings using rules for a predecided fitness test. If improvement has occurred in the overall properties, then it is adopted as one of the full set. If not better, then it is discarded. In this way the set of strings, and thus the total system capability, gradually improves toward an optimal state.

The operational aspects of such systems are beyond description here. Suffice to say that the technique is well established — but highly specialized — and is undergoing massive international research effort in hope of alleviating certain limitations.

The first limitation is that although it is generally agreed each adopted change for the better takes the overall system capability closer to the goal being sought, there is, as yet, no theory that can be applied to show the state of maximum optimization. GAs, therefore, always have doubt associated with their solutions as to how much more improvement might be possible.

The second limitation becomes obvious when the computational demands are considered in terms of the number of comparison operations needed to be run in the improvement process. This number can be truly huge, especially as the range of options rises with increase in string length. This kind of operation usually needs very large and fast computing power. In cases where there is plenty of time to determine an improvement, this is not a major limitation. An example of effective use would be finding how best to operate a wide range of functions in a complex system when the task is then fixed; this method was used to set up more optimal use of power supplies in a space vehicle. In cases where the variables in a code string are fast changing, the method may not be applicable.

When running this form of computer program, it is essential to have a measure of the speed — the dynamic behavior — at which this highly iterative process is moving toward the optimization goal: the user can then decide if it is a viable method for obtaining improvement.

With this background, it is probably obvious why GAs are sometimes used to set up the membership functions of fuzzy logic systems. As explained above, the membership functions each are part of a set of individual functions that form the mapping for a multisensor system. The choice of the membership function is largely heuristic and thus may not be the best function to use. By setting up several sets of functions it is then possible to apply GA computing methods iteratively to select a better set to use. Such

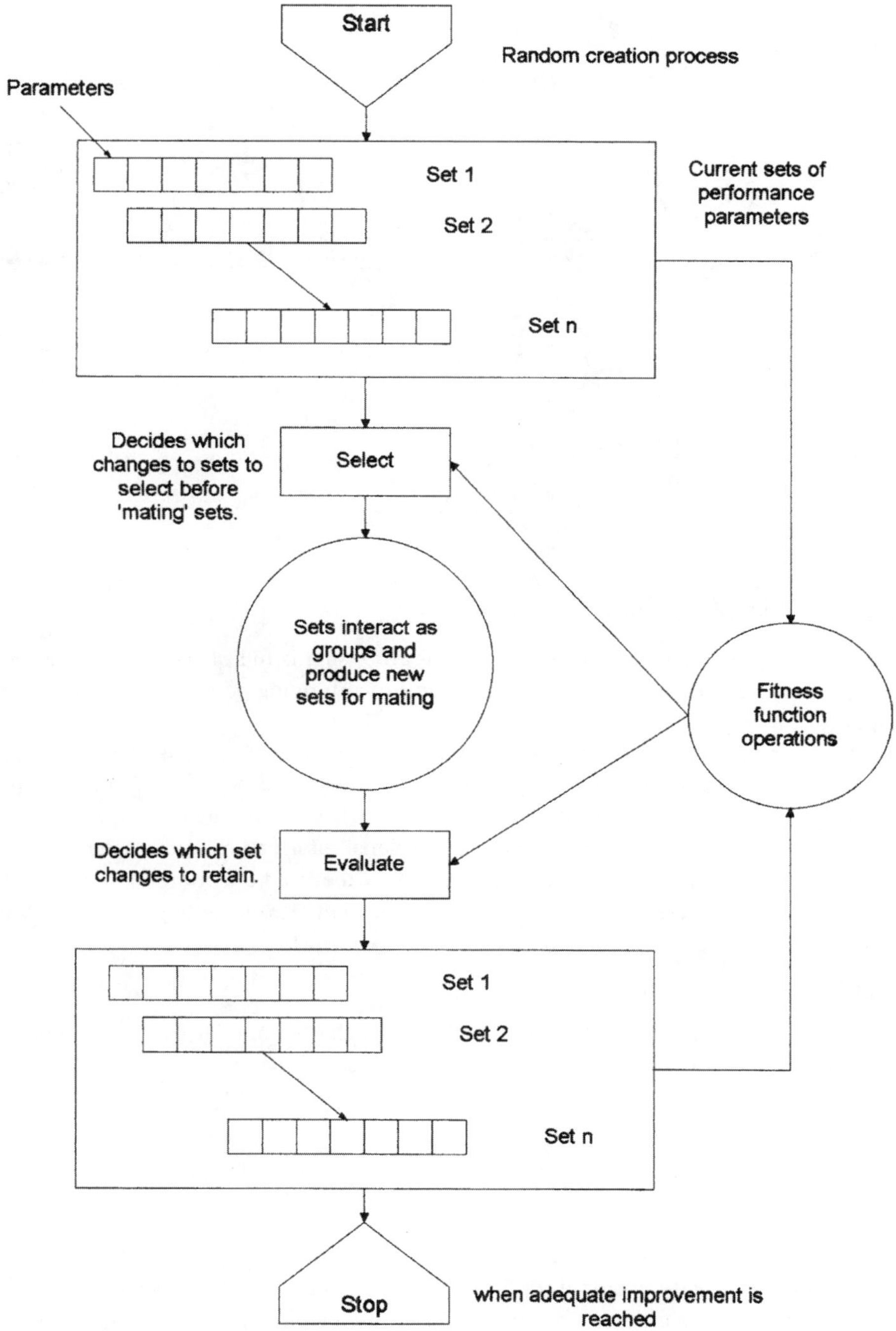

FIGURE 24.12 In the GA method sets of code strings are first modified by some form of genetic operations. They are then intercompared using fitness functions to select a better code to use in the subsequent set of strings. This iterative process is continued until some event disrupts it.

systems have been used where the time of currency of a function set is longer than the time needed to improve the functions. Obviously, use of GAs increases system set up and operational complexity, but once implemented may yield worthwhile gains.

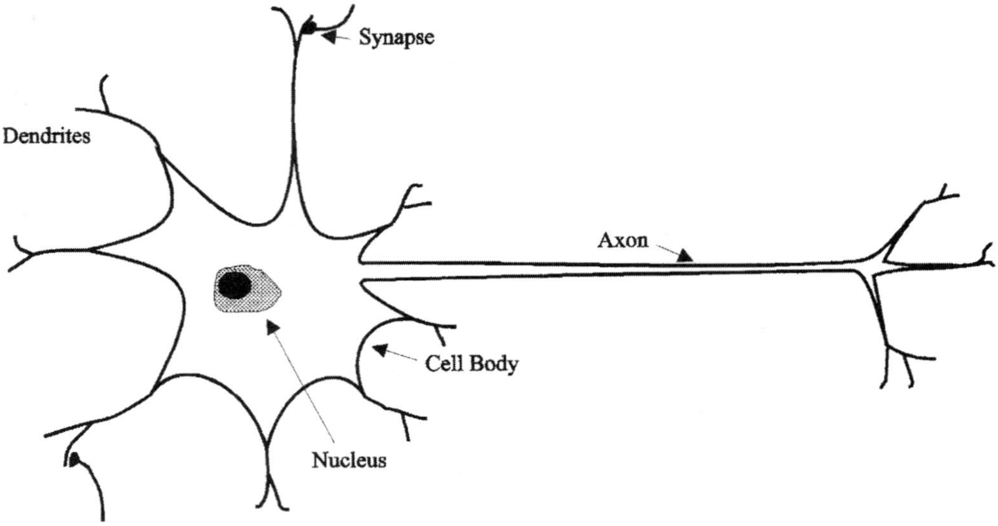

FIGURE 24.13 A biological neuron.

Artificial Neural Networks

The AI basket of potentially useful methods in signal processing is full of surprises. Attention is now directed to another method, which can learn, after an initial training period, how to better operate a given mapping process.

Neurocomputing or use of ANNs is another AI technique well suited to make a mapping processor in some circumstances. (These are sometimes called NNs but should not be used for human-devised systems as that leads to confusion with life sciences research on living neural networks.) ANNs are particularly useful when mapping complex multidimensional information to a simpler representation. Because of their ability to deal with nonlinear relationships, they find application in areas where traditional statistical techniques are of limited use. Some application areas include pattern recognition and classification, categorization, function approximation, and control.

Biological Neuron

The ANN has been inspired by the biological structure of the brain, and it is an attempt to mimic processes within the biological nervous system. The neuron is an information-processing cell in the brain, Figure 24.13. It consists of:

1. A body or *soma*
2. Input branches or *dendrites*
3. Output branch or *axon*

The neuron receives signals through its dendrites, and then transmits signals from its cell body along the axon. The neuron will generate an output (or fire) when the aggregation of the inputs reaches a threshold level. At the terminals of the axon are *synapses.* The synapse couples the signals from the axon of one neuron to the dendrites of another neuron. The strength with which these signals are transferred from the axon to the dendrite is controlled by the synapse and can be altered; hence the synapses can learn from experience.

Desirable characteristics of neural systems include:

- Massive parallelism
- Learning ability
- Ability to generalize

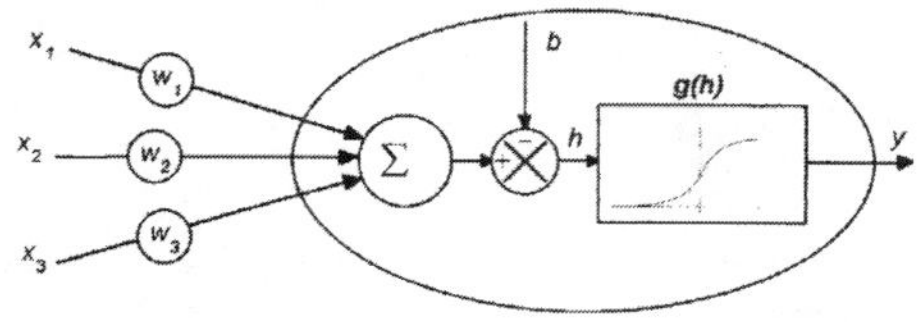

FIGURE 24.14 This neuron model is commonly used.

- Adaptability
- Fault tolerance

It is the attempt to construct machines that exhibit these characteristics that has led to the ANN methods of signal processing.

Artificial Neural Network

In 1943 McCulloch and Pitts proposed the first model of an artificial neuron. This has formed the basis for the generally accepted form of synthetic neuron model or processing element (PE); see Figure 24.14.

The output of the processing element y is given by

$$y = g\left[\left(\sum_i w_i \cdot x_i\right) - b\right]$$

where x_i are the PE inputs with weights (synaptic strengths) w_i, b the PE bias, and g the activation or transfer function. Many types of function have been proposed but the most popular is the sigmoid function, defined by

$$g(h) = \frac{1}{\left(1 + e^{(-\beta h)}\right)}$$

where β is the slope parameter.

By themselves the processing elements are very simple; however, when the individual elements are joined into large interconnected networks, complex relationships can be represented. Although a number of network architectures [2] exist, one of the most commonly discussed architectures found in the literature is the feed-forward multilayered perceptron. Figure 24.15 provides a simple illustration of how

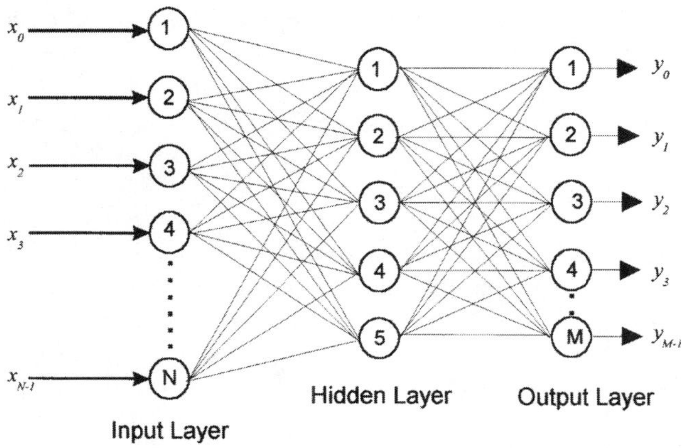

FIGURE 24.15 A typical three-layer neural network as is commonly used to create an optimal mapping from sensors to outputs.

a multilayered ANN maps a multidimensional input vector $x_0, \ldots x_{N-1}$ in an input space to a vector $y_0, \ldots y_{M-1}$ in an output space.

Learning

A fundamental characteristic of the ANN, once it has been set up as an operational tool in software form, is that it does not need to be programmed for the application. ANNs appear to learn rules from a representative set of examples, rather than having rules programmed in by an expert. The knowledge acquired by the system that controls how the system maps input to output is held within the connection weights of the network.

The focus of extensive ongoing research is the search for optimal training techniques. These techniques tend to fall into two broad categories — supervised and unsupervised learning.

In supervised learning, a representative set of inputs is presented to the network which then modifies its internal weights in order to achieve a desired output. With unsupervised learning, the input data only are presented to the network, following which the network organizes itself through self-modification of its internal weights so that it responds differently to each input stimulus.

It is beyond the scope of this text to review all the current learning techniques for ANNs. This is well documented in the literature [8]. One popular training algorithm — backpropagation — will be discussed as an example.

Backpropagation

Backpropagation is an example of a supervised learning paradigm commonly used with multilayer perceptron network architectures. Backpropagation follows the error-correction principle and uses the error signal {d (desired output) – y (actual output)} to modify the connection weights to reduce this error. The backpropagation algorithm is implemented as follows:

1. Initialize network weights to small random values.
2. Present an input vector $x_0, \ldots x_{N-1}$ and the corresponding desired output vector $d_0, \ldots d_{M-1}$.
3. Calculate the actual output vector $y_0, \ldots y_{M-1}$ by propagating the input through the network.
4. Use a recursive algorithm starting at the output layer and adjust the weights backward by

$$w_{ij}(t+1) = w_{ij}(t) + \eta\delta_j x_i'$$

where $w_{ij}(t)$ = the weight from an input to node j at time t
x_i' = either the output of node i or an input
η = a gain term ($0.0 < \eta < 1.0$)
δ_j = an error term for node j

For the output layer $l = L$ the error is calculated:

$$\delta_j^L = g'(h_j^L)[d_j - y_j]$$

where h_j^L = the net input to the jth unit in the L layer
g' = the derivative of the activation function g

For hidden layers $l = (L-1),\ldots,1.$ the error is calculated:

$$\delta_j^L = g'(h_j^l)\sum_l w_{ij}^{l+1}\delta_j^{l+1}$$

where h_j^l = the net input to the jth unit in the lth layer.
g' = the derivative of the activation function g

5. Return to Step 2 and repeat for the next pattern until the error reaches a predefined minimum level.

Unfortunately, no method exists that allows the ANN to create or learn information that is not contained in the training data; that is, the ANN can only reproduce based on experience. Under certain conditions, however, the network can generalize, that is, approximate, output values for data not contained in the training set.

The neural network can be considered as a universal approximator. During supervised learning, the output eventually approximates a target value based on training data. While this is a useful function, the ability to provide output data for test cases not in the training data is more desirable. Loosely, generalization can be viewed in terms of interpolation and extrapolation based on training data. If a test case is closely surrounded by training data, then (as with interpolation) the output accuracy is generally reliable. If, however, the test case is outside of, and not sufficiently close to, training data then (as with extrapolation) the accuracy is notoriously unreliable. Therefore, if the training cases are a sufficiently large sample of the total population of possible input data so that each test case is close to a training case, then the network will adequately generalize.

While multilayer feed-forward networks are finding increasing application in a wide range of products, many design issues such as determining an optimal number of layers, units, and training set for good generalization are research topics. Current theory provides loose guidelines and many of these design issues are resolved by trial and error. Another disadvantage of ANNs is the high demand that many training algorithms can put on computing resources because of their recursive nature.

The ANN, then, provides another alternative for development of suitable mapping models for measurement systems. They can be used to describe complex nonlinear relationships using a network of very simple processing elements. The attraction of the ANN lies in its ability to learn. As long as there exists a sufficiently representative sample of input-to-output data available, the mathematical relationship of the mapping function need not be known. It is effectively taught to the network during a learning process.

Again, there exist important limitations. The worst is the time it might take to adjust the system nodes and weights to a nearly final state. This can often require considerably more time than the time-varying properties of the inputs allow. In many potential applications, they take too long to learn and are not effective. Again computational speed and power are governing factors.

Despite their shortcomings in some applications, ANNs are now a commonly used procedure to set up sensor mapping systems. Examples are banknote image detection and the increased sensitivity of the detection of aluminum in water.

24.7 Problems in Calibration of AI Processing Methods

Calibration of a measurement system is the result of using an agreed upon, often legally binding, process by which it is proven to possess a declared level of accuracy in its measurement outcome. In conventional instrument terms this implies the system can be set up and compared with a measurement method of superior performance to give its error of accuracy plus its variance from the agreed value determined with a level of uncertainty. This kind of instrument system is then accepted to have a known behavior that could be explained by the laws of physics as causal and unique in performance. The prime example, the physical standard apparatus for a parameter, can be and is defined such that it will always give very closely the same outcome, even if built in different laboratories. It will have predictable behavior. At the heart of this acceptability is that it can be modeled in terms of an algorithm. All parts in it follow formal laws and have the same outcomes from implementation to implementation.

Most of this notion has to be put aside because, as explained above, AI-based instrument signal processors are built on a raft of transformations that convert subjective situations into objective ones or they carry out unexplained processes. It is, therefore, not hard to see that calibration is a major issue with this type of processor.

Processes make use of heuristics to at least start them going. The designer, when invoking any of the decision-making methods, will almost certainly not select the same rules, processes, and parameters that another designer will choose. There is a lack of consistency in AI processors. The outcomes are fuzzy, not crisp as in instrumentation that complies with a physical law. They act like humans do in that they provide a range of solutions to the same problem.

At first sight this seems to imply that we should ignore the AI possibilities for they cannot be calibrated according to long-standing metrological practices. However, their performance is often very worthy and will be workable where algorithmic methods are not. This calibration constraint must be considered in terms of human thinking, not so much in terms of physics and mathematical models.

At present, AI processing methods have been well proved in many fields, these tending to be fields in which performance does not need calibration with regard to the standards regime. Examples are the use of fuzzy logic in clothes washing machines to improve the wash by varying the wash cycle parameters, in neural network methods to aid the recognition of banknotes, and in rule-based controllers in industrial process plant controls. Genetic algorithms have been used to schedule power supply and usage in space shuttles — said to be an impossible task by any other known means. These all are typified as being needs where much experience can be devoted to ensuring they work satisfactorily.

They should normally not be the critical processing element in safety-critical situations. They can certainly be used in routine management situations where they can outperform algorithmic methods, but there they should be backed up with conventional alarms. They are often used in off-line plant control where the human and alarms are still the final arbiter. This, however, seems to be only a cautious step in our slow acceptance of new ideas.

The process of calibration here is more akin to that of conducting evaluation and validation. Does the system give the range of outcomes expected in given circumstances? Are the outcomes better than those without the processor? Could it be done as well or better by algorithmic-based processing? Is the speed it gives worth the problems it may bring? Problems in their testing, and thus calibration, are discussed by Sizemore [6]. The issues that need to be considered in the calibration of conventional instrumentation [5] are relevant to the calibration of AI-based processing but need much more care in their execution.

Such questions require consideration of the very same elements of decision theory upon which they are based to test them. They have been set up to think like humans so it is expected they will have to be calibrated and evaluated like humans — that is not at all easy.

At present, the calibration and validation of AI systems are not standardized well enough. This impedes acceptance, but standardization will improve as world knowledge of this relatively new method of processing develops to greater maturity inside instrumentation.

There will be opposition to the use of AI methods, but the performance gain they bring will ensure they are used. The forward-thinking academic measurement community is showing signs of addressing AI signal processing — but it will take time.

References

1. Baker, D., Kohler, D.C., Fleckenstein, W.O., Roden, C.E., and Sabia, R., Eds., *Physical Design of Electronic Systems,* Vol. 4, Prentice-Hall, Englewood Cliffs, NJ, 1972.
2. Jain, A.K. and Mao, J., Artificial neural networks: a tutorial, *IEEE Comput.,* pp. 31–44, 1996.
3. Kaufmann, A., *The Science of Decision Making,* Weidenfeld and Nicholson, London, 1968.
4. Mauris, G., Benoit, E., and Foulloy, L., Fuzzy sensors for the fusion of information, in *Proc. 13th IMEKO Conference,* Turin, pp. 1009–1014, 1994.
5. Nicholas, J., in *Handbook of Measurement and Control,* Vol. 3, Sydenham, P.H. and Thorn, R., Eds., John Wiley and Sons, Chichester, U.K., 1992.
6. Sizemore, N.L., Test techniques for knowledge-based systems, *ITEA J.,* XI (2), 34–43, 1990.
7. Tang, K.S., Man, K.F., Kwong, S., and He, Q., Genetic algorithms and their applications, *IEEE Signal Processing Mag.,* 13(6), 22–37, 1996.
8. Venmuri, *Artificial Neural Networks: Theoretical Concepts,* IEEE Computer Society Press, Los Angeles, CA, 1988.

25

Analog-to-Digital Converters

E. B. Loewenstein
National Instruments

25.1 Introduction

Almost every modern instrumentation system includes some form of digitizer, or *analog-to-digital converter (ADC)*. An ADC converts real-world signals (usually voltages) into digital numbers so that a computer or digital processor can (1) acquire signals automatically, (2) store and retrieve information about the signals, (3) process and analyze the information, and (4) display measurement results. A digitizing system can do these jobs with greater speed, reliability, repeatability, accuracy, and resolution than a purely analog system normally can.

The two main functions of an ADC are *sampling* and *quantization*. These two processes convert analog signals from the time and voltage continuums (respectively) into digital numbers having discrete amplitudes, at discrete times. To represent changing signals at every instant in time or at every possible voltage would take an infinite amount of storage. So for every system there is an appropriate *sampling rate* and degree of quantization (*resolution*) so that the system retains as much information as it needs about the input signals while keeping track of manageable amounts of data. Ultimately, the purpose of sampling and quantization is to reduce as much as possible the amount of information about a signal that a system must store in order to reconstruct or analyze it meaningfully.

25.2 Sampling

To prevent having to digitize an infinite amount of information, an analog signal must first be sampled. Sampling is the process of picking one value of a signal to represent the signal for some interval of time. Normally, digitizers take samples uniformly in time, e.g., every microsecond. It is not necessary to sample uniformly, but doing so has some interesting and convenient mathematical properties, which we will see later.

0-8493-1733-9/04/$0.00+$1.50

Sampling is done by a circuit called a *sample-and-hold* (*S/H*), which, at a sampling instant, transfers the input signal to the output and holds it steady, even though the input signal may still be changing. An S/H usually consists of a signal buffer followed by an electronic switch connected to a capacitor. At a sampling instant, the switch briefly connects the buffer to the capacitor, allowing the capacitor to charge to the input voltage. When the switch is disconnected, the capacitor retains its charge and thus keeps the sampled input voltage steady while the ADC that follows does its job. Quite often, sampling is actually done by a circuit called a *track-and-hold* (*T/H*), which differs from an S/H only slightly. Whereas the S/H holds the analog signal until the next sampling instant, the T/H holds the analog signal still only until the ADC has finished its conversion cycle. After the ADC is through, the T/H reconnects the buffer to the capacitor and follows the input signal until the next sampling instant. The result is more accurate sampling, because the buffer has more time to charge the capacitor and "catch up" with (track) the input signal, which has changed since the last sampling instant. Nearly every modern ADC chip has a built-in S/H or T/H, and virtually all data acquisition systems include them.

Of course, sampling necessarily throws away some information, so the art of sampling is in choosing the right sample rate so that enough of the input signal is preserved. The major pitfall of *undersampling* (sampling too slowly) is *aliasing*, which happens whenever the input signal has energy at frequencies greater than one half the sample rate. In Figure 25.1a, a signal (the fast sine wave) is sampled at a rate

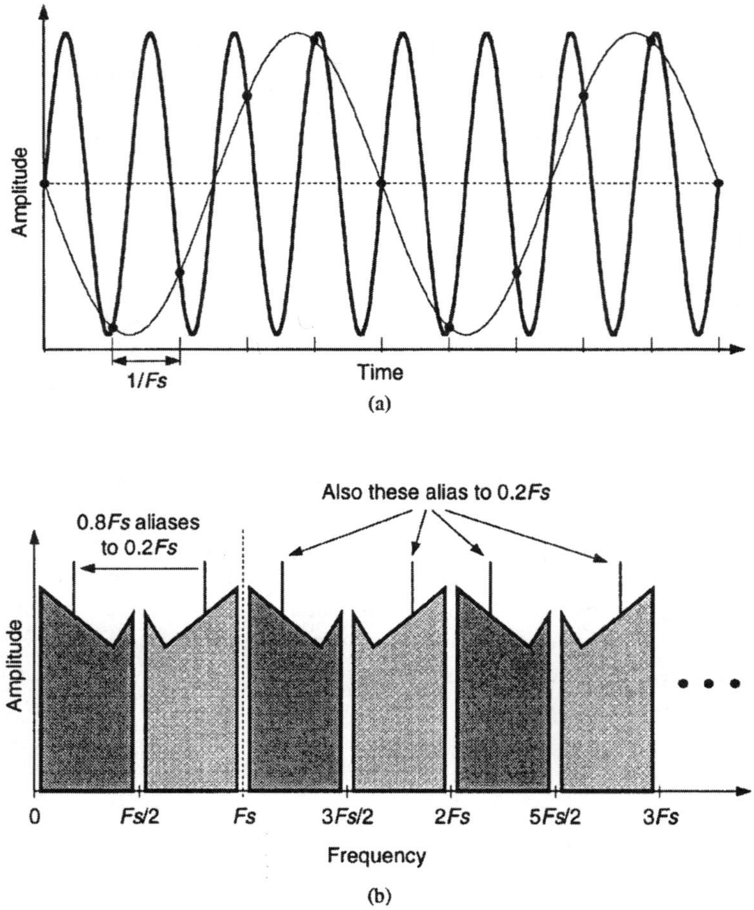

FIGURE 25.1 A demonstration of aliasing. An ADC sampling at rate *Fs* cannot distinguish between a 0.8 *Fs* sine wave and a 0.2 *Fs* sine wave. (a) A time-domain illustration. (b) A frequency-domain illustration. Theoretically, a sampler aliases an infinite number of 0.5 *Fs*-wide frequency bands into the baseband (0 to 0.5 *Fs*). Practically, finite analog bandwidth eventually limits how far out in frequency aliases can come from.

Fs, shown by the hash marks at the bottom of the graph. The sine wave has a frequency of 0.8 *Fs,* which is higher than one half the sample rate (0.5 *Fs*). Notice that sampling the lighter sine wave of 0.2 *Fs* produces the same set of samples. The resulting sampled data is ambiguous in that we cannot tell from the data what the frequency of the incoming sine wave actually is. In fact, even though the data set appears to represent a sine wave of 0.2 *Fs,* the actual signal could be any sine wave having a frequency of $(n)\ Fs \pm 0.2\ Fs$, where n is any integer, starting with 0. So the original signal could be 0.2 *Fs,* 0.8 *Fs,* 1.2 *Fs,* 1.8 *Fs,* 2.2 *Fs,* etc. (or even more than one of those). We say that 0.2 *Fs* is the *alias* of a signal that may actually be at another frequency entirely. During interpretation of sampled data, it is customary to treat signals as though they occurred in the baseband (0 to 0.5 *Fs*), whether or not that is the case. In general, in a system sampling at *Fs,* a signal at a frequency *F* will alias into the baseband at

$$Fa = \text{abs}\left[(n)Fs - F\right], \tag{25.1}$$

where abs denotes absolute value, $n \geq 0$, and $(n)Fs$ is the closest integer multiple of *Fs* to *F.*

Everyone has seen a demonstration of aliasing at the movies, in the form of "wagon-wheeling." As the stagecoach or wagon takes off, the wheels begin to turn, slowly at first, then faster. As the wagon speeds up, the spokes suddenly appear to be turning backward, even though the wagon is moving forward. Sometimes the spokes appear to be standing still. The reason for this is that a motion picture camera shooting film at 24 frames/s is a sampling system operating at 24 samples/s. The turning wagon wheel is a periodic signal that the camera undersamples. When the wheel begins turning just fast enough that one spoke travels at least half the distance to the next spoke in 1/24th of a second, the spokes begin to appear to move backward, and the system is aliasing. When the wheel is turning so that a spoke moves exactly the distance between two spokes in 1/24th of a second, the spokes appear to be standing still, since they all look the same to the camera.

It follows from Equation 25.1 that if we put into a sampler a signal with no energy at frequencies greater than one half the sample rate (0.5 *Fs*), then aliasing will not occur. This is the essence of the Shannon sampling theorem [1], which states that, with mathematical interpolation, the complete input waveform can be recovered *exactly* from the sampled data, at all times at and in between the sampling instants, as long as the sample rate is at least twice as high as the highest frequency content in the signal. Sometimes we refer to 0.5 *Fs* as the *Nyquist frequency,* because Nyquist was concerned with the maximum bandwidth of signals [2]. Similarly, twice the highest frequency content of a signal (i.e., the minimum nonaliasing sample rate) is sometimes called the *Nyquist rate.* Sample rates are specified in samples/s, or S/s, and it is also common to specify rates in kS/s, MS/s, and even GS/s.

It is not always necessary to worry about aliasing. When an instrument is measuring slow-moving dc signals or is gathering data for statistical analysis, for instance, getting frequencies right is not important. In those cases we choose the sample rate so that we can take enough data in a reasonable amount of time. On the other hand, if the instrument is a spectrum analyzer, where frequency does matter, or an oscilloscope, where fine time detail is needed, aliasing certainly is an issue. When aliased signals from beyond the frequency band of interest can interfere with measurement, an instrument needs to have an *antialias filter* before the S/H. An antialias filter is a low-pass filter with a gain of 1 throughout most of the frequency band of interest. As frequency increases, it begins to attenuate the signal; by the Nyquist frequency it must have enough attenuation to prevent higher-frequency signals from reaching the S/H with enough amplitude to disturb measurements. An efficient antialias filter must attenuate rapidly with frequency in order to make most of the baseband usable. Popular analog filters with rapid cutoff include elliptic and Chebyshev filters, which use zeros to achieve fast cutoff, and Butterworth filters (sixth order and above), which do not attenuate as aggressively, but have very flat passband response. A good book about filters is Reference 3.

Some ADCs do not need a S/H or T/H at all. If the ADC is converting a slow-moving or dc signal and precise timing isn't needed, the input may be stable enough during conversion that it is as good as sampled. There are also *integrating ADCs* (discussed later), which average the input signal over a period of time rather than sampling it. However, internally they actually sample the average.

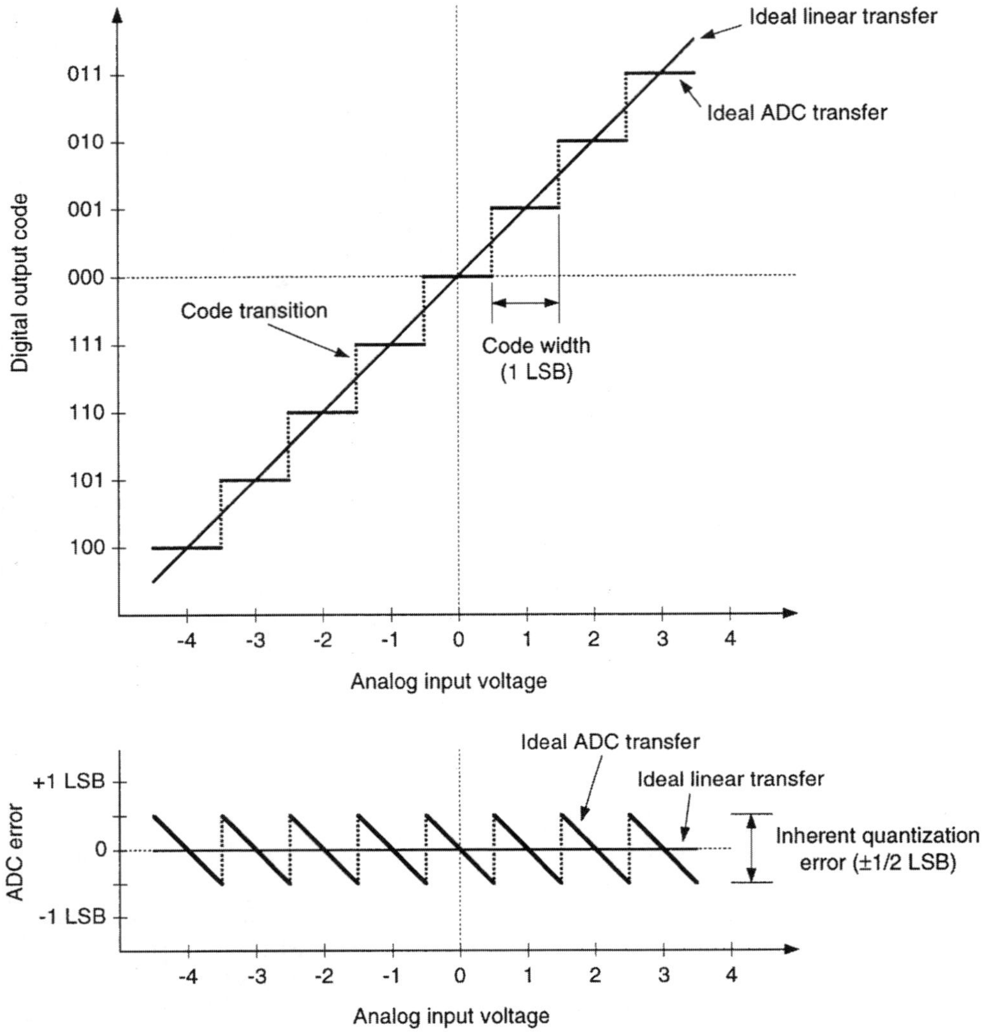

FIGURE 25.2 The ideal three-bit quantizer has eight possible digital outputs. The analog input-to-digital output transfer function is a uniform staircase with steps whose width and height are 1 LSB exactly. The bottom graph shows the ideal transfer function (a straight line) subtracted from the staircase transfer function.

25.3 Quantization

What sampling accomplishes in the time domain, quantization does in the amplitude domain. The process of digitization is not complete until the sampled signal, which is still in analog form, is reduced to digital information. An ADC quantizes a sampled signal by picking one integer value from a predetermined, finite list of integer values to represent each analog sample. Each integer value in the list represents a fraction of the total analog input range. Normally, an ADC chooses the value closest to the actual sample from a list of uniformly spaced values. This rule gives the *transfer function* of analog input-to-digital output a uniform "staircase" characteristic. Figure 25.2 represents a three-bit quantizer, which maps a continuum of analog input values to only eight (2^3) possible output values. Each step in the staircase has (ideally) the same width along the x-axis, which we call *code width* and define as 1 *LSB* (*least significant bit*). In this case 1 LSB is equal to 1 V. Each digital code corresponds to one of eight 1-LSB intervals making up the analog input range, which is 8 LSB (and also 8 V in this case).

Of course, we would like our measurement system to have a transfer function that is a straight line and has no steps at all. The bottom graph in Figure 25.2 is the ideal transfer function (a straight diagonal line) subtracted from the staircase function, or the *quantization error.* In an ideal ADC, the quantization error is bounded by ±½ LSB, and, over the input range, the average error is 0 LSB and the standard deviation of error is $1/\sqrt{12}$ LSB. As the bottom graph shows, the quantization error at any point is a deterministic function of the input signal.

25.4 ADC Specifications

Range and Resolution

The *input range* of an ADC is the span of voltages over which a conversion is valid. The end points at the bottom and the top of the range are called *–full-scale* and + *full-scale,* respectively. When –full-scale is 0 V the range is called *unipolar,* and when –full-scale is a negative voltage of the same magnitude as +full-scale the range is said to be *bipolar.* When the input voltage exceeds the input range, the conversion data are certain to be wrong, and most ADCs report the code at the end point of the range closest to the input voltage. This condition is called an *overrange.*

The *resolution* of an ADC is the smallest change in voltage the ADC can detect, which is inherently 1 LSB. It is customary to refer to the resolution of an ADC by the number of binary bits or decimal digits it produces; for example, "12 bits" means that the ADC can resolve one part in 2^{12} (= 4096). In the case of a digital voltmeter that reads decimal digits, we refer to the number of digits that it resolves. A "6-digit" voltmeter on a 1 V scale measures from –0.999999 V to +0.999999 V in 0.000001 V steps; it resolves one part in 2,000,000. It is also common to refer to a voltmeter that measures from –1.999999 to +1.999999 as a "6-digit" voltmeter. Figure 25.3 compares the resolutions of common word lengths for ADCs.

Coding Conventions

There are several different formats for ADC output data. An ADC using *binary* coding produces all 0s (e.g., 000 for the three-bit converter) at –full-scale and all 1s (e.g., 111) at +full-scale. If the range is bipolar, so that –full-scale is a negative voltage, binary coding is sometimes called *offset binary,* since the code 0 does not refer to 0 V. To make digital 0 correspond to 0 V, bipolar ADCs use *two's complement* coding, which is identical to offset binary coding except that the *most significant bit* (*MSB*) is inverted, so that 100 … 00 corresponds to –full-scale, 000 … 00 corresponds to 0 V (*midscale*), and 011 … 11 corresponds to +full-scale. All of the figures in this chapter depicting three-bit ADC transfer functions use two's complement coding.

Decimal-digit ADCs, such as those used in digital voltmeters, use a coding scheme called *binary-coded decimal* (*BCD*). BCD data consists of a string of four-bit groups of binary digits. Each four-bit group represents a decimal digit, where 0000 is 0, 0001 is 1, and so on, up to 1001 for 9. The other six combinations (1010 through 1111) are invalid, or can be used for special information, such as the sign of the conversion.

Linear Errors

Linear errors are the largest and most common errors in an ADC and are easily corrected by simple calibrations or by additions with and multiplications by correction constants. Linear errors do not distort the transfer function; they only change somewhat the input range over which the ADC operates.

Figure 25.4 shows the transfer function of an ideal three-bit ADC with some *offset error.* The straight line joining the centers of the code transitions is raised, or offset, by 0.6 LSB, and the bottom graph shows the resulting error. Figure 25.5 shows an ideal three-bit ADC with a +25% *gain error.* The slope of the line through the code transitions is 1.25 times the ideal slope of 1.00. If the slope of the line were 0.75 instead, the gain error would be –25%. The bottom graph shows the error resulting from excessive

Bits	Digits	Voltmeter "Digits"	Steps in FSR	Step Size, ppm	Theoretical Dynamic Range (dB)
30	8.730		1 073 741 824	0.001	182.379
28.575	8.301	8 1/2	400 000 000	0.003	173.802
28	8.128		268 435 456	0.004	170.338
27.575	8	8	200 000 000	0.005	167.782
26	7.526		67 108 864	0.015	158.297
25.253	7.301	7 1/2	40 000 000	0.025	153.802
24.253	7	7	20 000 000	0.05	147.782
• 24	6.924		16 777 216	0.060	146.255
22	6.322		4 194 304	0.238	134.214
21.932	6.301	6 1/2	4 000 000	0.25	133.802
20.932	6	6	2 000 000	0.5	127.782
• 20	5.720		1 048 576	0.954	122.173
18.610	5.301	5 1/2	400 000	2.5	113.802
18	5.118		262 144	3.815	110.132
17.610	5	5	200 000	5	107.782
• 16	4.515		65 536	15.259	98.091
15.288	4.301	4 1/2	40 000	25	93.802
14.288	4	4	20 000	50	87.782
14	3.913		16 384	61.035	86.049
• 12	3.311		4 096	244.141	74.008
11.966	3.301	3 1/2	4 000	250	73.802
10.966	3	3	2 000	500	67.782
10	2.709		1024	976.563	61.967
8.644	2.301	2 1/2	400	2500	53.802
• 8	2.107		256	3906.25	49.926
7.644	2	2	200	5000	47.782
6	1.505		64	15625	37.885

FIGURE 25.3 Comparison of theoretical resolutions of ADCs. "Bits" refers to binary word length, and "digits" refers to decimal word length. • denotes popular binary word lengths. FSR is full-scale range, and theoretical dynamic range is computed from the formula $1.7609 + 6.0206n$, where n is the number of bits (see discussion of dynamic range).

gain. Offset errors can be compensated for simply by adding a correcting voltage in the analog circuitry or by adding a constant to the digital data. Gain errors can be corrected by analog circuitry like potentiometers or voltage-controlled amplifiers or by multiplying the digital data by a correction constant.

Nonlinear Errors

Nonlinear errors are much harder to compensate for in either the digital or analog domain, and are best minimized by choosing well-designed, well-specified ADCs. Nonlinearities are characterized in two ways: differential nonlinearity and integral nonlinearity.

Differential nonlinearity (*DNL*) measures the irregularity in the code step widths by comparing their widths to the ideal value of 1 LSB. Figure 25.6 illustrates the three-bit ADC with some irregular code widths. Most of the codes have the proper width of 1 LSB and thus contribute no DNL, but one narrow code has a width of 0.6 LSB, producing a DNL of –0.4 LSB, and one wide code has a width of 1.8 LSB, producing a DNL of +0.8 LSB at that code. This converter would be consistent with a DNL specification of ±0.9 LSB, for example, which guarantees that all code widths are between 0.1 and 1.9 LSB. It is possible for a code not to appear at all in the transfer function. This happens when the code has a width of 0 LSB, in which case we call it a *missing code.* Its DNL is –1 LSB. If an ADC has a single missing code, the step size at that point in the transfer function is doubled, effectively reducing the local resolution of the ADC by a factor of two. For this reason it is important for an ADC specification to declare that the ADC has *no missing codes*, guaranteeing that every code has a width greater than 0 LSB. Even if an ADC has missing codes, no code can have a width less than 0 LSB, so the DNL can never be worse than –1 LSB.

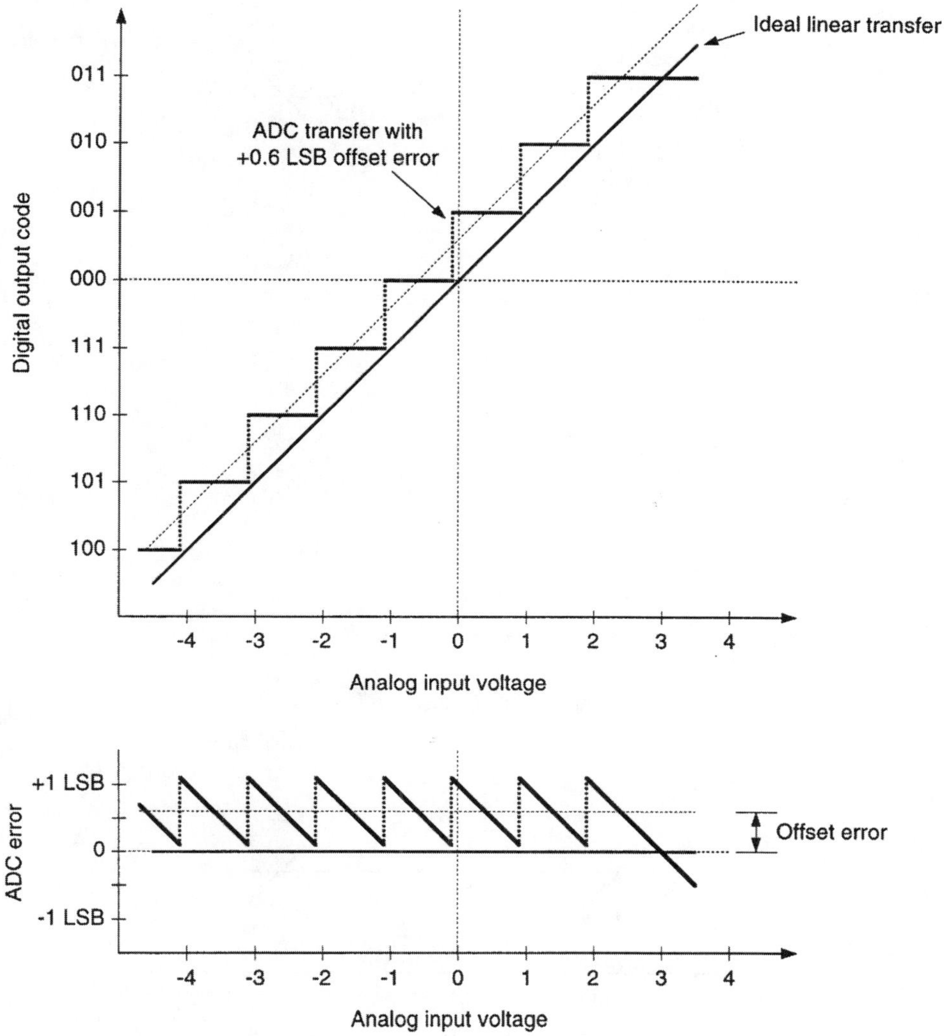

FIGURE 25.4 An ideal three-bit quantizer, only with +0.6 LSB of offset error.

Integral nonlinearity (*INL*) measures the deviation of the code transitions from the ideal straight line, providing that the linear errors (offset and gain) have been removed. Figure 25.7 depicts an ADC with an INL error of +0.7 LSB. The offset and gain errors have been calibrated at the end points of the transfer function.

Relative accuracy (*RA*) is a measure of nonlinearity related to INL, but more useful. It indicates not only how far away from ideal the code transitions are, but how far any part of the transfer function, including quantization "staircase" error, deviates from ideal (assuming offset and gain errors have been calibrated at the end points). In a noiseless ADC, the worst-case RA always exceeds the worst-cast INL by ±0.5 LSB, as demonstrated in Figure 25.7. In an ADC that has a little inherent noise or has noise (called *dither*) added at the input, the RA actually improves because the addition of noise to the quantizer tends to smooth the *averaged* transfer function. Figure 25.8 shows the average of the digital output data as a function of the input voltage when 0.1 LSB rms of Gaussian random noise is intentionally added to the input. The RA improves to ±0.3 LSB from ±0.5 LSB in the noiseless case. If about 0.5 LSB rms of Gaussian noise is added, the quantization staircase becomes nearly straight. This improvement in linearity comes at the expense of the random error in each individual conversion caused by the noise. Adding

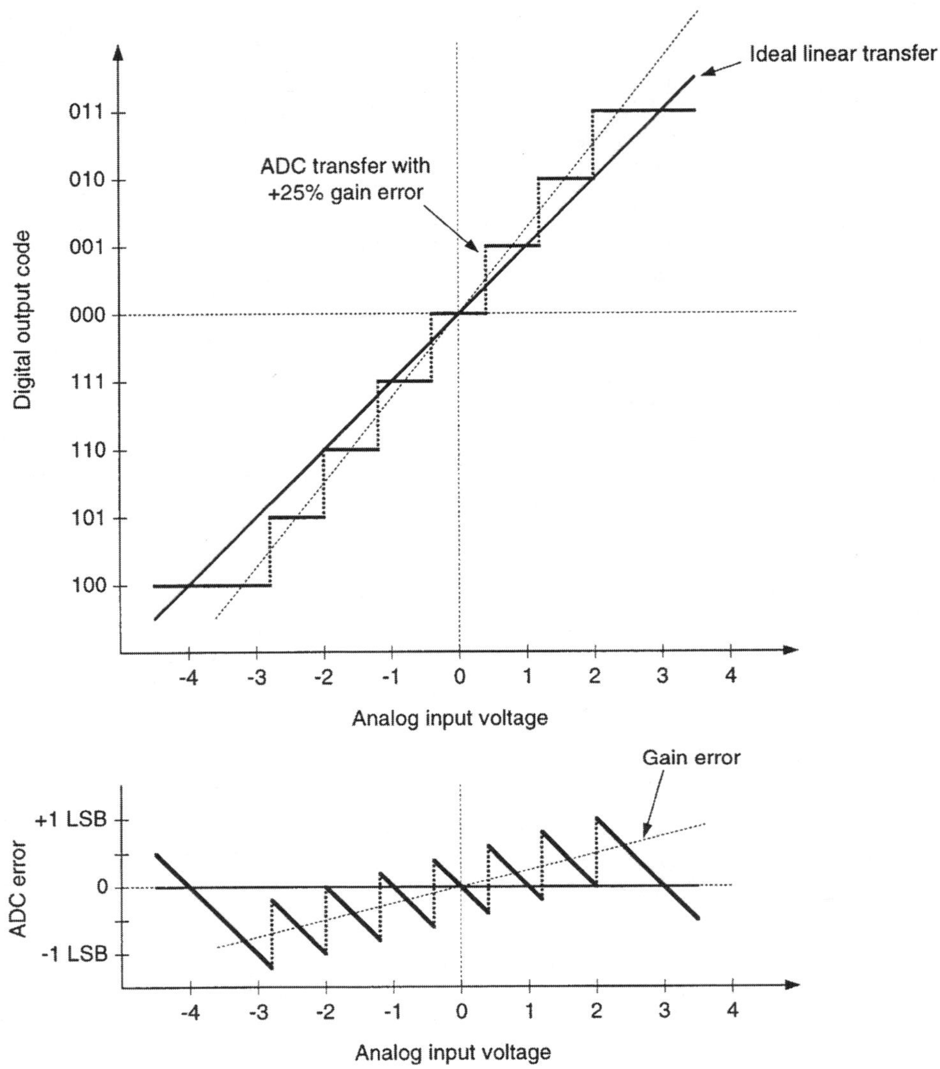

FIGURE 25.5 An ideal three-bit quantizer, only with a gain of 1.25 instead of 1.00. This represents a +25% gain error.

more noise to the ADC does not improve the average quantization error much more, but it does tend to smooth out local nonlinearities in the averaged transfer function. For a good discussion of noise and dither, see Reference 4.

Aperture Errors

Aperture errors have to do with the timing of analog-to-digital conversions, particularly of the S/H. *Aperture delay* characterizes the amount of time that lapses from when an ADC (S/H) receives a convert pulse to when the sample is held as a result of the pulse. Although aperture delay (sometimes called *aperture time*) is usually specified as a few nanoseconds for an ADC or S/H by itself, this delay is usually much more than negated by the group delay in any amplifiers that precede the S/H, so that the convert pulse arrives at the S/H quite some time before the analog signal does. For instance, a typical 1 MHz bandwidth amplifier has 160 ns of delay; if the ADC or S/H it was connected to had an aperture delay of 10 ns, the effective aperture delay for the system would be –150 ns.

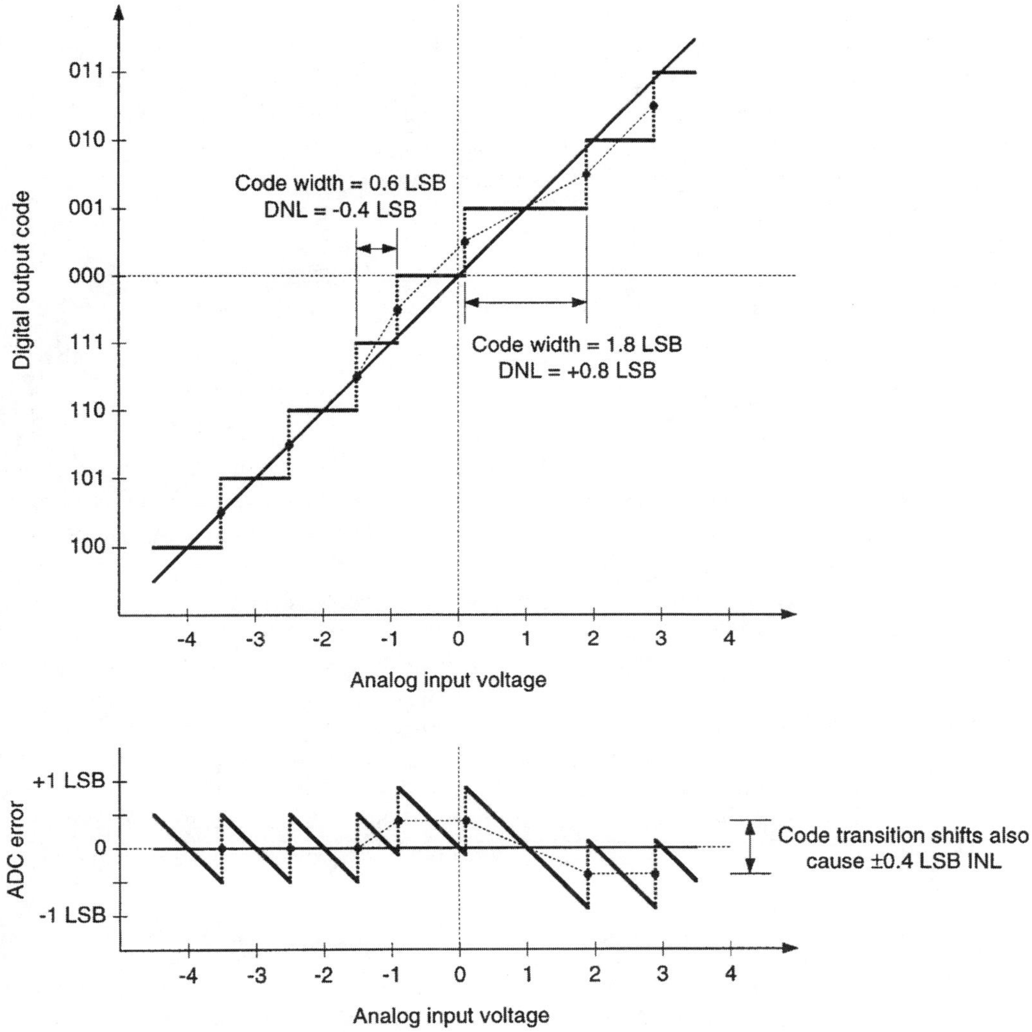

FIGURE 25.6 A three-bit quantizer with substantial DNL errors. The bottom graph illustrates the resulting INL errors.

Jitter (or *aperture jitter*) characterizes the irregularity in times at which samples are taken. If the nominal period between samples in an ADC is 1 μs, the actual time may vary from 1 μs by as much as a few hundred picoseconds or even as much as a nanosecond from cycle to cycle. Contributions to these variations can come from the crystal clock source (if included under the jitter specification), digital clock circuitry, or the S/H. Jitter is usually specified in picoseconds peak-to-peak or picoseconds rms.

Jitter interferes with measurements (particularly spectral analysis) by effectively frequency modulating the input signal by the jitter profile. A jittery ADC sampling a pure sine wave would scatter energy from the sine wave all throughout the spectrum, perhaps covering up useful spectral information. In a typical ADC, however, most of the interference from jitter tends to occur at frequencies very close to the main signal.

Noise

Noise, whether inherent in an ADC or introduced intentionally (see dither above), limits the resolution of an ADC by adding an interfering waveform to the input signal as the data is converted. Noise comes

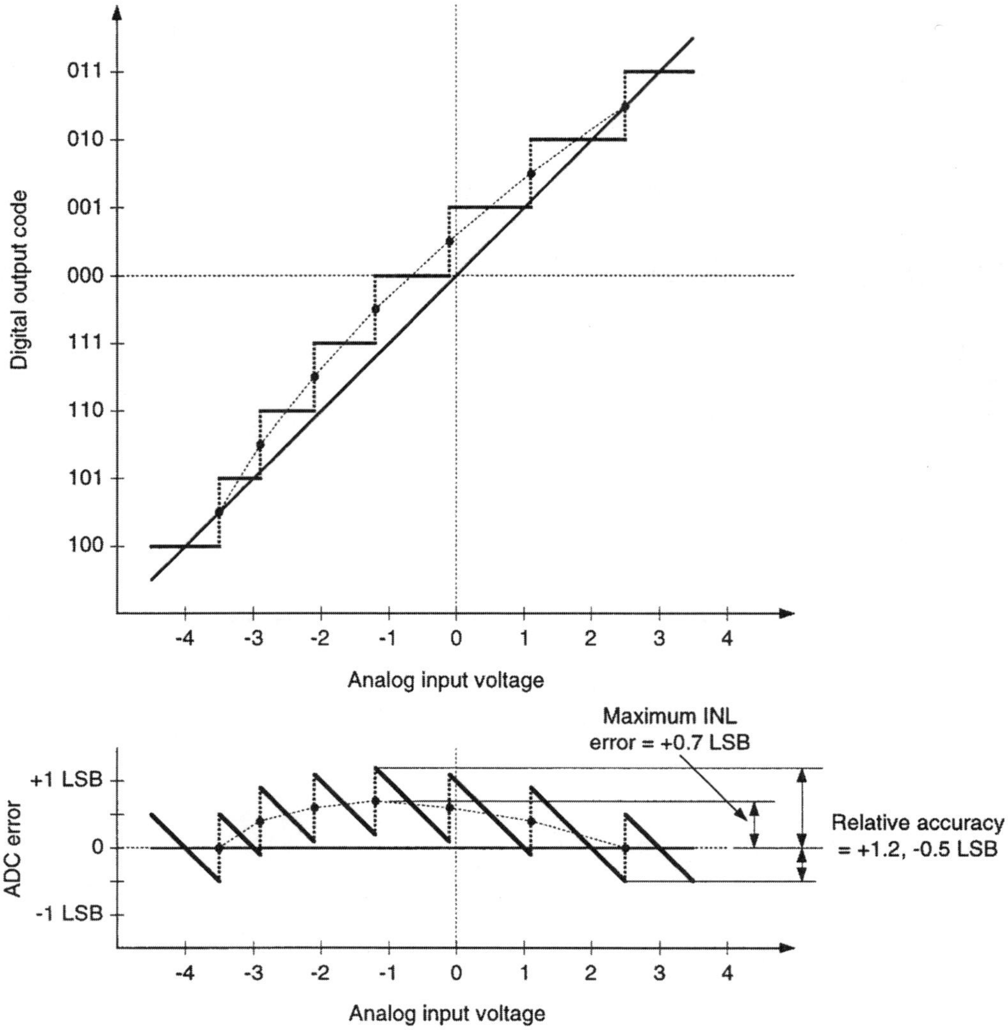

FIGURE 25.7 A three-bit quantizer with substantial INL errors. Here, the DNL error is still significant; but, for example, a 12-bit converter with 0.7 LSB of INL from a smooth error "bow" like the one above could have negligible DNL because it would have so many more steps over which to accumulate error.

from many places. The most common kind of noise is *thermal noise*, which is caused by the random nature of electric conduction in resistors and transistors. Thermal noise is worse at higher temperatures and higher resistances. Most other ADC noise is coupled electromagnetically from nearby circuitry, such as clock or logic circuits, or from routing of other input signals. Noise is usually specified in volts rms or peak-to-peak, or LSBs rms or peak-to-peak.

Quantization error (see above) can sometimes be thought of as *quantization noise*. Although quantization error is perfectly predictable with respect to the input signal, when a signal is fairly "busy" (i.e., busy enough that consecutive conversions do not tend to repeat data) the quantization error becomes chaotic, and it can be thought of as another source of random noise, whose statistical distribution is uniform from –0.5 LSB to +0.5 LSB and whose standard deviation is $1/\sqrt{12}$ LSB. This is sometimes the dominant source of noise in spectral analysis applications.

Once noise gets into an ADC, there are ways to process out the noise if it is independent of the signal. Acquisitions of dc signals can be quieted by collecting a number of points and averaging the collection. If the noise is *white random noise*, which has equal energy density at all frequencies, averaging can reduce

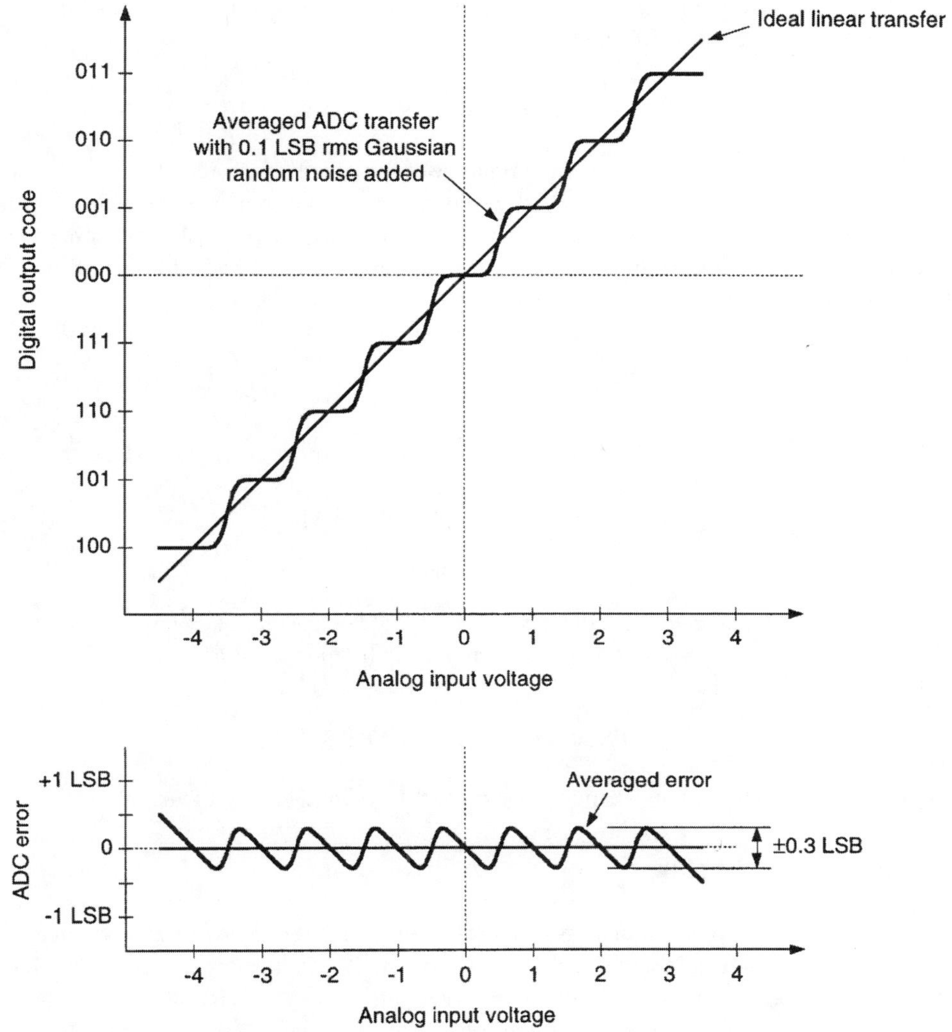

FIGURE 25.8 An ideal three-bit quantizer with 0.1 LSB rms Gaussian random noise (dither) added at the input. The relative accuracy has improved to ±0.3 LSB rms from the ±0.5 LSB expected from a noiseless quantizer. With the application of 0.5 LSB rms Gaussian noise, the transfer function becomes almost perfectly straight. Larger amounts of dither produce essentially no improvement in linearity.

the amount of noise by the square root of the number of samples averaged. The noise interfering with a repetitive waveform can be quieted by acquiring many waveforms using a level trigger and averaging the collection to produce an average waveform. Most digital oscilloscopes have waveform averaging. Quantization noise, as described above, cannot be averaged out unless other random noise is present.

The noise specifications for an ADC are for quiet, low-impedance signals at the input, such as a dead short. To preserve the noise performance of the ADCs, the user must carefully connect signals to the input with tidy cabling that keeps away from sources of electromagnetic noise. For more information on noise sources and treatment and prevention of noise, see References 5 and 6.

Dynamic Range

The *dynamic range* (*DR*) of an ADC is the ratio of the largest to the smallest signals the converter can represent. The largest signal is usually taken to be a full-scale sine wave, and the smallest signal is usually taken to be the background noise level of the ADC. It can be expressed simply as a ratio, but it is more common to express it in decibels (dB):

$$DR = 20\ \log(S/N), \tag{25.2}$$

where DR is dynamic range in dB, S is the rms amplitude of the largest signal, and N is the rms amplitude of the smallest signal, the noise. The noise must include the quantization noise of the ADC, which for a perfect, noiseless converter is $1/\sqrt{12}$ LSB rms. For an n-bit converter, a full-scale sine wave has a peak amplitude of 2^{n-1} LSB, which corresponds to an rms amplitude of $2^{n-1}/\sqrt{2}$ LSB, or $2^{n-1.5}$ LSB rms. Hence a perfect ADC had a dynamic range of

$$\begin{aligned} DR &= 20\ \log\left(2^{n-1.5}\sqrt{12}\right) \\ &= 20\ \log\left(2^{n}\right) + 20\ \log\left(2^{-1.5}\sqrt{12}\right) \\ &= (n)\left[20\ \log(2)\right] + 20\ \log\left(\sqrt{1.5}\right) \\ &\approx 6.0206n + 1.7609. \end{aligned} \tag{25.3}$$

Equation 25.3 can be used to determine the *effective number of bits* (*ENOB*) of an imperfect ADC. ENOB may take only noise into account, or it may include noise and harmonic distortion products of the input signal. It is computed as

$$ENOB = (DR - 1.7609)/6.0206. \tag{25.4}$$

For example, a 16-bit ADC with a dynamic range of 92 dB has an ENOB of 14.988 bits.

25.5 Types of ADCs

The fundamental building block of analog-to-digital conversion is the *comparator*. Every type of ADC has at least one comparator in it, and some ADCs have many. The comparator itself is a one-bit ADC; it has two analog voltage inputs and (usually) one digital output. If the voltage at the + input is greater than the voltage at the – input, the output of the comparator is a digital 1. If the voltage at the + input is less than the voltage at the – input, the output is a digital 0 (see Figure 25.9).

Another piece that all ADCs have in common is a linearity reference. This is what a comparator in an ADC compares the input signal with in the process of conversion. It directly determines the differential and integral nonlinearities of the ADC. Examples of linearity references include capacitors (in integrating ADCs) and DACs (found in successive-approximation ADCs).

The third piece that every ADC has is a voltage reference. The reference(s) determine the full-scale input range of the ADC and are usually part of or closely associated with the linearity reference.

Flash

Flash converters are the fastest ADCs, achieving speeds near 1 GS/s and resolutions of 10 bits and below. The flash converter with n bits of resolution has $2^n - 1$ high-speed comparators operating in parallel (see Figure 25.10). A string of resistors between two voltage references supplies a set of uniformly spaced voltages that span the input range, one for each comparator. The input voltage is compared with all of these voltages simultaneously, and the comparator outputs are 1 for all voltages below the input voltage and 0 for all the voltages above the input voltage. The resulting collection of digital outputs from the comparators is called a "thermometer code," because the transition between all 1s and all 0s floats up

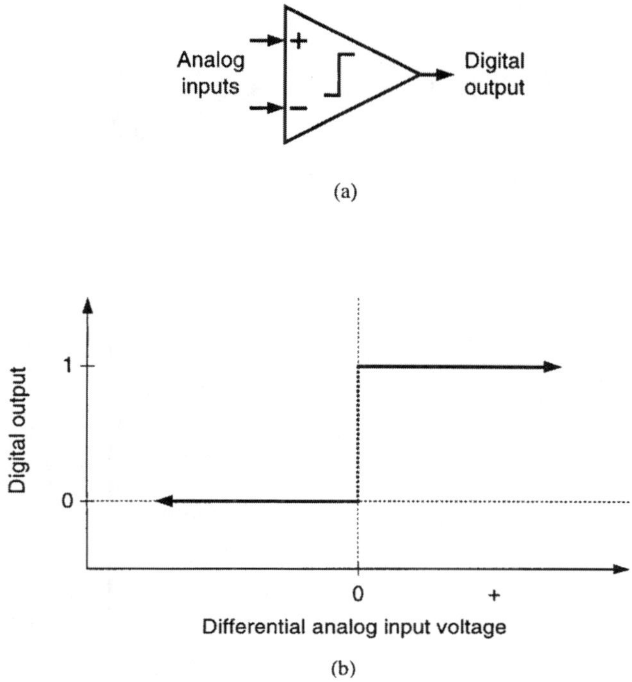

FIGURE 25.9 The comparator is the essential building block of all ADCs. (a) Comparator symbol. (b) Comparator input/output transfer function.

and down with the input voltage. Fast logic converts the thermometer codes to normal n-bit binary numbers.

Because of their simplicity, they are fast, but flash converters are limited to resolutions of 10 bits and below because the number of comparators and resistors goes up exponentially with resolution. Because the string resistor values typically vary only a few percent from each other in practical devices, the differential linearity of the flash ADC is quite good. But the same resistor variations can accumulate error across the input range and cause integral nonlinearity of a few LSB.

Successive-Approximation Register

Successive-approximation register (*SAR*) ADCs are the most common ADCs, having resolutions of 8 to 16 bits and speeds of 1 MS/s and below. They are generally low in cost, and they typically have very good integral linearity. The n-bit SAR ADC contains a high-speed n-bit DAC and comparator in a feedback loop (see Figure 25.11). The successive-approximation register sequences the DAC through a series of n "guesses," which are compared with the input voltage (Figure 25.12). As the conversion progresses, the register builds the n-bit binary conversion result out of the comparator outputs. By the end of the sequence the register has converged to the closest DAC value to the input voltage.

The speed of an SAR ADC is limited by the speed of the DAC inside the feedback loop. The DAC must settle n times to within $1/2^{-n}$ of full-scale within the conversion time of the ADC. Current SAR technology achieves 12-bit resolution at 1 MS/s and 16-bit resolution at 200 kS/s. Faster conversion at these resolutions requires multistage architectures.

Multistage

To achieve higher sample rates than SAR ADCs at resolutions of 10 to 16 bits, *multistage* ADCs (sometimes called *subranging* or *multipass* ADCs) use the iterative approach of SAR ADCs but reduce the number

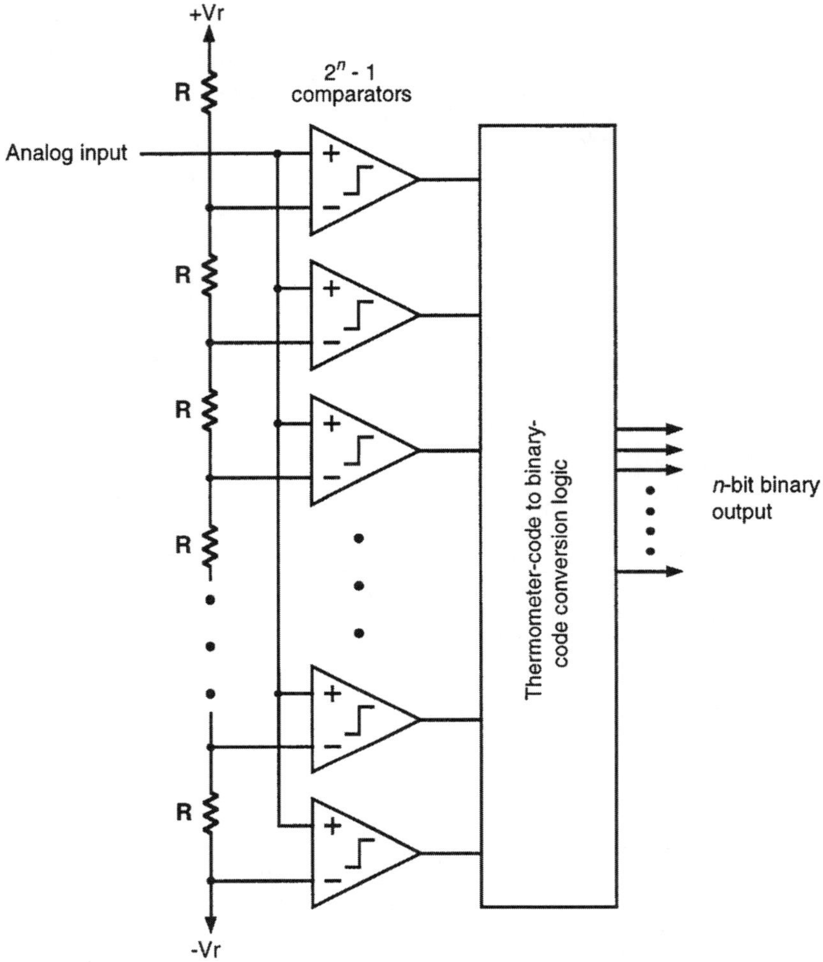

FIGURE 25.10 A flash converter has $2^n - 1$ comparators operating in parallel. It relies on the uniformity of the resistors for linearity.

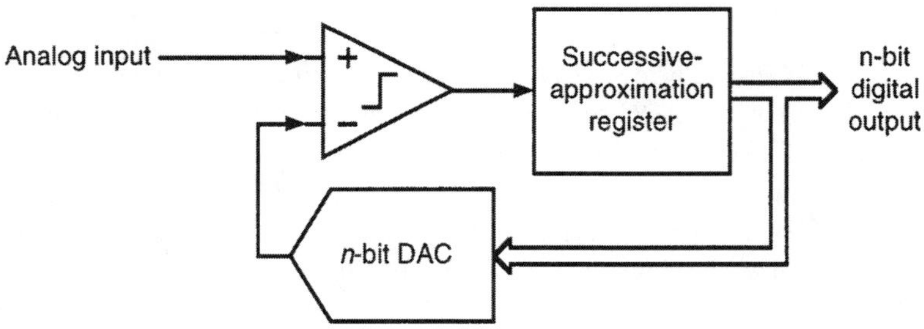

FIGURE 25.11 A successive-approximation converter has only one comparator and relies on an internal, precision DAC for linearity.

of iterations in a conversion. Instead of using just a comparator, the multistage ADC uses low-resolution flash converters (4 to 8 bits) as building blocks. Figure 25.13 illustrates an example of a 12-bit two-stage ADC built out of two flash ADCs and a fast DAC. The 6-bit flash ADC converts the residual error of the 8-bit flash ADC. The two digital outputs are combined to produce a 12-bit conversion result.

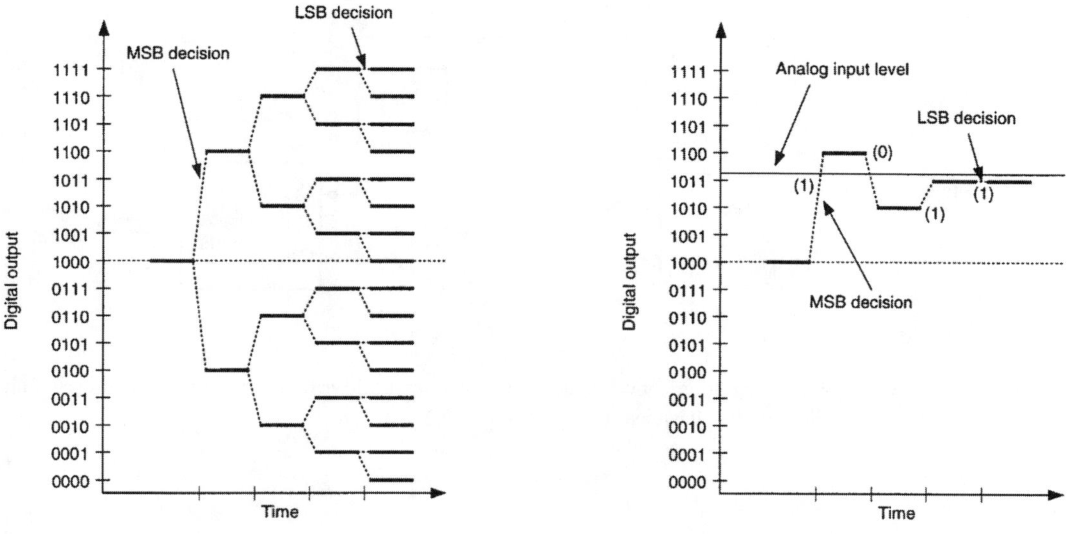

FIGURE 25.12 (a) Decision tree shows all the possible digital "guesses" of a four-bit successive-approximation converter over time. (b) Decision tree for conversion of four-bit code 1011.

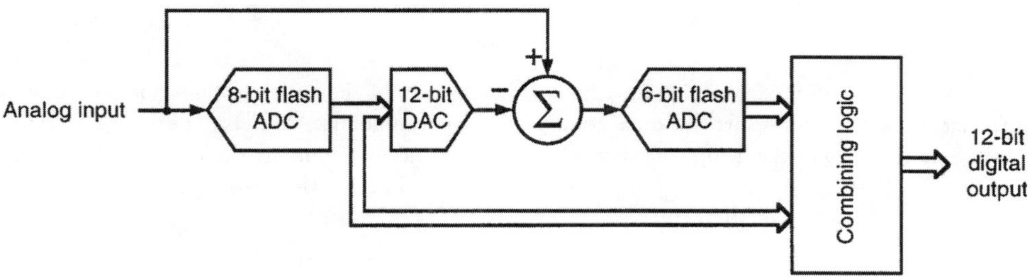

FIGURE 25.13 An example of a 12-bit multistage ADC built out of two flash ADCs and a fast DAC. The 8-bit flash ADC takes a first "guess" at the input signal and the 6-bit flash ADC converts the error in the guess, called the "residue." The 12-bit DAC actually needs to have only 8 bits, but it must be accurate to 12 bits. If the 8-bit flash ADC were perfect, the second flash ADC would only need 4 bits. But since the first flash actually may have some error, the second flash has 2 bits of "overlap."

If each flash ADC has a T/H at its input, then each stage can be converting the residual error from the previous stage while the previous stage is converting the next sample. The whole converter then can effectively operate at the sample rate of the slowest stage. Without the extra T/Hs, a new conversion cannot start until the residues have propagated through all the stages. This variation of the multistage ADC is called a *pipelined* ADC.

Integrating

Integrating converters are used for low-speed, high-resolution applications such as voltmeters. They are conceptually simple, consisting of an integrating amplifier, a comparator, a digital counter, and a very stable capacitor for accumulating charge (Figure 25.14). The most common integrating ADC in use is the dual-slope ADC, whose action is illustrated in Figure 25.15. Initially, the capacitor is discharged and so has no voltage across it. At time 0, the input to the integrator is switched to the analog input and the

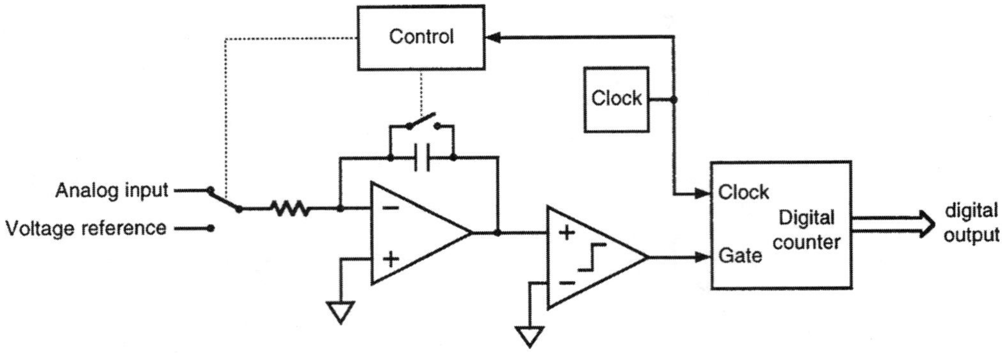

FIGURE 25.14 A dual-slope integrating converter uses a comparator to determine when the capacitor has fully discharged and relies on the capacitor for linearity.

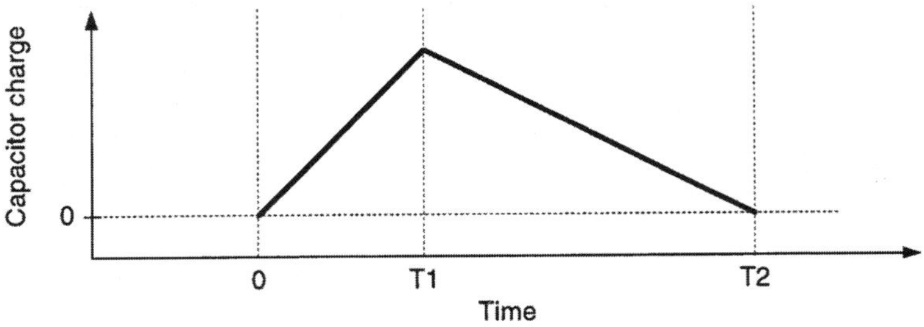

FIGURE 25.15 Charge on the integrating capacitor vs. time. At time 0, the input is switched to analog input and the switch across the capacitor opens. The capacitor integrates charge until fixed time T1. The input is then switched to the voltage reference to discharge the capacitor, and the counter begins counting a known clock. The comparator turns off the counter when the capacitor charge reaches 0 again, at time T2. The resulting count is proportional to the average input voltage over the time interval 0 to T1.

capacitor is allowed to charge for an amount of time, T1, which is always the same. Its rate of charging and thus its voltage at T1 are proportional to the input voltage. At time T1 the input switch flips over to the voltage reference, which has a negative value so that the capacitor will begin to discharge at a rate proportional to the reference. The counter measures how long it takes to discharge the capacitor completely. If the capacitor is of high quality, the ratio of the discharge time to the charge time is proportional to the ratio of the input voltage to the voltage reference, and so the counter output represents the analog input voltage.

An elaboration of the dual-slope ADC is the *multislope* integrating ADC. It achieves even higher resolution than the dual-slope ADC by discharging the capacitor at several progressively slower rates. At each rate, the counter is able to resolve finer increments of accumulated charge.

An important distinction between integrating converters and other ADCs is the way they sample the input voltage. Integrating converters do not sample the voltage itself; they *average* the voltage over the integration period and *then* they sample the average that is accumulated on the capacitor. This tends to reject noise that conventional sampling cannot, especially periodic noises. Most integrating ADCs operate with an integration period that is a multiple of one ac line period ($^1/_{60}$ or $^1/_{50}$ s) so that any potential interference from stray electric or magnetic fields caused by the power system is canceled.

Integrating converters are gradually being replaced in the marketplace with low-speed, high-resolution sigma–delta converters (see below). Sigma–delta converters are generally more flexible than integrating ADCs, and they are easier to use because they do not require an external charging capacitor. The resolution and speed of the two types are comparable, although integrating converters still have the highest linearity.

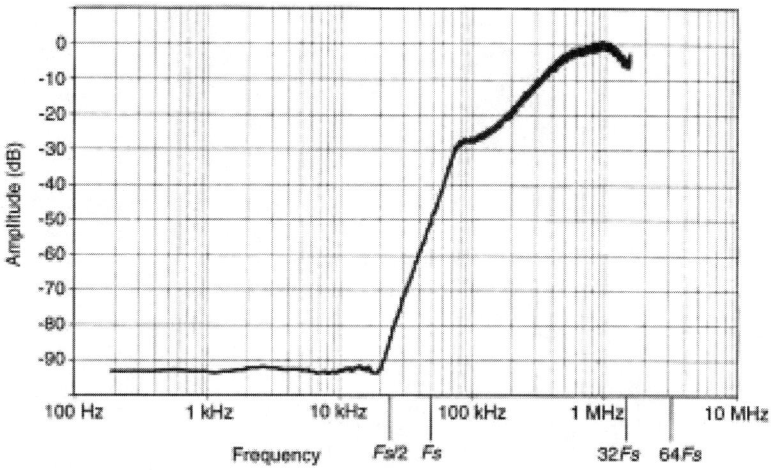

FIGURE 25.16 Spectrum of a 64-times oversampling SD ADC before the digital decimation filter. The modulator loop samples at 3.072 MS/s and the data comes out of the filter at 48 kS/s. The filter cuts off sharply at *Fs*/2, or 24 kHz, leaving only the small amount of noise left below 24 kHz.

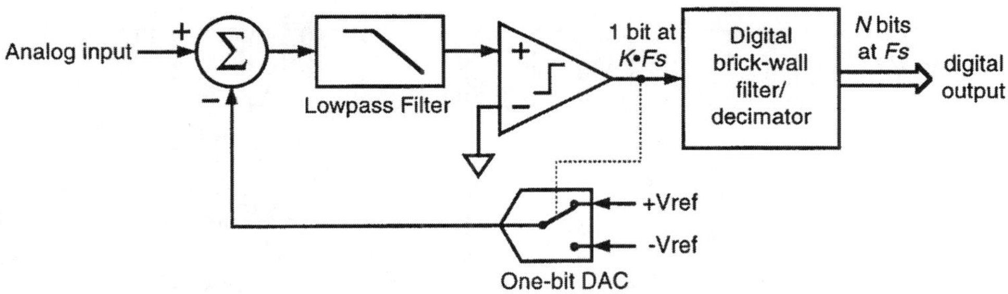

FIGURE 25.17 A SD modulating ADC uses a comparator simply as a one-bit quantizer. The linearity of a SD ADC is theoretically perfect because the one-bit DAC can only assume two values, and thus is linear by definition. Modern SD ADCs are made with switched-capacitor circuits which operate at *KFs*, where *Fs* is the output data sample rate and *K* is the oversampling ratio.

Sigma–Delta ADCs

The *sigma–delta* (*SD*) ADC is quickly becoming one of the most popular types of ADC. SD ADCs typically have resolutions of 16 to 24 bits and sample rates of 100 kS/s down to 10 S/s. Because of their high resolution at 48 kS/s, they are the most common type of converters in modern digital audio equipment. SD ADCs defy intuition by quantizing initially with *very* low resolution (often one bit) at very high rates, typically 64× to 128× the eventual sample rate (called *oversampling*). The high-rate, low-resolution quantizer operates inside a feedback loop with an analog lowpass filter and a DAC to force the large amount of otherwise unavoidable quantization error (noise) to frequencies higher than the band of interest. The resulting spectral redistribution of the quantization noise is called *noise shaping*, illustrated in Figure 25.16. The low-resolution digital output of the ADC loop is fed into a digital filter that increases the resolution from the resolution of the ADC loop to the output resolution, reduces the data rate from the rate of the ADC loop to the output sample rate, and applies a low-pass digital filter, leaving only the signals in the frequency band of interest and a little inherent electronic noise.

Figure 25.17 shows how a one-bit sigma–delta ADC works. The comparator *is* the ADC, and its output is processed digitally, so that no further analog errors can accumulate. The comparator is in a feedback

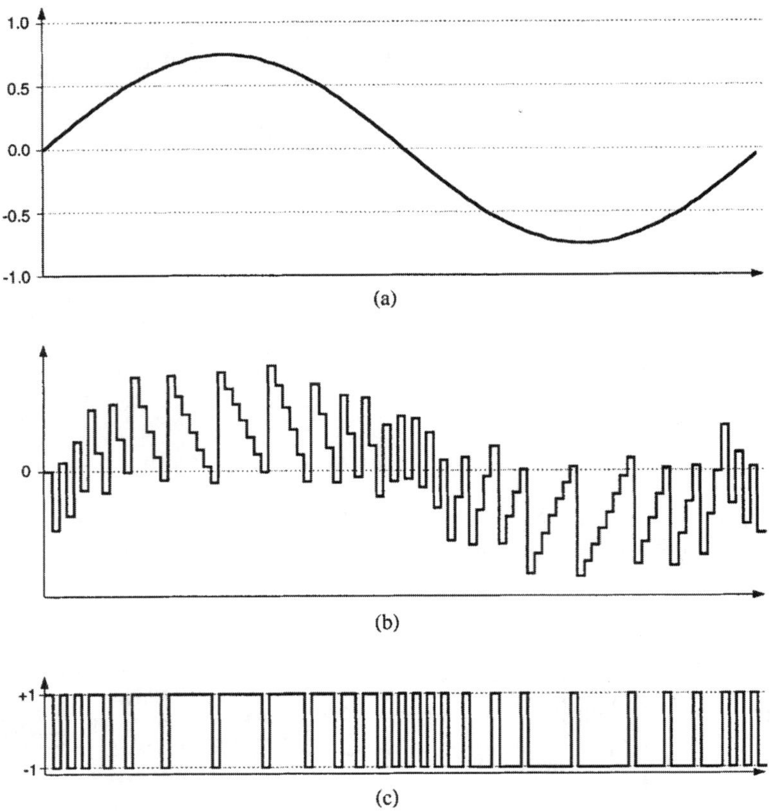

FIGURE 25.18 Behavior of a discrete-time (switched-capacitor) first-order SD modulator, where the low-pass filter is simply an integrator. In each graph, the x-axis represents time, and the y-axis represents signal level. (a) The input waveform. (b) Input to the comparator. (c) The one-bit digital comparator output. The duty cycle of this waveform corresponds to the input waveform. The digital filter and decimator recover the original waveform from this one bit.

loop with a low-pass filter (typically third to fifth order) and a one-bit DAC. The one-bit DAC can take on only one of two values, +full-scale and –full-scale, so it is perfectly linear. The low-pass filter causes the loop gain to be high at low frequencies (the signal band of interest) and low at high frequencies. Since the error in a feedback loop is low when the gain is high and high when the gain is low, the errors dominate at high frequencies and are low in the band of interest. The result is a one-bit output whose duty cycle is proportional to the input signal. Together, the elements of the feedback loop are called a *sigma–delta modulator.*

Figure 25.18 illustrates the operation of a simple discrete-time (switched-capacitor) SD ADC. In this first-order example, the low-pass filter is just an integrator. The loop tries to force the input to the comparator back to the baseline, and the figure shows how the duty cycle of the resulting digital output reflects input signal. The digital data here have undesirable patterns which tend to repeat, called *limit cycles.* They can appear in the band of interest and interfere with the signal. Higher-order loop filters (third and above) make the bit activity so chaotic that it has no substantial limit cycles.

The chief advantage of a SD converter is that it has a built-in antialias filter, and a darn good one at that. Most SD parts have a *finite-impulse response* (*FIR*) digital filter, which has an extremely flat frequency response in the passband and an extremely sharp cutoff, properties impossible to implement in analog filters. The ADC still needs an antialias filter to reject signals above one half the oversampling rate. But this filter is simple to build, since it has to be flat only up to one half the output sampling rate and has many octaves (all the way to near the oversampling rate) to fall off. The combination of the two filters provides watertight protection from aliases, often 96 dB of attenuation over the entire spectrum.

An important improvement of the one-bit SD is the multibit SD, wherein the comparator is replaced by a flash converter with as much as four bits of resolution. This improves the ENOB of the whole converter by several bits.

Voltage-to-Frequency Converters

Voltage-to-frequency converters (*VFCs*) are versatile, low-cost circuits that convert analog voltages to periodic waveforms whose frequency is proportional to the analog input voltage. A VFC is conceptually similar to an integrating converter (see above) except that the digital counter is missing and is replaced with a short-pulse generator that quickly discharges the capacitor. The voltage reference is not connected intermittently to the input; instead, it appears all the time at the minus input of the comparator instead of ground. The capacitor charges at a rate proportional to the input voltage until the voltage is equal to the voltage reference. Then the comparator trips the pulse generator, which quickly discharges the capacitor, and the cycle begins again. The periodic pulse at the comparator output can be used as the digital output.

The advantage of the VFC over conventional ADCs is that the one-bit output can be transmitted digitally, through isolation transformers, through fiber-optic cable, or through any other isolating, nonisolating, long-distance, or short-distance transmission medium. All that is needed at the receiving end to complete the analog-to-digital conversion is a digital counter, which does not need to be synchronized to the VFC itself. Sometimes, the digital conversion is not needed; a VFC can be used with an isolating transformer and a *frequency-to-voltage converter* (*FVC*) to create an isolation amplifier. For a good discussion of VFCs, see Reference 7.

25.6 Instrumentation and Components

Integrated Circuits

Table 25.1 lists several popular high-quality ADCs in integrated circuit form. The prices given are approximate for small quantities and for the lowest grade of part, as of mid-1996. By no means exhaustive, the list is a sampling of a few of the most popular or best-performing chips of each type of ADC. Table 25.2 contains addresses, phone numbers, and internet sites for the manufacturers in Table 25.1.

Instrumentation

Plug-in data acquisition cards are becoming increasingly popular as personal computer prices come down and processor performance goes up. These cards typically contain one or more ADCs (with S/H), instrumentation amplifiers with gain and differential input, and multiplexers to switch to different inputs. Some have DACs on-board, and some have digital data and timing functions as well. Once considered low performance and hard to use, data acquisition cards have improved dramatically, equaling and in some cases exceeding capabilities of stand-alone instruments. Most come with drivers that interface to user-friendly software packages for creating easy-to-use yet custom-built computer instrumentation. Table 25.3 lists a few popular plug-in data acquisition boards and Table 25.4 lists how their manufacturers may be contacted.

TABLE 25.1 ADC Integrated Circuits

Part	Type	Sample Rate	Resolution, bits	Manufacturer	Approx. Price, $
ADC160	Integrating	1 S/s	24	Thaler	225.00
AD7714	Sigma–delta	2.62 S/s	24	Analog Devices	22.00
MAX132	Integrating	6 S/s	19	MAXIM	15.11
CS5508	Sigma–delta	20 S/s	20	Crystal	21.50
HI7190	Sigma–delta	10 S/s	24	Harris	17.85
AD1879	Sigma–delta	50 kS/s	18	Analog Devices	46.00
CS5390	Sigma–delta	50 kS/s	20	Crystal	75.30
ADS7809	SAR	100 kS/s	16	Burr-Brown	41.54
CS5101A	SAR	100 kS/s	16	Crystal	67.20
AD7893	SAR	117 kS/s	12	Analog Devices	14.00
AD976	SAR	200 kS/s	16	Analog Devices	36.50
AD7722	Sigma–delta	200 kS/s	16	Analog Devices	39.80
LTC1278	SAR	500 kS/s	12	Linear Technology	17.08
AD1385	Multistage	500 kS/s	16	Analog Devices	1053.00
ADS7819	SAR	800 kS/s	12	Burr-Brown	31.90
AD9220	Multistage	10 MS/s	12	Analog Devices	22.95
AD775	Multistage	20 MS/s	8	Analog Devices	14.00
AD9050	Multistage	40 MS/s	10	Analog Devices	39.00
AD9066	Flash	60 MS/s	6	Analog Devices	7.00
HI1276	Flash	500 MS/s	8	Harris	338.58

TABLE 25.2 Companies That Manufacture ADC Integrated Circuits

Analog Devices, Inc. One Technology Way P.O. Box 9106 Norwood, MA 02062-9106 (617) 329-4700 http://www.analog.com	Harris Corp. Semiconductor Products Division P.O. Box 883 Melbourne, FL 37902 (407) 729-4984 http://www.semi.harris.com	Maxim Integrated Products, Inc. 120 San Gabriel Drive Sunnyvale, CA 94086 (408) 737-7600 http://www.maxim-ic.com
Burr-Brown Corporation P.O. Box 11400 Tucson, AZ 85734-1400 (520) 746-1111 http://www.burr-brown.com	Linear Technology Corporation 1630 McCarthy Blvd. Milpitas, CA 95035-7417 (408) 432-1900 http://www.linear-tech.com	Thaler Corporation 2015 N. Forbes Boulevard Tucson, AZ 85745 (520) 882-4000 http://www.thaler.com
Crystal Semiconductor Corporation P.O. Box 17847 Austin, TX 78760 (512) 445-7222 http://www.cirrus.com/prodtech/crystal.html		

TABLE 25.3 Plug-In Data Acquisition Boards

Part	Type	Sample Rate	Resolution, bits	Manufacturer	Approx. Price, $
AT-A2150	Sigma–delta	51.2 kS/s	16	National Instruments	1495
AT-MIO-16XE-50	SAR	20 kS/s	16	National Instruments	995
AT-MIO-16E-10	SAR	100 kS/s	12	National Instruments	995
CIO-DAS1600/12	SAR	160 kS/s	12	ComputerBoards, Inc.	599
AT-MIO-16XE-10	SAR	100 kS/s	16	National Instruments	1995
CIO-DAS1600/16	SAR	100 kS/s	16	ComputerBoards, Inc.	699
DT-3001	SAR	330 kS/s	12	Data Translation, Inc.	995
DAS-1800AO	SAR	333 kS/s	12	Keithley Metrabyte	1299
AT-MIO-16E-1	SAR	1 MS/s	12	National Instruments	1795
FAST16-1	Multistage	1 MS/s	16	Analogic	3895

TABLE 25.4 Companies That Manufacture Plug-In Data Acquisition Boards

Analogic Corporation 360 Audubon Road Wakefield, MA 01880 (508) 977-3000	Keithley Metrabyte 440 Myles Standish Blvd. Taunton MA 02780 (508) 880-3000 http://www.metrabyte.com
ComputerBoards, Inc. 125 High Street Mansfield, MA 02048 (508) 261-1123	National Instruments Corporation 6504 Bridge Point Parkway Austin, TX 78730 (512) 794-0100 http://www.natinst.com
Data Translation, Inc. 100 Locke Drive Marlboro, MA 01752-1192 (508) 481-3700 http://www.datx.com	

References

1. C.E. Shannon, Communication in the presence of noise, *Proc. IRE,* 37(1): 10–21, 1949.
2. H. Nyquist, Certain topics in telegraph transmission theory, *AIEE Transactions,* 617–644, April, 1928.
3. A.B. Williams and F.J. Taylor, *Electronic Filter Design Handbook: LC, Active, and Digital Filters,* 2nd ed., New York: McGraw-Hill, 1988.
4. S.P. Lipshitz, R.A. Wannamaker, and J. Vanderkooy, Quantization and dither: a theoretical survey, *J. Audio Eng. Soc.,* 40, 355–375, 1992.
5. H.W. Ott, *Noise Reduction Techniques in Electronic Systems,* 2nd ed., New York: John Wiley & Sons, 1988.
6. R. Morrison, *Grounding and Shielding Techniques in Instrumentation,* 3rd ed., New York: John Wiley & Sons, 1986.
7. J. Williams, Designs for High Performance Voltage-to-Frequency Converters, Application Note 14, Linear Technology Corporation, March 1986.

Further Information

M. Demler, *High-Speed Analog-to-Digital Conversion,* San Diego, CA: Academic Press, 1991.

B.M. Gordon, *The Analogic Data-Conversion Systems Digest,* Wakefield, MA: Analogic Corporation, 1981.

D.H. Sheingold, Ed., *Analog-Digital Conversion Handbook,* Englewood Cliffs, NJ: Prentice-Hall, 1986.

Crystal Semiconductor, *Delta Sigma A/D Conversion Technique Overview,* Application Note AN10, Austin, TX, 1989.

26

Computers

A. M. MacLeod
University of Abertay Dundee, Scotland

P. F. Martin
University of Abertay Dundee, Scotland

W. A. Gillespie
University of Abertay Dundee, Scotland

26.1 Introduction

Computers are an essential feature of most instrumentation systems because of their ability to supervise the collection of data and allow information to be processed, stored, and displayed. Many modern instruments are capable of providing a remote user with access to measurement information via standard computer networks.

26.2 Computer-Based Instrumentation Systems

The main features of a computer-based instrumentation system are shown in Figure 26.1. The actual implementation of such systems will depend on the application. Many commercially produced instruments such as spectrophotometers or digital storage oscilloscopes are themselves integrated computerbased measurement systems. These "stand-alone" instruments may be fitted with interfaces such as IEEE- 488 or RS-232 to allow them to be controlled from personal computers (PCs) and also to support the transfer of data to the PC for further processing and display. Alternatively, the instrumentation system may be based around a PC/workstation or an industrial bus system such as a VME or Multibus, allowing the user the ability to customize the computer to suit the application by selecting an appropriate set of add-on cards. Recently, the introduction of laptop and notebook PCs fitted with PCMCIA interfaces with input/output capability has provided opportunities for the development of highly portable instrumentation systems.

The Single-Board Computer

The simplest form of a computer is based around the single-board computer (SBC) which contains a microprocessor, memory, and interfaces for communicating with other electronic systems. The earliest

0-8493-1733-9/04/$0.00+$1.50

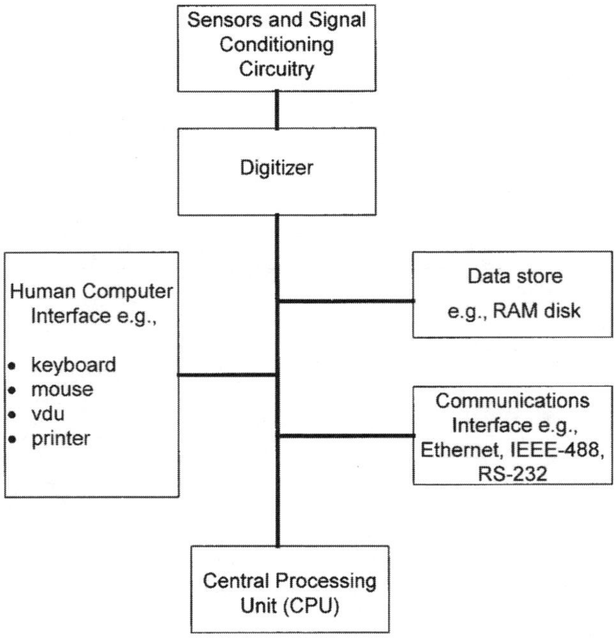

FIGURE 26.1 Elements of a computer-based instrumentation system.

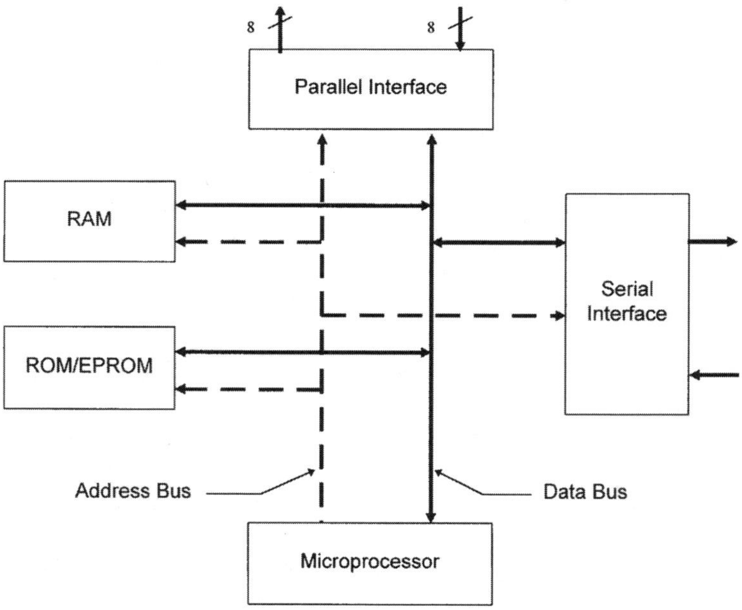

FIGURE 26.2 An overview of a single-board computer.

form of personal computers simply comprised an SBC together with a keyboard, display, disk drives, and a power supply unit. Today, the SBC still offers a solution for the addition of limited intelligence to instrumentation systems as well as forming an element of most computer systems, e.g., as the motherboard of a PC.

An overview of a simple SBC is shown in Figure 26.2. The microprocessor, which contains the central processor unit, is responsible for executing the computer program and for controlling the flow of data around the SBC. The random access memory (RAM) acts as a temporary (i.e., volatile) storage area for both program code and data. On the motherboard of a PC the read only memory (ROM) is largely used for storing the low-level code used to access the input/output hardware, e.g., the BIOS (basic input output system). The operating system and applications software are loaded into RAM from the disk unit. In small, dedicated systems such as an oscilloscope the ROM may be used to store the program code to avoid the need for a disk drive. For small production runs or development work erasable programmable ROM (EPROM) is used as an alternative to ROM, allowing the code to be upgraded without the high cost of commissioning a new ROM chip from the manufacturers.

Data are transferred around the SBC on its data bus, which will be typically 8, 16, or 32 bits wide (corresponding to the number of bits that can be transmitted at the same time). SBCs with 8-bit data buses are less complex and consequently of lower cost than 32-bit systems and may well be the optimum solution for those process control applications which require minimal processing of 8-bit data, e.g., from temperature and position sensors. However, the 32-bit bus width of most modern PCs and workstations is essential to ensure the fast operation of Windows-based applications software.

The address bus is used to identify the destination of the data on the data bus. Data transfer is usually between the microprocessor and the memory or interfaces. However, some SBCs support DMA (direct memory access) which allows data to be transferred directly between interfaces and memory without the need for the information to pass through the processor. DMA is inherently faster than program-driven data transfer and is used for moving large blocks of data, e.g., loading programs from disk or the transfer of digitized video images.

SBCs are fitted with interfaces to allow them to communicate with other circuits. Interfaces carry out two main functions. First, they serve to ensure that the signals on the internal buses of the SBC are not affected by the connection of peripheral devices. Second, they ensure that signals can pass into and out of the computer and that appropriate voltage levels and current loading conditions are met. A well-designed interface should also provide adequate electrical protection for the SBC from transients introduced via the external connection. Parallel interfaces allow data to be transferred, usually 8 or 16 bits at a time, and contain registers that act as temporary data stores. Serial interfaces must carry out the conversion of data from the parallel format of the internal SBC data bus to and from the serial format used by the interface standard (e.g., RS-232 or RS-422). Some SBCs may contain additional interfaces to support communication with a VDU, local area network (LAN), or a disk drive. Interfaces can range in complexity from a single parallel interface chip to a LAN controller which may require a significant fraction of the SBC board area.

Computer Bus Architectures

All but the simplest computer systems contain several circuit board cards which plug into a printed circuit board backplane. The cards will include at least one SBC and a number of "add-on" cards providing functions such as interfaces to peripherals (e.g., a LAN, graphics display, disk unit) or additional memory. The actual structure of bus architectures is quite variable but the simple model shown in Figure 26.3 contains the essential features. A group of tracks will carry the data and address information with a second group of tracks being used to control the flow of data and to ensure its reliable transfer. Other

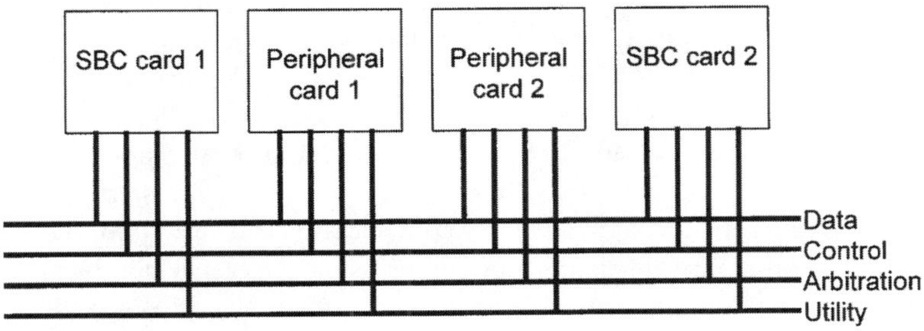

FIGURE 26.3 A simplified model of a computer bus.

TABLE 26.1 Typical Microprocessor Bus Standards

Bus Standard	STE	G96	VME	Multibus II
Data width (bits)	8	8/16	8/16/32	8/16/32
Max address (bytes)	1 M	32 M	4 G	4 G
Synchronous/asynchronous	Async	Async	Async	Sync
Connectors	64 pin	96 pin	96 pin	96 pin + 32 pin

tracks are reserved for the signals which provide arbitration between SBC cards to ensure that only one such card has control of the bus at any given moment. There will also be tracks which provide utility functions such as clock signals and the transmission of event or interrupt signals.

The main advantage of bus-based systems is that they help one build a complex computerized system using standard cards. By conforming to agreed standard buses (see Table 26.1 for typical examples), the system integrator can minimize problems of incompatibility and keep system costs to a minimum. The complexity of many bus systems is such that a significant fraction of the area of each card is dedicated to providing the logic required to interface to the bus, adding appreciably to the cost. In addition, data transfer between cards, even using DMA, is relatively slow compared with transfers between chips on the same SBC. There is, therefore, a trade-off between integrating functions on a single SBC and a lack of flexibility in customizing the system. An alternative to bus-based systems is to utilize processors such as transputers which are designed to be connected together directly into multiprocessor structures.

Most modern computerized systems adopt some form of distributed intelligence strategy in order to free the central processor unit to run the main application. Graphics controller cards generally employ a dedicated graphics processor able to handle the creation of the basic features such as line drawing or scaling the size of objects. LAN controller cards require intelligence to handle the network protocols and buffer the data flow to and from the network. Disk controller cards relieve the central processor of the need to handle the processes of reading and writing to the disk.

The use of a bus system with more than one SBC allows the designer to dedicate an SBC to the process of data acquisition and temporary data storage in order to ensure that the main SBC can concentrate on data processing and communication with the user. Such a strategy enables the data acquisition SBC to maintain a real-time response to making measurements while allowing the user access to processed data. In most systems one of the SBCs acts as a system controller, which has ultimate control over the use of the bus and will normally contain the bus arbitration hardware. Only one SBC at a time can obtain the authority of the controller to act as a bus master and initiate data communication. The other cards will be configured as slaves allowing them to respond to requests for information from the master. In some systems, e.g., Multibus II, bus arbitration is distributed throughout the intelligent cards.

Industrial Computers

Many instrumentation systems utilize standard PCs or workstations either fitted with add-on cards or linked to intelligent instruments via standard protocols such as IEEE-488 or RS-232. In many industrial environments there is a need to provide protection against hazards such as dust, damage by impact or vibration, and unauthorized access. The systems may have to fit more stringent electromagnetic compatibility (EMC) requirements. Industrial or ruggedized versions of desktop computers are available to meet this market. Cardframes designed to accommodate the standard industrial bus systems such as VME and Multibus can be readily customized to meet the demands of the industrial environment and ruggedized versions of PCs are also available. In designing industrial computers, care must be paid to the specification of adequate power supplies, both to provide sufficient current for the add-on cards and also to provide protection against fluctuations in the mains supply. It may also be necessary in safety-critical applications to use an uninterruptable power supply that will guarantee the operation of the system during short failures in the mains supply.

Software

All but the simplest computer systems require an operating system to support the operation of applications software. The operating system will allocate areas of memory for use by the applications programs and provide mechanisms to access system resources such as printers or displays. Some operating systems, such as MS DOS, are single-tasking systems, meaning that they will only support the operation of one program or task at a time. Instrumentation systems which simultaneously take measurements, process data, and allow the user to access information generally require a multitasking operating system, such as OS-9, UNIX, or Windows 95. In such instrumentation systems the applications software may well comprise a single program, but it may generate software processes or tasks each with their own local data and code. The multitasking operating system will allow the actual execution of these tasks to be scheduled and also provide mechanisms for communication of information between tasks and for the synchronization of tasks.

Instrumentation systems are real-time environments, i.e., their operation requires that tasks be carried out within specified time intervals; for example, the capture of data must occur at specified moments. Operating systems adopt a range of strategies for determining the scheduling of software tasks. Round-robin scheduling, for example, provides each task with execution time on a rota basis, with the amount of time being determined by the priority of the task. Operating systems which support preemptive scheduling allow high-priority tasks to become active if a specified event such as trigger signal occurs. The speed at which an operating system can switch from one task to another (i.e., context switch) is an important metric for real-time systems.

The extent to which an operating system can respond to a large number of events is extremely limited, as low-priority tasks will have a poor response time. If an instrumentation system needs to make a large number of measurements and also support other activities, such as information display, the solution is generally to use a distributed system with dedicated SBCs operating as microcontrollers to capture the data. These microcontrollers will not usually require an operating system and will each operate a single program which will be either downloaded from the master SBC or located in ROM.

Commercial packages (for examples, see Table 26.2) are readily available to support PC/workstation-based instrumentation systems. A typical system will provide a Windows-type environment to control measurements and store, process, and display data and increasingly make use of the virtual-instrument concept which allows the user to configure instrumentation systems from an on-screen menu. Most packages will allow stand-alone instruments to be controlled using interface cards supporting standards such as IEEE-488 or RS-232. Some packages also support the use of add-on cards for analog-to-digital conversion or a range of control functions such as channel switching and waveform generation; however, their main restriction is the limited range of hardware configurations supported by the software supplier. Each stand-alone instrument and add-on card requires a piece of code called a device driver so that the operating system can access the hardware resources of the card and instrument. Therefore, the development of device drivers requires an intimate knowledge of both the hardware and the operating system.

System Development

The software component of any computerized instrumentation system can form a significant percentage of its total cost. This is especially true of on-off systems which require software to be developed for a specific application, where the cost of the implementation of the software and its maintenance can considerably exceed the hardware costs. In such circumstances the ability to reuse existing code and the access to powerful development systems and debugging tools are of crucial importance. Some debug software only provides support for stepping through the operation of code written at the assembly code level. When developing device drivers, for example, access to a source-level debugger which can link the execution of code written in a high-level language to the contents of processor registers is useful.

Many SBCs that are intended for operation as stand-alone microcontrollers, instead of in bus-based systems, are often supported by a high-level language such as C or FORTH and a development environment. A common practice is to utilize a PC to develop the code which can be downloaded via a serial

TABLE 26.2 Typical Data Acquisition and Display Software Packages for PC-Based Instrumentation Systems

Product	Platforms	Manufacturer
DTV	PC (Windows 3.1 or 95)	Data Translation Inc. 101 Locke Drive Marlboro, MA 01752-1192 Tel: (U.S. and Canada) (800) 525-8528 @em@em(worldwide) +1 508-481-3700
Labtech Notebook/Notebook Pro	PC (DOS, Windows 3.1 or 95)	Laboratory Technologies Corporation 400 Research Drive Wilmington, MA 01887 Tel: (U.S. and Canada) 800-8799-5228 @em@em(worldwide) +1 508-658-9972
LabVIEW	PC, Mac, Sun, Power PC	National Instruments 6504 Bridge Point Parkway Austin, TX 78730-5039 Tel: +1 512-794-0100
LabWindows	PC (DOS)	National Instruments 6504 Bridge Point Parkway Austin, TX 78730-5039 Tel: +1 512-794-0100
LabWindows/CVI	PC (Windows 3.1 or 95) Sun	National Instruments 6504 Bridge Point Parkway Austin, TX 78730-5039 Tel: +1 512-794-0100
ORIGIN	PC (Windows 3.1 or 95)	MicroCal Software Inc. One Roundhouse Plaza Northampton, MA 01060 Tel: +1 413-586-2013

link to the microcontroller. Some debugging support for the execution of the code on the microcontroller is also provided. The development of powerful compact Intel 486 or Pentium microcontroller cards running operating systems such as MS DOS can greatly reduce software development time because of the ready access to PC-based software. The implementation of systems based on dedicated microcontrollers may require the use of a logic analyzer to view data on system buses.

26.3 Computer Buses

The VMEbus (IEEE P1014)

The VMEbus [1–3], utilizes a backplane fitted with two 96-pin connectors (called P1 and P2). The P1 connector provides access to all the bus signals with the exception of bits 16 to 31 of the data bus and bits 24 to 31 of the address bus, which are provided on the P2 connector. The location of cards in a VME system is important, as slot 1 must contain the system controller and the priority of the SBCs is determined by their proximity to the system controller SBC. As only 32 of the pins on the P2 connector are defined, the remaining 64 tracks on the P2 backplane may be specified by the user. VME cards may be single height (fitted with a P1 connector only) or double height (fitted with both P1 and P2 connectors). Single-height boards are 100 mm high and 160 mm deep and fit 3U height cardframes. Double-height boards are 233.35 mm high and 160 mm deep and fit 6U height cardframes.

VMEbus Signals

The VMEbus can be described as having four distinct groups of signals.

Data Transfer Bus (DTB).

The DTB provides 32-bit data transfer with 32-bit address information using a nonmultiplexed asynchronous approach. Transfer is supported by a simple handshake mechanism with the bus master

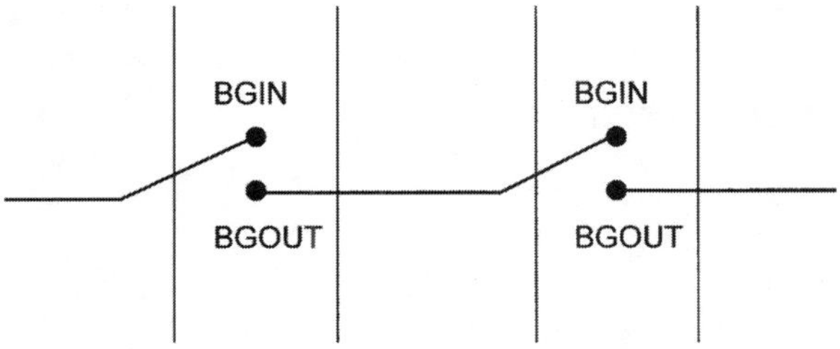

FIGURE 26.4 VMEbus arbitration daisy chain.

initiating the transfer providing an address strobe (AS) signal to tell the destination card to read the address information. The master also controls two data strobe (DS1 and DS2) lines which in a write operation indicate the presence of valid data on the data bus and in a read operation that the master is ready to receive data from the slave. The slave card uses the data acknowledge signal (DTACK) in a read operation to indicate that it has placed valid data on the bus and in a write operation to confirm that it has read the data. There is clearly a danger that the bus will hang up if the slave does not respond to a request for a data transfer, so in most VME systems a watchdog timer is provided to produce a signal on the bus error (BERR) line if a DTACK signal is not detected within the time-out period.

Revision C of the VME bus specification will support 8-,16-, and 32-bit data transfers, whereas revision D supports 64-bit transfers by making use of the address bus to transfer the additional 32 bits. Data may also be transferred in blocks of up to 256 bytes without the need to retransmit the address information. Most SBCs have both master and slave interfaces allowing their memory to be accessed from other SBCs in the system. The VME bus also supports three types of addressing mode, short (16 bit), standard (24 bit), and extended (32 bit), allowing the memory decoding circuitry of cards to be minimized for small systems or for input/output cards. Six address modifier lines (AM0 to AM5) are used to indicate the addressing mode and also to give information about the mode of transfer (e.g., block transfer and whether the transfer is in privileged, supervisor, or nonprivileged mode).

Arbitration Bus.

The arbitration bus provides a mechanism for deciding which master is allowed to gain control of the bus. The arbitration bus provides four bus request lines (BR0 to BR3) and four bus grant lines. The SBC in slot 1 of the VME bus acts as a bus controller which provides arbitration using a number of scheduling algorithms. In a simple priority-based algorithm each of the bus request lines is assigned a priority, whereas in a round-robin system access to the bus cycles around the bus request lines. Each of the four bus grant lines is in fact allocated two pins on the P1 connector (see Figure 26.4). The bus grant signal therefore passes through each card (called daisy chaining); thus, in Figure 26.4 the signal enters via BG1IN and exits via BG1OUT. This scheme allows a card to intercept the bus grant signal, giving cards nearer the system controller a higher priority. A consequence of daisy chaining is that any unused card slots must be fitted with jumpers between the bus grant pins.

Priority Interrupt Bus.

The priority interrupt bus provides seven interrupt request lines (IRQ1 to IRQ7) with IRQ7 having the highest priority. The interrupt requests may be handled by several SBCs, provided that all interrupts of the same level are handled by the same master. The interrupt acknowledge line is daisy chained through the cards in a similar way to the bus grant lines (the pins are called IACKIN and IACKOUT). Again, any unused card slots must be fitted with jumpers between these pins.

Utility Bus.

The utility bus provides the power supply tracks, the system reset, a 16 MHz clock and two system failure lines, SYSFAIL, which allows a card to indicate that it has suffered a failure, and ACFAIL, which is generated by the power supply monitor. Both failure lines are handled by the system controller SBC.

Associated with the VMEbus is VXI (*V*MEbus e*x*tensions for *i*nstrumentation). This blends the IEEE 1014 VMEbus standard with the IEEE-488 instrumentation bus using the uncommitted pins on the P2 connector to form the IEEE P1155 standard.

Multibus II (IEEE 1296)

Multibus II [3] is a synchronous bus system which is implemented on a single 96-track backplane. Cards conform to the Eurocard format of 220 × 233.65 mm; however, the bus requires only a single 96-pin DIN 41612-type connector. A second connector may be used to support a local bus or merely provide mechanical stability for a double-height card.

Unlike VME, the Multibus II standard has the following features:

1. A synchronous bus; i.e., data transfer is linked to the bus clock rather than the strobe/acknowledge control lines of the asynchronous VME system;
2. A 32-bit multiplexed address and data bus;
3. A message-passing strategy as an alternative to interrupt request lines;
4. Distributed bus arbitration rather than a central system controller (each card contains a message-passing controller, MPC, chip which participates in implementing the message-passing algorithm);
5. Only intelligent cards (i.e., fitted with microprocessors) that can interface to the main bus; simple input/output (I/O) cards should make use of a local bus (e.g., the iLBX bus) using the second DIN connector.

Multibus Signals

Multibus II may be described as having five distinct groups of signals.

Central Control.
The central services module (CSM) in slot 0 is responsible for the generation of this group of signals. The CSM may be implemented on a dedicated card or incorporated onto a SBC. The CSM produces the 10 MHz bus clock as well as a range of reset signals to support both a cold and warm start as well as recovery from power supply failure.

Address/Data.
Multibus II supports a 32-bit multiplexed address/data bus (AD0 to AD31) with four parity lines (PAR0 to PAR3) providing parity information for each data byte. As in all synchronous systems, data are only sampled on an edge of the system clock, a feature that enhances noise immunity.

System Control.
There are ten system control lines (SC0 to SC9) which provide basic handshake information to assist data transfer, e.g., requester is ready or replier is not ready. In addition, these lines are used to convey information such as data width, data error indication, and whether the bus is in the request or reply phase.

Exception Signals.
Two exception signals, bus error (BUSERR) and time-out (TIMOUT), are provided. Any card detecting a data integrity problem must report this to all other cards using BUSERR. The CSM generates a TIMOUT when it detects a data communications hang up on the bus.

Arbitration Group.
The arbitration signals, which determine which card gains control of the bus, consist of a single bus request line (BREQ) and six bus arbitration lines (ARB0 to ARB5). To request exclusive use of the bus, a card asserts the BREQ line and provides it arbitration ID on ARB0 to ARB4. It also uses ARB5 to indicate whether the request has high priority or whether “fairness mode” is acceptable. In this latter mode the card will not make another request until after all other requesters have used the bus. Each card has the same bus arbitration logic within its MPC chip. If several cards make high-priority requests, the order of access is determined by the numerical value of the arbitration ID.

Message Passing on Multibus II

Multibus II uses message passing to implement block data transfers and interrupt requests. Each MPC chip contains a first-in first-out (FIFO) buffer, which ensures that the full bandwidth of the bus can be utilized by storing data immediately before or after transfer. In "solicited" transfers, the MPCs cooperate by warning each other that a message is to be sent. These messages may be sent as 32-byte packets in a manner that is analogous to the block transfer mechanism in VME. "Unsolicited" packets not expected by the receiving MPC are used to set up a block transfer or to act as the equivalent of interrupt request signals.

System Configuration

Multibus II employs a software configuration approach in which information such as the base memory address and arbitration ID are sent down the bus rather than by the use of jumpers or dip switches. Some information such as card type and serial number are coded on the cards themselves.

Other System Buses for Small-Scale Systems

The G64/G96 and STE standards [3] are examples of buses well suited to small industrial real-time instrumentation and control systems because of their relatively simple and hence lower-cost bus interfaces, compared with VME and Multibus. Both buses support DMA, have multiprocessor bus arbitration, and use single-height Eurocard (100 × 160 mm) cards. Prototyping cards fitted with bus interfaces are readily available and may be used to develop custom designed I/O cards. Power is supplied on the buses at +5 and ±12 V. In addition, the real-time multitasking operating system OS-9 has been ported onto SBCs which operate with these buses. Development systems that support the use of MS-DOS are also available, thus providing access to a wide range of PC-based software especially for graphics applications.

The G64 bus, which was defined by the Swiss company Gespac in 1979, specifies 64 bus lines which are mapped to rows of a DIN41612 connector. The bus has a 16-bit nonmultiplexed data bus and a 16-bit address bus; 32-bit transfers may be achieved by multiplexing the upper 16 bits of the data bus with the address bus. The G96 standard adds a further 32 bus lines by making use of the third row of a DIN41612 connector to extend the address bus width to 24 bits and provide additional interrupt request lines. The STE bus provides an unmultiplexed 8-bit data and 20-bit address bus using 64-pin DIN41612 connectors.

26.4 Personal Computer Buses

There are three main bus standards for personal computers — industry standard architecture (ISA), extended ISA (EISA), and the microchannel architecture (MCA). In addition, the Personal Computer Memory Card International Association (PCMCIA) architecture has been developed primarily for use in laptop and notebook computers.

ISA and EISA are pin compatible and are both synchronous buses with a clock rate of 8 MHz regardless of the clock rate of the main processor, whereas the MCA bus is an asynchronous bus. Slow slave add-on cards can utilize the ISA and EISA buses by using an appropriate number of wait states. The MCA architecture will not be covered in this chapter. Further details of these bus architectures are given in References 4 through 6.

ISA Bus

The original IBM PC and its clones used the standard PC bus which supported 8 data bus and 20 address bus lines and employed a 62-pin printed circuit card edge connector. When IBM introduced the PC-AT, a second 36-pin connector was added to the motherboard backplane slots to provide a 16-bit data bus and increase the number of address lines to 24. This new bus subsequently became known as the ISA bus and is capable of supporting both cards designed for the original PC bus, and cards with two connectors providing full 16-bit data access. The bus master line allows a card to take over control of the bus; however, this is generally only suited to long-term takeovers. The ISA bus supports 8-bit and 16-bit DMA transfers allowing efficient transfer of blocks of data, e.g., while loading software from disk into RAM. The ISA bus supports the I/O addressing mode of the Intel 80 × 86 range of processors with an I/O address space of 768 locations in addition to the 256 locations reserved for the motherboard.

EISA Bus

The EISA bus provides full 32-bit data and 32-bit address bus lines. ISA cards can fit into EISA bus connectors; however, cards designed for the EISA standard have bilevel edge connectors, providing double the number of contacts as an ISA card. The presence of a notch in the EISA cards allows them to be inserted farther into the motherboard socket than the ISA cards and thus mate with the additional contacts. The EISA standard increases the maximum ISA data width in DMA transfer to 32 bits and also provides a much-enhanced bus master. The following features of EISA are worthy of note.

Bus Arbitration

All EISA systems have a central arbitration control (CAC) device on the motherboard. The CAC uses a multilevel rotating priority arbitration scheme. The top-priority level rotates around three customers, the DMA controller, the dynamic memory refresh controller, and, alternately, either the main CPU or one of the bus master cards. A device that does not make a request is skipped over in the rotation process. The bus masters take it in turns to gain access to the top-priority level. Whereas ISA supported a single bus request line, the EISA standard provides a dedicated pair of request lines (MREQ0 to 14) and acknowledge lines (MAK0 to 14) for each bus master. (*Note:* Although this allows 15 bus masters to be used, in many systems the chip set which implements the CAC supports bus masters in a limited number of the EISA sockets.) The CAC supports preemption, i.e., it allows a device making a request to capture control from another device if it is next in turn. A bus master card must release the bus within 64 bus clock cycles, whereas a DMA controller has 32 clock cycles to surrender the bus.

Input/Output

The EISA bus provides 768 additional I/O locations for each slot in addition to the 768 ISA I/O locations that may be accessed from any slot. EISA cards contain nonvolatile memory to store configuration information (see below), and a minimum of 340 bytes of the I/O address space of each slot is reserved for this purpose.

System Configuration

The EISA system provides support for automatic system configuration to replace the use of jumpers and dip switches to specify parameters such as the base memory address, interrupt request line number, or DMA channel number used by each add-on card. Information on the product type and manufacturer are stored on each EISA add-on card and is read by the CPU during system start-up, making it possible to identify the slots that are fitted with full EISA cards. Manufacturers of both ISA and EISA cards should supply a configuration file containing information on the programmable options that are available. When a system configuration program is run, an optimum configuration of the boards will be determined, and the configuration information written into the I/O space of each EISA card. For ISA cards, the user can be informed of the required jumper settings.

PCMCIA

The PCMCIA architecture was developed by the Personal Computer Memory Card International Association and the Japan Electronics Industry Development Association for removable add-on cards for laptop and notebook computers. Each PCMCIA socket has its own host bus adapter (HBA), which acts as an interface to the main computer bus. Cards may be plugged into PCMCIA sockets either before or after the computer has been powered up. There are three types of PC cards all with the same planar dimensions (54.00 × 85.6 mm) but with differing thicknesses, namely, 3.3 mm for type I, 5.0 mm for type II, and 10.5 mm for type III. Cards and sockets are keyed to prevent them from being inserted the wrong way around. The PCMCIA standard supports cards that operate at several possible voltage levels, i.e., 5.0 V cards, 3.3 V cards, or dual-voltage 5.0 V/3.3 V cards.

Configuration

PC cards contain configuration information called the card information structure (CIS) which is stored in nonvolatile memory. Cards may be configured either on system power-up or on insertion of the card into the socket (i.e., plug and play). Configuration is carried out using a form of device driver called an enabler. The enabler may make use of two additional software services called card services and socket services. Socket services, which may be contained in ROM on the PC or loaded from disk during power-up, provide function calls to allow the HBA to be configured to cooperate with the PC card. Card services, which may be an extension to the operating system or an installable device driver, act as a server for the enabler, which performs as a client. Card services provide a range of functions such as accessing the CIS of the card, requesting system resources required by the card, and telling the enabler that a card has been inserted or removed from the socket. Enablers may be classified as dedicated to one particular card or generic, i.e., designed for a range of cards. Note that early PCMCIA cards were not designed for use with card services.

The PCMCIA Socket Interface

The PCMCIA socket comprises a 68-pin connector with 26 address lines providing access to 64 MB of memory space and a 16-bit data bus. Two V_{cc} pins and four ground pins are supplied. The maximum current that can be supplied to the card is 1.0 A with a maximum of 0.5 A from each of the two power supply pins. Release 2.x sockets apply 5.0 V to the V_{cc} pins on power-up and reduce this to 3.3 V if the card has dual-voltage capability. Cards that operate at 3.3 V only are keyed so they cannot fit into this type of socket and only into low-voltage sockets. The supply voltage provided by low-voltage sockets depends on the logic state of its voltage sense inputs. A PCMCIA socket can be configured either as a *memory only socket* or as a *memory or I/O socket.* Initially, the socket acts as a memory only socket but is converted by the enabler to a memory or I/O socket if the card is required to support I/O functions. In this mode the card can generate an interrupt request via a single IRQ pin and support both 8-bit and 16-bit I/O data transfers. DMA may be supported but not by Release 2.x systems.

PC/104

The enormous popularity of PC architecture resulted in its use in embedded systems. A need then arose for a more compact implementation of the ISA bus that could accommodate the reduced space and power requirements of embedded applications. The PC/104 specification (1992) was adopted as the base for an IEEE draft standard called the P996.1 Standard for Compact Embedded PC Modules. The key features of the PC/104 are

- Size reduced to 90 × 96 mm (3.550 × 3.775 in.);
- Self-stacking bus allowing modules to be "piggy-backed" and eliminating backplanes or cardframes;
- Rugged 64-contact and 40-contact male and female headers replacing the PC edge connectors (64 + 40 = 104, hence PC/**104**);
- Lower power consumption (<2 W per module).

PC/104 CPU modules range from a basic 9.6 MHz, 8088 compatible XT with one serial port and a keyboard connector to a 100 MHz, 80486DX4 with four serial ports, parallel port, IDE disk controller, display controller, Ethernet adapter, keyboard port, and up to 64 MB of on-board RAM. Pentium-based systems are also available. Systems can be customized from a wide range of modules, including data acquisition boards, solid-state disk modules, and LAN support, all in the same 3.6 by 3.8 in. stackable format.

TABLE 26.3 Manufacturers/Suppliers of Bus-Based Systems and Industrial PCs

System	Manufacturer/Supplier	System	Manufacturer/Supplier
VME	Wordsworth Technology Ltd. 6 Enterprise Way Edenbridge, Kent TN8 6HF U.K. Tel: +44 (0) 1732 866988	STE	Arcom Control Systems Units 8-10 Clifton Road Cambridge CB1 4WH, U.K. Tel +44 (0) 1223 411200
	PEP Modular Computers, Inc. 750 Holiday Drive, Building 9 Pittsburg, PA 15220 Tel: (412) 921-3322	G64	Gespac SA 18 Chemin des Aulx 1228 Geneva, Switzerland Tel: +41 (22) 794 34 00
	BVM Ltd, Hobb Lane Hedge End Southampton, SO30 0GH, U.K. Tel: +44 (0) 1489 780144		Altek Microcomponents Ltd. Lifetrend House Heyes Lane Alderley Edge, Cheshire SK9 7LW, U.K. Tel: +44 (0) 1625 584804
	Motorola, Inc. Computer Group 2900 S Diablo Way Tempe, AZ 85282 Tel: (800) 759-1107	Industrial PC	Blue Chip Technology Ltd. Chowley Oak, Tattenhall Chester, Cheshire CH3 9EX, U.K. Tel: +44 (0) 1829 772 000
VXI	National Instruments 6504 Bridge Point Parkway Austin, TX 78730-5039 Tel: (512) 794-0100		Capax Industrial PC Systems Ltd. Airport House, Purley Way Croydon, Surrey, CR0 0XZ, U.K. Tel +44 (0) 181 667 9000
Multibus	Tadpole Technology, Inc. 2001 Gateway Place, Suite 550 West, San Jose, CA 95110 Tel: (408)441-7920	PC/104	ComputerBoards, Inc. 125 High Street Mansfield, MA 02048 Tel: (508) 261-1123
	Intel Corp. 3065 Bowers Avenue Santa Clara, CA 95051		Diamond Point International (Europe) Ltd. Unit 9, North Point Business Estate, Enterprise Close Rochester, Kent ME2 4LY, U.K. Tel +44 (0)1634 718100
	Syntel Microsystems Queens Mill Road Huddersfield, HD1 3PG, U.K. Tel: +44 (0) 1484 535101		

The wide availability of software development tools for the PC, the large number of software developers familiar with the PC environment, and the ease of transferring software developed on a conventional PC to the PC/104 make this an increasingly popular format. Table 26.3 lists some manufacturers and suppliers of bus-based systems and industrial computers.

26.5 Peripherals

Computer peripherals fall conveniently into two categories. The first category may be considered to be internal to the computer system and comprises cards plugged directly into a computer bus slot. The second category comprises instruments external to the computer but controlled by it. These external instruments are usually themselves "intelligent," being controlled by their own CPU, and are linked to the main computer by a serial (RS-232) line or the IEEE-488 bus (GPIB). Such instruments may normally be operated in a stand-alone mode in response to their front panel controls without the requirement of an external computer.

TABLE 26.4 Typical Internal Peripherals

Card Type	Facilities Offered
Display adapter	Provides video and graphics facilities
Serial communications adapter	Serial communication to a similar device using the RS232, RS422, or RS485 standard
IEEE-488 Adapter	Communication with intelligent instruments using the IEEE-488 bus (GPIB)
Digital input/output (I/O)	Individual I/O lines, normally grouped as an 8-bit byte, which may be used to provide logic high and low signals (output) or sense logic high or low signals (input); inputs and outputs may be optically isolated; outputs may drive relays
Counter/timer	Hardware counter timer allowing external pulses to be counted, digital waveforms generated or pulse widths measured; counter/timers are often available as an additional facility on digital I/O cards
Analog input	Converts an analog input voltage to an integer value which may be read by the computer; resolution typically 8, 10, 12, or 16 bits; input voltage range may be fixed or may be user selectable; conversion times vary from several seconds to tens of nanoseconds
Analog output	Produces an analog output voltage proportional to a digital input; resolution typically be 8, 10, 12, or 16 bits

Internal Cards

Internal cards are available to perform a wide range of functions. Table 26.4 provides a brief list of representative types. Note that both the serial interface and the IEEE-488 adapter required for the control of external "intelligent" peripherals will be fitted as internal cards. Almost all internal peripherals need to be configured before use to set up such parameters as the base address, the interrupts, and/or DMA channels used. This may involve setting jumpers or switches on the card or may be accomplished under software control using a configuration file supplied by the manufacturer. To operate the cards, bytes are written or read from appropriate addresses on the cards by the controlling SBC. The mechanism for doing this is discussed further in the section concerned with software for data acquisition. Further detail on some of the card types is given below.

Display Adapter

While graphics support is normally available as standard on a PC-based system, this is not the case with other bus-based systems such as VME or Multibus. These systems are provided with a serial port that may be connected to a terminal to provide a text-only dialogue with the operating system running on the SBC. In such circumstances the choice of display adapter will determined by the user requirements, taking into account the support for the device provided by any software packages that are to be used. If the user intends to write custom graphics software, it is essential to ensure that a graphics library is available from the vendor providing as a minimum line drawing, arc drawing, and block color fill facilities.

IEEE-488 Adapter

This device provides support for communications across the IEEE-488 bus or GPIB (general purpose-interface bus). The bus itself comprises eight data lines, five interface management lines and three handshake lines. Transfers are parallel, synchronous, and at rates up to 1.5 MB/s. The IEEE-488 bus and its applications are discussed further in the section on external instruments.

Serial Communications Adapter

This device provides support for serial communications using the RS-232, RS-422, or RS-485 standards. It is used to provide a text-based terminal for systems based on Multibus or VME and to communicate with "intelligent" instruments such as position controllers, multimeters, or storage oscilloscopes fitted with similar interfaces.

Serial adapters convert parallel data to and from a bit stream in which each data byte (5, 7, or 8 bits) is framed by a start bit, an optional parity bit, and one or more stop bits. The bit stream may be sent via a circuit consisting of only two wires at rates of up to 115,200 bits per second (commonly referred to as 115200 baud). Common bit rates are 300, 600, 1200, 2400, 4800, 9600, 19200, 28800, 38400, 57600,

115200 baud. Both the transmitting and receiving adapters must be configured, normally by software, for the same baud rate, number of data bits, stop bits, and parity. Some form of flow control to prevent a receiver from being overloaded with incoming data is essential. This is accomplished either by separate handshake lines (e.g., those denoted by RTS and CTS in the standard) or by software where the receiver sends a special control byte (XOFF) back to the transmitter telling it to stop sending until it receives a second control byte (XON) to reenable it. For historical reasons the RS-232 standard does not define a bidirectional handshake procedure, and manufacturers have been forced to implement their own schemes which are not always compatible with each other.

Serial communication between devices may be

Full duplex, where either device may transmit or receive data at any time;
Half duplex, where both devices are capable of transmission or reception but only one may transmit at any instant;
Simplex, where one device is a transmitter, one is a receiver, and data can only flow in a single direction.

RS-232, developed by the Electronics Industries Association (EIA), is the oldest standard, originally developed in the early 1960s, to allow mainframe computers to communicate with terminals via modems and telephone lines. This is the origin of the names of some of the connections (e.g., RI ring indicator, DCD data carrier detect), which have no relevance in the applications considered here. A related problem is that the standard expects that the devices being connected are data terminal equipment (DTE), at one end of the link, and data communication equipment (DCE) at the other. Computers and terminals are DTE, while modems are DCE. The most commonly used revision of the standard, RS-232-C (revisions D and E also exist), was made in 1969 and is still widely used. Serial communication is made using voltage levels in the region of ± 12 V, over distances up to 15 m (50 ft) at speeds up to 20000 baud.

RS-422, also developed by the EIA, is an enhancement of the RS-232 standard. Differential transmitters and receivers are employed which allow one transmitter to drive up to ten receivers, using a twisted-pair connection for each circuit, at bit rates up to 10 MBaud at distances up to 12 m (40 ft) or 100 kBaud at distances up to 1200 m (4000 ft). The RTS and CTS lines (defined in the standard) are used for flow control, while the RXD and TXD lines are used to transmit and receive data. Thus, a two-twisted-pair cable is required for duplex connection without hardware handshaking. A four-twisted-pair cable is required if hardware handshaking is used.

RS-485 is based on RS-422 and allows up to 32 driver/receiver pairs to be connected to a common data bus (two twisted pairs). Clearly, only one device can be allowed to transmit at any one time. The RTS circuit is used to disable the other transmitters connected to the bus if a device is required to transmit data. Handshaking is performed using software.

The serial interfaces on instruments are usually configured as DTE devices. We are faced with the problem of connecting one DTE device (the computer serial interface) to another (the instrument), which is not what the RS-232 standard was designed for. Furthermore, since the standard does not define a bidirectional handshake to control data flow, several incompatible handshaking schemes exist. A comprehensive survey of these is presented in Reference 7. A common solution to the DTE to DTE connection problem is the so-called *null modem*, which is nothing more than a specially wired cable. Figure 26.5 shows two such connection schemes. One requires the software handshaking procedure and the other implements a bidirectional hardware handshake. The reader should refer to Reference 7 for details of other schemes and for the definitions of the mnemonics used to label the connections.

Some common problems encountered in practice are

1. The received data are garbage. This is almost always due to the baud rate, parity, and number of stop bits not being the same at both ends of the link.
2. Data initially correct, but parts in the middle are missing. This is probably a handshake problem. The transmitting device is sending data faster than the receiver can process it.
3. No communications at all. Probably a handshake problem where the transmitter does not sense that the receiver is ready.

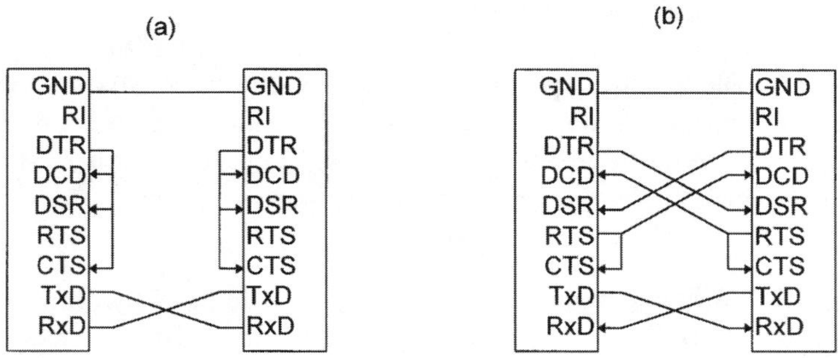

FIGURE 26.5 Two null modems for connecting DTE to DTE. In (a) all handshaking must be in software. The DTR line "fools" the serial interface that it is connected to the handshake lines of another device. Scheme (b) implements a hardware handshake. The DTR–DSR connection shows each device that the other is present. The RTS lines, connected to the DCD of the other device, which it can monitor, are used to control the flow of data in either direction.

Digital Input and Output

These cards provide I/O lines, normally in groups of eight, which may be used to sense or generate digital signals for devices outside the computer. A group of eight input lines is referred to as an (8-bit) input port and a group of eight output lines as an (8-bit) output port. Input and output levels vary from card to card and it is best to consult the appropriate data sheet. Typically, voltages between 2.5 and 5.0 V are considered as high logic levels, whereas voltages between 0.0 and 0.5 are considered as low logic levels. These levels are sometimes referred to loosely as TTL (transistor transistor logic) levels. Note that the actual logic levels used by the various TTL families differ from these slightly. The "high" and "low" ranges may be slightly different for input and output lines. Output lines often have limited current sourcing and sinking abilities compared with TTL, and it is therefore often necessary to buffer them. It is important that voltages exceeding the maximum rated values do not appear on inputs or outputs (e.g., attempting to switch an inductive load might produce a dangerously high transient voltage at an output); otherwise, the device may be damaged. Where this is likely to be a problem, inputs and outputs should be suitably buffered or even optically isolated, which provides protection up to a few kilovolts. Outputs may also drive appropriately connected relays. I/O cards with these facilities on board are readily available.

As a minimum, an I/O card may be expected to support a control register, two I/O ports each with an associated data register, and some handshake lines. Handshake lines may sometimes be used to generate interrupts on the controlling SBC. A byte written to the control register is used to configure the I/O ports, i.e., to determine if they are to behave as input or output ports as well as to select the function, if any, of the handshake lines. It may not be possible to select the direction of optically isolated or buffered ports. Writing a byte to an output port causes a pattern of high and low voltages to appear on the lines reflecting the pattern of zeros and ones in the binary representation of the byte written. Similarly, when a byte is read from an input port, the number read is specified in binary representation by the pattern of high and low logic levels on the input lines. The following example illustrates this.

The Intel 8255 Programmable Peripheral Interface (PPI) is commonly used in digital I/O cards for the PC. Data for this device are readily available [8]. The 8255 provides three ports, denoted A, B, and C, and a control register. Ports A and B may be designated as an 8-bit input port or an 8-bit output port. Port C may be considered as two independent 4-bit ports, which may be chosen independently as input or output. Port C may also provide handshake functions. There are three modes in which the chip may operate. The simplest, mode 0, which provides basic input and output without automatic handshaking, is used in our example. Note that the 8 bits of an I/O port are conventionally labeled bits 0 to 7. This is because bit 0 is weighted 2^0, bit 1 weighted 2^1, etc. in the binary representation of the number read from or written to the port.

To configure the PPI to operate in mode 0 with port A as an input port and port B as an output port the bit pattern 10011001 must be written to the control register. This number is equivalent to 99 in hexadecimal or 153 in decimal. Suppose now that switches connected to port A hold bits 2 and 7 high and the remaining bits low. The resulting binary pattern will be 10000100, which is equivalent to hexadecimal 84 or decimal 132. When port A is read, the number 132 (decimal) will therefore be obtained. To hold the lines connected to bits 2, 3, and 5 of port B high while leaving the remaining lines low, we see that the binary pattern 00101100 must appear at port B. 00101100 binary is equivalent to 2C hexadecimal or 44 decimal. We therefore write the decimal number 44 to port B.

Counter/Timers

These devices typically provide software-programmable event counting, pulse, and frequency measurement. As output devices, they may generate a single pulse (one-shot) when a programmable number of input pulses have been counted and produce square waves of arbitrary frequency and complex duty cycles. Frequencies generated are normally based on an on-board crystal clock to provide independence from the internal clock speed of the computer. Counter/timer cards commonly support at least three independent 16-bit counters.

Common applications include:

1. Alarms. The counter is in one-shot mode and generates a single pulse on timeout. This is connected to interrupt the computer and alert the user in the middle of the currently executing task.
2. Watchdog timer. This is used to detect problems, particularly in systems which are intended to operate without operator intervention. It is similar to the Alarm described earlier except that the interrupt is used to reset the computer. In normal operation this will never occur as all software tasks executing are designed to update the counter constantly so that it never reaches its terminal count. Only if a problem develops, e.g., a software "crash," will the counter time out and the system be reset.
3. The generation of complex waveforms, e.g., for pulse-width modulation. This application uses two counters in cascade, one (T1) to provide regular pulses at the carrier frequency triggering another (T2), in one-shot mode, to provide the variable duty cycle as shown in Figure 26.6.

Analog Input

An analog input card uses an analog-to-digital converter (ADC) that accepts an input voltage and supplies an integer proportional to that voltage to the computer. Many cards now are produced with on-board signal conditioning circuits that provide for variable gain either by means of switches or under program control. Cards with specialized signal conditioning circuits for common applications such as thermocouple linearization or interfacing to strain gages and other bridge sensors, are available. Signal conditioning to protect cards destined to be used in hostile electrical environments is also available. Cheaper cards may provide fixed gain and require additional signal conditioning circuits to be provided external to the computer. Many cards also feature multiplexed inputs where one of several inputs may be selected under program control to be fed to the ADC. Some important parameters to consider in selecting a analog input board are given in Table 26.5.

At rates above a few tens of kilohertz, interrupt-driven data capture is essential to maintain speed. Faster data rates require on-board memory to avoid degrading the performance of the controlling SBC. In this case DMA may be used to transfer data to main memory and increase performance further.

A timer function is often incorporated to allow samples to be taken at regular intervals independently of what the controlling SBC is doing. An interrupt is generated when the conversion is complete, and an interrupt service routine is then activated to read the result of the conversion into memory. The writing and installation of interrupt service routines is not a trivial task and is best left to those with an intimate knowledge of the operating system running on the SBC. Fortunately, most manufacturers supply software (device drivers) for this purpose. The simpler analog input boards may be driven by writing values directly to registers on them in a similar manner to the example given for digital I/O cards. Manufacturers now commonly provide a software library which may be called from a variety of high-level languages to allow the user to access the card in a more intuitive way. This is discussed further in the section on software.

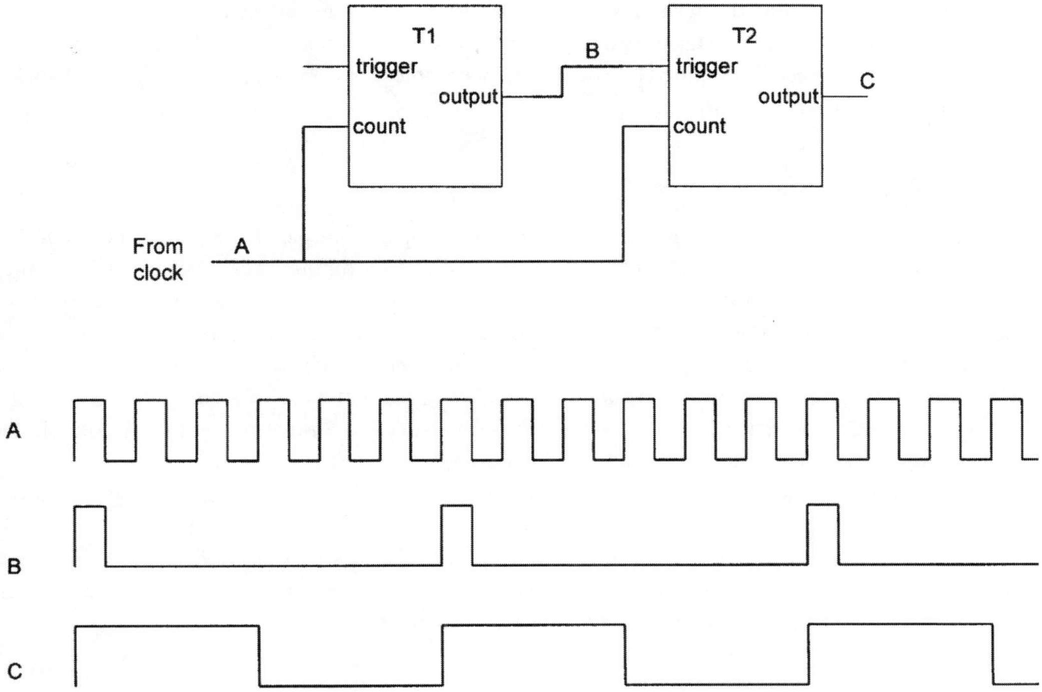

FIGURE 26.6 Using two timers to produce pulse-width modulation under software control. Both timers receive input pulses at a constant frequency from an external clock, as shown at A. Timer T1 operates in continuous mode. The trigger has no effect in this mode. The output of T1 is a single positive-going pulse when it has counted the specified (by software) number of input pulses, as shown at B. Timer T2 operates in one-shot mode. Each time it receives a trigger pulse, its output goes high for a specified (again by software) number of counts, as shown at C. In this way the frequency of the output at C is controlled by the count specified for T1 and the width of the positive-going part of C is controlled by the count specified for T2.

TABLE 26.5 Common ADC Parameters

Parameter	Description
Resolution	The smallest change in input detectable in the digital output; resolutions are commonly expressed in the number of significant bits in the digital output; hence, 8-bit resolution means 1 part in 256; 10-bit, 1 part in 1024, and 12-bit, 1 part in 4096
Linearity	The extent to which the output deviates from a linear relationship with the input; good devices will be linear to ±1 least significant bit; i.e., the output value is guaranteed to be within ±1 of an exactly linear conversion
Range	The maximum (and minimum) input voltages; inputs may be unipolar, e.g., 0–5 V or bipolar ±5 V; voltages are specified relative to ground unless the inputs are differential, e.g., those designed for bridge sensors
Conversion speed	The time taken to convert an input voltage into a digital output, typically 1 s to 1 μs; may also be quoted as a sample rate
Linearity	The extent to which the conversion is linear, e.g., a linearity of ±1 least-significant bit means that the output value is within ±1 of the ideal linear conversion
Input impedance	The impedance between the input terminal and ground or between differential inputs

Analog Output

Analog output cards are available as 8-, 10-, or 12-bit devices. Frequently, a card will support several channels of analog output with provision for delaying the updating of channels so that all can be updated simultaneously. Output voltages may be unipolar or bipolar and current outputs (4 to 20 mA) are also available. Signal conditioning (buffering) is necessary to drive loads drawing currents of more than a few milliamps. Special care should be taken with inductive loads, e.g., motors to avoid damage to the device by transient voltages. Specially designed position-control modules incorporating suitably buffered analog and digital I/O are available for this purpose.

External Peripherals

These are usually "intelligent" devices which can operate via their front panel controls without another computer but are additionally capable of being controlled by a computer. Many common laboratory instruments are available with such facilities, including power supplies, signal generators, storage oscilloscopes, voltmeters, spectrophotometers, and position controllers. A computer can coordinate the actions of several such instruments to gather then manipulate and display data in a way which enhances the power of the instrumentation system. An almost trivial example is the use of a computer to control a signal generator and a voltmeter in order to generate the frequency response of an amplifier automatically. Such a system has an obvious role in automatic testing rigs.

Two common methods are used to control such devices: a serial link or the IEEE-488 bus. In both cases the devices are controlled by sending messages consisting of sequences of ASCII characters. Usually the sequences are chosen to have an obvious meaning, as in the example that follows, but this is not always the case — particularly with older devices where user friendliness was often sacrificed as a result of limited memory and processing power! The message sequence

"FREQ10kHz"
"SINE"
"1.0VOLTRMS"

might be used to set a signal generator to produce a 10 kHz sinusoidal signal at 1 V rms. There is little standardization in the form of device messages used although the IEEE standard 488.2 goes some way in this direction. Responses from instruments are sent in the same way, i.e., as ASCII characters so that a voltmeter might respond to a command to make a measurement with the data

"AC2.01mV"

to indicate that it was on an ac voltage range and measured 2.1 mV. Again, there is little standardization in the format of responses. Large blocks of data may be send in a binary format where possible (8-bit serial links or IEEE-488 bus) to minimize the amount of data to be transferred.

Serial Devices

Serial control of devices is accomplished using links conforming to one of the serial standards (RS-232, RS-422, or RS-485) described elsewhere in this chapter. This is a relatively simple method of control and has the advantage that much of the preliminary testing and debugging of a system can be done using a terminal or a terminal emulator program such as the public domain KERMIT available from Kermit Distribution (Columbia University Academic Information Systems, 612 West 115th Street, New York, NY 10025, Tel: 212-854-3703). The writing of custom software that accesses the serial interface of the controlling computer is relatively easy under common operating systems including DOS, Windows, Windows 95, UNIX, and OS-9 using languages such as C, Pascal, or BASIC. It is increasingly common, particularly for DOS and Windows applications, for manufacturers to provide software support for their devices.

Disadvantages of serial transfers are the relative slowness when large amounts of data are transferred, the lack of standardization in device messages, and the limited control facilities available. Advantages are the ease of testing, the simplicity of the controlling software, the relative simplicity of the interconnection scheme, and — for remote instrumentation systems — the fact that with the use of modems data can be transferred over large distances using standard telephone lines or even a radio link.

IEEE 488 Devices

The IEEE standard 488 was developed in the 1970s and rapidly became an industry standard for the interconnection and control of test equipment. This standard was modified slightly in 1987 (IEEE standard 488.1) to allow for the considerable enhancements of IEEE standard 488.2 which was introduced at the same time [9,10]. The original IEEE standard 488 specifies the electrical characteristics of the bus, the mechanical characteristics of its connectors, and a set of messages to be passed between interfaces. It does not attempt to provide any syntax or structure for communicating these messages, to specify commonly used commands, or to establish a standard for device-specific messages. These issues are addressed in IEEE standard 488.2.

The bus itself supports synchronous parallel transfers of data using three groups of lines,

A bidirectional 8-bit data bus,
Five interface management lines, and
Three handshake lines,

over distances of up to 20 m and at data transfer rates of up to 1 MB/s.

Devices on the bus are classed as *talkers, listeners,* or *controllers.* In general, the computer system is the bus controller which can also talk (send data) or listen (receive data). Most devices are both talkers and listeners: for example, a digital voltmeter will be a listener when receiving instructions to set the voltage range prior to making a measurement but will be a talker when returning the result of the measurement to the controller, which is itself acting as a listener. Each device on the bus must be assigned a unique address which is a number between 0 and 30. This may be done from the front panel of the device or, less conveniently, by setting switches elsewhere on the device.

It is a difficult, time-consuming, and error-prone process to write software to drive an IEEE-488 card. Purchasers of new IEEE-488 interfaces are strongly advised to obtain a device driver from the manufacturer. Such device drivers are now readily available and integrate the card into the filing system of the operating system running on the controlling computer. This allows the interface to be accessed in a natural way from high-level languages running on the controller.

26.6 Software for Instrumentation Systems

The difficulty involved in writing software for instrumentation systems depends largely on the support available from the manufacturers of the subsystems, on the operating system (if any), and on the development tools available. On one extreme, one may be working in a virtual instrument environment, such as that provided by the National Instrument LabVIEW, where software development is entirely graphical and, for small projects at least, is readily undertaken by users with little or no prior experience. On the other extreme, one is faced with the problem of developing software, which is at the very least interrupt driven and probably multitasking, for a target SBC with no resident operating system; this requires considerable expertise in software design and development together with the availability of development tools such as cross compilers and source-level debuggers.

Virtual Instruments

Figure 26.7 shows a layer model of the software for a generalized instrumentation system. The application layer handles the data acquisition, analysis, and presentation. The instrument drivers provide a mechanism

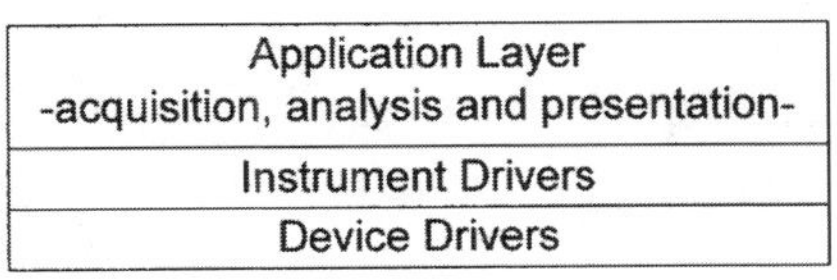

FIGURE 26.7 A layer model for instrumentation systems.

for communicating with the instruments in a standard way without requiring the user to know about the often cryptic data strings which need to be sent. For example, all digital multimeters will need the facility to chose a specific input voltage range. The range coding, resolution, etc. that have to be sent to the multimeter to achieve this will vary from instrument to instrument; however, the instrument driver allows the software writer programming in the application layer to call a procedure such as

SetVoltageRange(VoltageValue)

and this procedure call is the same for all multimeters. Although some manufacturers use the term slightly differently, the instrument driver is in effect the virtual instrument. Writing *instrument drivers* is a time-consuming but not too difficult task. Instrument drivers for proprietary instrumentation software design packages are readily available from instrument manufacturers. Device drivers integrate the controlling interface (e.g., IEEE-488, RS-232, or internal card) into the operating system of the computer. Writing a *device driver* requires a detailed knowledge of the device hardware and of the computer operating system. This is a difficult task and new interfaces should be purchased with a device driver appropriate for the operating system wherever possible.

A number of development environments which are based on the virtual instrument concept are now available. These free the user from the problems of writing conventional software to control instruments and handle the data produced. Instruments appear to the software developer as "front panels" drawn on the computer screen, complete with familiar buttons, knobs, and displays. Data flow is handled by linking instruments in a block diagram using a mouse in an environment that resembles an ordinary drawing package. The software developer is working only in the application layer.

While the graphical environment allows simple systems to be developed rapidly, experienced programmers may find it restrictive. There are software development systems available that give the programmer access libraries containing instrument drivers, data analysis routines, graphics functions, and data visualizations facilities in commonly used high-level languages such as C, Pascal, BASIC, and FORTRAN. Table 26.2 lists representative software packages.

Working Directly with Peripherals

It may occasionally be the case that the cost of software support for a virtual instrument development environment is not justified for a small application. Software must then be written to interface directly with the peripheral. The earlier section on digital input and output explained in principal what was necessary to program a simple interface chip. We now continue this example and show using the language C how this might be achieved.

The method of accessing the registers of peripheral cards depends on the microprocessor involved and may not even be a standard feature of the language being used. Where the peripheral forms part of the same address space as the computer memory, such as in the Motorola 680XX series, pointers can be used to read and write values in the registers. The Intel 80 × 86 series of processors often place peripherals in a separate address space which may not be accessed by pointers. In this case an extension to the language is required. Borland's Turbo C and C++ provide functions to read and write I/O mapped devices:

unsigned char inportb(int portid)
void outportb(int portid, unsigned char value)

These are used in the code fragment which implements the software for our earlier example. We assume that the 8255 PPI has base address 0 × 1b0 and that the program copies the value read from the input port (port A) directly to the output port (port B), until bit 0 of the input becomes zero

```
/ * define the addresses of the registers for the PPI */
#define BASE 0x1b0
#define PABASE
#define PBBASE+1
#define CONTROLBASE+3
```

```
unsigned char xin;/ * declare an 8-bit variable */
outportb(CONTROL, 153)/* set port A as input, port B as output */
do
{
          xin = inportb(PA);/ * read the input */
          outportb(PB,xin);/ * write the output */
}
while(xin & 1);/* loop if bit 0 is still 1 */
```

This fragment also illustrates that high-level languages generally only support input and output of bytes. Masking techniques must be used to access individual bits.

The Choice of Operating System

The software development support discussed so far is typically available under DOS, Windows 3.1, Windows 95, Windows NT, and UNIX. When a system is multitasking, it has to meet stringent real-time constraints; however, none of these operating systems is particularly appropriate. DOS does not support multitasking, and the others are not optimized for real-time systems which require speedy context switching and rapid interrupt response. The most fundamental requirement of real-time applications is that ability of the system to respond to external events with very short, bounded, and predictable delays. Table 26.6 lists some important real-time operating systems and kernels.

Real-time operating systems tend not to have the mature and powerful software development support available for conventional operating systems. It is not possible simply to develop the software on a familiar operating system and then transfer the working programs to the target system. Much of the debugging, testing, and system integration will have to be done on the target itself to access the hardware. A common solution is a development system in which a conventional workstation is linked to the target system. Software is developed on the workstation using familiar tools, e.g., a Windows-based editor, support for version control, and a powerful filing system. At any time, code can be cross-compiled (i.e., compiled for the processor on the target system) and downloaded to the target. The workstation may then monitor the execution of the software running on the target processor. Features that allow the user to single-step

TABLE 26.6 Some Important Real-Time Operating Systems and Kernels

System	OS/Kernel	Manufacturer
OS-9	OS	Microware Systems Corporation 1900 NW 114th Street, Des Moines, IA 50325
LynxOS	OS	Lynx Real-Time Systems, Inc. 16870 Lark Avenue Los Gatos, CA 95030-2315
VxWorks	OS	Wind River Systems 1010 Atlantic Avenue Alameda, CA 94501
VRTX/OS	OS	Microtec Research 2350 Mission College Blvd. Santa Clara, CA 95054
VRTX	Kernel	Microtec Research 2350 Mission College Blvd. Santa Clara, CA 95054
iRMX	Kernel	Intel Corp. 3065 Bowers Avenue Santa Clara, CA 95051

through the source code, seen in a workstation window, while viewing the status of key variables in another are available. It is also possible to set breakpoints and allow the processes to run until one is encountered. Manufacturers of real-time operating systems are often able to provide development support of this type for their product for a variety of workstations.

References

1. Micrology pbt, Inc., *The VMEbus Specification Manual: Revision C*1, PRINTEX, 1985.
2. S. Heath, *VMEbus: A Practical Companion*, Boston, MA: Butterworth Heinemann, 1993.
3. J. Di Giacomo, Ed., *Digital Bus Handbook*, New York: McGraw-Hill, 1990.
4. T. Shanley and D. Anderson, *ISA System Architecture,* 3rd ed., Reading, MA: Addison-Wesley, 1995.
5. T. Shanley and D. Anderson, *EISA System Architecture,* 2nd ed., Reading, MA: Addison-Wesley, 1995.
6. T. Shanley and D. Anderson, *PCMCIA System Architecture,* 2nd ed., Reading, MA: Addison-Wesley, 1995.
7. M.D. Seyer, *RS*232 *Made Easy*, Englewood Cliffs, NJ: Prentice-Hall, 1984.
8. Intel Corporation, *Intel Microsystems Component Handbook*, Vol. 2.
9. IEEE Std 488.1-1987, *IEEE Standard Digital Interface for Programmable Instrumentation.*
10. IEEE Std 488.2-1987, *IEEE Standard Codes, Formats, Protocols and Common Commands.*

27

Telemetry

Albert Lozano-Nieto
The Pennsylvania State University

27.1 Introduction

Telemetry is the science of gathering information at some remote location and transmitting the data to a convenient location to be examined and recorded. Telemetry can by done by different methods: optical, mechanical, hydraulic, electric, etc. The mechanical methods, either pneumatic or hydraulic have acceptable results for short distances and are used in environments that have a high level of electromagnetic interference and in those situations where, for security reasons, it is not possible to use electrical signals, for example, in explosive environments. More recently, use of optical fiber systems allows the measurement of broad bandwidth and high immunity to noise and interference. Other proposed telemetry systems are based on ultrasound, capacitive or magnetic coupling, and infrared radiation, although these methods are not routinely used. The discussion in this chapter will be limited to the most used systems: telemetry based on electric signals. The main advantage of electric over mechanical methods is that electrically based telemetry does not have practical limits regarding the distance between the measurement and the analysis areas, and can be easily adapted and upgraded in already existing infrastructures. Electric telemetry methods are further divided depending on the transmission channel that they use as wire telemetry and wireless (or radio) telemetry. Wire telemetry is technologically the simplest solution. The limitations of wire telemetry are the low bandwidth and low transmission speed that it can support. However, it is used when the transmission wires can use the already existing infrastructure, as, for example, in most electric power lines that are also used as wire telemetry carriers. Wireless telemetry is more complex than wire telemetry, as it requires a final radio frequency (RF) stage. Despite its complexity, it is widely used because it can transmit information over longer distances; thus, it is used in those applications in which the measurement area is not normally accessible. It can also transmit at higher speeds and have enough capacity to transmit several channels of information if necessary.

Figure 27.1 displays a generic telemetry system. It consists of (not all the blocks will be always present) (1) transducers to convert physical variables to be measured into electric signals that can be easily processed; (2) conditioning circuits to amplify the low-level signal from the transducer, limit its bandwidth, and adapt impedance levels; (3) a signal-processing circuit that sometimes can be integrated in the previous circuits; (4) a subcarrier oscillator whose signal will be modulated by the output of the different transducers once processed and adapted; (5) a codifier circuit, which can be a digital encoder, an analog modulator, or a digital modulator, that adapts the signal to the characteristics of the transmission channel, which is a wire or an antenna; (6) a radio transmitter, in wireless telemetry, modulated by the composite signal; (7) an impedance line adapter, in case of wire transmission, to adapt the characteristic

0-8493-1733-9/04/$0.00+$1.50

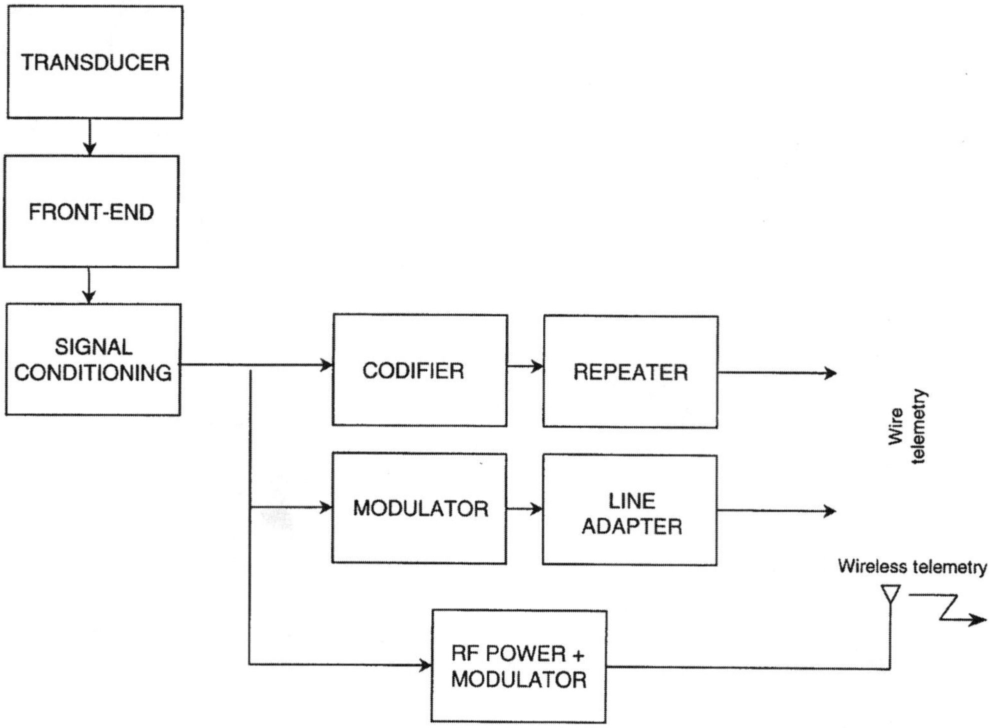

FIGURE 27.1 Block diagram for a telemetry system. Telemetry using wires can be performed in either base-band or by sending a modulated signal, while wireless telemetry uses an RF carrier and an antenna.

impedance of the line to the output impedance of the circuits connected to the adapter; and (8) for wireless communication, a transmitting antenna. The receiver end consists of similar modules. For wireless telemetry, these modules are (1) a receiving antenna designed for maximum efficiency in the RF band used; (2) a radio receiver with a demodulation scheme compatible with the modulation scheme; and (3) demodulation circuits for each of the transmitted channels. For wire telemetry, the antenna and the radio receiver are replaced by a generic front end to amplify the signal and adapt the line impedance to the input impedance of the circuits that follow. The transmission in telemetry systems, in particular wireless ones, is done by sending a signal whose analog variations in amplitude or frequency are a known function of the variations of the signals from the transducers. More recently, digital telemetry systems send data digitally as a finite set of symbols, each one representing one of the possible finite values of the composite signals at the time that it was sampled. The effective communication distance in a wireless system is limited by the power radiated by the transmitting antenna, the sensitivity of the receiver and the bandwidth of the RF signal. As the bandwidth increases, the contribution of noise to the total signal also increases, and consequently more transmitted power is needed to maintain the same signal-to-noise ratio (SNR). This is one of the principal limitations of wireless telemetry systems. In some applications, the transmission to the receiver is done on base band, after the conditioning circuits. The advantage of base-band telemetry systems is their simplicity, although because of the base-band transmission, they are normally limited to only one channel at low speeds.

Not uncommonly, a measurement system needs to acquire either different types of signals or the same type of data at different locations in the process that is being monitored. These different information signals can be transmitted using the same common carrier by multiplexing the data signals. Multiplexing allows different signals to share the same channel. Multiplexing techniques are usually considered either frequency division multiplexing (FDM) or time division multiplexing (TDM). In FDM, different subcarrier frequencies are modulated by the different measurement channel signals, which causes the information

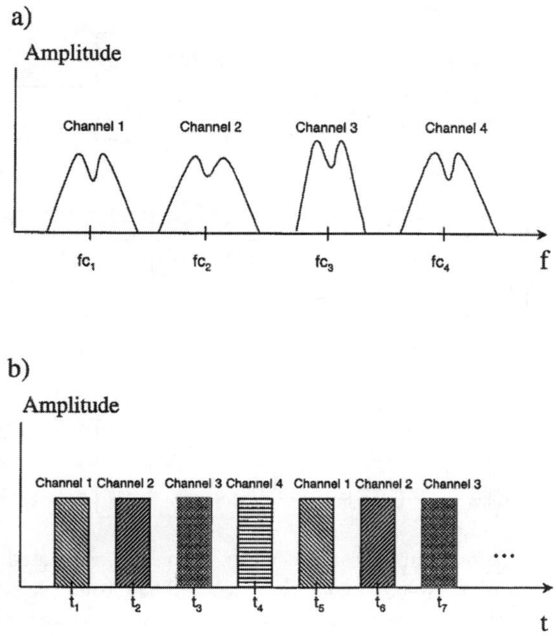

FIGURE 27.2 Basic characteristics of (a) FDM and (b) TDM signals. In FDM different channels are allocated at different subcarrier frequencies (f_{c1}, f_{c2}, …) while in TDM only one channel is transmitted at a given time. The remaining channels are transmitted sequentially.

spectrum to shift from base band to the subcarrier frequency. Then, the subcarrier frequencies modulate the RF carrier signal, which allows the transmission of all desired measurement channels simultaneously. In TDM, the whole channel is assigned entirely to each measurement channel, although only during a fraction of the time. TDM techniques use digital modulation to sample the different measurement channels at different times. Then, these samples are applied sequentially to modulate the RF carrier. Figure 27.2 illustrates these concepts by showing frequency and time graphs for FDM and TDM, respectively.

Almost all instrumentation and measurement situations are candidates for use of a telemetry link. Telemetry is widely used in space applications for either telemeasurement of a distant variable or telecommandment of actuators. In most of these types of applications, for example, in space telemetry, it is very important to design the telemetry systems to minimize the consumption of power [1]. Some land-mobile vehicles, such as trains, also use telemetry systems, either wireless or by using some of the existing power wires to transmit data to the central station and receive its commands [2]. In clinical practice, the telemetry of patients increases their quality of life and their mobility, as patients do not need to be connected to a measurement system to be monitored. Several medical applications are based on implanting a sensor in a patient and transmitting the data to be further analyzed and processed either by radio [3] or by adapted telephone lines [4] from the receiving station. Optical sensors and fiber-optic communications are used in industry to measure in environments where it is not desirable to have electric signals such as explosive atmospheres [5]. The designer of a telemetry system needs also to keep in mind the conditions in which the system will have to operate. In most of the applications, the telemetry systems must operate repeatedly without adjustment and calibration in a wide range of temperatures. Finally, as different telemetry systems are developed, the need to permit tests to be made interchangeable at all ranges increases, which require compatibility of transmitting, receiving, and signal-processing equipment at all ranges. For this reason, the Department of Defense Research and Development Squad created the Guided Missiles Committee, which formed the Working Group on Telemetry. This later became the Inter-Range Instrumentation Group (IRIG) that developed Telemetry Standards. Today, the IRIG Standard 106-96 is the primary Telemetry Standard used worldwide by both government and industry.

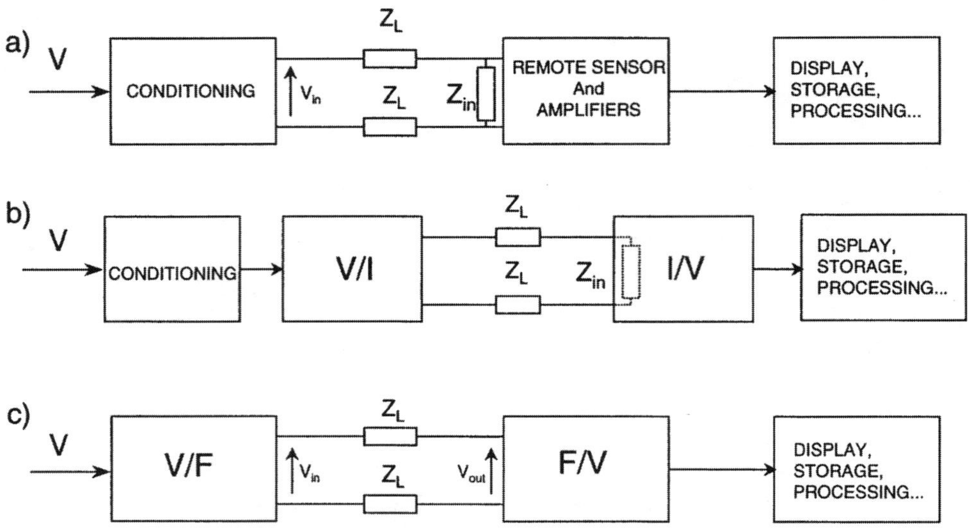

FIGURE 27.3 Different configurations for base-band telemetry. In voltage-based base-band telemetry (a) the information is transmitted as variations of a voltage signal. Current-based base-band telemetry (b) is based on sending a current signal instead of a voltage signal to neutralize the signal degradation due to the voltage divider made up by the input impedance of the receiver (Z_{in}) and the impedance of the lines (Z_L). In frequency-based base-band telemetry (c), the information is transmitted as variations of frequency which makes this system immune to noise and interference that affect the amplitude of the transmitted signal.

27.2 Base-Band Telemetry

Base-band telemetry uses a wire line to communicate the signal from the transducer after being processed and conditioned with the receiver. We will briefly describe telemetry systems based either on amplitude or frequency. More in-depth study of these base-band telemetry systems can be found in Reference 6.

Base-Band Telemetry Based on Amplitude

Voltage-Based Base-Band Telemetry

Figure 27.3a shows a simple voltage-based telemetry system. The signal from the transducer is amplified, normally to a voltage level between 1 and 15 V, and sent through a line consisting of two wires to the receiver. By making the low end of the scale 1 V, this system can detect short circuits [6]. The main problem of this configuration is the limitation on the transmission distance, which depends on the resistance of the line and the input resistance for the receiver. Also, the connecting wires form a loop that is very susceptible to interference from parasitic signals.

Current-Based Base-Band Telemetry

The limitation on transmission distance of the voltage-based system due to the impedance of the line are solved by using a current signal instead of a voltage, as is shown in Figure 27.3b. This requires an additional conversion module after the signal-processing circuits from voltage to current. At the receiver end, the signal is detected by measuring the voltage across a resistor. The most-used system in industry is the 4 to 20 mA loop. This means that 0 V is transmitted as 4 mA, while the highest voltage value is transmitted as a 20 mA current. The advantage of transmitting 4 mA for 0 V is the easy detection of an open circuit in the loop (0 mA). Other standard current values are 0 to 5, 0 to 20, 10 to 50, 1 to 5, and 2 to 10 mA. Also, voltage drops due to resistance of the wires do not affect the transmitted signal, which allows the use of thinner wires. Because this is a current mode, the parasitic voltages induced in the line do not affect the signal either. Current-based telemetry allows the use of grounded or floating transmitters with few modifications [6].

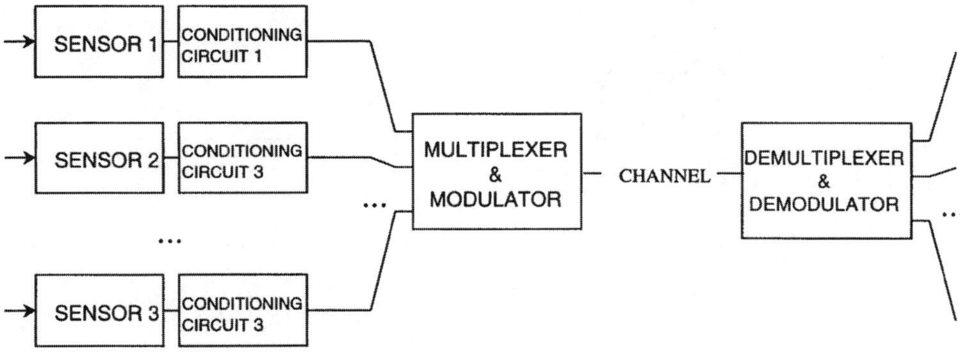

FIGURE 27.4 In multiple-channel telemetry a common transmission channel is used to transmit the measured signals from different channels using different sharing schemes.

Base-Band Telemetry Based on Frequency

Frequency-based transmission is known to have higher immunity to noise than amplitude-based transmission. Frequency-based telemetry, shown in Figure 27.3c, is used in the presence of inductive or capacitive interference due to its immunity to noise. It also offers the possibility of isolating the receiver from the transmitter. The signal at the output of the conditioning circuit modifies the frequency of the telemetry signal, normally using a voltage-to-frequency converter. In the receiver, a frequency-to-voltage converter performs the opposite function. A special case of frequency-based telemetry is pulse telemetry, in which the modulating signal changes some characteristics of a train of pulses. Because of its importance and widespread use, pulse telemetry will be analyzed in depth in the following sections.

27.3 Multiple-Channel Telemetry

Most of the industrial processes in which telemetry is used require the measurement of different physical variables to control the process, the measurement of only one physical variable at different locations, or normally a combination of both. In these multiple-channel measurements, base-band telemetry is not an option, as it would require building a different system for each channel. Multiple-channel telemetry is achieved by sharing a common resource (transmission channel), as is shown in Figure 27.4. The sharing of the transmission channel by all the measurement channels is designated by *multiplexing*. There are two basic multiplexing techniques: FDM and TDM. In FDM, different channels are assigned to different spectral bands and the composite signal is transmitted through the communication channel. In TDM, the information for different channels is transmitted sequentially through the communication channel.

Frequency Division Multiplexing

In FDM, shown in Figure 27.5a, each measurement channel modulates a sinusoidal signal of different frequency. These sinusoidal signals are called subcarriers. Each of the modulated signals is then low-pass-filtered to ensure that the bandwidth limits are observed. After the filtering stage, all the modulated signals are fed into a summing block, producing what is known as a base-band signal. A base-band signal indicates here that the final carrier has not yet been modulated. The spectrum of the base-band signal is shown in Figure 27.5b, where it is possible to see how each measurement channel spectrum signal is allocated its own frequency. This composite signal finally modulates a carrier signal whose frequency depends on the transmission medium that is used. The signal is then fed into a transmission wire (similar to TV-broadcasting systems by cable) or, more commonly, into an antenna in the case of wireless telemetry systems. In wireless telemetry, the frequency of the carrier cannot be chosen arbitrarily, but is chosen in accordance with international agreements on the use of the electromagnetic spectrum. In the U.S., the Federal Communications Commission (FCC) is the body that regulates the allocation of frequencies for

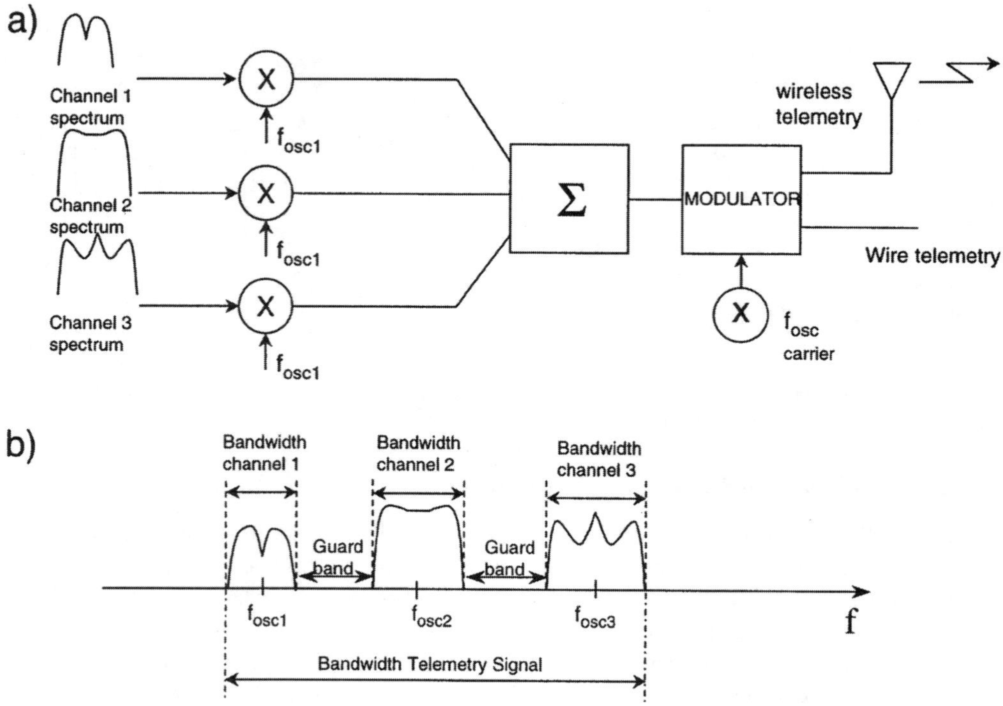

FIGURE 27.5 The different channels in an FDM system (a) are allocated at different subcarrier frequencies producing a composite signal shown in (b) that is later modulated by an RF frequency according to the transmission channel used. The guard bands limit the closeness of contiguous channels to avoid intermodulation and cross talk.

TABLE 27.1 Frequency Bands Allocated for Telemetry

Frequency Band, MHz	Uses	Notes
72–76	Biotelemetry	Low-power devices; restricted by Part 15 of FCC rules
88–108	Educational	Four frequencies in this band; Part 90 of FCC rules
154	Industry	Band in TV channels 7–13
174–216	Biotelemetry	Low-power operations restricted to hospitals
216–222	Multiple	BW < 200 kHz
450–470	General	Telemetry as secondary basis; limited to 2 W of RF
467	Industry	Business band; limited to 2 W of RF
458–468	Biotelemetry	Band in TV channels 21–29
512–566	Biotelemetry	Low-power operations restricted to hospitals
1427–1435	Fixed	Uses in land-mobile services (telemetering and telecommand)
1435–1535	Aeronautical	
2200–2290	Mobile	

different communication services. Table 27.1 shows the most common telemetry frequency bands and their intended use. Table 27.1 is for informational purposes only, and it is not a comprehensive guide to telemetry frequencies. To find the allowed telemetry frequencies for a specific application, the maximum power allowed, and other limitations, the reader should consult the applicable FCC documents [7,8]. The allocation of bands is a process subject to change. For example, in October 1997 the FCC assigned some of the TV channel bands for patient telemetry inside hospitals, with restricted power. The FCC publishes all changes that affect frequency bands or other technical characteristics for telemetry.

At the receiver end, the carrier demodulator detects and recovers the composite base-band signal. The next step is to separate each of the subcarriers, by feeding the signal into a bank of parallel passband filters. Each channel is further demodulated, recovering the information from the transducer. The main practical problem of FDM systems is the cross talk between channels. Cross talk appears due to the nonlinearities of the electronic devices, which originates when the signal for one channel partially modulates another subcarrier in addition to the one assigned to that channel. Cross talk also originates when the spectra for two adjacent channels overlap. To avoid this effect, the subcarriers have to be chosen so that there is a separation (guard band) between the spectra of two contiguous channels. By increasing the guard band, the possibility of cross talk decreases, but the effective bandwidth also increases. The effective bandwidth equals the sum of the bandwidth of all channels, plus the sum of all the guard bands.

There are three alternative methods for each of the two modulation processes: the modulation of the measurement channel signals and the modulation of the composite signal. These methods are amplitude modulation (AM), frequency modulation (FM), and phase modulation (PM). The usual combinations are FM/FM, FM/PM, or AM/FM [6]. Here, we will analyze only on the subcarrier modulation schemes, while the modulation for the RF signal is analyzed in Chapter 21.

Subcarrier Modulation Schemes for Frequency Division Multiplexing

Subcarrier Modulation of Amplitude.

In an AM subcarrier modulation scheme, the amplitude of a particular subcarrier signal is changed according to the value of the measured channel assigned to that frequency. The resulting AM signal is given by

$$v(t) = A_c\left[1 + m(t)\right]\cos(\omega_c t)$$

where A_c is the amplitude of the carrier, $m(t)$ the modulating signal, and ω_c the frequency of the carrier.

The advantage of this type of modulation is the simplicity of the circuits that perform the modulation and the circuits required for the demodulation, in order to recover the modulating signal that carries the desired information. The percentage of modulation denotes the extent to which a carrier has been amplitude modulated. Assuming for simplicity that the modulating signal is sinusoidal of frequency ω_m, such as

$$m(t) = m \times \cos(\omega_m t)$$

the percentage of modulation (P) can be found as

$$P = m \times 100(\%)$$

In a more general way, the percentage of modulation (P) is expressed as

$$\frac{P}{100\%} = \frac{A_{c(\max)} - A_{c(\min)}}{2A_c}$$

where $A_{c(\max)}$ and $A_{c(\min)}$ are the maximum and minimum values that the carrier signal achieves.

Figure 27.6 shows the spectrum of an amplitude-modulated signal, assuming that the modulating signal is a band-limited, nonperiodic signal of finite energy. Figure 27.6 shows that it consists of two sidebands that are symmetrical in reference to the subcarrier. Figure 27.6 shows the main disadvantages of AM schemes. First, the bandwidth of the modulated channel is two times the bandwidth of the modulating signal, due to the two similar sidebands that appear. This results in an inefficient use of the spectrum. Second, the analysis of power for each of the components in Figure 27.6 shows that at least 50% of the transmitted power is used in transmitting the subcarrier, which is independent of the measured signal, as it does not contain any information. The remaining power is split between the two sidebands, which results in a maximum efficiency that it is theoretically possible to achieve of below 25%. The third main problem of AM is the possibility of overmodulation, which occurs when $m > 1$. Once a signal is

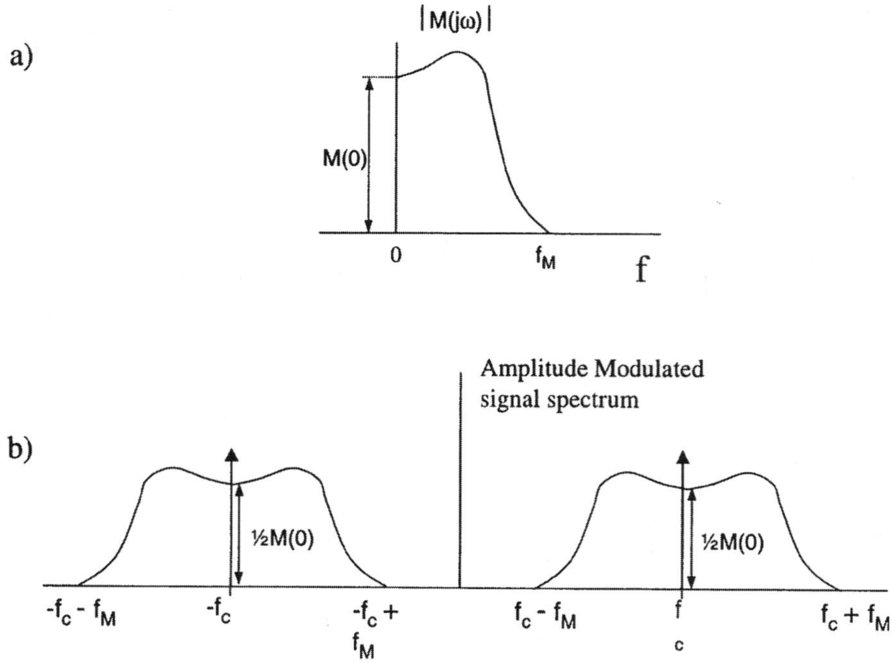

FIGURE 27.6 Resulting spectrum after amplitude modulation of a signal shown in (a). The resulting spectrum has doubled the required bandwidth, while only 0.25 of the total power is used in transmitting the desired information.

overmodulated, it is not possible to recover the modulating signal with the simple circuits that are widely used for AM telemetry transmission.

The limitations of AM subcarrier modulation can be overcome using more efficient modulation techniques such as double sideband (DSB), single sideband (SSB), and compatible single sideband (CSBB), which are also considered AM techniques. However, the complexity of these modulation systems and the cost associated with systems capable of recovering subcarrier signals modulated this way cause these not to be used in most commercial telemetry systems. Most of the available systems that use AM subcarrier techniques use the traditional AM that has been described here, because its simplicity overcomes the possible problems of its use.

Subcarrier Modulation of Frequency.

FM (or PM) is by far the most-used subcarrier modulation scheme in FDM telemetry systems. These angle modulations are inherently nonlinear, in contrast to AM. Angle modulation can be expressed as

$$v(t) = A\cos\left[\omega_c t + \phi(t)\right]$$

where $\phi(t)$ is the modulating signal, that is, the signal from the transducers after conditioning.

It is then possible to calculate the value of the instantaneous frequency as

$$f = \frac{1}{2\pi}\frac{d}{dt}\left[\omega_c t + \phi(t)\right] = \frac{\omega_c}{2\pi} + \frac{d}{dt}\phi(t)$$

This equation shows how the signal $v(t)$ is modulated in frequency. We can analyze two parameters that can be derived from the previous equations: frequency deviation and modulation index. Frequency deviation (f_m) is the maximum departure of the instantaneous frequency from the carrier frequency. The modulation index (β) is the maximum phase deviation. The following equations show how these parameters are related. The value of the instantaneous frequency (f) is [9]

$$f = \frac{\omega_c}{2\pi} + \frac{\beta\omega_m}{2\pi}\cos(\omega_m t) = f_c + \beta\, f_m \cos(\omega_m t)$$

The maximum frequency deviation is Δf and is given by

$$\Delta f = \beta f_m$$

Therefore, we can write the equation for the frequency modulated signal as

$$v(t) = A\cos\left[\omega_c t + \frac{\Delta f}{f_m}\sin(\omega_m t)\right]$$

The previous equation shows that the instantaneous frequency, f, lies in the range $f_c \pm \Delta f$. However, it does not mean that all the spectral components lie in this range. The spectrum of an angle-modulated waveform cannot be written as a simple equation. In the most simple case, when the modulating signal is a sinusoidal signal, a practical rule states that the bandwidth of an FM signal is twice the sum of the maximum frequency deviation and the modulating frequency. For modulating signals commonly found in measuring systems, the bandwidth is dependent upon the modulation index; that is, as the bandwidth allocated for each channel is limited, the modulation index will also be limited.

Frequency Division Multiplexing Telemetry Standards

IRIG Standard 106-96 is the most used for military and commercial telemetry, data acquisition, and recording systems by government and industry worldwide [10]. It recognizes two types of formats for FM in FDM systems: proportional-bandwidth modulation (PBW) and constant-bandwidth modulation (CBW). It also allows the combination of PBW and CBW channels. In PBW, the bandwidth for a channel is proportional to the subcarrier frequency. The standard recognizes three classes of subcarrier deviations: 7.5, 15, and 30%. There are 25 PBW channels with a deviation frequency of 7.5%, numbered 1 to 25. The lowest channel has a central frequency of 400 Hz, which means that the lower deviation frequency is 370 Hz and the upper deviation frequency is 430 Hz. The highest channel (channel 25) has a center frequency of 560,000 Hz (deviation from 518,000 to 602,000 Hz). The center frequencies have been chosen so that the ratio between the upper deviation limit for a given channel and the lower deviation limit for the next channel is around 1.2. There are 12 PBW channels with a deviation frequency of 15%, identified as A, B, … L. The center frequency for the lowest channel is 22,000 Hz (deviation from 18,700 Hz to 25,300 Hz), while the center frequency for the highest channel is 560,000 Hz (476,000 to 644,000 Hz), with a ratio for the center frequencies of adjacent channels being about 1.3. There are also 12 PBW channels for a deviation frequency of 30%, labeled from AA, BB, … to LL. The center frequency for these channels is the same as that for the 15% channels.

CBW channels keep the bandwidth constant and independent of its carrier frequency. There are eight possible maximum subcarrier frequency deviations labeled A (for 2 kHz deviation) to H (for 256 kHz deviation). The deviation frequency doubles from one group to the next. There are 22 A-channels, whose center frequency range from 8 to 176 kHz. The separation between adjacent channels is a constant of 8 kHz. Table 27.2 shows a summary of the characteristics of CBW channels.

IRIG Standard 106-96 gives in its appendix criteria for the use of the FDM Standards. It focuses on the limits, most of the time dependent on the hardware used, and performance trade-offs such as data accuracy for data bandwidth that may be required in the implementation of the system. The subcarrier deviation ratio determines the SNR for a channel. As a rule of thumb, the SNR varies as the three-halves power of the subcarrier deviation ratio. On the other hand, the number of subcarrier channels that can be used simultaneously to modulate an RF carrier is limited by the channel bandwidth of the RF carrier as well as considerations of SNR. Given a limited RF bandwidth, as more channels are added to the FDM system, it is necessary to reduce the deviation ratio for each channel, which reduces the SNR for each channel. It is then very important to evaluate the acceptable trade-off between the number of subcarrier

TABLE 27.2 Characteristics of Constant Bandwidth (CBW) Channels for FDM

Channel Denomination	Frequency Deviation, kHz	Lowest Channel Center Frequency, kHz	Highest Channel Center Frequency, kHz	No. of Channels	Separation between Channels, kHz
A	±2	8	176	22	8
B	±4	16	352	22	16
C	±8	32	704	22	32
D	±16	64	1408	22	64
E	±32	128	2816	22	128
F	±64	256	3840	15	256
G	±128	512	3584	7	512
H	±256	1024	3072	4	1024

channels and the acceptable SNR values. A general equation that might be used to estimate the thermal noise performance of an FM/FM channel is the following [11]:

$$\left(\frac{S}{N}\right)_{\mathrm{d}} = \left(\frac{S}{N}\right)_{\mathrm{c}} \left(\frac{3}{4}\right)^{1/2} \left(\frac{B_{\mathrm{c}}}{F_{\mathrm{ud}}}\right) \left(\frac{f_{\mathrm{dc}}}{f_{\mathrm{s}}}\right) \left(\frac{f_{\mathrm{ds}}}{F_{\mathrm{ud}}}\right)$$

where $(S/N)_{\mathrm{d}}$ represents the SNR at the discriminator output, $(S/N)_{\mathrm{c}}$ represents the SNR of the receiver, B_{c} is the intermediate-frequency bandwidth of the receiver, F_{ud} is the subcarrier discriminator output filter (at –3 dB), f_{s} is the subcarrier center frequency, f_{dc} is the carrier peak deviation for the subcarrier considered, and f_{ds} is the subcarrier peak deviation.

According to the Standard, the FM/FM composite FDM signal that is used to modulate an RF carrier can be of PBW format, CBW format, or a combination of both, with the only limitation that the guard bands between the channels used in the mixed format are equal to or greater than the guard bands for the same channels in an unmixed format.

Time Division Multiplexing

TDM is a transmission technique that divides the time into different slots, and assigns one slot to each measurement channel. In TDM, all the transmission bandwidth is assigned entirely to each measurement channel during a fraction of the time. After the signals from the measurement channels have been low-pass filtered, they are sequentially sampled by a digital switch that samples all the measurement channels in a period of time (T) that complies with the Nyquist criteria. Figure 27.7a shows a basic block diagram for an FDM system. The output of the sampler is a train of AM pulses that contains the individual samples for the channels framed periodically, as is shown in Figure 27.7b. Finally, the composite signal modulates an RF carrier. The set of samples from each one of the input channels is called a frame. For M measurement channels, the period between two consecutive pulses is $T_{\mathrm{s}}/M = 1/Mf_{\mathrm{s}}$, where T_{s} is the sampling period. The period between samples from the same channel is T_{s}. At the receiver end, by separating the digital signals into different channels by a synchronized demultipler and by low-pass filtering, it is possible to recover the original signal for each measurement channel.

TDM systems have advantages over FDM systems. First, FDM requires subcarrier modulators and demodulators for each channel, whereas in TDM only one multiplexer and demultiplexer are required. Second, TDM signals are resistant to the error sources that originate cross talk in FDM: nonideal filtering and cross modulation due to nonlinearities. In TDM, the separation between channels depends on the sampling system. However, because it is impossible in practice to produce perfectly square pulses, their rise and fall times are different from zero. It is then necessary to provide guard time between pulses, similar to the band guards in FDM systems. Cross talk in TDM can be easily estimated assuming that the pulse decay is exponential with a time constant (τ) approximately equal to

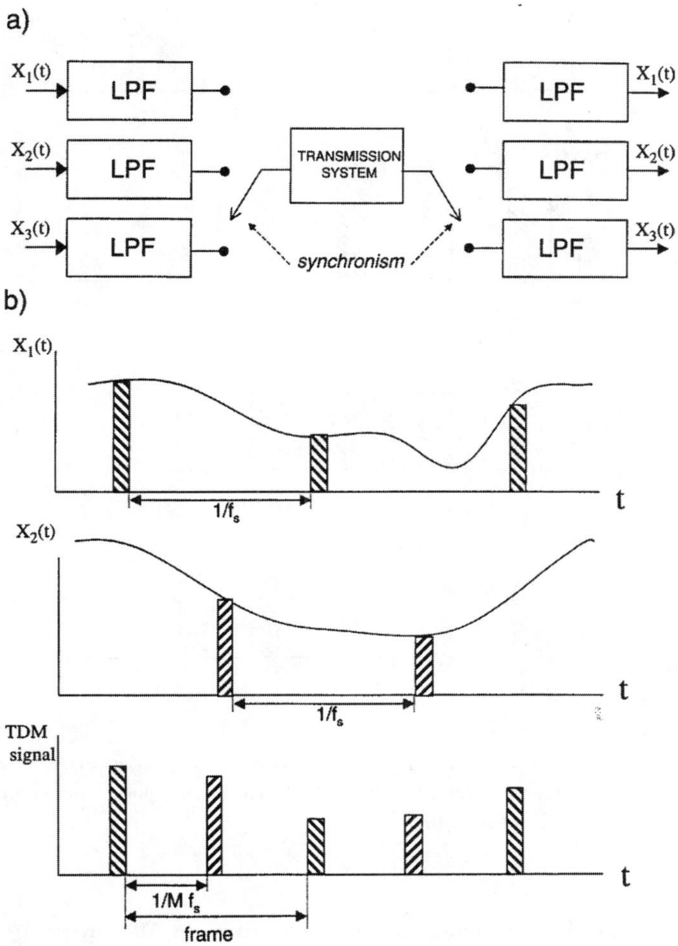

FIGURE 27.7 TDM systems (a) are based on sequentially sampling M different channels at a sampling frequency f_s, and sending the information for each channel sequentially (b). In TDM, the synchronism between transmitter and receiver is critical to recovery of the sampled signal. In this figure, the TDM signal is made of only two channels to increase readability. The blocks labeled LPF represent low-pass filters.

$$\tau = \frac{1}{2\pi B}$$

where B is the –3 dB channel bandwidth. The cross talk (k) between channels can be approximated as

$$k = -54.5T_g \text{ (dB)}$$

where T_g is the minimum time separation between channels, called guard time.

A common situation in measurement systems occurs when the M signals that need to be measured have very different speeds. The channel sampling rate is determined by the fastest signal, thus needing an M-input multiplexer capable of handling signals at that sampling frequency. A convenient solution is to feed several slow signals into one multiplexer, then combine its output with the fast signal in a second multiplexer [6].

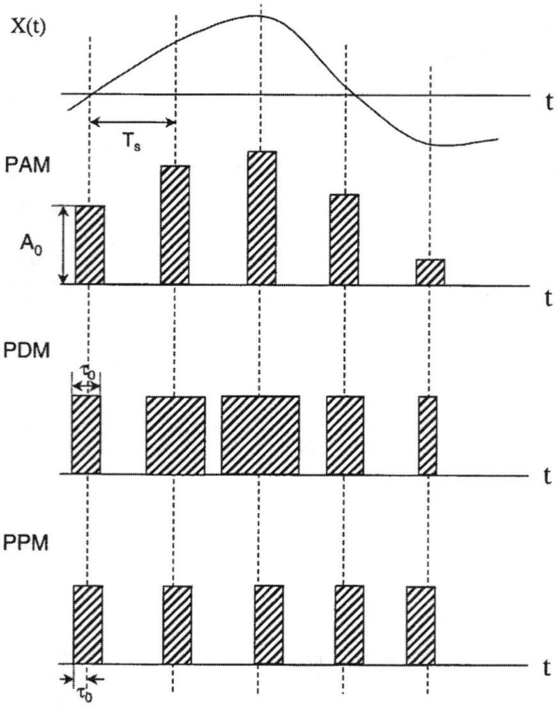

FIGURE 27.8 Different analog modulation schemes used in TDM. The variations in amplitude of the signal $x(t)$ are transmitted as amplitude variations of pulses (PAM), duration changes of pulses (PDM), or changes in the relative position of the pulses (PPM). In all the cases, the level 0 is transmitted by a pulse whose amplitude (A_0), duration (τ_0), or relative position (τ_0) is different from 0.

Analog Subcarrier Modulation Schemes for Time Division Multiplexing

In analog modulation for subcarriers the signal that results after the multiplexing and sampling process modulates a train of pulses. The most common methods for analog subcarrier modulation are pulse amplitude modulation (PAM), pulse duration modulation (PDM), and pulse position modulation (PPM). Figure 27.8 illustrates these three modulation schemes, where the pulses are shown square for simplicity. In analog modulation, the parameter that is modulated (amplitude, duration, or relative position) changes proportionally to the amplitude of the sampled signal. However, in PAM and PDM the values have an offset, so that when the value of the sample is zero, the pulse amplitude or the pulse width is different from zero. The reason for these offsets is to maintain the rate of the train of pulses constant, which is very important for synchronization purposes. The common characteristics of the different analog modulation schemes for pulses in TDM are (1) a modulated signal spectrum with a large low-frequency content, especially close to the sampling frequency; (2) the need to avoid overlaying between consecutive pulses in order to conserve the modulation parameters; and (3) the possibility of reconstructing the original samples from the modulated signal through low-pass filtering after demultiplexing. The reduction of noise depends on the bandwidth of the modulated signal, with this being the principal design criterion.

Pulse Amplitude Modulation.

PAM waveforms are made of unipolar, nonrectangular pulses whose amplitudes are proportional to the values of the samples. It is possible to define the modulation index using similar criteria as in analog AM. Similarly, in PAM the modulation index is limited to values less than 1.

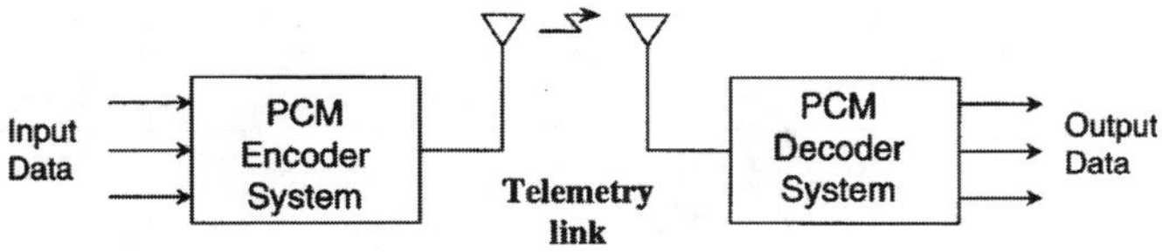

FIGURE 27.9 Block diagram showing a basic PCM link for telemetry.

Pulse Duration Modulation.
PDM is made of unipolar, rectangular pulses whose durations or widths depend on the values of the samples. The period between the center of two consecutive pulses is constant. The analysis of the resulting spectrum shows that it is possible to reconstruct the samples by low-pass filtering [9].

Pulse Position Modulation.
PPM is closely related to PDM, as PPM can be generated through PDM. In PPM the information resides on the time location of the pulses rather than in the pulses by themselves. It is then possible to transmit very narrow pulses to reduce the energy needed; this energy reduction is the most important advantage of PPM.

Pulse Code Modulation for Time Division Multiplexing.
All the previously analyzed subcarrier modulation schemes in telemetry systems are based on an analog signal that modulates either an analog carrier or a train of pulses. Pulse code modulation (PCM) is different: it is a digital modulation in which the measured signal is represented by a group of codified digital pulses. Two variations of PCM that are also often used are delta modulation (DM) and differential pulse code modulation (DPCM). In analog modulation schemes, the modulating signal from the transducer can take any value between the limits. If noise alters the modulating signal, it is impossible to decide its real value. Instead, if not all the values in the modulating signal are allowed, and the separation between the allowed levels is higher than the expected noise values, it is then possible to decide which were the values sent by the transmitter. This immunity against noise makes PCM systems one of the preferred alternatives for telemetry. Figure 27.9 shows the basic elements of a PCM telemetry system. A PCM encoder (or PCM commutator) converts the input data into a serial data format suitable for transmission through lines by wireless techniques. At the receiving end, a PCM decoder (or PCM decommutator) converts the serial data back into individual output data signals. PCM systems transmit data as a serial stream of digital words. The PCM encoder samples the input data and inserts the data words into a PCM frame. Words are assigned specific locations in the PCM frame, so the decoder can recover the data samples corresponding to each input signal. The simplest PCM frame consists of a frame synchronization word followed by a string of data words. The frame repeats continually to provide new data samples as the input data change. Frame synchronization enables the PCM decoder to locate the start of each frame easily.

Pulse Code Modulation Telemetry Standards.
IRIG Standard 106-96 also defines the characteristics of PCM transmission for telemetry purposes, in particular, the pulse train structure and system design characteristics. The PCM formats are divided into two classes for Standards purposes: class I and class II. The simpler types are class I, whereas the more complex types are class II. Some of the characteristics of class II systems are bit rates greater than 5 Mbit/s, word lengths in excess of 16 bits, fragmented words, unevenly spaced subcommutation, format changes, tagged data formats, asynchronous data transmission, and merger of multiple format types, among others. Table 27.3 provides a brief summary of relevant PCM specifications. Readers interested in the detailed specifications and descriptions should refer to Chapter 4 of the IRIG 106-96 Standard [10].

The following PCM codes, shown in Figure 27.10, are recognized by the IRG Standards: NRZ-L (nonreturn to zero — level), NRZ-M (nonreturn to zero — mark), NRZ-S (nonreturn to zero — space), BiØ-L (Biphase — level), BiØ-M (Bi-Phase — mark), and BiØ-S (Biphase — space). The Standard also

TABLE 27.3 Summary of the Most Relevant PCM Specifications According to IRIG 106-96

Specification	Class I	Class II
Class format support	Class I (simple formats) supported on all ranges	Class II (complex formats) requires concurrence of range involved
Primary bit representation (PCM codes)	NRZ-L, NRZ-M, NRZ-S, RNRZ-L, BiØ-L, BiØ-M, BiØ-S	Same as class II
Bit rate	10 bps to 5 Mbps	10 bps to > 5 Mbps
Bit rate accuracy and stability	0.1%	Same as class I
Bit jitter	0.1 bit	Same as class I
Bit numbering	MSB = bit number 1	Same as class I
Word length	4 to 16 bits	4 to 64 bits
Fragmented words	Not allowed	Up to 8 segments each; all segments in the same minor frame
Minor frame length	<8192 bits or <1024 words (includes synchro)	<16,384 bits (includes synchro)
Major frame length	<256 minor frames	Same as class I
Minor frame numbering	First minor frame in each major frame in number 1	Same as class I
Format change	Not allowed	Frame structure is specified by frame format identification (FFI) word in every minor frame

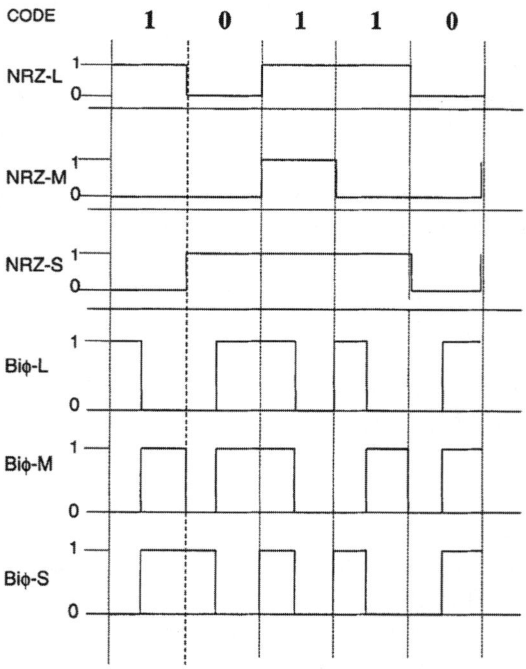

FIGURE 27.10 Different PCM codes. All lower levels in NRZ use a value different from zero. In biphase codes the information resides in the transitions rather than in the levels. In NRZ-L, a 1 is represented by the highest level, while a 0 is represented by a lower level. In NRZ-M, a 1 is represented by a change in level, while a 0 is represented by no change in level. In NRZ-S, a 1is represented by no change of level, while a 0 is represented by a change of level. In Biϕ-L, a 1 is represented by a transition to the lower level, while a 0 is represented by a transition to the higher level. In Biϕ-M, the 1 is represented by no change of level at the beginning of the bit period, while the 0 is represented by a change of level at the beginning of the bit period. In Biϕ-S, a 1 is represented by changing the level at the beginning of the bit period, while the 0 is represented by no change of level at the beginning of the bit period.

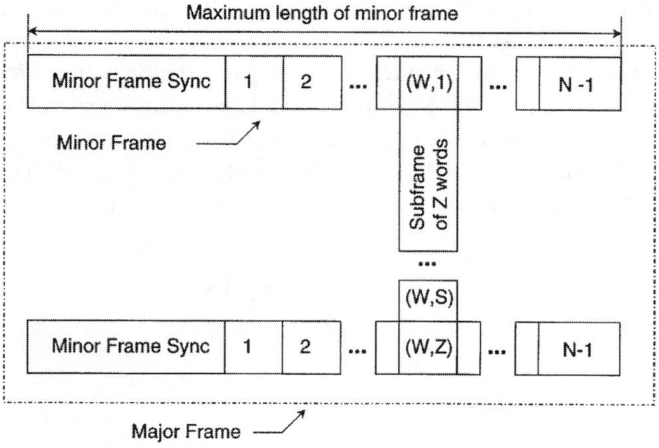

FIGURE 27.11 Structure of a PCM Frame. The maximum length of a minor frame is 8192 bits or 512 for class I and 16,284 bits for class II. A major frame contains $N \times Z$ words, where Z is the number of words in the maximum subframe, and N is the number of words in the minor frame. Regardless of its length, the minor frame synchronism is considered as one word. W is the word position in the minor frame, while S is the word position in the subframe.

recommends that the transmitted bit stream be continuous and contain sufficient transitions to ensure bit acquisition and continued bit synchronization. Bit rates should be at least 10 bits/s. If the bit rate is above 5 Mbit/s, the PCM system is classified as class II. In reference to the word formats, the Standard defines a fixed format as one that does not change during transmissions with regard to the frame structure, word length or location, commutation sequence, sample interval, or measurement list. Individual words may vary in length from 4 bits to not more than 16 bits in class I and not more than 64 bits in class II. Fragmented words, defined as a word divided into not more than eight segments and placed in various locations within a minor frame, are only allowed in class II. All word segments used to form a data word are constrained to the boundaries of a single minor frame. The Frame Structure allowed by the Standards for PCM telemetry specifies that data are formatted into fixed frame lengths, that contain a fixed number of equal-duration bit intervals. A minor frame is defined as the data structure in time sequence from the beginning of a minor frame synchronization pattern to the beginning of the next minor frame synchronization pattern. The minor frame length is the number of bit intervals from the beginning of the frame synchronization pattern to the beginning of the next synchronization pattern. The maximum length of a minor frame will not exceed 8192 bits nor 1024 words in class I and will not exceed 16,384 bits in class II. Minor frames consist of the synchronization pattern, data words, and subframe synchronization words if they are used. The Standard allows the use of words of different length if they are multiplexed in a single minor frame. Figure 27.11 shows a graphical representation of a PCM frame structure. Major frames contain the number of minor frames required to include one sample of every parameter in the format. Their length is defined as minor frame length multiplied by the number of minor frames contained in the major frame. The maximum number of minor frames per major frame is limited to 256.

Appendix C in the 106-96 IRIG Standard gives recommendations for maximal transmission efficiency in PCM telemetry. The intermediate-frequency (IF) bandwidth for PCM telemetry data receivers should be selected so that 90 to 99% of the transmitted power spectrum is within the receiver 3-dB bandwidth. The IF also has effects on the bit error probability (BEP) according to the following equation for NRZ-L PCM/FM [10]:

$$\mathrm{BEP} = 0.5e^{(k\,\mathrm{SNR})}$$

where $k \approx -0.7$ for IF bandwidth equal to bit rate
$k \approx -0.65$ for IF bandwidth equal to 1.2 times bit rate
$k \approx -0.55$ for IF bandwidth equal to 1.5 times bit rate

Other data codes and modulation techniques have different BEP vs. SNR performance characteristics, but in any case they will have similar trends.

The Standard also specifies the recommended frame synchronization patterns for general use in PCM telemetry. There are different lengths for synchronization patterns, but in all of them the 111 is the first bit sequence transmitted. The patterns for lengths 16 to 30 were selected in order to minimize the probability of false synchronization over the entire pattern overlap portion of the ground station frame synchronization [12]. The spectral density (S) for the NRZ and BiØ codes are:

$$\text{NRZ Codes} \qquad S = \frac{\sin^2(\pi f T)}{(\pi f T)^2}$$

$$\text{Biphase Codes} \qquad S = \frac{\sin^4(\pi f T/2)}{(\pi f T/2)^4}$$

The calculation of spectral densities allows the determination of the BEP for the previous type of codes assuming perfect bit synchronization. These calculations show that for the same SNR, the lowest BEP is achieved for NRZ-L and Bi codes, followed by NRZ and BiØ mark and space codes and finally for random NRZ-L codes (RNRZ-L).

Telemetry data are usually recorded onto magnetic tape for later analysis. When recording PCM data, it is important to ensure that the tape recorder provides sufficient frequency response to capture and reproduce the PCM signal. Useful rules to calculate the maximum bit rate for various PCM codes specify that for NRZ and RNRZ codes the maximum bit rate is 1.4 times the tape recorder frequency response, while for all biphase codes, the maximum rate is 0.7 times the tape recorder response. To limit the transmission bandwidth that PCM creates because it is a digital signal with sharp transitions, the PCM signal is usually passed through a premodulation filter before it is fed into the transmitter input. The filter cutoff frequency can be calculated as 0.7 times the PCM bit rate for NRZ and RNRZ codes and 1.4 times the PCM bit rate for all biphase codes.

Defining Terms

Bandwidth: The range of frequencies occupied by a signal.

Carrier: A frequency that is modulated by a signal containing information.

Channel: A subcarrier that carries information.

Constant bandwidth (CBW) channel: A channel whose bandwidth is independent of its carrier frequency.

Deviation ratio: The ratio of the maximum carrier frequency deviation to the maximum data frequency deviation.

Frequency deviation: The difference between the center frequency of a carrier and its upper or lower deviation limit.

Frequency division multiplexing (FDM): A composite signal consisting of a group of subcarriers arranged so that their frequencies do not overlap or interfere with each other.

Frequency response: The highest data frequency that can be carried by the channel.

IRIG: Inter-Range Instrumentation Group of the Range Commanders Council (RCC).

Proportional bandwidth (PBW) channel: A channel whose bandwidth is proportional to its carrier frequency.

Remote switching: Telemetry consisting only of yes/no or on/off orders.

Signaling: Telemetry consisting of binary information.

Subcarrier: A carrier combined with other carriers to create a composite signal.

Subcarrier bandwidth: The difference between the upper and lower frequencies of a modulated carrier.

References

1. B.P. Dagarin, R.K. Taenaka, and E.J. Stofel, Galileo probe battery system, *IEEE Aerospace Electron. Syst. Mag.*, 11 (6): 6–13, 1996.
2. M.W. Pollack, Communications-based signaling: advanced capability for mainline railroads, *IEEE Aerospace Electron. Syst. Mag.*, 11 (11): 13–18, 1996.
3. M.C. Shults, R.K. Rhodes, S.J. Updike, B.J. Gilligan, and W.N. Reining, A telemetry-instrumentation system for monitoring multiple subcutaneously implanted glucose sensors, *IEEE Trans. Biomed. Eng.*, 41: 937–942, 1994.
4. M. Rezazadeh and N.E. Evans, Multichannel physiological monitor plus simultaneous full-duplex speech channel using a dial-up telephone line, *IEEE Trans. Biomed. Eng.*, 37: 428–432, 1990.
5. Y.S. Trisno, P. Hsieh, and D. Wobschall, Optical pulses powered signal telemetry system for sensor network application, *IEEE Trans. Instrum. Meas.*, 39: 225–229, 1990.
6. R. Pallás-Areny and J.G. Webster, *Sensors and Signal Conditioning*, 2nd ed., New York: John Wiley & Sons, 2001, 352–379.
7. Part 90: Private land mobile radio services. Subpart J: Non voice and other specialized operations (Sec. 90.238: Telemetry), *Code of Federal Regulations*, Title 47, Telecommunication. Chapter I: Federal Communications Commission.
8. Part 15: Radio frequency devices, *Code of Federal Regulations*, Title 47, Telecommunication. Chapter I: Federal Communications Commission.
9. H. Taub and D. L. Schilling, *Principles of Communication Systems*, 2nd ed., New York: McGraw-Hill, 1986.
10. IRIG, *Telemetry Standard IRIG 106-96*, Range Commander Council, U.S. Army White Sands Missile Range, 1996.
11. K. M. Uglow, Noise and bandwidth in FM/FM radio telemetry, *IRE Trans. Telemetry Remote Control*, 19–22, 1957.
12. J.L. Maury and J. Styles, Development of optimum frame synchronization codes for Goddard space flight center PCM telemetry standards, *Proc. Natl. Telemetering Conf.*, 1964.

Further Information

E. H. Higman, Pneumatic instrumentation, in B.E. Noltingk, Ed., *Instrument Technology: Instrumentation Systems*, Vol. 4, London: Butterworths, 1987.

C.H. Hoeppner, Telemetry, in R.C. Dorf, Ed., *The Electrical Engineering Handbook*, Boca Raton, FL: CRC Press, 1993.

R.S. Mackay, *Bio-Medical Telemetry*, 2nd ed., Piscataway, NJ: IEEE Press, 1993.

Telemetry Group, *Test Methods for Telemetry Systems and Subsystems. Document 118-97*, Range Commanders Council, U.S. Army White Sands Missile Range, 1997.

Telemetry Group, *Telemetry Applications Handbook Document 119-88*, Range Commanders Council, U.S. Army White Sands Missile Range, 1988.

28

Sensor Networks and Communication

Robert M. Crovella

28.1 Introduction

What Is a Communication Network?

A communication network provides a system by which multiple users may share a single communication path (or medium) to exchange information. The telephone system is an example of a system containing many communication networks, which can be considered to be a single communication network as an abstract example. Communication networks are commonly used in various industries and applications to provide an economical means to allow multiple, geographically separated users to exchange information.

Ordinary Sensors vs. Networked Sensors

A definition of the function of a sensor is to map or convert one measured variable (e.g., spatial, mechanical, electromagnetic, etc.) into another — usually electric — variable or signal. This signal may then be passed to a measurement or processing system for capture and analysis, or as a direct input to some controlled process. In this case, the measured variable is represented as an electric signal. This signal must be handled individually by the measurement system, and it may also be subject to corruption from a variety of sources, such as electromagnetic interference in the case of an electric signal.

0-8493-1733-9/04/$0.00+$1.50

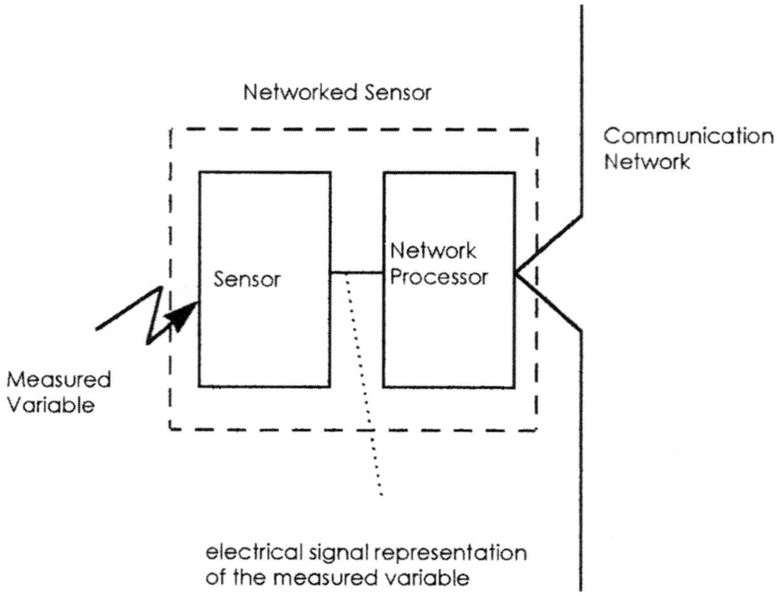

FIGURE 28.1 A networked sensor is an ordinary sensor with network communication components added.

In applications where a number of sensing devices are needed, and/or where the sensing devices are distributed geographically (or are distant from the measurement and analysis system), the application designer may wish to use a communication network to transmit sensor data from the measurement point to the measurement and analysis system. Such applications typically involve some sort of digital computing machinery at the measurement and analysis point, or as part of the control system. Figure 28.1 depicts a representative block diagram of a sensor/network system showing the relationship of the various components.

Networked sensors can be distinguished into two components: those performing the measurement function, and those components performing the communication function. In some cases, these two functions may be designed as a single unit, such that the "sensor" intrinsically includes communication capability. In other cases, an ordinary sensor may be connected to a conversion unit, which converts the output signal of the ordinary sensor into a form suitable for the network, and manages the delivery of this information on the network.

Why Use Networked Sensors?

Network communication combined with sensor technology can provide several benefits to an application, as well as to the sensor designer. The most obvious benefit of a network is the simplification of the wiring for the transmission of the signals from one place to another. For a system containing N users, the number of wires or cables T required to individually connect each user with each other user is given by Equation 28.1:

$$T = 2^{(N-1)} - 1 \tag{28.1}$$

assuming each wire or cable can carry information in both directions between the two users connected by that cable. For more than a few users, the number of cables required (T) to provide an individual connection between each pair of users is large. Sensors are often connected to a central measurement and analysis system, and may only need to communicate with the central system. In this case, the number of individual wires or cables needed is equal to the number of sensors ($N - 1$). Even with this smaller number of cables, the wiring for a large number of sensors in some applications can be quite complex. A network may be able to reduce the total number of cables required to a much smaller number. In fact, in a sensor network, all of the sensors and the central measurement and analysis system can be connected to a single cable.

An indirect benefit of networking may be in its handling of the sensor signal. Because most modern networks are digital in nature, an analog sensor signal typically must be digitized before it can be transmitted on a network. With a networked sensor, the digitization will typically be carried out by circuitry in relatively close proximity to the sensor. As a result, the analog signal will have traveled a short distance before being converted to a digital signal. This can be a benefit in two ways. The first is that the analog signal will not suffer as much attenuation or degradation due to electric losses associated with carrying a signal over a great distance. The second is that once in digital form, the "signal" can be made relatively immune to the effects of distortion or degradation due to electromagnetic interference (EMI). Although digital transmission of signals is still subject to EMI, modern protocols and transmission systems can be designed to be very robust, using signaling that is resistant to EMI as well as using error control techniques. As a result, the effect of attenuation and disturbances can be essentially eliminated by digital transmission of the signal.

Another benefit of networking is the ability to communicate a much wider range of information — in both directions — when compared with a single cable carrying a sensor signal. With many modern networks suitable for networked sensing applications, a microprocessor is used at the sensor to manage the handling of the sensor signal and its transmission on the network. But there is generally no need to limit the microprocessor to this one function alone. The combination of the network and the microprocessor provides a platform upon which many additional functions and features can be incorporated into the networked sensor. For example, the signal of a sensor may need a certain calibration or correction function applied to it before it can be used in calculations. It may be beneficial to load into the networked sensor (through the network) a set of correction parameters or coefficients, and then have the microprocessor correct or calibrate the output of the sensor before transmitting it to the network. Sensors can be easily designed to have multiple sensing functions, such as temperature and pressure, for example. Each signal can be handled separately and transmitted separately on the network, with no need for additional connections. Sensors may be designed to store certain types of information, such as the name of the manufacturer, or certain calibration parameters determined by the manufacturer at the time of manufacture. This information can then be read out over the network and used for a variety of purposes. A sensor can even be designed to have "intelligent" functions, such as the ability to sense its environment and determine when certain parameters have been exceeded (such as operating temperature range), or report a special message containing an "alarm" when the sensor signal level exceeds a certain threshold. The combination of the network and the microprocessor leads to an endless variety of functions and features that can be added to the basic sensor technology.

Potential Problems with Networked Sensors

Networked sensors will generally require more complex circuitry than equivalent, nonnetworked sensors. A drawback of analog-to-digital (A/D) conversion and digital transmission of signals is the time and level quantization effect that A/D conversion can have on the analog signal. These effects can be mitigated with modern, high-speed A/D converters (to minimize the effect of time quantization, or the sampling effect) with the ability to convert in high resolution (i.e., using a large number of digital bits to represent the analog signal level). These drawbacks are not unique to networked sensors but rather to digitized sensor values and digital control whether or not it uses a network. Finally, the capacity of the network to carry information (the bandwidth) must be considered in any communication system. Putting a large number of sensors on a single network may overload the information-carrying capability of the network, resulting in queuing delays in the reception of sensor signals and, in some cases, lost data.

28.2 Communication and Networking Concepts

In order to be able to select an appropriate network technology, it is necessary to understand some basic terminology so that the features and capabilities of various networks and technologies can be categorized and compared.

Station

A station represents a single communicating element on a network system. Each user of the network must access the communication capability of the network via a station. Each station will typically have some implementation of the open systems interconnection (OSI) network reference model as the means of utilizing the network system.

Media Access

Media access is the method by which individual stations determine when they are permitted to transmit, or "use" the media. Media access control (MAC) is a function that is usually performed in the data link layer of the OSI reference model. Some well-known methods of media access control include carrier sense multiple access with collision detection (CSMA/CD) and token passing. CSMA/CD systems (such as Ethernet) allow all stations on a network equal access. Each station must "listen" to the network to determine periods of inactivity before transmitting. Any station wishing to use the network may begin transmitting providing the network is inactive when it checks for activity. If multiple stations attempt to transmit simultaneously, a collision occurs. This is detected by all transmitting stations, which must all immediately stop transmitting and each wait a randomly determined period of time, before attempting to use the network again. Controller area network (CAN), for example, uses a variant of CSMA/CD for media access. Token-passing systems have a logical "token" which is exchanged among stations via network messaging. The station that holds the token has permission to transmit. All other stations are only permitted to receive messages. Stations wishing to transmit but not having the token must wait until the station holding the token passes it on. Another commonly used method of media access control is master–slave. In this method, one station on the network (designated the master) is generally in charge of, and originates, all communications. Slaves only respond to the master, and only respond when the master initiates communications with them via sending a message to the slave. Profibus-FMS (see below) is an example of a protocol which uses both token passing (in some cases) and master–slave (in some cases) to control media access.

Bandwidth

Bandwidth may have several different definitions. For digital communication systems, bandwidth describes the capacity of the system to transport digital data from one place to another. This term may be applied to the raw capability of the physical and data link layers to transport message data (*raw bandwidth*, closely related to the bit-rate concept) or it may be applied to the effective rate at which user-meaningful information is transported (*effective bandwidth*). The bandwidth of a given system is generally inversely proportional to the worst-case node-to-node distance. The smaller the network span, the higher its bandwidth can be.

Addressing

Addressing is a concept that assigns generally unique identifiers to each station in a network system. This identifier (the address) can then be used by the network for a variety of purposes, including identifying the origin and/or destination of messages, or arbitrating access to a shared communications medium. Another addressing or identifier concept assigns unique identifiers not to stations, but to unique pieces of data or signals that will be carried by the network. Stations then use an identifier according to what type of data they will be transmitting. Many, but not all networking methods require establishment of an explicit address for each network station.

Arbitration

Arbitration is a function closely related to MAC. Arbitration is used by some networks to define the procedure followed when multiple stations wish to use the network simultaneously.

Signaling

Signaling refers to the actual physical (e.g., electrical, optical, or other) representation of data as it is carried on the media. For example, in some networks, data elements may be represented by certain voltage levels or waveforms in the media. In other networks, data elements may be represented by the presence of certain wavelengths of light in the media. The association of all the representable data elements (e.g., 0/1 or on/off) with the corresponding signal representations in the media is the signaling scheme or method. An important signaling method where electric wires are used as the medium is differential signaling. Differential signaling represents a particular data element (1 or 0) as two different states on a pair of wires. Determining the data element requires measuring the voltage difference between the two wires, not the absolute level of the voltage on either wire. Different data elements are then represented by the (signed) voltage difference between the two wires. For example, RS-485 represents a digital 1 data element as a 5 V signal level on the first wire and a 0 V signal level on the second wire, and a digital 0 as a 0 V signal level on the first wire and 5 V signal level on the second wire. One of the principal benefits of differential signaling is that it is possible to determine the data being transmitted without knowing the ground reference potential of the transmitter. This allows the transmitter and receiver to operate reliably, even when they have different ground potentials (within limits), which is a common occurrence in communication systems.

Encoding

Encoding refers to the process of translating user-meaningful information into data elements or groups of data elements to be transported by the network system. A code book refers to the set of all relationships between user-meaningful information and data carried by the network. Encoding may occur at several levels within the OSI reference model, as user-meaningful information is transformed successively until it becomes an actual network message, produced by the data link layer. Decoding is the reverse process, whereby a network message is successively translated back into user-meaningful information.

Modulation

Modulation in a classical sense refers to a signaling technique by which data or information is used to control some combination of the frequency, phase, and/or amplitude of a carrier signal. The carrier signal carries the information to a remote receiver where it will be demodulated to retrieve the information. Modulated network systems are outside the scope of this chapter.

Message

A message is the fundamental, indivisible unit of information which is exchanged between stations. User-meaningful information will be grouped into one or more messages by the OSI network reference model.

Multiplexing

Multiplexing refers to the ability to use the media in a network to carry multiple messages or information streams "simultaneously." Multiplexed systems allow several communication channels to use the same physical wire or media. Each message or information stream may have different sources and destinations. Multiplexing may be accomplished using a variety of means. Time division multiplexing (TDM) involves breaking access to the media into a series of time quanta. During each time quantum, the media carries a separate message or information stream. The close arrangement of time quanta allows the network media to carry multiple messages "simultaneously." Code division multiplexing (CDM) involves the separation of the code book (see Encoding) into sections. Each section of the code book provides all of the messages that will be used for a particular information stream. Therefore, a particular information stream within the network media is distinguished by all of the messages that belong to the section of the code book for that stream. Frequency division multiplexing (FDM) divides an available bandwidth of a communication channel into several frequency ranges, and assigns one information stream to each frequency range.

Protocols

A protocol is a defined method of information exchange. Protocols typically are defined at several levels within the OSI network reference model, such as at the application layer and at the data link layer. Protocols are used to define how the services provided by a particular layer are to be exercised, and how the results of these services are to be interpreted.

Service

A service represents a specific function or operation that is supported by a particular layer in the OSI network reference model. For example, an application layer service might be provided for the reading of or writing to a data element contained in another device (or station) on the network. This service might make use of a data link layer service which might be provided for supporting the exchange of a message with another device (or station) on the network.

Topology

Topology refers to the physical or geographic layout or arrangement of a network. Certain types of canonical topologies are commonly discussed in the context of networks, such as trunkline/branchline, star (or hub), ring, and daisy chain.

Bit Rate

Bit rate refers to the speed at which binary pieces of information (bits) are transmitted on a particular network. The raw bit rate of a network generally refers to the actual speed of transmission of bits on the network. The effective bit rate — or throughput — generally refers to the speed at which user information is transmitted. This number is less than or equal to the raw bit rate, depending on what percentage of the bits transmitted is used for carrying user information. The bits not carrying user information are overhead, used to carry protocol, timing, or other network information.

Duplex (Half and Full Duplex)

Half duplex refers to a communication system in which a station can either transmit information or receive information, but not both simultaneously. A full duplex network allows a station to transmit information and receive information simultaneously.

Error Control

Many network systems provide mechanisms to control errors. Error control has four aspects: prevention, detection, correction, and isolation. Error prevention may simply be shielding for the media to minimize electromagnetic disturbances, or it may be more complicated, such as signal sampling control to optimize the probability that a signal will be in the correct state when sampled. Error detection generally depends on detecting violations of protocol rules at various network levels, or violations of computed data added to a message for error control purposes. Some examples of error detection techniques are parity and cyclic redundancy check (CRC). Both methods involve the computation of additional bits of information based on the data that is contained in a message, and appending these bits to the message. For example, a particular protocol may require that the data link layer compute and append a CRC to a message prior to transmission. The receiver of the message may then also compute the CRC and compare it to the CRC which has been appended to the message. If a mismatch exists, then it is assumed an error has occurred.

Error correction may take on a variety of forms. One of the simplest methods of error correction is to require that the data link layer of the transmitter retransmit a message which has been detected to have an error during transmission. This method is based on the assumption that the error was caused by a disturbance which is unlikely to occur again. Another method of error correction involves transmission of additional bits of information along with the user information in a message. These additional bits of information are computed by the transmitter to provide redundant information in the message. When fewer than a certain number of bit-level errors have occurred during the transmission of the message, the receiver is able to reconstruct the original user information accurately using the redundant information (bits) supplied within the message. Error isolation is a capability of some networks to localize the source of errors and isolate the sections of the network or the stations at which the errors have been localized. Error isolation allows the fault-free portions of the network to continue communicating even when other portions of the network have degraded to the point of generating errors.

Internetworking

There are occasions when communications between two or more points are best handled by multiple networks. This may be the case when a single network has limitations that prevent it from tying the points together (e.g., distance limits) or when multiple networks are required for other reasons (e.g., to carry different types of data). When multiple networks are used to provide communications, there may be a need to pass messages or information directly from one network to another.

A repeater may be used when the networks to be joined are logically identical, and the purpose is simply to extend the length of the network or extend its capabilities in some way. A repeater generally has no effect on messages, and simply carries all messages from one cable or port to another (i.e., a change of physical media). A repeater allows for connection of networks at the physical layer level.

A bridge is similar to a repeater, but allows for connection of networks at the data link layer level. Generally, a bridge will pass all messages from one network to another, by passing messages at the data link layer level.

A router usually has the function of partitioning similar networks. Two networks may be based on the same technologies and protocol, but may not be logically identical. In these cases, some, but not all, of the messages on one network may need to be carried or transported to the other network. The router has the function of determining which messages to pass back and forth based on certain rules. Functions to enable efficient, automatic routing of messages may be included in layer 3 (the network layer) of the OSI network reference model, and a router allows for connection of networks at the network layer level.

A gateway may have a function similar to a router, or it may have the function of joining dissimilar networks, i.e., networks based on dissimilar technologies and/or protocols. When functioning like a router, a gateway usually performs its discrimination at a higher protocol level than a router. When a gateway joins dissimilar networks, generally a more complex set of rules must be designed into the gateway so that message translation, mapping, and routing can occur within the gateway as it determines which messages to pass from one network to the other.

ISO/OSI Network Reference Model

The explosion in the use and types of communication networks over the last several decades has led to more precise descriptions and treatment of communication networks in general. The International Organization for Standardization (ISO) has recognized one such method of precise description of networks, called the OSI reference model [1]. As shown in Figure 28.2, this model decomposes an arbitrary communication network into a "stack" of seven "layers." At each layer, certain types of network communication functions are described. The user of the communication system — usually another system that needs to communicate on the network — interacts with layer 7, the highest layer. The actual transmission medium (e.g., copper cable, fiber optic, free space, etc.) is connected to layer 1, the lowest layer. Most communication networks do not implement all of the layers in the reference model. In this case, formal

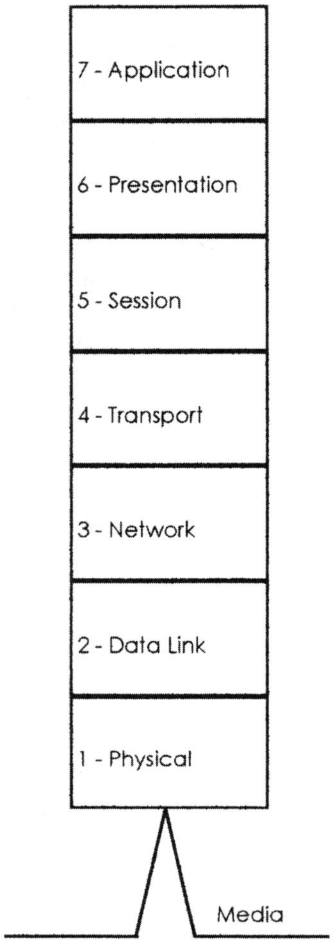

FIGURE 28.2 The ISO-OSI Seven-Layer Model provides a method for segmenting communication functions.

definition, treatment, or inclusion of certain layers of the model in the actual network design are omitted. Layers 1, 2, and 7 are typically present in all networks, but the other layers may only be explicitly included or identifiable when their function is an important part of the network communications. In many sensor communication networks, the functions performed by layers 3, 4, 5, and 6 are "collapsed" into vestigial additions to the functions of layer 7, the application layer.

Physical Layer

The physical layer is the lowest layer of the model. This layer is responsible for converting between the symbolic or data representation of the network messages and the actual physical representation of data in the network medium. This layer specifies the behavior of the electric circuits referred to as the transmitter and the receiver. It also defines physical structures for connectors.

Data Link Layer

The data link layer, or layer 2, is responsible for several functions. This layer manages access to the network medium (MAC), structures the bits of information into well-defined groups identified as "frames" or messages, handles identification of source and destination stations on the network, and provides for error-free transmission of a message from source to destination stations, all according to the data link layer protocol. A number of standard data link layer protocols exist, which act as the basis for many of the communication networks in wide use. Ethernet, or IEEE 802.3, for example, specifies a MAC sublayer that works with the IEEE 802.2 Logical Link Control layer to form the data link layer protocol used in the majority of office information networks [2].

Network Layer

The network layer encapsulates functions related to routing of messages, both within a single network and among multiple networks. This layer typically uses addressing in a variety of forms as a key part of the functions of directing and routing messages, and the search and usage of the available communication paths.

Transport Layer

The transport layer provides any additional data transfer functions not directly provided by the data link layer for end-to-end reliable messaging. For example, some data transfer functions between stations may require the use of multiple data link layer messages to accomplish a reliable message transfer. The generation of multiple messages and the sequential disassembly, delivery, and assembly of data is accomplished by the transport layer. The transport layer also recovers lost, duplicated, and misordered messages.

Session Layer

The session layer provides for a higher level of control and management of network usage and data flow than that provided at lower layers, including opening or building up a communication channel, maintaining the channel, and closing the channel. This layer is infrequently implemented in contemporary systems.

Presentation Layer

The presentation layer provides functions to transform data from formats that are transportable by the network to the user-accessible formats that are defined in the application layer and understood in the local station.

Application Layer

The application layer, or layer 7, provides communication services directly to the user application. The usage and formatting of these services is summarized in the application layer protocol. The user interacts with the network by invoking functions and services provided by the application layer and passing data to and from the network through these services.

28.3 Network Technologies

There is a wide range of technologies in various stages of development and standardization, which address virtually all levels or layers of the ISO/OSI network reference model. One or more of the available technologies will probably suit almost any networking need. An analysis of the available technologies and their limitations will also be beneficial if it is deemed that a networking method must be designed to meet a particular application. The selection and description of technologies is by no means complete or exhaustive. The technologies presented are selected from several industries which make common use of networking to communicate sensor data. Figure 28.4 provides a comparison of selected parameters for a set of networks.

RS-232

RS-232 (ANSI/EIA/TIA-232-E-91) is a widely used method of communication, which has been standardized in a variety of places including the Electronics Industry Association [3]. RS-232 represents elements of layer 1 of the OSI model, for communicating between two (and only two) stations. RS-232 provides a separate wire for transmission of data in each direction between the two stations, and gives the two stations different designations — data terminal equipment (DTE), and data communications equipment (DCE) — so that a method exists to distinguish which station will use which wire to transmit and receive. The signal levels for RS-232 represent a digital 1 bit as a voltage in the range of 5 to 12 V

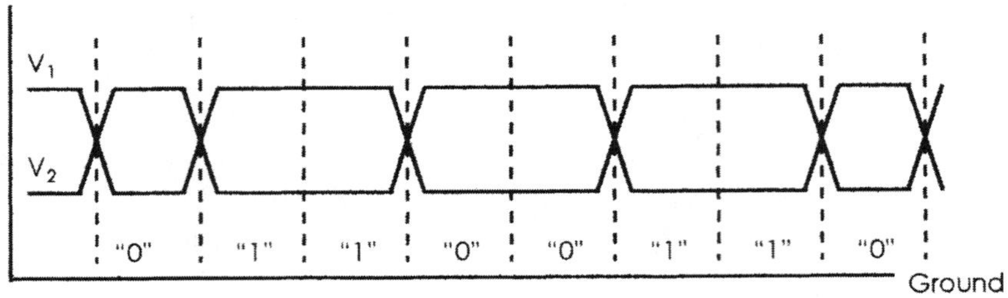

FIGURE 28.3 A sample RS-485 waveform showing voltages on differential wire pair (V_1, V_2) and superimposed bit intervals showing 0 and 1 bits. The ground reference is arbitrary within the defined signaling range.

	Length	Stations	Bit Rate	Wires	Media	Topology
RS-232	30 m	2	115 kb/s	2	TP	P-P
RS-485	1200 m	32	10 Mb/s	2	TP	D-C
Seriplex	1500 m	256	200 kb/s	4	2STP	D-C,Free
AS-i	100 m	32	167 kb/s	2	UP	T-B
Interbus-S	25.6 km*	64	500 kb/s	6	3STP	Ring
CAN	450 m	64	1 Mb/s	4	2STP	T-B
4-20 mA	1000 m	2	-	2	STP	P-P
HART	1000 m	2(15)	1200 b/s	2	STP	P-P(D-C)
Profibus	9600 m	126	12 Mb/s	2	STP	D-C
Found. Fieldbus	1900 m	32	2.5 Mb/s	2	STP	D-C
LonWorks	1400 m	64	1.2 Mb/s	2	STP	D-C,Free

FIGURE 28.4 A comparison of selected parameters (maximum values) for various network technologies. *Notes:* P-P = point to point; D-C = daisy-chain; T-B = trunkline-branchline; TP = twisted pair; STP = shielded twisted pair; UP = unshielded pair. * Maximum 400 m between stations. Maximum parameters for networks are not achievable simultaneously, and do not include repeaters, routers, or gateways. Maximum parameters are estimates based on available information.

on the wire, and a digital 0 bit as a voltage of negative 5 to 12 V on the wire. RS-232 is typically implemented in a full duplex fashion, since each station can transmit to the other simultaneously using separate wires. RS-232 can be made to operate at a variety of bit rates, but typically is used at bit rates from 300 bit/s up to 115,200 bit/s.

RS-485

EIA RS-485 was made a standard in 1983, derived from the RS-422 standard. RS-485 provides for differential transmission of data on a pair of wires among 32 or more stations. Like RS-232, the standard is a layer 1 specification. RS-485 provides for half duplex communication, since a station cannot simultaneously transmit and receive independent data streams. Each station in an RS-485 system can have either a transmitter or a receiver, or both (commonly called a transceiver). Most implementations provide a transceiver. When one transceiver is transmitting, all others should be receiving (i.e., not transmitting). Which station is allowed to transmit at which time is not specified in the standard, and must be covered by a higher layer protocol (e.g., Interbus-S, Profibus-DP). Figure 28.3 shows a sample RS-485 waveform, indicating the differential nature of the signaling.

Seriplex[1]

Seriplex® is a digital, serial multiplexing system developed by Automated Process Control, Inc., in Jackson, MS. Square D Corporation purchased Automated Process Control and the rights to Seriplex in 1995,

[1]Seriplex is a trademark of the Seriplex Technology Organization.

and subsequently launched Seriplex Technology Organization (STO) to manage the protocol. Seriplex is designed to be particularly efficient at handling large numbers of digital or on/off input and output points. Seriplex provides three communication wires, one for a clock signal, one for a data signal, and a ground reference. The system can be operated in two different modes (peer-to-peer and master–slave). In master–slave mode, one station is designated the master. The master synchronizes all data transmission among stations by driving a digital waveform on the clock line which all stations listen to and use for timing of transmit and receive operations. The master generates a repetitive pattern on the clock line which causes all stations to transmit and/or receive data on each cycle, or "scan" of the network. Each station is given an address, and uses the address along with the clock signal to determine when to drive the data line (in the case of an input point) or when to monitor the data line for valid output data (in the case of an output point). There are variations possible in implementation which allow for various clock speeds and bit rates (16, 100, and 200 kHz). Other protocol details allow for the handling of analog or multibit input and output points (by combining several bits on sequential scans together), bus fault detection, input redundancy, and communication error control using multiple scans of the network. Implementing the protocol in a sensor or other device typically requires using a Seriplex ASIC (Application Specific Integrated Circuit) which must be licensed from the STO [4].

AS-i

Actuator Sensor Interface, or AS-i, was developed by a consortium of primarily European companies interested in developing a low-cost, flexible method for connecting sensors and actuators at the lowest levels of industrial control systems. The system is managed by an independent worldwide organization [5]. The AS-i system provides a two-wire, nontwisted cable for interconnection of devices. Devices may draw current from the two wires (nominally at 24 V dc) for powering circuitry, and the data communications are modulated on top of the nominal dc level at a bit rate of 167 kHz, under the control of the master. A single parity bit per station is used for error detection. Similar to Seriplex, an AS-i device is typically implemented using a special ASIC which handles the communication.

Interbus-S

Interbus-S was developed by Phoenix Contact [6] and is controlled by the Interbus-S Club. The topology of the network is a ring, with data being sequentially shifted from point to point on the ring under the control of a network master. Each device in the ring acts as a shift register, transmitting and receiving data simultaneously at 500 kHz. The actual serial data transmission between stations conforms to RS-485. Interbus-S transmissions include a CRC for error detection. Interbus-S (Interbus-S Remote Bus) has also been extended to include a subprotocol called Interbus-Sensor Loop (or Interbus-S Local Bus). This subprotocol provides an alternate physical layer, with a single twisted pair carrying power and data on the same lines, and a reduction in the minimum size of the shift register in each station from 16 to 4 bits. Each Interbus sensor loop system can act as a single station on an Interbus-S network, or the sensor loop can be connected directly to a controller or master. Interbus-S devices are usually implemented with a special ASIC.

CAN

Controller Area Network (CAN) is a data link layer (layer 2) network technology developed by Robert Bosch Corporation [7], with an application target of onboard automotive networking. The technology is standardized in ISO 11898 [8], licensed to all major integrated circuit manufacturers, and is widely available — both as separate CAN controllers as well as CAN controllers integrated with microprocessors. As a result, CAN has been used in a variety of industries. As a data link layer technology, it is not a complete network definition. A number of physical layer options are usable with CAN (e.g., twisted pair, fiber optic, radio frequency wireless) and some have been subject to standardization (e.g., ISO 11898). Also, a number of application layer protocols have been developed for use with CAN, such as DeviceNet,

Smart Distributed System (SDS), CANOpen [9], and SAE J1939 [10]. Both DeviceNet [11] and Smart Distributed System [12] have developed systems for creating networks of industrial field devices for the factory floor, including sensors and actuators.

4 to 20 mA Current Loop

The 4 to 20 mA current loop is a widely used method for transferring information from one station (the transmitter) to another station (the receiver). Therefore, this system allows for only two stations. A typical current loop system assigns a sensing range (e.g., 0 to 100°C) to the current range between 4 and 20 mA. A loop exists (i.e., two wires) between the transmitter and receiver. The transmitter can impress a certain current in the loop (using a controlled current source) so that the receiver can measure the current in the loop (e.g., by placing a small resistor in series with the loop and measuring the voltage drop across the resistor). After measuring the current, the receiver can then determine the present level of the sensed signal within the defined sensing range. This method uses current signaling, instead of voltage signaling, and therefore is relatively unaffected by potential differences between the transmitter and the receiver. This is similar to the benefit of differential (voltage) signaling, which also requires two wires. Another characteristic of this method is that it is not primarily digital in nature, as many other sensor communication systems are. The measured value can vary continuously in the range of 4 to 20 mA, and therefore can easily represent an analog sensing range, rather than a set of digital signals. Also, the signal is continuously variable and available. Another characteristic of this method is that the integrity of the loop can be verified. As long as the loop is unbroken and the transmitter is in good working order, the current in the loop should never fall below 4 mA. If the current approaches 0 mA, then the receiver can determine that a fault exists — perhaps a broken cable. These systems are widely used in various process control industries (e.g., oil refining) for connecting sensors (transmitters) with control computers. Because one station is always the transmitter and one station is always the receiver, this is a unidirectional, half duplex communication system.

HART[2]

HART® is a protocol which builds upon 4 to 20 mA communication systems. The basic idea is that additional data (beyond the basic sensor signal being carried in the current loop) can be transmitted by modulating a signal on top of the current flowing in the loop. The actual modulation method conforms closely to the Bell 202 standard for analog modem communications on telephone lines at 1200 bit/s. Because a 4 to 20 mA current loop carries a relatively slowly varying signal, it is easy to separate the 4 to 20 mA signal from the digital signal using filters. The Bell 202 standard uses continuous-phase frequency shift keying between two frequencies at up to 1200 shifts/s to modulate digital ones and zeros onto the 4 to 20 mA current loop. This method allows for bidirectional, full duplex communication between the two stations, on top of the 4 to 20 mA signal. It is also possible to configure HART communications on a network that is not carrying a 4 to 20 mA signal, in which case up to 15 devices can be connected together on the network. HART was developed by Fisher-Rosemount Corporation, and has been transferred to an independent foundation for management [13]. Because HART is compatible with U.S. telephone systems, it can theoretically be run over the telephone line and is therefore capable of running over arbitrarily long distances.

Profibus

Profibus (PROcess FIeld BUS) is one of three networks standardized by a European standard [14]. Profibus is under the control of a global organization, PNO [15]. Profibus is an umbrella network standard

[2]HART is a trademark of the HART Communications Foundation.

which encompasses three subnetworks within the Profibus family. Profibus-DP (Distributed Periphery) is the variant which is designed specifically for communication with field devices (sensors and actuators) at the device I/O level. Profibus-PA (Process Automation) is a variant which has more capabilities designed to support the needs of device-level networking for process industries, such as oil refining. One of the capabilities of Profibus-PA is its ability to be installed in an intrinsically safe way, thus providing a higher degree of safety in environments which may be explosive or otherwise hazardous. Profibus-PA typically uses a special physical layer specification standardized under IEC 1158-2, which is used by several network systems for process automation applications. IEC 1158-2 specifies a two-wire twisted pair implementation carrying both power and data on the same two wires at 31.25 kbit/s. Profibus-FMS (Fieldbus Messaging Specification) represents the highest level implementation, which is used to link together controllers (not field or I/O devices) in a factory.

Profibus-DP systems are typically master–slave systems, where usually a single network master (the host controller) communicates with a number of slave devices (remote I/O blocks and other I/O devices). The protocol provides for cyclic exchange of I/O information as well as on-demand exchange of other types of information. Profibus-DP can be implemented on several different physical layers, including RS-485 and fiber optics, at various bit rates up to 12 Mbit/s. Profibus messages include a CRC for error detection.

Foundation Fieldbus

Foundation Fieldbus (FF) is a networking standard which has grown out of an effort within industry standards organizations, especially ISA-SP50 [16], and IEC SC65C/WG6 [17], to provide a replacement for the 4 to 20 mA analog sensor communication standard. FF provides two basic levels of networking: H1 and H2. H1 is a lower-speed system that can provide intrinsically safe (IS) operation and uses a single twisted pair to deliver both power and data communications to field devices, according to IEC 1158-2. Running at a bit rate of 31.25 kbit/s, H1 is very similar to Profibus-PA, when run on the IEC 1158-2 physical layer standard. The H1 system is designed to be able to connect hierarchically "upward" to an H2 system, which acts as the host. FF H2 can be run at either 1 or 2.5 Mbit/s on twisted-pair wires, and also provides an IS option at the 1 Mbit/s rate. The H2 system can act as a network backbone in a factory environment, carrying data among various H1 systems.

WorldFIP

WorldFIP [18] is another technology of the three that were standardized in the European standard EN 50 170, running on the IEC 1158-2 physical layer. Many of the proponents of WorldFIP have embraced FF, and contributed to the development of that standard. WorldFIP is a member of the FF, and FF has incorporated many of the capabilities of WorldFIP as a result. When run on the IEC 1158-2 physical layer, WorldFIP has similar capabilities to FF.

LonWorks[3]

LonWorks® is a networking technology developed and controlled by the Echelon Corporation [19]. LonWorks is designed to be a general-purpose networking technology suitable for a variety of industries. LonWorks has been applied extensively in the building automation and control industry, as well as a variety of other industries. The core LonWorks technology for devices is contained in special integrated circuits — called Neuron® chips — which combine several microprocessors to manage the network, communications, and provide a general-purpose control environment. These chips are available from Motorola, Inc., and the Toshiba Corporation, which are licensees of the LonWorks technology. Echelon

[3]LonWorks, LonTalk, and Neuron are trademarks of the Echelon Corporation.

has also announced the possibility to license the LonTalk® protocol to other manufacturers for implementation in other microprocessors. LonWorks networks can be implemented on a variety of physical layers, including twisted pair at several bit rates and wireless options at 4800 bit/s, but the most common is a differential twisted-pair system running at 78 kbit/s. Most of the networking details (the LonTalk protocol) are hidden from the user, and are encapsulated as functions within the general-purpose control environment. The user programs (using a language like the C programming language) the Neuron chip for each station to behave in a certain way and communicate various data items to other stations. Then, specialized tools are used to tie all of the stations together (handling addressing and other network details) to yield a functioning network. The system combines flexibility with a certain amount of ease of implementation, and can easily be applied to a variety of applications.

28.4 Applying Network Communications

Shielding

Many communication networks require shielding of the media (the cable). Shielding constitutes an electric conductor which completely encases the communication media (e.g., twisted pair) to provide protection against EMI. Shielding provides an electric conductive barrier to attenuate electromagnetic waves external to the shield, and provides a conduction path by which induced currents can be circulated and returned to the source, typically via a ground reference connection. Shields in communication systems are often grounded at only a single point. This single point of ground prevents the shield from participating in a "ground loop," which is an alternative path for current to flow between two points of potential difference connected to a common ground. Ground loops can lead to noise problems, and can be destructive if the stray currents are large enough, since a shield ground is usually not constructed to carry heavy currents.

Media

The most common media types for network systems fall into three categories: electric, optical, and electromagnetic. Electric media are based on conductors (e.g., copper wire), whereas optical media are based on optical waveguides, or fiber optics. Electromagnetic media consists of free space, or general electromagnetic wave-permeable materials, and are referred to as wireless systems. Within the category of electric media are a large variety of conductor configurations. The most common are unshielded pair, unshielded twisted pair (UTP), shielded twisted pair (STP), and coaxial (COAX). These conductor configurations have various properties which are significant for the transmission of electric signals, as well as varying degrees of immunity to EMI. As a rule of thumb, the quality of the transmission line characteristics (signal transmission and immunity to EMI) improves in the order listed. Twisted-pair systems are generally easier to install, whereas coaxial and fiber-optic systems generally require more specialized tools and termination methods. Of course, wireless systems are easy to install, but attention must still be paid to the media. The characteristics of the free space such as distance and amount of EMI present must be considered for reliable operation of the network.

Bit Rate

Some networks provide only one choice of bit rate, whereas others provide user-selectable options for bit rate. Bit-rate options may be dependent on the type of media that is installed. As a rule of thumb, the bit rate chosen should be the lowest possible bit rate that still supports the application requirements for speed of data transfer and overall bandwidth. This generally results in more reliable operation, and generally gives the network more immunity to minor degradations, specification violations, and EMI.

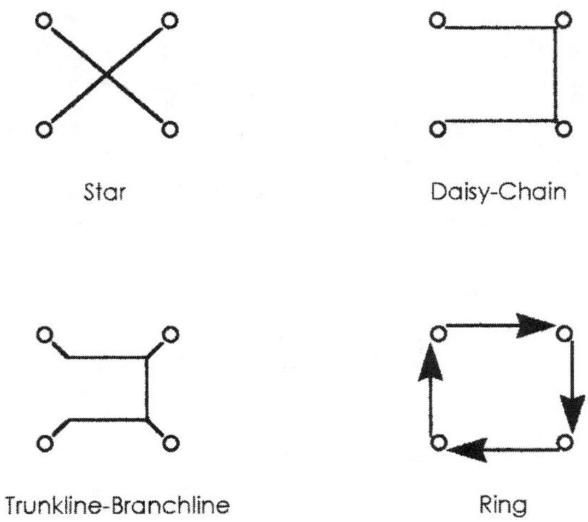

FIGURE 28.5 Some examples of the many possible network topologies using four stations.

Topologies

There are a variety of network topologies that are commonly used. Topology refers to the physical arrangement and interconnection of stations by the media. Some networks can be run using several different topologies; some can only be run with a certain topology (e.g., Interbus-S requires a ring topology). The most common topologies are daisy-chain, trunkline–branchline, ring, and star. Variations on these exist, and networks that incorporate or can be run on highly varied topologies are sometimes called free-form, tree, or free topology networks. Figure 28.5 depicts graphically several different types of topologies. In some cases, networks require certain topologies. Deviating from these can cause degradation in network behavior (e.g., corruption of messages) or network failure.

Configuration

Most networks involve some sort of configuration. Configuration is the process of connecting stations together and assigning certain programmable parameters to each station required for proper operation of the network. The most common configurable parameter in many networks is the station address. Some networks may require other parameters to be preset, such as the communication speed, or bit rate. Some networks have the capability to autoconfigure, which means to assign parameters automatically to stations as part of the network start-up process, without explicit user intervention (e.g., Interbus-S). Many networks define various tools, which may be computer based, to assign parameters to each station in order to configure the network. In other cases, the stations may incorporate switches or other manual means to configure the necessary parameters for network operation.

28.5 Advanced Topics

Wireless Technologies

The need for networking is present even in environments where an electrical or optical cable cannot be easily distributed. This may be due to various limitations, such as difficulty in running a new cable from one building to another, or connecting to sensors in motion or on vehicles. There are two general categories of wireless communications, based on electromagnetic frequency spectra. Various wireless technologies employ the infrared spectrum. These technologies generally have transmission limited to applications that have a direct line of sight between stations. Also, the distances are generally limited to

100 m or less. Because of these limitations, there are generally no legal restrictions in employing these frequency spectra, and infrared transceivers are now becoming available from a variety of manufacturers.

The other general category of wireless communications is based on radio frequency (RF) communications. In most countries, use of these spectra is tightly controlled by governmental agencies. As a result, employing wireless networking in most of these frequency ranges requires special licensing. However, a number of frequency ranges are reserved for low-power public communications. Within these frequency ranges, devices are allowed to communicate in an unlicensed fashion as long as they transmit according to certain rules about transmitted power output. RF-based wireless systems are generally not limited to line-of-sight applications, and can be designed to cover greater distances than infrared-based systems.

Wireless technologies can be viewed as simply another choice for the physical layer media, i.e., free space. As such, it is possible to consider, in some cases, a wireless media for implementation of a variety of protocols. For example, both CAN and LonWorks systems could be candidates for wireless networking.

Fiber Optics

Another physical layer media choice is fiber-optic media. Fiber-optic media employs pulses of light delivered along a tubular waveguide (glass or plastic fiber) to transmit information from one station to another. Fiber optics enjoy some benefits over traditional copper wiring. First, attenuation of light within fiber optics is generally about an order of magnitude less than attenuation of an electric signal within a copper wire. Second, fiber-optic transmission systems can be modulated (or pulsed) at much higher frequencies, yielding greater potential bandwidths and bit rates than copper media. Finally, fiber-optic systems are generally immune to the traditional sources of EMI that can cause trouble for copper media systems. There are also limitations in present implementations of fiber-optic systems. One of the limitations is that special tools and termination techniques must be used to connect a fiber-optic cable to a sensor or field device. Second, fiber-optic "taps" are not easily created. Therefore, most fiber-optic systems are implemented in a point-to-point fashion. When multiple devices are involved in a network, each device usually acts as an optical repeater, with a fiber-optic input and a fiber-optic output port.

Network Design Considerations

Designing a network communication system from the ground up can be a lengthy undertaking, and should not be considered unless a careful review of available technologies has yielded no solutions to the particular requirements of the application. The designers must take into account a number of fundamental questions to shape the capability of the network. One topic mentioned frequently in the area of networking for control applications is the subject of determinism. This refers to the ability of the network to behave in a predictable fashion under any given set of stimuli, or external conditions. Many networks do not exhibit this characteristic. Another question to be resolved is the subject of priority, and media access. The designers must determine the conditions under which any particular station is allowed to transmit, and if multiple stations are attempting to transmit, how it will be determined which station will be given priority to transmit first. Media access methods often impact the ability of a network to behave in a deterministic fashion.

Integrating Sensors with Communications — IEEE P1451

A recent interesting development in the area of sensor networks is an effort being sponsored by the IEEE [20] out of its TC-9 committee, called IEEE P1451. This activity is working toward the development of a standard to define sensor (or transducer) interfaces to networks generically. The first part of the proposed standard, IEEE P1451.1, includes definitions for the interface between the device and the network (refer to Figure 28.1). The second part, IEEE P1451.2, includes definitions for the interface between the transducer (or sensor) and the network interface block within the device. P1451.2 includes a definition for a transducer electronic data sheet, or TEDS, which defines a summary set of information pertinent to the sensor, allowing for standardized exchange of data on the network. The proposed standard has the potential benefits to make it easier to connect a sensor to a variety of networks, and to allow similar sensors from different manufacturers to be handled in a similar fashion on the network.

References

1. *ISO/IEC 7498-1:1994 Information Technology — Open Systems Interconnection — Basic Reference Model: The Basic Model,* International Organization for Standardization (ISO), 1, rue de Varembé, Case postale 56, CH-1211 Genève 20, Switzerland, [online]. Available http://www.iso.ch/.
2. *8802-3: 1996 (ISO/IEC) [ANSI/IEEE Std 802.3, 1996 Edition] Information Technology — Telecommunications and information exchange between systems — Local and metropolitan area networks — Specific requirements — Part 3: Carrier sense multiple access with collision detection (CSMA/CD) access method and physical layer specifications,* Institute of Electrical and Electronics Engineers, 445 Hoes Lane, Piscataway, NJ 08855-1331, [online]. Available http://www.ieee.org/.
3. *ANSI/EIA/TIA-232-E-91,* Electronic Industries Association, 2500 Wilson Boulevard, Arlington, VA 22201-3834, [online]. Available http://www.eia.org/.
4. *Distributed, Intelligent I/O for Industrial Control and Data Acquisition… The SERIPLEX Control Bus, Bulletin No. 8310PD9501R4/97,* Seriplex Technical Organization, P.O. Box 27446, Raleigh, NC 27611-7446, [online]. Available http://www.seriplex.org/.
5. AS-Interface U.S.A., 5010 East Shea Blvd., Suite C-226, Scottsdale, AZ 85254, [online]. Available http://www.as-interface.com/.
6. *Interbus-S Protocol Structure, Data Sheet 0005C,* Phoenix Contact, P.O. Box 4100, Harrisburg, PA 17111, [online]. Available http://www.ibsclub.com/.
7. *CAN Specification, Version 2.0, 1991,* Robert Bosch GmbH, Postfach 50, D-7000 Stuttgart 1, Germany.
8. *ISO 11898:1993 Road vehicles—Interchange of digital information—Controller area network (CAN) for high-speed communication,* International Organization for Standardization (ISO), 1, rue de Varembé, Case postale 56, CH-1211 Genève 20, Switzerland, [online]. Available http://www.iso.ch/.
9. *CiA Draft Standard 301 (Version 3.0), CANopen Communication Profile for Industrial Systems,* CiA Headquarters, Am Weichselgarten 26, D-91058 Erlangen, Germany, [online]. Available http://www.can-cia.de/.
10. *SAE J 1939 — Recommended Practice for Serial Control and Communications Vehicle Network,* Society of Automotive Engineers, 400 Commonwealth Drive, Warrendale, PA 15096, [online]. Available http://www.sae.org/.
11. *DeviceNet Specification v1.4,* Open DeviceNet Vendors Association, 8222 Wiles Road, Suite 287, Coral Springs, FL 33067, [online]. Available http://www.odva.org/.
12. *Smart Distributed System Application Layer Protocol Specification v2.0, 1996,* Honeywell MICRO SWITCH Division, 11 West Spring Street, Freeport, IL 61032, [online]. Available http://www.sensing. honeywell.com/sds/.
13. HART Communication Foundation, 9390 Research Boulevard, Suite I-350, Austin, TX 78759 [online]. Available http://www.ccsi.com/hart/hcfmain.html.
14. *EN 50 170 — Volume 2,* CENELEC Central Secretariat, 35, rue de Stassart, B-1050 Brussels, Belgium.
15. PROFIBUS Trade Organization U.S.A., 5010 East Shea Blvd., Suite C-226, Scottsdale, AZ 85254-4683, [online]. Available http://www.profibus.com/.
16. ISA, the International Society for Measurement & Control, P.O. Box 12277, Research Triangle Park, NC 27709, [online]. Available http://www.isa.org/.
17. *IEC 61158-2(1993-12), Fieldbus standard for use in industrial control systems — Part 2: Physical layer specification and service definition,* International Electrotechnical Commission, 3, rue de Varembé, PO Box 131, 1211 Geneva 20, Switzerland, [online]. Available http://www.iec.ch/.
18. WorldFIP Headquarters, 2, rue de Bone, 92160 Antony, France, [online]. Available http://www.world-fip.org/.

19. *005-0017-01 Rev C, LonTalk Protocol,* Echelon Corporation, 4015 Miranda Avenue, Palo Alto, CA 94304, [online]. Available http://www.echelon.com/.
20. *P1451.2, Draft Standard for a Smart Transducer Interface for Sensors and Actuators — Transducer to Microprocessor Communication Protocols and Transducer Electronic Data Sheet (TEDS) Formats,* Institute of Electrical and Electronics Engineers, 445 Hoes Lane Piscataway, NJ 08855-1331, [online]. Available http://www.ieee.org/.

Further Information

P.Z. Peebles, Jr., *Digital Communications Systems,* Englewood Cliffs, NJ: Prentice-Hall, 1987, provides a good general text on communication.

B. Svacina, *Understanding Device Level Buses*, Minneapolis, MN: TURCK Inc. (3000 Campus Dr., Minneapolis, MN 55441), 1996, is an in-depth study of the subject of communication networks for industrial field devices.

J.D. Gibson, Ed., *The Communications Handbook*, Boca Raton, FL: CRC Press, 1997, includes recent material on communication techniques.

29

Electromagnetic Compatibility

Daryl Gerke
Kimmel Gerke Associates, Ltd.

William Kimmel
Kimmel Gerke Associates, Ltd.

Jeffrey P. Mills
Illinois Institute of Technology

29.1 Grounding and Shielding

Daryl Gerke and William Kimmel

EMC (electromagnetic compatibility) is crucial to successful operation of industrial systems. Due to the increased electronic content of most industrial controls, electromagnetic interference (**EMI**) problems have increased dramatically in recent years. Two keys to EMC success are grounding and shielding. This section will briefly discuss how to implement these two crucial EMC strategies. It will also provide a general introduction to EMI problems in today's industrial electronic systems. The primary emphasis will be on practical insights and ideas gained in dealing with numerous industrial control problems.

Understanding EMI Problems

Here are three general observations on dealing with EMI problems in industrial electronics.

First, the industrial environment is harsh. The primary EMI threats are power disturbances, **RFI** (radio frequency interference), and **ESD** (electrostatic discharge). In addition, analog sensor circuits are often plagued with 50/60 Hz "ground loop" problems. Industrial electronics need more EMC care than most commercial electronics, and even more than many military systems.

Second, electronics often play a secondary role in electronics systems. Unlike a computer system, where electronics is the core technology, industrial electronics are often used to support another technology, such as chemical, mechanical, or process functions. This leads to EMC challenges when integrating the electronics to nonelectronic technologies.

Third, EMC rules and regulations are finally catching up to industrial electronics. For many years, industrial electronics were exempt from mandatory EMC rules, so unless there was an actual problem, EMC was often ignored. With the EMC directives of the European Union (**EU**) now in force, industrial electronics are no longer exempt.

Three Types of Problems

There are three aspects of the EMC problem: *emissions, susceptibility* (also known as immunity), and *self-compatibility.* **Emissions** originate within the equipment, and may upset other nearby equipment. On the other hand, external energy may upset equipment, leading to **susceptibility** (or a lack of immunity).

0-8493-1733-9/04/$0.00+$1.50

Finally, energy internal to the system may interfere with other internal circuits, resulting in a self-compatibility problem.

Problems with both emissions and susceptibility have led to EMC regulations. Two of the best known are the **FCC** (Federal Communications Commission) regulations for emissions in the U.S., and the EU regulations for both emissions and immunity in Europe. Industrial controls have always been exempt from the FCC regulations, but they are not exempt from the EU regulations which became mandatory in January 1996.

Four Major EMC Threats

Most industrial EMC problems fall into one of four key areas: *emissions, power disturbances, radio frequency interference,* and *electrostatic discharge.* In the past, industrial systems were usually only concerned with power disturbances. Today, all four threats must be considered.

Emissions.
Emissions refer to electric energy originating within equipment that can interfere with other equipment. The prime concern of this threat is jamming nearby television receivers, which is the basis for the now mandatory EU emissions regulations. Emissions problems between industrial electronic systems, however, are rare. While it is possible to interfere with any other nearby equipment, most industrial electronics generate only minute amounts of conducted and radiated interference, well below upset thresholds for digital or analog circuits.

Emissions are best addressed at the equipment design stage. Strategies include printed circuit board design techniques, high-frequency filtering on power and signal interfaces, shielded cables, and enclosure shielding. Fixes in the field are usually limited to shielded cables or enclosures, add-on filters, and ferrite clamps on cables.

Power Disturbances.
Power disturbances can take many forms, from short transients to long sags, surges, or complete power outages. The three most serious power threats to industrial controls are transients, voltage sags, and power outages. Stray 50/60 Hz currents can also cause problems with sensitive analog circuits, particularly due to ground loops (to be discussed later). Other power disturbances, like frequency or waveform variations, often have little effect on electronic systems.

Power disturbances are very common in industrial environments. As a result, most industrial systems are pretty robust against this threat, at least at low frequencies. High-frequency threats, such as fast transients or RF on the power lines, can still cause problems. The EU tests simulate these threats with the **EFT** (electrical fast transient) and injected RF tests.

Most power disturbances are caused by nearby equipment, rather than external sources. (One critical exception is lightning, which can result in some nasty voltage and current surges). Power disturbances solutions include grounding, power filters, transient protectors, and in extreme cases, uninterruptible power systems (UPS).

Radio Frequency Interference.
RFI deals with threats in the RF range. RFI is quite common in industrial environments, and will likely get worse with the proliferation of handheld radios and cellular telephones. It is expected that wireless LANs (local area networks) will also provide some interesting EMI challenges. There have been cases where handheld radios were banned from use due to repeated EMI problems with industrial electronics.

It turns out that the nearby handheld radio is a much bigger threat than a large commercial broadcast station several kilometers away. A key metric is electric field intensity, measured in "volts/meter." This is a function of both transmitter power, and distance from the antenna, and can be quickly predicted by the formula:

$$E(V/m) = 5.5\sqrt{PA}/d$$

where P = transmitter power in watts, A = antenna gain, and d = distance from the antenna in meters. For example, the electric field from a 1 W radio with a zero gain antenna at 1 m is about 5 V/m, while the electric field from a 10,000 W broadcast station at 1 km is about 0.5 V/m. Since unprotected equipment can fail in the 0.1 to 1 V/m range, problems can and do occur. The EU "heavy industrial" limits of 10 V/m are clearly aimed at protecting against the nearby handheld radio.

Solutions to RFI problems include high-frequency filtering on power and signal cables, shielded cables, and shielded enclosures. Analog circuits are particularly vulnerable to RFI, so they often need extra protection. Do not overlook banning radio transmitters in the immediate vicinity. Often, maintaining a 3 to 10 m distance is enough to solve the problem.

Electrostatic Discharge.
ESD refers to the sudden discharge that can occur after a gradual buildup of electric charge. ESD is most commonly associated with humans (touching controls or keyboards), but ESD can also be caused by internal arcing due to the movement of paper, plastic, etc. Internal ESD problems are increasing in industrial systems.

Although the static buildup can take a long time (seconds or even minutes), the discharge is almost instantaneous (nanoseconds or less). Furthermore, it is the sudden current, not the voltage, that is the culprit. The effect is a bit like having a dam burst — the ESD current is like water running down a mountain, destroying anything in its path. Fortunately, the current surge does not last too long, so the energy levels are not high. They are high enough, however, to damage or upset electronic devices.

The extremely fast discharge results in high frequencies well into the UHF range. At 1 ns, the transient bandwidth is over 300 MHz. As a result, it does not take a "direct ESD hit" to cause a problem. ESD upsets 5 to 10 m away are not uncommon, due to the intense electromagnetic fields associated with an ESD event. These problems are particularly insidious, since the ESD event may be occurring on a different piece of equipment.

Solutions to ESD problems include transient protection, high-frequency filtering, cable shielding, and enclosure shielding. Grounding is a very important factor in ESD protection, but it must be designed for high frequencies. Since many times ESD causes "reset" problems, extra attention to microprocessor reset circuits is beneficial.

Sources, Paths, and Receptors

A common EMI problem is gathering and organizing data. This is particularly important when troubleshooting EMI in the field. The "source–path–receptor" model is popular. Simply stated, three elements are necessary for any EMI problem:

1. There must be a source of energy;
2. There must be a receptor that is upset by that energy;
3. There must be a coupling path between the source and receptor.

All three elements must exist at the same time, and if any one is missing, there is no EMI problem. Sometimes one can identify all three, and, other times, one can only guess. While this may seem simple, it is a useful tool to organize EMI information.

Figure 29.1 illustrates this model, giving typical sources, paths, and receptors. Several possible sources have been discussed: emissions from digital circuits, ESD, RFI from communications transmitters, and power disturbances (including lightning). Several different receptors have also been suggested: communications receivers, analog electronics, and digital electronics. Note the two types of paths: radiated and conducted. In both cases, the object is to block unwanted energy from reaching a receptor, which is done with shielding (for the radiated path) and filtering (for the conducted path).

Grounding

Grounding is probably the most important, yet least understood, aspect of EMI control. Every circuit is connected to some sort of "ground," so every circuit is affected by EMI grounding issues.

Any interference problem can be broken down into
- the SOURCE of interference
- the RECEPTOR of interference
- the PATH coupling the source to the receptor

Sources
- Microprocessors
- Video Drivers
- ESD
- Transmitters
- RF Generators
- Power Disturbances
- Lightning

Paths
- Radiated
  - EM Fields
  - Crosstalk
    - Capacitive
    - Inductive
- Conducted
  - Signal
  - Power
  - Ground

Receptors
- Digital
  - Microprocessors
  - Reset
  - Other logic
- Low level analog
- Receivers

FIGURE 29.1 The source–path–receptor model for assessing EMI problems. All three elements must be necessary for an EMI problem to occur.

TABLE 29.1 A Ground May Work Over Wide Frequency Ranges

Type	Frequency	Typical Current Levels	Typical Duration
Power	50/60 Hz	10–1000 A	Seconds or minutes
Lightning	300 kHz	100,000 A	Tens of milliseconds
ESD	300 MHz	10–50 A	Tens of nanoseconds
EMI	Dc–Daylight	μA–A	Nanoseconds to years

What Is a Ground?

A major problem with the subject of grounding is the ambiguity of the term. Our favorite definition is one popular in the EMC community, which says that a *ground is simply a return path for current flow.* These currents can be intended, or unintended. The unintended currents are often referred to as "sneak grounds," and can cause many kinds of EMI problems. Finally, a physical connection is not even necessary at higher frequencies, where parasitic capacitance or inductance may form part of a ground path.

Different Types of Grounds

Grounds are used for many reasons, including power, safety, lightning, EMI, and ESD. Although they may share common functions, they may vary widely when it comes to frequencies and current amplitudes. Recognizing these key differences is key to understanding grounding issues.

Table 29.1 shows some frequency and amplitude requirements of several different types of grounds. Note that power and safety grounds must handle high currents, but only at low frequencies. Grounds for EMI and ESD, on the other hand, must often handle high frequencies at relatively low current levels. Lightning grounds must handle extremely high currents, but at moderate frequencies.

The frequency of transient events is calculated using the formula $f = 1/(\pi t_r)$, where f is the equivalent frequency, and t_r is the transient rise/fall time. This relationship can be derived using Fourier analysis. For example, ESD has an equivalent frequency of about 300 MHz based on a typical 1 ns rise time, and lightning has an equivalent frequency of about 300 kHz based on a 1 μs rise time.

Note that of all these types of grounds, only one actually needs an Earth connection — lightning. Other grounds may be connected to Earth by convention or for other safety reasons. For example, power neutrals are connected to Earth in many parts of the world to help provide lightning protection. On the other hand, in many other parts of the world, the power systems do not have Earth connections. When dealing with power grounding, the local safety codes will determine the proper Earth grounding methods.

TABLE 29.2 Impedance Parameters for 10-cm-Length Wires

Gage	Ω/m	μH/m	*Z* @ 10 kHz	*Z* @ 1 MHz	*Z* @ 100 MHz
10	0.0033	1.01	0.006	0.63	63
12	0.0052	1.05	0.007	0.66	66
14	0.0083	1.10	0.007	0.69	69
16	0.0132	1.15	0.007	0.72	72
18	0.0209	1.19	0.007	0.75	75
20	0.0333	1.24	0.008	0.78	78
22	0.0530	1.29	0.009	0.81	81
24	0.0842	1.33	0.010	0.84	84
26	0.1339	1.38	0.012	0.87	87
28	0.1688	1.40	0.019	0.88	88
30	0.2129	1.43	0.022	0.90	90

TABLE 29.3 Impedance Values for Ground Plane Impedance

	Thickness		
Frequency	0.1 mm	1 mm	10 mm
60 Hz	172 μΩ	17.2 μΩ	1.83 μΩ
1 kHz	172	17.5	11.6
10 kHz	172	33.5	36.9
100 kHz	175	116	116
1 MHz	335	369	369
10 MHz	1.16 mΩ	1.16 mΩ	1.16 mΩ
100 MHz	3.69	3.69	3.69
1000 MHz	11.6	11.6	11.6

Ground Impedances

A good ground must have a low enough impedance to minimize voltage drop in the ground system, and must provide the preferred path for current flow. The key to success is maintaining that low impedance over the entire frequency range of interest. We cannot overemphasize this point. Most EMI grounding problems are due to using the wrong approach for a given range.

The impedance of a ground conductor consists of both resistance and inductance ($Z = R + j\omega L$). For frequencies from dc through about 10 kHz, the resistance is the major factor, so heavy-gage wires are often used for low-frequency ground conductors. As the frequency increases, however, the inductance becomes the limiting factor for impedance. As a rule of thumb, the inductance for round wires is in the range of 10 nH/cm.

Table 29.2 gives the resistance, inductance, and inductive reactance for typical wire sizes used in instrumentation power and signal circuits. It is apparent that at power and audio frequencies (dc to 10 kHz), resistance is the dominant factor in ground impedance. Thus, at low frequencies, look for ways to reduce resistance, typically by using larger wires. At frequencies above the audio range (>10 kHz), inductance becomes the dominant factor in ground impedance. Thus, at higher frequencies, look for ways to reduce the inductance of the ground path. This is accomplished by using ground planes, grids, and straps to lower the inductance.

Table 29.3 gives the impedances for solid ground planes at various frequencies. In this case, the impedances are in "ohms-per-square," which is a measure of impedance across a diagonal surface. By comparing this with Table 29.2, one can see that at high frequencies (such as 100 MHz) the ground plane impedance may be several orders of magnitude below the impedance of a wire. Furthermore, at high frequencies the thickness is not a factor, since the impedance is limited by the skin effect.

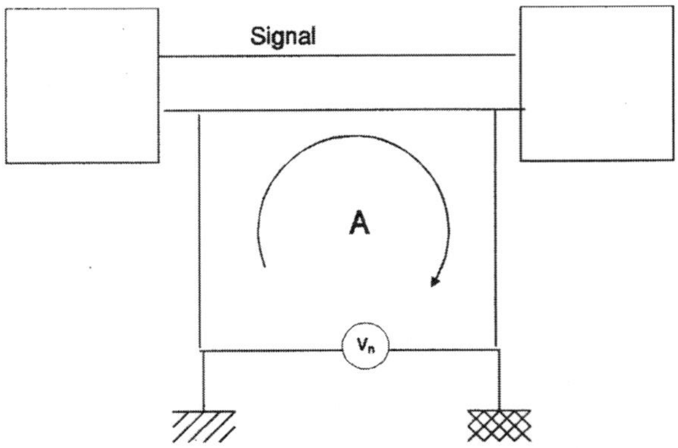

FIGURE 29.2 Typical industrial grounding situation, which also illustrates a ground loop.

Ground Topologies

Now that we have looked at ground impedance vs. frequency, we are ready to look at ground topologies vs. frequency. The impedance limitations yield two different grounding approaches, dependent on frequency. For low-frequency problems (dc to 10 kHz), single-point grounds are preferred, while at high frequencies (above 10 kHz), multipoint grounds with planes or grids become the preferred approach.

This dichotomy often causes confusion with industrial controls, but this can be minimized by determining the frequency of the EMI threat and then selecting the appropriate grounding approach. In many cases, both approaches may be necessary at the same time, leading to "hybrid" grounds, which use capacitors and inductors to alter the ground topology with frequency.

Single-Point Grounds.

At low frequencies, one can usually steer current via wires. Since the inductance is low, the limiting factor is the wire resistance itself. Furthermore, capacitive coupling from the ground wires to adjacent wires or surfaces is small, so virtually all the current follows the wiring path.

Figure 29.2 shows a typical industrial grounding scheme. Note what happens if the system is grounded at both ends. Any common noise current in the common ground path is now coupled into the circuit via the "common ground impedance." This results in the dreaded "ground loop," which will be discussed shortly. A single-point ground eliminates the ground loop, since there is no common impedance across which a common-mode voltage can be generated. Thus, single-point grounding is a very practical way to limit "ground noise" problems with the threat of low-frequency ground currents. This is very typical of 50/60 Hz currents getting into sensitive analog instrumentation circuits.

Multipoint Grounds.

Unfortunately, as the frequency increases, the inductive reactance of the wires increases. At the same time, parasitic capacitive reactance to adjacent wires or surfaces decreases, and soon it is no longer possible to maintain a true single-point ground, even if the system is wired that way. The only option left is to lower the ground path impedance, and that is accomplished with planes or grids. Furthermore, single-point connections to a grid or plane are usually not adequate because of transmission line effects, so multipoint grounds (combined with planes/grids) become the preferred approach above 10 kHz.

Ground grids have been used for years in computer facilities, and are seeing increasing use in industrial facilities. The recommended spacing for grids is no more than 1/20 of a wavelength at the highest frequency of concern. Computer room grids are often spaced about 0.7 m (about 2 ft), which meets this criteria from dc to about 25 MHz. This is very beneficial in addressing ground noise due to lightning and other power transients, which are usually in the 1 MHz range and below. But a 0.7 m grid does not help with VHF/UHF radio problems or ESD. In those cases, solid surfaces may be necessary.

Ground Loops.

Ground loops are a serious problem for sensitive analog circuits facing low-frequency threats. At high frequencies, ground loops generally do not pose serious threats if proper high-frequency precautions are taken when designing the ground system.

A ground loop exists whenever multiple ground paths exist. Unwanted currents can take unwanted paths, resulting in unwanted noise voltages at unwanted places. The problem is particularly acute with sensitive analog systems, where even a few microvolts can jam intended signals. A classic example is 60 Hz ground currents causing hum in an audio system.

Figure 29.2 shows a typical ground loop problem. Note that there must be the three conditions of any EMI problem: a source, a path, and a victim. In this case, the source can be circulating power currents, the path the common ground impedance, and the victim is often the sensitive analog circuit. With many systems problems, one cannot do anything about either the source or victim, so the solution is with the ground path. As we have already seen, single-point grounding is effective at low frequencies, and ground planes/grids are effective at higher frequencies.

If one cannot change the ground paths, one can still attack the ground loop by "breaking" it in other places. For example, transformers or optical isolators (or even fiber optics) can be used in cable connections, which will block common mode noise currents while passing intended differential mode signals. Balanced input/output (I/O) circuits can be used to "cancel" the noise through common-mode rejection. All of these are most effective at 50/60 Hz, and become less effective at higher frequencies due to parasitic capacitance.

Grounding Guidelines

By now it should be apparent that there is no magic solution for grounding. Rather, different methods and approaches are necessary for different circuits and operating conditions. Two key parameters are the threat frequency (low vs. high), and the circuit operating levels. Here are some guidelines, but keep in mind that even these may need to be modified for a particular situation.

Analog Circuits.

Since most analog circuits operate at low frequencies and are subject to low-frequency threats, single-point grounds are preferred. Typical threats are 50/60 Hz power return currents, stray switching power supply currents, and perhaps digital circuit return currents (if separate analog and digital power and grounds are not provided). Low-level analog circuits are the most vulnerable, since the signal levels are small.

Keep in mind that high-frequency threats (such as a VHF radio) to low-frequency circuits may require high-frequency grounding solutions, such as multipoint grounds. Often, this can be accomplished by using small high-frequency capacitors (1000 pF typical) which appear as a short circuit at 100 MHz, yet still appear as a high impedance at 50/60 Hz.

Digital Circuits.

Most digital circuits today operate at relatively high frequencies, so multipoint grounds and ground planes and grids are preferred. The connections between the circuits and their grounds need to be short, fast, and direct to minimize inductance.

Digital circuits, particularly I/O circuits, are vulnerable to external high-frequency threats like RF and ESD. They are also a key source of high-frequency emissions and internal problems like cross talk. For digital circuits, multilayer boards with internal ground planes are preferred. These ground planes typically are connected to a metallic enclosure through multiple low-inductance connections.

Pay particular attention to where digital and analog circuits meet. A single-point connection is usually preferred to minimize ground loops, but installing a small resistor (1 to 10 Ω typical) or inductor (1 to 100 μH typical) at that point is often helpful in providing additional isolation. One may need to experiment with this to determine the optimum solutions.

Power Safety Grounding.

Entire books have been written about this subject, and rightly so; this is an extremely important safety issue. The key concern here is human safety and prevention of electric shock. In most parts of the world, exposed metal on line-powered equipment must be bonded to a safety grounding conductor. Furthermore, the electric wiring codes (such as the National Electrical Code in the U.S.) give very specific guidelines on how power grounding must be accomplished.

These guidelines must be followed when wiring any industrial control system, and must never be compromised by "isolated" power grounds or other similar foolishness. *Finally, if there is ever a conflict between EMI and safety grounding, the safety issues must always prevail!*

Shielding

Many systems today require at least some **shielding** for proper operation or to meet radiated emission or immunity requirements. Many engineers consider shielding purely a mechanical issue, but nothing could be farther from the truth. EMI shielding needs both an electrical and a mechanical understanding of key issues to assure success.

Two of these key issues are selecting the right material and maintaining the shielding integrity over the desired frequency range. While most people worry more about the selection, shield integrity is usually much more important. We will soon see that even very thin metallic coatings can be effective shields, yet even very small holes or penetrations can completely destroy a shield. Like grounding, shielding cannot be left to chance, and must be properly designed and implemented.

How Shielding Works

EMI shielding involves two independent mechanisms: *reflection* and *absorption*. In reflection, an electromagnetic wave bounces off the surface, just like light off a mirror. In absorption, the electromagnetic wave penetrates the material and is absorbed as it passes through, much like heat loss through an insulating wall.

Shielding effectiveness is usually expressed as follows:

$$\text{SE}\left(\text{dB}\right) = R\left(\text{dB}\right) + A\left(\text{dB}\right)$$

where SE is the total shielding effectiveness in dB, and R and A are the reflection and absorption losses expressed in dB. Reflection is the primary mechanism for high-frequency shielding (emissions, RFI, ESD), while absorption is the key mechanism for low-frequency magnetic field shielding. The actual formulas for calculating reflection and absorption losses are a bit complex, and beyond the scope of this chapter, but several sources are included in the further information section.

Three Types of Fields

It is customary when dealing with shielding to use three types of "fields" to explain shielding. These three fields account for differences in shielding performance due to differences in frequency and circuit impedance levels. They also explain why the same shield can behave differently for different energy sources. These are plane waves, magnetic fields, and electric fields. Figure 29.3 shows typical shielding curves for copper, with references to each type of field.

Plane Wave Fields.

If one is located greater than about 1/6 wavelength from a point source, the wave impedance (ratio of electric field intensity to magnetic field intensity) is a constant 377 Ω in free space. This field is known as the "far field" or "radiation field," since real energy predominates here and propagates as a "plane wave." Since reflection losses are due to a mismatch between the wave impedance (377 Ω) and a metallic shield surface impedance (typically milliohms or less), shielding effectiveness is usually very high for plane wave sources.

At frequencies 30 MHz and above, once one is more than about 1 m away, one is in the plane wave region. Thus, even very thin shields work well for emissions, ESD, and RFI problems, with reflection as the prime shielding mechanism.

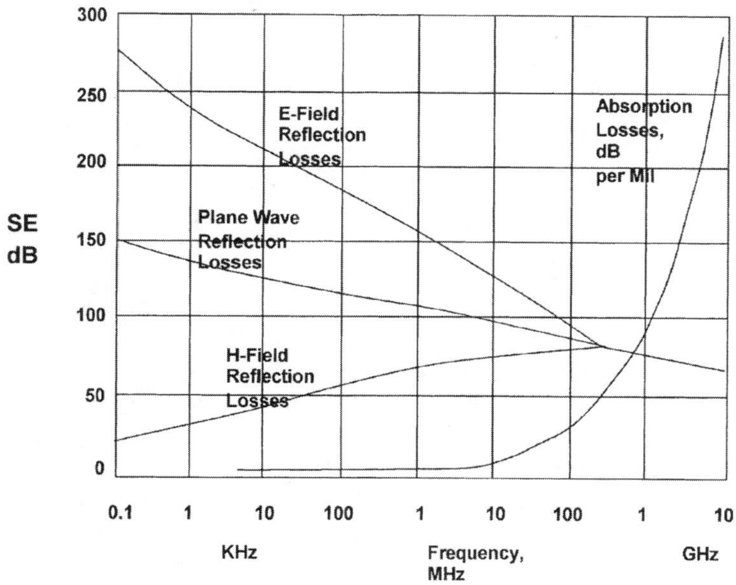

FIGURE 29.3 Typical shielding effectiveness curves for copper. Note two mechanisms (reflection and absorption) and three types of fields (electric, magnetic, and plane wave). Shielding for aluminum is almost the same as for copper.

Electric and Magnetic Fields.

If one is located less than about 1/6 wavelength from the source, then the wave impedance is dependent on the circuit impedance. This region is known as the "near field," since reactive energy predominates here. This region is further divided into "electric" and "magnetic" fields, both dependent on source circuit impedance. For high-impedance sources (electric fields), the reflection losses are still high, but for low-impedance sources (magnetic fields), the impedance can be quite low. In the latter case, the reflection losses can become minimal.

For power line frequencies, the near field almost always predominates. As a result, materials like aluminum or copper have no reflection losses and are virtually transparent to power line magnetic fields. (As a rule, remember that aluminum foil is transparent to 60 Hz magnetic fields.) To solve this problem, permeable materials are needed to boost the electric thickness for a given physical thickness. Steel or high-permeability mu-metals are usually used to absorb (not reflect) the magnetic fields. Even so, it can still be very difficult to shield for low-frequency magnetic fields.

Why Shielding Fails

While material selection is important, other factors must also be considered. For low-frequency/low-impedance threats (power supply or power line magnetic fields), steel or other high-permeability materials are needed. For high-frequency threats, however, even very thin materials like conductive paints provide high levels of shielding. Two problems at high frequencies, however, are shield openings and shield penetrations. Lack of attention to these areas can result in a loss of virtually all shielding effectiveness at high frequencies. Figure 29.4 illustrates these two high-frequency failure modes.

Intuition suggests that any opening in a shield can leak, much like an open window. The surprise is that for electromagnetic leakage it is not the area that is critical, but the longest dimension. For example, a 100 × 1 mm opening is about ten times more leaky than a 10 × 10 mm square hole. And that slot may not even be obvious. It could be a painted seam or a poorly fitting panel or door.

Slots act like small antennas. Because they are antennas, the longer they are, the better they radiate. While a half wavelength is very efficient, as a rule of thumb, we like to limit slots to 1/20 wavelength or less at the highest frequency of concern. For 100 MHz, this is 15 cm (about 6 in.); at 300 MHz (ESD

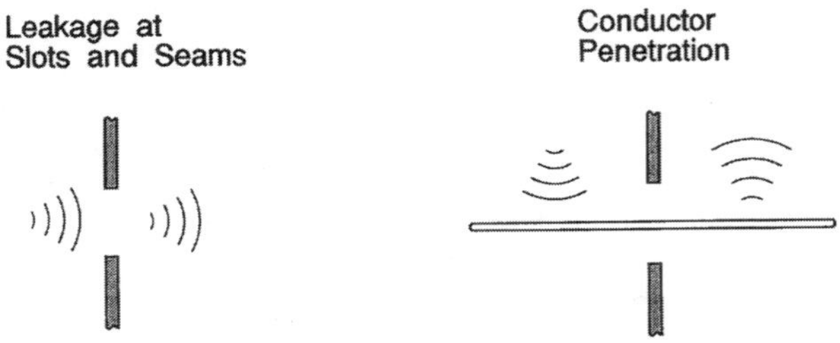

FIGURE 29.4 Two shielding failure modes, due to slots/seams and due to penetration of conductors. In both cases, the critical dimension is 1/20 wavelength for the highest frequency of concern.

frequencies), this drops to 5 cm (about 2 in.), and at 1000 MHz, it is only 1.5 cm (O in.). And even these dimensions may be too large, as they only assure 20 dB (tenfold) of shielding through the slot. Clearly, even small slots and other openings mean big shielding problems at high frequencies.

The other way to destroy a high-frequency shield is to pass unterminated metal through a shield. Hole dimensions do not matter here, and even a pinhole with an insulated wire passing through can carry large amounts across the shielding barrier. The dimension that does matter is how far the penetration extends on either side of the shield. Once again, the critical distance is 1/20 wavelength or more.

Shielding Guidelines

Now that we have looked at how shielding works (and fails), let us look at how to design good electromagnetic shields. Most of our focus will be on RF shielding in the 30 to 1000 MHz range, necessary for emissions, ESD, and RFI.

Material Selection.

We have already seen that for low-frequency magnetic interference problems, ferrous material like steel or mu-metals are necessary. Most instrumentation problems are either high impedance or high frequency in nature, so most of the time, thin conductive materials will work fine. For high frequencies, however, attention must be given to slots and penetrations.

Many enclosures today are made of plastic. For high-frequency shielding, conductive coatings also work quite well. Popular surface treatments include conductive paints, vacuum deposition, electroless plating, and even metal fabrics. Conductive plastics are also available, but they generally do not perform nearly as well as surface treatments for high frequencies.

Gasketing and Screening.

Large openings, such as ventilation ports or display areas, can be sealed with screening material. Seams or slots can be filled with conductive gaskets. In both cases, the secret is to provide complete and continuous metal-to-metal contact at all junctions. For high-frequency shielding, the connections must be almost watertight. Anything less is asking for problems.

For screening material, the smaller the openings, the higher the shielding. Window screen spacing is almost as effective as solid materials from dc to 1000 MHz, and even 5-mm (about 1/4 in.) openings are often acceptable at 1000 MHz. In any case, do not exceed 1/20 wavelength at the highest frequency of concern.

Cable Terminations and Filters.

Poor termination of shielded cables can cause big problems at high frequencies. If a shielded cable is not terminated directly at the shielding barrier, a lot of energy leaks, degrading both the cable and the enclosure shield. Pigtail connections, popular for terminating low-frequency cable shields, are particularly bothersome at high frequencies. In fact, this is a leading cause of EMI failures for RFI, emissions, and ESD. As a rule of thumb, pigtail connections should not be used on cable shields at frequencies above about 1 MHz.

Unshielded cables can also cause problems at high frequencies. In those cases, high-frequency filtering is needed directly at the interface to assure that the shield is not degraded at high frequencies. Common solutions are EMI filters on power and signal lines, or ferrite beads on the lines or cables. These must be installed as close to the shield penetration as possible. The best situation is to mount the filter directly in the shield itself, although this is not always necessary for moderate problems.

Internal Shields.
Finally, do not overlook using internal shields on critical circuits. Radio and television designers have been doing this for years, using selective shields on oscillators, power amplifiers, and the like. A classic example of this approach is the TV tuner, the most sensitive part of a television receiver. It is an inexpensive, yet highly effective shielding strategy.

Defining Terms

Conducted: Energy or interference that is propagated by a conductor, such as power, grounding, or signal interface wiring.

EFT: Electrical fast transient; a high-frequency burst of energy on power wiring.

EMC: Electromagnetic compatibility; the condition wherein electric and electronic equipment operate successfully in close proximity.

EMI: Electromagnetic interference; unwanted electric energy that may impair the function of electronic equipment.

Emissions: Electric energy emanating from an electronic source.

ESD: Electrostatic discharge; the rapid discharge that often follows a buildup of static charge.

EU: European union; formerly called the European Community.

FCC: Federal Communications Commission (U.S. government).

Ground: A return path for current.

Radiated: Energy or interference that is propagated by electromagnetic radiation through space.

RFI: Radio frequency interference; an older term for EMI, now usually used to describe interference caused by a nearby radio transmitter.

Shield: A metallic enclosure used to reduce electric or magnetic field levels.

Susceptibility: Vulnerability of electronic equipment to external sources of interference; often used interchangeably with immunity.

Further Information

D.D. Gerke and W.D. Kimmel, *EDN's Designer's Guide to Electromagnetic Compatibility*, Newton, MA: Cahners Publishing, 1994. Basic introduction on EMC issues written by the authors. Very good for non-EMC engineers.

W.D. Kimmel and D.D. Gerke, *Electromagnetic Compatibility in Medical Equipment,* New York: IEEE Press and Interpharm Press, 1995. Detailed introduction to EMC issues in medical instrumentation, written by the authors. Applicable to most industrial control instrumentation.

W.D. Kimmel and D.D. Gerke, Internet World Wide Web Home Page [online]. Available http://www.emi-guru.com, St. Paul, MN, 1996. Wide range of useful EMC information.

R. Morrison, *Grounding and Shielding Techniques in Instrumentation,* New York: John Wiley & Sons, 1986. Detailed analog coverage, with emphasis on low-level signal issues.

H.W. Ott, *Noise Reduction Techniques in Electronic Systems,* New York: John Wiley & Sons, 1988. Good introduction to EMC issues, particularly system and analog issues.

C.R. Paul, *Introduction to Electromagnetic Compatibility,* New York: John Wiley & Sons, 1992. Written as a college text, quite analytical with very good coverage of EMC issues.

29.2 EMI and EMC Test Methods

Jeffrey P. Mills

Electric and magnetic fields must be measured for a variety of reasons. A radio or TV broadcast station is licensed to provide reliable coverage over a specified geographic area, and any properly operating receiver must pick up the signal and properly respond to it. This can be assured only if the broadcast signal is of a guaranteed minimum strength. Also, the signal must not be so strong that it interferes with a distant station sharing the same frequency. The broadcast field must be measured over its geographic area of coverage to be sure that it satisfies both criteria.

Many electric devices unintentionally radiate electromagnetic fields. Examples include

- Oscillators in superhetrodyne radio or TV receivers
- Digital logic circuits
- Switching contacts, particularly if unsuppressed
- Automotive ignition systems

Stray fields (**emissions**) from these devices can interfere with other devices, or even with the radiating device itself. This process is known as **electromagnetic interference**, commonly abbreviated EMI. Interference between two devices is known as **intersystem** EMI, whereas if a device interferes with itself it is **intrasystem EMI**. Intrasystem EMI is usually easy to spot because the device itself does not operate correctly. Intersystem EMI is usually more difficult to isolate. Its result might be a simple annoyance, such as noise on a radio and TV receiver caused by an electric vacuum cleaner or a power drill. It could, however, be much more serious; a portable radio receiver might affect aircraft navigation or critical communications.

It is also possible for a device to be **susceptible** to fields intentionally generated by a licensed transmitter such as a broadcast or mobile-radio transmitter. Examples include

- Public-address systems
- Music (high-fidelity) systems
- Telephone lines and instruments
- Digital logic circuits

Again, the result may be only an annoyance, or it could be much more serious; aircraft control surfaces have been observed to move uncontrollably due to strong electromagnetic fields. Since the fields themselves cannot be eliminated in these cases, the devices must be made immune to electromagnetic fields.

In the above cases, the interference is usually through electric and/or magnetic fields in space, so the process is known as **radiated coupling**. Another coupling path exists if two devices share the same power source. One device may generate undesired high-frequency voltages on its power leads, which then appear on the power leads of the other. The second device may then malfunction because of this high-frequency voltage. This is known as **conducted coupling**. So we must consider both radiated and conducted noise.

It is not practical to eliminate all interfering fields completely, so a compromise must be reached. A stray field will not cause EMI if it is very weak compared with the desired field, which might be the field of a broadcast signal. The permissible strength of the stray field depends on the strength of the desired field; the stronger the desired field, the more stray field can be tolerated. It also depends on the device that is being interfered with (the *victim*); some receivers can reject undesired signals better than others. Since there are many combinations of interference sources and victims, a worst-case scenario is sought that will protect most real-life situations. This occurs where the weakest legal radio or TV signal (in its licensed area of coverage) is received by the poorest available receiver.

The maximum stray field strength that causes no EMI for this worst-case scenario is incorporated into government regulations. The field actually radiated by every device must then be measured to be sure that it does not exceed this level at the nearest practical distance from it, usually 10 or 30 m. To specify and measure these field strengths accurately, the nature of electric and magnetic fields must be understood.

Unlike most electrical engineering topics, EMI control is not very precise because of the complexity of practical hardware. It is virtually impossible to predict interference more precisely than within a factor of three, and usually the margin of error is even worse. Measurements can vary significantly between two supposedly identical samples, due to slight variations in physical dimensions. If one measures the EMI resulting from two different designs, the design that exhibits less EMI is probably better, but not always. An engineer can often judge if an EMI problem exists, but one must never rely on the accuracy normally expected in other branches of electrical engineering.

Nature of Electric and Magnetic Fields

An electric field is generated by a distribution of electric charge. If the distribution changes with time, then so will the electric field. A magnetic field may be generated by a permanent magnet or by an electric current. If the permanent magnet or the current path moves, or if the current magnitude varies with time, the magnetic field will vary with time. A time-varying electric field creates a magnetic field, and conversely.

Electric fields, designated E, are normally expressed in volts per meter (V/m). Magnetic fields are designated H and expressed in amperes per meter (A/m). More often, magnetic fields are perceived as magnetic flux density, which is designated B and expressed in webers per square meter (W/m^2), also known as *teslas* (T). A non-SI unit, sometimes found in older literature, is the *gauss*, equal to 10^{-4} T. Of course, any unit may be preceded by a scaling prefix such as micro or pico. In free space, B is equal to $\mu_0 H$, where, μ_0 is equal to 0.4π (approximately 1.257) μT·m/A (equivalent to μH/m).

Near a time-varying electric field source such as a charge distribution, the magnetic field is relatively weak, but it becomes stronger when observed from farther away. At a great enough distance, the ratio of E to H approaches $\sqrt{\mu/\epsilon}$, which in free space is equal to 120π (approximately 377) Ω. For a sinusoidal function of time with a frequency f, this occurs at any distance that is large compared with $\lambda/2\pi$ (approximately $\lambda/6$). Here, λ is the wavelength corresponding to f, equal to $3 \cdot 10^8/f$ m if f is specified in hertz. Distances much greater than $\lambda/2\pi$ are considered to be in the **far-field region**; nearer distances are in the **near-field region**. For a nonsinusoidal function of time, each Fourier frequency component must be considered separately, and the far-field region begins closer to the source for its higher frequency components.

Near a time-varying magnetic field source such as a current loop, the electric field is weak, becoming stronger when observed from a greater distance. At distances that are large compared with $\lambda/2\pi$ (the far-field region), the ratio of E to H again approaches 120π Ω.

Since $H = \sqrt{\epsilon/\mu}E$ and $B = \mu H = \sqrt{\mu\epsilon}E$ in the far-field region for either type of source, only E or B must be measured, and the other can easily be calculated from it. In free space, $\sqrt{\mu\epsilon} \approx 10^{-8}/3$ T·m/V (equivalent to s/m), so, if E is expressed in volts per meter, $B \approx 3.33E$ nT. By choice of a suitable antenna, either field can be measured. Far-field strengths are normally specified in terms of the E field, no matter whether the E or B field is measured.

Alternatively, the far-field strength may be specified in terms of **power density**, expressed in watts per square meter. This denotes the amount of radiated power passing through each square meter of a surface perpendicular to the direction away from the source. The *peak* power density P is equal to EH, and, for a sinusoidal source, the *average* power density is half this value. For a nonsinusoidal source, each frequency component must be considered separately, and the total average power is the sum of the average powers for all frequencies. Since $H = \sqrt{\epsilon/\mu}E$, it follows that $P = E^2/377$ Ω.

In regions other than the far field, the ratio of E to H varies greatly, approaching infinity for an electric field source or zero for a magnetic field source. A source may generate both electric and magnetic fields; for example, a charge moving between two electrodes causes a current to flow between them. Then the ratio of E to H may be any value at all. Therefore, at distances less than $\lambda/2\pi$ from a field source, both the E and B fields must be measured separately.

In the far-field region, both the electric and magnetic fields are perpendicular to the direction that an electromagnetic wave is propagating, and they are also perpendicular to each other. This still usually allows the fields to be oriented at many different angles with respect to the surface of the Earth. The direction of the electric field is called the **polarization** of the wave, which may be vertical, horizontal, or somewhere between. Or the wave may be **elliptically polarized**, which results from two waves that are not exactly in phase, one polarized vertically and the other horizontally. If the waves are equal in magnitude and exactly 90° out of phase, the wave is *circularly* polarized. To account for all these cases, all fields must be checked separately for vertically and horizontally polarized waves.

Measurement Antennas

Most electronic components and instruments are designed to respond to voltages or currents, not fields. To measure a field strength it is necessary to convert its effect to a voltage or a current. This is achieved by an antenna. Although many antennas are simple conductor shapes, they must be analyzed carefully if accurate quantitative measurements are desired.

A straight conductor immersed in a time-varying electric field will develop a current in it. If the conductor material is linear (the usual case), the current will be proportional to the applied electric field, so their ratio will be constant. This ratio, however, depends greatly on the geometric dimensions of the conductor and the frequency of the electric field. It must be known to calibrate the antenna.

Similarly, a closed conductive loop immersed in a time-varying magnetic field will develop a current in it. Again, if the conductor is linear, the ratio of the current to the magnetic field strength is constant but depends on the dimensions of the loop and the frequency of the magnetic field.

The easiest way to calibrate an antenna is to immerse it in a known electric or magnetic field and measure the current or voltage at the antenna terminals. The principal problem is generating the known field. To find its strength, one must use a "standard" antenna for which the current-to-field ratio can be calculated.

To calculate the required ratio, Maxwell's equations must be solved subject to the boundary conditions of the antenna conductor. For most antennas an exact closed-form solution is impossible. However, for a sinusoidally varying field encountering a straight cylindrical conductor called a **dipole antenna**, such a solution is possible, though difficult [1]. Once the solution is obtained, the required ratio becomes a simple expression if the antenna is *resonant* or *tuned*. This occurs for a precise length that is slightly less than one half the wavelength, λ, of the time-varying field. Obviously, the antenna will be resonant at only one frequency, so the ratio will be valid only for a field varying sinusoidally at that frequency. For nonsinusoidal fields, each Fourier frequency component must be measured separately, and the antenna length must be changed as different frequencies are measured. To simplify changing its length, two telescoping rods, mounted end to end, are normally used to make the dipole antenna. The measuring instrument is connected between these two rods via a transmission line.

For a given frequency, at any point on the antenna, there is a certain current I flowing in it, and there is also a certain voltage V on it with respect to ground. The ratio of these phasors, V/I, is known as the **driving-point impedance**. The precise resonant antenna length is that for which V and I are exactly in phase, i.e., for which the driving-point impedance is purely real. As mentioned above, this length is slightly less than half the wavelength, λ, and it also depends on the thickness of the telescoping rods (pp. 547–548 of Reference 1). For a rod thickness of $\lambda/400$, the resonant length is 0.476λ. The driving-point impedance of a dipole antenna of these dimensions is 64 Ω. If a voltage-measuring instrument such as a radio receiver or spectrum analyzer is connected to the antenna terminals via a transmission line, and is properly matched to the 64 Ω impedance, the measured voltage V_m will be equal to $0.148\lambda E$, where E is the applied field strength and λ is the wavelength at the frequency being measured. The ratio V_m/E, equal to 0.148λ, is known as the **effective length** (l_e) of the antenna, since it relates the field strength in volts per meter to the measured terminal voltage in volts. Obviously, it is not equal to the physical antenna length but is instead approximately one third of that value. With this ratio known, the electric field strength E that causes a certain terminal voltage V_m can easily be calculated.

To simplify calculations, E is often expressed in decibels with respect to a reference field of 1 μV/m and is designated E_d. Similarly, V_m is expressed in decibels with respect to a reference voltage of 1 μV and is designated V_d. The **antenna factor** (AF) is defined as the effective length expressed in negative decibels, or AF = –20 log(l_e). Then the multiplication becomes an addition, i.e., $E_d = V_d + \text{AF}$.

The above antenna factor assumes that the antenna is perfectly matched to the receiver, which implies maximum power transfer. A mismatch would change the antenna factor. Therefore, since the antenna driving-point impedance usually is not equal to the receiver input impedance, a matching circuit must be inserted between the antenna and receiver. Another essential consideration is antenna balance. Most receivers and spectrum analyzers have one input terminal grounded. If this grounded terminal is connected to one of the dipole antenna terminals, the impedances connected to the two antenna terminals will be unequal with respect to ground. This also will upset the antenna factor, since one side of the antenna will not be properly matched to the receiver. To prevent this, a balanced-to-unbalanced (**balun**) network must be inserted between the antenna and receiver. Such a circuit provides a high impedance *with respect to ground* for both input terminals, while providing the correct input impedance (such as 64 Ω) *between* its input terminals. Normally, a single network provides both the matching and balancing functions.

Unfortunately, unless the dipole antenna is precisely the correct length, its antenna factor is much more complicated. Even if the frequency being measured differs only a few percent from the antenna resonant frequency, the antenna factor becomes unpredictable and the driving-point impedance becomes complex. Thus, the electric field cannot be easily calculated from the measured terminal voltage. To achieve the simple antenna factor described above, the frequencies must be measured one at a time and the dipole antenna length properly adjusted for each frequency. It is impossible to sweep the spectrum rapidly, as when using a spectrum analyzer, unless the antenna length can somehow be varied also. This leads to mechanical difficulties and is usually impractical.

Other types of antennas, however, are less sensitive to frequency. Examples are the biconical antenna and the log-periodic antenna. A biconical antenna can perform acceptably over a range of 20 to 300 MHz, and a log-periodic antenna is useful from 300 to 1000 MHz. Their antenna factors are relatively constant, usually varying by no more than 20 dB, over their useful frequency ranges. The antenna factors are usually too difficult to calculate, but they may easily be measured simply by observing the terminal voltage resulting from a sinusoidally varying field of known strength. The known field is first measured using a tuned dipole antenna, for which the antenna factor can be calculated. The antenna factor is measured in this manner at several frequencies throughout its useful range, and the results are plotted for use with the antenna.

Unlike the tuned dipole, the biconical and log-periodic antennas do not exhibit constant driving-point impedances over their useful frequency range. Since the receiver input impedance cannot be made to follow the variation of driving-point impedance with frequency, an exact match is impossible. This affects the antenna factor just as it would for a mismatched tuned dipole. To compensate for this, the antenna factor must be measured with the antenna terminated into a known impedance, which must then be used for all measurements made with that antenna. Then the mismatch is accounted for in the antenna factor itself. The mismatch does cause the antenna to reradiate the received signal, but this effect may be minimized by performing the measurements in an open-field site, which will be discussed later.

Tuned dipole, biconical, and log-periodic antennas are *linearly polarized* antennas because they respond to only one polarization component of a propagating wave. If the antenna is oriented horizontally, only the horizontally polarized component of the wave will affect it. Similarly, only the vertically polarized component will affect a vertically oriented antenna. Thus, with two measurements, any linearly polarized antenna will detect any type of field polarization. Other types of antennas, such as the spiral antenna, are designed to detect a circularly polarized wave. They will detect vertically and horizontally polarized waves, but they could miss a wave that is circularly polarized in the reverse direction (counterclockwise instead of clockwise, for example). Consequently, circularly polarized antennas are forbidden for many types of field measurements.

All antennas discussed above respond to the electric field, *E*. As mentioned earlier, in the far-field region, the magnetic field, *B*, is simply 3.33 nT times the value of *E* expressed in volts per meter. In any other region, however, *B* is not so simply related to *E* and must be measured separately, using an antenna that responds to magnetic fields. A circular loop or coil of wire is such an antenna. The loop is cut at one point and the radio receiver or spectrum analyzer is connected between its two ends. For quantitative measurements, its antenna factor must be known. The factor can be measured by immersing the antenna in a known magnetic field and measuring its terminal voltage. To find the known magnetic field strength, the electric field is first measured, in the far-field region, using a tuned dipole antenna for which the factor is known. The magnetic field is then 3.33 nT times this value expressed in volts per meter. With the magnetic field thus determined, the antenna factor of the loop may be calculated, as required.

Measurement Environment

A major difficulty with electromagnetic field measurements is repeatability of results. Electromagnetic fields are affected by any materials in their vicinity, even by poor conductors and dielectrics. The measurement environment must therefore be carefully defined, and similar environments must be used for all comparable measurements.

The ideal environment would be one where (1) the only electromagnetic field source is the equipment under test (EUT) and (2) there is no “foreign” material at all that could affect the fields being measured. Unfortunately, the only natural location where this could be achieved is in outer space, since the Earth itself affects electromagnetic fields. Since this is impractical, attempts are made to simulate this environment on Earth.

A large outdoor open area simulates a hemisphere of free space. Such a test site is appropriately called an **open-field site**. If the conductivity, permittivity, and permeability of the Earth were constant, every open-field site would have the same effect on the electromagnetic fields radiating from the EUT. The Earth’s parameters do vary, however. To compensate for this variation, a large conductive floor, or ground plane, is laid under the EUT. This causes all electromagnetic waves to be totally reflected from the ground plane, so that the Earth’s properties have no effect. The ground plane must be large enough so that it appears infinite with respect to the EUT and the associated test equipment. Acceptable dimensions are $1.73d \times 2d$, where d is the distance between the measurement antenna and the EUT, normally 3 or 10 m. Radiated emissions must be measured in all directions from the EUT, and at various angles of inclination. This is most easily achieved by placing the EUT on a turntable, which is then rotated during the test. To allow measurement at various inclination angles, the receiving antenna height must be varied, and this is accomplished by mounting it on a halyard. A typical open-field site appears in Figure 29.5.

An open-field site provides repeatable data only if there are no nearby trees or structures that could cause undesirable reflections. Before it can be reliably used, it must be tested. This is done by generating a known electromagnetic field and measuring it. The field is normally generated by a radio frequency oscillator driving a tuned dipole antenna, for which the radiation can be calculated (pp. 237–238 of Reference 2). This radiation is then measured as though it were generated by a typical device being tested. The ratio of the voltage at the transmitting antenna terminals to that at the receiving antenna terminals is known as the **site attenuation**. If the site attenuation is within 3 dB of its calculated value, the test site is deemed acceptable.

Although an open-field site eliminates reflections, external field sources, such as licensed transmitters, still cause problems. Since electromagnetic radiation can travel thousands of miles, no open-field site will be completely free of electromagnetic fields. To eliminate the effects of these stray sources, testing must be performed inside a shielded enclosure. There, however, severe reflections occur, and measurements become inaccurate and unrepeatable.

An ideal test environment would be a shielded enclosure lined with material that does not reflect electromagnetic waves. Such an enclosure is called an *anechoic chamber*, with the understanding that the name refers to *electromagnetic* echoes. Until recently, such chambers were not practical except at very high frequencies, but improvements are constantly being made. Such an enclosure is acceptable for testing

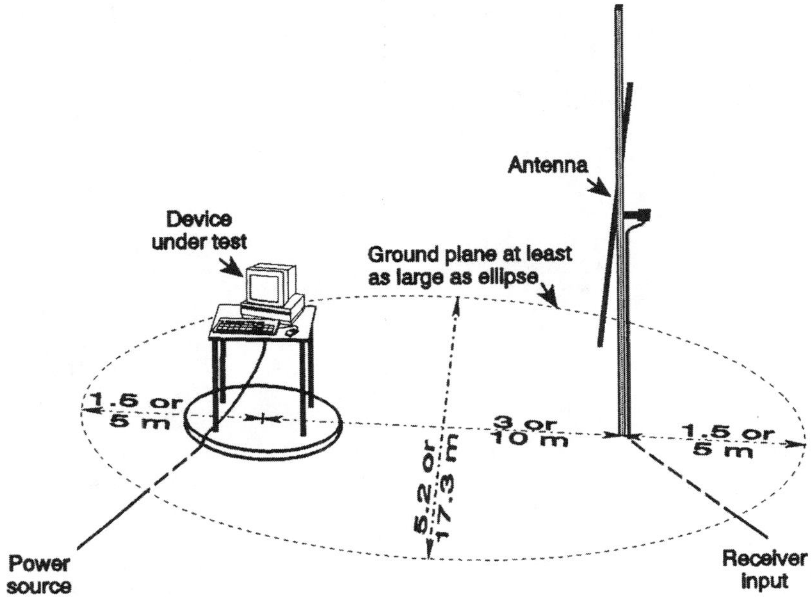

FIGURE 29.5 An open-field test site. Power and antenna cables are run under the ground plane so that they will not affect measured fields. The area outlined by the ellipse must be free of everything except the device under test, the table on which it rests, and the measuring antenna. To facilitate measuring radiation in all directions from the device, it is placed on a turntable. By rotating it during testing, and simultaneously varying the height of the receiving antenna, the direction of maximum radiation is found.

if it meets the site-attenuation requirements of a true open field. The site attenuation must be measured at several points inside the chamber, to assure that the proximity of the chamber walls has no effect. Unfortunately, such chambers are at present very expensive.

Another type of test chamber is the **transverse electromagnetic (TEM) cell**. This consists of an enlarged section of waveguide, in which the electromagnetic fields can be accurately predicted [3,4]. They are suitable only for testing small devices at relatively low frequencies. The TEM cell can be no larger than a wavelength at the frequency being tested. For example, to test at 200 MHz, the cell could be no larger than 1.5 m, or 5 ft, and the device itself must not exceed 1/6 of this value, or 10 in. For small devices, however, the TEM cell is very accurate and is unaffected by stray field sources.

If a suitable anechoic chamber is not available, a device may be tested in an ordinary shielded enclosure to learn what frequencies it emits. The field strengths will be inaccurate due to the internal reflections. Then the device is tested in a true open-field site, and the suspected frequencies are measured quantitatively. Any field that exceeds the acceptable limits is then observed while the device is shut off. If it does not disappear, it is obviously not being generated by the EUT. This procedure is acceptable, although not as simple as testing inside an anechoic chamber.

Preliminary measurements may even be performed in an ordinary room. They will not be comparable with similar measurements made anywhere else, because of the effects of nearby conductors and dielectrics. Here, also, the device must be shut off to decide if any emissions are from stray external sources instead of from the EUT. This procedure provides a rough estimate of the emissions from the EUT, and it usually saves time during any later testing at a true open-field site. The various measurement methods appear in Table 29.4.

Permissible emission levels appear in the *Code of Federal Regulations* [5]. These rules assume open-field measurements, which are the most accurate possible. Even there, variations of ±6 dB are typical. Therefore, a manufacturer should allow a safety factor when performing measurements intended to assure compliance with government regulations. Otherwise, a device may pass when tested by the manufacturer

TABLE 29.4 Comparison of EMI Measurement Methods

Method	Equipment Required	Space Required	Accuracy	Outside Influence	Cost	Comments
Ordinary room	Antenna and receiver	3 or 10 m radius around EUT	Medium, affected by structure, ±20 dB	May be severe, depending on location	Minimum	Usually acceptable for preliminary tests
Shielded room	Shielded room, antenna, and receiver	4 to 6 m radius around EUT	Poor, ±30–40 dB due to reflections	Usually none	Moderate	Use for preliminary tests in noisy areas
TEM cell	TEM cell and receiver	1 to 3 m^3	Very good, ±10 dB	Usually none	Moderate	Unusable for large EUT due to high-order modes
Open field	Antenna and receiver	17 × 20 m open field with no nearby structures	Excellent, usually ±6 dB	May be severe, depending on location	High, due to logistics of site (power, weather, etc.)	Standard test method
Shielded anechoic chamber	Anechoic chamber, antenna, and receiver	6 to 15 m radius around EUT	Very good, ±10 dB	Usually none	Very high	Use for accurate tests in noisy areas

but fail if later tested by the government using a supposedly identical test procedure. Since the government's measurements then prevail, the manufacturer's integrity could be questioned.

Further details on measurement techniques are available in the References 2 and 6.

Defining Terms

Antenna factor: Its effective length expressed in negative decibels.

Balun: An interface device used to isolate a dipole or other balanced antenna from the effects of a receiver having one grounded terminal.

Conducted coupling: Coupling due to voltages imposed on a shared power source.

Dipole antenna: An antenna consisting of two collinear rods with the feed line connected between them.

Driving-point impedance: The ratio of voltage to current at the driving point (normally the center) of an antenna.

Effective length: The ratio of the voltage observed at the driving-point of an antenna to the strength of its received electric field.

Electromagnetic compatibility (EMC): The capability of two or more electrical devices to operate simultaneously without mutual interference.

Electromagnetic interference (EMI): Any undesired effect of one electrical device upon another due to radiated electromagnetic fields or due to voltages imposed on a shared power source.

Elliptical polarization: Polarization of an electromagnetic wave consisting of two perpendicular electric fields of differing phase.

Emissions: Fields or conducted voltages generated by an electrical device.

Far-field region: Any location that is much farther than $\lambda/2\pi$ from an electric or magnetic field source, where λ is the wavelength at the frequency of concern.

Intersystem EMI: Electromagnetic interference between two or more systems.

Intrasystem EMI: Electromagnetic interference between two or more parts of the same system.

Near-field region: Any location that is much nearer than $\lambda/2\pi$ to an electric or magnetic field source, where λ is the wavelength at the frequency of concern.

Open-field site: A test location free of any conductors which would affect electromagnetic fields and taint the results.

Polarization: The direction of the electric field, *E*, of an electromagnetic wave.

Power density: Radiated power per unit of cross-sectional area.

Radiated coupling: Coupling due to radiated electric, magnetic, or electromagnetic fields.

Site attenuation: A measure of the degree to which electromagnetic fields at a test site are disturbed by environmental irregularities, obtained by comparing calculations with measured experimental results.

Susceptibility: The degree to which an electrical device is affected by externally generated fields or conducted voltages.

Transverse electromagnetic (TEM) cell: A relatively small test chamber in which fields can be accurately controlled by its geometric properties.

References

1. E.C. Jordan and K.G. Balmain, *Electromagnetic Waves and Radiating Systems,* 2nd ed., Englewood Cliffs, NJ: Prentice-Hall, 1968, 540–547.
2. J.P. Mills, *Electromagnetic Interference Reduction in Electronic Systems*, Englewood Cliffs, NJ: Prentice-Hall, 1993, 232–233.
3. M.L. Crawford, Generation of standard EM fields using TEM transmission cells, *IEEE Trans. Electromagn. Compat.*, EMC-16, 189–195, 1974.
4. M.L. Crawford and J.L. Workman, Predicting free-space radiated emissions from electronic equipment using TEM cell and open-field site measurements, *IEEE Int. Symp. Electromagn. Compat.*, Baltimore, MD, Oct 7–9, IEEE, Piscataway, NJ, 1980, 80–85.
5. Part 15 — Radio Frequency Devices, in *Code of Federal Regulations, Volume 47: Telecommunications,* Washington, D.C.: U.S. Government Printing Office, 1991, subpt. A-B.
6. Methods of measurement of radio-noise emissions from low-voltage electrical and electronic equipment (ANSI C63.4-1988), *American National Standards Institute (ANSI) Standards,* New York: The Institute of Electrical and Electronics Engineers, 1989.

Further Information

Guide for construction of open area test sites for performing radiated emission measurements (ANSI C63.7-1988), *American National Standards Institute (ANSI) Standards,* New York: The Institute of Electrical and Electronics Engineers, 1988.

H.W. Ott, *Noise Reduction Techniques in Electronic Systems,* 2nd ed., New York: John Wiley & Sons, 1988.

W.H. Hayt, *Engineering Electromagnetics,* 5th ed., New York: McGraw-Hill, 1989.

J.D. Kraus, *Electromagnetics,* 4th ed., New York, NY: McGraw-Hill, 1992.

S. Ramo, J.R. Whinnery, and T. Van Duzer, *Fields and Waves in Communication Electronics,* 2nd ed., New York: John Wiley & Sons, 1984.

III

Displays

30

Human Factors in Displays

Steven A. Murray
California State University

Barrett S. Caldwell
Purdue University

30.1 Introduction

The display system is the final link between the measuring process and the user. If the display is not easy to see and easy to understand, then that process is compromised. The user's sensory capabilities and cognitive characteristics, therefore, must both be addressed in display system selection. Furthermore, display technologies and performance capabilities are easier to evaluate in the context of their intended application. Consideration of the following issues can narrow the search for candidate systems, and can prevent needless frustration during system use:

Environment. Will the display be operated in sunlight or at night?
Application. Will the display present alphanumeric data, video images, graphics, or some combination?
Task scenario. Are portability, handheld operation, or group viewing required?
System characteristics. Weight, volume, power, maintenance, cost, etc.

This chapter begins with basic treatments of light and vision. It then proceeds to discussions of visual capabilities and the display characteristics that must be matched to them.

0-8493-1733-9/04/$0.00+$1.50

30.2 Fundamentals of Light Measurement

The foundation metric of light is *luminous flux,* which is the rate at which light energy is emitted from a source, and is expressed in lumens (lm). *Luminous intensity* is luminous flux per unit solid angle, and its unit of measurement is the candela (cd). This is distinguished from *illuminance,* or illumination, which is simply luminous flux per unit area, expressed as lux (lx). *Luminance* is a measure of the brightness, i.e., the amount of light, per unit area, either emitted by or reflected from a surface. Units of luminance measurement are candelas per square meter (cd/m^2) or nits. Finally, *reflectance* is a unitless ratio of the amount of light striking a surface to the amount of light leaving it:

$$R = \pi \times \frac{\text{luminance}}{\text{illuminance}} \tag{30.1}$$

High reflectance can create glare, dramatically reducing visual performance.

30.3 Fundamentals of Vision

The eye functions very much like a conventional camera. Light enters the eye through a transparent *cornea* and is modulated by the *pupil,* a variable aperture opening controlled by muscles of the *iris.* The pupil grows larger in dark surroundings and smaller in bright surroundings to control the range of light intensity to the eye. Light rays are then refracted by an adjustable *lens* and brought into focus on the *retina,* where neural imaging begins. The retina contains both *cones* and *rods,* two distinctly different types of photoreceptors. Cones are concentrated near the *fovea,* or the central 2° of the visual field, and decrease rapidly with distance from this region. In contrast, rods are essentially absent in the fovea and increase in density with distance.

The eye is sensitive to three characteristics of electromagnetic radiation: (1) *brightness* (the intensity of ambient or incident light, measured in lux), (2) *hue* (the wavelength of light, measured in nm), and (3) *saturation* (relative concentration of specific hues in light, measured as a dimensionless ratio from 0 to 1). Cones are differentially sensitive to wavelength, i.e., hue, and have greater resolving power than rods because of their one-to-one mapping onto visual nerves. Cones can be further divided into three types, each maximally sensitive to a different portion of the visible light spectrum: (1) red (670 nm peak), (2) green (470 nm peak), and (3) blue (400 nm peak). Rods are more sensitive to light than cones and have many-to-one connections with the nerves that exit the eye, a feature that permits neural summation of low light signals. Human ability to discriminate differences in levels of brightness, saturation, or hue is governed by a psychophysical function known as *Weber's law:*

$$K = \frac{\Delta I}{I} \tag{30.2}$$

where ΔI is the difference, or change, in intensity, I is the initial intensity, and K is the Weber fraction. Values of K have been experimentally determined for brightness (0.079), saturation (0.019 for red), and hue ($\approx$0.02 to 0.04, depending on the region of the visible spectrum).

Photopic vision occurs at light levels where both rods and cones are sensitive. The minimum light intensity required for photopic vision is approximately 2 lx; colors are seen in this region. As brightness decreases, a transition from photopic to scotopic vision takes place and color perception drops out gradually, a phenomenon that can impact the interpretation of color-coded information in poor light. Perception of blues and reds is lost first, then cyan and yellow-orange, and finally green, i.e., the wavelengths where the eye is most sensitive. The eye becomes most sensitive to wavelengths of about 550 nm (green) near the limit of photopic vision. *Scotopic vision* occurs at low light levels (2×10^{-7} lx to 2 lx) and primarily involves the rods; only achromatic shades of gray are seen. The transition from photopic to scotopic vision occurs slowly, requiring approximately 30 min for complete adjustment from photopic to scotopic visual sensitivity.

30.4 Visual Performance

Visual acuity is the ability to discriminate detail. The action of the lens, to change focus for objects at different distances, is called *accommodation. Minimum separable acuity,* the most common measure of discrimination, is determined by the smallest feature that the eye can detect, and is measured in terms of the reciprocal of the visual angle subtended at the eye by that feature. *Visual angle,* in minutes of arc, is calculated as

$$\mathrm{VA} = \frac{3438H}{D} \tag{30.3}$$

where H is the height of the object and D (in the same units) is the distance from the observer. The ability to distinguish an object from its surroundings is known as *visibility.* The term is related to visual acuity, but implicitly combines factors of object size, contrast (i.e., including differences in hue and saturation), and brightness that all interact to determine true visual detection performance. On a more functional level, *readability* or *legibility* describes the ability to distinguish meaningful groups of objects (e.g., words extracted from groups of letters on a display).

Other parameters affecting visual performance include viewing angle and viewing distance. *Viewing angle* at the eye is measured from a line through the visual axis to the point being viewed, and determines where an object will register on the retina. The best image resolution occurs at the fovea, directly on the line of gaze, and visual acuity degrades with increasing angle away from this axis. *Viewing angle* at the display is the angle, in degrees, between a line normal to the display surface and the user's visual axis. The best viewing angle is, of course, on the visual axis and normal to the display surface, as luminance falls off for most displays as the angle from normal increases. Luminance reduction with viewing angle can be calculated as

$$E = E_{\mathrm{m}} \cos^4 \theta \tag{30.4}$$

where E_{m} is the illuminance at the center of the display and θ is the viewing angle. Note that two viewing angles — at the eye and at the display — have been defined. *Viewing distance* is determined primarily by the minimum size requirements (i.e., visual angle) for objects that the user must see. A conventional reading distance is about 71 cm, although VDTs are frequently read at 46 cm. Most design criteria assume a viewing distance of between 50 and 70 cm.

Visual fatigue is an imprecise term, but one in common use, referring to the annoyance, irritation, or discomfort associated with visual tasks performed under poor conditions or for extended periods of time. A common cause of visual fatigue is *glare,* which can be due to a combination of high brightness, high reflectance, and specular (mirrorlike) reflections causing source light to reflect directly into the eye. Minimizing or eliminating glare is essential for effective display performance, and usually involves a thoughtful placement of the display, proper positioning of room lights, control of ambient light (e.g., window shades), or the use of display hoods.

30.5 Display Performance Considerations

Resolution is a measure of the smallest resolvable object on a display, and is expressed as display lines per millimeter or centimeter. Although *sharpness,* and its converse *blur,* are normally defined by the subjective reaction of the display user, sharpness has been formally measured as the ratio of the blurred border zone of letters to their stroke width [3]. Legibility is related to character quality, or readability, and depends on the sharpness of characters.

Contrast, or contrast ratio, is the measure of the luminance difference between an object and its background. While different definitions exist in the literature, luminance contrast as adopted by the International Lighting Commission (CIE) is given as

$$C_R = \frac{\text{luminance of brighter object} - \text{luminance of darker object}}{\text{luminance of brighter object}} \tag{30.5}$$

Lower luminance displays require greater contrast to achieve the same visibility of objects. The *contrast,* or *luminance ratio* between two surfaces in the central field of vision (e.g., a display and the desk on which it rests) should be around 3:1, while the ratio between the central field and surfaces farther away (e.g., around a room) can be as high as 10:1. Ratios greater than this can induce glare. The simplest methods for contrast enhancement are the use of hooded shades or displays that can be tilted away from the offending light. Contrast-enhancing *filters,* however, can be more effective. All filters involve reducing the amount of ambient light reflected back to the user, while leaving the emitted light from the display content as unchanged as possible. Several strategies for filtering exist, including *etching* or *frosting* the display surface to break up and scatter specular reflections. *Neutral density filters* increase contrast by reducing the amount of light passing through them; ambient light must pass through twice before reaching the user's eye, while display content must only pass through once. Micromesh filters placed on the display surface limit light penetration so only rays falling perpendicular to the mesh can penetrate; this stops both specular and diffuse reflections and increases contrast. *Circular polarizers* are neutral density filters that polarize incident light, which is then prevented from returning through the filter. *Quarter-wave thin film coatings* interfere with both specular and diffuse reflections.

Gray scale refers to the number of luminance levels, or shades of gray, available in a display. The common definition is a luminance ratio of 1.4 between levels, although the eye can discriminate changes as small as 1.03 (a Weber's K value of 0.03). The number of gray shades is useful for evaluating the capability of a display to render pictorial information or the range of luminance levels that can be used for coding. The highest luminance level is determined by display capabilities, but the lowest level is determined by the luminance of the display surface when no signal is present. Bright light incident on the display can elevate this minimum level and reduce the number of usable gray shades.

Flicker is the term for detectable changes in display luminance, and occurs when the frequency of those changes is below the integrating capability of the eye. The minimum frequency at which this occurs is the *critical flicker fusion* frequency, or CFF, which depends on the luminance level of the image, i.e., displays which do not flicker at high luminance levels may still flicker at low levels. The CFF is calculated as

$$CFF = a \log L_{\text{a}} + b \tag{30.6}$$

where $a = 12.5$ for high (photopic) ambient light levels and 1.5 for low (scotopic) levels, L_{a} is the average luminance of the image in cd/m^2, or nits, and $b = 37$. This is an empirical formula, and the values for a and b are only approximate. Because the eye cannot adapt to flicker fast enough to control the light on the retina, visual irritation usually occurs where flicker is present.

Many display parameters are stated in terms of the *pixel,* or picture element. The pixel is the smallest addressable element in an electronic display, or the smallest resolvable information element seen by the user. *Refresh rate* is the frequency with which display pixels are reilluminated. Refresh rates below 50 to 80 Hz may induce perceptible flicker. The *update rate* is the frequency with which the information content of display is changed.

Linearity is the deviation of a straight line from its exact form, expressed as a percentage of total line length. Pattern distortion is the deviation of any element of a pattern (e.g., a grid) from its exact location, expressed in dimensions of the total pattern. While no specific limits are associated with these parameters, interpretation of measurement data can obviously be affected if nonlinearities are observable on the display.

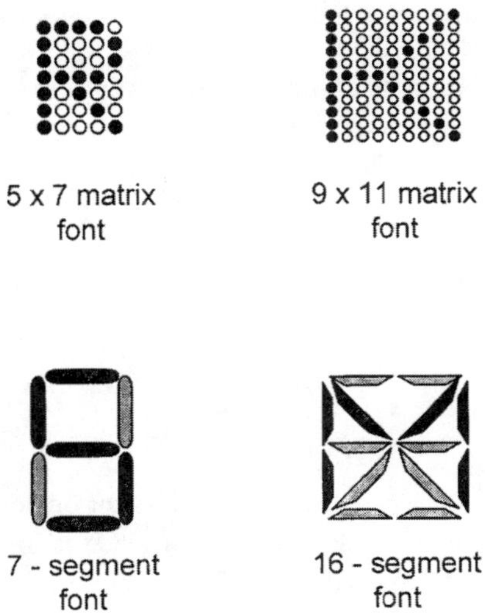

FIGURE 30.1 Examples of alphanumeric displays.

Several mathematical models have been developed to quantify display *image quality* as single "figures of merit." These models, while useful, are too involved for treatment here and the reader is referred to excellent summaries in the literature [1] for further information.

30.6 Display Content Considerations

Configurable software packages for scientific measurement (e.g., LabView™) allow great flexibility in the design of display formats. Human factors principles for display content, therefore, are as important to effective measurement as the electronic characteristics of the display itself. The following principles are an introduction to the kinds of issues that must be considered when designing display content. The interested reader is referred to Helander (1987) and Grandjean [3]) for information on human factors and design guidelines beyond those presented here.

Alphanumeric Displays

The size of a letter is its *pitch.* A general recommendation is that characters should subtend a minimum of about 12 min of arc at common reading distances. Alphanumeric displays are usually constructed of pixel arrays or segmented bars (Figure 30.1). A 5 × 7 pixel array is considered the minimum necessary to represent the ASCII character set. More pixels can increase legibility, but at higher cost for the control electronics. The seven-segment bar matrix is also a common design and has good performance, but up to 16-segment arrays are available for better appearance.

Stroke width is the ratio of the thickness of the stroke to the height of the letter. Recommendations for legibility are 1:6 to 8 or 1:8 to 10 [2]. As illumination is reduced, thick letters are easier to read than thin ones. With low illumination and low contrast, letters should be boldface type with low stroke width–height ratios (e.g., 1:5).

Quantitative and Status Displays

While numeric readout displays are easy to implement in software, analog pointer displays show an advantage when it is important to observe the direction or rate of change of the values presented [4]. If the measurement application involves "more or less" or "up or down" interpretations, a straight-line or thermometer scale is preferred because it shows the measurement in relation to zero. *Moving pointers* are better able to convey minor changes in readings than fixed pointers or numeric readouts. *Scale markings* should reflect the length of the smallest unit to be read. It is desirable to have a marker for each unit, so that no interpolation is required.

Check reading indicators are used to determine whether a condition is "normal." The normal criterion point, therefore, should be clearly coded. If several indicators are grouped together, they should be arranged so that the deviant reading stands out (e.g., by indicating a different column height or dial angle, etc.).

Color is an excellent method for organizing information on a display and for locating objects rapidly. Although display users can distinguish between many different colors, they usually cannot remember more than seven to ten of them, so the number should be limited if color is going to be used as a coding dimension.

Summary

The next sections address different display technologies in light of the principles discussed here. While display guidelines are available for essentially any parameter, it is important to remember that visual perception is an integrative process. No single guideline functions alone, and display quality is usually a product of interacting needs and trade-offs.

30.7 Cathode Ray Tube Displays

The cathode ray tube (CRT, see Chapter 31) is by far the most common display technology in use today, and its widespread use in televisions and computer monitors should guarantee its continued presence throughout the foreseeable future. Advantages of CRT-based displays include (1) versatility (the same CRT can be used for alphanumerics, pictures, or graphics), (2) high-resolution capability and high luminous efficiency, extremely fast dynamic response (which can be important for rapidly changing signals), (3) extensive commercial availability (e.g., 2.5 to 64 cm, diagonally), (4) high reliability, (5) long life, and (6) relatively low cost. CRT displays can function well in high ambient illumination if filtering is used. Potential disadvantages of CRT displays are bulk (the depth of conventional CRT tubes can match or exceed their diagonal dimension, although flat CRTs are available), and vulnerability to ambient reflections and high illumination. Light falling on the smooth display surface can produce a veiling illuminance that **washes out** screen contrast and reduces the number of colors that can be perceived.

CRT Performance

Many characteristics of CRT displays depend on the type of phosphor selected for the design. Phosphor materials vary widely in their luminous efficiency, their color, and their decay time. **Decay time** interacts with display refresh rate; a phosphor with a short decay time will require a higher refresh rate to avoid observable flicker effects. Selecting a CRT with a high phosphor decay time will also result in selecting a tube with a high average luminance. Resolution depends on the **spot size** of the energized phosphor (which is effectively the thickness of the raster line). Spot size will also depend on the acceleration voltage and beam current of the cathode gun, so manufacturer's data must be noted for the voltage and current where the measurements were recorded, and compared with expected operating conditions. CRT resolution is measured in two ways. *Television lines* are the number of the most closely spaced, discernible lines that can be distinguished on an EIA (Electronic Industries Association) test chart. *Shrinking raster lines* involve a process of reducing the spacing between lines of a test pattern until no lines can be seen. This point is then expressed as the number of lines per centimeter, and is a better metric for measurement display applications. Increasing numbers of scan lines improve symbol legibility and image quality. Most conventional CRT monitors have 525 scan lines, but up to 5000-line systems are available [5]. The primary method for achieving *color CRT* displays is the use of single or multiple electron beams to energize phosphors for three primary colors — red, blue, and green. A complete range of colors is obtained by selectively energizing appropriate combinations of these three basic phosphors. Beam efficiency is reduced in this process, which means that color CRT displays are not as bright as monochrome CRT systems.

Contrast ratio is diminished by high ambient light levels, and it is often necessary to compromise between light requirements for work tasks and light levels for optimum CRT visibility. In low ambient lighting conditions, a CRT contrast ratio of 10:1 is usually attainable. A ratio of 25:1 can be achieved with contrast-enhancing filters, but at the expense of brightness.

Types of CRT Displays

In addition to the conventional, raster-scanned CRTs used for computer monitors and workstations, two variants of CRT technology should be mentioned, the direct-view storage CRT and the flat CRT. *Direct-view storage CRTs* have been designed to get around the need to refresh phosphors constantly. Direct-view systems usually add an additional, electrically charged layer — the storage element — somewhere behind the phosphor layer and an additional electron gun to maintain this charge. Displayed information is retained until it is actively erased. *Flat CRTs* have been developed to answer the need for CRT performance in a smaller physical package. The basic design technique places the electron gun at right angles to the screen and, through additional focusing circuitry or a slightly angled phosphor screen, writes the raster pattern at a high angle. Additional detail on these and other CRT designs can be found in Sherr [1].

30.8 Liquid Crystal Displays

Liquid crystal displays (LCDs, see Chapter 32) belong to the class of nonemissive technologies, i.e., displays that do not generate light of their own but control the transmission or reflection of an external light source. LCDs alter the optical path of light when an electric field is placed across the liquid crystal (LC) material.

The principal advantages of LCDs include (1) very low power consumption (important for battery-operated and portable systems such as calculators), (2) a flat display package, (3) low cost of the basic materials, and (4) excellent contrast in high ambient illumination. Some LCDs, however, have slow dynamic response (i.e., for switching display elements on and off); 100 to 500 ms rise times, for example, are visually noticeable and such systems may be unacceptable for measurement applications. Low luminance is another drawback, and can make the display difficult to read in low-light conditions without an external light source. In addition, viewing angle is limited by inherent LC characteristics, and is usually less than 45° without special designs. Many LCD features, such as switching thresholds and response times, are temperature dependent.

LCD Performance

A full range of resolution capabilities is available, from simple alphanumeric displays to systems with 63 million pixels and resolutions of 47 lines/cm [6]. LCDs are primarily used in small display applications (e.g., calculators, watches, etc.), although 53 cm diagonal, full-color video-capable displays have been developed [7].

Contrast in polarized systems is determined by the *extinction ratio* of the polarizer, i.e., the ratio of light transmitted in the parallel polarizing orientation to light transmitted in the cross-polarizing orientation. Polarizers with good extinction ratios, however, also suffer high loss of light in the transmitting orientation, so maximum brightness is traded for contrast. Contrast ratios of up to 50:1 [6] have been achieved, although 20:1 is more common.

Color displays can be achieved by placing a color mosaic over the LCD and switching the cells behind the proper combinations of mosaic holes into their transmission states. This method reduces resolution and brightness, however, as the available pixels must be assigned to each of the three primary colors. The use of thin-film transistors (TFT) as a switching technology for LCDs is the latest approach to generating large, high-resolution displays and is the subject of extensive engineering research [8].

30.9 Plasma Displays

The simplicity and durability of plasma displays (see Chapter 33) makes this technology an attractive candidate for diverse measurement needs, especially where harsh environments are expected. In addition, the switching characteristics of plasma gases have not yet been fully exploited, and this technology offers excellent potential for future engineering improvements. Plasma methods are used extensively for alphanumeric displays in portable, laptop, and handheld computers, and for the display of video imagery. Advantages of plasma displays include enhanced memory capability, brightness, and luminous efficiency. It is also possible to retain pixels in the on state without continuous refresh signals, which means that increased brightness can be obtained for the same power and driving circuitry. This advantage also allows for excellent contrast ratios in high ambient illumination.

Plasma displays also exhibit long display life and ruggedness. It is not unusual for the display to outlast the life of the product in which it is installed, and the relatively simple panel construction can tolerate high shock and vibration environments, or extremes of temperature. Some disadvantages of plasma displays include high cost (relative to CRTs) and high power requirements (relative to LCDs). Other technologies can compete effectively with plasma devices for small alphanumeric displays in benign conditions.

Plasma Display Performance

Commercial plasma displays are available with resolutions of 40 lines/cm, and systems with almost 50 lines/cm are under development. Systems of 2 million pixels have been constructed [9]. Gray scale is achieved with dc displays by adjusting the discharge current in each cell. Displays using ac voltage can trade resolution for gray scale with spatial modulation methods, i.e., by controlling the number and location of activated pixels, rather than the level of pixel illumination. While plasma displays have good gray scale, their brightness is not yet equivalent to that of CRTs. Plasma displays show the color of the ionized gas, usually orange (i.e., where neon is used), although different gas mixtures or phosphors have been successfully used to expand the range of colors. The *hybrid ac-dc display* was designed to combine the memory capability of ac systems with the efficient matrix circuitry of dc devices [10]. The display uses both types of current to generate gas discharge; the dc component controls the pixel addressing, while the ac component controls the memory states of the cells. The *hybrid plasma–CRT display* attempts to use the gas discharge effect of the plasma panel as a matrix-addressable source of electrons for the CRT. The result is a full-color system with high brightness and good luminous efficiency.

30.10 Electroluminescent Displays

With the exception of light-emitting diodes (LEDs, see Chapter 35), electroluminescent (EL, see Chapter 34) technologies are not as prominent in the commercial arena as other types of display systems. EL materials are complex (i.e., driven and controlled by processes related to solid-state semiconductor physics) and are more difficult to work with than other display materials. Nevertheless, they offer great potential for high brightness and low cost that deserves consideration, especially as new designs become available. Matrix addressing is used for control of information display applications. EL materials are applied in two forms — powders (PEL) and thin films (TFEL) — and are controlled by both ac and dc voltages, generating four basic design approaches. Some advantages of EL displays are (1) high luminous efficiency (except ac powder designs), (2) readability in sunlight, (3) color capability, (4) compact, flat panel designs, and (5) significant potential for low-cost manufacture. A disadvantage of EL displays is that ac powder (ACPEL) systems have low luminance and contrast ratio. In addition, phosphors in powder designs scatter and reflect ambient light, reducing their contrast.

EL Display Performance

Contrast ratios of 50 to 150:1 have been demonstrated with monochrome thin-film (ACTFEL) systems, while 15 to 20:1 have been achieved with dc designs. ACTFEL designs also show excellent brightness, with demonstrated luminance levels of over 157 cd/m^2 (monochrome) and 26 cd/m^2 (color). DCPEL designs, representing newer technologies, have achieved over 100 cd/m^2 with monochrome designs, but may soon meet or exceed ac-based values. Resolution of EL displays is limited by the duty factor of matrix addressing; a finite amount of time is needed to energize each row, and a minimum luminance level is needed for adequate display performance, so the remaining variable becomes the number of addressable lines. A demonstrated ACTFEL display with 640 × 400 elements, with six colors and a resolution of 27 lines/cm, is typical of this technology.

30.11 Light-Emitting Diode Displays

Light-emitting diode (LED) displays (see Chapter 35) involve single-crystal phosphor materials, which distinguishes them from the polycrystal EL materials discussed in the previous section. The basic physics behind their operation is, however, quite similar. LED displays are highly versatile and well suited to a variety of measurement applications. Advantages of LED displays include high reliability and graceful degrades; individual LED elements can fail without affecting overall display performance. LEDs are rugged, for operation in harsh environments, and they are more tolerant of temperature extremes than other technologies. LEDs demonstrate better viewing angles than LCDs, and excellent brightness for visibility in sunlight. Unfortunately, LED displays also have high power consumption when packaged in large, flat panel displays, and the cost is high for the complex assembly. Optical cross talk between array elements can occur if transparent substrates are used. LEDs are the most restricted display in terms of color range (e.g., no blue device is commercially available).

LED Display Performance

LED devices have excellent brightness, but because display brightness is also a function of the filters or magnification lenses used over the LED elements, device luminance is not, by itself, a reliable measure of overall system performance. LED displays also show very good luminance contrast. *Chrominance contrast*, however — the color difference between the LED and its background — is a factor in evaluating LED performance that is not found in other technologies. Chrominance contrast is significant because of the high saturation of most LED phosphors. It is affected by display filters, and can have significantly more influence on display performance than luminance contrast.

LED displays with resolutions of 20 to 25 lines/cm have been constructed. Flat panel displays of 38,400 discrete elements have also been demonstrated with luminance levels of around 137 cd/m^2 (increasing to 240 cd/m^2 with reduced resolution), and at least one aircraft display with 49,000 elements has been built. CRT-equivalent displays with 600 × 400 elements have also been realized with engineering development models.

Defining Terms

Decay time: The time required for the peak brightness of a phosphor to drop to a defined fraction of peak luminance; a measure of how long the phosphor remains illuminated after being energized by the electron beam.

Duty cycle: The time spent addressing each pixel during a refresh cycle; inversely proportional to the number of pixels.

Font: Refers to the form in which alphanumerics and symbols are produced.

Spot size: The size of the illuminated spot from the electron beam; limits the size of the raster line.

Transillumination: Illumination from the side of a display surface, to highlight information on the surface itself, e.g., lighting for automobile or aircraft instruments.

Wash out: The loss of contrast (i.e., reduction in dynamic range) in an LED as the ambient light reflected off the background of the display surface approaches the light level of the active area.

References

1. S. Sherr, *Electronic Displays*, 2nd ed. New York: John Wiley & Sons, 1993.
2. M.S. Sanders and E.J. McCormick, *Human Factors in Engineering and Design,* 6th ed. New York: McGraw-Hill, 1987.
3. E. Grandjean, Design of VDT workstations, in G. Salvendy, Ed., *Handbook of Human Factors,* New York: John Wiley & Sons, 1987.
4. D.G. Payne, V.A. Lang, and J.M. Blackwell, Mixed vs. pure display format in integration and nonintegration visual display monitoring tasks, *Human Factors*, 37(3), 507–527, 1995.
5. N.H. Lehrer, The challenge of the cathode-ray tube, in L.E. Tannas, Jr., Ed., *Flat-Panel Displays and CRTs*, New York: Van Nostrand Reinhold, 1985.
6. M. Hartney, ARPA display program and the national flat-panel display initiative, in *Proc. 2nd Int. Workshop on Active Matrix Liquid Crystal Displays,* IEEE, 8–15, 1995.
7. T. Morita, An overview of active matrix LCDs in business and technology, in *Proc. 2nd Int. Workshop on Active Matrix Liquid Crystal Displays,* IEEE, 1–7, 1995.
8. I-W. Wu, High-definition displays and technology trends in TFT-LCDs, *J. Soc. Inf. Display*, 2(1), 1–14, 1991.
9. L. Weber, Plasma displays, in L.E. Tannas, Jr., Ed., *Flat-Panel Displays and CRTs*, New York: Van Nostrand Reinhold, 1985.
10. Y. Amano, A new hybrid ac–dc plasma display panel, *J. Soc. Inf. Display*, 2(1), 57–58, 1994.

Further Information

E. Grandjean, *Ergonomics in Computerized Offices*, London: Taylor & Francis, 1987, is an excellent all-around treatise on the principles of effective VDT selection and use. Summarizes a wide range of research literature. If this volume is difficult to obtain, a chapter by the same author is also included in the *Handbook of Human Factors* (Reference 2).

M.G. Helander, Design of visual displays, in G. Salvendy, Ed., *Handbook of Human Factors,* New York: John Wiley & Sons, 1987, is an excellent and concise review of major human factors principles for display design and use. Includes a critical review of the foundation literature in this area.

S. Sherr, *Electronic Displays*, 2nd ed., New York: John Wiley & Sons, 1993, offers clear presentations of all important display technologies, together with a good summary of performance measurement methods for display systems. Well illustrated with a variety of commercial products.

L.E. Tannas, Jr., Ed., *Flat-Panel Displays and CRTs*, New York: Van Nostrand Reinhold, 1985, provides a thorough, yet highly readable examination of the physical principles behind essentially every major display technology. Although the technology capabilities have become dated since publication, this is well worth review.

31

Cathode Ray Tube Displays

Christopher J. Sherman

31.1 Introduction

The cathode ray tube (CRT) is unequaled in its ability to produce dynamic, quality, high-information-content imagery at high resolution. Even more impressive is that it achieves this for a lower cost per pixel than any other comparable electronic display technology. For the instrument designer requiring a high-information-content display, it offers numerous advantages. As a raw image tube, it is commonly available as an off-the-shelf item with a broad infrastructure of vendors, integrators, support, and part suppliers. Interface standards are well established and as a complete system, ready to take a standard signal input, it is available worldwide in a variety of performance ranges. To meet different application requirements, it is available in diagonal sizes from 12 mm to over 1 m with resolution from thousands of pixels to over 5 million pixels per frame. Tube characteristics improve on a yearly basis, and prices continue to decrease. Stanford Resources has been tracking the CRT market for almost 20 years, and is an excellent source of information. In its latest report, it indicates that despite competition from other display technology, the CRT will remain the single largest market in the display industry. The worldwide market for CRT tubes in 1997 was 261 million units worth $26 billion U.S. This is expected to grow to 341 million units worth more than $34 billion by 2003 [1, 2]. Although there are many competing information display technologies, the CRT will be with us well into the 21st century.

0-8493-1733-9/04/$0.00+$1.50

31.2 History

The CRT has a rich and distinguished history. The roots of today's CRT technology extend back more than 100 years to the latter half of the 19th century. Eugene Goldstein first introduced the term *cathode rays;* John W. Hittorf, Heinrich Geissler, Julius Plücker, and others made important contributions specific to CRT technology [3]. Throughout the 19th century, researchers were interested in the nature of a luminous gas discharge and shadowy rays that occurred when a high voltage potential was applied between two electrodes in a vacuum. Sir William Crookes was an active experimentalist in this field, and early cathode ray devices came to be known as Crookes tubes. Crookes noted that a glow was generated by the surface that the rays struck; the rays themselves were not the source of light. By producing tubes with a vacuum of 1.3×10^{-4} Pa (10^{-6} torr), he eliminated the luminous gas discharge and worked directly with the cathode rays [4]. Crookes demonstrated the following: luminance depended directly upon the material properties of the surface the rays struck; a magnetic field would deflect the path of the rays; the deflection was proportional to the strength of the magnetic field; and a magnetic field could be used to focus the rays into a beam. He also suggested the rays emitted by the cathode were a stream of charged tiny particles, which were soon to be identified by Joseph John Thomson as electrons. Continuing CRT experimentation lead to the discovery of x rays by Wilhelm Konrad Röntgen in 1895.

Ferdinand Braun was the first person to envision the CRT as a tool for the display of information and is generally credited with the invention of the first device to be a direct forerunner to the modern CRT [5]. Braun in 1896 designed a CRT "indicator tube" for monitoring high frequencies in power-generating equipment. His design contained all of the same elements as today's CRTs. It utilized a cathode as an electron source, two control coils for vertical and horizontal deflection, an anode for electron acceleration and beam control, and a focusing slit. He also incorporated a phosphor screen normal to the beam for tracing its path. Although crude by later standards, this was the direct prototype for CRT oscilloscopes. In 1903 to 1905 Arthur Wehnelt added several very significant advances to Braun's design. Wehnelt developed and implemented a hot oxide–coated cathode and a beam control grid [5]. This lowered the voltage necessary to generate the electron stream and provided for much finer control of the beam current. These developments were the forerunner of the modern electron gun.

The CRT remained mostly a laboratory device until three important applications ushered it onto the center stage of information display: oscilloscopes, television, and military radar. By the early 1900s oscilloscopes were being widely used for the study of time-varying electric circuits in communication. Television was developed in the 1920s and in July of 1930 the National Broadcasting Company began the experimental broadcast of television in New York City. During World War II the CRT underwent a second wave of maturation with the advent of radar technology. By the end of the 1940s the TV-CRT was available as a consumer product. In the 1950s RCA perfected shadow mask technology and before long color became a new standard, gaining wide consumer acceptance in the 1960s. While the principal components of a CRT system have not fundamentally changed in many decades, the CRT has continued to improve.

31.3 Image Formation with a CRT

Although there are many different CRT designs, some of which are quite intricate, the process of forming an image using a CRT is straightforward. The procedure can be divided into four basic stages: beam formation, beam focusing, beam deflection, and energy conversion. These stages occur in four different regions of the CRT and follow in order from the rear of the tube (the neck) to the faceplate. As shown in Figure 31.1, the elements at the rear of the tube are collectively called the electron gun.

Beam Generation with the Electron Gun

The cathode generates the stream of electrons used to form the image-writing beam. The traditional cathode is a metal conductor such as nickel coated with a thin layer of oxide, typically a barium strontium

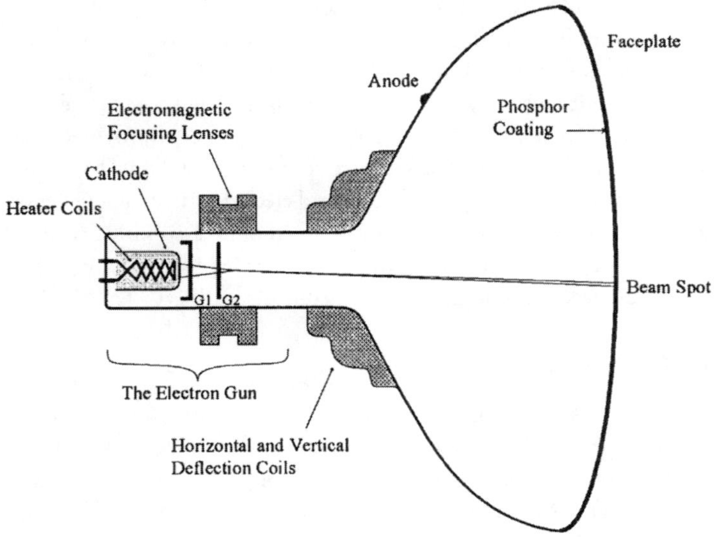

FIGURE 31.1 Example of a typical CRT with electromagnetic focus, and electromagnetic deflection. This type of tube design is usually used for demanding applications such as bright, high-resolution image projection. Tubes like this are available from a number of manufacturers such as Sony, Matsushita, and Thomson.

compound. To reduce the voltage required to generate electron emission, the cathode is heated to 700 to 1200°C. Applications that require high brightness often use more advanced and expensive dispenser cathode designs to increase the beam current while maintaining reasonable cathode life. These designs incorporate complex cathode structures and materials such as barium ceramics, molybdenum, rhenium, and tungsten. Readers interested in advanced cathode designs and additional information on CRT materials technology should consult References 6 through 10. The flow of electrons from the cathode is controlled by varying the potential between the cathode and a series of control grids commonly known as G1 (the control grid) and G2 (the acceleration grid). A voltage potential of 100 to 1000 V between G2 and the cathode creates the potential necessary to pull a stream of electrons off the cathode, forming the beam. The beam amplitude can be controlled and even completely shut off by varying the potential on G1. Thus, the voltage at G1 controls brightness because the brightness is proportional to beam current. The design of the cathode with respect to impedance and loading influences the maximum rate at which the beam can be modulated. The cathode and its associated control grids can be designed to produce a crossover flow of electrons or a laminar flow. In crossover gun designs, the emitted electrons converge to a point in front of the cathode. By using electron optics, this beam spot is imaged onto the phosphor screen. Due to inherent advantages, the crossover design is widely used. A crossover beam is narrow, making it easier to deflect than a thicker beam, and the spot can be very small, improving resolution at the screen. In theory, a laminar flow design provides for the possibility of higher beam current from a similar-sized cathode. In practice, the improvement is not usually advantageous enough to offset the added difficulty of controlling a wider beam.

Electron Beam Focusing

Beam focusing and beam current are critical in determining the final spot size and thus the resolution of the CRT. Focusing a beam of electrons is directly analogous to focusing a beam of light; the discipline is called electron optics. Concerns familiar to optical imaging such as magnification, spherical aberration, and astigmatism also confront electron optics. As CRTs become larger, and operate at higher deflection angles, spot control becomes critical. Beam focusing is achieved using either electrostatic focusing grids or electromagnetic focusing coils. Electrostatic focus is the most extensively used technique. It can be found in use in applications from television to desktop computer monitors. Electrostatic focus is achieved by applying a succession of potentials across a complex sequence of focusing grids built into the electron gun. As designers seek to improve performance further, grid designs have become intricate [11, 12]. Magnetic focus is the system of choice for all high-performance systems where resolution and brightness are design objectives. Magnetic lenses are better at producing a small spot with few aberrations. External coils in a yoke around the neck of the tube control the beam. Since it provides superior performance, electromagnetic focus is common on high-resolution commercial systems. Magnetic focus can also be achieved using permanent magnets and specialized hybrid electrostatic/magnetic focus components. Due to the tremendous impact focus has on resolution, tube suppliers continue to improve focus control [13, 14]. For an excellent and comprehensive treatment of electron physics in CRTs, beam control, detailed design discussions, and other aspects of CRT devices, consult Sol Sherr's textbook [15].

Electron Beam Deflection

The beam deflection system is responsible for controlling the position of the spot on the front face of the CRT. As with focusing, beam deflection techniques can be electromagnetic or electrostatic. A magnetic deflection system consists of two electromagnetic yolks on opposite sides of the CRT neck; a horizontal deflection coil and a vertical deflection coil. The position of the beam is easily controlled; the amount of beam deflection is directly proportional to the current in the coil. Deflection coil design also influences spot size and shape. Coil designs can incorporate correction for coma, convergence errors, and pincushion distortion [16]. Because of its low cost and efficiency, CRTs for both moderate- and high-resolution applications normally use magnetic deflection. Electrostatic deflection provides faster beam displacement but less spot size control. It is typically used in oscilloscope systems, where resolution requirements are moderate and deflection speed is paramount.

31.4 CRT Addressing: Raster Scanning vs. Stroke

Raster scanning is the most common CRT addressing technique. In raster scanning the electron beam writes each frame one line at a time. This means the horizontal deflection system requires a higher bandwidth than the vertical deflection system. High-performance raster CRTs have a horizontal bandwidth of 15 to 150 kHz, and a vertical deflection bandwidth of 30 to 180 Hz. The information is written onto the front face of the CRT from left to right and top to bottom, one line at a time. Raster scanning can be interlaced or progressive (noninterlaced). In interlace scanning, each frame of information is decomposed into two fields. The first field consists of all of the odd-numbered lines in the original frame, and the second field contains all of the even-numbered lines. Each field is scanned onto the CRT at twice the frame rate. Commercial television is interlaced. In the U.S., the frame rate is 30 Hz and the field rate is 60 Hz; this scheme is known as the NTSC standard. PAL and SECAM are two other well-known commercial formats in use today. There are numerous other standards, both analog and digital, in use worldwide, each with slightly different scan rates, resolution/addressing formats, luminance and color encoding, and timing protocols [17, 18]. Interlace scanning is an engineering compromise that conserves transmission bandwidth, electronics bandwidth, and CRT bandwidth, while maintaining acceptable performance with respect to moving imagery, resolution, and flicker. Computer monitors employ a progressive raster scan. Each frame has one field. The display is generated line by line, in order from the first pixel to the last pixel. Frame rates for the typical desktop monitor vary from 60 to 85 Hz to over 180 Hz for specialized systems.

Stroke is an alternative addressing technique that was once quite common. The name for the technique comes from the phrase "the stroke of a pen." Stroke is a point-to-point addressing system. The beam is directed to the starting point of a line, turned on, and then moved directly to the end of the line. Because there are no raster lines, stroke CRT resolution is independent of direction and is limited primarily by spot size. Stroke addressing is excellent for low-information-content screens with detailed graphic characters or simple vector graphics. It provides extremely high quality drawing, clean lines, precise detail,

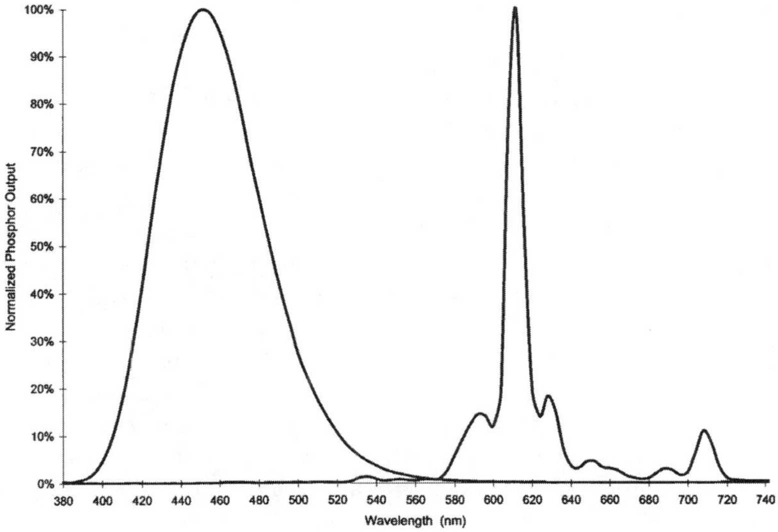

FIGURE 31.2 Phosphors can have a wide range of spectral characteristics. This graph illustrates the spectral profiles of two representative cases. The blue source is an example of a phosphor with a wide and smooth profile. The maximum output for this phosphor is in the blue at a wavelength of 452 nm. The red phosphor shows a sharper profile with several narrow peaks. The largest peak for this phosphor is at 612 nm.

and high speed while conserving power and bandwidth. Stroke systems address the CRT only where there is information, not the entire screen as in raster scanning. This technique is uncommon today because it is too slow and computationally intensive for systems that have high information content, gray scale, or imagery.

31.5 The Phosphor Screen

The phosphor functions as the CRT transducer. It converts the energy of the electron beam into light. This conversion process is called cathodoluminescence. CRT phosphors are inorganic crystalline materials doped with one or more impurities called activators and coactivators. Phosphors emit light in two ways, fluorescence and phosphorescence. Fluorescence is the emission of light by the phosphor material while it is under bombardment by the electron beam. The continued emission of light after the bombardment has ceased is called phosphorescence. The length of time phosphorescence lasts is known as persistence. Persistence can vary from tens of nanoseconds to many minutes, or even hours. CRTs take advantage of both forms of cathodoluminescence. Briefly, cathodoluminescence occurs when the electron beam excites the electrons of the phosphor into higher, unstable energy states available due to the presence of the activators. When the electrons transition back to their stable states, light is emitted. The choice of phosphor depends on the requirements of the application with respect to wavelength characteristics (narrow emission spectra or broadband emission), color, brightness, resolution, and persistence. Commercial television CRTs typically make use of the following phosphor powders: ZnS:Ag:Cl (blue), Zn(Cd)S:Cu:Al or ZnS:Cu:Au:Al (green), and Y_2O_2 S:Eu [19]. Television CRTs use moderately shortpersistence phosphors. This ensures a new frame does not exhibit blurring due to the previous frame. Traditionally, phosphors for radar displays, where the screen is refreshed infrequently, had a mix of short and long persistence. However, with the advent of today's digital systems, the use of long-persistence phosphors has declined. The Electronics Industries Association maintains an information database of commercial phosphors for CRTs. The interested reader should consult its publication TEP116-C [20]. Figure 31.2 illustrates the spectral characteristics of several typical phosphors.

31.6 Color CRTs Using Shadow Masks

Color has been a part of the CRT world since the early 1950s. A sound understanding of human color perception is important for a detailed understanding of color in CRTs. Readers interested in an excellent comprehensive treatment of color science should consult Wyszecki and Stiles [21] or Robertson and Fisher [22] for a brief overview. The most common method of introducing color to the CRT is the three-gun shadow mask technique. The shadow mask is a metal grid of many tiny apertures held in place immediately before the phosphor screen. The phosphor screen is an array of three different phosphor subpixels. Three electron guns are placed closely together in the neck of the tube, one gun for each one of the primary colors (red, green, and blue). The guns make a small angle with respect to each other. The shadow mask and beams are aligned so that each beam falls on the appropriate phosphor dot. The dots are placed closely together so the eye spatially integrates the subpixels into a continuous blend of color. There are two common configurations, delta and in-line. Delta, the traditional method, arranges the electron guns and the phosphor dots in a triad. This requires a shadow mask consisting of a grid of circular apertures. More recently, Sony introduced the Trinitron™ design. In the Trinitron tube the guns are placed in line, the shadow mask consists of vertical slots, and the phosphors are in vertical stripes. The design offers improved vertical resolution and is easier to converge, but the mask is slightly more difficult to support and is subject to thermal stress problems.

Trinitron continues to be an important and popular design [23]. The shadow mask technique of producing a color image has three main drawbacks. First, the shadow mask typically absorbs more than 75% of the electron beam energy, limiting brightness and causing undesirable thermal effects. Second, both require precise beam control to converge the three different images from the electron guns. Third, the resolution of the display is limited by the requirements of spatial color; three subpixels are needed to make each full-color pixel.

31.7 Alternative Techniques for Realizing Color Using CRTs

There are alternative approaches to producing color with CRTs. In its own way, each seeks to improve upon one or more of the enormously successful shadow mask designs. The three most noteworthy challengers are beam index tubes, penetration phosphor tubes, and field sequential color systems. All three have been around for decades, but only field sequential color systems are commercially available.

Beam Index and Penetration Phosphor Tube Designs

A beam index tube uses a slotted mask and vertical phosphor stripes. One electron beam is focused to less than the width of a phosphor stripe. A reflected ultraviolet signal from the mask provides a feedback signal. This signal is used to index the position of the beam very accurately. Color selection is achieved by precise beam positioning and rapid beam modulation. Although beam indexing offers perfect convergence, few guns, energy savings, and high resolution, practical problems in control and manufacturing have left this approach unrealized. Penetration tubes are potentially even more advantageous since color is produced with one electron gun and no mask is required. The phosphor screen consists of a layering of several different phosphors. There are many approaches to the layering structure, but the basic principle is that each layer produces a different color. To select among the different colors, the beam energy is varied, altering its layer penetration. In theory, multiple layers of phosphors could be used to produce a full-color display. Unfortunately, this design requires high switching voltages, and the color purity is poor because of the dilution caused by leakage from the unselected phosphor layers. In the past, a few two-color systems have been produced on a limited basis with stroke writer systems (p. 147 of Reference 4).

Field Sequential Color

Field sequential color (FSC) is a different approach to realizing color using a monochrome CRT and color filters [24]. The shadow mask relies on the ability of the eye to perform a *spatial integration on a group of color elements.* Field sequential color exploits the ability of the eye to perform a *temporal integration of color fields.* Field sequential color is not a new concept; it was an early design suggestion for commercial color television. To implement field sequential color, each full-color frame is decomposed into primary color fields (red, green, and blue). Each of these color fields is displayed sequentially on a monochrome CRT while simultaneously a color filter is placed over the front of the CRT. The filters must be rapidly changed to stay in sequence with the individual fields which are displayed at three times the desired frame rate. This sequence is so quick that the eye fuses the individual color fields into a steady blend of continuous color. Field sequential color systems have all of the advantages of a monochrome CRT, namely, superior resolution and simplicity. The major historical drawbacks to field sequential color have been the difficulty in controlling a bulky, spinning, mechanical filter wheel, and two thirds of the light energy of the phosphor is thrown away in each field. In the past, some viewers have reported the ability to perceive the individual color fields intermittently; this artifact is known as "color break up." Field sequential color systems have seen a commercial rebirth due to several factors. The luminous efficiency of several phosphors has improved to the point where the light loss penalty is no longer a severe price to pay. Awkward, mechanical filter wheels have been replaced by several different types of compact, low-power, liquid crystal shutters. The use of faster frame rates and digital sampling techniques has reduced color artifacts. And finally, monochrome CRTs using field sequential color meet the growing market need for small high-resolution displays.

31.8 Image Quality and Performance

For any display technology, the ultimate test of performance and quality is how the image looks to the viewer in a real-world environment. This is highly subjective, but critical none the less. Table 31.1

TABLE 31.1 Parameters That Influence CRT Performance and Image Quality

Resolution	Spot size and shape
	Focus accuracy
	Line width
	Addressing/scan format
	Shadow mask design
Luminance	Dynamic range
	Maximum brightness
	Gray scale in high and low ambient light
	Gray scale vs. display brightness
	Phosphor luminous efficiency
Contrast	Contrast under high ambient illumination
	Contrast under low ambient illumination
	Large area contrast
	Pixel-to-pixel contrast (small area)
Image fidelity	Frame rate
	Flicker/refresh
	Uniformity and linearity
	Aberrations (linearity, pincushion, barrel distortion, keystone, and focus)
Color	Phosphor selection
	Convergence accuracy
	Color gamut and saturation
	Color uniformity
	Color accuracy
Bandwidth	Video bandwidth (cathode design)
	Horizontal deflection design
	Vertical deflection design
	Bandwidth of the CRT supporting electronics
	Bandwidth of the signal delivery electronics

highlights some important parameters associated with performance and image quality. Resolution is the most critical characteristic of any CRT. A note of caution: resolution should not be confused with addressing. A system can have a precise addressing format, but if the spot is too large the CRT resolution capability will be poor despite the addressing format. Alternatively, if the spot is precise and the addressing format is coarse, the *image* will be low resolution, but the CRT resolution will be high. Resolution is affected by almost every system parameter: spot size, focus, deflection, raster format, the mask, and beam current. For a perfectly designed system, line width and spot size will ultimately determine the resolution. Spot diameter is limited by the quality of the electron optics, the particle size and thickness of the phosphor layer, and beam current. The cross-section energy distribution of the electron beam is approximately Gaussian. The higher the current, the larger the beam cross section. Scattering by individual phosphor particles also increases the spot size. Thus, high resolution is much more difficult to achieve for high-brightness applications. The champions of small spot sizes are the small 12 to 25 mm monochrome CRTs. Several commercial systems achieve ≤25 μm spot sizes, and sizes ≤15 μm have been reported [25]. Monochrome CRTs in the 53 cm class (21-in.) have spot diameters ≤150 μm. Color CRTs with shadow mask pitches ≤250 μm are available in sizes up to 21 in. at commodity prices, and this will continue to improve. Spot size does not tell the whole resolution story. In most systems something other than spot size limits performance. The scan format may be well below the performance of the CRT, the bandwidth of the support electronics is frequently less than the tube bandwidth, and, for color, the shadow mask will limit resolution more than spot size. Engineers can still design a CRT that maintains its performance even with state-of-the-art electronics. MTF (modulation transfer function) is the standard metric for resolution. MTF compares the modulation of the displayed output with the modulation of a known input signal. As the spatial frequency of the input increases, the performance of the output will show a steady roll-off in modulation depth and quality. Readers interested in an introduction to this complex topic should consult Infante's review [26]. Contrast and its related companion, gray scale, are also important to CRT image quality. Since contrast can be measured and stated in so many different terms, gray scale is probably a more practical indicator of performance from a user's standpoint. As it attempts to replace traditional film, the medical imaging community demands the most of the monochrome CRT. Medical CRTs can produce up to 256 shades of gray with 8-bit controllers. Systems are available that will do 10-bit monochrome and color. Although frequently quoted, *shades of gray* is not a true indication of CRT performance and should not be confused with true *gray scale* capability. Shades of gray is a characteristic of the image source and the capability of the electronics; gray scale reflects the capability of the CRT. Gray scale is typically defined as a series of steps in brightness in increments of $2^{1/2}$ from black to maximum CRT output. Thus, gray scale is a realistic indicator that combines CRT dynamic range, contrast ratio, halation, and ambient reflections. More than 12 steps of gray scale is considered good performance.

The subject of display image quality and evaluation is an important topic. In addition to resolution and gray scale, users and designers need to consider many other metrics, such as luminance, dimmability, contrast, readability, color, persistence, and convergence. Users may also be aware of visual distortions such as pincushion, barrel, and keystone. A discussion of these topics and the other issues raised by Table 31.1 is beyond the scope of a single chapter; indeed, it has been the subject of books. There are good sources of information in this area [15, 27]; in particular, Keller's [28] text is superb. It is essential to remember that no matter which metrics are used, the most single most important metric is the subjective determination of the end users. Although many factors will affect final display performance, the bottom line remains *how good does the display look* in the real-world application environment it was designed to meet.

31.9 CRT and Phosphor Lifetime

The lifetime of a CRT depends on how it is driven. The higher the beam current, the more rapidly its performance will degrade. In general, degradation is due to a failure of three parts of the CRT — the

TABLE 31.2 The Strengths and Weaknesses of CRT Technology

Advantage	Weakness
Highest resolution technology	High power consumption
Versatile addressing formats	Weight of glass
Excellent image fidelity	Size of footprint
High speed	Emits EM radiation
Bandwidth (_350 MHz)	Must be refreshed
Excellent contrast/gray scale	
Good selection of phosphors	
High luminous efficiency	
Bright display (up to 20,000 fL)	
Excellent color gamut	
Simplicity of interface	
Universal interface standards	
Good design flexibility	
Broad application base	
Long life and reliability	
Mature and broad knowledge base	
Worldwide sources of tubes and parts	
Inexpensive — low cost per resolution element	

cathode, the phosphor, and the glass. Brighter displays require higher beam current, and this, in turn, requires a higher level of cathode loading. Oxide cathodes can comfortably operate with a loading range of 0.1 to 10 A/cm^2. This will provide a typical lifetime in the range of 10,000 h. Dispenser cathode designs can improve cathode life 2 to 3 times, but again, individual cathode life depends on how much brightness the user regularly demands from the tube. Phosphor degradation is the second mechanism affecting CRT lifetime and performance. All phosphors degrade under constant electron bombardment, an effect called phosphor burning. The rate of degradation is related to beam current, anode voltage, and material parameters specific to each phosphor. Today, there is a broad range of good phosphor materials with greatly improved aging characteristics. The third mechanism is glass browning. This effect is a complicated interaction between the high energy of the electron beam and the molecular structure of the glass [10]. In recent years, glass suppliers have made great strides in providing CRT manufacturers with improved glass materials. However, phosphor aging and glass browning are still important concerns, especially in extremely bright (high beam current) applications, such as CRT projection.

31.10 CRT Strengths and Weaknesses

With a century of development behind it, the CRT has abundant advantages that make it an attractive display option for the system designer dealing with high information content or dynamic imagery. Table 31.2 lists some attributes, positive and negative, which the system designer should consider in evaluating whether or not to use CRT technology. The most notable traits of the CRT are flexibility and cost. CRTs can operate over a wide range of resolution formats from commercial television to the strict requirements of medical imaging and digital-to-film recording. One unit can easily be designed to accept a myriad of scan formats and resolutions from static text to dynamic imagery with megapixels at frame rates of hundreds of frames per second. This flexibility means the system can be upgraded later without having to redesign the display subsystem. Interfacing with the CRT is also straightforward; there are well-established standards and hardware for many scan formats from the television NTSC to the RGB standards of the personal computer industry. Finally, cost is important to emphasize; CRTs are inexpensive. The desktop CRT delivers its performance for ≤\$0.00005 per color resolution element. Currently, no other high-information-content display technology can match this cost.

31.11 CRT Selection

The design requirements for (1) resolution, (2) brightness, and (3) screen size will quickly steer the system designer to a particular class of CRTs. If the design requirements are close to either television or desktop computer monitors, the choice is easy. There are literally hundreds of vendors offering integrated CRTs and electronic drivers as off-the-shelf items. Color shadow mask systems with more than 40 elements/cm are commonly available. If higher resolution is required, then medical imaging and film recorder CRTs may meet the requirements. Physically large display requirements can be supplied by projection CRTs or by tiled CRTs using video wall processors. Small to miniature CRTs provide resolutions up to 500 to 600 elements/cm and portability. If the application falls out of the mainstream, the designer will need to speak directly with CRT vendors. Table 31.3 is a partial list of CRT suppliers. Since CRTs have been designed to satisfy a broad range of specialty applications, the chances are excellent that an acceptable design will already exist. If not, the CRT is so mature that vendors working from a proven experience base can design a tube directly to customer specifications with little in the way of design problems or surprises.

31.12 Future Trends in CRT Technology

CRT technology is mature, and the industry does not foresee major surprises on the horizon. The CRT will continue to evolve, resolution will improve, screens will be flatter, deflection angles will be larger, and footprints will be reduced. The computer world continues to demand more resolution. In 1987, the 36-cm (14-in.) VGA class monitor (640 × 480) with its 300,000 pixels and 16 colors was the new standard. In 1998, the 43-cm (17-in.) desktop monitor is now capable of 1600 × 1200 performance (1.9 megapixels) with 24-bit color. The 53-cm (21-in.) monitor class, once an expensive custom device, is also widely available. The trend of increasing resolution and performance for decreasing cost shows every sign of continuing. The medical community is quickly moving to monochrome systems with 12-bit gray scale at a resolution of 2560 × 2048 (5 megapixels). All of this means the instrument designer has better resolution and fidelity for less money.

There is a quiet revolution taking place in the information display industry, which may have a significant effect on CRTs and all information display technology. There is a design trend to decouple the resolution and fidelity of the display device from its physical size. There are a number of applications motivating this change: communications, military, portable computing, simulation, and virtual reality. This is why miniature CRT technology has been more active in recent years. This is also the engine driving some of the major competitors to the miniature CRT, such as on-chip field emission displays (FED), deformable mirror devices (DMD), ferroelectric liquid crystal integrated circuit displays, and on-chip active matrix liquid crystal display (LCD) technology. If realized, these promise to be low-cost solutions because they are built on the technology foundation of the silicon chip industry. These challengers have the very real potential of eliminating the miniature CRT, and in some application areas, the LCD panel as well. However, the CRT backed by its formidable 100 years of design evolution and maturity is not standing still; its assets and market remain impressive. Although it is one of the few vacuum tubes still in use today, the traditional CRT is not doomed for obsolescence any time in the immediate future.

TABLE 31.3 Companies That Manufacture Cathode Ray Tubes

Clinton Electronics Corporation
6701 Clinton Road
Rockford, Illinois 61111
Tel. 815-633-1444 Fax. 815-633-8712
Tel. 040-783749 Fax. 040-788399

Hitachi, Ltd.
New Marunouchi Bldg.
Marunouchi 1-chrome
Chiyoda-ku, Tyoko 100, Japan
Tel. 81-3-3212-1111 Fax. 81-3-3212-3857
U.S. address
Hitachi America, Ltd., Electron Tube Division
3850 Holcomb Bridge Road, Suite 300
Norcross, Georgia 30092-2202
Tel. 770-409-3000 Fax. 770-409-3028

Hughes Lexington, Inc.
A subsidiary of Hughes Electronics Company
1501 Newtown Pike
Lexington, Kentucky 40511
Tel. 606-243-5500 Fax. 606-243-5555

Image Systems Corporation
11595 K-tel Drive
Hopkins, Minnesota 55343
Tel. 612-935-1171 Fax 612-935-1386

Imaging & Sensing Technology Corporation
300 Westinghouse Circle
Horseheads, New York 14845-2299
Tel. 607-796-4400 Fax. 607-796-4482

ITPO Institute of Surface Engineering and Optoelectronics
Teslova 30
1000 Ljubljana, Slovenia
Tel. 386-61 1264 592/111 Fax. 386-61 1264 593

L. G. Electronics
20 Yoido-dong Youngdungpo-gu
Seoul, Korea
E-mail display2@www.goldstar.co.kr or
monitor@www.goldstar.co.kr
U.S. address
L G Electronics, Monitors Division
1000 Silvan Avenue
Englewood Cliffs, New Jersey 07632
Tel. 201-816-2000 Fax. 201-816-2188

Matsushita Electric Industrial Ltd.
Twin 21 National Tower
1-61, Shiromi, 2-Chome, Chuo-ku,
Osaka 540, Japan
Tel. (06)908-1121
U.S. address
Panasonic Industrial Company
Computers and Communications Division
2 Panasonic Way
Secaucus, New Jersey 07094
Tel. 201-392-4502 Fax. 201-392-4441

Phillips Netherlands, BV
Phillips Components and Semiconductors
Bldg. VB Postbus 90050
5600 PB Eindhoven The Netherlands
U.S. address
Discrete Products Division
2001 West Blue Heron Blvd.
Riviera Beach, Florida 33404
Tel. 407-881-3200 or 800-447-3762 Fax. 407-881-3300

Rank Brimar, Ltd.
Greenside Way
Middleton, Manchester M24 1SN, England
Tel. 0161-681 7072 Fax. 0161-682-3818
U.S. address
25358 Avenue Stanford
Valencia, California 91355-1214
Tel. 805-295-5770 Fax. 805-295-5087

Sony Corporation
Display Systems
16550 Via Esprillo
San Diego, California 92127
Tel. 619-487-8500

Thomas Electronics
100 Riverview Drive
Wayne, New Jersey 07470
Tel. 201-696-5200 Fax. 201-696-8298

Thomson Tubes Electronics
13, avenue Morane Saulnier
Bâtiment Chavez — Vélizy Espace
BP 121/F-78148 VELIZY CEDEX France
Tel. 33-1 30 70 35 00 Fax. 33-1 30 70 35 35
U.S. address
Thompson Components and Tubes Corporation
40 G Commerce Way
Totowa, New Jersey 07511
Tel. 201-812-9000 Fax. 201-812-9050

Toshiba Corporation
Electronic Components, Cathode Ray Tube Division
1-1 Shibaura 1-Chrome, Minato-KU
Tokyo 105, Japan
Tel. 03-457-3480 Fax. 03-456-1286
U.S. address
Toshiba America, Inc.
1220 Midas Way
Sunnyvale, California 94086-4020
Tel. 408-737-9844

References

1. G. Aboud, *Cathode Ray Tubes, 1997,* 2nd ed., San Jose, CA, Stanford Resources, 1997.
2. G. Aboud, *Cathode Ray Tubes, 1997,* Internet excerpts, available http://www.stanfordresources.com/sr/crt/crt.html, Stanford Resources, February 1998.
3. G. Shires, Ferdinand Braun and the Cathode Ray Tube, *Sci. Am.,* 230 (3): 92–101, March 1974.
4. N.H. Lehrer, The challenge of the cathode-ray tube, in L.E. Tannas, Jr., Ed., *Flat Panel Displays and CRTs,* New York: Van Nostrand Reinhold, 1985.
5. P. Keller, *The Cathode-Ray Tube, Technology, History, and Applications,* New York: Palisades Press, 1991.
6. D.C. Ketchum, CRTs: the continuing evolution, *Society for Information Display International Symposium, Conference Seminar M-3,* 1996.
7. L.R. Falce, CRT dispenser cathodes using molybdenum rhenium emitter surfaces, *Society for Information Display International Symposium Digest of Technical Papers,* 23: 331–333, 1992.
8. J.H. Lee, J.I. Jang, B.D. Ko, G.Y. Jung, W.H. Kim, K. Takechi, and H. Nakanishi, Dispenser cathodes for HDTV, *Society for Information Display International Symposium Digest of Technical Papers,* 27: 445–448, 1996.
9. T. Nakadaira, T. Kodama, Y. Hara, and M. Santoku, Temperature and cutoff stabilization of impregnated cathodes, *Society for Information Display International Symposium Digest of Technical Papers,* 27: 811–814, 1996.
10. W. Kohl, *Materials Technology for Electron Tubes,* New York, Reinhold Publishing, 1951.
11. S. Sugawara, J. Kimiya, E. Kamohara, and K. Fukuda, A new dynamic-focus electron gun for color CRTs with tri-quadrupole electron lens, *Society for Information Display International Symposium Digest of Technical Papers,* 26: 103–106, 1995.
12. J. Kimiya, S. Sugawara, T. Hasegawa, and H. Mori, A 22.5 mm neck color CRT electron gun with simplified dynamically activated quadrupole lens, *Society for Information Display International Symposium Digest of Technical Papers,* 27: 795–798, 1996.
13. D. Imabayashi, M. Santoku, and J. Karasawa, New pre-focus system structure for the trinitron gun, *Society for Information Display International Symposium Digest of Technical Papers,* 27: 807–810, 1996.
14. K. Kato, T. Sase, K. Sasaki, and M. Chiba, A high-resolution CRT monitor using built-in ultrasonic motors for focus adjustment, *Society for Information Display International Symposium Digest of Technical Papers,* 27: 63–66, 1996.
15. S. Sherr, *Electronic Displays,* 2nd ed., New York: John Wiley, 1993.
16. N. Azzi and O. Masson, Design of an NIS pin/coma-free 108° self-converging yoke for CRTs with super-flat faceplates, *Society for Information Display International Symposium Digest of Technical Papers,* 26: 183–186, 1995.
17. J.F. Fisher and R.G. Clapp, Waveforms and spectra of composite video signals, in K. Benson and J. Whitaker, *Television Engineering Handbook, Featuring HDTV Systems,* New York: McGraw-Hill Reinhold, 1992.
18. D. Pritchard, Standards and recommended practices, in K. Benson and J. Whitaker, *Television Engineering Handbook, Featuring HDTV Systems,* New York: McGraw-Hill Reinhold, 1992.
19. A. Vecht, Phosphors for color emissive displays, *Society for Information Display International Symposium Conference Seminar Notes F-2,* 1995.
20. *Optical Characteristics of Cathode Ray Tube Screens,* EIA publication TEP116-C, Feb., 1993.
21. G. Wyszecki and W.S. Stiles, *Color Science: Concepts and Methods, Quantitative Data and Formulae,* 2nd ed., New York: John Wiley & Sons, 1982.
22. A. Robertson and J. Fisher, Color vision, representation, and reproduction, in K. Benson and J. Whitaker, *Television Engineering Handbook, Featuring HDTV Systems,* New York: McGraw-Hill Reinhold, 1992.
23. M. Maeda, Trinitron technology: current status and future trends, *Society for Information Display International Symposium Digest of Technical Papers,* 27: 867–870, 1996.

24. C. Sherman, Field sequential color takes another step, *Inf. Display,* 11 (3): 12–15, March, 1995.
25. L. Ozawa, Helmet mounted 0.5 in. crt for SVGA images, *Society for Information Display International Symposium Digest of Technical Papers,* 26: 95–98, 1995.
26. C. Infante, CRT display measurements and quality, *Society for Information Display International Symposium Conference Seminar Notes M-3,* 1995.
27. J. Whitaker, *Electronic Displays, Technology, Design, and Applications,* New York: McGraw-Hill, 1994.
28. P. Keller, *Electronic Display Measurement, Concepts, Techniques, and Instrumentation,* New York: John Wiley & Sons, 1997.

Further Information

L. Ozawa, *Cathodoluminescence: Theory and Applications,* New York: Kodansha, 1990.

V.K. Zworykin and G.A. Morton, *Television: The Electronics of Image Transmission in Color and Monochrome,* New York: John Wiley & Sons, 1954.

B. Wandell, The foundations of color measurement and color perception, *Society for Information Display International Symposium, Conference Seminar M-1,* 1993. A nice brief introduction to color science (31 pages).

Electronic Industries Association (EIA), 2500 Wilson Blvd., Arlington, VA 22201 (Internet: www.eia.org). The Electronic Industries Association maintains a collection of over 1000 current engineering publications and standards. The EIA is an excellent source for information on CRT engineering, standards, phosphors, safety, market information, and electronics in general.

The Society for Information Display (SID), 1526 Brookhollow Dr., Suite 82, Santa Ana, CA 92705-5421 (Internet: www.display.org). The Society for Information Display is a good source of engineering research and development information on CRTs and information display technology in general.

Internet Resources

The following is a brief list of places to begin looking on the World Wide Web for information on CRTs and displays, standards, metrics, and current research. Also many of the manufacturers listed in Table 31.3 maintain Web sites with useful information.

The Society for Information Display	www.display.org
The Society of Motion Picture and Television Engineers	www.smpte.org
The Institute of Electrical and Electronics Engineers	www.ieee.org
The Electronic Industries Association	www.eia.org
National Information Display Laboratory	www.nta.org
The International Society for Optical Engineering	www.spie.org
The Optical Society of America	www.osa.org
Electronics & Electrical Engineering Laboratory	www.eeel.nist.gov
National Institute of Standards and Technology (NIST)	www.nist.gov
The Federal Communications Commission	www.fcc.gov

32

Liquid Crystal Displays

Kalluri R. Sarma
Honeywell, Inc.

32.1 Introduction

Liquid crystals (LC) are an important class of materials with applications ranging from display devices, optoelectronic devices, sensors, and biological and structural materials. The focus of this chapter will be on LCs for display applications. In general, most substances have a single melting point where a solid possessing a positional and orientational order changes upon melting to an isotropic liquid that has neither positional nor orientational order. However, some materials when melted from the solid state change into a cloudy liquid with orientational order at one temperature, and upon further heating change into an isotropic liquid that has no order, as shown in Figure 32.1. Thus, an LC is a mesophase existing between the melting temperature T_m, of a crystalline phase, and clearing point T_c, of the liquid phase; i.e., below T_m, the material has a crystalline phase, above T_c, it has a liquid (isotropic) phase, and between T_m and T_c, it has a liquid crystal phase. This type of LC in which the mesophase is defined by the temperature (between T_m and T_c) is called a *thermotropic* LC. When the mesophase is defined by a solvent concentration, it is called a *lyotropic* LC. Thermotropic LCs are used for display applications. The orientational order in LC materials results in important physical properties, such as birefringence, that make these materials useful for display devices. Because LCs have the attributes of low drive voltage, low power consumption, thin form factor (flat panel displays), light weight, full-color, gray scale with a wide dynamic range, full motion video, superior image quality, and high reliability, LC displays (LCDs) are the preferred approach for battery-powered (portable) applications ranging from wristwatch displays and handheld TVs to laptop computer displays. They are also replacing cathrode ray tubes (CRTs) in select applications such as avionic displays because of their high brightness and readability in sunlight. LCs are also being used in projection display devices for head-mounted display (HMD) and for large-screen display applications. The following will discuss the various types of LC materials, their properties,

0-8493-1733-9/04/$0.00+$1.50

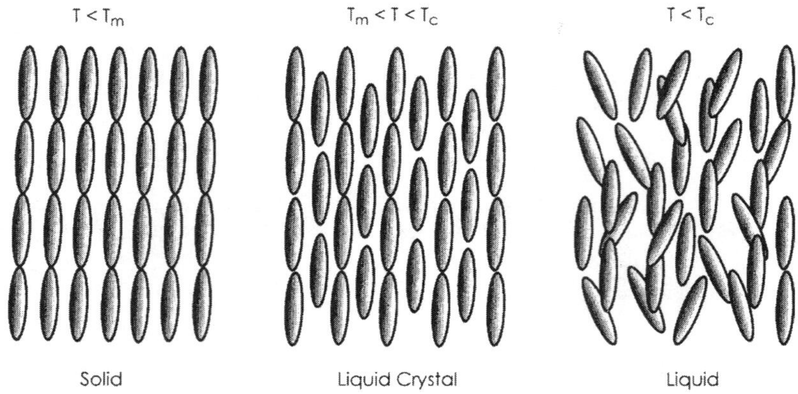

FIGURE 32.1 Illustration of a solid, an LC, and a liquid. A solid has an orientational as well as a positional order for the molecules. An LC has an orientational order only. A liquid phase is isotropic with neither positional nor orientational order.

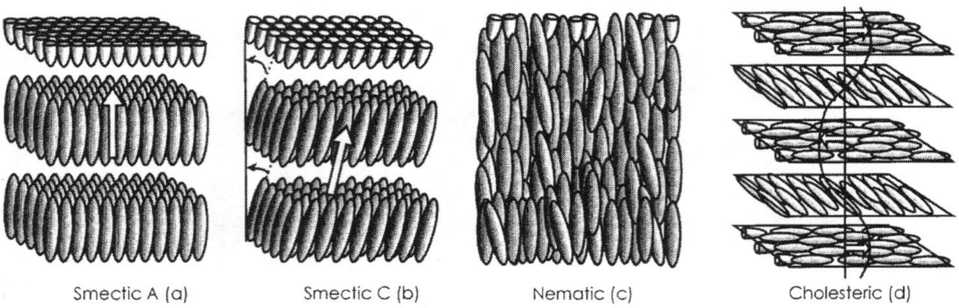

FIGURE 32.2 Molecular orientation in (a) smectic A, (b) smectic C, (c) nematic, and (d) cholesteric LC phases.

LCD materials and fabrication processes, various LCD modes, and display addressing methods. There are many good general references, for example, References 1 through 7, on LCs and LCDs. At the time of this writing, LCD technology is advancing very rapidly with respect to technology development for LCD products with improved viewing angles, improved image quality, lower power consumption, and larger display sizes. The purpose of this chapter, however, is to present the basic LCD principles and technologies, as opposed to reviewing the current state of the art.

32.2 Types of Liquid Crystal Materials

Most of the LC materials are organic compounds which consist of rod-shaped or disk-shaped molecules. For display applications, LC materials with rod-shaped molecules are the most commonly used. The LC materials are broadly classified into three types (phases) — smectic, nematic, and cholesteric — according to their molecular order, as shown in Figure 32.2. In a smectic liquid crystal, the rod-shaped molecules are arranged in layers with molecules parallel to each other. There are many different smectic phases, but smectic A and smectic C are the most common. In the smectic A LC, the molecular axis (director) is perpendicular to the layers as shown in Figure 32.2a, and in smectic C it is tilted at an angle from the layer normal as shown in Figure 32.2b. Also, in the nematic LC, the rod-shaped molecules are parallel to each other, but the individual molecules move relatively easily in the direction along their axis without a layer structure, as shown in Figure 32.2c. In the cholesteric LC, the molecules are arranged in a layered fashion as in smectic LC, but the molecular axis is in the plane of each layer as shown in Figure 32.2d. In addition, the cholesteric LC shows a helical structure in which the director $\boldsymbol{n}$ changes from layer to layer. The same LC material may have different LC phases at different temperatures. For example, the

LC material may have a smectic C phase at a lower temperature, and as the temperature increases it may change to a smectic A phase and then to a nematic phase, before changing to an isotropic liquid phase at T_c.

The nematic LC is the basis for most widely used active matrix-addressed twisted nematic (TN) LCDs, and passive matrix-addressed supertwisted nematic (STN) LCDs. An example of a smectic C LC display is a passive matrix-addressed ferroelectric LCD [8]. An example of a cholesteric display is a passive matrix-addressed stabilized cholesteric texture (SCT) display with bistability [9]. A classic example of a nematic LC material is *p*-azoxyanisole (PAA) with a nematic phase in the range of 117 to 136°C.

In PAA, the two benzene rings are nearly coplanar and the rigid rod is about 20 Å long and 5 Å wide. Another historical example of a nematic LC is *N*-(*p*-methoxybenzylidene-*p*-butylaniline) (MBBA) with a nematic phase in the range of 22 to 47°C.

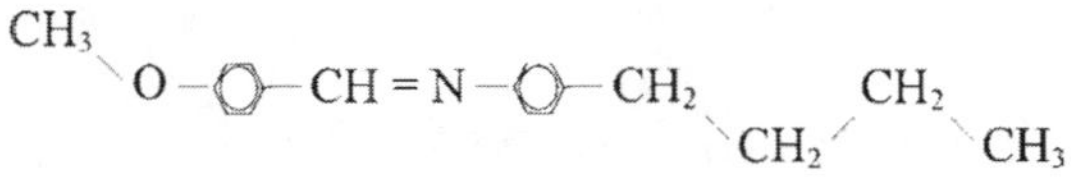

MBBA has a central group that connects the two ringlike cores firmly and serves to maintain the linearity of the entire LC molecule. The terminal groups and conjugated bonds in the core are largely responsible for the dielectric, optical, and other anisotropic properties of the material. Azoxy and Schiff's base compounds were among the materials used earliest in LCDs. Because of environmental stability problems, they were replaced by biphenyl materials. More recently, phenylcyclohexane, bicyclohexane, and estercyclohexane compounds are developed to satisfy the requirements of broad temperature operation and enhanced electro-optical characteristics. The LC materials used in current LCDs are highly developed mixtures of various compounds tailored to meet the requirements of environmental stability, wide operating temperature range, proper response to the applied electric field, high electrical resistivity for matrix addressing, and fast response time.

The degree of order in the LC is an important parameter, and is defined by the *order parameter, S,* given by

$$S = \frac{1}{2}\left\langle 3\cos^2\theta - 1 \right\rangle \tag{32.1}$$

where θ is the angle between the molecular axis and the predominant molecular orientation $\boldsymbol{n}$. The symbol < > represents averaging over the whole space. The predominant molecular orientation, which is known as the LC *director* $\boldsymbol{n}$ is defined as the average alignment direction of the long molecular axis. For a perfect orientational order, i.e., when all the molecules align parallel to the director, as in a perfect crystal, $\theta = 0$ and thus $S = 1$. For no orientational order, i.e., for a completely random molecular orientation, as in an isotropic liquid, $S = 0$. In a typical nematic LC, S is in the range of 0.3 to 0.7, and in a typical smectic LC it is in the range of 0.7 to 0.8, with higher values at lower temperatures. Figure 32.3 shows the temperature dependence of S for the LC material PAA [10] as an example. The order parameter decreases rapidly to a value of around 0.3, close to the clearing point T_c, and becomes zero in the isotropic state. The exact dependence of S on temperature T depends on the type of molecules considered. However, the following analytical expression derived from theory has been shown to be useful [11]:

$$S = \left(1 - yT/T_c\right)^{\beta} \tag{32.2}$$

where T_c corresponds to nematic–isotropic transition, y is of the order of 0.98, and β is an exponent in the range of 0.13 to 0.18 depending on the material in question.

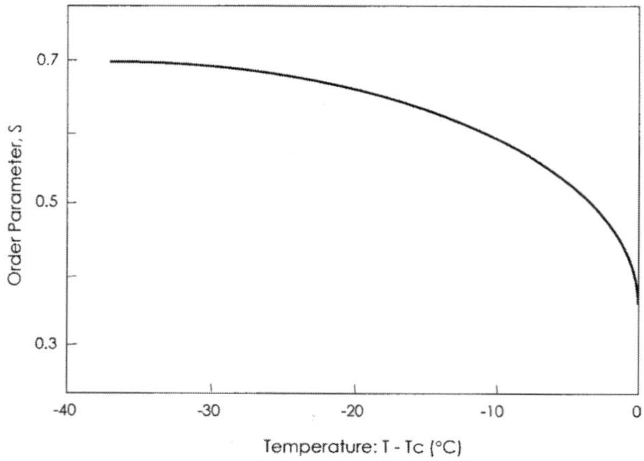

FIGURE 32.3 Temperature dependence of the order parameter S of the nematic LC P = azoxy anisole (PPA) [10]. S decreases with increasing temperature T, and rapidly approaches 0 near the clearing temperature, T_c.

32.3 Physical Properties of Liquid Crystals

Because of ordered structure with the molecules aligned with their long axis parallel to each other, LC molecules exhibit anisotropic properties. That is, various physical properties, such as dielectric constant ε, refractive index n, magnetic susceptibility χ, conductivity σ, and viscosity η, have different values in the direction parallel ($\|$) and perpendicular ($\perp$) to the molecular axis. The anisotropic physical properties, in conjunction with the ease of controlling the initial orientation (boundary condition) by surface alignment and the ease of reorienting the molecular axis by applying a voltage, is the basis for application of LCs for displays. The following will discuss the anisotropy of the dielectric constant and refractive index, elastic constants, and electro-optical characteristics which are important in the use of LCs for displays.

Dielectric Anisotropy

LC molecules exhibit dielectric anisotropy because of their permanent and induced dipoles. The dielectric anisotropy, $\Delta\varepsilon$, is expressed as

$$\Delta\varepsilon = \varepsilon_{\|} - \varepsilon_{\perp} \tag{32.3}$$

where $\varepsilon_{\|}$ and $\varepsilon_{\perp}$ are the dielectric constants measured parallel and perpendicular to the LC director. Materials that exhibit positive dielectric anisotropy ($\Delta\varepsilon > 0$) are referred to as p-type materials and the materials that exhibit negative dielectric anisotropy ($\Delta\varepsilon < 0$) are referred to as n-type materials. The p-type LC materials tend to align themselves with their molecular axis parallel to the applied electric field, whereas n-type materials align themselves with their molecular axis perpendicular to the applied field. Generally, the dielectric constant $\varepsilon_{\|}$ decreases with increasing frequency [12] due to the relaxation phenomenon. However, $\varepsilon_{\perp}$ is independent of frequency over a large range of frequencies. At the crossover frequency, f_c, where $\varepsilon_{\|} = \varepsilon_{\perp}$, the LC material becomes isotropic. Depending on the material, this frequency, f_c, falls in the range of 100 kHz to >1 MHz. The dielectric constants $\varepsilon_{\|}$ and $\varepsilon_{\perp}$ also change as a function of temperature [13], as shown in Figure 32.4 for the nematic LC cyanobiphenyl. The two dielectric constants rapidly converge as the temperature approaches T_c, where $\varepsilon_{\|} = \varepsilon_{\perp} = \varepsilon_{\text{isotropic}}$.

Refractive Index Anisotropy

An LC is birefringent with anisotropic refractive indices. It has two principal refractive indices, n_o and n_e, as shown in Figure 32.5. For the ordinary refractive index n_o, the electric field vector of the light beam

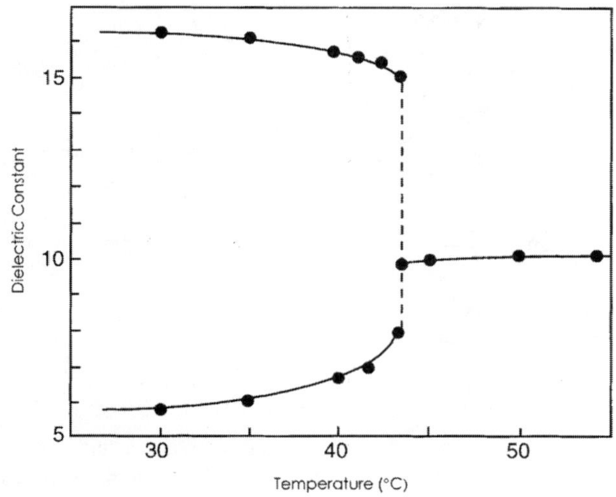

FIGURE 32.4 Temperature dependence of dielectric constant of the nematic LC cyanobiphenyl at a frequency of 100 KHz [13] exhibiting positive dielectric anisotropy ($\varepsilon_{\parallel} > \varepsilon_{\perp}$). $\varepsilon_{\parallel}$ decreases with temperature, whereas $\varepsilon_{\perp}$ increases with temperature until both are equal at the clearing temperature corresponding to the isotropic liquid.

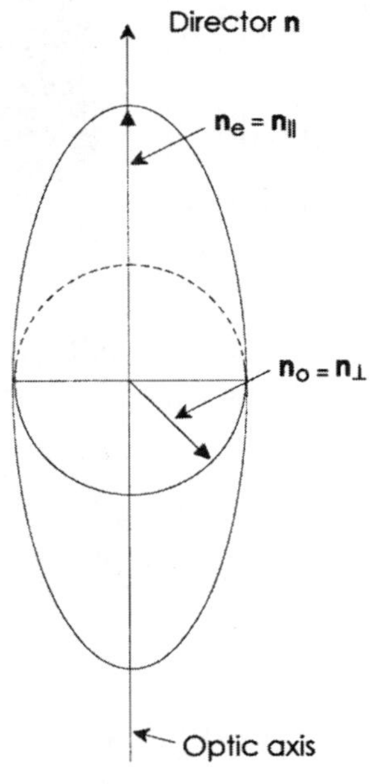

FIGURE 32.5 Refractive index anisotropy of uniaxial LC.

oscillates perpendicular to the optic axis, and for the extraordinary refractive index n_e, the electric field vector oscillates parallel to the optic axis. In the nematic and smectic LCs, the direction of the LC director $\boldsymbol{n}$ is the optic axis of the uniaxial crystal and therefore the refractive indices for light rays with oscillations in the directions parallel and perpendicular to the director are $n_{\parallel}$ and $n_{\perp}$, respectively, i.e., $n_o = n_{\perp}$, $n_e = n_{\parallel}$, and the optical anisotropy or birefringence, Δn, is given by

$$\Delta n = n_{\parallel} n_{\perp} = n_e - n_o \tag{32.4}$$

Figure 32.6a and b show the temperature dependence [2] and wavelength dependence [2] of the refractive indices of typical LC materials. The dependence of refractive index on wavelength λ is generally expressed by the so-called Cauchy equation:

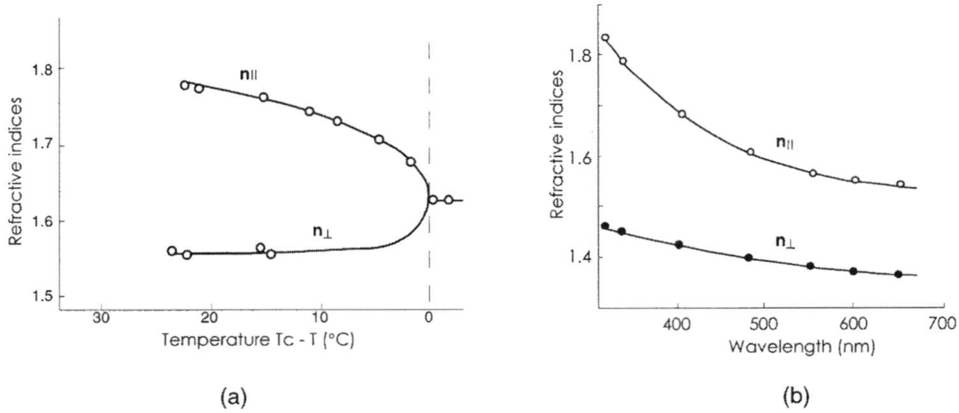

FIGURE 32.6 (a) Temperature dependence of the refractive indices $n_{\|}$ and $n_{\perp}$ for a nematic LC MBBA at $\lambda = 546$ nm; (b) wavelength dependence of refractive indices of the nematic LC 4-butoxyphenyl ester of 4′-hexyloxybenzoic acid at 80°C.

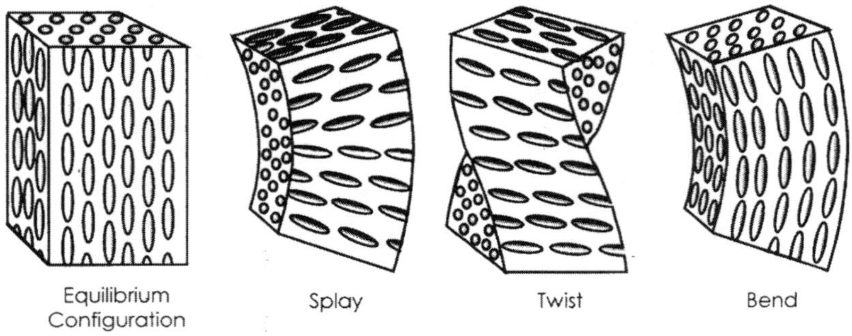

FIGURE 32.7 Deformation of nematic LC molecules from equilibrium configuration is shown in (a). Three types of deformations — (b) splay, (c) twist, and (d) bend can describe all possible types of deformations.

$$n_{o,e} = n_{\infty} + \alpha_{o,e} / \lambda^2 \tag{32.5}$$

where n_{∞} is the refractive index extrapolated to infinite wavelength and α is a material-specific coefficient.

Elastic Constants

In uniaxial LCs, the preferred or equilibrium orientation of the LC molecule is given by the director ***n***, which may be imposed by the surface treatments at the boundary conditions or by an external field. When the LC is perturbed from an equilibrium condition by application or removal of an external field, the elastic and electrical forces determine the static deformation pattern of the LC. The transition of the director from one direction to the other induces curvature strain in the medium. Frank [14] showed that an arbitrary deformation state can be envisaged as the combination of three basic operations; *Splay*, *Twist*, and *Bend*, denoted by the elastic constants K_{11}, K_{22}, and K_{33} following the notation of the Oseen–Frank theory. Figure 32.7 illustrates the *Splay*, *Twist*, and *Bend* deformations. The elastic part of the internal free energy, *F*, of a perturbed liquid crystal is given by the equation:

$$F = \frac{1}{2}\left[K_{11}(\nabla \cdot n)^2 + K_{22}(n \cdot \nabla \times n)^2 + K_{33}(n \times \nabla \times n)^2\right] \tag{32.6}$$

The free energy density is thus a quadratic function of the curvature strains with the elastic constants appearing as constants of proportionality. The elastic constants K_{11}, K_{22}, and K_{33} are temperature dependent, and decrease with increase in temperature. The magnitudes of the elastic constants, K_{ii}, can be approximated [15] by

$$K_{ii} \alpha S^2 \tag{32.7}$$

Electro-Optical Characteristics

When an electric field is applied to an LC with an initial molecular (director) orientation, it will change to a new molecular orientation due to the dielectric anisotropy ($\Delta\varepsilon = \varepsilon_{\parallel} - \varepsilon_{\perp}$) of the LC. This change in molecular orientation is accompanied by a change in the optical transmission/reflection characteristics of the LC which forms the basis for LCDs. This phenomenon of an electrically driven optical modulation is known as the electro-optic effect of the LC. When an electric field E is applied to an LC, it produces an electric energy, f_e, given by

$$f_e = -\frac{1}{2}\varepsilon_{\perp} E^2 - \frac{1}{2}\Delta\varepsilon (n \cdot E)^2 \tag{32.8}$$

The initial molecular orientation (boundary condition achieved by surface alignment) of the LC molecules (director $\boldsymbol{n}$) is either parallel (for $\Delta\varepsilon > 0$) or perpendicular (for $\Delta\varepsilon < 0$) to the plane of the two parallel electrodes of the display. When a field is applied across the parallel electrodes with the LC material in between, the director $\boldsymbol{n}$ orients parallel to the electric field E in *+ve* $\Delta\varepsilon$ materials, and it orients perpendicular to the field in *−ve* $\Delta\varepsilon$ materials. The total free energy, F, of the LC when the initial undeformed molecular orientation undergoes deformation due to the applied field is given by the sum of the electric energy f_e and the elastic energy. This transition from an undeformed state, known as *Freedericksz transition*, occurs as the field is increased to a critical field E_c. The Freedericksz transition is simply a transition from a uniform director configuration to a deformed director configuration; i.e., at any point in the LC, the order of the molecules relative to one another remains the same. The threshold electric field, E_c, is calculated by a free energy minimization technique [3], and it is given by

$$E_c = (\pi/d)\left(K_{ii}/|\Delta\varepsilon|\right)^{1/2} \tag{32.9}$$

Thus the threshold voltage V_{th} of the LC electro-optic effect is given by

$$V_{th} = E_c d = \pi\left(K_{ii}/|\Delta\varepsilon|\right)^{1/2} \tag{32.10}$$

In Equations 32.9 and 32.10, d is the thickness of the LC and k_{ii} is the appropriate elastic constant. When the field is perpendicular to the initially homogeneous orientation of the director, $K_{ii} = K_{11}$ or K_{22}. When the field is parallel to the initially homogeneous orientation, $K_{ii} = K_{33}$. In the case of a twisted orientation, $K_{ii} = K_{11} + (K_{33} - 2K_{22})/4$.

32.4 LCD Materials and Fabrication Processes

There are several types of LCDs utilizing different LC materials and LCD modes which are discussed in the next section. However, the general display assembly processes and materials are very similar for all these LCD modes. Figure 32.8a and b show the plan and cross-sectional view of a passive matrix-addressed LCD. The display fabrication can be broadly divided into three parts: (1) lower and upper glass fabrication processes, (2) cell assembly processes, and (3) polarizer and driver attachment and module assembly

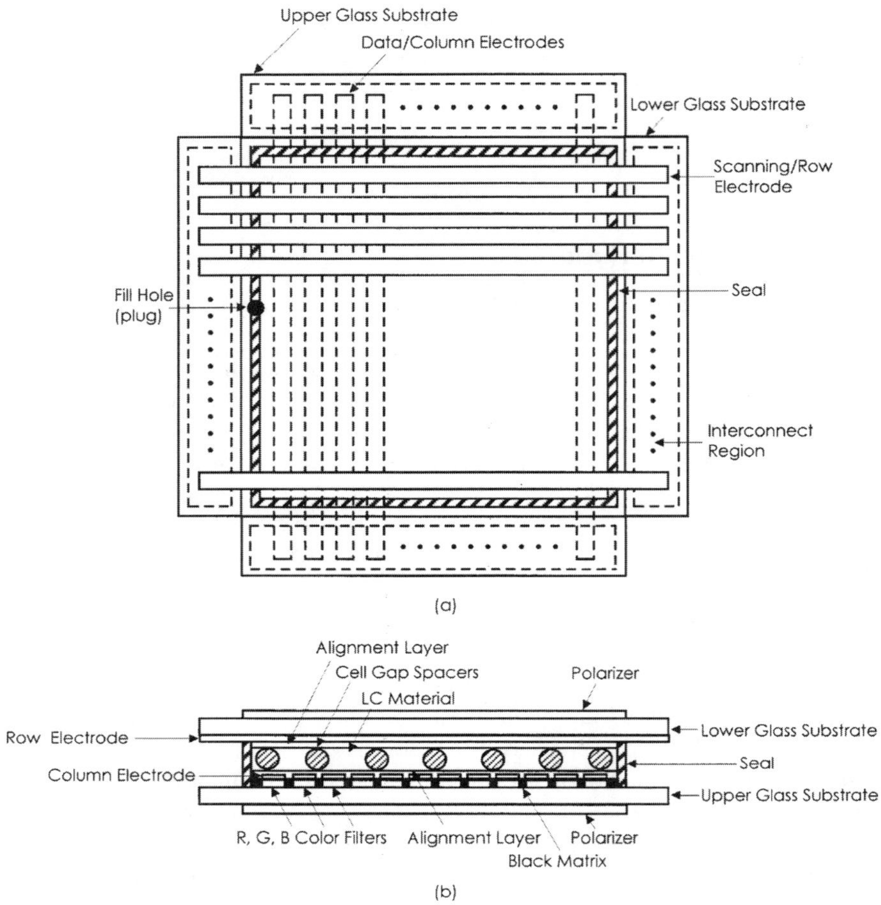

FIGURE 32.8 Plan (a) and cross-sectional view (b) of a passive matrix-addressed LCD.

processes as illustrated in Figure 32.9. The following will describe the various display materials and the assembly processes.

Glass Substrate

The quality of the glass substrate is important with regard to its chemical compatibility with the LC materials, surface flatness, defects, and dimensional stability under processing temperatures associated with various display fabrication steps. With a typical LCD cell gap in the range of 5 to 10 μm, the importance of the glass flatness and surface quality is clear. Glass substrate defects such as voids, scratches, streaks, and attached particles can cause electrode defects and hinder uniform LC cell spacing. Therefore, depending on the type of display, glass substrates are sometimes polished to achieve the required surface quality. Typical display glass materials include borosilicate (e.g., Corning 7059) and aluminosilicate glasses (e.g., Corning 1737), with a thickness of 0.7 or 1.1 mm.

Color Filters

In full-color LCDs, color most often is generated by use of red, green, and blue (R, G, B) color filters fabricated at each pixel, as shown in Figure 32.8b, using a white backlight system. The color filter requirements include proper spectral transmission characteristics and chemical, thermal, and dimensional stability. The display color gamut is a function of the spectral characteristics of the backlight used

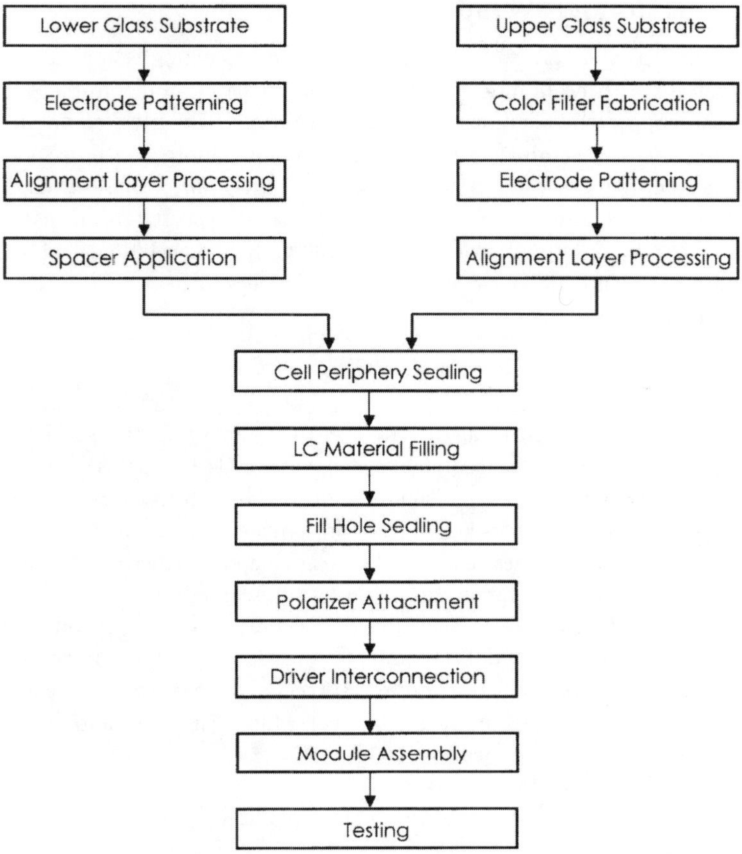

FIGURE 32.9 LCD assembly flowchart.

and the color filter transmission characteristics. By a suitable choice of these parameters, an LCD can achieve a color gamut comparable to that of a high-quality CRT. However, trade-offs are sometimes made between color purity and transmission (brightness) characteristics of the LCD. Typical color filter thickness is about 2 μm. Color filter materials include dye and pigment dispersed polyimides and photoresists. The color filter materials are applied on the display glass by various processes, such as spin coating, printing, electrodeposition, and photolithography. First, color filter material of the first color is applied and photolithographically patterned. Then, the color filter material for the second color is processed, and then the third. A black matrix material is also applied and patterned between the color filters to block the light transmission from the interpixel regions. In some cases, a passivation layer such as low-temperature SiO_2 dielectric is deposited on the color filters to act as a barrier for impurities and to achieve a smooth surface for the subsequent transparent electrode deposition.

Transparent Electrodes

Most often, indium tin oxide (ITO) with a typical concentration of 90% In_2O_3 and 10% SnO_2 is used as the transparent conducting electrode material. The typical transmission of ITO is about 90%. It is generally deposited by e-beam evaporation or sputtering in an oxygen-containing atmosphere. The film thickness is typically 50 to 300 nm depending on the required sheet resistance and transmission. Typical resistivity is in the range of $2 \times 10^{-4}\ \Omega \cdot cm$. Light transmission through ITO is not linearly proportional to the film thickness because of light interference.

Alignment Materials and Techniques

After the transparent electrodes are patterned, an LC alignment coating is applied. Depending on the display mode of interest, either a homogeneous (parallel), tilted, or a homeotropic (vertical) alignment is achieved using an appropriate alignment material and process. The most commonly used TN LCD requires a tilted alignment to eliminate reverse tilt disclinations. Inorganic films such as an obliquely evaporated SiO, as well as organic materials such as polyimide, can be used as alignment materials. Polyimide is most commonly used as an alignment layer. It is spin-coated or printed to achieve a layer thickness of about 50 nm, and cured around 150 to 200°C. It is then buffed (rubbed) using a roller covered with cotton or a synthetic fabric. For TN LCD, the alignment process is selected to achieve a pretilt angle of around 2 to 5°.

Cell Spacing and Sealing

The cell assembly process starts after the alignment layer treatment of the two substrates by rubbing. To control the LC cell spacing (thickness) accurately, spacing materials are applied in the active area of the display as well as in the peripheral seal area, where the two glass substrates are sealed. Typical spacing materials include glass or plastic fibers (cylinders) or spheres (balls), with a tightly controlled size (diameter) distribution. A sealing adhesive material such as an epoxy is then applied at the seal area with an opening (fill hole) left for LC material injection after the sealing process. The seal is typically 1 to 3 mm wide and is a few millimeters from the active display area. The requirement for the seal material is that it must not chemically react with (contaminate) the LC material and must be a barrier against moisture and contamination from outside. After the spacers and seal materials are applied on the first (active) substrate, it is precisely aligned to the second (color filter) substrate, and the seal is cured either by heating it to a higher temperature, typically 100 to 150°C, or by ultraviolet exposure depending on the type of epoxy seal material used.

LC Material Filling

After the empty cell is fabricated, it is filled with the LC material. Because the cell thickness (spacing) is small (~5 to 10 μm), it is filled using special techniques. The most popular filling method is by evacuating the cell in a vacuum chamber, dipping the fill hole into a vessel containing the LC material, and increasing the pressure in the chamber. As the chamber pressure is raised, the cell gets filled by capillary action. After the cell is completely filled, the fill hole is capped by using an epoxy adhesive that is chemically compatible with the LC material.

External Components

The external components of an LCD include polarizers, reflectors, display drivers, and a backlight assembly. A reflective LCD uses a reflector at the back of the display, works by modulating the ambient light, and does not require backlighting. In the most commonly used transmissive mode TN LCDs, a polarizer is attached on the front as well as back surfaces of the LCD after the cell assembly is complete. Also, in the most commonly used normally white mode TN LCD, the polarizers are attached with their polarization axis crossed and along the rubbing directions of the alignment layers. The polarizer is a three-layer composite film with a stretched iodine doped polyvinyl alcohol (PVA) polarizing film in the center, and two outer films of triacetyl cellulose (TAC) for protecting the PVA film from the ambient (moisture, temperature, and harsh environment) conditions. A typical transmission range of a polarizer is 41 to 45%, with a polarization efficiency in the range of 99.9 to 99.99%.

The display row and column IC drivers are attached to the row and column bond pads of the display either by TAB (tape-automated bonding) using an ACA (anisotropic conductive adhesive), or chip on glass (COG) approaches. For backlighting transmissive LCDs, a fluorescent lamp is generally used. The R, G, B emission spectrum of the backlight and transmission spectrum of the R, G, B color filters are tuned together to achieve the desired color coordinates for the primary colors. Also, a diffuser is used to achieve uniform backlighting of the display. The backlight system may also use brightness enhancement films to tailor the light intensity distribution in the viewing cone. In addition to the above components, LCDs for specialized applications requiring enhanced performance may use a cover glass with EMI and antireflection coatings at the front, and a heater glass at the back side (between the backlight and the LCD) which facilitates low-temperature operation.

32.5 Liquid Crystal Display Modes

LC displays based on many different modes of operation have been developed. Historically, the phase change (PC) effect was discovered first in 1968 by Wysoki et al. [16]. The same year, dynamic scattering (DS) mode [17] and guest–host (GH) mode [18] were announced by Heilmeier et al. Then in 1971, the TN mode [19] was reported by Schadt and Helfrich and electrically controlled birefringence (ECB) was reported by Schiekel and Fahrenschon [20] and Hareng et al. [21]. The physical effects and the various display modes based on these effects include

Current effects:	• DS effect
Electric field effects:	• TN effect
	• STN effect
	• ECB
	• GH effect
	• Phase change effect
Thermal effects:	• Smectic effect

The DS effect is based on the anisotropy of the conductivity. Because of higher voltage operation and higher power consumption, the DS mode is not currently used. The TN and STN effects are most widely used among all the LCDs. The following will discuss various display modes.

Twisted Nematic Effect

Figure 32.10 shows a schematic of a display based on the TN effect. It consists of nematic LC material with a positive dielectric anisotropy ($\Delta\varepsilon > 0$) with a layer thickness of about 5 μm, sandwiched between two transparent substrates with transparent electrodes. The surfaces of the transparent electrodes are

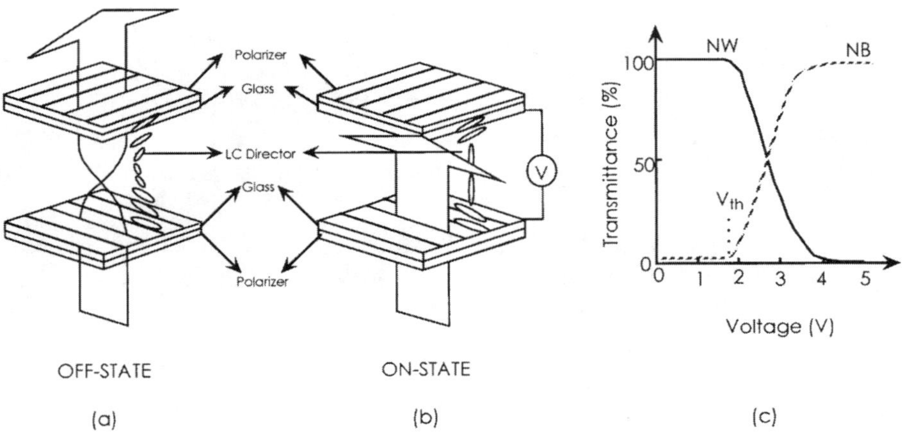

FIGURE 32.10 Illustration of TN effect: (a) in the off-state, the incident polarized light is transmitted through the entrance polarizer, the 90° TN LC, and the crossed exit polarizer; (b) in the on-state; (c) the solid line shows the voltage-transmission behavior for the NW configuration, shown in (a) and (b), with crossed polarizers. The dashed line is for an NB configuration with parallel polarizers.

coated with a polyimide alignment layer and rubbed to orient LC molecules at the substrate surfaces along the rubbing direction with a small (~3°) pretilt angle. The molecules on the two substrates are oriented 90° from each other as shown in Figure 32.10a; i.e., the LC molecular axis rotates (twists) continuously through 90° from the first substrate to the second substrate. The TN display can be fabricated to operate in a normally black (NB) or normally white (NW) mode based on how the polarizers are attached to the outer surface of the two glass substrates. Figure 32.10a and b shows the on- and off-state of a NW mode TN LCD with crossed (orthogonal) polarizers attached with their polarization direction parallel to the LC director orientation on that substrate. Since the pitch of the twist is sufficiently large compared with the wavelength of the visible light, the direction of polarization of linearly polarized light incident normally on one surface of the display rotates through 90° by the twist of the LC molecules as it propagates through the cell and exits through the second polarizer. When a voltage is applied to the TN cell, the molecules align parallel to the direction of the field as shown in Figure 32.10b, and the 90° optical rotatory power is eliminated. Thus, the incident polarized light from the first polarizer is not rotated as it goes through the LC cell and gets blocked by the crossed exit polarizer.

Figure 32.10c shows the voltage-transmission characteristics of a TN cell for NB and NW modes of operation. For an NW mode, when a sufficiently high voltage is applied, LC molecules are aligned homeotropically (parallel to the field) and there is no rotation of the electrical field vector of the polarized light. This results in complete suppression of transmission regardless of the wavelength of light. The transmission in the on-state is, however, wavelength dependent, but this does not have a significant effect on the contrast ratio, although it can influence the color balance. In the NB mode of operation, the transmission is suppressed to zero [22] for the off-state only for a monochromatic light of wavelength. Therefore, in a practical display using a broadband backlight, there is a small amount of light leakage which lowers the display contrast ratio. In a color display, the cell gaps for the R, G, B pixels can be optimized to eliminate the light leakage and improve contrast [22].

The threshold voltage of a TN mode LCD is given by

$$V_{\mathrm{th}} = \pi \cdot \sqrt{\left\{\left[K_{11} + \left(K_{33} - 2K_{22}\right)/4\right]/\varepsilon_0 \cdot \Delta\varepsilon\right\}} \tag{32.11}$$

V_{th} depends on the dielectric anisotropy and elastic constants, and is generally in the range of 2 to 3 V, with the maximum operating voltage being in the range of 5 to 8 V. This low voltage driving, coupled with low current due to high resistivity of LC materials, contributes to the very low power consumption (~1 μW/cm^2) of LCDs. The response times measured by the rise and decay times of the display τ_r, τ_f, are given by [23]:

$$\tau_{\mathrm{d}} = \gamma d^2 / \Delta\varepsilon\left(V^2 - V_{\mathrm{th}}^2\right) \tag{32.12}$$

$$\tau_{\mathrm{d}} = \gamma d^2 / \left(\Delta\varepsilon \cdot V_{\mathrm{th}}^2\right) \tag{32.13}$$

where γ is the rotational viscosity coefficient. The above equations show that rise time can be improved by using a thinner cell gap d, a higher $\Delta\varepsilon$, and a higher drive voltage V. Similarly, the decay time can be improved by reducing the cell gap. The turn-on time τ_r is usually shorter than the turn-off time τ_d. At room temperature, these times are of the order of 10 ms, which is adequate for many common applications such as computer and TV displays. In a TN LCD, gray scale is generated by varying the voltage using the electro-optic curve shown in Figure 32.10c. The shallow slope of the electro-optic curve works well for the gray scale generation in active matrix-addressed displays (see next section). However, in the case of passive matrix addressing, the shallow slope greatly limits the multiplexibility (number of addressable rows) of the display, which led to the development of the STN effect.

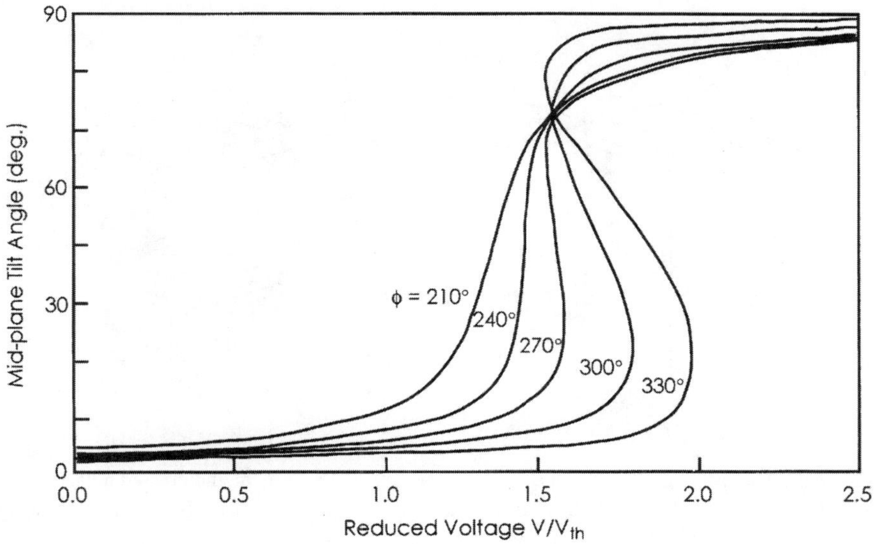

FIGURE 32.11 Calculated curves of tilt angle of local directors in the midplane of an STN cell as a function of reduced voltage V/V_{th}, where V_{th} is the Freedericksz threshold voltage of a nontwisted layer with a zero pretilt angle. The steepness of the curves increase as the twist angle, Φ, is increased, and bistability is achieved when $\Phi > 240°$.

STN Effect

In a passive matrix-addressed display, the addressability or the maximum number of addressable rows N is given by the Alt and Pleshko [24] limit:

$$V_{on}/V_{off} = \sqrt{\left(N^{1/2}+1\right)/\left(N^{1/2}-1\right)} \tag{32.14}$$

where V_{on} and V_{off} are the rms voltages at the select and nonselect pixels. Equation 32.14 shows that as N increases, V_{off} approaches V_{on} and the contrast ratio becomes 1, which makes the display not viewable. For $N = 100$, $V_{on} = 1.11V_{off}$; i.e., select voltage is only 11% higher than the nonselect voltage. This will result in a very low contrast ratio when using a TN mode with a shallow turn-on curve (Figure 32.10c). STN displays have been developed [25, 26] to achieve a steep electro-optic curve, so that large numbers of rows can be multiplexed. The STN effect uses a twist angle of 180° to about 270° with a relatively high pretilt angle alignment. Figure 32.11 [25] illustrates the STN effect. The figure shows the voltage dependence of the midplane director tilt of a chiral nematic layer with a pretilt of 28° at both substrates. Bistability is achieved when a twist angle, ϕ, greater than 245° is used. In highly multiplexed displays twist angles in the range of 240° to 275° and tilt angles in the range of 5° to 30° are generally used. High pretilts ensure that competing distortional structure which has 180° less twist is eliminated. For a 270° left-handed twist, optimum results are achieved when the front polarizer is oriented with its polarization axis at 30° with the LC director and the rear polarizer is oriented at an angle of 60° with the projection of the director at the rear substrate. Due to interference of the optical normal modes propagating in the LC layer, the display has a yellow birefringence color in the nonselect state (yellow mode). Rotation of one of the polarizers by 90° results in a complementary image with a bright colorless state, and a blue nonselect state (blue mode). White-mode STN displays are made using retardation films. The response time of a typical STN display is on the order of 150 ms. These displays typically have a lower contrast ratio and a narrow viewing angle.

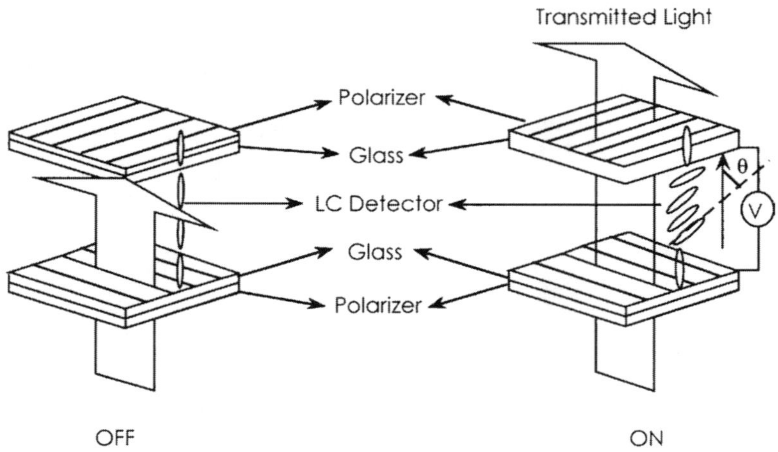

FIGURE 32.12 Illustration of an ECB display: (a) with homeotropic alignment and crossed polarizers, the off-state is black; (b) in the on-state, the output light through the LC is elliptically polarized due to the LC birefringence and the light is transmitted through the crossed polarizer.

Electrically Controlled Birefringence (ECB)

This display technique is based on controlling the birefringence of the LC cell by application of an electric field. There are a number of types of this display depending on the molecular orientation of the LC cell used; examples include DAP type, homogeneous type, HAN (hybrid aligned nematic) type, and IPS (in-plane switching) type. A DAP type (homeotropic orientation) is made using an LC with a negative $\Delta\varepsilon$, sandwiched between transparent electrode substrates and placed between crossed polarizers, as shown in Figure 32.12. In the off-state (with no electric field), the incident polarized light does not see birefringence when passing through the cell, and thus gets blocked by the crossed exit polarizer. In the on-state (when a voltage is applied), the molecular axis of the LC is inclined at an angle θ (as shown in Figure 32.12), so the linearly polarized light becomes elliptically polarized as it passes through the cell due to birefringence. Hence a portion of the light passes through the crossed exit polarizer; the intensity I of the transmitted light through the cell is given by [20]

$$I = I_o \sin^2 2\theta \cdot \sin^2\left(\pi d \Delta n(V)/\lambda\right) \tag{32.15}$$

where I_0 is the intensity of the incident light, θ is the angle between the direction of polarization of the incident light and the direction of oscillation of the ordinary light within the cell, d is the cell spacing, $\Delta n\ (V)$ is the birefringence of the cell, $d\Delta n$ is the optical phase difference, and λ is the wavelength of the incident light. The equation shows that I depends on the applied voltage and λ.

In case of the homogeneous technique, the LC cell uses positive material with a homogeneous orientation. With this method, the threshold voltage is obtained by replacing the bend elastic coefficient K_{33}, with splay elastic coefficient K_{11}. The HAN cell is characterized by a hybrid orientation cell in which the molecules are aligned perpendicular to one substrate, but parallel to the second substrate. In this mode, both positive and negative $\Delta\varepsilon$ materials can be used, and since there is no clear threshold voltage, it has the advantage of a very low drive voltage. Recently, an ECB mode based on an IPS type display has been developed to produce LCDs with extremely wide viewing angles. In the IPS mode displays, LC is homogeneously aligned and switched between on- and off-states using interdigitated electrodes fabricated on one of the display substrates [27]. The IPS mode displays use an NB mode with either positive or negative $\Delta\varepsilon$ LC materials.

Guest–Host Type

Some organic dyes show anisotropy of light absorption; i.e., they absorb more light in a specific wavelength band when the **E** vector of the light is parallel to the optic axis of the dye molecules, than they do when it is perpendicular. LCDs based on this principle are called guest–host (GH) displays. In these displays a small amount of a dichroic dye (guest) is mixed in the LC material (host). These dye molecules get aligned to the LC molecules; hence, their orientation can be changed (by changing the orientation of the LC molecules) by application of an electric field. When an LC material with a positive $\Delta\varepsilon$ is used, in the off-state, the **E** vector of the polarized light coincides with the light absorption axis of the dichroic dye; hence, light is absorbed and transmitted light is colored. When a voltage is applied for the on-state, the **E** vector of the polarized light is orthogonal to the absorption axis of the dye; hence, no absorption takes place, and transmitted light is not colored (white). GH LCD requires only one polarizer. Further, because the optical effect is based on absorption, the display provides a better viewing angle than a TN mode LCD.

Phase-Change Type

The phase change type of display is based on a change in molecular orientation from a helical cholesteric phase to a homeotropic nematic phase, and vice versa. For this technique, a cholesteric LC with a long helical pitch with a positive or negative $\Delta\varepsilon$ is used. No polarizers are used in this display. In the off-state of this display, the incident light passing through the cholesteric cell with a focal conic orientation is optically dispersed (scattered), and the cell looks cloudy. However, when a voltage is applied, helical structure of the cholesteric phase changes to a nematic phase with a homeotropic orientation, and the cell becomes transparent.

Thermal Effects

Thermal effect is based on a change in the electro-optical behavior due to a change in the molecular orientation of the LC when it is heated or cooled. This effect is utilized with smectic LCs with a homeotropic alignment. When this cell is heated until the isotropic phase is reached and cooled, then if the cooling is sudden, the cell becomes cloudy, whereas if the cooling is gradual, the cell becomes transparent. These cloudy and transparent states correspond, respectively, to the focal conic and homeotropic orientations of the smectic A LC. This effect is used for large-size memory type displays, in which a laser beam is used to write the image thermally. The heating can also be accomplished by one of the transparent electrodes, while the other transparent electrode is used as a signal electrode.

32.6 Display Addressing

Display addressing (driving) techniques have a major influence on the LCD image quality. The addressing techniques can be classified in three essential types, namely, direct (static) addressing, passive matrix addressing, and active matrix addressing. In the case of low-information-content displays such as numeric displays, bar graph displays, and other fixed pattern displays, using segmented electrodes, direct addressing is used. A common example of direct-addressed displays is a numeric display using seven segmented electrodes for each digit. Each of these segmented electrodes on the front substrate and the common electrode on the back substrate are directly connected to drive signals. A voltage is selectively applied to each of the segments so that any of the digits between 0 and 9 can be displayed. For high-information-content displays, this approach becomes impractical because of the huge number of interconnects, and, hence, either passive matrix or active matrix addressing is used.

Passive Matrix Addressing

A passive matrix (PM) display comprising an LC between a matrix of transparent conducting row and column electrodes (Figure 32.8) is the simplest and least expensive matrix-addressed LCD to manufacture.

An example of a PM LCD is a color (R,G,B) VGA (video graphics array) display using the STN effect, with a pixel format of 640 (× 3 = 1920) H × 480 V, with 1920 columns and 480 rows, for a total of 2400 interconnects used for addressing a display containing 921,600 pixels. In PM addressing, the row voltages are scanned in succession with a voltage, V_r, while all the columns in a given row are driven in parallel, during the row time, with a voltage of $\pm V_c$ depending on whether the pixel is selected to be ON or OFF. As discussed above under the STN effect, the contrast ratio of PMLCDs is influenced by the Alt and Pleshko [24] addressability limitation. To enhance the operating margin for improved contrast ratio, DSTN (dual-scan STN) configuration is used in higher-information-content displays. In a DSTN, the display is separated into two halves, and the rows in each half are scanned simultaneously and synchronously, to essentially double the duty ratio of the ON pixels to increase the contrast ratio. One of the major shortcomings of the passive matrix-addressed STN display is the slow response time of the LC, which is of the order of 150 ms. This slow response time is not adequate for video applications and is barely fast enough for the graphical interface of a computer. The response time of the STN LCDs can be improved by active addressing or multiline addressing techniques [28, 29]. These techniques involve simultaneous addressing of several rows of a display to suppress the frame response problems of conventional STN LCDs.

Active Matrix Addressing

Active matrix (AM) addressing removes the multiplexing limitations [24] of the PM LCDs by incorporating a nonlinear control element in series with each pixel, and provides 100% duty ratio for the pixel using the charge stored at the pixel during the row addressing time. Figure 32.13a illustrates an active matrix array with row and column drivers and the associated display module electronics. Figure 32.13b shows a magnified view of the active matrix array in the AM LCD panel. In the figure, C_{LC} and C_S represent the pixel capacitance and the pixel storage capacitance. Typically, a storage capacitor, C_S, is incorporated at each pixel to reduce the pixel voltage offset (see Equation 32.16 below) and for a broad temperature operation. Figure 32.14 shows the cross section through an AM LCD illustrating various elements of the display. Figure 32.15 shows a typical AM LCD pixel, showing the gate and data busses, thin-film transistor (TFT), ITO pixel electrode, and the storage capacitor. Fabrication of the active matrix substrate is one of the major aspects of AM LCD manufacturing. Both two-terminal devices such as back-to-back diodes, and metal-insulator-metal (MIM) diodes as well as three-terminal TFTs are developed for active matrix addressing. While two-terminal devices are simple to fabricate and cost less, their limitations include lack of uniform device performance (breakdown voltage/threshold voltage) over a large display area, and lack of total isolation of the pixel when neighboring pixels are addressed. For a superior image quality AM LCDs use TFT for the active matrix device, which provides a complete isolation of the pixel from the neighboring pixels. Large-area AM LCDs use amorphous silicon (a-Si) TFTs [6], while polysilicon TFTs with integrated row and column drivers are used in small high-resolution LCDs [30].

Figure 32.16 shows the electric equivalent of a TFT-LCD pixel, display drive waveforms, and the resulting pixel voltage. As in most matrix-addressed displays with line-at-a-time addressing, the rows (gates) are scanned with a select gate pulse $V_{g,sel}$, during the frame time t_f, while all the pixels in a row are addressed simultaneously with the data voltage $\pm V_d$ during the row time t_r (= t_f/N). During the row time the select gate voltage, $V_{g,sel}$, "turns on" the TFT and charges the pixel and the storage capacitor to the data voltage V_d. After the row time, the TFT is "switched off" by application of the nonselect gate voltage, $V_{g,non\text{-}sel}$; hence, the voltage (charge) at this pixel is isolated from the rest of the matrix structure until it is time to charge the pixel during the next frame time. Note that the LC pixel must be driven in an ac fashion with $+V_d$ and $-V_d$, during alternate frame periods, with no net dc across the pixel. A net dc voltage across the pixel results in flicker and image sticking effects [33], resulting from LC conductivity. Large and sustained dc voltages also degrade the LC material due to electrolysis. The shift in pixel voltage, ΔV_p shown in Figure 32.16, at the end of the row time is due to the parasitic gate-to-drain capacitance, C_{gd}, of the TFT. When the gate voltage is switched, the distribution of the charge from the TFT gate dielectric causes the pixel voltage shift ΔV_p, given by

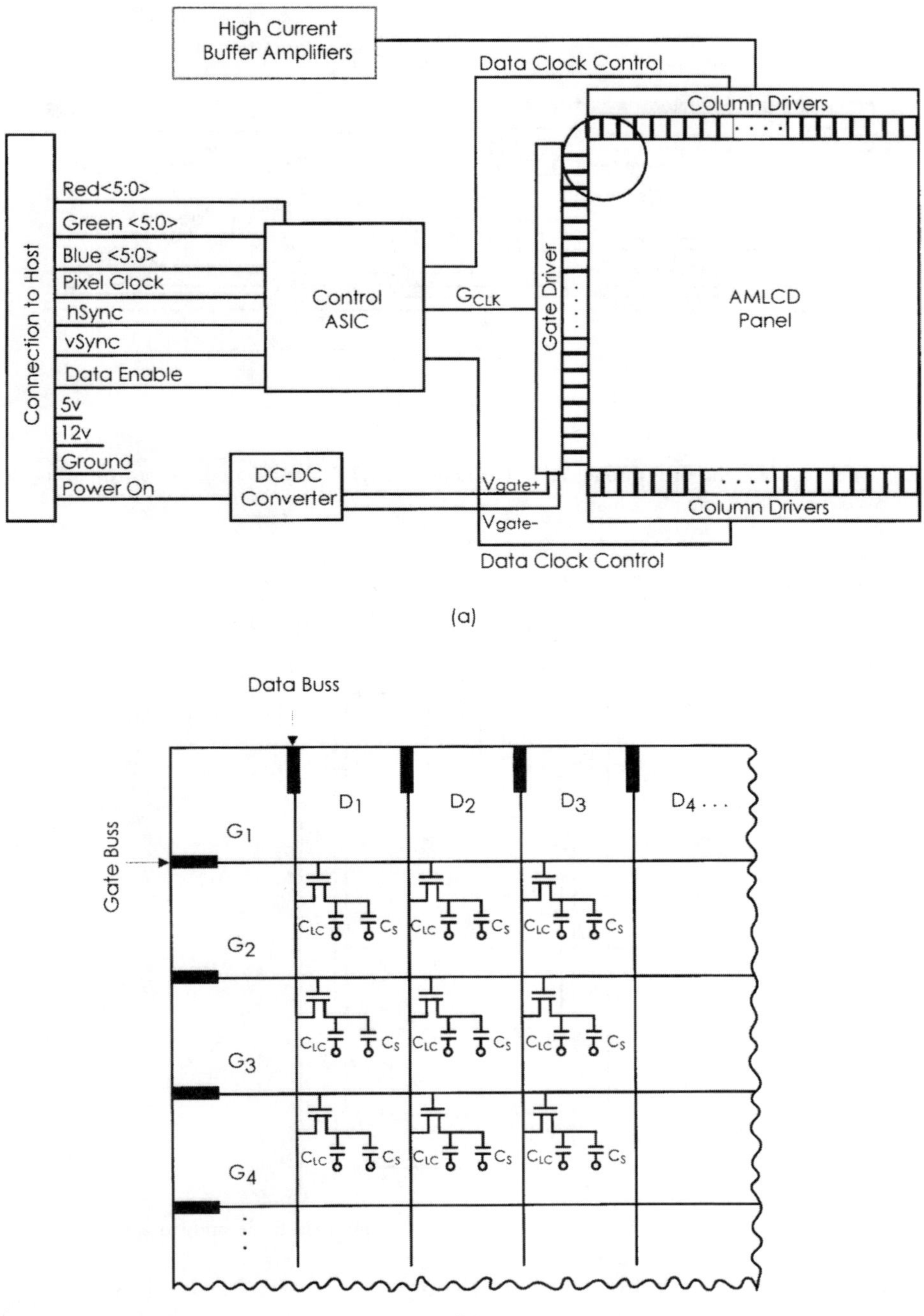

FIGURE 32.13 (a) AM LCD module electronics block diagram; (b) magnified view of the region shown in (a) illustrating active matrix TFT array.

$$\Delta V_p = (\Delta V_g) C_{gd} / (C_{gd} + C_{lc} + C_s) \tag{32.16}$$

For the *n*-channel enhancement mode a-Si TFT, this voltage shift ΔV_p is negative for both the positive and negative frames; thus, it helps pixel charging in the negative frame and hinders it in the positive frame. Further, due to increased gate bias during the negative frame, the pixel attains the data voltage

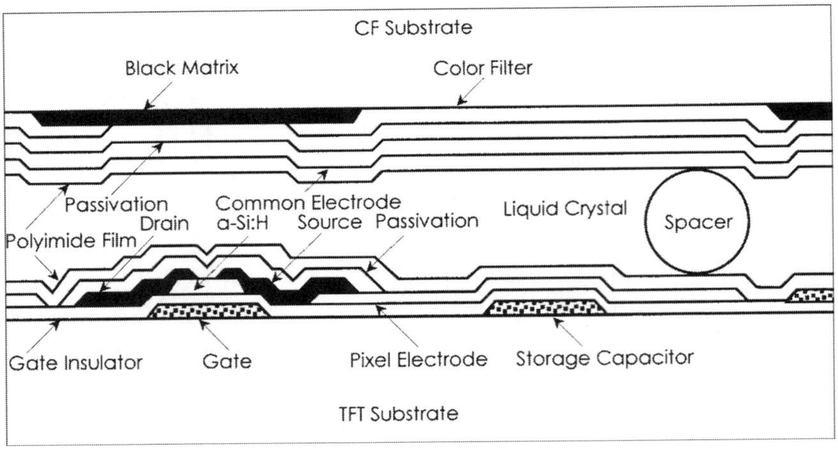

FIGURE 32.14 Cross-sectional view through an AM LCD showing TFT, pixel electrode, storage capacitor, polyimide alignment layers, color filter, and black matrix.

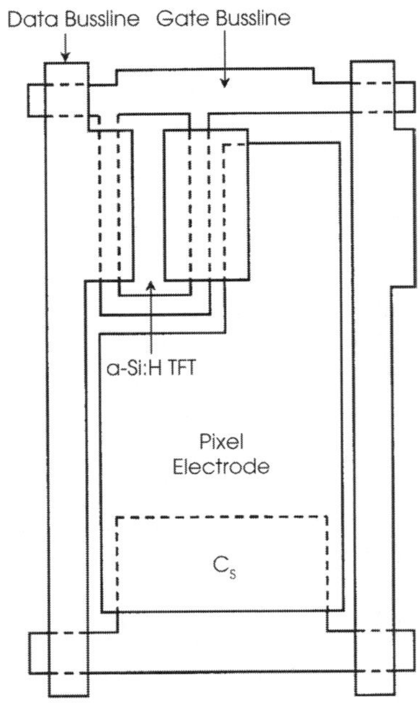

FIGURE 32.15 Layout of a pixel in a TFT-LCD showing the TFT, pixel electrode, and storage capacitor.

much more rapidly during the addressing period. Hence, the TFT is designed for the worst-case positive frame conditions. ΔV_p is reduced by minimizing C_{gd} by decreasing the source drain overlap area of the TFT and by using a storage capacitor. Further, ΔV_p is compensated by adjusting the common electrode voltage V_{com} as shown in Figure 32.16. Note that C_{lc} is a function of the V_p (V_{lc}) due to the dielectric anisotropy of the LC; and hence, adjustment to V_{com} alone does not eliminate dc for all gray levels, and modification of the gray scale voltages is required to compensate for the dielectric anisotropy of the LC.

Display Module Electronics

Figure 32.13 shows a block diagram for an AM LCD module electronics. The control block and power supply generation means are separately mounted on a PC board and connected to the row and column

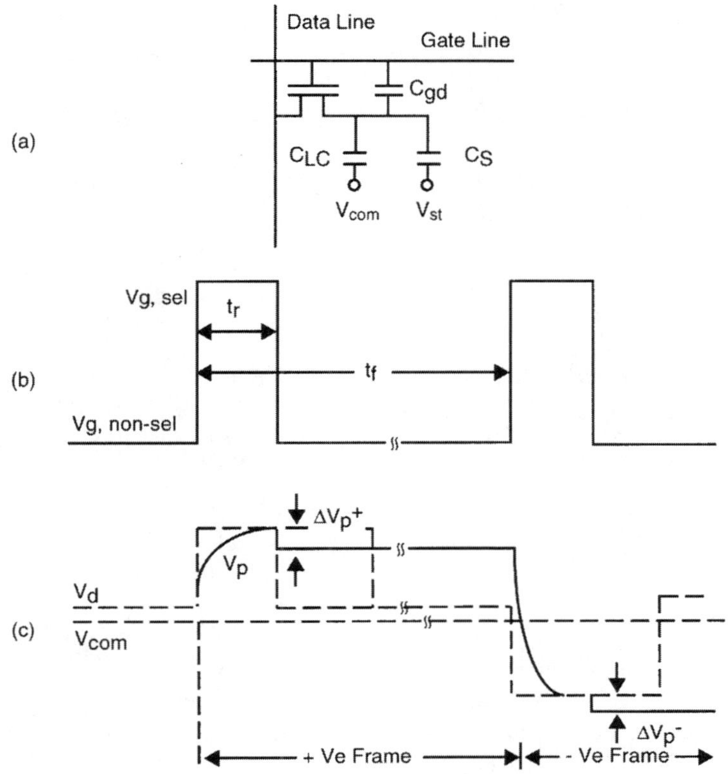

FIGURE 32.16 (a) Electric equivalent of a TFT-LCD pixel; (b) gate drive voltage waveform; (c) data voltage waveform and pixel charging behavior in the positive and negative frames.

drivers of the LCD on one side and to the host controller on the other. The control block may include level shifters, timing generators, and analog functions in some cases; the control block takes in digital data from the host system, which is typically a graphics controller chip, and converts it into timing and signal levels required by the row and column drivers. The architecture and design of the module electronics encompassing row and column drivers have a significant impact on not only the display system cost and power consumption, but also the image quality. The LC material typically requires about 5 V to achieve optical saturation (see Figure 32.10c). Considering the need for an ac drive, the required voltage swing across the LC material is about 10 V. To achieve this 10 V swing across the LC material, the column drivers typically use 12 V power supplies. Column driver voltage can be reduced by using a V_{com} modulation drive method. In this method, the V_{com} node (which is connected to all pixels in the display) is driven above and below a 5 V range of the column drivers. Each and every row time, the V_{com} node is alternated between a voltage above and a voltage below the 5 V output range of the column drivers. This achieves 10 V across the LC material using 5 V column drivers. This method requires additional components and consumes additional power due to the oscillation of the V_{com} node. In addition, to avoid capacitive injection problems, the row drivers usually have their negative supply modulated with the same frequency as the V_{com} node. Note, however, that compared to 10 V column drivers, 5 V column drivers consume less power, and are simpler to design and fabricate using small-geometry CMOS. The V_{com} modulation drive method can be used with a row (polarity) inversion scheme only (for elimination of pixel flicker) which results in some horizontal cross talk. However, column inversion and pixel inversion schemes provide better image quality with much-reduced cross talk, but they cannot be used with the V_{com} modulation drive.

References

1. P.J. Collings, *Liquid Crystals: Nature's Delicate Phase Matter*, Princeton University Press, Princeton, NJ, 1990.
2. L.M. Blinov and V.G. Chigrinov, *Electro-optical Effects in Liquid Crystal Materials,* Springer, New York, 1996.
3. P.G. De Gennes and J. Prost, *The Physics of Liquid Crystals,* 2nd ed., Clarendon Press, Oxford, 1993.
4. B. Bahadur, Ed., *Liquid Crystals: Applications and Uses,* World Scientific, Singapore, 1990.
5. E. Kaneko, *Liquid Crystal TV Displays: Principles and Applications of Liquid Crystal Displays,* D. Reidel Publishing, Boston, 1987.
6. T. Tsukada, *TFT/LCD: Liquid Crystal Display Addressed by Thin Film Transistors*, Gordon and Breach, Canada, 1996.
7. E. Lueder, Fundamentals of passive and active addressed liquid crystal displays, *Short Course S-1, SID Conference*, San Diego, CA, May 12, 1996.
8. P.W. Ross, L.K.M. Chan, and P.W. Surguy, Ferroelectric LCD: simplicity and versatility, *SID '94 Digest*, 147–150, 1996.
9. J.W. Doane, D.K. Yang, and Z. Yaniv, Front-lit flat panel display for polymer stabilized cholesteric textures, *Proc. 12th International Display Conference*, Japan Display '92, Oct. 12–14, 73–76, 1992.
10. J.C. Rowell, W.D. Phillips, L.R. Melby, and M. Panar, NMR studies of some liquid crystal systems, *J. Chem. Phys.*, 43, 3442–3454, 1965.
11. L. Pohl and E. Merck, Physical properties of liquid crystals, in *Liquid Crystals: Applications and Uses,* B. Bahadur, Ed., World Scientific, Singapore, 1990, 139–170.
12. P.R. Kishore, N.V.S. Rao, P.B.K. Sarma, T.F.S. Raj, M.N. Avadhanlu, and C.R.K. Murty, Field and frequency effects in nematic mixtures of negative and positive dielectric anisotropy, *Mol. Cryst. Liq. Cryst.*, 45, 3/4, 231–241, 1978.
13. D. Lippens, J.P. Parneix, and A. Chapoton, Study of 4-heptyl 4′-cyanobiphenyl using the analysis of its dielectric properties, *J. Phys.*, 38, 1465, 1977.
14. F.C. Frank, On the theory of liquid crystals, *Discuss. Faraday Soc.,* 25, 19, 1958.
15. W.H. de Jeu, *Physical Properties of Liquid Crystalline Materials,* Gordon and Breach, New York, 1980.
16. J.J. Wysocki, A. Adams, and W. Haas, Electric-field-induced phase change in cholesteric liquid crystals, *Phys. Rev. Lett.*, 20, 1024, 1968.
17. G.H. Heilmeier, L.A. Zanoni, and L.A. Barton, Dynamic scattering: a new electo-optic effect in certain classes of nematic liquid crystals, *Proc. IEEE*, 56, 1162, 1968.
18. G.H. Heilmeier and L.A. Zanoni, Guest-host interaction in nematic liquid cyrstal — a new electro-optic effect, *Appl. Lett.,* 13, 91, 1968.
19. M. Schadt and W. Helfrich, Voltage-dependent optical activity of a twisted nematic liquid crystal, *Appl. Phys. Lett.*, 18, 127, 1971.
20. M.F. Schiekel and K. Fahrenschon, Deformation of nematic liquid crystal with vertical orientation in electric fields, *Appl. Phys. Lett.,* 19, 391, 1971.
21. M. Hareng, G. Assouline, and E. Leiba, Liquid crystal matrix display by electrically controlled birefringence, *Proc. IEEE,* 60, 913, 1972.
22. C.H. Gooch and H.A. Tarry, The optical properties of twisted nematic liquid crystal structures with twist angles $\leq = 90°$, *J. Phys. D Appl. Phys.*, 8, 1575–1584, 1975.
23. E. Jakeman and E.P. Raynes, Electro-optical response times in liquid crystals, *Phys. Lett.,* 39A, 69–70, 972.
24. P.M. Alt and P. Pleshko, Scanning limitations of liquid-crystal displays, *IEEE Trans. Electron Dev.*, ED-21, 146, 1974.
25. T.J. Scheffer and J. Nerring, A new highly multiplexable liquid crystal display, *Appl. Phys. Lett.,* 45, 1021, 1984.
26. T.J. Scheffer, *Super Twisted Nematic (STN) LCDs, SID '95 Seminar Notes*, Vol. I, M-2, 1995.

27. M. Oh-e, M. Ohta, S. Aratani, and K. Kondo, Principles and characteristics of electro-optical behavior of in-plane switching mode, in *Proc. 15th International Display Research Conf.*, Asia Display '95, 577–580, 1995.
28. T. Scheffer and B. Clifton, Active addressing method for high-contrast video-rate STN displays, *SID'92 Digest*, 228, 1992.
29. H. Muraji et al., A 9.4-in. color VGA F-STN display with fast response time and high contrast ratio by using MLS method, *SID '94 Dig.*, 61, 1994.
30. Higashi et al., A 1.8-in poly-Si TFT-LCD for HDTV projectors with 5V fully integrated driver, *SID '95 Dig.*, 81, 1995.
31. Nano et al., Characterization of sticking effects in TFT-LCD, *SID '90 Dig.*, 404, 1990.

33

Plasma-Driven Flat Panel Displays

Robert T. McGrath
The Pennsylvania State University

Ramanapathy Veerasingam
The Pennsylvania State University

William C. Moffatt
Sandia National Laboratories

Robert B. Campbell
Sandia National Laboratories

33.1 An Introduction to Plasma-Driven Flat Panel Displays

Development History and Present Status

Plasma-driven flat panel display pixels were invented by Bitzer and Slottow at the University of Illinois in 1966 [1-3]. Figure 33.1 shows one of the inventors' early designs and demonstrates its simplicity. Parallel sets of thin conducting wires are deposited on two glass substrates which are then mounted with the conductor sets perpendicular to one another as shown in the Figure 33.1. A spacer, in this case a perforated glass dielectric, is used to maintain a gap separation of about 100 μm between the glass plates. The gap region then is filled with an inert gas, typically at a pressure of half an atmosphere. Individual pixels formed by the intersection of two conductor wires are aligned with the perforations. Pixels are illuminated by applying a voltage between two intersecting wires sufficient to initiate gas breakdown. Over the years, this basic pixel design has undergone a multitude of refinements and improvements, but the fundamental concept is still widely used.

Throughout the 1980s, plasma display products on the market were monochrome and operated with neon-based gases, directly producing within the discharge volume the red-orange (585 to 640 nm) visible photons that are characteristic of the quantum energy level structure of the neon atom. Dot matrix displays of the type shown in Figure 33.2 were widely used [3,4]. Early work by Owens-Illinois led to improvements in glass sealing and spacer supports [5,6], and work by IBM led to improved understanding

0-8493-1733-9/04/$0.00+$1.50

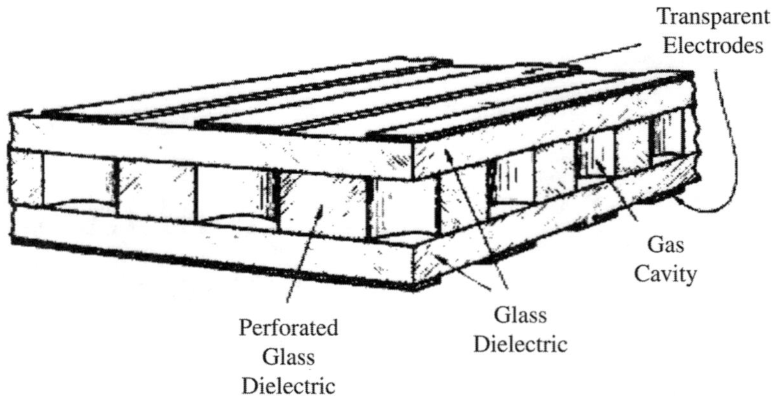

FIGURE 33.1 Structure of the ac plasma display invented at the University of Illinois. (From Bitzer, D.L. and Slottow, H.G., *AFIPS Conf. Proc.*, Vol. 29, p. 541, 1966. With permission.)

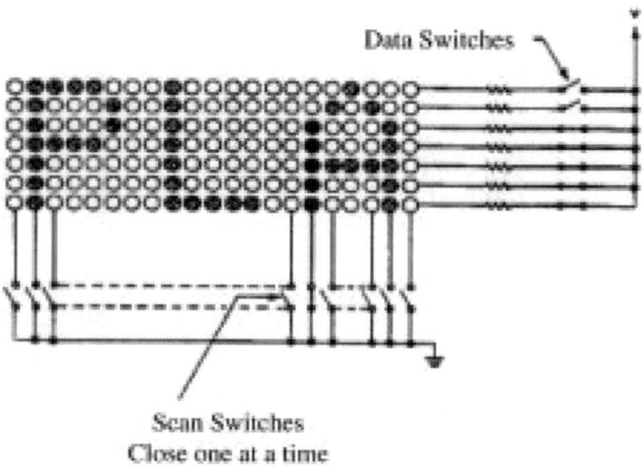

FIGURE 33.2 A simple dot matrix plasma display and data scanning switches. (From Weber, L.F., in *Flat Panel Displays and CRTs*, L.E. Tannas, Jr., Ed., Van Nostrand Reinhold, New York, 1985. With permission.)

and control of the discharge [7-15]. These advances ultimately paved the way for manufacture of large-area, high-resolution monochrome displays. The largest area plasma display panels ever manufactured were produced by Photonics Imaging. These monochrome displays had a 1 m diagonal dimension and contained over 2 million pixels with a pixel pitch of 20 pixels/cm (50 lines/in.) [4].

Advances in lithography, patterning, and phosphors have enabled continued improvement of plasma display performance and resolution. Today, many companies offer full-color plasma flat panel displays. Table 33.1 presents a summary list compiled by the National Research Institute of display panel specifications for some of the major companies investing in plasma flat panel manufacturing [16]. According to Stanford Research, Inc., sales for plasma display panels in 1995 totaled $230 million, but projected sales for 2002 are $4.1 billion [17]. NEC projects a more aggressive market growth reaching $2.0 billion by the year 2000 and $7.0 billion by 2002 [17]. The production capacities listed in Table 33.1 represent investments committed to manufacturing as of January 1997. As these production facilities come online, color plasma flat panel display production will grow to nearly 40,000 units per month by the end of 1998, and to over 100,000 units per month by early in 2000. In 1993 Fujitsu was the first to market a high-information-content full-color plasma flat panel display, a 21 in. diagonal, ac-driven system with a 640 × 480 pixel array [17,18]. Two examples of more recent market entries are shown in Figures 33.3 and 33.4. The first example is the NHK full-color, 102 cm (40 in.) diagonal, high-definition television

TABLE 33.1 Plasma Flat Panel Display Specifications and Manufacturer's Business Plans

	Product Specification			Efficiency Specification				Plan			
Company	Inch	Aspect	Pixels	Luminecence (cd/m^2)	Contrast	lm/W	Power (W)	Factory	Capital Cost ($M)	Product Ability unit/month	Target Region
Fujitsu	42	16:9	852 × 480	300	70:1	0.7	350(set) 300(panel)	Miyazaki	20	10,000	Europe (Philips), Japan
NEC	33	4:3	640 × 480	200	150:1	1.2	270(set) 190(panel)	Tamagawa, Kagoshima	5	2,000	Japan
Pioneer	40	4:3	640 × 480	400	150:1	1.2	350(set)	Kofu	5	10,000	Japan
Mitsubishi	40	4:3	640 × 480	350	200:1	0.8	350(set) 300(panel)	Kyoto	14.8	10,000	U.S.
MEC	42	16:9	852 × 480	450	150:1	10	300(panel)	Kyoto	10	5,000	Japan, U.S.
Photonics	21	5:4	1280 × 1024	100	50:1	—	300(panel)	Ohio	—	—	—
Hitachi	25	4:3	1024 × 768	150	50:1	—	250(set)	Yokohama	3	1,000	—
NHK	40	16:9	1344 × 800	93	80:1	—	—	—	—	—	—

Source: Wakabayshi, H., paper presented at *Imaging 2001: The U.S. Display Consortium Business Conference,* January 28, San Jose, CA, 1997. With permission.

FIGURE 33.3 The 40-in. diagonal dc-driven plasma display from NHK. (From Mikoshiba, S., *Inf. Display,* 10(10), 21, 1994. With permission.)

(HDTV) [19-22]. The system comprises 1,075,000 full-color pixels (1344 × 800) with a pixel pitch of 0.65 mm in both horizontal and vertical directions (15.4 pixels/cm). This pulsed, dc-driven display has a peak luminance of 93 cd/m^2, a contrast ratio of 80 to 1, and produces 256 gray levels. The display has an overall thickness of only 8 cm and weighs only 8 kg. The dimensions of the display panel itself are 87.5 × 52.0 cm with a width of only 6 mm. Shown in Figure 33.4 is the 76 cm (30 in.) diagonal, full-color ac Plasma Display manufactured by Photonics Imaging [23-24]. The display contains an array of 1024 × 768 full-color pixels. At 16.8 pixels/cm (pixel pitch = 0.59 mm) this is the highest resolution full-color, plasma display manufactured to date. This unit has 64 gray levels per color channel and an average area (white) luminance greater than 103 cd/m^2 (30 fL).

Dc and Ac Plasma Pixels

As indicated above, plasma display pixels can be designed for either ac or dc operation. Figure 33.5 shows schematic diagrams for the simplest dc and ac pixel designs. In either case, sets of parallel conductor wires are deposited on glass substrates. In most cases display costs are kept low by utilizing ordinary soda-lime float glass. The two glass plates are then mounted with a separation of about 100 μm and with the conductor wire sets perpendicular to one another. The gap region between the glass plates is filled with an inert gas, which discharges and illuminates the pixels when sufficient voltage is applied across two intersecting wires.

For dc pixels, shown in Figure 33.5a, the working gas is in direct contact with the electrodes. Electrons produced within the discharge volume flow rapidly to the anode, while ions produced flow more slowly toward the cathode. At 53.3 kPa (400 torr), a gas gap of 100 μm, and an applied voltage of 200 V, the electron and ion transit times across the gap are roughly 0.2 and 20 ns, respectively. Once breakdown is initiated, the electrical resistance of the discharge is negligible. Consequently, dc operation requires that external resistors in series with each pixel be included in the circuit in order to limit the current amplitude. Often, dc pixels are operated in pulsed discharge mode with frequency modulation used to define the pixel brightness. For either ac or dc pixels, a base firing frequency of 50 kHz is typical. This frequency is too fast for the human eye to detect any *on–off* flicker, but allows sufficient flexibility for intensity and refresh control. In reviewing the literature on plasma displays, it is easy to confuse dc and ac pixels since dc pixels are often operated in pulsed mode and with electrode polarity reversal which distributes sputter damage over both electrode surfaces. The dc pixels are readily identified by conducting electrodes in direct contact with the discharge gas and the inclusion of a current-limiting resistor in the circuit for each pixel. While polarity reversal is optional for dc pixel operation, it is inherently required for ac pixel

FIGURE 33.4 The 40-in. diagonal ac-driven plasma display from Photonics Imaging. (From Friedman, P.S., *Inf. Display,* 11(10), October 1995. With permission.)

operation as discussed below. Drive electronics, current limiting, gray scale, and other aspects of both dc and ac pixel operation are discussed in greater detail in subsequent sections.

Figure 33.5b shows a schematic representation of an ac plasma pixel configuration. One can see that the differences between ac and dc pixel geometry are slight; however, the resulting operational differences are significant. In the ac pixel, the conductor wires are covered with a dielectric film. Typically, lead oxide (PbO), which has a dielectric constant of about 15, is deposited at a film thickness of about 25 μm. Most ac pixels are made with a thin film (50 to 200 nm) of magnesium oxide (MgO) dielectric coating covering the PbO and in contact with the working gas. This dual material dielectric film serves two principal functions, charge storage and secondary electron emission.

The exact voltage required for gas breakdown depends upon the gap width, the gas pressure, the gas composition, and MgO surface conditioning. For the pixel parameters shown in Figure 33.5b, an externally applied voltage of about 120 to 180 V is required to initiate a discharge. In the ac pixel, once the discharge is initiated, electrons and ions flow toward the anode and cathode, respectively, as in the dc pixel. However, in the ac case, charge carriers are unable to reach the conductor wires and instead collect as a surface charge on the dielectric coating. The electric field within the gas gap is always the sum of that produced by the externally applied voltage and that produced by the surface charge. During pixel firing, if the externally applied voltage is held constant for only a few microseconds, the net electric field within the gas gap very quickly decreases (~100 to 200 ns). The gap potential drop produced by the surface charge shields out that produced by the externally applied voltage. Eventually, the gap electric field is insufficient to sustain the discharge and the pixel turns *off*. Thus, each ac pixel is inherently

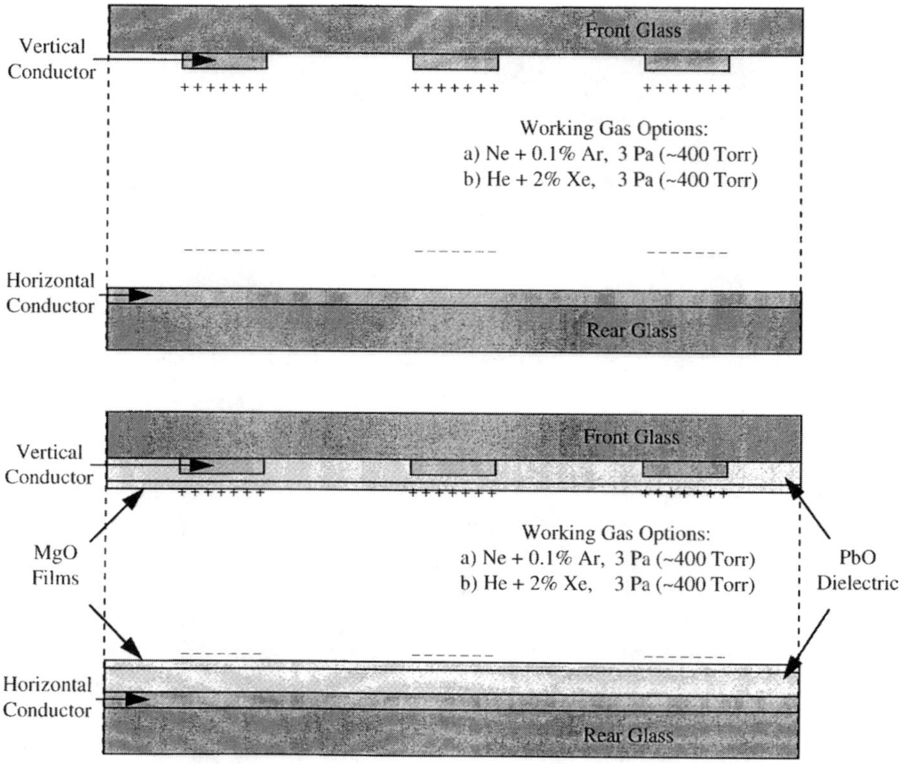

FIGURE 33.5 Schematic diagrams of (a) dc and (b) ac opposed electrode plasma pixels.

self-current-limiting and, unlike the dc pixel, requires no external resistance in series with it. At the start of the next ac half cycle, the externally applied voltage is reversed. When this occurs, the voltage across the gas gap is the sum of the external voltage and the voltage produced by the surface charge established during the previous discharge. If a sufficient surface charge is present, a new discharge pulse can be initiated by application of an external voltage, which by itself would be insufficient to break down the gas. Within the new discharge, charge carriers flow quickly to reverse the polarity of the surface charge concentrations. Once again, the field within the gap is diminished and the discharge turns *off.* Storage of surface charge make ac pixels easily controllable and provides them with their inherent memory properties. The presence or absence of surface charge determines whether or not a given pixel will discharge at the onset of the next ac half cycle of the externally applied voltage. The details of how these discharge dynamics are used to write, erase, and sustain each pixel are discussed in subsequent sections, along with drive mechanisms for gray scale and for pixel array refresh.

General Attributes of Plasma Displays

Plasma-driven flat panel displays offer a number of advantages over competing display technologies. The highly nonlinear electrical behavior of each pixel, with inherent memory properties, can be used to advantage in design of the drive electronics required to refresh and to update the pixel array of the display. The simplicity of the pixel design makes large-area manufacturing problems, such as alignment and film thickness uniformity, somewhat more manageable. Relative to color active matrix liquid crystal displays (AMLCDs) which use a thin-film transistor (TFT) to control each pixel, less complicated manufacturing and less complicated drive electronics give plasma flat panel displays advantage for large-area applications. On the other hand, plasma displays require more robust drive electronics with voltages of 100 to 275 V. Plasma displays are also not well suited for portable applications since power consumption is high relative to other display technologies, but not restrictive for office or domestic use. The 76 cm (30 in.) diagonal

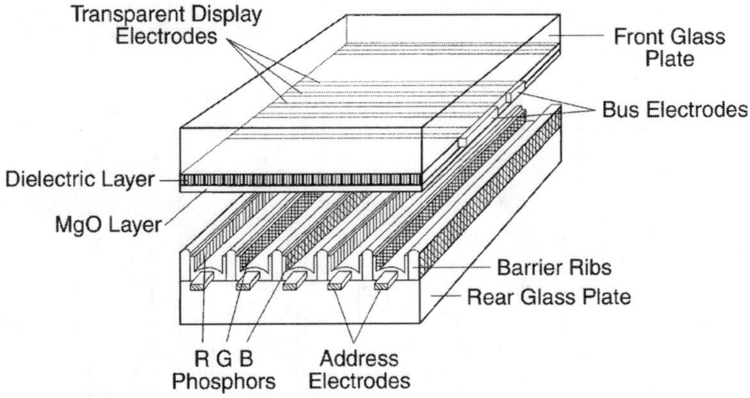

FIGURE 33.6 Structure of the ac color plasma display manufactured by Fujitsu. (From Mikoshiba, S., *Inf. Display,* 10(10), 21, 1994. With permission.)

color display manufactured by Photonics Imaging shown in Figure 33.4 has a peak power consumption of only 300 W [23]. At high power levels, plasma-driven flat panel displays are bright enough to be readable in sunlight. The displays are also easily adjusted to a low-ambient-light condition by discharge amplitude or frequency modulation.

Plasma flat panel displays are well suited for large-area (0.5 to 5 m) applications such as videoconferencing, large meeting room displays, outdoor displays, and simulators requiring large viewing areas. Thin, high-resolution, large-area, color plasma displays are also very attractive for desktop workstation or personal computer applications requiring high-resolution graphics. Note, too, that plasma flat panel displays have very large viewing angles, greater than 160° in many designs [22-24]. For displays using metal electrodes, one often finds that the best viewing angle is slightly off normal since the front electrode wire blocks out a portion of the pixel emission. This occurs both for monochrome pixels producing visible emissions within the discharge and for color plasma displays where the viewer sees visible red, green, and blue (RGB) emissions from vacuum ultraviolet (VUV) photon-stimulated phosphors. Some manufacturers have investigated use of transparent electrodes, such as indium-tin oxide (ITO), but there is a trade-off with power consumption since the conductivity of ITO is less than that of metal electrodes [18]. In contemporary designs, the metal conductor width is thin (~20 μm) and its opacity does not present a major problem.

For color pixels, the discharge gas mixture is modified to produce emissions in the VUV. In all other respects, the operational principals of the plasma discharge by the pixel are identical for color and for monochrome displays. Ideally in color plasma displays, no visible emissions are produced within the discharge itself and VUV-photostimulated luminous phosphors are used to produce the required RGB visible light. The ac color pixel design concept shown in Figure 33.6 is that utilized by Fujitsu [18]. Long, straight barrier structures, each about 100 μm tall, are constructed parallel to and between each of the vertically oriented conductor wires on the rear glass plate. The sidewalls of these barriers are alternately coated with red, green, and blue photostimulated phosphors. Note that the Fujitsu panel employs a three-electrode, ac-driven surface discharge pixel design which is slightly more complicated than the opposed electrode ac design shown in Figure 33.5b. This chapter will return to surface discharge configurations and other aspects of color pixel design and operation after reviewing fundamentals of the discharge physics and electrical behavior governing pixel operation.

33.2 Fundamentals of Plasma Pixel Operation

Atomic Physics Processes

Although simplistic in design, the plasma display pixel is a rich environment for study of basic atomic physics, electron collisional processes, photon production and transport, and plasma–surface interactions.

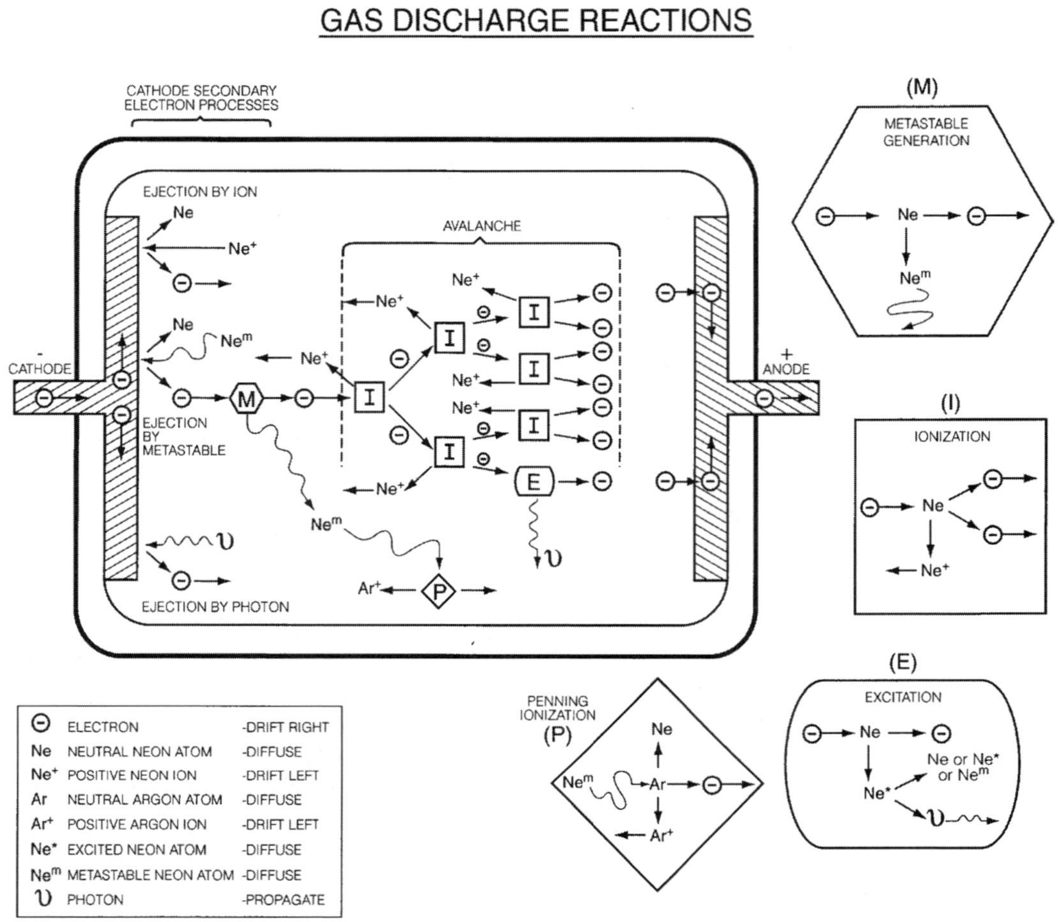

FIGURE 33.7 Collisional and surface interactions in a gas discharge. (From Weber, L.F., in *Flat Panel Displays and CRTs*, L.E. Tannas, Jr., Ed., Van Nostrand Reinhold, New York, 1985. With permission.)

The coupling of these processes for a neon–argon monochrome pixel discharge was nicely summarized in the diagram from Weber which is reproduced here as Figure 33.7 [4]. The reader interested in additional information on fundamental discharge physics is directed to one of the excellent textbooks in this field [25–27].

The physical processes governing the behavior of the pixel discharge are closely coupled and form a closed-loop system. The discussion begins by assuming that a seed electron is resident within the gas gap and is subjected to an electric field which results from application of an externally applied voltage to the two conductors forming that pixel. Some of the gas and surface processes for production of the seed electrons will become evident as the discussion progresses. In order to ensure reliable discharge initiation, seed particles, which are either electrons or electron-producing photons or metastable atoms, are often provided by a controlled source which may be external to the pixel being fired. Some display panels include electrodes for production of seed particles at the edges of the panel outside the field of view or hidden behind opaque conductor wires. Other display panels use well-controlled temporal sequencing to ensure that nearest-neighbor pixels provide seed particles for one another [4,19,28]. Pixel addressing sequences are discussed further later in this chapter.

The transport of electrons or ions across the gas gap is a balance between field acceleration and collisional energy loss. In the example of Figure 33.7, the gas is mostly neon (98 to 99.9%) and field-accelerated electrons will predominantly collide with Ne atoms. The quantum energy level diagram for excitation of the Ne atom is shown schematically in Figure 33.7 [29]. Note that the lowest-lying excited state is 16.6 eV above the ground state, while the ionization energy is 21.6 eV. This means that electrons with energies less than 16.6 eV can only experience elastic collisions with the Ne atoms. When an electron is field-accelerated to an energy in excess of 16.6 eV, inelastic collisions which transfer energy from the incident electron to one of the outer-shell electrons in the Ne atom can take place. Incident electrons with kinetic energies in excess of 21.6 eV can drive ionization reactions:

$$\mathrm{Ne} + \mathrm{e}^- \rightarrow \mathrm{Ne}^+ + 2\mathrm{e}^- \tag{33.1}$$

Excitation and ionization collisions transfer energy from the electron population to the neutral atoms in the gas. At the same time, the electron population available to ionize the Ne further is increased with every ionizing event. The result is the discharge avalanche schematically shown in Figure 33.7, which manifests itself experimentally as a rapid increase in electric current flowing in the pixel gas gap. In dc panels, an external resistor of about $R = 500\ \mathrm{k}\Omega$ is placed in series with each pixel. The amplitude of the externally applied voltage provided by the driving electronics, V_a, is held constant and the total voltage across the gas gap, $V_g = V_a - IR$, decreases as the circuit current, I, increases. Very quickly, a steady-state dc current in the gas gap and in the circuit is established. Brightness and gray scale are controlled by frequency modulation of the pulsed dc pixel firing using a base frequency of about 50 kHz. In ac pixel discharges, electrons and ions are driven by the applied field to the dielectric-covered anode and cathode, respectively. The buildup of charge on the dielectric surfaces shields the gap region from the field produced by the externally applied voltage. Eventually, the electric field in the gap drops below a level sufficient to sustain the discharge and the pixel turns *off.*

For electron energies greater than 16.6 eV, collisions with Ne atoms can excite outer-shell electrons in the atom to one of the numerous excited energy states shown in Figure 33.7.

$$\mathrm{Ne} + \mathrm{e}^- \rightarrow \mathrm{Ne}^{\mathrm{ex}} + \mathrm{e}^- \tag{33.2a}$$

$$\mathrm{Ne}^{\mathrm{ex}} \rightarrow \mathrm{Ne}^* + h\nu \tag{33.2b}$$

Most of these excited states have short lifetimes ranging from fractions to tens of nanoseconds [30] and quickly decay to lower-lying atomic quantum states accompanied by the emission of a characteristic photon, indicated in Equation 33.2 by $h\nu$, the product of Planck's constant times the photon frequency. As can be seen in Figure 33.8, the characteristic red-orange Ne gas emissions result from electron transitions within the atom from higher-energy 2p quantum states to lower-lying 1s energy levels [30,31]. Two of the four 1s energy levels radiate to ground-emitting VUV photons with wavelengths of 74.4 and 73.6 nm. Due to quantum mechanical exclusion principles, electron decay from the other two 1s levels is more complex and depends upon fine details of the electronic wave function and upon very small perturbing interactions [31]. Consequently, decay lifetimes for these so-called metastable states are measured in seconds, which is very long relative to other dynamic physical processes governing pixel discharge behavior, such as charge or neutral particle transport. An Ne atom with an electron trapped in one of these metastable levels harbors 16.6 eV of latent energy. The metastable atom, Ne*, is unable to dissipate its stored energy in collisions with ground-state Ne atoms, yet readily liberates its energy whenever a lower-lying energy configuration can be accessed. The principal channels in this system to lower energy configurations are Ne* collisions with Ar or Ne* incidence onto pixel interior surfaces.

Figure 33.8 shows simplified energy-level diagrams for several inert gases. The relative positioning of the allowable energy levels provides insight into the energy exchange that occurs in collisional coupling. The ionization energy of the Ar atom is 15.8 eV and lies 0.8 eV below the metastable-state Ne*. Consequently, the Ne* has sufficient stored energy to ionize the Ar atom:

$$\mathrm{Ne}^* + \mathrm{Ar} \rightarrow \mathrm{Ne} + \mathrm{Ar}^+ + \mathrm{e}^- \tag{33.3}$$

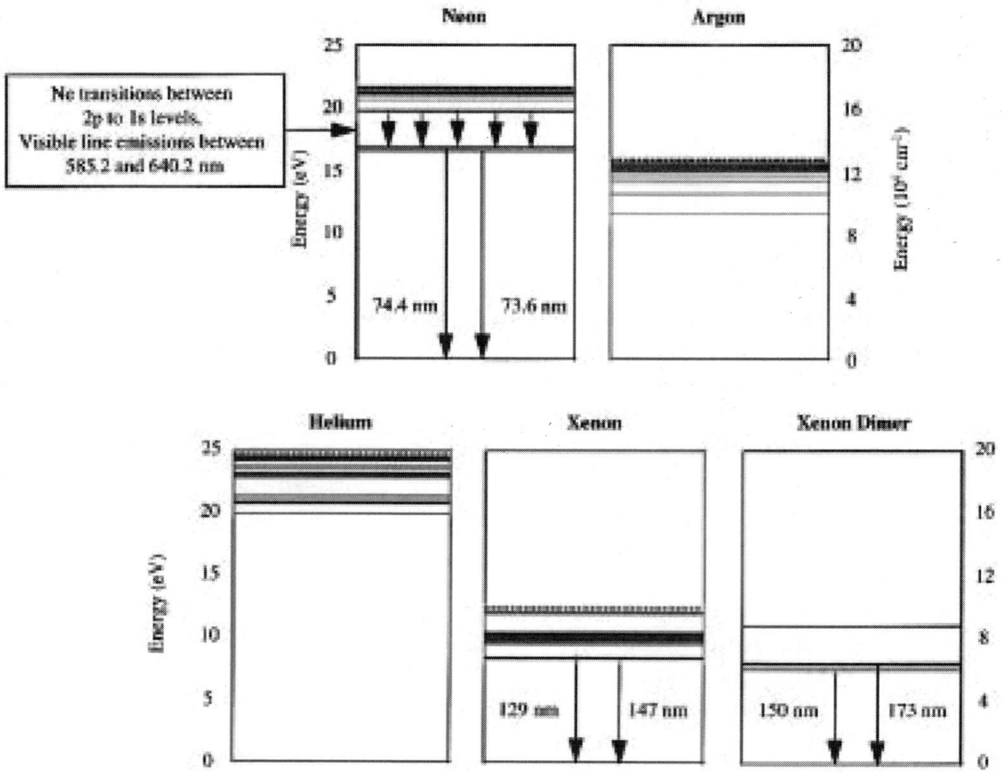

FIGURE 33.8 Quantum energy level diagrams for He, Ne, Ar, Xe, and the Xe^{2*} dimer.

Ionizing reactions of this type are called Penning reactions, and gas mixtures that rely on metastable states of the majority gas constituent (Ne) for ionization of the minority gas constituent (Ar) are referred to as Penning gas mixtures [25,26,32]. Figure 33.9 shows the efficiency with which charge pairs are produced through ionization within Ne/Ar Penning gases containing various fractions of Ar. The curves show that for any given pressure, ion pair production per volt applied is optimal at low Ar gas fractions (0 to 10%) except for very large values of E/P, greater than 75 V/m/Pa (100 V/cm/torr), where E is the electric field strength and P is the gas pressure. Penning gas mixtures have been studied for many years. Figure 33.9 shows the original data on Ne/Ar gas breakdown published by Kruithof and Penning in 1937 [32]. An extensive volume of literature has been published on inert gas Penning processes since then, and the interested reader is referred to the excellent texts which have recently been re-released through the American Vacuum Society and MIT Press [25,26].

Plasma display pixels usually operate at pressures near 53.3 kPa (400 torr) in order to achieve sufficient photon production and brightness. Typical pixel fields are roughly 100 MV/m. Consequently, plasma pixels operate with E/P values near 18.8 V/m/Pa (25 V/cm/torr). Both charge pair production and luminous efficiency are then optimized with Ar gas fractions between 0.1 and 10%, depending upon the specifics of the pixel gas pressure, gap width, and driving voltage. For a given applied voltage, the product of the gas pressure (P) and the gas gap dimension (d) provides a measure of the balance between electron acceleration by the electric field and electron energy loss due to collisions with the background gas. Paschen curves, which plot the gas breakdown voltage vs. the Pd product, for several inert gas mixtures are shown in Figure 33.10 [26,33,45]. In each case, minimum voltage for breakdown occurs at a value of the Pd product which is dependent upon the ionization levels, collisionality, and energy channels within the gas. For example, in Ne atomic excitation and ionization processes dominate, while in air much of the energy absorbed by the gas goes into nonionizing molecular vibration, rotation, and dissociation. For fixed pressure, the Paschen curves show that increased gap dimension lowers the electric

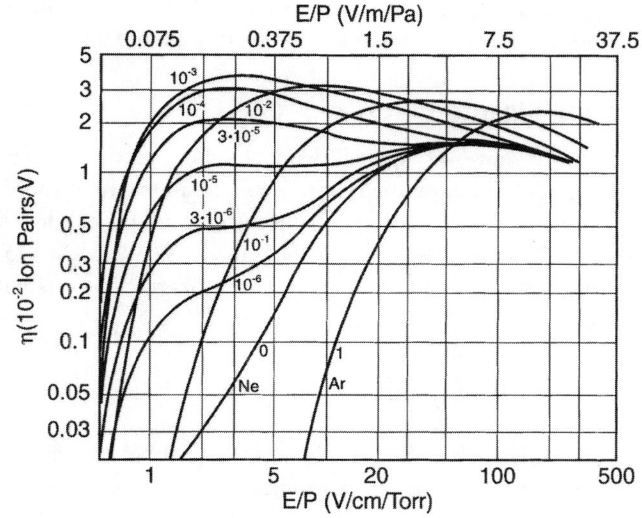

FIGURE 33.9 Ionizing collisions plotted vs. electric field strength divided by pressure. The numbers on each curve indicate the ratio of the Ar partial pressure to the total gas pressure. (From Brown, S., *Basic Data of Plasma Physics — The Fundamental Data on Electrical Discharges in Gas,* American Institute of Physics Press, New York, 1993. With permission.)

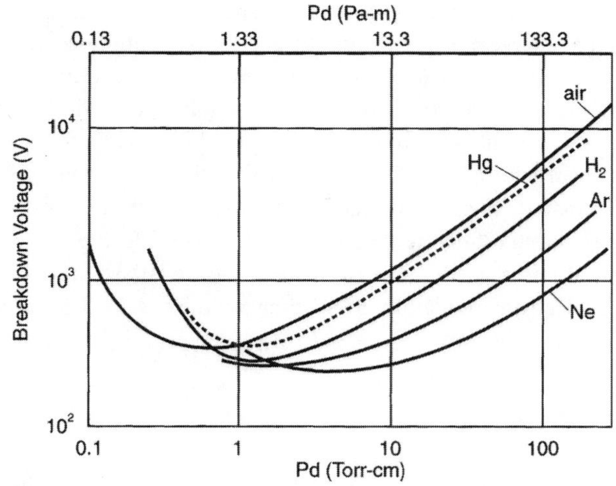

FIGURE 33.10 Breakdown voltage as a function of pressure — gas gap length product for various gases. (From Brown, S., *Basic Data of Plasma Physics — The Fundamental Data on Electrical Discharges in Gas,* American Institute of Physics Press, New York, 1993. With permission.)

field strength per volt applied and a large voltage is required for breakdown. On the other hand, if d is reduced for a given pressure, the electric field strength can be large, but electrons transit the gap without initiating a sufficient number of collisions to drive the type of discharge avalanche shown in Figure 33.7. If the gas gap, d, is held fixed while pressure is varied, the shapes of the Paschen curves are again explained by electron acceleration and collisional processes. For high pressures, the mean free paths between electron collisions with the background gas atoms are short and electrons are unable to accelerate to energies sufficient to initiate ionization unless the electric field is especially strong. At low pressures, the electrons may be accelerated by the field to energies sufficient to initiate ionization, but few collisions with the background gas occur and, again, the avalanche is difficult to initiate. Penning processes are especially efficient at driving ionization. Introduction of 0.1% Ar into the neon gas lowers the minimum breakdown voltage from the value near 250 V shown in Figure 33.10, to about 150 V. The minimum breakdown voltage occurs at a *Pd* product of 40 Pa·m (30 torr·cm) for this gas mixture.

Discharge Physics for Plasma Pixels

Within any discharge, electrons move very quickly, while the more massive ions move relatively slowly in comparison. In a charge-neutral plasma that is subjected to an externally applied electric field, the mobile electrons quickly respond to the applied field and rush toward the anode. The inertia-laden ions, in a much slower fashion, begin their motion toward the cathode. Very quickly, a local charge imbalance is established as the electrons reach the anode faster than the rate of arrival of ions at the cathode. Poisson's equation

$$\nabla \cdot E(x) = 4\pi\rho(x) = 4\pi e\big(n_{\mathrm{i}}(x) - n_{\mathrm{e}}(x)\big) \tag{33.4}$$

shows that a local electric field is established in response to the net positive charge density, $\rho(x)$, in the plasma region. Here, $n_{\mathrm{i}}(x)$ and $n_{\mathrm{e}}(x)$ are the spatial profiles of the ion and electron densities, respectively, and e is the electron charge. The field established retards the rate at which electrons flow out of any volume within the plasma column and forces them to follow the net ion motion. The ion drift motion is correspondingly accelerated, but this acceleration is smaller by a factor proportional to the mass ratio of the electron to the ion. The net ion/electron motion is called ambipolar flow and is described in detail in many basic plasma physics texts [25-27].

In steady-state dc plasma pixel discharges, the amplitude of the current flowing in the circuit and in the gas gap is defined by the value of the applied voltage and the resistor in series with the pixel. Steady-state operation dictates that charge buildup within the gap region cannot occur. The rate at which charge particle pairs arrive at the electrodes must equal their rate of production due to ionization. At the same time, the rates at which ions and electrons leave the plasma volume, arriving at the cathode and anode, respectively, must be equal. Equilibrium is sustained by establishment of the spatial potential profile within the gas gap shown in Figure 33.11a. Due to the high electron mobility, the plasma is extremely efficient in shielding out externally applied electric fields. As a result, the potential profile is flat across the gas gap of a pixel sustaining a fully developed discharge. The entire potential drop is localized in a small zone called the sheath adjacent to each electrode. The spatial extent of the sheath is determined by the effectiveness of the electron population in shielding out the electric fields produced by the electrode potentials. The Debye length,

$$\lambda_{\mathrm{D}} = \sqrt{kT_{\mathrm{e}} / 4\pi e^2 n_{\mathrm{e}}(x)} \tag{33.5}$$

provides a measure of the shielding distance. The expression for λ_{D} implies that the sheath thickness increases with increasing electron temperature, T_{e}, and decreases as the electron density, n_{e}, increases. For fully developed plasma pixel discharges, the product of Boltzmann's constant and the electron temperature, kT_{e}, is at most a few electron volts, and n_{e} is of order $10^{16}/\mathrm{m}^3$. Thus, the sheath thickness is roughly 5 μm. The potential within the plasma region adjusts, V_{P}, within the discharge volume rises to a value just above that of the applied voltage at the anode. Consequently, only the most energetic electrons can overcome the potential barrier at the anode which adjusts to a potential such that the rate of electron loss at the anode equals the rate of ion loss at the cathode. For ac plasma pixels, a similar potential profile is established, but changes dynamically as the pixel pulse evolves. Charge pairs incident upon the anode and cathode in ac pixels are trapped there by the dielectric film covering the conductor wires. Consequently, the potential at the discharge boundary is diminished as surface charge collects at each electrode, as shown in Figure 33.11b. Ultimately, the discharge terminates as the electric field produced by the surface charge cancels that produced by the externally applied voltage. As the density of charge carriers is reduced near the termination of an ac pixel discharge pulse, the effectiveness of the

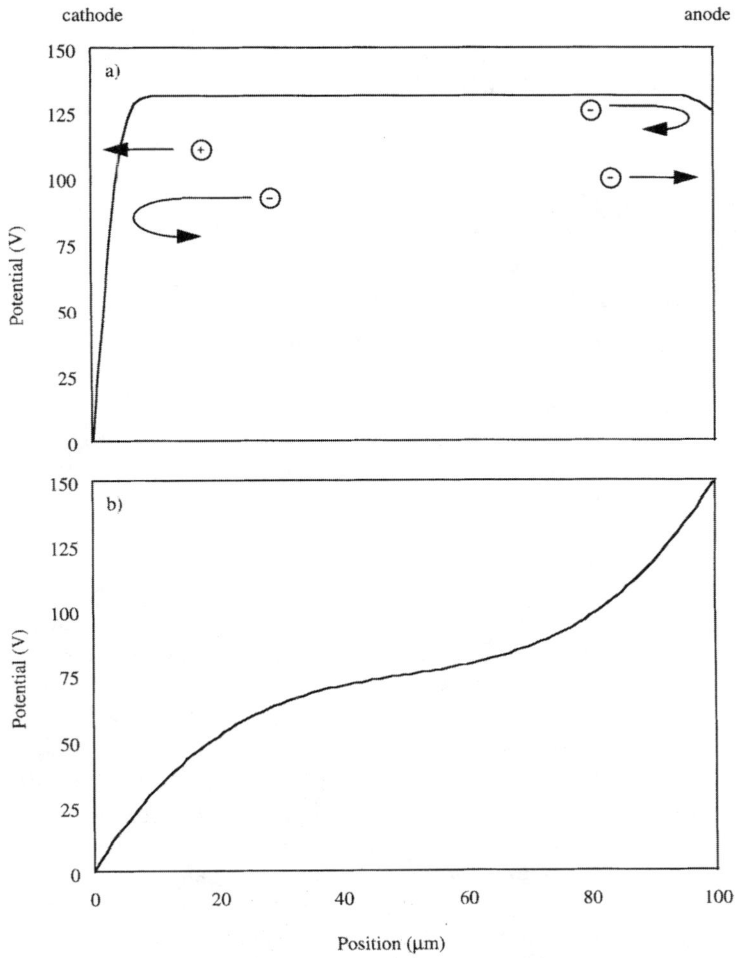

FIGURE 33.11 Potential profiles in the pixel gap region for (a) high-electron-density and (b) low-electron-density discharges.

electrons to shield out electric fields within the gap is diminished. In this situation, the sheath potential drop is small but the sheath region occupies a large fraction of the gap [36,37].

Plasma Surface Interactions

Ion-Induced Secondary Electron Emission

Ions arriving at the cathode sheath are accelerated by the sheath potential drop. Incident ions strike the cathode with kinetic energies equal to the plasma potential, V_p, which is just over 200 V in the example shown in Figure 33.12a. Ions incident on the cathode quickly capture an electron, additionally depositing on the cathode surface an energy equal to the recombination or ionization energy for that atom. Energy deposition on the cathode surface drives two important processes for plasma pixels — ion-induced secondary electron emission and sputtering. The first process significantly enhances the luminous efficiency of plasma pixels. The second shortens their operational lifetime as is discussed in subsequent sections.

Ion-induced secondary electron emission occurs when ion energy deposition on the surface results in electron ejection. Secondary electrons are exceptionally effective at driving discharge ionization since they gain large amounts of kinetic energy as they are accelerated across the cathode sheath and because

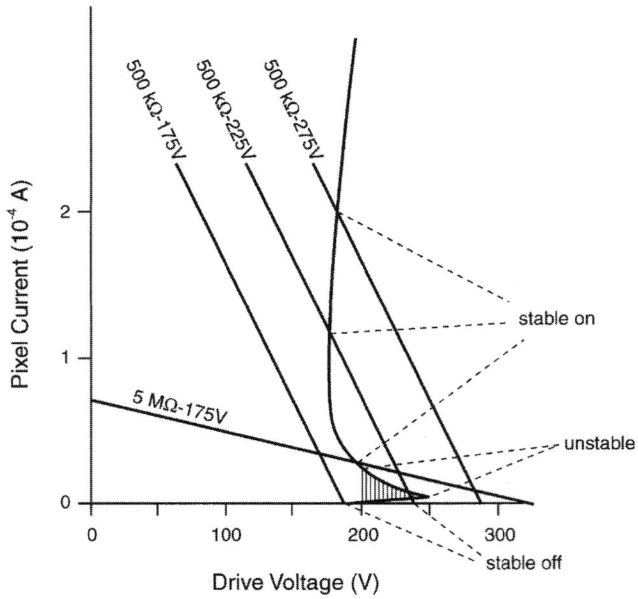

FIGURE 33.12 A representative current–voltage characteristic for gas breakdown. Load lines representative of plasma pixel operation are also shown. (From Weber, L. F., in *Flat Panel Displays and CRTs*, Tannas, L. E. Jr., Ed., Van Nostrand Reinhold, New York, 1985. With permission.)

they have ample opportunities for ionizing collisions as they traverse the entire width of the gas gap. The secondary electron emission coefficient, γ, is defined as the number of electrons ejected per incident ion [25,26]. As one would expect, γ varies with incident ion energy and with cathode material. Most ac plasma display panels take advantage of the strong secondary electron emission of MgO, which is also a good insulating material as required for surface charge storage in ac operation. Measurement of the MgO γ value is difficult, especially for low-energy ion incidence (<500 eV), and is complicated by charge buildup on the samples during the measurements [38]. Most often, relative values of secondary electron yields for different materials are deduced from discharge intensity measurements [11,12,39-42]. Chou directly measured the ion-induced secondary electron emission coefficient for MgO using a pulsed ion beam with sample surface neutralization between pulses. For ion incidence at 200 eV, he found $\gamma = 0.45$ and $\gamma = 0.05$ for Ne^+ and Ar^+, respectively [39]. Note, too, that photons and metastable atoms incident on the electrode surfaces are also capable of initiating secondary electron emission, as shown in Figure 33.7. Since neither photons nor metastables are influenced by the electric fields within the gas gap, they propagate isotropically throughout the gas volume and are often utilized as seed particles.

Sputtering

Ions accelerated across the sheath deposit energy on the cathode surface. This often initiates sputtering, whereby an atom within the cathode material is ejected from the surface. Sputtering processes erode the cathode surface and degrade pixel performance. Contamination of the discharge by sputtered surface impurities can lead to reduction in luminous efficiency due to visible emissions from the contaminant atoms or molecules which compromise the color purity of the pixel. Unwanted surface coatings from redeposited materials can also degrade the electrical characteristics of the pixel or, in color applications, shield the phosphors from VUV photons, further degrading luminous efficiency. For argon ion, Ar^+, bombardment of MgO surfaces at 2 keV, the measured sputtering yield is slightly greater than one ejected atom per incident ion [43]. Data on sputtering yields at lower energy ion incidence are difficult to obtain. Because yields are small, large incident ion currents are required to obtain measurable signals and sample charging is once again a problem. In spite of the lack of detailed data on low-energy MgO sputtering, manufacturers of ac plasma panels have been able to demonstrate display lifetimes well in excess of

10,000 h [18,23]. Shone et al. [44] have demonstrated that Rutherford backscattering of high-energy (2.8 MeV) alpha particle can be used to measure the thickness of MgO film on a PbO substrate. The film thickness accuracy obtained was ±1.5 nm. Because the technique requires a large (and expensive) particle accelerator, this technique is a very nice research tool but is ill suited for any fabrication line measurements.

33.3 Pixel Electrical Properties

Electrical Properties of Dc Pixels

Figure 33.5 shows schematic diagrams and circuit models for dc and ac pixels. In the dc case, the pixel gas gap functions electrically as a variable impedance resistor. Prior to gas breakdown, the resistance is large and the pixel represents an open-circuit element. Once breakdown is initiated, the plasma is an excellent conductor and offers only modest resistance, R_p, to current flow. Since $R \gg R_p$, the circuit equation simplifies to

$$V_a = I(R + R_p) \approx IR \tag{33.6}$$

and the circuit current, *I*, is defined by the amplitude of the applied voltage and the size of the circuit series resistor, *R*. The externally applied voltage, V_a, is typically a 50 kHz square wave with a fast voltage rise time (~50 ns). The dc driving voltages range from 175 to 275 V and a typical value for the series resistor is $R = 500$ kΩ. Pixel currents then range from 0.35 to 0.55 mA. Note that without a large resistance in series with the pixel, the current is limited by some physical failure such as melting of the pixel electrodes.

Figure 33.12 shows the characteristic *I–V* behavior of a dc pixel which has a breakdown voltage of 250 V [4]. Only a very small current due to a few stray charge carriers flows across the gas gap as the voltage increases from 0 to 250 V and the pixel remains in the *off* state. At the breakdown voltage, the situation is dynamic with the current growing rapidly and the voltage across the gas gap dropping as a result. The steady-state operating point achieved is identified by the intersection of the load line, $V_a = IR$, with the discharge *I–V* characteristic as shown in Figure 33.12. For an applied voltage of 175 V the pixel is always *off*, while for $V_a = 275$ V the pixel is always *on*. For a line resistance of 500 kΩ, the bimodal operation and memory of the dc pixel at $V = 225$ V is evident in the figure. If an applied voltage of 225 V is approached from the low-voltage direction, the pixel remains *off*. If, on the other hand, a large voltage is applied and subsequently lowered to $V_a = 225$ V, then the pixel will be in an *on* state. Note that the region where the 225 V/500 kΩ load line intersects the negative resistance portion of the *I–V* characteristic is unstable. The pixel discharge will quickly transition to either the stable *on* or stable *off* operating point. As a practical matter, one should note that the negative resistance region of the *I–V* characteristic curve cannot be experimentally measured in a pixel circuit operating with a 500 kΩ series resistance. Instead, as shown in the figure, a much larger series resistor, $R = 5$ MΩ, provides a load line with slope small enough to produce stable operation in the negative resistance regime.

Electrical Properties of Ac Pixels

The physical design of an opposed electrode ac pixel is shown in Figure 33.5b. Electrically, the pixel functions as a breakdown capacitor and is described by the circuit equation:

$$V_a(t) = I(t)R + \frac{1}{C}\int_0^t I(t')\mathrm{d}t' = I(t)R + Q(t)/C \tag{33.7}$$

where V_a is the externally applied voltage, *I* the circuit current, *C* the pixel capacitance, and *Q* the charge collected. For ac pixels the line resistance, *R*, is minimized in order to minimize power consumption and Equation 33.7 simplifies to

$$V_a(t) = \frac{1}{C}\int_0^t I(t')dt' = Q(t)/C \tag{33.8}$$

The capacitance for each pixel is the series summation of the capacitance for each dielectric film and for the gas gap:

$$\frac{1}{C} = \frac{1}{C_{PbO}} + \frac{1}{C_{MgO}} + \frac{1}{C_{gas}} \tag{33.9}$$

In each case,

$$C_i = \frac{e_i A}{d_i} \tag{33.10}$$

where i is the material index and the surface area, A, is roughly equal to the square of the conductor wire width. As shown in Figure 33.5b, an ac pixel is typically constructed with a PbO film of thickness d = 25 μm, while the thin-film MgO has thickness d = 50 to 200 nm. The lead oxide has a dielectric constant of roughly $\varepsilon_{PbO} = 15\varepsilon_0$, while that for MgO is $\varepsilon_{MgO} = 6\varepsilon_0$ with exact values dependent upon the film purity and microstructure [45]. Note that the MgO contribution to the total capacitance is negligible and that this material is incorporated into the design because of its excellent secondary electron emission properties. Prior to gas breakdown, the capacitance of the pixel is attributed largely to the gas gap. For 20 μm-thick conductor wires the capacitance of a pixel gas gap prior to breakdown is about 500 pF. The time derivative of Equation 33.8 gives the circuit current:

$$I(t) = C\frac{dV(t)}{dt} \tag{33.11}$$

This charge displacement current appears as the initial large amplitude current peak in Figure 33.13, which shows the temporal current response of a 45 × 45 ac pixel array to a single pulse within a 50 kHz square wave applied voltage pulse train. The electrical measurement shown was made using a simple induction loop probe to measure the current and a high impedance voltage probe (1 MΩ, 3 pF) to monitor the applied voltage. The signals were captured using a high-speed (300 MHz) oscilloscope.

If the applied voltage amplitude is below the gas breakdown threshold, only the capacitor charging displacement current, defined by Equation 33.11, is observed as shown in Figure 33.13a. If the voltage for gas breakdown is exceeded, a second current pulse due to the plasma discharge current within the gas gap is observed in the circuit, Figure 33.13b. The plasma pulse is accompanied, of course, by strong photon emission from the gas gap region. The total charge displacement in the discharge pulse as a function of amplitude of the square wave–applied voltage is plotted in Figure 33.14 for a helium–xenon (2%) Penning gas mixture [35]. The hysteresis or inherent memory property of the ac pixel is apparent. As the applied voltage amplitude is increased from zero to 180 V, no measurable current flows across the pixel gas gap. When no surface charge is present, below 180 V the electric field within the gap region is insufficient to drive the electron collisions into the avalanche regime. For any voltage amplitude in excess of 180 V, a gas discharge is initiated and the pixel turns *on*. If a pixel is subjected to a single voltage pulse with amplitude less than 135 V, the pixel turns *off* even if a surface charge is present.

In ac pixels, charge pairs produced during one discharge pulse collect on the surfaces of the dielectric films at the boundaries of the gas gap and are available to assist formation of the next discharge pulse in the sequence. In a fully developed ac pixel discharge, the surface charge accumulation on the dielectric film produces an electric field within the gas gap, which cancels the gap field produced by the externally applied voltage. This is shown in Figure 33.15, which is a composite representation of experimental current measurements and computational model predictions of the surface charge accumulation producing the surface or wall voltage [34,36]. When the polarity of the applied voltage is reversed, the

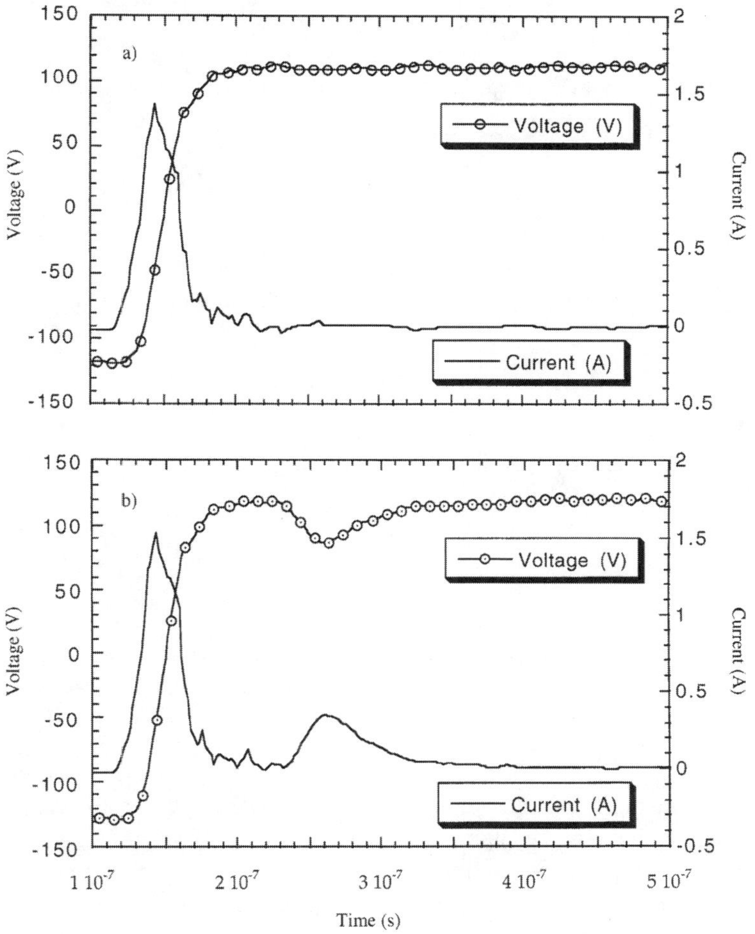

FIGURE 33.13 Voltage and current traces for a 45 × 45 array of ac plasma pixels in the (a) *on* and (b) *off* states. Drive voltage amplitudes were 117 and 127 V, respectively.

potential drop due to the surface charge and that due to the applied voltage suddenly are additive as shown in the figure. The gas gap is momentarily subjected to an intense electric field which results from a potential drop roughly equal to twice the applied voltage. The presence or absence of surface charge results in the bimodal current–voltage behavior shown in Figure 33.14.

Addressing of ac pixels is easily accomplished by taking advantage of the inherent memory of the pixel that results from this bimodal *I–V* behavior. For the pixel electrical properties shown in Figure 33.14, each pixel would be continuously supplied with an ac square wave applied voltage pulse train with an amplitude of 160 V, called the sustain voltage, $V_{sustain}$. If the pixel is initially in an *off* state, it will remain so indefinitely since no surface charge is available to enhance the field produced by the sustain voltage. To turn the pixel *on*, a single high-amplitude voltage pulse, called an address (or write) pulse is delivered across the pixel electrodes. In this example, an address pulse of 200 V initiates a discharge whose charge pairs collect on the internal dielectric surfaces of the pixel. The self-limiting nature of the ac pixel is such that the surface charge concentration produced for a fully developed pixel discharge completely shields the gap region from the externally applied field. When the next sustain polarity reversal occurs, the pixel gas gap experiences a voltage equal to the sum of the sustain voltage (160 V) plus the voltage due to the surface charge produced by the previous pulse, $V_{surface}$ = 200 V in this case. The new gap voltage of 360 V is more than sufficient to initiate a second discharge and to establish a new surface charge whose polarity is opposite that of the preceding pulse. Once again, the surface charge adjusts to produce a

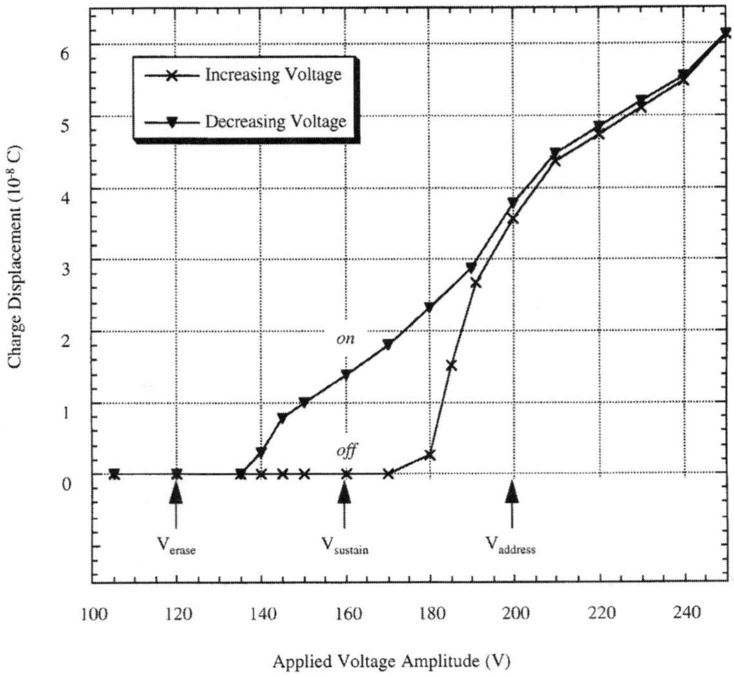

FIGURE 33.14 Discharge charge displacement for operation of a 45 × 45 array of ac-opposed electrode pixels with an He – Xe (2%) gas mixture at 53.3 kPa (400 torr).

voltage exactly canceling the field of the applied voltage. For this pulse, V_{wall} = 160 V, and the next sustain voltage polarity reversal subjects the gap to a potential difference of $V_{gap} = V_{sustain} + V_{address}$ = 160 V + 160 V = 320 V, which is again sufficient to initiate a new discharge pulse. Consequently, the pixel remains in the *on* state until action is taken to eliminate or diminish the surface charge buildup accompanying each discharge. This is accomplished by application of a single low-voltage pulse called an erase pulse with amplitude V_{erase} = 120 V for the example shown in Figures 33.14 and 33.15. Application of the erase pulse produces a potential drop across the gas gap of $V_{gap} = V_{applied} + V_{surface} = V_{erase} + V_{surface}$ = 120 V + 160 V = 280 V. The erase pulse produces a discharge of lower intensity which is insufficient to reestablish a reversed polarity surface charge. Consequently, the erase discharge diminishes the concentration of the surface charge so that no discharge is initiated during the next pulse in the sustain applied voltage train. Ideally, the erase pulse drives the surface charge concentration identically to zero, but this rarely occurs in practice and is not essential for ac pixel operation, as can be seen in Figure 33.15. Very low intensity discharges with negligible photon production drive the pixel to its ideal *off* state within a few ac cycles. Fortunately, these minor deviations from the ideal *off* condition have little effect on subsequent write pulses for frequency-modulated ac operation, and therefore do not affect the timing of pixel addressing and refresh which is covered in the next subsection.

33.4 Display Priming, Addressing, Refresh, and Gray Scale

Pixel priming is necessary to provide the initial source of electrons, or the priming current, required to initiate a discharge avalanche. Metastable atoms or photons can also be used as priming particles since these produce electrons via ionization of the background gas. Pilot cell priming and self-priming are two options used in currently available commercial products. In pilot cell priming, a separate cell which generates electrons is located near the pixel to be addressed. Pilot cells are often located on the periphery of the display outside the viewing area, yet can produce seed electrons throughout the display. In self-priming, an auxiliary discharge created within each pixel provides the priming electron source for the

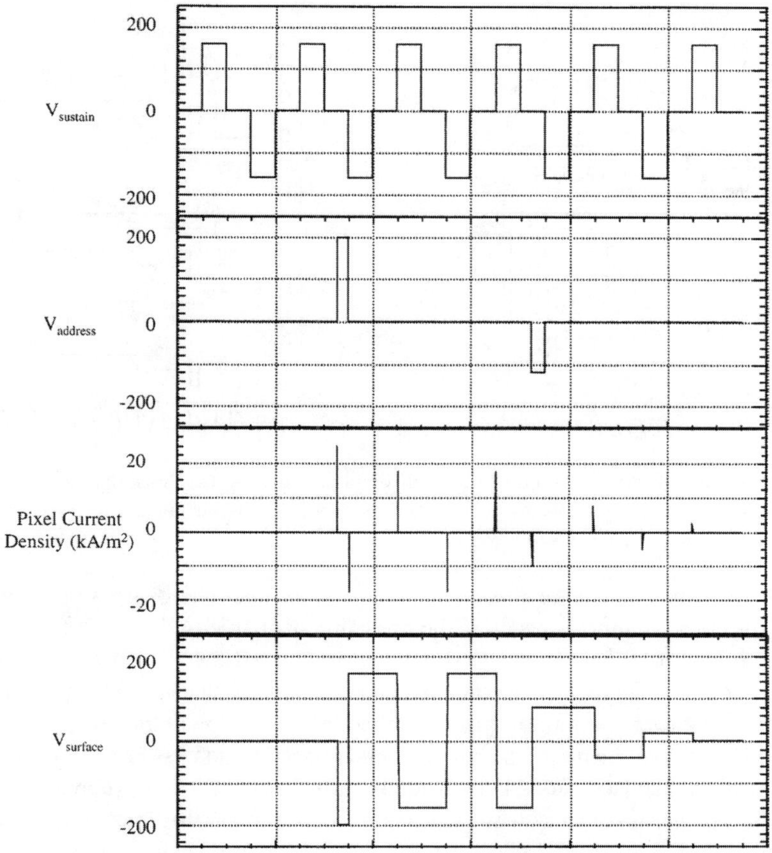

FIGURE 33.15 Sustain and address voltage waveforms for ac-driven plasma pixels. The amplitude of the pixel current density and wall voltage resulting from the surface charge buildup provide a measure of the discharge intensity.

main pixel discharge. These priming discharges are often hidden from view by positioning them behind opaque conductor wires. Introducing a trace amount of radioactive Kr^{85} into the gas mixture provides a passive priming option. The ionizing radiation from Kr^{85} generates priming electrons uniformly throughout the display interior. Because the required Kr concentration is low and because the beta radiation produced cannot penetrate the glass enclosure of the display, the radiation exposure risk to the display user is negligible. However, display manufacture using radioactive seeding involves potential health hazards associated with radioactive material handling. Consequently, this seeding approach, while very effective, is not at present employed in commercial products.

A simplistic scanning scheme for pixel illumination is shown in Figure 33.2, reproduced here from Reference 4. The scan switches on one axis open and close sequentially in a repetitive fashion, while the data switches on the other axis determine if the pixel is fired on a given scan. This simplistic refresh and data update method fails to take advantage of the discharge properties or inherent memory functions available with plasma pixels. High-resolution dynamic displays utilizing this address scheme would not be cost-competitive since display drivers constitute a significant portion of the total cost of plasma displays. Driver circuit costs also increase with required output voltage. Thus, it is desirable to design plasma displays with operating voltages as low as possible and which require the fewest number of driver chips. Designers strive then to maximize the number of pixels driven by a single chip.

For nonmemory dc pixels, one option for reducing the number of external drive switches required is to sweep the firing of priming discharges repetitively across each pixel row, such as in the self-scan circuitry developed by Burroughs [46,47]. More recently, NHK has developed a pulse memory drive scheme for its 102 cm (40 in.) diagonal dc HDTV plasma display, which is being widely used [48]. Sustain

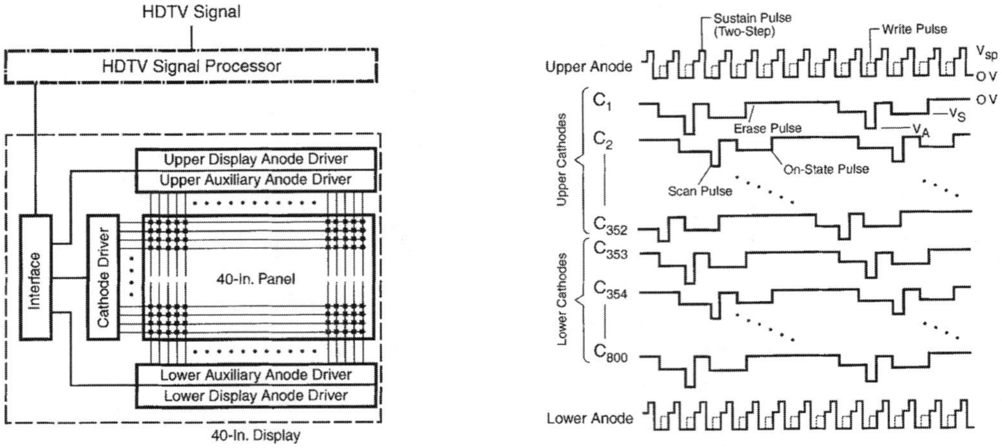

FIGURE 33.16 Pulsed memory operation of the NHK dc plasma display. (a) Block diagram of the driver system and pixel array. (b) Temporal sequences for pulsed memory operation. (From Yamamoto, T. et al., *SID '93 Sympos. Dig.*, 165–168, 1993. With permission.)

operation at high frequency is used to take advantage of residual charge pairs and metastable atoms present in the pixel gas volume as result of the preceding discharge [49]. In this fashion, each pixel is self-seeding, with seed particle populations dependent upon the time elapsed since the termination of the preceding discharge. The high-frequency operation is fast enough to take advantage of the short duration memory characteristic of the dc pixel. As the sustain voltage pulse train is applied to the electrode of a pixel, it will remain in the *on* or *off* state indefinitely until an address or erase pulse is supplied. In the NHK scheme, an auxiliary anode is used to assist in the *address* access operations. Figure 33.16a shows the block diagram of such a system, while Figure 33.16b shows the time sequences for the scheme [48]. Note that the pulses are dc and that *on* state pulses have larger gap voltages than erase pulses. The timing sequence is critical to address a pixel selectively within the matrix. Implementation of this scheme requires (1) display anode drivers, (2) auxiliary anode drivers, (3) cathode drivers, and (4) interfaces to decode the HDTV signals provided to the drivers.

For ac displays with memory, drivers need to provide (1) *address* (or *write*) pulses, (2) *sustain* pulses, and (3) *erase* pulses. A complex refresh scan signal is not required since a continuously supplied sustain signal, coupled with the ac pixels inherent memory, trivially maintains each pixel in either an *on* or *off* state. Pixels respond to address, sustain, and erase pulses as described in the preceding section. Similar to dc pixel dynamic control discussed above, ac pixel addressing requires well-timed application of voltage waveforms to the rows and columns of the display matrix so that the proper voltage appears across the electrodes for the pixel of interest without modifying the state of adjacent pixels. A more-detailed discussion of ac pixel addressing can be found in Reference 4 or 47.

Gray scale is achieved for dc or ac plasma displays either by modulation of the discharge current or by duty cycle modulation with fixed current. Modulating the applied voltage amplitude to vary the discharge current is not widely used because of practical limitations in effectively controlling the nonlinear response of the discharge current. However, duty cycle modulation is a viable technique both for pulse memory-driven dc displays and for ac memory displays. In either case, duty cycle modulation requires complex circuit design for the well-timed delivery of *on* and *off* pulses. Gray scale is achieved by varying the time a pixel is *on* compared with *off* during each refresh cycle. In 50 kHz operation, a sustain half cycle is 10 μs. VUV photon emission occurs usually in less than 1 μs. For color displays the visible light emission persists much longer, with the fastest phosphors having 10% persistence times of about 5 μs. More typical phosphors have 10% persistence times in the 5 to 10 ms range [50]. If the image is updated every 40 ms, corresponding to a refresh rate of 25 images per second, then a 1/8-level brightness is achieved by having the pixel *on* for 5 ms and *off* for 35 ms during that refresh cycle. The time *on* is

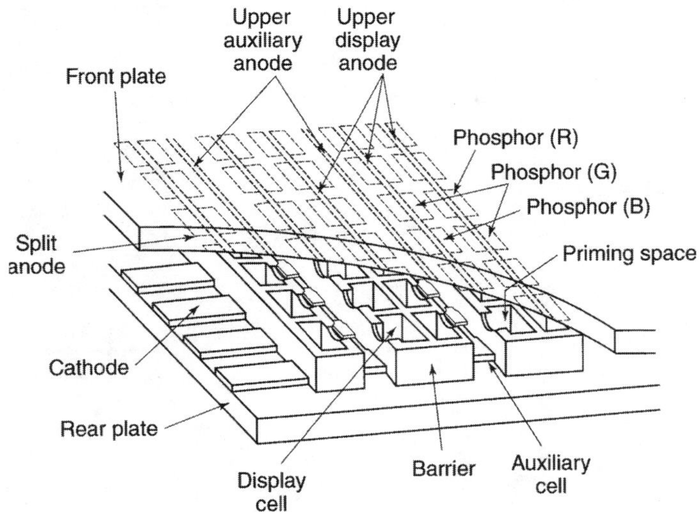

FIGURE 33.17 Structure of the 40 in. color display manufactured by NHK. (From Yamamoto, T. et al., *SID '93 Sympos. Dig.*, 165–168, 1993. With permission.)

interspersed throughout the 40 ms refresh period by appropriate timing circuit design. For example, the NHK 102 cm (40 in.) display has a 2^8 or 256 levels of gray scale per color, providing a total of 16 million (256^3) color scale levels [48].

33.5 Color Plasma Flat Panel Displays

Color Pixel Structures

In color plasma display panels, photoluminescent phosphors provide the primary RGB optical emissions required for full-color image display. In this case, visible emissions from the discharge itself must be suppressed in order to avoid color contamination. A common approach is to utilize xenon as the minority species constituent in the Penning gas mixture of the panel. The structure and phosphor layout of the 102 cm (40 in.) diagonal color dc plasma display available from NHK is shown in Figure 33.17, while that of the Fujitsu 53 cm (21 in.) diagonal ac color display is shown in Figure 33.6. Each uses screen printing and hard mask or abrasive-resistant lithographic processes for conductor wire deposition, barrier structure definition, and phosphor deposition [51]. In the NHK design, the fourth section within the honeycomb color pixel structure houses a redundant green phosphor subpixel to compensate for the lower photoluminance of green phosphors relative to that of either red or blue phosphors. In a similar honeycomb dc color pixel structure, Panasonic instead incorporates a series resistor in this fourth subpixel position [20]. Printing the series resistor for each pixel on the display glass substrate complicates panel manufacturing but simplifies design requirements for the drive electronics. In the Fujitsu, the opposed electrode ac color pixel structure shown in Figure 33.6, barrier or separation rib structures running between and parallel to each conductor wire are fabricated on the rear glass substrate. The barrier rib heights are typically 100 to 150 μm. ac barrier rib structures and dc pixel honeycomb lattice structures are usually composed of the same PbO thick-film dielectric used to cover the conductor wires.

VUV Photon Production and Utilization for Color Plasma Flat Panel Displays

Color plasma display gas mixtures utilizing xenon as the minority species are optimized for production of VUV emissions which are used to excite RGB photoluminescent phosphors. Both neon–xenon and

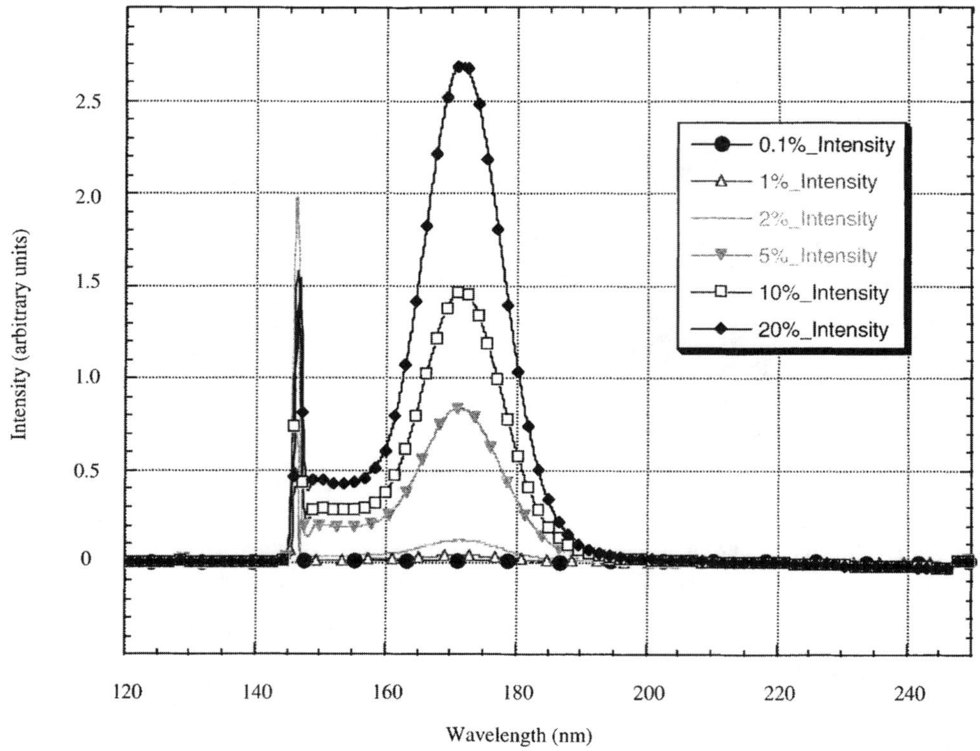

FIGURE 33.18 VUV emission spectra from opposed electrode ac plasma pixel discharges in He/Xe gas mixtures. Each spectrum was collected near the minimum sustain (or first *on*) voltage for that gas mixture, which ranged from 150 V for 0.1% Xe to 350 V for 20% Xe.

helium–xenon combinations are popular. The ionization energy of xenon at 12.3 eV lies below the lowest excited atomic states of either neon or helium, as shown in Figure 33.8. Consequently, electrons accelerated by electric fields within the pixel volume preferentially impart their kinetic energy to the xenon atoms. In addition, the excited states of He or Ne produced readily transfer stored energy to the xenon atoms through ionizing Penning collision processes. Consequently, the red-orange visible emissions typical of Ne discharges are suppressed as Xe concentration is increased. Fujitsu utilizes an Ne–10% Xe working gas in its color display [18], while Photonics Imaging prefers to use an He-based background gas in its panel [23] where suppression of unwanted optical emissions from the discharge can be accomplished at somewhat lower xenon concentrations.

The tendency of xenon to fill its outermost electronic shell results in the formation of the xenon dimer molecule, Xe_2^*, whose energy states are also shown in Figure 33.8 [52,53]. Radiative dissociation of the dimer produces photons with wavelengths near 173 and 150 nm. Figure 33.18 shows how the dimer emissions dominate the VUV spectra from He–Xe gas pixel discharges as the fraction of Xe is increased. Since VUV photons are completely absorbed by glass, the spectra shown in the figure were measured by mounting opposed electrode pixels inside of a vacuum chamber filled with the gas mix of interest. The boundaries of the panel glass were not sealed which then allowed on-edge viewing of the pixel discharges with a McPherson 0.2 m monochromator operating in conjunction with a Princeton Instrument CCD optical multichannel analyzer [34]. The background gas mix was varied to obtain the various Xe concentrations in He shown while maintaining a total gas pressure of 53.3 kPa (400 torr). At low Xe concentrations, photons from the atomic Xe $1s^4$ and $1s^5$ states dominate the emission spectra producing lines at 147 nm and, with much less intensity, at 129 nm. The Tachibana laser-induced spectroscopic measurements show the spatial and temporal evolution of the $1s^4$ Xe atomic state in He/Xe plasma display discharges [54]. Both of these atomic lines experience significant resonant absorption and reemission.

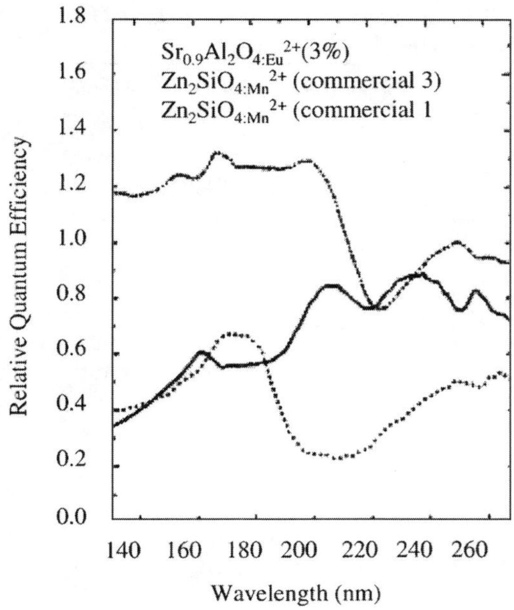

FIGURE 33.19 Relative quantum efficiencies of a Tb-activated lanthanum phosphate compared to that of yttrium and gadolinium phosphate prepared by Sarnoff Research Center. (From Yocum, N. et al., *J. SID*, 4/3, 169–172, 1996. With permission.)

Thus, the measured line intensities are strong functions of photon path length traveled and of Xe partial pressure in the background gas [55]. For the emission spectra shown in Figure 33.18, the lithium fluoride (LiF) entrance window to the evacuated spectrometer chamber was positioned between 100 and 150 nm from the nearest pixel discharges, which is roughly the location of the phosphors relative to the discharge in an opposed electrode ac color display panel; for an example, see Figure 33.5.

Recall that the optimal charge pair production per volt applied in Penning gas discharges occurs at minority species concentrations as low as 0.1%; see Figure 33.9. However, color plasma pixels must optimize usable photon production per watt while maintaining stringent color purity requirements. Consequently, color plasma pixels typically operate with xenon concentrations ranging between 2 and 10%. Figure 33.18 shows that increased xenon concentration results in significant dimer formation and radiative emission from dimer dissociation. Since the dimer dissociation is a three-body process involving a photon and two xenon atoms, the momentum and energy conservation equations do not demand unique solutions. Consequently, emissions lines produced cover a broad spectral range spanning several tens of nanometers. Increased dimer emission is accompanied by the suppression of xenon atomic emission as energy within the atomic manifolds continues to flow toward the lowest available atomic levels; see Figure 33.8. Note, too, that the dimer emission lines are not subject to resonant absorption. Therefore, the measured intensities shown reflect dimer emission from all pixel rows within the line of site of the spectrometer (four for the data of Figure 33.18). In contrast, due to the strong resonance absorption of the atomic lines, more than 90% of the measured intensity of the 147 nm line is produced in the pixel row adjacent to the spectrometer window [34,55]. Care must be taken to account for these large variations in photon mean free paths when analyzing emission data.

Phosphor Excitation and Emission for Color Plasma Flat Panels

A variety of photoluminous phosphors are commercially available. Efficiencies for conversion of VUV photons to visible emissions has a complex dependence on excitation photon wavelength as can be seen in Figure 33.19, which shows quantum conversion efficiencies relative to a sodium salicylate standard for some of the available green phosphors [50]. Conversion efficiencies for red, blue, and other green

TABLE 33.2 Relative Quantum Efficiencies (QE) and Chromaticity Coordinates for Selected Phosphors [50,57]

Phosphor	Rel. QE[a] (174 nm)	Rel. QE[a] (170 nm)	Lifetime (10%) ms	*x*	*y*
NTSC green				0.21	0.71
$(La_{0.87}Tb_{0.13})PO_4$	1.1	1.4	13	0.34	0.57
$(La_{0.6}Ce_{0.27}Tb_{0.13})PO_4$	1.1	1.5	12	0.33	0.59
$(Y_{0.85}Ce_{0.1}Tb_{0.05})PO_4$	1.1	1.1	—	—	—
$(Y_{0.6}Ce_{0.27}Tb_{0.13})PO_4$	1.35	1.35	—	—	—
$(Gd_{0.87}Ce_{0.1}Tb_{0.03})PO_4$	1.0	1.1	10	0.34	0.58
$(Gd_{0.6}Ce_{0.27}Tb_{0.13})PO_4$	1.35	1.45	—	—	—
$(Ce,Tb)MgAl_{11}O_{19}$	0.9	1.4	—	—	—
$Sr_{0.9}Al_2O_4$:Eu^{2+} (3%)	0.4	0.7	0.01	0.26	0.59
Zn_2SiO_4:Mn	1.2	1.3	12.5	0.21	0.72
Zn_2SiO_4:Mn	1.1	1.5	9.8	—	—
Zn_2SiO_4:Mn	0.45	0.55	5.4	—	—
Zn_2SiO_4:Mn	1.0	1.1	5	—	—
Zn_3SiO_4:Mn	1.0			0.21	0.72
$BaAl_{12}O_{19}$:Mn	1.1			0.16	0.74
$BaMgAl_{14}O_{23}$:Mn	0.92			0.15	0.73
$SrAl_{12}O_{19}$:Mn	0.62			0.16	0.75
$ZnAl_{12}O_{19}$:Mn	0.54			0.17	0.74
$CaAl_{12}O_{19}$:Mn	0.34			0.15	0.75
YBO_3:Tb	1.1			0.33	0.61
$LuBO_8$:Tb	1.1			0.33	0.61
$GdBO_3$:Tb	0.53			0.33	0.61
$ScBO_8$:Tb	0.36			0.35	0.60
$Sr_4\ Si_8\ O_8\ Cl_4$:Eu	1.3			0.14	0.33
NTSC red				0.67	0.33
Y_2O_3:Eu	0.67			0.65	0.34
Y_2Si_{O5}:Eu	0.62			0.66	0.34
$Y_3A_{l5}O_{12}$:Eu	0.47			0.63	0.37
$Zn_8(PO_4)_2$:Mn	0.34			0.67	0.33
YBO_3:Eu	1.0			0.65	0.35
$(Y,Gd)BO_8$:Eu	1.2			0.65	0.35
$GbBO_3$:Eu	0.94			0.64	0.36
$ScBO_3$:Eu	0.94			0.61	0.39
$LuBO_8$:Eu	0.74			0.63	0.37
NTSC blue				0.14	0.08
$CaWO_4$:Pb	0.74			0.17	0.17
Y_2SiO_5:Ce	1.1			0.16	0.09
$BaMgAl_{14}O_{23}$:Mn	1.6			0.14	0.09

[a] QEs above double rule are relative to sodium salicylate, those below relative to Zn_2SiO_4:Mn.

phosphors can be found in References 56 and 57. Table 33.2 provides the compositions of some selected commercially available phosphors and lists their relative quantum efficiencies for the principal emission lines of xenon discharges. Note that quantum efficiencies listed are relative values and that phosphors that convert 8.4 eV photons to visible photons near 2.3 eV have absolute efficiencies of only 27%. In principle, it is possible to produce two or more visible photons from a single high energy photon, but to date no such phosphors have been developed [58]. Table 33.2 also lists the chromaticity diagram coordinates which provide a measure of the color purity of the visible RGB emission spectra produced. The chromaticity diagram can be found in many references including Reference 59. Another consideration in the plasma display phosphor selection is persistence. Most of the phosphors listed in Table 33.2 require 5 to 13 ms for the emission intensity to decay to 10% of maximum value. For ac pixel operation at

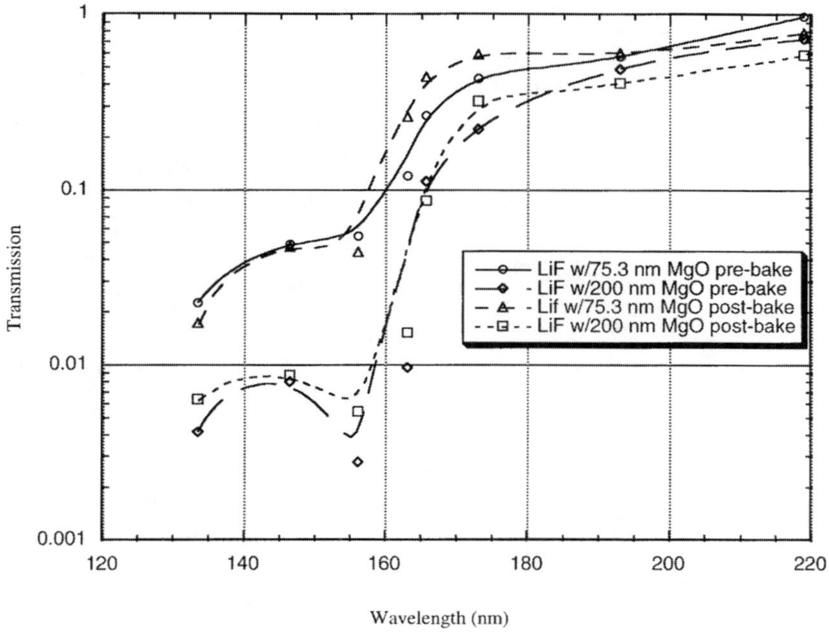

FIGURE 33.20 Photon transmission through MgO films before and after bake-out to remove water vapor.

50 kHz, each sustain voltage half cycle lasts only 10 μs while the discharge produces VUV emissions for only a small fraction of that time. Efforts are continuing for development of phosphors with faster response times. For example, Eu_{2+} green phosphors with 10% decay times of only 5 to 10 μs and with good quantum efficiencies near 173 nm have been developed [50].

Color Plasma Display Lifetime Considerations

Phosphors for plasma flat panel displays must be tolerant of the harsh environment produced by the pixel discharge. Photoluminous phosphor degradation mechanisms are at present not well understood. Contamination of the discharge by phosphor materials is a serious concern. Discharge modeling indicates that damage results principally from the accumulated fluence of photon and metastable bombardment, although fringe electric fields and prolonged surface charge accumulations could also result in ion bombardment [37]. Most ac plasma displays take advantage of MgO for enhancement of the discharge intensity by coating dielectric surfaces above the electrodes with an additional thin film of MgO. For ease of fabrication, the MgO is most often deposited using electron beam evaporation as one of the final manufacturing steps before glass seal and gas fill. If no mask is used, the MgO can also cover the phosphors. While providing the phosphors with a protective coating, the MgO film also attenuates the intensity of the VUV photon flux available to excite the phosphors. Figure 33.20 shows VUV photon transmission as a function of wavelength through MgO films [34,60]. The measurements show that the primary atomic Xe emission lines at 129 and 147 nm are nearly 95% attenuated by an MgO film only 75 nm thick. In contrast, the dimer emission lines centered near 173 nm are much less attenuated with transmission factors of 30 to 50%, respectively, for MgO films of 200 and 75 nm. Consequently, ac plasma display designers must often achieve a balance among discharge enhancement due to MgO secondary electron emission, discharge degradation due to MgO photon absorption, and fabrication complexity associated with MgO deposition, which impacts the final cost of the display. Additionally, the designer must consider the aging or brightness degradation with time of the display, which is influenced in part by the rate of MgO sputter erosion discussed briefly below.

TABLE 33.3 Equipment for Noncontact Profiling and Chemical Characterization of Thin Films

	CyberOptics Point Range Sensor	Zygo NewView 200	Leitz FTM400
Physical principle	Laser triangulation	Scanning white light interferometry	Laser interferometry
Maximum thickness/step height	150 μm (high resolution head)	100 μm (standard) (to 5 mm with z-drive)	70 μm
Vertical resolution	0.38 μm	100 pm	150 pm or 0.5%
Spatial resolution	>0.5 μm	0.22 μm	0.5 μm
Spot diameter/FOV	5.1 μm	140 × 110 μm	20 μm
Scan rate	10 points/s	2 or 4 μm/s	75 mm/s max.
Maximum sample size	Based on Table	10 cm × 15 cm	47 cm × 37 cm
Maximum sample weight	Based on Table	4.5 kg	4.5 kg (including holder)
Elemental detection	N.A.	N.A.	N.A.

33.6 Inspection and Metrology

Plasma display panel manufacturing uses some process steps typically found in semiconductor device manufacturing. Although the feature sizes are much larger than those required in semiconductor manufacturing, the dimensions over which feature integrity must be assured are also much larger. Confirmation that a display manufacturing process is under control or rapid quantitative characterization of the deviation from acceptable process limits is essential to producing high-quality displays with high yield. Information on a variety of length scales is required. Blanket deposition of materials onto substrates should be uniform; large deviations in thickness of the deposited material from one point on the substrate to a widely separated point should be avoided. Similarly, roughness of the deposited layers must be kept within defined limits. Finally, chemical compositions of deposited layers and their spatial homogeneity need to be measured and controlled.

The three commercial profile measuring devices, the CyberOptics, Zygo, and Leitz units, are based on laser or white light triangulation or interferometry; see Table 33.3 [61-63]. These machines are well suited for determining the profile of metallization after patterning. They do not appear to be useful for measuring the thickness of opaque layers and cannot, for example, measure the thickness of the blanket metallization before patterning. The CyberOptics unit is in wide use in the electronic packaging industry because of the low noise in its signal. Improvements in vertical resolution would increase its value for display manufacturing process control. The Zygo instrument has better vertical resolution, but has no real scanning capability. The Leitz unit is designed specifically for metrology of unfinished liquid crystal display components, and is presumably optimized for that application. Both the CyberOptics and Zygo units have a variety of heads with differing resolutions and fields of view. However, the vertical distance over which the interference can take place in the Zygo unit is very limited, so it may not be suitable for measuring features with large vertical dimensions.

Depending on system requirements, a useful laser metrology system can be purchased for as little as $20,000. Full-featured systems can easily cost $100,000. The bulk of the cost is, however, in the precision transport tables used for moving samples in a precisely controlled manner beneath stationary measurement heads. Since precision transport tables meeting the needs of the metrology equipment may already be part of the display production process or would require only minor upgrades to be brought in line with requirements, the cost to bring the metrology capability online may be much smaller than the figures mentioned above. Since laser head costs are low, it may also be desirable to use multiple fixed heads to speed up inspection and metrology.

Listed in Table 33.4 are the characteristics of two advanced measurement techniques for noncontact profiling and chemical analysis. To the authors' knowledge, systems of this type are not yet commercially available. They are being utilized, however, by researchers at Sandia National Laboratories and show

TABLE 33.4 Advanced Techniques for Noncontact Profiling/Chemical Characterization of Thin Films

Physical Principle	β-Backscatter	X-Ray Microfluorescence
Maximum thickness/step height	A few mm	50 μm
Vertical resolution	About 0.1 nm	0.1 μm
Spatial resolution	Depends on pinhole size	50 μm–2 mm
Spot diameter/FOV	Depends on pinhole size	100 μm
Scan rate	<1 min/measurement	1 spot/min
Maximum sample size	No restriction	24 cm × 21.5 cm
Maximum sample weight	No restriction	4.5 kg
Elemental detection	N.A.	<10–300 ppm

promise for commercial usage. In beta backscattering measurements, the energy spectra of the backscattered beta particles provides an accurate measure of the elemental composition of a surface. This technique has been used to measure the thin-film thickness of trace-deposited metals on large-area surfaces (several square meters) in tokamaks [64]. With a depth resolution of about 100 pm, beta backscatter could serve as the physical basis for a high-sensitivity surface-profiling device. X-ray fluorescence measurements, routinely performed in air, can give information on film thickness and composition, and performance parameters for this technique are also listed in Table 33.4. Vacuum must be used when studying light elements like carbon, a potential undesired product of phosphor binder burnout, because the x rays produced are "soft". Although x-ray fluorescence equipment may have good lateral resolution, a larger spot size may prove useful when more averaged information such as film thickness is needed. Finally, eddy current measurements may provide a fast and reliable method for determining the thickness of the opaque, blanket metallization before patterning. This technique should be capable of providing 10% thickness measurement accuracy for conductor layers as thin as 1 μm.

Many of the devices for quantitative characterization of the thickness and chemistry of the deposited layers suffer from the fact that they have a rather limited field of view. Characterization techniques are needed that will allow rapid identification of visual defects in the blanket-deposited layers and in the patterned layers produced from them. Visual inspection equipment is available for defect identification in liquid crystal displays. Manufacturers include Minato Electronics, Advantest, ADS, Photon Dynamics, and Teradyne. All use CCD devices and special algorithms for the identification of line and area defects. The defects found during high-volume plasma display manufacturing may be sufficiently similar in appearance to those found in liquid crystal display manufacturing that this equipment will prove useful with appropriate adjustments. These devices, however, are expensive.

Although researchers and equipment manufacturers believe that the equipment and techniques described above will be suitable for online process control during plasma display manufacturing, all agree that more development is required. Real parts will need to be characterized extensively on specific commercial equipment in production environments before conclusions can be drawn about the suitability of the equipment and techniques for the intended application. This is particularly true for more difficult measurements, like that of the thickness of dielectric above conductor lines. It is clear that there is no single instrument that will meet all film and feature dimensional measurement requirements and those for chemical characterization. A number of instruments will be needed to measure confidently the parameters needed for inspection, and characterize and control the manufacturing processes for plasma displays.

References

1. D.L. Bitzer and H.G. Slottow, The plasma display panel — a digitally addressable display with inherent memory, *1966 Fall Joint Computer Conf.*, Washington, D.C., *AFIPS Conf. Proc.*, 29, 541, 1966.
2. H.G. Slottow, Plasma displays, *IEEE Trans. Electron. Devices*, 23, 760–772, 1976.

3. T.N. Criscimagna and P. Pleshko, Ac plasma display. chap. 3, in *Topics in Applied Physics,* Vol. 40, *Display Devices,* Springer-Verlag, Berlin, 91–150, 1980.
4. L.F. Weber, Plasma displays, chap. 10, in *Flat Panel Displays and CRTs,* L.E. Tannas, Jr., Ed., Van Nostrand Reinhold, New York, 1985.
5. J. Nolan, Gas discharge display panel, *1969 International Electron Devices Meeting,* Washington, D.C., 1969.
6. H.J. Hoehn and R.A. Martel, A 60 line per inch plasma display panel, *IEEE Trans. Electron Devices,* 18, 659–663, 1971.
7. O. Sahni and W.P. Jones, Spatial distribution of wall charge density in ac plasma display panels, *IEEE Trans. Electron Devices,* 25, 223–226, 1979.
8. O. Sahni and C. Lanza, Origin of the bistable voltage margin in the ac plasma display panel, *IEEE Trans. Electron. Devices,* 24, 853–859, 1977.
9. C. Lanza and O. Sahni, Numerical calculation of the characteristics of an isolated ac gas discharge display panel cell, *IBM J. Res. Dev.,* 22, 641–646, 1978.
10. O. Sahni and M.O. Aboelfotoh, The pressure dependence of the bistable voltage margin of an ac plasma panel cell, *Proc. SID,* 22, 212–218, 1981.
11. O. Sahni and C. Lanza, Importance of the secondary electron emission coefficient on E/Po for Paschen breakdown curves in ac plasma panels, *J. Appl. Phys.,* 47, 1337, 1976.
12. O. Sahni and C. Lanza, Influence of the secondary electron emission coefficient of argon on Paschen breakdown curves in ac plasma panels for neon + 0.1% argon mixture, *J. Appl. Phys.,* 47, 5107, 1976.
13. C. Lanza, Analysis of an ac gas display panel, *IBM J. Res. Dev.,* 18, 232–243, 1974.
14. O. Sahni, C. Lanza, and W.E. Howard, One-dimensional numerical simulation of ac discharges in a high-pressure mixture of Ne + 0.1% Ar confined to a narrow gap between insulated metal electrodes, *J. Appl. Phys.,* 49, 2365, 1978.
15. O. Sahni and C. Lanza, Failure of Paschen's scaling law for Ne –0.1% Ar mixtures at high pressures, *J. Appl. Phys.,* 52, 196, 1981.
16. H. Wakabayshi, Display market projections report from Namora Research Institute, paper presented at *Imaging 2001: The U.S. Display Consortium Business Conference,* January 28, 1997, San Jose, CA.
17. K. Werner, Plasma hits the ground running, *Inf. Display,* 12(12), 30–34, 1996.
18. *Nikkei Materials & Technology,* [135], November, 1993, pp. 22–23.
19. H. Murakami and R. Toyonaga, A pulse discharge panel display for producing a color TV picture with high luminance and luminous efficiency, *IEEE Trans. Electron Devices,* 29, 988, 1982.
20. M. Ushirozawa, Y. Motoyama, T. Sakai, K. Wani, and K. Takahashi, Color DC-PDP with an improved resistor design in each cell, paper 33.2, pp. 719–722, *SID '94 Symp. Dig.,* 1994.
21. S. Mikoshiba, Color plasma displays: where are we now?, *Inf. Display,* 10(10), 21, 1994.
22. Y. Takano et al., A 40-in, DC-PDP with new pulse memory drive scheme, *SID '97 Dig.,* 1997.
23. P.S. Friedman, Are plasma display panels a low-cost technology?, *Inf. Display,* 11(10), 22–25, 1995.
24. *Inf. Display,* 10(7 & 8), 28, 1994.
25. A. von Engle, *Ionized Gases,* AIP Press, New York, 1994.
26. S. Brown, *Basic Data of Plasma Physics — The Fundamental Data on Electrical Discharges in Gas,* American Institute of Physics Press, New York, 1993.
27. F.F. Chen, *Introduction to Plasma Physics,* Plenum Press, New York, 1977.
28. D. Miller, J. Ogle, R. Cola, B. Caras, and T. Maloney, An improved performance self-scan I panel design, *Proc. SID,* 22, 159–163, 1981.
29. M.H. Miller and R.A. Roig, Transition probabilities of XeI and XeII, *Phys. Rev. A,* 8, 480, July 1973.
30. *Atomic Data and Nuclear Data Tables,* Vol. 21, No. 6, Academic Press, New York, 1978.
31. R.S. Van Dyck, C.E. Johnson, and H.A. Shugart, Lifetime lower limits for the $_3P_0$ and $_3P_2$ metastable states of neon, argon and krypton, *Phys. Rev. A,* 5, 991–993, 1972.
32. A.A. Kruithof and F.M. Penning, The Townsend ionization coefficient and some elementary processes in neon with small admixtures of argon, *Physica,* 4(6), 450, 1937.

33. F.L. Jones and W.R. Galloway, The sparking potential of mercury vapor, *Proc. Phys. Soc. Lond.*, 50, 207–212, 1938.
34. R.T. McGrath, R. Veerasingam, J.A. Hunter, P.D. Rockett, and R.B. Campbell, Measurements and simulations of VUV emissions from plasma flat panel display pixel microdischarges, *IEEE Trans. on Plasma Sci.*, 26(5), 1998.
35. P. Hines, Spectroscopic and electrical measurements on opposed electrode ac plasma display pixels, Honors engineering senior thesis, The Pennsylvania State University, University Park, 1997.
36. R. Veerasingam, R.B. Campbell, and R.T. McGrath, One-dimensional single and multipulse simulations of the ON/OFF voltages and bistable margin for He, Xe, and He/Xe filled plasma display pixels, *IEEE Trans. Plasma Sci.*, 24(6), 1399–1410, 1996.
37. R. Veerasingam, R.B. Campbell, and R.T. McGrath, Two-dimensional simulations of plasma flow and charge spreading in ac plasma displays, *IEEE Trans. Plasma Sci.*, 24, 1411–1421, 1996.
38. M. Scott, Los Alamos National Laboratory, Private communication, 1997.
39. H. Uchike, K. Miura, N. Nakayama, T. Shinoda, and Y. Fukushima, Secondary electron emission characteristics of dielectric materials in ac-operated plasma display panels, *IEEE Trans. Electron. Devices*, 23, 1211–1217, 1976.
40. N.J. Chou and O. Sahni, Comments on "Secondary Emission Characteristics of Dielectric Materials in ac Operated Plasma Display Panels," *IEEE Trans. Electron. Devices*, 25, 60–62, 1978.
41. T. Shinoda, H. Uchike, and S. Andoh, Low-voltage operated ac plasma-display panels, *IEEE Trans. Electron. Devices*, 26, 1163–1167, 1979.
42. M. Aboelfotoh and J.A. Lorenzen, Influence of secondary-electron emission from MgO surfaces on voltage-breakdown curves in Penning mixtures for insulated-electrode discharges, *J. Appl. Phys.*, 48, 4754–4759, 1977.
43. T. Nenadovic, B. Perraillon, Z. Bogdanov, Z. Djordjevic, and M. Milic, Sputtering and surface topography of oxides, *Nucl. Instrum. Methods Phys. Res.*, B48, 538–543, 1990.
44. Schone, D. Walsh, R.T. McGrath and J.H. Burkhart, Microbeam Rutherford backscattering measurements of flat panel display thin film erosion, *Nuclear Instru. Methods Phys. Rev.-B*, 130(1–4), 543–550, 1998.
45. *CRC Handbook of Chemistry and Physics, 48th ed.*, R.C. Weast and S. M. Selby, Eds., The Chemical Rubber Co., Cleveland, OH, 1967.
46. S. Matsumoto, *Electronic Displays Devices*, John Wiley & Sons, New York, 1990.
47. G.E. Holz, The primed gas discharge cell — a cost and capability improvement for gas discharge matrix displays, *Proc. SID*, 13(2), 1972.
48. T. Yamamoto, T. Kuriyama, M. Seki, T. Katoh, H. Murakami, K. Shimada, and H. Ishiga, A 40 in. diagonal HDTV plasma display, *SID' 93 Symp. Dig.*, 165–168, 1993.
49. J. Deschamps and H. Doyeus, Plasma displays, *Phys. World*, June 1977.
50. N. Yocum, R.S. Meltzer, K.W. Jang, and M. Grimm, New green phosphors for plasma displays, *J. SID*, 4/3, 169–172, 1996.
51. H. Fujii, H. Tanabe, H. Ishiga, M. Harayama, and M. Oka, A sandblasting process for fabrication of color PDP phosphor screens, paper 37.5, 728–731, *SID '92 Symp. Dig.*, 1992.
52. M.C. Castex, Experimental determination of the lowest excited Xe_2 molecular states from VUV absorption spectra, *J. Chem. Phys.*, 74, 759–771, 1981.
53. M.R. Flannery, K.J. McCann, and N.W. Winter, Cross sections for electron impact ionization of metastable rare gas excimers (He_2^*, Kr_2^*, Xe_2^*), *J. Phys. B Mol. Phys.*, 14, 3789–3796, 1981.
54. K. Tachibana, Spatio-temperal measurement of excited $Xe(1s_4)$ atoms in a discharge cell of a plasma display panel by laser induced spectroscopic microscopy, *Appl. Phys.* Lett., 65, 935–937, 1994.
55. T. Holstein, Imprisonment of resonance radiation in gases, *Phys. Rev.*, 72, 1212, 1947.
56. De Husk and S.E. Schnatterly, Quantum efficiency and linearity of 16 phosphors in the soft x-ray regime, *J. Opt. Soc. Am. B*, 9, 660–663, 1992.
57. J. Koike, T. Kojima, and R. Toyonaga, New tricolor phosphors for gas discharge display, *J. Electrochem. Soc. Solid-State Sci. Tech.*, 1008, June 1979.

58. P.N. Yocum, Future requirements of display phosphors from an historical perspective, *J. SID*, 4/3, 149, 1996.
59. L.E. Tannas, Jr., Ed., *Flat Panel Displays and CRTs*, Van Hostrand Reinhold, New York, 1985.
60. S.G. McLean and W.W. Duley, VUV absorption in thin MgO films, *J. Phys. Chem. Solids*, 45, 223, 1984.
61. R. Kuntz, CyberOptics, Private communication, 1-800-746-6315.
62. Zygo, Private communication, 1-860-347-8506, www.zygo.com.
63. A. Machura, Leica, Private communication, 49 (0) 6441 29-2316, Andreas.Machura@lmw.leica.com.
64. B.E. Mills, D.A. Buchenauer, A.E. Pontau, and M. Ulrickson, Characterization of deposition and erosion on the TFTR bumper limiter and wall, *J. Nucl. Mater.*, 162–164, 343–349, 1989.

34

Electroluminescent Displays

William A. Barrow
Planar Systems

34.1 Introduction

Electroluminescence (EL) is the nonthermal generation of light resulting from the application of an electric field to a substance. The light-emitting substance is generally a luminescent crystal. Most commercially available monochrome EL displays utilize ZnS:Mn as the luminescent material. EL displays have become very important in certain display markets. These include medical instrumentation, industrial control equipment, portable instrumentation, and military vehicles. The attributes of EL displays that make them attractive in these types of applications are mechanical ruggedness, relative insensitivity to ambient temperature, rapid display response, essentially unlimited viewing angle, compactness, and light weight.

There are four types of EL devices: ac thin film, ac powder, dc thin film, and dc powder. The ac thin-film display is by far the dominant device type. Some liquid crystal displays use ac powder EL for backlights. There is currently little or no commercial application of dc EL devices, either thin film or powder [1]. The focus here is on ac thin-film EL (ACTFEL) devices. While there are no widely established standard measurement techniques for the other EL device types, measurements similar to those described herein for ACTFEL devices could be applied with appropriate modifications to the other methods of device excitation.

0-8493-1733-9/04/$0.00+$1.50

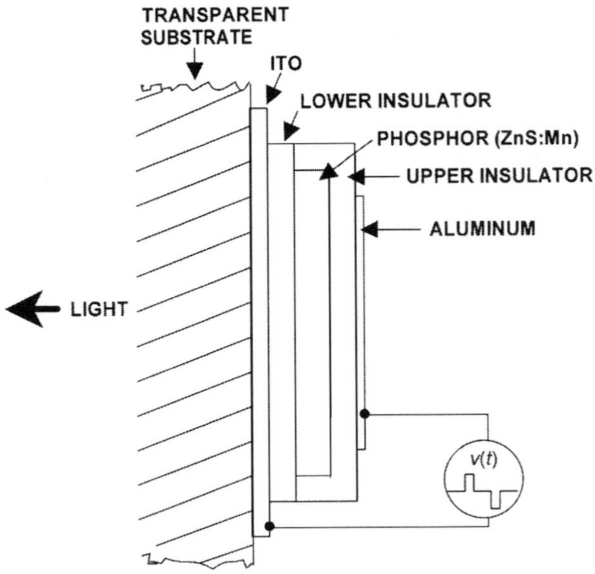

FIGURE 34.1 Schematic, cross-sectional diagram of the basic ACTFEL device structure.

34.2 Device Structure and Operation

A schematic, cross-sectional representation of the basic ACTFEL device structure [2] is shown in Figure 34.1. The supporting substrate is usually made of very low sodium glass. If a suitable ion barrier layer is deposited between the substrate and lower electrode of the EL device, soda-lime glass can be used. The lower electrode, usually a transparent conductor such as indium tin oxide (ITO) is deposited next. The ITO is usually between 350 and 120 nm in thickness, providing sheet resistance in the range of 5 to 15 Ω/square. On top of the ITO electrode a lower insulator is deposited. This layer is typically SiON or aluminum titanium oxide (ATO). The thickness is normally around 200 nm. The phosphor layer, typically ZnS:Mn, is deposited between the lower insulator and a similar upper insulator. The phosphor thickness is typically in the range 200 to 1000 nm thickness, depending on the application. The upper electrode, typically aluminum, is deposited on top of the upper insulator. The aluminum is generally 100 to 200 nm thick.

A matrix-addressed monochrome display is created by dividing the upper and lower electrodes into orthogonal arrays of electrode stripes. The EL device is then excited locally by applying a voltage between a pair of crossing electrodes, causing an electric field to exist between them, which excites the phosphor. A color display can be created by dividing the phosphor into stripes of red-, green-, and blue-emitting phosphors which are aligned with one of the sets of electrode stripes [3,4]. This is shown schematically in Figure 34.2. Color displays can also be created by using an unpatterned phosphor which emits a broad spectrum including red, green, and blue and filtering the emission using either a patterned color filter which is aligned with the electrode stripes or a frame sequential color filter.

34.3 Device Fabrication

Thin-Film Deposition Methods

A wide variety of deposition techniques are used by various manufacturers and researchers in the fabrication of ACTFEL devices. The electrode materials ITO and aluminum are usually deposited by physical vapor deposition (PVD) techniques. ITO is almost universally deposited by dc magnetron

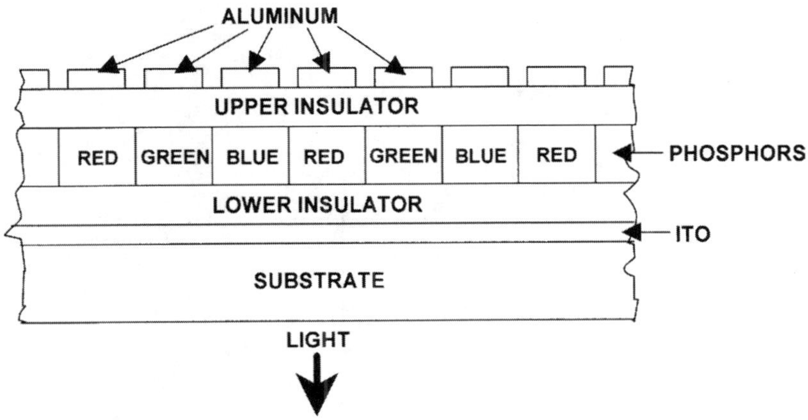

FIGURE 34.2 Schematic, cross-sectional diagram of a color ACTFEL device with the red, green, and blue primary colors produced by patterned color phosphor stripes. The color phosphor stripes are end on in this view.

sputtering from either a conductive ITO target or from a metal alloy target [5]. Optimum ITO conductivity is obtained by postdeposition annealing in a very low oxygen environment. Al is deposited either by dc magnetron sputtering from an Al metal target or by electron beam evaporation of Al metal. The insulator and phosphor layers are deposited by chemical vapor deposition (CVD) as well as PVD. Phosphor layers have been deposited by thermal as well as e-beam evaporation, sputtering, metal–organic CVD (MOCVD), and atomic layer epitaxy (ALE). Insulator layers have been deposited by e-beam evaporation, radio frequency (RF) sputtering, plasma-enhanced CVD (PECVD), and ALE.

Thin-Film Patterning Methods

Patterning of the electrodes is generally accomplished either through etching or liftoff, although some early workers patterned the upper aluminum electrodes by evaporating through a shadow mask. Aluminum is easily etched wet or dry. There are commercially available wet etches for Al. Dry etching of Al can be carried out using standard chlorine chemistries, e.g., Cl_2/BCl_3. Al can also be patterned by evaporating onto a reversed photoresist pattern and lifting off excess metal. ITO can be etched wet or dry. ITO wet etches generally consist of mixtures of HCl and HNO_3. ITO can be dry-etched in HI or HBr. Patterning of the phosphor layers is more problematic. Phosphor etches exist for most ACTFEL phosphor materials, but they are generally proprietary. Thin-film phosphors are difficult to etch because they are usually water sensitive, and some color phosphors contain heavy metals which are difficult to volatilize in a dry-etch process.

34.4 Device Operation

Luminescence Mechanisms

As the device structure shown in Figure 34.1 indicates, ACTFEL devices are capacitively coupled. Since only displacement current can flow across the insulator layers, the drive signal must be an alternating polarity waveform. A typical alternating polarity, trapezoidal waveform is shown in Figure 34.3. If the peak voltage, V_P, in Figure 34.3 is larger than the **threshold voltage** of the device, V_{th}, then, when the positive pulse of the waveform is applied between the Al and ITO electrodes of the device structure shown in Figure 34.1, the energy band diagram of the ACTFEL device will be as shown in Figure 34.4. Electrons, which are the majority carriers in ACTFEL devices, tunnel out of the interface states on the left and into the conduction band. Once in the conduction band the electrons are accelerated by the large electric field, which is approximately 1 MV/cm = 100 kV/μm. The conduction electrons drift across the ZnS:Mn layer until they impact excite an Mn^{2+} center, transferring some energy to one of its electrons and causing

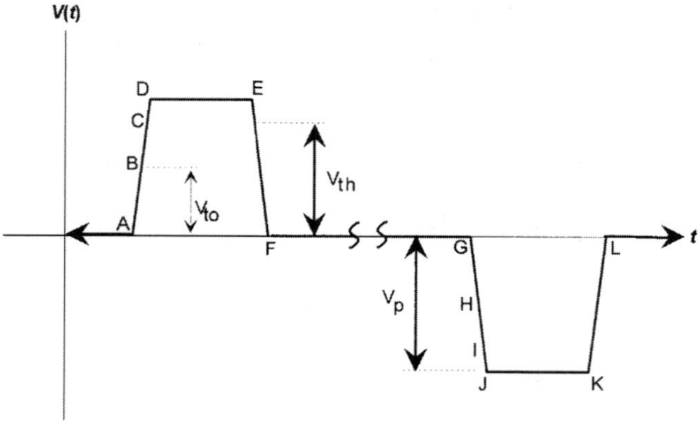

FIGURE 34.3 Typical alternating polarity, pulse drive waveform. Letters A–L mark points on the drive waveform which are referenced later in Figures 34.5 and 34.6. The pulse width would generally be about 30 μs and the frequency would be between 60 and 500 Hz for a passive matrix-addressed display.

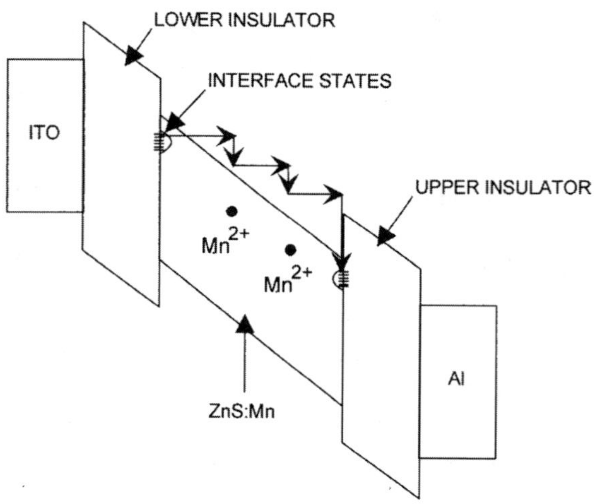

FIGURE 34.4 Energy band diagram of an ACTFEL device during the peak of the applied voltage pulse. Electrons tunnel out of insulator/phosphor interface states into the conduction band, are swept across the phosphor layer, and impact exciting emission centers as they go until they are finally thermalized and trapped at the opposite interface.

it to undergo a transition to an excited state. The conduction electron may undergo additional collisions, eventually reaching the right interface and getting trapped in interface states there until the next voltage pulse, which is of the opposite polarity. This pulse causes the electrons to tunnel out and drift back across the ZnS:Mn layer, transferring energy to Mn^{2+} centers along the way, until eventually they are trapped again at the left interface. This transfer of charge back and forth between the interface states continues as long as the alternating polarity drive signal with a peak amplitude above the threshold voltage of the device continues to be applied. Light emission occurs when the Mn^{2+} centers, which have been impact excited by the hot electrons, relax [6-8]. The light emission thus results from transitions of the electrons within the Mn^{2+} centers, rather than electron–hole pair recombination near a *pn* junction as occurs in a light-emitting diode (LED).

Description of Charge Flow

If the external charge, *Q*, flowing into the ACTFEL device is plotted vs. the externally applied voltage, *V*, the resulting curve is called a ***QV* loop** [9,10]. If the amplitude of the applied voltage pulses is less than

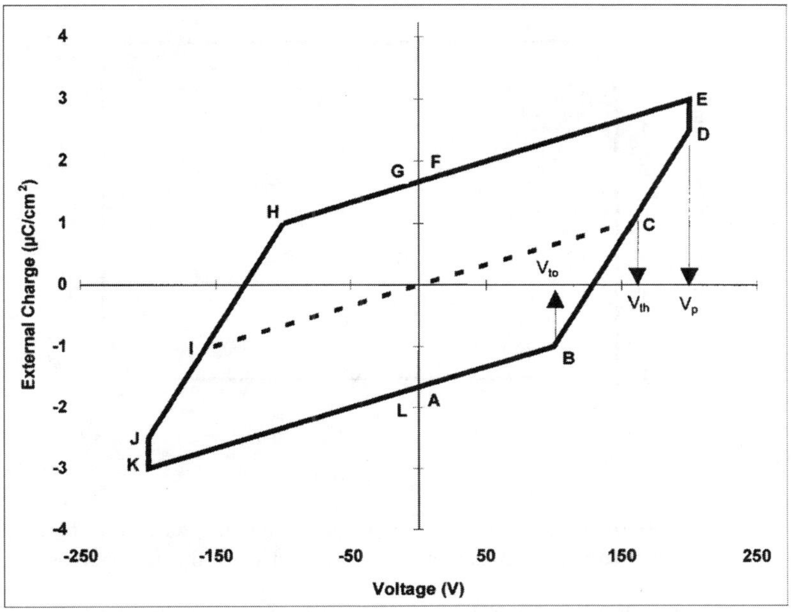

FIGURE 34.5 Idealized *QV* loop with no charge leakage from interface states between drive pulses. Letters A–L mark points on the *QV* loop which are coincident in time with the points labeled A–L on the drive waveform in Figure 34.3. The dashed line is the *QV* loop for the case just below threshold. The solid line is the open loop above threshold. The area of the *QV* loop represents the energy dissipated by the device per cycle of the drive waveform.

the threshold voltage of the device, the *QV* loop is simply a straight line with slope equal to the total capacitance of the insulator/phosphor/insulator stack. If the amplitude of the applied voltage pulses is greater than the threshold voltage of the device, the *QV* loop opens up. *QV* loops below and above threshold are shown in Figure 34.5. Above threshold, power is dissipated in the ACTFEL device. The area encompassed by the *QV* loop is equal to the energy delivered to the device per period of the drive waveform.

The *QV* loop is measured directly. A theoretical extension of the *QV* loop that is sometimes used by researchers studying the physics of ACTFEL devices is a plot of the actual charge flow across the phosphor layer, Q_p, vs. the electric field across the phosphor layer, F_p. The quantities required for a plot of Q_pF_p can be calculated from the *QV* data if the thicknesses and dielectric constants of the insulator layers and phosphor layer are known. The actual charge flow across the phosphor is calculated by subtracting the reactive charge from the total charge. The field in the phosphor layer is calculated by adding the externally applied field to the internal **polarization field** due to the actual flow of charges across the phosphor layer. A Q_pF_p loop corresponding to the above threshold *QV* loop of Figure 34.5 is shown in Figure 34.6. The Q_pF_p loop is useful because it expresses the internal electrical characteristics of the phosphor layer during device operation.

Device Excitation

Whether the device under test is a test dot or a matrix display, the drive waveform must be ac coupled. Passively addressed matrix displays are scanned one row at a time. Data are loaded into the column drivers for a single row of pixels and the selected pixels in the row are all turned on simultaneously. The row pulse brings the voltage across each pixel in the row to a level just below threshold. The columns of selected pixels are then pulsed to bring the voltage across selected pixels to a level above threshold and the pixels are turned on. During the following frame the voltage polarities across the pixels are reversed. Each individual pixel is subjected to a drive signal similar to that shown in Figure 34.3. To activate a pixel fully requires the voltage to be held above V_{th} for 10 to 20 μs. Since each row must be scanned in sequence,

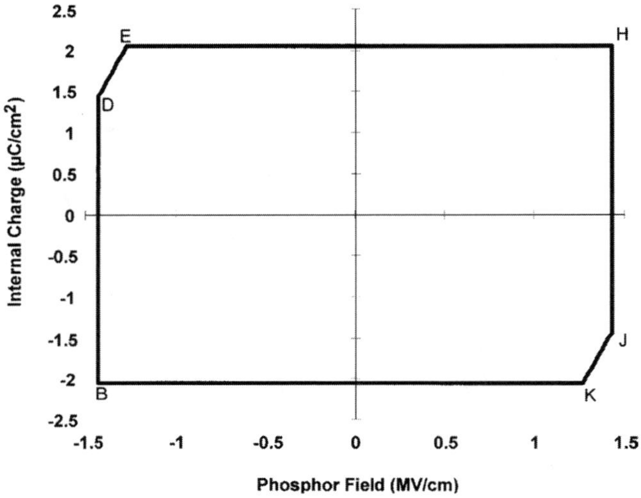

FIGURE 34.6 Q_p F_p loop corresponding to the above threshold *QV* loop in Figure 34.5. This loop represents the charge flow across the phosphor layer as a function of the electric field across the phosphor layer.

a typical display with approximately 500 rows requires at least 5 ms to scan one frame. The maximum frame rate is thus approximately 200 Hz. Displays that are addressed by an active matrix of transistors are not scanned a line at a time as passively addressed displays are, but instead are bulk driven. They have one unpatterned common electrode, usually an upper layer of ITO, to which an ac drive signal is applied. Each pixel is connected to ground through a transistor which drops a portion of the applied drive voltage when the pixel is not selected. In this type of display the drive waveform can be any ac signal of the appropriate voltage. Sine, trapezoid, and triangle waveforms have been used. An active matrix display can be driven with a frame rate higher than a passive matrix-addressed display by a factor equal to the number of rows in the display, since row-at-a-time scanning is not required. Since a light pulse is emitted for each voltage pulse, the average luminance is proportional to the frame rate, resulting in much higher luminance capability for active matrix-addressed displays.

34.5 Standard Measurements

Measurable quantities of interest include luminance, luminous efficiency, emission spectrum, latent image, and defects. The luminance of an ACTFEL display is a function of the peak voltage and frequency of the drive waveform, the intrinsic efficiency of the insulator/phosphor/insulator stack, and the operating history (aging). The efficiency is a function of the drive waveform shape and frequency as well as various device parameters. The emission spectrum is primarily determined by the phosphor host material and activators, although it is also affected by deposition and anneal conditions and in some cases by the drive voltage and frequency. **Latent image** is the burning in of a permanent image of a fixed pattern which is displayed for long periods. It can appear as a faint dark image superimposed on a bright background or as a faint bright image superimposed on a dark background. Formation of latent image is affected primarily by drive waveform symmetry and device processing parameters. Display defects include pixel, line, and mura defects.

34.6 Time-Resolved Measurements

Measurements of luminance, efficiency, and emission spectra as introduced so far involve time-averaged light emission. There are two fairly common time-resolved measurements of ACTFEL light emission: light emission decay time and time-resolved light emission spectroscopy. The **light emission decay time,** τ, is the time it takes for the light emission from one excitation pulse to fall to $1/e$ times its initial value.

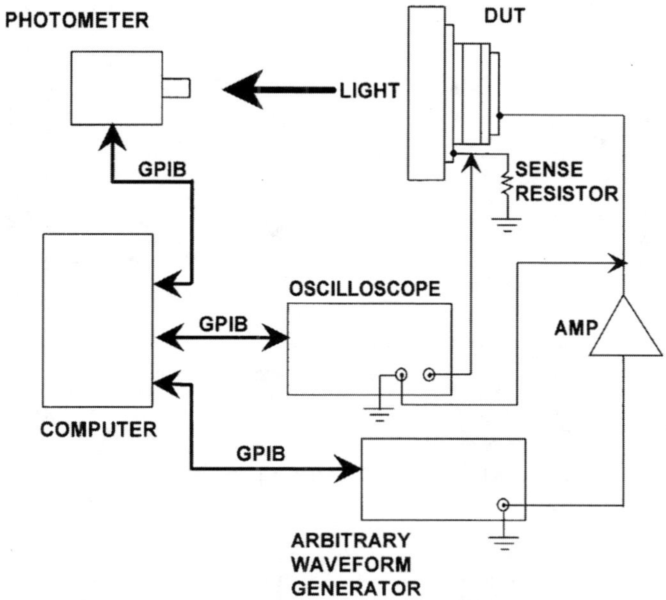

FIGURE 34.7 Schematic diagram of a system for measuring luminance and efficiency as functions of the peak drive voltage. The voltage waveform, $V(t)$, is measured by the oscilloscope at the Al electrode of the device under test. The current waveform, $I(t)$, is measured by the oscilloscope as the voltage across the sense resistor divided by its resistance. The luminance is measured by the photometer.

The measurement of τ is started just after the trailing edge of the excitation pulse. This is necessary in order that the measured value of τ not be affected by the continuing excitation of additional emission centers, so that it represents the intrinsic relaxation time of the emission center. In devices with evaporated ZnS:Mn phosphor, τ is a strong function of the Mn concentration. It can thus be used as an analytical technique to determine the Mn concentration. Time-resolved spectroscopy is the measurement of the emission spectrum occurring during specific portions of the excitation and emission process. An example of the application of this technique is the study of the separate light pulses emitted during the leading and trailing edges of the excitation pulse with the blue-emitting phosphor SrS:Ce. Both of these techniques are frequently used to help elucidate the excitation and emission mechanisms in ACTFEL phosphors.

34.7 Test Dot Characterization

Luminance and Efficiency

A schematic representation of a measurement system for collecting luminance and efficiency vs. voltage data on a test dot is shown in Figure 34.7. An arbitrary waveform generator provides a drive signal with the waveshape, frequency, and peak voltage determined by a control computer. The waveform generator output signal is amplified from a ±5 V range to a ±300 V range. A photometer, also under computer control, measures the luminance, L, of the test dot. An oscilloscope measures the voltage across the test dot and the voltage across the sense resistor. The current is calculated by dividing the voltage across the sense resistor by its resistance. The control computer can thus adjust the peak voltage up and down and collect the luminance data, as well as the waveforms representing the voltage across the test dot and the current through it. The energy dissipated, E_p, during each drive pulse is calculated as

$$E_p = \int_D I(t) \times V(t) \mathrm{d}t \tag{34.1}$$

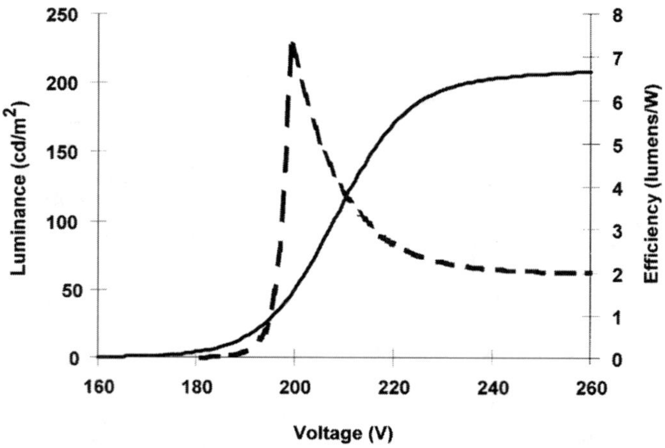

FIGURE 34.8 Plots of luminance and efficiency as functions of peak drive voltage. The solid line is the luminance and the dashed line is the efficiency.

where $I(t)$ and $V(t)$ are the current and voltage waveforms, respectively, and D is the duration of either the positive or negative drive pulse. This integration can be carried out by most oscilloscopes or the $I(t)$ and $V(t)$ data can be transferred to the control computer for the calculation, although this approach is generally slower. The average energy dissipated per period of the drive waveform, E, is the average of E_P for a positive pulse and a negative pulse. The average power dissipated, P, can be calculated by multiplying E by the frequency of the drive waveform. The efficiency, η, is calculated as follows:

$$\eta = \frac{\pi L A}{P} \tag{34.2}$$

where L is the luminance in cd/m^2, A is the area of the test dot in m^2, and P is the average power in watts. Values of L and η are collected for peak voltages ranging from 10 V below threshold to 40 or 50 V above threshold. Plots of L and η vs. peak voltage for a typical device are shown in Figure 34.8.

Charge Flow and Electric Field

The QV loop is measured using a circuit identical to that in Figure 34.7, except that the sense resistor is replaced by a sense capacitor and the photometer is not used. The sense capacitor value is chosen to be much larger than the capacitance of the ACTFEL device so that the voltage dropped by the sense capacitor is small. Since the sense capacitor and the ACTFEL device are in series, the charges stored on them are equal. The charge, Q, on the ACTFEL device is thus

$$Q = C_S V_S \tag{34.3}$$

where C_S is the capacitance of the sense capacitor and V_S is the voltage appearing across it. When $Q(t)$ is plotted vs. $v(t)$, the QV loop results. In the idealized QV loop shown in Figure 34.5, V_{th} (C) is the threshold voltage, V_{to} (B) is the turn on voltage, and V_p (D,E) is the peak voltage of the drive waveform. Threshold voltage is the voltage at which the first knee in the LV curve occurs. If C_T is the total capacitance of the device below threshold, this is also the voltage at which the line $Q = C_T V$ intersects the open, above-threshold QV loop (C). In practice, it is sometimes defined as the voltage at which a certain luminance value occurs at a given frequency, e.g., the voltage at which the luminance is 1 cd/m^2 at 60 Hz. V_{to} is the voltage at which the slope of the QV loop changes from C_T to C_I, where C_I is the capacitance of the insulator layers.

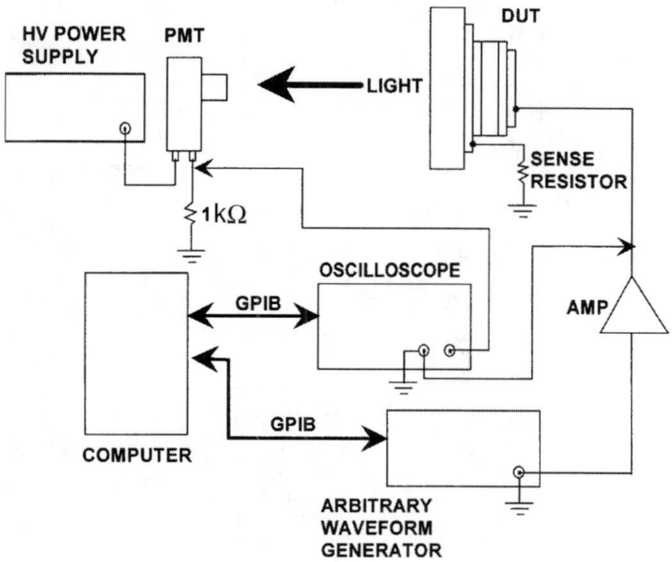

FIGURE 34.9 Apparatus for measuring the luminescent decay time of the ACTFEL phosphor. The light signal is measured by the oscilloscope across the sense resistor on the output of the PMT. The oscilloscope is triggered on the drive pulse.

Since a differential element of energy delivered to the ACTFEL device is $dE = V(t)dQ$, then

$$E = \int V(t)\mathrm{d}Q \tag{34.4}$$

The energy delivered per period of the drive waveform is thus equal to the area encompassed by the *QV* loop. The power dissipated is just the energy per period multiplied by the frequency. In practice, the area of the *QV* loop is measured by numerical integration. The calculation can be carried out on the oscilloscope if it has analysis capabilities or the data can be transferred to the control computer for integration.

Generation of the $Q_P F_P$ loop does not require any electrical measurements other than those required for the *QV* loop. Q_P is the charge separation across the phosphor layer and F_P is the electric field across the phosphor layer. If the thicknesses and dielectric constants of the insulator and phosphor layers are known, Q_P and F_P can be calculated from the values of $Q(t)$ and $V(t)$ on the *QV* loop. This is accomplished by applying the following equations [11,12]:

$$Q_P(t) = \frac{C_i + C_P}{C_i} Q(t) - C_P V(t) \tag{34.5}$$

and

$$F_P(t) = \frac{1}{d_P}\left(\frac{Q(t)}{C_i} - V(t)\right) \tag{34.6}$$

where C_i is the capacitance of the insulators, C_P is the capacitance of the phosphor layer, and d_P is the thickness of the phosphor layer.

Time-Resolved Measurements

The apparatus for measuring τ is shown in Figure 34.9. A photomultiplier tube (PMT) is used to detect the light emission as a function of time. The drive system is set to provide relatively narrow drive pulses,

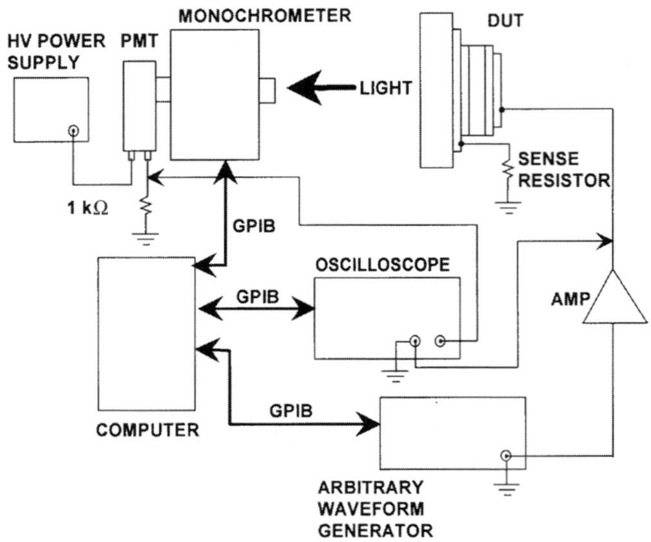

FIGURE 34.10 System for measuring the time-resolved emission spectrum of an ACTFEL device. The oscilloscope is set to integrate the light signal from the PMT during a selected time window. The monochrometer wavelength is scanned over the entire spectral range and the emission intensity data is collected from the oscilloscope.

typically 10 μs pulse width, at relatively low frequency, typically 60 Hz. This approach works well for phosphors with relatively long decay times, on the order of 100 μs to a few milliseconds. This is the case for many common ACTFEL phosphors such as ZnS:Mn and ZnS:Tb. Phosphors such as SrS:Ce, however, have very fast decay times and cannot be measured in this manner. In such cases the photoluminescent decay time must be measured using a pulsed laser to excite the phosphor and appropriately low *RC* response time of the light detection system.

The general setup for measuring time-resolved emission spectra from ACTFEL devices is shown schematically in Figure 34.10. The oscilloscope is triggered on the drive waveform and the signal integration period is set to the region of interest. A boxcar integrator can also be used to integrate the light signal during the desired time window. The monochrometer wavelength is scanned and the emission spectrum is collected for the selected portion of the emission process.

Aging

ACTFEL devices tend to stabilize after a few tens of hours of burn-in, but can exhibit complex aging behavior during the burn-in process. The luminance vs. voltage curves for ZnS:Mn devices in which the phosphor layer is deposited by evaporation, for example, tend to shift to slightly higher voltage during burn-in [2]. Luminance vs. voltage curves for devices in which the ZnS:Mn is deposited by ALE tend to shift to slightly lower voltage [13]. Aging data is collected by measuring luminance vs. voltage at selected time intervals during aging. The measurement is carried out as described earlier for luminance vs. voltage measurements. The aging is done by continuously operating the device at a fixed voltage or at a fixed voltage above threshold. The aging process can be accelerated by operating the device at higher frequency.

34.8 Characterization of Matrix-Addressed Displays

The characterization of matrix-addressed displays differs from characterization of test dots because less control of the drive waveform is readily available. The row and column drivers do not provide great flexibility, although some control can be exercised by varying the composite drive waveform and the control signals to the driver chips. These types of modifications, however, require detailed knowledge of the addressing and control electronics involved and are best left to the original display manufacturer. Measurements that are more accessible and of more general interest involve characterization of the luminance, chromaticity, uniformity, display life, latent image, and defects. For these measurements the display under test is controlled by a computer through a standard video output or a custom video display interface provided with the display. The display operating voltage and frequency are fixed. The display is placed in a dark room or enclosure. Luminance is measured with a photometer. Chromaticity is measured with a spectrophotometer or a photometer equipped with tristimulus filters. Uniformity can be measured by mounting either the display or the photometer/spectrophotometer on a translation system and collecting data at points in a regular array of locations throughout the display surface. Display life is characterized by operating the display in a full-on pattern, cycling through checkerboard patterns, etc., and making luminance measurements at exponentially increasing time intervals. Latent image is formed by displaying small blocks in various fixed locations and various fixed gray levels (if available) continuously for long periods. It is characterized by setting the entire screen to each gray level available and measuring the luminance in the locations which had blocks displayed during aging as well as unaged areas nearby. Latent image is the percent luminance difference between aged blocks and nearby unaged areas. Latent image is typically measured after aging the block pattern for 1000 h. Defects are classified into pixel, line, and mura defects. They are characterized by visual inspection with the aid of an eye loupe or by an automated flat panel inspection system.

34.9 Excitation and Measurement Equipment

Excitation of Test Dots

Test dots can be excited by any signal source that provides bipolar pulses up to a peak voltage of 300 V and sufficient current sourcing and sinking capability to charge and discharge the device capacitance. Bipolar pulse drivers with sufficient voltage and current output can be built with commercially available components. A hybrid circuit op amp, produced by Apex Microtechnology, provides sufficient voltage output and frequency response. A current boosting stage can be added to the output if the DUT capacitance is too large to drive directly. This amplifier approach is very flexible since any waveform that can be generated by the arbitrary waveform generator can be used. There are also some commercial amplifiers available that are effective for driving test dots for some measurements. They tend to have limited bandwidth and must be used with caution in situations in which the measurement is sensitive to the exact shape of the drive pulse. This would be the case, for example, in *QV* measurements and often in efficiency measurements.

Excitation of Matrix-Addressed Displays

Excitation of matrix-addressed displays is straightforward since the drive electronics are integrated with the display. Generally, only a standard ac power outlet and a computer with an appropriate video card are required. Test patterns for measurement can be created using simple computer programs.

34.10 Measurement Instruments

Measurement of Drive Voltage and Current

Digital oscilloscopes are used for all of the electrical measurements described in this chapter. Suitable instruments are available from several manufacturers, including Tektronix and Hewlett-Packard. A bandwidth of 100 MHz is more than sufficient. Waveform analysis capabilities are very helpful but not absolutely necessary. RS 232 or IEEE 488 interfaces are required for computer control and data transfer.

TABLE 34.1 Light Measurement Instruments

Instrument Type	Model	Manufacturer
Photometer	PR880	Photo Research
Photometer/spectroradiometer	PR650	Photo Research
	Pritchard 1980B	Photo Research
	GS-1280 RadOMAcam	Gamma Scientific (EG&G)
Photomultiplier tubes	—	Oriel Corp.
	—	Hamamatsu
Photodiode	PIN 10AP	UDT Sensors, Inc.
Flat panel inspection system	FIS 250	Photon Dynamics, Inc.

TABLE 34.2 Manufacturers of Light Measurement Instruments

Photo Research 9330 DeSoto Avenue, P.O. Box 2192 Chatsworth, CA 91313-2192 (818) 341-5151	Hamamatsu Photonics Systems Corp. 360-T Foothill Road, P.O. Box 6910 Bridgewater, NJ 08807-0910
Gamma Scientific (EG&G) 8581 Aero Dr. San Diego, CA 92123-1876 (619) 279-8034	UDT Sensors, Inc. 12525 Chadron Ave. Hawthorne, CA 90250 (310) 978-0516
Oriel Corp. 252 Long Beach Blvd., P.O. Box 872 Stratford, CT 06497-0872 (203) 380-4200	Photon Dynamics, Inc. 6325 San Ignacio Ave. San Jose, CA 95119 (408) 226-9900

Measurement of Emitted Light

Several types of instruments are used for measuring light emission from ACTFEL displays. Photometers are used for luminance measurements. Spectrophotometers are used for measuring the emission spectrum and with suitable software can also provide luminance measurements. Time-resolved measurements are accomplished by using photomultiplier tubes or photodiode detectors. Table 34.1 lists some examples of photometers, spectroradiometers, photomultipliers, and photodiodes along with the names of the companies that manufacture them. A relatively new development for characterizing the light emission characteristics of flat panel displays, including ACTFEL displays, is the flat panel inspection system. This is a large measurement system comprising a CCD camera detector, light-tight enclosure, control computer, image processor, and specialized software. This type of system images an entire flat panel display on the CCD camera and measures luminance, chromaticity, and various defects by analyzing the image. These systems are intended for high throughput manufacturing environments and cost several hundred thousand dollars. An example of this type of system is also included in Table 34.1. Contact information for the companies listed in Table 34.1 is provided in Table 34.2.

Defining of Terms

Electroluminescence: The nonthermal generation of light resulting from the application of an electric field to a substance, usually a luminescent crystal.

Latent image: The ghost image of a previously displayed pattern which can sometimes be seen in a full field on an electronic display screen.

Light emission decay time: The time it takes for the light emission from one excitation pulse to fall to $1/e$ times its initial value.

Polarization charge: The charge trapped at the phosphor/insulator interface following the application of a drive pulse.

Polarization field: The field across the phosphor layer resulting from the polarization charge.

Q_pF_p loop: The closed curve which results from plotting the internal charge flow across the phosphor layer (Q_p) vs. the electric field across the phosphor layer (F_p).

***QV* loop:** The closed curve which results from plotting the external charge (Q) flowing into a TFEL device vs. the externally applied voltage (V).

Threshold voltage: The voltage amplitude of the drive waveform above which current flows across the phosphor layer and light is emitted from a TFEL device.

Turn-on voltage: The voltage corresponding to the first knee in the *QV* loop of a TFEL device. This is the voltage at which charge begins to flow across the phosphor layer. This voltage is generally less than the threshold voltage because the internal field across the phosphor layer is enhanced by the polarization field once the polarization charge has built up in the steady state.

References

1. Y.A. Ono, *Electroluminescent Displays,* Singapore: World Scientific, 1995.
2. T. Inoguchi, M. Takeda, Y. Kakihara, Y. Nakata, and M. Yoshida, "Stable high-brightness thin-film electroluminescent panels," *Dig. 1974 SID International Symposium,* 84, 1974.
3. C.N. King, R.E. Coovert, and W.A. Barrow, "Full-color 320 × 240 TFEL display panel," *Eurodisplay '87,* 14, 1987.
4. W.A. Barrow, R.E. Coovert, C.N. King, and M.J. Ziuchkovski, "Matrix-addressed full-color TFEL display," *Dig. 1988 SID International Symposium,* 284, 1988.
5. R. Tueta and M. Braguier, "Fabrication and characterization of indium tin oxide thin films for electroluminescent applications," *Thin Solid Films,* 80: 143, 1981.
6. J.D. Davidson, J.F. Wager, and I. Khormaei, "Electrical characterization and SPICE modeling of ZnS:Mn ACTFEL devices," *Dig. 1991 SID International Symposium,* 77, 1991.
7. A.A. Douglas and J.F. Wager, "ACTFEL device response to systematically varied pulse waveforms," in *Electroluminescence — Proceedings of the Sixth International Workshop on Electroluminescence,* V.P. Singh and J.C. McClure, Eds., El Paso, TX: Cinco Puntos Press, 1992.
8. D.H. Smith, "Modeling AC thin-film electroluminescent devices," *J. Lumin.*, 23: 209, 1981.
9. P.M. Alt, "Thin-film electroluminescent displays: device characteristics and performance," *Proc. SID,* 25: 123, 1984.
10. Y.A. Ono, H. Kawakami, M. Fuyama, and K. Onisawa, "Transferred charge in the active layer and EL device characteristics of TFEL cells," *Jpn. J. Appl. Phys.,* 26: 1482, 1987.
11. E. Bringuier, "Charge transfer in ZnS-type electroluminescence," *J. Appl. Phys.,* 66: 1314, 1989.
12. A. Abu-Dayah, S. Kobayashi, and J. F. Wager, "Internal charge-phosphor field characteristics of alternating-current thin-film electroluminescent devices," *Appl. Phys. Lett.*, 62: 744, 1993.
13. A. Mikami, K. Terada, K. Okibayashi, K. Tanaka, M. Yoshida, and S. Nakajima, "Aging characteristics of ZnS:Mn electroluminescent films grown by a chemical vapor deposition technique," *J. Appl. Phys.,* 72: 773, 1992.

35

Light-Emitting Diode Displays

Mohammad A. Karim
City College of New York

A light-emitting diode (LED) is a particular solid-state *p–n* junction diode that gives out light upon the application of a bias voltage. The luminescence process in this case is electroluminescence, which is associated with emission wavelengths in the visible and infrared regions of the spectrum. When a forward bias is applied to the *p–n* junction diode, carriers are injected into the depletion region in large numbers. Because of their physical proximity, the electron–hole pairs undergo a recombination that is associated with the emission of energy. Depending on the semiconductor band-gap characteristics, this emitted energy can be in the form of heat (as phonons) or light (as photons).

The solution of the Schrödinger equation for a typical crystal reveals the existence of Brillouin zones. A plot between the energy E of an electron in a solid and its wave vector $\mathbf{k}$ represents the allowed energy bands. It may be noted that the lattice structure affects the motion of an electron when k is close to $n\pi/l$ (where n is any integer and l is the crystal periodicity) and the effect of this constraint is to introduce an energy band gap between the allowed energy bands. Figure 35.1a shows portions of two E vs. k curves for neighboring energy bands within the regions $k = \pi/l$ and $k = -\pi/l$ (also known as the reduced zone).

While the upper band of Figure 35.1 represents the energy of conduction band electrons, the curvature of the lower band can be associated with electrons having negative effective mass. The concept of negative effective mass can readily be identified with the concept of holes in the valence band. While the majority of the electrons are identified with the minima of the upper E–k curve, the majority of the holes are identified with the maxima of the lower E–k curve. The minimum value of the conduction band and the maximum value of the valence band in Figure 35.1a both have identical k values. A semiconductor having such a characteristic is said to have a direct band gap, and the associated recombination in such a semiconductor is referred to as direct.

The *direct recombination* of an electron–hole pair always results in the emission of a photon. In a direct band-gap semiconductor, the emitted photon is not associated with any change in momentum (given by $hk/2\pi$) since $\Delta k = 0$. However, for some semiconducting materials, the E vs. k curve may be somewhat different, as shown in Figure 35.1b. While the minimum conduction band energy can have a nonzero k, the maximum valence band energy can have $k = 0$. The electron–hole recombination in such a semiconductor is referred to as indirect.

An *indirect recombination* process involves a momentum adjustment. Most of the emission energy is thus expended in the form of heat (as phonons). Very little energy is left for the purpose of photon emission, which in most cases is a very slow process. Furthermore, since both photons and phonons are involved in this energy exchange, such transitions are less likely to occur. The interband recombination rate is basically given by

0-8493-1733-9/04/$0.00+$1.50

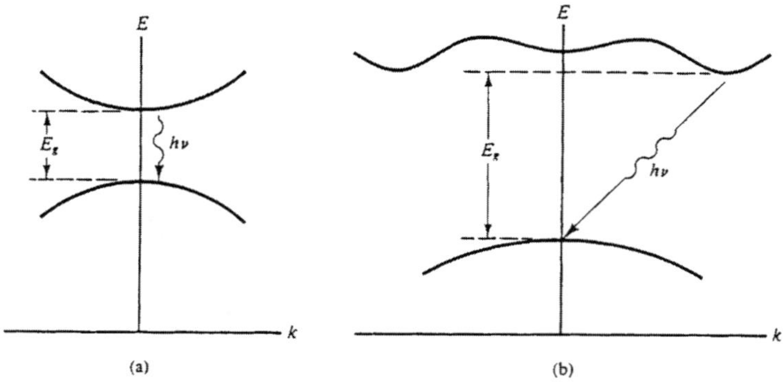

FIGURE 35.1 *E* vs. *k* for semiconductors having (a) a direct band gap and (b) an indirect band gap.

$$\mathrm{d}n/\mathrm{d}t = B_r np \tag{35.1}$$

where B_r is a recombination-dependent constant which for a direct band-gap semiconductor is ~10^6 times larger than that for an indirect band-gap semiconductor. For direct recombination, B_r value ranges from 0.46×10^{-10} to 7.2×10^{-10} cm^3/s.

All semiconductor crystal lattices are alike, being dissimilar only in terms of their band characteristics. Si and Ge both have indirect band transitions, whereas GaAs, for example, is a semiconductor that has a direct band transition. Thus, while Si and Ge are preferred for fabrication of transistors and integrated circuits, GaAs is preferred for the fabrication of LEDs.

The direct recombination (when k = constant) results in a photon emission whose wavelength (in micrometers) is given by

$$\lambda = hc/E_g = 1.24/E_g(\mathrm{eV}) \tag{35.2}$$

where E_g is the band-gap energy. The LEDs under proper forward-biased conditions can operate in the ultraviolet, visible, and infrared regions. For the visible region, however, the spectral luminous efficiency curves of Figure 35.2, which account for the fact that the visual response to any emission is a function of wavelength, should be of concern. It is unfortunate that there is not a single-element semiconductor suitable for fabrication of LEDs, but there are many binary and ternary compounds that can be used for fabrication of LEDs. Table 35.1 lists some of these binary semiconductor materials. The ternary semiconductors include GaAlAs, $CdGeP_2$, and $ZnGeP_2$ for infrared region operation, $CuGaS_2$ and $AgInS_2$ for visible region operation, and $CuAlS_2$ for ultraviolet region operation. Ternary semiconductors are used because their energy gaps can be tuned to a desired emission wavelength by picking appropriate composition.

Of the ternary compounds, gallium arsenide–phosphide (written as $GaAs_{1-x}P_x$) is an example that is basically a combination of two binary semiconductors, namely, GaAs and GaP. The corresponding bandgap energy of the semiconductor can be varied by changing the value of x. For example, when $x = 0$, $E_g = 1.43$ eV. E_g increases with increasing x until $x = 0.44$ and $E_g = 1.977$ eV, as shown in Figure 35.3. However for $x \geq 0.45$, the band gap is indirect. The most common composition of $GaAs_{1-x}P_x$ used in LEDs has $x = 0.4$ and E_g. 1.3 eV. This band-gap energy corresponds to an emission of red light. Calculators and watches often use this particular composition of $GaAs_{1-x}P_x$.

Interestingly, the indirect band gap of $GaAs_{1-x}P_x$ (with $1 \geq x \geq 0.45$) can be used to output light ranging from yellow through green provided the semiconductor is doped with impurities such as nitrogen. The dopants introduced in the semiconductor replace phosphorus atoms which, in turn, introduce electron trap levels very near the conduction band. For example, if $x = 0.5$, the doping of nitrogen increases the LED efficiency from 0.01 to 1%, as shown in Figure 35.4. It must be noted, however, that nitrogen doping

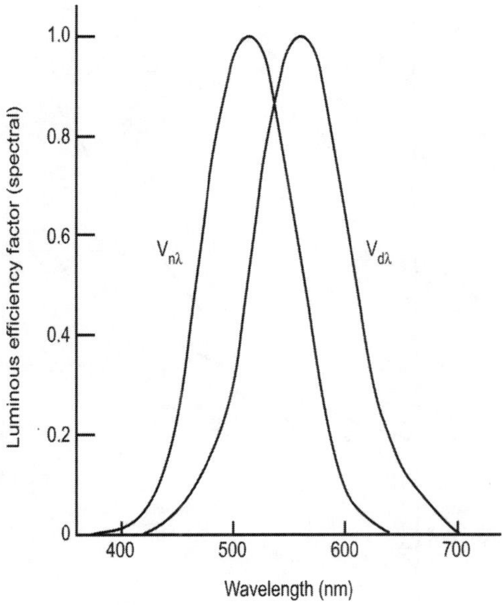

FIGURE 35.2 Spectral luminous efficiency curves. The photopic curve $V_{d\lambda}$ corresponds to the daylight-adapted case while the scotopic curve $V_{n\lambda}$ corresponds to the night-adapted case. (From Boyd, R.W., *Radiometry and the Detection of Optical Radiation,* John Wiley & Sons, New York, 1983. With permission.)

TABLE 35.1 Binary Semiconductors Suitable for LED Fabrication

	Material	E_g (eV)	Emission Type
III–V	GaN	3.5	UV
II–VI	ZnS	3.8	UV
II–VI	SnO_2	3.5	UV
II–VI	ZnO	3.2	UV
III–VII	CuCl	3.1	UV
II–VI	BeTe	2.8	UV
III–VII	CuBr	2.9	UV — visible
II–VI	ZnSe	2.7	Visible
III–VI	In_2O_3	2.7	Visible
II–VI	CdS	2.52	Visible
II–VI	ZnTe	2.3	Visible
III–V	GaAs	1.45	IR
II–VI	CdSe	1.75	IR — Visible
II–VI	CdTe	1.5	IR
III–VI	GaSe	2.1	Visible

shifts the peak emission wavelength toward the red. The shift is comparatively larger at and around $x =$ 0.05 than $x = 1.0$. The energy emission in nitrogen-doped $GaAs_{1-x}P_x$ devices is a function of both x and the nitrogen concentration.

Nitrogen is a different type of impurity from those commonly encountered in extrinsic semiconductors. Nitrogen, like arsenic and phosphorus, has five valence electrons, but it introduces no net charge carriers in the lattice. It provides active radiative recombination centers in the indirect band-gap materials. For an electron, a recombination center is an empty state in the band gap into which an electron falls and, thereafter, falls into the valence band by recombining with a hole. For example, while a GaP LED emits green light (2.23 eV), a nitrogen-doped GaP LED emits yellowish green light (2.19 eV), and a heavily nitrogen-doped GaP LED emits yellow light (2.1 eV).

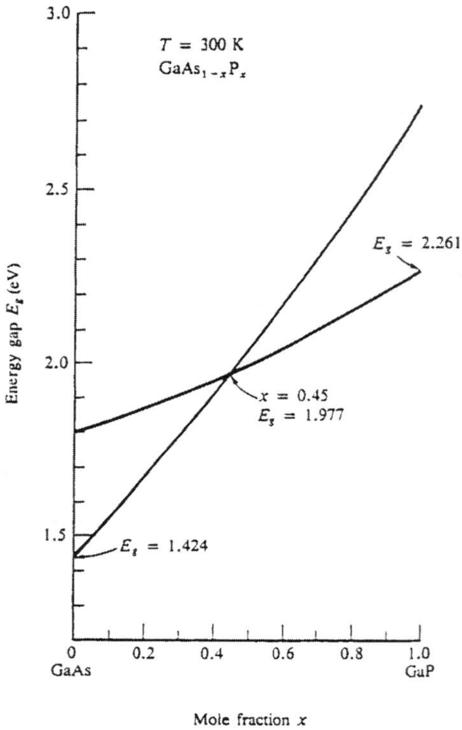

FIGURE 35.3 Band-gap energy vs. x in $GaAs_{1-x}P_x$. (From Casey, H.J., Jr. and Parish, M.B., Eds., *Heterostructure Lasers*, Academic Press, New York, 1978. With permission.)

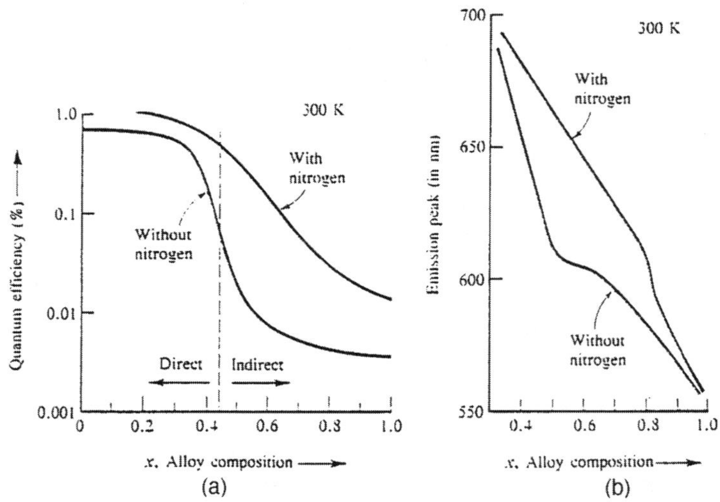

FIGURE 35.4 The effects of nitrogen doping in $GaAs_{1-x}P_x$: (a) quantum efficiency vs. x and (b) peak emission wavelength vs. x.

The injected excess carriers in a semiconductor may recombine either radiatively or nonradiatively. Whereas nonradiative recombination generates phonons, radiative recombination produces photons. Consequently, the internal quantum efficiency η, defined as the ratio of the radiative recombination rate R_r to the total recombination rate, is given by

$$\eta = R_r/(R_r + R_{nr}) \tag{35.3}$$

where R_{nr} is the nonradiative recombination rate. However, the injected excess carrier densities return to their value exponentially as

$$\Delta p = \Delta n = \Delta n_0 e^{-t/\tau} \tag{35.4}$$

where τ is the carrier lifetime and Δn_0 is the excess electron density at equilibrium. Since $\Delta n/R_r$ and $\Delta n/R_{nr}$ are, respectively, equivalent to the radiative recombination lifetime τ_r and the nonradiative recombination lifetime τ_{nr}, we can obtain the effective minority carrier bulk recombination time τ as

$$(1/\tau) = (1/\tau_r) + (1/\tau_{nr}) \tag{35.5}$$

such that $\eta = \tau/\tau_r$. The reason that a fast recombination time is crucial is that the longer the carrier remains in an excited state, the larger the probability that it will give out energy nonradiatively. In order for the internal quantum efficiency to be high, the radiative lifetime τ_r needs to be small. For indirect band-gap semiconductors, $\tau_r \gg \tau_{nr}$ so that very little light is generated, and for direct band-gap semiconductors, τ_r increases with temperature so that the internal quantum efficiency deteriorates with the temperature.

As long as the LEDs are used as display devices, it is not too important to have fast response characteristics. However, LEDs are also used for the purpose of optical communications, and for those applications it is appropriate to study their time response characteristics. For example, an LED can be used in conjunction with a photodetector for transmitting optical information between two points. The LED light output can be modulated to convey optical information by varying the diode current. Most often, the transmission of optical signals is facilitated by introducing an optical fiber between the LED and the photodetector.

There can be two different types of capacitances in diodes that can influence the behavior of the minority carriers. One of these is the *junction capacitance*, which is caused by the variation of majority charge in the depletion layer. While it is inversely proportional to the square root of bias voltage in the case of an abrupt junction, it is inversely proportional to the cube root of bias voltage in the case of a linearly graded junction. The second type of capacitance, known as the *diffusion capacitance*, is caused by the minority carriers.

Consider an LED that is forward biased with a dc voltage. Consider further that the bias is perturbed by a small sinusoidal signal. When the bias is withdrawn or reduced, charge begins to diffuse from the junction as a result of recombination until an equilibrium condition is achieved. Consequently, as a response to the signal voltage, the minority carrier distribution contributes to a signal current.

Consider a one-dimensional *p*-type semiconducting material of cross-sectional area *A* whose excess minority carrier density is given by

$$\delta \Delta n_p/\delta t = D_n \delta^2 \Delta n_p/\delta x^2 - \Delta n_p/\tau \tag{35.6}$$

As a direct consequence of the applied sinusoidal signal, the excess electron distribution fluctuates about its dc value. In fact, we may assume excess minority carrier density to have a time-varying component as described by

$$\Delta n_p(x,t) = \langle \Delta n_p(x) \rangle + n_p'(x) e^{j\omega t} \tag{35.7}$$

where $\langle \Delta n_p(x) \rangle$ is a time-invariant quantity. By introducing Equation 35.7 into Equation 35.6, we get two separate differential equations:

$$\delta^2/\delta x^2 \left(\langle \Delta n_p(x) \rangle \right) = \langle \Delta n_p(x) \rangle / (L_n)^2 \tag{35.8a}$$

and

$$\delta^2/\delta x^2\left[\Delta n_p'(x)\right] = \Delta n_p'(x)/\left[L_n^*\right]^2 \tag{35.8b}$$

where

$$L_n^* = L_n/(1 + j\omega\tau)^{1/2} \tag{35.9a}$$

and

$$L_n = (D_n\tau)^{1/2} \tag{35.9b}$$

The dc solution of Equation 35.8a is well known. Again, the form of Equation 35.8b is similar to that of Equation 35.8a and, therefore, its solution is given by

$$\Delta n_p'(x) = \Delta n_p'(0)e^{-x/L} \tag{35.10}$$

Since the frequency-dependent current $I(\omega)$ is simply a product of eAD_n and the concentration gradient, we find that

$$\begin{aligned} I(\omega) &= \left|eAD_n \mathrm{d}n_p'(x)/\mathrm{d}x\right|_{x=0} \\ &= I(0)/\left(1 + \omega^2\tau^2\right)^{1/2} \end{aligned} \tag{35.11}$$

where $I(0)$ is the intensity emitted at zero modulation frequency. We can determine the admittance next by dividing the current by the perturbing voltage. The real part of the admittance, in this case, will be equivalent to the diode conductance, whereas its imaginary part will correspond to the diffusion capacitive susceptance.

The modulation response as given by Equation 35.11 is, however, limited by the carrier recombination time. Often an LED is characterized by its modulation bandwidth, which is defined as the frequency band over which signal power (proportional to $I^2(\omega)$) is half of that at $\omega = 0$. Using Equation 35.11, the 3-dB modulation bandwidth is given by

$$\Delta\omega \approx 1/\tau_r \tag{35.12}$$

where the bulk lifetime has been approximated by the radiative lifetime. Sometimes the 3-dB bandwidth of the LED is given by $I(\omega) = ½I(0)$, but this simplification contributes to an erroneous increase in the bandwidth by a factor of 1.732.

Under conditions of thermal equilibrium, the recombination rate is proportional to the product of initial carrier concentrations, n_0 and p_0. Then, under nonequilibrium conditions, additional carriers $\Delta n = \Delta p$ are injected into the material. Consequently, the recombination rate of injected excess carrier densities is given by initial carrier concentrations and injected carrier densities as

$$\begin{aligned} R_{\Delta r} &= \left[B_r(n_o + \Delta n)(p_o + \Delta p) - B_r n_o p_o\right] \\ &= B_r(n_o + p_o + \Delta n)\Delta n \end{aligned} \tag{35.13}$$

where B_r is the same constant introduced in Equation 35.1. For p-type GaAs, for example, $B_r = 1.7 \times 10^{-10}$ cm^3/s when $p_0 = 2.4 \times 10^{18}$ holes/cm^3. Equation 35.13 is used to define the radiative carrier recombination lifetime by

$$\tau_r = \Delta n / R_{\Delta r} = \left[B_r\left(n_o + p_o + \Delta n\right)\right]^{-1} \quad (35.14)$$

In the steady-state condition, the excess carrier density can be calculated in terms of the active region width *d* by

$$\Delta n = J\tau_r / ed \quad (35.15)$$

where *J* is the injection current density.

The radiative recombination lifetime is found by solving Equation 35.14 after having eliminated Δn from it using Equation 35.15:

$$\tau_r = \left[\left\{\left(n_o + p_o\right)^2 + \left(4J/B_r ed\right)\right\}^{1/2} - \left(n_o + p_o\right)\right] \Big/ \left(2J/ed\right) \quad (35.16)$$

Thus, while for the low carrier injection (i.e., $n_o + p_o \gg \Delta n$), Equation 35.16 reduces to

$$\tau_r \approx \left[B_r\left(n_o + p_o\right)\right]^{1/2} \quad (35.17a)$$

for the high carrier injection (i.e., $n_o + p_o \ll \Delta n$), it reduces to

$$\tau_r \approx \left(ed/JB_r\right)^{1/2} \quad (35.17b)$$

Equation 35.17a indicates that in highly doped semiconductors, τ_r is small. But the doping process has its own problem, since in many of the binary LED compounds higher doping may introduce nonradiative traps just below the conduction band, thus nullifying Equation 35.12. In comparison to Equation 35.17a, Equation 35.17b provides a better alternative whereby τ_r can be reduced by decreasing the active region width or by increasing the current density. For the case of *p*-type GaAs, the radiative lifetimes vary between 2.6 and 0.35 ns, respectively, when p_0 varies between 1.0×10^{18} holes/cm^3 and 1.5×10^{19} holes/cm^3.

Usually, LEDs are operated at low current ($\gg$10 mA) and low voltages ($\gg$1.5 V), and they can be switched on and off in the order of 10 ns. In addition, because of their small sizes, they can be reasonably treated as point sources. It is, therefore, not surprising that they are highly preferred over other light sources for applications in fiber-optic data links.

Two particular LED designs are popular: *surface emitters* and *edge emitters*. They are shown in Figure 35.5. In the former, the direction of major emission is normal to the plane of the active region, whereas in the latter the direction of major emission is in the plane of the active region. The emission pattern of the surface emitters is very much isotropic, whereas that of the edge emitters is highly directional.

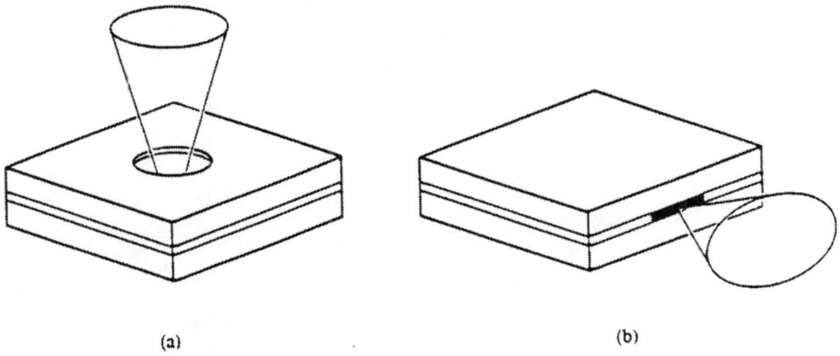

FIGURE 35.5 LED type: (a) surface emitter and (b) edge emitter.

As the LED light originating from a medium of refractive index n_1 goes to another medium of refractive index $n_2(n_2 < n_1)$, only a portion of incident light is transmitted. In particular, the portion of the emitted light corresponds to only that which originates from within a cone of semiapex angle θ_c, such that

$$\theta_c = \sin^{-1}(n_2/n_1) \tag{35.18}$$

In the case of an LED, n_1 corresponds to the refractive index of the LED medium and n_2 corresponds to that of air (or vacuum). Light originating from *beyond* angle θ_c undergoes a total internal reflection. However, the light directed from *within* the cone of the semiapex angle θ_c will be subjected to Fresnels loss. Thus, the overall transmittance T is given by

$$T = 1 - \{(n_1 - n_2)/(n_1 + n_2)\}^2 \tag{35.19}$$

Accordingly, the total electrical-to-optical conversion efficiency in LEDs is given by

$$\begin{aligned}
\eta_{\text{LED}} &= T[(\text{solid angle within the cone})/(4\pi)] \\
&= (T/2)(1 - \cos\theta_c) \\
&= (T/4)\sin^2\theta_c \\
&= (1/4)(n_2/n_1)^2\left[1 - \{(n_1 - n_2)/(n_1 + n_2)\}^2\right]
\end{aligned} \tag{35.20}$$

Only two schemes increase the electrical-to-optical conversion efficiency in an LED. The first technique involves guaranteeing that most of the incident rays strike the glass-to-air interface at angles less than θ_c. It is accomplished by making the semiconductor–air interface hemispherical. The second method involves schemes whereby the LED is encapsulated in an almost transparent medium of high refractive index. The latter means is comparatively less expensive. If a glass of refractive index 1.5 is used for encapsulation, the LED efficiency can be increased by a factor of 3. Two of the possible encapsulation arrangements and the corresponding radiation patterns are illustrated in Figure 35.6.

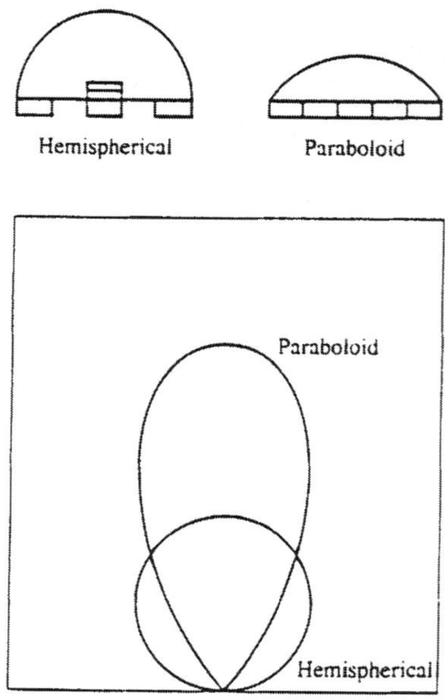

FIGURE 35.6 LED encapsulation geometries and their radiation patterns.

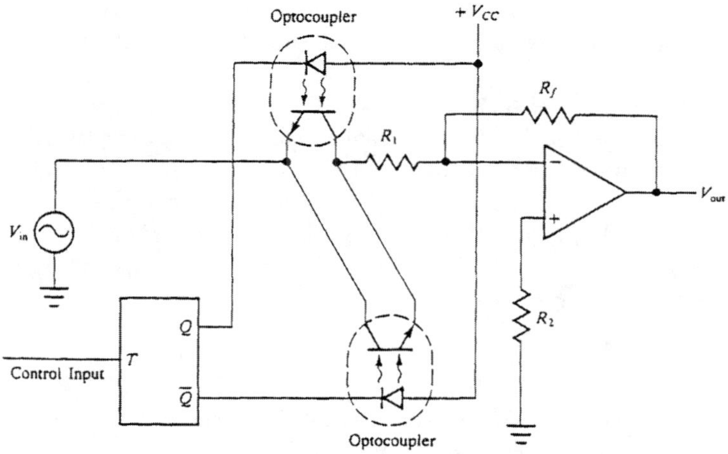

FIGURE 35.7 A chopping circuit with an amplifier.

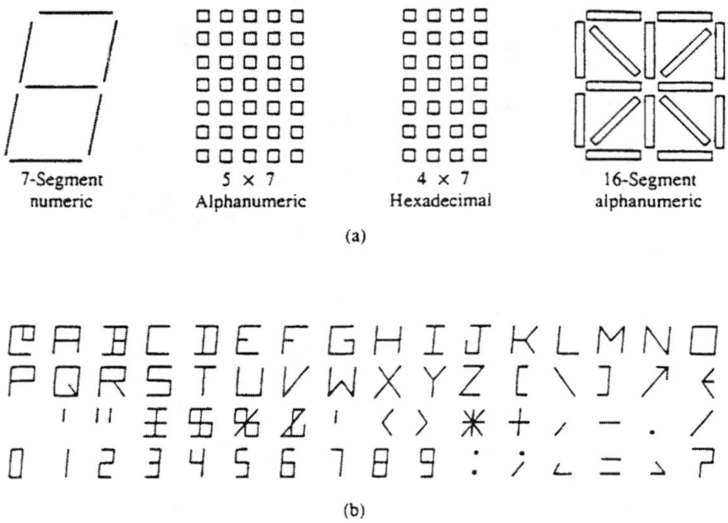

FIGURE 35.8 (a) LED display formats; and (b) displayed alphanumeric characters using 16-segment displays.

LEDs are often used in conjunction with a phototransistor to function as an optocoupler. The optocouplers are used in circumstances when it is desirable to have a transmission of signals between electrically isolated circuits. They are used to achieve noise separation by eliminating the necessity of having a common ground between the two systems. Depending on the type of coupling material, these miniature devices can provide both noise isolation as well as high voltage isolation. Figure 35.7 shows a typical case where two optocouplers are used to attain a chopper circuit. The two optocouplers chop either the positive or the negative portion of the input signals with a frequency of one half that of the control signal that is introduced at the *T* flip-flop. The operational amplifier provides an amplified version of the chopped output waveform. In comparison, a chopper circuit that uses simple bipolar transistors produces noise spikes in the output because of its inherent capacitive coupling.

The visible LEDs are best known for their uses in displays and indicator lamps. In applications where more than a single source of light is required, an LED array can be utilized. An LED array is a device consisting of a row of discrete LEDs connected together within or without a common reflector cavity. Figure 35.8a shows different LED arrangements for displaying hexadecimal numeric and alphanumeric characters, whereas Figure 35.8b shows, for example, the possible alphanumeric characters using 16-segment displays. In digital systems, the binary codes equivalent to these characters are usually decoded and, consequently, a specific combination of LED segments are turned on to display the desired alphanumeric character.

The dot matrix display provides the most desirable display font. It gives more flexibility in shaping characters and has a lower probability of being misinterpreted in case of a display failure. However, these displays involve a large number of wires and increased circuit complexity. LED displays, in general, have an excellent viewing angle, high resonance speed ($\gg$10 ns), long life, and superior interface capability with electronics with almost no duty cycle limitation. LEDs with blue emission are not available commercially. When compared with passive displays, LED displays consume more power and involve complicated wiring with at least one wire per display element.

References

M.A. Karim, *Electro-Optical Devices and Systems,* Boston: PWS-Kent Publishing, 1990.

M.A. Karim (Ed.), *Electro-Optical Displays,* New York: Marcel Dekker, 1992.

L.E. Tannas, Jr. (Ed.), *Flat-Panel Displays and CRTs,* New York: Van Nostrand Reinhold, 1985.

T. Uchida, Multicolored Liquid Crystal Displays, *Opt. Eng.,* 23, 247–252, 1984.

J. Wilson and J.F.B. Hawkes, *Optoelectronics: An Introduction,* Englewood Cliffs, NJ: Prentice-Hall International, 1985.

36

Reading/Recording Devices

Herman Vermariën
Vrije Universiteit Brussel

Edward McConnell
National Instruments

Yufeng Li
Samsung Information Systems America

36.1 Graphic Recorders

Herman Vermariën

A **graphic recorder** is essentially a measuring apparatus that is able to produce in real time a hard copy of a set of time functions with the purpose of immediate and/or later visual inspection. The curves are mostly drawn on a (long) strip of paper (from a roll or Z-fold); as such, the instrument is indicated as a strip chart recorder. The independent variable time (t) then corresponds to the strip length axis and the physical variables measured are related to the chart width. Tracings are obtained by a writing process at sites on the chart short axis (y) corresponding to the physical variables magnitudes with the strip being moved at constant velocity to generate the time axis. Graphs cannot be interpreted if essential information is absent; scales and reference levels for each physical variable recorded and for time are all necessities. Additional information concerning the experimental conditions of the recording is also necessary and is preferably printed by the apparatus (data, investigated item, type of experiment, etc.). The capacity of the graphic recorder is thus determined by its measuring accuracy, its ability to report additional information and its graphical quality, including the sharpness of tracings, the discriminability of tracings (e.g., by different colors), and the stability of quality with respect to long-term storage. Simple chart recorders only produce tracings on calibrated paper; more-advanced graphic recorders generate tracings and calibration lines, display additional information in alphanumeric form on charts, store instrument settings and recorded data in memory (which can be reproduced on charts in diverse modes), have a built-in waveform monitor screen, and can communicate with a PC via standard serial interfacing. The borderlines between these types of intelligent graphic recorders and, on the one hand, digital storage oscilloscopes equipped with a hard copy unit and, on the other hand, PC-based data acquisition systems

0-8493-1733-9/04/$0.00+$1.50

with a laser printer or a plotter, become very unclear. The property of producing the hard copy in real time is probably the most discriminating factor between the graphic recorder and other measuring systems that produce hard copies. Graphic recorders are used for test and measurement applications in laboratory and field conditions and for industrial process monitoring and control. Graphic recorders are intensively used in biomedical measurement applications [1].

Whereas the time axis is generated by moving the chart at constant velocity, the ordinate can be marked in an analog or a digital manner. Analog recorder transducers generate a physical displacement of the writing device, e.g., a pen or a printhead. With digital transducers, moving parts are absent and the writing device is a stationary rectilinear array of equidistant writing points covering the complete width of the chart; the writing act then consists in activating the point situated at the site corresponding to the signal magnitude and putting a dot on the paper. Analog recorders thus can produce continuous lines, whereas digital recorders generate dotted lines. If ordinate and time axis resolutions are sufficient, digital recordings have excellent graphic quality regarding visual impression of continuity. Analog transducers can be used in a discontinuous mode and thus handle a set of slowly varying signals; during the scanning cycle a dot is set on the paper by the moving writing device at the sites corresponding to the magnitudes of the signal. A single digital transducer and a single analog transducer applied in the scanning mode can handle a set of signals; a single analog transducer can process only one signal in the continuous mode. For a digital transducer the number of signals recorded is essentially unlimited; it is thus programmed to draw the necessary calibration lines. With analog transducers calibrated paper is used. In this case, generally ink writing is applied and different colors ensure excellent tracing identifiability. With the digital array, dot printing can be more or less intensified for different channels or more than one adjacent points can be activated, resulting in more or less increased line blackness and thickness. However, tracing identification is usually performed by alphanumeric annotations.

In **analog graphic recorders,** the transducer can be designed in a direct mode or in a servo mode. In the direct mode the signal magnitude is directly transduced to a position of the writing device (e.g., the simple galvanometric type). In the servo mode (also called *feedback* or *compensation*) the position of the writing device is measured by a specific sensor system and the difference between the measured value and the value of the signal to be traced is applied to the transducer motor, resulting in a movement tending to zero the difference value and thus to correct the position [2,3]. In both methods the moving parts set a limit to the system frequency bandwidth. Moreover, in the feedback mode velocity and acceleration limitations may be present; thus linear system theory description of the apparatus behavior with respect to signal frequency may not be applicable. As such, the bandwidths of servo systems can be dependent on the writing width. Movement of the writing device can be generated by a rotation or by a translation. In the latter case the writing part is mechanically guided; primarily, the servo method is applied [3]. A rotation is generally obtained with a galvanometric motortype [3,4]; the galvanometer may rotate a pen, an ink jet, a light beam. The inertia of the moving part is the major parameter determining the bandwidth of the system. Translational devices allow a bandwidth of a few hertz. Higher bandwidths can be obtained with galvanometric pen types (about 100 Hz), ink jets (up to 1 kHz), and optical systems (up to 10 kHz) [1], but, being replaced by dot array recorders or data acquisition systems, these types are disappearing from the market. Major reasons are inherent errors and limitations of these analog types, maintenance needs of moving parts and ink devices, cost of photographic paper, and the lack of the possibilities of digital types.

Whereas moving parts restrict the analog recorder bandwidth, a corresponding capacity of **digital graphic recorders** is determined by the sampling frequency and the writing frequency. According to the sampling criterion the sampling frequency should be twice the highest signal frequency. This implies two samples to display a complete sine wave period, which can hardly be called a good graphic representation. Ten samples may be a minimum. The sampling frequency is a pure electronic matter; the maximal writing frequency of the dot array is the limiting factor in real time. Alternatively, if the signal spectrum exceeds the real-time bandwidth of the recorder, data can be stored at a sufficient sampling rate in memory and reproduced off-line at a slower rate which can be handled by the apparatus. Most digital recorders have this memory facility; some recorders are specifically referred to as "memory" recorders when their offline capabilities largely exceed their online performance.

Real-time recording is primarily performed as a function of time (*t–y* recorders). On the other hand, *x–y* recording is another way of representing the data. In this case the relation between two physical variables is traced and the independent variable time is excluded (apart from the fact that, if a dashed line is used, each dash can represent a fixed time interval). In standard analog *x–y* recorders the chart is stationary (e.g., electrostatically fixed to the writing table); two similar analog writing transducers with identical input circuitry are assembled in the recorder. The first transducer (y) translates a long arm covering the width of the paper at which the second transducer (x) carrying the pen is moved. Evidently, recorders with memory facilities and appropriate software may produce an *x–y* recording in off-line mode by setting dots while the paper progresses. Recording accuracy can be formulated in similar terms as for any measuring instrument [5]. This accuracy is determined, on the one hand, by the input signal conditioning, similar for digital as well as for analog types and, on the other hand, by the accuracy of the recording transducer and its driver electronics. In the digital type, digitization bit accuracy, sampling frequency, dot array resolution, and dot writing frequency are major parameters. In the analog type, typical inconveniences of analog transducer systems can be found (such as static nonlinearity, noise and drift, dead zone and hysteresis, limited dynamic behavior); servo systems are known to be more accurate as compared with direct systems. For example, drift in analog recording can be the result of a small shift of the calibrated chart along the y-axis; the latter is excluded if the recorder draws its own calibration lines.

With respect to graphic quality, clarity and sharpness of the tracings are important (within a large range of writing velocities). Tracing quality depends on the writing velocity, i.e., the velocity of the writing device with respect to the moving paper. Evidently, the flow of writing medium (e.g., ink or heat) should be more or less adapted to this writing velocity to prevent poorly visible tracings at high velocities and thick lines at low velocities. Good identifiability of overlapping curves is essential. Sufficient dot resolution (with adequate interpolation techniques) is important in discontinuous types for easy visual inspection. A graphic recorder can be designed as a single-channel instrument or can have a multichannel input. Inputs can be standard or modular, so that the user can choose the specific signal conditioners for the application and the number of channels. A recorder can be called "general purpose" or can be assembled in a specific measuring apparatus (e.g., in biomedical applications such as electrocardiography and electroencephalography). The recorder can be portable for operation in the field or can be mounted in a laboratory rack or in a control panel. The paper can be moved vertically or horizontally on a writing table ("flat bed" recorder) and can be supplied from a roll or in Z-fold. Besides strip chart recorders and *x–y* recorders, circular chart recorders exist. In this case the chart rotates, one rotation corresponds to a complete measurement interval, and the chart is provided with appropriate calibration lines adapted to movement of the pens writing on it.

Apart from the low-bandwidth translational pen devices there is a decreasing interest in analog graphic recorders. They are being replaced by thermal dot array recorders or by data acquisition systems. Nevertheless, as some of them may still be manufactured and a number of apparatus may still be in use, different techniques are mentioned. A description of analog recorder principles and performances can be found in Reference 1. Galvanometric recorders apply rotational transducers. The direct as well as the servo principles are used. The direct type makes use of the d'Arsonval movement as applied in ordinary galvanometers [1, 2, 4]. Dynamically, the galvanometer acts as a mechanical resonant system and the bandwidth is thus determined by its resonant frequency, the latter being dependent on the inertia of the moving parts. Evidently, rotation gives rise to inherent errors in graphic recorders. If pen tips (perpendicular to the pen arm, thermal or ink) are used, the rotation of the pen arm fixed to the galvanometer coil occurs in a plane parallel to the chart plane, so the recording is curvilinear instead of rectilinear, introducing an error with respect to the time axis and to the ordinate axis (the ordinate value being proportional to the tangent of the rotation angle). Calibrated paper with curvilinear coordinate lines may solve this problem; nevertheless, the tracing is deformed and zero offset is critical. Rectilinear recording can be realized with pen systems, ink jets, and light beams. Rectilinear pen recording can be approximated with pen tips in case of "long-arm" pens and by mechanical rectilinearization; alternatively "knife-edge" recording is a solution [1]. In the case of ink jet and light beam recorders the rotation plane and the chart plane do not have to be parallel; writing then occurs at the intersecting line of both planes and is thus essentially rectilinear. In ink jet recording a miniature nozzle through which ink is pumped is mounted in a direction perpendicular to the axis of the galvanometer. In optical recording a sharp light beam is reflected by a small mirror connected to the galvanometric moving coil toward the photosensitive paper. In these methods miniaturization of the moving parts gives rise to higher resonant frequencies and thus higher bandwidths. Whereas a typical bandwidth for a galvanometric pen system is 100 Hz, the bandwidth for an ink jet system can be 1000 Hz and for an optical system 10 kHz may be reached. In the fiber-optic cathode ray tube (FO-CRT) no mechanical moving parts are present and thus there are no mechanical limits on bandwidth. The FO-CRT is essentially a one-dimensional CRT. A phosphor layer at the inside of the screen converts the cathode ray into ultraviolet (UV) light. This UV light is guided by an outer faceplate composed of glass fibers onto the photosensitive chart. As in ordinary oscilloscopes, the deflection of the spot is directly proportional to the signal applied at the input of the deflection unit. The bandwidth is determined by the driving electronics. The system can be used in scanning mode as the beam intensity is easily controlled. In the following paragraphs further details will be given on translational pen recorders and thermal dot array recorders.

Translational Pen Recorders

In **translational pen recorders** the writing device is usually a fiber-tip pen with an ink cartridge. In discontinuous applications the writing device can be a printhead with different color styli or with a colored ribbon. A manual or automatic pen lift facility is included. During recording, the writing device is translated along the *y*-axis as it is linked to a mechanical guidance and a closed-loop wire system. A motor and wheels system pulls the wire and thus the writing device. In some designs a motor and screw system is applied. Translational recorders are primarily designed as a servo type. The position of the pen is accurately measured, and the difference voltage between the input signal and the position magnitude (following appropriate amplification and conditioning) drives the servomotor. Servo motors can be dc or stepper types; servo electronics can be analog or digital. Position sensing can be potentiometric ("potentiometric" recorders): the pen carriage is equipped with a sliding contact on the resistor (wire wound or thick film) which covers the complete width of the paper. More recently developed methods use optical or ultrasonic principles for position sensing; with these methods contacts are absent resulting in less maintenance and longer lifetime. For example, in the ultrasonic method the pen position is sensed by a detector coil from the propagation time of an ultrasound pulse, which is imparted by a piezoelectric transducer to a magnetostrictive strip covering the chart width. Accordingly, brushless dc-motors are used in some apparatuses. In the servo system accuracy is determined for the larger part by the quality of the sensing system. A poor contact with the resistor can give rise to noise; there may be a mechanical backlash between pen tip and the sliding contact on the potentiometer. The velocity of the pen carriage is limited, about 0.5 to 2 m/s dependent on motor and mechanics design. This results in a bandwidth of the recorder depending on the amplitude of the tracing: the –3 dB frequency fits in the range from 1 to 5 Hz for a full-scale width of 200 to 250 mm. Alternatively, the pen response time to a full-scale step input is given (5 to 95% of full-scale tracing): 0.1 to 0.5 s. Overshoot of the pen step response is extremely small in accurate designs.

In most pen recorders each tracing can cover the complete width. As such, pens must have the possibility to pass each other resulting in a small shift between adjacent pens (a few millimeters) along the time axis. In some apparatus, tracings can be synchronized with a POC-system ("pen offset compensation"); signals are digitized, stored in memory, and reproduced after a time delay correcting for the physical displacement of the pen. If immediate visual inspection is required, applying POC can be inconvenient as a consequence of this time delay. In process monitoring, slowly varying signals such as temperature, pressure, flow, etc. are followed. These signals can be handled by a single transducer in a discontinuous way; all input signals are scanned during the scanning cycle and for each signal a dot is

TABLE 36.1 Pen Recorders

Designation	Description	Manufacturer	Approximate Price (U.S.$)
LR8100	Test, meas.; 4,6,8 c. ch.; POC; printer; display; memory; analysis; alarm; interface	Yokogawa E. C.	11,500 (8 ch.)
LR102	Test, meas.; 1,2 c. ch.	Yokogawa E. C.	1,800 (2 ch.)
LR122	Test, meas.; 1,2 c. ch.; *x–y;* alarm; interface	Yokogawa E. C.	2,300 (2 ch.)
MC1000	Test, meas.; 4,6,8,12 c. ch.; POC; printer; waveform display; analysis; alarm; interface	Graphtec C.	20,200 (12 ch.)
BD112	Test, meas.; 2 c. ch.; POC	Kipp Z.	2,700 (2 ch.)
BD200	Test, meas.; 4,6,8 c. ch.; POC; display; *x–y;* alarm; interface	Kipp Z.	12,200 (8 ch.)
L250	Test, meas.; 1,2 c. ch.; POC	Linseis	2,000 (2 ch.)
L2066	Test, meas.; 1 to 6 c. ch.; POC; *x–y;* interface	Linseis	8,100 (6 ch.)
MCR560	Test, meas.; 2,4,6 c. ch.	W+W	3,800 (2 ch.)
DCR540	Test, meas.; 1 to 4 c. ch.; POC; display; *x–y;* interface	W+W	6,600 (4 ch.)
PCR500SP	Test, meas.; 2,4,6,8 c. ch.; POC; display; *x–y;* analysis; alarm; interface; transient option	W+W	15,900 (8 ch.)
Omega640	Test, meas.; 1,2,3 c. ch.	Omega	4,300 (3 ch.)
Omega600A	Test, meas.; 1 to 6 c. ch.; POC; printer; *x–y;* memory; interface	Omega	23,700 (6 ch.)
µR1000	Process mon.; 1,2,3,4 c. ch., 6 s. ch.; POC; printer; display; analysis; alarm; interface	Yokogawa E. C.	4,000 (4 c. ch.)
µR1800	Process mon.; 1,2,3,4 c. ch., 6,12,18,24 s. ch.; POC; printer; display; analysis; alarm; interface	Yokogawa E. C.	5,600 (4 c. ch.)
DR240	Process mon.; 30 s. ch.; printer; display; analysis; alarm; interface	Yokogawa E. C.	6,500 (30 s. ch.)
RL100	Process mon.; 1,2 c. ch., 6 s. ch.; printer; alarm	Honeywell	1,500 (2 c. ch.)
DPR100C/D	Process mon.; 1,2,3 c. ch., 6 s. ch.; POC; printer; display; alarm; analysis; interface	Honeywell	3,000 (6 s. ch.)
DPR3000	Process mon.; 4 to 32 s. ch.; printer; display; alarm; analysis; interface	Honeywell	7,700 (32 s. ch.)
4101	Process mon.; 1 to 4 c. ch., 6 s. ch.; POC; printer; alarm	Eurotherm	2,400 (4 c. ch.)
4180 G	Process mon.; 8,16,24,32 s. ch.; printer; waveform display; alarm; analysis; interface	Eurotherm	9,800 (32 s. ch.)
Sirec L	Process mon.; 1,2,3 c. ch.	Siemens	1,200 (3 c. ch.)

Note: c. ch. = continuous channel; s. ch. = scanned channel.

printed ("multipoint" recorder). The minimum scanning time is dependent on the moving writing device. For chart progression, dc and stepper motors are used. Calibrated paper is pulled by sprocket wheels seizing in equidistant perforations at both sides of the chart. Translational pen recorders range from simple purely analog design to intelligent microprocessor-controlled types handling a large number of channels with a broad range of control and monitor facilities (e.g., printing of a report after alarm).

Table 36.1 displays a set of translational pen recorders; some of them are equipped with a printhead. Under "Description" the major application is given: test and measurement or process monitoring. Furthermore the following are indicated: the number of continuous (c. ch.) and scanned (s. ch.) channels; the availability of POC, a printer (for additional information or for trace printing), a display for alphanumeric information (such as calibration values for each channel), or even a waveform display, data memory (allowing memory recorder functioning), *x–y* recording facility, alarm generation (after reaching thresholds of recorded variables), and standard serial interface options allowing communication with a PC (introduction of recorder settings, storage, and processing of recorded data, etc.). Table 36.2 gives a summary of pen recorder specifications (multipoint types also included).

Thermal Dot Array Recorders

In **thermal dot array recorders,** apart from the chart-pulling system, no moving parts are present; the writing transducer is essentially a rectilinear array of equidistal writing points which covers the total

TABLE 36.2 Pen Recorder Specifications

Type (Test, Measurement)	Recording Width (mm)	Chart Velocity max. (mm/min)	Chart Velocity min. (mm/h)	Pen Velocity, max. (m/s)	Pen Step Response Time(s)	Bandwidth (–3 dB) (Hz)	Number of Continuous Channels, max.
LR8100	250	1200	10	1.6		5	8
LR102	200	600	10	0.4	0.5	1.5	2
LR122	200	600	10	0.4	0.5	1.5	2
MC1000	250	1200	7.5	1.6			12
BD112	200	1200	6		0.2		2
BD200	250	1200	5		0.25		8
L250	250	1200	6	1	0.12	3.6	2
L2066	250	3000	1	1	0.3	2	6
MCR560	250	600	10		0.3	1.5	6
DCR540	250	600	10		0.3	1.5	4
PCR500SP	250	1200	10	2	0.15	4.5	8
Omega640	250	600	30	0.5			3
Omega600A	250	600	10	0.1			6

Type (Process Monitor)	Recording Width (mm)	Chart Velocity max. (mm/min)	Chart Velocity min. (mm/h)	Pen Step Response Time (s)	Printhead Scanning Cycle (s)	Number of Channels (max.)[a]
μR1000	100	200	5	1	10	4c, 6s
μR1800	180	200	5	1.5	10 (6s)	4c, 24s
DR240	250	25	1	—	2	30s
RL100	100	8	10	3.2	5	2c, 6s
DPR100C/D	100	100	1	1	0.6	3c, 6s
DPR3000	250	25	1	—	5	32s
4101	100	25	1	2	5	4c, 6s
4180 G	180	25	1	—	3	32s
Sirec L	100	20	1		—	3c

[a] c = continuous; s = scanned.

width of the paper. Although some apparatuses apply an electrostatic method [1], the thermal dot array and thermosensitive paper are generally used. In this array the writing styli consist of miniature electrically heated resistances; thermal properties of the resistances (in close contact with the chart paper) and the electric activating pulse form determine the maximal writing frequency. The latter ranges in real-time recorders from 1 to 6.4 kHz. Heating of the thermosensitive paper results in a black dot with good long-term stability. The heating pulse is controlled in relation to the chart velocity in order to obtain sufficient blackness at high velocities. Tracing blackness or line thickness is seldom used for curve identification; alphanumeric annotation is mostly applied. With the dot array a theoretically unlimited number of waveforms can be processed; the apparatus is thus programmed to draw its own calibration lines. Different types of grid patterns can be selected by the user. Moreover, alphanumeric information can be printed for indicating experimental conditions.

Ordinate axis resolution is determined by the dot array: primarily, 8 dots/mm; exceptionally, 12 dots/mm (as in standard laser printers). The resolution along the abscissa depends on the thermal array limitations and programming. Generally, a higher resolution is used (mostly 32 dots/mm, maximally 64 dots/mm) except for the highest chart velocities (100, 200, 500 mm/s). At these high velocities and consequently short chart contact times, dots become less sharp and less black. Most of the dot array instruments are intended for high-signal-frequency applications: per channel sampling frequencies of 100, 200, and even 500 kHz are used in real time. These sampling frequencies largely exceed the writing frequencies; during the writing cycle, data are stored in memory and for each channel within each writing interval a dotted vertical line is printed between the minimal and the maximal value. For example, a sine wave with a frequency largely exceeding the writing frequency is represented as a black band with a width equal to

TABLE 36.3 Thermal Dot Array Recorders

Designation	Description	Manufacturer	Approximate Price (U.S.$)
WR 5000	8 a. ch.; memory	Graphtec C.	17,100 (8 ch.)
WR 9000	4,8,16 a. ch.; monitor; memory; *x–y*; analysis; FFT	Graphtec C.	11,300 (4 ch.)
Mark 12	4 to 52 a. ch., 4 to 52 d. ch.; monitor; memory	W. Graphtec	30,300 (16 a. ch.)
MA 6000	2 to 16 a. ch.; monitor; memory; *x–y*; analysis; FFT	Graphtec C.	23,500 (8 ch.)
ORP 1200	4,8 a. ch., 16 d. ch.; monitor; memory; *x–y*	Yokogawa E. C.	11,600 (8 a. ch.)
ORP 1300	16 a. ch., 16 d. ch.; monitor; memory; *x–y*	Yokogawa E. C.	18,000 (16 a. ch.)
ORM 1200	4,8 a. ch., 16 d. ch.; monitor; memory; *x–y*	Yokogawa E. C.	14,100 (8 a. ch.)
ORM 1300	16 a. ch., 16 d. ch.; monitor; memory; *x–y*	Yokogawa E. C.	21,900 (16 a. ch.)
OR 1400	8 a. ch., 16 d. ch.; monitor; memory; *x–y*	Yokogawa E. C.	16,100 (8 a. ch.)
TA 240	1 to 4 a. ch.	Gould I. S.	8,500 (4 ch.)
TA 11	4,8,16 a. ch.; monitor; memory	Gould I. S.	18,900 (16 ch.)
TA 6000	8 to 64 a. ch., 8 to 32 d. ch.; monitor; memory	Gould I. S.	33,900 (16 a. ch.)
Windograf	2 to 4 a. ch.; monitor	Gould I. S.	10,200 (4 ch.)
Dash 10	10,20,30 a. ch.; monitor; memory	Astro-Med	22,500 (10 ch.)
MT95K2	8 to 32 a. ch., 32 d. ch.; monitor; memory; *x–y*; analysis	Astro-Med	32,600 (8 a. ch.)
8852	4 a. ch., 24 d. ch.; monitor; memory; *x–y*; analysis; FFT	Hioki E. E. C.	22,300 (4 a. ch.)
8815	4 a. ch., 32 d. ch.; memory; *x–y*	Hioki E. E. C.	4,500 (4 a. ch.)
8825	16 a. ch., 32 d. ch.; monitor; memory; *x–y*; analysis	Hioki E. E. C.	28,800 (8 a. ch.)

Note: a. ch. = analog channel; d. ch. = digital channel.

the sine amplitude. In this way the graphs indicate the presence of a phenomenon with a frequency content exceeding the writing frequency. As data are stored in memory they can be reproduced at a lower rate thus revealing the actual high-frequency waveform captured. Some apparatuses use a much lower sampling rate in real time and only perform off-line: in this case the apparatus is indicated as a "memory" recorder. Digitization accuracy ranges from 8 to 16 bit, whereas the largest number of dots full scale is 4800. In this way the useful signal may be superposed on a large dc-offset: it can be written or reproduced with excellent graphic quality with the offset digitally removed and the scale adapted.

In a high-performance recorder a waveform display is extremely useful to avoid paper spoiling, in real-time and in off-line recording as well. The display is also used for apparatus settings. Signals can be calibrated and real physical values and units can be printed at the calibration lines. Via memory *x–y* plots can be obtained. Some apparatuses allow application of mathematical functions for waveform processing and analysis: original and processed waveforms can be drawn together off-line. A few types are equipped with FFT software. Computer interfacing, a large set of triggering modes (including recording at increased velocity after a specific trigger), event channels, etc. are standard facilities. Table 36.3 shows a set of thermal dot array recorders (under "Description" : number of analog channels (a. ch.) and digital channels (d. ch.); waveform monitor; signal data memory, *x–y* facility, mathematical analysis, FFT) and Table 36.4 gives specifications.

Concluding Remarks

Table 36.5 gives addresses and fax and phone numbers of manufacturers of recorders mentioned in Tables 36.1 and 36.3. It should be remarked that prices mentioned in these tables hold for purchasing a complete functioning apparatus (number of channels indicated) from firms in Belgium representing the manufacturers and having provided the data sheets from which specifications were derived. With the expression "a complete functioning" apparatus a standard system is meant, thus including simple input couplers (in case of a modular design), standard RAM and analysis software, no specific options. Obviously, the list of manufacturers is incomplete. It should be mentioned that the number of manufacturers of graphic recorders is decreasing; a significant and increasing amount of applications has been taken over by "paperless" recorders, i.e., data acquisition systems. Nevertheless, the possibility of generating graphs in real time remains an important feature, e.g., to provide evidence of the presence of a specific

TABLE 36.4 Thermal Dot Array Recorder Specifications

Type	Recording Width (mm)	Thermal Array Resolution (dots/mm)	Chart Velocity		Maximal Writing Frequency (dots/s)	Time Axis Resolution	
			max. (mm/s)	min. (mm/h)		max. (dots/mm)	min. (dots/mm)
WR 5000	384	8	200	1	1600	64	8
WR 9000V	200	8	100	1		32	
WR 9000M	200	8	100	1		32	
Mark 12	384	8	200	1	1600	64	8
MA 6000	205	8	100	1		40	
ORP 1200	201	8	100	10	1600	32	16
ORP 1300	201	8	100	10	1600	32	16
ORM 1200	201	8	100	10	1600	32	16
ORM 1300	201	8	100	10	1600	32	16
OR 1400	201	8	250	10	6400	32	25.6
TA 240	104	8	125	36	1000	32	8
TA 11	264	8	200	36	1600	16	8
TA 6000	370	8	200	36	1600	16	8
Windograf	104	8	100	36	800	32	8
Dash 10	256	12	200	60	1200	12	6
MT95K2	400	12	500	1	2000	48	4
8852	100	8	25	10	200	16	8
8815	104	6	8	10	50	12	6
8825	256	8	20	10	200	10	10

Type	Sampling Frequency Real-Time, max. (kHz)	Bit Accuracy, max. (bits)	No., max., Channels[a]	Display Dimensions (mm)	Display Resolution (pixels)	Sampling Frequency Memory, max. (kHz)	Samples Stored/ Channel, max.
WR 5000	64	14	8a	—	—	64	32 k
WR 9000V	250	12	8a	192 × 120	640 × 400	250	512 k
WR 9000M	50	14	8a	192 × 120	640 × 400	50	512 k
Mark 12	200	16	52a, 52d	97 × 77	256 × 320	200	2 M
MA 6000	500	16	16a	192 × 120	640 × 400	500	512 k
ORP 1200	100	14	8a, 16d	127[a]	320 × 240	100	32 k
ORP 1300	100	14	16a, 16d	127[a]	320 × 240	100	32 k
ORM 1200	100	14	8a, 16d	127[a]	320 × 240	100	128 k
ORM 1300	100	14	16a, 16d	127[a]	320 × 240	100	128 k
OR 1400	100	16	8a, 16d	127[a]	320 × 240	100	256 k
TA 240	5	12	4a	—	—	—	—
TA 11	250	12	16a	198 × 66	640 × 200	250	500 k
TA 6000	250	12	64a, 32d	224 × 96	640 × 200	250	500 k
Windograf	10	12	4a	178[a]	800 × 350	—	—
Dash 10	250	12	30a		256 × 64	250	512 k
MT95K2	200	12	32a, 32d			200	500 k
8852	1.6	8	4a, 24d	178[a]		100×10^3	1 M
8815	12.5	8	4a, 32d	—	—	500	30 k
8825	8	12	16a, 32d	254[a]	640 × 480	200	500 k

[a] Diagonal.
a = analog; d = digital.

phenomenon. In recent years analog recorders have become less used and manufactured (apart from the translational pen types, especially in industrial process monitoring). Thermal array recorders have become more important: the quality and long-term stability of thermal paper have improved and cost levels are comparable with calibrated paper for ink recording. In new designs, recorders provide more capabilities

TABLE 36.5 Companies that Make Graphic Recorders

Astro-Med, Inc. Astro-Med Industrial Park, West Warwick, RI 02893, USA Tel: (401) 828-4000 Fax: (401) 822 - 2430	Linseis GMBH Postfach 1404, Vielitzer Strasse 43, D-8672 Selb, Germany Tel: 09287/880-0 Fax: 09287/70488
Eurotherm Recorders Ltd. Dominion Way, Worthing, West Sussex BN148QL, Great Britain Tel: 01903-205222 Fax: 01903-203767	Omega Engineering, Inc. P.O. Box 4047, Stamford, CT 06907-0047, USA Tel: (203) 359-1660 Fax: (203) 359-7700
Gould Instrument Systems, Inc. 8333 Rockside Road, Valley View, OH 44125-6100, USA Tel: (216) 328-7000 Fax: (216) 328-7400	Siemens AG, Bereich Automatisierungstechnik Geschäftsgebiet Processgeräte, AUT 34, D-76181 Karlsruhe, Germany Tel: 0721/595-2058 Fax: 0721/595-6885
Graphtec Corporation 503-10 Shinano-cho, Totsuka-ku, Yokohama 244, Japan Tel: (045) 825-6250 Fax: (045) 825-6396	Western Graphtec, Inc. 11 Vanderbilt, Irvine, CA 92718-2067, USA Tel: (800) 854-8385 Fax: (714) 770-6010
Hioki E. E. Corporation 81 Koizumi, Ueda, Nagano, 386-11, Japan Tel: 0268-28-0562 Fax: 0268-28-0568	W+W Instruments AG Frankfurt-Strasse 78, CH-4142 Münchenstein, Switzerland Tel: +41 (0) 614116477 Fax: +41 (0) 6141166685
Honeywell Industrial Automation and Control 16404 North Black Canyon Hwy., Phoenix, AZ 85023, USA Tel: (800) 343-0228	Yokogawa Electric Corporation Shinjuku-Nomura Bldg. 1-26-2 Nishi-Shinjuku, Shinjuku-ku, Tokyo 163-05, Japan Tel: 81-3-3349-1015 Fax: 81-3-3349-1017
Kipp & Zonen, Delft BV Mercuriusweg 1, P.O. Box 507, NL-2600 AM Delft, The Netherlands Tel: 015-561000 Fax: 015-620351	

and appear more intelligent, obviously leading to increased complications with respect to instrument settings and thus increased need for training and experience in the use of the instrument.

Defining Terms

Analog graphic recorder: A graphic recorder that makes use of an analog transducer system (e.g., a moving pen).

Analog recorder bandwidth: The largest frequency that can be processed by the analog recorder (–3 dB limit).

Digital graphic recorder: A graphic recorder that makes use of a digital transducer system (e.g., a fixed dot array).

Graphic recorder: A measuring apparatus that produces in real time a hard copy of a set of time-dependent variables.

Maximal sampling frequency: Maximal number of data points sampled by the digital recorder per time unit (totally or per channel).

Maximal writing frequency: Maximal number of writing (or printing) acts executed by the digital recorder per time unit.

Thermal dot array recorder: A digital recorder applying a fixed thermal dot array perpendicular to the time axis.

Translational pen recorder: An analog recorder with one or several pens being translated perpendicularly to the time axis.

References

1. H. Vermariën, Recorders, graphic, in J.G. Webster, Ed., *Encyclopedia of Medical Devices and Instrumentation,* New York: John Wiley & Sons, 1988.
2. D.A. Bell, *Electronic Instrumentation and Measurements,* 2nd ed., Englewood Cliffs, NJ: Prentice-Hall, 1994.
3. A. Miller, O.S. Talle, and C.D. Mee, Recorders, in B.M. Oliver and J.M. Cage, Eds., *Electronic Measurementsand Instrumentation,* New York: McGraw-Hill, 427–479, 1975.
4. R.J. Smith, *Circuits, Devices and Systems: A First Course in Electrical Engineering,* New York: John Wiley & Sons, 1976.
5. W.H. Olson, Basic concepts in instrumentation, in J.G. Webster, Ed., *Medical Instrumentation: Application and Design, 3rd ed.,* New York: John Wiley & Sons, 1998.

36.2 Data Acquisition Systems

Edward McConnell

The fundamental task of a **data acquisition system** is the measurement or generation of real-world physical signals. Before a physical signal can be measured by a computer-based system, a sensor or transducer is used to convert the physical signal into an electrical signal, such as voltage or current. Often only a plug-in data acquisition (DAQ) board is considered the data acquisition system; however, a board is only one of the components in the system. A complete DAQ system consists of sensors, signal conditioning, interface hardware, and software. Unlike stand-alone instruments, signals often cannot be directly connected to the DAQ board. The signals may need to be conditioned by some signal-conditioning accessory before they are converted to digital information by the plug-in DAQ board. Software controls the data acquisition system — acquiring the raw data, analyzing the data, and presenting the results. The components are shown in Figure 36.1.

Signals

Signals are physical events whose magnitude or time variation contains information. DAQ systems measure various aspects of a signal in order to monitor and control the physical events. Users of DAQ

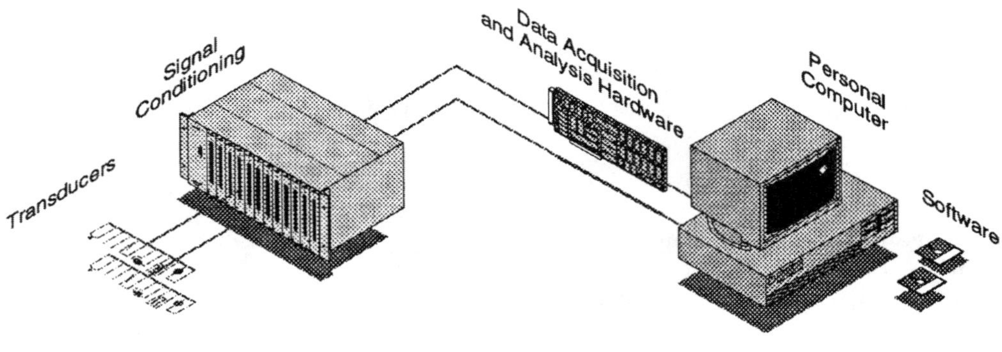

FIGURE 36.1 Components of a DAQ system.

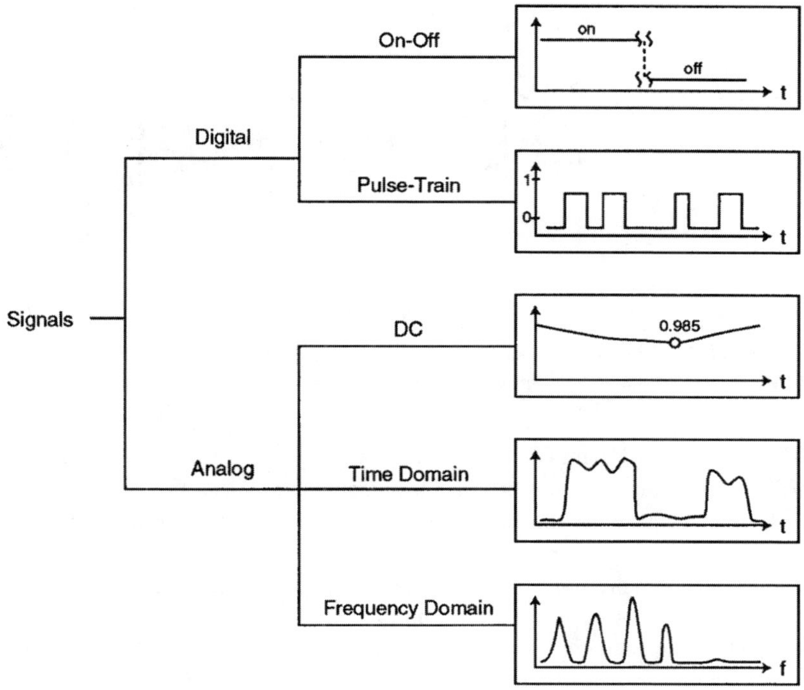

FIGURE 36.2 Classes of signals.

systems need to know the relation of the signal to the physical event and what information is available in the signal. Generally, information is conveyed by a signal through one or more of the following signal parameters: state, rate, level, shape, or frequency content. The physical characteristics of the measured signals and the related information help determine the design of a DAQ system.

All signals are, fundamentally, analog, time-varying signals. For the purpose of discussing the methods of signal measurement using a plug-in DAQ board, a given signal should be classified as one of five signal types. Because the method of signal measurement is determined by the way the signal conveys the needed information, a classification based on this criterion is useful in understanding the fundamental building blocks of a DAQ system.

As shown in Figure 36.2, any signal can generally be classified as analog or digital. A digital, or binary, signal has only two possible discrete levels of interest — a high (on) level and a low (off) level. The two digital signal types are on–off signals and pulse train signals. An analog signal, on the other hand, contains information in the continuous variation of the signal with time. Analog signals are described in the time or frequency domains depending upon the information of interest. A dc type signal is a low-frequency signal, and if the phase information of a signal is presented with the frequency information, then there is no difference between the time or frequency domain representations. The category to which a signal belongs depends on the characteristic of the signal to be measured. The five types of signals can be closely paralleled with the five basic types of signal information — state, rate, level, shape, and frequency content. Basic understanding of the signal representing the physical event being measured and controlled assists in the selection of the appropriate DAQ system.

Plug-In DAQ Boards

The fundamental component of a DAQ system is the plug-in DAQ board. These boards plug directly into a slot in a PC and are available with analog, digital, and timing inputs and outputs (I/O). The most versatile of the plug-in DAQ boards is the multifunction I/O board. As the name implies, this board typically contains various combinations of analog-to-digital converters (ADCs), digital-to-analog converters

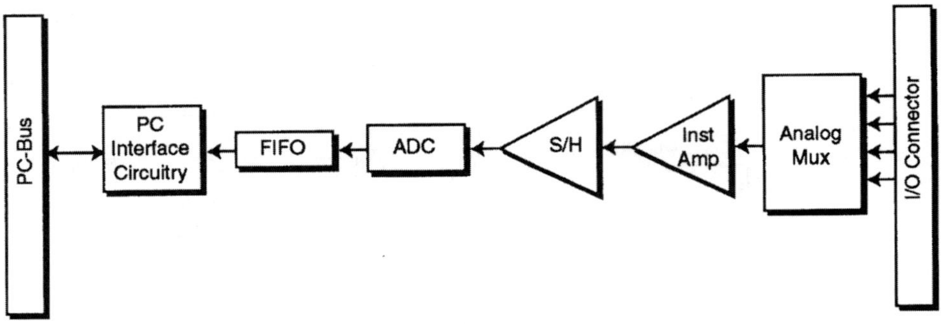

FIGURE 36.3 Analog input section of a plug-in DAQ board. *Note:* FIFO = first-in first-out buffer, S/H = sample-and-hold, Inst. Amp = instrumentation amplifier, and Mux = analog multiplexer.

(DACs), digital I/O lines, and counters/timers. ADCs and DACs measure and generate analog voltage signals, respectively. The digital I/O lines sense and control digital signals. Counters/timers measure pulse rates, widths, delays, and generate timing signals. These many features make the multifunction DAQ board useful for a wide range of applications.

Multifunction boards are commonly used to measure analog signals. This is done by the ADC, which converts the analog voltage level into a digital number that the computer can interpret. The analog multiplexer (MUX), the instrumentation amplifier, the sample-and-hold (S/H) circuitry, and the ADC compose the analog input section of a multifunction board (see Figure 36.3).

Typically, multifunction DAQ boards have one ADC. Multiplexing is a common technique for measuring multiple channels (generally 16 single-ended or 8 differential) with a single ADC. The analog MUX switches between channels and passes the signal to the instrumentation amplifier and the S/H circuitry. The MUX architecture is the most common approach taken with plug-in DAQ boards. While plug-in boards typically include up to only 16 single-ended or 8 differential inputs, the number of analog input channels can be further expanded with external MUX accessories.

Instrumentation amplifiers typically provide a differential input and selectable gain by jumpers or software. The differential input rejects small common-mode voltages. The gain is often software programmable. In addition, many DAQ boards also include the capability to change the amplifier gain while scanning channels at high rates. Therefore, one can easily monitor signals with different ranges of amplitudes. The output of the amplifier is sampled, or held at a constant voltage, by the S/H device at measurement time so that voltage does not change during digitization.

The ADC transforms the analog signal into a digital value which is ultimately sent to computer memory. There are several important parameters of A/D conversion. The fundamental parameter of an ADC is the number of bits. The number of bits of an A/D determines the range of values for the binary output of the ADC conversion. For example, many ADCs are 12-bit, so a voltage within the input range of the ADC will produce a binary value that has one of $2^{12} = 4096$ different values. The more bits that an ADC has, the higher the resolution of the measurement. The resolution determines the smallest amount of change that can be detected by the ADC. Resolution is expressed as the number of digits of a voltmeter or dynamic range in decibels, rather than with bits. Table 36.6 shows the relation among bits, number of digits, and dynamic range in decibels.

The resolution of the A/D conversion is also determined by the input range of the ADC and the gain. DAQ boards usually include an instrumentation amplifier that amplifies the analog signal by a gain factor prior to the conversion. This gain amplifies low-level signals so that more accurate measurements can be made.

Together, the input range of the ADC, the gain, and the number of bits of the board determine the minimum resolution of the measurement. For example, suppose a low-level ±30 mV signal is acquired using a 12-bit ADC that has a ±5 V input range. If the system includes an amplifier with a gain of 100, the resulting resolution of the measurement will be $\text{range}/(\text{gain} * 2^{\text{bits}}) = \text{resolution}$, or $10\ \text{V}/(100 * 2^{12}) = 0.0244$ mV.

TABLE 36.6 Relation Among Bits, Number of Digits, and Dynamic Range (dB)

Bits	Digits	dB
20	6.0	120
16	4.5	96
12	3.5	72
8	2.5	48

Finally, an important parameter of digitization is the rate at which A/D conversions are made, referred to as the sampling rate. The A/D system must be able to sample the input signal fast enough to measure the important waveform attributes accurately. In order to meet this criterion, the ADC must be able to convert the analog signal to digital form quickly enough.

When scanning multiple channels with a multiplexing DAQ system, other factors can affect the throughput of the system. Specifically, the instrumentation amplifier must be able to settle to the needed accuracy before the A/D conversion occurs. With multiplexed signals, multiple signals are being switched into one instrumentation amplifier. Most amplifiers, especially when amplifying the signals with larger gains, will not be able to settle to the full accuracy of the ADC when scanning channels at high rates. To avoid this situation, consult the specified settling times of the DAQ board for the gains and sampling rates required by the application.

Types of ADCs

Different DAQ boards use different types of ADCs to digitize the signal. The most popular type of ADC on plug-in DAQ boards is the successive approximation ADC, because it offers high speed and high resolution at a modest cost.

Subranging (also called half-flash) ADCs offer very high speed conversion with sampling speeds up to several million samples per second.

The state-of-the-art technology in ADCs is sigma–delta modulating ADCs. These ADCs sample at high rates, are able to achieve high resolution, and offer the best linearity of all ADCs.

Integrating and flash ADCs are mature technologies still used on DAQ boards today. Integrating ADCs are able to digitize with high resolution but must sacrifice sampling speed to obtain it. Flash ADCs are able to achieve the highest sampling rate (gigahertz) but are available only with low resolution. The different types of ADCs are summarized in Table 36.7.

TABLE 36.7 Types of ADCs

Type of ADC	Advantages	Features
Successive approximation	High resolution High speed Easily multiplexed	1.25 MS/s sampling rate 12-bit resolution 200 kS/s sampling rate 16-bit resolution
Subranging	Higher speed	1 MHz sampling rate 12-bit resolution
Sigma–delta	High resolution Excellent linearity Built-in antialiasing State-of-the-art technology	48 kHz sampling rate 16-bit resolution
Integrated	High resolution Good noise rejection Mature technology	15 kHz sampling rate
Flash	Highest speed Mature technology	125 MHz sampling rate

Analog Input Architecture

With a typical DAQ board, the multiplexer switches among analog input channels. The analog signal on the channel selected by the multiplexer then passes to the programmable gain instrumentation amplifier (PGIA), which amplifies the signal. After the signal is amplified, the sample and hold (S/H) keeps the analog signal constant so that the ADC can determine the digital representation of the analog signal. A good DAQ board will then place the digital signal in a first-in first-out (FIFO) buffer, so that no data will be lost if the sample cannot transfer immediately over the PC I/O channel to computer memory. Having a FIFO becomes especially important when the board is run under operating systems that have large interrupt latencies, such as Microsoft Windows.

Basic Analog Specifications

Almost every DAQ board data sheet specifies the number of channels, the maximum sampling rate, the resolution, and the input range and gain.

The number of channels, which is determined by the multiplexer, is usually specified in two forms — differential and single ended. Differential inputs are inputs that have different reference points for each channel, none of which is grounded by the board. Differential inputs are the best way to connect signals to the DAQ board because they provide the best noise immunity.

Single-ended inputs are inputs that are referenced to a common ground point. Because single-ended inputs are referenced to a common ground, they are not as good as differential inputs for rejecting noise. They do have a larger number of channels, however. Single-ended inputs are used when the input signals are high level (greater than 1 V), the leads from the signal source to the analog input hardware are short (less than 5 m), and all input signals share a common reference.

Some boards have pseudodifferential inputs which have all inputs referenced to the same common — like single-ended inputs — but the common is not referenced to ground. These boards have the benefit of a large number of input channels, like single-ended inputs, and the ability to remove some common-mode noise, especially if the common-mode noise is consistent across all channels. Differential inputs are still preferable to pseudodifferential, however, because differential is more immune to magnetic noise.

Sampling rate determines how fast the analog signal is converted to a digital signal. When measuring ac signals, sample at least two times faster than the highest frequency of the input signal. Even when measuring dc signals, oversample and average the data to increase the accuracy of the signal by reducing the effects of noise.

If the physical event consists of multiple dc-class signals, a DAQ board with interval scanning should be used. With interval scanning, all channels are scanned at one sample interval (usually the fastest rate of the board), with a second interval (usually slow) determining the time before repeating the scan. Interval scanning gives the effects of simultaneously sampling for slowly varying signals without requiring the additional cost of input circuitry for true simultaneous sampling.

Resolution is the number of bits that are used to represent the analog signal. The higher the resolution, the higher the number of divisions the input range is broken into, and therefore the smaller the possible detectable voltage. Unfortunately, some DAQ specifications are misleading when they specify the resolution associated with the DAQ board. Many DAQ board specifications state the resolution of the ADC without stating the linearities and noise, and therefore do not give the information needed to determine the resolution of the entire board. Resolution of the ADC, combined with the settling time, **integral nonlinearity** (INL), **differential nonlinearity** (DNL), and noise will give an understanding of the accuracy of the board.

Input range and gain determine the level of signal that should be connected to the board. Usually, the range and gain are specified separately, so the two must be combined to determine the actual signal input range as

$$\text{signal input range} = \text{range/gain}$$

For example, a board using an input range of ±10 V with a gain of 2 will have a signal input range of ±5 V. The closer the signal input range is to the range of the signal, the more accurate the readings from the DAQ board will be. If the signals have different input ranges, use a DAQ board with the feature of different gains per channel.

Data Acquisition Software

The software is often the most critical component of the DAQ system. Users of DAQ systems usually program the hardware in one of two ways — through register programming or through high-level device drivers.

Board Register-Level Programming

The first option is not to use vendor-supplied software and program the DAQ board at the hardware level. DAQ boards are typically register based; that is, they include a number of digital registers that control the operation of the board. The developer may use any standard programming language, such as C, C++, or Visual BASIC, to write a series of binary codes to the DAQ board to control its operation. Although this method affords the highest level of flexibility, it is also the most difficult and time-consuming, especially for the inexperienced programmer. The programmer must know the details of programming all hardware, including the board, the PC interrupt controller, the DMA controller, and PC memory.

Driver Software

Driver software typically consists of a library of function calls usable from a standard programming language. These function calls provide a high-level interface to control the standard functions of the plug-in board. For example, a function called SCAN_OP may configure, initiate, and complete a multiple-channel scanning DAQ operation of a predetermined number of points. The function call would include parameters to indicate the channels to be scanned, the amplifier gains to be used, the sampling rate, and the total number of data points to be collected. The driver responds to this one function call by programming the plug-in board, the DMA controller, the interrupt controller, and CPU to scan the channels as requested.

What Is Digital Sampling?

Every DAQ system has the task of gathering information about analog signals. To do this, the system captures a series of instantaneous "snapshots" or samples of the signal at definite time intervals. Each sample contains information about the signal at a specific instant. Knowing the exact time of each conversion and the value of the sample, one can reconstruct, analyze, and display the digitized waveform.

Real-Time Sampling Techniques

In real-time sampling, the DAQ board digitizes consecutive samples along the signal (Figure 36.4). According to the **Nyquist sampling theorem**, the ADC must sample at least twice the rate of the maximum frequency component in that signal to prevent aliasing. **Aliasing** is a false lower-frequency component that appears in sampled data acquired at too low a sampling rate. The frequency at one half the sampling frequency is referred to as the Nyquist frequency. Theoretically, it is possible to recover information about those signals with frequencies at or below the Nyquist frequency. Frequencies above the Nyquist frequency will alias to appear between dc and the Nyquist frequency.

For example, assume the sampling frequency, f_s, is 100 Hz. Also assume the input signal to be sampled contains the following frequencies — 25, 70, 160, and 510 Hz. Figure 36.5 shows a spectral representation of the input signal.

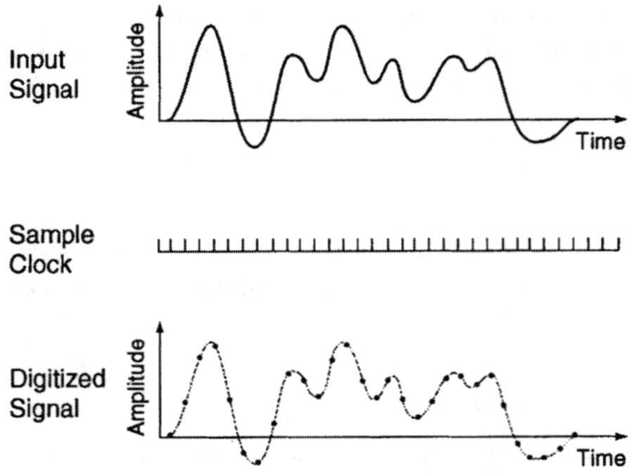

FIGURE 36.4 Consecutive discrete samples recreate the input signal.

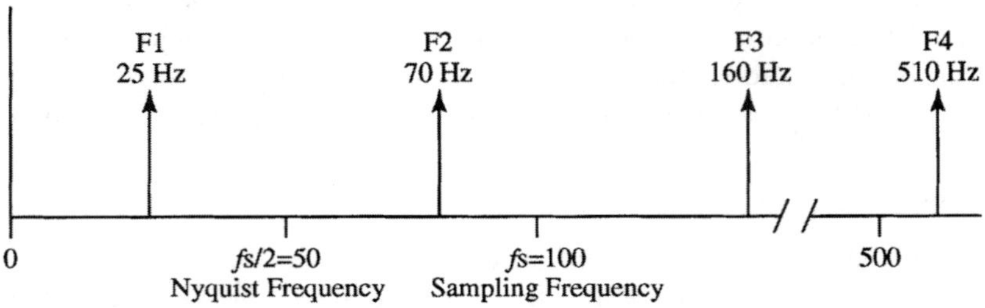

FIGURE 36.5 Spectral of signal with multiple frequencies.

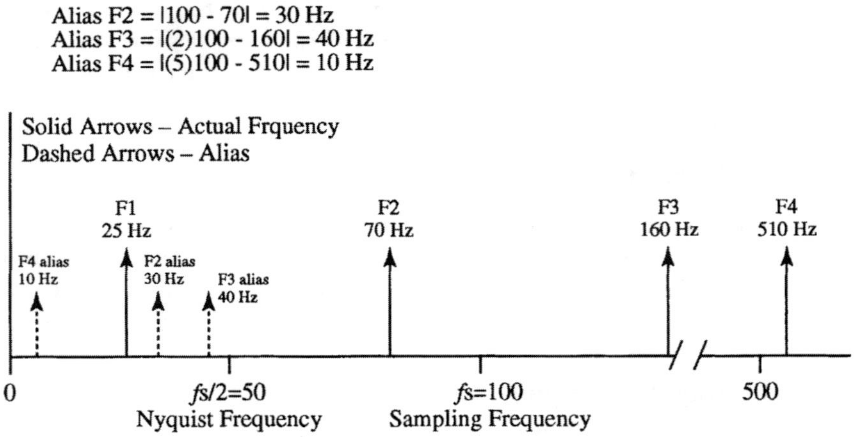

FIGURE 36.6 Spectral of signal with multiple frequencies after sampling at $f_s = 100$ Hz.

The mathematics of sampling theory show us that a sampled signal is shifted in the frequency domain by an amount equal to integer multiples of the sampling frequency, f_s. Figure 36.6 shows the spectral content of the input signal after sampling. Frequencies below 50 Hz, the Nyquist frequency ($f_s/2$), appear correctly. However, frequencies above the Nyquist appear as aliases below the Nyquist frequency. For example, F1 appears correctly; however, F2, F3, and F4 have aliases at 30, 40, and 10 Hz, respectively.

The resulting frequency of aliased signals can be calculated with the following formula:

$$\text{Apparent (Alias) Freq.} = \text{ABS (Closest Integer Multiple of Sampling Freq.} - \text{Input Freq.)}$$

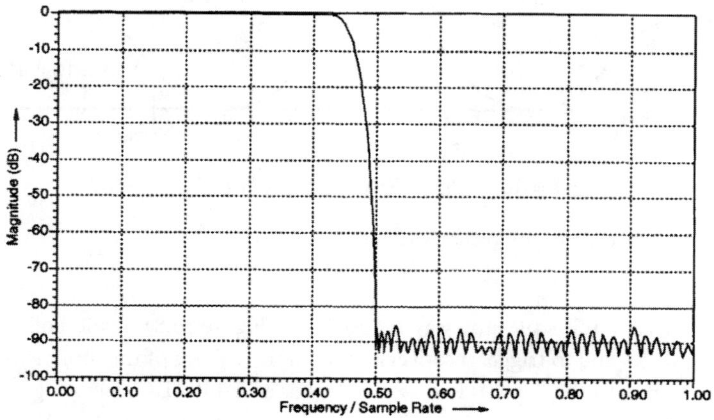

FIGURE 36.7 Magnitude portion of transfer function of an antialiasing filter.

For the example of Figures 36.5 and 36.6:

$$\text{Alias F2} = |100 - 70| = 30 \text{ Hz}$$

$$\text{Alias F3} = |(2)100 - 160| = 40 \text{ Hz}$$

$$\text{Alias F4} = |(5)100 - 510| = 10 \text{ Hz}$$

Preventing Aliasing

Aliasing can be prevented by using filters on the front end of the DAQ system. These antialiasing filters are set to cut off any frequencies above the Nyquist frequency (half the sampling rate). The perfect filter would reject all frequencies above the Nyquist; however, because perfect filters exist only in textbooks, one must compromise between sampling rate and selecting filters. In many applications, one- or two-pole passive filters are satisfactory. The rule of thumb is to oversample (5 to 10 times) and use these antialiasing filters when frequency information is crucial.

Alternatively, active antialiasing filters with programmable cutoff frequencies and very sharp attenuation of frequencies above the cutoff can be used. Because these filters exhibit a very steep roll-off, the DAQ system can sample at two to three times the filter cutoff frequency. Figure 36.7 shows a transfer function of a high-quality antialiasing filter.

The computer uses digital values to recreate or to analyze the waveform. Because the signal could be anything between each sample, the DAQ board may be unaware of any changes in the signal between samples. There are several sampling methods optimized for the different classes of data; they include software polling, external sampling, continuous scanning, multirate scanning, simultaneous sampling, interval scanning, and seamless changing of the sample rate.

Software Polling

A software loop polls a timing signal and starts the A/D conversion via a software command when the edge of the timing signal is detected. The timing signal may originate from the internal clock of the computer or from a clock on the DAQ board. Software polling is useful in simple, low-speed applications, such as temperature measurements.

The software loop must be fast enough to detect the timing signal and trigger a conversion. Otherwise, a window of uncertainty, also known as jitter, will exist between two successive samples. Within the window of uncertainty, the input waveform could change enough to reduce the accuracy of the ADC drastically.

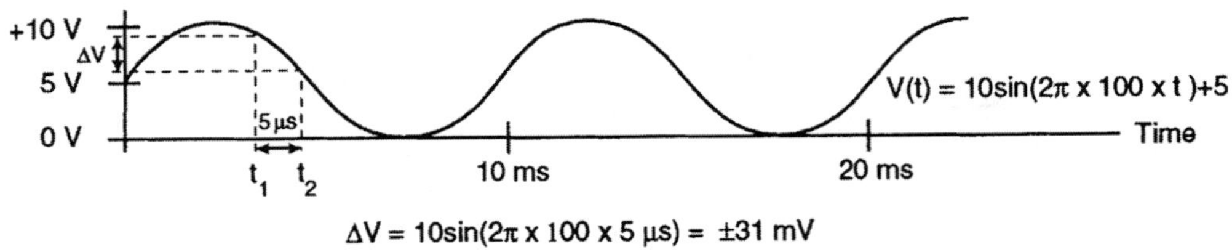

FIGURE 36.8 Jitter reduces the effective accuracy of the DAQ board.

Suppose a 100-Hz, 10-V full-scale sine wave is digitized (Figure 36.8). If the polling loop takes 5 ms to detect the timing signal and to trigger a conversion, then the voltage of the input sine wave will change as much as 31 mV, [$\Delta V = 10 \sin(2\pi \times 100 \times 5 \times 10^{-6})$]. For a 12-bit ADC operating over an input range of 10 V and a gain of 1, one least significant bit (LSB) of error represents 2.44 mV:

$$\left(\frac{\text{Input range}}{\text{gain} \times 2^n}\right) = \left(\frac{10\text{ V}}{1 \times 2^{12}}\right) = 2.44\text{ mV}$$

But because the voltage error due to jitter is 31 mV, the accuracy error is 13 LSB.

$$\left(\frac{31\text{ mV}}{2.44\text{ mV}}\right)$$

This represents uncertainty in the last 4 bits of a 12-bit ADC. Thus, the effective accuracy of the system is no longer 12 bits but rather 8 bits.

External Sampling

Some DAQ applications must perform a conversion based on another physical event that triggers the data conversion. The event could be a pulse from an optical encoder measuring the rotation of a cylinder. A sample would be taken every time the encoder generates a pulse corresponding to n degrees of rotation. External triggering is advantageous when trying to measure signals whose occurrence is relative to another physical phenomenon.

Continuous Scanning

When a DAQ board acquires data, several components on the board convert the analog signal to a digital value. These components include the analog MUX, the instrumentation amplifier, the S/H circuitry, and the ADC. When acquiring data from several input channels, the analog MUX connects each signal to the ADC at a constant rate. This method, known as continuous scanning, is significantly less expensive than having a separate amplifier and ADC for each input channel.

Continuous scanning is advantageous because it eliminates jitter and is easy to implement. However, it is not possible to sample multiple channels simultaneously. Because the MUX switches between channels, a time skew occurs between any two successive channel samples. Continuous scanning is appropriate for applications where the time relationship between each sampled point is unimportant or where the skew is relatively negligible compared with the speed of the channel scan.

If samples from two signals are used to generate a third value, then continuous scanning can lead to significant errors if the time skew is large. In Figure 36.9, two channels are continuously sampled and added together to produce a third value. Because the two sine waves are 90° out-of-phase, the sum of the signals should always be zero. But because of the skew time between the samples, an erroneous sawtooth signal results.

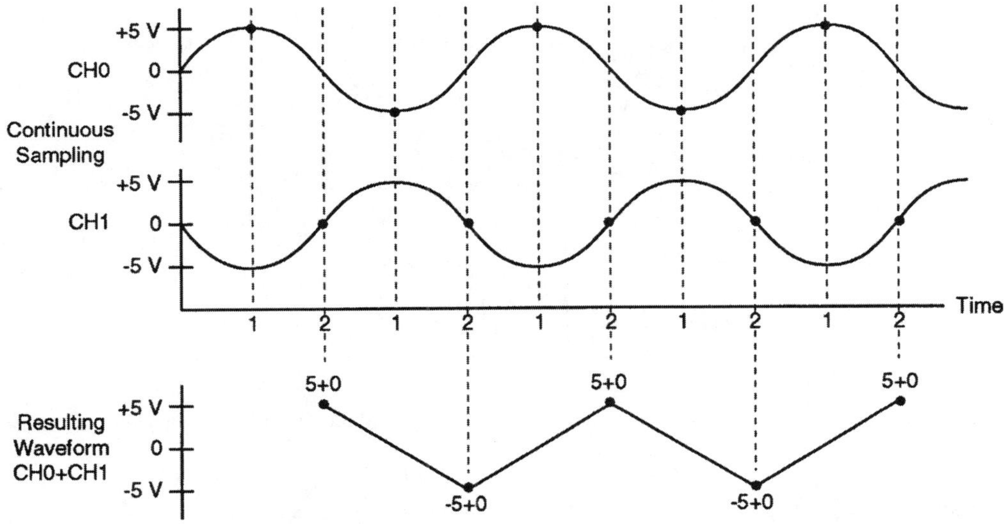

FIGURE 36.9 If the channel skew is large compared with the signal, then erroneous conclusions may result.

Multirate Scanning

Multirate scanning, a method that scans multiple channels at different scan rates, is a special case of continuous scanning. Applications that digitize multiple signals with a variety of frequencies use multirate scanning to minimize the amount of buffer space needed to store the sampled signals. Channel-independent ADCs are used to implement hardware multirate scanning; however, this method is extremely expensive. Instead of multiple ADCs, only one ADC is used. A channel/gain configuration register stores the scan rate per channel and software divides down the scan clock based on the per-channel scan rate. Software-controlled multirate scanning works by sampling each input channel at a rate that is a fraction of the specified scan rate.

Suppose the system scans channels 0 through 3 at 10 kS/s, channel 4 at 5 kS/s, and channels 5 through 7 at 1 kS/s. A base scan rate of 10 kS/s should be used. Channels 0 through 3 are acquired at the base scan rate. Software and hardware divide the base scan rate by 2 to sample channel 4 at 5 kS/s, and by 10 to sample channels 5 through 7 at 1 kS/s.

Simultaneous Sampling

For applications where the time relationship between the input signals is important, such as phase analysis of ac signals, simultaneous sampling must be used. DAQ boards capable of simultaneous sampling typically use independent instrumentation amplifiers and S/H circuitry for each input channel, along with an analog MUX, which routes the input signals to the ADC for conversion (as shown in Figure 36.10).

To demonstrate the need for a simultaneous-sampling DAQ board, consider a system consisting of four 50 kHz input signals sampled at 200 kS/s. If the DAQ board uses continuous scanning, the skew between each channel is 5 μs (1S/200 kS/s) which represents a 270° [(15 μs/20 μs) × 360°] shift in phase between the first channel and fourth channel. Alternatively, with a simultaneous-sampling board with a maximum 5 ns interchannel time offset, the phase shift is only 0.09° [(5 μs/20 μs) × 360°]. This phenomenon is illustrated in Figure 36.11.

Interval Scanning

For low-frequency signals, interval scanning creates the effect of simultaneous sampling, yet maintains the cost benefits of a continuous-scanning system. This method scans the input channels at one rate and uses a second rate to control when the next scan begins. If the input channels are scanned at the fastest

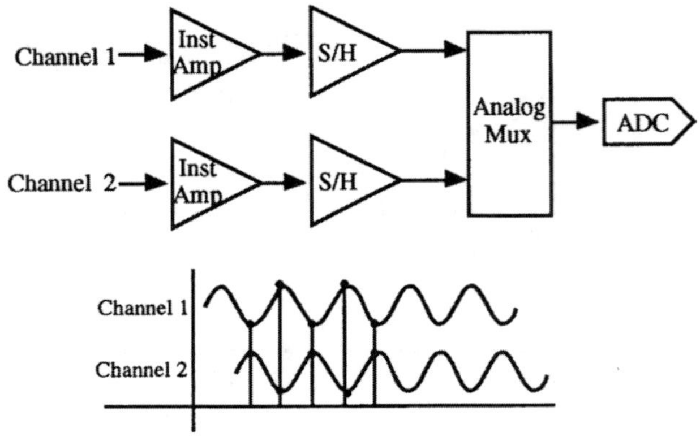

FIGURE 36.10 Block diagram of DAQ components used to sample multiple channels simultaneously.

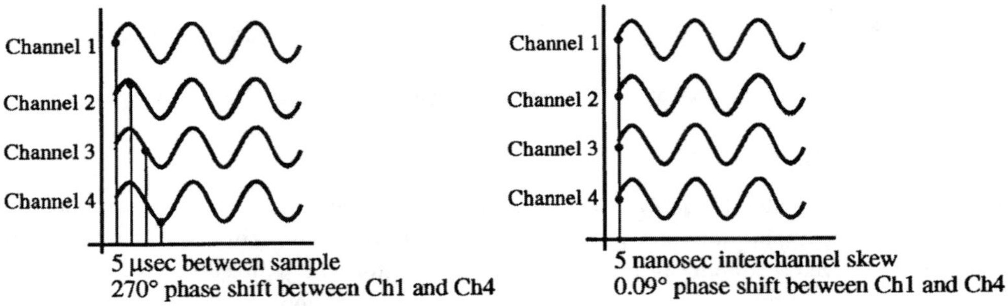

FIGURE 36.11 Comparison of continuous scanning and simultaneous sampling.

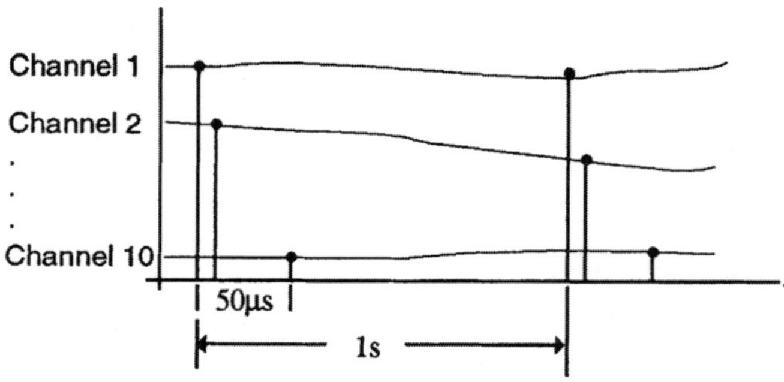

FIGURE 36.12 Interval scanning — all ten channels are scanned within 45 μs; this is insignificant relative to the overall acquisition rate of 1 S/s.

rate of the ADC, the effect of simultaneously sampling the channels is created. Interval scanning is appropriate for slow-moving signals, such as temperature and pressure. Interval scanning results in a jitter-free sample rate and minimal skew time between channel samples. For example, consider a DAQ system with ten temperature signals. By using interval scanning, a DAQ board can be set up to scan all channels with an interchannel delay of 5 μs, then repeat the scan every second. This method creates the effect of simultaneously sampling ten channels at 1 S/s, as shown in Figure 36.12.

To illustrate the difference between continuous and interval scanning, consider an application that monitors the torque and RPMs of an automobile engine and computes the engine horsepower. Two signals, proportional to torque and RPM, are easily sampled by a DAQ board at a rate of 1000 S/s. The values are multiplied together to determine the horsepower as a function of time.

A continuously scanning DAQ board must sample at an aggregate rate of 2000 S/s. The time between which the torque signal is sampled and the RPM signal is sampled will always be 0.5 ms (1/2000). If either signal changes within 0.5 ms, then the calculated horsepower is incorrect. But using interval scanning at a rate of 1000 S/s, the DAQ board samples the torque signal every 1 ms, and the RPM signal is sampled as quickly as possible after the torque is sampled. If a 5-μs interchannel delay exists between the torque and RPM samples, then the time skew is reduced by 99% [(0.5 ms – 5 μs)/0.5 ms], and the chance of an incorrect calculation is reduced.

Factors Influencing the Accuracy of Measurements

How does one determine if a plug-in DAQ will deliver the required measurement results? With a sophisticated measuring device like a plug-in DAQ board, significantly different accuracies can be obtained depending on the type of board used. For example, one can purchase DAQ products on the market today with 16-bit ADCs and get less than 12 bits of useful data, or one can purchase a product with a 16-bit ADC and actually get 16 bits of useful data. This difference in accuracies causes confusion in the PC industry where everyone is used to switching out PCs, video cards, printers, and so on, and experiencing similar results between equipment.

The most important thing to do is to scrutinize more specifications than the resolution of the ADC that is used on the DAQ board. For dc-class measurements, one should at least consider the settling time of the instrumentation amplifier, DNL, **relative accuracy**, INL, and noise. If the manufacturer of the board under consideration does not supply these specifications in the data sheets, ask the vendor to provide them or run tests to determine these specifications.

Defining Terms

Alias: A false lower frequency component that appears in sampled data acquired at too low a sampling rate.

Asynchronous: (1) Hardware — A property of an event that occurs at an arbitrary time, without synchronization to a reference clock. (2) Software — A property of a function that begins an operation and returns prior to the completion or termination of the operation.

Conversion time: The time required, in an analog input or output system, from the moment a channel is interrogated (such as with a read instruction) to the moment that accurate data are available.

DAQ (data acquisition): (1) Collecting and measuring electric signals from sensors, transducers, and test probes or fixtures and inputting them to a computer for processing: (2) Collecting and measuring the same kinds of electric signals with ADC and/or DIO boards plugged into a PC, and possibly generating control signals with DAC and/or DIO boards in the same PC.

DNL (differential nonlinearity): A measure in LSB of the worst-case deviation of code widths from their ideal value of 1 LSB.

INL (integral nonlinearity): A measure in LSB of the worst-case deviation from the ideal A/D or D/A transfer characteristic of the analog I/O circuitry.

Nyquist sampling theorem: A law of sampling theory stating that if a continuous bandwidth-limited signal contains no frequency components higher than half the frequency at which it is sampled, then the original signal can be recovered without distortion.

Relative accuracy: A measure in LSB of the accuracy of an ADC. It includes all nonlinearity and quantization errors. It does not include offset and gain errors of the circuitry feeding the ADC.

Further Information

House, R., "Understanding Important DA Specifications," *Sensors*, 10(10), June 1993.

House, R., "Understanding Inaccuracies Due to Settling Time, Linearity, and Noise," *National Instruments European User Symposium Proceedings*, November 10–11, 1994, pp. 11–12.

McConnell, E., "PC-Based Data Acquisition Users Face Numerous Challenges," *ECN*, August 1994.

McConnell, E., "Choosing a Data-Acquisition Method," *Electronic Design*, 43(6), 147, 1995.

McConnell, E. and Jernigan, D., "Data Acquisition," in *The Electronics Handbook*, J.C. Whitaker (Ed.), Boca Raton, FL: CRC Press, 1996, 1795–1822.

Potter, D. and Razdan, A., "Fundamentals of PC-Based Data Acquisition," *Sensors*, 11(2), 12–20, February 1994.

Potter, D., "Sensor to PC — Avoiding Some Common Pitfalls," *Sensors Expo Proceedings*, September 20, 1994.

Potter, D., "Signal Conditioners Expand DAQ System Capabilities," *I&CS*, 25–33, August 1995.

Johnson, G. W., *LabVIEW Graphical Programming*, New York: McGraw-Hill, 1994.

McConnell, E., "New Achievements in Counter/Timer Data Acquisition Technology," *MessComp 1994 Proceedings*, September 13–15, 1994, 492–498.

McConnell, E., "Equivalent Time Sampling Extends DA Performance," *Sensors Data Acquisition*, Special Issue, June, 13, 1995.

36.3 Magnetic and Optical Recorders

Yufeng Li

The heart of recording technology is for the process of information storage and retrieval. In addition to its obvious importance in different branches of science and engineering, it has become indispensable to our daily life. When we make a bank transaction, reserve an airplane ticket, use a credit card, watch a movie from a video tape, or listen to music from a CD, we are using the technology of recording. The general requirements for recording are information integrity, fast access, and low cost. Among the different techniques, the most popularly used ones are magnetic and optical recording.

Typical recording equipment consists of a read/write head, a medium, a coding/decoding system, a data access system, and some auxiliary mechanical and electronic components. The head and medium are for data storage and retrieval purposes, and the coding/decoding system is for data error correction. The data access system changes the relative position between the head and the medium, usually with a servo mechanism for data track following and a spinning mechanism for on-track moving. While the data access system and the auxiliary components are important to recording equipment, they are not considered essential in this chapter to the understanding of recording technology, and will not be covered. Interested readers are referred to Reference 1.

Magnetic Recording

At present, magnetic recording technology dominates the recording industry. It is used in the forms of hard disk, floppy disk, removable disk, and tape with either digital or analog mode. In its simplest form, it consists of a magnetic head and a magnetic medium, as shown in Figure 36.13. The head is made of a piece of magnetic material in a ring shape (core), with a small gap facing the medium and a coil away from the medium. The head records (writes) and reproduces (reads) information, while the medium stores the information. The recording process is based on the phenomenon that an electric current i generates a magnetic flux ϕ as described by Ampere's law. The flux ϕ leaks out of the head core at the gap, and magnetizes the magnetic medium which moves from left to right with a velocity V under the head gap. Depending on the direction of the electric current i, the medium is magnetized with magnetization M pointing either left or right. This pattern of magnetization is retained in the memory of the medium even after the head moves away.

Two types of head may be used for reproducing. One, termed the *inductive head,* senses magnetic flux change rate, and the other, named the *magnetoresistive* (MR) *head,* senses the magnetic flux. When an inductive head is used, the reproducing process is just the reverse of the recording process. The flux coming out of the magnetized medium surface is picked up by the head core. Because the medium magnetization under the head gap changes its magnitude and direction as the medium moves, an electric

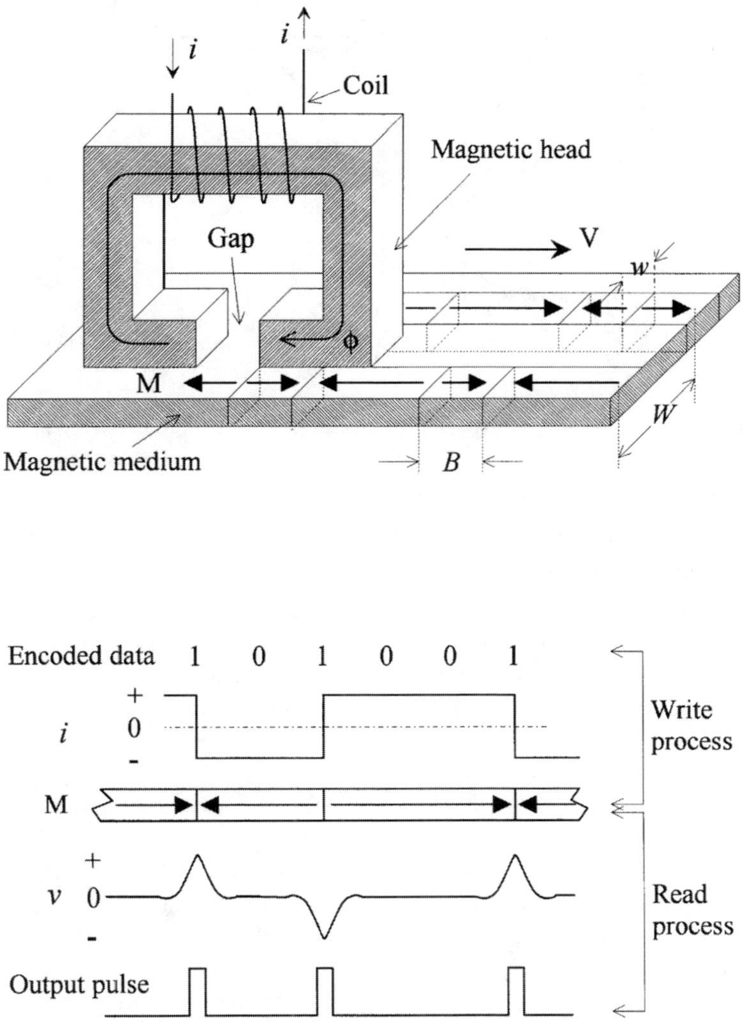

FIGURE 36.13 Conceptual diagrams illustrating the magnetic recording principle (a), and recording/reproducing process (b).

voltage is generated in the coil. This process is governed by Faraday's law. Figure 36.13b schematically shows the digital recording/reproducing process. First, all user data are encoded into a binary format — a serial of 1s and 0s. Then a write current *i* is sent to the coil. This current changes its direction whenever a 1 is being written. Correspondingly, a change of magnetization, termed a *transition,* is recorded in the medium for each 1 in the encoded data. During the reproducing process, the electric voltage induced in the head coil reaches a peak whenever there is a transition in the medium. A pulse detector generates a pulse for each transition. These pulses are decoded to yield the user data.

The minimum distance between two transitions in the medium is the flux change length *B*, and the distance between two adjacent signal tracks is the track pitch *W*, which is wider than the signal track width *w*. The flux change length can be directly converted into bit length with the proper code information. The reciprocal of the bit length is called *linear density,* and the reciprocal of the track pitch is termed *track density.* The information storage areal density in the medium is the product of the linear density and the track density. This areal density roughly determines how much information a user can store in a unit surface area of storage medium, and is a figure of merit for a recording technique. Much effort has been expended to increase the areal density. For example, it has been increased 50 times during the

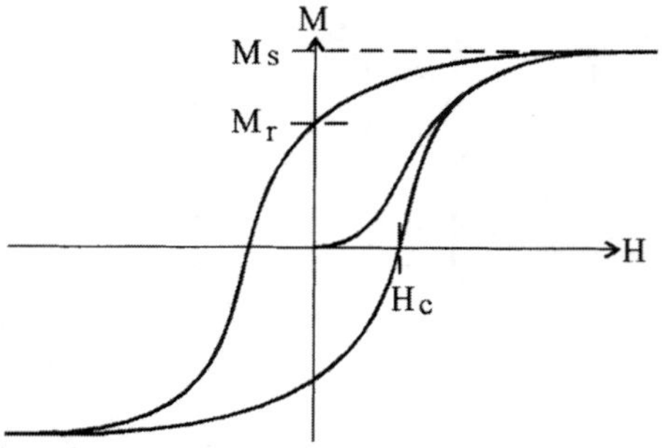

FIGURE 36.14 Hysteresis loop of a magnetic material shows the nonlinear relationship between M and H which results in magnetic memory.

last decade in hard disk drives, and is expected to continue increasing 60% per year in the foreseeable future. At present, state-of-the-art hard disk products feature areal densities of more than 7 Mbits/mm^2 ($B < 0.1$ μm and $W < 1.5$ μm). This gives a total storage capacity of up to 6 Gbytes for a disk of 95 mm diameter.

Magnetism and Hysteresis Loop

Magnetism is the result of uncompensated electron spin motions in an atom. Only transition elements exhibit this property, and nearly all practical interest in magnetism centers on the first transition group of elements (Mn, Cr, Fe, Ni, and Co) and their alloys. The strength of magnetism is represented by magnetization *M*, and is related to magnetic field *H* and magnetic flux density *B* by

$$B = \mu_0 (H + M) \tag{36.1}$$

where μ_0 is the permeability of vacuum. Since *M* is a property of a magnetic material, it does not exist outside the magnetic material. *H* represents the strength acting on a magnetic material from a magnetic field which is generated either by a magnetic material or by an electric current. *B* is the flux density which determines the induced electric voltage in a coil. The ratio of *B* with and without a magnetic material is the relative permeability μ of that magnetic material.

When a magnetic field *H* is applied to a piece of demagnetized magnetic material, the magnetization *M* starts increasing with *H* from zero. The rate of increase gradually slows down and *M* asymptotically approaches a value M_s at high *H*. If *H* is reduced to zero, then *M* is reduced to a lower value M_r. Continuous reduction of *H* to a very high negative value will magnetize the material to $-M_s$. In order to bring the material to demagnetized state, a positive field H_c is required. Further increase in the *H* field will bring the trace of *M* to a closed loop. This loop is the major hysteresis loop, as shown in Figure 36.14. The hysteresis loop shows that a magnetic material has memory. It is this memory that is used in the medium for storing information. H_c is the coercivity, indicating the strength of magnetic field required to erase the memory of a magnetic material. Magnetic materials with high H_c are "hard" magnets, and are suitable for medium applications if they have high M_r. On the other hand, magnetic materials with low H_c are "soft" magnets, and are candidates for head core materials if they have high M_s and high μ. M_r and M_s are the remanent and saturation magnetization, respectively, and their ratio is the remanent squareness. The flux density corresponding to M_s is B_s.

Magnetic Media

Magnetic media are used to store information in a magnetic recording system. In order to increase the areal density, we need to reduce flux change length *B* and track width *w*. Since *B* is limited by the term $M_r\delta/H_c$, where δ is the magnetic layer thickness, we can reduce *B* by either decreasing $M_r\delta$ or increasing H_c. However, the amplitude of the magnetic signal available for reproducing head is proportional to the term $M_r\delta w$. If we reduce track width *w* to increase areal density, we must increase $M_r\delta$ to avoid signal

TABLE 36.8 Remanence (M_r) and Coercivity (H_c) Values of Some Commonly Used Magnetic Media (some values are from Reference 5)

Group	Material	M_r (kA/m)	H_c (kA/m)	Application
Particulate	γ-Fe_2O_3	56–140	23–32	Floppy disk, audio, video, and instrumentation tapes
	γ-Fe_2O_3+Co	60–140	44–74	Floppy disk, audio, video, and instrumentation tapes
	CrO_2	110–140	38–58	Floppy disk, audio, video, and instrumentation tapes
	$BaFe_{12}O_{19}$	56	58	Floppy disk
Thin film	Co–Ni	600–1100	30–85	Hard disk
	Co–Fe	1100–1500	60–150	Hard disk
	Co–P	600–1000	36–120	Hard disk
	Co–Ni–Pt	600–1100	60–175	Hard disk
	Co–Cr–Ta	350–900	55–190	Hard disk
	Co–Cr–Pt	300–750	56–200	Hard disk

deterioration. In addition, if the magnetic layer is so thin that it causes thickness nonuniformity, more noise will appear in the reproducing process. Therefore, the major requirements for magnetic layer are high H_c, high M_r, and ease of making a uniform thin layer. Additional requirements include good magnetic and mechanical stability.

There are two groups of magnetic media. The first group is called particulate media because the magnetic materials are in the form of particles. This group includes iron oxide (γ-Fe_2O_3), cobalt-modified iron oxide (γ-Fe_2O_3+Co), chromium dioxide (CrO_2), metal particles, and barium ferrite ($BaFe_{12}O_{19}$). Some of these have been used in the magnetic recording for several decades. More recently, another group of media has been developed largely due to the ever-increasing demand for higher storage capacity in the computer industry. This group of media is the thin-film media, where the magnetic layer can be made as a continuous thin film. Most materials in this group are cobalt-based metal alloys. Compared with particulate media, the thin-film media usually have a higher coercivity H_c, a higher remanence M_r, and can be deposited in a very thin continuous film. Table 36.8 lists H_c and M_r for some of the most popularly used particulate and thin-film media. Note that magnetic properties are affected by the fabrication process and film structure. Therefore, their values can be out of the ranges of Table 36.8 if different processes are used.

Magnetic media can be classified into three general forms of applications. Tape is the oldest form and remains an important medium today. It is central to most audio, video, and instrumentation recording, although it is also used in the computer industry for archival storage. Tape is economical and can hold a large capacity, but suffers slow access time. Hard disk is primarily used as the storage inside a computer, providing fast data access for the user, but having poor transportability. Flexible disk is designed for easy data transportation, but is limited in capacity. Besides these three general forms of applications, a hybrid of flexible and hard disk is being gradually accepted. It is a removable rigid disk capable of holding up to several gigabytes of digital data. In addition, magnetic stripes are getting wide use in different forms of cards.

The magnetic layer alone cannot be used as a medium. It needs additional components to improve its chemical and mechanical durability. Typical cross sections of a particulate magnetic tape and a thin-film hard disk are shown in Figure 36.15. In the case of tape application, iron particles with typical size of 0.5 μm long and 0.1 μm wide are dispersed in a polymeric binder, together with solvents, lubricants, and other fillers to improve magnetic and mechanical stability. This dispersed material is then coated on an abiaxially oriented polyethylene terephthalate substrate. An optional back coat may also be applied to the other side of the substrate. The cross section of a hard disk is more complex. A high-purity aluminum–magnesium (5 wt%) substrate is diamond turned to a fine surface finish, and then electrolessly plated with a nonmagnetic nickel–phosphorus (10 wt%) undercoat. This layer is used to increase the

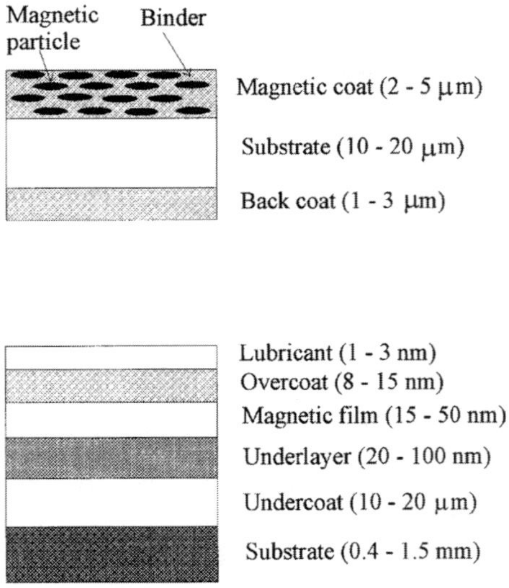

FIGURE 36.15 Cross-sectional views of a particulate magnetic tape (top) and a thin film hard disk (bottom).

TABLE 36.9 Relative Permeability (μ), Saturation Flux Density (B_s), Coercivity (H_c) and Resistivity (ρ) Values of Some Commonly Used Magnetic Head Materials at Low Frequency (some values are from Reference 5)

Material	μ	B_s (T)	H_c (A/m)	ρ (μΩ·cm)	Application
Ni–Fe–Mo	11000	0.8	2.0	100	Audio tape
Ni–Zn	300–1500	0.4–0.46	11.8–27.6	10^{11}	Floppy and hard disk drives, video and instrumentation tapes
Mn–Zn	3000–10000	0.4–0.6	11.8–15.8	10^6	Floppy and hard disk drives, video and instrumentation tapes
Fe–Si–Al	8000	1.0	2.0	85	Floppy and hard disk drives, video and instrumentation tapes
Ni–Fe	2000–4000	1.0	<10	20	Hard disk drives

hardness, reduce the defects, and improve the finish of the Al–Mg alloy, and is polished to a super surface finish. Next, an underlayer of chromium is sputtered to control the properties of the magnetic film, followed by sputtering the magnetic film. Finally, a layer of hydrogenated or nitrogenated carbon is overcoated on the magnetic film, and an ultrathin layer of perfluorinated hydrocarbon liquid lubricant is applied on top. The carbon and lubricant layers are used to improve the corrosion and mechanical resistance of the disk. For a 95 mm disk the finished product should have a surface flatness better than 10 μm and a tightly control surface roughness. In some applications, an arithmetic average roughness (R_a) of less than 0.5 nm is required.

Magnetic Heads

Magnetic heads have three functions: recording, reproducing, and erasing. Usually for stationary head applications such as tape drives, multiple heads are used to perform these functions. For moving head applications such as disk drives, a single head is employed because of the requirements of simple connections and small head mass for fast data access. Most of these heads are the inductive type, where the fundamental design is an inductive coil and a magnetic core. The general requirements for the core materials are high relative permeability μ, high saturation flux density B_s, low coercivity H_c, high electric resistivity ρ, and low magnetostriction coefficient λ. Some of the properties for the commonly used core materials are listed in Table 36.9.

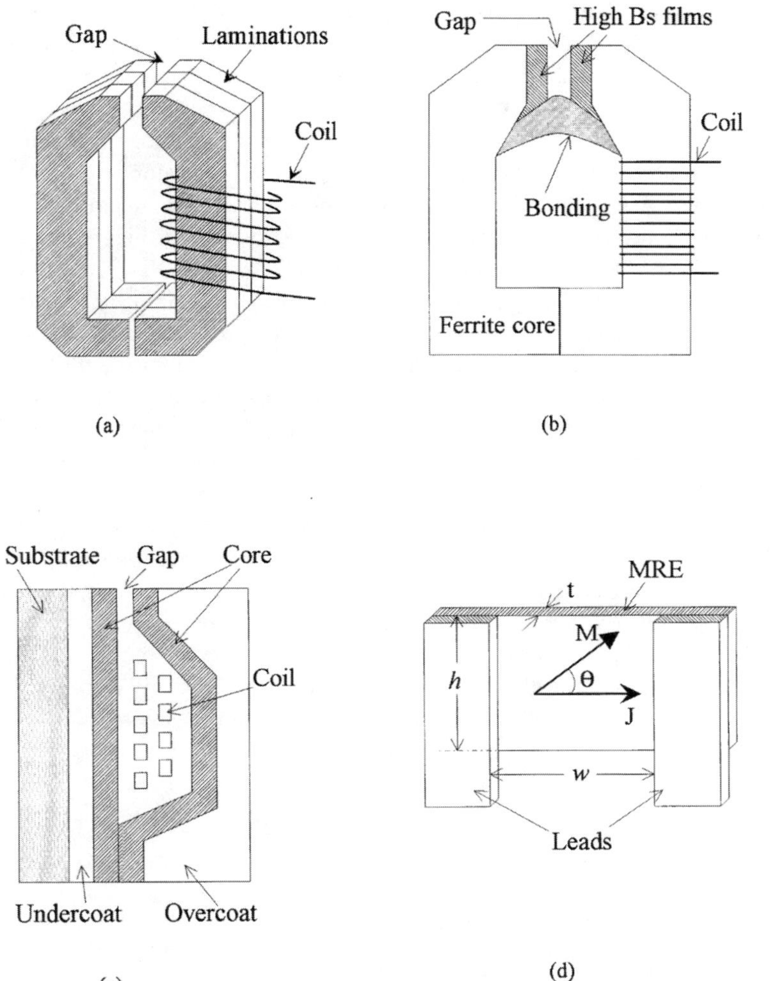

FIGURE 36.16 Schematic illustrations of (a) a laminated head, (b) cross-section of an MIG head, (c) cross-section of a thin film head, and (d) an MR sensor with leads.

The evolution of the magnetic head follows the selection of core materials, as shown in Figure 36.16. Early heads used laminated molybdebum Permalloy (Ni-Fe-Mo, 79-17-4 wt%). These heads are inexpensive to make, and have low H_c and high μ and B_s. The primary drawbacks are frequency limitation, gap dimension inaccuracy, and mechanical softness. Frequency limitation is caused by the difficulty of making the lamination layer thinner than 25 μm. Eddy current loss, which is proportional to layer thickness and square root of frequency, reduces the effective permeability. As a result, laminated heads are seldom used for applications exceeding 10 MHz. Gap dimension inaccuracy is associated with the head fabrication process, and makes it unsuitable for high areal density applications. Lack of mechanical hardness reduces its usable life.

One way to reduce eddy current loss is to increase core material electric resistivity. Two types of ferrite material have high resistivity (four to nine orders higher than Permalloy) and reasonable magnetic properties: Ni–Zn and Mn–Zn. These materials are also very hard, elongating head life during head/medium contacts. The major deficiency of ferrite materials is their low B_s values. In order to record in high H_c media, high flux density B is needed in the head core. When the flux density in the core material reaches its saturation B_s, it will not increase despite the increase of recording current or coil turns. This saturation starts from the corners of the gap due to its geometry. To remedy this deficiency, a layer of metallic alloy material with much higher B_s is deposited on the gap faces. This type of head is called the metal-in-gap (MIG) head. Sendust (Fe–Si–Al, 85–9.6–5.4 wt%) is one of the materials used for the deposition. MIG heads are capable of recording up to 100 MHz frequency and 180 kA/m medium coercivity.

Thin-film heads capitalize on semiconductor-like processing technology to reduce the customized fabrication steps for individual heads. The core, coil, gap, and insulator layers are all fabricated by electroplating, sputtering, or evaporation. Due to the nature of the semiconductor process, the fabrication is accurate for small dimensions. Small gap dimensions are suitable for high linear and track density, and small core dimensions allow the use of high B_s Permalloy material (Ni–Fe, 80–20 wt%) as core with low inductance for high data rate applications. Thin-film heads are used for high medium H_c, high areal density applications. The high cost of the semiconductor-like process is offset by high throughput: a 150 × 150 mm wafer can produce 16,000 nanoslider heads. One disadvantage is the limited-band recording capability because the small pole length limits low-frequency response and introduces undershoots. A second disadvantage is the Barkhausen noise, which is caused by the relatively small number of magnetic domains in the core. At present, thin-film heads are used up to frequencies of 80 MHz and medium coercivity of 200 kA/m. MIG thin-film heads are also being used for high-coercivity applications.

An inductive head is often used for both recording and reproducing. The optimal performance cannot be achieved because recording and reproducing have contradictory requirements for head design. To solve this problem, the MR head has been developed. The MR head is for reproducing only, and an inductive head is used for recording. As schematically shown in Figure 36.16, an MR head has a magnetoresistive element (MRE) and two electric leads. The MRE is a Permalloy stripe (Ni–Fe, 80–20 wt%), with thickness t, width w, and height h. An electric current, with density J, passes through the MRE through the leads. The electric resistivity of the MRE is a function of the angle θ between J and MRE magnetization M:

$$\rho_\theta = \rho\left(1 + \frac{\Delta\rho}{\rho}\cos^2\theta\right) \tag{36.2}$$

where ρ_θ is the resistivity at θ, ρ is the resistivity at $\theta = 90°$, and $\Delta\rho$ is the resistivity difference between $\theta = 0°$ and $\theta = 90°$. $\Delta\rho/\rho$ is the anisotropic MR ratio (AMR) of the MRE. Usually a transverse magnetic field is applied to the MRE so that $\theta = \theta_0$ when the head is not reading a magnetic signal. Assume that a magnetic signal from the medium rotates M from θ_0 to θ, then an electric voltage change v will be detected across the MRE as the output signal:

$$v = Jw\rho\frac{\Delta\rho}{\rho}\left(\sin^2\theta_0 - \sin^2\theta\right) \tag{36.3}$$

where θ_0 is the bias angle and is set to 45° for good linearity. In practice, a longitudinal bias is also used along the MRE width direction to stabilize the magnetic domain and reduce large Barkhausen noise. To compare the output between an MR head and an inductive head, we write the inductive head output using Faraday's equation:

$$v = -nV\frac{d\phi}{dx} \tag{36.4}$$

where n is the number of the coil turns, V is the medium velocity, ϕ is the magnetic flux, and x is the coordinate axis fixed on the medium surface. Equations 36.3 and 36.4 tell us that while inductive head output is proportional to medium velocity and not suitable for low-velocity applications, the MR head can be used for either high- or low-velocity applications.

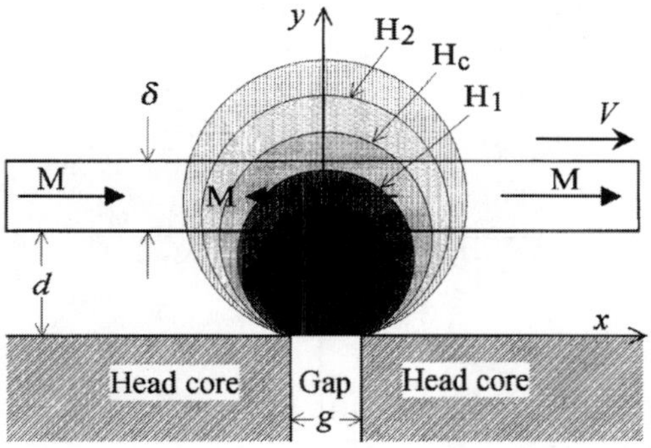

FIGURE 36.17 The constant horizontal fields of Karlqvist approximation are circles resting on the gap corners of a head, and the change of magnetization in the medium is gradual.

Recording Process

We can imagine the recording process in two steps. First, the magnetic flux flowing in the head core generates a fringe magnetic field around the gap. Then the magnetic field magnetizes the magnetic medium and leaves a magnetization transition in it. Partly due to the nonlinear nature of the hysteresis loop, the recording process is so complex that there has been no rigorous explanation. However, we can still obtain significant insights into the recording process by using simple models if we keep in mind the limitations.

If we set the origin of a coordinate system at the center of the gap with x axis on the head surface and y axis pointing away from the head, then the longitudinal magnetic field H_x and perpendicular magnetic field H_y of this head can be expressed by the Karlqvist approximation [2]:

$$H_x = \frac{ni}{\pi\left(g + lA_g/\mu A_c\right)} \tan^{-1}\left[\frac{yg}{x^2 + y^2 - \left(g^2/4\right)}\right] \tag{36.5}$$

$$H_y = \frac{ni}{2\pi\left(g + lA_g/\mu A_c\right)} \ln\left[\frac{\left(x - g/2\right)^2 + y^2}{\left(x + g/2\right)^2 + y^2}\right] \tag{36.6}$$

where n is the number of coil turns, i is the current in the coil, g is the gap length, l is the core length, A_g is the core cross-sectional area, μ is the relative permeability of core material, and A_c is the gap crosssectional area. Both Equations 36.5 and 36.6 give accurate results for points $0.25g$ away from gap corners. Since longitudinal recording mode dominates the magnetic recording industry, we will focus on the field H_x. Equation 36.5 shows that the contours of constant H_x field are circles nesting on the two gap corners, as shown in Figure 36.17. The greater the diameter of the circle, the weaker the magnetic field. Assume a magnetic medium, moving from left to right with a distance d above the head, has a thickness δ and a magnetization M pointing to the right. At some instant the recording current turns on and generates the magnetic field H_x above the gap as depicted in Figure 36.17. On the circumference of $H_x = H_c$, half of the medium material has its magnetization reversed and half remains the same, resulting in a zero total magnetization. Since H_x has a gradient, the medium closer to the gap (inside a smaller circle) gets its magnetization reversed more completely than the medium farther away from the gap (outside a bigger circle). Therefore, magnetic transition is gradual in the medium even if the change of recording current follows a step function. Assume the original magnetization is M_r and the completely reversed magnetization is $-M_r$, this gradual change of magnetization for an isolated transition can be modeled by [3]:

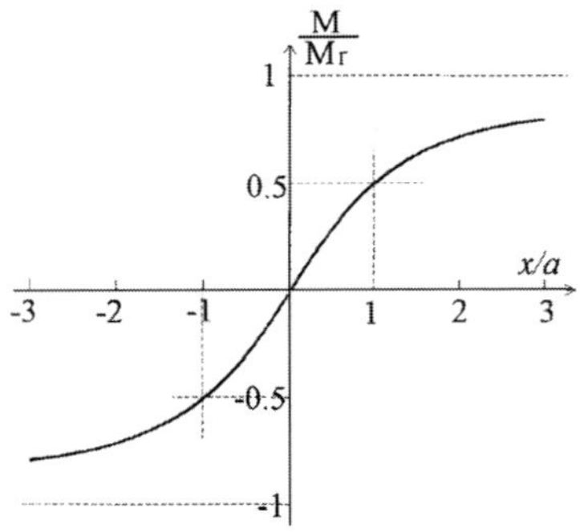

FIGURE 36.18 An isolated arctangent magnetization transition from negative M_r to positive M_r.

$$M = \frac{2}{\pi} M_r \tan^{-1} \frac{x}{a} \tag{36.7}$$

where x is the distance from the center of transition and a is a parameter characterizing the sharpness of the transition as shown in Figure 36.18. Assuming a thin-film medium and using the Karlqvist approximation for the head field, a is found to be [4–6]:

$$a = \frac{(1-S^*)(d+\delta/2)}{\pi Q} + \sqrt{\left[\frac{(1-S^*)(d+\delta/2)}{\pi Q}\right]^2 + \frac{M_r\delta(d+\delta/2)}{\pi Q H_c}} \tag{36.8}$$

where S^* is the medium loop squareness and Q is the head-field gradient factor. For a reasonably well designed head, $Q \approx 0.8$. It is obvious that we want to make parameter a as small as possible so that we can record more transitions for a unit medium length. If the head gap length g and medium thickness δ are small compared with head/medium separation d, and the medium has a squareness of one, then the minimum possible value of a is [7]

$$a_m = \begin{cases} \dfrac{M_r\delta}{2\pi H_c} & \dfrac{M_r\delta}{4\pi H_c d} \geq 1 \\ \sqrt{\dfrac{M_r\delta d}{\pi H_c}} & \dfrac{M_r\delta}{4\pi H_c d} < 1 \end{cases} \tag{36.9}$$

In order to decrease the value of a and therefore increase areal density, we need to reduce medium remanence M_r, thickness δ, head/medium separation d, and to increase coercivity H_c.

Reproducing Process

In contrast to the recording process, the reproducing process is well understood. The flux density induced in the head core is on the order of a few millitesia, yielding a linear process for easier mathematical treatment. The head fringe field is the Karlqvist approximation (Equation 36.5) and the foundation is the reciprocity theorem. For an isolated transition (Figure 36.18) with a thin magnetic layer $\delta \ll d$, the induced electric voltage v for an inductive head is [7]:

$$v(x) = \frac{-2\mu_0 V w M_r \delta n}{\pi\left(g + lA_g/\mu A_c\right)} \left[\tan^{-1}\left(\frac{g/2+x}{a+d}\right) + \tan^{-1}\left(\frac{g/2-x}{a+d}\right)\right] \tag{36.10}$$

where μ_0 is the permeability of vacuum, μ is the relative permeability of the core, V is the medium velocity, w is the track width, n is the number of coil turns, g is the head gap length, d is the head/medium separation, and x is the distance between the center of the medium transition and the center of the head

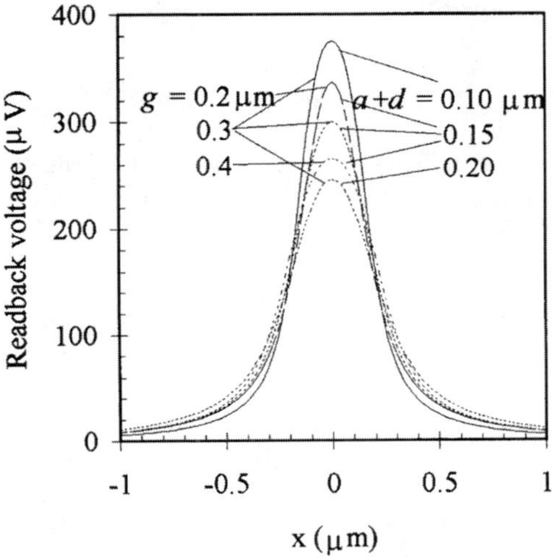

FIGURE 36.19 The reproducing voltage of an inductive head over an isolated arctangent transition shows the effects of gap length *g*, parameter *a*, and head/medium separation *d*.

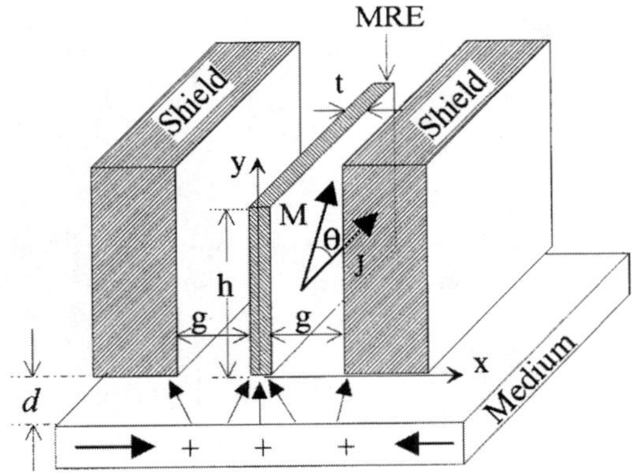

FIGURE 36.20 Schematic diagram of a shielded MR head with a shield to MRE distance *g*.

gap. The term $lA_g/\mu A_c$ is closely related to g for head efficiency. When a transition passes under the head, its voltage starts with a very low value, reaches a peak, then falls off again, as shown in Figure 36.19, where the following typical values for a hard disk drive are used: $V = 20$ m/s, $w = 3.5$ μm, $M_r = 450$ kA/m, $\delta = 50$ nm, $n = 50$, $lA_g/\mu A_c = 0.1g$. The effects of g and $a + d$ are shown in Figure 36.19. Since a greater peak voltage and a narrower spatial response are desired for the reproducing process, smaller g and $a + d$ values are helpful.

When an MR head is used for reproducing, the MRE is usually sandwiched between two magnetic shields to increase its spatial resolution to medium signals, as shown in Figure 36.20. Since the MR head is flux sensitive, the incident flux ϕ_i on the bottom surface of the MRE should be derived as a function of the distance (x) between the center of MRE and the center of the transition [7, 8]:

$$\phi_i(x) = \frac{2\mu_0 w M_r \delta(a+d)}{\pi g}\left\{ f\left[\frac{x+(g+t)/2}{a+d}\right] - f\left[\frac{(x+t)/2a+d}{}\right] + f\left[\frac{x-(g-t)/2}{a+d}\right] - f\left[\frac{(x-t)/2}{a+d}\right]\right\} \tag{36.11}$$

where g is the distance between the MRE and the shield, t is the MRE thickness, and

$$f(\beta) = \beta \tan^{-1}\beta - \ln\sqrt{1+\beta^2} \tag{36.12}$$

The angle between magnetization and current varies along the MRE height h. To find out the variation, we need to calculate the signal flux decay as a function of y by

$$\phi_s(y) = \phi_i \frac{\sinh[(h-y)/l_c]}{\sinh(h/l_c)} \tag{36.13}$$

where

$$l_c = \sqrt{\mu g t/2} \tag{36.14}$$

Then the bias angle θ_0 and signal angle θ can be calculated by

$$\sin\theta_0(y) = \frac{\phi_b(y)}{M_s t} \tag{36.15}$$

and

$$\sin\theta(y) = \frac{\phi_s(y)+\phi_b(y)}{M_s t} \tag{36.16}$$

where ϕ_b is the biasing flux in the MRE and M_s is the saturation magnetization of the MRE. Application of Equations 36.15 and 36.16 to Equation 36.3 and integration over height h lead to

$$v = Jw\rho \frac{\Delta\rho}{\rho}\frac{1}{h}\left[\int_0^h \sin^2\theta_0(y)\,dy - \int_0^h \sin^2\theta(y)\,dy\right] \tag{36.17}$$

For an MR head with a 45° bias at the center and small height $h \ll l_c$, the peak voltage is [6]

$$v_p \approx \frac{9\Delta\rho J w M_r \delta(g+t)}{8\sqrt{2}\,tgM_s}\tan^{-1}\left[\frac{g}{2(a+d)}\right] \tag{36.18}$$

The general shape of the reproducing voltage from an MR head is similar to that in Figure 36.19.

The study of an isolated transition reveals many intrinsic features of the reproducing process. However, transitions are usually recorded closely in a magnetic medium to achieve high linear density. In this case, the magnetization variation in the medium approaches a sinusoidal wave. That is,

$$M(x) = M_r \sin\frac{2\pi}{\lambda}x \tag{36.19}$$

where λ is the wavelength. The reproducing voltage in an inductive head becomes [9, 10]

$$v(x) = \frac{-\mu_0 V w M_r n g}{g + lA_g/\mu A_c}\left(e^{-2\pi d/\lambda}\right)\left(1-e^{-2\pi\delta/\lambda}\right)\left(\frac{\sin\frac{\pi g}{\lambda}}{\pi g/\lambda}\right)\cos\frac{2\pi}{\lambda}x \tag{36.20}$$

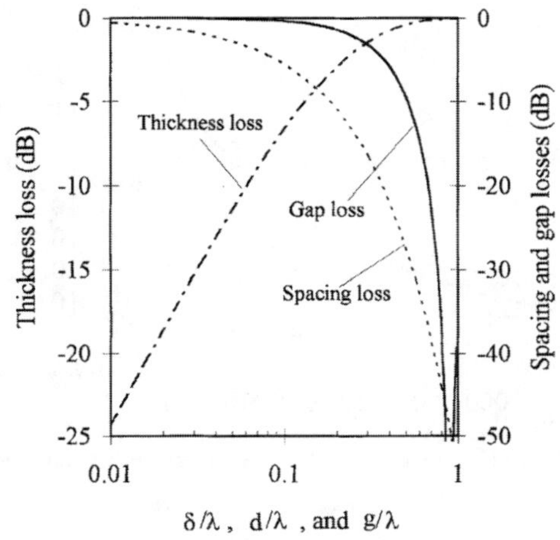

FIGURE 36.21 Spacing, thickness, and gap losses of an inductive head vs. frequency for the reproducing of a sinusoidal medium magnetization.

This equation presents all the important features of the high-linear-density reproducing process. The term $\exp(-2\pi d/\lambda)$ is the spacing loss. It shows that the reproducing voltage falls exponentially with the ratio of head/medium spacing to wavelength. The second term $1 - \exp(-2\pi\delta/\lambda)$ is the thickness loss. The name of this term is misleading because its value increases with a greater medium thickness. However, the rate of increase diminishes for thicker medium. In fact, 80% of the maximum possible value is achieved by a medium thickness of 0.25λ. The last term $\sin(\pi g/\lambda)/(\pi g/\lambda)$ is the gap loss. This term is based on the Karlqvist approximation. If a more accurate head fringe field is used, this term is modified to $\sin(\pi g/\lambda)/(\pi g/\lambda)\cdot(1.25g^2 - \lambda^2)/(g^2 - \lambda^2)$ [11]. It shows a gap null at $\lambda = 1.12g$, and limits the shortest wavelength producible. These three terms are plotted in Figure 36.21. The most significant loss comes from the spacing loss term, which is 54.6 dB for $d = \lambda$. Therefore, one of the biggest efforts spent on magnetic recording is to reduce the head/medium spacing as much as possible without causing mechanical reliability issues. For an MR head, the reproducing voltage is [11]

$$v \propto \frac{4M_r i\Delta\pi w\lambda}{ht}\left(e^{-2\pi d/\lambda}\right)\left(1-e^{-2\pi\delta/\lambda}\right)\left(\frac{\sin\dfrac{\pi g}{\lambda}}{\pi g/\lambda}\right)\sin\frac{\pi(g+t)}{\lambda}\cos\frac{2\pi}{\lambda}x \qquad (36.21)$$

Digital vs. Analog Recording

Due to the nonlinearity of the hysteresis loop, magnetic recording is intrinsically suitable for digital recording, where only two states (1 and 0) are to be recognized. Many physical quantities, however, are received in analog form before they can be recorded, such as in consumer audio and instrumentation recording. In order to perform such recording, we need to either digitize the information or use the analog recording technique. In the case of digitization, we use an analog-to-digital converter to change a continuous signal into binary numbers. The process can be explained by using the example shown in Figure 36.22. An electric signal V, normalized to the range between 0 and 1, is to be digitized into three bits. The signal is sampled at time $t = 1, 2, \ldots, 6$. At each sampling point, the first bit is assigned a 1 if the value of the continuous signal is in the top half (>0.5), otherwise assigned a 0. The second bit is assigned a 1 if the value of the continuous signal is in the top half of each half ($0.25 \le V < 0.5$, or >0.75), otherwise assigned a 0. The third bit is assigned similarly. The first bit is the most significant bit (MSB), and the last bit is the least significant bit (LSB). The converted binary numbers are listed below each sampling point in Figure 36.22. This process of digitization is termed *quantization*. In general, the final quantization interval is

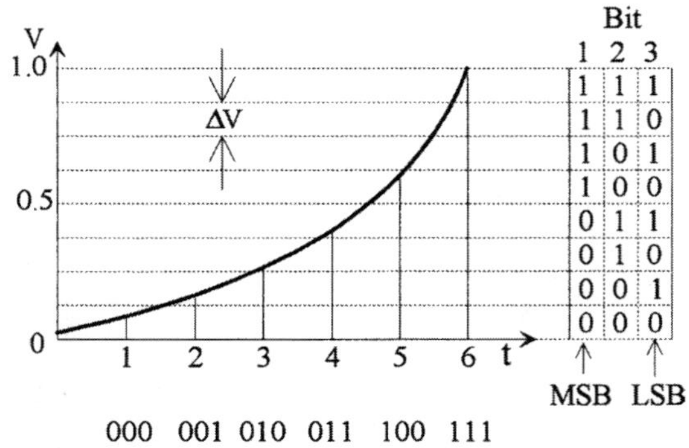

FIGURE 36.22 Schematic illustration of the quantization of a continuous signal to three bits.

$$\Delta V = \frac{V}{2^N} \tag{36.22}$$

Where V is the total voltage range and N is the number of bits. Because we use a finite number of bits, statistically there is a difference between the continuous signal and the quantized signal at the sampling points. This is the quantization error. It leads to a signal-to-quantization-noise ratio (SNR) [12]:

$$\text{SNR} = \frac{12P_s}{\Delta V^2} \tag{36.23}$$

where P_s is the mean square average signal power. For a signal with uniform distribution over its full range, this yields

$$\text{SNR} = 2^{2N} \tag{36.24}$$

For a sinusoidal signal, it changes to

$$\text{SNR} = 1.5 \times 2^{2N} \tag{36.25}$$

The SNR can be improved by using more bits. This improvement, however, is limited by the SNR of the incoming continuous signal. The quantized signal is then pulse code modulated (PCM) for recording.

For analog recording, a linear relationship between the medium magnetization and the recording signal is required. This is achieved through the anhysteretic magnetization process. If we apply an alternating magnetic field and a unidirectional magnetic field to a previously demagnetized medium, and then reduce the amplitude of the alternating field to zero before we remove the unidirectional field, the remanent magnetization shows a pseudolinear relationship with the unidirectional field strength H_u up to some level. Figure 36.23 shows such an anhysteretic curve. The linearity deteriorates as H_u gets greater. In applications, the recording signal current is biased with an ac current of greater amplitude and higher frequency. Therefore, it is also termed ac-biased recording. Analog recording is easy to implement, at the price of a lowered SNR because remanent magnetization is limited to about 30% of the maximum possible M_r to achieve good linearity.

Recording Codes

PCM is a scheme of modifying input binary data to make them more suitable for a recording and reproducing channel. These schemes are intended to achieve some of the following goals: (1) reducing the dc component, (2) increasing linear density, (3) providing self-clocking, (4) limiting error propagation, and (5) achieving error-free detection. There are numerous code schemes; only three of the ones developed early are shown in Figure 36.24. The earliest and most straightforward one is the return-to-zero (RZ)

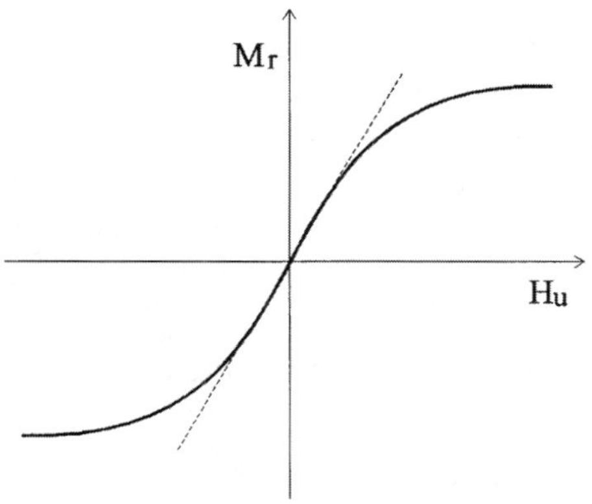

FIGURE 36.23 An anhysteretic remanent magnetization shows a pseudolinear relationship with the applied unidirectional magnetic field to some H_u level.

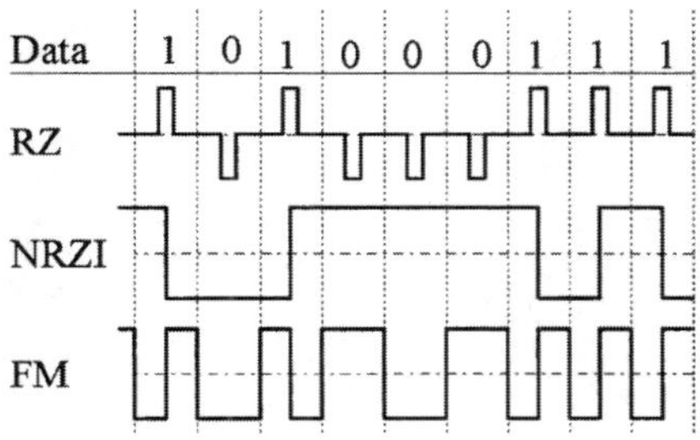

FIGURE 36.24 Comparison of some early developed codes.

code. In this scheme a positive and negative pulse is used to represent each 1 and 0, respectively, of the data. The main drawback is that direct recording over old data is not possible due to the existence of zero recording current between two data. It also generates two transitions for each bit, therefore reducing the linear density. In addition, it only uses half of the available recording current range for a transition. The non-return-to-zero-invert (NRZI) method was developed to alleviate some of these problems. It changes the recording current from one direction to the other for each 1 of the data, while making no changes for all 0s. However, it has a strong dc component and may lose reproducing synchronization if there is a long string of 0s in the input data. In addition, reproducing circuits are usually not designed for dc signal processing. In frequency modulation (FM) code there is always a transition at the bit–cell boundary which acts as a clock. There is also an additional transition at the bit–cell center for each 1 and no transition for 0s. It reduces the dc component significantly. The primary deficiency is the reduction of linear density since there are two transitions for each 1 in the data.

The most popularly used codes for magnetic recording are the run-length-limited (RLL) codes. They have the general form of $m/n(d, k)$. In these schemes, data are encoded in groups. Each input group has m bits. After encoding, each group contains n bits. In some schemes multiple groups are coded together. d and k are the minimum and maximum 0s, respectively, between two consecutive 1s in the encoded sequence. While d is used to limit the highest transition density and intersymbol interference, k is employed to ensure adequate transition frequency for reproducing clock synchronization. The encoding is carried out by using a lookup table, such as Table 36.10 for a 1/2(2,7) code [13].

Head/Medium Interface Tribology

As expressed in Equation 36.20, the most effective way to increase signal amplitude, therefore areal density, is to reduce head/medium spacing d. However, wear occurs when a moving surface is in close proximity to another surface. The amount of wear is described by Archard's equation:

TABLE 36.10 1/2(2,7) Code

Before Coding	After Coding
10	0100
11	1000
000	000100
010	100100
011	001000
0010	00100100
0011	00001000

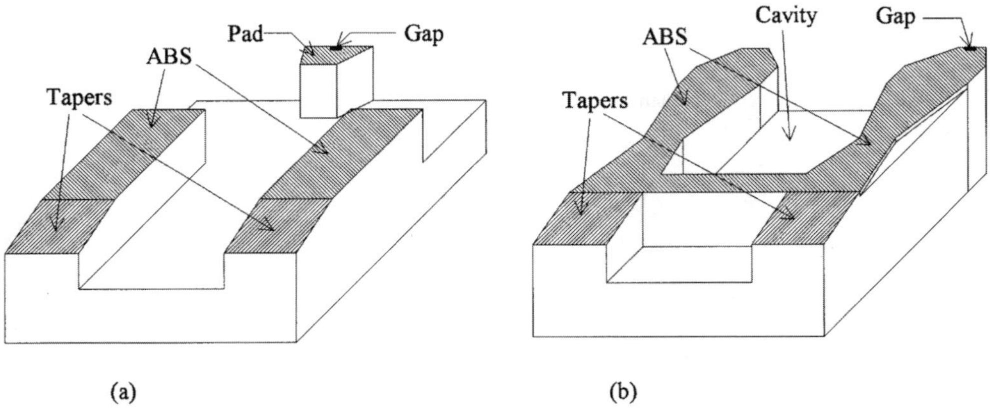

FIGURE 36.25 The ABS of (a) a tri-pad slider for pseudocontact recording and (b) a SAP slider for conventional flying recording.

$$V = k\frac{Ws}{H} \tag{36.26}$$

where V is the volume worn away, W is the normal load, s is the sliding distance, H is the hardness of the surface being worn away, and k is a wear coefficient. In order to increase medium hardness H, hard Al_2O_3 particles are dispersed in particulate media and a thin layer of hard carbon (≈10 nm) with either hydrogenation or natrogenation is coated on thin-film media of hard disks. A layer of liquid lubricant, typically perfluoropolyethers with various end groups and additives, is applied on top of the carbon film to reduce the wear coefficient k. Load is minimized to reduce wear while keeping adequate head/medium dynamic stability. For applications where the sliding distance s is modest over the lifetime of the products such as floppy disk drives and consumer tapes drives, the head physically contacts the medium during operations. In the case of hard disk application, heads are separated nominally from the media by a layer of air cushion. The head is carried on a slider, and the slider uses air-bearing surfaces (ABS) to create the air film based on hydrodynamic lubrication theory. Figure 36.25 shows two commonly used ABS. Tapers are used to help the slider take off and maintain flying stability. ABS generates higher-than-ambient pressure to lift the slider above the medium surface during operations. The tripad slider is for pseudo-contact applications while the subambient-pressure (SAP) slider is for flying (such as MR head) applications. Because the relative linear velocity between the slider and the medium changes when the head moves to different radii, a cavity region is used in the SAP slider to generate suction force to reduce flying height variation. The ABS is designed based on the modified Reynolds equation:

$$\frac{\partial}{\partial X}\left(PH^3Q\frac{\partial P}{\partial X}\right) + \frac{\partial}{\partial Y}\left(PH^3Q\frac{\partial P}{\partial Y}\right) = \Lambda_x\frac{\partial(PH)}{\partial X} + \Lambda_y\frac{\partial(PH)}{\partial Y} + \sigma\left(\frac{\partial(PH)}{\partial T}\right) \tag{36.27}$$

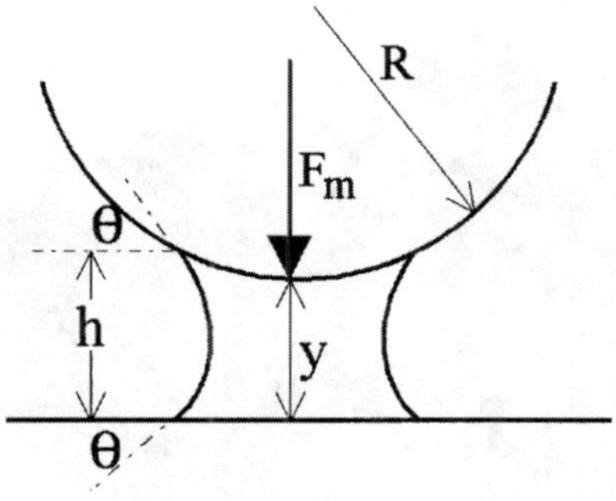

FIGURE 36.26 Formation of meniscus between a sphere tip and a flat surface.

where X and Y are coordinates in the slider longitudinal and transverse directions normalized by the slider length and width, respectively, P is the hydrodynamic pressure normalized by the ambient pressure, H is the distance between the ABS and medium surface normalized by the minimum flying height, Q is the molecular slip factor, T is time normalized by the characteristic squeeze period, Λ_x and Λ_y are the bearing numbers in the x and y directions, respectively, and σ is the squeeze number. A full derivation and explanation of the Reynolds equation can be found in Reference 14. At present, high-end hard disk drives feature a flying height on the order of 20 to 50 nm.

When power is turned off, the slider in the popularly used Winchester-type drives rests on the medium surface. Although the ABS and medium surface look flat and smooth, they really consist of microscopic peaks and valleys. If we model an ABS/medium contact by a flat surface and a sphere tip, the liquid lubricant on the medium surface causes a meniscus force F_m as depicted in Figure 36.26 [15]:

$$F_m = \frac{4\pi R\gamma\cos\theta}{1 + y/(h-y)} \tag{36.28}$$

where R is the radius of the sphere, γ is the surface tension of the lubricant, θ is the contact angle between the lubricant and the surfaces, y is the sphere to flat surface distance, and h is the lubricant thickness. Detailed analysis [16] shows that the static friction F at a head/medium interface is a function of several parameters:

$$F = f(h, R, A, \eta, \gamma, \theta, E', \phi, \sigma, s) \tag{36.29}$$

where A is the ABS area, η is the peak density, E' is the effective modulus of elasticity, ϕ is the peak height distribution, σ is the rms peak roughness, and s is the solid-to-solid shear strength. If friction F is too large, either the drive cannot be started or the head/medium interface is damaged. While friction can be reduced practically by reducing A, γ, and increasing θ, the most effective ways are to control h, σ, η, and ϕ. Too thin a lubricant layer will cause wear, and too thick will induce high friction. This limits h to the range of 1 to 3 nm. σ is controlled by surface texture. Historically, texture is created by mechanical techniques using either free or fixed abrasives. This leaves a surface with a random feature and is unsuitable for controlling η and ϕ. Recently, people started to use lasers [17]. This technique generates a surface texture with well-defined η and ϕ to improve wear and friction performance. Figure 36.27 shows AFM images of a mechanical and a laser texture.

Optical Recording

The major obstacle to achieving higher areal density in magnetic recording is the spacing loss term. It is a great engineering challenge to keep heads and media in close proximity while maintaining the head/medium interface reliable and durable. Care must also be taken in handling magnetic media since even

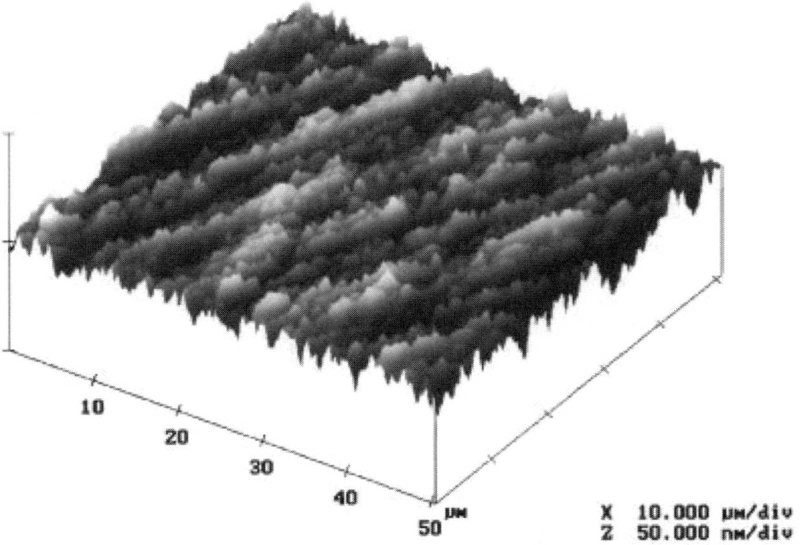

(a)

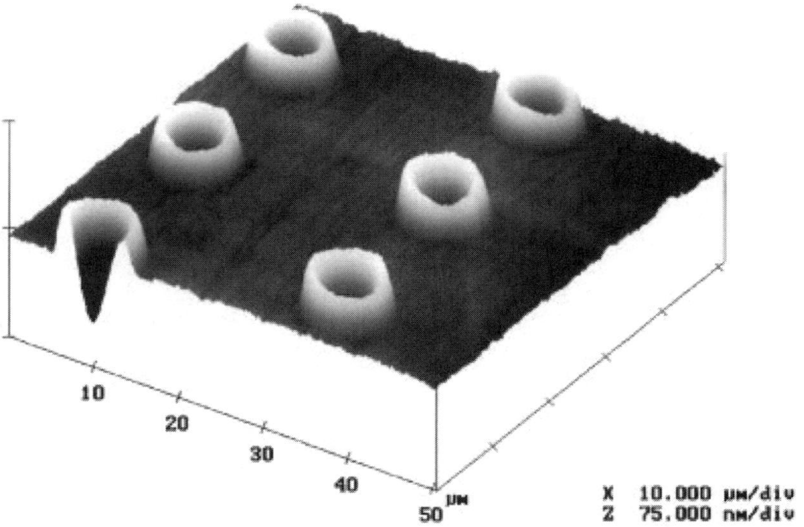

(b)

FIGURE 36.27 AFM images of (a) a mechanical texture and (b) a laser texture. (Courtesy of J. Xuan.)

minute contamination or scratches can destroy the recorded information. In addition, the servo technique of using magnetic patterns limits the track density to about one order lower than the linear density. Optical recording, on the other hand, promises to address all these concerns.

Optical recording can be categorized into three groups. In the first group, information is stored in the media during manufacturing. Users can reproduce the information, but cannot change or record new information. CD-ROM (compact disk–read only memory) belongs to this group. The second group is WORM (write once read many times). Instead of recording information during manufacturing, it leaves this step to the user. This is usually achieved by creating physical holes or blisters in the media during the recording process. Once it is recorded, however, the medium behaves like the first group: no further recording is possible. The third group is similar to magnetic recording. Recording can be performed infinitely on the media by changing phase or magnetization of the media. The most noticeable example

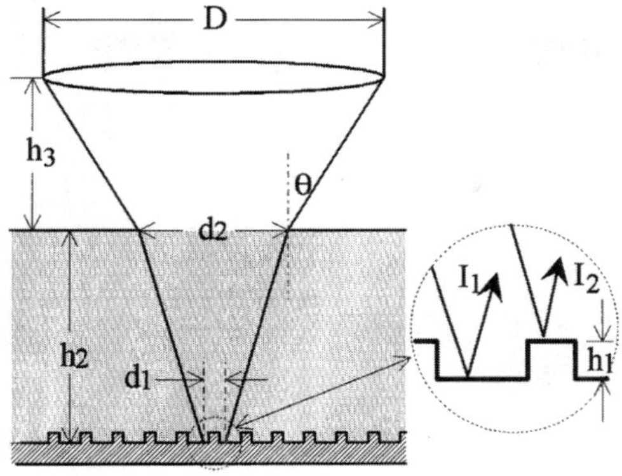

FIGURE 36.28 Schematic representation of the CDROM reproducing principle.

in this group is the magneto-optic (MO) technique [18]. Only CD-ROM and the magneto-optic recording are described in the following.

CD-ROM

Figure 36.28 shows the CD-ROM reproducing principle. Data are pressed as physical pits and lands on one surface of a plastic substrate, usually polycarbonate. A layer of aluminum is deposited on this surface to yield it reflective. Lacquer is then coated to protect the aluminum layer. During the reproducing process, an optical lens is used to focus a laser beam on the reflective pit and land surface. The diameter of the lens is D, the distance between the lens and the substrate is h_3, and the substrate thickness is h_2. The diameter of the laser beam is d_2 when entering the substrate, and becomes d_1 when focused on the reflective surface. The width of the pits are designed smaller than d_1. The reflected light consists of two portions: I_1 from the land and I_2 from the pit. According to the theory of interference, the intensity of the reflected light is

$$I = I_1 + I_2 + 2\sqrt{I_1 I_2}\cos\frac{4\pi h_1}{\lambda} \tag{36.30}$$

where λ is the wavelength of the laser and $h_1 \approx \lambda/4$ is the pit height. This leads to

$$I = \begin{cases} I_1 + I_2 - 2\sqrt{I_1 I_2} & \text{if there is a pit } \left(h_1 = \lambda/4\right) \\ I_1 + I_2 + 2\sqrt{I_1 I_2} & \text{if there is no pit } \left(h_1 = 0\right) \end{cases} \tag{36.31}$$

This change of light intensity is detected and decoded to yield the recorded data. The reflected light is also used for focusing and track following.

The fundamental limit on optical recording density is the focused beam diameter d_1. For a Gaussian (TEM_{00}) laser, this is the diffraction-limited diameter at which the light intensity is reduced to $1/e_2$ of the peak intensity:

$$d_1 \approx \frac{2\lambda}{\pi\theta} \tag{36.32}$$

where θ is the aperture angle. The following values are typical for a CD-ROM system: λ (gallium arsenide laser) = 780 nm, $\theta = 27°$, $h_2 = 1.2$ mm, $D = 5$ mm, $h_3 = 4.2$ mm. This yields $d_1 \approx 1.0$ μm and $d_2 \approx 0.7$ mm, and sets the areal density limit of optical (including magneto-optic) recording to about 1 Mbit/mm^2. For most CD-ROM applications, the areal density is smaller than this limit, and a disk with 120 mm diameter holds about 600 Mbyte information. In order to increase areal density, we can either reduce light wavelength or increase numerical aperture. Much of the effort has been to adopt a new light source with

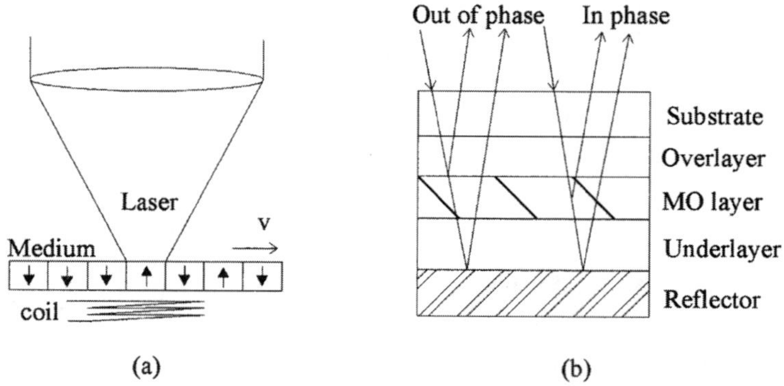

FIGURE 36.29 Schematic illustrations of (a) MO recording/reproducing and (b) quadrilayer medium cross section.

short wavelength such as a blue laser. Increasing numerical aperture is more difficult because increasing lens diameter is cost prohibitive and reducing h_2 or h_3 is reliability limited. Note that although the beam size on the focus plane is on the order of 1 μm (d_1), it is two to three orders greater at the air/substrate interface (d_2). This means that optical recording can tolerate disk surface contamination and scratches much better than magnetic recording. However, the performance of optical recording does not match magnetic recording in general. The data transfer rate of CD-ROM drives is expressed as multiple (×) of 150 kB/s. Even for a 12× CD-ROM drive, the data access time and data transfer rate are still on the order of 100 ms and 1.8 MB/s, respectively, while for a high-performance rigid disk drive these values are less than 10 ms and greater than 30 MB/s, respectively.

Magnetooptic Recording

The primary drawback of a CD-ROM to an end user is its inability to record. This deficiency is remedied by MO recording technology, as depicted in Figure 36.29. A linearly polarized laser beam is focused on a layer of magnetic material, and a coil provides a dc magnetic field on the other side of the medium. This dc magnetic field is too weak to affect the medium magnetization at normal temperature. The recording process utilizes the thermomagnetic property of the medium, and the reproducing process is achieved by using the Kerr effect. During recording, the medium is initially magnetized vertically in one direction, and the dc magnetic field is in the opposite direction. The laser heats up the medium to its Curie temperature, at which the coercivity becomes zero. During the cooling process, the dc magnetic field aligns the medium magnetization of the heated region to the magnetic field direction. In the process of reproducing, the same laser is used with a smaller intensity. The medium is heated up to its compensation temperature, at which the coercivity becomes extremely high. Depending on the direction of the magnetization, the polarization of the reflected light is rotated either clockwise or counterclockwise (Kerr rotation). This rotation of polarization is detected and decoded to get the data. The main disadvantage of MO recording is that a separate erasing process is needed to magnetize the medium in one direction before recording. Recently some technologies have been developed to eliminate this separate erasing process at the cost of complexity.

The medium used in MO recording must have a reasonable low Curie temperature (<300°C). The materials having this property are rare earth transition metal alloys, such as $Tb_{23}Fe_{77}$ and $Tb_{21}Co_{79}$. Unfortunately, the properties of these materials deteriorate in an oxygen and moisture environment. To protect them from air and humidity, they are sandwiched between an overlayer and a underlayer, such as SiO, AlN, SiN, and TiO_2. Another issue with the rare earth transition metal alloys is their small Kerr rotation, about 0.3°. To increase this Kerr rotation, multiple layers are used. In the so-called quadrilayer structure (Figure 36.29b), the overlayer is about a half-wavelength thick and the underlayer is about a quarter-wavelength thick [18]. The MO layer is very thin (≈3 nm). Light reflected from the reflector is out-of-phase with the light reflected from the surface of the MO layer, and is in-phase with the light reflected from the inside of the MO layer. As a result, the effective Kerr rotation is increased several times.

TABLE 36.11 Digital Magnetic and Optical Storage Devices

Description	Manufacturers	Approximate Price, $
Thin-film head for hard disk drive	AMC, Read-Rite, SAE	6.00–9.00
MR head for hard disk drive	AMC, Read-Rite, SAE, Seagate	8.00–12.00
Thin-film hard disk	Akashic, HMT, Komag, MCC, Stormedia	7.00–10.00
Hard disk drive	IBM, Maxtor, Quantum, Samsung, Seagate, WD	0.02–0.20/Mbytes
Floppy drive	Panasonic, Sony	20.00–40.00
Floppy disk	3M, Fuji, Memorex, Sony	0.15–0.50
Removable rigid disk drive	Iomega, Syquest	100.00–400.00
Removable rigid disk	Iomega, Maxell, Sony	5.00–20.00/100 Mbytes
Tape drive	Exabyte, HP, Seagate	100.00–400.00
Backup tape	3M, Sony, Verbatim	4.00–25.00/Gbytes
8 × CD-ROM drive	Goldstar, Panasonic	100.00–200.00
Recordable optical drive	JVC, Philips	300.00–500.00
Recordable optical disk	3M, Maxell, Memorex	3.00–15.00/650 Mbytes

Compared with magnetic recording, optical recording has the intrinsic advantages of superior reliability and portability. However, its performance is inferior due to slower data access time and transfer rate. Another advantage of optical recording, higher areal density, has been disappearing or even reversing to magnetic recording. Both magnetic and optical recording will be continuously improved in the near future, probably toward different applications. Currently there are some emerging techniques that try to combine the magnetic and optical recording techniques. Table 36.11 is a short list of representative magnetic and optical devices for digital recording.

References

1. C.D. Mee and E.D. Daniel, Eds., *Magnetic Storage Handbook*, 2nd ed., New York: McGraw-Hill, 1996.
2. O. Karlqvist, Calculation of the magnetic field in the ferromagnetic layer of a magnetic drum, *Trans. R. Inst. Technol. (Stockholm)*, 86, 3–28, 1954.
3. J.J. Miyata and R.R. Tartel, The recording and reproduction of signals on magnetic medium using saturation-type recording, *IRE Trans. Elec. Comp.*, EC-8, 159–169, 1959.
4. M.L. Williams and R.L. Comstock, An analytical model of the write process in digital magnetic recording, *A.I.P. Conf. Proc.*, 5, 738–742, 1972.
5. V.A.J. Maller and B.K. Middleton, A simplified model of the writing process in saturation magnetic recording, *IERE Conf. Proc.*, 26, 137–147, 1973.
6. H.N. Bertram, *Theory of Magnetic Recording*, Cambridge, UK: Cambridge University Press, 1994.
7. C.D. Mee and E.D. Daniel, Eds., *Magnetic Recording Technology*, 2nd ed., New York: McGraw-Hill, 1996.
8. R. Potter, Digital magnetic recording theory, *IEEE Trans. Magn.*, MAG-10, 502–508, 1974.
9. W.K. Westmijze, Studies on magnetic recording, *Philips Res. Rep.*, 8, 161–183, 1953.
10. G.J. Fan, A study of the playback process of a magnetic ring head, *IBM J. Res. Dev.*, 5, 321–325, 1961.
11. J.C. Mallinson, *The Foundations of Magnetic Recording*, 2nd ed., San Diego, CA: Academic Press, 1993.
12. M.S. Roden, *Analog and Digital Communication Systems*, 4th ed., Upper Saddle River, NJ: Prentice-Hall, 1996.
13. P.A. Franaszek, *Run-length-limited variable length coding with error propagation limitation,* U.S. Patent No. 3, 689, 899, 1972.
14. W.A. Gross, L. Matsch, V. Castelli, A. Eshel, T. Vohr, and M. Wilamann, *Fluid Film Lubrication*, New York: Wiley, 1980.
15. J.N. Israelachvili, *Intermolecular and Surface Forces*, London: Academic Press, 1985, 224.

16. Y. Li and F.E. Talke, A model for the effect of humidity on stiction of the head/disk interface, *Tribology Mech. Magn. Storage Syst.*, SP-27, 79–84, 1990.
17. R. Ranjan, D.N. Lambeth, M. Tromel, P. Goglia, and Y. Li, Laser texturing for low-flying-height media, *J. Appl. Phys.*, 68, 5745–5747, 1991.
18. J. Watkinson, *The Art of Data Recording*, Oxford U.K.: Focal Press, 1994.
19. G.A.N. Connell and R. Allen, Magnetization reversal in amorphous rare-earth transition-metal alloys: TbFe, *Proc. 4th Int. Conf. on Rapidly Quenched Metals,* Sendai, Japan, 1981.

Index

A

B

C

D

E

F

G

M

N

O

P

S

T

U

V

W

X

Z